Modern Spectrum Analysis, II

OTHER IEEE PRESS BOOKS

The Calculus Tutoring Book, *By C. Ash and R. Ash*
Imaging Technology, *Edited by H. Lee and G. Wade*
Phase-Locked Loops, *Edited by W. C. Lindsey and Chak M. Chie*
VLSI Circuit Layout: Theory and Design, *Edited by T. C. Hu and E. S. Kuh*
Monolithic Microwave Integrated Circuits, *Edited by R. A. Pucel*
Next Generation Computers, *Edited by E. A. Torrero*
Kalman Filtering: Theory and Application, *Edited by H. W. Sorenson*
Spectrum Management and Engineering, *Edited by F. Matos*
Digital VLSI Systems, *Edited by M. T. Elmasry*
Introduction to Magnetic Recording, *Edited by R. M. White*
Insights Into Personal Computers, *Edited by A. Gupta and H. D. Toong*
Television Technology Today, *Edited by T. S. Rzeszewski*
The Space Station: An Idea Whose Time Has Come, *By T. R. Simpson*
Marketing Technical Ideas and Products Successfully! *Edited by L. K. Moore and D. L. Plung*
The Making of a Profession: A Century of Electrical Engineering in America, *By A. M. McMahon*
Power Transistors: Device Design and Applications, *Edited by B. J. Baliga and D. Y. Chen*
VLSI: Technology Design, *Edited by O. G. Folberth and W. D. Grobman*
General and Industrial Management, *By H. Fayol; revised by I. Gray*
A Century of Honors, *An IEEE Centennial Directory*
MOS Switched-Capacitor Filters: Analysis and Design, *Edited by G. S. Moschytz*
Distributed Computing: Concepts and Implementations, *Edited by P. L. McEntire, J. G. O'Reilly, and R. E. Larson*
Engineers and Electrons, *By J. D. Ryder and D. G. Fink*
Land-Mobile Communications Engineering, *Edited by D. Bodson, G. F. McClure, and S. R. McConoughey*
Frequency Stability: Fundamentals and Measurement, *Edited by V. F. Kroupa*
Electronic Displays, *Edited by H. I. Refioglu*
Spread-Spectrum Communications, *Edited by C. E. Cook, F. W. Ellersick, L. B. Milstein, and D. L. Schilling*
Color Television, *Edited by T. Rzeszewski*
Advanced Microprocessors, *Edited by A. Gupta and H. D. Toong*
Biological Effects of Electromagnetic Radiation, *Edited by J. M. Osepchuk*
Engineering Contributions to Biophysical Electrocardiography, *Edited by T. C. Pilkington and R. Plonsey*
The World of Large Scale Systems, *Edited by J. D. Palmer and R. Saeks*
Electronic Switching: Digital Central Systems of the World, *Edited by A. E. Joel, Jr.*
A Guide for Writing Better Technical Papers, *Edited by C. Harkins and D. L. Plung*
Low-Noise Microwave Transistors and Amplifiers, *Edited by H. Fukui*
Digital MOS Integrated Circuits, *Edited by M. I. Elmasry*
Geometric Theory of Diffraction, *Edited by R. C. Hansen*
Modern Active Filter Design, *Edited by R. Schaumann, M. A. Soderstrand, and K. B. Laker*
Adjustable Speed AC Drive Systems, *Edited by B. K. Bose*
Optical Fiber Technology, II, *Edited by C. K. Kao*
Protective Relaying for Power Systems, *Edited by S. H. Horowitz*
Analog MOS Integrated Circuits, *Edited by P. R. Gray, D. A. Hodges, and R. W. Broderson*
Interference Analysis of Communication Systems, *Edited by P. Stavroulakis*
Integrated Injection Logic, *Edited by J. E. Smith*
Sensory Aids for the Hearing Impaired, *Edited by H. Levitt, J. M. Pickett, and R. A. Houde*
Data Conversion Integrated Circuits, *Edited by D. J. Dooley*
Semiconductor Injection Lasers, *Edited by J. K. Butler*
Satellite Communications, *Edited by H. L. Van Trees*
Frequency-Response Methods in Control Systems, *Edited by A. G. J. MacFarlane*
Programs for Digital Signal Processing, *Edited by the Digital Signal Processing Committee, IEEE*
Automatic Speech & Speaker Recognition, *Edited by N. R. Dixon and T. B. Martin*
Speech Analysis, *Edited by R. W. Schafer and J. D. Markel*
The Engineer in Transition to Management, *By I. Gray*
Multidimensional Systems: Theory & Applications, *Edited by N. K. Bose*
Analog Integrated Circuits, *Edited by A. B. Grebene*
Integrated-Circuit Operational Amplifiers, *Edited by R. G. Meyer*
Modern Spectrum Analysis, *Edited by D. G. Childers*
Digital Image Processing for Remote Sensing, *Edited by R. Bernstein*

Modern Spectrum Analysis, II

Edited by
Stanislav B. Kesler
Associate Professor
Drexel University

A volume in the IEEE PRESS Selected Reprint Series, prepared under the sponsorship of the IEEE Acoustics, Speech, and Signal Processing Society.

IEEE PRESS

The Institute of Electrical and Electronics Engineers, Inc., New York.

PRINTED IN THE UNITED STATES OF AMERICA

IEEE Order Number: PC01958

Library of Congress Cataloging-in-Publication Data

Modern spectrum analysis, II.

(IEEE Press selected reprint series)
Continues: Modern spectrum analysis. 1978.
Includes bibliographical references and index.
1. Power spectra—Addresses, essays, lectures.
2. Spectrum analysis—Addresses, essays, lectures.
I. Kesler, Stanislav B., 1942– II. Modern spectrum analysis. III. Title: Modern spectrum analysis, 2. IV. Title: Modern spectrum analysis, two.
TA348.M63 1986 620′.0042 86-322

ISBN 0-87942-203-3

Contents

Acknowledgment

The editor gratefully acknowledges the comments and suggestions from the members of the Spectrum Estimation Committee of the Acoustics, Speech, and Signal Processing Society, including J. Allen (former chairman of the society's Publications Board), J. Cadzow, T. Durrani, B. Friedlander, W. Gabriel, V. Jain, M. Kaveh (present chairman of the society's Publications Board), S. Kay, J. Makhoul, L. Marple, J. McClellan, C. Nikias, A. Nuttall, D. Tufts, and J. Woods. The help and encouragement from H. Leander of the IEEE PRESS is also gratefully acknowledged.

Preface

THE past decade witnessed the emergence of the field of power spectrum estimation as a rather independent subfield of digital signal processing. Advances in very large scale integration technology have had a major impact on the technical areas to which spectrum estimation techniques are being applied. Research in power spectrum estimation has led to a variety of parametric and nonparametric techniques, extensions to multidimensional, multichannel, and spatio-temporal processing algorithms, and to computationally fast procedures. Concurrently, various techniques have been developed for estimation of signal parameters, like frequencies of sinusoids and poles and zeros. In some cases, parameter estimation was an intermediate step in some parametric spectrum estimation procedures, e.g., the determination of prediction error filter coefficients in the maximum entropy method. In general, however, parameter estimation techniques have grown to become a research area in its own right, although closely related to spectrum estimation. Extensive research has been conducted on the statistical properties of estimation techniques. The number of important applications have constantly been increased.

Rapid developments in the field have naturally led to an expansion of the technical literature. In 1978, in order to make a number of the important papers in the field easily accessible, the IEEE PRESS published the book, *Modern Spectrum Analysis*, edited by D. Childers [1]. Since the publication of that first reprint book, there have been many important developments in the field with regard to theoretical approaches, extensions, and applications. These have been particularly significant in the area of parameter estimation. As a consequence of such a rapid development, a number of publications entirely devoted to power spectrum estimation have appeared in print. They include the proceedings of four spectrum estimation workshops, initiated by the Rome Air Development Center [2], [3] and later sponsored by the IEEE Acoustics, Speech, and Signal Processing Society [4], [5]. A special issue of the *Proceedings of the IEEE* was published in 1982 [6], and one of the *IEE Proceedings*, part F, in 1983 [7]. These publications provide a chronology of spectral estimation research since 1978.

This second volume of selected reprints on power spectrum estimation complements the first. In selecting the papers to be included in this reprint book, an attempt was made to cover different topics. Each paper included contains at least one important aspect of spectral analysis not covered by other papers in the collection. On the other hand, every effort was made to reduce the inevitable overlap of the material to a minimum. In deciding among a number of papers treating a similar subject, we inclined toward selecting those which were the best combination of representation, length, and tutorial value. Therefore, in some cases, when a good review or tutorial paper was available, it was included in place of the original source paper.

Within the last ten years or so, an approximate time period covered in this collection, the papers on spectral analysis have appeared in a wide variety of technical journals published both within and outside the IEEE. Within the IEEE itself, the Acoustics, Speech, and Signal Processing Society has had the strongest interest in the area. However, a wide range of applications, as well as the interest in some theoretical aspects, have caused a significant number of papers in spectral analysis to appear in, at least, ten other IEEE Groups' and Societies' Publications. Therefore, it was possible to draw most of the selections for this reprint book from the IEEE sources, and still retain the generality in presenting the important topics. However, a number of important papers from other sources are also included.

Due to the page limitation, a number of excellent papers could not be included in spite of their appropriate content and/or tutorial value. In particular, most of the September 1982 special issue of the *Proceedings of the IEEE* [6] fits into this category. Therefore, it was decided not to include any paper which was published in [6], and treat that special issue as a companion volume.

The material in this book is divided into six parts. The division is somewhat arbitrary since a number of papers treat more than one specific topic. Part I, "Introduction", contains one paper, which became classic within a few years after being published. It gives a clear and thorough overview of up-to-date developments in the area of spectrum estimation. Due to the clear presentation of the subject matter, this paper represents excellent introductory reading for newcomers to the field. Part II, "Parametric Methods", deals with estimation procedures which are based on some *a priori* assumptions about the signal under analysis. The assumptions are made about the probabilistic mechanism (mostly, the second-order statistics) that governs the signal generating sources, resulting in assigning a particular parametric model to the signal. The most common models are autoregressive (AR), moving average (MA), and mixed autoregressive-moving average (ARMA) models. It is the discovery of fast computational methods in this class that initiated a rapid development of the field of spectrum estimation nearly two decades ago. Part III, ''Nonparametric Methods'', treats methods which do not assume *a priori* information about the signals. They are based on either the classical Fourier decomposition or some other orthogonal decomposition of signals. Since no model is imposed upon the signal, a method from the nonparametric class is theoretically, at least, applicable to a wider variety of signals than is a parametric method. However, the latter performs better on signals which *are known* to fit the assumed model.

Part IV, ''Multidimensional, Multichannel, and Spatial Spectral Analysis'', deals with processing of either the signals which are functions of more than one independent variable (multidimensional), or a set of signals which are vector processes of one independent variable (multichannel). Spatial-temporal spectral analysis of data received by an array of sensors is a typical example of the combination of the above cases. Array processing finds applications in virtually all areas where temporal-frequency spectral analysis is used. Parametric and nonparametric methods can be implemented using various computational algorithms. Some of them are given in Part V, ''Algorithms and Adaptive Techniques''. A majority of the earlier algorithms are based on the global optimization (e.g., in a least mean square sense) of some system parameters. These are recently being superceded by fast algorithms which are recursive and/or adaptive. Development of such algorithms enables faster processing of large amounts of input data. Important by-products of spectral analysis are signal detection and estimation of signal parameters. They are treated in Part VI, ''Statistics and Detection'', together with the statistical properties of the analysis techniques. An important issue in detection of multiple signals is the signal resolvability, and a few papers that deal with it are included in this part.

Since there are several papers in the companion volume [6] that should have been included in this reprint collection, we will list the contents of [6] with a brief description of each paper.

The first paper entitled ''A historical perspective of spectrum estimation'', by E. A. Robinson presents a detailed overview of the developments in spectral analysis from Newton to the present.

Nine of the remaining 12 papers that follow, are concerned with various aspects of parametric methods. J. A. Cadzow's paper, ''Spectral estimation: An overdetermined rational model equation approach,'' and E. T. Jaynes' paper, ''On the rationale of maximum-entropy methods,'' treat spectral analysis methods, based on rational modeling of time series, from two viewpoints. The former discusses parameter hypersensitivity when their estimates are obtained by using a minimal set of Yule–Walker equations, i.e., when the number of equations are equal to the number of parameters. It suggests counteracting this hypersensitivity by using more than minimal number of equations. The latter paper treats various methods from the information theoretic point of view.

The next paper, entitled ''Maximum-entropy spectral analysis of radar clutter'' by S. Haykin, B. W. Currie, and S. B. Kesler, describes the digital processor for classifying the different forms of radar clutter as encountered in an air traffic control environment. The Doppler-based features are obtained by the multisegment maximum-entropy procedure. The paper by J. P. Burg, D. G. Luenberger, and D. L. Wenger, entitled ''Estimation of structured covariance matrices'' approaches the subject through the maximum likelihood estimate of the covariance matrix, rather then the power spectrum density estimate. The underlying process is assumed to be zero-mean Gaussian, and the covariance matrix of the special structure is sought. When the maximum-entropy estimate is found from the covariance matrix obtained in this way, there is no line splitting effect.

In their paper, ''Estimation of frequencies of multiple sinusoids: Making linear prediction perform like maximum likelihood'', D. W. Tufts and R. Kumaresan suggest a modification of the least-squares linear prediction method, by replacing the usual covariance matrix estimate with the least squares approximation matrix having the lower rank. The estimation performance for short data records and low signal-to-noise ratio is significantly improved. This paper gives somewhat more detailed account of the subject than the paper [9] in Part II. Combination of forward and backward linear prediction is a part of a number of parametric spectral analysis procedures. It is naturally parametrized by lattice filter structures. A comprehensive summary of lattice algorithms for estimating model parameters is given in B. Friedlander's paper ''Lattice methods for spectral estimation''. It also shows the methods of computation of various model parameters from lattice parameters.

The equivalence between the problem of estimating the principal frequency components in the time series and the problem of determining the bearing of a radiating source with an array of sensors is presented in D. H. Johnson's paper ''The application of spectral estimation methods to bearing estimation problems''. This treatment is similar to the one given in the paper [6] in Part IV. J. H. McClellan's paper ''Multidimensional spectral estimation'' gives a thorough review of the subject and discusses several types of estimators including Fourier, MLM, MEM, and Pisarenko estimators.

The last paper in the section on parametric methods, entitled ''Spectral approach to geophysical inversion by Lorentz, Fourier, and Radon transforms'', written by E. A. Robinson, deals with the application of the spectral analysis to one-dimensional (1-D) and two-dimensional

(2-D) geophysical inversion problem. The objective of the inversion is to determine the structure of the earth from the seismic data obtained at the surface. The earth is modeled as being composed of horizontal layers with different propagation parameters, and the seismic ray-paths are taken to be in either vertical (1-D) or slanted (2-D) direction.

First of the three papers in nonparametric section, ''Spectrum estimation and harmonic analysis'', by D. J. Thomson treats the estimation problem through the solution of an integral equation that defines a Fourier transform of the time series. The solution is given in terms of orthogonal data windows (discrete prolate spheroidal sequences). The method shows a good performance for time series with both narrow-band and wide-band spectral components. The following paper, entitled ''Robust-resistant spectrum estimation'', by R. D. Martin and D. J. Thomson deals with the preprocessing of time series which contain local perturbations, such as missing data points or non-Gaussian additive noise. These perturbations normally cause bias and variance increases in estimated spectra. To prevent this, authors suggest ''data cleaning'' by either one-sided or two-sided perturbation interpolators based on autoregressive approximations, prior to spectrum estimation.

The last paper in the nonparametric section, ''Spectral estimation using combined time and lag weighting'', by A. H. Nuttall and G. C. Carter, presents a computationally efficient spectral estimation method with good statistical properties. Also, their procedure yields classical methods, such as the Blackman–Tukey and the Welch method, as special cases. The procedure is described in the papers [4], [5] in Part III.

References

[1] D. G. Childers, Ed., *Modern Spectrum Analysis* New York: IEEE PRESS, 1978.

[2] Proceedings of the First RADC Spectrum Estimation Workshop, Rome Air Development Center, Griffiss Air Force Base, NY, 1978.

[3] Proceedings of the Second RADC Spectrum Estimation Workshop, Rome Air Development Center, Griffiss Air Force Base, NY, 1979.

[4] Proceedings of the First ASSP Workshop on Spectral Estimation, McMaster University, Hamilton, ON, Canada, 1981.

[5] Proceedings of the ASSP Spectrum Estimation Workshop II, Tampa, FL, 1983.

[6] *Proceedings of the IEEE*, Special Issue on Spectral Estimation, vol. 70, no. 9, Sept. 1982.

[7] *IEE Proceedings*, part F, Special Issue on Spectral Analysis, vol. 130, part F, no. 3, Apr. 1983.

Part I
Introduction

THIS introductory section consists of a single paper, entitled "Spectrum analysis—A modern perspective" by S. M. Kay and S. L. Marple, Jr. This paper is a comprehensive review of the spectrum analysis techniques developed up to the time of its publication in 1981. In addition to the classical Blackman–Tukey and periodogram methods, the paper treats modern techniques by first examining the modeling and parameter identification approaches, on which these techniques are based. Modern techniques discussed include three rational modeling approaches namely, moving average (MA), autoregressive (AR), and autoregressive-moving average (ARMA), then the minimum variance distortionless unbiased technique (also known as Capon's maximum likelihood method), and the Pisarenko and Prony techniques. The comparative overview of all methods gives a significant tutorial value to the paper, so that it represents an excellent introductory reading for the newcomer to the field.

Spectrum Analysis—A Modern Perspective

STEVEN M. KAY, MEMBER, IEEE, AND STANLEY LAWRENCE MARPLE, JR., MEMBER, IEEE

Abstract—A summary of many of the new techniques developed in the last two decades for spectrum analysis of discrete time series is presented in this tutorial. An examination of the underlying time series model assumed by each technique serves as the common basis for understanding the differences among the various spectrum analysis approaches. Techniques discussed include the classical periodogram, classical Blackman-Tukey, autoregressive (maximum entropy), moving average, autoregressive-moving average, maximum likelihood, Prony, and Pisarenko methods. A summary table in the text provides a concise overview for all methods, including key references and appropriate equations for computation of each spectral estimate.

I. INTRODUCTION

ESTIMATION of the power spectral density (PSD), or simply the spectrum, of discretely sampled deterministic and stochastic processes is usually based on procedures employing the fast Fourier transform (FFT). This approach to spectrum analysis is computationally efficient and produces reasonable results for a large class of signal processes. In spite of these advantages, there are several inherent performance limitations of the FFT approach. The most prominent limitation is that of frequency resolution, i.e., the ability to distinguish the spectral responses of two or more signals. The frequency resolution in hertz is roughly the reciprocal of the time interval in seconds over which sampled data is available. A second limitation is due to the implicit windowing of the data that occurs when processing with the FFT. Windowing manifests itself as "leakage" in the spectral domain, i.e., energy in the main lobe of a spectral response "leaks" into the sidelobes, obscuring and distorting other spectral responses that are present. In fact, weak signal spectral responses can be masked by higher sidelobes from stronger spectral responses. Skillful selection of tapered data windows can reduce the sidelobe leakage, but always at the expense of reduced resolution.

These two performance limitations of the FFT approach are particularly troublesome when analyzing short data records. Short data records occur frequently in practice because many measured processes are brief in duration or have slowly time-varying spectra that may be considered constant only for short record lengths. In radar, for example, only a few data samples are available from each received radar pulse. In sonar, the motion of targets results in a time-varying spectral response due to Doppler effects.

In an attempt to alleviate the inherent limitations of the FFT approach, many alternative spectral estimation procedures have been proposed within the last decade. A comparison of the spectral estimates shown in Fig. 1 illustrate the improvement that may be obtained with nontraditional approaches. The three spectra illustrated were computed using the first nine autocorrelation lags[1] of a process consisting of two equi-amplitude sinusoids at 3 and 4 Hz in additive white noise. The conventional spectral estimate based on the nine known lags $R_{xx}(0), \cdots, R_{xx}(8)$ is shown in Fig. 1(a). The spectrum is a plot of 512 values obtained by application of a 512-point FFT to the nine lags, zero-padded with 503 zeros. This spectrum, often termed the Blackman–Tukey (BT) estimate of the PSD, is characterized by sidelobes, some of which produce negative values for the PSD, and by an inability to distinguish the two sinusoidal responses.

Fig. 1(b) shows the spectral response of the autoregressive (AR) method based on the same nine lags. The improvement in resolution over that shown in Fig. 1(a) has contributed to the popularity of this alternative spectral estimate. Although the AR spectral estimate was originally developed for geophysical data processing, where it was termed the maximum entropy method (MEM) [16], [37]–[39], [50], [84], [136], [138], [158], [221], [231], [246], [247], it has been used for applications in radar [75], [92], [99], [116], [125], [126], [216], sonar [122], [198], imaging [98], radio astronomy [162], [264], [265], biomedicine [71], [74], oceanography [96], ecological systems [88], and direction finding [70], [128], [233]. The AR approach to spectrum analysis is closely related to linear prediction coding (LPC) techniques used in speech processing [80], [130], [143], [145]. The AR PSD estimator fits an AR model to the data. The origin of AR models may be found in economic time series forecasting [31], [276] and statistical estimation [189]–[191]. The MEM approach makes different assumptions about the lags, but for practical purposes, the MEM and AR spectral estimators are identical for one-dimensional analysis of wide sense stationary, Gaussian processes.

The ultimate resolution of the two sinusoidal signals into two delta function responses in a uniform spectral floor, representing the white noise PSD level, is achieved with the Pisarenko harmonic decomposition (PHD) method shown in Fig. 1(c). This technique yields the most accurate estimate of the spectrum of sinusoids in noise, at least when the autocorrelation lags are known.

As evidenced by the spectrum examples of Fig. 1, the development of alternative spectral estimates in widely different application areas has led to a confusion of conflicting terminology and different algorithm development viewpoints. Thus

Manuscript received September 8, 1980; revised June 2, 1981. The submission of this paper was encouraged after the review of an advanced proposal.

S. M. Kay is with the Department of Electrical Engineering, University of Rhode Island, Kingston, RI 02881.

S. L. Marple, Jr. is with the Analytic Sciences Corporation, McLean Operation, 8301 Greensboro Drive, Suite 1200, McLean, VA 22102.

[1] The autocorrelation function $R_{xx}(k)$ of a stochastic wide sense stationary discrete process x_n at lag k is defined in this paper as the expectation of the product $x_{n+k}x_n^*$, or $R_{xx}(k) = E[x_{n+k}x_n^*]$, where x_n is assumed to have zero mean. The * denotes complex conjugate, since complex processes are assumed in general, and $E(\ )$ denotes the expectation operator.

Reprinted from *Proc. IEEE*, vol. 69, pp. 1380–1419, Nov. 1981.

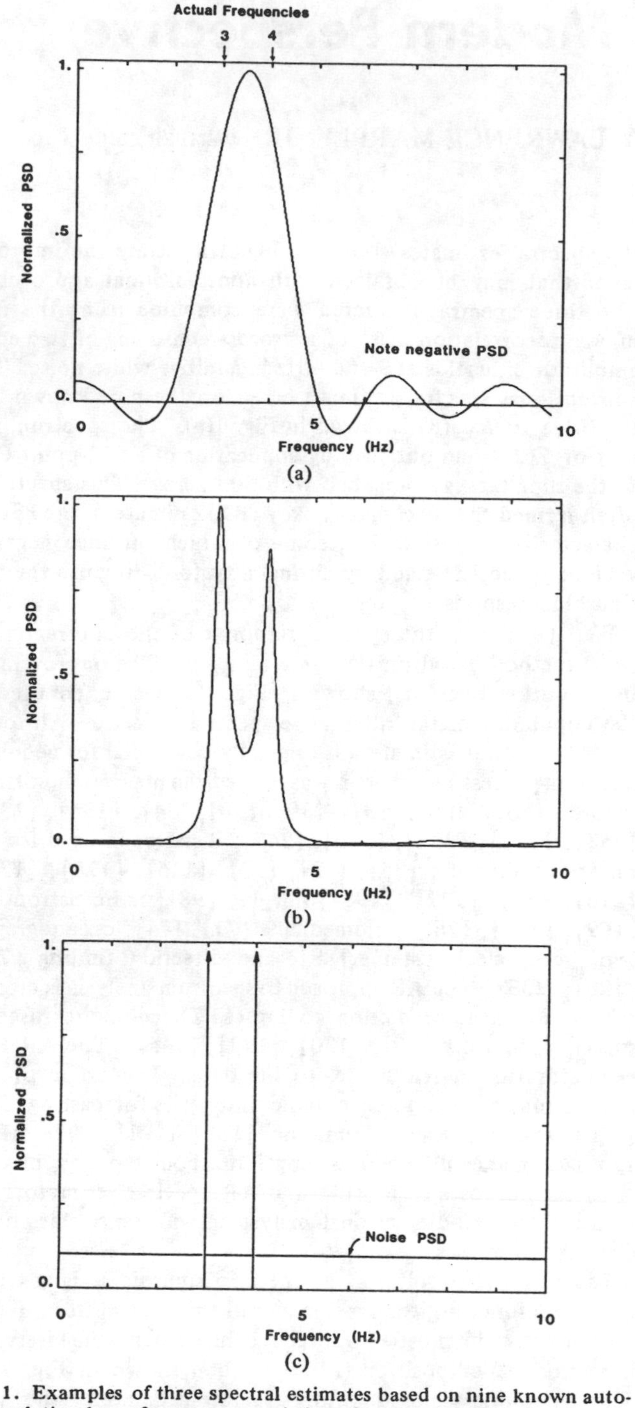

Fig. 1. Examples of three spectral estimates based on nine known autocorrelation lags of a process consisting of two equi-amplitude sinusoids in additive white noise (the variance of the noise is 10 percent of the sinusoid power). (a) BT PSD. (b) Autoregressive PSD. (c) Pisarenko harmonic decomposition PSD.

two purposes of this review are 1) to establish a common framework of terminology and symbols and 2) to unify the various approaches and algorithm developments that have evolved in various disciplines.

Claims have been made concerning the degree of improvement obtained in the spectral resolution and the signal detectability when AR and Pisarenko techniques are applied to sampled data [36], [206], [250], [251]. These performance advantages, though, strongly depend upon the signal-to-noise ratio (SNR), as might be expected. In fact, for low enough SNR's the modern spectral estimates are often no better than those obtained with conventional FFT processing [122], [150]. Even in those cases where improved spectral fidelity is achieved by use of an alternative spectral estimation procedure, the computational requirements of that alternative method may be significantly higher than FFT processing. This may make some modern spectral estimators unattractive for real-time implementation. Thus a third objective of this paper is to present tradeoffs among the various techniques. In particular, the performance advantages and disadvantages will be highlighted for each method, the computational complexity will be summarized, and criteria will be presented for determining if the selected spectral estimator is appropriate for the process being analyzed.

Some historical perspective is instructive for an appreciation of the basis for modern spectral estimation. The illustrious history of the Fourier transform can be traced back over 200 years [34], [223]. The advent of spectrum analysis based on Fourier analysis can be traced to Schuster, who was the first to coin the term "periodogram" [218], [219]. Schuster made a Fourier series fit to the variation in sun-spot numbers in an attempt to find "hidden periodicities" in the measured data. The next pioneering step was described in Norbert Wiener's classic paper on "generalized harmonic analysis" [269]. This work established the theoretical framework for the treatment of stochastic processes by using a Fourier transform approach. A major result was the introduction of the autocorrelation function of a random process and its Fourier transform relationship with the power spectral density. Khinchin [127] defined a similar relationship independently of Wiener.

Blackman and Tukey, in a classical publication in 1958 [25], provided a practical implementation of Wiener's autocorrelation approach to power spectrum estimation when using sampled data sequences. The method first estimates the autocorrelation lags from the measured data, windows (or tapers) the autocorrelation estimates in an appropriate manner, and then Fourier transforms the windowed lag estimates to obtain the PSD estimate. The BT approach was the most popular spectral estimation technique until the introduction of the FFT algorithm in 1965, generally credited to Cooley and Tukey [53]. This computationally efficient algorithm renewed an interest in the periodogram approach to PSD estimation. The periodogram spectral estimate is obtained as the squared magnitude of the output values from an FFT performed directly on the data set (data may be weighted). Currently, the periodogram is the most popular PSD estimator [17], [24], [32], [105]–[107], [109].

Conventional FFT spectral estimation is based on a Fourier series model of the data, that is, the process is assumed to be composed of a set of harmonically related sinusoids. Other time series models have been used in nonengineering fields for many years. Yule [276] and Walker [258] both used AR models to forecast trends in economic time series. Baron de Prony [202] devised a simple procedure for fitting exponential models to data obtained from an experiment in gas chemistry. Other models have arisen in the statistical and numerical analysis fields. The modern spectral estimators have their roots in these nonengineering fields of time series modeling.

The use of nontraditional spectral estimation techniques in a significant manner began in the 1960's. Parzen [189], in 1968, formally proposed AR spectral estimation. Independently in 1967, Burg [37] introduced the maximum entropy method, motivated by his work with linear prediction filtering in geoseismological applications. The one-dimensional MEM was shown formally by Van den Bos [255] to be equivalent to the AR PSD estimator. Prony's method also bears some mathematical similarities to the AR estimation algorithms. An area of current research is that of autoregressive-moving average (ARMA) models. The ARMA model is a generalization of the AR model. It appears that methods based upon these may provide even better resolution and performance than AR methods. The PHD [194], [195] is one example of a spectral estimation technique based upon a special case ARMA model.

The unifying approach employed in this paper is to view each spectral estimation technique as being based on the fitting of measured data to an assumed model. The variations in performance among the various spectral estimates may often be attributed to how well the assumed model matches the process under analysis [173]. Different models may yield similar results, but one may require fewer model parameters and is therefore more efficient in its representation of the process. Spectral estimates of various techniques computed from samples of a process consisting of sinusoids in colored Gaussian noise are presented in Section III to illustrate these variations. The process has both narrow-band and broad-band components. This process helps to illustrate how some spectral estimates tend to better estimate the narrow-band components while other spectral estimates better estimate the broad-band components of the spectra. This example process emphasizes the need to understand the underlying model before passing judgement on a spectral estimation method.

This tutorial is divided into five sections. Section II is the largest section. It contains a tutorial review of all the methods considered in this paper. Section III provides a summary table and illustration that highlights and compares the various modern spectral estimation methods. Section IV briefly examines other application areas that utilize the spectral estimation methods discussed in this paper.

A table of contents of these three sections is included below to enable the reader to quickly locate topics of interest.

No discussion of band-limited extrapolation techniques for spectral estimation is presented here since a good tutorial is already available [103]. The conclusion, Section V, makes observations concerning trends in research and application of modern spectral estimation.

II. Review of Spectral Estimation Techniques

A. Spectral Density Definitions and Basics

Traditional spectrum estimation, as currently implemented using the FFT, is characterized by many tradeoffs in an effort to produce statistically reliable spectral estimates. There are tradeoffs in windowing, time-domain averaging, and frequency-domain averaging of sampled data obtained from random processes in order to balance the needs to reduce sidelobes, to perform effective ensemble averaging, and to ensure adequate spectral resolution. To summarize the basics of conventional spectrum analysis, consider first the case of a deterministic analog waveform $x(t)$, that is a continuous function of time. For generality, $x(t)$ will be considered complex-valued in this paper. If $x(t)$ is absolute integrable, i.e., the signal energy $\mathcal{E}$ is finite

$$\mathcal{E} = \int_{-\infty}^{\infty} |x(t)|^2 \, dt < \infty \tag{2.1}$$

then the continuous Fourier transform (CFT) $X(f)$ of $x(t)$ exists and is given by

$$X(f) = \int_{-\infty}^{\infty} x(t) \exp(-j2\pi ft) \, dt. \tag{2.2}$$

(Note that (2.1) is a sufficient, but not a necessary condition for the existence of a Fourier transform [33].) The squared modulus of the Fourier transform is often termed the spectrum, $\mathcal{S}(f)$, of $x(t)$,

$$\mathcal{S}(f) = |X(f)|^2. \tag{2.3}$$

Parseval's energy theorem, expressed as

$$\int_{-\infty}^{\infty} |x(t)|^2 \, dt = \int_{-\infty}^{\infty} |X(f)|^2 \, df \tag{2.4}$$

is a statement of the conservation of energy; the energy of the time domain signal is equal to the energy of the frequency domain transform, $\int_{-\infty}^{\infty} \mathcal{S}(f)\, df$. Thus $\mathcal{S}(f)$ is an *energy spectral density* (ESD) in that it represents the distribution of energy as a function of frequency. If the signal $x(t)$ is sampled at equispaced intervals of Δt s to produce a discrete sequence $x_n = x(n\Delta t)$ for $-\infty < n < \infty$, then the sampled sequence can be represented as the product of the original time function $x(t)$ and an infinite set of equispaced Dirac delta functions $\delta(t)$. The Fourier transform of this product may be written, using distribution theory [33], as

$$X'(f) = \int_{-\infty}^{\infty} \left[\sum_{n=-\infty}^{\infty} x(t)\delta(t - n\Delta t)\Delta t \right] \exp(-j2\pi ft) \, dt$$
$$= \Delta t \sum_{n=-\infty}^{\infty} x_n \exp(-j2\pi fn\Delta t). \tag{2.5}$$

Expression (2.5) corresponds to a rectangular integration approximation of (2.2); the factor Δt ensures conservation of integrated area between (2.2) and (2.5) as $\Delta t \to 0$. Expression (2.5) will be identical in value to the transform $X(f)$ of (2.2) over the interval $-1/(2\Delta t) \leqslant f \leqslant 1/(2\Delta t)$ Hz, as long as $x(t)$ is band limited and all frequency components are in this interval. Thus the continuous energy spectral density

$$\mathcal{S}'(f) = |X'(f)|^2 \tag{2.6}$$

for data sampled from a band-limited process is identical to that of (2.3).

If a) the data sequence is available from only a finite time window over $n = 0$ to $n = N - 1$, and b) the transform is discretized also for N values by taking samples at the frequencies $f = m\Delta f$ for $m = 0, 1, \cdots, N - 1$ where $\Delta f = 1/N\Delta t$, then one can develop the familiar discrete Fourier transform (DFT) [33] from (2.5),[2]

$$\begin{aligned} X_m &= \Delta t \sum_{n=0}^{N-1} x_n \exp(-j2\pi m \Delta f n \Delta t) \\ &= \Delta t \sum_{n=0}^{N-1} x_n \exp(-j2\pi mn/N) \end{aligned}$$

$$\text{for } m = 0, \cdots, N - 1. \tag{2.7}$$

Both (2.7) and its associated inverse transform are cyclic with period N. Thus by using (2.7), we have forced a periodic extension to both the discretized data and the discretized transform values, even though the original continuous data may not have been periodic. A discrete ESD may then be defined as

$$\mathcal{S}_m = |X_m|^2 \tag{2.8}$$

also for $0 \leqslant m \leqslant N - 1$. Both the discrete $\mathcal{S}_m$ and the continuous $\mathcal{S}'(f)$ have been termed *periodogram* spectral estimates. Note however that $\mathcal{S}_m$ and $\mathcal{S}'(f)$, when evaluated at $f = m/N\Delta t$ for $m = 0, \cdots, N - 1$, do not yield identical values. $\mathcal{S}_m$ is, in effect, a sampled version of a spectrum determined from the convolution of $X(f)$ with the transform of the rectangular window that contains the data samples. Thus the discrete spectrum $\mathcal{S}_m$ based on a finite data set is a distorted version of the continuous spectrum $\mathcal{S}'(f)$ based on an infinite data set.

A different viewpoint must be taken when the process $x(t)$ is a wide sense stationary, stochastic process rather than a deterministic, finite-energy waveform. The energy of such processes are usually infinite, so that the quantity of interest is the power (time average of energy) distribution with frequency. Also, integrals such as (2.2) normally do not exist for a stochastic process. For the case of stationary random processes, the autocorrelation function

$$R_{xx}(\tau) = E[x(t + \tau)x^*(t)] \tag{2.9}$$

provides the basis for spectrum analysis, rather than the random process $x(t)$ itself. The Wiener–Khinchin theorem relates $R_{xx}(\tau)$ via the Fourier transform to $\mathcal{P}(f)$, the PSD,

$$\mathcal{P}(f) = \int_{-\infty}^{\infty} R_{xx}(\tau) \exp(-j2\pi f \tau)\, d\tau. \tag{2.10}$$

[2] The inverse transform is given by $x_n = \Delta f \sum_{m=0}^{N-1} X_m \exp(+j2\pi mn/N)$ and the energy theorem is

$$\sum_{n=0}^{N-1} |x_n|^2 \Delta t = \sum_{n=0}^{N-1} |X_m|^2 \Delta f.$$

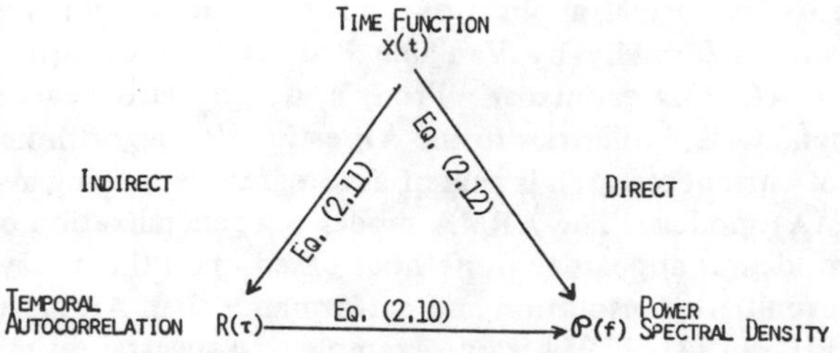

Fig. 2. Direct and indirect methods to obtain PSD (stationary and ergodic properties assumed).

As a practical matter, one does not usually know the statistical autocorrelation function. Thus an additional assumption often made is that the random process is ergodic in the first and second moments. This property permits the substitution of time averages for ensemble averages. For an ergodic process, then, the statistical autocorrelation function may be equated to

$$R_{xx}(\tau) = \lim_{T\to\infty} \frac{1}{2T} \int_{-T}^{T} x(t + \tau)x^*(t)\, dt. \tag{2.11}$$

It is possible to show [107], [132], [187], with the use of (2.11), that (2.10) may be equivalently expressed as

$$\mathcal{P}(f) = \lim_{T\to\infty} E\left\{\frac{1}{2T}\left|\int_{-T}^{T} x(t) \exp(-j2\pi f t)\, dt\right|^2\right\}. \tag{2.12}$$

The expectation operator is required since the ergodic property of $R_{xx}(\tau)$ does not couple through the Fourier transform; that is, the limit in (2.12) without the expected value does not converge in any statistical sense. Fig. 2 depicts the direct and indirect approaches to obtain the PSD from the signal $x(t)$, based on the formal relationships (2.10), (2.11), and (2.12).

Difficulties may arise if (2.12) is applied to finite data sets without regard to the expectation and limiting operations. Statistically inconsistent (unstable) estimates result if no statistical averaging is performed; i.e., the variance of the PSD estimate will not tend to zero as T increases without bound [183].

B. Traditional Methods

Two spectral estimation techniques based on Fourier transform operations have evolved. The PSD estimate based on the indirect approach via an autocorrelation estimate was popularized by Blackman and Tukey [14]. The other PSD estimate, based on the direct approach via an FFT operation on the data, is the one typically referred to as the periodogram.

With a finite data sequence, only a finite number of discrete autocorrelation function values, or lags, may by estimated. Blackman and Tukey proposed the spectral estimate

$$\hat{\mathcal{P}}_{BT}(f) = \Delta t \sum_{n=-M}^{M} \hat{R}_{xx}(m) \exp(-j2\pi f m \Delta t) \tag{2.13}$$

based on the available autocorrelation lag estimates $\hat{R}_{xx}(m)$, where $-1/(2\Delta t) \leqslant f \leqslant 1/(2\Delta t)$ and ^ denotes an estimate. This spectral estimate is the discrete-time version of the Wiener–Khinchin expression (2.10). An obvious companion autocorrelation estimate, based on (2.11), is the unbiased estimator

$$\hat{R}_{xx}(m) = \frac{1}{N - m} \sum_{n=0}^{N-m-1} x_{n+m} x_n^* \tag{2.14}$$

for $m = 0, \cdots, M$, where $M \leq N - 1$. The negative lag estimates are determined from the positive lag estimates as follows:

$$\hat{R}_{xx}(-m) = \hat{R}^*_{xx}(m) \tag{2.15}$$

in accordance with the conjugate symmetric property of the autocorrelation function of a stationary process. Instead of (2.14), both Jenkins–Watts [107] and Parzen [188], [189] provide arguments for the use of the autocorrelation estimate

$$\hat{R}'_{xx}(m) = \frac{1}{N} \sum_{n=0}^{N-m-1} x_{n+m} x_n^* \tag{2.16}$$

defined for $m = 0, \cdots, M$, since it tends to have less mean-square error than (2.14) for many finite data sets. $\hat{R}'_{xx}(m)$ is a biased estimator since $E[\hat{R}'_{xx}(m)] = [(N - m)/N]\, R_{xx}(m)$. The mean value is a triangular window weighting (sometimes called a Bartlett weighting) of the true autocorrelation function.

The direct method of spectrum analysis is the modern version of Schuster's periodogram. A sampled data version of expression (2.12), for which measured data is available only for samples $x_0, \cdots, x_{N-1}$, is

$$\hat{\mathcal{P}}_{\mathrm{PER}}(f) = \frac{1}{N\Delta t} \left| \Delta t \sum_{n=0}^{N-1} x_n \exp(-j2\pi f n \Delta t) \right|^2 \tag{2.17}$$

also defined for the frequency interval $-1/(2\Delta t) \leq f \leq 1/(2\Delta t)$. Note that the expectation operation in (2.12) has been ignored for the moment. Use of the fast Fourier transform (FFT) will permit evaluation of (2.17) at the discrete set of N equally spaced frequencies $f_m = m\Delta f$ Hz, for $m = 0, 1, \cdots, N - 1$ and $\Delta f = 1/N\Delta t$,

$$\hat{\mathcal{P}}_m = \hat{\mathcal{P}}_{\mathrm{PER}}(f_m) = \frac{1}{N\Delta t} |X_m|^2 \tag{2.18}$$

where X_m is the DFT of (2.7). $\hat{\mathcal{P}}_m$ is identical to the energy spectral density $\mathcal{S}_m$ of (2.8) except for the division by the time interval of $N\Delta t$ seconds required to make $\hat{\mathcal{P}}_m$ a power spectral density. The total power in the process, which is assumed periodic due to the DFT property, is

$$\text{Power} = \sum_{m=0}^{N-1} \hat{\mathcal{P}}_m \Delta f \tag{2.19}$$

based on rectangular integration approximation of $\hat{\mathcal{P}}_{\mathrm{PER}}$. If the Δf factor is incorporated into $\hat{\mathcal{P}}_m$, then

$$\hat{\hat{\mathcal{P}}}_m = \hat{\mathcal{P}}_m \Delta f = \frac{1}{(N\Delta t)^2} |X_m|^2 = \left| \frac{1}{N} \sum_{n=0}^{N-1} x_n \exp(-j2\pi mn/N) \right|^2. \tag{2.20}$$

This is the quantity often computed as the periodogram, but it is not scaled appropriately as a PSD. Using (2.20), it is the *peak* in the PSD plot, rather than the *area* under the plot, that is equal to the power of the assumed periodic signal. The computational economy of the FFT algorithm has made this approach a popular one.

Often a periodogram of N data samples is computed using (2.18) when the measured process has deterministic components imbedded in random noise. As pointed out earlier, care must be taken since statistically inconsistent results can occur if (2.18) is used literally without regard to the expectation operation. This need for some sort of ensemble averaging, or smoothing of the sample spectrum, is illustrated by Oppenheim and Schafer [183, p. 546] and Otnes and Enochson [184, p. 328] with examples of a white-noise process in which the variance of the spectral estimate does not decrease, even though longer and longer data sequences are used. Bartlett [18] had recognized the statistical problems with (2.18) and suggested splitting the data into segments, computing $\hat{\mathcal{P}}_m$ for each segment, and averaging the periodograms of all segments. Welch [262], [263] suggested a special digital procedure with the FFT that involves averaging periodograms.

Other mechanisms for approximating an ensemble average make use of windows in the time or frequency domain, or both [183], [184]. Overlapped weighted segment averaging is advocated by Nuttall and Carter [43], [44], [175], [179], [181] to give stability and to minimize the impact of window sidelobes.

Other references for the FFT and its application for PSD estimation may be found in Bergland [19], Bertram [22], [23], Brigham and Morrow [32], [33], Cochran *et al.* [52], Cooley *et al.* [54], [55], Glisson *et al.* [77], Nuttall and Carter [43], [181], Richards [207], Rife and Vincent [208], Webb [259], and Yuen [274], [275].

In general, the spectral estimates $\hat{\mathcal{P}}_{\mathrm{BT}}(f)$ and $\hat{\mathcal{P}}_{\mathrm{PER}}(f)$ are not identical. However, if the biased autocorrelation estimate (2.16) is used and as many lags as data samples ($M = N - 1$) are computed, then the BT estimate and the periodogram estimate yield identical numerical results [184]. Thus the periodogram can be viewed as a special case of the BT procedure. It is for this reason that the BT and periodogram estimates are occasionally termed taper and transform (TT) approaches [116].

Many of the problems of the periodogram PSD estimation technique can be traced to the assumptions made about the data outside the measurement interval. The finite data sequence may be viewed as being obtained by windowing an infinite length sample sequence with a boxcar function. The use of only this data implicitly assumes the unmeasured data to be zero, which is usually not the case. This multiplication of the actual time series by a window function means the overall transform is the convolution of the desired transform with the transform of the window function. If the true power of a signal is concentrated in a narrow bandwidth, this convolution operation will spread that power into adjacent frequency regions. This phenomena, termed leakage, is a consequence of the tacit windowing inherent in the computation of the periodogram.

In addition to the distorting effects of leakage on the spectral estimate, leakage has a detrimental impact on power estimation and detectability of sinusoidal components [201], [224], [234]. Sidelobes from adjacent frequency cells add in a constructive or destructive manner to the main lobe of a response in another frequency cell of the spectrum, affecting the estimate of power in that cell. In extreme cases, the sidelobes from strong frequency components can mask the main lobe of weak frequency components in adjacent cells, as illustrated in Fig. 3. Sidelobes characteristic of the $\sin \pi f/\pi f$ function (the transform of a rectangular time-domain window) are evident in this illustration.

Data windowing is also the fundamental factor that determines the frequency resolution of the periodogram. The convolution of the window transform with that of the actual signal transform means that the most narrow spectral response of the resultant transform is limited to that of the main-lobe width of

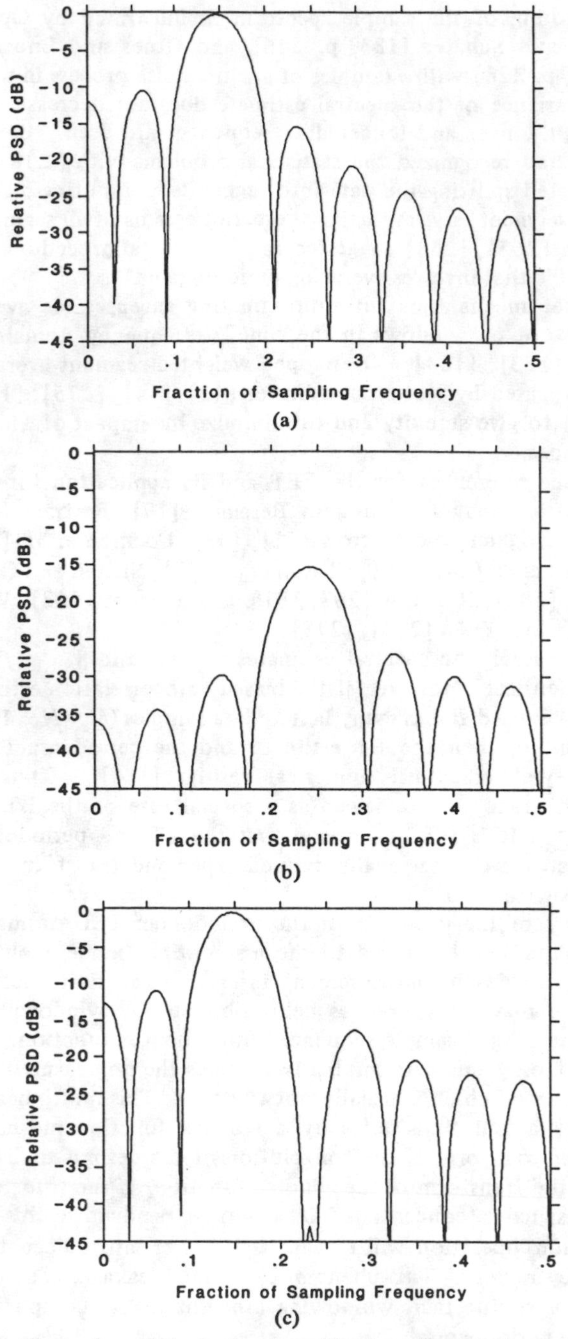

Fig. 3. Illustration of weak signal masking by adjacent strong signals when using periodogram spectrum analysis. Number of samples used in spectra (a)-(c) was 16. (a) PSD of a single sinusoid of amplitude unity, fractional sampling frequency 0.15, and initial phase 45°. (b) PSD (relative to previous PSD level) of a single sinusoid of amplitude 0.19, fractional sampling frequency 0.24, and initial phase 162°. (c) Combined signal PSD—note that there is little response at the weaker signal frequency location.

the window transform, independent of the data. For a rectangular window, the main-lobe width between 3-dB levels (and therefore, the resolution) of the resulting $(\sin \pi f)/\pi f$ transform is approximately the inverse of the observation time of $N\Delta t$ seconds. Other windows may be used, but the resolution will always be proportional to $1/N\Delta t$ Hz. Leakage effects due to data windowing can be reduced by the selection of windows with nonuniform weighting. Harris [90] has provided a good summary of the merits of various windows. Nuttall [182] provides a correction to the sidelobe behavior for some of the windows described by Harris. Other references are [12], [61], [161], [224], [260], [273]. No attempt is made here to summarize the relative merits of various window functions. The price paid for a reduction in the sidelobes is always a broadening in the main lobe of the window transform, which in turn means a decrease in the resolution of the spectral estimate.

There is a common misconception that zero-padding the data sequence before transforming will improve the resolution of the periodogram. Transforming a data set with zeros only serves to interpolate additional PSD values within the frequency interval $-1/(2\Delta t) \leqslant f \leqslant 1/(2\Delta t)$ Hz between those that would be obtained with a non-zero-padded transform. Fig. 4 shows periodograms of an N-point data set with no zero padding, data padded with N zeros, $7N$ zeros, and $31N$ zeros. In each case, the additional values of the periodogram, computed by an FFT applied to the zero-padded data set, fill in the shape of the continuous-frequency periodogram as defined by expression (2.17). In no case of zero padding, however, is there an improvement in the fundamental frequency resolution (reciprocal of the measurement interval). Zero padding is useful for 1) smoothing the appearance of the periodogram estimate via interpolation, 2) resolving potential ambiguities as illustrated in Fig. 4, and 3) reducing the "quantization" error in the accuracy of estimating the frequencies of spectral peaks. Mathematically, a $2N$-point DFT of a $2N$-point sequence $x_0, \cdots, x_{2N-1}$ is

$$X_m = \Delta t \sum_{n=0}^{2N-1} x_n \exp(-j2\pi mn/2N) \tag{2.21}$$

for $m = 0, 1, \cdots, 2N-1$.. If the data set had been zero padded with N zeros, $x_n = 0$ for $n = N, \cdots, 2N-1$, then (2.21) becomes

$$X_m = \Delta t \sum_{n=0}^{N-1} x_n \exp\left(-j2\pi \left[\frac{m}{2}\right] n/N\right) \tag{2.22}$$

which is the same as the N-point transform (2.7), but evaluated over the interval $-1/(2\Delta t) \leqslant f \leqslant 1/(2\Delta t)$ at twice as many frequencies as (2.7). By eliminating the operations on zeros introduced by zero padding, or pruning as it is called, a more efficient FFT algorithm is possible [148].

In summary, the conventional BT and periodogram approaches to spectral estimation have the following advantages: 1) computationally efficient if only a few lags are needed (BT) or if the FFT is used (periodogram), 2) PSD estimate directly proportional to the power for sinusoid processes, and 3) a good model for some applications (to be explained in more detail in the next section). The disadvantages of these techniques are: 1) suppression of weak signal main-lobe responses by strong signal sidelobes, 2) frequency resolution limited by the available data record duration, independent of the characteristics of the data or its SNR, 3) introduction of distortion in the spectrum due to sidelobe leakage, 4) need for some sort of pseudo ensemble averaging to obtain statistically consistent periodogram spectra, and 5) the appearance of negative PSD values with the BT approach when some autocorrelation sequence estimates are used.

C. Modeling and the Parameter Identification Approach

At this point, we shall depart from the traditional perspective of spectrum analysis as presented in the last section. The conventional approach used FFT operations on either windowed data or windowed lag estimates. Windowing of data or

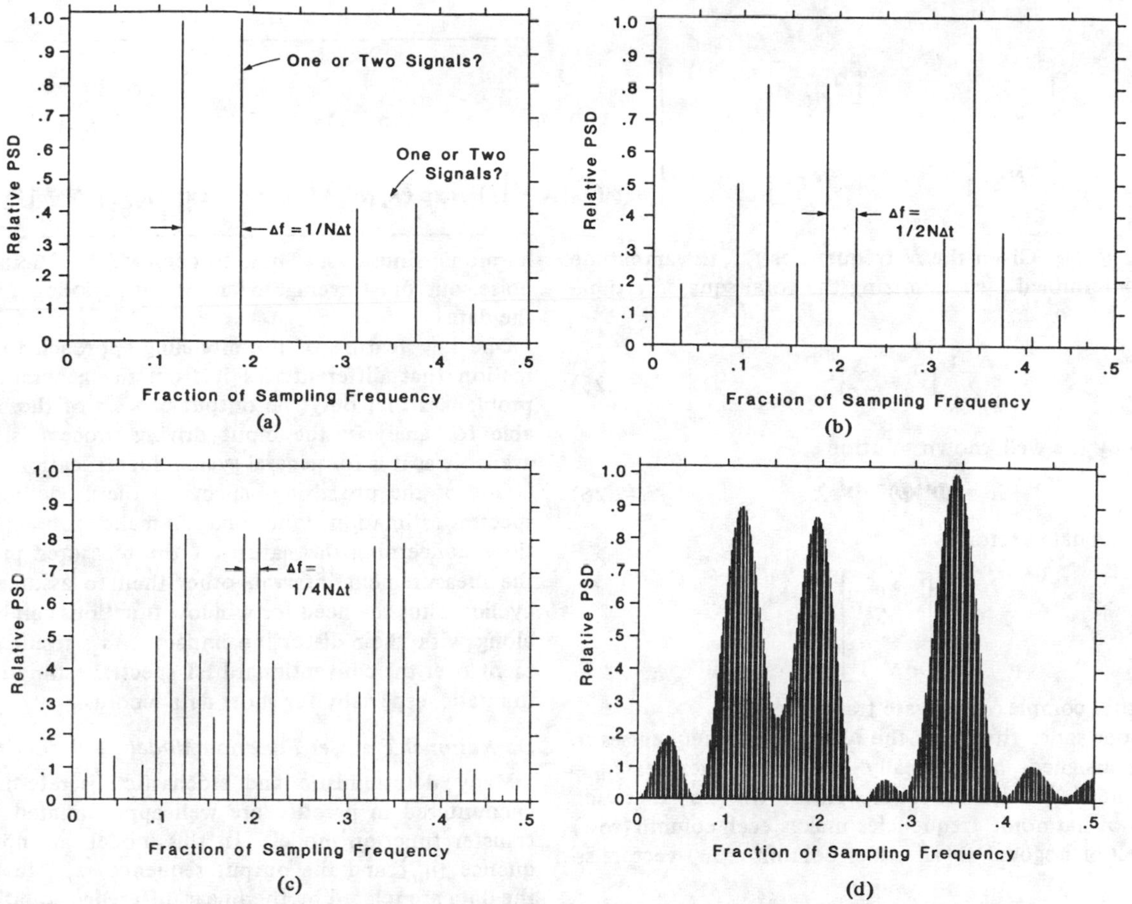

Fig. 4. Impact of zero padding the periodogram to interpolate the spectral shape and to resolve ambiguities. The spectra were estimated using the same 16 samples of a process consisting of three sinusoids of fractional sampling frequencies 0.1335, 0.1875, 0.3375 and initial phases 0°, 90°, 0°, respectively. (a) No zero padding; ambiguities are present in the spectrum. (b) Double padding; ambiguities resolved. (c) Quadruple padding; smoothest spectrum seen. (d) 32-times padding; envelope is approximation to continuous Fourier transform.

lags makes the implicit assumption that the unobserved data or lag values outside the window are zero, which is normally an unrealistic assumption. A smeared spectral estimate is a consequence of the windowing.

Often one has more knowledge about the process from which the data samples are taken, or at least is able to make a more reasonable assumption other than to assume the data is zero outside the window. Use of *a priori* information (or assumptions) may permit selection of an exact model for the process that generated the data samples, or at least a model that is a good approximation to the actual underlying process. It is then usually possible to obtain a better spectral estimate based on the model by determining the parameters of the model from the observations. Thus spectrum analysis, in the context of modeling, becomes a three step procedure. The first step is to select a time series model. The second step is to estimate the parameters of the assumed model using either the available data samples or autocorrelation lags (either known or estimated from the data). The third step is to obtain the spectral estimate by substituting the estimated model parameters into the theoretical PSD implied by the model. One major motivation for the current interest in the modeling approach to spectral estimation is the higher frequency resolution achievable with these modern techniques over that achievable with the traditional techniques previously discussed. The degree of improvement in resolution and spectral fidelity, if any, will be determined by the ability to fit an assumed model with a few parameters to the measured data. The selection of a model for the spectral estimate is intimately tied to estimation and identification techniques employed in linear system theory [271], [272].

To illustrate the modeling viewpoint of spectral estimation, the discrete periodogram PSD estimate (2.18) will be shown to be equivalent to a least squares fit of the data to a harmonic model, namely the discrete Fourier series. The least squares fit to a Fourier series is well known [28]. Only the essential ideas are presented here. If N samples $x_0, \cdots, x_{N-1}$ of a continuous-time process $x(t)$ are modeled by a discrete sequence $\hat{x}_n$ composed of N complex sinusoids of arbitrary frequencies $f_0, \cdots, f_{N-1}$, then

$$\hat{x}(n\Delta t) = \hat{x}_n = \sum_{m=0}^{N-1} a_m \exp(j2\pi f_m n\Delta t) \qquad (2.23)$$

for $n = 0, \cdots, N-1$. Thus the signal $x(t)$ over the interval $N\Delta t$ seconds is represented with periodic functions, whether or not $x(t)$ is itself cyclic. The weights a_m, to be determined, are assumed to be complex-valued for generality. The N terms of (2.23) can be expressed in matrix form as

$$\hat{X} = \Phi A \qquad (2.24)$$

where

$$\hat{X} = \begin{bmatrix} \hat{x}_0 \\ \vdots \\ \hat{x}_{N-1} \end{bmatrix}, \quad A = \begin{bmatrix} a_0 \\ \vdots \\ a_{N-1} \end{bmatrix}, \quad \Phi = \begin{bmatrix} 1 & 1 & \cdots & 1 \\ \exp(\lambda_0) & \exp(\lambda_1) & \cdots & \exp(\lambda_{N-1}) \\ \vdots & \vdots & & \vdots \\ \exp(\lambda_0[N-1]) & \exp(\lambda_1[N-1]) & \cdots & \exp(\lambda_{N-1}[N-1]) \end{bmatrix}$$

and $\lambda_i = j2\pi f_i \Delta t$. Given the N frequencies f_m, the amplitude vector A determined by minimizing the total squared estimation error,

$$\sum_{n=0}^{N-1} |x_n - \hat{x}_n|^2 \tag{2.25}$$

is provided by the well-known solution

$$A = (\Phi^H \Phi)^{-1} \Phi^H X \tag{2.26}$$

where X is the data vector

$$X = \begin{bmatrix} x_0 \\ \vdots \\ x_{N-1} \end{bmatrix}$$

and H denotes complex conjugate transpose.

The Fourier series fit selects the N sinusoidal frequencies to be the preassigned, harmonically related frequencies $f_m = m\Delta f$ Hz, where $\Delta f = 1/N\Delta t$. It is well known [28] that such a selection of harmonic frequencies makes each column (row) vector of Φ orthogonal to all other column (row) vectors so that

$$(\Phi^H \Phi)^{-1} = \frac{1}{N} I \tag{2.27}$$

where I is the identity matrix. The amplitude vector A is then given by

$$A = \frac{1}{N} \Phi^H X \tag{2.28}$$

or

$$a_m = \frac{1}{N} \sum_{n=0}^{N-1} x_n \exp(-j2\pi mn/N)$$

for $m = 0, \cdots, N-1$. The power of the sinusoidal component at the preassigned frequency f_m is

$$|a_m|^2 = \left| \frac{1}{N} \sum_{n=0}^{N-1} x_n \exp(-j2\pi mn/N) \right|^2 \tag{2.29}$$

which is identical to expression (2.20). Thus the discrete periodogram spectral estimate may be viewed as a least squares fit of a harmonic set of complex sinusoids to the data.

The case where frequencies are not preselected and are not necessarily harmonic is treated in Section II-J, which provides a discussion of Prony's method. The harmonic model preassigned the frequencies and number of sinusoids so that only estimation of the sinusoidal powers was necessary. The nonharmonic model of Prony's method will require estimation of not only the powers, but also the number of sinusoids present and their frequencies. Another aspect of the harmonic model that is noteworthy is the fact that noise is not accounted for in the model. Any noise present must also be modeled by the harmonic sinusoids. Thus, to decrease the fluctuations due to noise, one must average over a set of periodograms made from the data.

One key feature of the modeling approach to spectral estimation that differentiates it from the general identification problem is that only the output process of the model is available for analysis; the input driving process is not assumed available as it is for general system identification.

One of the promising aspects of the modeling approach to spectral estimation is that one can make more realistic assumptions concerning the nature of the measured process outside the measurement interval, other than to assume it is zero or cyclic. Thus the need for window functions can be eliminated, along with their distorting impact. As a result, the improvement over the conventional FFT spectral estimate can be quite dramatic, especially for short data records.

D. Rational Transfer Function Modeling Methods

Many deterministic and stochastic discrete-time processes encountered in practice are well approximated by a rational transfer function model. In this model, an input driving sequence $\{n_n\}$ and the output sequence $\{x_n\}$ that is to model the data are related by the linear difference equation,

$$x_n = \sum_{l=0}^{q} b_l n_{n-l} - \sum_{k=1}^{p} a_k x_{n-k}. \tag{2.30}$$

This most general linear model is termed an ARMA model. The interest in these models stems from their relationship to linear filters with rational transfer functions.

The system function $H(z)$ between the input n_n and output x_n for the ARMA process of (2.30) is the rational expression

$$H(z) = \frac{B(z)}{A(z)} \tag{2.31}$$

where

$$A(z) = z\text{ - transform of AR branch} = \sum_{m=0}^{p} a_m z^{-m}$$

$$B(z) = z\text{ - transform of MA branch} = \sum_{m=0}^{q} b_m z^{-m}.$$

It is well known that the power spectrum at the output of a linear filter, $P_x(z)$, is related to the power spectrum of the input stochastic process, $P_n(z)$, as follows:

$$P_x(z) = H(z)H^*(1/z^*)P_n(z) = \frac{B(z)B^*(1/z^*)}{A(z)A^*(1/z^*)} \cdot P_n(z). \tag{2.32}$$

Expression (2.32) is normally evaluated along the unit circle, $z = \exp(j2\pi f\Delta t)$ for $-1/(2\Delta t) \leqslant f \leqslant 1/(2\Delta t)$. Often the driving process is assumed to be a white-noise sequence of zero mean and variance σ^2. The PSD of the noise is then $\sigma^2 \Delta t$. (Note that we have included the Δt factor in the expression for power spectral density of the noise so that $P_x(\exp[j2\pi f\Delta t])$, when integrated over $-1/2\Delta t \leq f \leq 1/2\Delta t$, yields the true power of an analog signal). The PSD of the ARMA output process is then

$$\mathcal{P}_{\text{ARMA}}(f) = \mathcal{P}_x(f) = \sigma^2 \Delta t \, |\mathcal{B}(f)/\mathcal{A}(f)|^2 \tag{2.33}$$

where $\mathcal{A}(f) = A(\exp[j2\pi f\Delta t])$ and $\mathcal{B}(f) = B(\exp[j2\pi f\Delta t])$. Specification of the parameters $\{a_k\}$ (termed the autoregressive coefficients), the parameters $\{b_k\}$ (termed the moving-average coefficients), and σ^2 is equivalent to specifying the spectrum of the process $\{x_n\}$. Without loss of generality, one can assume $a_0 = 1$ and $b_0 = 1$ since any filter gain can be incorporated into σ^2.

If all the $\{a_k\}$ terms except $a_0 = 1$ vanish, then

$$x_n = \sum_{l=0}^{q} b_l n_{n-l} \tag{2.34}$$

and the process is strictly a moving average of order q, and

$$\mathcal{P}_{\text{MA}}(f) = \sigma^2 \Delta t \, |\mathcal{B}(f)|^2. \tag{2.35}$$

This model is sometimes termed an all-zero model [266].

If all the $\{b_i\}$, except $b_0 = 1$, are zero, then

$$x_n = -\sum_{k=1}^{p} a_k x_{n-k} + n_n \tag{2.36}$$

and the process is strictly an autoregression of order p. The process is termed AR in that the sequence x_n is a linear regression on itself with n_n representing the error. With this model, the present value of the process is expressed as a weighted sum of past values plus a noise term. The PSD is

$$\mathcal{P}_{\text{AR}}(f) = \frac{\sigma^2 \Delta t}{|\mathcal{A}(f)|^2}. \tag{2.37}$$

This model is sometimes termed an all-pole model.

The Wold decomposition theorem [266] relates the ARMA, MA, and AR models. Basically, the theorem asserts that any stationary ARMA or MA process of finite variance can be represented as a unique AR model of possibly infinite order; likewise, any ARMA or AR process can be represented as a MA process of possibly infinite order. This theorem is important because if we choose the wrong model among the three, we may still obtain a reasonable approximation by using a high order. Thus an ARMA model can be approximated by an AR model of higher order. Since the estimation of parameters for an AR model results in linear equations, as will be shown, it has a computational advantage over ARMA and MA parameter estimation techniques. The largest portion of research effort on rational transfer function modeling has therefore been concerned with the AR model.

E. Autoregressive PSD Estimation

Introduction: Since this section is detailed, reflecting the extensive research on this PSD estimation method, it is worthwhile to briefly outline the material to be presented. The Yule–Walker equations are first derived. They describe the linear relationship between the AR parameters and the autocorrelation function. The solution of these equations is provided by the computationally efficient Levinson–Durbin algorithm, the details of which reveal some fundamental properties of AR processes.

Next, the relationship between AR modeling, linear prediction theory, and maximum entropy spectral estimation (MESE) is examined. Selection of the AR model order and the associated AR parameters is then addressed. Included in the discussion are batch estimation techniques based upon linear prediction theory and sequential estimation methods based on recursive least squares and adaptive algorithms. Tradeoffs between the various approaches are noted.

Finally, some limitations of AR spectral estimation that reduce its applicability in practice are described. These involve the degrading effect of observation noise, spurious peaks, and some anomolous effects which occur when the data are dominated by sinusoidal components. Some techniques for reducing these effects are presented.

Yule–Walker Equations: If an autoregression is a reasonable model for the data, then the AR power spectral density estimate based on (2.37) may be rewritten as

$$\mathcal{P}_{\text{AR}}(f) = |H(\exp[j2\pi f\Delta t])|^2 \mathcal{P}_n(f) = \frac{\sigma^2 \Delta t}{\left|1 + \sum_{k=1}^{p} a_k \exp(-j2\pi f k\Delta t)\right|^2}. \tag{2.38}$$

Thus, to estimate the PSD one need only estimate $\{a_1, a_2, \cdots, a_p, \sigma^2\}$. To do this, a relationship between the AR parameters and the autocorrelation function (known or estimated) of x_n is now presented. This relationship is known as the Yule–Walker equations [31]. The derivation proceeds as follows:

$$R_{xx}(k) = E[x_{n+k}x_n^*] = E\left[x_n^*\left(-\sum_{l=1}^{p} a_l x_{n-l+k} + n_{n+k}\right)\right]$$

$$= -\sum_{l=1}^{p} a_l R_{xx}(k-l) + E[n_{n+k}x_n^*].$$

Since $H(z)$ is assumed to be a stable, causal filter, we have

$$E(n_{n+k}x_n^*) = E\left[n_{n+k}\sum_{l=0}^{\infty} h_l^* n_{n-l}^*\right]$$

$$= \sum_{l=0}^{\infty} h_l^* \sigma^2 \delta_{k+l}$$

$$= \sigma^2 h_{-k}^*$$

$$= \begin{cases} 0, & \text{for } k > 0 \\ h_0^* \sigma^2, & \text{for } k = 0. \end{cases}$$

Note that δ_m is the discrete delta function, i.e., $\delta_m = 1$ if $m = 0$ or 0 if $m \neq 0$. But $h_0 = \lim_{z\to\infty} H(z) = 1$, and therefore,

$$R_{xx}(k) = \begin{cases} -\sum_{l=1}^{p} a_l R_{xx}(k-l), & \text{for } k > 0 \\ -\sum_{l=1}^{p} a_l R_{xx}(-l) + \sigma^2, & \text{for } k = 0. \end{cases} \tag{2.39}$$

Expression (2.39) is the Yule–Walker equations. To determine the AR parameters, one need only choose p equations from (2.39) for $k > 0$, solve for $\{a_1, a_2, \cdots, a_p\}$, and then find σ^2 from (2.39) for $k = 0$. The set of equations which require the fewest lags of the autocorrelation function is the selection

$k = 1, 2, \cdots, p$. They can be expressed in matrix form as

$$\begin{bmatrix} R_{xx}(0) & R_{xx}(-1) & \cdots & R_{xx}(-(p-1)) \\ R_{xx}(1) & R_{xx}(0) & \cdots & R_{xx}(-(p-2)) \\ \vdots & \vdots & & \vdots \\ R_{xx}(p-1) & R_{xx}(p-2) & \cdots & R_{xx}(0) \end{bmatrix} \begin{bmatrix} a_1 \\ a_2 \\ \vdots \\ a_p \end{bmatrix} = - \begin{bmatrix} R_{xx}(1) \\ R_{xx}(2) \\ \vdots \\ R_{xx}(p) \end{bmatrix} \quad (2.40)$$

Note that the above autocorrelation matrix, R_{xx}, is Hermitian ($R_{xx}^H = R_{xx}$) and it is Toeplitz since the elements along any diagonal are identical. Also, the matrix is positive definite (assuming x_n is not purely harmonic) which follows from the positive definite property of the autocorrelation function [41], [81], [92].

It should be noted that (2.40) can also be augmented to incorporate the σ^2 equation, yielding

$$\begin{bmatrix} R_{xx}(0) & R_{xx}(-1) & \cdots & R_{xx}(-p) \\ R_{xx}(1) & R_{xx}(0) & \cdots & R_{xx}(-(p-1)) \\ & & & \vdots \\ R_{xx}(p) & R_{xx}(p-1) & \cdots & R_{xx}(0) \end{bmatrix} \begin{bmatrix} 1 \\ a_1 \\ \vdots \\ a_p \end{bmatrix} = \begin{bmatrix} \sigma^2 \\ 0 \\ \vdots \\ 0 \end{bmatrix} \quad (2.41)$$

which follows from (2.39). This form will be useful later. Thus, to determine the AR parameters and σ^2, one must solve (2.41) with the $p+1$ estimated autocorrelation lags $R_{xx}(0), \cdots, R_{xx}(p)$ and use $R_{xx}(-m) = R_{xx}^*(m)$.

Levinson–Durbin Algorithm [56], [60], [142], [270]: The Levinson–Durbin algorithm provides an efficient solution for (2.41). The algorithm requires only order p^2 operations, denoted $o(p^2)$, as opposed to $o(p^3)$ for Gaussian elimination. Although appearing at first to be just an efficient algorithm, it reveals fundamental properties of AR processes. The algorithm proceeds recursively to compute the parameter sets $\{a_{11}, \sigma_1^2\}, \{a_{21}, a_{22}, \sigma_2^2\}, \cdots, \{a_{p1}, a_{p2}, \cdots, a_{pp}, \sigma_p^2\}$. Note that an additional subscript has been added to the AR coefficients to denote the order. The final set at order p is the desired solution. In particular, the recursive algorithm is initialized by

$$a_{11} = -R_{xx}(1)/R_{xx}(0) \quad (2.42)$$

$$\sigma_1^2 = (1 - |a_{11}|^2) R_{xx}(0) \quad (2.43)$$

with the recursion for $k = 2, 3, \cdots, p$ given by

$$a_{kk} = -\left[R_{xx}(k) + \sum_{l=1}^{k-1} a_{k-1,l} R_{xx}(k-l) \right] \Big/ \sigma_{k-1}^2 \quad (2.44)$$

$$a_{ki} = a_{k-1,i} + a_{kk} a_{k-1,k-i}^* \quad (2.45)$$

$$\sigma_k^2 = (1 - |a_{kk}|^2) \sigma_{k-1}^2. \quad (2.46)$$

It is important to note that $\{a_{k1}, a_{k2}, \cdots, a_{kk}, \sigma_k^2\}$, as obtained above, is the same as would be obtained by using (2.41) for $p = k$. Thus the Levinson–Durbin algorithm also provides the AR parameters for all the lower order AR model fits to the data. This is a useful property when one does not know *a priori* the correct model order, since one can use (2.42)–(2.46) to generate successively higher order models until the modeling error σ_k^2 is reduced to a desired value. In particular, if a process is actually an AR(p) process (an AR process of order p), then $a_{p+1,k} = a_{pk}$ for $k = 1, 2, \cdots, p$ and hence $a_{p+1,p+1} = 0$. In general for an AR(p) process, $a_{kk} = 0$ and $\sigma_k^2 = \sigma_p^2$ for $k > p$. Hence, the variance of the excitation noise is a constant for a model order equal to or greater than the correct order. Thus the point at which σ_k^2 does not change would appear to be a good indicator of the correct model order. It can be shown that $|a_{kk}| \leq 1$, so that $\sigma_{k+1}^2 \leq \sigma_k^2$ [9], [41]. This means that σ_k^2 first reaches its minimum at the correct model order. This point is discussed further under the topic of model order determination.

TABLE I

SUMMARY OF AR PROCESS PROPERTIES (EXCLUDING PURELY HARMONIC PROCESSES)

- Autocorrelation matrix is positive definite: $X^H R_{xx} X > 0$ for all X vectors
- Reflection coefficient sequence satisfies $|K_i| < 1$ for $i=1,2,\ldots,p$
- Zeros of A(z) lie within unit circle: $|z_i| < 1$ for $i=1,2,\ldots,p$
- Prediction error powers monotically decrease: $\sigma_1^2 \geq \sigma_2^2 \geq \ldots \geq \sigma_p^2 \geq 0$

The parameters $\{a_{11}, a_{22}, \cdots, a_{pp}\}$ are often called the reflection coefficients and are designated as $\{K_1, K_2, \cdots, K_p\}$. They have the property that for $\{R_{xx}(0), R_{xx}(1), \cdots, R_{xx}(p)\}$ to be a valid autocorrelation sequence, i.e., the autocorrelation matrix is positive semidefinite, then it is necessary and sufficient that $|a_{kk}| = |K_k| \leq 1$ for $k = 1, 2, \cdots, p$ [41]. Furthermore, a necessary and sufficient condition for the poles of $A(z)$ to be on or within the unit circle of the z plane is $|K_k| \leq 1$ for $k = 1, 2, \cdots, p$ [41], [140]. It should be noted that if $|K_k| = 1$ for some k, then the recursion (2.42)–(2.45) must terminate since $\sigma_k^2 = 0$. The process in this case is purely harmonic (consists only of sinusoids). These properties are summarized in Table I.

The problem of AR parameter identification is closely related to the theory of linear prediction. Assume x_n is an AR(p) process. If one wishes to predict x_n on the basis of the previous p samples [92],

$$\hat{x}_n = -\sum_{k=1}^{p} \alpha_k x_{n-k} \quad (2.47)$$

then $\{\alpha_1, \alpha_2, \cdots, \alpha_p\}$ can be chosen to minimize the prediction error power Q_p where

$$Q_p = E[|x_n - \hat{x}_n|^2]. \quad (2.48)$$

By the orthogonality principle [187]

$$E[(x_n - \hat{x}_n) x_k^*] = 0, \qquad \text{for } k = n-1, \cdots, n-p$$

or

$$R_{xx}(k) = -\sum_{l=1}^{p} \alpha_l R_{xx}(k-l), \quad \text{for } k = 1, 2, \cdots, p. \quad (2.49)$$

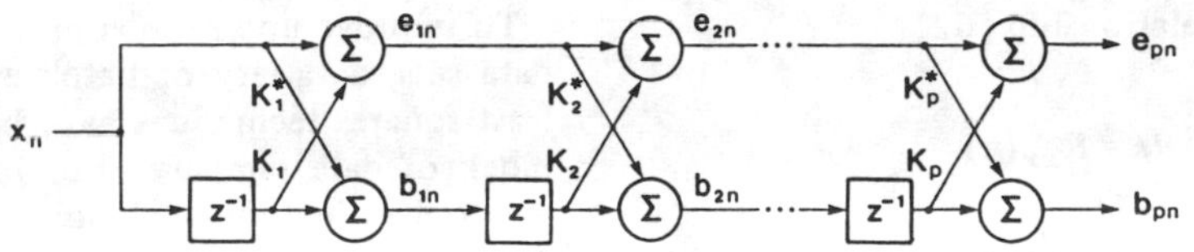

Fig. 5. Lattice formulation of prediction error (whitening, or inverse) filter.

The minimum prediction error power is

$$Q_{p_{\min}} = E[(x_n - \hat{x}_n)x_n^*] = R_{xx}(0) + \sum_{k=1}^{p} \alpha_k R_{xx}(-k). \quad (2.50)$$

These equations are identical to (2.39). Thus it must be true that $\alpha_k = a_{pk}$ for $k = 1, \cdots, p$ and $Q_{p_{\min}} = \sigma_p^2$, so that the best linear predictor is just $\hat{x}_n = -\sum_{k=1}^{p} a_{pk} x_{n-k}$. The error sequence, although uncorrelated with the linear estimate, is not necessarily a white process (it will be if x_n is a AR(p) process). In the limit as $p \to \infty$, the error sequence becomes white.

Note that $\{a_{k1}, a_{k2}, \cdots, a_{kk}\}$ and σ_k^2 constitute the parameters for the optimum kth-order linear predictor and the corresponding minimum prediction error power, respectively. Therefore, AR parameter identification and linear prediction of an AR(p) process yield identical results and the theory of one is applicable to the other.

The theory of linear prediction lends an important interpretation to the Levinson–Durbin algorithm. Denote the prediction error for a pth-order linear predictor as e_{pn}. Then, using (2.45)

$$e_{pn} = x_n + \sum_{k=1}^{p} a_{pk} x_{n-k} \quad (2.51)$$

$$= x_n + \sum_{k=1}^{p-1} (a_{p-1,k} + K_p a_{p-1,p-k}^*) x_{n-k} + K_p x_{n-p} \quad (2.52)$$

$$= e_{p-1,n} + K_p b_{p-1,n-1}$$

where

$$b_{pn} = x_{n-p} + \sum_{k=1}^{p} a_{pk}^* x_{n-p+k}. \quad (2.53)$$

The term b_{pn} is the backward prediction error, i.e., the error when one attempts to "predict" x_{n-p} on the basis of samples $x_{n-p+1}, \cdots, x_n$. It is seen that the predictor coefficients for the backwards predictor are complex conjugates of those of the forward predictor, which is a consequence of the stationary autocorrelation function assumption [41]. Similarly, it can be shown

$$b_{pn} = b_{p-1,n-1} + K_p^* e_{p-1,n}. \quad (2.54)$$

The relationships of (2.52) and (2.54) give rise to the so called lattice filter structure shown in Fig. 5 [146]. Note that the transfer function of the entire filter is just

$$A(z) = 1 + \sum_{k=1}^{p} a_{pk} z^{-k}$$

which is the inverse of $H(z) = 1/A(z)$. This follows from (2.51). This filter is termed either the "inverse" filter or "prediction error" filter. The lattice filter interpretation of the Levinson–Durbin algorithm leads to an important time recursive formulation for estimating the AR parameters. By updating the reflection coefficients using (2.52) and (2.54), the AR parameters are obtained via the order recursion (2.45). More details are presented under the topic of sequential estimation of AR parameters.

If one minimizes $E(|e_{pn}|^2)$, as given by (2.52), with respect to K_p, one obtains

$$K_p = \frac{-E(e_{p-1,n} b_{p-1,n-1}^*)}{E(|b_{p-1,n-1}|^2)}. \quad (2.55)$$

As stated previously, both e_{pn} and b_{pn} have the same statistical properties, so that (2.55) can be written as

$$K_p = \frac{-E(e_{p-1,n} b_{p-1,n-1}^*)}{\sqrt{E(|e_{p-1,n}|^2) E(|b_{p-1,n-1}|^2)}}. \quad (2.56)$$

Thus K_p is the negative of the normalized correlation coefficient between $e_{p-1,n}$ and $b_{p-1,n-1}$, so we must have $|K_p| \leq 1$. In the statistical literature, $-K_p$ is known as a partial correlation coefficient since it is the normalized correlation between x_n and x_{n-p} with the correlation of $x_{n-1}, x_{n-2}, \cdots, x_{n-p+1}$ removed [31].

Maximum Entropy Spectral Estimation [41], [62], [78], [171], [172], [205], [215], [248], [249]: MESE is based upon an extrapolation of a segment of a known autocorrelation function for lags which are not known. In this way the characteristic smearing of the estimated PSD due to the truncation of the autocorrelation function can be removed. If we assume $\{R_{xx}(0), R_{xx}(1), \cdots, R_{xx}(p)\}$ are known, the question arises as to how $\{R_{xx}(p+1), R_{xx}(p+2), \cdots\}$ should be specified in order to guarantee that the entire autocorrelation sequence is positive semi-definite. In general, there are an infinite number of possible extrapolations, all of which yield valid autocorrelation functions. Burg [37], [41] argued that the extrapolation should be made so that the time series characterized by the extrapolated autocorrelation function has maximum entropy. The time series will then be the most random one which has the known autocorrelation lags for its first $p+1$ lags. Alternately, the power spectral density is the one with the flattest (whitest) spectrum of all spectra for which $\{R_{xx}(0), R_{xx}(1), \cdots, R_{xx}(p)\}$ is equal to the known lags. The resultant spectral estimate is termed the maximum entropy spectral estimate. The rationale for the choice of the maximum entropy criterion is that it imposes the fewest constraints on the unknown time series by maximizing its randomness, thereby producing a minimum bias solution.

In particular, if one assumes a Gaussian random process, then the entropy per sample is proportional to

$$\int_{-1/2\Delta t}^{1/2\Delta t} \ln \mathcal{P}_x(f)\, df \quad (2.57)$$

where $\mathcal{P}_x(f)$ is the PSD of x_n. $\mathcal{P}_x(f)$ is found by maximizing (2.57) subject to the constraints that the $(p+1)$ known lags satisfy the Wiener–Khinchin relationship [62],

$$\int_{-1/2\Delta t}^{1/2\Delta t} \mathcal{P}_x(f) \exp(-j2\pi f n \Delta t)\, df = R_{xx}(n), \qquad \text{for } n = 0, 1, \cdots, p. \tag{2.58}$$

The solution is found by the Lagrange multiplier technique and is

$$\mathcal{P}_x(f) = \frac{\sigma_p^2 \Delta t}{\left| 1 + \sum_{k=1}^{p} a_{pk} \exp(-j2\pi f k \Delta t) \right|^2} \tag{2.59}$$

where $\{a_{p1}, \cdots, a_{pp}\}$ and σ_p^2 are just the pth-order predictor parameters and prediction error power, respectively, With knowledge of $\{R_{xx}(0), R_{xx}(1), \cdots, R_{xx}(p)\}$, the MESE will be equivalent to an AR PSD, with $\mathcal{P}_x(f)$ given by (2.59) and based on parameters found by solving (2.41). The maximum entropy relationship to AR PSD analysis is only valid for Gaussian random processes and known autocorrelation lags.

AR Autocorrelation Extension: An alternative representation for (2.38) is [59]

$$\mathcal{P}_{AR}(f) = \frac{\sigma_p^2 \Delta t}{\left| 1 + \sum_{k=1}^{p} a_{pk} \exp(-j2\pi f k \Delta t) \right|^2} = \Delta t \sum_{n=-\infty}^{\infty} r_{xx}(n) \exp(-j2\pi f n \Delta t) \tag{2.60}$$

where

$$r_{xx}(n) = \begin{cases} R_{xx}(n), & \text{for } |n| \leq p \\ -\sum_{k=1}^{p} a_{pk}\, r_{xx}(n-k), & \text{for } |n| > p. \end{cases} \tag{2.61}$$

From this, it is easy to see that the AR PSD preserves the known lags and recursively extends the lags beyond the window of known lags. The AR PSD function (2.60) summation is identical to the BT PSD function (2.13) up to lag p, but continues with an infinite extrapolation of the autocorrelation function rather than windowing it to zero. Thus AR spectra do not exhibit sidelobes due to windowing. Also, it is the implied extrapolation given by (2.61) that is responsible for the high resolution property of the AR spectral estimator.

AR Parameter Estimation: In most practical situations, one has data samples rather than known autocorrelation lags available for the spectral estimation procedure. To obtain reliable estimates of the AR parameters, standard statistical estimation theory can be used. The usual estimator for a nonrandom set of parameters is the maximum likelihood estimator (MLE). However, the exact solution of the MLE for the parameters of an AR(p) process is difficult to obtain [31]. If $N \gg p$, an approximate MLE can be found which amounts to nothing more than solving the Yule–Walker equation with the autocorrelation function replaced by a suitable estimate. For long data records, this AR parameter estimator produces good spectral estimates. For short data records, which are more commonly encountered in practice (and for which the periodogram spectral estimate has the poorest resolution), the use of the Yule–Walker approach produces poor resolution spectral estimates.

To improve upon the approximate MLE approach for short data sets, a variety of batch estimation techniques based on least squares techniques have been proposed that operate on a block of data samples. For longer data sets, a variety of sequential estimation techniques are available for updating the AR estimates as new data are received. These techniques are especially useful for tracking processes that slowly vary with time. The next two topic areas cover batch and sequential AR estimation methods.

Many additional references on the statistical properties of AR spectral estimates may be found in [13], [100], [116], [186], [212]. Some references dealing with frequency estimation accuracy of AR spectral estimates are [124], [212], [239]. It has been empirically shown by Sakai [212] that the frequency variance of an AR spectral estimate is inversely proportional to both the data length and the square of the SNR; Keeler [124] has empirical evidence that the variance is inverse to both the length and the SNR (rather than the square of SNR).

Batch Estimation of the AR Parameters: Although the Yule–Walker equations could be solved using lag estimates, several least squares estimation procedures are available that operate directly on the data to yield better AR parameter estimates. These techniques often produce better AR spectra than that achievable with the Yule–Walker approach. Two types of least squares estimators will be considered. The first type utilizes forward linear prediction for the estimate, while the second type employs a combination of forward and backward linear prediction.

Assume the data sequence $x_0, \cdots, x_{N-1}$ is used to find the pth-order AR parameter estimates. The foward linear predictor will have the usual form

$$\hat{x}_n = -\sum_{k=1}^{p} a_{pk} x_{n-k}. \tag{2.62}$$

The prediction is forward in the sense that the prediction for the current data sample is a weighted sum of p previous samples. The forward linear prediction error is

$$e_{pn} = x_n - \hat{x}_n = \sum_{k=0}^{p} a_{pk} x_{n-k} \tag{2.63}$$

where

$$a_{p0} = 1.$$

We may compute e_{pn} for $n = 0$ to $n = N + p - 1$ if one assumes the terms outside the measurements are zero, i.e., $x_n = 0$ for $n < 0$ and $n > N - 1$. There is an implied windowing of the data sequence in order to extend the index range for e_{pn} from 0 to $N + p - 1$. Using a matrix formulation for (2.63),

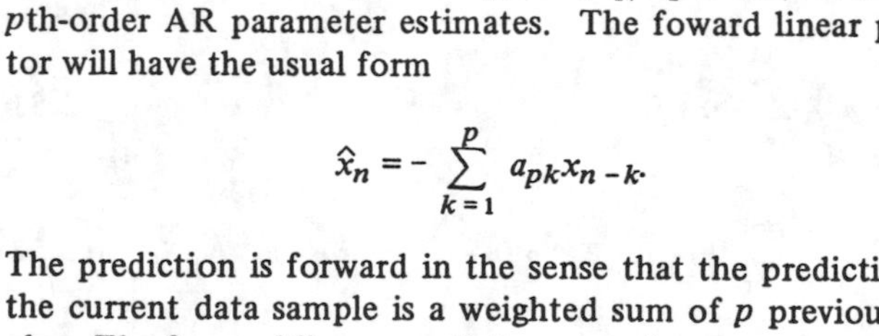

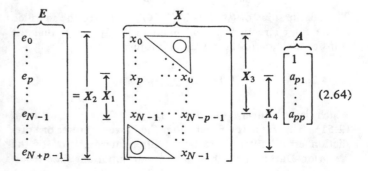

(2.64)

or

$$E = XA. \tag{2.65}$$

The prediction error energy is simply

$$\mathcal{E}_p = \sum_n |e_{pn}|^2 = \sum_n \left| \sum_{k=0}^{p} a_{pk} x_{n-k} \right|^2. \tag{2.66}$$

The summation range for $\mathcal{E}_p$ is purposely not specified for the moment. To minimize $\mathcal{E}_p$, the derivatives of $\mathcal{E}_p$ with respect to the $\{a_{pk}\}$ are set to zero and the resultant equations solved for the AR parameters. The result is

$$\sum_{k=0}^{p} a_{pk} \left(\sum_n x_{n-k} x_{n-i}^* \right) = 0, \quad \text{for } 1 \leqslant i \leqslant p \tag{2.67}$$

with minimum error energy,

$$\mathcal{E}_p = \sum_{k=0}^{p} a_{pk} \left(\sum_n x_{n-k} x_n^* \right). \tag{2.68}$$

Expressions (2.67) and (2.68) can be reformulated in normal equation form

$$(X_l^H X_l) A = \begin{pmatrix} \mathcal{E}_p \\ 0 \\ \vdots \\ 0 \end{pmatrix} \tag{2.69}$$

for which four special indexing ranges $l = 1, 2, 3, 4$ are selected, as indicated in (2.64). Note that (2.69) has the same structure as (2.41); however, the data matrix product $(X_l^H X_l)$ is not necessarily Toeplitz as are the Yule–Walker equations.

If the data matrix X_1 is selected, the normal (2.69) are termed the covariance equations, often encountered in LPC of speech (see Makhoul [144]). If the data matrix X_2 is selected, the resulting normal equations are called autocorrelation equations since the product matrix $(X_2^H X_2)/N$ reduces exactly to the Yule–Walker equations, for which the biased autocorrelation estimator (2.16) has been used in lieu of the known autocorrelation function. Note that a data window has been assumed for this case. This data window reduces the resolution of AR spectra estimated with data matrix X_2, as will be illustrated in Section III. If the data matrix X_3 is selected, the normal equations are termed the prewindowed normal equations due to the zero value assumptions made for the missing data prior to x_0. If the data matrix X_4 is selected, the normal equations are termed the postwindowed normal equations since a zero data assumption is made for the data beyond x_{N-1}.

It would appear that only the data matrix X_2 will yield normal equations with Toeplitz structure to permit an efficient recursive solution (namely, the Levinson recursion); as outlined in Fig. 6. However, even though the product matrix $(X_i^H X_i)$ may not be Toeplitz, each of the four data matrices X_i have Toeplitz structure. This property allows one to develop recursive algorithms with $o(p^2)$ operations in each of the four cases. Morf *et al.* have provided the details of the recursive algorithms for the covariance case [163], [164], [167], [169] and for the prewindowed case [66], [168]. The interested reader may consult these references for details of these algorithms. They are not discussed in detail here because the forward and backward prediction approaches, to be discussed next, yield better spectral estimates in most cases.

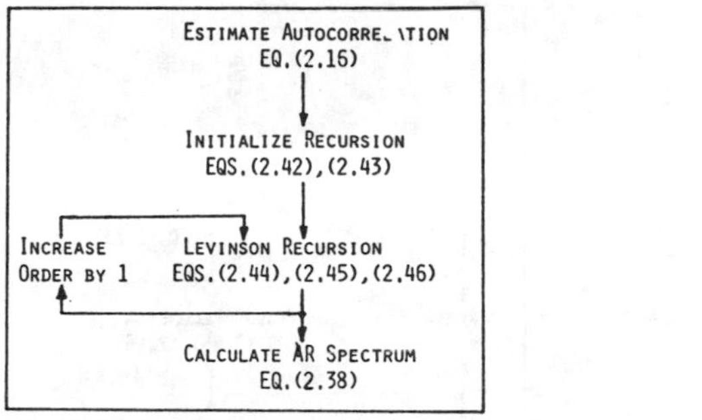

Fig. 6. Summary of Levinson recursion algorithm for AR spectral estimation.

Many problems with the forward only prediction approach to AR spectral estimation exist. The autocorrelation matrix $(X_2^H X_2)$ solution yields AR spectra with the least resolution among these four AR least squares estimates when the data sequence is short. The decrease in resolution is due to the inherent windowing in the data matrix X_2. The covariance matrix $(X_1^H X_1)$ solution produces AR parameters whose resulting spectra have been observed by several authors [227], [228], [251] to have more false peaks and greater perturbations of spectral peaks from their correct frequency locations than other AR estimation approaches. The covariance normal equations, which are also used in the Prony method (discussed in Section II-J), lead to AR parameter estimates with greater sensitivity to noise. Spectral line splitting, the placement of two or more closely spaced peaks in the spectrum where only one should be present, has been observed in all four forward prediction cases considered here. The reasons for line splitting have been documented by Kay and Marple [120].

If the process is wide sense stationary, the coefficients of the optimum backward prediction error filter are identical to the coefficients of the optimum forward prediction error filter, but conjugated and reversed in time [41]. The use of the backward prediction errors when estimating the AR parameters was introduced by Burg [37], [41].

The most popular approach for AR parameter estimation with N data samples was introduced by Burg in 1967. The Burg algorithm, separate and distinct from the maximum entropy viewpoint discussed earlier, may be viewed as a constrained least squares minimization. Assuming a wide sense stationary process, the forward linear prediction error of (2.63) is defined for $p \leqslant n \leqslant N-1$ (p is the predictor order) and the backward linear prediction error is given by

$$b_{pn} = \sum_{k=0}^{p} a_{pk}^* x_{n-p+k} \tag{2.70}$$

also for $p \leqslant n \leqslant N-1$. Recall that a_{po} is defined as unity, the a_{pk} are the predictor parameters at order p, the x_n are the data samples, and the nonwindowed prediction error range has been assumed.

To obtain estimates of the predictor (or AR) parameters, Burg minimized the sum of the forward and backward prediction error energies,

$$\mathcal{E}_p = \sum_{n=p}^{N-1} |e_{pn}|^2 + \sum_{n=p}^{N-1} |b_{pn}|^2 \tag{2.71}$$

subject to the constraint that the AR parameters satisfy the

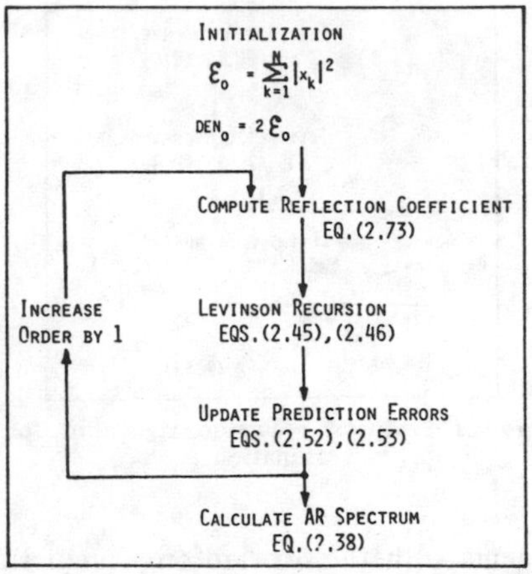

Fig. 7. Summary of Burg algorithm for AR spectral estimation.

Levinson recursion

$$a_{pk} = a_{p-1,k} + a_{pp}a^*_{p-1,p-k} \tag{2.72}$$

for all orders from 1 to p. This constraint was motivated by Burg's desire to ensure a stable AR filter (poles within the unit circle). By substituting the lattice recursion expressions (2.52) and (2.54) into (2.71), $\mathcal{E}_p$ becomes a function of only the unknown reflection coefficient a_{pp} and the prediction errors at order $p-1$, which are assumed known. Thus one need only estimate a_{ii} for $i = 1, 2, \cdots, p$ if (2.72) is used. Setting the derivative of $\mathcal{E}_i$ with respect to a_{ii} to zero then yields

$$a_{ii} = \hat{K}_i = \frac{-2\sum_{k=i}^{N-1} b^*_{i-1,k-1}e_{i-1,k}}{\sum_{k=i}^{N-1}(|b_{i-1,k-1}|^2 + |e_{i-1,k}|^2)}. \tag{2.73}$$

Note that $|a_{ii}| \leqslant 1$, which may be easily shown using (2.73). Thus (2.72) and (2.73) together will guarantee a stable all-pole filter. A recursive relationship for the denominator of (2.73) was found by Anderson [10]

$$\begin{aligned} \text{DEN}(i) &= \sum_{k=i}^{N-1}(|b_{i-1,k-1}|^2 + |e_{i-1,k}|^2) \\ &= \text{DEN}(i-1)\,[1 - |a_{i-1,i-1}|^2] \\ &\quad - |b_{i-1,N-i}|^2 - |e_{i-1,i}|^2. \end{aligned} \tag{2.74}$$

Fig. 7 is an outline of the Burg procedure for AR spectral estimation. A computational complexity analysis indicates that $3Np - p^2 - 2N - p$ complex adds, $3Np - p^2 - N + 3p$ complex multiplications, and p real divisions are required. Storage of $3N + p + 2$ complex values is also required [9], [57], [91], [92]. Multichannel versions of Burg's algorithm may be found in [166], [177], [178], [180], [210], [225].

The Burg algorithm has several problems associated with it, including spectral line splitting and biases in the frequency estimate. If one minimizes $\mathcal{E}_p$ in (2.71) with respect to all the a_{pk} for $k = 1, \cdots, p$, then these problems can be mitigated. Ulrych and Clayton [251] and Nuttall [176] independently suggested this least squares procedure for forward and backward prediction in which the Levinson recursion constraint imposed by Burg is removed.

To obtain the p normal equations for the LS (also called forward–backward) algorithm, determine the minimum of $\mathcal{E}_p$ by setting the derivatives of $\mathcal{E}_p$ with respect to all the AR parameters a_{p1} through a_{pp} to zero. This yields

$$\frac{\partial \mathcal{E}_p}{\partial a_{pi}} = 2\sum_{j=0}^{p} a_{pj} r_p(i,j) = 0 \tag{2.75}$$

for $i = 1, \cdots, p$, where $a_{po} = 1$ by definition and

$$r_p(i,j) = \sum_{k=0}^{N-p-1} (x_{k+p-j}x^*_{k+p-i} + x_{k+i}x^*_{k+j}) \tag{2.76}$$

for $0 \leqslant i, j \leqslant p$. The minimum prediction error energy may be determined to be

$$\mathcal{E}_p = \sum_{j=0}^{p} a_{pj} r_p(0,j). \tag{2.77}$$

Expressions (2.75) and (2.77) can be combined into a single $(p+1)$ by $(p+1)$ matrix expression

$$R_p A_p = E_p \tag{2.78}$$

where

$$A_p = \begin{bmatrix} 1 \\ a_{p1} \\ \vdots \\ a_{pp} \end{bmatrix}, E_p = \begin{bmatrix} \mathcal{E}_p \\ 0 \\ \vdots \\ 0 \end{bmatrix}, R_p = \begin{bmatrix} r_p(0,0) & \cdots & r_p(0,p) \\ \vdots & & \vdots \\ r_p(p,0) & \cdots & r_p(p,p) \end{bmatrix}. \tag{2.79}$$

Direct solution of (2.78) by Gaussian elimination requires a computational complexity proportional to $o(p^3)$. Barrodale and Erickson [16] discuss such a solution and its numerical stability. However, expression (2.79) has a structure that can be exploited to generate an algorithm of $o(p^2)$ operations. Basically, R_p can be expressed as sums and products of Toeplitz and Hankel matrices. This structure enabled a recursive algorithm of $o(p^2)$ to be developed (see Marple [154]). A flowchart of the algorithm may be found in the same reference and is not provided here due to its complexity. The LS algorithm is almost as computationally efficient as the Burg algorithm, requiring typically about 20 percent more computations. The improvement obtained from the LS approach over the Burg algorithm is well worth the slight additional computation. These improvements include less bias in the frequency estimate of spectral components, reduced variance in frequency estimation, and absence of observed spectral line splitting [141], [154], [227].

Sequential Estimation of AR Parameters: Three major approaches to time updating the AR parameter estimates on a data sample by data sample basis are available. These are the recursive least squares method, the gradient adaptive approach, and sequential identification using a lattice filter.

The recursive least squares method [14], [20], [211], [220] in appearance resembles a Kalman filtering procedure. By eliminating the $\mathcal{E}_p$ term from the normal equations (2.69), the least squares solution for the AR parameter estimate vector $\hat{A}_m$ for order p is

$$\hat{A}_m = [\mathcal{X}^H_m \mathcal{X}_m]^{-1} \mathcal{X}^H_m Y_m \tag{2.80}$$

where an m subscript has been added to all the vectors and matrices to indicate that the time samples up to index m are included. The vector Y_m is composed of data samples,

$$Y_m = \begin{bmatrix} x_0 \\ \vdots \\ x_m \end{bmatrix}$$

and $\mathcal{X}_m$ is a modified version of data matrices X_1 or X_3,

$$\mathcal{X}_m = \begin{bmatrix} x_0 & & \bigcirc \\ \vdots & \ddots & \\ \vdots & & x_0 \\ \vdots & & \vdots \\ x_{m-1} & \cdots & x_{m-p} \end{bmatrix} \quad \text{or} \quad \begin{bmatrix} x_{p-1} & \cdots & x_0 \\ \vdots & & \vdots \\ \vdots & & \vdots \\ x_{m-1} & \cdots & x_{m-p} \end{bmatrix}$$

corresponding to the covariance or prewindowed cases discussed under batch processing methods.

The addition of a new time sample x_{m+1} can be accounted for by partitioning Y_{m+1} and $\mathcal{X}_{m+1}$ as follows:

$$Y_{m+1} = \begin{bmatrix} Y_m \\ \hline x_{m+1} \end{bmatrix}, \quad {}_{m+1} = \begin{bmatrix} \mathcal{X}_m \\ \hline H_{m+1} \end{bmatrix}$$

where $H_{m+1} = [x_m \cdots x_{m-p+1}]$. If we define $P_m = [\mathcal{X}_m^H \mathcal{X}_m]^{-1}$ and substitute the partitioned Y_{m+1} and $\mathcal{X}_{m+1}$, then

$$\begin{aligned} \hat{A}_{m+1} &= P_{m+1} \mathcal{X}_{m+1}^H Y_{m+1} \\ &= P_{m+1} [\mathcal{X}_m^H Y_m + H_{m+1}^H x_{m+1}] \\ &= P_{m+1} [P_m^{-1} P_m \mathcal{X}_m^H Y_m + H_{m+1}^H x_{m+1}]. \end{aligned} \tag{2.81}$$

Noting that $P_m \mathcal{X}_m^H Y_m = \hat{A}_m$ and $P_m^{-1} = P_{m+1}^{-1} - H_{m+1}^H H_{m+1}$, then

$$\begin{aligned} \hat{A}_{m+1} &= P_{m+1} [(P_{m+1}^{-1} - H_{m+1}^H H_{m+1}) \hat{A}_m + H_{m+1}^H x_{m+1}] \\ &= \hat{A}_m + P_{m+1} H_{m+1}^H (x_{m+1} - H_{m+1} \hat{A}_m). \end{aligned} \tag{2.82}$$

Using the matrix inversion lemma, $(A + BCD)^{-1} = A^{-1} - A^{-1}B (C^{-1} + DA^{-1}B)^{-1} DA^{-1}$, then an alternative recursive formulation for P_{m+1} is

$$\begin{aligned} P_{m+1} &= (P_m^{-1} + H_{m+1}^H H_{m+1})^{-1} \\ &= P_m - P_m H_{m+1}^H (1 + H_{m+1} P_m H_{m+1}^H)^{-1} H_{m+1} P_m \\ &= (I - K_{m+1} H_{m+1}) P_m \end{aligned} \tag{2.83}$$

where by definition,

$$K_{m+1} = P_m H_{m+1}^H (1 + H_{m+1} P_m H_{m+1}^H)^{-1}. \tag{2.84}$$

The sequential recursion (2.82) then reduces to

$$\begin{aligned} \hat{A}_{m+1} &= \hat{A}_m + K_{m+1} (x_{m+1} - H_{m+1} \hat{A}_m) \\ &= \hat{A}_m + K_{m+1} e_{p,m+1}. \end{aligned} \tag{2.85}$$

Equations (2.13)–(2.85) are similar in structure to a Kalman filter in which the data vector H_m and data matrix P_m are similar to the covariance vector and matrix assumed available in the Kalman formulation. In fact, the least squares formulation presented here could be modified to incorporate *a priori* knowledge of any statistics available concerning the linear prediction error noise statistics.

The recursive least squares technique can be applied to other parameter estimation problems other than this AR application, of on-line spectral estimation, as long as the model is linear in the parameters [170]. In order to start the recursion, $\hat{A}_0$ and P_0 must be specified. If P_0^{-1} is other than an all zero matrix, then the selection of $\hat{A}_0$ must be made carefully since the $\hat{A}_m$ estimate will be biased toward $\hat{A}_0$. This bias can be removed by setting P_0^{-1} to an all zero matrix, but this requires an alternative formulation of (2.84) and (2.85) in terms of P_0^{-1} rather than P_0.

The update relationships in the least squares recursion require $o(p^3)$ operations with each new data point, which may be formidable. An alternative approach that requires $o(p)$ operations for each sequential update is the adaptive linear prediction filter approach [8], [83], [85], [243]. The adaptive approach recursively estimates the AR parameter vector using a gradient technique

$$\hat{A}_{m+1} = \hat{A}_m - \mu \nabla E(|e_{pm}|^2) \tag{2.86}$$

where μ is the step size and ∇ denotes the gradient. By substituting

$$e_{pm} = x_m + \sum_{k=1}^{p} a_{pk} x_{m-k}$$

and taking expectations, then

$$E(|e_{pm}|^2) = R_{xx}(0) + \hat{A}_m^H R_{xx} \hat{A}_m + 2 \operatorname{Re} (\hat{A}_m^H r_{xx})$$

where R_{xx} is the autocorrelation matrix and r_{xx} is the autocorrelation vector $r_{xx} = (R_{xx}(1) \cdots R_{xx}(p))^T$. The gradient of $E(|e_{pm}|^2)$ is

$$\nabla E(|e_{pm}|^2) = 2 r_{xx} + 2 R_{xx} \hat{A}_m. \tag{2.87}$$

When the gradient technique converges, $\nabla E(|e_{pm}|^2) = 0$ and

$$r_{xx} = -R_{xx} A$$

which is the same as the Yule–Walker equations (2.40). In practice, r_{xx} and R_{xx} are unavailable, so instantaneous estimates are substituted. This yields

$$r_{xx} \to x_m X_{m-1}^* \qquad \text{where} \quad X_{m-1}^* = \begin{bmatrix} x_{m-1}^* \\ \vdots \\ x_{m-p}^* \end{bmatrix}$$

$$R_{xx} \to X_{m-1}^* X_{m-1}^T.$$

Noting that $e_{pm} = x_m + X_{m-1}^T \hat{A}_m$ and using (2.87), then (2.86) becomes

$$\hat{A}_{m+1} = \hat{A}_m - 2\mu e_{pm} X_{m-1}^*. \tag{2.88}$$

The above gradient formulation is called the least mean square (LMS) algorithm. It has a similar structure to (2.85), except μ is fixed whereas the gain K_{m+1} is variable.

Convergence is guaranteed as long as the constant μ is selected between 0 and $1/\lambda_{\max}$, where $\lambda_{\max}$ is the maximum eigenvalue of R_{xx}. The choice of μ involves a tradeoff between rate of convergence of $E[\hat{A}_m]$ to A_m and the amount of steady-state variance (sometimes termed misadjustment) to be tolerated once convergence is achieved. Thus the price paid for a much reduced computational burden over the recursive least squares approach is a slow convergence requiring the need for a longer data record to achieve reliable AR parameter estimates.

In an attempt to achieve the accuracy of the recursive least squares technique and the computational savings of the adaptive gradient techniques, a third method has been advocated based on the lattice filter. However, the lattice recursive relationships update only the reflection coefficients with $o(p)$ operations. If the AR parameter estimates are to be updated with each new data sample, then the Levinson recursion (2.45) will need to be used, which requires $o(p^2)$ operations. Thus computational savings with the lattice technique are achievable only if the AR parameter estimates are updated infrequently rather than with each new sample.

A sequential algorithm based on a lattice structure may be developed for each of the least squares techniques presented under the batch estimation topic [164], [168]. An illustration of one such algorithm based on Burg's algorithm is given here. The reflection coefficient K_{mn} for order m and time index n was determined in the Burg algorithm by computing

$$K_{mn} = \frac{-2 \sum_{i=m}^{n} e_{m-1,i} b^*_{m-1,i-1}}{\sum_{i=m}^{n} (|b_{m-1,i-1}|^2 + |e_{m-1,i}|^2)}. \tag{2.89}$$

A time update recursive formulation for (2.89) is given by [222]

$$K_{m+1,n+1} = K_{m+1,n} - \frac{[K_{m+1,n}(|e_{mn}|^2 + |b_{m,n-1}|^2) + 2e_{mn}b^*_{m,n-1}]}{\sum_{i=m}^{n} [|e_{mi}|^2 + |b_{m,i-1}|^2]}. \tag{2.90}$$

Thus, (2.90) in combination with (2.52) and (2.54) for $m = 1, \cdots, p$ and with initial conditions $e_{0n} = b_{0n} = x_n$, form a sequential time-update algorithm for the reflection coefficients.

AR Spectral Power Estimation: It has been shown [136] that, unlike conventional Fourier spectral estimates, the peak amplitudes in AR spectral estimates are not linearly proportional to the power when the input process consists of sinusoids in noise. Lacoss [136] has shown that for high SNR, the peak is proportional to the square of the power, although the area under the peak is proportional to power. One method for obtaining an estimate of the actual power of real sinusoids from an AR spectrum was suggested by Andersen and Johnsen [108]. The method works best for high SNR components in the process. The AR PSD in z-transform notation is

$$P_{AR}(z) = \frac{\sigma^2 \Delta t}{A(z)A^*(1/z^*)} \tag{2.91}$$

where

$$A(z) = 1 + \sum_{k=1}^{p} a_{pk} z^{-k}.$$

If a peak is at f_i in the AR spectral estimate, i.e., $z_i = \exp(j2\pi f_i \Delta t)$, then the estimated power is approximately

$$\text{Power}(f_i) = 2 \times \text{Real}\left\{\text{Residue of } \frac{P_{AR}(z)}{z} \text{ at } z_i\right\} \tag{2.92}$$

where

$$\text{Residue } \frac{P_{AR}(z)}{z} = (z - z_i) \left| \frac{P_{AR}(z_i)}{z} \right|_{z=z_i}$$

$$z_i = \text{Root of } A(z) = \exp(\alpha_i + j2\pi f_i)\Delta t.$$

Note that it is assumed that the peak occurs at the angular location of the pole z_i. Negative power estimates can occur with this technique if peaks are very close together. This approach is closely related to the power estimation procedure utilized in the Prony method presented in Section II-J.

Model Order Selection [21], [100], [110], [114], [133], [235], [237], [238]: Since the best choice of filter order p is not generally known *a priori*, it is usually necessary in practice to postulate several model orders. Based on these, one then computes some error criterion that indicates which model order to choose. Too low a guess for model order results in a highly smoothed spectral estimate. Too high an order introduces spurious detail into the spectrum. One intuitive approach would be to construct AR models of increasing order until the computed prediction error power reaches a minimum. However, all the least squares estimation procedures discussed in this paper have prediction error powers that decrease monotonically with increasing order p. For example, the Burg algorithm and Yule–Walker equations involve the relationship

$$\mathcal{E}_i = \mathcal{E}_{i-1}[1 - |a_{ii}|^2]. \tag{2.93}$$

As long as $|a_{ii}|^2$ is nonzero (it must be ≤ 1), the prediction error power decreases. Thus the prediction error power alone is not sufficient to indicate when to terminate the search.

Several criteria have been introduced as objective bases for selection of the AR model order. Akaike [1]–[7] has provided two criteria. His first criterion is the final prediction error (FPE). This criterion selects the order of the AR process so that the average error for a one step prediction is minimized. He considers the error to be the sum of the power in the unpredictable (or innovation) part of the process and a quantity representing the inaccuracies in estimating the AR parameters. The FPE for an AR process is defined as

$$\text{FPE}_p = \mathcal{E}_p \left(\frac{N + p + 1}{N - p - 1}\right) \tag{2.94}$$

where N is the number of data samples. Note that (2.94) assumes one has subtracted the sample mean from the data. The term in parentheses increases the FPE as p approaches N, reflecting the increase in the uncertainty of the estimate $\mathcal{E}_p$ of the prediction error power. The order p selected is the one for which the FPE is minimum. The FPE has been studied for application by Gersch and Sharpe [73], Jones [111], Fryer *et al.* [69], and Ulrych and Bishop [250]. For AR processes, the FPE works fairly well. However, when processing actual geophysical data, both Jones [111] and Berryman [21] found the order selected tended to be too low.

Akaike suggested a second-order selection criterion using a maximum likelihood approach to derive a criterion termed the Akaike information criterion (AIC). The AIC determines the model order by minimizing an information theoretic function. Assuming the process has Gaussian statistics, the AIC is

$$\text{AIC}_p = \ln(\mathcal{E}_p) + 2(p + 1)/N. \tag{2.95}$$

The term $(p + 1)$ in (2.95) is sometimes replaced by p, since $2/N$ is only an additive constant which accounts for the subtraction of the sample mean. The second term in (2.95) represents the penalty for the use of extra AR coefficients that do not result in a substantial reduction in the prediction error power. Again, the order p selected is the one that minimizes the AIC. As $N \to \infty$, the AIC and FPE are equivalent. Kashyap [115] claims the AIC is statistically inconsistent in that the probability of error in choosing the correct order does not tend to zero as $N \to \infty$.

A third method was proposed by Parzen [192] and is termed the criterion autoregressive transfer (CAT) function. The order p is selected to be the one in which the estimate of the difference of the mean-square errors between the true prediction error filter (which may be of infinite length) and the estimated filter is a minimum. Parzen showed that this difference can be calculated, without explicitly knowing the true prediction error filter, by

$$\mathrm{CAT}_p = \left(\frac{1}{N}\sum_{j=1}^{p}\frac{1}{\hat{\mathcal{E}}_j}\right) - \frac{1}{\hat{\mathcal{E}}_p} \tag{2.96}$$

where $\hat{\mathcal{E}}_j = (N/(N-j))\mathcal{E}_j$. Again p is chosen to minimize CAT_p.

The results of spectra using the FPE, AIC, and CAT have been mixed, particularly against actual data rather than simulated AR processes. Ulrych and Clayton [251] have found that for short data segments, none of the criteria work well. For harmonic processes in noise, the FPE and AIC also tend to underestimate the order if the SNR is high [92], [138]. Ulrych and Ooe [92] suggest in the case of short data segments that an order selection between $N/3$ to $N/2$ often produces satisfactory results. In the final analysis, more subjective judgment is still required in the selection of order for data from actual processes than that required for controlled simulated computer processes.

Anomalies of and Patches for the AR Spectral Estimator: Several anomalies of the AR spectral estimator have been observed by researchers. When the model order is chosen to be too large relative to the number of data points, the AR spectral estimate exhibits spurious peaks [212], [250]. Ideally, if the autocorrelation lags, or equivalently, the reflection coefficents, were estimated without error, then the estimated AR parameters for an AR(p) model would be

$$\hat{a}_{pi} = \begin{cases} a_{pi}, & \text{for } i = 1, 2, \cdots, p \\ 0, & \text{for } i = p+1, \cdots, n \end{cases} \tag{2.97}$$

where a_{pi} are the AR(p) parameters. However, when estimation errors are present, then $a_{pi} \neq 0$ in general for $i > p$. Correspondingly, there will be n-p "extra" poles. When the estimated extra poles occur near the unit circle, spurious spectral peaks result. It is this possibility of spurious peaks that is the basis of the recommendation that the maximum model order should be no greater than $N/2$, where N is the data record length [251].

It has been observed for a process consisting of a sinusoid in noise that the peak location in the AR spectral estimate depends critically on the phase of the sinusoid [45], [229]. Also, it has been observed that the spectral estimate sometimes exhibits two closely spaced peaks, falsely indicating a second sinusoid. The latter phenomenon is known as spectral line splitting (SLS) [64].

The phase dependence of the AR spectral estimate decreases as the data record length increases. The amount of phase dependence varies for the dfiferent AR estimation procedures. For the Burg algorithm, the shift in peak location can be as much as 16 percent [230]. The forward–backward prediction error approach is least dependent on phase [154], [251]. Two techniques have been proposed to reduce this effect. In the first approach, the phase dependence is attributed to the interaction between the positive and negative frequency components of the real sinusoid, much in the same way as the peak of the periodogram depends upon phase [119], [231], [234]. Based on this premise, the solution is then to replace the real valued signal by the analytic signal. The analytic signal process is then down sampled by two and the AR spectral estimate for complex data used. The model order for complex data need only be half as large as for real data since the complex conjugate pole pairs in the real case are not required in the complex approach. Using this approach, the phase dependence of the Burg spectral estimate can be decreased [119].

The other alternative procedure (for the Burg spectral estimate) is to employ the estimator

$$\hat{K}_i = -2\sum_{n=i}^{N-1} v_n e_{i-1,n} b^*_{i-1,n-1} \Bigg/ \sum_{n=i}^{N-1} v_n \left[|e_{i-1,n}|^2 + |b_{i-1,n-1}|^2\right] \tag{2.98}$$

which weights the reflection coefficient terms with the real sequence $\{v_n\}$. This windowing of the residual time series, suggested by Swingler [228], has the effect of reducing the end effects of the short data record. Simulations indicate the phase dependence is also reduced by this method.

The problem of spectral line splitting in AR spectra produced by the Burg algorithm was first documented by Fougere *et al.* [64]. They noted that spectral line splitting was most likely to occur when 1) the SNR is high, 2) the initial phase of sinusoidal components is some odd multiple of 45°, 3) the time duration of the data sequence is such that sinusoidal components have an odd number of quarter cycles, and 4) the number of AR parameters estimated is a large percentage of the number of data values used for the estimation. Many spurious spectral peaks often accompany spectra that exhibit line splitting. Again the phenomenon is associated with short data records since it tends to disappear as the data record increases. SLS has been observed in the Burg and Yule–Walker spectral estimates for multiple sinusoids in white noise [120]. A solution to the problem has been proposed by Fougere [65]. He attributes the splitting to the fact that the prediction error power is not truly minimized using Burg's estimate for the reflection coefficients. His technique minimizes the prediction error power by varying all reflection coefficients simultaneously. From his simulation, the technique appears to eliminate SLS for at least one sinusoid. Another method of eliminating SLS for one sinusoid is to use the analytic signal approach [94], [120]. The analytic signal approach yields the true reflection coefficients for the Burg spectral estimate when noise is negligible, the condition where SLS is most likely to be observed. For the Yule–Walker spectral estimate, if the analytic signal approach and the unbiased autocorrelation estimate are used, the true autocorrelation function is obtained. Thus, in either case, SLS is eliminated. For multiple sinusoids, the performance of Fougere's algorithm is undocumented and the use of complex data in conjunction with Burg's reflection coefficient estimate can still exhibit SLS [94]. Using the forward–backward LS approach in conjunction with the recursive algorithm of Marple, SLS has not been observed [154].

Besides phase, signals with large dc levels or a linear trend have also been found to corrupt AR spectra [113], particularly the low-frequency end of the spectral estimate. These components should be removed before applying AR spectral analysis techniques.

A very important problem with the AR spectral estimator, which limits its utility, is its sensitivity to the addition of observation noise to the time series [186]. An example is given in Fig. 8. It is seen that the spectral peaks are broadened and dis-

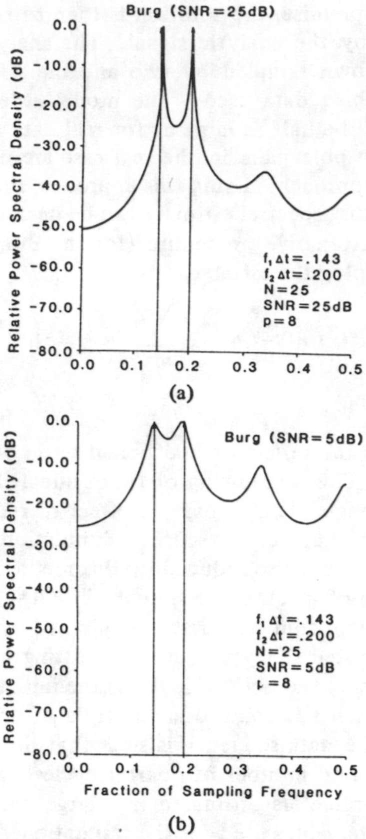

Fig. 8. Spectral estimates for two sinusoids in white Gaussian noise. (a) High SNR. (b) Low SNR.

placed from their true positions (indicated by arrows). In particular, it has been shown that the resolution of the AR spectral estimate for two equiamplitude sinusoids in white noise decreases as the SNR decreases [149], [152]. For low SNR, the resolution is no better than that of the periodogram. The reason for the degradation is that the all-pole model assumed in AR spectral analysis is no longer valid when observation noise is present. To see this, assume y_n denotes the noise corrupted AR process, x_n. Thus

$$y_n = x_n + w_n \tag{2.99}$$

where w_n is the observation noise. If w_n is white noise with variance σ_w^2 and is uncorrelated with x_n,

$$\begin{aligned} P_y(z) &= \frac{\sigma^2 \Delta t}{A(z)A^*(1/z^*)} + \sigma_w^2 \Delta t \\ &= \frac{[\sigma^2 + \sigma_w^2 A(z)A^*(1/z^*)]\,\Delta t}{A(z)A^*(1/z^*)}. \end{aligned} \tag{2.100}$$

Thus the PSD of y_n is characterized by poles and zeros, i.e., y_n is an ARMA (p, p) process. The inconsistency of the AR model for a noise corrupted AR process leads to the degradation observed in Fig. 8 [121]. The phenomenon is explained as follows. The effect of noise is to reduce the dynamic range of the PSD of x_n. Since the prediction error filter $\hat{A}(z)$ attempts to whiten the PSD, it is not surprising that for low SNR, the zeros of $\hat{A}(z)$ are located near the origin of the z plane, i.e., $\hat{A}(z) \approx 1$. This is because the PSD of y_n is already relatively flat due to noise so that subsequent filtering operations, i.e., the use of a prediction error filter, will not significantly whiten the PSD further.

To reduce the degradation of the AR spectral estimate in the presence of noise, four general approaches have been proposed. One can

1) use the ARMA spectral estimate;
2) filter the data to reduce the noise;
3) use a large order AR model;
4) compensate either the autocorrelation function estimates or the reflection coefficient estimates for the noise effects.

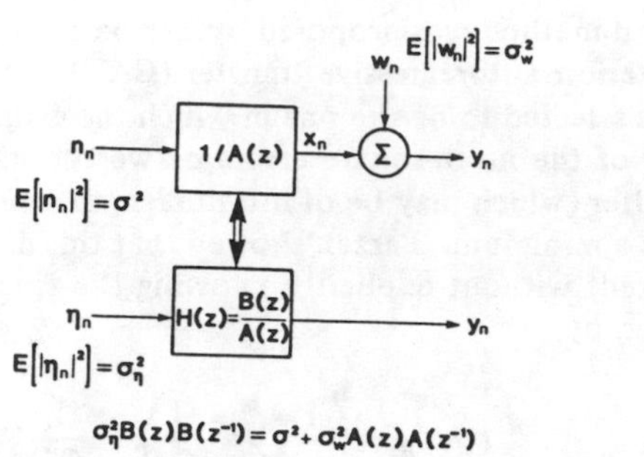

Fig. 9. ARMA model for AR process in white noise.

The ARMA approach assumes that the noise corrupted AR process is a general ARMA (p, p) process even though the AR and MA parameters are related by (2.100). The most common approach has been the use of the modified Yule–Walker equations as described in Section II-F. [58], [72]. For an ARMA (p, p) process, this means solving the set of equations

$$R_{yy}(k) = -\sum_{l=1}^{p} a_l R_{yy}(k-l) \tag{2.101}$$

for $k = p+1, p+2, \cdots, 2p$ in order to obtain the AR parameters. Although simple to implement, this approach has met with only moderate success. Reasonable results are obtained for long data records and/or high SNR's. A more suitable solution to the noise problem is to use the maximum likelihood ARMA estimate. However, this procedure leads to a set of highly nonlinear equations [31]. In the case for which the maximum likelihood equations are determined specifically for an AR process in white noise, a suboptimal solution to these equations leads to an iterative filtering scheme as described in [143]. Other ARMA filtering schemes can be found in references [118], [151], [159], [172]. All the methods rely on a boot-strapping approach to design the filter since the power spectral density of x_n, which is what we are attempting to estimate, is unknown.

Another technique to combat noise is to employ an AR model with a model order larger than the true AR model. This is because an ARMA(p, p) process is equivalent to an AR(∞) model, as guaranteed by the Wold decomposition. Using (2.100), let

$$\sigma^2 + \sigma_w^2 A(z)A^*(1/z^*) = \sigma_\eta^2 B(z)B^*(1/z^*)$$

where

$$B(z) = 1 + \sum_{k=1}^{p} b_{pk} z^{-k}$$

so that y_n can be represented as the output of a pole-zero filter, $H(z)$, driven by white noise (with variance σ_η^2) as shown in Fig. 9. If we divide $A(z)$ by $B(z)$, we have

$$H(z) = 1/(A(z)/B(z)) = 1/C(z)$$

where

$$C(z) = \frac{A(z)}{B(z)} = 1 + \sum_{k=1}^{\infty} c_k z^{-k}.$$

Thus, y_n can be modeled by an AR(∞) process with parameters $\{c_k\}$. Clearly, as the assumed AR model order increases, the estimated AR PSD will approach the true PSD of y_n. This property is also evident from the maximum entropy formulation since it is shown there that

$$\hat{R}_{yy}(k) = R_{yy}(k), \qquad \text{for } |k| \leqslant p$$

where $\hat{R}_{yy}(k)$ is the autocorrelation function corresponding to the AR(p) model and $R_{yy}(k)$ is the true autocorrelation function of y_n. Thus, as $p \to \infty$, the autocorrelation function of the model matches the true autocorrelation function. Hence, the spectra must also match each other.

It would seem that a model order as large as possible should be used. However, due to the spurious peak problem, one should limit the maximum model order to no more than one half the number of data points, as discussed previously.

In practice a larger order model will be needed when the zeros of $B(z)$ are near the unit circle of the z plane. In this case, the c_k sequence will die out slowly. Since the zeros $B(z)$ move outward as the SNR decreases [121], increasing the model order will be necessary as the SNR decreases. To quantify the effect of model order on the AR spectral estimate for an AR process in white noise, consider two equiamplitude sinusoids in white noise. It has been shown [149] for this case that the resolution δf in hertz of the AR spectral estimate, assuming a known autocorrelation function, is approximately

$$\delta f \approx \frac{[\rho(p+1)]^{0.31}}{6.471 \times 2\pi p \Delta t} \tag{2.102}$$

where ρ is the SNR. As expected, the resolution increases with increasing model order. An example of this behavior is shown in Fig. 10. Note that the extra poles are approximately uniformly spaced within the unit circle, producing an equiripple approximation to the flat noise spectrum.

Many noise cancellation schemes that compensate the autocorrelation lags for the noise are available [123], [151], [159], [214], [240], [268]. Details may be found in these references. The PHD is a special case of these schemes. In general, these noise cancellations schemes can reduce the bias, but will increase the variance of the spectral estimate. A serious deficiency is that, in general, one does not know how much noise power to remove. Thus, if σ_w^2 is too large, the estimated AR spectrum will exhibit sharper peaks than the true spectrum. Thus one must be careful in applying these techniques.

F. Moving Average PSD Estimation

As presented in the introduction to Section II, a MA process is a stochastic process obtained from the output of a filter whose transfer function contains only zeros, and whose input is a white noise process, i.e.,

$$x_n = \sum_{m=0}^{q} b_m n_{n-m} \tag{2.103}$$

with

$$E[n_n] = 0 \qquad E[n_{n+m} n_n^*] = \sigma^2 \delta_m$$

where δ_m is 1 for $m = 0$, and 0 otherwise. Based on (2.103),

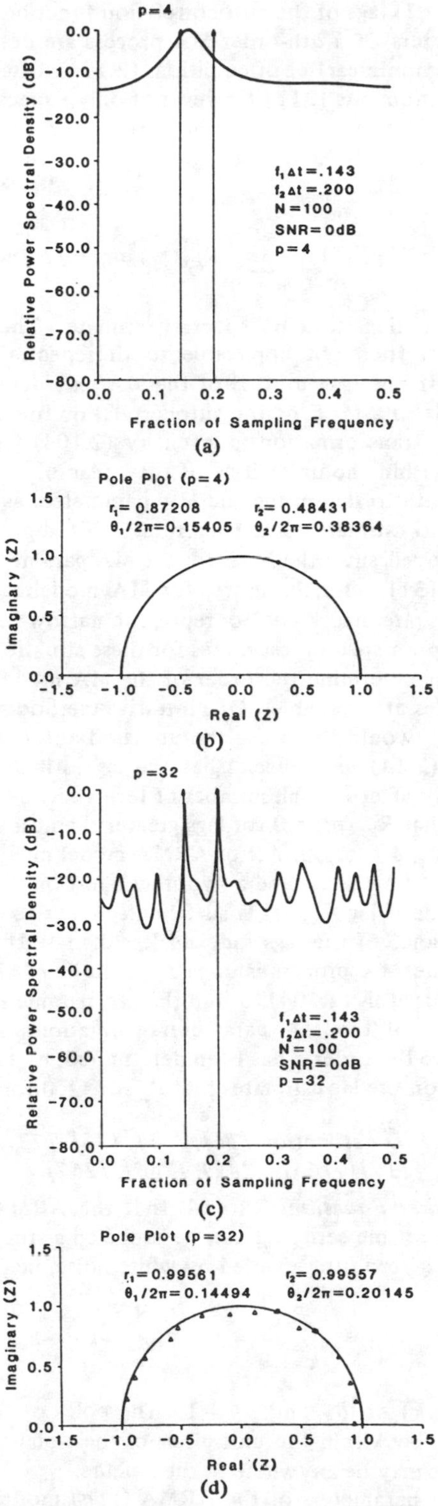

Fig. 10. Burg spectral estimate for two sinusoids in white noise. Effect of model order. (a) $p = 4$. (b) Pole plot for $p = 4$. (c) $p = 32$. (d) Pole plot for $p = 32$.

the autocorrelation function of a MA process of order q is

$$R_{xx}(k) = \begin{cases} \sigma^2 \sum_{i=0}^{q-k} b_i^* b_{i+k}, & \text{for } k = 0, 1, \cdots, q. \\ 0, & \text{for } k > q \end{cases} \tag{2.104}$$

Thus, if $(q+1)$ lags of the autocorrelation function are known, the parameters of a qth-order MA process are determined by solving the nonlinear set of equations (2.104), often called the method of moments [31]. However, if only a spectral estimate is desired, then there is no need to solve for the MA parameters, but only to determine the autocorrelation function, since

$$\mathcal{P}_{MA}(f) = \sum_{m=-q}^{q} R_{xx}(m) \exp(-j2\pi fm\Delta t)$$

which is identical to a BT spectral estimate. The method of moments is then not applicable to the spectral estimation problem. If one uses a MLE of the MA parameters, this also corresponds to a MLE of the autocorrelation function, since a one-to-one transformation is given by (2.104) (assuming the zeros are within the unit circle of the z plane). In this case, it is appropriate to determine the MA parameters as an intermediate step to estimating the spectrum. This approach has not been employed since the MLE for the MA parameters is highly nonlinear [31]. Furthermore, for MA modeling, too many coefficients are necessary to represent narrow-band spectra, leading to poor spectral estimates for these situations.

One must determine the order of the MA model when only data samples are available. One intuitive method suggested by Chow [48] would be to use the unbiased autocorrelation lag estimator (2.14) and check that the lag estimates approach zero rapidly after a small number of terms, since from (2.104) we know that $R_{xx}(m) = 0$ for lags greater than the order of the MA process. If not, an AR or ARMA model may be more appropriate. Chow suggested a hypothesis test on successive lags to determine if lag $R_{xx}(q)$ is sufficiently close to zero relative to the variance of the lags indexed less than q. If so, then the order of the MA process is considered to be q. The lag estimates are used in (2.104) to find the MA parameters. Further refinements of the MA parameter estimation can be made [236], once the order q has been determined, by enforcing the constraint on the lag estimates that $R_{xx}(m) = 0$ for $m > q$.

G. ARMA PSD Estimation [11], [31], [35], [72], [79], [86], [87], [89], [117], [165], [242], [266], [267]

Yule–Walker Equations: Recall that the ARMA model assumes that a time series x_n can be modeled as the output of a p pole and q zero filter excited by white noise, i.e.,

$$x_n = -\sum_{k=1}^{p} a_k x_{n-k} + \sum_{k=0}^{q} b_k n_{n-k} \tag{2.105}$$

where $R_{nn}(k) = \sigma^2 \delta_k$ and $b_0 \equiv 1$. The poles of the filter are assumed to be within the unit circle of the z-plane. The zeros of the filter may lie anywhere in the z-plane.

Once the parameters of the ARMA (p, q) model are identified, the spectral estimate is obtained as

$$\mathcal{P}_x(f) = |H(\exp[j2\pi f\Delta t])|^2 \, \mathcal{P}_n(f)$$

$$= \frac{\sigma^2 \Delta t \left|1 + \sum_{k=1}^{q} b_k \exp(-j2\pi fk\Delta t)\right|^2}{\left|1 + \sum_{k=1}^{p} a_k \exp(-j2\pi fk\Delta t)\right|^2} \tag{2.106}$$

The relationship of the ARMA parameters to the autocorrelation function is easily found as follows. Multiply (2.105) by x^*_{n-l} and take the expectation to yield

$$R_{xx}(l) = -\sum_{k=1}^{p} a_k R_{xx}(l-k) + \sum_{k=0}^{q} b_k R_{nx}(l-k) \tag{2.107}$$

where

$$R_{nx}(k) = E(n_n x^*_{n-k}).$$

But $R_{nx}(k) = 0$ for $k > 0$ since a future input to a causal, stable filter cannot affect the present output and n_n is white noise. Therefore,

$$R_{xx}(l) = \begin{cases} -\sum_{k=1}^{p} a_k R_{xx}(l-k) + \sum_{k=0}^{q} b_k R_{nx}(l-k), & \text{for } l = 0, \cdots, q \\ -\sum_{k=1}^{p} a_k R_{xx}(l-k), & \text{for } l = q+1, q+2, \cdots. \end{cases}$$

From the derivation of the Yule–Walker equations, it was shown

$$R_{nx}(k) = \sigma^2 h^*_{-k}$$

and therefore

$$R_{xx}(l) = \begin{cases} -\sum_{k=1}^{p} a_k R_{xx}(l-k) + \sigma^2 \sum_{k=l}^{q} b_k h^*_{k-l}, & \text{for } l = 0, 1, \cdots, q \\ -\sum_{k=1}^{p} a_k R_{xx}(l-k), & \text{for } l = q+1, q+2, \cdots. \end{cases} \tag{2.108}$$

These normal equations for an ARMA process are analogous to the Yule–Walker equations for an AR process.

Estimation of ARMA Parameters: Many ARMA parameter estimation techniques have been formulated theoretically, which usually involve many matrix computations and/or iterative optimization techniques. These approaches are normally not practical for real-time processing. Suboptimum techniques have therefore been developed to make the computational load more manageable. These techniques are usually based on a least squares error criterion and require solutions of linear equations. These methods generally estimate the AR and MA parameters separately rather than jointly as required for optimal parameter estimation. The AR parameters can be estimated independently of the MA parameters first if one uses the Yule–Walker equations as given by (2.108). A final point in favor of the suboptimal linear approaches is that iterative optimization techniques are not guaranteed to converge or may converge to the wrong solution. The nonlinearity of the equations encountered is typified by (2.108). Since the impulse response is a function of $a_1, \cdots, a_p, b_1, \cdots, b_q$, the equations given by (2.108) are nonlinear in the ARMA parameters. As an example, consider an ARMA (1, 1) process. In this case,

$$h_k = (-a_1)^k u(k) + b_1(-a_1)^{k-1} u(k-1)$$

$$R_{xx}(0) = -a_1 R_{xx}(-1) + \sigma^2(1 + |b_1|^2 - a_1^* b_1)$$

$$R_{xx}(1) = -a_1 R_{xx}(0) + \sigma^2 b_1$$

$$R_{xx}(l) = -a_1 R_{xx}(l-1), \quad \text{for } l \geq 2 \tag{2.109}$$

where $u(k)$ is the unit step function. Although numerous researchers have proposed means of solving these equations, there appear to be few successful applications of these approaches.

A more popular approach to this problem is to use (2.108) for $l > q$ to find $(a_1, a_2, \cdots, a_p)$ and then to apply some appropriate technique to find $(b_1, b_2, \cdots, b_q)$ or an equivalent parameter set. For example, to find the AR parameters, using (2.108) and $l = q + 1, q + 2, \cdots, q + p$, we solve the following matrix expression [72]:

$$\underbrace{\begin{bmatrix} R_{xx}(q) & R_{xx}(q-1) & \cdots & R_{xx}(q-p+1) \\ R_{xx}(q+1) & R_{xx}(q) & \cdots & R_{xx}(q-p+2) \\ \vdots & \vdots & & \vdots \\ R_{xx}(q+p-1) & R_{xx}(q+p-2) & \cdots & R_{xx}(q) \end{bmatrix}}_{R'_{xx}} \begin{bmatrix} a_1 \\ a_2 \\ \vdots \\ a_p \end{bmatrix} = - \begin{bmatrix} R_{xx}(q+1) \\ R_{xx}(q+2) \\ \vdots \\ R_{xx}(q+p) \end{bmatrix}. \tag{2.110}$$

These equations have been called the extended, or modified, Yule–Walker equations. The matrix is Toeplitz, although not symmetric, and is therefore not guaranteed to be either positive-definite or nonsingular. An algorithm requiring $o(p^2)$ operations has been developed by Zohar [278] for solving (2.110).

In order to choose an appropriate model order p for the AR portion of the ARMA model, the property [48]

$$|\boldsymbol{R}'_{xx}| = 0$$

for dimension of $\boldsymbol{R}'_{xx}$ greater than the AR order p can be used. Here $|\boldsymbol{R}'_{xx}|$ denotes the determinant of the matrix $\boldsymbol{R}'_{xx}$. This means that one need only monitor the determinant, $|\boldsymbol{R}'_{xx}|$, for $i = 1, 2, \cdots$ until it becomes sufficiently small. Once the AR parameter estimates $\{\hat{a}_k\}$ have been found, the MA parameters may be found by filtering the data with the all-zero filter $\hat{A}(z)$, where

$$\hat{A}(z) = 1 + \sum_{k=1}^{p} \hat{a}_k z^{-k}$$

to yield a purely MA process. Having performed this operation, the techniques of Section II-F for MA processes can be applied. A spectral factorization is required to determine the MA parameters. To avoid the spectral factorization, note that for spectral estimation one is only concerned with finding $A(z)A^*(1/z^*)$ and $B(z)B^*(1/z^*)$, since the spectral estimate is [117], [129]

$$\mathcal{P}_x(f) = \left. \frac{\sigma^2 \Delta t B(z) B^*(1/z^*)}{A(z)A^*(1/z^*)} \right|_{z = \exp(j2\pi f \Delta t)}.$$

If

$$B_k = Z^{-1}[\sigma^2 \Delta t B(z) B^*(1/z^*)]$$

where Z and Z^{-1} denote the z-transform and inverse z-transform, respectively, then the spectral estimate is

$$\mathcal{P}_x(f) = \frac{\sum_{k=-q}^{q} B_k \exp(-j2\pi f k \Delta t)}{|A(\exp[j2\pi f \Delta t])|^2}$$

where $B_{-k} = B_k^*$. To obtain B_k, observe that

$$B_k = Z^{-1}[A(z)A^*(1/z^*) P_x(z)].$$

Letting $A_k = Z^{-1}[A(z)A^*(1/z^*)]$, which is known, then

$$B_k = \sum_{n=-p}^{p} A_n R_{xx}(k-n), \quad \text{for } k = 0, 1, \cdots, q.$$

To determine B_k requires knowledge of $\{R_{xx}(0), R_{xx}(1), \cdots, R_{xx}(p+q)\}$. To insure a nonnegative spectral estimate, B_k must be a positive-semidefinite sequence.

The performance of the modified Yule–Walker approach as applied to ARMA modeling varies greatly. For some processes, the estimates of the ARMA parameters obtained will be quite accurate. However, for some processes, this will not be the case. As an example, consider the asymptotic variance (as $N \to \infty$) of the AR parameter estimate for a real ARMA (1, 1) process. The estimate as given by (2.110)

$$\hat{a}_1 = -\hat{R}_{xx}(2)/\hat{R}_{xx}(1)$$

can be shown to have a variance [72]

$$\text{Var}(\hat{a}_1) = \frac{\sigma^2}{N} \frac{(1 + b_1^2) R_{xx}(0) + 2 b_1 R_{xx}(1)}{R_{xx}^2(1)}. \tag{2.111}$$

But

$$R_{xx}(0) = \sigma^2 \left[\frac{1 + b_1^2 - 2a_1 b_1}{1 - a_1^2} \right] \quad \text{and} \quad R_{xx}(1) = -a_1 R_{xx}(0) + \sigma^2 b_1$$

so (2.111) may be rewritten as

$$\text{Var}(\hat{a}_1) = \frac{1}{N} \frac{\left[1 + \frac{(a_1 - b_1)^2}{(1 - a_1^2)}\right]^2 + \frac{2b_1^2}{(1 - a_1^2)}}{\left(\frac{1}{1 - a_1^2}\right) \left[b_1 - a_1 \left(1 + \frac{(a_1 - b_1)^2}{(1 - a_1^2)}\right)\right]^2}.$$

For a reasonably accurate estimate of a_1, one might require the rms error to be no greater than 0.1,

$$\sqrt{\text{Var}(\hat{a}_1)} \leq 0.1.$$

To meet this requirement for $b_1 = 0.5$, the minimum number of samples versus a_1 is shown in Fig. 11. It may be seen that the statistical fluctuation will vary greatly for a given N, depending on the spectral shape. As $N \to \infty$, the ARMA(1, 1) model is seen to be inappropriate as $a_1 \to 0.5$, since the pole and zero cancel resulting in a white noise process. This example illustrates that care must be taken when using the modified Yule–Walker approach, especially when the model order p is unknown. This is because the modified Yule–Walker matrix $\boldsymbol{R}'_{xx}$ formed from exactly known lags will be singular for a dimension in excess of the true model order.

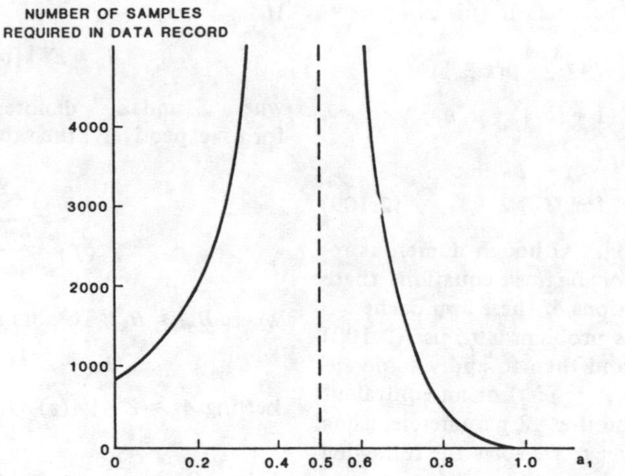

Fig. 11. Required number of samples in data record for accurate estimation of AR parameters for ARMA (1, 1) process using modified Yule-Walker equations.

A second technique for estimating the ARMA parameters utilizes the identify

$$\frac{B(z)}{A(z)} = \frac{1}{C(z)}$$

where

$$C(z) = 1 + \sum_{k=1}^{\infty} c_k z^{-k}$$

to equate an ARMA model to an infinite order AR model. The $\{c_k\}$ may be estimated using AR techniques only and then related to the ARMA parameters. Specifically, let $\hat{C}(z) = 1 + \sum_{k=1}^{M} \hat{c}_k z^{-k}$ be the estimated AR parameters, where $M \geqslant p + q$. Assuming $p > q$, then [79]

$$\frac{\hat{B}(z)}{\hat{A}(z)} = \frac{1}{\hat{C}(z)} \qquad \text{or}$$

$$\sum_{k=0}^{q} \hat{b}_k \hat{c}_{n-k} = \hat{a}_n, \qquad \text{for } n = 1, 2, \cdots.$$

where $\hat{b}_0 = 1$. Since a_n should be equal zero for $n > p$, set

$$\sum_{k=0}^{q} \hat{b}_k \hat{c}_{n-k} = 0, \qquad \text{for } n = p+1, p+2, \cdots, p+q.$$

This expression may be written in matrix form,

$$\begin{bmatrix} \hat{c}_p & \hat{c}_{p-1} & \cdots & \hat{c}_{p+1-q} \\ \hat{c}_{p+1} & \hat{c}_p & \cdots & \hat{c}_{p+2-q} \\ \vdots & \ddots & \vdots & \vdots \\ \hat{c}_{p+q-1} & \hat{c}_{p+q-2} & \cdots & \hat{c}_p \end{bmatrix} \begin{bmatrix} \hat{b}_1 \\ \hat{b}_2 \\ \vdots \\ \hat{b}_q \end{bmatrix} = - \begin{bmatrix} \hat{c}_{p+1} \\ \hat{c}_{p+2} \\ \vdots \\ \hat{c}_{p+q} \end{bmatrix} \tag{2.112}$$

Once $\{\hat{b}_1, \hat{b}_2, \cdots, \hat{b}_q\}$ are found, then $\{\hat{a}_1, \hat{a}_2, \cdots, \hat{a}_p\}$ may be found by solving

$$\sum_{k=0}^{q} \hat{b}_k \hat{c}_{n-k} = \hat{a}_n, \qquad \text{for } n = 1, 2, \cdots, p \tag{2.113}$$

where

$$\hat{c}_0 \equiv 1.$$

In matrix form, this is

$$\begin{bmatrix} \hat{a}_1 \\ \hat{a}_2 \\ \vdots \\ \hat{a}_p \end{bmatrix} = \begin{bmatrix} \hat{c}_1 & 1 & 0 & \cdots & 0 \\ \hat{c}_2 & \hat{c}_1 & 1 & \cdots & 0 \\ \vdots & \vdots & \vdots & & \vdots \\ \hat{c}_p & \hat{c}_{p-1} & \hat{c}_{p-2} & \cdots & \hat{c}_{p-q} \end{bmatrix} \begin{bmatrix} 1 \\ \hat{b}_1 \\ \vdots \\ \hat{b}_q \end{bmatrix}. \tag{2.114}$$

Since $C(z) = A(z)/B(z)$, a very large-order AR model must be used when the zeros of $B(z)$ are near the unit circle. In this case, the c_k sequence will not die out rapidly. This will usually be the case of interest, for if the zeros of $B(z)$ are near the origin, they will have negligible effect upon the PSD. In this case, an AR model would suffice. Nevertheless, some promising results have been obtained with this method [79], [197].

A third technique based upon least squares input–output identification has also been proposed [131], [155]. From (2.107) it may be seen that the nonlinear character of the normal equations is due to the unknown cross correlation between the input and output. If n_n is unobservable, then $R_{nx}(k)$ cannot be estimated. If, however, n_n were known, so that $R_{nx}(k)$ could be estimated, then the ARMA parameters could be found as the solutions of a set of linear equations. In practice, n_n is estimated from x_n in a boot-strap approach to be discussed later. To set up the linear equations, rewrite x_n from (2.105) as

$$x_n = -\sum_{k=1}^{p} a_k x_{n-k} + \sum_{k=1}^{q} b_k n_{n-k} + n_n, \qquad \text{for } n = 0, \cdots, N-1. \tag{2.115}$$

From (2.115), one can observe that

$$\boldsymbol{z} = \boldsymbol{H\theta} + \boldsymbol{v} \tag{2.116}$$

where

$$\boldsymbol{z} = [x_0 \quad x_1 \quad \cdots \quad x_{N-1}]^T$$

$$\boldsymbol{v} = [n_0 \quad n_1 \quad \cdots n_{N-1}]^T$$

$$\boldsymbol{\theta} = [-a_1 \quad -a_2 \quad \cdots \quad -a_p \quad b_1 \quad b_2 \quad \cdots \quad b_q]^T$$

and

$$H = \begin{bmatrix} x_{-1} & x_{-2} & \cdots & x_{-p} & n_{-1} & n_{-2} & \cdots & n_{-q} \\ x_0 & x_{-1} & \cdots & x_{-p+1} & n_0 & n_{-1} & \cdots & n_{-q+1} \\ \vdots & \vdots & & \vdots & \vdots & \vdots & & \vdots \\ x_{N-1} & x_{N-3} & \cdots & x_{N-p-1} & n_{N-2} & n_{N-3} & \cdots & n_{N-q-1} \end{bmatrix} = \text{Input/Output Data Matrix.}$$

Note that H has dimensions $N \times (p+q)$. Equation (2.115) is the standard form for a linear least squares problem, for which the solution is

$$\hat{\theta} = (H^H H)^{-1} H^H z. \tag{2.117}$$

This approach is similar to the least squares approach for AR parameter estimation of pure AR processes. The correlation matrix $H^H H$ of the ARMA process involves the estimate of the cross-correlation function $R_{nx}(k)$. The initial conditions $\{x_{-p}, x_{-p+1}, \cdots, x_{-1}, n_{-q}, n_{-q+1}, \cdots n_{-1}\}$ need to be specified, or assumed equal to zero (in a similar fashion to least squares estimation of AR parameters). To estimate n_n [155], one models x_n by a large AR model and sets n_n equal to the prediction error time series. The ARMA parameter estimates are then further improved by an iterative procedure. The technique works well only if the zeros of $B(z)$ are well within the unit circle. It should be observed that (2.117) is the least squares solution corresponding to (2.107), in which the exactly known statistical correlations have been replaced by their estimates. The least squares approach does not utilize the additional information that $R_{nx}(k) = 0$ for $k > 0$.

ARMA parameter estimation continues to be an active area of research as no one method seems to stand out over another method in terms of its performance and/or lower computational complexity.

H. Pisarenko Harmonic Decomposition

If a stochastic process consists solely of sinusoids in additive white noise, then it is possible to model it as a special case ARMA process. Unlike the model for the periodogram, this model assumes the sinusoids are, in general, nonharmonically related. The mathematical properties of this special ARMA process leads to an eigenanalysis for the estimation of its parameters. Hence, a separate treatment from the general ARMA model discussion is given to this process.

Sinusoids in additive noise is a frequently used test process for evaluating spectrum analysis techniques. To motivate the selection of an ARMA process as the appropriate model for sinusoids in white noise, consider the following trigonometric identity:

$$\sin(\Omega n) = 2\cos\Omega \sin(\Omega[n-1]) - \sin(\Omega[n-2]) \tag{2.118}$$

for $-\pi < \Omega \leqslant \pi$. By letting $\Omega = 2\pi f \Delta t$, where $-1/2\Delta t < f \leqslant 1/2\Delta t$, $\sin \Omega n$ represents a sinusoid sampled at increments of Δt s. By setting $x_n = \sin(\Omega n)$, (2.118) may be rewritten as a second-order difference equation

$$x_n = (2\cos\Omega)x_{n-1} - x_{n-2} \tag{2.119}$$

permitting the current sinusoid value to be recursively computed from the two previous values x_{n-1} and x_{n-2}. If the z transform of (2.119) is taken, then

$$X(z)[1 - 2\cos\Omega z^{-1} + z^{-2}] = D(z) \tag{2.120}$$

where $D(z)$ is a polynomial of second degree that reflects the initial conditions. It has the characteristic polynomial $1 - 2\cos\Omega z^{-1} + z^{-2}$, or equivalently $z^2 - 2\cos\Omega z + 1$, with roots $z_1 = \exp(j2\pi f\Delta t)$ and $z_2 = z_1^* = \exp(-j2\pi f \Delta t)$. The roots are of unit modulus, $|z_1| = |z_2| = 1$, and the sinusoidal frequency in hertz is determined from the roots as follows:

$$f_i = [\tan^{-1}(\mathrm{Im}\{z_i\}/\mathrm{Re}\{z_i\})]/2\pi\Delta t, \quad \text{for } i = 1, 2. \tag{2.121}$$

Note that $f_1 = -f_2$. Observe that (2.119) is the limiting case of an AR(2) process in which the driving noise variance tends to zero and the poles tend to the unit circle. Also, with only two coefficients and knowledge of two samples, (2.119) makes it possible to perfectly predict the sinusoidal process for all time.

In general, a $2p$th-order difference equation of real coefficients of the form

$$x_n = -\sum_{m=1}^{2p} a_m x_{n-m} \tag{2.122}$$

can represent a deterministic process consisting of p real sinusoids of the form $\sin(2\pi f_i \Delta t)$. In this case, the $\{a_m\}$ are coefficients of the polynomial

$$z^{2p} + a_1 z^{2p-1} + \cdots + a_{p-1}z^{p+1} + a_p z^p + a_{p+1} z^{p-1} + \cdots$$
$$+ a_{2p-1} z + a_{2p} = \prod_{i=1}^{p} (z - z_i)(z - z_i^*) = 0 \tag{2.123}$$

with unit modulus roots that occur in complex conjugate pairs of the form $z_i = \exp(j2\pi f_i \Delta t)$, where the f_i are arbitrary frequencies such that $-1/2\Delta t \leqslant f_i < 1/2\Delta t$, and $i = 1, \cdots, p$. For this purely harmonic process, it can be shown that $a_i = a_{2p-i}$ for $i = 0, \cdots, p$.

For sinusoids in additive white noise w_n, the observed process is

$$y_n = x_n + w_n = -\sum_{m=1}^{2p} a_m x_{n-m} + w_n \tag{2.124}$$

where $E[w_n w_{n+k}] = \sigma_w^2 \delta_k$, $E[w_n] = 0$, and $E[x_n w_m] = 0$ since the noise is assumed to be uncorrelated with the sinusoids. Substituting $x_{n-m} = y_{n-m} - w_{n-m}$ into (2.124), it is possible to rewrite (2.124) as

$$\sum_{m=0}^{2p} a_m y_{n-m} = \sum_{m=0}^{2p} a_m w_{n-m} \tag{2.125}$$

where $a_0 = 1$ by definition. Expression (2.125), first developed by Ulrych and Clayton [251], represents the sinusoids in white-noise process in terms of the noise w_n and the noisy observations y_n; it has the structure of an ARMA(p, p). However, this ARMA has a special symmetry in which the AR parameters are identical to the parameters of the MA portion of the model.

If the autocorrelation function of y_n is known, the ARMA parameters can be found as the solution to an eigenequation, as is now shown. An equivalent matrix expression for (2.125) is [67]

$$Y^T A = W^T A \tag{2.126}$$

where

$$Y^T = [y_n \;\; y_{n-1} \;\; \cdots \;\; y_{n-2p}]$$
$$A^T = [1 \;\; a_1 \;\; \cdots \;\; a_{2p-1} a_{2p}]$$
$$W^T = [w_n \;\; w_{n-1} \;\; \cdots \;\; w_{n-2p}].$$

Premultiplying both sides of (2.148) by the vector Y and taking the expectation yields

$$E[YY^T]A = E[YW^T]A. \tag{2.127}$$

Defining

$$X^T = [x_n \cdots x_{n-2p}]$$

then

$$E[YY^T] = R_{yy} = \begin{bmatrix} R_{yy}(0) & \cdots & R_{yy}(-2p) \\ \vdots & \ddots & \vdots \\ R_{yy}(2p) & \cdots & R_{yy}(0) \end{bmatrix} \tag{2.128}$$

$$E[YW^T] = E[(X + W)W^T] = E[WW^T]$$
$$= \sigma_w^2 I. \tag{2.129}$$

R_{yy} is the Toeplitz autocorrelation matrix for the observed process and I is the identity matrix. The fact that $E[XW^T] = 0$ follows from the assumption that the sinusoids are uncorrelated with the noise. Expression (2.127) is then rewritten as

$$R_{yy}A = \sigma_w^2 A \tag{2.130}$$

which is an eigenequation where the noise variance (σ_w^2) is an eigenvalue of the autocorrelation matrix R_{yy}. The ARMA parameter vector A is the eigenvector associated with the eigenvalue σ_w^2, scaled so that the first element is unity. Equation (2.130) will yield the ARMA parameters when the lags are known. Knowledge of the noise variance σ_w^2 is not required. It may be shown [149, app. C] for a process consisting of p real sinusoids in additive white noise that σ_w^2 corresponds to the *minimum* eigenvalue of R_{yy} when the dimension of R_{yy} is $(2p + 1) \times (2p + 1)$ or greater (the minimum eigenvalue is repeated if the dimension is greater than $2p + 1$).

Equation (2.130) forms the basis of a harmonic decomposition procedure developed by Pisarenko [195]. This procedure gives the exact frequencies and powers of p real sinusoids in white noise assuming exact knowledge of $2p + 1$ autocorrelation lags, including the zero lag. Since only the autocorrelation lags are assumed known, phase information about each sinusoid is lost. Pisarenko noted the applicability of a trigonometric theorem of Caratheodory [82] for developing a method to find not only the frequencies $\Omega_i = 2\pi f_i \Delta t$, but also the powers $P_i = A_i^2/2$ and the noise PSD $\sigma_w^2 \Delta t$, from only knowledge of $2p + 1$ values of the autocorrelation function. For sinusoids in white noise, the autocorrelation function is

$$R_{yy}(0) = \sigma_w^2 + \sum_{i=1}^{p} P_i$$

$$R_{yy}(k) = \sum_{i=1}^{p} P_i \cos(2\pi f_i k \Delta t), \qquad \text{for } k \neq 0. \tag{2.131}$$

Noting that white noise only affects the zero lag term, Pisarenko was led to the eigenequation (2.130) by the approach of Caratheodoy's theorem, rather than by the approach presented here.

Once the ARMA coefficients a_i are found, the roots z_n of the polynomial

$$z^{2p} + a_1 z^{2p-1} + \cdots + a_{2p-1} z + a_{2p} = 0 \tag{2.132}$$

formed from the coefficients will yield the sinusoid frequencies, since the roots are of unit modulus with

$$z_n = \exp(j 2\pi f_n \Delta t). \tag{2.133}$$

See the discussion leading to (2.123). Due to the structure of (2.128), it turns out that (2.132) must be symmetrical, that is, $a_i = a_{2p-i}$.

No recursive technique is known for solving the eigenequation (2.130) for order p based on knowledge of the solution for order $(p - 1)$. If the number of sinusoids is unknown, but the autocorrelation lags are exactly known, then independent solutions of (2.130) for successively higher orders must be computed until a point is reached where the minimum eigenvalue does not change from one order to the next higher order. This is an indication that the correct order has been reached. At this point, the minimum eigenvalue is the noise variance. In practice, only autocorrelation estimates are available, so that one must choose the number of sinusoids p as that order in which the minimum eigenvalue of (2.130) changes little from the minimum eigenvalue at order $p - 1$. The computational requirements for solution of (2.130) can be reduced somewhat by utilizing the Toeplitz structure of the matrix R_{yy}. The minimum eigenvalue and associated eigenvectors may be found by the classical power method in which the sequence of vectors

$$A(k + 1) = R_{yy}^{-1} A(k), \qquad \text{for } k = 0, 1, \cdots \tag{2.134}$$

converges in the limit to the eigenvector of the minimum eigenvalue, for some initial guess $A(0)$. Equation (2.134) can be rewritten as

$$R_{yy} A(k + 1) = A(k) \tag{2.135}$$

which can be solved for the unknown vector $A(k + 1)$, given $A(k)$. Gaussian elimination type techniques require $o(p^3)$ operations. However, algorithms are available [63], [277] that use the Toeplitz structure of R_{yy} to solve (2.135) in $o(p^2)$ operations. A good starting vector is the all unity vector $A^T(0) = [1, \cdots, 1]$, and the eigenvector $A(\infty)$ is usually obtained after only a few iterations. Once A is found, the minimum eigenvalue λ_{min} (and therefore the noise variance estimate) is given by

$$\lambda_{min} = \sigma_w^2 = \frac{A^T R_{yy} A}{A^T A}. \tag{2.136}$$

Note that $A(k)$ is rescaled for use in (2.134) each time by the Rayleigh quotient of (2.136) until convergence to λ_{min} is achieved.

Once the frequencies have been determined from the polynomial rooting of A, the sinusoid powers can be determined. The autocorrelation lags $R_{yy}(1)$ to $R_{yy}(p)$ may be expressed in matrix form, based on (2.131), as

$$FP = r \tag{2.137}$$

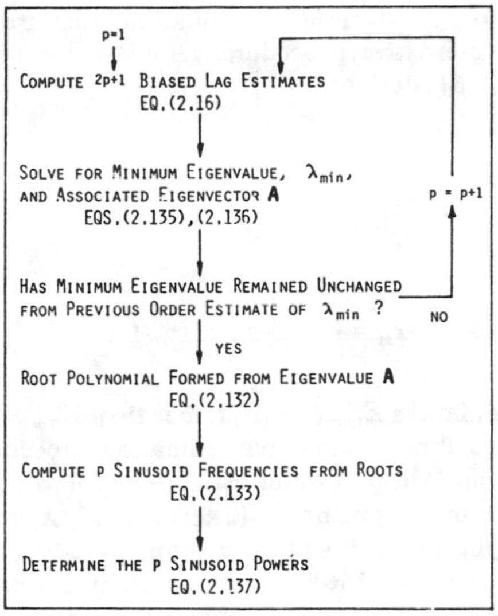

Fig. 12. Summary of Pisarenko spectral line decomposition procedure.

where

$$F = \begin{bmatrix} \cos(2\pi f_1 \Delta t) & \cdots & \cos(2\pi f_p \Delta t) \\ \vdots & & \vdots \\ \cos(2\pi f_1 p\Delta t) & \cdots & \cos(2\pi f_p p\Delta t) \end{bmatrix}$$

$$P = \begin{bmatrix} P_1 \\ \vdots \\ P_p \end{bmatrix} \quad \text{and} \quad r = \begin{bmatrix} R_{yy}(1) \\ \vdots \\ R_{yy}(p) \end{bmatrix}.$$

The matrix F is composed of terms that depend upon the sinusoid frequencies as determined from polynomial rooting. The sinusoid powers are found by solving the simultaneous equation set (2.137) for the power vector P. The noise power can also be determined from

$$\sigma_w^2 = R_{yy}(0) - \sum_{i=1}^{p} P_i.$$

Fig. 12 is a summary of the PHD technique.

Since the order is usually unknown, determining order by checking the minimum eigenvalue involves several solutions of eigenequation (2.130). This is not only computationally expensive, but it is also not often clear when the minimum eigenvalue has been reached since estimated lags, rather than known lags, are normally used. If the selected order is too high, p sinusoids for eigenequation order $2p$ will be computed, even though fewer than p sinusoids really exist. Thus spurious components will be introduced. If the order is too low, then the spectral components that are found tend to appear at incorrect frequencies. The use of the biased autocorrelation lag estimates guarantees a positive-definite Toeplitz autocorrelation matrix. However, the implied triangular windowing of the biased autocorrelation estimate, as discussed in Section II-B, yields significantly inaccurate frequency and power estimates for actual signals present. It will also introduce spurious components into the spectral decomposition. Unbiased lag estimates like (2.14) could be used, but the autocorrelation matrix is not guaranteed to be positive definite, as required in order to perform the Pisarenko decomposition. This can lead to negative eigenvalues and meaningless frequency estimates. Non-Toeplitz positive-definite autocorrelation matrix forms such as $X_1^H X_1$ from Section II-E have been tried in place of R_{yy} in (2.130), but they produce nonunit modulus roots and do not seem to improve the results. For colored noise that contributes to a finite number of lags beyond lag zero, a modification of the Pisarenko technique can be used [214].

The eigenanalysis approach of the Pisarenko technique can be generalized to the idea of extracting the most information concerning a signal by processing for the largest eigenvalues and corresponding eigenvectors of an estimated correlation matrix [135], [185], [252].

J. Prony's Energy Spectral Density [36], [49], [101], [102], [156], [160], [199], [200], [217], [253], [261]

Prony's method, a technique for modeling data of equally spaced samples by a linear combination of exponentials, is not a spectral estimation technique in the usual sense, but a spectral interpretation is provided in this section. Gaspard Riche, Baron de Prony [202], was led to believe that laws governing expansion of various gases could be represented by sums of exponentials. He proposed a method for providing interpolated data points in his measurements by fitting an exponential model to the measured points and computing the interpolated values by evaluation of the exponential model at these points. The modern version of Prony's method bears little resemblance to his original approach due to evolutionary changes that have been made. The original procedure exactly fitted an exponential curve having p exponential terms (each term has two parameters—an amplitude A_i and an exponent α_i where $A_i \exp(\alpha_i t)$) to $2p$ data measurements. This approach is discussed in Hildebrand [95]. For the case where only an *approximate* fit with p exponentials to a data set of N samples is desired, such that $N > 2p$, a least squares estimation procedure is used. This procedure is called the extended Prony method.

The model assumed in the extended Prony method is a set of p exponentials of arbitrary amplitude, phase, frequency, and damping factor. The discrete-time function

$$\hat{x}_n = \sum_{m=1}^{p} b_m z_m^n, \quad \text{for } n = 0, \cdots, N-1 \tag{2.138}$$

is the model to be used for approximating the measured data $x_0, \cdots, x_{N-1}$. For generality, b_m and z_m are assumed complex and

$$b_m = A_m \exp(j\theta_m)$$
$$z_m = \exp[(\alpha_m + j2\pi f_m)\Delta t] \tag{2.139}$$

where A_m is the amplitude, θ_m is the phase in radians, α_m is a damping factor, f_m is the oscillation frequency in hertz, and Δt represents the sample interval in seconds. Finding $\{A_m, \theta_m, \alpha_m, f_m\}$ and p that minimize the squared error

$$\mathcal{E} = \sum_{n=0}^{N-1} |x_n - \hat{x}_n|^2 \tag{2.140}$$

is a difficult nonlinear least squares problem. The solution involves an iterative process in which an initial guess of the unknown parameters is successively improved. McDonough and Huggins [157] and Holtz [97] provide such iterative schemes for the solution of (2.140). An alternative suboptimum solution that does not minimize (2.140) but still provides satisfactory results, is based on Prony's technique. Prony's method solves two sequential sets of linear equations with an intermediate polynomial rooting step that concentrates the nonlinearity of the problem in the polynomial rooting procedure.

The key to the Prony technique is to recognize that (2.138) is the homogeneous solution to a constant coefficient linear difference equation, the form of which is found as follows. Define the polynomial $\Psi(z)$ as

$$\Psi(z) = \prod_{k=1}^{p} (z - z_k) = \sum_{i=0}^{p} a_i z^{p-i}, \quad a_0 = 1. \tag{2.141}$$

Thus $\Psi(z)$ has the complex exponentials z_k of (2.139) as its roots and complex coefficients a_i when multiplied out. Based on (2.138), one way of expressing $\hat{x}_{n-m}$ is

$$\hat{x}_{n-m} = \sum_{l=1}^{p} b_l z_l^{n-m} \tag{2.142}$$

for $0 \leqslant n - m \leqslant N - 1$. Multiplying (2.142) by a_m and summing over the past $p + 1$ products yields

$$\sum_{m=0}^{p} a_m \hat{x}_{n-m} = \sum_{l=1}^{p} b_l \sum_{m=0}^{p} a_m z_l^{n-m} \tag{2.143}$$

defined for $p \leqslant n \leqslant N - 1$. If in (2.143) the substitution $z_l^{n-m} = z_l^{n-p} z_l^{p-m}$ is made, then

$$\sum_{m=0}^{p} a_m \hat{x}_{n-m} = \sum_{l=1}^{p} b_l z_l^{n-p} \sum_{m=0}^{p} a_m z_l^{p-m} = 0. \tag{2.144}$$

The zero result in (2.144) follows by recognizing that the final summation above is just the polynomial $\Psi(z_l)$ of (2.141), evaluated at one of its roots. Expression (2.144) then yields the recursive difference equation

$$\hat{x}_n = -\sum_{m=1}^{p} a_m \hat{x}_{n-m} \tag{2.145}$$

defined for $p \leqslant n \leqslant N - 1$. Compare this with (2.122) of the PHD procedure. Thus the exponential parameters are found by rooting polynomial (2.141) using the a_m coefficients.

To set up the extended Prony method, first define the difference between the actual measured data x_n and the approximation $\hat{x}_n$ to be e_n, so that

$$x_n = \hat{x}_n + e_n \tag{2.146}$$

defined for $0 \leqslant n \leqslant N - 1$. Substituting (2.145),

$$x_n = -\sum_{m=1}^{p} a_m \hat{x}_{n-m} + e_n$$

$$= -\sum_{m=1}^{p} a_m x_{n-m} + \sum_{m=0}^{p} a_m e_{n-m} \tag{2.147}$$

defined for $p \leqslant n \leqslant N - 1$, where $\hat{x}_{n-m} = x_{n-m} - e_{n-m}$ has been used. Based on (2.147), an alternative model to the sum of exponentials plus additive noise model is that of an ARMA model with identical AR and MA parameters driven by the noise process e_n. Unlike the Pisarenko technique, the a_i coefficients are not constrained to produce polynomial roots of unit modulus (no damping). Although the true least squares estimate of the parameters is obtained by minimizing

$$\sum_{n=p}^{N-1} |e_n|^2$$

this leads to a set of nonlinear equations that are difficult to solve. An alternative procedure, termed the extended Prony approach [266], defines

$$\epsilon_n = \sum_{m=0}^{p} a_m e_{n-m}, \qquad \text{for } n = p, \cdots, N - 1 \tag{2.148}$$

so that

$$x_n = -\sum_{m=1}^{p} a_m x_{n-m} + \epsilon_n. \tag{2.149}$$

One then minimizes $\sum_{n=p}^{N-1} |\epsilon_n|^2$, rather than $\sum_{n=p}^{N-1} |e_n|^2$. Thus the extended Prony parameter estimation procedure reduces to that of an AR parameter estimation for which the least square covariance algorithm of (2.69) with $X_1^H X_1$ may be used. Note that the nonwhite random input process ϵ_n is derived from a MA process driven by the approximation error e_n, as indicated by (2.148). Also, ϵ_n is the difference between x_n and its linear prediction based on p past data samples, whereas e_k is the difference between x_n and its exponential approximation. The number of exponentials p is determined using the AR order selection techniques discussed in Section II-E. An alternate scheme to determine p involves an eigenanalysis of the $X_1^H X_1$ matrix [252] and bears a close relationship to the non-Toeplitz Pisarenko eigenanalysis discussed in Section II-H.

Once the z_i have been determined from the polynomial rooting, expression (2.138) reduces to a set of linear equations in the unknown b_m parameters, expressible in matrix form as

$$\mathbf{\Phi} \, \boldsymbol{B} = \hat{\boldsymbol{X}} \tag{2.150}$$

where

$$\mathbf{\Phi} = \begin{bmatrix} 1 & 1 & \cdots & 1 \\ z_1 & z_2 & \cdots & z_p \\ \vdots & \vdots & & \vdots \\ z_1^{N-1} & z_2^{N-1} & \cdots & z_p^{N-1} \end{bmatrix}$$

$$\boldsymbol{B} = [b_1 \ \cdots \ b_p]^T$$

$$\hat{\boldsymbol{X}} = [\hat{x}_0 \ \cdots \ x_{N-1}]^T.$$

Note that $\mathbf{\Phi}$ is a Van der Monde matrix similar to (2.24), except that the z_i terms have damping and arbitrary frequency assignments instead of a harmonic relationship. A least squares minimization of $\Sigma(x - \hat{x})^2$ yields the well-known solution

$$\boldsymbol{B} = [\mathbf{\Phi}^H \mathbf{\Phi}]^{-1} \mathbf{\Phi}^H \boldsymbol{X}. \tag{2.151}$$

A useful relationship that reduces the computational burden of (2.151) is

$$\mathbf{\Phi}^H \mathbf{\Phi} = \begin{bmatrix} \gamma_{11} & \cdots & \gamma_{1p} \\ \vdots & & \vdots \\ \gamma_{p1} & \cdots & \gamma_{pp} \end{bmatrix}$$

where

$$\gamma_{ij} = \frac{(z_i^* z_j)^N - 1}{(z_i z_j) - 1}. \tag{2.152}$$

Determining the a_i parameters by a least squares estimation, rooting the polynomial, and then solving for the b_j parameters (or residues) constitute the extended Prony method. To obtain the amplitude A_i, phase θ_i, damping factor α_i, and fre-

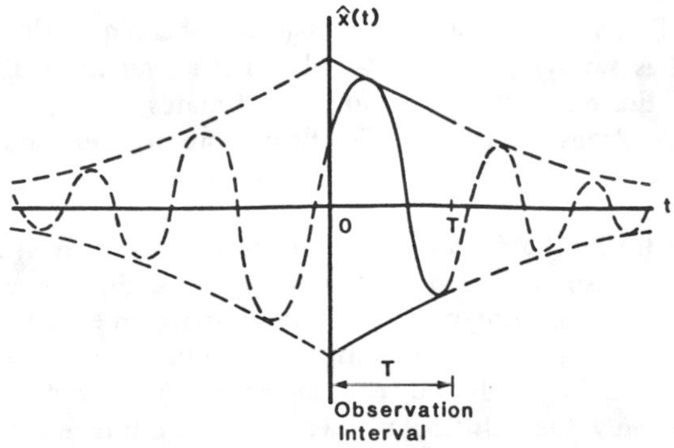

Fig. 13. Symmetric envelope exponential model.

quency f_i from the z_i and b_i estimates, simply compute

$$
\begin{aligned}
A_i &= |b_i| \\
\theta_i &= \tan^{-1} [\mathrm{Im}\,(b_i)/\mathrm{Re}\,(b_i)] \\
\alpha_i &= \ln |z_i|/\Delta t \\
f_i &= \tan^{-1} [\mathrm{Im}\,(z_i)/\mathrm{Re}\,(z_i)]/2\pi\Delta t.
\end{aligned} \quad (2.153)
$$

Normally, the Prony method is completed with the computation of the exponential parameters given in (2.153). As such, Prony's method has found most of its application in transient analysis, such as finding resonant modes in electromagnetic pulse problems [196]. However, a "spectrum analysis" can be performed in the following manner. Although many different spectra could be defined, one "spectrum" found to be useful makes the assumption that the model of the process has symmetry as illustrated for one damped real sinusoid in Fig. 13. The assumed approximation function becomes

$$\hat{x}(t) = \sum_{m=1}^{p} A_m \exp(\alpha_m |t|) \exp(j[2\pi f_m t + \theta_m]) \quad (2.154)$$

defined for $-\infty < t < \infty$. For $x(t)$ real, complex conjugate pairs like $\exp j(2\pi f_m t + \theta_m)$ and $\exp -j(2\pi f_m + \theta_m)$ in (2.154) are required. It is further assumed that all the damping factors are negative, so that decaying exponentials are obtained. One motivation for the selection of a symmetric envelope is that for $\alpha = 0$, $\hat{x}(t)$ will have undamped sinusoidal components which are defined over $-\infty < t < \infty$. As a result, unwindowed sinusoids are accurately modeled by this approach.

Since (2.154) is a finite energy, deterministic expression, its ESD based on the Fourier transform of (2.154) is

$$\hat{S}_{\mathrm{PRONY}}(f) = |\hat{X}(f)|^2 \quad (2.155)$$

where

$$\hat{X}(f) = \sum_{m=1}^{p} A_m \exp(j\theta_m) \frac{2\alpha_m}{[\alpha_m^2 + (2\pi[f - f_m])^2]}. \quad (2.156)$$

This tnen constitutes one possible Prony "spectrum." Note that the spectral estimate (2.155) maintains peaks that are linearly proportional to the energy, unlike AR spectra peaks, which are nonlinearly related to power [136]. The Prony spectrum has the ability to produce narrow-band or wide-band spectral shapes, the shapes being a function of the size of the damping factor (illustrated in Fig. 14). The bell shaped curves have bandwidths (to the -3 dB points) of α/π Hz, so resolution varies as a function of damping. Note that for selection of model order p, the Prony spectrum requires $2p$ parameters to characterize the spectrum, which is twice that required for the

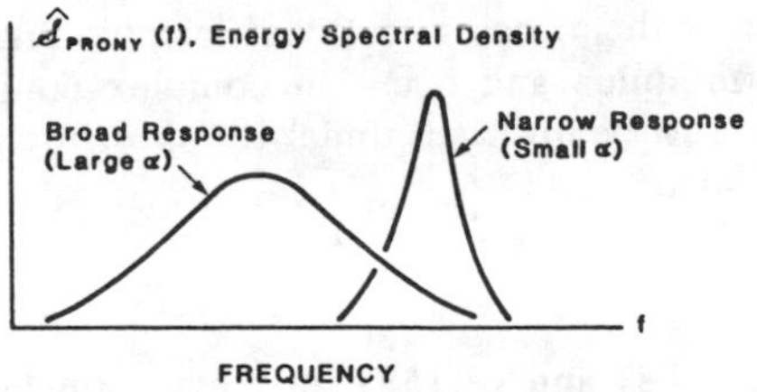

Fig. 14. Narrow-band and wide-band Prony spectral responses.

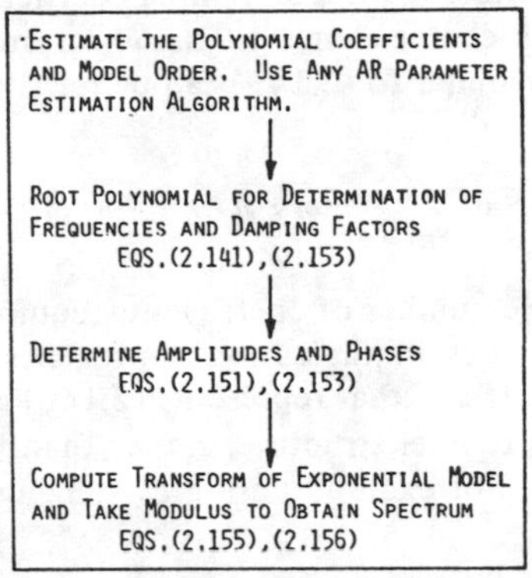

Fig. 15. Prony spectrum estimation procedure.

AR spectral estimate. Also, the Prony method yields phase information not available with AR spectral estimation. Fig. 15 summarizes the Prony spectrum estimation procedure.

The Prony technique is a data adaptive procedure in the sense that it adjusts the parameters of a damped exponential model of varying frequency, phase, amplitude, and damping to fit the data. The periodogram, in contrast, uses a fixed number of undamped sinusoids of fixed frequencies.

There are several problems of which one should be aware when applying Prony's method. The problem of determinating the number of exponential terms is similar to the problem of model order selection in AR estimation, so that the same considerations apply. However, since $2p$ parameters are computed, the maximum order is limited to be $p \leqslant N/2$, whereas $p > N/2$ is possible with AR spectral estimation (although not advisable). Noise impacts the accuracy of the Prony pole estimates greatly in some situations [252]–[254]. Noise also can cause the damping factors to be too large.

K. Prony Spectral Line Estimation

For a process consisting of p real undamped ($\alpha = 0$) sinusoids in noise, a special variant of Prony's method has been developed. The basic approach was described by Hildebrand [95]. In this case, (2.138) may be expressed as

$$\hat{x}_n = \sum_{m=1}^{p} [b_m z_m^n + b_m^* z_m^{*n}] = \sum_{m=1}^{p} A_m \cos(2\pi f_m n\Delta t + \theta_m) \quad (2.157)$$

where $b_m = A_m \exp(j\theta_m)/2$ and $z_m = \exp(j2\pi f_m \Delta t)$. Note that the z_m are roots of unit modulus with arbitrary frequencies and occur in complex conjugate pairs as long as $f_m \neq 0$ or $1/2\Delta t$. Thus one must solve (2.141) for the roots of the polynomial

$$\Psi(z) = \prod_{i=1}^{p} (z - z_i)(z - z_i^*) = \sum_{k=0}^{2p} a_k z^{2p-k} = 0 \quad (2.158)$$

with $a_0 = 1$ and the a_i being real coefficients. Since the roots are of unit modulus and occur in complex conjugate pairs, then (2.158) must be invariant under the substitution z^{-1} for z,

$$z^{2p}\Psi\left(\frac{1}{z}\right) = z^{2p} \sum_{k=0}^{2p} a_k z^{k-2p} = \sum_{k=0}^{2p} a_k z^k = 0. \quad (2.159)$$

Comparing (2.158) and (2.159), one may conclude that $a_j = a_{2p-j}$ for $j = 0$ to p, with $a_0 = a_{2p} = 1$. Thus the requirement for complex conjugate root pairs of unit modulus is implemented by constraining the polynomial coefficients to be symmetric about the center element. Based on order $2p$, a linear prediction error similar to (2.149) can be rewritten as

$$\epsilon_n = \sum_{m=0}^{p} a_m (x_{n+m} + x_{n-m}) \quad (2.160)$$

which reduces the number of coefficients required by one-half. All the least squares minimization approaches apply, except now the data matrix $\boldsymbol{X}$ that represents (2.160) is a data matrix of Toeplitz plus Hankel structure, rather than just Toeplitz as in the AR case. For example, using ϵ_n ranging from $n = p$ to N, we have

$$\begin{bmatrix} \epsilon_{p+1} \\ \vdots \\ \epsilon_{N-p} \end{bmatrix} = \boldsymbol{XA} \quad (2.161)$$

where

Toeplitz Data Matrix

$$\boldsymbol{A} = \begin{bmatrix} 1 \\ a_1 \\ \vdots \\ a_{p-1} \\ \frac{1}{2}a_p \end{bmatrix}, \quad \boldsymbol{X} = \boldsymbol{T} + \boldsymbol{H}, \quad \boldsymbol{T} = \begin{bmatrix} x_{p+1} & \cdots & x_1 \\ \vdots & \ddots & \vdots \\ x_{N-2p} & & x_{p+1} \\ \vdots & \ddots & \vdots \\ x_{N-p} & \cdots & x_{N-2p} \end{bmatrix}$$

Hankel Data Matrix

$$\boldsymbol{H} = \begin{bmatrix} x_{p+1} & \cdots & x_{2p+1} \\ \vdots & \cdot^{\cdot^{\cdot}} & \vdots \\ x_{2p+1} & & x_{N-p} \\ \vdots & \cdot^{\cdot^{\cdot}} & \vdots \\ x_{N-p} & \cdots & x_N \end{bmatrix}$$

so that the minimum of the squared error $\sum_{n=p+1}^{N-p} |\epsilon_n|^2$ determines the real coefficients $a_1, \cdots, a_{p-1}, a_p/2$ analogous to those solutions in AR batch estimation. Note that the last coefficient is $a_p/2$ rather than a_p. The factor of half is due to a symmetry in $\boldsymbol{X}$ that counts the last factor a_p twice. These coefficients are used to set up the order $2p + 1$ symmetrical coefficient polynomial (2.158). Although the unit modulus roots give rise to a symmetrical polynomial, the converse is not necessarily true. Symmetric coefficients only guarantee that if a root z_i occurs, then so does its reciprocal z_i^{-1}; to have $|z_i| = 1$ is not required. In practice [226], nonunit modulus roots are only observed rarely, but when they do occur the roots are usually at $f_i = 0$ or $f_i = 1/2\Delta t$. The algorithm is completed with the determination of amplitude and phase as given by (2.151), which can be reduced in size by one-half by combining related complex pairs. The spectrum will then consist of delta functions, representing the sinusoids, and damped exponentials for those rare cases of nonunit modulus root pairs.

The Prony harmonic decomposition technique described above has several performance advantages over the PHD procedure. For one, autocorrelation lag estimates are not required with the Prony method. The Prony method appears from experiments to yield fewer spurious spectral lines than the Pisarenko approach since the order can be better determined by monitoring the residual squared error of the special Prony method. Also, the frequency and power estimates are less biased than those obtained from the Pisarenko method [153], [226]. See Fig. 16 in the summary section for a comparison of the spectral lines given by each approach. The Prony method requires only the solution of two sets of simultaneous linear equations and a polynomial rooting. The Pisarenko approach requires a more computationally complex eigenequation solution.

L. Maximum Likelihood (Capon) Spectral Estimation [42], [137], [203]

In maximum likelihood spectral estimation (MLSE), originally developed for seismic array frequency-wave number analysis [42], one estimates the PSD by effectively measuring the power out of a set of narrow-band filters [136]. MLSE is actually a misnomer in that the spectral estimate is not a true maximum likelihood estimate of PSD. MLSE is sometimes referenced as the Capon spectral estimate [92]. The name MLSE is retained here only for historic reasons. The difference between MLSE and conventional BT/periodogram spectral estimation is that the shape of the narrow-band filters in MLSE are, in general, different for each frequency whereas they are fixed with the BT/periodogram procedures. The filters adapt to the process for which the PSD is sought. In particular, the filters are finite impulse response (FIR) types with p weights (taps),

$$\boldsymbol{A} = [a_0 a_1 \cdots a_{p-1}]^T. \quad (2.162)$$

The coefficients are chosen so that at the frequency under consideration, f_0, the frequency response of the filter is unity (i.e., an input sinusoid at that frequency would be undistorted at the filter output) and the variance of the output process is minimized. Thus the filter should adjust itself to reject components of the spectrum not near f_0 so that the output power is due mainly to frequency components close to f_0. To obtain the filter, one minimizes the output variance σ^2, given by

$$\sigma^2 = \boldsymbol{A}^H \boldsymbol{R}_{xx} \boldsymbol{A} \quad (2.163)$$

subject to the unity frequency response constraint (so that the sinusoid of frequency f_0 is filtered without distortion)

$$\boldsymbol{E}^H \boldsymbol{A} = 1 \quad (2.164)$$

where $\boldsymbol{R}_{xx}$ is the covariance matrix of x_n, and $\boldsymbol{E}$ is the vector

$$\boldsymbol{E} = [1 \exp(j2\pi f_0 \Delta t) \cdots \exp(j2\pi[p-1] f_0 \Delta t)]^T$$

and H denotes the complex conugate transpose. The solution for the filter weights is easily shown to be [203]

$$\boldsymbol{A}_{\text{OPT}} = \frac{\boldsymbol{R}_{xx}^{-1}\boldsymbol{E}}{\boldsymbol{E}^H \boldsymbol{R}_{xx}^{-1} \boldsymbol{E}} \quad (2.165)$$

and the minimum output variance is then

$$\sigma^2_{\text{MIN}} = \frac{1}{\boldsymbol{E}^H \boldsymbol{R}_{xx}^{-1} \boldsymbol{E}}. \quad (2.166)$$

It is seen that the frequency response of the optimum filter is unity at $f = f_0$ and that the filter characteristics change as a function of the underlying autocorrelation function. Since the minimum output variance is due to frequency components near f_0, then $\sigma^2_{\text{MIN}}\Delta t$ can be interpreted as a PSD estimate. Thus, the MLSE PSD is defined as

$$\hat{\mathcal{P}}_{\text{ML}}(f_0) = \frac{\Delta t}{E^H R_{xx}^{-1} E}. \tag{2.167}$$

To compute the spectral estimate, one only needs an estimate of the autocorrelation matrix.

In practice, the MLSE exhibits more resolution than the periodogram and BT spectral estimators, but less than an AR spectral estimator [136]. When the autocorrelation function must be estimated, it has been observed and verified analytically for large data records that the MLSE exhibits less variance than the AR spectral estimate [13]. It should be noted that for a narrow-band process, in which the autocorrelation function is known, the peak of the AR spectrum is proportional to the square of the power of the process, while for the MLSE the peak is proportional to the power [136], [137]. Also, the AR spectral estimate power can be found by determining the area under the peak, while the area under a MLSE peak is proportional to the square root of the power.

The MLSE and ARSE have been related analytically as follows [40]:

$$\frac{1}{\hat{\mathcal{P}}_{\text{ML}}(f)} = \frac{1}{p} \sum_{m=1}^{p} \frac{1}{\hat{\mathcal{P}}_{\text{AR}}^{(m)}(f)} \tag{2.168}$$

where $\hat{\mathcal{P}}_{\text{AR}}^{(m)}(f)$ is the AR PSD for an mth order model and $\hat{\mathcal{P}}_{\text{ML}}(f)$ is the MLSE PSD, both based upon a known autocorrelation matrix of order p [40], [203]. Thus the lower resolution of the MLSE can be explained by the "parallel resistor network averaging" effect of combining the low-order AR spectra of least resolution with the high order AR spectra of highest resolution. Also of interest is the fact that the inverse Fourier transform of $\hat{\mathcal{P}}_{\text{ML}}(f)$, which yields the estimated autocorrelation function, is not identical to the autocorrelation function used to obtain the PSD. The inverse Fourier transform of the AR PSD, on the other hand, yields the identical autocorrelation functions over the known range of lag values, as indicated by (2.61).

III. Summary of Techniques

Table II provides a summary of eleven of the more commonly used spectral estimation techniques presented in this paper. A brief overview of key properties, equation references for computing each spectral estimate, and a list of key references will aid the reader to readily implement any of the techniques.

Fig. 16 illustrates typical spectra of the eleven techniques described in Table II. Each spectral estimate is based on the same 64-point real sample sequence from a process consisting of three sinusoids and a colored noise process obtained by filtering a white Gaussian process. Table III is a list of the data samples used. The true PSD is shown in Fig. 16(a). The frequency axis ranges from 0.0 to 0.5 and represents the fraction of the sampling frequency. The three sinusoids are at fractional frequencies of 0.10, 0.20, and 0.21 and have SNR's of +10, +30, and +30 dB, respectively, where SNR is defined as the ratio of the sinusoid power to the total power in the passband noise process. The noise process passband is centered at 0.35. This particular signal was selected to demonstrate how each spectral estimation technique performs against both narrow-band and wide-band processes. Fig. 16 is intended to illustrate properties of each technique, especially for short data records, rather than to serve as a basis for comparing relative performance among the techniques.

A periodogram based on the 64 data samples of Table III is shown in Fig. 16(b). The periodogram was generated with an FFT that had been double padded with 64 zeros. The nominal resolution in Hz of a 64-point sequence is 0.015 625 times the sampling frequency, so that the sinusoids at 0.20 and 0.21 are closer than the resolution width. Indeed, the periodogram shown here is unable to resolve these two sinusoidal components. The weaker sinusoid can be seen among the sidelobes (no data windowing was used). The presence of the colored noise is also indicated by the discrete spectral lines in the upper part of the frequency band. Fig. 16(c) illustrates the BT spectrum, based on 16 autocorrelation lag estimates. The number of lags was around 20 percent of the number of data samples, as recommended by Blackman and Tukey.

Several AR PSD estimates are pictured in Figs. 16(d)–(f). Although all are AR spectral estimates, differing only in the manner that the AR coefficients are estimated, the resulting AR spectra are quite different. Using the 64 samples, sixteen coefficients were computed for the AR and all the remaining techniques to be discussed. The Yule–Walker AR approach, which requires estimation of lags, does not resolve the two closely spaced sinusoids (it has the least resolution of all AR methods) and does not give much insight into the spectrum on either side of the main response. The AR PSD estimate based on the Burg algorithm shown in Fig. 16(e) provides sharp responses at the three sinusoid frequencies, although the one at .1 is barely visible on the scale shown. It also shows power is present at the high frequency end of the spectrum, although it is not a smooth, broad spectrum as it should be. This illustrates the "peaky" nature of AR spectra. A more accurate response for the three sinusoids frequencies is obtained with the forward–backwards (or least squares) technique for AR spectral estimation, as shown in Fig. 16(f). Otherwise, AR spectra Figs. 16(e) and (f) are comparable.

The MA PSD estimate is depicted in Fig. 16(g). It is identical to the BT spectrum since only autocorrelation lag estimates were used. The broad-band response of the MA spectrum stands in contrast to the sharp narrow-band response of the AR spectra. It is unable to resolve the two close sinusoids; the response around 0.1 is as broad as the response at the high frequency end of the spectrum, making it difficult to detect narrow-band components in a wide-band response. One ARMA (8, 8) PSD estimate is illustrated in Fig. 16(h), based on the modified Yule–Walker approach with biased lag estimates computed from the data samples. It is not a very good spectral estimate, although an ARMA (16, 16) spectrum not shown here was able to separate the three sinusoid components.

The Pisarenko spectral line decomposition of Fig. 16(i) is, like the FFT periodogram, a discrete spectrum. The two close sinusoids are resolved, but the frequencies and powers are grossly inaccurate. The lower level sinusoid at 0.10 has a spectral line near this frequency, but there are many other spectral lines, making selection of actual signals from spurious components difficult. The broad portion of the spectrum has been modeled by placing several spectral lines in the area of the broad-band process spectrum. Thus, the Pisarenko method does not model the broad-band processes well, though it shows there is power in this frequency region.

The energy spectral density based on the extended Prony method yields the spectrum shown in Fig. 16(j). The three

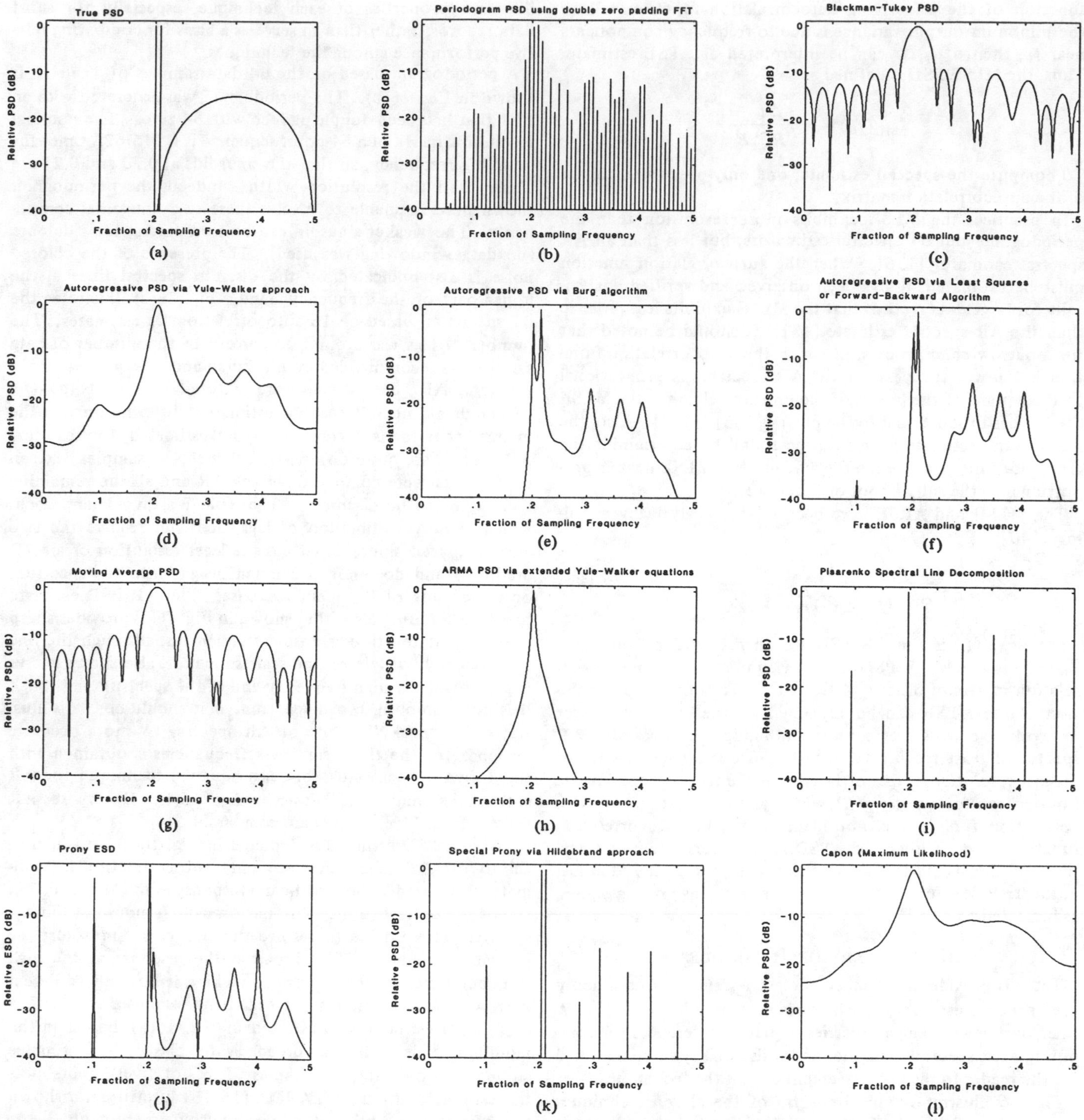

Fig. 16. Illustration of various spectra for the same 64-point sample sequence.

sinusoid components have very sharp responses at the sinusoid frequencies, with a broad response at the higher end of the spectrum. Table IV shows the actual parameter estimates obtained with the Prony method. The actual amplitudes for the sinusoids of frequencies 0.1, 0.2, and 0.21 were 0.1, 1., and 1., respectively. The most accurate estimates of the three sinusoid powers and frequencies is provided by the spectral line decomposition variant of the Prony method, pictured in Fig. 16(k). This is no surprise since this technique is the least squares approach that assumes a sinusoidal model. It is a discrete spectrum so that the broad-band process is not well modeled, although several lines are present to indicate spectral power in this region. Table V lists the actual parameter estimates obtained with this procedure.

The maximum likelihood spectrum, shown in Fig. 16(l) has a smooth spectrum. It cannot resolve the two closely spaced sinusoidal components. The smooth nature of the MLSE spectrum, being the equivalent of an average of all the AR spectra from order 1 to 16, is typical of this method.

If more accurate frequency estimation of noisy sinusoids

TABLE II
SUMMARY OF MODERN SPECTRAL ESTIMATION METHODS

TECHNIQUE	KEY REFS.	DISCRETE LINE OR CONTINUOUS SPECTRUM	ALGORITHM PROCEDURE	ROUGH COMPUTATIONAL COMPLEXITY (ADDS / MULTS)	MODEL(S)	ADVANTAGES AND DISADVANTAGES	REMARKS
Periodogram** FFT Version	24 46 207 218 262	D	Eqn. (2.18)	$N \log_2 N$	Sum of harmonically related sinusoids	Output directly proportional to power Most computationally efficient Resolution roughly the reciprocal of the observation interval Performance poor for short data records Leakage distorts spectrum and masks weak signals	Harmonic least squares fit Requires some type of frequency domain statistical averaging to stabilize spectrum (e.g., Welch method) Windowing can reduce sidelobes at expense of resolution
Blackman-** Tukey (BT)	25 33 204	C	Eqns. (2.13) & (2.16)	Lag Ests.: NM PSD Est.*: MS	Identical to MA with windowing of the lags	Most computationally efficient if M<<N Resolution roughly the reciprocal of the observation interval Leakage distorts spectrum and masks weak signals	Negative PSD values in spectrum may result with some window weightings and autocorrelation estimates (e.g., unbiased)
Autoregressive (AR) Yule-Walker Version	59 108 250 276	C	Fig. 6**	Lag. Ests.: NM AR Coeffs.: M^2 PSD Est.*: MS	Autoregressive (all-pole) process	Model order must be selected Better resolution than FFT or BT, but not as good as other AR methods Spectral line splitting occurs Implied windowing distorts spectrum No sidelobes	Model applicable to seismic, speech, radar clutter data Minimum-phase (stable) linear prediction filter guaranteed if biased lag estimates computed AR related to linear prediction analysis and adaptive filtering Models peaks in spectrum better than valleys
Autoregressive (AR) Burg-Algorithm Version	* 9 15 37 41 57 64 65	C	Fig. 7	AR Coeffs.: $NM + M^2$ PSD Est.*: MS	Autoregressive (all-pole) process	High resolution for low noise levels Good spectral fidelity for short data records Spectral line splitting can occur Bias in the frequency estimates of peaks No implied windowing No sidelobes Must determine order	Stable linear prediction filter guaranteed Adaptive filtering applicable Uses constrained recursive least squares approach
Autoregressive (AR) Least squares or forward-backward linear prediction version	154 227	C	See Ref. 154 for flowchart	AR Coeffs.: $MN + M^2$ PSD Est.*: MS	Autoregressive (all-pole) process	Sharper response for narrowband processes than other AR estimates No spectral line splitting observed Bias reduced in the frequency estimates Must determine order No sidelobes	Stable linear prediction filter not guaranteed, though stable filter results in most instances Based on exact recursive least squares solution with no constraint
Moving Average (MA)	47 236	C	Eqns. (2.104) & (2.35)	MA Coeffs.: Nonlinear Simult.Eqn.Set Lag Ests.: NM PSD Est.*: MS	Moving average (all-zero) process	Broad spectral responses (low resolution) Must determine order Has sidelobes	Generalized form of BT technique
ARMA (Yule Walker Version)	31 97	C	High Order YW Eqns. (2.108)	Lag Ests.: NM Coeff. Computation: M^3 PSD Est.*: MS	ARMA process (Rational Transfer Function) (MA order ≠ AR order)	Must determine AR & MA orders	Models all rational transfer function processes Requires accurate lag estimates to obtain good results
Pisarenko Harmonic Decomposition (PHD)	195 251	D	Fig. 12	Lag Ests.: NM Eigen eqn.: M^2 to M^3 Poly. Rooting: Dependent on Root Algorithm Powers: M^3	Special ARMA with equal MA and AR coefficients Sum of nonharmonically undamped sinusoids in additive white noise	Must determine order Does not work well in high noise levels Eigenequation and rooting are computationally inefficient	Requires accurate lag estimates to obtain good results Spurious spectral lines if order selected too high
Prony's Method (Extended)	95 97 252	C	Fig. 15	AR Coeffs.: $M^2 + NM$ Poly. Rooting: Dependent on Root Algorithm Amp. Coeffs.: M^3 PSD Est.*: MS	Sum of nonharmonically related damped exponentials ARMA with equal MA and AR coefficients and equal orders (p=q)	Must determine order Output linearly proportional to power Requires a polynominal rooting Resolution as good as AR techniques, sometimes better No sidelobes	Uses least squares estimates to obtain exponential parameters First step same as AR least sqaures estimation
Prony Spectral Line Decomposition	95 226	D	Eqns. (2.161) (2.158) (2.151) (2.153)	Coeffs.: M^3 Rooting: Function of root algorithm used Amp. Coeffs.: M^3	Sum of nonharmonically related sinusoids	Must determine order Output linearly proportional to power Requires a polynominal rooting Resolution as good as AR techqniues, sometimes better No sidelobes	Uses least squares estimation
Capon Maximum Likelihood (MLSE)	40 136 193 203	C	Eqns. (2.167) & (2.16)	Lag Ests.: NM Matrix Inversion M^3 PSD Ests.*: MS	Forms an optimal bandpass filter for each spectral component	Resolution better than BT; not as good as AR Statistically less variability in MLSE spectra than AR spectra	MLSE is related to AR spectra (see Eqn. (2.168))

** Computer programs may be found in Programs for Digital Signal Processing, edited by Digital Signal Processing Committee of the IEEE ASSP Society, IEEE Press, 1979.
* FFT could be used to generate $S = 2^{\nu}$ values of the PSD.
N=Number of data samples
S=Number of Spectral Samples Computed (usually S>>M)
M=Order of Model (or number of Autocorrelation Lags)

TABLE III
LIST OF DATA SAMPLES

X(1)=	1.291061	X(33)=	0.309840
X(2)=	-2.086368	X(34)=	1.212892
X(3)=	-1.691316	X(35)=	-0.119905
X(4)=	1.243138	X(36)=	-0.441686
X(5)=	1.641072	X(37)=	-0.879733
X(6)=	-0.008688	X(38)=	0.306181
X(7)=	-1.659390	X(39)=	0.795431
X(8)=	-1.111467	X(40)=	0.189598
X(9)=	0.985908	X(41)=	-0.342332
X(10)=	1.991979	X(42)=	-0.328700
X(11)=	-0.046613	X(43)=	0.197881
X(12)=	-1.649269	X(44)=	0.071179
X(13)=	-1.040810	X(45)=	0.185931
X(14)=	1.054665	X(46)=	-0.324595
X(15)=	1.855816	X(47)=	-0.366092
X(16)=	-0.951182	X(48)=	0.368467
X(17)=	-1.476495	X(49)=	-0.191935
X(18)=	-0.212242	X(50)=	0.519116
X(19)=	0.780202	X(51)=	0.008320
X(20)=	1.416003	X(52)=	-0.425946
X(21)=	0.199202	X(53)=	0.651470
X(22)=	-2.027026	X(54)=	-0.639978
X(23)=	-0.483577	X(55)=	-0.344389
X(24)=	1.664913	X(56)=	0.814130
X(25)=	0.614114	X(57)=	-0.385168
X(26)=	-0.791469	X(58)=	0.064218
X(27)=	-1.195311	X(59)=	-0.380008
X(28)=	0.119801	X(60)=	-0.163008
X(29)=	0.807635	X(61)=	1.180961
X(30)=	0.895236	X(62)=	0.114206
X(31)=	-0.012734	X(63)=	-0.667626
X(32)=	-1.763842	X(64)=	-0.814997

TABLE IV
LIST OF PRONY METHOD PARAMETER ESTIMATES

Signal No.	Parameter Estimates							
1	AMP=	0.0924044	PHASE=	127.8651	DAMP=	0.0001263	FREQ=	0.100033
2	AMP=	1.2756225	PHASE=	150.5105	DAMP=	0.0006901	FREQ=	0.201151
3	AMP=	0.8447961	PHASE=	190.7096	DAMP=	0.0092935	FREQ=	0.208949
4	AMP=	0.2363850	PHASE=	288.9569	DAMP=	-0.0592385	FREQ=	0.275976
5	AMP=	0.5834477	PHASE=	51.5229	DAMP=	-0.0383019	FREQ=	0.310200
6	AMP=	0.2562685	PHASE=	101.3553	DAMP=	-0.0320243	FREQ=	0.357724
7	AMP=	0.2442874	PHASE=	103.9528	DAMP=	-0.0119484	FREQ=	0.402038
8	AMP=	0.1313714	PHASE=	51.7046	DAMP=	-0.0738351	FREQ=	0.451616

TABLE V
LIST OF PRONY SPECTRAL LINE METHOD PARAMETER ESTIMATES

Signal No.	Parameter Estimates							
1	AMP=	0.1033464	PHASE=	125.2777	DAMP=	0.0000000	FREQ=	0.100039
2	AMP=	1.0255654	PHASE=	156.4714	DAMP=	0.0000000	FREQ=	0.200267
3	AMP=	1.0396414	PHASE=	170.5962	DAMP=	0.0000000	FREQ=	0.209599
4	AMP=	0.0564410	PHASE=	36.3485	DAMP=	0.0000000	FREQ=	0.270887
5	AMP=	0.2108561	PHASE=	65.7856	DAMP=	0.0000000	FREQ=	0.309460
6	AMP=	0.1186532	PHASE=	174.4037	DAMP=	0.0000000	FREQ=	0.358735
7	AMP=	0.1971185	PHASE=	95.5934	DAMP=	0.0000000	FREQ=	0.402381
8	AMP=	0.0278794	PHASE=	32.9251	DAMP=	0.0000000	FREQ=	0.450810

and also improved resolution are the most important aspects of spectral estimation, rather than spectral shape, then some recent research by Tufts and Kumaresan [244], [245], [134], [135] has addressed this problem. They consider improvements in linear prediction, eigen-analysis, and maximum likelihood approaches to reduce the frequency estimation variance and increase the resolution. However, these deal strictly with the sinusoids in noise process.

IV. Other Applications of Spectral Estimation Methods

A. Introduction

The preceding sections have discussed the theory and application of modern spectral estimation. Much of the underlying theory presented, however, has been applied to areas other than spectral estimation. Since these further applications are of sufficient interest to researchers in many fields, this section summarizes some of these applications. The topics to be discussed are not meant to be an all inclusive listing of these additional applications, but only a representative sampling of the more common areas.

B. Time Series Extrapolation and Interpolation

The theoretical foundations of modern spectral estimation have led to other applications. An obvious one is that of extrapolation of a time series of unknown PSD. If the time series is an AR(p) process, for example, then the optimum linear predictor parameters are the AR parameters. The latter are estimated from the data as discussed in Section II-E. If the

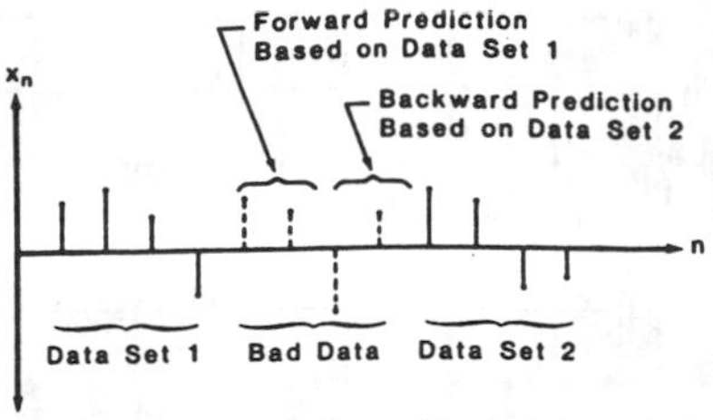

Fig. 17. Interpolation of bad data points.

process is not an AR process, but an AR model is used, then the number of linear prediction parameters of the optimal predictor is, in general, infinite. Theoretically, as the number of predictor parameters increases, the extrapolation error will decrease. When one is limited to a finite data set from which to estimate the predictor parameters, the prediction error power will be minimized by choosing a predictor with as large an order as possible, subject to the constraint that the predictor parameters can be accurately estimated [51].

Although the techniques described in Section II-E can only yield a one step predictor, one can use the predicted sample as if it were part of the original data set and continue the extrapolation to the next sample [27], [29]. It has even been proposed to use the enlarged set of original and extrapolated data with a conventional periodogram or a BT spectral estimator to improve the resolution [68]. In addition to extrapolation, interpolation may be performed by using a forward and backward predictor as shown in Fig. 17. This is valuable for replacing bad data points [176].

C. Prewhitening Filters

A prewhitening filter is a natural use of the parameters obtained from a spectral estimate. For example, in AR spectral analysis $\hat{A}(z)$ of (2.91) is a whitening filter. The output of the filter (the prediction error) is white noise if the observed process is AR(p) and the predictor coefficients estimated are the actual AR parameters. In the event that the time series is not an AR process, the output time series power spectral density will still be flatter than the input and "approximately" white. This property is particularly valuable in the design of detectors for signals in colored noise of unknown spectral shape. The detection of target returns in a background of clutter is an example. The optimal detector is a prewhitener followed by a matched filter, matched to the signal at the prewhitener output [256]. Since the clutter spectrum usually is time varying, the whitening filter parameters and matched filter must be updated. The success of the prewhitening scheme will depend upon the time variation of the clutter spectrum and the ability to estimate the parameters of the spectrum before they change [30].

A closely related concept is that of prewhitening a time series to reduce the bias of conventional spectral estimators. It may be shown that [107]

$$E[\hat{\mathcal{P}}_x(f)] = \int_{-1/2\Delta t}^{1/2\Delta t} \mathcal{P}_x(\nu)\, W(f-\nu)\, d\nu \tag{4.1}$$

where

$\hat{\mathcal{P}}_x(f)$ is a BT type spectral estimate,
$\mathcal{P}_x(f)$ is the true PSD,
$W(f)$ is a spectral window required to reduce the variance of the estimate $[\int_{-1/2\Delta t}^{1/2\Delta t} W(f)\, df = 1]$.

If $\mathcal{P}_x(f)$ is nearly constant with frequency, then

$$E[\hat{\mathcal{P}}_x(f)] \approx \mathcal{P}_x(f) \int_{-1/2\Delta t}^{1/2\Delta t} W(f-\nu)\, d\nu = \mathcal{P}_x(f). \tag{4.2}$$

Thus, to reduce the bias, one should attempt to prewhiten the data [213], [232]. Following the whitening, a BT estimate $\hat{\mathcal{P}}_e(f)$ of the filtered time series e_n is found. Since e_n is the output of the prewhitening filter $\hat{A}(z)$, its PSD is $\mathcal{P}_e(f) = |\hat{A}(\exp[j2\pi f\Delta t])|^2\, \mathcal{P}_x(f)$. The spectral estimate of x_n is then given as

$$\hat{\mathcal{P}}_x(f) = \frac{\hat{\mathcal{P}}_e(f)}{|\hat{A}(\exp[j2\pi f\Delta t])|^2} \tag{4.3}$$

where an all-zero prewhitener is assumed. It is interesting to note that this approach yields a spectral estimate that is the standard AR spectral estimate, with the white noise PSD $\hat{\sigma}_n^2\Delta t$ replaced by the PSD estimate of the residual time series, $\hat{\mathcal{P}}_e(f)$.

D. Bandwidth Compression

An important problem in speech research is that of bandwidth compression. If the redundancy of speech can be reduced, then more speech signals can be transmitted through a fixed bandwidth channel or stored in some mass storage. One common technique is differential pulse code modulation (DPCM) [104]. The basis of DPCM is to transmit only information that cannot be predicted, often termed the innovations of the process [112]. In fact, if the speech waveform were perfectly predictable from a set of previous samples, then the receiver, once it had those samples, could perfectly reconstruct the entire waveform (assuming no channel noise). Transmission could be halted! In practice using DPCM, speech samples are analyzed at the transmitter to determine the predictor parameters. Then, only the prediction error time series and the predictor parameters are transmitted. The speech signal is reconstructed at the receiver. The bandwidth reduction is possible because the variance of the prediction error time series is less than that of the speech waveform [251], i.e.,

$$\sigma_e^2 = R_{xx}(0) \prod_{i=1}^{p} (1 - K_i^2) \leqslant R_{xx}(0). \tag{4.4}$$

Thus fewer quantizer levels are necessary to code the residual time series. Note that for maximum bandwidth compression, the predictor parameters must be continually updated as the statistical character of speech changes, i.e., voiced to unvoiced and vice versa.

The most dramatic technique for bandwidth reduction is linear predictor coding (LPC), in which AR modeling is used to represent the speech waveform [145]. Assuming speech can be accurately modeled as the output of an all-pole filter driven by white noise for unvoiced speech, or driven by an impulse train for voiced speech, the speech waveform may be reduced to a small set of parameters. Thus, only the model parameters and the period of the impulse train need be transmitted or stored. Speech synthesis is then accomplished by employing the appropriate model for each speech sound.

E. Spectral Smoothing

Conventional periodogram and BT analysis lead to spectral estimates that are characterized by many "hills and valleys," since the Fourier transform of a zero mean random process is itself a zero mean random process. Autocorrelation lag windowing or spectral window smoothing will substantially reduce the fluctuations but not eliminate them. An AR spectral estimator can be used to smooth these fluctuations since a pth-order AR spectral estimate is constrained to have p or less peaks (or troughs). For p small, a smoothed spectral estimate will result. It is now shown that the AR spectral model accurately represents the peaks of a periodogram but not the valleys [26], [76], [145]. Consider the estimate of the AR parameters found by minimizing the error criterion ((2.64) with the use of $X_2^H X_2$, the autocorrelation normal equations)

$$\mathcal{E} = \frac{1}{N} \sum_{n=-\infty}^{\infty} |e_n|^2 \tag{4.5}$$

where it is assumed x_n, $n = 0, 1, \cdots, N-1$, is available and $x_n = 0$ outside this interval. Then, by Parseval's theorem

$$\mathcal{E} = \frac{1}{N\Delta t} \int_{-1/2\Delta t}^{1/2\Delta t} |E(\exp[j2\pi f\Delta t])|^2 \, df$$

where

$$E(z) = \Delta t \sum_{n=-\infty}^{\infty} e_n z^{-n}$$

Since

$$X(z) = \Delta t \sum_{n=-\infty}^{\infty} x_n z^{-n}$$

then E becomes

$$E = \frac{1}{N\Delta t} \int_{-1/2\Delta t}^{1/2\Delta t} |X(\exp[j2\pi f\Delta t])|^2 \cdot |\hat{A}(\exp[j2\pi f\Delta t])|^2 \, df$$

$$= \hat{\sigma}^2 \Delta t \int_{-1/2\Delta t}^{1/2\Delta t} \frac{1}{N\Delta t} |X(\exp[j2\pi f\Delta t])|^2 \Big/ \hat{\mathcal{P}}_x(f) \, df \tag{4.6}$$

where $|X(\exp[j2\pi f\Delta t])|^2/N\Delta t$ is the periodogram and $\hat{\mathcal{P}}_x(f) = \hat{\sigma}^2\Delta t/|\hat{A}(\exp[j2\pi f\Delta t])|^2$. Thus when

$$\frac{1}{N\Delta t} |X(\exp[j2\pi f\Delta t])|^2$$

is large, $\hat{\mathcal{P}}_x(f)$ should match the periodogram to reduce $\mathcal{E}$. For $|X(\exp[j2\pi f\Delta t])|^2/N\Delta t$ small, there is only a small contribution to the error, so that matching is not necessary. The result is that the AR spectral estimate matches the peaks, but not the valleys, of the periodogram. If one wishes to represent the peaks of a spectrum, then one need only take an inverse Fourier transform to find the first $p+1$ autocorrelation lags, which are then used to find the AR(p) model. An example is shown in Fig. 18 for the periodogram of a speech signal.

F. Beamforming

In beamforming, one is interested in obtaining an estimate of the spatial structure of a random spatial field. If one samples in space a random field using a line (linear) array of sensors, then a vector time series, $\{x(t, 0), x(t, \Delta x), \cdots, x(t, [M-1]\Delta x)\}$, is obtained, where $x(t, i\Delta x)$ is the continuous waveform at the ith sensor, $0 \leq i \leq M-1$, and M is the number of sensors as shown in Fig. 19. The field can be expressed as [42]

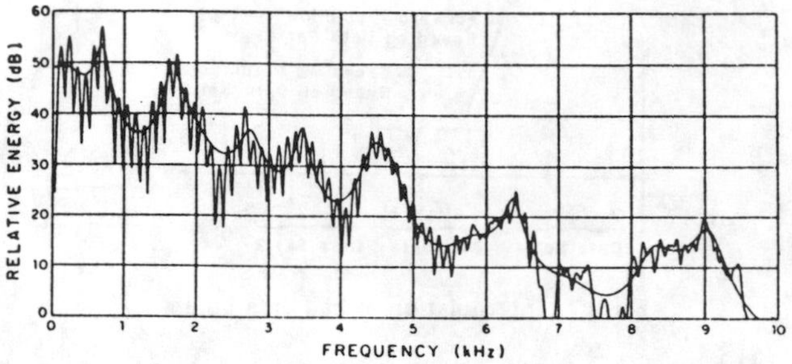

Fig. 18. Periodogram PSD and smoothed AR(28) PSD estimates of speech data (from Makhoul [144]).

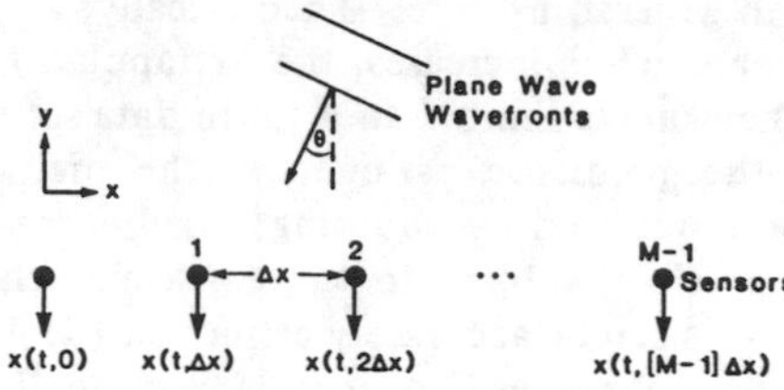

Fig. 19. Line array geometry.

$$x(t, i\Delta x) = \iint \Psi(f, k_x) \exp[j2\pi(ft - k_x i\Delta x)] \, df \, dk_x \tag{4.7}$$

which represents the field as the sum of an infinite number of monochromatic plane waves with random amplitudes $\Psi(f, k_x)$. The temporal frequency is denoted by f, while the spatial frequency along the x direction is denoted by k_x. The wavenumber component k_x is the reciprocal of the wavelength of a monochromatic plane wave along the x direction. Since $k_x = (f/c)\sin\theta$, where θ is the angle indicated in Fig. 19, then $E[|\Psi(f, k_x)|^2]$ is the power at frequency f arriving from the θ direction. From (4.7) the inverse Fourier transform is

$$\Psi(f, k_x) = \sum_{i=0}^{M-1} \left(\int x(t, i\Delta x) \exp(-j2\pi ft) \, dt \right) \exp(j2\pi k_x i\Delta x)$$

$$= \sum_{i=0}^{M-1} X_f(i\Delta x) \exp(j2\pi k_x i\Delta x). \tag{4.8}$$

Expression (4.8) is a Fourier transform relationship between a spatial "time series" $X_f(i\Delta x)$, where Δx is the distance between the samples, and its spectrum $\Psi(f, k_x)$. If the spatial field is assumed homogeneous, i.e.,

$$E[x(t, i\Delta x)\, x^*(t, j\Delta x)] = f(t, [i-j]\Delta x) \tag{4.9}$$

then the spatial "time series" is wide sense stationary and the estimation of $E(|\Psi(f, k_x)|^2)$ for all k_x at a given temporal frequency f is analogous to the one-dimensional temporal power spectral estimation. Any of the techniques described in this paper are then applicable if the time data record is replaced by the spatial data record $\{x(t_0, 0), x(t_0, \Delta x), \cdots, x(t_0, [M-1]\Delta x)\}$ at some time t_0. Note that some extra averaging is afforded in the spatial case that is not available in the temporal case. For instance, the spatial "autocorrelation" estimate could be chosen to be

$$\hat{R}_x(k\Delta x) = \frac{1}{N(M-k)} \sum_{n=0}^{N-1} \sum_{i=0}^{M-k-1} x^*(n\Delta t, i\Delta x) \cdot x(n\Delta t, (i+k)\Delta x) \qquad k \geqslant 0 \tag{4.10}$$

where $x(n\Delta t, i\Delta x)$ is assumed to be temporally stationary over the interval $0 \leqslant n\Delta t \leqslant (N-1)\Delta t$. This estimate includes an extra time averaging operation.

G. Lattice Filters

The minimum phase lattice filter described in Section II-E has the property that its coefficients are bounded by one in magnitude. This is very desirable when one must quantize the coefficients for transmission or storage [257]. The lattice structure may be used to synthesize stable minimum-phase FIR filters, stable all-pole IIR filters, or stable pole-zero IIR filters that have zeros within the unit circle of the z-plane.

H. Other Applications

Other applications of the techniques described in this paper are

1) equalization for digital communications [147]
2) transient analysis [252], [254]
3) digital filter deisgn [183]
4) predictive deconvolution [209]
5) cepstral analysis [139].

The interested reader may consult the references for further details.

V. Conclusions

Modern spectral estimation techniques are based upon modeling of the data by a small set of parameters. When the model is an accurate representation of the data, spectral estimates can be obtained whose performance exceed that of the classical periodogram or BT spectral estimators. The improvement in performance is manifested by higher resolution and a lack of sidelobes. It should also be emphasized that in addition to an accurate model of the data, one must base the spectral estimator on a good estimator of the model parameters. Usually, this entails a maximum likelihood parameter estimator. If the model is inappropriate, as in the case of an AR model for an AR process with additive observation noise, poor (biased) spectral estimates will result. If the model is accurate but a poor statistical estimator of the parameters is employed, as in the case of the ARMA spectral estimate using the modified Yule-Walker equations, poor (inflated variance) spectral estimates will also result.

Computationally efficient procedures exist for maximum-likelihood AR spectral estimation. These techniques generally do not require substantially more computation than conventional Fourier spectral estimators. However, maximum likelihood ARMA spectral estimation involves the solution of nonlinear equations so that no efficient computational procedures now exist. Since ARMA spectral estimators are more desirable than AR spectral estimators when the data characteristics are unknown, due to their robustness, future research is being directed at computationally efficient maximum likelihood ARMA spectral estimation.

A multitude of modern spectral estimation algorithms have been proposed, with only a small but representative subset described in this tutorial. Unfortunately few algorithms, if any, have been analyzed statistically for finite data records. Comparisons among various competing algorithms have been based on limited computer simulations, which can be misleading. Therefore, future research should also be directed at providing more complete statistical descriptions of modern spectral estimators.

In summary, modern spectral estimation techniques, when used properly, are extremely valuable for data analysis. It has been the intent of the authors to present the various techniques in a unified modeling framework and with common nomenclature. Hopefully, this approach will aid users in the selection of the spectral estimation method appropriate to their application.

References

References marked with * may be found in *Modern Spectrum Analysis*, D. G. Childers, Ed., IEEE Press selected Reprint Series, New York, 1978.

[1] H. Akaike, "Fitting autoregressive models for prediction," *Ann. Inst. Statist. Math.*, vol. 21, pp. 243-247, 1969.

[2] —, "Power spectrum estimation through autoregression model fitting," *Ann. Inst. Statist. Math.*, vol. 21, pp. 407-419, 1969.

[3] —, "On a semi-automatic power spectrum estimation procedure," in *Proc. 3rd Hawaii Int. Conf. System Sciences*, Part 2, pp. 974-977, 1970.

[4] —, "Statistical predictor identification," *Ann. Inst. Statist. Math.*, vol. 22, pp. 203-217, 1970.

[5] —, "Autoregressive model fitting for control," *Ann. Inst. Statist. Math.*, vol. 23, pp. 163-180, 1971.

[6] —, "Use of an information theoretic quantity for statistical model identification," in *Proc. 5th Hawaii Int. Conf. System Sciences*, pp. 249-250, Jan. 11-13, 1972.

*[7] —, "A new look at the statistical model identification," *IEEE Trans. Autom. Contr.*, vol. AC-19, pp. 716-723, Dec. 1974.

*[8] M. A. Alam, "Adaptive spectral estimation," in *Proc. Joint Automat. Contr. Conf.*, vol. 1, San Francisco, CA, June 22-24, 1977, pp. 105-112.

*[9] N. O. Andersen, "On the calculation of filter coefficients for maximum entropy spectral analysis," *Geophys.*, vol. 39, pp. 69-72, Feb. 1974.

[10] —, "Comments on the performance of maximum entropy algorithms," *Proc. IEEE*, vol. 66, pp. 1581-1582, Nov. 1978.

[11] T. W. Anderson, "Estimation by maximum likelihood in autoregressive moving average models in the time and frequency domains," Dep. Statistics, Tech. Rep. 20, Contract NR-042-034, Stanford University, Stanford, CA, June 1975.

[12] H. Babic and G. C. Temes, "Optimum low-order windows for discrete Fourier transform systems," *IEEE Trans. Acoustics, Speech, Signal Process.*, vol. ASSP-24, pp. 512-517, Dec. 1976.

*[13] A. B. Baggeroer, "Confidence intervals for regression (MEM) spectral estimates," *IEEE Trans. Inform. Theory*, vol. IT-22, pp. 534-545, Sept. 1976.

[14] J. F. Banas, Comments on "Harmonic analysis via Kalman filtering technique," *Proc. IEEE*, vol. 61, pp. 1759-1760, Dec. 1973.

[15] T. E. Barnard, "The maximum entropy spectrum and the Burg technique," Tech. Rep. 1, Advanced Signal Processing, Texas Instruments, prepared for Office of Naval Research, ALEX (03)-TR-75-01, June 25, 1975.

[16] I. Barrodale and R. E. Erickson, "Algorithms for least-squares linear prediction and maximum entropy spectral analysis—Part I: Theory and Part II: FORTRAN Program," *Geophys.*, vol. 45, pp. 420-446, Mar. 1980.

[17] M. S. Bartlett, "Periodogram analysis and continuous spectra," *Biometrika*, vol. 37, pp. 1-16, June 1950.

[18] M. S. Bartlett and J. Medhi, "On the efficiency of procedures for smoothing periodograms from time series with continuous spectra," *Biometrika*, vol. 42, pp. 143-150, 1955.

[19] G. D. Bergland, "A guided tour of the fast Fourier transform," *IEEE Spectrum*, vol. 6, pp. 41-52, July 1969.

[20] A. J. Berkhout and P. R. Zaanen, "A comparison between Weiner filtering, Kalman filtering, and deterministic least squares estimation," *Geophysical Prospecting*, vol. 24, pp. 141-197, Mar. 1976.

[21] J. G. Berryman, "Choice of operator length for maximum entropy spectral analysis," *Geophys.*, vol. 43, pp. 1384-1391, Dec. 1978.

[22] S. Bertram, "On the derivation of the fast Fourier transform," *IEEE Trans. Audio Electroacoust.*, vol. AU-18, pp. 55-58, Mar. 1970.
[23] —, "Frequency analysis using the discrete Fourier transform," *IEEE Trans. Audio Electroacoust.*, vol. AU-18, pp. 495-500, Dec. 1970.
*[24] C. Bingham, M. D. Godfrey, and J. W. Tukey, "Modern techniques of power spectrum estimation," *IEEE Trans. Audio Electroacoust.*, vol. AU-15, pp. 56-66, June 1967.
[25] R. B. Blackman and J. W. Tukey, *The Measurement of Power Spectra From the Point of View of Communications Engineering*. New York: Dover, 1959.
[26] S. F. Boll, "*A priori* digital speech analysis," Ph.D. dissertation, Dep. Elec. Eng., Utah Univ., Mar. 1973.
[27] S. B. Bowling, "Linear prediction and maximum entropy spectral analysis for radar applications," M.I.T. Lincoln Lab., Project Rep. RMP-122 (ESD-TR-77-113), May 24, 1977.
[28] P. Bloomfield, *Fourier Analysis of Time Series: An Introduction*. New York: Wiley, 1976.
[29] S. B. Bowling and S. Lai, "The use of linear prediction for the interpolation and extrapolation of missing data prior to spectral analysis," in *Proc. RADC Workshop on Spectrum Estimation*, pp. 39-50, Oct. 3-5, 1979.
[30] D. E. Bowyer, P. K. Rajasekaran, and W. W. Gebhart, "Adaptive clutter filtering using autoregressive spectral estimation," *IEEE Trans. Aerospace Elec. Syst.*, vol. AES-15, pp. 538-546, July 1979.
[31] G.E.P. Box and G. M. Jenkins, *Time Series Analysis: Forecasting and Control*. San Francisco, CA: Holden-Day, 1970.
[32] E. O. Brigham and R. E. Morrow, "The fast Fourier transform," *IEEE Spectrum*, vol. 4, pp. 63-70, Dec. 1967.
[33] E. Oran Brigham, *The Fast Fourier Transform*. Englewood Cliffs, NJ: Prentice-Hall, 1974.
[34] D. R. Brillinger, "Fourier analysis of stationary processes," *Proc. IEEE*, vol. 62, pp. 1628-1643, Dec. 1974.
[35] S. Bruzzone and M. Kaveh, "On some suboptimum ARMA spectral estimates," *IEEE Trans. Acoustics, Speech, Signal Process.*, vol. ASSP-28, pp. 753-755, Dec. 1980.
[36] H. P. Bucker, "Comparison of FFT and Prony algorithms for bearing estimation of narrow-band signals in a realistic ocean environment," *J. Acoust. Soc. Amer.*, vol. 61, pp. 756-762, Mar. 1977.
*[37] J. P. Burg, "Maximum entropy spectral analysis," in *Proc. 37th Meeting Society of Exploration Geophysicists* (Oklahoma City, OK), Oct. 31, 1967.
*[38] —, "A new analysis technique for time series data," NATO Advanced Study Institute on Signal Processing with Emphasis on Underwater Acoustics, Enschede, The Netherlands, Aug. 12-23, 1968.
[39] —, "New concepts in power spectral estimation," in *Proc. 40th Annu. Int. Society of Exploration Geophysicists (SEG) Meeting* (New Orleans, LA) Nov. 11, 1970.
*[40] —, "The relationship between maximum entropy and maximum likelihood spectra," *Geophys.*, vol. 37, pp. 375-376, Apr. 1972.
[41] —, "Maximum entropy spectral analysis," Ph.D. dissertation, Dep. Geophysics, Stanford Univ., Stanford, CA, May 1975.
*[42] J. Capon, "High-resolution frequency-wavenumber spectrum analysis," in *Proc. IEEE*, vol. 57, pp. 1408-1418, Aug. 1969.
[43] G. C. Carter and A. H. Nuttall, "On the weighted overlapped segment averaging method for power spectral estimation," *Proc. IEEE*, vol. 68, pp. 1352-1354, Oct. 1980.
[44] —, "A brief summary of a generalized framework for power spectral estimation," *Signal Processing*, vol. 2, pp. 387-390, Oct. 1980.
[45] W. Y. Chen and G. R. Stegen, "Experiments with maximun entropy power spectra of sinusoids," *J. Geophysical Res.*, vol. 79, pp. 3019-3022, July 10, 1974.
[46] D. Childers and A. Durling, *Digital Filtering and Signal Processing*. St. Paul, MN: West Publishing, 1975.
[47] J. C. Chow, "On the estimation of the order of a moving-average process," *IEEE Trans. Automat. Contr.*, vol. AC-17, pp. 386-387, June 1972.
[48] —, "On estimating the orders of an autoregressive moving-average process with uncertain observations," *IEEE Trans. Automat. Contr.*, vol. AC-17, pp. 707-709, Oct. 1972.
[49] C. W. Chuang and D. L. Moffatt, "Natural resonances of radar targets via Prony's method and target discrimination," *IEEE Trans. Aerospace Electron. Syst.*, vol. AES-12, pp. 583-589, Sept. 1976.
[50] J. F. Claerbout, *Fundamentals of Geophysical Data Processing*. New York: McGraw-Hill: 1976.
[51] W. S. Cleveland, "Fitting time series models for prediction," *Technometrics*, pp. 713-723, Nov. 1971.
[52] W. T. Cochran *et al.*, "What is the fast Fourier transform?," *IEEE Trans. Audio Electroacoust.*, vol. AU-15, pp. 45-55, June 1967.
[53] J. W. Cooley and J. W. Tukey, "An algorithm for machine calculation of complex Fourier series," *Math. Comput.*, vol. 19, pp. 297-301, Apr. 1965.
[54] J. W. Cooley, P.A.W. Lewis, and P. D. Welch, "The fast Fourier transform algorithm: Programming considerations in the calculation of sine, cosine, and Laplace transforms," *J. Sound Vibration*, vol. 12, pp. 315-337, July 1970.
[55] —, "The application of the fast Fourier transform algorithm to the estimation of spectra and cross-spectra," *J. Sound Vibration*, vol. 12, pp. 339-352, July 1970.
[56] G. Cybenko, "Round-off error propagation in Durbin's, Levinson's, and Trench's Algorithms," *Rec. 1979 IEEE Int. Conf. Acoustics, Speech, and Signal Processing*, pp. 498-501.
[57] A. K. Datta, Comments on "The complex form of the maximum entropy method for spectral estimation," *Proc. IEEE*, vol. 65, pp. 1219-1220, Aug. 1977.
[58] W. J. Done, "Estimation of the parameters of an autoregressive process in the presence of additive white noise," Computer Science Dep., Univ. Utah, Rep. UTEC-CSC-79-021, Dec. 1978 (order NTIS no. ADA 068749).
*[59] R. E. Dubroff, "The effective autocorrelation function of maximum entropy spectra," *Proc. IEEE*, vol. 63, pp. 1622-1623, Nov. 1975.
[60] J. Durbin, "The fitting of time series models," *Rev. Inst. Int. de Stat.*, vol. 28, pp. 233-244, 1960.
[61] A. Eberhard, "An optimal discrete window for the calculation of power spectra," *IEEE Trans. Audio Electroacoust.*, vol. AU-21, pp. 37-43, Feb. 1973.
*[62] J. A. Edward and M. M. Fitelson, "Notes on maximum entropy processing," *IEEE Trans. Inform. Theory*, vol. IT-19, pp. 232-234, Mar. 1973.
[63] D. C. Farden, "Solution of a Toeplitz set of linear equations," *IEEE Trans. Antennas Propagat.*, vol. AP-24, pp. 906-908, Nov. 1976.
[64] P. F. Fougere, E. J. Zawalick, and H. R. Radoski, "Spontaneous line splitting in maximum entropy power spectrum analysis," *Physics Earth Planetary Interiors*, vol. 12, pp. 201-207, Aug. 1976.
[65] P. F. Fougere, "A solution to the problem of spontaneous line splitting in maximum entropy power spectrum analysis," *J. Geophysical Res.*, vol. 82, pp. 1051-1054, Mar. 1, 1977.
[66] B. Friedlander, M. Morf, T. Kailath, and L. Ljung, "New inversion formulas for matrices classified in terms of their distance from Toeplitz matrices," *Linear Algebra and Its Applications*, vol. 27, pp. 31-60, 1979.
[67] O. L. Frost, "Power spectrum estimation," in *Proc. 1976 NATO Advanced Study Institute on Signal Processing with Emphasis on Underwater Acoustics*, Portovener (LaSpezia), Italy, Aug. 30-Sept. 11, 1976.
[68] O. L. Frost and T. M. Sullivan, "High resolution two dimensional spectral analysis," in *Conf. Rec. 1979 ICASSP*, pp. 673-676, 1979.
[69] J. Fryer, M. E. Odegard, and G. H. Sutton, "Deconvolution and spectral estimation using final prediction error," *Geophys.*, vol. 40, pp. 411-425, June 1975.
[70] W. F. Gabriel, "Spectral analysis and adaptive array superresolution techniques," *Proc. IEEE*, vol. 68, pp. 654-666, June 1980.
[71] W. Gersch, "Spectral analysis of EEG's by autoregressive decomposition of time series," *Math. Biosci.*, vol. 7, pp. 205-222, 1970.
[72] —, "Estimation of the autoregressive parameters of a mixed autoregressive moving-average time series," *IEEE Trans. Automat. Contr.*, vol. AC-15, pp. 583-588, Oct. 1970.
[73] W. Gersch and D. R. Sharpe, "Estimation of power spectra with finite-order autoregressive models," *IEEE Trans. Automat. Contr.*, vol. AC-18, pp. 367-369, Aug. 1973.
[74] W. Gersch and J. Yonemoto, "Automatic classification of EEG's: A parametric model new features for classification approach," in *Proc. 1977 Joint Automatic Control Conf.* (San Francisco, CA), June 22-24, 1977, pp. 762-769.
[75] J. Gibson, S. Haykin, and S. B. Kesler, "Maximum entropy (adaptive) filtering applied to radar clutter," in *Rec. 1979 IEEE Int. Conf. Acoustics, Speech, and Signal Processing*, pp. 166-169.
[76] J. D. Gibson *et al.*, "Digital speech analysis using sequential estimation techniques," *IEEE Trans. Acoustics, Speech, Signal Process.*, pp. 362-369, Aug. 1975.
[77] T. H. Glisson, C. I. Black, and A. P. Sage, "The digital computation of discrete spectra using the fast Fourier transform," *IEEE Trans. Audio Electroacoust.*, vol. AU-18, pp. 271-287, Sept. 1970.
[78] J. Grandell, M. Hamrud, and P. Toll, "A remark on the correspondence between the maximum entropy method and the autoregressive models," *IEEE Trans. Inform. Theory*, vol. IT-26, pp. 750-751, Nov. 1980.
[79] D. Graupe, D. J. Krause, and J. B. Moore, "Identification of

autoregressive moving-average parameters of time series," *IEEE Trans. Automat. Contr.*, vol. AC-20, pp. 104-107, Feb. 1975.

[80] A. H. Gray, Jr. and D. Y. Wong, "The Burg algorithm for LPC speech analysis/synthesis," *IEEE Trans. Acoustics, Speech, Signal Process.*, vol. ASSP-28, pp. 609-615, Dec. 1980.

[81] R. M. Gray, "Toeplitz and circulant matrices: A review," Information Systems Laboratory, Center for Systems Research, Stanford University, Tech. Rep. No. 6502-1, June 1971.

[82] O. Grenander and G. Szego, *Toeplitz Forms and Their Applications.* Berkeley, CA: Univ. California Press, 1958.

*[83] L. J. Griffiths, "Rapid measurement of digital instantaneous frequency," *IEEE Trans. Acoustics, Speech, Signal Process.*, vol. ASSP-23, pp. 207-222, Apr. 1975.

[84] L. J. Griffiths and R. Prieto-Diaz, "Spectral analysis of natural seismic events using autoregressive techniques," *IEEE Trans. Geosci. Electron.*, vol. GE-15, pp. 13-25, Jan. 1977.

[85] L. J. Griffiths, "Adaptive structures for multiple-input noise canceling applications," in *Conf. Rec. ICASSP 79*, pp. 925-928.

[86] P. R. Gutowski, E. A. Robinson, and S. Treitel, "Novel aspects of spectral estimation," in *Proc. 1977 Joint Automatic Control Conf.*, vol. 1 (San Francisco, CA), June 22-24, 1977, pp. 99-104.

*[87] —, "Spectral estimation: Fact or fiction?," *IEEE Trans. Geosci. Electron.*, vol. GE-16, pp. 80-84, Apr. 1978.

[88] C. S. Hacker, "Autoregressive and transfer function models of mosquito populations," in *Time Series and Ecological Processes*, H. M. Shugat, Ed. Boulder, CO: SIAM, 1978, pp. 294-303.

[89] E. Hannan, *Multiple Time Series Analysis.* New York: Wiley, 1970.

[90] F. J. Harris, "On the use of windows for harmonic analysis with the discrete Fourier transform," *Proc. IEEE*, vol. 66, pp. 51-83, Jan. 1978.

[91] S. Haykin and S. Kesler, "The complex form of the maximum entropy method for spectral estimation," *Proc. IEEE*, vol. 64, pp. 822-823, May 1976.

[92] S. S. Haykin, Ed., *Nonlinear Methods of Spectral Analysis.* New York: Springer-Verlag, 1979.

[93] S. Haykin and J. Reilly, "Mixed autoregressive-moving average modeling of the response of a linear array antenna to incident plane waves," *Proc. IEEE*, vol. 68, pp. 622-623, May 1980.

[94] R. W. Herring, "The cause of line splitting in Burg maximum-entropy spectral analysis," *IEEE Trans. Acoustics, Speech, Signal Process.*, vol. ASSP-28, pp. 692-701, Dec. 1980.

[95] F. B. Hildebrand, *Introduction to Numerical Analysis.* New York: McGraw-Hill, 1956, ch. 9.

[96] S. Holm and J. M. Hovem, "Estimation of scalar ocean wave spectra by the maximum entropy method," *IEEE J. Ocean. Eng.*, vol. OE-4, pp. 76-83, July 1979.

[97] H. Holtz, "Prony's method and related approaches to exponential approximation," Aerospace Corp., Rep. ATR-73(9990)-5, June 15, 1973.

[98] M. Hsu, "Maximum entropy principle and its application to spectral analysis and image reconstruction," Ph.D. dissertation, Ohio State Univ., 1975.

[99] F. M. Hsu and A. A. Giordano, "Line tracking using autoregressive spectral estimates," *IEEE Trans. Acoustics, Speech, Signal Process.*, vol. ASSP-25, pp. 510-519, Dec. 1977.

[100] M. Huzii, "On a spectral estimate obtained by an autoregressive model fitting," *Ann. Inst. Statist. Math.*, vol. 29, pp. 415-431, 1977.

[101] L. B. Jackson and F. K. Soong, "Observations on linear estimation," in *Rec. 1978 IEEE Int. Conf. Acoustics, Speech, and Signal Processing*, pp. 203-207.

[102] L. B. Jackson, D. W. Tufts, F. K. Soong, and R. M. Rao, "Frequency estimation by linear prediction," in *Rec. 1978 IEEE Int. Conf. Acoustics, Speech, and Signal Processing*, pp. 352-356.

[103] A. K. Jain and S. Ranganath, "Extrapolation algorithms for discrete signals with a application in spectral estimation," *IEEE Trans. Acoust., Speech, Signal Processing*, vol. ASSP-29, pp. 830-845, Aug. 1981.

[104] N. S. Jayant, "Digital coding of speech waveforms: PCM, DPCM, and DM quantizers," *Proc. IEEE*, pp. 611-632, May 1974.

[105] G. M. Jenkins, "General considerations in the analysis of spectra," *Technometrics*, vol. 3, pp. 133-166, May 1961.

[106] —, "A survey of spectral analysis," *Appl. Stat.*, vol. 14, pp. 2-32, 1965.

[107] G. M. Jenkins and D. G. Watts, *Spectral Analysis and Its Applications.* San Francisco, CA: Holden-Day, 1968.

[108] S. J. Johnsen and N. Andersen, "On power estimation in maximum entropy spectral analysis," *Geophys.*, vol. 43, pp. 681-690, June 1978.

[109] R. H. Jones, "A reappraisal of the periodogram in spectral analysis," *Technometrics*, vol. 7, pp. 531-542, November 1965.

*[110] —, "Identification and autoregressive spectrum estimation," *IEEE Trans. Automat. Contr.*, vol. AC-19, pp. 894-898, Dec. 1974.

[111] —, "Autoregression order selection," *Geophys.*, vol. 41, pp. 771-773, Aug. 1976.

[112] T. Kailath, "A view of three decades of linear filtering theory," *IEEE Trans. Informa. Theory*, vol. IT-20, pp. 146-181, Mar. 1974.

[113] R. P. Kane and N. B. Trivedi, "Effects of linear trend and mean value on maximum entropy spectral analysis," Institute de Pesquisas Espaciais, Rep. INPE-1568-RPE/069, Sao Jose dos Compos, Brasil (Order NTIS no. N79-33949).

[114] R. P. Kane, "Maximum entropy spectral analysis of some artificial samples," *J. Geophys. Res.*, vol. 84, pp. 965-966, Mar. 1, 1979.

[115] R. L. Kashyap, "Inconsistency of the AIC rule for estimating the order of autoregressive models," *IEEE Trans. Automa. Contr.*, vol. AC-25, pp. 996-998, Oct. 1980.

[116] M. Kaveh and G. R. Cooper, "An empirical investigation of the properties of the autoregressive spectral estimator," *IEEE Trans. Informa. Theory*, vol. IT-22, pp. 313-323, May 1976.

[117] M. Kaveh, "High resolution spectral estimation for noisy signals," *IEEE Trans. Acoust., Speech, Signal Process.*, vol. ASSP-27, pp. 286-287, June 1979.

[118] S. M. Kay, "Improvement of autoregressive spectral estimates in the presence of noise," in *Rec. 1978 Int. Conf. Acoustics, Speech, and Signal Processing*, pp. 357-360.

[119] —, "Maximum entropy spectral estimation using the analytical signal," *IEEE Trans. Acoust., Speech Signal Process.*, vol. ASSP-26, pp. 467-469, Oct. 1978.

[120] S. M. Kay and S. L. Marple, Jr., "Sources of and remedies for spectral line splitting in autoregressive spectrum analysis," in *Rec. 1979 Int. Conf. Acoustics, Speech, and Signal Processing*, pp. 151-154.

[121] S. M. Kay, "The effects of noise on the autoregressive spectral estimator," *IEEE Trans. Acoust., Speech, Signal Process.*, vol. ASSP-27, pp. 478-485, Oct. 1979.

[122] —, "Autoregressive spectral analysis of narrowband processes in white noise with application to sonar," Ph.D. dissertation, Georgia Institute of Technology, Mar. 1980.

[123] —, "Noise compensation for autoregressive spectral estimates," *IEEE Trans. Acoust., Speech, Signal Process.*, vol. ASSP-28, pp. 292-303, June 1980.

[124] R. J. Keeler, "Uncertainties in adaptive maximum entropy frequency estimators," NOAA Tech. Rep. ERL-105-WPL53, Feb. 1979 (Order NTIS no. N79-32489).

[125] S. B. Kesler and S. S. Haykin, "The maximum entropy method applied to the spectral analysis of radar clutter," *IEEE Trans. Inform. Theory*, vol. IT-24, pp. 269-272, Mar. 1978.

[126] —, "Maximum entropy estimation of radar clutter spectra," in *Natl. Telecommunications Conf. Rec.* (Birmingham, AL), pp. 18.5.1-18.5.5, Dec. 3-6, 1978.

[127] A. Ya. Khinchin, "Korrelationstheorie der Stationären Stochastischen Prozesse," *Math. Annalen*, vol. 109, pp. 604-615, 1934.

[128] W. R. King, "Maximum entropy spectral analysis in the spatial domain," Rome Air Development Center, Tech. Rep. RADC-TR-78-160, July 1978 (Order NTIS no. ADA 068558).

[129] J. F. Kinkel *et al.*, "A note on covariance-invariant digital filter design and autoregressive moving average spectrum analysis," *IEEE Trans. Acoust., Speech, Signal Process.*, pp. 200-202, Apr. 1979.

[130] H. Kobatake *et al.*, "Linear predictive coding of speech signals in a high ambient noise environment," in *Conf. Rec. 1978 ICASSP*, pp. 472-475.

[131] I. S. Konvalenka *et al.*, "Simultaneous estimation of poles and zeros in speech analysis and ITIF—Iterative inverse fitting algorithm," *IEEE Trans. Acoust., Speech, Signal Process.*, pp. 485-491, Oct. 1979.

[132] L. H. Koopmans, *The Spectral Analysis of Time Series.* New York: Academic Press, 1974.

[133] F. Kozin and F. Nakajima, "The order determination problem for linear time-varying AR models," *IEEE Trans. Automat. Contr.*, vol. AC-25, pp. 250-257, Apr. 1980.

[134] R. Kumaresan and D. W. Tufts, "Improved spectral resolution III: Efficient realization," *Proc. IEEE*, vol. 68, pp. 1354-1355, Oct. 1980.

[135] —, "Data-adaptive principal component signal processing," in *Proc. 19th IEEE Conf. Decision and Control* (Albuquerque, NM), Dec. 10-12, 1980, pp. 949-954.

*[136] R. T. Lacoss, "Data adaptive spectral analysis methods," *Geophysics*, vol. 36. pp. 661-675, Aug. 1971.

[137] —, "Autoregressive and maximum likelihood spectral analysis methods," in *Proc. 1976 NATO Conf. Signal Processing.*

*[138] T. E. Landers and R. T. Lacoss, "Some geophysical applications of autoregressive spectral estimates," *IEEE Trans. Geosci. Electron.*, vol. GE-15, pp. 26-32, Jan. 1977.

[139] T. Landers, "Maximum entropy cepstral analysis," in *Proc. 1978 RADC Spectral Estimation Workshop*, pp. 245-258.

[140] S. W. Lang and J. H. McClellan, "A simple proof of stability for all-pole linear prediction models," *Proc. IEEE*, vol. 67, pp. 860-861, May 1979.

[141] —, "Frequency estimation with maximum entropy spectral

estimators," *IEEE Trans. Acoust., Speech, Signal Processing*, vol. ASSP-28, pp. 716–724, Dec. 1980.
[142] N. Levinson, "The Wiener (root mean square) error criterion in filter design and prediction," *J. Math. Phys.*, vol. 25, pp. 261–278, 1947.
[143] J. S. Lim, "All pole modeling of degraded speech," *IEEE Trans. Acoust., Speech, Signal Processing*, vol. ASSP-26, pp. 197–209, June 1978.
*[144] J. Makhoul, "Linear prediction: A tutorial review," *Proc. IEEE*, vol. 63, pp. 561–580, Apr. 1975.
[145] —, "Spectral linear prediction: Properties and applications," *IEEE Trans. Acoust., Speech, Signal Processing*, vol. ASSP-23, pp. 283–296, June 1975.
*[146] —, "Stable and efficient lattice methods for linear prediction," *IEEE Trans. Acoust., Speech, Signal Processing*, vol. ASSP-25, pp. 423–428, Oct. 1977.
[147] J. Makhoul and R. Viswanathan, "Adaptive lattice methods for linear prediction," in *Conf. Rec. 1978 ICASSP*, pp. 83–86.
[148] J. D. Markel, "FFT pruning," *IEEE Trans. Audio Electroacoust.*, vol. AU-19, pp. 305–311, Dec. 1971.
[149] S. L. Marple, Jr., "Conventional Fourier, autoregressive, and special ARMA methods of spectrum analysis," Engineer's Dissertation, Stanford Univ., Stanford, CA, Dep. Elec. Eng., Dec. 1976.
[150] —, "Resolution of conventional Fourier, autoregressive and special ARMA methods of spectral analysis," in *Rec. 1977 Int. Conf. Acoustics, Speech and Signal Processing* (Hartford, CT), May 9–11, 1977, pp. 74–77.
[151] —, "High resolution autoregressive spectrum analysis using noise power cancellation," in *Rec. 1978 IEEE Int. Conf. Acoustics, Speech and Signal Processing*., pp. 345–348.
[152] —, "Frequency resolution of high-resolution spectrum analysis techniques," in *Proc. 1978 RADC Spectrum Estimation Workshop*, pp. 19–35.
[153] —, "Spectral line analysis by Pisarenko and Prony methods," in *Rec. 1979 IEEE Int. Conf. Acoustics, Speech, and Signal Processing*, pp. 159–161.
[154] —, "A new autoregressive spectrum analysis algorithm," *IEEE Trans. Acoust., Speech, Signal Process.*, vol. ASSP-28, pp. 441–454, Aug. 1980.
[155] D. Q. Mayne *et al.*, "Linear estimation of ARMA systems," Dep. Computing and Control, Imperial College of Science and Technology, London, England, Research Rep. 77113.
[156] R. N. McDonough, "Representations and analysis of signals. Part XV. Matched exponents for the representation of signals," Ph.D. dissertation, Dep. Elec. Eng., Johns Hopkins Univ., Baltimore, MD, Apr. 30, 1963.
[157] R. N. McDonough and W. H. Huggins, "Best least-squares representation of signals by exponentials," *IEEE Trans. Automat. Contr.*, vol. AC-13, pp. 408–412, Aug. 1968.
*[158] R. N. McDonough, "Maximum entropy spatial processing of array data," *Geophysics*, vol. 39, pp. 843–851, Dec. 1974.
[159] J. M. Melsa and J. D. Tomcik, "Linear predictive coding with additive noise for application to speech digitization," in *Proc. 14th Allerton Conf. Circuits and Systems Theory*, pp. 500–508, Sept. 27, 1976.
[160] R. Mittra and L. W. Pearson, "A variational method for efficient determination of SEM poles," *IEEE Trans. Antennas Propagat.*, vol. AP-26, pp. 354–358, Mar. 1978.
[161] A. Mohammed and R. G. Smith, "Data windowing in spectral analysis," Defense Research Establishment Atlantic, Dartsmouth, Nova Scotia, Rep. 7512, June 1975.
[162] D. R. Moorcroft, "Maximum entropy spectral analysis of radio-auroral signals," Department of Physics and Centre for Radio Science, Univ. Western Ontario, London, Ontario, Canada.
[163] M. Morf, "Fast algorithms for multivariable systems," Ph.D. dissertation, Stanford University, Stanford, CA, Dep. Elec. Eng., Aug. 1974.
[164] M. Morf, T. Kailath, and L. Ljung, "Fast algorithms for recursive identification," in *Proc. 1976 IEEE Conf. Decision and Control* (Clearwater, FL), Dec. 1–3, 1976, pp. 916–921.
*[165] M. Morf, D. T. Lee, J. R. Nickolls, and A. Vieira, "A classification of algorithms for ARMA models and ladder realizations," in *Rec. 1977 IEEE Int. Conf. Acoustics, Speech, and Signal Processing*. (Hartford, CT), May 9–11, 1977, pp. 13–19.
[166] M. Morf, A. Vieira, D. T. Lee, and T. Kailath, "Recursive multichannel maximum entropy method," in *Proc. 1977 Joint Automatic Control Conf.* (San Francisco, CA), pp. 113–117, June 22–24, 1977.
*[167] M. Morf, B. Dickinson, T. Kailath, and A. Vieira, "Efficient solution of covariance equations for linear prediction," *IEEE Trans. Acoust., Speech, Signal Process.*, vol. ASSP-25, pp. 429–433, Oct. 1977.
[168] M. Morf, A. Vieira, and D. T. Lee, "Ladder forms for identification and speech processing," in *Proc. 1977 Conf. Decision and Control* (New Orleans, LA), pp. 1074–1078, Dec. 1977.
*[169] M. Morf, A. Vieira, D. T. Lee, and T. Kailath, "Recursive multichannel maximum entropy spectral estimation," *IEEE Trans. Geosci. Electron.*, vol. GE-16, pp. 85–94, Apr. 1978.
[170] P. O. Neudorfer, "Alternate methods of harmonic analysis," *Proc. IEEE*, vol. 61, pp. 1661–1662, Nov. 1973.
*[171] W. I. Newman, "Extension to the maximum entropy method," *IEEE Trans. Inform. Theory*, vol. IT-23, pp. 89–93, Jan. 1977.
[172] —, "Extension to the maximum entropy method II," *IEEE Trans. Inform. Theory*, vol. IT-25, pp. 705–708, Nov. 1979.
[173] R. Nitzberg, "Spectral estimation: An impossibility?" *Proc. IEEE*, vol. 67, pp. 437–439, Mar. 1979.
[174] B. Noble and J. W. Daniel, *Applied Linear Algebra*, 2nd ed. Englewood Cliffs, NJ: Prentice-Hall, 1977.
[175] A. H. Nuttall, "Spectral estimation by means of overlapped fast Fourier transform processing of windowed data," NUSC Tech. Rep. 4169, New London, CT, Oct. 13, 1971.
[176] —, "Spectral analysis of a univariate process with bad data points, via maximum entropy, and linear predictive techniques," Naval Underwater Systems Center, Tech. Rep. 5303, New London, CT, Mar. 26, 1976.
[177] —, "FORTRAN program for multivariate linear predictive spectral analysis, employing forward and backward averaging," Naval Underwater Systems Center, Tech. Document 5419, New London, CT, May 19, 1976.
[178] —, "Multivariate linear predictive spectral analysis employing weighted forward and backward averaging: A generalization of Burg's algorithm," Naval Underwater Systems Center, Tech. Rep. 5501, New London, CT, Oct. 13, 1976.
[179] —, "Probability distribution of spectral estimates obtained via overlapped FFT processing of windowed data," Naval Underwater Systems Center, Tech. Rep. 5529, New London, CT, Dec. 3, 1976.
[180] —, "Positive definite spectral estimate and stable correlation recursion for multivariate linear predictive spectral analysis," NUSC Tech. Rep. 5729, Naval Underwater Systems Center, New London, CT, Nov. 14, 1977.
[181] A. H. Nuttall and G. C. Carter, "A generalized framework for power spectral estimation," *IEEE Trans. Acoust., Speech, Signal Process.*, vol. ASSP-28, pp. 334–335, June 1980.
[182] A. H. Nuttall, "Some windows with very good sidelobe behavior," *IEEE Trans. Acoust., Speech, Signal Process.*, vol. ASSP-29, pp. 84–89, Feb. 1981.
[183] A. V. Oppenheim and R. W. Schafer, *Digital Signal Processing*. Englewood Cliffs, NJ: Prentice-Hall, 1975.
[184] R. K. Otnes and L. Enochson, *Digital Time Series Analysis*. New York: Wiley, 1972.
[185] N. L. Owsley, "Adaptive data orthogonalization," in *Proc. 1978 IEEE ICASSP* (Tulsa, OK), pp. 109–112, April 1978.
[186] M. Pagano, "Estimation of models of autoregressive signal plus white noise," *Ann. Statistics*, vol. 2, pp. 99–108, 1974.
[187] A. Papoulis, *Probability, Random Variables, and Stochastic Processes*. New York: McGraw-Hill, 1965.
[188] E. Parzen, "Mathematical considerations in the estimation of spectra," *Technometrics*, vol. 3, pp. 167–190, May 1961.
[189] —, "Statistical spectral analysis (single channel case) in 1968," Dep. Statistics, Stanford Univ., Stanford, CA, Tech. Rep. 11, June 10, 1968.
[190] —, "Multiple time series modelling," Dep. Statistics, Stanford, Univ., Stanford, CA, Tech. Rep. 12, July 8, 1968 (Contract Nonr-225-80).
[191] —, "Multiple time series modelling," *Multivariate Analysis II*, P. R. Krishnaiah, Ed. New York: Academic Press, pp. 389–409, 1969.
*[192] —, "Some recent advances in time series modeling," *IEEE Trans. Automat. Contr.*, vol. AC-19, pp. 723–730, Dec. 1974.
[193] J. V. Pendrel and D. E. Smylie, "The relationship between maximum entropy and maximum likelihood spectra," *Geophysics*, vol. 44, pp. 1738–1739, Oct. 1979.
[194] V. F. Pisarenko, "On the estimation of spectra by means of nonlinear functions of the covariance matrix," *Geophysical J. Royal Astronomical Soc.*, vol. 28, pp. 511–531, 1972.
[195] —, "The retrieval of harmonics from a covariance function," *Geophysical J. Royal Astronomical Soc.*, vol. 33, pp. 347–366, 1973.
[196] A. J. Poggio, M. L. Van Blaricum, E. K. Miller, and R. Mittra, "Evaluation for a processing technique for transient data," *IEEE Trans. Antennas Propagat.*, vol. AP-26, pp. 165–173, Jan. 1978.
[197] J. Ponnusarny *et al.*, "Identification of complex autoregressive processes," in *Rec. 1979 ICASSP*, pp. 384–387.
[198] G. Prado and P. Moroney, "Linear predictive spectral analysis for Doppler sonar applications," Charles Stark Draper Laboratory, Tech. Rep. R-1109, Sept. 1977.
[199] —, "The application of linear predictive methods of spectral analysis to frequency estimation," in *Rec. 1978 NAECON*.
[200] —, "The accuracy of center frequency estimators using linear predictive methods," in *Rec. 1978 IEEE Int. Conf. Acoustics, Speech and Signal Processing*, pp. 361–364.
[201] J. F. Prewitt, "Amplitude bias in the Fourier transforms of noisy

signals," *IEEE Trans. Antennas Propagat.*, vol. AP-26, pp. 730–731, Sept. 1978.

[202] G.R.B. Prony, "Essai experimental et analytique, etc.," *Paris, J. de L'Ecole Polytechnique*, vol. 1, cahier 2, pp. 24–76, 1795.

[203] L. C. Pusey, "High resolution spectral estimates," Lincoln Laboratory, M.I.T., Tech. Note 1975-7, Jan. 21, 1975.

[204] C. M. Rader, "An improved algorithm for high speed autocorrelation with applications to spectral estimation," *IEEE Trans. Audio Electroacoust.*, vol. AU-18, pp. 439–442, Dec. 1970.

[205] H. R. Radoski, P. F. Fougere, and E. J. Zawalick, "A comparison of power spectral estimates and applications of the maximum entropy method," *J. Geophysical Res.*, vol. 80, pp. 619–625, Feb. 1, 1975.

*[206] H. R. Radoski, E. J. Zawalick, and P. F. Fougere, "The superiority of maximum entropy power spectrum techniques applied to geomagnetic micropulsations," *Phys. Earth Planetary Interiors*, vol. 12, pp. 208–216, Aug. 1976.

[207] P. I. Richards, "Computing reliable power spectra," *IEEE Spectrum*, vol. 4, pp. 83–90, Jan. 1967.

[208] D. C. Rife and G. A. Vincent, "Use of the discrete Fourier transform in the measurement of frequencies and levels of tones," *Bell Syst. Tech. J.*, vol. 49, pp. 197–228, Feb. 1970.

[209] E. A. Robinson, "Predictive decomposition of time series with application to seismic exploration," *Geophysics*, vol. 32, pp. 418–484, June 1967.

[210] ——, *Multichannel Time Series Analysis with Digital Computer Programs.* San Francisco, CA: Holden-Day, 1967.

[211] A. P. Sage and J. L. Melsa, *Estimation Theory with Applications to Communication and Control.* New York: McGraw-Hill, 1971, ch. 6.

[212] H. Sakai, "Statistical properties of AR spectral analysis," *IEEE Trans. Acoust., Speech, Signal Processing*, vol. ASSP-27, pp. 402–409, Aug. 1979.

[213] M. R. Sambur, "A preprocessing filter for enhancing LPC analysis/synthesis of noisy speech," in *Rec. 1979 ICASSP*, pp. 971–974.

[214] E. H. Satorius and J. T. Alexander, "High resolution spectral analysis of sinusoids in correlated noise," in *Rec. 1978 ICASSP* (Tulsa, OK), pp. 349–351, Apr. 10–12, 1978.

[215] E. H. Satorius and J. R. Zeidler, "Maximum entropy spectral analysis of multiple sinusoids in noise," *Geophysics*, vol. 43, pp. 1111–1118, Oct. 1978.

[216] J. H. Sawyers, "Applying the maximum entropy method to adaptive digital filtering," in *Rec. 12th Asilomar Conf. Circuits, Systems and Computers*, pp. 198–202.

[217] D. H. Scahubert, "Application of Prony's method to time-domain reflectometer data and equivalent circuit synthesis," *IEEE Trans. Antennas Propagat.*, vol. AP-27, pp. 180–184, Mar. 1979.

[218] A. Schuster, "On the investigation of hidden periodicities with application to a supposed 26 day period of meteorological phenomena," *Terrestrial Magnetism*, vol. 3, pp. 13–41, Mar. 1898.

[219] ——, "The periodogram of magnetic declination as obtained from the records of the Greenwich Observatory during the years 1871–1895," *Trans. Cambridge Philosophical Soc.*, vol. 18, pp. 107–135, 1899.

[220] K.L.S. Sharma and A. K. Mahalanabis, "Harmonic analysis via Kalman filtering technique," *Proc. IEEE*, pp. 391–392, Mar. 1973.

[221] D. E. Smylie, G.K.C. Clarke, and T. J. Ulrych, "Analysis of irregularities in the earth's rotation," in *Methods in Computational Physics*, vol. 13. B. A. Bolt, Ed. New York: Academic Press, 1973, pp. 391–430.

[222] M. D. Srinath and M. M. Viswanathan, "Sequential algorithm for identification of parameters of an autoregressive process," *IEEE Trans. Automat. Contr.*, vol. AC-20, pp. 542–546, Aug. 1975.

[223] G. G. Stokes, "Note on the search for periodic inequalities," *Proc. Royal Soc. London*, vol. 29, pp. 122–123, May 29, 1879.

[224] N. R. Strader, II, "Effects of subharmonic frequencies on DFT coefficients," *Proc. IEEE*, vol. 68, pp. 285–286, Feb. 1980.

[225] O. N. Strand, "Multichannel maximum entropy spectral analysis," *IEEE Trans. Automat. Contr.*, vol. AC-22, pp. 634–640, Aug. 1977.

[226] T. M. Sullivan, O. L. Frost, and J. R. Treichler, "High resolution signal estimation," ARGO Systems, Internal Rep. June 16, 1978.

[227] D. N. Swingler, "A comparison between Burg's maximum entropy method and a nonrecursive technique for the spectral analysis of deterministic signals," *J. Geophysical Res.*, vol. 84, pp. 679–685, Feb. 10, 1979.

[228] ——, "A modified Burg algorithm for maximum entropy spectral analysis," *Proc. IEEE*, vol. 67, pp. 1368–1369, Sept. 1979.

[229] ——, "Frequency errors in MEM processing," *IEEE Trans. Acoust., Speech, Signal Processing*, vol. ASSP-28, pp. 257–259, Apr. 1980.

[230] ——, "Comments on maximum entropy spectral estimation using the analytical signal," *IEEE Trans. Acoust., Speech, Signal Processing*, vol. ASSP-28, pp. 259–260, Apr. 1980.

[231] D. J. Thomson, M. F. Robbins, C. G. Maclennan, and L. J. Lanzerotti, "Spectral and windowing techniques in power spectral analyses of geomagnetic data," *Phys. Earth Planetary Interiors*, vol. 12, pp. 217–231, Aug. 1976.

[232] D. J. Thomson, "Spectrum estimation techniques for characterization and development of WT4 waveguide-I," *Bell Syst. Tech. J.*, vol. 56, pp. 1769–1815, Nov. 1977.

[233] T. Thorvaldsen, A. T. Waterman Jr., and R. W. Lee, "Maximum entropy angular response patterns of microwave transhorizon signals," *IEEE Trans. Antennas Propagat.*, vol. AP-28, pp. 722–724, Sept. 1980.

[234] K. Toman, "The spectral shift of truncated sinusoids," *J. Geophysical Res.*, vol. 70, pp. 1749–1750, Apr. 1, 1965.

*[235] H. Tong, "Autoregressive model fitting with noisy data by Akaike's information criterion," *IEEE Trans. Inform. Theory*, vol. IT-21, pp. 476–480, July 1975.

[236] ——, "Fitting a smooth moving average to noisy data," *IEEE Trans. Inform. Theory*, vol. IT-22, pp. 493–496, July 1976.

*[237] ——, "More on autoregressive model fitting with noisy data by Akaike's information criterion," *IEEE Trans. Inform. Theory*, vol. IT-23, pp. 409–410, May 1977.

[238] H. Tong, "Final prediction error and final interpolation error: A paradox?" *IEEE Trans. Inform. Theory*, vol. IT-25, pp. 758–759, Nov. 1979.

[239] J. P. Toomey, "High-resolution frequency measurement by linear prediction," *IEEE Trans. Aerospace Electron. Syst.*, vol. AES-16, pp. 517–525, July 1980.

[240] J. Treichler, "γ-LMS and its use in a noise-compensating adaptive spectral analysis technique," in *Rec. 1979 ICASSP*, pp. 933–936.

[241] W. Trench, "An algorithm for the inversion of finite Hankel matrices," *J. Soc. Indust. Appl. Math.*, vol. 13, no. 4, pp. 1102–1107, Dec. 1965.

[242] S. A. Tretter and K. Steiglitz, "Power spectrum identification in terms of rational models," *IEEE Trans. Automat. Contr.*, vol. AC-12, pp. 185–188, Apr. 1967.

[243] D. W. Tufts, "Adaptive line enhancement and spectrum analysis," *Proc. IEEE*, vol. 65, pp. 169–173, Jan. 1977.

[244] D. W. Tufts and R. Kumaresan, "Improved spectral resolution," *Proc. IEEE*, vol. 68, pp. 419–420, Mar. 1980.

[245] ——, "Improved spectral resolution II," in *Proc. 1980 ICASSP* (Denver, CO), vol. 2, pp. 592–597, Apr. 9–11, 1980.

[246] T. J. Ulrych, "Maximum entropy power spectrum of long period geomagnetic reversals," *Nature*, vol. 235, pp. 218–219, Jan. 28, 1972.

[247] ——, "Maximum entropy power spectrum of truncated sinusoids," *J. Geophysical Research*, vol. 77, pp. 1396–1400, March 10, 1972.

[248] T. J. Ulrych, D. E. Smylie, O. G. Jensen, and G.K.C. Clarke, "Predictive filtering and smoothing of short records by using maximum entropy," *J. Geophysical Res.*, vol. 78, pp. 4959–4964, Aug. 10, 1973.

[249] T. J. Ulrych and O. G. Jensen, "Cross-spectral analysis using maximum entropy," *Geophysics*, vol. 39, pp. 353–354, June 1974.

*[250] T. J. Ulrych and T. N. Bishop, "Maximum entropy spectral analysis and autoregressive decomposition," *Rev. Geophysics Space Phys.*, vol. 13, pp. 183–200, Feb. 1975.

[251] T. J. Ulrych and R. W. Clayton, "Time series modelling and maximum entropy," *Phys. Earth Planetary Interiors*, vol. 12, pp. 188–200, Aug. 1976.

[252] M. L. Van Blaricum and R. Mittra, "Problems and solutions associated with Prony's method for processing transient data," *IEEE Trans. Antennas Propagat.*, vol. AP-26, pp. 174–182, Jan. 1978.

[253] M. L. Van Blaricum, "A review of Prony's method techniques for parameter estimation," in *Proc. Rome Air Development Center Spectrum Estimation Workshop*, Griffis Air Force Base, May 24–26, 1978. pp. 125–139.

[254] M. L. Van Blaricum and R. Mittra, Correction to "Problems and solutions associated with Prony's method for processing transient data," *IEEE Trans. Antennas Propagat.*, vol. AP-28, p. 949, Nov. 1980.

*[255] A. VanDenBos, "Alternative interpretation of maximum entropy spectral analysis," *IEEE Trans. Inform. Theory*, vol. IT-17, pp. 493–494, July 1971.

[256] H. L. Van Trees, *Detection, Estimation, and Modulation Theory.* New York: McGraw-Hill, 1968, vol. I.

[257] R. Viswanathan and J. Makhoul, "Quantization properties of transmission parameters in linear predictive systems," *IEEE Trans. Acoust., Speech, Signal Processing*, vol. ASSP-23, pp. 309–321, June 1975.

[258] G. Walker, "On periodicity in series of related terms," *Proc. Roy. Soc. London, Series A*, vol. 131, pp. 518–532, 1931.

[259] C. Webb, "Practical use of the fast Fourier transform (FFT) algorithm in time series analysis," ARL, Univ. Texas, Austin, TX, ARL-TR-70-22, June 22, 1970.

[260] R. J. Webster, "Leakage regulation in the discrete Fourier transform spectrum," *Proc. IEEE*, vol. 68, pp. 1339-1341, Oct. 1980.

[261] L. Weiss and R. N. McDonough, "Prony's method, Z-transforms, and Pade approximation," *SIAM Rev.*, vol. 5, pp. 145-149, Apr. 1963.

*[262] P. D. Welch, "The use of fast Fourier transform for the estimation of power spectra: A method based on time averaging over short, modified periodograms," *IEEE Trans. Audio Electroacoust.*, vol. AU-15, pp. 70-73, June 1967.

[263] P. D. Welch, "On the variance of time and frequency averages over modified periodograms," in *Rec. 1977 IEEE Int. Conf. Acoustics, Speech, and Signal Processing* (Hartford, CT), May 9-11, 1977, pp. 58-62.

*[264] S. J. Wernecke and L. R. D'Addario, "Maximum entropy image reconstruction," *IEEE Trans. Comput.*, vol. C-26, pp. 351-364, Apr. 1977.

[265] S. J. Wernecke, "Two-dimensional maximum entropy reconstruction of radio brightness," *Radio Sci.*, vol. 12, pp. 831-844, Sept.-Oct. 1977.

[266] E. C. Whitman, "The spectral analysis of discrete time series in terms of linear regressive models," Naval Ordnance Labs Rep. NOLTR-70-109, White Oak, MD, June 23, 1974.

[267] P. Whittle, "On the fitting of multivariate autoregressions, and the approximate canonical factorization of a spectral density matrix," *Biometrika*, vol. 50, pp. 129-134, 1963.

[268] B. Widrow *et al.*, "Adaptive noise cancelling: Principles and application," *Proc. IEEE*, vol. 63, pp. 1692-1716, Dec. 1975.

[269] N. Wiener, "Generalized harmonic analysis," *Acta Mathematica*, vol. 55, pp. 117-258, 1930.

[270] R. A. Wiggins and E. A. Robinson, "Recursive solution to the multi-channel filtering problem," *J. Geophysical Res.*, vol. 70, pp. 1885-1891, Apr. 1965.

[271] A. S. Willsky, "Relationship between digital signal processing and control and estimation theory," *Proc. IEEE*, vol. 66, pp. 996-1017, Sept. 1978.

[272] ——, *Digital Signal Processing and Control and Estimation Theory: Points if Tangency, Areas of Intersection, and Parallel Directions.* Cambridge, MA: M.I.T. Press, 1980.

[273] T. H. Wonnacott, "Spectral analysis combining a Bartlett window with an associated inner window," *Technometrics*, vol. 3, pp. 235-243, May 1961.

[274] C. K. Yuen, "A comparison of five methods for computing the power spectrum of a random process using data segmentation," *Proc. IEEE*, vol. 65, pp. 984-986, June 1977.

[275] ——, "Comments on modern methods for spectral estimation," *IEEE Trans. Acoust., Speech, Signal Processing*, vol. ASSP-27, pp. 298-299, June 1979.

[276] G. U. Yule, "On a method of investigating periodicities in disturbed series, with special reference to Wolfer's sunspot numbers," *Philosophical Trans. Royal Soc. London, Series A*, vol. 226, pp. 267-298, July 1927.

[277] S. Zohar, "The solution of a Toeplitz set of linear equations," *J. Assoc. Comput. Mach.*, vol. 21, pp. 272-276, Apr. 1974.

[278] ——, "FORTRAN subroutines for the solution of Toeplitz sets of linear equations," *IEEE Trans. Acoust., Speech, Signal Processing*, vol. ASSP-27, pp. 656-658, Dec. 1979.

Corrections to "Spectrum Analysis—A Modern Perspective"

S. LAWRENCE MARPLE, JR.

In the above titled paper[1] the following errors should be corrected:

Page 1389.
Table I. "$\sigma_p^2 \geqslant 0$" should read "$\sigma_p^2 > 0$".

Manuscript received March 15, 1982.

The author is with the Analytic Sciences Corp., Washington Systems Engineering, McLean, VA 22102.

[1] S. M. Kay and S. L. Marple, Jr., *Proc. IEEE*, vol. 69, no. 11, pp. 1380-1419, Nov. 1981.

Page 1393.
Fig. 7. The index range for $\mathcal{E}_0$ should be 0 to $N-1$ rather than 1 to N.

Page 1394.
Equation above (2.81) should read

$$\mathfrak{X}_{m+1} = \begin{bmatrix} \mathfrak{X}_m \\ \hline H_{m+1} \end{bmatrix}.$$

Sentence below (2.85) should read "(2.83)-(2.85)" rather than "(2.13)-(2.85)."
In upper right corner, the "$0(p^3)$" should be "$0(p^2)$."
Paragraph beginning with "Convergence is guaranteed . . ." should read "Convergence of $E[\hat{A}_m]$ to A_m is guaranteed . . . ."

Page 1395.
Equation below (2.92) should be

$$\text{Residue } \frac{P_{AR}(z)}{z} = (z - z_i) \left. \frac{P_{AR}(z)}{z} \right|_{z=z_i}$$

Page 1397.
Fig. 9. The equation "$\sigma_\eta^2 B(z)\, B(z^{-1}) = \sigma^2 + \sigma_w^2 A(z)\, A(z^{-1})$" should be "$\sigma_\eta^2 B(z)\, B^*(1/z^*) = \sigma^2 + \sigma_w^2 A(z)\, A^*(1/z^*)$."

Page 1398.
Equation (2.102) should be

$$\delta f \approx \frac{6.471[\rho(p+1)]^{-0.31}}{2\pi p \Delta t}.$$

Page 1399.
The following three references were inadvertently omitted in the heading for Section G: [1], [2], [3].

Page 1400.
The sentence "As $N \to \infty$, the ARMA (1, 1) model . . . ," should read "As $a_1 \to 0.5$, the ARMA (1, 1) model is seen to be inappropriate, since . . . ."

Page 1402.
In the $\boldsymbol{H}$ matrix at the top of the page, the bottom left-hand term should be x_{N-2} rather than x_{N-1}. The sentence below that which reads "Equation (2.115) is the standard . . ." should be "Equation (2.116) is the standard . . . ."

Page 1405.
The vector $\hat{\boldsymbol{X}}$ below (2.150) should be "$\hat{\boldsymbol{X}} = [\hat{x}_0 \cdots \hat{x}_{N-1}]^T$."

Page 1407.
The sentence above (2.161) should read "For example, using ϵ_n ranging from $n = p + 1$ to $N - p$, we have . . . ."

Page 1410.
Table II. The following reference should be added to the AR Yule-walker Technique: 176.
The following references should be added to the AR Least Squares Technique: 176, 251.

Page 1413.
The E above (4.6) should be $\mathcal{E}$, as follows "then $\mathcal{E}$ becomes $\mathcal{E}$ = ..."
In equation (4.8), a Δx term should be placed to the left of each Σ.

Page 1417.
The page numbers for reference [182] should be pp. 84-91.

References

[1] J. A. Cadzow, "High performance spectral estimation—A new ARMA method," *IEEE Trans. Acoust. Speech, Signal Processing.*, vol. ASSP-28, pp. 524-529, Oct. 1980.

[2] ——, "Autoregressive moving average spectral estimation: A model equation error procedure," *IEEE Trans. Geosci. Remote Sensing*, vol. GE-19, pp. 24-28, Jan. 1981.

[3] D. Lee, B. Friedlander, and M. Morf, "Recursive ladder algorithms for ARMA modeling," *IEEE Trans. Automat. Contr.*, vol. AC-27, no. 3, pp. 753-764, Aug. 1982.

Reprinted from *Proc. IEEE*, vol. 70, pp. 1238, Oct. 1982.

Part II
Parametric Methods

MODELING of time series by rational transfer functions dates back to 1920's. One of the pioneering works on the subjects was that of G. U. Yule, which is reprinted here. It is an interesting historical reference that gave us the autoregressive (AR) approximation for sinusoids. The work was motivated by the analysis of Wolfer's sunspot number, a time series which has become one of the standards for testing various signal processing techniques. The papers by A. Papoulis and by Lang and McClellan present an excellent review of the application of the maximum entropy principle to spectral estimation with the former one putting an emphasis to the information theoretic aspect of entropy. Another approach to analysis of multiple sinusoids by maximum entropy method (MEM) is given in paper by Satorius and Zeidler. They used the method of undetermined coefficients to reduce the number of normal equations.

For some signals autoregressive modeling is not satisfactory. In fact, even a simple test signal, containing sinusoids in noise, must be modeled by a transfer function containing *zeros* as well as poles, if the noise power is of the same order of magnitude as the powers of deterministic components. This is certainly true in cases of more complicated signals. In such cases, spectral analysis is based on the autoregressive-moving average (ARMA) modeling of signals. The paper by Kaveh and Bruzzone gives a comparison overview of a number of ARMA techniques. Comparison is based on the computational and statistical efficiencies of the method. It also presents good practical guidelines for those applying these techniques.

The minimum cross-entropy method (MCEM) is treated in the paper by J. E. Shore. It includes the prior information in shaping the spectral estimates. The various shaped estimates correspond to various underlying probability density functions (pdf's). The resulting spectral estimate is obtained by minimizing the cross-entropy of the pdf's.

The next three papers present the methods of spectral/parameter estimation which are based on eigenstructure of the covariance matrix. Pisarenko's harmonic decomposition consists of finding the minimum eigenvalue and the corresponding eigenvector of the exact covariance matrix. The method is based on the assumption that signal contains only sinusoids in white noise and thus have the covariance matrix exactly known. Schmidt's MUSIC algorithm utilizes all of the noise subspace eigenvectors instead of only one, as in the Pisarenko decomposition. This averaging makes the former method less sensitive to the assumption of the exactness of the covariance matrix. The MUSIC algorithm is presented in terms of spatial signals arriving at an array of sensors, but the reformulation to temporal domain is straightforward. The third paper in this group, authored by Tufts and Kumaresan, presents the method which belongs to the class of least-squares methods. Instead of the eigen-decomposition of the covariance matrix, a singular value decomposition (SVD) of the data matrix is performed, and the singular values corresponding to the noise subspace are set to zero. In the last paper of the section, Kay and Demeure point out the potential danger of using visual examination as the only criterion for assessing resolution capabilities of spectral analysis methods.

VII. *On a Method of Investigating Periodicities in Disturbed Series, with special reference to Wolfer's Sunspot Numbers.*

By G. UDNY YULE, *C.B.E., M.A., F.R.S., University Lecturer in Statistics, and Fellow of St. John's College, Cambridge.*

(Received December 11, 1926.—Revised February 17, 1927.)

CONTENTS.

I. INTRODUCTORY: SUPERPOSED FLUCTUATIONS AND DISTURBANCES.

If we take a curve representing a simple harmonic function of the time, and superpose on the ordinates *small* random errors, the only effect is to make the graph somewhat irregular, leaving the suggestion of periodicity still quite clear to the eye. Fig. 1 (*a*) shows such a curve, the random errors having been determined by the throws of dice. If the errors are increased in magnitude, as in fig. 1 (*b*), the graph becomes more irregular, the suggestion of periodicity more obscure, and we have only sufficiently to increase the "errors" to mask completely any appearance of periodicity. But, however large the errors, periodogram analysis is applicable to such a curve, and, given a sufficient number of periods, should yield a close approximation to the period and amplitude of the underlying harmonic function.

Reprinted with permission from *Phil. Trans. Roy. Soc. London*, vol. 226, Series A, pp. 267–298, July 1927.

When periodogram analysis is applied to data respecting any physical phenomenon in the expectation of eliciting one or more true periodicities, there is usually, as it seems to me, a tendency to start from the initial hypothesis that the periodicity or periodicities are masked solely by such more or less random *superposed fluctuations*—fluctuations which do not in any way disturb the steady course of the underlying periodic function or functions. It is true that the periodogram itself will indicate the truth or otherwise of the hypothesis made, but there seems no reason for assuming it to be the hypothesis most likely *a priori*.

If we observe at short equal intervals of time the departures of a simple harmonic pendulum from its position of rest, errors of observation will cause superposed fluctuations of the kind supposed in fig. 1. But by improvement of apparatus and automatic methods of recording, let us say, errors of observation are practically eliminated. The recording

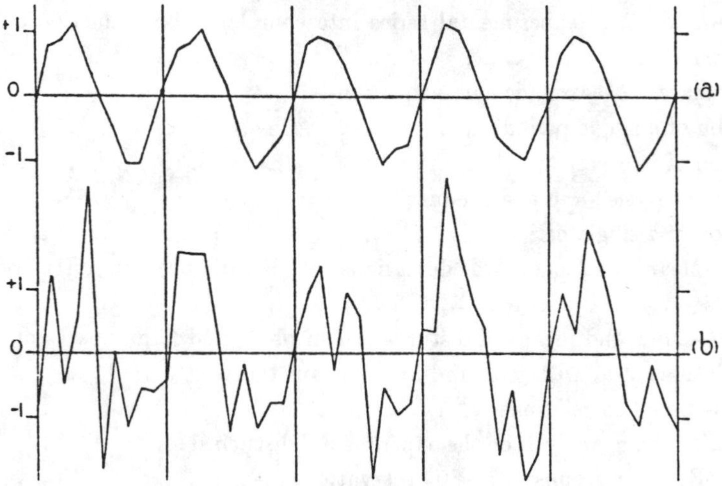

FIG. 1.—Graphs of simple harmonic functions of unit amplitude with superposed random fluctuations: (*a*) smaller fluctuations, (*b*) larger fluctuations.

apparatus is left to itself, and unfortunately boys get into the room and start pelting the pendulum with peas, sometimes from one side and sometimes from the other. The motion is now affected, not by *superposed fluctuations* but by true *disturbances*, and the effect on the graph will be of an entirely different kind. The graph will remain surprisingly smooth, but amplitude and phase will vary continually.

Working with finite in lieu of infinitesimal intervals, we may construct an approximation to a curve of the kind supposed. Let the terms of the trigonometrical series be

$$\left.\begin{aligned} u_0 &= \mathrm{A} \sin 2\pi \frac{t+\tau}{\mathrm{T}} \\ u_1 &= \mathrm{A} \sin 2\pi \frac{t+\tau+h}{\mathrm{T}} \\ u_2 &= \mathrm{A} \sin 2\pi \frac{t+\tau+2h}{\mathrm{T}} \end{aligned}\right\}, \qquad (1)$$

etc., where A is the amplitude, t the time, T the period, h the interval between successive terms, and τ gives the phase. Then, with a little reduction, we have for the second difference

$$\Delta^2 (u_0) = -\left(4 \sin^2 \pi \frac{h}{T}\right) u_1 = - \mu u_1, \tag{2}$$

where, if θ is the angle corresponding to the interval,

$$\mu = 4 \sin^2 \pi \frac{h}{T} = 2 (1 - \cos \theta). \tag{3}$$

But, in terms of the u's, (2) gives

$$u_2 = (2 - \mu) u_1 - u_0, \tag{4}$$

where it may be noted

$$2 - \mu = 2 \cos \theta. \tag{5}$$

If there are no disturbances, (4) gives u_x generally in terms of u_{x-1} and u_{x-2}. Provided the interval were infinitesimal, (4) would still give u_x correctly, even if the velocity in the interval $x - 2$ to $x - 1$ were affected by an impulse, so long as the interval $x - 1$ to x were undisturbed. But if a disturbance occurred also in the latter interval, we would have, say,

$$u_x = (2 - \mu) u_{x-1} - u_{x-2} + \varepsilon, \tag{6}$$

where ε is an "error" varying with the impulse or disturbance.

Fig. 2 shows a graph constructed in the following way from this equation:—The period was taken as 10 intervals, and the first two ordinates as 0 and sin 36° (0·588). Thereafter all the ordinates were calculated in succession by (6), the errors or "disturbances" ε being given by dice-throwing. Four dice were thrown together: the divergence of the sum of the pips from the mean number (that is, 14) was divided by 20, and this was taken as the value of ε. The values so determined fluctuate round zero with a standard deviation 0·1708, and are thus fairly considerable, ranging up to $\pm$ 0·5. Inspection of the figure shows that there are now no abrupt variations in the graph, but the amplitude varies within wide limits, and the phase is continually shifting. Increasing the magnitude of the disturbances simply increases the amplitude: the graph remains smooth. At one point, in fact, the "disturbance" was inadvertently magnified by an error of calculation, but there was no appreciable kink in the graph to direct attention to the blunder. It is, of course, true that the graph may be made to pass through any assigned series of points, however irregular, but to introduce such irregularities appropriate large and erratic disturbances must be given: abrupt irregularities do not naturally occur with random disturbances.

It is of interest to look a little more closely into the question why the graph does present such a smooth appearance. An undisturbed harmonic function may be regarded as the solution of the difference equation

$$\Delta^2 (u_x) + \mu (u_{x+1}) = 0. \tag{7}$$

2 o 2

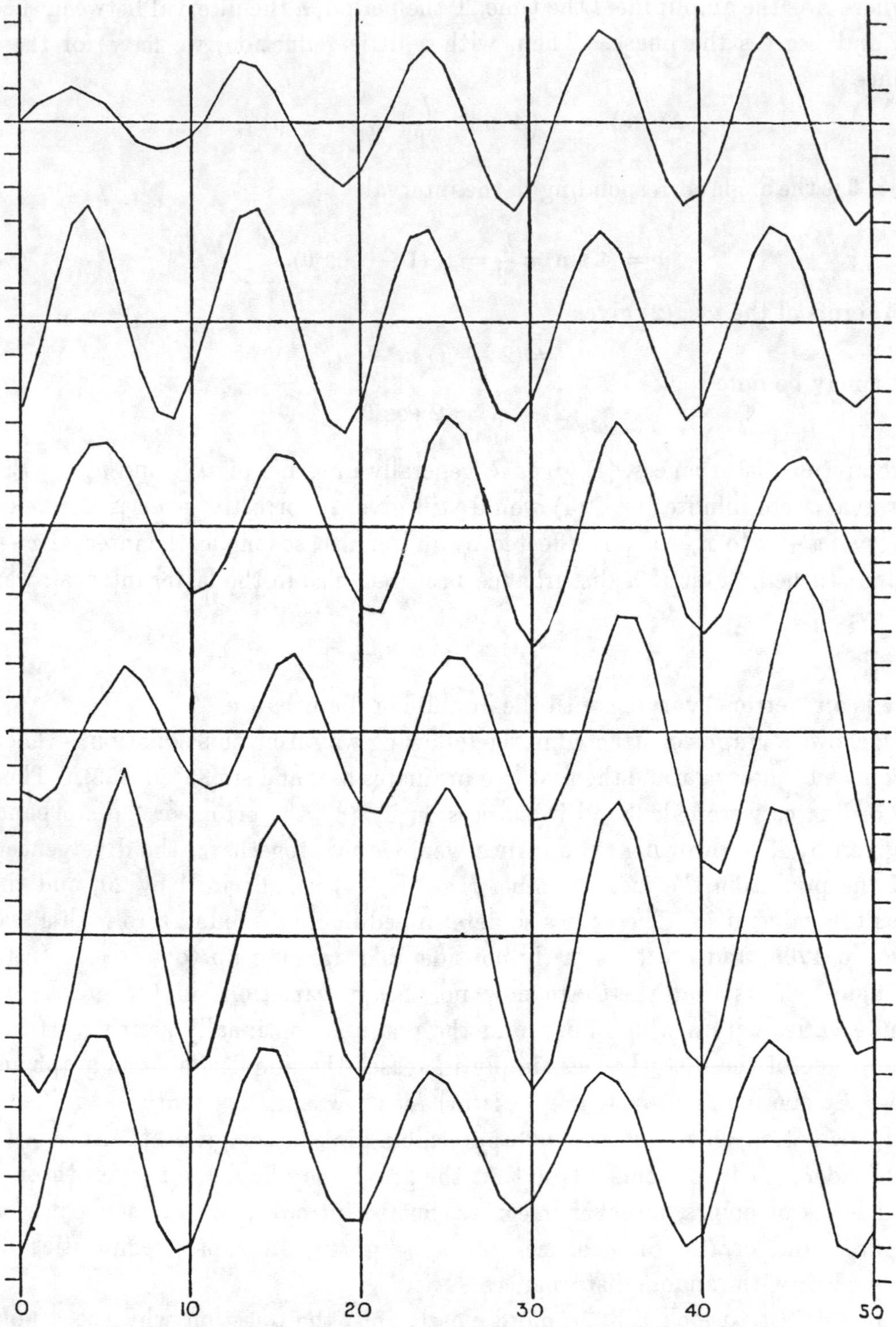

FIG. 2.—Graph of a disturbed harmonic function, experimental series.

If the motion is disturbed, on the other hand, we have, say,

$$\Delta^2 (u_x) + \mu (u_{x+1}) = \phi (t), \tag{8}$$

where $\phi(t)$ may be regarded as a "disturbance function." Looking at the matter from this standpoint, the simple harmonic function is the complementary function in the solution of (8), and the difference between the simple harmonic function and the oscillatory function of the time represented by the graph is the particular integral.

Consider the formation of the series by equation (6), given initial terms u_0 and u_1 and the disturbances ε_2, ε_3, ε_4, etc., appropriate to u_2 onwards. Writing for brevity

$$2 - \mu = k, \qquad (9)$$

the series will be

0. u_0
1. u_1
2. $ku_1 - u_0 + \varepsilon_2$
3. $(k^2 - 1)\, u_1 - ku_0 + k\varepsilon_2 + \varepsilon_3$
4. $\{k(k^2 - 1) - k\}\, u_1 - (k^2 - 1)\, u_0 + (k^2 - 1)\, \varepsilon_2 + k\varepsilon_3 + \varepsilon_4$
5. $\{k[k(k^2 - 1) - k] - (k^2 - 1)\}\, u_1 - \{k(k^2 - 1) - k\}\, u_0 + \{k(k^2 - 1) - k\}\, \varepsilon_2 + (k^2 - 1)\, \varepsilon_3 + k\varepsilon_4 + \varepsilon_5,$

etc. Examining the ε terms, we see that the coefficients 1, k, $k^2 - 1$, $k(k^2 - 1) - k$, etc., are related by an equation of the form

$$A_m = k(A_{m-1}) - A_{m-2}. \qquad (10)$$

But this is simply an equation of the form (4), and the coefficients of the ε's are therefore the terms of a sine series, of the same period as the complementary function, with initial terms 1 and k. For the experimental series they are

$$\begin{array}{r} +\,1 \\ +\,1\cdot618034 \\ +\,1\cdot618034 \\ +\,1 \\ 0 \\ -\,1 \\ -\,1\cdot618034 \end{array}$$

etc. Table I shows the first 30 terms of the experimental series, the complementary function, the particular integral, and the values of ε. Thus, on line 10 of the table we have

$$\begin{aligned} -\,0\cdot54272 = &+ 1\,(-\,0\cdot10) \\ &+ 1\cdot618034\,(-\,0\cdot10) \\ &+ 1\cdot618034\,(+\,0\cdot20) \\ &+ 1\,(-\,0\cdot05) \\ &+ 0 \\ &- 1\,(+\,0\cdot30) \\ &- 1\cdot618034\,(-\,0\cdot10) \\ &- 1\cdot618034\,(+\,0\cdot35) \\ &- 1\,(-\,0\cdot15) \end{aligned}$$

TABLE I.—Analysis of first 30 terms of experimental series used for fig. 2 into complementary function (simple harmonic function) and particular integral (function of the disturbances alone).

Term.	Observed series.	Complementary function.	Particular integral.	Disturbance ε
0	0	0	0	0
1	+ 0·58779	+ 0·58779	0	0
2	+ 0·80106	+ 0·95106	− 0·15	− 0·15
3	+ 1·05835	+ 0·95106	+ 0·10729	+ 0·35
4	+ 0·81139	+ 0·58779	+ 0·22360	− 0·10
5	+ 0·55451	0	+ 0·55451	+ 0·30
6	− 0·01417	− 0·58779	+ 0·57362	− 0·10
7	− 0·62744	− 0·95106	+ 0·32362	− 0·05
8	− 0·80105	− 0·95106	+ 0·15001	+ 0·20
9	− 0·76869	− 0·58779	− 0·18090	− 0·10
10	− 0·54272	0	− 0·54272	− 0·10
11	+ 0·14155	+ 0·58779	− 0·44624	+ 0·25
12	+ 1·12175	+ 0·95106	+ 0·17069	+ 0·35
13	+ 1·87348	+ 0·95106	+ 0·92242	+ 0·20
14	+ 1·65960	+ 0·58779	+ 1·07181	− 0·25
15	+ 0·96181	0	+ 0·96181	+ 0·15
16	− 0·10336	− 0·58779	+ 0·48443	0
17	− 0·92905	− 0·95106	+ 0·02201	+ 0·20
18	− 1·54987	− 0·95106	− 0·59881	− 0·15
19	− 1·92869	− 0·58779	− 1·34090	− 0·35
20	− 1·47082	0	− 1·47082	+ 0·10
21	− 0·50115	+ 0·58779	− 1·08894	− 0·05
22	+ 0·55994	+ 0·95106	− 0·39112	− 0·10
23	+ 1·75715	+ 0·95106	+ 0·80609	+ 0·35
24	+ 2·23319	+ 0·58779	+ 1·64540	− 0·05
25	+ 1·70623	0	+ 1·70623	− 0·15
26	+ 0·37755	− 0·58779	+ 0·96534	− 0·15
27	− 1·09534	− 0·95106	− 0·14428	0
28	− 2·24985	− 0·95106	− 1·29879	− 0·10
29	− 2·69499	− 0·58779	− 2·10720	− 0·15
30	− 1·86074	0	− 1·86074	+ 0·25

The series tends to be oscillatory, since, if we take adjacent terms, most of the periodic coefficients of the ε's are of the same sign, and consequently the adjacent terms are positively correlated; whereas if we take terms, say, 5 places apart, the periodic coefficients of the ε's are of opposite sign, and therefore the terms are negatively correlated. The series also tends to be smooth—*i.e.*, adjacent terms highly correlated—since adjacent terms represent simply differently weighted sums of ε's, all but one of which are the same. [*Added, February* 17.—It may be noted that if, in constructing an empirical series by equation (6), we put $u_0 = u_1 = 0$, there would be no true harmonic component, and the series would reduce to the particular integral alone; but the graph would present to the eye an appearance hardly different from that of fig. 2. The case would correspond to that of a pendulum initially at rest, but started into movement by the disturbances.]

It is evident that the problem of determining with any precision the period of the fundamental undisturbed function from the data of such a graph as fig. 2 is a much more difficult one than that of determining the period when we have only to deal with superposed fluctuations. It is doubtful if any method can give a result that is not subject to an unpleasantly large margin of error if our data are available for no more than, say, 10 or 15 periods. Determining the epochs of minima on the graph of fig. 2 by interpolation, I find periods for individual waves ranging from 8·57 to 11·29 intervals, the true period being 10. The first 15 waves give an average of 10·185, the second 15 an average of 9·850, and it takes the whole 30 to give an average, 10·026, near the truth. The question of the applicability of periodogram analysis to a series of this type is further discussed in Section IV below; but from mere inspection of the graph it is, I think, clear that it must give results subject to a much larger margin of error than is usually supposed—results, consequently, which must be interpreted with the greatest caution, and that if applied to data covering only a few periods it may easily give results which are apparently absurd or highly paradoxical.

Inspection of a graph of WOLFER's annual sunspot numbers, the upper curve in fig. 8 (p. 296) suggests quite definitely to my eye that we have to deal with a graph of the type of fig. 2, not of the type of fig. 1, at least as regards its principal features. It is true that there are minor irregularities, which may represent superposed fluctuations, probably in part of the nature of errors of observation; for the sunspot numbers can only be taken as more or less approximate "index-numbers" to sunspot activity. But in the main the graph is wonderfully smooth, and its departures from true periodicity, which have troubled all previous analysts of the data, are precisely those found in fig. 2—great variation in amplitude and continual changes of phase.

If this interpretation is correct, it seems desirable to break away from the periodogram method: the problem is, in fact, no longer one merely of determining the period, but also of determining the values of ϵ, the "disturbances," as I term them for short. It is natural, then, to approach the problem from the standpoint of the equation relating u_x to u_{x-1}, u_{x-2}, etc. Starting also, as I did, with the conception of periodogram analysis and harmonic periodicities in my mind, it was natural to assume an equation of the form (6), and an equation of corresponding form for two periodicities. This gave the first method tried.

It only occurred to me later that the method started from an unnecessarily limited assumption; that it would be better simply to find the linear regression equation of u_x on u_{x-1}, u_{x-2}, and more terms if necessary, and solve this as a finite difference equation. This gave my second method. As the results of the first method were interesting, I give both methods below.

On doing the work I was puzzled by the fact that the equation first found suggested a period obviously too short. A little consideration suggested that this was probably due to the presence of superposed fluctuations: as already noted, the graph of sunspot numbers suggests the presence of minor irregularities due to this cause, notwithstanding that a certain amount of smoothing has already been introduced by employing the annual average and not the monthly numbers. I therefore desired to repeat the work on graduated figures, assuming that graduation would largely eliminate such irregularities, and used the following method. The u's were first summed in overlapping sets of three, thus:—

$$\left.\begin{aligned} w_1 &= u_0 + u_1 + u_2 \\ w_2 &= u_1 + u_2 + u_3 \\ w_3 &= u_2 + u_3 + u_4 \end{aligned}\right\} . \tag{11}$$

Here $w_2/3$ is evidently a first approximation to a graduated value of u_2. As the second approximation was taken the corrected value

$$\bar{u}_2 = \frac{w_2}{3} - \frac{1}{9}\Delta^2 (w_1). \tag{12}$$

The results were not very good, as will be seen from the graph, the second curve in fig. 8 (p. 296), and from a discussion below in Section III, p. 282. A better result could probably have been obtained by graphic smoothing on a large-scale chart; but a "mechanical" method of graduation gave directly figures for calculation, and they served the purpose of comparison. The graduated figures, together with WOLFER's numbers,* are given in Table A at the end of the paper. A test of the graduation gives for the mean difference (actual number less graduated number) $-0 \cdot 04$; standard deviation of differences, $6 \cdot 04$. Over 90 per cent. of the differences lie within $\pm$ 10 points.

II. First Method of Analysis: Harmonic Curve Equation.

A.—*Assumption of a Single Period only.*—In equation (6) the average value of ε is assumed to be zero. Hence, if we form an equation

$$u_x = ku_{x-1} - u_{x-2}, \tag{13}$$

and determine k by the method of least squares, we have

$$k = 2 - \mu = 2 \cos \theta \tag{14}$$

by (5). But

$$S(u_x - \overline{ku_{x-1} - u_{x-2}})^2 = S(\overline{u_x + u_{x-2}} - ku_{x-1})^2,$$

and hence we can most readily determine k by finding the correlation between $u_x + u_{x-2}$ and u_{x-1}, and forming the regression equation of the former on the latter. In the case of the sunspot numbers we would have to work, of course, with the deviations of the u's from the general mean, and subsequently transform to zero as origin, so that there would be a constant on the right of (13).

Since I could see no valid method of determining probable errors in cases of the present

* 'Terrestrial Magnetism and Atmospheric Electricity,' Baltimore (June, 1925).

kind, where we are not dealing with random samples in the ordinary sense but with samples from series all the terms of which are highly correlated with one another,* a practical test of the method on the data of fig. 2 seemed to be of interest. The series of 300 terms was divided into two series of 150 terms each, which gave the following results—accents to the u's denote deviations :—

Empirical Series.

First 150 terms :

$$u'_x = 1\cdot 62438\, u'_{x-1} - u'_{x-2},$$

$$\cos\theta = 0\cdot 81219\,;\quad \theta = 35^\circ\cdot 69\,;\quad \text{period} = 10\cdot 087.$$

Second 150 terms :

$$u'_x = 1\cdot 60636\, u'_{x-1} - u'_{x-2}$$

$$\cos\theta = 0\cdot 80318\,;\quad \theta = 36^\circ\cdot 56\,;\quad \text{period} = 9\cdot 845.$$

The periods thus found are not far from those obtained from the interpolated minima (p. 273), viz., 10·185 and 9·850 : the coefficient of u'_{x-1} for a period 10 should be 1·61803. The respective equations give values of the disturbances ε having correlations $+0\cdot 997$ and $+0\cdot 987$ with the true disturbances, and even in the latter case give quite a fair picture of the true state of affairs. On the whole, I think the result may be regarded as reasonably satisfactory.

Turning now to WOLFER's sunspot numbers, the series was used as a whole (1749–1924) : the deviations of the individual numbers from the general mean were written down to the nearest unit, and the correlation worked without further grouping. The results were :—

WOLFER's *Sunspot Numbers*, 1749–1924 :

s.d. of whole series = 34·66 points

$$u_x = 1\cdot 62373\, u_{x-1} - u_{x-2} + 16\cdot 99 \tag{15}$$

$$\cos\theta = 0\cdot 81187\,;\quad \theta = 35^\circ\cdot 72\,;\quad \text{period} = 10\cdot 08 \text{ years.}$$

s.d. of disturbances = 17·05 points.

As, in view of the subsequent work, I judge that the disturbances calculated from (15) have no special importance, it has not been thought worth while to tabulate them, but they are shown in the third graph in fig. 8 : the graph is to double the scale of the graphs of sunspot numbers, and the line drawn gives quinquennial averages. It will be seen that the disturbances are very variable, running up to over ± 50 points. But the course of affairs is rather curious. From 1751 to 1792, or thereabouts, the disturbances are mainly positive and highly erratic ; from 1793 to 1834 or thereabouts, when the

* *Cf.* the general discussion of the nature of time-series in "Why do we sometimes get nonsense correlations between time-series ? A study in sampling and the nature of time-series." Presidential Address, G. U. YULE, 'Journ. Stat. Soc.,' vol. 89 (1926).

sunspot curve was depressed, they are mainly negative and very much less scattered; from 1835 to 1875, or thereabouts, they are again mainly positive and highly erratic; and finally, from 1876 to 1915, or thereabouts, once more mainly negative and much less erratic. It looks as if the "disturbance function" had itself a period of somewhere about 80 to 84 years, alternate intervals of 40 to 42 years being highly disturbed and relatively quiet. This characteristic appears in whatever way the disturbances are calculated, whether from the graduated or ungraduated numbers, and is returned to below (p. 283).

But it is evident that the period, 10·08 years, given by equation (15) is markedly too low: it ought, one would expect, to be in fair agreement with the usual estimate of rather over 11 years—11·125 years (SCHUSTER*) or 11·21 years (LARMOR and YAMAGA†). As already mentioned, the divergence might be due to the presence of superposed fluctuations. If such fluctuations are present, our two variables $u_x + u_{x-2}$ and u_{x-1} are, as it were, affected by errors of observation, which would have the effect of reducing the correlation and also the regression. Reducing the regression means reducing the value of cos θ—that is, increasing θ or reducing the apparent period. It was therefore attempted to eliminate superposed fluctuations by graduating the numbers, using the method already described (p. 274), and doing the work again on these graduated figures, which will be found in Table A at the end of the paper. The results were as follows :—

Graduated Sunspot Numbers, 1753–1920:

s.d. of whole series = 33·75 points.

$$u_x = 1\cdot68426\, u_{x-1} - u_{x-2} + 14\cdot13 \qquad (16)$$

$$\cos\theta = 0\cdot84213;\ \theta = 32^\circ\cdot63;\ \text{period} = 11\cdot03 \text{ years.}$$

s.d. of disturbances = 11·43 points.

The estimate of the period is now much closer to that usually given, and I think it may be concluded that the reason assigned for the low value obtained from the ungraduated numbers is correct.

For lack of space on the plate a graph has not been given of the disturbances as calculated from (16). The scatter is greatly reduced (s.d. of disturbances 11·43 against 17·05), but the general course of affairs is very similar to that shown by the graph for the ungraduated numbers.

The graphic test was applied to see whether the regression of $u_x + u_{x-2}$ on u_{x-1} was, in fact, appreciably linear or no. Figs. 3 and 4 show dot-diagrams for the ungraduated and the graduated numbers respectively. It will be seen that over the greater part of the range the regression is effectively linear. But for the larger negative deviations (low values of the sunspot numbers) there is an appreciable, though small, divergence

* 'Phil. Trans.,' A, vol. 206, pp. 69–100 (1906).

† 'Roy. Soc. Proc.,' A, vol. 93, pp. 493–506 (1917).

from linearity affecting some 10 per cent. of the points. On the whole, however, divergence from linearity does not look as if it would be a serious trouble.

If we ask ourselves the question how much of the variance of u_x has been accounted for by u_{x-1} and u_{x-2}, the answer is not afforded directly in the usual way by the

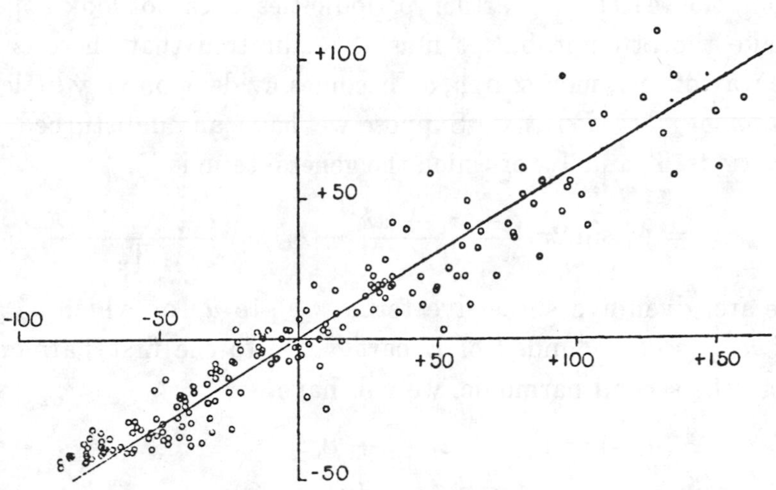

FIG. 3.—Graph to test approximation to linearity of regression of $u'_x + u'_{x-2}$ (horizontal) on u'_{x-1} (vertical): WOLFER'S sunspot numbers.

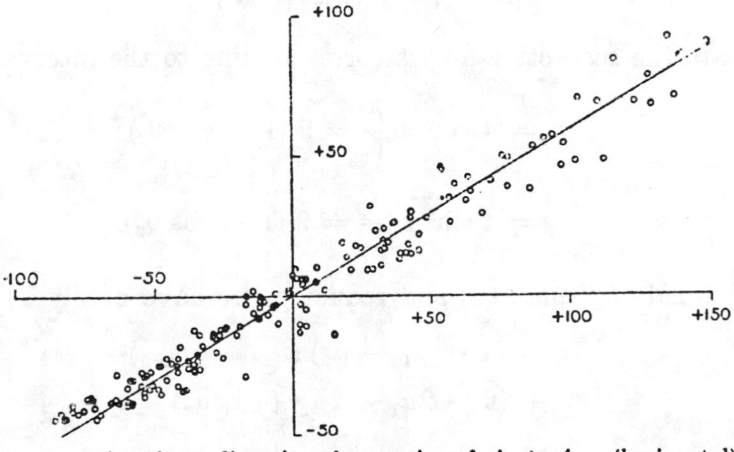

FIG. 4.—Graph to test approximation to linearity of regression of $u'_x + u'_{x-2}$ (horizontal) on u'_{x-1} (vertical): graduated sunspot numbers.

correlation calculated, which is the correlation between $u_x + u_{x-2}$ and u_{x-1}, not the correlation between the right and left-hand sides of (13). We must take the standard deviations given. We have

WOLFER'S numbers.		Graduated numbers.	
$34 \cdot 66^2$	1201·3	$33 \cdot 75^2$	1139·1
$17 \cdot 05^2$	290·7	$11 \cdot 43^2$	130·6
	910·6		1008·5

and have therefore accounted for some 76 per cent. of the variance in the case of WOLFER'S numbers and some 89 per cent. in the case of the graduated figures. On the present lines we cannot account for superposed fluctuations, and, bearing in mind the look of the chart and its suggestion to the eye that a great part of the variation of the disturbances is merely random, the search for further periodicities does not look hopeful. It seems desirable to make the attempt, but it must be admitted that there is a very serious difficulty in the way of any such search, as becomes evident on very little consideration.

B.—*Assumption of Two Periods.*—Suppose we have an undisturbed periodic series involving two periods, T_1 and T_2, of which the general term is

$$u_x = A_1 \sin 2\pi \frac{t + \tau_1 + xh}{T_1} + A_2 \sin \frac{t + \tau_2 + xh}{T_2}, \tag{17}$$

and suppose we are given five successive terms, u_{x-4} to u_x, of which u_{x-2} is the central term. Then if u_{x-2} is compounded of a part a due to the first harmonic and a part $(u_{x-2} - a)$ due to the second harmonic, we will have

$$\begin{aligned} \Delta^2 (u_{x-3}) &= u_{x-1} - 2u_{x-2} + u_{x-3} \\ &= - \mu_1 a - \mu_2 (u_{x-2} - a) \end{aligned} \tag{18}$$

$$\begin{aligned} \Delta^4 (u_{x-4}) &= u_x - 4u_{x-1} + 6u_{x-2} - 4u_{x-3} + u_{x-4} \\ &= \mu_1^2 a + \mu_2^2 (u_{x-2} - a), \end{aligned} \tag{19}$$

where, if θ_1, θ_2 are the respective angles corresponding to the interval h.

$$\mu_1 = 4 \sin^2 \pi \frac{h}{T_1} = 2 (1 - \cos \theta_1) \tag{20}$$

$$\mu_2 = 4 \sin^2 \pi \frac{h}{T_2} = 2 (1 - \cos \theta_2). \tag{21}$$

Using (18) to eliminate a from (19) and reducing, we have finally

$$\begin{aligned} u_x = &(4 - \mu_1 - \mu_2) (u_{x-1} + u_{x-3}) \\ &- (6 - 2\mu_1 - 2\mu_2 + \mu_1 \mu_2) u_{x-2} \\ &- u_{x-4}. \end{aligned} \tag{22}$$

If the series is completely undisturbed, (22) can be used to calculate in succession all the following terms, when the first four are given.

But now comes the problem. What happens if the series is " disturbed " in the sense of the previous work? Can we, exactly as before, assume

$$u_x = k_1 (u_{x-1} + u_{x-3}) - k_2 u_{x-2} - u_{x-4} + \varepsilon, \tag{23}$$

where ε is a deviation, of the nature of an error of observation, varying with the disturbance? It seems clear that we cannot legitimately make any such assumption. It was going far enough to treat the single interval as if it were infinitesimal: we cannot possibly stretch the assumption to cover three intervals. If, admitting this, we nevertheless assume a relation of the form (23), and proceed to determine k_1 and k_2 by the method of least squares, regarding $u_x + u_{x-4}$, $u_{x-1} + u_{x-3}$, and u_{x-2} as our three variables, and forming the regression equation for the first on the last two, can this give us any useful information ? I think it can. The results may afford a certain criterion as between the respective conceptions of the curve being affected by superposed fluctuations or by disturbances. If there are no *disturbances* in the sense in which the term is here used, the application of the suggested method is perfectly legitimate, and should bring out any secondary period that exists. To put the matter in a rather different way : *disturbances* occurring in every interval imply an element of unpredictability very rapidly increasing with the time. *Superposed fluctuations* imply an element of unpredictability which is no greater for several years than for one year. If, then, there is a secondary period in the data, and we might well expect a period of relatively small amplitude—if only a sub-multiple of the fundamental period—equation (23) should certainly bring out this period, *provided that we have only to do with superposed fluctuations and not disturbances.*

I accordingly worked out the results, which proved rather unexpected. I simply give the equations and the resulting values of the μ's and θ's. Accented u's denote deviations as before : it was not worth while working out the constant term in view of the results.

WOLFER'S *Sunspot Numbers*, 1749–1924 :

$$u'_x = 1\cdot16051\,(u'_{x-1} + u'_{x-3}) - 1\cdot01486\,u'_{x-2} - u'_{x-4} \tag{24}$$

$$\mu_1 = 2\cdot56945 \qquad \mu_2 = 0\cdot27003$$

$$\cos\theta_1 = -0\cdot28473 \qquad \cos\theta_2 = +0\cdot86498$$

$$\theta_1 = 106^\circ\cdot54 \text{ or } 253^\circ\cdot46$$

First period = 3·37 or 1·42 years.

$$\theta_2 = 30^\circ\cdot12$$

Second period = 11·95 years.

s.d. of disturbances = 21·95 points.

Graduated Numbers, 1753–1920 :

$$u'_x = 1\cdot65539\,(u'_{x-1} + u'_{x-3}) - 1\cdot83955\,u'_{x-2} - u'_{x-4} \tag{25}$$

$$\mu_1 = 2\cdot09183 \qquad \mu_2 = 0\cdot25278$$

$$\cos\theta_1 = -0\cdot04592 \qquad \cos\theta_2 = +0\cdot87361$$

$$\theta_1 = 92^\circ\cdot63 \text{ or } 267^\circ\cdot37$$

First period = 3·89 or 1·35 years.

$$\theta_2 = 29^\circ\cdot12$$

Second period = 12·36 years.

s.d. of disturbances = 17·47 points.

Since the values of the μ's give $\cos\theta$ and not θ itself, the value of θ is not strictly determinate: the longer period is naturally taken as approximate to the fundamental, but the choice of the shorter period is quite uncertain. So far as the results go then, they at first sight suggest the existence of two periods, one a year or more longer than the value which anyone, on a mere inspection of the graph, would be inclined to take for the fundamental, and the other much shorter. On the face of it the result looks odd, and the last figures given for the ungraduated and graduated numbers respectively show that it is really of no meaning. *The standard deviations found for the disturbances are in both cases larger than when we assumed the existence of a single period only:* 21·95 *against* 17·05, *and* 17·47 *against* 11·43. So far from having improved matters by the assumption of a second period, we have made them very appreciably worse: we get a worse and not a better estimate of u_x when u_{x-3} and u_{x-4} are brought into account than when we confine ourselves to u_{x-1} and u_{x-2} alone. To put it moderately, there is at least no evidence that any secondary period exists—a conclusion in entire accord with that of LARMOR and YAMAGA (*loc. cit.*). The result also bears out the assumption that it is disturbances rather than superposed fluctuations which are the main cause of the irregularity, the element of unpredictability, in the data.

The fact that we get a worse and not a better estimate, although that estimate is based on a larger number of variables, will naturally seem paradoxical to those who are accustomed to the ordinary theory of correlation. It is simply due to the fact that we have insisted on the regression equation being of a particular form, the coefficients of u_{x-1} and u_{x-3} being identical, and the coefficient of u_{x-4} unity. The result tells us merely that, if we insist on this, such and such values of the coefficients are the best, but even so they cannot give as good a result as the equation of form (13) with only two terms on the right.

III. SECOND METHOD OF ANALYSIS: REGRESSION EQUATION.

A.—*Assumption of a Single Period only.*—We form the ordinary regression equation for u_x on u_{x-1} and u_{x-2}:

$$u_x = b_1 u_{x-1} - b_2 u_{x-2}. \tag{26}$$

With a curve that fluctuates round zero as base-line there will be no constant term on the right; with the sunspot numbers there must be a constant as before, but this is immaterial. If the curve is of periodic form, the roots of the equation*

$$E^2 - b_1 E + b_2 = 0 \tag{27}$$

must be imaginary. Let the roots be

$$\alpha \pm i\beta,$$

* I follow BOOLE, 'Finite Differences,' 2nd edition, chap. xi, pp. 208–212.

and let
$$\alpha^2 + \beta^2 = b_2 = e^{2\lambda} \tag{28}$$
$$\tan\theta = \beta/\alpha. \tag{29}$$

Then the general solution of the difference equation (26) is of the form

$$u_x = e^{\lambda x}(\mathrm{A}\cos\theta x + \mathrm{B}\sin\theta x). \tag{30}$$

Here θ is, as before, the angle corresponding to the interval h. For a real physical phenomenon one would in general expect λ to be negative, the solution representing damped harmonic vibrations; or zero, the solution being simple harmonic vibrations. The condition for the latter solution is that b_2 shall be unity.

This method also was tested on the empirical data of fig. 2, using two series of 150 terms each as before. The correlation between u_{x-1} and u_{x-2} was assumed to be the same as that between u_x and u_{x-1}, both representing correlations between adjacent terms: we cannot get correlations between terms one apart, two apart, etc., that involve precisely the same terms, and results are, in so far, approximate; but the closeness of approximation will be the greater the longer the series.*

Empirical Series.

First 150 terms.

$$u_x = 1\cdot6117\, u_{x-1} - 0\cdot9867\, u_{x-2}.$$

Roots of (27): $0\cdot80585 \pm 0\cdot58078\, i$.

$\tan\theta = 0\cdot72070$: $\theta = 35^\circ\cdot78$: Period $= 10\cdot06$: $\lambda = -0\cdot0067$.

Second 150 terms.

$$u_x = 1\cdot5975\, u_{x-1} - 0\cdot9875\, u_{x-2}$$

Roots of (27): $0\cdot79875 \pm 0\cdot59119\, i$.

$\tan\theta = 0\cdot74014$: $\theta = 36^\circ\cdot51$: Period $= 9\cdot86$: $\lambda = -0\cdot0063$.

The values found for the period, 10·06 and 9·86, are close to the values given by the harmonic curve equation, viz., 10·087 and 9·845. The values found for λ, which should be zero, are, in fact, numerically less than 0·01 in each case. The agreement seems quite satisfactory.

Proceeding now to the work on the sunspot numbers, the following are the results:

WOLFER'S *Sunspot Numbers*, 1749–1924:

$$u_x = 1\cdot34254\, u_{x-1} - 0\cdot65504\, u_{x-2} + 13\cdot854. \tag{31}$$

Roots of (27): $0\cdot67127 \pm 0\cdot45215\, i$.

$\tan\theta = 0\cdot67358$: $\theta = 33^\circ\cdot963$: Period $= 10\cdot600$ years:

$\lambda = -\,0\cdot21154$:

s.d. of disturbances $= 15\cdot41$ points.

* The correlations required are the first two *serial correlations*, as I have termed them. *Cf. Address*, already cited, 'Journ. Stat. Soc.,' vol. 89 (1926).

Graduated Sunspot Numbers, **1753–1920**:

$$u_x = 1 \cdot 51527\, u_{x-1} - 0 \cdot 80245\, u_{x-2} + 12 \cdot 854. \tag{32}$$

Roots of (27): $0 \cdot 75764 \pm 0 \cdot 47795\, i$.

$\tan\theta = 0 \cdot 63085$: $\theta = 32^\circ \cdot 246$: Period $= 11 \cdot 164$ years.

$\lambda = -0 \cdot 11004$:

s.d. of disturbances $= 10 \cdot 79$ points.

The magnitudes of the disturbances as calculated from equations (31) and (32) respectively are given in Table A at the end of the paper: the regressions were cut down to three decimal places for these calculations. The disturbances are also shown, together with the quinquennial averages as indicated by the lines, in the fourth and fifth graphs in fig. 8 (p. 296).

The period derived from WOLFER's numbers is now higher than that given by the harmonic formula (10·60 against 10·08), but still too low; that derived from the graduated data is also a little higher (11·16 against 11·03), and now lies between the values suggested by SCHUSTER, and by LARMOR and YAMAGA, respectively.

But the solution of both equations is a *heavily damped* and not a simple harmonic movement. The damping given by the graduated data is, however, only about half that given by WOLFER's numbers: (31) gives a vibration reduced to 0·106 of the original amplitude in the duration of a period, (32) a vibration reduced to 0·293 only. This is at first sight a very puzzling result, and precisely the reverse of what was to be expected by the elimination or reduction of superposed fluctuations. For let x_1, x_2, x_3 be three variables with the same standard deviation σ, and let the correlations between x_1 and x_2 and between x_2 and x_3 be r_1, and the correlation between x_1 and x_3 be r_2. Then, with a little reduction, we have for the partial regressions in the usual notation,

$$\left.\begin{aligned} b_{12\cdot3} &= \frac{r_1(1-r_2)}{1-r_1^2} \\ b_{13\cdot2} &= \frac{r_2-r_1^2}{1-r_1^2} \end{aligned}\right\} \tag{33}$$

Now suppose all three variables to have random errors of the same standard deviation—errors completely uncorrelated with each other—superposed on them. Then both correlations will be reduced in the same proportion and become, say, pr_1 and pr_2. Whence, for the partial regressions in this case, we have

$$\left.\begin{aligned} b'_{12\cdot3} &= \frac{pr_1(1-pr_2)}{1-p^2r_1^2} \\ b'_{13\cdot2} &= \frac{pr_2-p^2r_1^2}{1-p^2r_1^2} \end{aligned}\right\}. \tag{34}$$

The respective ratios of the second coefficient to the first are

$$\left.\begin{aligned}\frac{b_{13\cdot 2}}{b_{12\cdot 3}} &= \frac{r_2 - r_1^2}{r_1(1 - r_2)}\\ \frac{b'_{13\cdot 2}}{b'_{12\cdot 3}} &= \frac{r_2 - pr_1^2}{r_1(1 - pr_2)}\end{aligned}\right\}. \qquad (35)$$

The condition that the second ratio shall be greater than the first (r_2 not equal to r_1) reduces simply to $p < 1$, which is necessarily true. That is to say, where superposed random errors occur, we would expect the ratio of the second partial regression to the first to be greater than when such errors are eliminated or reduced. We have found precisely the contrary, for the ratio is greater for the graduated than for the ungraduated numbers.

An examination of the chart and of the figures suggests that the explanation may lie in an unexpected and unintended effect of the graduation. The occurrence of a damping term in the solution of the empirical finite difference equation may conceivably be due to an attempt of the equation to represent the asymmetry of the waves in the sunspot curve, which is a marked feature of the sunspot curve in waves of large amplitude. A careful inspection of the graphs and of the figures of Table A suggests that the graduation has tended to lessen this asymmetry, owing presumably to second differences only having been taken into account. As definite features in Table A, it may be noted that graduation has pushed forward the maximum from 1769 to 1770, has greatly lessened the difference between the ordinates at 1778 and 1779, and has advanced the maximum again from 1787 to 1788, from 1870 to 1871, from 1905 to 1906, and from 1917 to 1918. If we take two damped sine curves with the above respective values of λ, but, for fair comparison, the same period—say, 11·164 years—the first with the greater damping factor would have its first maximum at 2·15 years, the second with the lower damping factor at 2·44 years—*i.e.*, the maximum would be advanced by roundly 0·3 of a year. The graduation seems to have had, unintentionally, the effect of producing an average advance of this order of magnitude, and therefore of reducing the apparent damping.

The question whether in fact the damping factor represents a physical reality or merely an attempt of the empirical formula to adjust itself to the asymmetry of the waves is for the present postponed. In the first place, a more detailed examination of the disturbances is desirable.

First let us examine more closely the apparent alternation of disturbed and quiet periods, each some 40 to 42 years in duration, which was already noted in the graph of the disturbances as calculated from the harmonic formula and is equally evident in the present graphs. It is not possible to assign the beginnings and ends of such periods with precision; and, as it happens, doing the work independently, I did not take precisely the same years for WOLFER'S and the graduated numbers. There is also a difficulty at the commencement of the data. As I judge it, the magnitude of the disturbance in 1751 indicates this year as within a disturbed period, and it is the first year for which we can calculate a disturbance, so that it must be taken as the beginning of the period for WOLFER's numbers. But 1753 is the first year for which we can calculate a disturbance for the graduated numbers, and this must be taken as the opening of the period. Table II gives the mean values of the disturbances for the periods finally adopted, and also the standard deviations. It will be seen that they completely confirm the impression given by the graph. Alternate periods give positive and negative mean values* of the disturbance: the periods with positive mean give a high value of the standard deviation, the periods with negative mean a low value of the standard deviation. At

TABLE II.—Means and Standard Deviations of disturbances in disturbed and quiet periods of 40 to 42 years. Disturbances from Table A at the end of the paper.

Period.	Number of years.	Mean disturbance.	Standard deviation.
WOLFER's numbers.			
1751–1792	42	+ 6·74	17·80
1793–1834	42	− 5·80	7·32
1835–1875	41	+ 4·43	17·85
1876–1915	40	− 2·61	10·59
Graduated numbers.			
1753–1793	41	+ 4·41	11·62
1794–1834	41	− 5·42	5·30
1835–1875	41	+ 3·95	12·49
1876–1915	40	− 2·95	6·61

the same time the last quiet period, taken as 1876 to 1915 inclusive, was more disturbed than the very conspicuously quiet period from 1793 to 1834 or thereabouts. While a much longer experience will be necessary to confirm the result, this alternation seems a rather conspicuous feature of the existing data.

Further inspection of the graphs suggests another feature which is at least not obvious in the first graph for the disturbances calculated from the harmonic curve formula. In the lines showing quinquennial averages, on the two lower graphs, there are distinct "humps" more or less consilient with the waves in the sunspot graph, but a little earlier

* It may be noted that positive or negative disturbances as calculated from (31) or (32) are what is meant. But the constant term in each of these equations may be understood as a steady positive "disturbance," and if added to the disturbances of Table A, from which the graphs are plotted, would render positive the bulk of the negative disturbances there given.

in phase. Examination of the figures of Table A suggests, in fact, that positive disturbances tend to begin at or just after the minimum, and continue till the maximum or a year or two before, disturbances from the maximum to the minimum, or a year or two after the minimum, being preponderantly negative. Preceding the maximum there is often a group of two or more large positive disturbances. Table III gives a summary

TABLE III.—Sums and means of disturbances in rising and preponderantly falling parts of the graph of sunspot numbers: "rising" implying from the minimum or a year beyond to the maximum or a year or two before.

WOLFER's numbers.				Graduated numbers.			
Years.	Rise + or fall −	Sum of disturbances.	Mean.	Years.	Rise + or fall −	Sum of disturbances.	Mean.
1751–1756	−	− 37·5	− 6·2	1753–1755	−	− 10·8	− 3·6
1757–1761	+	+ 36·0	+ 7·2	1756–1760	+	+ 13·2	+ 2·6
1762–1766	−	− 38·4	− 7·7	1761–1765	−	− 16·3	− 3·3
1767–1769	+	+ 58·2	+ 19·4	1766–1768	+	+ 36·9	+ 12·3
1770–1775	−	− 25·3	− 4·2	1769–1774	−	− 7·6	− 1·3
1776–1778	+	+102·5	+ 34·2	1775–1778	+	+ 71·4	+ 17·8
1779–1784	−	− 27·5	− 4·6	1779–1784	−	− 4·8	− 0·8
1785–1787	+	+ 77·5	+ 25·8	1785–1787	+	+ 59·5	+ 19·8
1788–1799	−	− 10·7	− 0·9	1788–1799	−	− 0·2	− 0·0
1800–1802	+	− 5·7	− 1·9	1800–1802	+	− 9·0	− 3·0
1803–1810	−	− 64·5	− 8·1	1803–1810	−	− 56·7	− 7·1
1811–1816	+	− 37·7	− 6·3	1811–1816	+	− 36·9	− 6·1
1817–1823	−	− 68·2	− 9·7	1817–1823	−	− 58·9	− 8·4
1824–1830	+	+ 6·9	+ 1·0	1824–1829	+	− 6·4	− 1·1
1831–1833	−	− 32·3	− 10·8	1830–1833	−	− 23·4	− 5·8
1834–1836	+	+ 76·7	+ 25·6	1834–1836	+	+ 61·7	+ 20·6
1837–1843	−	− 7·7	− 1·1	1837–1843	−	+ 3·5	+ 0·5
1844–1848	+	+ 66·0	+ 13·2	1844–1848	+	+ 39·6	+ 7·9
1849–1856	−	− 23·5	− 2·9	1849–1855	−	− 1·4	− 0·2
1857–1859	+	+ 42·0	+ 14·0	1856–1859	+	+ 25·6	+ 6·4
1860–1867	−	− 27·5	− 3·4	1860–1866	−	− 15·9	− 2·3
1868–1870	+	+ 89·4	+ 29·8	1867–1871	+	+ 62·0	+ 12·4
1871–1879	−	− 53·5	− 5·9	1872–1879	−	− 24·7	− 3·1
1880–1884	+	+ 16·1	+ 3·2	1880–1884	+	+ 3·1	+ 0·6
1885–1890	−	− 49·8	− 8·3	1885–1889	−	− 32·7	− 6·5
1891–1892	+	+ 32·3	+ 16·1	1890–1892	+	+ 17·8	+ 5·9
1893–1901	−	− 48·2	− 5·4	1893–1902	−	− 42·9	− 4·3
1902–1907	+	+ 6·9	+ 1·1	1903–1908	+	+ 0·5	+ 0·1
1908–1913	−	− 54·0	− 9·0	1909–1913	−	− 41·4	− 8·3
1914–1917	+	+ 48·1	+ 12·0	1914–1918	+	+ 20·9	+ 4·2
1918–1923	−	− 51·4	− 8·6	1919–1922	−	− 19·6	− 4·9

of the disturbances over such alternate rising or preponderantly falling periods. Owing to the small shift noted as a secondary effect of the graduation, the years taken are not quite the same for WOLFER's numbers and for the graduated numbers. In both cases the tendency to alternation in sign is clear, though the closer for the ungraduated data. During the earlier part of the "very quiet" period that ended with 1834, all the sums of disturbances are negative, whether for a rising or falling part of the graph, but the sums tend to be higher during a fall than during a rise. During the rather abnormal long fall from 1788 to 1799 there is also an irregularity. The numbers seem to have been maintained during these years by a succession of positive disturbances up to 1792 or 1793, and the negative disturbances of the following years only just overbalance these and leave a small negative total.

This distribution of the disturbances seems to me to have some bearing on the question whether we may perhaps tentatively regard the damped harmonic formula at which we have empirically arrived as being something more than merely empirical, and representing some physical reality. As it seems to me, the disturbances do occur just in the kind of way that would be necessary to maintain a damped vibration, and this suggests that broadly the conception fits the facts.*

Clearly, however, a simple damped vibration, varying round zero, is not quite what is wanted. One would rather expect a function of the form of the square of a damped harmonic vibration, say,

$$y = Ae^{-at}(1 - \cos \theta t). \tag{36}$$

The form of this function is shown in fig. 5, and it would look as though a train of such functions superposed on each other would give a graph not unlike that of the sunspot numbers (*cf.* below, fig. 6). But the difference equation of this function is of the third order, and would therefore have to be extended to include u_{x-3}. This raises again the serious theoretical difficulties briefly mentioned in Section II, B, p. 278. Even if the difference equation is in fact of the form supposed, it is doubtful if it can be determined. The question does render it necessary, however, to examine the correlations of u_x with u_{x-3} and more distant terms, and see what information they give us.

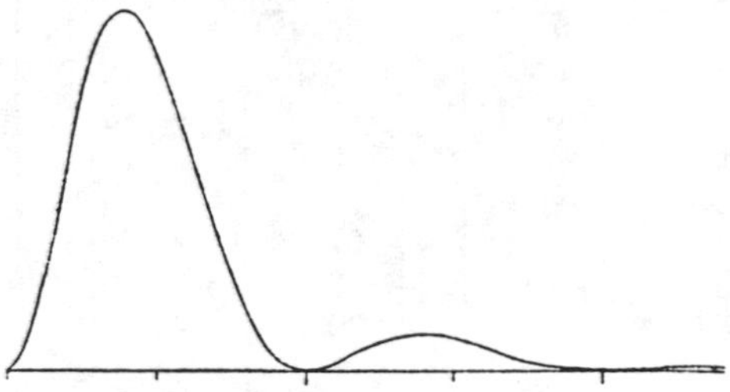

FIG. 5.—Graph of the function equation (36).

B.—*The Correlations of u_x with u_{x-3} and more Distant Terms, and the Information given thereby.*—On the left of Table IV are given the *serial correlations*, as I have termed them, for WOLFER's sunspot numbers and the graduated numbers respectively: r_1 is the correlation between u_x and u_{x-1}, r_2 the correlation between u_x and u_{x-2}, and so on. From these all the partial correlations are calculated on the assumption that the series is indefinitely long, so that we may assume that the correlation between u_{x-1} and u_{x-2} is the same as that between u_x and u_{x-1}, and so forth—an assumption which implies

* But I fail to find any relation between the disturbances and TURNER's dates of discontinuity of phase ('Monthly Notices,' R.A.S., vol. 74, p. 82).

corresponding equalities between partial correlations. The serial correlations are given as far as r_5, which is the negative maximum, r_6 being slightly smaller numerically.

TABLE IV.—Serial correlations for WOLFER'S and the graduated sunspot numbers, and the deduced partial correlations, etc. In the serial correlations, 1 denotes the correlation between u_x and u_{x-1}, 2 the correlation between u_x and u_{x-2}, and so on. In the partial correlations 13·2 denotes the correlation between u_x and u_{x-2}, u_{x-1} constant, and so on.

Serial correlations.		Partial correlations.		$1 - r^2$.	Continued product of $1 - r^2$.
WOLFER'S sunspot numbers.					
1	+ 0·811180	12	+ 0·811180	0·341987	0·341987
2	+ 0·433998	13·2	− 0·655040	0·570923	0·195248
3	+ 0·031574	14·23	− 0·101043	0·989790	0·193255
4	− 0·264463	15·234	+ 0·013531	0·999817	0·193219
5	− 0·404119	16·2345	− 0·050001	0·997500	0·192736
Graduated sunspot numbers.					
1	+ 0·840670	12	+ 0·840670	0·293274	0·293274
2	+ 0·471388	13·2	− 0·802451	0·356072	0·104427
3	+ 0·047038	14·23	+ 0·037840	0·998568	0·104277
4	− 0·264147	15·234	+ 0·351917	0·876154	0·091363
5	− 0·404327	16·2345	+ 0·325556	0·894013	0·081680

In the case of the partial correlations, $r_{13\cdot 2}$ denotes the correlation between u_x and u_{x-2}, u_{x-1} constant; $r_{14\cdot 23}$ the correlation between u_x and u_{x-3}, u_{x-1} and u_{x-2} constant; and so on. Only those partial correlations are given which are necessary to show how far we can improve the estimate of u_x by taking into account the successive terms beyond u_{x-1}. The continued products of $(1 - r^2)$ are given in the last column on the right, and we may fix our attention on these, considering first the figures for WOLFER'S numbers. It will be seen that after the first two terms all the correlations are so small that the continued product of $(1 - r^2)$ hardly falls at all, the variance of the disturbances—that is, the errors made in estimating u_x from the preceding terms—only falling from some 19·5 per cent. to some 19·3 per cent. of the total variance of the numbers themselves. It seems quite clear that in the case of the ungraduated numbers it would be an entire waste of time to take into account any terms more distant from u_x than u_{x-2} for purposes of estimation. As regards the idea suggested that the difference equation should be of the form required for such a function as (36), it may be noted that $r_{14\cdot 23}$ is of the wrong sign: a positive correlation would be required. The correlations give no evidence at all of any periodicity other than the fundamental, nor of any other exponential function. They strongly emphasise the increase of the element of unpredictability with the time.

When we turn to the correlations for the graduated numbers, matters are not altogether so clear. The last two partial correlations given in Table IV rise to markedly higher values than are found for the ungraduated figures, both exceeding $+0 \cdot 3$. The total correlations are based on 167 to 171 observations, and if we calculated the standard error by the ordinary formula, it would be under $0 \cdot 08$, and we would reckon both correlations as significant. Here, however, we have to do with a correlated series, not a random sample; the standard error is probably higher, and personally I am inclined to doubt whether either correlation is really significant. The discrepancies of sign as compared with the partial correlations for the ungraduated numbers in the case of the third and fifth partials may alone suffice to raise doubts. In any case, the effect of these correlations on the continued product of $(1 - r^2)$ is extremely small, only reducing the variance of disturbances from $10 \cdot 4$ to $8 \cdot 2$ per cent. of the total. In this case also there is very little to be gained by taking into account any terms beyond u_{x-2}, even if that little be significant.

This result, that terms beyond u_{x-2} hardly come into account if we attempt to estimate u_x by means of the preceding terms is, as it seems to me, what ought to be expected if the series is in fact "disturbed." But, in a sense, it is rather disappointing. Fig. 6

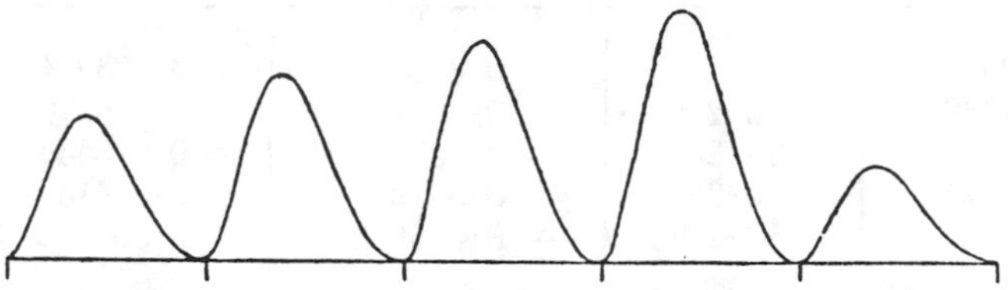

FIG. 6.—Graph of a series of superposed functions of the form of fig. 5, each one starting when the one before reaches its first minimum.

shows a graph formed by superposing a series of functions of the form (36), or fig. 5, of varying amplitudes, a new one starting when the one before reaches its first minimum. It will be seen that the graph is very like that of the sunspot numbers. It may be that this is, or is a close approximation to, the actual function, but when disturbances affect the movement, it does not seem possible to determine the constants, at least by the present method. The road seems to be blocked at the first approximation given by the regression equations of the second order (31) and (32).

The objection may be raised that the suggested function is not "anti-symmetrical," as required by the result of LARMOR and YAMAGA'S investigation, so that $F(t) = -F(-t)$ when the origin is taken on the periodic curve at the mean height, and the Fourier series consists of sine terms only. But the divergence of the suggested function from anti-symmetry is quite small, and LARMOR and YAMAGA only state that the cosine terms found for the mean sunspot wave had amplitudes less than unity. To get a comparison with their results I have worked out the harmonic analysis of the first period of the function (36) with the following numerical values, being those used for fig. 5:

$$y = 100e^{-0.23026t}\left(1 - \cos\frac{2\pi}{10}t\right), \tag{37}$$

where $0 \cdot 23026 = 1/10 \log e$. Integration has first to be effected with zero as time-origin, and when the sine and cosine amplitudes have been determined the origin can be shifted to the epoch at which the function attains its mean value, $t = 1 \cdot 6802$. The final result is

$$\left.\begin{aligned} y = 34 \cdot 459 + 36 \cdot 10 \sin 0x + 4 \cdot 07 \sin 2\theta x + 1 \cdot 06 \sin 3\theta x \\ + 0 \cdot 63 \cos \theta x - 1 \cdot 61 \cos 2\theta x + 0 \cdot 47 \cos 3\theta x \end{aligned}\right\}. \tag{38}$$

The expansion LARMOR and YAMAGA find for the unmodified sunspot curve is

$$y = 44 \cdot 5 + 35 \cdot 4 \sin kt + 6 \cdot 6 \sin 2kt, \tag{39}$$

and for the same curve as modified by equalising the amplitudes of all the waves in the sunspot curve over the period considered

$$y = 45 + 37 \cdot 8 \sin k't + 6 \cdot 5 \sin 2k't + 1 \cdot 4 \sin 3k't. \tag{40}$$

I chose the amplitude in (37) so as to make the amplitude in the first sine term approximately the same in (38) as in (39) and (40), so that the three expansions are fairly comparable. It will be seen that in (38) two of the three cosine amplitudes are less than unity, and would have been ignored on LARMOR and YAMAGA's criterion: the third is only $1 \cdot 61$. It does not seem to me that (38) differs very materially from (39) or (40): the main difference seems to lie in the relatively low value obtained in (38) for the amplitude of the second sine term, rather than in the amplitudes of the cosine terms, and the value taken for the damping coefficient, $a = 0 \cdot 23026$, is high. In the above case a/θ is $0 \cdot 37$ roundly: from the position of the maximum in LARMOR and YAMAGA's fig. 4 I should estimate a at about $0 \cdot 146$, a/θ (the year as unit) at about $0 \cdot 26$.

IV. SOME TRIALS OF PERIODOGRAM ANALYSIS ON THE EXPERIMENTAL DISTURBED SERIES.

The opinion was expressed in Section I that the application of periodogram analysis to "disturbed" functions must yield results subject to a much larger margin of error than is usually supposed. I can see no direct way of tackling the problem and finding the standard error of the amplitude of a period found from any number n of observations, given the standard deviation of the disturbances, even assuming for simplicity that these are random, as in my experimental case. For the u's—*i.e.*, the terms of the observed series—are not terms of a random but of an oscillatory series, in which u_n is correlated with u_{n+x} and the sign of the correlation changes as x increases.* The problem is too complex for my abilities at least. Hence I thought it worth while to carry out some trials on the data used for fig. 2, and for testing the methods developed in Sections II and III. To avoid decimal places in the analysis, the original figures were cut down to two decimals and then multiplied by 100, so that one unit in fig. 2 corresponds to an amplitude of 100, or an intensity of 10,000.

A.—For the first test four groups of observations were used, covering roughly the first, second, third and fourth quarter of the observations respectively. For each of these groups the intensities of the periods 8, 9, 10, 11 and 12 alone were determined in the first instance. Subsequently, periods 9·5 and 10·5 were added. Since for these periods it was necessary to take observations covering an even number of periods, 8 periods were used, so that 84 observations were required for the period 10·5. The original intention was to use only 70 observations or as few more as could be helped; hence only 70 observations were employed to get the intensity of the period 10. The figures, however, are reasonably comparable, and are given in Table V: the second (italicised) line of figures for each group gives the calculated intensity for a simple harmonic function of period 10 and the intensity shown in the table.

It will be seen that while in every group the intensity for period 10 is the greatest, the relative intensities of the respective periods in the different groups vary largely. In Group II the intensity of 10·5 is nearly equal to the intensity of 10, while the intensity of 9·5 is less than a third of the intensity of 10. In Group I the intensities of 9·5 and 10·5 are not far from equal, while in Groups III and IV the intensity of 9·5 is much greater than that of 10·5. In Groups I and II the intensity of 9·5 is less than the calculated figure, in Groups III and IV substantially greater, while for the period 10·5 matters go just the other way.

If we look at the figures for periods diverging more largely from the fundamental, the variation is almost more striking. In the case of periods 8 and 9, observed intensities vary roundly from one-fourth to four times the calculated intensity for a simple harmonic function, or more. For period 11 the range is from under a fourth to about three times the calculated figure. For period 12 Group IV shows an intensity well over four times the calculated figure: Group III gives an almost vanishingly small intensity against a calculated intensity of nearly 3,000.

* Any series that is worth analysing at all must be an oscillatory series in this sense, the sense in which the term is used in the *Address* already cited ('Journ. Stat. Soc.' (1926)). SCHUSTER's exponential formula renders great service by enabling us to exclude at once terms which might arise even in the analysis of a random series; but fluctuations of sampling in intensities based on samples from an oscillatory series may be much larger than the corresponding fluctuations in samples from a random series, and at present one is thrown back on empirical tests of significance in such cases. It is true that they may also be lower, but SCHUSTER's formula then leaves one on the safe side.

TABLE V.—Periodogram analysis, for seven periods only, of four groups of 70 to 80 observations each in the experimental series. The first line in the table for each group gives the intensity found, the second line in italic type the calculated intensity for a simple harmonic function of period 10 and the intensity shown. Original ordinates multiplied by 100.

Group.	Period : intervals.						
	8	9	9·5	10	10·5	11	12
I	880 *673*	9,139 *3,410*	35,606 *35,708*	62,341 —	40,233 *35,708*	12,480 *8,437*	708 *1,516*
II	2,344 *659*	771 *3,338*	18,030 *34,949*	61,015 —	60,047 *34,949*	26,399 *8,258*	245 *1,483*
III	5,649 *1,255*	23,961 *6,356*	92,193 *66,562*	116,206 —	51,695 *66,562*	14,511 *15,727*	54 *2,825*
IV	3,738 *982*	6,591 *4,974*	65,395 *52,082*	90,927 —	33,457 *52,082*	2,912 *12,306*	9,476 *2,210*
Periods covered by analysis.	9	8	8	7	8	7	6

As might be expected from an examination of the graph, fig. 2, the phase of the fundamental period varies largely from group to group. The relative phases are

Group I—310°
II—276°
III—252°
IV—295°

It is evident that, for curves of this type, identity of phase in successive sections of the observations cannot serve as an empirical test of the reality of a period. The periodicity —or, rather, the fundamental tendency to the given period—may be absolutely real, but phase may shift backwards and forwards over quite a large fraction of the period.

B.—The second test carried out was a detailed periodogram analysis of the series as a whole (273 to 327 observations) between periods 9 and 11. The results are shown in Table VI and the graph fig. 7.

One effect of the shifting phase of the fundamental period is immediately noticeable. The intensity of period 10 comes out at 67,600 roundly, a figure little higher than the intensities for Groups I and II in Table V, and much less than those shown by Groups III and IV. The average for Groups I to IV would be 82,600 ; but even this figure is

TABLE VI.—Periodogram analysis of experimental series (273–327 observations) for periods between 9 and 11 intervals. The values of A and B for periods 9 and 11 have been adjusted to observation 1 as origin. Original ordinates multiplied by 100.

Period: intervals.	Observations used.	A.—Sine amplitude.	B.—Cosine amplitude.	Intensity: $I = A^2 + B^2$	Calculated I for harmonic function.
9·0	2 — 298	− 16·58	− 23·60	832	0
9·1	1 — 273	− 56·37	− 27·26	3,921	615
$9\frac{1}{8}$	1 — 275	− 10·79	− 52·63	2,886	1,096
9·2	1 — 276	+ 15·07	− 43·46	2,116	1,076
9·3	1 — 279	+ 14·33	+ 33·57	2,398	148
$9\frac{1}{3}$	1 — 280	− 17·66	+ 45·50	2,382	0
9·4	1 — 282	− 85·16	+ 9·00	7,333	730
9·5	1 — 285	− 51·52	− 117·14	16,376	3,045
9·6	1 — 288	+ 94·98	− 84·57	16,173	1,644
$9\frac{2}{3}$	1 — 290	+ 99·77	+ 24·52	10,555	0
9·7	1 — 291	+ 58·12	+ 67·24	7,899	808
9·8	1 — 294	− 113·40	+ 14·63	13,074	17,215
9·9	1 — 297	− 55·21	− 193·85	40,618	49,829
10·0	*1 — 300*	*+ 190·71*	*− 176·79*	*67,625*	*67,625*
10·1	1 — 303	+ 238·48	+ 57·14	60,138	49,829
10·2	1 — 306	+ 77·83	+ 156·29	30,484	17,215
10·3	1 — 309	− 5·73	+ 86·89	7,583	808
$10\frac{1}{3}$	1 — 310	− 4·92	+ 67·45	4,574	0
10·4	1 — 312	+ 2·91	+ 58·58	3,440	1,644
10·5	1 — 315	− 22·01	+ 62·44	4,383	3,045
10·6	1 — 318	− 33·74	+ 21·46	1,599	730
$10\frac{2}{3}$	1 — 320	− 12·88	+ 5·55	14	0
10·7	1 — 321	− 3·97	+ 9·43	10	148
10·8	1 — 324	− 17·27	+ 31·30	1,278	1,076
$10\frac{5}{8}$	1 — 325	− 29·41	+ 27·62	1,628	1,096
10·9	1 — 327	− 38·06	− 0·67	1,449	615
11·0	2 — 298	− 6·51	− 8·70	118	0

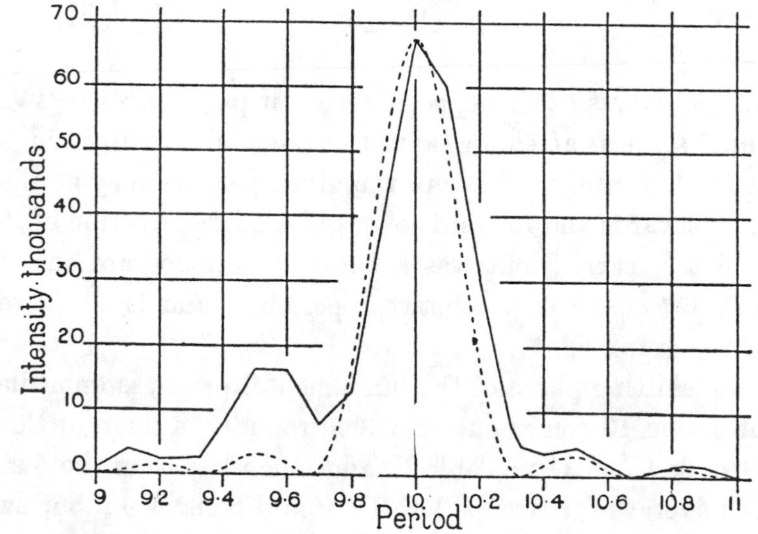

FIG. 7.—Periodogram for the experimental disturbed series on which fig. 2 is based: data in Table VI.

misleadingly low, for the figure for each group is also lowered by the shifting phase. If the graph represented a simple harmonic function, the intensity would be given by $A^2 = 2\sigma^2$, where σ is the standard deviation of the series about the zero base-line, and for the series in question this would show an intensity of 92,700. The periodogram intensity is therefore too low by some 25 per cent.

It is clear, then, that periodogram analysis applied to functions of the present kind tends to give much too low* an intensity for the fundamental. But this at once raises the question, what would be the magnitude of such reduction if the series were indefinitely long? If we refer back to Section I and the analysis of the series into its component parts, the complementary function which is a simple harmonic function and the particular integral which is a fluctuating series, the answer is, I think, clear. The intensity of the complementary function alone would remain at its true value: the intensity of the particular integral would vanish. But if disturbances are sufficiently large, the particular integral contributes nearly the whole of the intensity: even in the present case it contributes roundly eight-ninths. Hence in a heavily disturbed series the intensity would in the long run tend almost to vanish, and the result given by the periodogram might tend to mislead the incautious interpreter. Probably, however, the result would not mislead in practice, because only a modest number of periods is usually available.

Turning now either to the last two columns of the table or to the graph, the main general effect of the varying phase is evidently to broaden the band that would be given by a simple harmonic function of the same period. At the points where the intensity should drop to zero for a simple harmonic function, the observed intensities show a tendency to drop to minima but remain above zero. The two sides of the periodogram are, however, strikingly unlike each other: the graph is markedly asymmetrical, and one cannot help wondering whether the same characteristic would have remained had the series been longer—3,000 observations, say, instead of 300. Outside limits of about 9·7 to 10·3, intensities are much higher for periods below the fundamental than for periods correspondingly above it.

It is of interest to compare fig. 7 with Graph A in fig. 2 of SCHUSTER's paper, in which the periodogram for the sunspot numbers between 1826 and 1900 is compared with the calculated periodogram for a simple harmonic function. It shows something of the same characteristics, the intensities of shorter periods being much higher than the calculated values, while the intensities of longer periods, round about 14 years, are actually *lower*. The graph of the periodogram for the entire series, from 1750 to 1900, SCHUSTER's fig. 1, somewhat similarly but even more markedly shows much higher intensities for periods from 8·25 to 10·25 years than for periods from 12·25 to 14·25, the average intensities over these ranges being 1,061 and 394 respectively.

* [*Added, February* 17.—Too low, I hope it is clear, only in the sense in which the whole of the intensity may be regarded as due to the tendency to follow the fundamental period. Actually the intensity is much higher than that of the complementary function, which is the strictly periodic component.]

It must be borne in mind that this section has been concerned solely with the effect of a certain series of random disturbances. In data referring to natural phenomena it is unlikely that the disturbances will be strictly random with respect to time, and the effect on the per iodogram may be much more marked.

V. Summary and Conclusions.

The sunspot numbers, it is suggested, should be regarded as analogous to the data that would be given by observations of a disturbed periodic movement, such as that of a pendulum subjected to successive small random impulses.

The graph of such a movement presents the principal features of the graph of sunspot numbers, viz., a surprising degree of smoothness accompanied by a continual change of amplitude and shift of phase.

It is suggested that in this case the application of the periodogram method gives results subject to a large margin of error, and may be misleading. Trial on an empirical disturbed series (Section IV) showed, in fact, that with only 7 or 8 periods results are highly erratic; with a larger number of periods, about 30, the main effect is a broadening of the band due to the fundamental and a reduction of the apparent intensity.

The problem of determining the period and the disturbances, in the case of the sunspot numbers, was attacked in the first instance (Section II) by finding the best (least square) linear equation relating $u_x + u_{x-2}$ to u_{x-1}, this giving the form of difference equation required for a simple harmonic function. The equation gave a period which was obviously too low. It is suggested that this result is due to the presence of superposed fluctuations in addition to disturbances, a suggestion borne out by applying the same method to graduated values of the numbers. This yielded a much closer approximation to the period suggested by the graph.

Applying an extension of the same method in an endeavour to determine whether or no there was any secondary period in addition to the fundamental, the paradoxical result was reached that (with the particular form of equation corresponding to two simple harmonic functions) u_x cannot be so closely estimated in terms of u_{x-1} to u_{x-4} as in terms of u_{x-1} and u_{x-2} alone. There is thus no evidence of the existence of any secondary period. The result also suggests the existence of disturbances (as distinct from superposed fluctuations), since only disturbances can give the required element of unpredictability rapidly increasing with the time.

The better and more general method was then applied (Section III) of determining the regression equation for u_x on u_{x-1} and u_{x-2}, and solving this as a finite difference equation. The solution is a rapidly damped harmonic function.

The "disturbances" deduced from the equation show two principal features: (1) a tendency to give preponderantly positive and highly variable disturbances, and preponderantly negative and less variable disturbances, in alternate intervals of 40 to 42 years (*cf.* Table II); (2) a tendency for positive disturbances during the approach to the maximum of the sunspot numbers, negative disturbances during the approach to minimum (*cf.* Table III). It is suggested that the second feature accords with the necessity for maintaining a damped periodic function.

A damped harmonic function of the time is, however, clearly not the mathematical form required: a form that would suit the data very well would be the square of a damped harmonic function. The difference equation of this function is of the third degree, and would therefore require u_{x-3} to be brought into the regression equation as well as u_{x-1} and u_{x-2}.

But investigation of the correlations (Table IV) shows that for the ungraduated numbers it would be no use whatever, for the graduated numbers very little use, to bring in further terms beyond u_{x-2}, once more emphasising the rapid increase of the element of unpredictability with the time.

Further work on this line seems, therefore, to be blocked—a rather disappointing conclusion, since the form of function suggested otherwise looks hopeful.

The correlations, like the method of Section II, equally fail to suggest the presence of any period other than the fundamental, a conclusion entirely in accord with the work of LARMOR and YAMAGA.

I do not put forward the methods used in the present paper as necessarily the best, nor even in all cases applicable. I was attacking a problem which, to me at least, was a new one, and used the methods that seemed best at the moment; but experience may suggest better methods. With the present experience, indeed, it seems clear that the method of Section III is not a good method for determining the period, for it tends to give too low a value when superposed fluctuations are present in addition to disturbances, as in all probability they nearly always are. It might be better, for example, to determine the period first by the simple and straightforward method of taking the interval between the first and last maxima (or minima), and dividing by the number of intervening periods, leaving only the damping factor to be found by least squares or otherwise. But while this is quite a possible method for the sunspot numbers, it would not be possible with a variable largely affected by superposed fluctuations: maxima and minima would be too indefinite. Variables affected largely both by disturbances and by superposed fluctuations present a very difficult problem for analysis.

And I would like in conclusion to suggest that many series which have been or might be subjected to periodogram analysis may be subject to "disturbance" in the sense in which the term is here used,* and that this may possibly be the source of some rather odd results which have been reached. Disturbance will always arise if the value of the variable is affected by external circumstance and the oscillatory variation with time is wholly or partly self-determined, owing to the value of the variable at any one time being a function of the immediately preceding values. Disturbance, as it seems to me,

* [*Added, February* 17.—A number of the graphs in HEDGES and MYERS' 'The Problem of Physico-Chemical Periodicity' (Arnold, 1926) look obviously of the "disturbed" type, the waves being very smooth, but varying in phase and amplitude.]

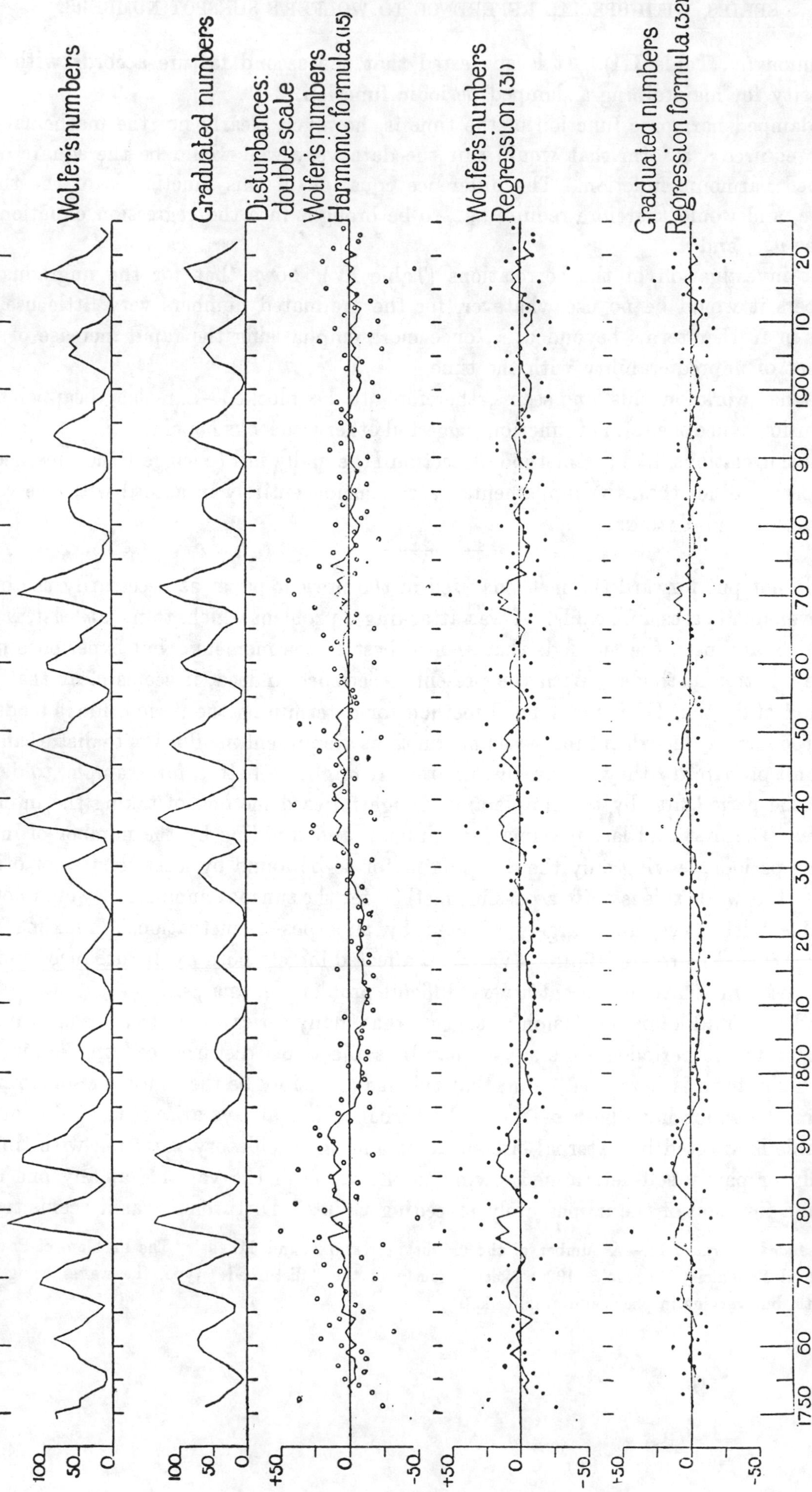

Fig. 8.—Graphs of Wolfer's numbers and the graduated numbers, and of the disturbances by equations (15), (31) and (32): the lines on the graphs of disturbances show quinquennial averages.

can only be excluded if either (1) the variable is quite unaffected by external circumstance, or (2) we are dealing with a forced vibration and the external circumstances producing this forced vibration are themselves undisturbed.

TABLE A.—WOLFER's sunspot numbers and graduated numbers, and the disturbances calculated by equations (31) and (32) respectively. The graduated numbers are given solely because they were used for part of the preceding work: they are not a good graduation (*cf.* text, pp. 282–283).

Year.	WOLFER's number.	Disturbance.	Graduated number.	Disturbance.	Year.	WOLFER's number.	Disturbance.	Graduated number.	Disturbance.
1749	80·9	—	—	—	1793	46·9	− 3·9	50·8	+ 9·8
1750	83·4	—	—	—	1794	41·0	+ 3·5	35·5	− 9·5
1751	47·7	− 25·2	61·8	—	1795	21·3	− 16·9	26·5	+ 0·6
1752	47·8	+ 24·5	40·2	—	1796	16·0	+ 0·4	12·6	− 11·9
1753	30·7	− 16·1	30·5	+ 6·3	1797	6·4	− 15·0	7·9	− 2·8
1754	12·2	− 11·6	15·5	− 11·3	1798	4·1	− 7·9	3·8	− 10·9
1755	9·6	− 0·5	6·1	− 5·8	1799	6·8	− 8·4	6·0	− 6·3
1756	10·2	− 8·6	15·4	+ 5·7	1800	14·5	− 5·8	17·5	− 1·4
1757	32·4	+ 11·1	29·4	− 1·9	1801	34·0	+ 5·1	32·2	− 2·4
1758	47·6	− 3·1	46·1	+ 1·1	1802	45·0	− 5·0	42·4	− 5·2
1759	54·0	− 2·6	54·0	− 5·1	1803	43·1	− 8·9	47·0	− 4·3
1760	62·9	+ 7·7	71·1	+ 13·4	1804	47·5	+ 5·2	45·6	− 4·5
1761	85·9	+ 22·9	72·8	− 4·5	1805	42·2	− 7·2	41·8	− 2·4
1762	61·2	− 26·8	67·6	+ 1·5	1806	28·1	− 11·3	26·4	− 13·2
1763	45·1	+ 5·3	46·5	− 10·4	1807	10·1	− 13·9	14·5	− 4·8
1764	36·4	+ 2·1	33·4	+ 4·3	1808	8·1	− 0·9	5·2	− 8·4
1765	20·9	− 12·3	19·0	− 7·2	1809	2·5	− 15·6	3·2	− 5·9
1766	11·4	− 6·7	18·1	+ 3·3	1810	0·0	− 11·9	0·3	− 13·2
1767	37·8	+ 22·3	34·6	+ 9·6	1811	1·4	− 10·8	1·1	− 9·6
1768	69·8	+ 12·6	74·8	+ 24·0	1812	5·0	− 10·7	6·2	− 8·1
1769	106·1	+ 23·3	97·9	− 0·5	1813	12·2	− 7·5	8·4	− 13·0
1770	100·8	− 9·8	101·9	+ 0·7	1814	13·9	− 13·1	20·1	− 0·5
1771	81·6	+ 1·9	85·9	− 2·8	1815	35·4	+ 10·9	32·4	− 4·2
1772	66·5	+ 9·1	59·3	− 2·0	1816	45·8	− 6·5	44·3	− 1·5
1773	34·8	− 14·9	44·9	+ 11·1	1817	41·1	− 11·1	41·0	− 13·0
1774	30·6	+ 13·6	19·2	− 14·1	1818	30·4	− 8·7	32·3	− 7·1
1775	7·0	− 25·2	10·6	+ 4·7	1819	23·9	− 3·9	23·2	− 5·7
1776	19·8	+ 16·6	30·3	+ 16·8	1820	15·7	− 10·3	15·0	− 7·1
1777	92·5	+ 56·6	93·5	+ 43·2	1821	6·6	− 12·7	8·1	− 8·9
1778	154·4	+ 29·3	136·9	+ 6·7	1822	4·0	− 8·4	2·4	− 10·7
1779	125·9	− 34·7	130·4	− 14·9	1823	1·8	− 13·1	3·6	− 6·4
1780	84·8	+ 3·0	93·1	− 7·5	1824	8·5	− 5·2	6·5	− 9·9
1781	68·1	+ 22·8	61·0	+ 11·7	1825	16·6	− 7·5	19·7	− 0·1
1782	38·5	− 11·3	42·7	+ 12·1	1826	36·3	+ 5·7	33·7	− 3·8
1783	22·8	+ 1·8	19·0	− 9·6	1827	49·7	− 2·0	51·2	+ 3·1
1784	10·2	− 9·1	10·8	+ 3·4	1828	62·5	+ 5·7	60·8	− 2·6
1785	24·1	+ 11·5	32·2	+ 18·2	1829	67·0	+ 1·8	70·8	+ 6·9
1786	82·9	+ 43·4	81·3	+ 28·3	1830	71·0	+ 8·4	64·7	− 6·7
1787	132·0	+ 22·6	123·2	+ 13·0	1831	47·8	− 17·5	51·3	− 2·8
1788	130·9	− 5·9	135·6	+ 1·3	1832	27·5	− 3·8	24·8	− 13·9
1789	118·1	+ 15·2	115·4	− 4·1	1833	8·5	− 11·0	9·3	0·0
1790	89·9	+ 3·2	90·8	+ 11·9	1834	13·2	+ 5·9	16·9	+ 9·9
1791	66·6	+ 9·4	70·5	+ 12·6	1835	56·9	+ 30·9	62·5	+ 31·5
1792	60·0	+ 15·6	55·9	+ 9·1	1836	121·5	+ 39·9	114·3	+ 20·3

TABLE A (continued).

Year.	WOLFER'S number.	Disturbance.	Graduated number.	Disturbance.	Year.	WOLFER'S number.	Disturbance.	Graduated number.	Disturbance.
1837	138·3	− 1·5	130·1	− 5·8	1881	54·3	+ 1·0	51·2	+ 0·4
1838	103·2	− 16·8	113·5	− 4·8	1882	59·7	− 5·9	61·7	− 4·2
1839	85·8	+ 23·9	83·1	+ 2·6	1883	63·7	+ 5·2	64·2	− 1·1
1840	63·2	+ 1·7	61·4	+ 13·7	1884	63·5	+ 3·2	63·2	+ 2·6
1841	36·8	− 5·7	40·4	+ 1·2	1885	52·2	− 5·2	48·4	− 8·7
1842	24·2	+ 2·3	20·5	− 4·3	1886	25·4	− 17·0	29·7	− 5·8
1843	10·7	− 11·6	12·4	+ 0·9	1887	13·1	− 0·7	12·2	− 6·8
1844	15·0	+ 2·6	18·1	+ 2·8	1888	6·8	− 8·0	7·3	− 0·2
1845	40·1	+ 13·1	35·2	+ 4·9	1889	6·3	− 8·1	2·9	− 11·2
1846	61·5	+ 3·6	66·6	+ 14·9	1890	7·1	− 10·8	12·1	+ 0·7
1847	98·5	+ 28·3	100·3	+ 14·8	1891	35·6	+ 16·3	37·3	+ 8·4
1848	124·3	+ 18·4	113·6	+ 2·2	1892	73·0	+ 16·0	68·4	+ 8·7
1849	95·9	− 20·4	98·7	− 5·8	1893	84·9	− 3·7	84·3	− 2·3
1850	66·5	+ 5·3	73·6	+ 2·3	1894	78·0	− 2·1	79·4	− 6·3
1851	64·5	+ 24·2	60·2	+ 15·0	1895	64·0	+ 1·0	62·2	− 3·3
1852	54·2	− 2·7	54·4	+ 9·4	1896	41·8	− 6·9	42·4	− 1·0
1853	39·0	− 5·4	38·3	− 8·7	1897	26·2	− 1·9	30·7	+ 3·5
1854	20·6	− 10·1	20·7	− 6·5	1898	26·7	+ 5·0	20·2	− 5·2
1855	6·7	− 9·3	6·4	− 7·1	1899	12·1	− 20·5	16·9	− 1·9
1856	4·3	− 5·1	6·2	+ 0·3	1900	9·5	− 3·1	6·2	− 16·1
1857	22·8	+ 7·6	22·7	+ 5·6	1901	2·7	− 16·0	3·3	− 5·4
1858	54·8	+ 13·1	59·0	+ 16·7	1902	5·0	− 6·3	8·0	− 4·9
1859	93·8	+ 21·3	87·0	+ 3·0	1903	24·4	+ 5·6	21·7	− 0·6
1860	95·7	− 8·2	95·2	− 2·1	1904	42·0	− 1·3	46·5	+ 7·2
1861	77·2	− 3·7	79·2	− 8·1	1905	63·5	+ 9·2	54·1	− 11·8
1862	59·1	+ 4·2	57·7	+ 1·2	1906	53·8	− 17·8	63·7	+ 6·2
1863	44·0	+ 1·3	49·9	+ 13·2	1907	62·0	+ 17·5	54·2	− 11·8
1864	47·0	+ 12·8	40·4	− 1·8	1908	48·5	− 13·4	55·2	+ 11·3
1865	30·5	− 17·7	32·6	− 1·4	1909	43·9	+ 5·5	36·9	− 16·1
1866	16·3	− 7·7	12·9	− 16·9	1910	18·6	− 22·4	22·5	− 2·0
1867	7·3	− 8·5	14·7	+ 8·5	1911	5·7	− 4·4	6·7	− 10·6
1868	37·3	+ 24·3	31·2	+ 6·4	1912	3·6	− 5·7	1·2	− 3·8
1869	73·9	+ 14·7	89·9	+ 41·6	1913	1·4	− 13·6	0·4	− 8·9
1870	139·1	+ 50·4	113·2	− 10·8	1914	9·6	− 3·8	18·1	+ 5·6
1871	111·2	− 41·1	128·5	+ 16·3	1915	47·4	+ 21·6	33·7	− 6·3
1872	101·7	+ 29·6	92·4	− 24·3	1916	57·1	− 14·1	76·3	+ 26·9
1873	66·3	− 11·3	72·9	+ 23·1	1917	103·9	+ 44·4	83·5	− 17·9
1874	44·7	+ 8·4	39·4	− 9·8	1918	80·6	− 35·4	90·8	+ 12·6
1875	17·1	− 13·4	21·9	+ 7·8	1919	63·6	+ 9·6	59·3	− 24·1
1876	11·3	+ 3·8	11·5	− 2·9	1920	37·6	− 8·9	41·9	+ 12·0
1877	12·3	− 5·5	8·1	− 4·6	1921	26·1	+ 3·4	24·0	− 4·8
1878	3·4	− 19·6	4·4	− 11·5	1922	14·2	− 10·1	12·9	− 2·7
1879	6·0	− 4·4	10·5	− 2·5	1923	5·8	− 10·0	—	—
1880	32·3	+ 12·6	30·6	+ 5·4	1924	16·7	+ 4·4	—	—

Maximum Entropy and Spectral Estimation: A Review

ATHANASIOS PAPOULIS, FELLOW, IEEE

***Abstract*—The method of maximum entropy is reviewed with emphasis on its relationship to entropy rate, Wiener filters, autoregressive processes, extrapolation, the Levinson algorithm, lattice, all-pole and all-pass filters, and stability.**

I. Introduction

In the last decade, several papers have been published discussing a method of spectral estimation based on the principle of maximum entropy [1]-[4] and the relationship of this method to entropy rate [5], the Wiener theory of prediction [6], [7], autoregressive processes, the Levinson algorithm [8], lattice filters [9], all-pole and all-pass filters, and stability. However, it appears that no single publication in the open literature explains simply the interconnection of these topics. The purpose of this paper is an attempt to do so starting from first principles [10]. The effectiveness of the method in the solution of specific problems will not be considered here. In the Appendix we comment briefly on its conceptual justification. The material is developed with some originality; however, the paper is essentially tutorial.

The entire development is based on the orthogonality principle [11]: in the estimation of a random variable (RV) y by a linear combination

$$\hat{y} = a_1 x_1 + \cdots + a_N x_N \tag{1}$$

of the N random variables $x_1, \cdots, x_N$ (data), the MS error

$$P = E\{(y - \hat{y})^2\} \tag{2}$$

is minimum if the estimation error

$$e = y - \hat{y} \tag{3}$$

is orthogonal to the data x_k, that is, if

$$E\{e x_k\} = 0 \qquad k = 1, \cdots, N. \tag{4}$$

Manuscript received October 20, 1980; revised May 1, 1981. This work was supported by the Advanced Research Projects Agency of the Department of Defense and was monitored by the Office of Naval Research under Contract N00014-76-C-0144, and in part by the Joint Services Technical Advisory Committee under Contract F49620-80-C-0077.

The author is with the Polytechnic Institute of New York, Farmingdale, NY 11743.

Reprinted from *IEEE Trans. Acoust., Speech, Signal Processing*, vol. ASSP-29, pp. 1176–1186, Dec. 1981.

The resulting MS error P is then given by

$$P = E\{e^2\} = E\{ey\}. \tag{5}$$

We state also, for later use, the following results from the theory of linear systems with stochastic inputs [11]. Suppose that the input to a discrete linear system is a stationary process $x[n]$ with autocorrelation

$$R_{xx}[m] = E\{x[n+m]\, x[n]\} \tag{6}$$

and power spectrum

$$S_{xx}(z) = \sum_{m=-\infty}^{\infty} R_{xx}[m]\, z^{-m}. \tag{7}$$

If $h[n]$ is the delta response and $H(z)$ the system function of the system, then the power spectrum of the resulting output $y[n] = x[n] * h[n]$ is given by

$$S_{yy}(z) = S_{xx}(z)\, H(z)\, H(1/z). \tag{8}$$

In the above we assumed, as we shall throughout the paper, that all processes and systems are real. With trivial modifications, the results hold also for complex processes.

The spectral estimation problem has two parts:

1) *Deterministic:* Estimate the power spectrum $S(z)$ of a process $s[n]$ in terms of the $N+1$ values $R[0], R[1], \cdots, R[N]$ of its autocorrelation.

2) *Random:* Estimate the power spectrum $S(z)$ of a process $s[n]$ in terms of the N_0 values $s[1], s[2], \cdots, s[N_0]$ of a single realization of $s[n]$.

As we show in the paper, the maximum entropy solution of part 1) can be presented as a recursive modification of the Wiener prediction filter. The modification is based on the Levinson algorithm expressed in terms of forward and backward predictors. The solution of part 2) is given by an estimator whose various parameters satisfy the same equations as in the deterministic case, with the only difference that in the evaluation of the recursion coefficient Γ_N [see (53)], all ensemble averages are replaced by suitable time averages.

II. Prediction

We wish to estimate the future value $s[n+1]$ of a random signal $s[n]$ in terms of the sum

$$\hat{s}_N[n+1] = a_1^N s[n] + \cdots + a_N^N s[n-N+1]$$

$$= \sum_{k=1}^{N} a_k^N s[n-k+1] \tag{9}$$

involving its N most recent values $s[n-k]$. The set of weights a_k^N that minimizes the MS value of the prediction error defines an FIR filter of order N (Fig. 1) called the *forward predictor* (one-step) of $s[n]$. The superscript N in a_k^N specifies the order of the predictor. Since $s[n]$ is stationary, the optimum weights a_k^N are independent of n. We can give, therefore, to the variable n in (9) any value. With

$$\hat{e}_N[n] = s[n] - \hat{s}_N[n] \tag{10}$$

the *forward predictor* error, we have [see (4)]

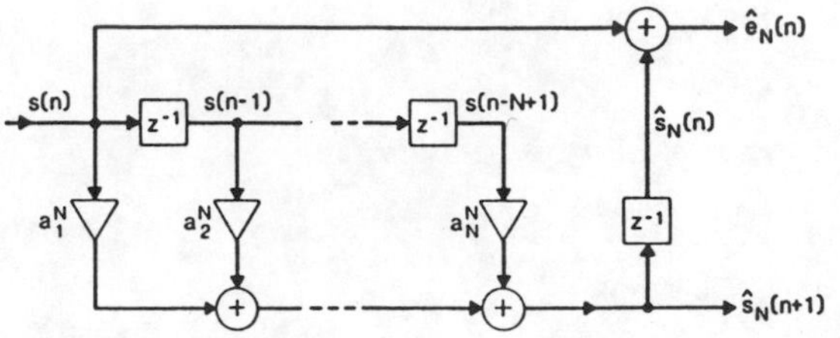

Fig. 1. Forward predictor filter $\hat{H}_N(z)$. $\hat{s}_N[n]$: predictor of $s[n]$, $\hat{e}_N[n]$: predictor error.

$$E\{\hat{e}_N[n]\, s[n-k]\} = 0 \qquad 1 \leq k \leq N. \tag{11}$$

This yields the system

$$\begin{aligned} R[0]\, a_1^N + R[1]\, a_2^N + \cdots + R[N-1]\, a_N^N &= R[1] \\ R[1]\, a_1^N + R[0]\, a_2^N + \cdots + R[N-2]\, a_N^N &= R[2] \\ &\vdots \\ R[N-1]\, a_1^N + R[N-2]\, a_2^N + \cdots + R[0]\, a_N^N &= R[N] \end{aligned} \tag{12}$$

expressing the predictor coefficients a_k^N in terms of the $N+1$ values $R[0], \cdots, R[N]$ of the autocorrelation $R[m]$ of $s[n]$. In the next section, we discuss a recursion method for solving this system.

Applying (5) to our estimator, we conclude that the MS estimation error $\hat{P}_N$ is given by

$$\hat{P}_N = E\{\hat{e}_N^2[n]\} = E\{\hat{e}_N[n]\, s[n]\} = R[0] - \sum_{k=1}^{N} a_k^N R[k]. \tag{13}$$

Note: Consider two processes $s[n]$ and $s_0[n]$ with autocorrelations $R[m]$ and $R_0[m]$, respectively. From (12) it follows that if

$$R[m] = R_0[m] \qquad \text{for } 0 \leq m \leq N \tag{14}$$

then the Nth order predictors of $s[n]$ and $s_0[n]$ are identical. Conversely, if the Nth order predictors of $s[n]$ and $s_0[n]$ are identical, then (12) shows that $R[m] = cR_0[m]$ $0 \leq m \leq N$. The proportionality constant c equals 1 if $R[m]$ and $R_0[m]$ satisfy one additional equation, for example, if the Nth order MS errors are equal or if $R_0[0] = R[0]$, that is, if the two processes have the same average power

$$P_0 = E\{s^2[n]\} = R[0]. \tag{15}$$

The prediction error [see (10)]

$$\hat{e}_N[n] = s[n] - \sum_{k=1}^{N} a_k^N s[n-k] \tag{16}$$

is the output of a system (Fig. 1) with input $s[n]$ and system function

$$\hat{H}_N(z) = 1 - \sum_{k=1}^{N} a_k^N z^{-k}. \tag{17}$$

This system will be called the *forward error filter*.

The Backward Predictor: We shall now estimate the process $s[n]$ in terms of the *backward predictor*

$$\check{s}_N[n] = \sum_{k=1}^{N} b_k^N s[n+k]$$

involving its N closest future values $s[n+k]$. With

$$\check{e}_N[n] = s[n] - \check{s}_N[n] \tag{18}$$

the *backward predictor* error, we have as in (11)

$$E\{\check{e}_N[n]\, s[n+k]\} = 0 \qquad 1 \leqslant k \leqslant N. \tag{19}$$

This yields the system

$$\sum_{r=1}^{N} b_r^N R[r-k] = R[k] \qquad 1 \leqslant k \leqslant N$$

which is identical to the system (12). Hence, $b_k^N = a_k^N$, that is, the backward predictor of $s[n]$ is the sum

$$\check{s}_N[n] = a_1^N s[n+1] + \cdots + a_N^N s[n+N]. \tag{20}$$

The predictor error is thus given by

$$\check{e}_N[n] = s[n] - \sum_{k=1}^{N} a_k^N s[n+k]. \tag{21}$$

Denoting by $\check{P}_N$ its MS value, we conclude, as in (13), that

$$\check{P}_N = E\{\check{e}_N[n]\, s[n]\} = R[0] - \sum_{k=1}^{N} a_k^N R[k].$$

In other words, the forward and backward MS predictor errors are equal:

$$\check{P}_N = \hat{P}_N = P_N. \tag{22}$$

Clearly, $\check{e}_N[n]$ is the output of a system with input $s[n]$ and system function

$$\check{H}_n(z) = 1 - \sum_{k=1}^{N} a_k^N z^k. \tag{23}$$

This system will be called the *backward error filter.* Comparing with (17), we conclude that

$$\check{H}_N(z) = \hat{H}_N(1/z) \tag{24}$$

In the above, we assumed that $s[n]$ is a real process. The results hold also for complex processes subject to the following modifications.

$$R[-m] = R^*[m] \quad b_k^N = (a_k^N)^* \quad \check{H}_N(z) = \hat{H}_N^*(1/z^*).$$

Autoregressive Processes: An autoregressive process (AR) of order M is a random signal $s[n]$ satisfying the recursion equation

$$s[n] - c_1 s[n-1] - \cdots - c_M s[n-M] = \zeta[n] \tag{25}$$

where $\zeta[n]$ is stationary white noise with

$$R_{\zeta\zeta}[m] = \sigma^2 \delta[m] \qquad S_{\zeta\zeta}(z) = \sigma^2. \tag{26}$$

From the definition it follows that $s[n]$ is the output of a linear system with input $\zeta[n]$ and system function

$$T(z) = \frac{1}{1 - \sum_{k=1}^{M} c_k z^{-k}} \tag{27}$$

If this system is stable, then $s[n]$ is a stationary process given by

$$s[n] = \sum_{r=0}^{\infty} h[r]\, \zeta[n-r] \tag{28}$$

where $h[n]$ is the causal inverse transform [12] of $T(z)$. This shows that for any $k \geqslant 1$, the random variable $s[n-k]$ is a linear combination of only the past values of $\zeta[n]$; hence

$$E\{s[n-k]\, \zeta[n]\} = 0 \qquad k \geqslant 1 \tag{29}$$

because $\zeta[n]$ is white noise by assumption.

We maintain that the predictor $\hat{s}_N[n]$ of $s[n]$ of order $N \geqslant M$ is the sum

$$\hat{s}[n] = c_1 s[n-1] + \cdots + c_M s[n-M] \tag{30}$$

and the predictor error $\hat{e}_N[n] = s[n] - \hat{s}_N[n]$ equals the process $\zeta[n]$:

$$\hat{e}_N[n] = \zeta[n]. \tag{31}$$

Indeed, as we see from (25), $s[n] - \hat{s}_N[n] = \zeta[n]$. Hence, the orthogonality condition (11) follows from (29). The resulting MS error is given by

$$P_N = E\{\hat{e}_N^2[n]\} = E\{\zeta^2[n]\} = \sigma^2.$$

We note that the sum in (30) is of the form (9) where

$$a_k^N = \begin{cases} c_k & 1 \leqslant k \leqslant M \\ 0 & M < k \leqslant N. \end{cases} \tag{32}$$

With $\hat{H}_N(z)$ the error filter as defined in (17), it follows from (27) and (32) that the system function $T(z)$ of the system (AR filter) specified by (25) is given by

$$T(z) = \frac{1}{\hat{H}_N(z)}. \tag{33}$$

Clearly, $s[n]$ is the output of $T(z)$ with input $\zeta[n]$. Denoting by $S(z)$ its power spectrum, we conclude from (8) and (33) that

$$S(z) = \frac{S_{\zeta\zeta}(z)}{\hat{H}_N(z)\, \hat{H}_N(1/z)}. \tag{34}$$

Since $S_{\zeta\zeta}(z) = \sigma^2$, the above yields

$$\bar{S}(\omega) \equiv S(e^{j\omega T}) = \frac{\sigma^2}{\left|1 - \sum_{k=1}^{M} c_k e^{-jk\omega T}\right|^2}. \tag{35}$$

Extrapolation: The inverse transform of $S(z)$ is the autocorrelation $R[m]$ of $s[n]$. From (12) and (32) it follows that

$$R[N] = \sum_{k=1}^{M} c_k R[N-k] \qquad \text{for every } N \geqslant M. \tag{36}$$

Setting $N = M$, we obtain a system of M equations expressing the M coefficients c_k in terms of the first $M+1$ values of $R[m]$. With c_k so determined, we can use (36) to evaluate successively $R[N]$ for every $N > M$. Thus, (36) is an *extrapolation* formula for $R[N]$.

III. The Levinson Algorithm

The solution of the system (12) involves the inversion of the matrix

$$\begin{bmatrix} R[0] & R[1] & \cdots R[N-1] \\ R[1] & R[0] & \cdots R[N-2] \\ \vdots & \vdots & \vdots \\ R[N-1] & R[N-2] & \cdots R[0] \end{bmatrix} \tag{37}$$

This matrix has a special form (Toeplitz [13]) and can be inverted easily by a simple iteration known as Levinson's algorithm [8]. We shall present the result as a recursion involving directly the predictor coefficients.

Theorem: The forward predictor $\hat{s}_N[n]$ can be written as a sum

$$\hat{s}_N[n] = \hat{s}_{N-1}[n] + \Gamma_N(\hat{s}[n-N] - \check{s}_{N-1}[n-N]) \tag{38}$$

where

$$\hat{s}_{N-1}[n] = \sum_{k=1}^{N-1} a_k^{N-1} s[n-k]$$
$$\check{s}_{N-1}[n-N] = \sum_{k=1}^{N-1} a_k^{N-1} s[n-N+k] \tag{39}$$

are the forward and backward predictors of $s[n]$ and $s[n-N]$, respectively, and the coefficient Γ_N is a constant to be determined. Equation (38) can be expressed in terms of the forward and backward predictor errors

$$\hat{e}_{N-1}[n] = s[n] - \hat{s}_{N-1}[n]$$
$$\check{e}_{N-1}[n-N] = s[n-N] - \check{s}_{N-1}[n-N].$$

Indeed, subtracting both sides from $s[n]$, we obtain

$$\hat{e}_N[n] = \hat{e}_{N-1}[n] - \Gamma_N \check{e}_{N-1}[n-N]. \tag{40}$$

It suffices, therefore, to prove (40).

Proof by Induction: Clearly, $\hat{s}_N[n]$ is a linear combination of the N most recent values $s[n-k]$ of the signal. It suffices, therefore, to show that $\hat{e}_N$ satisfies the orthogonality condition. By the induction hypothesis, we assume that the sequences $\hat{e}_{N-1}$ and $\check{e}_{N-1}$ are the predictor errors of order $N-1$, that is [see (11) and (19)]

$$E\{\hat{e}_{N-1}[n]\, s[n-k]\} = 0 \qquad 1 \leqslant k \leqslant N-1$$
$$E\{\check{e}_{N-1}[n-N]\, s[n-k]\} = 0 \qquad 1 \leqslant k \leqslant N-1. \tag{41}$$

We shall show that if $\hat{e}_N$ is given by (40), then it is the Nth order predictor error. As we know, this is true if

$$E\{\hat{e}_N[n]\, s[n-k]\} = 0 \qquad 1 \leqslant k \leqslant N. \tag{42}$$

From (41) it follows that (42) holds for $1 \leqslant k \leqslant N-1$. It suffices, therefore, to select Γ_N such as to satisfy (42) for $k = N$. This yields $E\{\hat{e}_n[n]\, s[n-N]\} = 0$. Inserting (40) into the above, we obtain

$$E\{\hat{e}_{N-1}[n]\, s[n-N]\} = \Gamma_N E\{\check{e}_{N-1}[n-N]\, s[n-N]\} \tag{43}$$

and since [see (39)]

$$E\{\hat{e}_{N-1}[n]\, s[n-N]\}$$
$$= E\left\{\left(s[n] - \sum_{k=1}^{N-1} a_k^{N-1} s[n-k]\right) s[n-N]\right\}$$
$$= R[N] - \sum_{k=1}^{N-1} a_k^{N-1} R[N-k]$$

and

$$E\{\check{e}_{N-1}[n-N]\, s[n-N]\} = P_{N-1}$$

(43) yields

$$P_{N-1}\Gamma_N = R[N] - \sum_{k=1}^{N-1} a_k^{N-1} R[N-k]. \tag{44}$$

We have, thus, expressed Γ_N in terms of the coefficients a_k^{N-1} of the predictor of order $N-1$ and the corresponding MS error P_{N-1}. This error is given also by [see (22)]

$$P_{N-1} = E\{\hat{e}_{N-1}[n]\, s[n]\}. \tag{45}$$

With this choice of Γ_N, (42) holds for every k from 1 to N.

Using (38), we can express a_k^N in terms of a_k^{N-1} and the constant Γ_N. Indeed, with $\hat{s}_N[n]$ as in (9), we obtain, equating coefficients of both sides of (38), the recursive equation

$$a_k^N = a_k^{N-1} - \Gamma_N a_{N-k}^{N-1} \qquad 1 \leqslant k \leqslant N-1 \quad a_N^N = \Gamma_N \tag{46}$$

where Γ_N is determined from (44). Since this equation involves the MS error P_{N-1}, to complete the induction, we must determine its Nth order value

$$P_N = E\{\hat{e}_N[n]\, s[n]\}. \tag{47}$$

We maintain that

$$P_N = (1 - \Gamma_N^2) P_{N-1}. \tag{48}$$

To show this, we insert $\hat{e}_N[n]$, as given by (40), into (47) and use (45) and the fact that

$$E\{\check{e}_{N-1}[n-N]\, s[n]\}$$
$$= E\left\{\left(s[n-N] - \sum_{k=1}^{N-1} a_k^{N-1} s[n-N+k]\right) s[n]\right\}$$
$$= R[N] - \sum_{k=1}^{N-1} a_k^{N-1} R[N-k] = \Gamma_N P_{N-1}.$$

The result is (48).

The induction starts with $\Gamma_0 = 0, P_0 = E\{s^2[n]\} = R[0]$ and for $N = 1$ it yields $P_0\Gamma_1 = R[1], P_1 = P_0(1 - \Gamma_1^2)$.

The recursion (38) and its equivalent (40) hold also for the backward predictors. Reasoning similarly, we obtain

$$\check{s}_N[n-N] = \check{s}_{N-1}[n-N] + \Gamma_N(s[n] - \hat{s}_{N-1}[n]) \tag{49}$$

$$\check{e}_N[n-N] = \check{e}_{N-1}[n-N] - \Gamma_N \hat{e}_{N-1}[n]. \tag{50}$$

Lattice: Equations (40) and (50) can be given the following graphical interpretation [9], [14]. In Fig. 2, we show N lattice sections connected in cascade. Each section consists of one delay element and two multipliers. The input to the system so formed equals $s[n] = \hat{e}_0[n] = \check{e}_0[n]$ and the two outputs equal the forward and backward predictor errors.

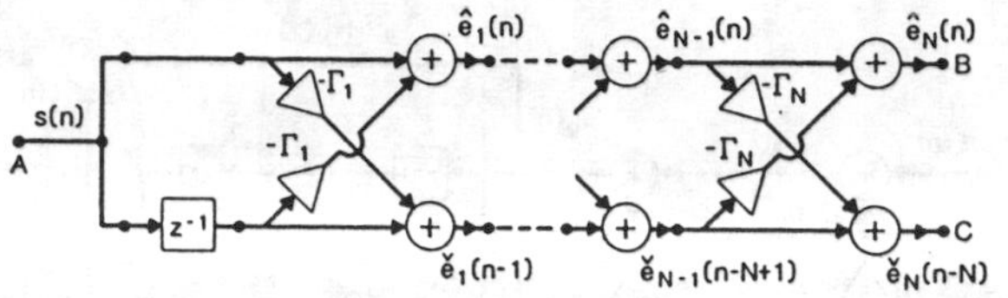

Fig. 2. Lattice filter. $\hat{e}_N[n]$: forward error, $\check{e}_N[n]$: backward error.

We note that the transfer functions from the input A to the two outputs B and C equal $\hat{H}_N(z)$ and

$$z^{-N}\check{H}_N(z) = z^{-N}\hat{H}_N(1/z)$$

respectively, where $\hat{H}_N(z)$ is the forward error filter and $\check{H}_N(z)$ is the backward error filter.

Note: We derive next, for later use, a modified form of (44). Clearly [see (41)]

$$P_N = E\{e_N^2[n]\} = E\{(\hat{e}_{N-1}[n] - \Gamma_N \check{e}_{N-1}[n-N])^2\}. \quad (51)$$

Since the coefficients of the predictor minimize P_N, we must have

$$\frac{\partial P_N}{\partial \Gamma_N} = 0 = E\{-2(\hat{e}_{N-1}[n] - \Gamma_N \check{e}_{N-1}[n-N])\,\check{e}_{N-1}[n-N]\}.$$

Hence

$$P_{N-1}\Gamma_N = E\{\hat{e}_{N-1}[n]\,\check{e}_{N-1}[n-N]\}. \quad (52)$$

The above can be written in the symmetrical form

$$\Gamma_N = \frac{E\{\hat{e}_{N-1}[n]\,\check{e}_{N-1}[n-N]\}}{\frac{1}{2}E\{\hat{e}_{N-1}^2[n] + \check{e}_{N-1}^2[n-N]\}}. \quad (53)$$

This is a consequence of the fact that the forward and backward MS errors are equal. It can also be derived by writing P_N in the symmetrical form

$$P_N = \tfrac{1}{2}E\{\hat{e}_N^2[n] + \check{e}_N^2[n-N]\}$$
$$= \tfrac{1}{2}E\{(\hat{e}_{N-1}[n] - \Gamma_N\check{e}_{N-1}[n-N])^2 + (\check{e}_{N-1}[n-N] - \Gamma_N\hat{e}_{N-1}[n])^2\} \quad (54)$$

and minimizing with respect to Γ_N.

Stability: We have shown that the Nth order MS error P_N is given by $P_N = (1 - \Gamma_N^2)\,P_{N-1}$. From this it follows that

$$|\Gamma_N| \leq 1 \quad (55)$$

with equality iff $P_N = P_{N-1}$. We shall use this result to show that the forward error filter

$$\hat{H}_N(z) = 1 - \sum_{k=1}^{N} a_k^N z^{-k} \quad (56)$$

is a Hurwitz polynomial, i.e., all its roots z_i are inside the unit circle [14]

$$|z_i| \leq 1. \quad (57)$$

From this it will follow that all roots of the backward error filter $\check{H}_N(z)$ are outside the unit circle because

$$\check{H}_N(z) = \hat{H}_N(1/z). \quad (58)$$

Proof by Induction:[1] Clearly, $\hat{H}_1(z) = 1 - \Gamma_1 z^{-1}$; hence, $|z_1| = |\Gamma_1| \leq 1$. Suppose that (57) is true for all orders up to $N-1$. We shall show that it is true for order N. From (40) and (58), it follows that

$$\hat{H}_N(z) = \hat{H}_{N-1}(z) - \Gamma_N z^{-N}\check{H}_{N-1}(z)$$
$$= \hat{H}_{N-1}(z) - \Gamma_N z^{-N}\hat{H}_{N-1}(1/z). \quad (59)$$

If z_i is a root of $\hat{H}_{N-1}(z)$, then, by the induction hypothesis, $|z_i| \leq 1$. And, since $1/z_i$ is a root of $\hat{H}_{N-1}(1/z)$, we conclude that the ratio

$$H_a(z) = \frac{z^{-N}\hat{H}_{N-1}(1/z)}{\hat{H}_{N-1}(z)} \quad (60)$$

is an all-pass system; hence, it can be factored into a product

$$H_a(z) = \frac{zz_1^* - 1}{z - z_1} \cdots \frac{zz_{N-1}^* - 1}{z - z_{N-1}}.$$

From this it follows that

$$|H_a(z)| \begin{cases} <1 & |z| > 1 \\ =1 & |z| = 1 \\ >1 & |z| < 1 \end{cases} \quad (61)$$

because this holds for every bilinear term of $H_a(z)$. To complete the proof, we shall show that if z_0 is a root of $\hat{H}_N(z)$, then $|z_0| \leq 1$. Setting $z = z_0$ in (59), we obtain

$$\hat{H}_N(z_0) = 0 = \hat{H}_{N-1}(z_0) - \Gamma_N z_0^{-N}\hat{H}_{N-1}(1/z_0).$$

Hence

$$\frac{1}{\Gamma_N} = \frac{z_0^{-N}\hat{H}_N(1/z_0)}{\hat{H}_{N-1}(z_0)} = H_a(z_0).$$

But $|\Gamma_N| \leq 1$; therefore, $|H_a(z_0)| \geq 1$ and (61) yields $|z_0| \leq 1$.

IV. Spectral Estimation

We shall now relate the preceding results to the method of maximum entropy.

Deterministic Case: We are given the first $N+1$ values of the autocorrelation $R[m]$ of a random process $s[n]$ and we wish to estimate its power spectrum

$$S(z) = \sum_{m=-\infty}^{\infty} R[m]\, z^{-m}. \quad (65)$$

For this purpose, we shall construct an AR process $s_0[n]$ of order N with autocorrelation $R_0[m]$ such that $R_0[m] = R[m]$ for $|m| \leq N$. The power spectrum $S_0(z)$ of this process will be used as the estimate of $S(z)$.

[1] This proof was suggested to the author by Th. Andrikos.

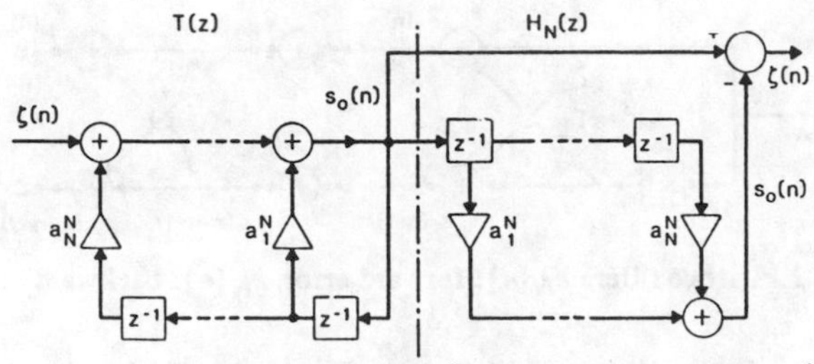

Fig. 3. Cascade of AR filter $T(z) = 1/\hat{H}_N(z)$ and predictor filter $\hat{H}_N(z)$.

The construction of $s_0[n]$ is based on the determination of the Nth order prediction

$$\hat{s}_N[n] = \sum_{k=1}^{N} a_k^N s[n-k] \tag{66}$$

of $s[n]$. As we have shown, the coefficients a_k^N of the predictor can be determined by solving the system (12) or, equivalently, from the recursion equations

$$a_k^N = a_k^{N-1} + \Gamma_N a_{N-k}^{N-1} \qquad 1 \leqslant k \leqslant N-1$$

$$P_{N-1}\Gamma_N = R[N] - \sum_{k=1}^{N} a_k^{N-1} R[N-k] \qquad N \geqslant 1$$

$$a_N^N = \Gamma_N \quad P_N = (1-\Gamma_N^2)P_{N-1} \tag{67}$$

with the initial condition $P_0 = R[0]$. In either case, the solution is uniquely determined in terms of the known values of $R[m]$. We next form the AR filter of Fig. 3 with system function

$$T(z) = \frac{1}{\hat{H}_N(z)} \quad \text{where} \quad \hat{H}_N(z) = 1 - \sum_{k=1}^{N} a_k^N z^{-k} \tag{68}$$

is the error filter of the predictor $\hat{s}_N[n]$ of $s[n]$ [see (17)]. As input to this system we use a stationary white noise process $\zeta[n]$ with average power P_N. Denoting by $s_0[n]$ the resulting output, we conclude that

$$s_0[n] - \sum_{k=1}^{N} a_k^N s_0[n-k] = \zeta[n]$$

$$R_{\zeta\zeta}[m] = P_N \delta[m]. \tag{69}$$

The system $T(z)$ is stable because $\hat{H}_N(z)$ is a Hurwitz polynomial. Therefore, its output $s_0[n]$ is stationary and since it satisfies (69), it is AR. From this and (32), it follows that the Nth order predictor $\hat{s}_0[n]$ of $s_0[n]$ is given by

$$\hat{s}_0[n] = \sum_{k=1}^{N} a_k^N s_0[n-k]. \tag{70}$$

This shows that the process $s_0[n]$ of Fig. 3 and the original process $s[n]$ have identical predictors; therefore [see (14)], their corresponding autocorrelations $R_0[m]$ and $R[m]$ are equal for $|m| \leqslant N$ within a factor. We maintain that this factor is one. Indeed, with $\hat{e}_0[n] = s_0[n] - \hat{s}_0[n]$ the prediction error of $s_0[n]$, it follows from (69) and (70) that

$$E\{\hat{e}_0^2[n]\} = E\{\zeta^2[n]\} = R_{\zeta\zeta}[0] = P_N.$$

Hence (see Note, Section II)

$$R_0[m] = R[m] \qquad |m| \leqslant N. \tag{71}$$

This shows that if we use as the estimate of the unknown spectrum $S(z)$ of $s[n]$ the spectrum $S_0(z)$ of the AR process $s_0[n]$, its inverse transform will agree with the given values of $R[m]$. Since $S_{\zeta\zeta}(z) = P_N$, it follows from (34) that

$$S_0(z) = \frac{P_N}{\hat{H}_N(z)\hat{H}_N(1/z)} \tag{72}$$

and on the unit circle

$$\bar{S}_0(\omega) = S_0(e^{j\omega T}) = \frac{P_N}{\left|1 - \sum_{k=1}^{N} a_k^N e^{-jk\omega T}\right|^2}. \tag{73}$$

This is the maximum entropy estimate of the unknown spectrum $\bar{S}(\omega)$. The numerator P_N and the coefficients a_k^N are determined from (67).

Random Case: We are given the N_0 samples (data) $s[1]$, $s[2], \cdots, s[N_0]$ of a single realization of a process $s[n]$ and we wish to estimate its power spectrum $\bar{S}(\omega)$. The maximum entropy estimator $\hat{\bar{S}}(\omega)$ of $\bar{S}(\omega)$ is an all-pole function as in (73). However, unlike the deterministic case, the value of N is not specified. The problem now is to select first N and then to estimate the coefficients a_k^N. Suppose that we have somehow decided on the value of N. We then proceed as in the deterministic case using the recursion equations (67). These equations specify a_k^N in terms of the constants Γ_N and the initial condition $P_0 = R[0]$. It suffices, therefore, to determine the estimates $\hat{\Gamma}_N$ and $\hat{P}_0$ of these constants by appropriate time averages involving the given data.

For the estimate of P_0, we use the sum

$$\hat{P}_0 = \frac{1}{N_0}\sum_{n=1}^{N} s^2[n]. \tag{74}$$

For the estimate of Γ_N, we use the time-average version of (44) or (53). As we have shown, these equations are equivalent; however, because of end effects the corresponding time averages are not equivalent. We shall use the latter because, unlike (44), it leads to an estimate $\hat{\Gamma}_N$ that satisfies the stability condition

$$|\hat{\Gamma}_N| \leqslant 1. \tag{75}$$

Our problem thus is reduced to the determination of the time-average form of the equation

$$\Gamma_N = \frac{E\{\hat{e}_{N-1}[n]\,\check{e}_{N-1}[n-N]\}}{\frac{1}{2}E\{\hat{e}^2_{N-1}[n] + \check{e}^2_{N-1}[n-N]\}} \tag{76}$$

where

$$\hat{e}_{N-1}[n] = s[n] - (a_1^N s[n-1] + \cdots + a_{N-1}^{N-1} s[n-N+1])$$

$$\check{e}_{N-1}[n-N] = s[n-N] - (a_1^N s[n-N+1] + \cdots + a_{N-1}^{N-1} s[n-1]).$$

The above involves all samples of $s[n]$ from n to $n-N$. And since the data are available only from $n=1$ to $n=N_0$, to avoid overflow in the time-average form of (76), we must limit the values of n from $N+1$ to N_0. This interval has $N_0 - N - 1$ points; hence,

$$\hat{\Gamma}_N = \frac{\frac{1}{N_0 - N - 1}\sum_{n=N+1}^{N_0} \hat{e}_{N-1}[n]\,\check{e}_{N-1}[n-N]}{\frac{1}{2(N_0 - N - 1)}\sum_{n=N+1}^{N_0} (\hat{e}^2_{N-1}[n] + \check{e}^2_{N-1}[n-N])}. \tag{77}$$

The above ratio satisfies (75) because (Schwarz inequality)

$$\left|\sum \hat{e}_{N-1}[n]\,\check{e}_{N-1}[n-N]\right|^2 \le \sum \hat{e}^2_{N-1}[n] \sum \check{e}^2_{N-1}[n-N]$$

and

$$\sqrt{|xy|} \le \tfrac{1}{2}(|x| + |y|).$$

With $\hat{\Gamma}_N$ so determined, the Nth order estimate of the unknown spectrum is given by

$$\hat{\bar{S}}_N(\omega) = \frac{\hat{P}_N}{\left|1 - \sum_{k=1}^{N} \hat{a}_k^N e^{-jk\omega T}\right|^2} \tag{78}$$

where the coefficients are determined recursively as in (67):

$$\hat{a}_k^N = \hat{a}_k^{N-1} + \hat{\Gamma}_N \hat{a}_{N-k}^{N-1} \qquad \hat{a}_N^N = \hat{\Gamma}_N \tag{79}$$

$$\hat{P}_N = (1 - \hat{\Gamma}_N)\hat{P}_{N-1}. \tag{80}$$

The recursion starts with the estimate (74) of P_0.

We conclude with a brief comment on the choice of N. This choice is dictated by two conflicting requirements. For a satisfactory approximation of the unknown spectrum $\bar{S}(\omega)$ by an all-pole function $\bar{S}_0(\omega)$, N should be as large as possible. However, in the estimate (77) of $\hat{\Gamma}_N$, the number of terms in the time-average equals $N_0 - N - 1$, and as we know, this number should be large for the variance of the estimate to be small. Various schemes have been suggested for selecting N but they will not be discussed [15], [16].

The estimate (79) of Γ_N can be obtained by minimizing the time-average form

$$I_N = \frac{1}{2(N_0 - N - 1)}\sum_{m=N+1}^{N_0} [(\hat{e}_{N-1}[n] - \Gamma_N \check{e}_{N-1}[n-N])^2 + (\check{e}_{N-1}[n-N] - \Gamma_N \hat{e}_{N-1}[n])^2]$$

of the MS error P_N as given by (54). Setting $\partial I_N/\partial \Gamma_N = 0$, we obtain (77). However, the resulting value of I_N does not equal the estimate $\hat{P}_N$ of P_N obtained recursively from (80).

V. Maximum Entropy

We shall finally show that the all-pole model is a consequence of the principle of maximum entropy. The required background is discussed in the Appendix. We repeat the problem: we are given the $N+1$ values $R[0], \cdots, R[N]$ of the autocorrelation $R[m]$ of a random process $s[n]$ and we wish to estimate its power spectrum $\bar{S}(\omega)$. The statistics of $s[n]$ are determined in terms of the joint density of the random values $s[n], s[n-1], \cdots, s[n-r]$. Hence, to apply the method of maximum entropy, we must determine the unknown values of $R[m]$ so as to maximize the entropy $H(s_0, \cdots, s_r)$ of these RV and to find the limit as $r \to \infty$. This is equivalent to the maximization of the *entropy rate* H_s of $s[n]$ [see (A12)], subject to the given constraints. We shall show that H_s is maximum if $s[n]$ is a normal process with power spectrum as in (73).

We give three proofs. The first two involve the maximization of H_s. In the third, we find $R[N+1]$ by maximizing $H(s_0, \cdots, s_{N+1})$, and, with $R[N+1]$ so determined, we continue the process. This method can be questioned because it does not yield the maximum of $H(s_0, \cdots, s_N, \cdots, s_{N+k})$ subject to the given constraints. However, the result is corrected in the limit as $k \to \infty$.

Method 1: We form the Nth order predictor $\hat{s}_N[n]$ of $s[n]$ and the predictor error

$$s[n] - \sum_{k=1}^{N} a_k^N s[n-k] = \hat{e}_N[n]. \tag{81}$$

Clearly, $\hat{e}_N[n]$ is the output of the error filter $\hat{H}_N(z)$ [see (17)] with input $s[n]$. Hence [see (A13)], its entropy rate $H_{\hat{e}}$ is given by

$$H_{\hat{e}} = H_s + \frac{1}{4\omega_0}\int_{-\omega_0}^{\omega_0} \ln|\hat{H}_N(e^{j\omega T})|^2\, d\omega. \tag{82}$$

To maximize H_s, it suffices, therefore, to maximize $H_{\hat{e}}$ because the integral is specified in terms of the given values of $R[m]$. As we know, the average power of $e_N^2[n]$ is given by

$$E\{e_N^2[n]\} = P_N = R[0] - \sum_{k=1}^{N} a_k^N R[k]. \tag{83}$$

Therefore, $H_{\hat{e}}$ is maximum if the process $\hat{e}_N[n]$ is *normal white noise* (see the Appendix). And since $\hat{e}_N[n]$ is the right side of the recursion equation (81), we conclude that the optimum $s[n]$ is an AR process of order N; hence, its power spectrum is all-pole as in (73).

We note that the optimum $s[n]$ is a normal process because it is the output of the stable linear system $T(z) = 1/\hat{H}_N(z)$ whose input is the normal process $\hat{e}_N[n]$.

Method 2: In the following reasoning, we assume that the process $s[n]$ is normal. As we have just shown, this assumption is not restrictive. From the normality of $s[n]$ it follows that, within a constant, its entropy rate is given by [see (A20)]

$$H_s = \frac{1}{4\omega_0}\int_{-\omega_0}^{\omega_0} \ln \bar{S}(\omega)\, d\omega \qquad \omega_0 = \frac{\pi}{T} \tag{84}$$

where

$$\bar{S}(\omega) = \sum_{m=-\infty}^{\infty} R[m]\, e^{-jm\omega T}. \tag{85}$$

Since $R[m]$ is specified for $|m| \leqslant N$, the above integral depends on the values of $R[m]$ for $|m| > N$ and it is maximum if

$$\frac{\partial H_s}{\partial R[m]} = 0 = \frac{1}{4\omega_0}\int_{-\omega_0}^{\omega_0} \frac{e^{-jm\omega T}}{\bar{S}(\omega)}\, d\omega \qquad |m| > N. \tag{86}$$

This shows that the Fourier series coefficients of the function $1/\bar{S}(\omega)$ are zero for $|m| > N$. Hence

$$\frac{1}{\bar{S}(\omega)} = \sum_{k=-N}^{N} c_k e^{-jk\omega T} \tag{87}$$

and since $\bar{S}(\omega) \geqslant 0$, it follows from the Fejér-Riess theorem [12] that the above sum can be written as a square. This yields

$$\bar{S}(\omega) = \frac{1}{\left|\sum_{k=0}^{N} b_k e^{-jk\omega T}\right|^2}. \tag{88}$$

We have thus shown that $\bar{S}(\omega)$ is an all-pole function as in (73), where $P_N = 1/|b_0|^2$ and $a_k^N = b_k/b_0$.

Method 3: This method is iterative. In the first iteration, we determine $R[N+1]$ so as to maximize the entropy H of the RV $s[n], s[n-1], \cdots, s[n-N-1]$. For this purpose, we start with the assumption that $R[N+1]$ is specified and we determine the joint density of the above RV for maximum H. As we show in the Appendix, H is maximum if these RV are jointly normal with zero mean. In this case [see (A24)]

$$H = \ln \sqrt{(2\pi e)^{N+1}\Delta} \tag{89}$$

where

$$\Delta \equiv \begin{vmatrix} R[0] & R[1] \cdots R[N+1] \\ R[1] & R[0] \cdots R[N] \\ \vdots & \vdots \\ R[N+1] & R[N] \cdots R[0] \end{vmatrix}.$$

The above determinant is a nonnegative quadratic in $R[N+1]$ and it is maximum if

$$R[N+1] = \sum_{k=1}^{N} a_k^N R[N+1-k] \tag{90}$$

where the coefficients a_k^N satisfy (12). With $R[N+1]$ so determined, we continue the iteration, and at the rth step we determine $R[N+r]$ so as to maximize the entropy of the RV

$$s[n], \cdots, s[n-N-r]. \tag{91}$$

This yields the extrapolation formula

$$R[N+r] = \sum_{k=1}^{N} a_k^N R[N+r-k]. \tag{92}$$

The coefficients a_k^{N+r} of the predictor of $s[n]$ of order $N+r$ satisfy again the system (12), where now N is replaced by $N+r$. From this and (92) it follows that

$$a_k^{N+r} = 0 \qquad \text{for } N < k \leqslant N+r.$$

This shows that the Nth order predictor is also the predictor of any higher order; hence, $s[n]$ is an AR process of order N and its power spectrum is an all-pole function as in (88) [see also (35) and (36)].

Appendix

Entropy and Entropy Rate

We present next for easy reference the relevant concepts from the theory of entropy [23].

Consider a probability space S and a *partition* A of S, that is, a *countable* collection of mutually exclusive events A_i whose union equals S.

Definition: The entropy $H(A)$ of A is the sum

$$H(A) = -\sum_{i=1}^{N} p_i \ln p_i \qquad \text{where } p_i = P(A_i). \tag{A1}$$

Thus, entropy is a number associated to each partition of a probability space. This number has the following significance. As we know, if the experiment is performed n times and the event A_i occurs n_i times, then "almost certainly"

$$p_i \simeq n_i/n \tag{A2}$$

provided that n is "sufficiently large." This heuristic statement is the basis for the use of probability in real problem. It can be given a precise interpretation in the context of the law of large numbers [11].

We shall call each sequence of the forms $t = \{A_i \text{ occurs } n_i \simeq np_i \text{ times in a specific order}\}$ typical. The union of all such sequences will be denoted by T. Clearly

$$P(T) \simeq 1 \tag{A3}$$

because, according to (A2), the typical sequences occur "almost certainly." Each typical sequence is an event in the product space $S^n = S \times \cdots \times S^n$ and

$$P(t) = p_1^{n_1} p_2^{n_2} \cdots p_N^{n_N}. \tag{A4}$$

Since $n_i \cong np_i$ and $p_i = e^{\ln p_i}$, it follows from (A1) that

$$P(t) \cong e^{np_1 \ln p_1} \cdots e^{np_N \ln p_N} = e^{-nH(A)}. \tag{A5}$$

Hence, the total number N_T of typical sequences is given by [5]

$$N_T \simeq e^{nH(A)}. \tag{A6}$$

It follows readily from (A1) that $H(A) \leqslant \ln N$ with equality iff $p_i = 1/N$. And since the total number of sequences in S^n equals N^n, we conclude that if all p_i's are not equal, then

$$H(A) < \ln N \qquad N_T \ll N^n$$

for large n. Thus, although $P(T) \simeq 1$, the number N_T of sequences in T is small compared with the total number N^n of all possible sequences. It is this result that forms the basis for the applications of entropy. We shall use it to establish the conceptual equivalence between maximum entropy and the classical definition of probability.

Suppose first that we wish to determine p_i in the absence of any prior information (no constraints). In this case, all sequences in S^n are equally likely; hence, N_T must be nearly equal to N^n because $P(T) \simeq 1$. From this and (A6), it follows that $H(A)$ must equal its maximum $\ln N$.

Suppose next that prior information is available in the form of inequality constraints, or expected values. Such information leads to the condition that only certain sequences in the space S^n are admissible, forming the subset S_c^n. All typical sequences are now in S_c^n, and since $P(T) \simeq 1$ and the sequences in S_c^n are equally likely, N_T must contain most of them, i.e., $H(A)$ must be maximum subject to the given constraints.

The above argument is imprecise in the same sense as (A2); however, as in that case, it can be given a precise interpretation as a limit theorem.

A consequence of the conceptual equivalence between maximum entropy and classical definition is the conclusion that the former is subject to the same critique as the latter. We should note in support of maximum entropy that, in most problems involving prior constraints, the classical definition must be applied not to the original space S, but to the vastly more complex space S^n, whereas the maximum entropy deals only with quantities in S. This simplification is the primary reason for using maximum entropy. However, it is in such cases that the results are least reliable. We shall illustrate with the die experiment. In the absence of prior information, we reach the reasonable conclusion that $p_i = 1/6$. If we know, on the other hand, that the expected value of the zero-one RV associated with the event "one" equals 0.1998, say, then the conclusion is that $p_1 = 0.1998$, $p_2 = p_3 = \cdots = p_6 = 0.16004$. Unlike the fair-die case, our trust in the correctness of these values is not great, although we have no other reasonable alternative.

These observations are relevant, we believe, in the application of the method to spectral estimation problem. In our view, the method is popular not because it leads to an all-pole model as a logical imperative, but rather because the model is numerically simple, and, unlike earlier methods, it can detect sharp peaks in the unknown spectra.

Random Variables: Consider a discrete-type RV x taking the values x_i with probability p_i. Clearly, the events $\{x = x_i\}$ form a partition A_x. The entropy of the RV x is by definition the entropy of this partition

$$H(x) = H(A_x) = -\sum_i p_i \ln p_i. \tag{A7}$$

The entropy of a continuous-type RV x cannot be so defined because the events $\{x = x_i\}$ do not form a partition (they are not countable). In this case a limiting argument is used: The RV x is approximated by a discrete-type RV x_δ taking the values $x_i = i\delta$ with probability $p_i = f(x_i)\delta$ where $f(x)$ is the density of x. As we see from (A7)

$$H(x_\delta) = -\sum_i \delta f(x_i) \ln f(x_i) - \ln \delta \sum_i \delta f(x_i).$$

Hence, $H(x_\delta) \to \infty$ as $\delta \to 0$. This is so because of the underlying assumption that the various values of x can be recognized as distinct no matter how close they are. However,

$$H(x_\delta) + \ln \delta \to -\int_{-\infty}^{\infty} f(x) \ln f(x)\, dx \qquad \text{as } \delta \to 0$$

and it is this limit that is used as the entropy of x

$$H(x) = -\int_{-\infty}^{\infty} f(x) \ln f(x)\, dx = -E\{\ln f(x)\}. \tag{A8}$$

The addition of the term $\ln \delta$ is a recognition of the fact that, in real problems, only values of x whose difference exceeds a certain level can be considered as distinct.

The joint entropy of the vector RV $x = (x_1, \cdots, x_r)$ with density $f(x_1, \cdots, x_r)$ is defined similarly

$$H(x_1, \cdots, x_r) = -E\{\ln f(x_1, \cdots, x_r)\}. \tag{A9}$$

As we know [11], if y: $(y_1, \cdots, y_r)$ is a linear transformation of x, that is, if $y = Ax$ where A is an r by r nonsingular matrix, then

$$f(y_1, \cdots, y_r) = \frac{1}{|A|} f(x_1, \cdots, x_r).$$

Hence

$$H(y_1, \cdots, y_r) = -E\{\ln f(x_1, \cdots, x_r)\} + \ln |A|. \tag{A10}$$

Entropy Rate: We shall finally define the entropy rate of a discrete stationary process $x[n]$. From the stationarity of $x[n]$ it follows that the joint entropy of the RV $x[n], \cdots, x[n-r+1]$ is independent of n. The entropy rate H_x of the process $x[n]$ is, by definition, the limit

$$H_x = \frac{1}{r} H(x_1, \cdots, x_r) \qquad \text{as } r \to \infty. \tag{A12}$$

Suppose that $x[n]$ is the input to a stable causal system with delta response $h[n]$ and system function $H(z)$. If $x[n]$ is applied at $n = -\infty$, then the resulting response $y[n]$ is stationary with entropy rate H_y.

Theorem 1: If $H(z)$ is minimum phase, then

$$H_y = H_x + \frac{1}{4\omega_0} \int_{-\omega_0}^{\omega_0} \ln |H(e^{j\omega T})|^2\, d\omega \qquad \omega_0 = \frac{\pi}{T}. \tag{A13}$$

Proof: We can assume, introducing if necessary a change in sign and an appropriate shift of the time-origin, that $h[n] > 0$. If $x[n]$ is applied at $n = 0$, then the resulting response

$$\bar{y}[n] = \sum_{k=0}^{n} x[n-k]\, h[k] \tag{A14}$$

is not stationary. However, it tends to the stationary process $y[n]$ as $n \to \infty$. Clearly, (A14) is a linear transformation of the RV $x_0 = x[0], \cdots, x_n = x[n]$ into the RV $y_0 = \bar{y}[0], \cdots, y_n = \bar{y}[n]$ of the form $y = Ax$ where

$$A = \begin{bmatrix} h[0] & 0 & \cdots & 0 \\ h[1] & h[0] & \cdots & 0 \\ \vdots & \vdots & & \vdots \\ h[n] & h[n-1] & \cdots & h[0] \end{bmatrix} \qquad |A| = h^{n+1}[0].$$

Hence [see (A10)]

$$H(y_0, \cdots, y_n) = H(x_0, \cdots, x_n) + (n+1) \ln h[0].$$

Dividing by $n+1$ and making $n \to \infty$, we obtain

$$H_y = H_x + \ln h[0]. \tag{15}$$

Therefore, to complete the proof of the theorem it suffices to show that the term $\ln h[0]$ in (A15) equals the integral in (A13). Since $|H(e^{j\omega T})|^2 = H(e^{j\omega T}) H(e^{-j\omega T})$, we conclude with $z = e^{j\omega T}$ that

$$jT \int_{-\omega_0}^{\omega_0} \ln |H(e^{j\omega T})|^2 \, d\omega = \oint \frac{1}{z} \ln [H(z) H(1/z)] \, dz$$

where the line integral is along the unit circle. But

$$\oint \frac{1}{z} \ln H(z) \, dz = \oint \frac{1}{z} \ln H(1/z) \, dz;$$

hence, it suffices to show that

$$\ln h[0] = \frac{1}{2\pi j} \oint \frac{1}{z} \ln H(z) \, dz. \tag{A16}$$

From the assumption that $H(z)$ is minimum-phase, it follows that the integrand in (A16) is analytic for $|z| \geqslant 1$, hence, the circle of integration can be made arbitrarily large. And since $h[0] = \ln H(z)$ as $z \to \infty$, we conclude that the integral equals

$$\ln h[0] \oint \frac{dz}{z} = 2\pi j \ln h[0]$$

and (A16) results.

Normal Processes: If x is a normal RV with

$$f(x) = \frac{1}{\sigma\sqrt{2\pi}} e^{-x^2/2\sigma^2}$$

then $H(x) = -E\{\ln f(x)\} = \ln \sigma\sqrt{2\pi e}$.

If $\nu[n]$ is normal white noise with $E\{\nu^2[n]\} = \sigma^2$, then

$$f(\nu_1, \cdots, \nu_r) = f(\nu_1) \cdots f(\nu_r).$$

Hence

$$H(\nu_1, \cdots, \nu_r) = -E\{\ln [f(\nu_1) \cdots f(\nu_r)]\} = r \ln \sigma\sqrt{2\pi e}. \tag{A17}$$

From this and (A12), it follows that if $\nu[n]$ is white noise, then

$$H_\nu = \ln \sigma\sqrt{2\pi e}. \tag{A18}$$

Theorem 2: If $x[n]$ is a normal process with power spectrum $\bar{S}(\omega)$ such that

$$\int_{-\omega_0}^{\omega_0} \ln \bar{S}(\omega) \, d\omega < \infty \tag{A19}$$

then its entropy rate H_x is given by

$$H_x = \ln \sqrt{2\pi e} + \frac{1}{4\omega_0} \int_{-\omega_0}^{\omega_0} \ln \bar{S}(\omega) \, d\omega. \tag{A20}$$

Proof: Since $x[n]$ is normal, all its statistical properties, including its entropy rate [see (A12)], can be expressed in terms of its autocorrelation $R[m]$. From this it follows that if another process $y[n]$ has the same autocorrelation $R[m]$, then its entropy rate H_y will equal H_x. Since $\bar{S}(\omega)$ is an even, positive function, and it satisfies the discrete form (A19) of the Paley–Wiener condition [22], it can be factored into a product [12]

$$\bar{S}(\omega) = H(e^{j\omega T}) H(e^{-j\omega T}) \tag{A21}$$

where $H(z)$ is the system function of a real causal minimum phase system. Using, as input to this system, a white-noise normal process with zero mean and variance one, we obtain as output a normal process $s[n]$ with entropy rate [see (A13) and (A18)]

$$H_s = \ln \sqrt{2\pi e} + \frac{1}{4\omega_0} \int_{-\omega_0}^{\omega_0} \ln |H(e^{j\omega T})|^2 \, d\omega$$

and (A20) follows from (A21).

Maximum Entropy with Constraints: The solution of problems involving maximum entropy with constraints in the form of expected values is a simple consequence of the following inequality: if $f(x)$ and $g(x)$ are two arbitrary density functions, then

$$-\int_{-\infty}^{\infty} f(x) \ln g(x) \, dx \geqslant -\int_{-\infty}^{\infty} f(x) \ln f(x) \, dx \tag{A22}$$

with inequality iff $f(x) = g(x)$. Indeed, as it is easy to see, $\ln y \leqslant 1 - y$ and (A22) follows readily with $y = g(x)/f(x)$. The above holds also if $f(x)$ and $g(x)$ are replaced by joint densities of any order.

Using (A22), we shall determine the density $f(x)$ of an RV x so as to maximize its entropy $H(x)$ subject to the constraint

$$E\{x^2\} = \int_{-\infty}^{\infty} x^2 f(x) \, dx = \sigma^2.$$

With

$$g(x) = \frac{1}{\sigma\sqrt{2\pi}} e^{-x^2/2\sigma^2}$$

it follows from (A22) that

$$-\int_{-\infty}^{\infty} f(x) \ln f(x) \, dx \leqslant -\int_{-\infty}^{\infty} f(x) \left(-\frac{x^2}{2\sigma^2} - \ln \sigma\sqrt{2\pi}\right) dx$$

$$= \tfrac{1}{2} + \ln \sigma\sqrt{2\pi}.$$

The left side equals $H(x)$ and the right side is specified; hence, $H(x)$ is maximum iff $f(x) = g(x)$, that is, if x is normal with zero mean.

We now wish to find the joint density $f(x_1, \cdots, x_r)$ of the RV $x_1, \cdots, x_r$ so as to maximize their entropy $H(x_1, \cdots, x_r)$ subject to the constraints

$$E\{x_i x_j\} = \mu_{ij} \qquad i, j = 1, \cdots, r. \tag{A23}$$

With

$$x = (x_1, \cdots, x_r) \quad x_t = \begin{bmatrix} x_1 \\ \vdots \\ x_r \end{bmatrix} \quad \mu = \begin{bmatrix} \mu_{11}, \cdots, \mu_{1r} \\ \vdots \qquad \vdots \\ \mu_{r1}, \quad , \mu_{rr} \end{bmatrix}$$

and

$$g(x) = \frac{1}{\sqrt{(2\pi)^r |\mu|}} e^{-(1/2) x \mu^{-1} x_t}$$

we conclude applying the multidimensional form of (A22) that $f(x) = g(x)$ and

$$H(x) = \ln \sqrt{(2\pi e)^r |\mu|}. \tag{A24}$$

We similarly conclude that if μ_{ij} is specified for $i = j$ only, then $H(x)$ is maximum if the RV x_i are normal independent with zero mean and variance μ_{ii}.

From the above and (A12) it follows that if $x[n]$ is a random process with given average power $E\{x^2[n]\} = \sigma^2$, then its entropy rate H_x is maximum if $x[n]$ is normal white noise.

References

[1] E. T. Jaynes, "Prior probabilities," *IEEE Trans. Syst. Sci., Cybern.*, vol. SSC-4, 1968.

[2] J. P. Burg, "Maximum entropy spectral analysis," presented at the Int. Mccting Soc. Explor. Geophys., Orlando, FL, 1967.

[3] R. T. Lacoss, "Data adaptive spectral analysis methods," *Geophysics*, vol. 36, pp. 661–675, 1971.

[4] T. J. Ulrych and T. N. Bishop, "Maximum entropy spectral analysis and autoregressive decomposition," *Rev. Geophys. Space Phys.*, vol. 13, Feb. 1975.

[5] C. E. Shannon and W. Weaver, *The Mathematical Theory of Communication.* Urbana, IL: Univ. of Illinois Press, 1979.

[6] J. Makhoul, "Linear prediction: A tutorial review," *Proc. IEEE*, vol. 63, pp. 561–580, Apr. 1975.

[7] A. H. Nuttall and G. C. Carter, "A generalized framework for power spectral estimation," *IEEE Trans. Acoust., Speech, Signal Processing*, vol. ASSP-28, pp. 334–335, June 1980.

[8] N. Levinson, "The Wiener RMS error criterion in filter design and prediction," *J. Math. Phys.*, vol. 25, no. 4, 1947.

[9] J. Makhoul, "Stable and efficient lattice methods for linear prediction," *IEEE Trans. Acoust., Speech, Signal Processing*, vol. ASSP-26, Oct. 1978.

[10] A. H. Nuttall, Naval Underwater Syst. Center, New London, CT, Tech. Rep. 5303, Mar. 1976.

[11] A. Papoulis, *Probability, Random Variables, and Stochastic Processes.* New York: McGraw-Hill, 1965.

[12] —, *Signal Analysis.* New York: McGraw-Hill, 1977.

[13] F. Itakura and S. Saito, "Digital filtering techniques for speech analysis and synthesis," presented at the 7th Int. Cong. Acoust., Budapest, Hungary, 1971.

[14] U. Grenander and G. Szego, *Toeplitz Forms and Their Applications.* Berkeley, CA: Univ. of California Press, 1958.

[15] L. D. Davisson, "A theory of adaptive filtering," *IEEE Trans. Inform. Theory*, vol. IT-12, pp. 97–102, Apr. 1966.

[16] H. Akaike, "A new look at the statistical model identification," *IEEE Trans. Automat. Contr.*, vol. AC-19, pp. 716–723, Dec. 1974.

[17] D. E. Smylie, G. K. C. Clarice, and T. J. Ulrych, "Analysis of irregularities in the earth's rotation," in *Methods of Computational Physics*, vol. 13. New York: Academic, 1973, pp. 391–430.

[18] A. Van Den Bos, "Alternative interpretation of maximum entropy analysis," *IEEE Trans. Inform. Theory*, vol. IT-17, July 1971.

[19] R. N. McDonough, "Maximum entropy, spatial processing of array data," *Geophysics*, vol. 39, Dec. 1974.

[20] J. E. Shore and R. W. Johnson, "Axiamatic derivation of the principle of maximum entropy and the principle of minimum cross-entropy," *IEEE Trans. Inform. Theory*, vol. IT-26, pp. 26–37, Jan. 1980.

[21] J. E. Shore, "Minimum cross-entropy spectral analysis," *IEEE Trans. Acoust., Speech, Signal Processing*, vol. ASSP-29, pp. 230–237, Apr. 1981.

[22] A. Papoulis, *The Fourier Integral and Its Applications.* New York: McGraw-Hill, 1962.

[23] A. Papoulis, "Entropy: From first principles to spectral estimation," in *Proc. ASSP Workshop Spectral Estimation*, McMaster Univ., Hamilton, Ont., Canada, Aug. 1981.

Frequency Estimation with Maximum Entropy Spectral Estimators

STEPHEN W. LANG AND JAMES H. McCLELLAN, SENIOR MEMBER, IEEE

Abstract—The ability of a modified covariance method "maximum entropy" spectral estimator to estimate the frequencies of several sinusoids in additive white Gaussian noise is studied. Analytical expressions for the variance of the spectral estimate peak positions at high signal-to-noise ratios are derived. The calculated variance is compared to the Cramer-Rao lower bound and to the results of similar variance calculations for the more familiar covariance method. It is shown that performance approaching the Cramer-Rao bound can be obtained. Simulations demonstrate substantial agreement with the analytical results over a wide range of signal-to-noise ratios.

I. Introduction

THE statistical behavior of various spectral estimates of zero-mean stationary random processes has been treated in several papers [1]-[5]. The problem of estimating an unknown but nonrandom signal has been studied less frequently, although it is often a good model for situations in which spectral estimators are actually used. This paper is concerned with the behavior of maximum entropy method (MEM) spectral estimators when the observed data comes from a random process of the form

$$x(n) = \sum_{l=0}^{L-1} b_l e^{j\theta_l} e^{j\omega_l n} + w(n) \qquad (n = 0, \cdots, N-1) \qquad (1)$$

where the real positive amplitudes $\{b_l\}$, the initial phases $\{\theta_l\}$, and the frequencies $\{\omega_l\}$ are fixed but unknown parameters and $w(n)$ is complex white Gaussian noise of semivariance $\sigma^2/2$. The behavior of MEM estimators in this context has been studied in [6]-[11]. These papers present either empirical studies of the estimator's behavior on synthetic data or analytical studies in which the true process autocorrelations are used to form the MEM estimate.

This paper is concerned with the problem of estimating the frequencies ω_l in (1). The frequencies are determined from the peak positions of a MEM spectral estimate computed from a finite segment of the random process (1). The performance of a spectral estimator in such a parameter estimation problem is one way in which to judge how "good" it is; performance can be characterized by a few numbers such as bias and variance, which then form a basis for comparison to other spectral estimators as well as to theoretical bounds on performance. Furthermore, the case of estimation from a finite segment of data has not been treated in previous analytical work. Finally, the preceding frequency estimation problem, while restrictive, is important in its own right. Applications in geophysics, Doppler radar, and azimuth estimation with a linear antenna array are but a few examples of its widespread use.

The paper is structured in the following way. Sections II and III concern the general frequency estimation problem. The Cramer-Rao bound and estimation by the method of maximum likelihood (ML) are reviewed. MEM spectral estimators are considered for use as suboptimal frequency esimators, and the mathematics of the modified covariance method of linear prediction [7] are presented.

Calculation of the variance of such a frequency estimate is discussed in Sections IV-VI. The peak position error is approximated as a linear function of the noise samples. Simple variance expressions are obtained in two special cases. In the case of a signal frequency, or multiple well-separated frequencies, results indicate that the peak position variance can approach the Cramer-Rao bound when the model order is about $\frac{1}{3}$ of the data length. In the case of two closely spaced frequencies, the peak position variance has a dependence on frequency separation and phase difference which is similar to the Cramer-Rao bound. Again, it appears that for model orders comparable to the data length, the variance approaches the bound.

The results derived for the complex random process of (1) are applied to a real-valued process consisting of sinusoids in white noise in Section VII. Simulation results are presented to support the analytical results. Several cases in which the linear approximation breaks down are evident in the simulations. The dependence of this breakdown on noise level, model order, and data length are discussed in Section VIII.

II. The Frequency Estimation Problem

Before considering MEM spectral estimators, the general frequency estimation problem will be discussed from a theoretical standpoint. The Cramer-Rao bound [12]-[15] and the ML estimate [13]-[15] will be reviewed.

A. The Cramer-Rao Bound

For convenience, relabel the unknown parameters of the random process as

$$\begin{aligned} \alpha_{3l} &= \omega_l, \\ \alpha_{3l+1} &= \theta_l, \qquad (l = 0, \cdots, L-1) \\ \alpha_{3l+2} &= b_l, \end{aligned} \qquad (2)$$

Manuscript received August 28, 1979; revised May 21, 1980. This work was supported in part by a Fannie and John Hertz Foundation Fellowship and in part by the National Science Foundation under Grant ENG76-24117.

The authors are with the Department of Electrical Engineering and Computer Science, and the Research Laboratory of Electronics, Massachusetts Institute of Technology, Cambridge, MA 02139.

Reprinted from *IEEE Trans. Acoust., Speech, Signal Processing*, vol. ASSP-28, pp. 716–724, Dec. 1980.

and define the vectors

$$x = (x(0), \cdots, x(N-1))$$
$$\alpha = (\alpha_0, \cdots, \alpha_{3L-1}). \tag{3}$$

Then, from (1), the probability density function for the observed data conditioned on the unknown parameter vector can be written as

$$P(x|\alpha) = (\pi\sigma^2)^{-N} \cdot \exp\left[-\frac{1}{\sigma^2}\sum_{n=0}^{N-1}\left|x(n) - \sum_{l=0}^{L-1} b_l e^{j\theta_l} e^{j\omega_l n}\right|^2\right]. \tag{4}$$

The Cramer-Rao bound states that for any unbiased estimate $\hat{\alpha}_i$ of α_i,

$$\text{var}(\hat{\alpha}_i) \geq [J^{-1}(\alpha)]_{ii} \tag{5}$$

where $J(\alpha)$ is the $3L \times 3L$ Fischer information matrix with elements

$$[J(\alpha)]_{ij} = -E\left[\frac{\partial^2 \ln p(x|\alpha)}{\partial\alpha_i \partial\alpha_j}\,\Big|\alpha\right]. \tag{6}$$

For the case $L = 1$, $J(\alpha)$ can be inverted to obtain

$$\text{var}(\hat{\omega}_0) \geq 6\sigma^2/b_0^2 N(N^2 - 1). \tag{7}$$

A simple result such as (7) is not obtained for $L > 1$. However, the following general behavior has been shown.

1) The bound does not depend on absolute values of the frequencies and phases, but only on frequency differences and phase differences. This is true for any L, as shown by the invariance of the Cramer-Rao bound under any change of frequency or phase origin. In a similar fashion, it can be shown that the bound is invariant under a transformation which changes any frequency or phase by a multiple of 2π; thus, the frequency and phase dependence of the bound is only through quantities like

$$(\omega_l - \omega_k) \bmod 2\pi \quad \text{or} \quad (\theta_l - \theta_k) \bmod 2\pi \tag{8}$$

which can be taken to lie in the interval $[-\pi, \pi]$. The natural measure of frequency (phase) separation is then

$$|(\omega_l - \omega_k) \bmod 2\pi|$$

which will be denoted as $|\omega_l - \omega_k|$ for short. If ω_l and ω_k are thought of as points in the complex plane corresponding to $e^{j\omega_l}$ and $e^{j\omega_k}$, then $|\omega_l - \omega_k|$ is the shortest angular distance between them.

2) The bounds on the lth frequency are independent of the amplitudes of the other exponentials [13]. This can be shown by factoring J into the form DKD, where D is diagonal; b_l appears only in the elements D_{ii} $(i = 3l, 3l + 1, 3l + 2)$, and K is independent of the amplitudes.

3) When the frequencies are "well-separated"

$$(N|\omega_l - \omega_k| \gg 1, \quad \forall l \neq k)$$

then the bounds approach the $L = 1$ case of (7). This property is shown by observing that the factor K becomes nearly block diagonal in this case, decoupling the bounds for each frequency [13].

Some further observations, based on numerical studies, can be made for the case of two "closely spaced"

$$(N|\omega_0 - \omega_1| \ll 1)$$

frequencies. In this case, the bound is proportional to

$$1/|\omega_0 - \omega_1|^2$$

except for initial phases such that $\theta_0 - \theta_1 \simeq n\pi$. For these phase differences, the bound is proportional to $1/|\omega_0 - \omega_1|^4$, and is much worse.

B. The Maximum Likelihood (ML) Estimate

The ML estimate is defined as that parameter set $\hat{\alpha}$ which maximizes the conditional probability density function $p(x|\hat{\alpha})$ for the observed data x. In the $L = 1$ case, it only amounts to finding the peak of the periodogram, however, when $L > 1$, the maximization involves a difficult nonlinear least squares problem. Thus, although the variance of the ML estimate, asymptotically, approaches the Cramer-Rao bound, it may be unattractive in many applications because of the computational burden. The approach of using an easy to implement suboptimal estimator will be pursued in this paper. The frequency estimation procedure will be to form a spectral estimate from the observed data and then to choose the positions of the L highest peaks in the spectral estimate as the frequency estimates.

III. MEM Spectral Estimation

MEM spectral estimators have been called "high resolution" estimators because of their ability to resolve closely spaced frequencies. This property suggests their use as frequency estimators, particularly when high resolution is required from short data segments.

The MEM models the spectrum as all-pole:

$$S_{\text{MEM}}(\omega) = P_E/|A(\omega)|^2$$
$$A(\omega) = \sum_{k=0}^{P} a(k)\, e^{-j\omega k}$$
$$a(0) = 1. \tag{9}$$

There are several variations of the MEM which differ in the manner by which the model coefficients $\{P_E, a(1), \cdots, a(P)\}$ are computed. The autocorrelation method [16] and Burg's technique [17] are widely used, but were not considered for use as frequency estimators because they have been reported to behave badly even at high signal-to-noise ratio (SNR), exhibiting frequency shifts of the spectral peaks and "peak splitting" [8], [18], [19]. This behavior gets worse as the SNR increases, behavior opposite to what one would expect from any nearly optimal estimator, making them unsuitable in the high SNR limit.

Proposed by Ulrych and Clayton [7] and called by them the "least squares" estimator, the modified covariance technique obtains the model coefficients by minimizing the sum of the forward and backward prediction error energies:

$$P_E = \frac{1}{2(N-P)} \cdot \left(\sum_{n=P}^{N-1} |x(n) * a(n)|^2 + \sum_{n=0}^{N-P-1} |x(n) * a^*(-n)|^2\right). \tag{10}$$

An argument for using $S_{MEM}(\omega)$, as determined by the modified covariance technique, as a spectral estimate is that both $A(\omega)$ and $A^*(\omega)$ should act as whitening filters [20]. If the error sequences

$$e_f(n) = x(n) * a(n)$$

and

$$e_b(n) = x(n) * a^*(-n) \tag{11}$$

were exactly white, and P_E were exactly their variance, then $S_{MEM}(\omega)$ would be exactly the power spectrum.

The $a(k)$'s determined by the modified covariance method are those which minimize the Euclidean length of the error vector

$$\begin{bmatrix} e_f(P) \\ \vdots \\ e_f(N-1) \\ e_b^*(0) \\ \vdots \\ e_b^*(N-P-1) \end{bmatrix} = \begin{bmatrix} x(P) \\ \vdots \\ x(N-1) \\ x^*(0) \\ \vdots \\ x^*(N-P-1) \end{bmatrix} + \begin{bmatrix} x(P-1) & \cdots & x(0) \\ \vdots & & \vdots \\ x(N-2) & \cdots & x(N-P-1) \\ x^*(1) & \cdots & x^*(P) \\ \vdots & & \vdots \\ x^*(N-P) & \cdots & x^*(N-1) \end{bmatrix} \cdot \begin{bmatrix} a(1) \\ \vdots \\ a(P) \end{bmatrix}. \tag{12}$$

It can be shown that:

1) There is always at least one solution to this least squares problem which attains the minimum error energy P_E.

2) There may be more than one solution. In particular, if $P > 2N/3$, then the $2(N-P) \times P$ matrix in (12) is certainly singular; from one solution, another may be obtained by adding to it any vector in the nullspace of the matrix. This places a limit on the model order for which a unique spectral estimate can be obtained.

The error minimized by the modified covariance method has an important consequence for the frequency estimation problem. If the data consists of L complex exponentials with no noise added, and if $P \geqslant L$ then $A(\omega)$ will be zero at each of the exponentials' frequencies, since this solution drives the error energy P_E to zero. Therefore, L of the P zeros of $A(\omega)$ are fixed exactly at the exponential frequencies, and a frequency estimator which looks for peaks in $S_{MEM}(\omega)$ should obtain exact estimates of these frequencies. The other $P - L$ zeros may be placed arbitrarily. This last degeneracy is removed when noise is added to the problem; then these zeros will also be uniquely determined, although their locations may be very sensitive to the exact noise realization. As long as these $P - L$ zeros do not wander too close to the unit circle, and the L zeros near the exponential frequencies are not perturbed much by the noise, then the frequency estimates should not be greatly affected.

There are many algorithms for solving the least squares problem posed in (12), e.g., [21], [22]. One way is to derive the normal equations; P linear equations in P unknowns.

$$\sum_{k=1}^{P} \phi(l,k)\, a(k) = -\phi(l,0), \qquad (l = 1, \cdots, P) \tag{13}$$

where

$$\phi(l,k) = \frac{1}{2(N-P)} \cdot \left(\sum_{n=P}^{N-1} x^*(n-l)\, x(n-k) + \sum_{n=0}^{N-P-1} x(n+l)\, x^*(n+k) \right). \tag{14}$$

This system of equations may be inverted to obtain the $a(k)$'s (assuming a unique solution).

The more familiar covariance technique [16] involves minimizing only the forward prediction error energy. The least squares problem corresponds to the upper $N - P$ rows of (12). The normal equations (13) apply with the second sum in (14) deleted. The covariance technique has the same desirable behavior with respect to data composed of exponentials as does the modified covariance method: placing zeros exactly at the exponentials' frequencies in the absence of noise. The reduced number of rows in (12), however, makes the solution to the least squares problem relatively less stable with respect to additive noise in the data, as will become obvious in the following results.

IV. The Linear Approximation

The primary new result presented in this paper is the approximation of the peak position error of the MEM spectral estimate as a linear function of the noise samples in (1). A plausibility argument and some experimental evidence can be given to support the validity of this linear approximation.

Considering the peak position error as some smooth function of the noise samples, it possesses a Taylor series expansion with a finite region of convergence. For sufficiently high SNR, the peak position should be well approximated by truncating the series after the linear term. Once this approximation is made, it is a simple matter to calculate the peak position variance for any noise process, given knowledge of the noise covariance matrix. The validity of such a linear approximation is supported by the simulation results shown in Fig. 1. For the sake of simplicity, real data were used; $x(n)$ was of the form $\sin(\pi n/2) + w(n)$, where $w(n)$ was white Gaussian noise of variance σ^2. Fig. 1 shows the peak position variance plotted versus the SNR ($10 \log(1/2\sigma^2)$) for a data length of 32 and a model order of eight; the results of 200 trials were used to calculate the sample variances.

Over a wide range of SNR, the peak position variance is seen to be proportional to σ^2, just the result expected if the peak position error is a linear function of the noise samples. Also evident in Fig. 1 is that the linear approximation breaks

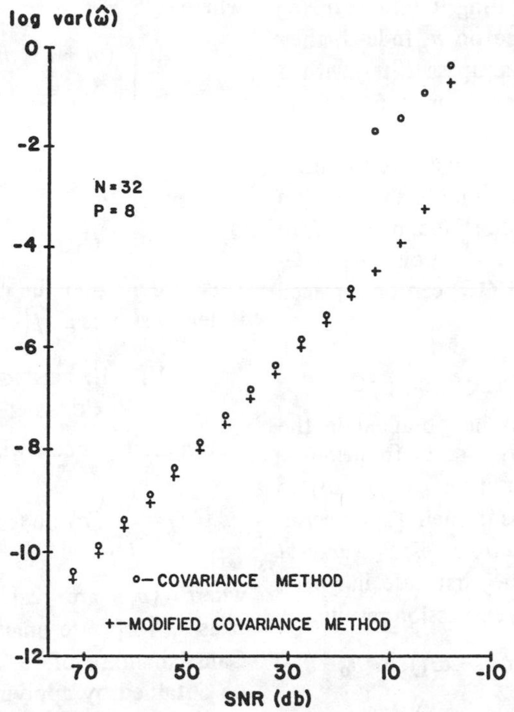

Fig. 1. The peak position variance depends linearly on the noise variance over a wide range of SNR (sinusoid frequency $\omega_0 = \pi/2$).

down at low SNR. This effect is due to spurious peaks appearing in the spectral estimates. If the highest peak in a particular spectral estimate is one of these spurious peaks, then its position is used as the frequency estimate; a very high estimation error can occur in this way. The presence of a "threshold" SNR below which the estimate variance increases dramatically is characteristic of a nonlinear estimation problem [23]. The region in which the linear approximation breaks down is not indicated by the calculations contained in this paper, although some comments are made in Section VIII, based on the simulation results.

A linear approximation can be derived in two steps. First, it is assumed that the effect of the noise is to perturb only slightly the solution to the least squares problem from a particular solution to the noise-free problem. The selection of a particular solution to the noise-free problem avoids having to deal with the fact that this problem often has more than one solution in the absence of noise. Second, the peak position is obtained by an approximate minimization of the prediction error energy. The calculation is sketched out in the Appendix; more details are available in [24].

V. Well-Separated Frequencies

When the condition $P|\omega_l - \omega_k| \gg 1$, $\forall l \neq k$ is met (this condition is actually a bit more stringent than the "well-separated" condition defined in Section II-A since $P < N$) then each frequency may be treated individually. The calculation in the Appendix results in the following linear expression for the peak position error in terms of the noise:

$$\hat{\omega}_l - \omega_l \simeq \sum_{n=0}^{N-1} [h_R(n) \operatorname{Re} w(n) + h_I(n) \operatorname{Im} w(n)] \tag{15}$$

where

$$h_R(n) = \begin{cases} -\sin(n\omega_l + \theta_l)/b_l P(N-P), & 0 \leq n \leq P-1 \\ 0, & P \leq n \leq N-P-1 \\ \sin(n\omega_l + \theta_l)/b_l P(N-P), & N-P \leq n \leq N-1 \end{cases}$$

$$h_I(n) = \begin{cases} \cos(n\omega_l + \theta_l)/b_l P(N-P), & 0 \leq n \leq P-1 \\ 0, & P \leq n \leq N-P-1 \\ -\cos(n\omega_l + \theta_l)/b_l P(N-P), & N-P \leq n \leq N-1. \end{cases} \tag{16}$$

Thus, for well-separated frequencies in additive white noise of semivariance $\sigma^2/2$:

$$\operatorname{var}(\hat{\omega}_l) \simeq \sigma^2/b_l^2 P(N-P)^2. \tag{17}$$

This expression has a broad minimum when the model order is $\frac{1}{3}$ the data length, at which the theoretical variance is $\frac{9}{8}$ times the Cramer-Rao bound of (7).

A similar analysis for the covariance method [24] results in the following variance expression for well-separated frequencies in additive white noise of semivariance $\sigma^2/2$:

$$\operatorname{var}(\hat{\omega}_l) \simeq \frac{2(2P+1)\sigma^2}{3P(P+1)(N-P)^2 b_l^2}. \tag{18}$$

This expression shows a broad minimum around a model order equal to $\frac{1}{3}$ the data length. The theoretical variance at that model order is about $\frac{3}{2}$ the Cramer-Rao bound, somewhat worse than the variance for the modified covariance method predicted by (17).

Returning to (16) note that the weighting functions $h_R(n)$ and $h_I(n)$ have a sinusoidal dependence on n, indeed, they resemble the impulse response of a bandpass filter with a bandwidth proportional to $1/P$, centered around $\omega = \omega_l$. Thus, if the noise is not white, but has a spectral density function which is "smooth," then the dominant contribution to the variance of a peak position at ω_l is due to that portion of the noise spectrum in the $1/P$ neighborhood of ω_l. If, in addition, the real and imaginary part of the noise are independent, then the σ^2 term in (17) and (18) can be replaced with the noise spectral density at ω_l.

VI. Two Closely Spaced Frequencies

Relatively simple expressions can also be obtained in the case of two closely spaced ($N|\omega_0 - \omega_1| \ll 1$) frequencies; the closely spaced assumption allows functions of $\omega_0 - \omega_1$ to be approximated by the low order terms in their Taylor series expansions. The problem separates into two cases: $\theta_0 - \theta_1$ *not* near $n\pi$, and $\theta_0 - \theta_1$ near $n\pi$. Only the first case has been fully worked out in [24]. The following expression results:

$$\hat{\omega}_0 - \omega_0 \simeq [6/b_0(N-P)(P+1)(P+2)(\omega_0 - \omega_1)\sin(\theta_0 - \theta_1)] \times \sum_{m=0}^{N-1} U_4(m)\left[\sin\left(\frac{\omega_0+\omega_1}{2} m + \theta_1\right) \operatorname{Re} w(m) + \cos\left(\frac{\omega_0+\omega_1}{2} m + \theta_1\right) \operatorname{Im} w(m)\right] \tag{19}$$

where

$$U_4(m) = \begin{cases} m[1/P - 3(P+1)/P(P-1)] + [2/P+1], & 0 \leq m \leq P-1 \\ 0, & P \leq m \leq N-P-1 \\ -U_4(m-(N-P)), & N-P \leq m \leq N-1. \end{cases} \tag{20}$$

Thus, with white noise of semivariance $\sigma^2/2$

$$\operatorname{var}(\hat{\omega}_l) \simeq \frac{18\sigma^2 \sum_{m=0}^{N-1} U_4^2(m)}{b_l^2(N-P)^2(P+1)^2(P+2)^2(\omega_0-\omega_1)^2 \sin^2(\theta_0-\theta_1)}. \tag{21}$$

This expression demonstrates the same $1/|\omega_0 - \omega_1|^2$ dependence as the Cramer-Rao bound for $\theta_0 - \theta_1$ not near $n\pi$. The increasing variance as $\theta_0 - \theta_1$ nears $n\pi$ suggests that this case will show a substantially higher variance. In fact, numerical results show that the variance in this second case is proportional to $1/|\omega_0 - \omega_1|^4$; thus, the modified covariance method demonstrates the same frequency dependence as the Cramer-Rao bound in both cases.

A similar analysis performed for the covariance method [24] results in the following variance expression for two closely spaced frequencies in additive white noise of semivariance $\sigma^2/2$:

$$\operatorname{var}(\hat{\omega}_l) \simeq \frac{10368\sigma^2 \sum_{m=0}^{N-1} U_3^2(m)}{b_l^2[(N-P)^2-1]^2(N-P)^2(\omega_0-\omega_1)^4(P+1)^2(P+2)^2} \tag{22}$$

where

$$U_3(m) = \begin{cases} (m+1)\{m^2 + m[2P - 3(N+P-1)/2] + (N+P-1)(P-1)\}/P(P-1), & 0 \leq m \leq P-1 \\ 0, & P \leq m \leq N-P-1 \\ -U_3(m-(N-P)), & N-P \leq m \leq N-1. \end{cases} \tag{23}$$

It is evident that the covariance method performs badly since it demonstrates a $1/|\omega_0 - \omega_1|^4$ behavior for all phases.

VII. Extension to Real-Valued Random Processes and Simulation Results

A random process which is the real-valued analog of (1) is

$$x(n) = \sum_{l=0}^{L-1} b_l \sin(\omega_l n + \phi_l) + w(n) \tag{24}$$

where $\{b_l\}$ are real, positive amplitudes, $\{\phi_l\}$ are initial phases, $\{\omega_l\}$ are angular frequencies, and $w(n)$ is real white Gaussian noise of variance σ^2. The Cramer-Rao bound can be obtained by applying (5) and (6). The results on the peak position variance in the MEM spectrum can be used by expressing the sine functions as sums of complex exponentials.

$$b_l \sin(\omega_l n + \phi_l) = \frac{b_l}{2} e^{j[\phi_l - (\pi/2)]} e^{j\omega_l n} + \frac{b_l}{2} e^{-j[\phi_l - (\pi/2)]} e^{-j\omega_l n}. \tag{25}$$

If the frequencies $\{-\omega_{L-1}, \cdots, -\omega_0, \omega_0, \cdots, \omega_{L-1}\}$ are well separated, then (17) for the modified covariance method becomes

$$\operatorname{var}(\hat{\omega}_l) \simeq 4\sigma^2/b_l^2 P(N-P)^2. \tag{26}$$

For the covariance method (18) becomes

$$\operatorname{var}(\hat{\omega}_l) \simeq 16\sigma^2/3b_l^2 P(N-P)^2. \tag{27}$$

Similarly, if $L = 1$ and $N|\omega_0| \ll 1$ (a single low frequency sinusoid) then (21) [for the modified covariance method] becomes

$$\operatorname{var}(\hat{\omega}_0) \simeq \frac{9\sigma^2 \sum_{m=0}^{N-1} U_4^2(m)}{b_0^2(N-P)^2(P+1)^2(P+2)^2\omega_0^2 \sin^2(\phi_0)} \tag{28}$$

and (22) [for the covariance method] becomes

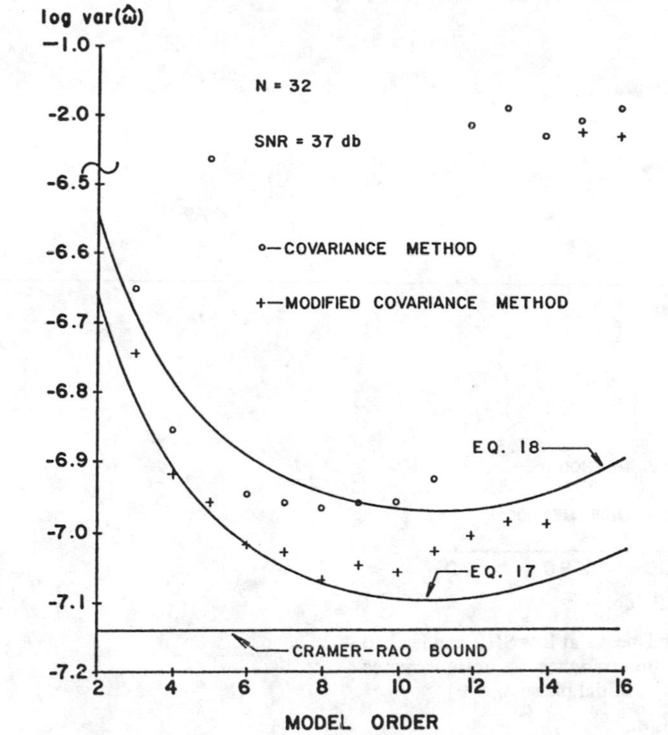

Fig. 2. The dependence of peak position variance on model order is compared to theoretical predictions and to the Cramer-Rao bound in the case of well-separated frequencies (sinusoid frequency $\omega_0 = \pi/2$).

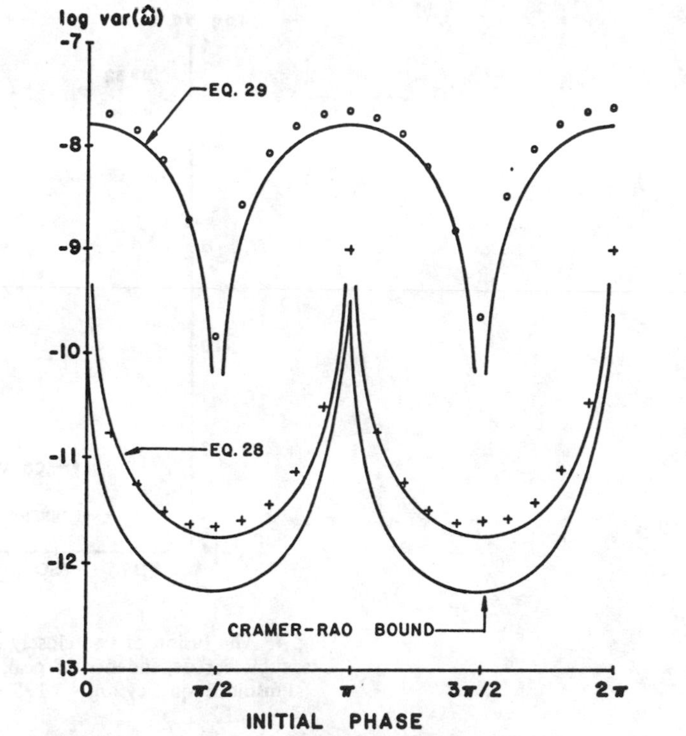

Fig. 3. The dependence of peak position variance on initial phase is compared to theoretical predictions and to the Cramer-Rao bound in the case of closely spaced frequencies (sinusoid frequency $\omega_0 = 3.125 \times 10^{-3}$, data length $N = 32$, model order $P = 8$, SNR = 117 dB).

$$\text{var}(\hat{\omega}_0) \simeq \frac{5184\sigma^2 \cos^2(\phi_0) \sum_{m=0}^{N-1} U_3^2(m)}{b_0^2 [(N-P)^2 - 1]^2 (N-P)^2 \omega_0^4 (P+1)^2 (P+2)^2}. \tag{29}$$

There is a phase dependence in this last equation, although (22) was phase independent because the noise now has only a real part.

Simulations were performed for the real data case. First, to test the well-separated frequencies case, data of the form $x(n) = \cos(\pi n/2) + w(n)$ were used, where $w(n)$ was white Gaussian noise of variance $\sigma^2 = 10^{-4}$ (SNR = 37 dB). Fig. 2 shows the resulting sample peak position variance, var ($\hat{\omega}_0$), for 1000 trials plotted versus the model order P for a data length of $N = 32$. Also plotted are the theoretical variances from (26) and (27), and the Cramer-Rao bound. The simulation results confirm the presence of a null in the peak position variance for a model order of about $\frac{1}{3}$ the data length. Effects due to spurious peaks are also present. The covariance method had a high variance for $P = 5$ and $P \geqslant 12$, while the modified covariance method had a high variance for $P \geqslant 15$. The simulations show the modified covariance method to be better than the covariance method, both because it had a lower variance when the linear approximation held, and because it was less susceptible to spurious peaks.

A low frequency sinusoid was used to test the closely spaced frequencies case; $x(n) = \sin(3.125 \times 10^{-3}\, n + \phi) + w(n)$ where $w(n)$ was white Gaussian noise of variance 10^{-12} (SNR = 117 dB). The data length was $N = 32$ and the model order was $P = 8$. The sample peak position variance, var ($\hat{\omega}_0$) for 200 trials is plotted in Fig. 3 versus the initial phase ϕ. The theoretical variances from (28) and (29) are also plotted, as well as the Cramer-Rao bound which was calculated numerically. The noise variance and model order were chosen low enough so that spurious peaks were not a problem; the simulation results are in good agreement with the linear theory. The $1/|\omega_0|^4$ dependence of the covariance method versus $1/|\omega_0|^2$ for the modified covariance method is evident. The particular phases where both the Cramer-Rao bound and the modified covariance method variance have a $1/|\omega_0|^4$ dependence show up as sharp peaks.

VIII. Breakdown of the Analysis

There are several regions in which the linear analysis may not provide an accurate description of what is going on. A high noise level is the primary cause of these effects, but just how much noise can be tolerated before this analysis breaks down is a complicated function of the signal parameters and the model order.

When the noise level is very high, spurious peaks may appear in the spectral estimate. If one of these peaks is chosen as the location of a sought-after tone instead of the correct peak, a very high estimation error can result. (In practice, one could use *a priori* information to restrict the search for ω_0, thus rejecting spurious peaks and lowering the estimate variance.) Just how much noise is needed to make the probability of this event significant depends on the model order P. If, for the same observed data, a sequence of spectral estimates is constructed using different model orders, spurious peaks may be

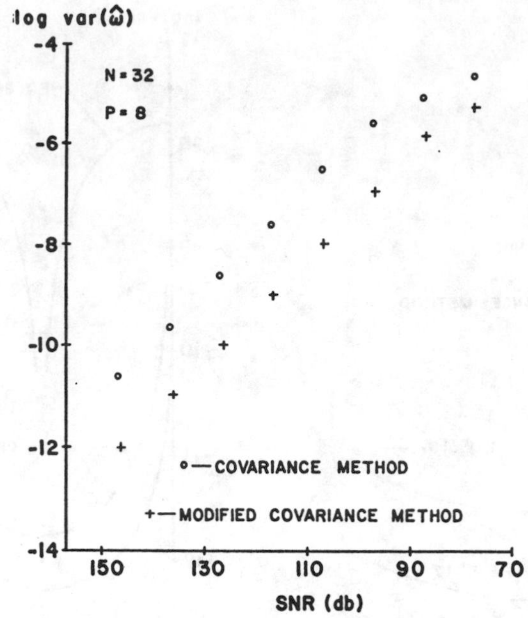

Fig. 4. The fusion of two closely spaced peaks at low SNR can lead to a sublinear dependence of peak position variance on noise variance (sinusoid frequency $\omega_0 = 3.125 \times 10^{-3}$, initial phase $\phi_0 = 0$).

present in some estimates, but not in others. Generally, higher model orders lead to a less stable least squares problem and, thus, to a higher probability of spurious peaks. Thus, it is not surprising that in Fig. 1, a threshold effect is present at high noise levels and in Fig. 2, a threshold effect is seen at high model orders. A mathematical analysis of the occurrence of these spurious peaks would be very interesting; although such work has been done for adaptive MEM [25], none is available for the methods considered in this paper.

Even when the noise level is not high enough to create spurious peaks of significant amplitude, it can still adversely affect the spectral estimate. For instance, in the closely spaced frequencies case, a little noise might cause only one peak to appear instead of two closely spaced peaks. This is confirmed by the simulation results of Fig. 4 where peak position variance is plotted versus SNR for a low frequency sinusoid. At low SNR the peak position variance is smaller than the linear analysis predicts because in a large percentage of the trials only one peak is observed at $\omega = 0$.

Another such effect results from the linear approximation used. Recall from (16) that only $2P$ of the N noise samples contributed to the shift in the peak position. The other noise samples must still contribute, though perhaps not linearly. Their nonlinear contribution may be evident when $N - 2P \gg P$ or when the noise variance is sufficiently high, resulting in a peak position variance which is not proportional to σ^2. Indeed, Sakai [26] has shown that in the case of fixed SNR and model order, as the data length $N \to \infty$, the peak position variance is proportional to σ^4.

IX. Summary and Conclusions

The use of a modified covariance technique maximum entropy spectral estimator for frequency estimation has been studied. A linear approximation has resulted in an expression for the peak position variance in the high signal-to-noise region. Simulations have verified the theoretical results for high SNR, and have shown where the approximations break down. Comparisons to the Cramer-Rao bound have shown that in the high SNR case, the peak positions are nearly optimal estimates of the frequencies; thus, this method of frequency estimation is an attractive alternative to the true maximum likelihood estimate with its high computational burden. MEM spectral estimators have more resolution and lower bias (in simulations, the measured bias was always negligible in comparison to the standard deviation) than periodogram or FFT based frequency estimators, the other popular alternatives to the maximum likelihood estimate. MEM spectral estimates based on the modified covariance method have proved better than those based on the covariance method; the increased numerical stability of the former method leads to fewer spurious peaks in the spectral estimate.

Appendix

The linear approximation is derived in two steps: a solution to the noise-free problem, and a perturbation of this solution to account for the effects of the noise. This process is sketched out below; a more detailed derivation can be found in [24]. For $P \geq L$ let

$$a_0(k) = \begin{cases} 1, & k = 0 \\ -(1/P) \sum_{q=0}^{L-1} d(q)\, e^{j\omega_q k}, & 1 \leq k \leq P \\ 0, & \text{else} \end{cases} \tag{A1}$$

where the $d(q)$'s solve the $L \times L$ system of equations:

$$\sum_{q=0}^{L-1} e^{-j(P+1)(\omega_p - \omega_q)/2} \frac{\sin [P(\omega_p - \omega_q)/2]}{P \sin [(\omega_p - \omega_q)/2]} d(q) = 1 \quad (p = 0, \cdots, L-1). \tag{A2}$$

Since $A_0(\omega_l) = \Sigma_{k=0}^{P} a_0(k)\, e^{-j\omega_l k} = 0$, the $a_0(k)$'s are a particular solution to the normal equations (13) in the case of no noise.

Assume that, for any given noise realization, $a(k)$ can be expressed as $a(k) = a_0(k) + a_1(k)$, where $a_1(k) = 0$ unless $1 \leqslant k \leqslant P$. The peak positions can be related to the perturbation $A_1(\omega) = \Sigma_{k=1}^{P} a_1(k)\, e^{-j\omega k}$. If $|A(\omega)|^2$ is approximated near ω_l by its truncated Taylor series expansion:

$$|A(\omega)|^2 \simeq |A(\omega_l)|^2 + (d|A(\omega)|^2/d\omega)|\,\omega_l(\omega - \omega_l) + \tfrac{1}{2}(d^2|A(\omega)|^2/d\omega^2)|\,\omega_l(\omega - \omega_l)^2 \tag{A3}$$

then the position of the null $\hat{\omega}_l$ in $|A(\omega)|^2$ can be found by setting $d|A(\omega)|^2/d\omega$ to zero, resulting in

$$\hat{\omega}_l - \omega_l \simeq -\frac{\mathrm{Re}\,(d|A(\omega)|^2/d\omega)|\,\omega_l}{(d^2|A(\omega)|^2/d\omega^2)|\,\omega_l}. \tag{A4}$$

Calculating these derivatives and neglecting terms involving A_1 in favor of terms involving A_0 results in

$$\hat{\omega}_l - \omega_l \simeq -\frac{\mathrm{Re}\,[A_1(\omega_l)\,(dA_0^*(\omega)/d\omega)|\,\omega_l]}{|(dA_0(\omega)/d\omega)|\,\omega_l|^2}. \tag{A5}$$

Thus, the peak position $\hat{\omega}_l$ for a given noise realization can be found from knowledge of $A_1(\omega_l)$. $A_1(\omega_l)$ can be found by an approximate minimization of the prediction error energy. Since $A_0(\omega_l) = 0$:

$$e_f(n) = w(n) * a_0(n) + \sum_{l=0}^{L-1} b_l e^{j\theta_l} A_1(\omega_l)\, e^{j\omega_l n} + w(n) * a_1(n)$$

and

$$e_b(n) = w(n) * a_0^*(-n) + \sum_{l=0}^{L-1} b_l e^{j\theta_l} A_1^*(\omega_l)\, e^{j\omega_l n} + w(n) * a_1^*(-n). \tag{A6}$$

Neglecting the last term in each sum, since they contain a product of two small quantities, a and w, results in

$$\begin{aligned}\sum_{n=P}^{N-1} |e_f(n)|^2 + \sum_{n=0}^{N-P-1} |e_b(n)|^2 &\simeq (N-P) \sum_{l,k} b_l A_1^*(\omega_l)\,[c_1(\omega_l - \omega_k) + c_2(\omega_l - \omega_k)] \\ &\cdot b_k A_1(\omega_k) + 2\,\mathrm{Re} \sum_{m=0}^{N-1} \\ &\cdot \Bigg[w^*(m) \sum_l e^{j\theta_l} \left(\sum_{n=P}^{N-1} a_0^*(n-m)\, e^{j\omega_l n}\right) b_l A_1(\omega_l) \\ &+ w(m) \sum_l e^{-j\theta_l} \left(\sum_{n=0}^{N-P-1} a_0^*(m-n)\, e^{-j\omega_l n}\right) b_l A_1(\omega_l)\Bigg] \\ &+ \text{terms which do not depend on } A_1 \end{aligned} \tag{A7}$$

where

$$c_1(\omega_l - \omega_k) = \frac{1}{N-P}\, e^{-j(\theta_l - \theta_k)} \sum_{n=P}^{N-1} e^{-j(\omega_l - \omega_k)n}$$

$$c_2(\omega_l - \omega_k) = \frac{1}{N-P}\, e^{j(\theta_l - \theta_k)} \sum_{n=0}^{N-P-1} e^{j(\omega_l - \omega_k)n}. \tag{A8}$$

Choosing the $A_1(\omega_k)$'s to minimize the error energy results in the $L \times L$ system of equations:

$$\begin{aligned}(N-P) \sum_{k=0}^{L-1} [c_1(\omega_l - \omega_k) + c_2(\omega_l - \omega_k)]\, b_k A_1(\omega_k) &= -\sum_{m=0}^{N-1} \Bigg[w(m)\, e^{-j\theta_l} \sum_{n=P}^{N-1} a_0(n-m)\, e^{-j\omega_l n} \\ &+ w^*(m)\, e^{j\theta_l} \sum_{n=0}^{N-P-1} a_0(m-n)\, e^{j\omega_l n}\Bigg] \\ (l = 0, \cdots, L-1). \end{aligned} \tag{A9}$$

(A9) can be combined with (A5) to give a linear relationship between the real and imaginary parts of the noise samples $w(m)$ and the peak shift $\hat{\omega}_l - \omega_l$. The calculation of these shifts requires inverting (A2) to obtain the d's, using the d's in (A1) to find $A_0(\omega)$, and inverting (A9) to obtain the $A_1(\omega_l)$'s. In this form, they are suitable for machine computation; further approximations must be made to obtain simple expressions. For well-separated frequencies, the two $L \times L$ systems of equations (A2) and (A9) are approximated as diagonal. For two closely spaced frequencies, the elements of the 2×2 matrices are approximated by truncated Taylor series in $\omega_0 - \omega_1$. (A9) can be applied to the covariance method by deleting c_2 and the $w^*(m)$ term. The preceding discussion is of necessity very brief. More details are available in [24].

ACKNOWLEDGMENT

The authors would like to thank one of the reviewers for his many helpful suggestions.

REFERENCES

[1] R. B. Blackman and J. W. Tukey, *The Measurement of Power Spectra.* New York: Dover, 1959.
[2] P. D. Welch, "The use of the fast Fourier transform for the estimation of power spectra," *IEEE Trans. Audio Electroacoust.*, vol. AU-15, pp. 70-73, June 1967.
[3] H. Akaike, "Power spectrum estimation through autoregressive model fitting," *Ann. Inst. Statist. Math.*, Tokyo, 1969, pp. 407–419.
[4] K. Berk, "Consistent autoregressive spectral estimates," *Ann. Statist.*, vol. 2, pp. 489-502, 1974.
[5] A. B. Baggeroer, "Confidence intervals for regression (MEM) spectral analysis," *IEEE Trans. Inform. Theory*, vol. IT-22, pp. 534-545, Sept. 1976.
[6] T. J. Ulrych, "Maximum entropy power spectrum of truncated sinusoids," *J. Geophys. Res.*, vol. 77, pp. 1396-1400, Mar. 1972.
[7] T. J. Ulrych and R. W. Clayton, "Time series modelling and maximum entropy," *Phys. Earth Planet. Inter.*, vol. 12, pp. 188–200, 1976.
[8] W. Y. Chen and G. R. Stegun, "Experiments with maximum entropy power spectra of sinusoids," *J. Geophys. Res.*, vol. 79, pp. 3019-3022, July 1974.
[9] L. C. Pusey, "High resolution spectral estimates," Lincoln Lab. Tech. Rep. 1975-7, Jan. 1975.

[10] R. T. Lacoss, "Data adaptive spectral analysis methods," *Geophys.*, vol. 36, pp. 661–675, Aug. 1971.
[11] E. H. Satorius and J. R. Zeidler, "Maximum entropy spectral analysis of multiple sinusoids in noise," *Geophys.*, vol. 43, pp. 1111–1118, Oct. 1978.
[12] J. R. Sklar and F. C. Schweppe, "The angular resolution of multiple targets," Lincoln Lab. Group Rep. 1964-2, Jan. 1964.
[13] D. C. Rife, "Digital tone parameter estimation in the presence of Gaussian noise," Ph.D. dissertation, Polytechnic Inst. Brooklyn, Brooklyn, NY, June 1973.
[14] D. C. Rife and R. R. Boorstyn, "Single tone parameter estimation from discrete time observations," *IEEE Trans. Inform. Theory*, vol. IT-20, pp. 591–598, Sept. 1974.
[15] R. Birgenheier, "Parameter estimation of multiple signals," Ph.D. dissertation, Univ. California, Los Angeles, 1972.
[16] J. Makhoul, "Linear prediction: A tutorial review," *Proc. IEEE*, vol. 63, pp. 561–580, Apr. 1975.
[17] J. P. Burg, "Maximum entropy spectral analysis," Ph.D. dissertation, Stanford University, Stanford, CA, May 1975.
[18] P. F. Fougere, E. J. Zawalick, and H. R. Radoski, "Spontaneous line splitting in maximum entropy power spectrum analysis," *Phys. Earth Planet. Inter.*, vol. 12, 1976.
[19] S. Kay and L. Marple, "Sources of and remedies for spectral line splitting in autoregressive spectrum analysis," in *Conf. Rec., IEEE Int. Conf. Acoust., Speech, Signal Processing*, 1979, pp. 151–154.
[20] J. P. Burg, "Maximum entropy spectral analysis," in *Proc. 37th Meet. Soc. Explor. Geophysicists*, 1967.
[21] C. L. Lawson and R. J. Hanson, *Solving Least Squares Problems.* Englewood Cliffs, NJ: Prentice-Hall, 1974.
[22] L. Marple, "A new autoregressive spectrum analysis algorithm," in *Conf. Rec. 12th Asilomar Conf. Circuits, Syst., Comput.*, pp. 277–281, 1978.
[23] J. M. Wozencraft and I. M. Jacobs, *Principles of Communication Engineering.* New York: Wiley, 1965.
[24] S. W. Lang, "Performance of maximum entropy spectral estimators," S.M. thesis, Massachusetts Inst. Technol., Cambridge, MA, May, 1979.
[25] R. Keeler, "Uncertainties in adaptive maximum entropy frequency estimators," *IEEE Trans. Acoust., Speech, Signal Processing*, pp. 469–471, Oct. 1978.
[26] H. Sakai, "Statistical properties of AR spectral analysis," *IEEE Trans. Acoust., Speech, Signal Processing*, pp. 402–409, Aug. 1979.

Maximum entropy spectral analysis of multiple sinusoids in noise

E. H. SATORIUS* AND J. R. ZEIDLER*

An analytical technique based on the method of undetermined coefficients is applied to the problem of computing the theoretical spectral estimate by the maximum entropy method (MEM) when the autocorrelation function of the data is known exactly and corresponds to N sinusoids in additive white noise and to N sinusoids in additive 1-pole, low-pass noise. For the white noise case, the L prediction filter coefficients are expanded directly in terms of the input sinusoids. This expansion leads to a transformation of the $L \times L$ normal equations for the prediction filter coefficients to a set of $2N \times 2N$ equations. The transformed equations are a smaller set of equations to be solved whenever $L > 2N$ and provide a convenient description of the interaction between the various frequency components of the sinusoids which occurs in the MEM estimate. Further, for certain cases where there is little interaction between some of the frequency components of the sinusoids, the solution of the $2N \times 2N$ equations may be approximated (to zeroth order) by the solution of a smaller set of coupled equations. A better approximation to the exact solution of the $2N \times 2N$ equations can then be obtained from a perturbation expansion of the exact solution about the zeroth order approximation.

For the case of N sinusoids in 1-pole, low-pass noise, the L prediction filter coefficients are expanded in terms of the input sinusoids as well as two delta functions which occur at the beginning and end of the filter. This expansion also leads to a set of $2N \times 2N$ equations. For this case the values of the MEM estimate evaluated at the frequencies of the sinusoids are shown to be a function of the frequencies of the sinusoids. This result is reasonable since the signal-to-noise ratio per unit bandwidth is also a function of frequency.

INTRODUCTION

The maximum entropy method (MEM) of spectral analysis, which was originally proposed by Burg (1967, 1975), has been widely applied in geophysical data processing. The correspondence between MEM and linear prediction filtering, as discussed by van den Bos (1971), has allowed the application of a large body of literature on autoregressive (AR) time series analysis to MEM. Ulrych and Bishop (1975) and Ulrych and Clayton (1976) give a thorough review and discussion of MEM and AR analysis.

An important application of MEM is to the spectrum analysis of data containing multiple sinusoids in noise. Even though the proper time series model for this case is not an AR model (as will be discussed in more detail in the next sections), MEM can still provide excellent sinusoid resolution (especially for large signal-to-noise ratios). The improved resolution offered by MEM over the more conventional Fourier spectral estimation methods for this case has been well documented in the literature (e.g., Ulrych, 1972; Ulrych and Bishop, 1975). For purposes of comparing the performance of MEM with other spectral estimation techniques, Lacoss (1971) and more recently Frost (1977) and Marple (1976) have examined the theoretical MEM spectral estimate when the autocorrelation function of the data is known exactly and consists of sinusoids in white noise. The use of an exact matrix inverse identity for computing the theoretical MEM spectral estimate for the cases of one and two sinusoids in white noise was discussed by Lacoss (1971) and Frost (1977). However, when there are more than two sinusoids, the use of the inverse identity becomes tedious. Even for two sinusoids, the analytical form of the theoretical MEM spectral estimate which is obtained from the repeated application

Manuscript received by the Editor April 1, 1977; revised manuscript received February 3, 1978.
*Naval Ocean Systems Center, Code 632 (Bayside), San Diego, CA 92152.

Reprinted with permission from *Geophysics*, vol. 43, pp. 111–1118, Oct. 1978.

of the inverse identity provides little insight into the interaction between the positive and negative frequency components of the sinusoids which occurs in the MEM estimate (Marple, 1976). In this paper an alternative approach based on the method of undetermined coefficients will be used to compute the theoretical MEM spectral estimate when the autocorrelation function of the data is known exactly and corresponds to multiple sinusoids in white noise and to multiple sinusoids in 1-pole low-pass noise. It will be shown that the method of undetermined coefficients provides a convenient description of the interaction between the various frequency components of the sinusoids which occurs in the theoretical MEM estimate. Further, useful approximations for the theoretical MEM spectral estimate may be obtained for cases in which the interaction between certain sinusoidal frequency components is small but not negligible. For the case of 1-pole low-pass noise, it will be shown that the MEM spectral resolution is a function of the frequency and becomes worse in the spectral regions where the noise power increases.

MEM SPECTRAL ANALYSIS OF MULTIPLE SINUSOIDS IN WHITE NOISE

In this section we will consider the case when the autocorrelation function of the data is known at uniform intervals of lag. Further, we will assume that the sampled autocorrelation function, $\rho(\ell)$, $\ell = 0, 1, \ldots, L$, can be written as follows

$$\rho(\ell) = \sigma_0^2 \delta(\ell) + \sum_{n=1}^{V} \sigma_n^2 \cos 2\pi f_n \ell, \quad (1)$$

where $\delta(\ell)$ is the Kronecker delta function, σ_n^2 is the power in the nth sinusoid, f_n represents the frequencies of the sinusoids (which are normalized to the sample frequency), and σ_0^2 is the white noise power. The MEM spectral estimate $S(f)$ may be written (Ulrych and Bishop, 1975)

$$S(f) = \Delta \sigma_r^2 \, Q(f), \quad (2)$$

where Δ is the sampling interval, and

$$Q(f) = \left|1 - \sum_{k=0}^{L-1} g(k)\, e^{-2\pi jf(k+1)}\right|^{-2}, \quad (3)$$

where f is normalized to the sample frequency. The L prediction coefficients, $g(k), k = 0, \ldots, L-1$, are obtained from the normal equations

$$\sum_{k=0}^{L-1} \rho(\ell - k) g(k) = \rho(\ell + 1), \quad \ell = 0, \ldots, L-1, \quad (4)$$

and the constant σ_r^2 is obtained from the $g(k)$ by

$$\sigma_r^2 = \rho(0) - \sum_{k=0}^{L-1} g(k)\, \rho(k+1). \quad (5)$$

Lacoss (1971) and Marple (1976) treat the problem of computing $S(f)$ either through the direct numerical solution of equations (4)–(5) or through the use of a well-known matrix inversion identity which is sometimes referred to as Woodbury's identity (Zielke, 1968). As noted above, the application of Woodbury's identity becomes quite tedious and leads to very extensive analytic expressions for $g(k)$ and $S(f)$ if N is larger than 2.

An alternative technique of examining $S(f)$ analytically and numerically is the method of undetermined coefficients. This method consists of substituting a solution for $g(k)$, which is expressed in terms of unknown constants, into equation (4). This substitution then leads to a set of equations for the unknown constants. This technique was originally applied to the continuous analog of equation (4) by Zadeh and Ragazzini (1950) and more recently has been applied directly to equation (4) by Satorius and Zeidler (1977) when the spectral density of the data contains both poles and zeroes.

The form of the assumed solution for $g(k)$, when $\rho(\ell)$ is given by equation (1), is

$$g(k) = \sum_{n=1}^{2N} A_n e^{2\pi jf_n k}, \quad (6)$$

where we define (for notational convenience) $f_{n+N} \equiv -f_n (n = 1, 2, \ldots, N)$, and the A_n are to be determined. This particular choice for the assumed solution leads to precisely $2N$ equations for the A_n. Substituting equation (6) into (4) with $\rho(\ell)$ given by equation (1), and equating coefficients of exp $(2\pi jf_r \ell)$ (for $r = 1, 2, \ldots, 2N$) in the resulting equation, leads to $2N$ equations for the amplitude A_r of each sinusoid,

$$A_r + \sum_{\substack{n=1 \\ n \neq r}}^{2N} \gamma_{rn} A_n = \frac{e^{2\pi jf_r}}{L + 2\sigma_0^2/\sigma_r^2}, \quad r = 1, 2, \ldots, 2N, \quad (7)$$

where in (7) we have defined $\sigma_{n+N}^2 \equiv \sigma_n^2 (n = 1, 2, \ldots, N)$ and γ_{rn} is given by

$$\gamma_{rn} = \frac{\phi_L(f_n - f_r)}{L + 2\sigma_0^2/\sigma_r^2}, \quad (8)$$

where

$$\phi_L(f) \equiv \sum_{k=0}^{L-1} e^{2\pi jfk}. \quad (9)$$

It is noted that the net effect of the method of undetermined coefficients is to transform the original $L \times L$ equations [equation (4)] to the set of $2N \times 2N$ equations [equation (7)]. This transformation yields a smaller set of equations to be solved whenever $L > 2N$.

Equations (6) to (8) show that the prediction coefficients $g(k)$ can be expressed as a sum of the positive and negative frequency components of the input sinusoids and that the amplitude of each sinusoid A_r is coupled to the amplitude of all the other sinusoids through coupling coefficients γ_{rn}. The coupling coefficients vanish if $|f_n - f_r|$ is some integral multiple of $1/L$. Also, as the factor, $L + 2\sigma_0^2/\sigma_r^2$, becomes large, the γ_{rn} approach zero. As the γ_{rn} approach zero, equation (7) decouples and the A_r are given to a good approximation by

$$A_r \simeq \frac{e^{2\pi jf_r}}{L + 2\sigma_0^2/\sigma_r^2}, r = 1, 2, \ldots, 2N. \quad (10)$$

Using equations (6) to (8), we may express all quantities of interest in terms of the A_n. From (5) we have for σ_r^2

$$\sigma_r^2 = \sigma_0^2 \left\{ 1 + \sum_{n=1}^{2N} A_n e^{-2\pi jf_n} \right\}. \quad (11)$$

As the γ_{rn} approach zero, σ_r^2 is given to a good approximation by

$$\sigma_r^2 \simeq \sigma_0^2 \left\{ 1 + \sum_{n=1}^{2N} (L + 2\sigma_0^2/\sigma_n^2)^{-1} \right\}. \quad (12)$$

From equation (3), we have for $Q(f)$

$$Q(f) = |1 - e^{-2\pi jf} \sum_{n=1}^{2N} A_n \phi_L(f_n - f)|^{-2}. \quad (13)$$

It is interesting to note that $Q(f)$ evaluated at the frequency of the rth sinusoid can be simply expressed in terms of A_r and σ_0^2/σ_r^2. The result for $Q(f_r)$ is

$$Q(f_r) = |A_r|^{-2}(\sigma_r^2/2\sigma_0^2)^2, r = 1, \ldots, N. \quad (14)$$

Equation (14) is valid, regardless of the frequency separation between the N sinusoids. [Of course, as the frequency separations approach zero, the maxima of $Q(f)$ will not necessarily be equal to the $Q(f_r)$.] As the γ_{rn} approach zero, $Q(f_r)$ is given approximately by

$$Q(f_r) \simeq (\sigma_r^2/2\sigma_0^2)^2(L + 2\sigma_0^2/\sigma_r^2)^2, \quad r = 1, \ldots, N. \quad (15)$$

When $L \gg 2\sigma_0^2/\sigma_r^2$, equation (15) becomes

$$Q(f_r) \simeq (\sigma_r^2/2\sigma_0^2)^2 L^2, \; r = 1, \ldots, N. \quad (16)$$

A result similar to equation (16) was also obtained by Lacoss (1971) for the theoretical peak values of the MEM estimate of a real sinusoid in white noise.

As pointed out in the recent paper by Ulrych and Clayton (1976), the proper time series model for N sinusoids in white noise is an autoregressive-moving average (ARMA) model which contains $2N$ autoregressive (AR) terms and $2N$ moving average (MA) terms. Therefore, since an infinite order AR model is required to model a finite-order ARMA process (Gersch and Sharpe, 1973), it is expected that as $L \to \infty$, $S(f)$ will converge to $\sigma_0^2 \Delta$ (for $f \neq f_n$; $n = 1, \ldots, N$). This can be seen quite simply by noting from equation (10) that

$$\lim_{L \to \infty} A_n \phi_L(f_n - f) = 0,$$

provided $f \neq f_n (n = 1, \ldots, N)$. Therefore, from equations (2), (11), and (13), $S(f)$ converges pointwise to $\sigma_0^2 \Delta$ as $L \to \infty$ when $f \neq f_n$ and from equation (16) it is seen that provided the autocorrelation function is known exactly, the resolution capabilities of $S(f)$ improve without limit as $L \to \infty$ regardless of the value of σ_0^2. Of course, in reality the autocorrelation function is never known exactly but must be estimated from a finite amount of data and, therefore, a practical limit is imposed on the value of L which provides a tradeoff between MEM resolution and the confidence in the estimates of the autocorrelation function (as well as the increased inaccuracies which occur when computing a larger number of prediction filter coefficients). The criteria which have been most frequently applied to the problem of determining the optimum value of L are the Akaike final prediction error (FPE) criterion (Akaike, 1969; 1970) and the Akaike information theoretic (AIC) criterion (Akaike, 1972; 1974). The application of these different criteria has been considered by a number of authors (e.g., Gersch and Sharpe, 1973; Akaike, 1974; Jones, 1976). The problem of determining the optimal value of L is further discussed by Ulrych and Bishop (1975) and Ulrych and Clayton (1976).

APPROXIMATIONS TO THE THEORETICAL MEM SPECTRAL ESTIMATE FOR SINUSOIDS IN WHITE NOISE

Note that when some of the γ_{rn} are negligible, an approximate solution (zeroth order) for the A_r may be obtained by setting the negligible γ_{rn} to zero in equation (7). A better approximation to the true A_r

may be obtained by expanding equation (7) in a perturbation expansion about the zeroth order approximation. In particular, let $\mathscr{B}$ be the matrix with elements B_{kj} given by

$$B_{kj} = \begin{cases} 1; \text{ if } k = j \\ \gamma_{kj}; \text{ if } \gamma_{kj} \text{ is not negligible in zeroth order.} \\ 0; \text{ otherwise} \end{cases}$$

Further, let $\mathscr{G}$ be the matrix with elements G_{kj} given by

$$G_{kj} = \begin{cases} 0; \text{ if } k = j \\ \gamma_{kj}; \text{ if } \gamma_{kj} \text{ is negligible in zeroth order} \\ 0; \text{ otherwise} \end{cases}$$

Equation (7) may now be expressed in the equivalent matrix form

$$(\mathscr{I} + \mathscr{B}^{-1} \cdot \mathscr{G}) \cdot \mathbf{A} = \mathbf{A}^{(0)}, \tag{17}$$

where $\mathscr{I}$ is the $2N \times 2N$ identity matrix; $\mathbf{A}$ is a column vector with kth element given by A_k; and $\mathbf{A}^{(0)} = \mathscr{B}^{-1} \cdot \mathbf{F}$, where $\mathbf{F}$ is a column vector with kth element given by $\exp(2\pi j f_k)/(L + 2\sigma_0^2/\sigma_k^2)$. $\mathbf{A}^{(0)}$ is the zeroth order approximation to $\mathbf{A}$. Equation (17) can be further expanded in a perturbation expansion for $\mathbf{A}$ in terms of the matrix $\mathscr{G}$. The result is

$$\mathbf{A} = \sum_{p=0}^{\infty} \mathbf{A}^{(p)}, \tag{18}$$

where

$$\mathbf{A}^{(p)} = (-\mathscr{B}^{-1} \cdot \mathscr{G})^p \cdot \mathbf{A}^{(0)}, \tag{19}$$

Equation (18) is the desired perturbation expansion for $\mathbf{A}$ in powers of $\mathscr{G}$. The convergence of equation (18) is discussed in the Appendix.

As a specific example, consider the case of two sinusoids in white noise in which there is little interaction between the positive and negative frequency components of the sinusoids, but there is appreciable interaction between the two positive frequencies (and, therefore, the two negative frequencies). For this case, the 8 coefficients γ_{13}, γ_{31}, γ_{41}, γ_{14}, γ_{23}, γ_{32}, γ_{24}, and γ_{42} may be neglected in zeroth order, and, therefore, $\mathbf{A}^{(0)}$ is obtained by solving the two independent sets of 2×2 equations involving only the nonnegligible coefficients γ_{12}, γ_{21}, γ_{34}, and γ_{43}. From equations (18) and (19), it is straightforward to obtain approximations for $Q(f)$. To zeroth order, $Q(f)$ is given by

$$\begin{aligned} Q(f) \simeq Q^{(0)}(f) \equiv |1 - e^{-2\pi jf} \cdot & \\ \cdot (A_1^{(0)} \phi_L(f_1 - f) & \\ + A_2^{(0)} \phi_L(f_2 - f))|^{-2}. & \end{aligned} \tag{20}$$

In equation (20), we have neglected $A_3^{(0)}$ and $A_4^{(0)}$ which is equivalent to neglecting the negative part of the frequency spectrum in $Q(f)$ and is consistent with neglection of the interaction between the positive and negative frequency components in the zeroth order approximation. The approximations in obtaining equation (20) are equivalent to applying Woodbury's identity to the correlation matrix formed by the positive frequency components of the two sinusoids as suggested by Lacoss (1971). Equation (20) will give a good approximation to $Q(f)$ for f sufficiently far from zero and will also give the zeroth order approximation to $Q(f_r)$, i.e.,

$$Q^{(0)}(f_r) = \frac{1}{4}\left(\frac{\sigma_r^2}{\sigma_0^2}\right)^2 |A_r^{(0)}|^{-2}, \; r = 1, 2. \tag{21}$$

However, for certain applications (e.g., Frost, 1977), the errors introduced by neglecting the negative frequency components in $Q(f)$ can be appreciable. To include these effects (to first order) we have from equations (18) and (19)

$$\begin{aligned} Q(f) \simeq Q^{(1)}(f) \equiv |1 - e^{-2\pi jf} \sum_{n=1}^{4} \cdot & \\ (A_n^{(0)} + A_n^{(1)}) \cdot \phi_L(f_n - f)|^{-2}. & \end{aligned} \tag{22}$$

Equation (22) will provide a better approximation to $Q(f)$ over a wider range of frequencies than $Q^{(0)}(f)$; however, $Q^{(1)}(f_r)$ will only be correct to zeroth order, i.e. [from (19) and (22)],

$$\begin{aligned} Q^{(1)}(f_r) = \frac{1}{4}\left(\frac{\sigma_r^2}{\sigma_0^2}\right)^2 |(A_r^{(0)} + A_r^{(1)}) & \\ - \frac{\sigma_r^2}{2\sigma_0^2} \sum_{n=3}^{4} A_n^{(1)} \phi_L(f_n - f_r)|^{-2} & \\ r = 1, 2. & \end{aligned} \tag{23}$$

Note that for small values of $\sigma_r^2/\sigma_0^2 (r = 1, 2)$, $Q^{(1)}(f_r)$ will be approximately correct to first order [as can be seen from equation (23)]. Higher order approximations, $Q^{(p)}(f)$ [correct to order $p - 1$ at $f = f_1$ or f_2], may also be obtained as in equation (22). It should be noted that such approximations to $Q(f)$ which include the interaction between the positive and negative frequencies would be difficult to obtain using Woodbury's identity. This is because the identity would have to be applied to both the positive and negative frequencies and would result in a complicated expression for $Q(f)$ which would provide little insight into the interaction between the various frequency components of the sinusoids.

As a numerical example to illustrate the difference between the approximations for $Q(f)$ given by (20) and (22) and the exact expression given by (13), consider the case when $f_1 = .25$; $f_2 = .26$; $L = 9$; and $\sigma_1^2 = \sigma_2^2 = .1\sigma_0^2$. For this case, the magnitudes of the coupling coefficients between the positive and negative frequencies are [from (8)]: $|\gamma_{24}| = |\gamma_{42}| = .029$; $|\gamma_{23}| = |\gamma_{32}| = |\gamma_{14}| = |\gamma_{41}| = .033$; $|\gamma_{13}| = |\gamma_{31}| = .034$, and the magnitude of the coupling coefficients between the closely spaced frequencies is: $|\gamma_{12}| = |\gamma_{21}| = |\gamma_{34}| = |\gamma_{43}| = .306$. As seen from equation (A–4) given in the Appendix, equation (18) converges for this case. In Figure 1 plots of $Q(f)$ (exact), $Q^{(0)}(f)$, and $Q^{(1)}(f)$ are presented. As shown, $Q^{(1)}(f)$ provides a better approximation to $Q(f)$ than does $Q^{(0)}(f)$. It should be pointed out that the value of L used in this example ($L = 9$) was only chosen for purposes of comparing the different approximations for $Q(f)$ and does not represent an optimal choice for L in this example. Indeed, as previously discussed, when the autocorrelation function is known exactly, there is no cut-off value for L and the theoretical MEM resolution improves without limit as L is increased.

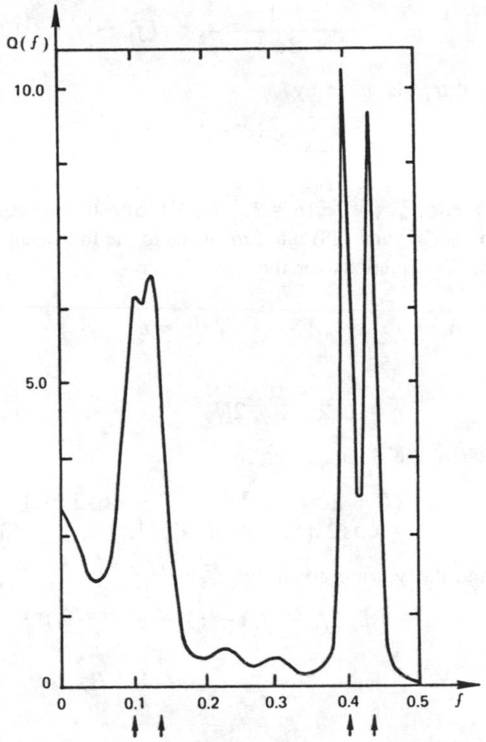

FIG. 1. Comparison between the exact expression and the zeroth order and first order approximations for $Q(f)$ for two sinusoids in white noise. For this case, $L = 9$; $f_1 = .25$; $f_2 = .26$; and $\sigma_1^2 = \sigma_2^2 = .1\sigma_0^2$.

MEM SPECTRAL ANALYSIS OF MULTIPLE SINUSOIDS IN 1-POLE, LOW-PASS NOISE

In this section we will consider the case when the sampled autocorrelation function is known exactly and can be written

$$\rho(\ell) = \sigma_0^2 e^{-\alpha|\ell|} + \sum_{n=1}^{N} \sigma_n^2 \cos 2\pi f_n \ell. \quad (24)$$

Equation (24) corresponds to the sum of N sinusoids in 1-pole, low-pass noise. A determination of $S(f)$ for this case indicates the resolution properties of MEM in a background noise with variable spectral density. For this case we expand the prediction coefficients as follows

$$g(k) = C_1\delta(k) + C_2\delta(k - L + 1) + \sum_{n=1}^{2N} A_n e^{2\pi j f_n k}, \quad (25)$$

where, as in equation (6), $f_{n+N} \equiv -f_n (n = 1, 2, \ldots, N)$. This particular choice for the assumed solution is similar to the solution of equation (4) for the more general case when the spectral density of the data contains both poles and zeroes (Satorius and Zeidler, 1977) and leads to precisely $2N + 2$ equations for the A_n, C_1, and C_2. Substituting (25) into (4) with $\rho(\ell)$ given by (24) and equating coefficients of $\exp(2\pi j f_r \ell)$, $r = 1, 2, \ldots, 2N$, and $\exp(\pm\alpha\ell)$ in the resulting equation leads to the following $2N + 2$ equations for the A_r, C_1, and C_2:

$$A_r + \sum_{\substack{n=1 \\ n\neq r}}^{2N} \mu_{rn} A_n = \frac{e^{2\pi j f_r} - C_1 - C_2 e^{-2\pi j f_r (L-1)}}{L + 2\sigma_0^2/\sigma'^2_r}$$

$$r = 1, 2, \ldots, 2N, \quad (26)$$

$$C_1 = e^{-\alpha} - \sum_{n=1}^{2N} \frac{A_n}{1 - e^{\alpha + 2\pi j f_n}} \quad (27)$$

and

$$C_2 = \sum_{n=1}^{2N} \frac{A_n e^{2\pi j f_n L}}{e^{\alpha} - e^{2\pi j f_n}}. \quad (28)$$

In equation (26), μ_{rn} is given by

$$\mu_{rn} = \frac{1}{L + 2\sigma_0^2/\sigma_r'^2}\phi_L(f_n - f_r), \quad (29)$$

and $\sigma_r'^2$ is given by

$$\sigma_r'^2 = \frac{1 - \cos 2\pi f_r/\cosh\alpha}{\tanh\alpha}\sigma_r^2, \quad (30)$$

where $\sigma_{n+N}^2 \equiv \sigma_n^2 (n = 1, \ldots, N)$. Substituting equations (27) and (28) into (26) leads to the following set of $2N$ equations for the A_r:

$$A_r + \sum_{\substack{n=1\\ n\neq r}}^{2N} \gamma_{rn} A_n = (e^{2\pi j f_r} - e^{-\alpha})\beta_r,$$
$$r = 1, 2, \ldots, 2N, \quad (31)$$

where the β_r are given by

$$\beta_r = \{L + 2\sigma_0^2/\sigma_r'^2 + (e^{-\alpha} - \cos 2\pi f_r)/(\cos 2\pi f_r - \cosh\alpha)\}^{-1}, \quad (32)$$

and the γ_{rn} are given by

$$\gamma_{rn} = \left\{\phi_L(f_n - f_r) - (1 - e^{\alpha + 2\pi j f_n})^{-1} + \frac{e^{-\alpha + 2\pi j((f_n - f_r)L + f_r)}}{1 - e^{-\alpha + 2\pi j f_n}}\right\}\beta_r. \quad (33)$$

Equation (31) is similar in structure to (7) and becomes identical to (7) as $\alpha \rightarrow \infty$, as it should. Although the γ_{rn} in equation (33) are considerably more complicated than those for the white noise case [equation (8)], they still approach zero as $L \rightarrow \infty$. As the γ_{rn} approach zero, equation (31) decouples and the A_r are given to a good approximation by

$$A_r \simeq (e^{2\pi j f_r} - e^{-\alpha})\beta_r. \quad (34)$$

Using equations (25) to (30), we may express all quantities of interest in terms of A_r, and C_1, and C_2. From (5) we have for σ_v^2 (after considerable simplification),

$$\sigma_v^2 = \sigma_0^2(1 - e^{-2\alpha}) \cdot \cdot\left\{1 + \sum_{n=1}^{2n} \frac{A_n}{e^{2\pi j f_n} - e^{-\alpha}}\right\}. \quad (35)$$

From (3) and (25), we have for $Q(f)$

$$Q(f) = |1 - C_1 e^{-2\pi j f} - C_2 e^{-2\pi j f L} - e^{-2\pi j f}\sum_{n=1}^{2N} A_n \phi_L(f_n - f)|^{-2}. \quad (36)$$

As in the case of white noise [equation (14)], $Q(f)$ evaluated at the frequency of the rth sinusoid can be simply expressed in terms of A_r and $\sigma_0^2/\sigma_r'^2$. The result for $Q(f_r)$ is

$$Q(f_r) = \frac{1}{4}\left(\frac{\sigma_r'^2}{\sigma_0^2}\right)^2 |A_r|^{-2}, \; r = 1, 2, \ldots, N. \quad (37)$$

As the γ_{rn} approach zero and for $L \gg \beta_r^{-1} - L$, $Q(f_r)$ is given to a good approximation by [from equation (34)]

$$Q(f_r) \simeq \frac{e^\alpha}{8}\left(\frac{\sigma_r^2}{\sigma_0^2}\right)^2 L^2 \frac{\cosh\alpha - \cos 2\pi f_r}{\sinh^2\alpha},$$
$$r = 1, 2, \ldots, N. \quad (38)$$

Equation (38) shows that for large L the values $Q(f_r)$ are a function of f_r as well as σ_r^2/σ_0^2. This result is reasonable since the signal-to-noise ratio (SNR) per unit bandwidth is a function of frequency. This indicates that the resolution capabilities of the MEM spectral estimate will depend upon where the sinusoids are located in the noise spectrum. For sinusoids located in the low SNR regions, the MEM resolution will be worse than for sinusoids located in the high SNR region. This is indicated in Figure 2 where $Q(f)$ is plotted for 4 sinusoids of equal amplitudes. Two of the sinusoids are located near the low SNR region and

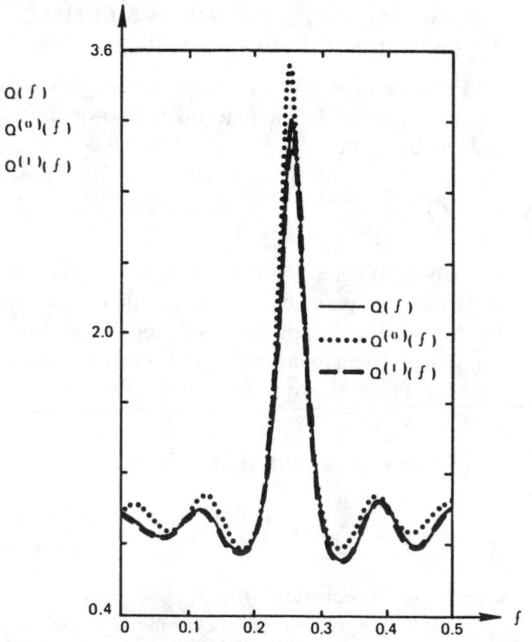

FIG. 2. A plot of $Q(f)$ for 4 sinusoids in 1-pole low-pass noise. For this case, $L = 14$; $\alpha/2\pi = .1$; $f_1 = .1$; $f_2 = .136$; $f_3 = .4$; $f = .436$; and $\sigma_1^2 = \sigma_2^2 = \sigma_3^2 = \sigma_4^2 = \sigma_0^2/2.25$.

the other two sinusoids are located near the high SNR region. As is seen, the two sinusoids in the low SNR region are poorly resolved, whereas the two sinusoids in the high SNR region are well resolved (arrows in Figure 2 indicate the correct location of the frequencies of the sinusoids). As in the case of Figure 1, the value of L used in Figure 2 ($L = 14$) was only chosen for purposes of illustrating the variation of two sinusoid MEM resolution versus frequency and does not represent any optimal value for L.

It should be noted that in complete analogy with the white noise case, the proper time series model for N sinusoids in additive 1-pole, low-pass noise is an ARMA model which contains $2N + 1$ AR terms and $2N$ MA terms. (The additional AR term is due to the 1-pole structure of the additive noise.) Therefore, one expects that as $L \to \infty$, $S(f)$ will converge to the power spectral density of the 1-pole noise (for $f \neq f_n$; $n = 1, \ldots, N$). This can easily be seen from the development presented in this section by noting from equation (34) that

$$\lim_{L \to \infty} A_n \phi_L(f_n - f) = 0,$$

provided $f \neq f_n (n = 1, \ldots, N)$. Further from equations (27) and (28), it is seen that $C_1 \to e^{-\alpha}$ and $C_2 \to 0$ as $L \to \infty$. Therefore, from equations (2), (35), and (36) we have

$$\lim_{L \to \infty} S(f) = \frac{\Delta \sigma_0^2 \sinh\alpha}{\cosh\alpha - \cos 2\pi f}, \quad f \neq f_n (n = 1, \ldots, N). \qquad (39)$$

Equation (39) is the expression corresponding to the power spectral density of the 1-pole, low-pass noise. Therefore, from equations (38) and (39) it is seen that $S(f)$ provides an increasingly accurate pointwise approximation to the background noise spectrum as well as precise sinusoid resolution capabilities as $L \to \infty$. However, as noted in the previous section, this only applies when the sampled autocorrelation function is known exactly.

CONCLUSIONS

In this paper, the theoretical MEM spectral estimate of multiple sinusoids in noise has been examined by the method of undetermined coefficients. For the case of sinusoids in white noise, the prediction filter coefficients were expanded directly in terms of the input sinusoids [equation (6)]. This expansion leads to a transformation of the original $L \times L$ equations for the prediction coefficients [equation (4)] to a set of $2N \times 2N$ equations [equation (7)]. The transformed equations are a smaller set of linear coupled equations than equation (4) when $L > 2N$ and are particularly useful for purposes of computing the theoretical MEM spectral estimate for large L (high-resolution limit). Further, the reduced equations [equation (7)] have been shown to provide additional insight into the analytical structure of the MEM spectral estimate. Also, for certain cases where there is little interaction between some of the frequency components of the input sinusoids, the $2N \times 2N$ equations may be approximated by a smaller set of coupled equations.

For the case of N sinusoids in 1-pole, low-pass noise, the prediction filter coefficients were expanded in terms of the input sinusoids as well as two delta functions which occur at the beginning and end of the prediction filter [equation (25)]. This expansion also leads to a set of $2N \times 2N$ equations [equation (31)]. For this case, the values of the MEM spectral estimate evaluated at the input sinusoid frequencies were shown to be a function of the frequency of the sinusoids. This result is reasonable since the SNR per unit bandwidth is also a function of the frequency.

The results derived in this paper give a further understanding of the MEM spectral estimate of sinusoids in noise when the autocorrelation function is estimated from the data. As the variance in the estimates of the autocorrelation function becomes appreciable, deviations in the MEM spectral estimate will begin to appear, especially at its peaks (e.g., Lacoss, 1971; Baggeroer, 1976).

ACKNOWLEDGMENTS

The authors wish to express their thanks to S. Alexander and P. Reeves of the Naval Ocean Systems Center for many helpful comments throughout the course of this work.

REFERENCES

Akaike, H., 1969, Power spectrum estimation through autoregressive model fitting: Ann. Inst. Stat. Math., v. 21, p. 407–419.

——— 1970, Statistical predictor identification: Ann. Inst. Stat. Math., v. 22, p. 203–217.

——— 1972, Use of an information theoretic quantity for statistical model identification: Proc. 5th Hawaii Int. Conf. on System Sciences, p. 99–101.

——— 1974, A new look at the statistical model identification: IEEE Trans. Automat. Control, v. AC-19, p. 716–723.

Baggeroer, A. B., 1976, Confidence intervals for regression (MEM) spectral estimates: IEEE Trans. on Information Theory, v. IT-22, no. 5, p. 534–545.

Burg, J. P., 1967, Maximum entropy spectral analysis: Presented at 37th Annual International SEG Meeting, October 31, Oklahoma City.

——— 1975, Maximum entropy spectral analysis: Ph.D. thesis, 123 p., Stanford University.

Frost, O. L., 1977, Power spectrum estimation, *in* Aspects of signal processing with emphasis on underwater acoustics, part 1: G. Tacconi, ed., Boston, D. Reidel Publishing Co., p. 125–162.

Gersch, W., and Sharpe, D. R., 1973, Estimation of power spectra with finite order autoregressive models: IEEE Trans. Automat. Control, v. AC-18, p. 367–369.

Jones, R. H., 1976, Autoregression order selection: Geophysics, v. 41, p. 771–773.

Lacoss, R. T., 1971, Data adaptive spectral analysis methods: Geophysics, v. 36, p. 661–675.

Marple, S. L., 1976, Conventional Fourier, autoregressive, and special ARMA methods of spectrum analysis: Engineer thesis, Dept. of Electrical Engineering, Stanford University.

Satorius, E. H., and Zeidler, J. R., 1977, Least mean square, finite length, predictive digital filters: Conference record, 1977 IEEE Intl. Confer. on Acoustics, Speech, and Signal Processing, May 9–11, Hartford, Conn.

Stewart, G. W., 1973, Introduction to matrix computations: Academic Press, New York, 441 p.

Ulrych, T. J., 1972, Maximum entropy power spectrum of truncated sinusoids: J. Geoph. Res., v. 77, p. 1396–1400.

Ulrych, T. J., and Bishop, T. N., 1975, Maximum entropy spectral analysis and autoregressive decomposition: Rev. Geophys. and Space Phys., v. 33, p. 183–200.

Ulrych, T. J., and Clayton, R. W., 1976, Time series modelling and maximum entropy: Physics of the Earth and Planetary Interior, v. 12, p. 188–200.

van den Bos, A., 1971, Alternative interpretation of maximum entropy spectral analysis: IEEE Trans. on Information Theory, v. IT-17, p. 493–494.

Zadeh, L. A., and Ragazzini, J. R., 1950, An extension of Wiener's theory of prediction: J. Appl. Phys., v. 21, p. 645–655.

Zielke, G., 1968, Inversion of modified symmetric matrices: J. Assoc. Comp. Machinery, v. 15, p. 402–408.

APPENDIX

It is the purpose of this Appendix to establish sufficient conditions for the convergence of the perturbation expansion expressed by equation (18). We make use of the following basic result (e.g., Stewart, 1973): The perturbation series $\mathcal{I} + \mathcal{P} + \mathcal{P}^2 + \ldots$ converges to $(\mathcal{I} - \mathcal{P})^{-1}$ if $\|\mathcal{P}\| < 1$, where $\|\cdot\|$ denotes any matrix norm such that $\|\mathcal{A} \cdot \mathcal{C}\| \leq \|\mathcal{A}\| \, \|\mathcal{C}\|$.

The particular norm of which we make use and which is relatively simple to calculate is the ∞ – norm which is denoted by $\|\cdot\|_\infty$. The ∞ – norm of a $p \times p$ matrix $\mathcal{A}$ is defined as follows (Stewart, 1973):

$$\|\mathcal{A}\|_\infty = \max\left(\sum_{j=1}^{p} |A_{ij}| : i = 1, \ldots, p\right), \quad \text{(A–1)}$$

where A_{ij} are the elements of $\mathcal{A}$. Therefore, a sufficient condition for the convergence of equation (18) is

$$\|\mathcal{B}^{-1} \cdot \mathcal{G}\|_\infty < 1. \quad \text{(A–2)}$$

Since $\|\mathcal{A}\mathcal{C}\|_\infty \leq \|\mathcal{A}\|_\infty \|\mathcal{C}\|_\infty$, then a simpler sufficient condition to check is

$$\|\mathcal{B}^{-1}\|_\infty \|\mathcal{G}\|_\infty < 1. \quad \text{(A–3)}$$

For the special case of two sinusoids of equal power in white noise, in which there is little interaction between the positive and negative frequency components of the sinusoids, condition (A–3) reduces to

$$\frac{1}{1 - |\gamma_{12}|} \cdot \max(|\gamma_{13}| + |\gamma_{14}|, |\gamma_{23}| + |\gamma_{24}|) < 1, \quad \text{(A–4)}$$

where the γ_{ij} are computed from equation (8).

A COMPARATIVE OVERVIEW OF ARMA SPECTRAL ESTIMATION

M. Kaveh and S. P. Bruzzone
University of Minnesota
Department of Electrical Engineering
123 Church Street S.E.
Minneapolis, Minnesota 55455

Summary

This paper gives an overview of ARMA spectral estimation. Optimum and some suboptimum techniques are reviewed and an algorithm for a complex unconditional least squares estimator is introduced. The main objective of the paper is to present several methods in a comparative fashion with respect to computational efficiency, statistical efficiency, and their applications to specific classes of signals such as wideband and narrowband stochastic signals and sinusoids in the presence of noise.

Introduction

Spectral estimation via parametric modeling of the signals has gained great popularity in recent years. A useful model is one with a rational spectrum. In this case, the observed signal is considered as the output of a linear system with a rational transfer function, driven by white noise. In the statistical literature such a signal is said to fit an autoregressive-moving-average (ARMA) model.

A subclass of the ARMA model, the autoregressive (AR, all pole) scheme has been widely used in spectral estimation. This is due to the fact that AR parameters can be estimated in a variety of computationally efficient manners with reasonably good statistical properties. Furthermore, model order determination is far more straightforward in the AR case. This is due to the fact that 1) many algorithms for the estimation of AR parameters are recursive, and can therefore be updated with an increase in the model order and 2) the search for the optimum order for the AR model is one dimensional, whereas for an ARMA process various combinations of the number of poles and zeros have to be considered.

Questions may, therefore, be posed as to the practicality of the ARMA spectral estimator and as to the reason for renewed interest in it. The reason is the principle of parsimony; spectral estimation using parametric models is really based on this principle. In other words, one should, if possible, use the minimum number of parameters that best describe a signal. Thus, if a signal spectrum contains a zero, it is best, theoretically, to fit an ARMA model to the signal rather than a large number of poles to account for the zero.

Although the impetus exists for ARMA modeling in signal processing applications in general, and spectral estimation in particular, this model has traditionally not been popular. This is for the most part due to the complexity of both the algorithms and statistical analysis of the estimators involved. However, there have been a number of methods suggested recently for ARMA parameter identification for spectral estimation. This paper is an attempt to place some of the more popular algorithms on a common framework and give a comparative view of their performance. The paper is organized as follows. The ARMA model is first reviewed, its statistics are developed and the likelihood function and its approximations are discussed. Subsequently, a statistical classification of ARMA spectral estimators as well as some algorithms in the classes are given. Finally, the applicability of these methods is discussed and some comparative examples are presented. In this manuscript most derivations are referenced and only the required equations for the discussion of the algorithms are given.

The ARMA Model

A complex zero mean stationary ARMA time series is given by

$$x_t = \sum_{i=1}^{L} a_i x_{t-i} + \sum_{i=1}^{M} b_i u_{t-i} + u_t \tag{1}$$

where a_i and b_i are the AR and MA parameters, L and M are the AR and MA orders and u_t is an orthogonal sequence. Alternatively, x_t can be considered as the output of a discrete time filter, H(z), driven by u_t, where H(z) is given by

$$H(z) = \frac{1 + \sum_{i=1}^{M} b_i z^{-i}}{1 - \sum_{i=1}^{L} a_i z^{-i}} \tag{2}$$

The relation between the autocorrelation function, r_τ, and the coefficients of an ARMA process can be obtained from equation (1) by multiplying both sides by $x^*_{t-\ell}$ and taking the expectation. The following results:

$$\sum_{i=1}^{L} a_i r_{\ell-i} + \gamma^*_{-\ell} = r_\ell \quad , \ \ell = 0,1,\ldots \tag{3}$$

where $r_\ell = E[x^*_t x_{t+\ell}]$ and γ_ℓ is the cross-correlation between x_t and the moving average residuals ε_t. That is,

$$\varepsilon_t = u_t + \sum_{i=1}^{M} b_i u_{t-i} \ \text{and} \ \gamma_\ell = E[\varepsilon^*_t x_{t+\ell}] \tag{4}$$

The relationships in (4) show that for $\ell \leq M$ equation (3) is highly non-linear in a_i and b_i. For $\ell > M$ however $\gamma_{-\ell} = 0$ and (3) is the familiar modified Yule-Walker (YW) equations that recursively generate r_τ from a_i

It is also worth noting the relation between the autocorrelation function of ε_t, c_τ, and the model parameters. From the definition of ε_t, one obtains:

Supported in part by Air Force Office of Scientific Research Grant #AFOSR-78-3628.

Reprinted with permission from *Proc. of the 1st ASSP Workshop on Spectral Estimation*, vol. 1, 1981, pp. 2.4.1–2.4.8.

$$c_\tau = \sum_{i=0}^{M-\tau} b_i^* b_{i+\tau} \sigma_u^2 \;, \qquad \tau = 0,\ldots,M$$

$$c_\tau = 0 \;, \quad \tau > M \;, \qquad b_o = 1 \tag{5}$$

Furthermore, matching the spectrum of the process based on the model and as the discrete Fourier transform of r_τ leads to[2]

$$c_\tau = r_\tau - \sum_{j=1}^{L} [a_j r_{\tau-j} + a_j^* r_{\tau+j}] + \sum_{i=1}^{L} \sum_{j=1}^{L} a_i a_j^* r_{\tau-i+j} \;, \quad \tau = 0,\ldots,M \tag{6}$$

The power spectrum of the ARMA process is then simply given by the following equivalent expressions, all depending on a finite number of parameters:

$$S_x(f) = \frac{S_\varepsilon(f)}{|DFT[a_i']|^2} \tag{7}$$

or explicitly in terms of c_τ as

$$S_x(f) = \frac{DFT[c_\tau]}{|DFT[a_i']|^2} \tag{8}$$

or

$$S_x(f) = \frac{|DFT[b_i]|^2 \sigma_u^2}{|DFT[a_i']|^2} \tag{9}$$

where $a_o' = 1$, $a_i' = -a_i$, $b_o = 1$

Therefore the ARMA spectral estimator has to obtain estimates of a_i along with those of b_i or c_τ or some other spectral estimate of $S_\varepsilon(f)$. In order to consider estimators of the various paremeters we will first review the likelihood function for an ARMA process and its practical approximations.[1]

The Likelihood Function

Let u_t be a zero-mean, uncorrelated Gaussian process. The functional dependence of u_t on the ARMA parameters and the observations $x_t, t=1,\ldots,N$ is given by equation (1). Therefore, the likelihood function can be found by substituting for u_t in terms of the data and the model parameters in the multinomial normal probability density function for u_t. The complicating problem is that the generation of $u_t, t=1,\ldots,N$, requires starting values of $\tilde{u}_t$, $t=1-M,\ldots,0$ and $\tilde{x}_t, t=1-L,\ldots,0$ which are unavailable. Newbold[3] used the relation between the starting values and the model parameters to derive an exact likelihood function for an ARMA process. The resulting log-likelihood function is given by

$$L(x_t;a_i,b_i) = g(a_i,b_i) - \sum_{i=1}^{N} |u_t|^2 \tag{10}$$

with initial values for u_t and x_t given as a linear transformation on the data. Unfortunately, the function $g(a_i,b_i)$ and the transformation kernel for the initial values are both highly non-linear in the parameters. In fact their values only for the L=M=1 case have been derived[3] and, even for this simple model, the order of the non-linearities is well in excess of the number of data samples.

As a practical approximation to the maximum likelihood estimates, Box and Jenkins[1] proposed the use of the unconditional least-squares method. In this technique the second term in (10), i.e., the sum of the residual energies, is minimized subject to appropriate starting values. They suggested back forecasting at every iteration of model identification for both $\tilde{u}_t$, $t \leq 0$ and $\tilde{x}_t$, $t \leq 0$. Most available implementations of this method, however, assume zero starting values, thus obtaining conditional least-squares estimates. In the next section a classification of some ARMA spectral estimators is given including a brief description of a complex unconditional least-squares estimator of the parameters.

A Statistical Classification of ARMA Spectral Estimators

The above discussions point to three general classes of ARMA spectral estimators. In the minimal sufficient (MS) class, the AR and MA parameters are adjusted simultaneously to give a least squares or approximately least squares fit to the observed data vector. The name "minimal sufficient" acknowledges that no reduction of the data is a sufficient statistic for stationary Gaussian time series having zeros in the power spectral density[4]. The resulting estimates, as we have seen, are approximately maximum likelihood (ML) for Gaussian data. We denote such estimates as optimum. The equations for the least squares solution are highly nonlinear, and thus estimators in this class are typified by nonlinear optimization algorithms and other iterative approaches and their concomitant pitfalls, i.e., nonconvergence or convergence to local rather than global extrema, the need for preliminary ARMA parameter estimates to start the iterations, and large computer memory and time requirements. Nevertheless, with the steady reduction in computation costs and with the possibility of implementing these algorithms in a distributed fashion, this category has to be ultimately preferred.

Examples of work in the MS class include those reported in references [1] and [5-9]. All of these techniques basically use the least-squares method discussed above. The predominant effort, therefore, has been in the development of numerical algorithms for the non-linear minimization process, and specifically in obtaining expressions and/or algorithms for the gradient and Hessian of the objective function with respect to the parameters. In the following, we briefly describe a generalization of Box and Jenkins' algorithm to complex data. This algorithm can be used in conjunction with the many available standard non-linear minimization routines for real data.

Let $x_t, t=1,\ldots,N$ be complex ARMA(L,M).

If the data record is short, back forecasting such as proposed in[1] should be used to initialize the generation of $u_t, t=1,\ldots,N$. Care should be taken here, for the effects of initial transients when poles and zeros are near the unit circle.

Most non-linear minimization algorithms require an objective function and its gradient with respect to the parameters. Some, however, obtain numerical estimates of the gradients adaptively.

In order to obtain an expression for the gradient of the objective function for complex data, the following vector notation is chosen:

$$\underline{A}^T = \begin{bmatrix} 1 \\ -a_1 \\ \cdot \\ \cdot \\ \cdot \\ -a_L \end{bmatrix}, \quad \underline{B}^T = \begin{bmatrix} b_1 \\ \cdot \\ \cdot \\ \cdot \\ b_M \end{bmatrix}, \quad \underline{U}^T = \begin{bmatrix} u_1 \\ \cdot \\ \cdot \\ \cdot \\ u_N \end{bmatrix}$$

$$\underline{X} = \begin{bmatrix} x_1, x_0, \ldots, x, \ldots, x_{1-L} \\ x_2, x_1, \ldots, \quad x_{2-L} \\ \cdot \\ \cdot \\ \cdot \\ x_N, x_{N-1}, \ldots, \quad x_{N-L} \end{bmatrix}, \quad \underline{V} = \begin{bmatrix} u_o, u_{-1}, \ldots, \quad u_{1-M} \\ u_1, u_o, \ldots, \quad u_{2-M} \\ \cdot \\ \cdot \\ \cdot \\ u_{N-1}, u_{N-2}, \ldots, u_{N-M} \end{bmatrix}$$

Denoting the real and imaginary parts of the various quantities by the subscripts 'r' and 'i', respectively, define the parameter vector $\underline{P}$ as:

$$\underset{1x2(M+L)}{\underline{P}} = [a_{1r}, a_{2r}, \ldots, a_{Lr}, a_{1i}, a_{2i}, \ldots, a_{Li}, \quad b_{1r}, \ldots, b_{Mr}, b_{1L}, \ldots, b_{Mi}]$$

The residual vector is given by

$$\underline{U}^T = \underline{X}\,\underline{A}^T - \underline{V}\,\underline{B}^T \tag{11}$$

and the objective function is:

$$f = \underline{U}^*\,\underline{U}^T = \underline{U}_r\,\underline{U}_r^T + \underline{U}_i\,\underline{U}_i^T \tag{12}$$

The gradient of f with respect to the 2(M+L) element vector $\underline{P}$ is given by:

$$\nabla f = 2[J_{\underline{U}_r}\,\underline{U}_r^T + J_{\underline{U}_i}\,\underline{U}_i^T] \tag{13}$$

where $J_{\underline{U}_r}$ ($J_{\underline{U}_i}$) is the Jacobian of $\underline{U}_r$ ($\underline{U}_i$) with respect to the parameter vector $\underline{P}$. For example,

$$[J_{\underline{U}_r}]_{k\ell} = \frac{\partial\, u_{\ell r}}{\partial\, P_k} \tag{14}$$

It is shown in the appendix that the elements of the Jacobians can be obtained recursively, from the data and current estimates of the coefficients. It should be mentioned that such an algorithm can be extended to estimates based on least-squares forward and backward error energies. This is done by simply entering the data backwards, conjugating the model parameters and averaging the forward and backward objectives.

The second and third categories of estimators are based on the equations (3) - (6). These methods make use of data reduction in the form of estimates of the autocorrelation function to solve for the model parameters. This approach, of course, is quite common in pure AR(L) modeling as $r_\tau, \tau=0,\ldots,L$ are approximately the sufficient statistics for the model.

The difference between the second and third classes is in the manner that $\hat{r}_\tau$ is used. In the second class, equations (3) are used for a simultaneous estimation of a_i and b_i. An algorithm for complex ARMA processes using this approach is given in[10]. Because of the non-linearity of equations (3) for $\tau < M$, however, the numerical complexities associated with this method are similar to the optimum case. We will, therefore, not dwell any further on this class.

The third category is also based on estimates of the autocorrelation function and equations (3). The methods in this category, however, make explicit use of equations (3) for $\tau > M$. These equations are theoretically independent of b_i and are linear in a_i. Therefore, the AR coefficients can be easily estimated. Based on these estimates, estimates of c_τ, b_i or $s_\varepsilon(f)$ can be obtained by respectively:

- Using equation (6).
- Solving equation (5) via a complex Newton-Raphson type algorithm[11].
- Generating residuals $\varepsilon_t = x_t - \sum_{i=1}^{L} \hat{a}_i x_{t-i}$ and estimating $s_\varepsilon(f)$ using any traditional spectral estimation method.

Once these quantities are obtained, formula (7), (8) or (9) can be used to evaluate the ARMA spectrum. We stress that all of the estimators in this class are suboptimum due to the data reduction through the truncated sample autocorrelation function. Further, an optimum fitting requires that all parameters be adjusted simultaneously. Suboptimality, then, is the price paid for the gains in computational simplification and relaxed memory requirements of estimators in this class. Examples of this class of estimators are given in references [2,12-19]. The method proposed by Cadzow in[20] does not explicitly use estimates of the autocorrelation function. However, because it estimates a_i separately from b_i using the model in equations (3) for $\tau > M$, it is included in this class.

Due to the computational efficiency of the third category some of its methods are discussed further. The main computational advantage of this class, as we have seen, lies in its separate estimation of the AR parameters from the modified YW equations. Thus, given N data points, equation (3) leads to a set of N-M-1 linear equations

$$\sum_{i=1}^{L} \hat{a}_i \hat{r}^*_{\ell-i} = \hat{r}^*_\ell, \quad \ell = M+1,\ldots,N \tag{15}$$

where $\hat{r}_\tau$ is some estimate of the autocorrelation function, usually taken as

$$\hat{r}_\tau = \frac{1}{N-\tau} \sum_{i=1}^{N-\tau} x_i x^*_{i+\tau} \tag{16}$$

or

$$\hat{r}_\tau = \frac{1}{N} \sum_{i=1}^{N-\tau} x_i x^*_{i+\tau} \tag{17}$$

Normally, N-M > L. Therefore the set of equations in (15) may be solved in a variety of ways. The most straightforward and computationally efficient method is to solve any set of L linear equations for a unique solution of $\hat{a}_i$. Because the variance of $\hat{r}_\tau$ increases as a function of τ, however, the first L equations, that is, for ℓ=M+1,M+L, are usually used (e.g., [2,12]). On the other hand, for highly resonant systems, sinusoidal signals in the extreme, $\hat{r}_\tau$ for larger values of τ contain substantial information. Therefore, one may obtain a least-squares solution of $\hat{a}_i$ based on all the available equations (15)[11].

Define the coefficient matrix for all the equations in (15) by $\underset{\sim}{Q}_M$, the right-hand side vector of autocorrelation functions by $\underset{\sim}{q}_M$ and the vector of the L AR coefficients by $\underline{\hat{A}}'$. The weighted least-squares solution of (15) is given by:

$$\underline{\hat{A}}' = (\underset{\sim}{Q}{}^{*T}_M \underline{W} \underset{\sim}{Q}_M)^{-1} \underset{\sim}{Q}{}^{*T}_M \underline{W} \underset{\sim}{q}_M \tag{18}$$

where $\underline{W}$ is a diagonal positive definite real matrix. Cadzow[20] proposed a method similar in essence, to the above, that does not use the data reduction in terms of the autocorrelation function estimates. He suggested using a $\underline{W}$ with decreasing diagonal elements to account for the reduced confidence in the higher lag estimates of the autocorrelation function[20,21]. It is worth emphasizing that the least-squares solution of the overdetermined system (15) can also be used in conjunction with pure AR parameter estimation of a process with infinite order AR representation.

Thus far, we have discussed the estimation of a_i. Since, in this class, the estimates of b_i are based directly on those of a_i, the accuracy of $\hat{a}_i$ greatly affects the usefulness of the estimate of the numerator spectrum. In applications that involve spectra with narrow peaks, however, the AR parameters determine the dominant, i.e., resonant, structure of the spectrum and the numerator spectrum does not play a significant role in terms of resolution. For spectra with dominant zeros, however, special care must be taken to assure accurate $\hat{b}_i$. In any case, when the computing power is available, these sub-optimum estimates should be used as preliminary estimates for optimum schemes[1].

It is also noted that none of the methods discussed here guarantee the stability and/or invertibility of the identified system. This, we believe, is not a serious concern in the context of spectrum estimation. Nevertheless, one can include such conditions on, say, the least-squares algorithm and solve a constrained minimization problem. Since the constraints are on the locations of poles and zeros, the resulting algorithms will be substantially more complex.

Discussions and Examples

In this section, some cases where ARMA spectral estimation may be used are discussed and examples of the estimates using the methods described earlier are given. Much of the recent interest in spectral estimation stems from the desire to obtain "high resolution" spectral estimates from short data records. This is where the traditional taper and transform and periodogram methods fail. The AR spectral estimator promised computationally efficient and "high resolution" estimates. The problem, however, is that even when the AR(L) model is the true description of a signal, it may be far from being an accurate model in the presence of noise.

Figure 1 shows the bandwidth expansion factor (BEF) for an AR(2) model + white noise as a function of the signal-to-noise ratio and pole-pair distance from the origin, ρ_m[22]. BEF is defined as the ratio of the 3-dB bandwidths of a second-order AR model of the AR(2) process in the presence of white noise to that of the noiseless case. It can be seen that the spectral accuracy deteriorates rapidly with noise for highly resonant models and less so for broad AR spectra.

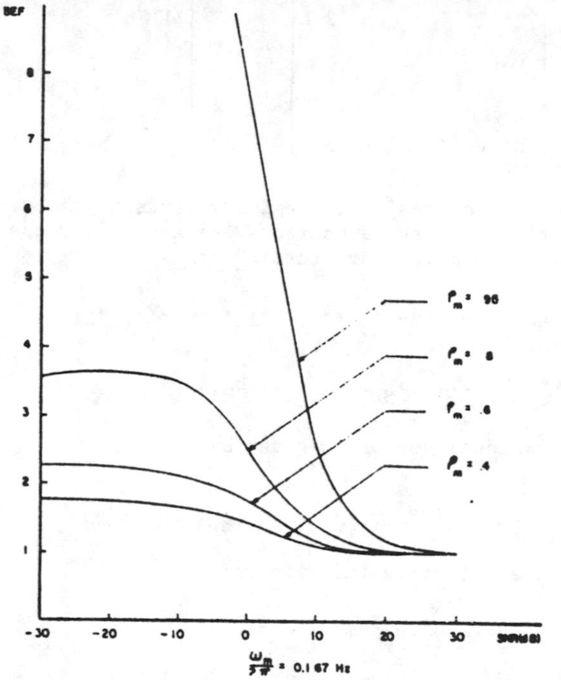

Fig. 1 Bandwidth expansion factor for noisy AR(2) process.

It can easily be shown[23] that the sum of several AR processes and white noise results in an ARMA process with equal AR and MA orders. Therefore, this situation is a good candidate for ARMA spectral estimation, especially since search of only a single order is needed, i.e., L = M. Much of the recent reports on ARMA spectral estimators have been directed to this application, with the special case of sinusoids in noise receiving special attention.

An examination of equation (3) shows that for ℓ > M, r_o does not appear in the equations for a_i. But, theoretically, r_o is the only sample of the autocorrelation function with a contribution from the white noise. Therefore, the suboptimum techniques, using the modified YW equations for ℓ > M should theoretically give the noiseless estimates of a_i resulting in high resolution spectral estimates. For short data records, however, the estimates $\hat{r}_\tau$ do not exactly decouple the noise from $\hat{a}_i$. But, the residual effect of the noise on $\hat{r}_\tau$ decreases at higher lags. Therefore, techniques such as the least-square extended YW estimates or that of Cadzow should improve resolution, at the expense of higher variance. These techniques, as mentioned earlier, are only useful in the highly resonant conditions where substantial signal information is contained in higher lags of the autocorrelation function. This has been demonstrated rigorously from a statistical information theoretic point of view [24].

This point can also be demonstrated heuristically from the structure of the autocorrelation function for an ARMA process. For an ARMA(L,M) process, r_τ, for $\tau>M$, is a mixture of geometrically damped sinusoids. Therefore $\hat{r}_\tau$ contains components due to the signal structure, i.e. the damped sinusoids, contaminated by the data-size dependent estimation errors in $\hat{r}_\tau$. Thus, if the model contains highly damped poles, the $\hat{r}_\tau$ for larger lags is dominated by noise and little additional information is gained by using the higher lags in the estimates. It is conjectured that there are different optimum algorithm for $\hat{a}_1$, within class 3, for models with highly, moderately or slightly damped sets of poles and zeros.

In the following several numerically generated examples are given to further illustrate the above mentioned points.

i) In the first set of examples we consider an ARMA(2,2) model obtained as the sum of an AR(2) process and white noise. Two cases are examined: one with the poles near the unit circle (i.e. small damping) and one with larger damping. Both examples have the same signal-to-noise ratio. The numerators for these cases have smooth spectra. Therefore Welche's periodogram as suggested by Cadzow [20] is used to estimate $S_\varepsilon(f)$. Figures 2 and 3 show the two examples as estimated by the modified Yule-Walker (MYW) method using the first 2 equations (15) and least-squares extend Yule-Walker (LSEYW) technique using all of the equations (15). The biased $\hat{r}_\tau$ is used in Fig. (2) and the unbiased one in Fig. (3). The exact spectra are shown as dashed lines.

Figure 2 confirms that the use of higher lags of $\hat{r}_\tau$ actually improves the spectral estimates in the highly resonant case (although exaggerating the peaks). Figure 3 on the other hand demonstrates the lack of improvement by using LSEYW, due to the highly damped poles. In fact, as expected, an increase in the variance of the estimates is observed.

ii) The second example deals with a signal containing a dominant zero. This example shows the problem with the separate estimation of the poles and zeros. Figure 4 presents the spectrum of an ARMA(2,2) process containing a narrow peak and a null. The least-squares (optimum) scheme gives excellent estimates of the spectrum. The class 3 methods, however, have difficulty determining the null. The main reason is that the previously used periodogram does not have the resolution needed for the numerator spectrum, for a low MA order (e.g. true order 2). The spectral estimates in Figs. 3(b) and (c) use a higher order MA order in order to improve numerator resolution, thereby compromising the statistical stability of the estimates.

iii) The final example (same as that in 20) is that of closely spaced sinusoids in noise, with estimates shown in Figure 5. This is an example of an extreme case where substantial information is contained in larger lags of $\hat{r}_\tau$. Furthermore the numerator spectrum is of no real consequence. Therefore, methods such as the LSEYW or Cadzow's [20] are able to resolve adjacent peaks where the MYW is not. Note the similarity of estimates based on the LSEYW and Cadzow's methods. Also, the price paid by choosing a large order (15,15) is the presence of many extraneous peaks in the estimates.

It is instructive to observe, here, that the resolution obtained in Figs. 5(a) and (b) is apparently due to the use of higher lags of $\hat{r}_\tau$ to estimate the

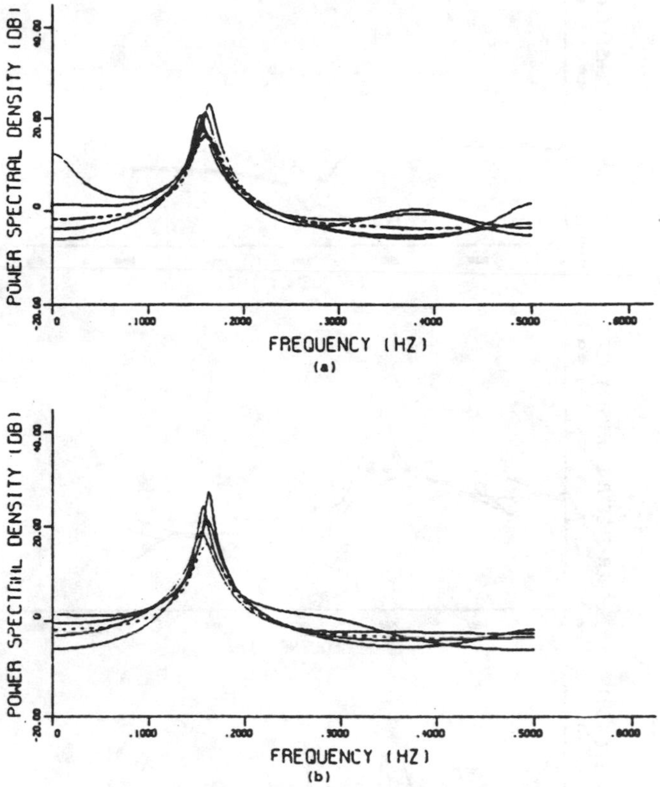

Fig. 2 Spectral estimates for $x_t = 1.027x_{t-1} - .902x_{t-2} + u_t + .64u_{t-1} - .36u_{t-2}$, N=40, (a) MYW estimates; (b) LSEYW estimates, uniform weight.

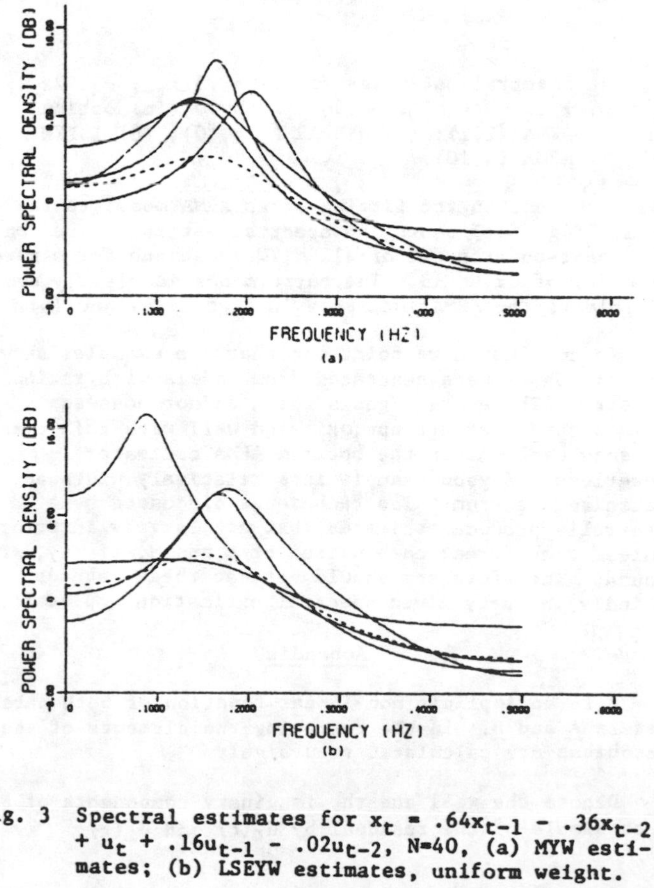

Fig. 3 Spectral estimates for $x_t = .64x_{t-1} - .36x_{t-2} + u_t + .16u_{t-1} - .02u_{t-2}$, N=40, (a) MYW estimates; (b) LSEYW estimates, uniform weight.

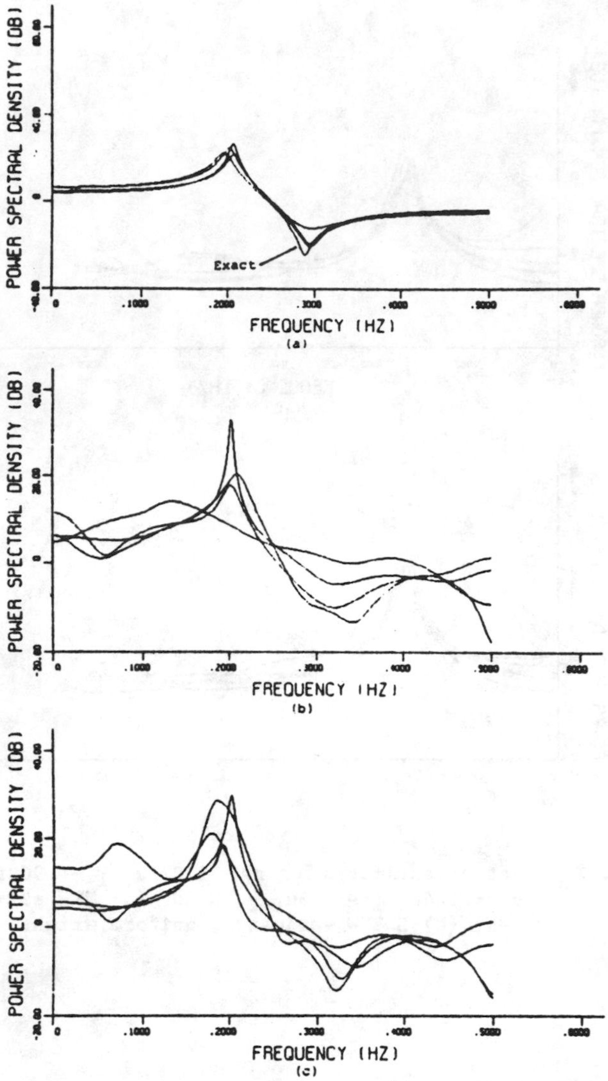

Fig. 4 Spectral estimates for $x_t = .51x_{t-1} - .92x_{t-2} + u_t - .44u_{t-1} - .94u_{t-2}$, N=40, (a) optimum ARMA (2,2); (b) MYW ARMA (4,10); (c) LSEYW ARMA (4,10).

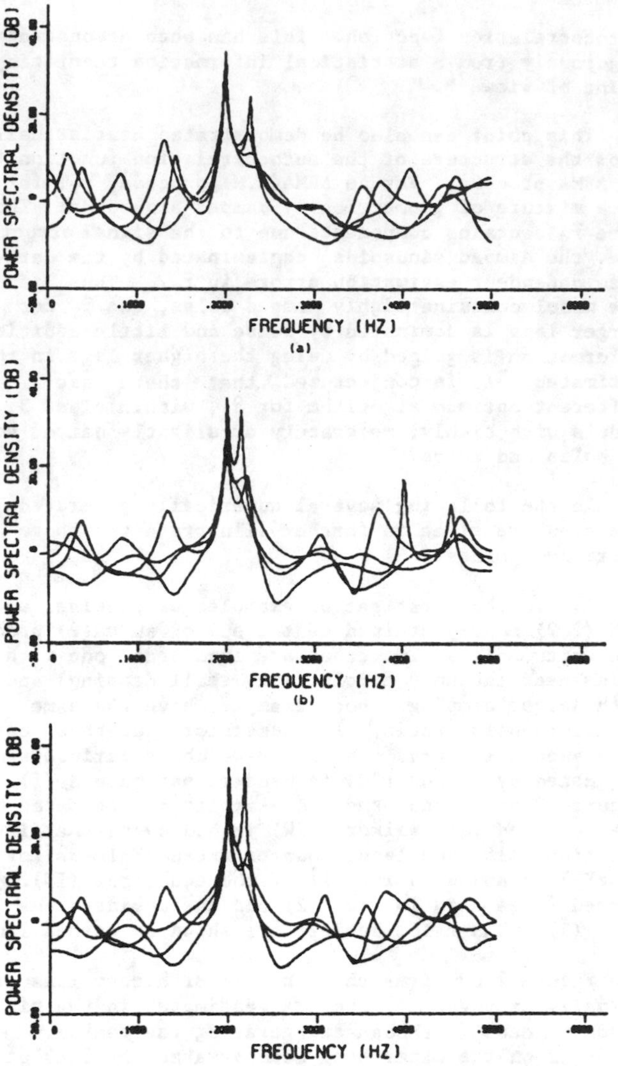

Fig. 5 Spectral estimates for $x_t = \sqrt{20}\cos(0.4\pi t) + \sqrt{2}\cos(0.426\pi t) + \omega_t$, ω_t AWGN unit variance, N=60, (a) LSEYW ARMA (15,15); (b) Cadzow's method ARMA (15,15); (c) LSYW AR(15).

poles and not on the fitting of an ARMA model to the data. Fig. 5(c) shows the spectral estimate based on the least-squares fit of all N YW equations for a pure AR model of order 15. The performance is clearly comparable to the ARMA estimates in Figs. 5(a) and (b).

In conclusion we point out that the examples shown in this paper were generated from models with rational spectra. There are signals which do not possess spectra that are approximated well with this class. In such cases, even the optimum ARMA estimator is powerless. A good example is a relatively wideband Gaussian spectrum. The techniques discussed here generally produce estimates that are largely in error, unless the correct combination of L and M, if any, are found. Therefore one should not use these methods blindly in every given spectral estimation application.

Appendix

<u>U</u> is an implicit non-linear function of both parameters <u>A</u> and <u>B</u>. In the following the elements of the Jacobians are calculated recursively.

Denote the real and the imaginary components of the t-th sample of the residual by $U_r(t)$ and $U_i(t)$, t=1,...,N. From (11), $U_r(t)$ and $U_i(t)$ are given by

$$U_r(t) = x_r(t) - \sum_{i=1}^{L} a_r(i)x_r(t-i) + \sum_{i=1}^{L} a_i(i)x_i(t-i) - \sum_{i=1}^{M} b_r(i)U_r(t-i) + \sum_{i=1}^{M} b_i(i)U_i(t-i) \quad \text{(A-1)}$$

$$U_i(t) = x_i(t) - \sum_{i=1}^{L} a_i x_r(t-i) - \sum_{i-1}^{L} a_r(i)x_i(t-i) - \sum_{i=1}^{M} b_r(i)U_i(t-i) - \sum_{i=1}^{M} b_i(i)U_r(t-i) \quad \text{(A-2)}$$

where for convenience, the notations a(i), b(i), x(i), etc. are used for the previously defined subscripted quantities a_i, b_i, x_t, etc.

The elements of the Jacobians, that is the partial derivatives of $U_r(t)$ and $U_i(t)$ with respect to the real and imaginary components of the parameters, are given by:

$$\frac{\partial U_r(t)}{\partial a_r(k)} = -x_r(t-k) - \sum_{i=1}^{M} b_r(i) \frac{\partial U_t(t-i)}{\partial a_r(k)} + \sum_{i=1}^{M} b_i(i) \frac{\partial U_i(t-i)}{\partial a_r(k)} \quad \text{(A-3)}$$

$$\frac{\partial U_r(t)}{\partial a_i(k)} = x_i(t-k) - \sum_{i=1}^{M} b_r(i) \frac{\partial U_r(t-i)}{\partial a_i(k)} + \sum_{i=1}^{M} b_i(i) \frac{\partial U_r(t-i)}{\partial a_i(k)} \quad \text{(A-4)}$$

$$\frac{\partial U_r(t)}{\partial b_r(k)} = -U_r(t-k) - \sum_{i \neq k}^{M} b_r(i) \frac{\partial U_r(t-i)}{\partial b_r(k)} + \sum_{i=1}^{M} b_i(i) \frac{\partial U_i(t-i)}{\partial b_r(k)} \quad \text{(A-5)}$$

$$\frac{\partial U_r(t)}{\partial b_i(k)} = U_i(t-k) - \sum_{i=1}^{M} b_r(i) \frac{\partial U_r(t-i)}{\partial b_i(k)} + \sum_{i \neq k}^{M} b_i(i) \frac{\partial U_i(t-i)}{\partial b_i(k)} \quad \text{(A-6)}$$

Similarly,

$$\frac{\partial U_i(t)}{\partial a_r(k)} = -x_i(t-k) - \sum_{i=1}^{M} b_r(i) \frac{\partial U_i(t-i)}{\partial a_r(k)} - \sum_{i=1}^{M} b_i(i) \frac{\partial U_r(t-i)}{\partial a_r(k)} \quad \text{(A-7)}$$

$$\frac{\partial U_i(t)}{\partial a_i(k)} = -x_r(t-k) - \sum_{i=1}^{M} b_r(i) \frac{\partial U_i(t-i)}{\partial a_i(k)} - \sum_{i=1}^{M} b_i(i) \frac{\partial U_r(t-i)}{\partial a_i(k)} \quad \text{(A-8)}$$

$$\frac{\partial U_i(t)}{\partial b_r(k)} = -U_i(t-k) - \sum_{i \neq k}^{M} b_r(i) \frac{\partial U_r(t-i)}{\partial b_r(k)} - \sum_{i=1}^{M} b_i(i) \frac{\partial U_r(t-i)}{\partial b_r(k)} \quad \text{(A-9)}$$

$$\frac{\partial U_i(t)}{\partial b_i(k)} = -U_r(t-k) - \sum_{i \neq k}^{M} b_i(i) \frac{\partial U_r(t-i)}{\partial b_i(k)} - \sum_{i=1}^{M} b_r(i) \frac{\partial U_i(t-i)}{\partial b_i(k)} \quad \text{(A-10)}$$

In the above derivations the back-forcasted values of x(t) are assumed fixed. It can be seen that (A-3) - (A-10) are a set of coupled recursive equations for $\frac{\partial U_r}{\partial p_k}$ and $\frac{\partial U_i}{\partial p_k}$ and can be calculated as such. Therefore the steps in the calculation of these partial derivatives may be summarized as follows:

1. Use the current values of the parameters, U(t) and the back-forcasted $\tilde{x}(t)$.

2. Calculate the partial derivatives recursively, i.e. $\{\frac{\partial U_r(1)}{\partial a_r(k)}, \frac{\partial U_r(1)}{\partial a_i(k)}, \frac{\partial U_r(1)}{\partial b_r(k)}, \frac{\partial U_r(1)}{\partial b_i(k)}\}$ and the same for $U_i(1)$ first. This implies calculating the columns of $\underline{J}_{U_i}$ and $\underline{J}_{U_i}$ one at a time starting with column one.

3. Based on values of $\frac{\partial U(t-i)}{\partial p_k}$, i=k,...,t-1 find $\frac{\partial U(t)}{\partial p_k}$.

References

1. G. E. P. Box and G. M. Jenkins, Time Series, Forecasting and Control, San Francisco; Holden-Day, 1970.

2. M. Kaveh, "High resolution spectral estimation for noisy signals," IEEE Trans. Acoust., Speech, Signal Processing, Vol. ASSP-27, pp. 286-287, June 1979.

3. P. Newbold, "The exact likelihood function for a mixed autoregressive-moving average process," Biometrika, Vol. 61, pp. 423-426, March 1974.

4. M. Arato, "On the sufficient statistics for stationary Gaussian random processes," Theo. Prob. Appl., Vol. 6, No. 2, pp. 199-201, 1961.

5. S. A. Tretter and K. Steiglitz, "Power spectrum identification in terms of rational models," IEEE Trans. Automat. Contr., pp. 185-188, April 1967.

6. E. J. Hannan, "The estimation of mixed moving average autoregressive systems," Biometrika, 56, 3, pp. 579-593, 1969.

7. H. Akaike, "Maximum likelihood identification of Gaussian autoregressive moving-average models," Biometrika 60, 2, pp. 255-265, 1973.

8. I. S. Konvalinka and M. R. Matausek, "Simultaneous estimation of poles and zeros in speech analysis and ITIF-iterative inverse filtering algorithm," IEEE Trans. Acoust., Speech, Signal Processing, Vol. ASSP-27, No. 5, pp. 485-492, October 1979.

9. P. Shaman, "On the inverse of the covariance matrix for an autoregressive-moving average process," Biometrika, Vol. 60, pp. 193-196, 1973.

10. M. Ooe, "An optimal complex ARMA model of the Chandler Wobble," Geophys. J. R. Astr. Soc., Vol. 53, pp. 445-457, 1978.

11. M. Kaveh, Internal Memos, M.I.T. Lincoln Laboratory, 1979-80.

12. J. F. Kinkel, J. Perl, L. L. Scharf and A. R. Stubberud, "A note on covariance-invariant digital filter design and autoregressive moving average spectrum analysis," IEEE Trans. Acoust., Speech, Signal Processing, Vol. ASSP-27, April 1979.

13. A. M. Walker, "Large-sample estimation of parameters for autoregressive processes with moving-average residuals," Biometrika, 49, pp. 117-131, 1962.

14. T. C. Hsia and D. A. Landgrebe, "On a method for estimating power spectra," IEEE Trans. Instrumentation and Measurement, Vol. IM-16, No. 3, pp. 255-257, September 1967.

15. D. Graupe, D. J. Krause, and J. B. Moore, "Identification of autogressive moving-average parameters of time series," IEEE Trans. Automat. Contr., pp. 104-107, February 1975.

16. H. Sakai and M. Arase, "Recursive parameter estimation of an autoregressive process disturbed by white noise," Int. J. Control, Vol. 30, No. 6, pp. 949-966, 1979.

17. E. H. Satorius and S. T. Alexander, "High resolution spectral analysis of sinusoids in correlated noise," 1978 IEEE ICASSP, Tulsa, OK, April 10-12, 1978.

18. S. M. Kay, "A new ARMA spectral estimator," IEEE Trans. Acoust., Speech, Signal Processing, Vol. ASSP-28, pp. 585-588, October 1980.

19. S. Bruzzone and M. Kaveh, "On some suboptimum ARMA spectral estimators," IEEE Trans. Acoust., Speech, Signal Processing, Vol. ASSP-28, pp. 753-755, December 1980.

20. J. A. Cadzow, "ARMA spectral estimation, a model equation error procedure," Proceedings of the 1980 Int. Conf. Acoust., Speech, Signal Processing, Denver, CO, April 14-16, 1980.

21. J. A. Cadzow and R. Moses, "A superresolution method of ARMA spectral estimation," Proceedings of the 1981 Int. Conf. Acoust., Sppech, Signal Processing, Atlanta, GA, March 30-April 1, 1981.

22. J. Haddad, "Spectral estimation of noisy signals using a parallel resonator model and state-variable techniques," M.S. paper, University of Minnesota, Minneapolis, MN, 1980.

23. M. Kaveh, "Resolution enhancement in the autoregressive spectral estimation of noisy signals," presented at the International Symposium on Information Theory, Ithaca, NY, October 10-14, 1977.

24. S. Bruzzone, "Statistical analysis of some suboptimum ARMA spectral estimators," Ph.D. Dissertation, University of Minnesota, Minneapolis, MN, September 1981.

Minimum Cross-Entropy Spectral Analysis

JOHN E. SHORE, SENIOR MEMBER, IEEE

Abstract—**The principle of minimum cross-entropy (minimum directed divergence, minimum discrimination information, minimum relative entropy) is summarized, discussed, and applied to the classical problem of estimating power spectra given values of the autocorrelation function. This new method differs from previous methods in its explicit inclusion of a prior estimate of the power spectrum, and it reduces to maximum entropy spectral analysis as a special case. The prior estimate can be viewed as a means of shaping the spectral estimator. Cross-entropy minimization yields a family of shaped spectral estimators consistent with known autocorrelations. Results are derived in two equivalent ways: once by minimizing the cross-entropy of underlying probability densities, and once by arguments concerning the cross-entropy between the input and output of linear filters. Several example minimum cross-entropy spectra are included.**

I. INTRODUCTION

THIS paper presents an information-theoretic method of estimating a power spectrum based on a prior estimate and on new information in the form of autocorrelation function values. The new method differs from previous methods in its explicit inclusion of a prior estimate of the power spectrum. This new approach, which avoids certain technical problems of maximum entropy spectral analysis (MESA) [1]-[5], reduces to MESA as a special case.

Our approach exploits properties of the cross-entropy between any two probability density functions q, p, defined by

$$H(q,p) = \int dx\, q(x) \log (q(x)/p(x)). \tag{1}$$

In particular, the approach is based on cross-entropy having unique properties as a measure of information dissimilarity [6]-[8] and on cross-entropy minimization having unique properties as an inference method [9], [10]. There have been many applications of cross-entropy minimization (for a list of references, see [9]). Recently, the combined properties of cross-entropy and cross-entropy minimization have been shown to be useful in speech processing. In particular, one formulation of the standard linear prediction coding (LPC) equations is based on minimizing a distortion measure introduced by Itakura and Saito [11]. In [12], it was shown that the Itakura-Saito distortion measure is a special case of asymptotic cross-entropy, and in [13] it was shown that the standard LPC equations can be obtained directly by cross-entropy minimization. The newly developed technique of speech coding by vector quantization [14], [15] was also derived in [13] directly by cross-entropy minimization. Furthermore, the derivation of vector quantization in [15] was carried out by exploiting properties of the Itakura-Saito distortion measure—for example, a triangle equality—that turn out to be special cases of properties of cross-entropy minimization [10], [16]. These properties have since been used in refining Kullback's classification method [6, p. 83], yielding a method that is optimal in a precise information-theoretic sense and computationally efficient [17].

Manuscript received May 31, 1970; revised May 19, 1980.

The author is with the Naval Research Laboratory, Washington, DC 20375.

Because LPC analysis fits an all-pole spectral model to the data, the work in [13] is an example of the use of cross-entropy minimization to perform spectral estimation. The results presented in this paper differ in two ways. First, they incorporate a prior estimate of the power spectrum that has an arbitrary form, rather than the all-pole (autoregressive) form assumed in [13]. Second, the derivation is carried out in the frequency domain rather than in the time domain as in [13].

Section II contains a brief discussion of spectral analysis from the point of view of extrapolating the autocorrelation function, and it summarizes the motivation and mathematics of cross-entropy minimization. In Section III, we derive the minimum cross-entropy spectral estimator. In Section IV, we discuss the connections with MESA and LPC. Some numerical examples are contained in Section V, and a concluding discussion follows in Section VI. For a more detailed discussion of minimum cross-entropy spectral analysis, see [18].

II. BACKGROUND

A. Spectral Analysis and Autocorrelation Function Extrapolation

Because the power spectrum $A(f)$ of a band-limited, stationary process is related to its autocorrelation function $R(t)$ by a Fourier transform, and because it is relatively easy to measure the autocorrelation function, many spectral analysis techniques start with values of the autocorrelation function. The classical approach uses spectral window functions [19]. In this approach one takes the Fourier transform of the product $R(t)W(t)$, where $R(t)$ is the measured autocorrelation function in the range $|t| < T$, and where $W(t)$ is a known window function with $W(t) = 0$ for $|t| > T$. One then estimates the unknown power spectrum $A(f)$ by exploiting the convolution theorem, which states that the Fourier transform of the product of two time domain functions is equal to the convolution in the frequency domain of their Fourier transforms. Although mathematically elegant, the classical procedure distorts the values of $R(t)$ in the known region $|t| < T$ and assumes that $R(t) = 0$ in the unknown region $|t| > T$, despite the fact that $R(t)$ cannot in general be zero everywhere in this region. An alternative approach is to extend $R(t)$ so as to take on reasonable values in the unknown region $|t| > T$ and to estimate

Reprinted from *IEEE Trans. Acoust., Speech, Signal Processing*, vol. ASSP-29, pp. 230-237, Apr. 1981.

$A(f)$ by taking the Fourier transform of the resulting extended function. As a general approach this seems more reasonable than the classical approach, but it leaves open the question of how to extend the measured portion of $R(t)$.

In proposing MESA, Burg suggested that $R(t)$ be extended in a manner that maximizes the entropy of the underlying stationary process [1], [2]. Specifically, he proposed that the power spectrum $A(f)$ be estimated by maximizing

$$\int_0^W df \log(A(f)) \tag{2}$$

subject to the known constraints

$$R(t_k) = \int_{-W}^{W} df A(f) \exp(2\pi i t_k f) \tag{3}$$

where W is the bandwidth, and where $R(t_k)$, $k = 1, 2, \cdots, m$, are known values of the autocorrelation function. This approach is motivated by expression (2) being the entropy gain in a stochastic process that is passed through a linear filter with characteristic function $Y(f)$, where $S(f) = |Y(f)|^2$ (see [4, pp. 412–414], [20, pp. 93–95], [21, p. 243]). If the input process is white noise, then the output process has spectral power density $A(f)$. This suggests that the process entropy can be maximized by maximizing the entropy gain of the filter that produces the process. Thus, (2) is maximized subject to the constraints (3). Such derivations of MESA suffer from several disadvantages: for continuous densities, entropy is mathematically ill-behaved [22, p. 31–32]; in comparison to cross-entropy, solutions to entropy extremum problems exist less generally and the extremum is less generally defined. Furthermore, there are logical problems in applying entropy maximization in the continuous case [23]. Finally, information theoretic justifications based on the principal of maximum entropy [24], [25] apply indirectly in terms of filtering rather than directly in terms of underlying probability densities.

Minimum cross-entropy spectral estimation can be viewed as an alternative means of extending $R(t)$–one that extends $R(t)$ in a manner that accounts for a prior estimate. Instead of maximizing the entropy of the underlying stationary process as in MESA, one minimizes the cross-entropy between the underlying process and the prior estimate. If the prior estimate of the power spectrum is flat, the results are equivalent to MESA. The derivation of minimum cross-entropy spectral analysis is carried out in terms of underlying probability densities and also avoids the other problems of the MESA derivations.

B. Cross-Entropy Minimization

Let x denote a single state of some system that has a set of D possible system states and a probability density $q^\dagger(x)$ of states. Let $\mathfrak{D}$ be the set of all probability densities q on D such that $q(x) \geqslant 0$ for $x \in D$ and

$$\int_D dx\, q(x) = 1. \tag{4}$$

We assume that the existence of $q^\dagger \in \mathfrak{D}$ is known but that $q^\dagger$ itself is unknown. The density $q^\dagger$ is sometimes known as a "true" density.

Suppose $p \in \mathfrak{D}$ is a *prior* density that is our current estimate of $q^\dagger$, and suppose we gain new information about $q^\dagger$ in the form of a set of expected values

$$\int_D dx\, q^\dagger(x) g_r(x) = \bar{g}_r \tag{5}$$

for a known set of functions $g_r(x)$ and numbers $\bar{g}_r$, $r = 0, \cdots, M$. Now, because the *constraints* (5) do not determine $q^\dagger$ completely, they are satisfied not only by $q^\dagger$ but by some subset of densities $\mathfrak{I} \subset \mathfrak{D}$. Which single density should we choose from this subset to be our new estimate of $q^\dagger$, and how should we use the prior p and the new information (5) in making this choice?

The principle of minimum cross-entropy provides a general solution to this inference problem [9]. The principle states that, of all the densities that satisfy the constraints, you should choose the posterior q with the least cross-entropy (1) with respect to the prior p. That is, the *posterior* density q satisfies

$$H(q, p) = \min_{q' \in \mathfrak{I}} H(q', p)$$

where $\mathfrak{I} \subset \mathfrak{D}$ comprises all of the densities that satisfy the constraints (5). As a general method of statistical inference, cross-entropy minimization was first introduced by Kullback [6] and has been advocated in various forms by others [8], [23], [26], [27]. The name cross-entropy is due to Good [26]. Other names include expected weight of evidence [28, p. 72], directed divergence [6, p. 7], discrimination information [6, p. 37], and the entropy of one distribution relative to another [7, p. 19]. The principle of maximum entropy [24], [25] is a special case of cross-entropy minimization under appropriate conditions [9], [13].

Given a positive prior probability density p, if there exists a posterior that minimizes the cross-entropy (1) and satisfies the constraints (4)-(5), then it has the form

$$q(x) = p(x) \exp\left(-\lambda - \sum_{k=0}^{M} \beta_k g_k(x)\right), \tag{6}$$

with the possible exception of a set of points on which the constraints imply that q vanishes [6, p. 38], [16]. In (6), λ and β_k are Lagrangian multipliers whose values are determined by the constraints (4) and (5), respectively. Conversely, if one can find values for β_k and λ in (6) such that the constraints are satisfied, then the solution exists and is given by (6) [16]. Conditions for the existence of solutions are discussed by Csiszár [16], who treats the general problem of cross-entropy minimization in a more general, measure-theoretic context. Usually, it is difficult or impossible to solve for the β_k explicitly in order to obtain a closed-form solution expressed directly in terms of the known expected values $\bar{g}_k$ rather than in terms of the Lagrangian multipliers. Computational methods for finding approximate solutions are, however, available [10], [29].

In what sense does cross-entropy minimization yield the best estimate of $q^\dagger$? To answer this question, it is useful to view cross-entropy minimization as one implementation of an abstract information operator $\circ$ that takes two arguments—a prior and new information—and yields a posterior. Thus, we write the posterior q as $q = p \circ I$, where I stands for the constraints (4)-(5). What would happen if other functionals besides cross-entropy (1) were used in implementing the information operator $\circ$? Recent work has shown that, if the operator $\circ$ is required to satisfy certain axioms of consistent inference, and if $\circ$ is implemented by means of functional minimization, then the principle of minimum cross-entropy follows necessarily [9]. Informally, the axioms state that different ways of taking the information I into account—for example, in different coordinate systems—should lead to consistent results. In terms of these axioms, the principle of cross-entropy minimization is correct in the following sense: given a prior probability density and new information in the form of constraints on expected values, there is only one posterior density satisfying these constraints that can be chosen by functional minimization in a manner that satisfies the axioms; this unique posterior can be obtained by minimizing cross-entropy.

An additional interpretation of the sense in which $q = p \circ I$ is the best estimate of $q^\dagger$ rests on cross-entropy's well-known [6] and unique [8] properties as an information measure. Informally speaking, $H[q,p]$ is a measure of the "information divergence" or "information disimilarity" between q and p. In these terms, one can interpret the principle of minimum cross-entropy as follows. Since $q = p \circ I$ minimizes $H[q,p]$, the posterior hypothesis for $q^\dagger$ is as close as possible in an information-measure sense to the prior hypothesis while at the same time satisfying the new constraints I. Furthermore, in the context of cross-entropy minimization, cross-entropy satisfies the triangle equality [10], [16]

$$H[q^\dagger, p] = H[q^\dagger, p \circ I] + H[p \circ I, p].$$

Thus, the minimum cross-entropy posterior estimate of $q^\dagger$ is not only logically consistent, but also closer to $q^\dagger$, in the cross-entropy sense, than is the prior p. Moreover, the difference $H[q^\dagger,p] - H[q^\dagger, p \circ I]$ is exactly the cross-entropy $H[p \circ I, p]$ between the posterior and the prior. Hence, $H[p \circ I, p]$ can be interpreted as the amount of information provided by I that is not inherent in p. Stated differently, $H[p \circ I, p]$ is the amount of additional distortion introduced if p is used instead of $p \circ I$. Since, for any density r there exist constraints I_r such that $r = p \circ I_r$ for any prior p, $H[r,p]$ is in general the amount of information needed to determine r when given p, or the amount of additional distortion introduced if p is used instead of r [10]. The foregoing triangle equality is a generalization of a similar property that holds for the Itakura-Saito distortion measure [15].

Yet another justification for using cross-entropy as a distortion measure in the context of cross-entropy minimization is provided by the "expectation matching" property [10], which states that, for an arbitrary density q^* and a density q of the general form (6), $H[q^*, q]$ is smallest when the expectations of q match those of q^*. In particular, it follows that $q = p \circ I$ is not only the density that minimizes $H[q,p]$, as already discussed, but also is the density of the form (6) that minimizes $H[q^\dagger, q]$. Hence, $p \circ I$ is not only closer to $q^\dagger$ than is p, as already discussed, but it is the closest possible density of the form (6). The expectation matching property is a generalization of a property of orthogonal polynomials [30, p. 12] that is called the "correlation matching property" in the speech literature [12], [31, ch. 2].

III. Minimum Cross-Entropy Spectral Analysis

We assume that the time-domain signal is a stationary, random process $g(t)$. Since any stationary random process can be obtained as the limit of a sequence of processes with discrete spectra [32, p. 36], we consider time-domain signals of the form

$$s(t) = \sum_{k=1}^{N} a_k \cos(2\pi f_k t) + b_k \sin(2\pi f_k t) \tag{7}$$

where the a_k and b_k are random variables and where the f_k are nonzero frequencies that need not be uniformly spaced. By a suitable choice of frequencies and random variables, the mean square error $E(|g(t) - s(t)|^2)$ can be made arbitrarily small. Since the power at each frequency is given by the variables x_k,

$$x_k = \tfrac{1}{2}(a_k^2 + b_k^2),$$

we will describe the random process (7) in terms of a joint probability density $q(x)$, where we write x for $x_1, x_2, \cdots, x_N$. Instead of constantly referring to $q(x)$ as the spectral power probability density of a stationary, random process, we will refer informally to $q(x)$ as a "process."

Let S_k, $k = 1, \cdots, N$, comprise a prior estimate of the power spectrum of an unknown process $q^\dagger$. In this case we assume a prior estimate $p(x)$ of $q^\dagger(x)$ to have the exponential form

$$p(x) = \prod_{k=1}^{N} (1/S_k) \exp[-x_k/S_k]. \tag{8}$$

Note that the expected spectral power at frequency f_k is just the prior estimate S_k, since $\int dx\, x_k p(x) = S_k$. Expressed in terms of the variables a_k and b_k, (8) becomes

$$p(a, b) = \prod_{k=1}^{N} (1/2\pi S_k) \exp[-(a_k + b_k)^2/2S_k]$$

which shows that the assumed prior (8) is equivalent to a Gaussian assumption for each of the amplitudes a_k, b_k. Such an assumption is generally considered to be reasonable and it is also convenient mathematically. An additional justification is that (8) itself is approximately a minimum cross-entropy posterior given a uniform prior density and expected value constraints $\int dx\, x_k q^\dagger(x) = S_k$. A uniform prior estimate of $q^\dagger$ in the absence of spectral shape information is reasonable since in any real situation there will be a physical limit on the magnitude of the x_k, and one may therefore assume that the domain of x is large but bounded. For more discussion about the assumed form (8), see [18]. We shall refer to signals of the form (8) as Gaussian.

Now, let T_k be the actual spectral powers of $q^\dagger(x)$, i.e.,

$$T_k = \int dx\, x_k q^\dagger(x) \tag{9}$$

and suppose that we obtain information about $q^\dagger$ in the form of $M + 1$ values of the autocorrelation function $R(t_r)$,

$$R_r = R(t_r) = \sum_{k=-N}^{+N} T_k \exp(2\pi i t_r f_k)$$

where $r = 0, \cdots, M$ and $t_0 = 0$, which we prefer to express in the noncomplex form

$$R_r = \sum_{k=1}^{N} T_k c_{rk}$$

where $c_{rk} = 2\cos(2\pi t_r f_k)$. Using (9), we write this as

$$R_r = \int dx \left(\sum_{k=1}^{N} x_k c_{rk} \right) q^\dagger(x) \tag{10}$$

which has the form of expected value constraints (5).

Now given the prior (8) and the new information (10), the minimum cross-entropy posterior estimate of $q^\dagger$ is given by (6),

$$q(x) = p(x) \exp\left(-\lambda - \sum_{r=0}^{M} \beta_r \sum_{k=1}^{N} x_k c_{rk} \right) \tag{11}$$

where the β_k are Lagrangian multipliers determined by the constraints (10) and λ is determined by the normalization constraint (4). By substituting (8) into (11) and eliminating λ by means of (4), it is easy to show that (11) becomes

$$q(x) = \prod_{k=1}^{N} \left(u_k + \frac{1}{S_k} \right) \exp\left(-\left(u_k + \frac{1}{S_k} \right) x_k \right) \tag{12}$$

where we have defined

$$u_k = \sum_{r=0}^{M} \beta_r c_{rk}.$$

Hence, the posterior is also a Gaussian process. The posterior estimate of the power spectrum is just

$$T_k = \int dx\, x_k q(x)$$

$$= \frac{S_k}{1 + S_k u_k} \tag{13}$$

or

$$T_k = \frac{1}{\frac{1}{S_k} + \sum_r \beta_r c_{rk}} \tag{14}$$

where the multipliers β_r in u_k are chosen so that the T_k satisfy the autocorrelation constraints

$$R_r = \sum_{k=1}^{N} T_k c_{rk}. \tag{15}$$

The result (13) has an interesting interpretation in terms of linear filtering: if a process with the prior probability density $p(x)$ is passed through a linear filter with characteristic function $Y(f)$, then the magnitude of each power spectrum component is changed by the factor $A_k = |Y(f_k)|^2$ from S_k to $S_k A_k$. Hence, (13) shows that the posterior estimate $q(x)$ can be produced from the prior estimate $p(x)$ by passing the prior estimate through a linear filter with $A_k = (1 + S_k u_k)^{-1}$. It follows that choosing the minimum cross-entropy posterior estimate (12) is equivalent to designing a linear filter whose output $q(x)$ satisfies the constraints (15) and has the smallest cross-entropy with respect to the fixed input $p(x)$. Since the posterior (12) has the form $q(x) = \pi_k (1/T_k) \exp[-1/T_k]$ and the prior has the form (8), it is easy to show by direct calculation that the normalized cross-entropy $H^*(q, p) = H[q, p]/N$ is

$$H^*(q, p) = \frac{1}{N} \sum_{k=1}^{N} \left(\frac{T_k}{S_k} - \log \frac{T_k}{S_k} - 1 \right) \tag{16}$$

which may be recognized as a discrete approximation to the Itakuro-Saito distortion measure [11], [12]. Hence, for a Gaussian prior (8), it follows that choosing the minimum cross-entropy posterior estimate (12) is equivalent to designing a linear filter whose output $q(x)$ satisfies the constraints (14) and has the smallest Itakuru-Saito distortion with respect to the fixed Gaussian input $p(x)$.

This filtering result can be seen in another way as follows: if $q(x)$ is the probability density of the signal that results from passing $p(x)$ through the filter, then the input $p(x)$ and the output $q(x)$ are related by

$$q(x) = \frac{p(x_1/A_1, x_2/A_2, \cdots, x_N/A_N)}{A_1 A_2 \cdots A_N}$$

where $A_k = |Y(f_k)|^2$, since the magnitude x_k of each power spectrum component is changed by the factor A_k. The filter has the effect of a linear coordinate transformation. The cross-entropy between the input and the output is

$$H(q, p) = \int dx\, q(x) \log\left(\frac{q(x)}{p(x)} \right)$$

$$= \int dy\, p(y_1, \cdots, y_N) \log\left(\frac{p(y_1, \cdots, y_N)}{p(y_1 A_1, \cdots, y_N A_N)} \right) - \sum_{k=1}^{N} \log(A_k) \tag{17}$$

where $y_k = x_k/S_k$. This is a general result for any input signal $p(x)$. When the input signal is Gaussian with the form (8), the cross-entropy (17) becomes

$$H(q, p) = -\sum_k \log(A_k) - \int dy\, p(y) \sum_k (y_k - y_k A_k)/S_k$$

or

$$H(q,p) = \sum_k (A_k - \log(A_k) - 1) \tag{18}$$

since $\int dy\, y_k p(y) = S_k$. Notice that this result is independent of the particular S_k values. Suppose the A_k are chosen so as to minimize (18) subject to the constraints (15), which we rewrite as

$$R_r = \sum_{k=1}^{N} S_k A_k c_{rk}.$$

The result is $A_k = (1 + S_k u_k)^{-1}$, so that $T_k = S_k A_k$ yields (14) again. Notice also that (18) is just another form of (16).

IV. Relation to Maximum Entropy Spectral Analysis and Linear Prediction

As mentioned in Section III, one way to reflect the absence of prior spectral shape information is to use a uniform prior density $p(x)$. When the filter input probability density $p(x)$ is uniform, the first term in (17) is zero, and the cross-entropy between the filter input and output is

$$H(q,p) = -\sum_k \log(S_k). \tag{19}$$

Except for sign, this is the discrete form of the expression (2); in the limit, minimizing (19) is equivalent to maximizing (2). This result reflects the general property that cross-entropy minimization reduces to entropy maximization when the prior is uniform.

Another way to reflect the absence of prior spectral shape information is to use a nonuniform prior of the Gaussian form (8), but one that is "white" in the sense that the prior estimate of the power sepctrum is flat: $S_k = S$ for $k = 1, \cdots, N$. In this case (14) applies. Since we have assumed a zero lag value $t_0 = 0$, the posterior estimate (14) becomes

$$T_k = \frac{1}{\sum_{r=0}^{M} \beta'_r c_{rk}} \tag{20}$$

where the β'_r are modified Lagrangian multipliers satisfying

$$\beta'_0 = \beta_0 + \frac{1}{S}$$

$$\beta'_r = \beta_r \quad (r = 1, \cdots, M).$$

If we assume equally spaced autocorrelation lags $t_r = r\Delta t$, we can rewrite (20) in the form

$$T(f) = \frac{1}{\sum_{r=-M}^{+M} \beta'_r z^{-r}} \tag{21}$$

where $z = \exp(-2\pi i f \Delta t)$ and where we have dropped the subscript k. Now, the Fejér–Riesz theorem states [33, p. 231] that (21) can also be written in the form

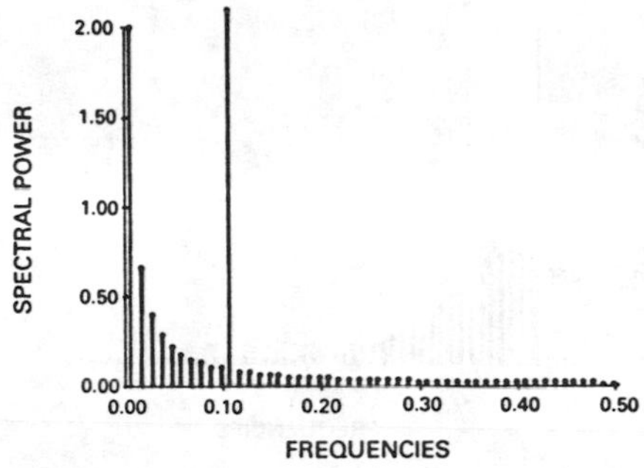

Fig. 1. Original spectrum for first example.

$$T(f) = \frac{\sigma^2}{\left|\sum_{r=0}^{M} a_r z^{-r}\right|^2} \tag{22}$$

where $a_0 = 1$. This may be recognized as the usual maximum entropy estimate [2], and is also the familiar form of standard LPC analysis [31]: the a_r are the inverse filter sample coefficients and σ^2 is the gain. By equating (21) and (22) one can show that the coefficients a_r are related to the modified Lagrangian multipliers β'_k by

$$\beta'_r = \frac{1}{\sigma^2} \sum_{k=0}^{M-r} a_{k+r} a_k^*.$$

V. Examples

Next, we consider three numerical examples in which conventional maximum-entropy spectral estimates are compared with minimum cross-entropy estimates that take into account prior information about the spectrum. In the first two examples, autocorrelations at a small number of equally spaced lags were obtained from the Fourier transform of an assumed "true" spectrum; then maximum-entropy and minimum cross-entropy spectra were computed from the autocorrelations. In the third example, the autocorrelations were estimated from samples of a noise-corrupted time-domain sequence.

The original spectrum in all the examples is the sum of a known "background" $G^{(b)}$ and an unknown "signal" $G^{(s)}$. In the first example, the background approximates $1/f$ noise, and the signal is a single sinusoid at a fixed frequency. Specifically, the background term is

$$G^{(b)}_{\pm k} = 0.01/f_k \qquad (k = 1, \cdots, 50)$$

for 50 equally spaced frequencies $f_k = (0.005, 0.015, \cdots, 0.495)$ between 0 and 0.5 (which is the Nyquist frequency: we take the spacing between autocorrelation lags to be unity). The signal term is

$$G^{(s)}_{\pm k} = \begin{cases} 2 & (f_k = 0.105) \\ 0 & \text{otherwise.} \end{cases}$$

The sum is shown in Fig. 1; for lags $t_r = 0, 1, 2, 3, 4, 5$, the corresponding autocorrelations R_r are $R_r = 15.7511, 11.6149, 7.8699, 4.5411, 2.0145, 1.1413$. The maximum entropy spectrum computed from these six autocorrelations is shown in Fig. 2. For the minimum cross-entropy calculation, the background term $G^{(b)}$ has been used as the prior spectral estimate;

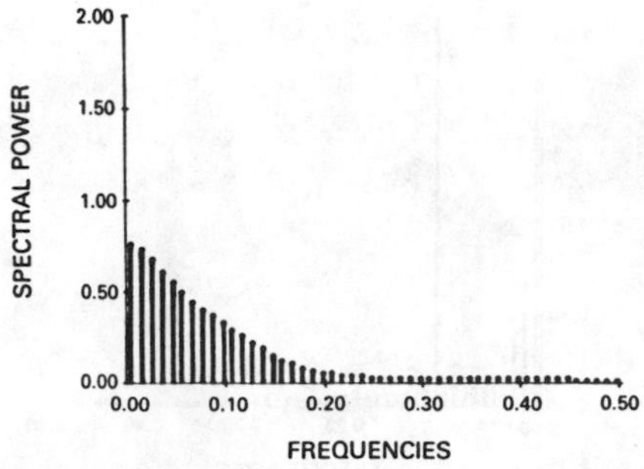

Fig. 2. Maximum entropy spectrum for first example.

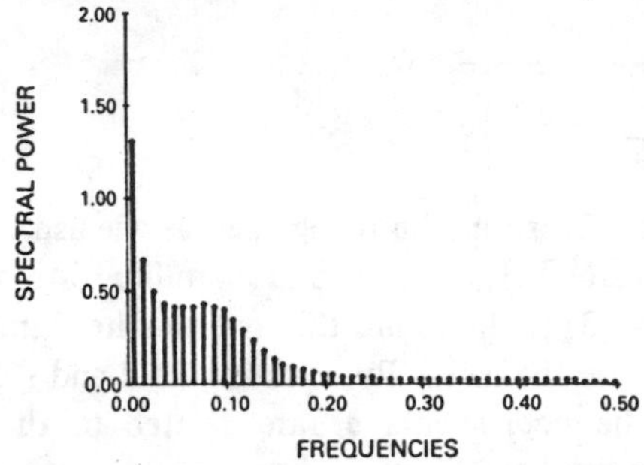

Fig. 3. Minimum cross-entropy spectrum for first example.

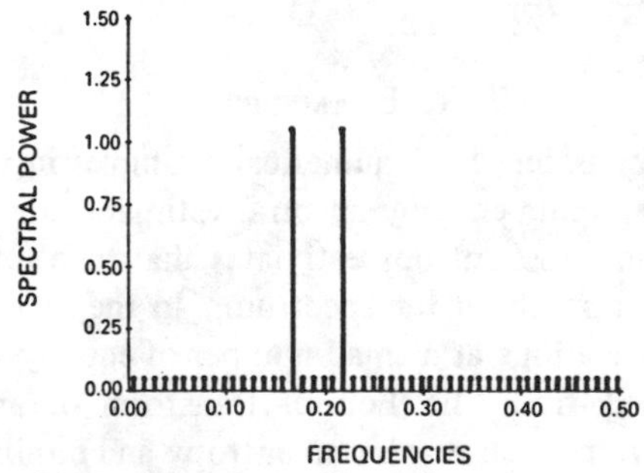

Fig. 4. Original spectrum for second and third examples.

the resulting posterior is shown in Fig. 3. As one might expect, the $1/f$ background is considerably better estimated in Fig. 3 than in Fig. 2. More important, however, is the clearly discernible peak in Fig. 3 corresponding to the sinusoidal signal at frequency 0.105; no such peak is evident in Fig. 2.

In the second example, spectral powers are shown at the same frequencies as for the first, and autocorrelations are computed for the same lags. The background consists of white noise plus a peak corresponding to a sinusoid at frequency 0.215:

$$G^{(b)}_{\pm k} = \begin{cases} 1.05 & (f_k = 0.215) \\ 0.05 & \text{otherwise.} \end{cases} \tag{23}$$

The signal term consists of a nearby, similar peak at frequency 0.165:

$$G^{(s)}_{\pm k} = \begin{cases} 1 & (f_k = 0.165) \\ 0 & \text{otherwise.} \end{cases}$$

The original spectrum is shown in Fig. 4, the autocorrelations are $R_r = 9.0000, 1.4544, -2.7732, -3.2248, 0.2032, 2.6900,$

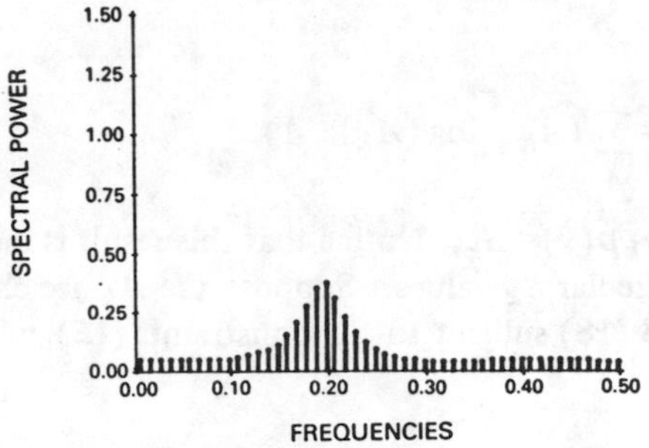

Fig. 5. Maximum entropy spectrum for second example.

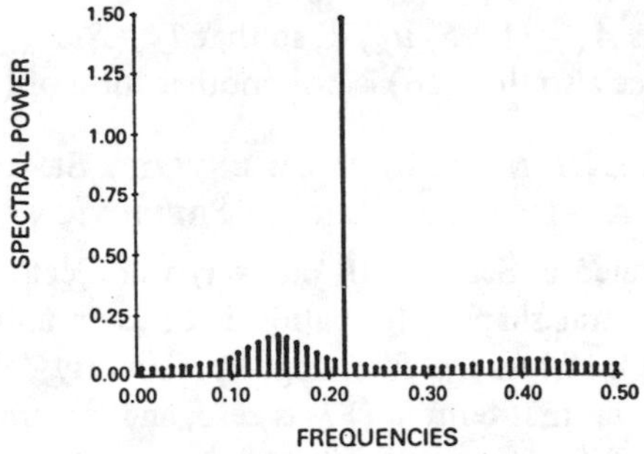

Fig. 6. Minimum cross-entropy spectrum for second example.

and the maximum-entropy spectrum is shown in Fig. 5. For the minimum cross-entropy calculation, the background term $G^{(b)}$ has again been taken as a prior spectral estimate. The posterior estimate is shown in Fig. 6. The information in the prior has permitted the resolution of the "expected" peak at frequency 0.215 from the "unexpected" peak at frequency 0.165. In the maximum-entropy estimate, by contrast, the two peaks are coalesced into a single peak at about the center frequency, 0.190.

The third example has the same "true" spectrum as the second example (Fig. 4). Instead of using exact autocorrelations, however, autocorrelations were estimated from noise-corrupted time-domain sequences. In particular, the following procedure was used.

1) Twenty-five samples were generated from the series

$$y_j = 2 \sin (r_1 + 2\pi f_1 t_j) + 2 \sin (r_2 + 2\pi f_2 t_j)$$

where $f_1 = 0.165$, $f_2 = 0.215$, $t_j = 0, 1, \cdots, 24$, and where the r_i are random initial phases. (The factor of two is to account for negative frequencies.)

2) Noise was added to each sample according to $z_j = y_j + e_j$, where the e_j are independent, zero mean, Gaussian random numbers with variance $5^{1/2}$. This variance accounts for the 0.05 level flat background in (23), including the contributions from negative frequencies.

3) Autocorrelations R_r, $r = 0, 1, \cdots, 5$, were estimated according to

$$R_r = \frac{1}{25} \sum_{j=1}^{25-r} z_{r+j} z_j.$$

(This is a biased estimate but it guarantees positive definiteness.)

4) Maximum entropy and minimum cross-entropy spectra were computed from the estimated autocorrelations. For the

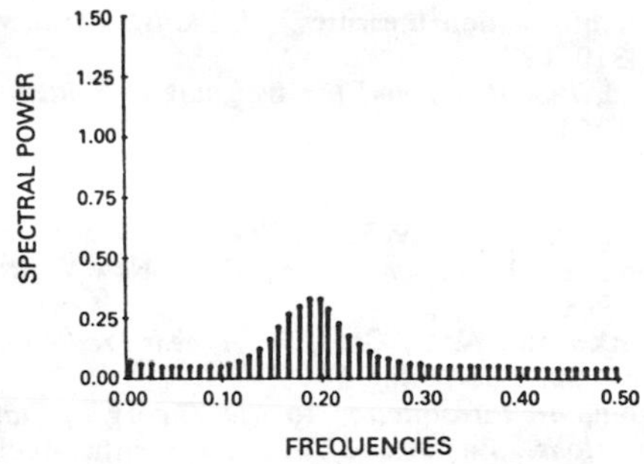

Fig. 7. Maximum entropy spectrum for third example (average of ten spectra).

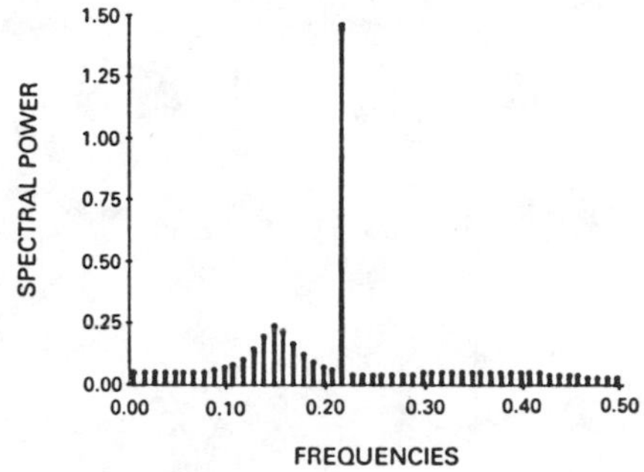

Fig. 8. Minimum cross-entropy spectrum for third example (average of ten spectra).

minimum cross-entropy calculation, the background (23) was used as the prior.

The foregoing procedure was repeated ten times. The average of the maximum entropy spectra is shown in Fig. 7 and the average of the minimum cross-entropy spectra is shown in Fig. 8. The results are similar to those in Figs. 6 and 7, which were obtained for exact autocorrelations.

VI. Discussion

Minimum cross-entropy spectral analysis is an information-theoretic method of estimating a power spectrum based on a prior estimate and on new information in the form of autocorrelation function samples. When the prior estimate of the power spectrum is flat, the results are the same as those for maximum entropy spectral analysis.

A minimum cross-entropy spectral estimate can be viewed as a means of extrapolating or extending the known autocorrelation values to values at arbitrary lags. That is, the discrete Fourier transform of (14) is equal to R_r at lags t_r, $r = 0, \cdots, M$. Since different priors result in different spectral estimates and therefore in different extensions of R_r, one can view the prior as a means of shaping the spectral estimate $\{T_k: k = 1, \cdots, N\}$ in a manner that is consistent with the known autocorrelation. A small value for S_k will depress the corresponding T_k and a large value for S_k will enhance T_k, but the set $\{T_k: k = 1, \cdots, N\}$ will always be consistent with $\{R_r: r = 0, \cdots, M\}$. Thus, cross-entropy minimization yields a family of shaped spectral estimators consistent with known autocorrelations. In this respect, our results are analogous and possibly equivalent to the flat echo estimator described recently by Youla [34].

In Section V we showed that the use of nonflat prior estimates can lead to better estimates than those of maximum entropy spectral analysis. This is hardly surprising, since the minimum cross-entropy estimate can exploit more information, but additional, more systematic studies are needed to establish the strengths and weaknesses of the new approach.

Acknowledgment

These results were obtained during joint work with R. W. Johnson. I thank him for many helpful discussions, for writing some of the computer programs used to obtain the examples, and for helping me to prepare parts of Section V. I also thank R. M. Gray for several helpful discussions and for his comments on a draft of Section III.

References

[1] J. P. Burg, "Maximum entropy spectral analysis," presented at the 37th Ann. Meeting Soc. Exploration Geophysicists, Oklahoma City, OK, 1967.

[2] —, "Maximum entropy spectral analysis," Ph.D. dissertation, Stanford Univ., Stanford, CA, 1975 (Univ. Microfilms No. 75-25, 499).

[3] R. T. Lacoss, "Data adaptive spectral analysis methods," *Geophys.*, vol. 36, pp. 661–675, 1971.

[4] D. E. Smylie *et al.*, "Analysis of irregularities in the earth's rotation," in *Methods in Computational Physics*, vol. 13. New York: Academic, 1973, pp. 391–431.

[5] T. J. Ulrych and T. N. Bishop, "Maximum entropy spectral analysis and autoregressive decomposition," *Rev. Geophys. Space Phys.*, vol. 43, no. 1, pp. 183–200, 1975.

[6] S. Kullback, *Information Theory and Statistics.* New York: Dover, 1969.

[7] M. S. Pinsker, *Information and Information Stability of Random Variables and Processes.* San Francisco, CA: Holden-Day, 1964.

[8] R. W. Johnson, "Axiomatic characterization of the directed divergences and their linear combinations," *IEEE Trans. Inform. Theory*, vol. IT-25, pp. 709–716, Nov. 1979.

[9] J. E. Shore and R. W. Johnson, "Axiomatic derivation of the principle of maximum entropy and the principle of minimum cross-entropy," *IEEE Trans. Inform. Theory*, vol. IT-26, pp. 26–37, Jan. 1980.

[10] —, "Properties of cross-entropy minimization," *IEEE Trans. Inform. Theory*, to be published.

[11] F. Itakura and S. Saito, "Analysis synthesis telephone based upon the maximum lilkihood method," *Rep. 6th Int. Cong. Acoust.*, Y. Yonasi, Ed., Tokyo, 1968.

[12] R. M. Gray, A. Buzo, A. H. Gray, Jr., and Y. Matsuyama, "Distortion measures for speech processing," *IEEE Trans. Acoust., Speech, Signal Processing.*, vol. ASSP-28, pp. 367–376, Aug. 1980.

[13] R. M. Gray, A. H. Gray, Jr., G. Rebolledo, and J. E. Shore, "Rate-distortion speech coding with a minimum discrimination information distortion measure," *IEEE Trans. Inform. Theory*, to be published.

[14] A. Buzo, A. H. Gray, Jr., R. M. Gray, and J. D. Markel, "Speech coding based upon vector quantization," in *Proc. 1980 IEEE Int. Conf. Acous., Speech, Signal Processing*, IEEE Cat. 80CH1559-4, pp. 15–17.

[15] —, "Speech coding based upon vector quantization," *IEEE Trans. Acoust., Speech, Signal Processing*, vol. ASSP-28, pp. 562–574, Oct. 1980.

[16] I. Csiszàr, "I-divergence geometry of probability distributions and minimization problems," *Ann. Prob.*, vol. 3, no. 1, pp. 146–58, 1975.

[17] J. E. Shore and R. M. Gray, "Minimum cross-entropy pattern classification and cluster analysis," *IEEE Trans. Pattern Anal. Machine Intell.*, to be published.

[18] J. E. Shore, "Minimum cross-entropy spectral analysis," Naval Res. Lab., Washington, DC, NRL Memo. Rep. 3921, Jan. 1979.

[19] R. B. Blackman and J. W. Tukey, *The Measurement of Power Spectra.* New York: Dover, 1959.

[20] C. E. Shannon and W. Weaver, *The Mathematical Theory of Communication*. Chicago, IL: Univ. Illinois Press, 1949.
[21] M. S. Bartlett, *An Introduction to Stochastic Processes*. Cambridge, England: Cambridge Univ. Press, 1966.
[22] R. G. Gallagher, *Information Theory and Reliable Communication*. New York: Wiley, 1968.
[23] E. T. Jaynes, "Prior probabilities," *IEEE Trans. Syst. Sci. Cybern.*, vol. SSC-4, pp. 227-241, 1968.
[24] W. M. Elsasser, "On quantum measurements and the role of the uncertainty relations in statistical mechanics," *Phys. Rev.*, vol. 52, pp. 987-999, Nov. 1937.
[25] E. T. Jaynes, "Information theory and statistical mechanics I," *Phys. Rev.*, vol. 108, pp. 171-190, 1957.
[26] I. J. Good, "Maximum entropy for hypothesis formulation, especially for multidimensional contingency tables," *Ann. Math. Statist.*, vol. 34, pp. 911-934, 1963.
[27] A. Hobson and B. Cheng, "A comparison of the Shannon and Kullback information measures," *J. Statist. Phys.*, vol. 7, no. 4, pp. 301-310, 1973.
[28] I. J. Good, *Probability and the Weighing of Evidence*. London: Griffen, 1950.
[29] R. W. Johnson, "Determining probability distributions by maximum entropy and minimum cross-entropy," *Proc. APL79* (ACM 0-89791-005), pp. 24-29, May 1979.
[30] L. Geronimus, *Orthogonal Polynomials*. New York: Consultants Bureau, 1961.
[31] J. D. Markel and A. H. Gray, Jr., *Linear Prediction of Speech*. New York: Springer-Verlag, 1976.
[32] Y. Yaglom, *An Introduction to the Theory of Stationary Random Functions*. Englewood Cliffs, NJ: Prentice-Hall, 1962.
[33] A. Papoulis, *Signal Analysis*. New York: McGraw-Hill, 1977.
[34] D. C. Youla, "The flat echo estimator: A new tunable high-resolution spectral estimator," Polytech. Inst. New York, Tech. Note 3, Contract F30602-78-C-0048.

The Retrieval of Harmonics from a Covariance Function

V. F. Pisarenko

(Received 1973 January 29)*

Summary

A new method for retrieving harmonics from a covariance function is introduced. The method is based on a theorem of Caratheodory about the trigonometrical moment problem. The relation between this method and the 'maximum entropy' spectral estimator is discussed, and the effect of a small addition of a noise component is investigated. A numerical example is discussed.

Introduction

Several new methods for estimating the spectra of stationary stochastic processes and noise fields have recently been developed:

1. The 'maximum likelihood' estimator (Capon 1969);
2. The 'maximum entropy' or 'autoregression' estimator (which we will refer to in this paper as MEE) due to Burg (Burg 1972; see also Ulrych 1972 and Lacoss 1971). This estimator was treated from the point of view of autoregression processes by Parzen (1969).
3. Spectral estimators based on the eigenvalues and eigenvectors of the covariance matrix, suggested by the present author (Pisarenko 1972).

These methods are particularly useful when good resolution of spectral peaks is desired, and they have been applied in various scientific fields, in geophysics, for example, in studying free oscillations of the Earth, seismic surface waves, spectra of microseisms, etc.

In this paper we introduce yet another method of detecting spectral peaks, based on a theorem of Caratheodory concerning the trigonometrical moment problem, and we show that MEE is in fact a smoothed version of our new method. In the last section of this paper we investigate the effect on the estimator of the addition of a small amount of noise to the record.

1. On the restitution of a spectrum from a finite segment of the covariance function

Let $x(t)$ be a complex stationary stochastic process, stationary in the wide sense, which depends on discrete time (see, for example, Doob 1953). We assume the mean value of the process to be zero. The spectral function $F(\lambda)$ of the process $x(t)$ is related to its covariance function $B(\kappa)$ by:

$$B(\kappa) = (2\pi)^{-\frac{1}{2}} \int_{-\pi}^{\pi} \exp(i\lambda\kappa)\, dF(\lambda) \quad -\infty \leqslant \kappa \leqslant \infty. \tag{1}$$

* Received in original form 1972 June 29

Reprinted with permission from *Geophys. J. Roy. Astron. Soc.*, vol. 33, pp. 347–366, 1973.

When $x(t)$ is assumed to be a real process we shall say so explicitly.

If $F(\lambda)$ is absolutely continuous its derivative $F'(\lambda)$ is called the spectral density $P(\lambda)$:

$$P(\lambda) = F'(\lambda). \tag{2}$$

If $F(\lambda)$ has a jump ρ_j at frequency λ_j, i.e. if $F(\lambda_j+0)-F(\lambda_j-0) = \rho_j$, then (2) is still valid if we introduce δ-functions; in this case $P(\lambda)$ contains a component $\rho_j\delta(\lambda-\lambda_j)$ and the covariance function $B(\kappa)$ contains a harmonic $\rho_j \exp(i\lambda_j\kappa)$.

Suppose we are given a segment of the covariance function of length $2m$, so that we have

$$B(\kappa) \quad -m \leqslant \kappa \leqslant m. \tag{3}$$

What can we say of the true spectral function $F(\lambda)$? To start with, we assume that the values $B(\lambda)$ are known absolutely, without any noise or error.

Assertion 1. Let the segment of $B(\kappa)$ be such that the covariance matrix B (whose order is $m+1$) is non-degenerate:

$$B = \{B(\kappa-j)\}_{\kappa, j=0,1,\ldots m} \tag{4}$$

Then there exist infinitely many spectral functions $F(\lambda)$ which are related by equation (1) to the given function $B(\kappa)$ when $|\kappa| \leqslant m$.

To prove this assertion, consider the MEE $P_E(\lambda)$:

$$P_E(\lambda) = \beta_{00}\left|\sum_{\kappa=0}^{m} \beta_{0\kappa}\exp(-i\lambda\kappa)\right|^{-2} \qquad -\pi < \lambda \leqslant \pi \tag{5}$$

where $\{\beta_{\kappa j}\}_{\kappa, j=0,1,\ldots,m}$ is the inverse of the matrix B. This estimator has been called the 'maximum entropy' estimator, since it maximizes the entropy integral (see Shannon 1948)

$$(2\pi)^{-1}\int_{-\pi}^{\pi} \log P(\lambda)\, d\lambda$$

among all the spectral density functions which satisfy the relations

$$(2\pi)^{-1}\int_{-\pi}^{\pi} \exp(i\lambda\kappa)\, P(\lambda)\, d\lambda = B(\kappa) \qquad |\kappa| \leqslant m.$$

Note that $P_E(\lambda)$ is just the spectral density of an autoregressive process of order m (see Jenkins & Watts 1970), of which the first m values of the covariance function are equal to the given values of the segment of $B(\kappa)$.

Since B is assumed to be positive definite, the matrix $R(\mu) = B-\mu I$ will also be positive definite for sufficiently small μ (here I denotes the identity matrix of order $m+1$). Thus for μ small enough the matrix B can be represented as the sum of two positive definite matrices

$$B = R(\mu)+\mu I. \tag{6}$$

Denoting the inverse of $R(\mu)$ by $\{r_{\kappa j}\}_{\kappa, j=0,1,\ldots,m}$ we get a family of spectral densities

$$r_{00}\left|\sum_{\kappa=0}^{m} r_{0\kappa}\exp(-i\lambda\kappa)\right|^{-2} + \mu \tag{7}$$

which correspond exactly to the given segment of the covariance function $B(\kappa)$. We notice that the second term in (7) represents the δ-functions $\mu\delta(\kappa)$ in the covariance function. The assertion is proved.

It follows from a theorem of Caratheodory (see Section 2 below) that for any λ_0 such that $-\pi < \lambda_0 \leqslant \pi$ there exists a spectral function $F(\lambda)$ which corresponds to the given segment of $B(\kappa)$ and has a jump at the frequency λ_0; moreover, it is possible to choose a function $F(\lambda)$ which has jumps at any prescribed finite set of frequencies $\lambda_0, \lambda_1, \ldots, \lambda_n$ (although the amplitudes of these jumps cannot be chosen arbitrarily, and these amplitudes will generally tend to zero as n increases).

Thus if a spectral estimate of a function of discrete time is based on a finite segment of the covariance function, and the corresponding covariance matrix is non-degenerate, then it must be taken into account that even if the covariance function values are known exactly there exist infinitely many corresponding spectral functions. Also, there is no upper limit for the spectral density at any frequency λ, since it can always be arranged that a δ-function is present at the frequency λ. Nevertheless, in such situations some upper and lower limits can be obtained for spectral functions which have been smoothed by some spectral window. In particular, the following theorem is true.

Theorem. Let $w(\lambda)$ be a spectral window, that is, a real periodic function of λ, having period 2π, differentiable r times (where $r \geqslant 1$), and

$$|w^{(r)}(\lambda)| \leqslant K$$

where K is some constant. We denote by $B(\kappa)$ the covariance function related to the spectral function $F(\lambda)$ by equation (1). Then for the smoothed spectrum

$$(2\pi)^{-1} \int_{-\pi}^{\pi} w(\lambda-\lambda_0)\, dF(\lambda)$$

the following upper and lower limits exist, based on the finite segment (3) of the covariance function:

$$\sum_{\kappa=-m}^{m} w_\kappa \exp(-i\lambda_0 \kappa)\, B^*(\kappa) - KB(0)\, m^{-r}.\,[(4/\pi^2)\log(m+1)+4]$$

$$\leqslant (2\pi)^{-1} \int_{-\pi}^{\pi} w(\lambda-\lambda_0)\, dF(\lambda)$$

$$\leqslant \sum_{\kappa=-m}^{m} w_\kappa \exp(-i\lambda_0 \kappa)\, B^*(\kappa) + KB(0)\, m^{-r}.\,[(4/\pi^2)\log(m+1)+4] \qquad (8)$$

where w_κ are the Fourier coefficients of the function $w(\lambda)$:

$$w_\kappa = (2\pi)^{-1} \int_{-\pi}^{\pi} \exp(-i\lambda\kappa)\, w(\lambda)\, d\lambda.$$

The proof is given in the Appendix. By this theorem we can estimate the power contribution of both the continuous and the discrete spectral components cut out by the spectral window $w(\lambda-\lambda_0)$.

For example, take the Parzen window of width Δ (see Jenkins & Watts 1969):

$$w(\lambda) = \begin{cases} 1-6(\lambda/\Delta)^2+6|\lambda/\Delta|^3 & |\lambda| \leqslant \Delta/2 \\ 2(1-|\lambda/\Delta|)^3 & \Delta/2 \leqslant |\lambda| \leqslant \Delta \\ 0 & \Delta \leqslant |\lambda| \leqslant \pi. \end{cases} \qquad (9)$$

It is easy to verify that the second derivative of $W(\lambda)$ exists and is less than $12/\Delta^2$, and that (Jenkins & Watts 1969)

$$w_\kappa = \frac{3\Delta}{8\pi}\left[\frac{\sin(\kappa\Delta/4)}{\kappa\Delta/4}\right]^4.$$

For $B(\kappa)$ we take the autoregression covariance:

$$B(\kappa) = \alpha^{|\kappa|} \qquad |\alpha| \leqslant 1$$

$$F'(\lambda) = (1-\alpha^2)(1+\alpha^2-2\cos\alpha\lambda)^{-1}$$

and we assume $\Delta = 1$, $\alpha = 0{\cdot}5$, $\lambda_0 = \frac{1}{2}\pi$, and $m = 50$. Then to two significant figures the integral

$$(2\pi)^{-1}\int_{-\pi}^{\pi} w(\lambda-\lambda_0)\,dF(\lambda) = (2\pi)^{-1}\int_{-\pi}^{\pi}\frac{(1-\alpha^2)\,w(\lambda-\lambda_0)}{1+\alpha^2-2\alpha\cos\lambda}\,d\lambda \tag{10}$$

is equal to 0·076. To the same accuracy, the sum

$$\sum_{\kappa=m}^{m} w_\kappa \exp(-i\lambda_0\kappa)\,B^*(\kappa)$$

is 0·077, and the term

$$KB(0)\,m^{-r}[(4/\pi^2)\log(m+1)+4] = 12{\cdot}50^{-2}[(4/\pi^2)\log 51+4]$$

is 0·027. Thus for the true value 0·076 of the integral (10) we find the limits $0{\cdot}077 \pm 0{\cdot}027$. We see that the limits (9) give rather rough estimates for (11) even for comparatively large m and a rather broad width Δ of the spectral window. Naturally, as m and Δ decrease the estimators (8) become worse.

Thus when only a finite segment of the covariance function is known, and there is no *a priori* information about the true spectrum, we should not speak of the true spectrum but rather of the whole family of true spectra.

2. A new method of retrieving harmonics from a finite segment of the covariance function

In this section we make use of the following theorem of Caratheodory (for a proof due to Szegö, see Grenander & Szegö 1958, Chapter 4).

Theorem. Let $c_1, \ldots, c_m$ be complex numbers (not all zero) and assume $m > 1$. Then there exists some integer r, $1 \leqslant r \leqslant m$ and some real numbers ρ_j and ω_j $(j = 1, \ldots, r)$ such that $\rho_j > 0$, $-\pi < \omega_j \leqslant \pi$, and $\omega_j \neq \omega_\kappa$ when $j \neq \kappa$, and such that the following representation of the sequence of numbers c_κ is true:

$$c_\kappa = \sum_{j=1}^{r} \rho_j \exp(i\omega_j\kappa) \qquad \kappa = 1, \ldots, m. \tag{11}$$

The constants ρ_j and ω_j are determined uniquely.

We now consider the segment (3) of the covariance function $B(\kappa)$ and put in Caratheodory's theorem $c_\kappa = B(\kappa)$, $\kappa = 1, \ldots, m$. Then (11) gives the following representation for $B(\kappa)$:

$$B(\kappa) = \sum_{j=1}^{r} \rho_j \exp(i\omega_j\kappa) \qquad \kappa = 1, \ldots, m;\ r \leqslant m. \tag{12}$$

When $\kappa = 0$, then generally

$$B(0) \geqslant \sum_{j=1}^{r} \rho_j. \tag{13}$$

It follows from Szegö's proof that the equality sign in (13) holds when and only when the matrix B is degenerate. If B is non-degenerate, the strict inequality holds. In this case we denote by ρ_0 the difference between the left side and the right side of (13):

$$\rho_0 = B(0) - \sum_{j=1}^{r} \rho_j. \tag{14}$$

Since $B(\kappa) = B^*(-\kappa)$, we have the following representation for $B(\kappa)$:

$$B(\kappa) = \rho_0\,\delta(\kappa) + \sum_{j=1}^{r} \rho_j \exp(i\omega_j \kappa) \qquad |\kappa| \leqslant m \tag{15}$$

where $\delta(0) = 1$, $\delta(\kappa) = 0$ when $\kappa \neq 0$; $\rho_0 > 0$ for non-degenerate matrices B, and $\rho_0 = 0$ for degenerate matrices B.

Equation (15) provides a representation for the segment of the covariance function $B(\kappa)$ in terms of a sum of harmonics and possibly a δ-function corresponding to white noise.

It should be noted that if the true covariance function consists of r harmonics (where $r \leqslant m$) and possibly a δ-function as well, then the representation (15) produces the true frequencies ω_j and amplitudes ρ_j, since according to Caratheodory's theorem these quantities are determined uniquely. But if $r > m$ or if $B(\kappa)$ contains some non-white spectral components, then (15) gives numbers ω_j and ρ_j which may not be the true frequencies and amplitudes. We will comment below on the relation between these numbers and the true spectrum.

If $B(\kappa)$ is real, the complex harmonics in (15) must be even in number and occur in complex conjugate pairs to produce real cosine harmonics:

$$\rho_{j1} = \rho_{j2} \qquad \omega_{j1} = -\omega_{j2} \tag{16}$$

$$\rho_{j1} \exp(i\omega_{j1}\kappa) + \rho_{j2} \exp(i\omega_{j2}\kappa) = 2\rho_{j1} \cos(\omega_{j1}\kappa).$$

Thus for real $B(\kappa)$ the representation (15) can be written in the form:

$$B(\kappa) = \rho_0\,\delta(\kappa) + 2\sum_{j=1}^{p} \rho_j \cos(\omega_j \kappa) \qquad |\kappa| \leqslant m \tag{17}$$

where $p \leqslant m/2$. Thus if the true number of cosine harmonics in a real covariance function is not more than $m/2$, these harmonics can be recovered exactly from a segment of length $2m$ of the covariance function.

We now show how the numbers r, ρ_j, and ω_j are derived; the argument follows from Grenander & Szegö (1958).

Algorithm for determining r, ρ_j, and ω_j. The algorithm consists of four steps:

1. We find the minimal eigenvalue μ_0 of the matrix B of order $m+1$; its multiplicity is denoted by ν. The number μ_0 is equal to the amplitude ρ_0 for the δ-function term in (15). If $\rho_0 = 0$, then the δ-function term in (15) vanishes. Two situations are possible: $\nu = 1$ and $\nu > 1$. If $\nu = 1$, we consider the matrix $B - \mu_0 I$ in what follows; if $\nu > 1$, we consider the principal minor of the matrix $B - \mu_0 I$, whose order is $r+1$

(where $r = m-v$). We denote this minor by:

$$(B-\mu_0 I)_r = \begin{bmatrix} B(0)-\mu_0 & B(1) & \dots & B(r) \\ B(-1) & B(0)-\mu_0 & \dots & B(r-1) \\ \vdots & \vdots & & \vdots \\ B(-r) & B(-r+1) & \dots & B(0)-\mu_0 \end{bmatrix}. \tag{18}$$

Thus the rank of the matrix $(B-\mu_0 I)_r$ is r and its order is $(r+1)$.

2. We find the eigenvector of $(B-\mu_0 I)_r$ corresponding to its unique zero eigenvalue, and denote its components by $p_0, p_1, \dots, p_r$.

3. We evaluate the roots of the polynomial

$$p_0+p_1 z+\dots+p_r z^r = 0 \tag{19}$$

which roots we denote by $z_1, \dots, z_r$. These roots are all different and of modulus unity, so they can be written uniquely in the form

$$z_j = \exp(i\omega_j) \qquad -\pi < \omega_j \leqslant \pi \qquad j = 1, \dots, r.$$

The numbers $\omega_1, \dots, \omega_r$ are the frequencies of the required harmonics.

4. As shown by Grenander & Szegö (1958), the amplitudes ρ_j can be expressed in terms of the $z_1, \dots, z_r$. We give here another derivation of expressions for ρ_j which may be more convenient for numerical calculation.

The amplitudes ρ_j satisfy the system of linear equations:

$$\sum_{j=1}^{r} \rho_j \exp(i\omega_j \kappa) = B(\kappa)-\mu_0 \delta(\kappa) \qquad |\kappa| \leqslant m. \tag{20}$$

We now show how to derive from (20) a system of real linear equations which determine $\rho_1, \dots, \rho_r$. First suppose that the absolute values $|\omega_1|, \dots, |\omega_r|$ are all different. Then from difference equation theory (Gelfond 1959) it can be shown that:

$$\det \begin{bmatrix} \sin\omega_1 & \sin\omega_2 & \dots & \sin\omega_r \\ \sin 2\omega_1 & \sin 2\omega_2 & \dots & \sin 2\omega_r \\ \vdots & \vdots & & \vdots \\ \sin r\omega_1 & \sin r\omega_2 & \dots & \sin r\omega_r \end{bmatrix} \neq 0 \tag{21}$$

when $\omega_j \neq 0$ and $w_j \neq \pi$.

$$\det \begin{bmatrix} 1 & 1 & & 1 \\ \cos\omega_1 & \cos\omega_2 & \dots & \cos\omega_r \\ \vdots & \vdots & & \vdots \\ \cos(r-1)\omega_1 & \cos(r-1)\omega_2 & \dots & \cos(r-1)\omega_r \end{bmatrix} \neq 0. \tag{22}$$

If none of the $\omega_1, \dots, \omega_r$ is equal to zero or π we can take the imaginary part of equations (20) for $\kappa = 0, \dots, r-1$:

$$\sum_{j=1}^{r} \rho_j \sin\omega_j \kappa = \operatorname{Im} B(\kappa) \qquad \kappa = 1, \dots, r. \tag{23}$$

Because of (21), the system (23) determines the amplitudes $\rho_1, \dots, \rho_r$ uniquely. If some $\omega_j = 0$ or $\omega_j = \pi$, the real parts of equations (20) can be taken for $k = 0, 1, \ldots, r-1$:

$$\sum_{j=1}^{r} \rho_j \cos \omega_j \kappa = \operatorname{Re} B(\kappa) - \mu_0 \delta(\kappa) \qquad \kappa = 0, \ldots, r-1. \tag{24}$$

The determinant of this system is the expression (22), and hence is non-zero.

Now assume that among the $\omega_1, \ldots, \omega_r$ there is one pair with identical modulus, say $\omega_1 = -\omega_2$, and the others, $|\omega_3|, \ldots, |\omega_r|$ are all different. Taking the terms with ω_1 and ω_2 together in (23) and (24), we have:

$$\rho_1 \sin \omega_1 \kappa + \rho_2 \sin \omega_2 \kappa = (\rho_2 - \rho_1) \sin \omega_2 \kappa$$

$$\rho_1 \cos \omega_1 \kappa + \rho_2 \cos \omega_2 \kappa = (\rho_2 + \rho_1) \cos \omega_2 \kappa.$$

We denote

$$\rho_2 - \rho_1 = \tilde{\rho}_2 \qquad \rho_2 + \rho_1 = \tilde{\tilde{\rho}}_2.$$

Now consider instead of (23) and (24) the truncated system:

$$\tilde{\rho}_2 \sin \omega_2 \kappa + \sum_{j=3}^{r} \rho_j \sin \omega_j \kappa = \operatorname{Im} B(\kappa) \qquad \kappa = 1, \ldots, r-1 \tag{25}$$

$$\tilde{\tilde{\rho}}_2 \cos \omega_2 \kappa + \sum_{j=3}^{r} \rho_j \cos \omega_j \kappa = \operatorname{Re} B(\kappa) - \mu_0 \delta(\kappa) \qquad \kappa = 0, 1, \ldots, r-2. \tag{26}$$

In these systems all the moduli $|\omega_2|, |\omega_3|, \ldots, |\omega_r|$ are different, so that using systems (23) and (24) we can determine $\tilde{\rho}_2, \tilde{\tilde{\rho}}_2, \rho_3, \ldots, \rho_r$. From these we get:

$$\rho_1 = \tfrac{1}{2}(\tilde{\tilde{\rho}}_2 - \tilde{\rho}_2) \qquad \rho_2 = \tfrac{1}{2}(\tilde{\tilde{\rho}}_2 + \tilde{\rho}_2). \tag{27}$$

This same method is applicable when there are several pairs of frequencies which differ only in sign. Having grouped together corresponding pairs in (23) and (24), one solves truncated systems and then uses (27) to find amplitudes. The terms with $\omega_j = 0$ or $\omega_j = \pi$ are always determined from (24).

Note that for a real covariance function we have instead of (20) the system of equations (17):

$$2 \sum_{j=1}^{p} \rho_j \cos \omega_j \kappa = B(\kappa) - \mu_0 \delta(\kappa) \qquad |\kappa| \leqslant m. \tag{28}$$

The moduli $|\omega_1|, \ldots, |\omega_p|$ are all different, so we can take equations (28) for $\kappa = 0, 1, \ldots, p-1$. Because of (22) the determinant of this system is non-zero, so $\rho_1, \ldots, \rho_p$ are uniquely determined.

The derivation of the numbers ω_j, ρ_j is thus complete. It is clear that this algorithm can be used in practice on digital computers for values of m up to several dozen.

3. On the connection between $P_E(\lambda)$ and the harmonics determined from a segment of the covariance function

We now assume that the minimal eigenvalue μ_0 of the matrix B is simple ($v = 1$) and that $\mu_0 > 0$. Then $r = m$ and it is easy to show that apart from a constant factor the polynomial (19) can be written as the determinant:

$$\det\begin{bmatrix} 1 & z & z^2 & \dots & z^m \\ B(-1) & B(0)-\mu_0 & B(1) & \dots & B(m-1) \\ \vdots & \vdots & \vdots & & \vdots \\ B(-m) & B(-m+1) & B(-m+2) & \dots & B(0)-\mu_0 \end{bmatrix} = 0. \tag{29}$$

For $\mu < \mu_0$ we denote the maximum entropy estimator (MEE) for the matrix $B-\mu I$ by $P_E^{(\mu)}(\lambda)$: the elements of the first row of the matrix $(B-\mu I)^{-1}$ are denoted by $\beta_{0\kappa}(\mu)$ and the corresponding signed minors by $B_{0\kappa}(\mu)$:

$$\beta_{0\kappa}(\mu) = \frac{B_{0\kappa}(\mu)}{\det(B-\mu I)} \qquad \kappa = 0, 1, \dots, m.$$

We see that the coefficients of the polynomial (29) are equal to $B_{0\kappa}(\mu)$. But from equation (5), the coefficients $\beta_{0\kappa}$ in $P_E(\lambda)$ can be written

$$\beta_{0\kappa} = \beta_{0\kappa}(0) = \frac{B_{0\kappa}(0)}{\det B} \qquad \kappa = 0, 1, \dots, m.$$

It is known (Grenander & Szegö 1958, Chapter 2) that if $\det B > 0$, then all the roots of the polynomial

$$\sum_{\kappa=0}^{m} \beta_{0\kappa} z^\kappa = \frac{1}{\det B} \sum_{\kappa=0}^{m} B_{0\kappa}(0) z^\kappa = 0.$$

lie inside the unit circle in the complex z-plane. Denoting the roots of the polynomial

$$\sum_{\kappa=0}^{m} \beta_{0\kappa}(\mu) z^\kappa = \frac{1}{\det(B-\mu I)} \sum_{\kappa=0}^{m} B_{0\kappa}(\mu) z^\kappa = 0 \tag{30}$$

by $z_1(\mu), \dots, z_m(\mu)$, it follows that for $\mu < \mu_0$:

$$|z_j(\mu)| < 1 \qquad j = 1, 2, \dots, m.$$

From (30) we see that the roots $z_j(\mu)$ tend to the roots of the polynomial (29) as $\mu \uparrow \mu_0$, i.e.

$$z_j(\mu) \to \exp(i\omega_j) \quad \text{when} \quad \mu \uparrow \mu_0, \qquad j = 1, \dots, m. \tag{31}$$

It follows from the definition (5) of the MEE $P_E(\lambda)$ that

$$(2\pi)^{-1} \int_{-\pi}^{\pi} \exp(i\lambda\kappa)\, P_E(\lambda)\, d\lambda$$

$$= (2\pi)^{-1} \int_{-\pi}^{\pi} \frac{\beta_{00} \exp(i\lambda\kappa)\, d\lambda}{\left|\sum_{j=0}^{m} \beta_{0j} \exp(-i\lambda j)\right|^2} = B(\kappa) \qquad |\kappa| \leqslant m. \tag{32}$$

Substituting the variable

$$\zeta = \exp(i\lambda)$$

in the integral (32), we find

$$\frac{\beta_{00}}{2\pi}\oint \frac{\zeta^{\kappa-1}\,d\zeta}{\left|\sum_{j=0}^{m}\beta_{0j}\zeta^j\right|^2} = B(\kappa) \qquad |\kappa| \leqslant m \tag{33}$$

where the path of integration is around the unit circle in the complex ζ-plane. Similarly, for the matrix $B - \mu I$:

$$\frac{\beta_{00}(\mu)}{2\pi}\oint \frac{\zeta^{\kappa-1}\,d\zeta}{\left|\sum_{j=0}^{m}\beta_{0j}(\mu)\zeta^j\right|^2} = B(\mu)-\mu\delta(\kappa) \qquad |\kappa| \leqslant m. \tag{34}$$

Assuming that all the roots $z_j(\mu)$ in the denominator of (34) are distinct, we have from the theory of complex residues:

$$B(\kappa)-\mu\delta(\kappa) = \sum_{j=1}^{m}\rho_j(\mu)\,z_j^{\kappa}(\mu) \qquad \kappa = 0, 1, \ldots, m. \tag{35}$$

For $\kappa < 0$ the equalities (35) are changed to the corresponding complex conjugate expressions. From (31) and (35) we have as $\mu \uparrow \mu_0$:

$$B(\kappa)-\mu_0\,\delta(\kappa) = \sum_{j=1}^{m}\rho_j(\mu_0)z_j^{\kappa}(\mu_0) \qquad \kappa = 0, 1, \ldots, m. \tag{36}$$

Comparing (36) and (15) we conclude that $\rho_j(\mu_0) = \rho_j$. Thus

$$\sum_{j=1}^{m}\rho_j(\mu)z_j^{\kappa}(\mu) \to \sum_{j=1}^{m}\rho_j\exp(i\omega_j\kappa) \quad \kappa \geqslant 0 \quad \text{when } \mu\uparrow\mu_0$$

or in spectral form,

$$P_E^{(\mu)}(\lambda) \to \sum_{j=1}^{m}\rho_j\delta(\lambda-\omega_j) \quad -\pi < \lambda \leqslant \pi \qquad \text{when } \mu\uparrow\mu_0. \tag{37}$$

The last equation makes clear some of the amplitude properties of this estimator (compare Lacoss 1971), and shows why the MEE $P_E(\lambda)$ sometimes does well in distinguishing harmonics in a white noise background. It is clear that subtraction of some diagonal matrix $\mu_0 I$ from the matrix B would result in exact determination of the harmonics in Lacoss's examples.

Before taking the limit $\mu\uparrow\mu_0$, the integrand in (34) can be decomposed into simple fractions of the form

$$|e^{i\lambda}-z_j(\mu)|^{-2}$$

which can be considered as 'smoothed' versions of the δ-functions $\delta(\lambda-\omega_j)$ in (37). Thus $P_E(\lambda)$ can be considered as a 'smoothed' version of the function

$$\mu_0+\sum_{j=1}^{m}\rho_j\delta(\lambda-\omega_j).$$

As μ increases toward μ_0, the effect of this smoothing diminishes, and in the limit when $\mu = \mu_0$, $P_E^{(\mu)}(\lambda)$ is decomposed into pure harmonics.

If the true spectrum is continuous, then the harmonics $\rho_j \delta(\lambda-\omega_j)$ in its 'linear' spectral representation (37) will 'outline' the true spectral density. The greater m is, the more detailed this outline will be.

Sometimes the linear spectral representation (37) may provide an economic spectral parameterization, which may be useful in practical computations.

4. Small random perturbations of the original data

Now we consider the situation when the covariance function is disturbed by additive random noise. This implies that in equation (15) the frequencies ω_j and amplitudes ρ_j are also random variables. Since they are non-linear functions of the noise and cannot be expressed generally as explicit functions of the noise, it is very difficult to obtain their statistical characteristics. Therefore we consider noise amplitudes small enough that the perturbation method is applicable (see, for example, Gelfand 1966).

We assume that we have a segment of the disturbed covariance function $\hat{B}(\kappa)$:

$$\hat{B}(\kappa) = B(\kappa)+\varepsilon\xi(\kappa) \qquad |\kappa| \leqslant m$$

where $B(\kappa)$ is the true covariance function, $\xi(\kappa)$ is a random noise such that $\xi(\kappa) = \xi^*(\kappa)$, and ε is a small real parameter. According to our assumptions, $\hat{B}(\kappa) = \hat{B}^*(-\kappa)$. We further assume that

$$\det B > 0 \tag{38}$$

and that with probability one

$$\det \hat{B} > 0. \tag{39}$$

We assume that the first two moments of the noise $\xi(\kappa)$ exist. We introduce separate notation for the real and imaginary parts of these moments:

$$\left.\begin{aligned} &E \operatorname{Re}\xi(\kappa) = a_\kappa && \kappa = 0, 1, \ldots, m \\ &E \operatorname{Im}\xi(\kappa) = a_{m+\kappa} && \kappa = 1, \ldots, m \end{aligned}\right\} \tag{40}$$

$$\left.\begin{aligned} &E[\operatorname{Re}\xi(\kappa)-a_\kappa][\operatorname{Re}\xi(j)-a_j] = \sigma(\kappa, j) && \kappa, j = 0, 1, \ldots, m \\ &E[\operatorname{Re}\xi(\kappa)-a_\kappa][\operatorname{Im}\xi(j)-a_{m+j}] = \sigma(\kappa, m+j) && \kappa = 0, \ldots, m \\ & && j = 1, \ldots, m \\ &E[\operatorname{Im}\xi(\kappa)-a_{m+\kappa}][\operatorname{Im}\xi(j)-a_{m+j}] = \sigma(m+\kappa, m+j) && \kappa, j = 1, \ldots, m \end{aligned}\right\}. \tag{41}$$

The column vector of quantities (40) is denoted by **a**, and the symmetrical covariance matrix $\sigma(\kappa, j)_{\kappa, j = 0, \ldots, 2m}$ of order $(2m+1)$ is denoted by Σ.

We assume in the representation (15) for $B(\kappa)$ that $r = m$, i.e.

$$\hat{B}(\kappa) = \rho_0 \delta(\kappa)+ \sum_{j=1}^{m} \rho_j \exp(i\omega_j \kappa) \qquad |\kappa| \leqslant m. \tag{42}$$

Then for sufficiently small ε, the covariance function $\hat{B}(\kappa)$ will be represented with probability arbitrarily close to unity by the similar form:

$$\hat{B}(\kappa) = \rho_0(\varepsilon) \delta(\kappa)+ \sum_{j=1}^{m} \rho_j(\varepsilon) \exp[i\omega_j(\varepsilon) \kappa] \qquad |\kappa| \leqslant m \tag{43}$$

where $\rho_j(\varepsilon)$ and $\omega_j(\varepsilon)$ are random functions of ε, and

$$\left.\begin{array}{ll}\rho_j(\varepsilon) \to \rho_j & j = 0, 1, \ldots, m \\ \omega_j(\varepsilon) \to \omega_j & j = 1, \ldots, m\end{array}\right\} \text{ when } \varepsilon \to 0. \tag{44}$$

In the neighbourhood of the point $\varepsilon = 0$ the functions $\rho_j(\varepsilon)$ and $\omega_j(\varepsilon)$ are differentiable and can be represented in the form:

$$\rho_j(\varepsilon) = \rho_j + \varepsilon\tilde{\rho}_j + \varepsilon^2 \tilde{\tilde{\rho}}_j + \ldots \tag{45}$$

$$\omega_j(\varepsilon) = \omega_j + \varepsilon\tilde{\omega}_j + \varepsilon^2 \tilde{\tilde{\omega}}_j + \ldots . \tag{46}$$

We wish to find the principal linear parts of the functions $\rho_j(\varepsilon)$ and $\omega_j(\varepsilon)$: for sufficiently small ε we have with probability close to one:

$$\begin{aligned}\rho_j(\varepsilon) &\approx \rho_j + \varepsilon\tilde{\rho}_j \\ \omega_j(\varepsilon) &\approx \omega_j + \varepsilon\tilde{\omega}_j.\end{aligned} \tag{47}$$

Inserting (45) and (46) into (43), expanding exponents in powers of ε, and equating coefficients of ε we get:

$$\tilde{\rho}_0\,\delta(\kappa) + \sum_{j=1}^{m} (\tilde{\rho}_j + \rho_j\,\tilde{\omega}_j\,i\kappa) \exp(i\omega_j\,\kappa) = \xi(\kappa) \qquad |\kappa| \leqslant m. \tag{48}$$

Taking separately the real and imaginary parts of (48) we obtain a system of $2m+1$ linear real equations for the $2m+1$ real unknown parameters $\tilde{\rho}_0, \ldots, \tilde{\rho}_m, \tilde{\omega}_1, \ldots, \tilde{\omega}_m$. For $\kappa = 0$ we get one real equation (48), and for $\kappa = 1, \ldots, m$ we get two real equations from each complex equation (48), making $2m+1$ equations in all. For $\kappa = -1, \ldots, -m$, the equations (48) provide no additional information, since these cases are complex conjugates of those for positive κ.

We now show that the homogeneous system corresponding to (48), i.e. $\xi(\kappa) \equiv 0$, has only a zero solution. Since the number of equations (48) coincides with the number of unknown parameters, it follows that (48) has a non-degenerate matrix, and its solution is unique.

Thus we have to prove that if

$$\xi(\kappa) \equiv 0 \qquad |\kappa| \leqslant m \tag{49}$$

then

$$\tilde{\rho}_0 = 0 \quad \tilde{\rho}_j = \tilde{\omega}_j = 0 \qquad j = 1, \ldots, m. \tag{50}$$

As shown in Section 3 above, the parameters ρ_0 and $\rho_0(\varepsilon)$ are the minimal eigenvalues of the matrices B and $\hat{B}$ respectively. It is known (Gelfand 1966) that

$$\tilde{\rho}_0 = \sum_{\kappa, j=0}^{m} p_\kappa\, p_j^*\, \xi(\kappa - j) \tag{51}$$

where $p_0, \ldots, p_m$ are components of the normalized eigenvector of B corresponding to the eigenvalue ρ. Hence if $\xi(\kappa) \equiv 0$ we get from (51) that $\rho_0 = 0$. The system (48) now has the form

$$\sum_{j=1}^{m} \tilde{\rho}_j \exp(i\omega_j\,\kappa) + i\kappa \sum_{j=1}^{m} \tilde{\omega}_j\,\rho_j \exp(i\omega_j\,\kappa) = 0 \qquad |\kappa| \leqslant m \tag{52}$$

and the $2m$ functions

$$\left.\begin{aligned} f_j(\kappa) &= \kappa \exp(i\omega_j \kappa) \\ g_j(\kappa) &= \exp(i\omega_j \kappa) \end{aligned}\right\} \quad j = 1, \ldots, m \tag{53}$$

are linearly independent solutions of a linear difference equation of order $2m$ with constant coefficients (see Gelfond 1959):

$$L(S)f(\kappa) = 0 \qquad |\kappa| \leqslant m \tag{54}$$

where S is the shift operator

$$Sf(\kappa) = f(\kappa+1)$$

and $L(S)$ is the difference operator of order $2m$:

$$L(S) = \prod_{j=1}^{m} [S - \exp(i\omega_j)]^2. \tag{55}$$

Since the functions (53) are linearly independent and the number of equations in (52) is larger than the order of the difference equation, it follows that (52) has only the zero solution (50)—which was what we set out to prove.

Now we divide equations (48) into real and imaginary parts for $\kappa = 0, 1, \ldots, m$:

$$\tilde{\rho}_0 \delta(\kappa) + \sum_{j=1}^{m} \tilde{\rho}_j \cos(\omega_j \kappa) - \kappa \sum_{j=1}^{m} \tilde{\omega}_j \rho_j \sin(\omega_j \kappa) = \operatorname{Re} \xi(\kappa) \qquad \kappa = 0, 1, \ldots, m$$

$$\sum_{j=1}^{m} \tilde{\rho}_j \sin(\omega_j \kappa) + \kappa \sum_{j=1}^{m} \tilde{\omega}_j \rho_j \cos(\omega_j \kappa) = \operatorname{Im} \xi(\kappa) \qquad \kappa = 1, \ldots, m.$$

In order to write this in vector form, we introduce a column vector $\mathbf{x}$ of the unknown parameters. The components of $\mathbf{x}$ are:

$$\left.\begin{aligned} x_0 &= \tilde{\rho}_0, \ldots, x_m = \tilde{\rho}_m \\ x_{m+1} &= \tilde{\omega}_1, \ldots, x_{2m} = \tilde{\omega}_m. \end{aligned}\right\} \tag{57}$$

Similarly, we introduce the column vector $\boldsymbol{\xi}$ of the right-hand parts of (56), having components

$$\left.\begin{aligned} \xi_0 = \xi(0),\ \xi_1 &= \operatorname{Re} \xi(1), \ldots, \xi_m = \operatorname{Re} \xi(m) \\ \xi_{m+1} &= \operatorname{Im} \xi(1), \ldots, \xi_{2m} = \operatorname{Im} \xi(m) \end{aligned}\right\} \tag{58}$$

The matrix of the system (58) is denoted by D:

$$D = \left[\begin{array}{c|ccc|ccc} 1 & 1 & \ldots & 1 & 0 & \ldots & 0 \\ \hline 0 & \cos \omega_1 & \ldots & \cos \omega_m & -\rho_1 \sin \omega_1 & \ldots & -\rho_m \sin \omega_m \\ \vdots & \vdots & & \vdots & \vdots & & \vdots \\ 0 & \cos m\omega_1 & \ldots & \cos m\omega_m & -m\rho_1 \sin m\omega_1 & \ldots & -m\rho_m \sin m\omega_m \\ \hline 0 & \sin \omega_1 & \ldots & \sin \omega_m & \rho_1 \cos \omega_1 & \ldots & \rho_m \cos \omega_m \\ \vdots & \vdots & & \vdots & \vdots & & \vdots \\ 0 & \sin m\omega_1 & \ldots & \sin m\omega_m & m\rho_1 \cos m\omega_1 & \ldots & m\rho_m \cos m\omega_m \end{array}\right] \tag{59}$$

The system (56) can thus be written:

$$D\mathbf{x} = \xi. \tag{60}$$

From what has been already proved, det $D \neq 0$. The solutions of equations (60) can thus be written:

$$\mathbf{x} = D^{-1}\,\xi \tag{61}$$

where D^{-1} is the inverse of D. From (61) we can find the statistical moments of the components of $\mathbf{x}$, in particular their mean value vector $\mathbf{b}$ and their covariance matrix W:

$$\mathbf{b} = E\mathbf{x} = D^{-1}\,E\xi = D^{-1}\,\mathbf{a},$$

where

$$b_0 = E\rho_0 \qquad \begin{aligned} b_j &= E\tilde{\rho}_j \\ & \qquad\qquad j = 1, \ldots, m \\ b_{m+j} &= E\tilde{\omega}_j \end{aligned} \tag{62}$$

$$W = E(\mathbf{x}-\mathbf{b})(\mathbf{x}-\mathbf{b})^* = ED^{-1}(\xi-\mathbf{a})(\xi-\mathbf{a})^*\,(D^{-1})^* = D^{-1}\,\Sigma(D^{-1})^* \tag{63}$$

where the asterisk denotes the complex conjugate matrix or vector, and $\mathbf{a}$ and Σ were defined in (40) and (41). Thus we have the following approximate mean values of the disturbed parameters:

$$\left.\begin{aligned} E\rho_j(\varepsilon) &\approx \rho_j + \varepsilon E\tilde{\rho}_j = \rho_j + \varepsilon b_j & j &= 0, \ldots, m \\ E\omega_j(\varepsilon) &\approx \omega_j + \varepsilon E\tilde{\omega}_j = \omega_j + \varepsilon b_{m+j} & j &= 1, \ldots, m. \end{aligned}\right\} \tag{64}$$

The covariance matrix of the disturbed parameters is $\varepsilon^2\,W$.

For practical use it is important to know how large the neglected terms of order ε^2 were in equations (45) and (46). Substituting (45) and (46) into (43) and equating coefficients of ε^2, we have the system of equations for the quadratic corrections $\tilde{\tilde{\rho}}_j$ and $\tilde{\tilde{\omega}}_j$:

$$\begin{aligned} &\tilde{\tilde{\rho}}_0\,\delta(\kappa) + \sum_{j=1}^{m} \tilde{\tilde{\rho}}_j \exp(i\omega_j\kappa) + i\kappa \sum_{j=1}^{m} \tilde{\tilde{\omega}}_j\,\rho_j \exp(i\omega_j\kappa) \\ &\qquad = \sum_{j=1}^{m} \left[\tfrac{1}{2}\rho_j(\tilde{\omega}_j\kappa)^2 - i\kappa\tilde{\rho}_j\tilde{\omega}_j\right] \exp(i\omega_j\kappa) \qquad |\kappa| \leqslant m. \end{aligned} \tag{65}$$

This differs from (48) by the right-hand parts only, so that the system (65) always has a unique solution. We see that $\tilde{\tilde{\rho}}_j$ and $\tilde{\tilde{\omega}}_j$ are quadratic functionals of the noise $\xi(\kappa)$. If the third and fourth moments of $\xi(\kappa)$ are known, then along the same lines as in the derivation of (63) and (64) it is possible to find $E(\tilde{\tilde{\rho}}_j)^2$ and $E(\tilde{\tilde{\omega}}_j)^2$. If it happens that

$$\left.\begin{aligned} E(\varepsilon^2\,\tilde{\tilde{\rho}}_j)^2 &\ll E(\varepsilon\tilde{\rho}_j)^2 & j &= 0, 1, \ldots, m \\ E(\varepsilon^2\,\tilde{\tilde{\omega}}_j)^2 &\ll E(\varepsilon\tilde{\omega}_j)^2 & j &= 1, \ldots, m \end{aligned}\right\} \tag{66}$$

then in (45) and (46) the quadratic corrections are negligible and the linear approximation (47) is justified.

When $r < m$ in the representation (15) of $B(\kappa)$, the corresponding representation for the disturbed function $\hat{B}(\kappa)$ will nevertheless generally contain m harmonics; i.e. $m-r$ false harmonics will appear with fictitious amplitudes and frequencies, the amplitudes being of order ε. Statistical analysis of the disturbances is much more difficult in this case, since even for the $\tilde{\rho}_j$ and $\tilde{\omega}_j$ the system of equations is non-linear.

For this situation a rough simplified procedure has merit: if we find in (15) that $m-r$ amplitudes $\rho_j(\varepsilon)$ are of order of magnitude ε, then it is possible to take a segment of the disturbed covariance function of length $2r$ rather than the original length $2m$ and carry out the analysis of the disturbances as above.

We note that in order to use (63) and (64) in practice, we have to know the matrix D, which depends on the true unknown parameters ρ_j and ω_j. In practice we usually put into D the disturbed parameters $\rho_j(\varepsilon)$ and $\omega_j(\varepsilon)$ obtained from data. The justification of this procedure lies in the fact that the disturbed matrix $D(\varepsilon)$, i.e. the matrix that contains $\rho_j(\varepsilon)$ and $\omega_j(\varepsilon)$ in place of ρ_j and ω_j, differs from the true matrix D only by a matrix whose elements are of magnitude ε (for sufficiently small ε). Of course, care must be taken, particularly when some frequencies ω_i, ω_j are close together or when the noise is not small. It would be possible to estimate the term linear in ε for $D(\varepsilon)$,

$$D(\varepsilon) = D+\varepsilon\tilde{D}+\varepsilon^2\,\tilde{\tilde{D}}+\ldots$$

and check that this term is negligible, but we omit such considerations here.

Now we write down the equations analogous to (48), (59), and (65) for the case of real covariance functions. These equations differ slightly from the complex expressions, but all the proofs and discussion are almost the same as in the more general case.

Suppose that in (17), $p = m/2$, $\omega_j \neq 0$, and $|\omega_j| < \pi$. For (48) we have:

$$\tilde{\rho}_0\,\delta(\kappa)+2\sum_{j=1}^{m}\tilde{\rho}_j\cos(\omega_j\kappa)-2\kappa\sum_{j=1}^{p}\tilde{\omega}_j\rho_j\sin(\omega_j\kappa) = \xi(\kappa) \qquad |\kappa| \leqslant m = 2p. \tag{67}$$

The matrix D has the form:

(Shown on p. 361)

For (58) we have

$$\xi_\kappa = \xi(\kappa) \qquad \kappa = 0, 1, \ldots, 2p. \tag{69}$$

For (40) and (41) we have

$$E\xi(\kappa) = a_\kappa \qquad \kappa = 0, 1, \ldots, 2p \tag{70}$$

$$E[\xi(\kappa)-a_\kappa][\xi(j)-a_j] = \sigma(\kappa, j) \qquad \kappa, j = 0, 1, \ldots, 2p. \tag{71}$$

The matrix Σ has elements $\sigma(\kappa, j)$, $\kappa, j = 0, \ldots, 2p$. With these changes, the equations (60)–(64) are valid. For (65) we have the system:

$$\sum_{j=1}^{p}[\tilde{\beta}_j\cos(\omega_j\kappa)-\kappa\tilde{\omega}_j\rho_j\sin(\omega_j\kappa)] = \sum_{j=1}^{p}[\tfrac{1}{2}\rho_j(\kappa\tilde{\omega}_j)^2\cos(\omega_j\kappa)+\tilde{\rho}_j\tilde{\omega}_j\kappa\sin(\omega_j\kappa)]$$
$$\kappa = 0, 1, \ldots, 2p. \tag{72}$$

Example of application of the perturbation method. We suppose that:

$$B(\kappa) = \rho_0\,\delta(\kappa)+\rho_1\exp(i\omega_1\kappa) \qquad |\kappa| \leqslant 1. \tag{73}$$

$$D = \begin{bmatrix} 1 & 2 & \cdots & 2 & 0 & \cdots & 0 \\ 0 & 2\cos\omega_1 & \cdots & 2\cos\omega_p & -2\rho_1 \sin\omega_1 & \cdots & -2\rho_p \sin\omega_p \\ \vdots & \vdots & & \vdots & \vdots & & \vdots \\ 0 & 2\cos p\omega_1 & \cdots & 2\cos p\omega_p & -2p\rho_1 \sin p\omega_1 & \cdots & -2p\rho_1 \sin p\omega_p \\ 0 & 2\cos(p+1)\omega_1 & \cdots & 2\cos(p+1)\omega_p & -2(p+1)\rho_1 \sin(p+1)\omega_1 & \cdots & -2(p+1)\rho_p \sin(p+1)\omega_p \\ \vdots & \vdots & & \vdots & \vdots & & \\ 0 & 2\cos 2p\omega_1 & \cdots & 2\cos 2p\omega_p & -4p\rho_1 \sin 2p\omega_1 & \cdots & -4p\rho_p \sin 2p\omega_p \end{bmatrix} \quad (68)$$

The matrices D and D^{-1} are:

$$D = \begin{bmatrix} 1 & 1 & 0 \\ 0 & \cos\omega_1 & -\rho_1 \sin\omega_1 \\ 0 & \sin\omega_1 & \rho_1 \cos\omega_1 \end{bmatrix}$$

$$D^{-1} = \begin{bmatrix} 1 & -\cos\omega_1 & -\sin\omega_1 \\ 0 & \cos\omega_1 & \sin\omega_1 \\ 0 & -\rho_1^{-1}\sin\omega_1 & \rho_1^{-1}\cos\omega_1 \end{bmatrix} \tag{74}$$

We further assume that $\xi(\kappa)$ is a complex Gaussian process, and:

$$E\xi(0) = E\xi(1) = 0$$

$$E\xi^2(0) = \sigma^2 \qquad E\xi(0)\,\xi^*(1) = 0$$

$$E[\operatorname{Re}\xi(1)]^2 = E[\operatorname{Im}\xi(1)]^2 = \sigma^2$$

$$\Sigma = \begin{bmatrix} \sigma^2 & 0 & 0 \\ 0 & \sigma^2 & 0 \\ 0 & 0 & \sigma^2 \end{bmatrix}.$$

From (62) we have

$$E\tilde{\rho}_0 = E\tilde{\rho}_1 = E\tilde{\omega}_1 = 0$$

and from (63) we have

$$\varepsilon^2 W = \varepsilon^2 D^{-1}\,\Sigma(D^{-1})^* = \varepsilon^2\sigma^2 \begin{bmatrix} 2 & -1 & 0 \\ -1 & 1 & 0 \\ 0 & 0 & \rho_1^{-2} \end{bmatrix}. \tag{75}$$

We now put in the following numerical values for the parameters ρ_0, ρ_1, and ω_1 : $\rho_0 = 1$, $\rho_1 = 10$, $\omega_1 = \pi/4$, $B(0) = 11$, $B(1) = 10(1+i)/2^{\frac{1}{2}} = 7{\cdot}07+i.7{\cdot}07$. We assume that $\varepsilon^2\sigma^2 = 0{\cdot}01$ and take the noise values

$$\varepsilon\xi(0) = -0{\cdot}1 \qquad \varepsilon\xi(1) = 0{\cdot}03-0{\cdot}07i.$$

Then the observed disturbed covariances $\hat{B}(\kappa)$ are:

$$\hat{B}(0) = 10{\cdot}9 \qquad \hat{B}(1) = 7{\cdot}1+7i$$

and using the algorithm of Section 2 above we find that

$$\rho_0(\varepsilon) = 10{\cdot}9-\surd(99{\cdot}41) \approx 0{\cdot}9295$$

$$\exp[i\omega_1(\varepsilon)] = \frac{7{\cdot}1+7i}{99{\cdot}41} \quad \omega_1(\varepsilon) \approx \tfrac{1}{4}\pi-0{\cdot}007$$

$$\rho_1(\varepsilon) = \surd(99{\cdot}41) \approx 9{\cdot}97.$$

From (75) we have:

$$E[\rho_0(\varepsilon)-\rho_0]^2 = 2\varepsilon^2\sigma^2 = 0{\cdot}02$$

$$E[\rho_1(\varepsilon)-\rho_1]^2 = \varepsilon^2\sigma^2 = 0{\cdot}01$$

$$E[\omega_1(\varepsilon)-\omega_1]^2 = \frac{\varepsilon^2\sigma^2}{\rho_1^2} \approx \frac{\varepsilon^2\sigma^2}{\rho_1^2(\varepsilon)} = 10^{-4}$$

We now estimate the quadratic corrections $\varepsilon^2 \tilde{\tilde{\rho}}_0$, $\varepsilon^2 \tilde{\tilde{\rho}}_1$, and $\varepsilon^2 \tilde{\tilde{\omega}}_1$. Equations (65) have the form

$$\tilde{\tilde{\rho}}_0+\tilde{\tilde{\rho}}_1 = 0$$

$$\tilde{\tilde{\rho}}_1 \cos\omega_1-\tilde{\tilde{\omega}}_1 \rho_1 \sin\omega_1 = \tfrac{1}{2}\rho_1 \tilde{\omega}_1{}^2 \cos\omega_1+\tilde{\rho}_1 \tilde{\omega}_1 \sin\omega_1$$

$$\tilde{\tilde{\rho}}_1 \sin\omega_1+\tilde{\tilde{\omega}}_1 \rho_1 \cos\omega_1 = \tfrac{1}{2}\rho_1 \tilde{\omega}_1{}^2 \sin\omega_1-\tilde{\rho}_1 \tilde{\omega}_1 \cos\omega_1$$

and hence we find

$$\tilde{\tilde{\rho}}_0 = -\tfrac{1}{2}\rho_1 \tilde{\omega}_1{}^2 \quad \tilde{\tilde{\rho}}_1 = \tfrac{1}{2}\rho_1 \tilde{\omega}_1{}^2 \quad \tilde{\tilde{\omega}}_1 = \frac{-\tilde{\rho}_1 \tilde{\omega}_1}{\rho_1}.$$

Further from (75) we have

$$\left.\begin{aligned} E(\varepsilon^2 \tilde{\tilde{\rho}})^2 &= \tfrac{1}{4}\varepsilon^4 E\tilde{\omega}_1{}^4 = \frac{3\varepsilon^4 \sigma^4}{4}\rho_1{}^{-4} \approx 7{\cdot}5.10^{-9} \\ E(\varepsilon^2 \tilde{\tilde{\rho}}_1)^2 &= 7{\cdot}5.10^{-9} \\ E(\varepsilon^2 \tilde{\tilde{\omega}}_1)^2 &= \frac{\varepsilon^4 \sigma^4}{\rho_1{}^4} = 10^{-8} \end{aligned}\right\} \qquad (76)$$

and

$$\left.\begin{aligned} E(\varepsilon\tilde{\rho}_0)^2 &= 2\varepsilon^2 \sigma^2 = 0{\cdot}02 \\ E(\varepsilon\tilde{\rho}_1)^2 &= \varepsilon^2 \sigma^2 = 0{\cdot}01 \\ E(\varepsilon\tilde{\omega}_1)^2 &= \frac{\varepsilon^2 \sigma^2}{\rho_1{}^2} = 10^{-4}. \end{aligned}\right\} \qquad (77)$$

Comparing (77) to (76) we see that in this case the quadratic corrections $\varepsilon^2 \tilde{\tilde{\rho}}_0$, $\varepsilon^2 \tilde{\tilde{\rho}}_1$, and $\varepsilon^2 \tilde{\tilde{\omega}}_1$ are much smaller than the linear corrections $\varepsilon\tilde{\rho}_0$, $\varepsilon\tilde{\rho}_1$, and $\varepsilon\tilde{\omega}_1$, and hence are negligible.

Conclusions

The method we have described for retrieving harmonics from a finite segment of the covariance function is applicable to the case of non-ergodic processes whose spectral functions contain jumps or discontinuities. The amplitudes ρ_j of the harmonics in the covariance functions are the *ensemble-averaged* squared amplitudes of the harmonics of the process itself. However, the method is also applicable to a different situation which is often encountered in practice, namely, the case when the observed time series $x(\kappa)$ contains some deterministic harmonics with unknown (but not random) amplitudes and frequencies, in addition to a random component having a continuous spectrum. In this case the sample covariance function based on a single realization of the process will contain the same harmonics but with some additions caused by the finite length of the realization. These additions can be regarded as noise, since they decrease as the length of the realization increases.

We are thus led to suggest the use of the method described here in the problem of 'hidden periodicities' (see, for example, Whittle 1952 or Serebrennikov & Pervozvansky 1965). This problem is the following: a time series $x(\kappa)$ is observed:

$$x(\kappa) = \sum_{j=1}^{m} \gamma_j \exp\,[i\omega_j \kappa+i\phi_j]+n(\kappa) \qquad |\kappa| \leqslant T$$

where $n(\kappa)$ is a noise component. The problem is to estimate the amplitudes γ_j and frequencies ω_j, and possibly also the phases ϕ_j of the harmonics. The application of the present method to this problem would be to compute a sample covariance function:

$$\hat{B}(\kappa) = (2T+1)^{-1} \sum_{j=-T+\kappa}^{T} x(j)\, x^*(j-\kappa) \qquad |\kappa| \leqslant m,\ m \ll T$$

and then use a sample of $\hat{B}(\kappa)$ of length $2m$ as described above. In using $\hat{B}(\kappa)$ rather than $x(\kappa)$ we of course lose the phase information, but the phase ϕ_j can easily be found from $x(\kappa)$ by means of standard least-squares methods after determining the frequencies ω_j.

The problem of hidden periodicities can, of course, be treated by itself without regard to covariance function analysis (for discussion, see Pisarenko 1973). However, the power of the method proposed here is that it allows the determination of several tens of harmonics at once, using high-speed digital computers.

The procedure described here may also be useful for resolving peaks in continuous spectra; for that problem it is not necessary to go to the limit of discrete harmonics, but our new procedure does offer a way to decrease the ‘ smoothing ’ which is caused by the maximum entropy spectral estimator.

Acknowledgments

The author thanks Professor E. A. Flinn for many useful suggestions which have considerably improved this paper.

Institute of Physics of the Earth,
Academy of Sciences of the USSR,
Moscow

References

Achiezer, N., 1965. *Lessons on approximation theory*, Nauka, Moscow (in Russian).

Burg, J. P., 1972. The relationship between maximum entropy spectra and maximum likelihood spectra, *Geophysics*, **37**, 375–376.

Capon, J., 1969. High-resolution frequency-wavenumber analysis, *Proc. IEEE*, **57**, 1408–1418.

Doob, J., 1953. *Stochastic processes*, John Wiley and Sons, New York.

Gelfand, I., 1966. *Lessons on linear algebra*, Nauka, Moscow (third edition; in Russian),

Gelfond, A., 1959. *Calculus of differences*, Fizmatgiz, Moscow (in Russian).

Grenander, U. & Szegö, G., 1958. *Toeplitz forms and their application*, Berkeley, University of California Press.

Jenkins, G. & Watts, D., 1969. *Spectral analysis and its applications*, Holden-Day, San Francisco.

Lacoss, R., 1971. Data-adaptive spectral analysis methods, *Geophysics*, **37**, 661–675.

Parzen, E., 1969. Multiple time series modeling, 389–409, in *Multivariate Analysis* II, ed. P. R. Krishnaiah, Academic Press, New York.

Pisarenko, V. F., 1972. On the estimation of spectra by means of non-linear functions of the covariance matrix, *Geophys. J. R. astr. Soc.*, **28**, 511–531.

Pisarenko, V. F., 1973. On the determination of hidden periodicities, *Vychislitel’naya Seismologiya*, vol. 6, in press.

Serebrennikov, M. & Pervozvansky, A., 1965. *The determination of hidden periodicities*, Nauka, Moscow (in Russian).
Shannon, C., 1948. A mathematical theory of communication, *Bell. Syst. Tech. J.*, **27**, 379–423, and 623–656.
Ulrych, T. J., 1972. Maximum entropy power spectrum of truncated sinusoids, *J. geophys. Res.*, **77**, 1396–1400.
Whittle, P., 1952. The simultaneous estimation of a time series harmonic components and covariance structure, *Trabajos de Estadistica*, **3**, 43–57.

Appendix

Here we prove the theorem stated in Section 1 above. Consider the function $\phi_m(\lambda-\lambda_0)$:

$$\phi_m(\lambda-\lambda_0) = \sum_{\kappa,j=0}^{m} g_{\kappa-j} \exp[i(\lambda-\lambda_0)(\kappa-j)]$$

$$= \sum_{\kappa=-m}^{m} g_\kappa(m+1-|\kappa|) \exp[i(\lambda-\lambda_0)\kappa] \tag{78}$$

where $g_\kappa = g^*_{-\kappa}$ are some complex numbers specified below. We have:

$$(2\pi)^{-1}\int_{-\pi}^{\pi} w(\lambda-\lambda_0)\,dF(\lambda) = (2\pi)^{-1}\int_{-\pi}^{\pi}[w(\lambda-\lambda_0)-\phi_m(\lambda-\lambda_0)]\,dF(\lambda)$$

$$+(2\pi)^{-1}\int_{-\pi}^{\pi}\phi_m(\lambda-\lambda_0)\,dF(\lambda). \tag{79}$$

First we evaluate the first term in the right-hand side of (79); to do this we set:

$$g_\kappa = \frac{w_\kappa}{m+1-|\kappa|} \qquad |\kappa| \leqslant m \tag{80}$$

where w_κ are the Fourier coefficients of the function $w(\lambda)$:

$$w_\kappa = (2\pi)^{-1}\int_{-\pi}^{\pi} w(\kappa\lambda)\exp(-i\kappa\lambda)\,d\lambda. \tag{81}$$

It is seen from (78) that $\phi_m(\lambda-\lambda_0)$ represents a finite Fourier sum of order m for the function $w(\lambda-\lambda_0)$. Since we assumed that r derivatives of $w(\lambda)$ exist (where $r \geqslant 1$), and that $|w^{(r)}(\lambda)| \leqslant K$, we have the following estimate for the deviation of the Fourier sum $\phi_m(\lambda-\lambda_0)$ from its generating function $w(\lambda-\lambda_0)$:

$$\max_\lambda |w(\lambda-\lambda_0)-\phi_m(\lambda-\lambda_0)| \leqslant Km^{-r}[(4/\pi^2)\log(m+1)+4] \tag{82}$$

(see Achiezer 1965). Now if $m \to \infty$ the first term on the right-hand side of (79) tends to zero. Hence the second term tends to the integral on the left side, i.e. as $m \to \infty$,

$$(2\pi)^{-1}\int_{-\pi}^{\pi}\phi_m(\lambda-\lambda_0)\,dF(\lambda) \to (2\pi)^{-1}\int_{-\pi}^{\pi} w(\lambda-\lambda_0)\,dF(\lambda) \quad \text{as} \quad m\to\infty. \tag{83}$$

The integral on the left-hand side of (83) can be written:

$$(2\pi)^{-1}\int_{-\pi}^{\pi}\phi_m(\lambda-\lambda_0)\,dF(\lambda) = (2\pi)^{-1}\sum_{\kappa=-m}^{m} w_\kappa \int_{-\pi}^{\pi}\exp\,[i(\lambda-\lambda_0)\,\kappa]\,dF(\lambda)$$

$$= \sum_{\kappa=-m}^{m} w_\kappa \exp\,(-i\lambda_0\,\kappa)\,B^*(\kappa). \qquad (84)$$

Substituting (84) into (79) and using (82) we get finally:

$$\left|(2\pi)^{-1}\int_{-\pi}^{\pi} w(\lambda-\lambda_0)\,dF(\lambda) - \sum_{\kappa=-m}^{m} w_\kappa \exp\,(-i\lambda_0\,\kappa)\,B^*(\kappa)\right|$$

$$\leqslant KB(0)\,m^{-r}[(4/\pi^2)\log\,(m+1)+4]$$

which was to be proved.

MULTIPLE EMITTER LOCATION AND SIGNAL PARAMETER ESTIMATION

RALPH SCHMIDT

ESL, Incorporated
495 Java Drive
Sunnyvale, CA 94086

Abstract

Processing the signals received on an array of sensors for the location of the emitter is of great enough interest to have been treated under many special case assumptions.

The general problem considers sensors with arbitrary locations and arbitrary directional characteristics (gain/phase/polarization) in a noise/interference environment of arbitrary covariance matrix.

This report is concerned first with the multiple emitter aspect of this problem and second with the generality of solution. A description is given of the Multiple Signal Classification (MUSIC) algorithm, which provides asymptotically unbiased estimates of

1. number of incident wavefronts present
2. directions-of-arrival (or emitter locations)
3. strengths and cross-correlations among the incident waveforms
4. noise/interference strength.

Examples and comparisons with methods based on Maximum Likelihood and Maximum Entropy, as well as conventional beam forming are included. An example of its use as a multiple frequency estimator operating on time series is included.

Reprinted with permission from *Proc. of the RADC Spectrum Estimation Workshop*, 1979, pp. 243–258.
This paper has been corrected by the author and is scheduled to be published in the *IEEE Trans. Antennas Propagat.*

Introduction

The term multiple signal classification (MUSIC) is used to describe experimental and theoretical techniques involved in determining the parameters of multiple wavefronts arriving at an antenna array from measurements made on the signals received at the array elements.

The general problem considers antennas with arbitrary locations and arbitrary directional characteristics (gain/phase/polarization) in a noise/interference environment of arbitrary covariance matrix. The Multiple Signal Classification (MUSIC) approach is described; it can be implemented as an algorithm to provide asymptotically unbiased estimates of

1. Number of signals
2. Directions-of-arrival
3. Strengths and cross-correlations among the directional waveforms
4. Polarizations
5. Strength of noise/interference.

These techniques are very general and of wide application. Special cases of MUSIC are

1. Conventional Interferometry
2. Monopulse DF, i.e., using multiple colocated antennas
3. Multiple Frequency Estimation.

The Data Model

The waveforms received at the M array elements are linear combinations of the D incident wavefronts and noise. Thus, the multiple signal classification (MUSIC) approach begins with the following model for characterizing the received M vector X as in

$$\begin{bmatrix} X_1 \\ X_2 \\ \vdots \\ X_M \end{bmatrix} = \begin{bmatrix} a(\theta_1) & a(\theta_2) & \cdots & a(\theta_D) \end{bmatrix} \begin{bmatrix} F_1 \\ F_2 \\ \vdots \\ F_D \end{bmatrix} + \begin{bmatrix} W_1 \\ W_2 \\ \vdots \\ W_M \end{bmatrix}$$

or

$$X = AF + W \tag{1}$$

The incident signals are represented in amplitude and phase at some arbitrary reference point (for instance the origin of the coordinate system) by the complex quantities $F_1, F_2, \ldots, F_D$. The noise, whether "sensed" along with the signals or generated internal to the instrumentation, appears as the complex vector W.

The elements of X and A are also complex in general. The a_{ij} are known functions of the signal arrival angles and the array element locations. That is, a_{ij} depends on the ith array element, its position relative to the origin of the coordinate system, and its response to a signal incident from the direction of the jth signal. The jth column of A is a "mode" vector $a(\theta_j)$ of responses to the direction-of-arrival θ_j of the jth signal. Knowing the mode vector $a(\theta_1)$ is tantamount to knowing θ_1 (unless $a(\theta_1) = a(\theta_2)$ with $\theta_1 \neq \theta_2$, an unresolvable situation, a type I ambiguity).

In geometrical language, the measured X vector can be visualized as a vector in M dimensional space. The directional mode vectors $a(\theta_j) = a_{ij}$ for i = 1, 2, ..., M, i.e., the columns of A, can also be so visualized. Equation (1) states that X is a particular linear combination of the mode vectors; the elements of F are the coefficients of the combination. Note that the X vector is confined to the range space of A. That is, if A has 2 columns, the range space is no more than a 2-dimensional subspace within the M space and X necessarily lies in the subspace. Also note that $a(\theta)$, the continuum of all possible mode vectors lies within the M space but is quite nonlinear. For help in visualizing this, see Figure 1. For example, in an azimuth-only DF system, θ will consist of a single parameter. In an azimuth/elevation/range system, θ will be replaced by θ, ϕ, r for example. In any case, $a(\theta)$ is a vector continuum such as a "snake" (azimuth only) or a "sheet" (AZ/EL) twisting and winding through the M space. (In practice, the procedure by which the $a(\theta)$ continuum is measured or otherwise established corresponds to calibrating the array.)

In these geometrical terms (see Figure 1), the problem of solving for the directions-of-arrival of multiple incident wavefronts consists of locating the intersections of the $a(\theta)$

**THE SIGNAL SUBSPACE
(DETERMINED BY THE DATA)**

**SIGNAL SPACE
EIGENVECTOR**

$a(\theta_2)$

$a(\theta_1)$

**SIGNAL SPACE
EIGENVECTOR**

**NOISE SPACE
EIGENVECTOR**

**THE a(θ) CONTINUUM OF
DIRECTION-OF-ARRIVAL
VECTORS (DETERMINED
BY ARRAY GEOMETRY
AND CHARACTERISTICS)**

$\bullet_1, \bullet_2, \bullet_{MIN}$ ARE THE EIGENVECTORS OF S CORRESPONDING
TO EIGENVALUES $\lambda_1 > \lambda_2 > \lambda_{MIN} > 0$

$\bullet_1, \bullet_2$ SPAN THE SIGNAL SUBSPACE

$a(\theta_1), a(\theta_2)$ ARE THE INCIDENT SIGNAL MODE VECTORS

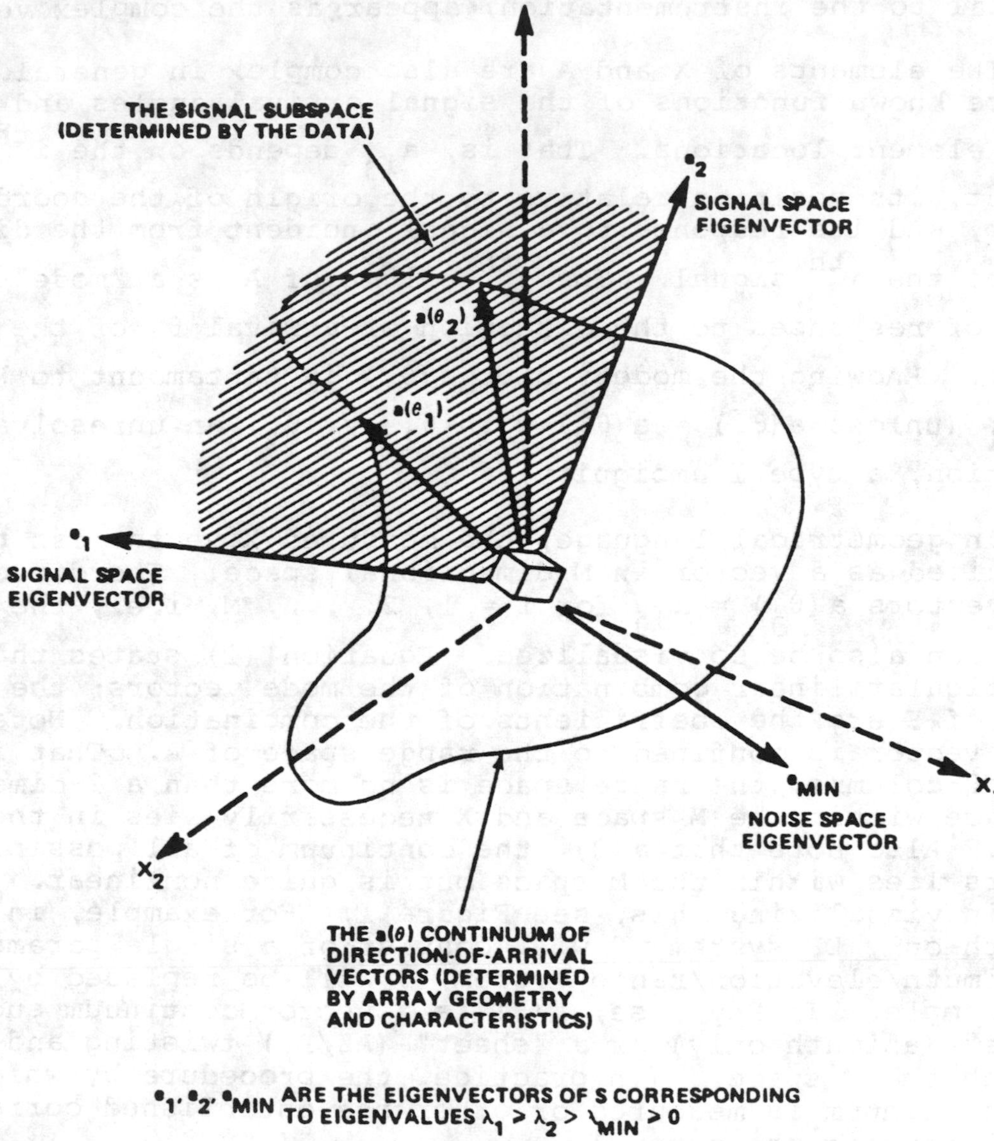

FIGURE 1. Geometric Portrayal for Three-Antenna Case

continuum with the range space of A. The range space of A is, of course, obtained from the measured data. The means of obtaining the range space and, necessarily, its dimensionality (the number D of incident signals) follows.

The S Matrix

The MxM covariance matrix of the X vector is

$$S \triangleq \overline{XX^*} = A\,\overline{FF^*}A^* + \overline{WW^*}$$

or

$$S = APA^* + \lambda S_o \qquad (2)$$

under the basic assumption that the incident signals and the noise are uncorrelated. Note that the incident waveforms represented by the elements of F may be uncorrelated (the DxD matrix $P \triangleq \overline{FF^*}$ is diagonal) or may contain completely correlated pairs (P is singular). In general, P will be "merely" positive definite which reflects the arbitrary degrees of pairwise correlations occurring between the incident waveforms.

When the number of incident wavefronts D is less than the number of array elements M, then APA* is singular; it has a rank less than M. Therefore

$$|APA^*| = |S - \lambda S_o| = 0 \qquad (3)$$

This equation is only satisfied with λ equal to one of the eigenvalues of S in the metric of S_o. But, for A full rank and P positive definite, APA* must be nonnegative definite. Therefore λ can only be the minimum eigenvalue λ_{min}. Therefore, any measured $S = \overline{XX^*}$ matrix can be written

$$S = APA^* + \lambda_{min} S_o, \quad \lambda_{min} \geq 0 \qquad (4)$$

where λ_{min} is the smallest solution to $|S - \lambda S_o| = 0$. Note the special case wherein the elements of the noise vector W are mean zero, variance σ^2; in which case, $\lambda_{min} S_o = \sigma^2 I$.

Calculating a Solution

The rank of APA* is D and can be determined directly from the eigenvalues of S in the metric of S_o. That is, in the complete set of eigenvalues of S in the metric of S_o, λ_{min} will not always be simple. In fact, it occurs repeated N = M-D times.

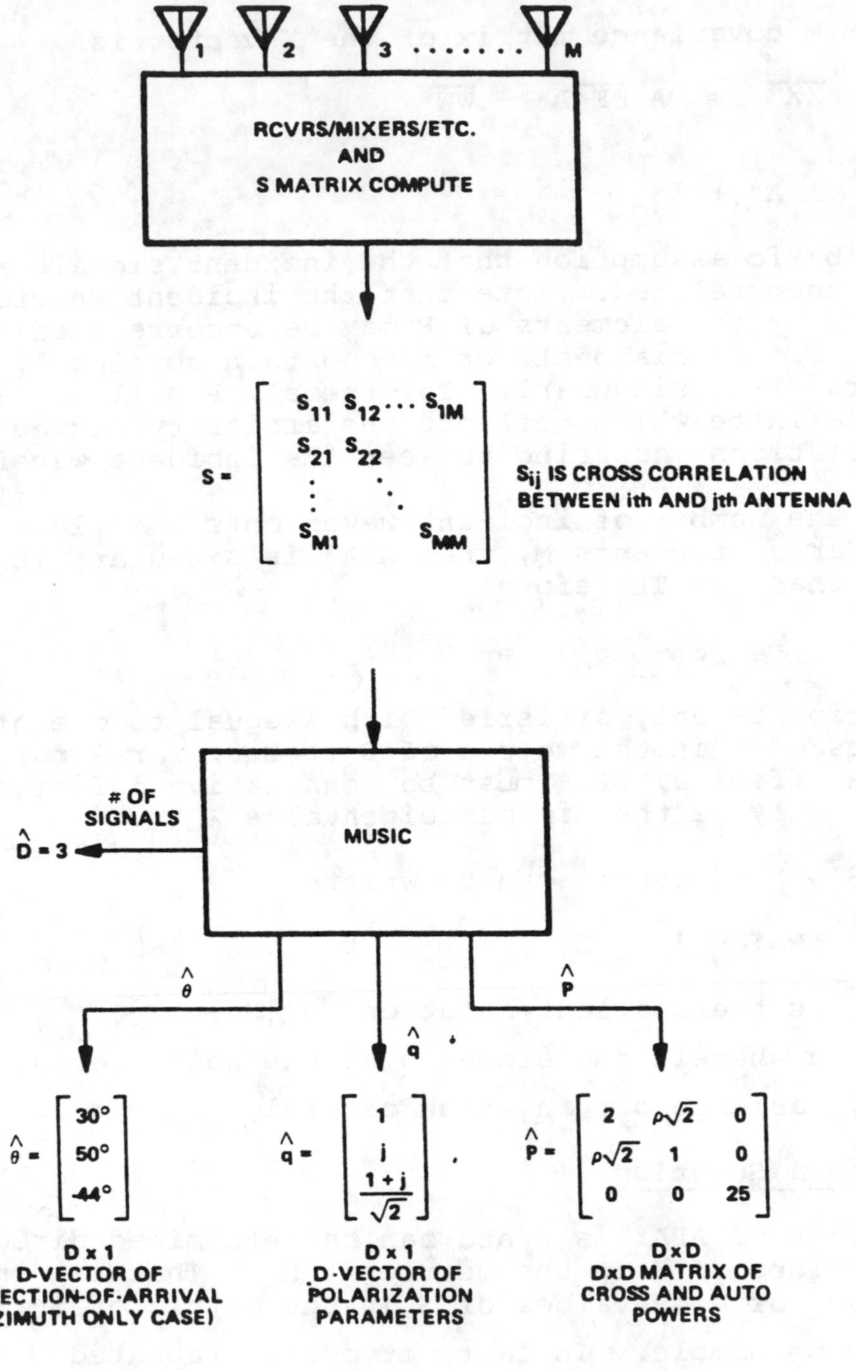

FIGURE 2. Block Diagram for Multiple Signal Classification

This is true because the eigenvalues of S and those of $S - \lambda_{min}S_o = APA^*$ differ by λ_{min} in all cases. Since the minimum eigenvalue of APA^* is zero (being singular), λ_{min} must occur repeated N times. Therefore, the number of incident signals estimator is

$$\hat{D} = M - \hat{N} \tag{5}$$

where $\hat{N}$ = the multiplicity of $\lambda_{min}(S,S_o)$ and $\lambda_{min}(S,S_o)$ is read "λ_{min} of S in the metric of S_o." (In practice, one can expect that the multiple λ_{min}'s will occur in a cluster rather than all precisely equal. The "spread" on this cluster decreases as more data is processed.)

<u>The Signal and Noise Subspaces</u>

The M eigenvectors of S in the metric of S_o must satisfy $Se_i = \lambda_i S_o e_i$, i = 1, 2, ..., M. Since $S = APA^* + \lambda_{min}S_o$, we have $APA^*e_i = (\lambda_i - \lambda_{min})S_o e_i$. Clearly, for each of the λ_i that is equal to λ_{min} - there are N - we must have $APA^*e_i = 0$ or $A^*e_i = 0$. That is, the eigenvectors associated with $\lambda_{min}(S,S_o)$ are orthogonal to the space spanned by the columns of A; the incident signal mode vectors!

Thus we may justifiably refer to the N dimensional subspace spanned by the N noise eigenvectors as the noise subspace and the D dimensional subspace spanned by the incident signal mode vectors as the signal subspace; they are disjoint.

<u>The Algorithm</u>

We now have the means to solve for the incident signal mode vectors. If E_N is defined to be the MxN matrix whose columns are the N noise eigenvectors, and the ordinary Euclidean distance (squared) from a vector Y to the signal subspace is $d^2 = Y^*E_N E_N^* Y$, we can plot $1/d^2$ for points along the $a(\theta)$ continuum as a function of θ. That is,

$$P_{MU}(\theta) = \frac{1}{a^*(\theta)E_N E_N a(\theta)} \tag{6}$$

(However, the $a(\theta)$ continuum may intersect the D dimensional signal subspace more than D times; another unresolvable situation occurring only for the case of multiple incident signals - a type II ambiguity.) It is clear from the expression that MUSIC is asymptotically unbiased even for multiple incident wavefronts because S is asymptotically perfectly measured so that E_N is also. $a(\theta)$ does not depend on the data.

Once the directions-of-arrival of the D incident signals have been found, the A matrix becomes available and may be used to compute the parameters of the incident signals. The solution for the P matrix is direct† and can be expressed in terms of $(S - \lambda_{min} S_o)$ and A. That is, since $APA^* = S - \lambda_{min} S_o$,

$$P = (A^*A)^{-1} A^* (S - \lambda_{min} S_o) A (A^*A)^{-1} \qquad (7)$$

Including Polarization

Consider a signal arriving from a specific direction θ_o. Assume that the array is not diverse in polarization; i.e., all elements are identically polarized, say, vertically. Certainly the DF system will be most sensitive to vertically polarized energy, completely insensitive to horizontal and partially sensitive to arbitrarily polarized energy. The array is only sensitive to the vertically polarized component of the arriving energy.

For a general or polarizationally diverse array, the mode vector corresponding to the direction θ_o depends on the signal polarization. A vertically polarized signal will induce one mode vector and horizontal another, and right hand circular (RHC) still another.

Recall that signal polarization can be completely characterized by a single complex number q. We can "observe" how the mode vector changes as the polarization parameter q for the emitter changes at the specific direction θ_o. It can be proven that as q changes through all possible polarizations, the mode vector sweeps out a two-dimensional "polarization subspace." Thus, only two independent mode vectors spanning the polarization subspace for the direction θ_o are needed to represent any emitter polarization q at direction θ_o. The practical embodiment of this is that only the mode vectors of two emitter polarizations need be calculated or kept in store for direction θ_o in

† [added in reprint] Equation(7) is true if S_0, the noise covariance matrix, is the identity matrix. In general, although there are many estimators of P, the least squares estimate based on $X = AF + W$ with $\overline{WW^*} = \lambda_{min} S_0$ requires *whitening* which leads to

$$P = (A^* S_0^{-1} A)^{-1} A^* S_0^{-1} (S - \lambda_{min} S_0) S_0^{-1} A (A^* S_0^{-1} A)^{-1} \qquad (7)$$

order to solve for emitter polarizations where only one was needed to solve for DOA in a system with an array that was not polarizationally diverse.

These arguments lead to an equation similar to Equation (6) for $P(\theta)$ but including the effects of polarization diversity among the array elements.

$$P_{MU}(\theta) = \frac{1}{\lambda_{min}\left(\begin{bmatrix} a_x^*(\theta) \\ a_y^*(\theta) \end{bmatrix} E_N E_N^* \begin{bmatrix} a_x(\theta) & a_y(\theta) \end{bmatrix}\right)} \tag{8}$$

where $a_x(\theta)$ and $a_y(\theta)$ are the two continua corresponding to, for example, separately taken x and y linear incident wavefront polarizations. The eigenvector corresponding to λ_{min} in Equation (8) provides the polarization parameter q since it is of the form $[1\ q]^T$.

The Algorithm

In summary, the steps of the algorithm are

Step 0: Collect data, form S

Step 1: Calculate Eigenstructure of S in metric of S_o

Step 2: Decide number of signals D; Equation (5)

Step 3: Evaluate $P_{MU}(\theta)$ vs. θ; Equation (6) or (8)

Step 4: Pick D peaks of $P_{MU}(\theta)$

Step 5: Calculate remaining parameters; Equation (7).

The above steps have been implemented in several forms to verify and evaluate the principles and basic performance. Field tests have been conducted using actual receivers, arrays, and multiple transmitters. The results of these tests have demonstrated the potential of this approach for handling multiple signals in practical situations. Performance results are being prepared for presentation in another paper.

Comparison With Other Methods

In comparing MUSIC with ordinary beamforming (BF) Maximum Likelihood (ML) and Maximum Entropy (ME), the following expressions were used. See Figures 3 and 4.

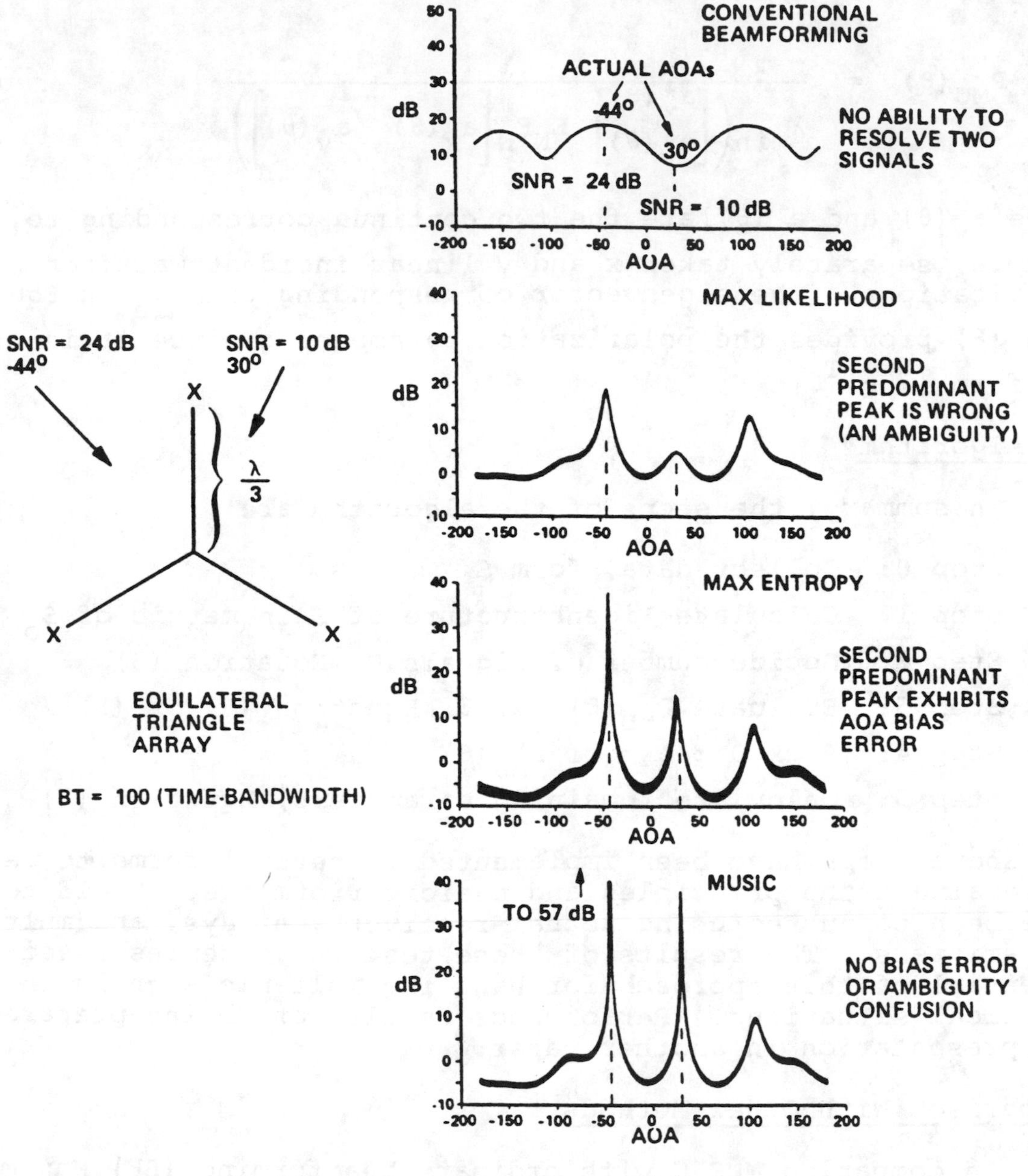

FIGURE 3. Example of Azimuth-Only DF Performance

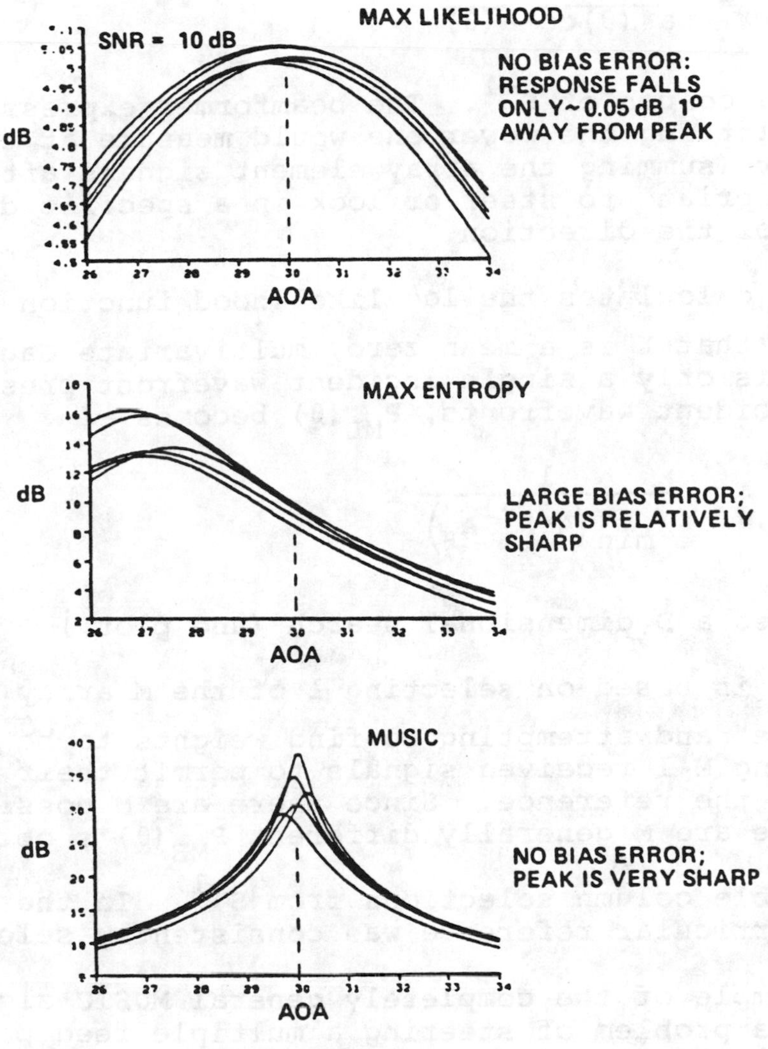

FIGURE 4. Example of Azimuth - Only DF Performance (Scale Expanded About Weaker Signal at 30°)

$$P_{BF}(\theta) = a^*(\theta)\ S\ a(\theta)$$

$$P_{ML}(\theta) = \frac{1}{a^*(\theta)S^{-1}a(\theta)}$$

$$P_{ME}(\theta) = \frac{1}{a^*(\theta)cc^*a(\theta)}$$

where c is a column of S^{-1}. The beamformer expression calculates for plotting the power one would measure at the output of a beamformer (summing the array element signals after inserting delays appropriat to steer or look in a specific direction) as a function of the direction.

$P_{ML}(\theta)$ calculates the lo likelihood function under the assumptions that X is a mean zero, multivariate Gaussian and that there is only a single incident wavefront present. For multiple incident wavefronts, $P_{ML}(\theta)$ becomes

$$P_{ML}(\theta) = \frac{1}{\lambda_{min}\left(A_\theta^* S^{-1} A_\theta\right)}$$

which implies a D dimensional search (and plot!)

$P_{ME}(\theta)$ is based on selecting 1 of the M array elements as a "reference" and attempting to find weights to be applied to the remaining M-1 received signals to permit their sum with a MMSE fit to the reference. Since there are M possible references, there are M generally different $P_{ME}(\theta)$'s obtained from the M possible column selections from S^{-1}. In the comparison plots, a particular reference was consistently selected.

An example of the completely general MUSIC algorithm applied to a problem of steering a multiple feed parabolic dish antenna is shown in Figure 5. $\frac{\sin x}{x}$ pencil beamshapes skewed slightly off boresight are assumed for the element patterns. Since the six antennas are essentially colocated, the DF capacity arises out of the antenna beam pattern diversity. The computer was used to simulate the "noisy" S matrix that would arise in practice for the conditions desired and then to subject

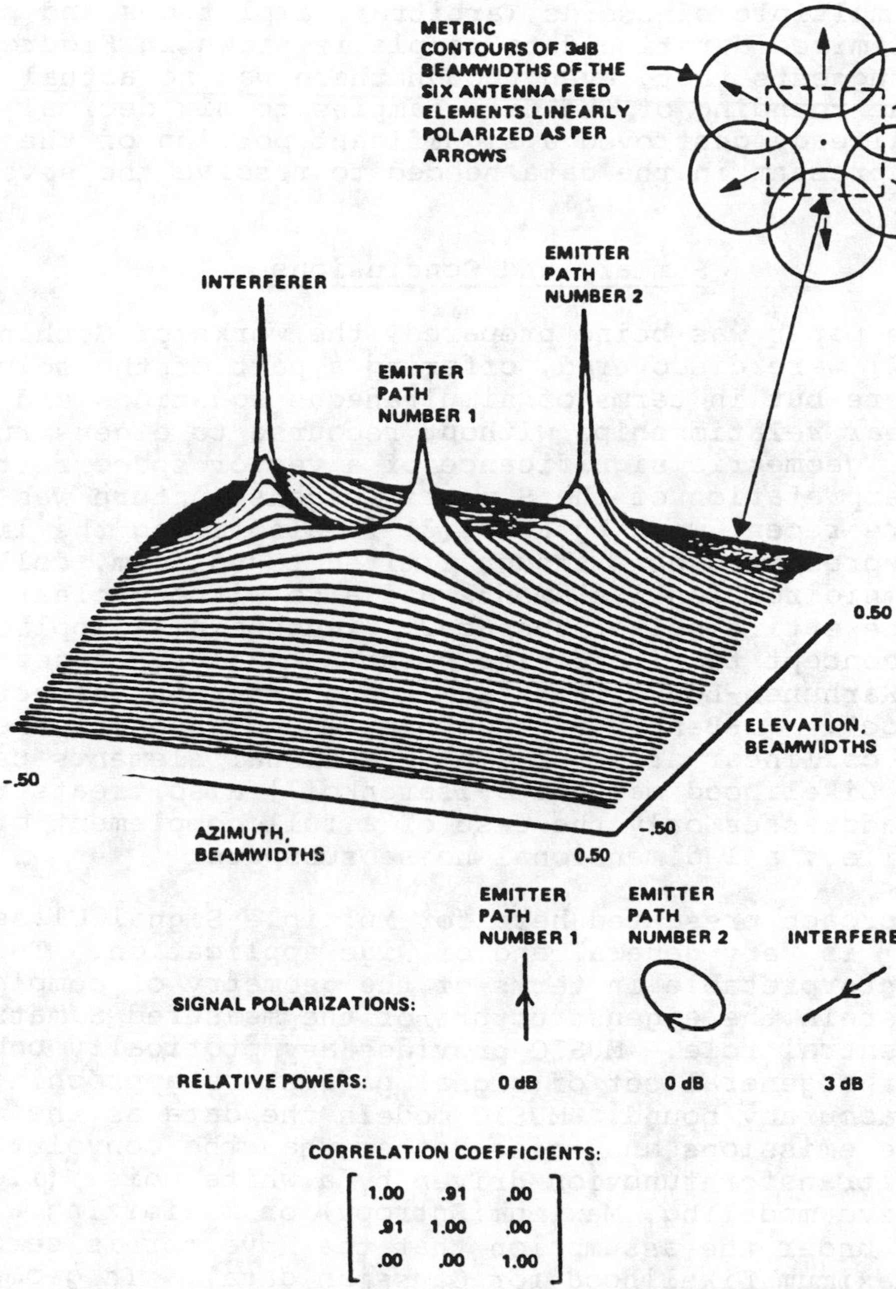

FIGURE 5. MUSIC Applied to a Multiple Feed, Parabolic Dish Antenna System

it to the MUSIC algorithm. Figure 5 shows how three directional signals are distinguished and their polarizations estimated even though two of the arriving signals are highly similar (90% correlated).

The application of MUSIC to the estimation of the frequencies of multiple sinusoids (arbitrary amplitudes and phases) for a very limited duration data sample is shown in Figure 6. The figure suggests that, even though there was no actual noise included, the rounding of the data samples to six decimal digits has already destroyed a significant portion of the information present in the data needed to resolve the several frequencies.

Summary and Conclusions

As this paper was being prepared, the works of Gething[1] and Davies[2] were discovered, offering a part of the solution discussed here but in terms of simultaneous equations and special linear relationships without recourse to eigenstructure. However, the geometric significance of a vector space setting and the interpretation of the S matrix eigenstructure was missed. More recent work by Reddi[3] is also along the lines of the work presented here though limited to uniform, collinear arrays of omnidirectional elements and also without clear utilization of the entire noise subspace. Ziegenbein[4] applied the same basic concept to time series spectral analysis referring to it as a Karhunen-Loeve Transform though treating aspects of it as "ad hoc". El-Behery and MacPhie[5] and Capon[6] treat the uniform collinear array of omnidirectional elements using the Maximum Likelihood method. Pisarenko[7] also treats time series and addresses only the case of a full complement of sinusoids; i.e., a 1 dimensional noise subspace.

The approach presented here for Multiple Signal Classification (MUSIC) is very general and of wide application. The method is interpretable in terms of the geometry of complex M spaces wherein the eigenstructure of the measured S matrix plays the central role. MUSIC provides asymptotically unbiased estimates of a general set of signal parameters approaching the Cramer-Rao accuracy bound. MUSIC models the data as the sum of point source emissions and noise rather than the convolution of an all pole transfer function driven by a white noise (i.e., autoregressive modeling, Maximum Entropy) or maximizing a probability under the assumption that the X vector is zero mean, Gaussian (Maximum Likelihood for Gaussian data). In geometric

DATA: 16 COMPLEX TIME SAMPLES OF 6 CISOIDS

FREQ (Hz)	REL AMP (dB)
78.1	-62.33
134.1	-11.10
138.6	0.0
142.9	-11.10
152.9	-40.0
165.3	-26.02

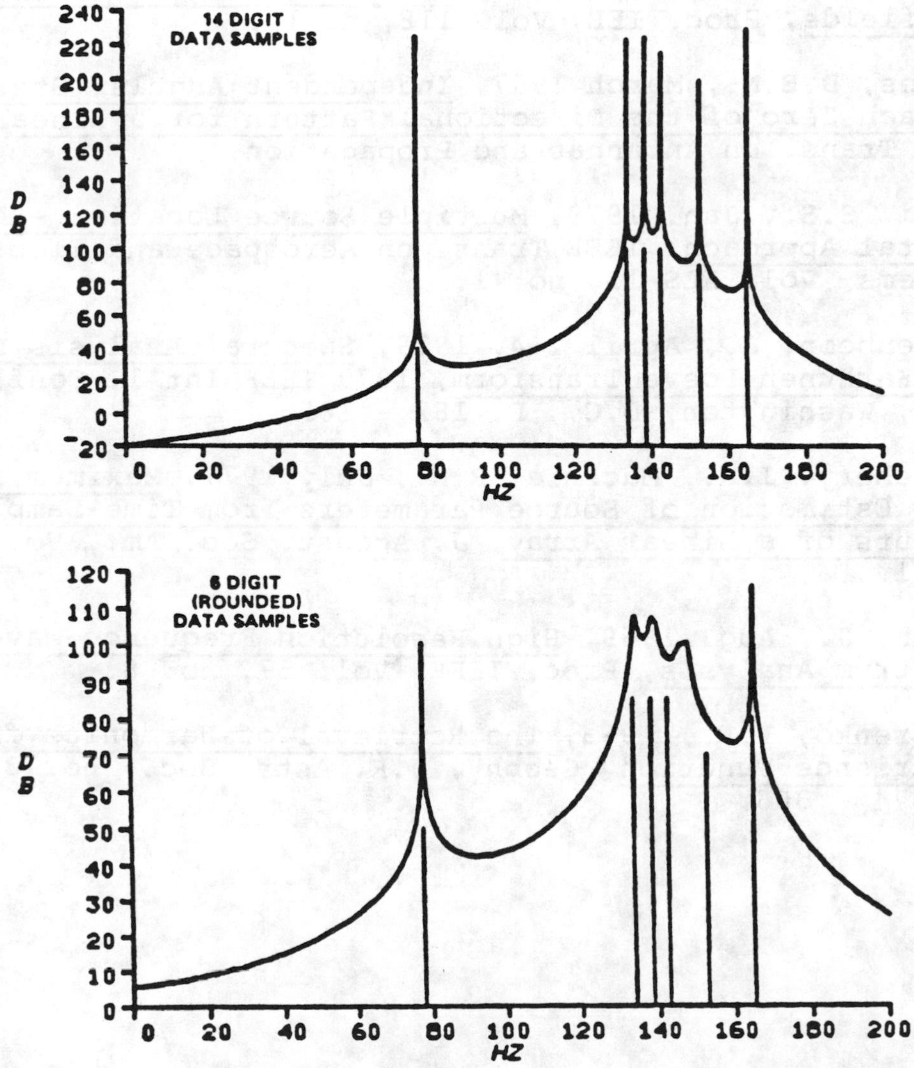

FIGURE 6. Example of MUSIC Used for Frequency Estimation

terms MUSIC minimizes the distance from the a(θ) continuum to the signal subspace whereas Maximum Likelihood minimizes a weighted combination all component distances.

No assumptions nave been made about array geometry. The array elements may be arranged in a regular or irregular pattern and may differ or be identical in directional characteristics (amplitude/phase) provided their polarization characteristics are all identical. The extension to include general polarizationally diverse antenna arrays will be more completely described in a separate paper.

References

1. Gething, P.J.D., Oct. 1971, Analysis of Multicomponent Wavefields, Proc. IEE, Vol. 118, no 10.

2. Davies, D.E.N., March 1967, Independent Angular Steering of Each Zero of the Directional Pattern for a Linear Array, IEEE Trans. on Antennas and Propagation.

3. Reddi, S.S., Jan. 1979, Multiple Source Location - A Digital Approach, IEEE Trans. on Aerospace and Electronic Systems, Vol. AES-15, no. 1.

4. Ziegenbein, J., April 2-4, 1979, Spectral Analysis Using the Karhunen-Loeve Transform, 1979 IEEE Int'l. Conf. on ASSP, Washington, D.C., P. 182 - 185.

5. El-Behery, I.N., MacPhie, R.H., July 1977, Maximum Likelihood Estimation of Source Parameters from Time-Sampled Outputs of a Linear Array, J. Aconst. Soc. Am., Vol. 62, no. 1.

6. Capon, J., Aug. 1969, High Resolution Frequency-Wavenumber Spectrum Analysis, Proc. IEEE. Vol. 57, no. 8.

7. Pisarenko, V.F., 1973, The Retrieval of Harmonics from a Covariance Function, Geophy. J.R. Astr. Soc., no. 33, P. 374 - 366.

Singular Value Decomposition and Improved Frequency Estimation Using Linear Prediction

DONALD W. TUFTS AND RAMDAS KUMARESAN

Abstract—Linear-prediction-based (LP) methods for fitting multiple-sinusoid signal models to observed data, such as the forward-backward (FBLP) method of Nuttall [5] and Ulrych and Clayton [6], are very ill-conditioned. The locations of estimated spectral peaks can be greatly affected by a small amount of noise because of the appearance of outliers. LP estimation of frequencies can be greatly improved at low SNR by singular value decomposition (SVD) of the LP data matrix. The improved performance at low SNR is also better than that obtained by using the eigenvector corresponding to the minimum eigenvalue of the correlation matrix, as is done in Pisarenko's method and its variants.

I. INTRODUCTION

In previous related work, we presented results concerning the estimation of closely spaced frequencies of multiple sinusoids in noise using maximum likelihood (ML) and linear prediction (LP) methods [1], [2]. We have found that the threshold SNR, where the variance of the frequency estimation error departs significantly from the Cramer–Rao (CR) bound, is much lower for ML frequency estimation than for LP frequency estimation [1], [2]. This is a manifestation of the ill-conditioned nature of the LP frequency estimation [3].

In [2] we showed that the ill-conditioning of LP frequency estimation is not intrinsic. Good frequency estimates were obtained below the LP threshold SNR by extraction of the principal eigenvectors of the estimated correlation matrix which appears in the linear prediction normal equations.

Further, we showed in [3] and [4] that for purposes of estimating closely spaced frequencies, better LP coefficients can be obtained from the principal eigenvectors of the estimated correlation matrix than can be obtained by conventional LP calculations using the total matrix [5], [6].

In this correspondence, we wish to provide some new insights into our improvements to frequency estimation by linear prediction. We demonstrate in more detail the ill-conditioned nature of conventional LP methods, and we discuss the source of this ill conditioning. We also discuss the use of the singular value decomposition (SVD) of the data matrix. Henderson's work [7] is closely related to our work reported here.

II. SINGULAR VALUE DECOMPOSITION AND LINEAR PREDICTION

Let us assume that we want to approximate some complex valued data $x(k)$ by L sinusoids in complex valued form,

$$s(k) = \sum_{m=1}^{L} a_m z_m^k \qquad k = 0, 1, 2, \cdots N-1, \tag{1}$$

in which each of the complex numbers z_m has unit magnitude. And we assume that $x(k)$ consists of $s(k)$ plus broad-band complex valued noise $w(k)$:

$$x(k) = s(k) + w(k) \qquad k = 0, 1, 2 \cdots N-1. \tag{2}$$

In an LP method for estimating the z_m values or their angles, a coefficient vector $\boldsymbol{c}$ is sought which will approximately satisfy the linear prediction equations.

$$\begin{bmatrix} x(0) & x(1) & \cdots x(P-1) \\ x(1) & x(2) & \cdots x(P) \\ \vdots & \vdots & \vdots \\ x(N-P-1) & x(N-P) & \cdots x(N-1) \end{bmatrix} \begin{bmatrix} c(P) \\ c(P-1) \\ \vdots \\ c(1) \end{bmatrix} \approx \begin{bmatrix} x(P) \\ x(P+1) \\ \vdots \\ x(N-1) \end{bmatrix} \tag{3a}$$

or

$$A\,\boldsymbol{c} \simeq \boldsymbol{b}. \tag{3b}$$

The matrix A and vector $\boldsymbol{b}$ are slightly different if the data are used in both forward and backward directions [5], [6].

More specifically, of all the coefficient vectors which minimize the sum of squared magnitudes

$$\|\boldsymbol{b} - A\,\boldsymbol{c}\|^2 = \sum_{k=P}^{N-1} \left| x(k) - \sum_{n=1}^{P} c(n) \times (k-n) \right|^2 \tag{4}$$

we wish to find a vector $\boldsymbol{c}$ of minimum norm. The norm is defined by

$$\|c\|^2 = \boldsymbol{c}^* \boldsymbol{c} = \sum_{n=1}^{L} |c(n)|^2 \tag{5}$$

in which the asterisk stands for complex conjugate transpose.

The minimum norm solution to this minimum prediction error problem can be compactly written in terms of the pseudoinverse [8] A^I of the data matrix A. The solution is

$$\boldsymbol{c} = A^I \boldsymbol{b}. \tag{6}$$

We made use of this solution to obtain computationally simple improvements to earlier LP methods [9].

In the conventional LP methods [5], [6], [10]–[12], arrangements are made such that there is likely to be one vector $\boldsymbol{c}$ which minimizes the prediction error energy of (4). For this special case, we can use the following formula for the pseudoinverse [8]:

$$A^I = (A^*A)^{-1} A^*. \tag{7}$$

Manuscript received June 10, 1981; revised January 4, 1982. This work was supported by the Office of Naval Research under Grant N0014-81-k-0144.

The authors are with the Department of Electrical Engineering, University of Rhode Island, Kingston, RI 02881.

Reprinted from *IEEE Trans. Acoust., Speech, Signal Processing*, vol. ASSP-30, pp. 671–675, Aug. 1982.

TABLE I
STANDARD DEVIATION OF FREQUENCY ESTIMATION ERROR (FOR $f_2 = 0.5$ Hz).
500 INDEPENDENT TRIALS WERE USED. σ_{CR} IS THE CRAMER–RAO BOUND.
PE: PRINCIPAL EIGENVECTORS METHOD. FBLP: FORWARD–BACKWARD
LINEAR PREDICTION METHOD. P = PREDICTION FILTER ORDER.
25 DATA SAMPLES WERE USED.

P	SNR=70dB, σ_{CR}=.311x10^{-5} Standard Deviation		SNR=30dB, σ_{CR}=.311x10^{-3} Standard Deviation		SNR=12dB, σ_{CR}=.276x10^{-2} Standard Deviation		SNR=7dB, σ_{CR}=.491x10^{-2} Standard Deviation	
	PE	FBLP	PE	FBLP	PE	FBLP	PE	FBLP
**2	.529x10^{-4}	.529x10^{-4}	.629x10^{-2}	.629x10^{-2}	*	*	*	*
4	.110x10^{-4}	.112x10^{-4}	.191x10^{-2}	.196x10^{-2}	*	*	*	*
6	.676x10^{-5}	.694x10^{-5}	.680x10^{-3}	.691x10^{-3}	*	*	*	*
8	.510x10^{-5}	.513x10^{-5}	.514x10^{-3}	.506x10^{-3}	.473x10^{-2}	*	*	*
10	.410x10^{-5}	.448x10^{-5}	.424x10^{-3}	*	.345x10^{-2}	*	*	*
12	.400x10^{-5}	.426x10^{-5}	.403x10^{-3}	*	.313x10^{-2}	*	*	*
14	.583x10^{-5}	*	.394x10^{-3}	*	.309x10^{-2}	*	.484x10^{-2}	*
16	.348x10^{-5}	*	.355x10^{-3}	*	.219x10^{-2}	*	.433x10^{-2}	*
18	.346x10^{-5}	*	.345x10^{-3}	*	.279x10	*	.417x10^{-2}	*
20	.332x10^{-5}	*	.338x10^{-3}	*	.264x10^{-2}	*	*	*
22	.362x10^{-5}	.347x10^{-5}	.353x10^{-3}	*	.275x10^{-2}	*	*	*
23	.408x10^{-5}	.400x10^{-5}	.406x10^{-3}	*	.443x10^{-2}	*	*	*
***24 $\frac{(N-L)}{2}$	.514x10^{-5}	.514x10^{-5}	.517x10^{-3}	.517x10^{-3}	.670x10^{-2}	.670x10^{-2}	*	*

* For these cases the frequency estimation error was large either due to a spurious zero of the prediction-error filter wandering too close to the unit circle (at large P) or due to merging of two zeros close to the signalfrequency locations.

** This corresponds to the Prony's method with the modification in which the data is used in both directions. PE and FBLP give identical results

*** This corresponds to the Kumaresan-Prony case. PE and FBLP (min. Norm Solution) give identical results.

Or, equivalently, for this special case we can solve the normal equations

$$A^*A\, \boldsymbol{c} = \cdot A^* \boldsymbol{b} \tag{8}$$

in which A^*A is proportional to an estimated correlation matrix. As examples of such arrangements, Ulrych and Clayton [6] and Lang and McClellan [12] recommend that, for a fixed number of data samples N, the number of coefficients P should be approximately $N/3$ (or in between $N/3$ and $N/2$) for best frequency estimation.

How can singular value decomposition (SVD) be used to improve LP frequency estimation? To answer this question, we first present some well-known [8] SVD results. Then we apply them first to a noiseless case and then to the case of noisy data. More details are available in [3], [13], [14].

The SVD of a rectangular matrix A which has real or complex entries is given by the formula [8], [15]

$$\underset{(N-P\times P)}{A} = \underset{(N-P\times N-P)}{U} \; \underset{(N-P\times P)}{\Sigma} \; \underset{(P\times P)}{V^*}. \tag{8}$$

The dimensions of each matrix are written below the matrix. The square matrices U and V are unitary and Σ is a rectangular diagonal matrix of the same size as A with real nonnegative diagonal entries. These diagonal entries, called the singular values of A, are ordered in decreasing order with the largest in the upper left-hand corner. These singular values are the square roots of the eigenvalues of an estimated correlation matrix A^*A or AA^*. The asterisk is used to denote complex conjugate transpose. The columns of U and V are the eigenvectors of AA^* and A^*A, respectively.

The pseudoinverse A^I of (6) above is related to the SVD of A by the formula

$$A^I = V\, \Sigma^I\, U^* \tag{9}$$

in which Σ^I is obtained from Σ by replacing each positive diagonal entry by its reciprocal.

In the absence of noise, or at very high SNR, the solution of (6) is good. This is because the right-hand side vector $\boldsymbol{b}$ in (6) lies primarily along the L principal eigenvectors of A^*A. It can be shown that [14], if the number of prediction coefficients P satisfies the inequality

$$L \leqslant P \leqslant N - L, \tag{10}$$

then the signal-related zeros of the prediction error filter polynomial [16] derived from the prediction coefficient vector of (6) will be in their correct or approximately correct locations, and any extra zeros (if $P > L$) will be located inside the unit circle in the complex plane and hence separable from the signal zeros. We have assumed that the values of P and $N - P$ are both greater than L. When there is no noise so that $x(k) = s(k)$ in (2), the rank of A is L and there will be only L nonzero elements in the diagonal matrices Σ and Σ^I. The introduction of a small amount of noise tends to change the situation. The matrix A tends to take on its full possible rank [15]. The matrix Σ then has some (or likely all) of its formerly zero singular values become small nonzero values. These small diagonal values of Σ, introduced by noise, become large diagonal values of Σ^I in (9), and this is a cause for large perturbations in the prediction coefficient vector of (6).

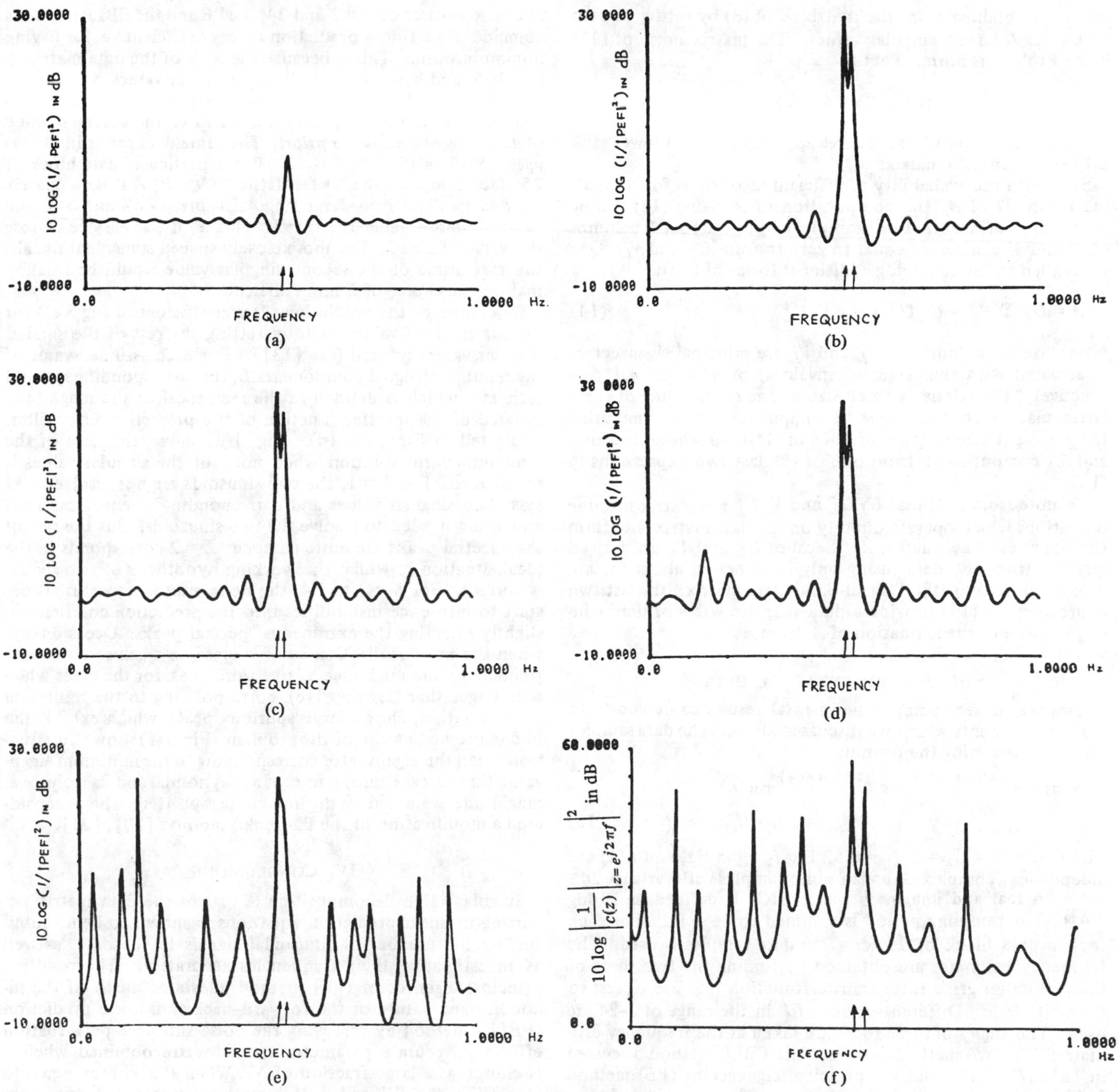

Fig. 1. Reciprocal of the prediction-error filter transfer function (PEF) using four different values L as working hypothesis. (a) $L = 1$, (b) $L = 2$, (c) $L = 3$, (d) $L = 4$. In (e) the minimum norm solution for the linear prediction equations was used. $P = 18$ and $N = 25$. (f) Reciprocal of the magnitude of a polynomial $c(z)$ (evaluated on the unit circle) constructed with the $L + 1$ elements of the eigenvector corresponding to the minimum eigenvalue of A'^*A'. A' is the augmented matrix $(-\boldsymbol{b}, A)$.

As shown by the work of Wilkinson [17], the perturbations of the L originally nonzero values of Σ and the corresponding first L columns of U and V are relatively small. The main perturbation effects of a small amount of noise are due to the increase in rank of A over its rank in the noiseless case and the associated large perturbations in the directions of the rest of the columns of U and V. This major effect can be alleviated by replacing A in the case of noisy data by a lower rank approximation $\hat{A}$ prior to calculation of the vector $\boldsymbol{c}$ of prediction coefficients. In 1936, Eckart and Young developed the following procedure, based on the SVD for finding the best lower rank approximation to a given matrix [18].

Let the rank of A be r, and let $s(L)$ be the set of all $(N - P) \times P$ matrices of rank $L < r$. Then for all matrices B in $S(L)$

$$\|A - \hat{A}\| \leqslant \|A - B\| \tag{11}$$

where

$$\hat{A} = U \hat{\Sigma} V^* \tag{12}$$

and $\hat{\Sigma}$ is obtained from the matrix Σ of (8) by setting to zero all but its L largest singular values. The matrix norm of (11) is the Frobenius norm. That is,

$$\|A - B\|^2 = \text{trace}\,[(A - B)^*(A - B)]. \tag{13}$$

Hence, $\hat{A}$ is the best least squares approximation of lower rank L to the given data matrix.

Even with the availability of efficient algorithms for computing the SVD of A, the computation of $\hat{A}$ using (12) can be made more efficient. For example, all but the first L columns of U and V can be set equal to zero to form U_L and V_L. One is then led to the following additional forms of (12):

$$\hat{A} = U_L \hat{\Sigma} V_L^* = U_L U_L^* A = A V_L V_L^*. \tag{14}$$

Because the columns of U_L and V_L are principal eigenvectors (associated with the larger eigenvalues) of A^*A and AA^*, respectively, the relatively small size of one or the other of these latter matrices can lead one to compute $\hat{A}$ by: 1) computing the principal eigenvectors of AA^* or A^*A, whichever is easier; and 2) computing $\hat{A}$ from one of the last two expressions in (14).

We note from (14) that $U_L U_L^*$ and $V_L V_L^*$ are matrix filtering operations which operate directly on the data matrix A to form the signal estimate matrix $\hat{A}$. Because U_L and V_L are derived directly from the data (using only information about the approximate rank of the "signal-only" data matrix), the last two expressions in (14) provide a data-adaptive Wiener-filtering interpretation of the formation of $\hat{A}$ from A.

III. Experimental Results

Here we present some experimental results to demonstrate the improvements which we discussed above. The data samples are generated using the formula

$$x(n) = e^{j(2\pi f_1 n + \phi_1)} + e^{j(2\pi f_2 n + \phi_2)} + w(n), \qquad n = 0, 1, \cdots N - 1 \tag{15}$$

where $\phi_1 = \pi/4$, $\phi_2 = 0$, $f_1 = 0.52$ Hz, $f_2 = 0.5$ Hz, and $w(n)$ are independent complex Gaussian noise samples with variance $2\sigma^2$ for each real and imaginary part. SNR is defined as 10 log $\frac{1}{2}\sigma^2$. The sampling period is assumed to be 1 Hz. For each trial, a data block of 25 ($N = 25$) data samples is used. The frequency estimates are obtained by finding the two roots of the prediction error filter transfer function that are closest to the unit circle. Different values of P in the range of 2–24 are used. The angles of these roots are taken as the frequency estimates. The two methods used are the FBLP method discussed in [5], [6], [12] and the principal eigenvector (PE) method [13], [14] using SVD discussed above. The standard deviation of the frequency estimation error for f_2 is computed for 500 independent trials. The estimated standard deviations are tabulated in Table I for different SNR and P values. The corresponding CR bound values [19] are also given. The estimation bias was negligible in all cases except at 7 dB. The biases for the three values of $P = 14, 16, 18$ were about a third of the respective standard deviations. The main point in Table I is that the FBLP method is primarily useful only at very high SNR values, whereas the PE method can be used at much lower SNR values. Also, by choosing the P value to be about $3N/4$ in the PE method which is not useful in the FBLP method, the PE method practically achieves the CR bound. The special case of $P = 2$ is a variant of Prony's method with the data being used in both the forward and backward directions.

The case of $P = 24$, called the Kumaresan–Prony (KP) case [3], [9], is simple to implement. It does not require SVD. It has superior performance at lower SNR values compared to the FBLP method at conventional values of P [6], [12]. For the two special cases of $P = 2$ and 24, the PE and the FBLP methods coincide if we find a prediction filter coefficient vector having minimum norm. This is because the rank of the data matrix A is only 2, and hence it has only two singular values.

In the second part of the experiment, we demonstrate the insensitivity of the frequency estimates to the assumed value of L if L is not known *a priori*. The same data set as in (15) is used. SNR is 10 dB. P is 18. For a particular data block of 25 data samples, the SVD of the $2(N - P) \times P$ data matrix showed two relatively large singular values, 4.83 and 3.8, and the rest were smaller than 0.95. Hence, it was easy to choose the value of L as 2. For more closely spaced sinusoidal signals, the magnitude of the second singular value would be smaller, making the choice of L more difficult.

We computed the prediction filter coefficients using SVD for various assumed values of L by setting the rest of the smaller singular values to zero (see [13]). For each assumed value of the number of signal components L, the corresponding spectral estimate, which is defined as the reciprocal of the magnitude squared of the transfer function of the prediction error filter, is plotted in Fig. 1(a)–(d). Fig. 1(e) shows the case of the minimum norm solution when none of the singular values is set to zero. For $L = 1$, the two sinusoids are not resolved. At least two singular values and corresponding eigenvectors in U and V are needed to resolve the two sinusoids. But the rest of the spectral peaks are quite damped. $L = 2$ corresponds to the ideal situation in which the working hypothesis of two signals is correct. For $L = 3$ and 4, the noise subspace perturbations start to introduce instabilities into the prediction coefficients, slightly effecting the extraneous spectral peaks. Occasionally, when the noise realization itself is close to a sinusoid of some frequency, one might see a large third peak for the cases when L is larger than 2. Fig. 1(e), corresponding to the minimum norm solution, shows large spurious peaks which exhibit the ill-conditioned nature of the problem. Fig. 1(f) shows the situation when the eigenvector corresponding to the minimum eigenvalue (in this case zero) is used as a polynomial and its reciprocal magnitude evaluated on the unit circle is plotted. This is considered a modification of the Pisarenko method [20], [21].

IV. Conclusions

Singular value decomposition (SVD) of the data matrix occurring in linear prediction equations is shown to be useful in finding the number of sinusoidal signals in the data, as well as in estimating their frequencies accurately. The resulting principal eigenvector (PE) method alleviates much of the ill-conditioned nature of the forward–backward linear prediction (FBLP) method by removing the noise subspace perturbation effects. Accurate parameter estimates are obtained when P is chosen as a large fraction of N. When P is chosen equal to $(N - L/2)$, the PE and FBLP methods (when the minimum norm prediction filter coefficients are found) are the same, and we called this case the Kumaresan–Prony (KP) case. This case does not require SVD calculations.

References

[1] D. W. Tufts and R. Kumaresan, "Improved spectral resolution," *Proc. IEEE*, vol. 61, pp. 419–420, Mar. 1980.

[2] —, "Improved spectral resolution II," in *Proc. ICASSP*, Apr. 1980, pp. 592–597.

[3] —, "Frequency estimation of multiple sinusoids: Making linear prediction perform like maximum likelihood," *Proc. IEEE*, to be published.

[4] R. Kumaresan and D. W. Tufts, "Data adaptive principal component signal processing," in *Proc. 19th IEEE Int. Conf. Decision Contr.*, Albuquerque, NM, Feb. 10–12, 1982.

[5] A. H. Nuttall, "Spectral analysis of a univariate process with bad data points via maximum entropy and linear predictive techniques," in *NUSC Scientific and Engineering Studies, Spectral*

Estimation, NUSC, New London, CT, Mar. 1976.

[6] T. J. Ulrych and R. W. Clayton, "Time series modelling and maximum entropy," *Phys. Earth and Planetary Interiors*, vol. 12, pp. 188-200, Aug. 1976.

[7] T. L. Henderson, "Geometric methods for determining system poles from transient response," *IEEE Trans. Acoust., Speech, Signal Processing*, Oct. 1981.

[8] C. L. Lawson and R. J. Hanson, *Solving Least-Squares Problems*. Englewood Cliffs, NJ: Prentice-Hall, 1974.

[9] R. Kumaresan and D. W. Tufts, "Improved spectral resolution III: Efficient realization," *Proc. IEEE*, vol. 68, Oct. 1980.

[10] F. B. Hildebrand, *Introduction to Numerical Analysis*. New York: McGraw-Hill, 1956.

[11] C. Lanczos, *Applied Analysis*. Englewood Cliffs, NJ: Prentice-Hall, 1956.

[12] S. W. Lang and J. H. McClellan, "Frequency estimation with maximum entropy spectral estimators," *IEEE Trans. Acoust., Speech, Signal Processing*, vol. 28, pp. 716-724, Dec. 1980.

[13] R. Kumaresan and D. W. Tufts, "Singular value decomposition and spectral analysis," in *Proc. 1st ASSP Workshop Spectral Estimation*, Hamilton, Ont., Canada, Aug. 17, 18, 1981, pp. 6.4.1-6.4.12.

[14] R. Kumaresan, "On the zeros of the linear prediction error filter for deterministic signals," *IEEE Trans. Acoust., Speech, Signal Processing*, submitted for publication.

[15] V. C. Klemma and A. J. Laub, "The singular value decomposition: Its computation and some applications," *IEEE Trans. Automat. Contr.*, vol. AC-25, pp. 164-176, Apr. 1980.

[16] S. M. Kay and S. L. Marple, "Spectrum analysis–A modern perspective," *Proc. IEEE*, vol. 69, pp. 1380-1419, Nov. 1981.

[17] J. H. Wilkinson, *The Algebraic Eigenvalue Problem*. Oxford, England: Clarendon, 1965.

[18] C. Eckart and G. Young, "The approximation of one matrix by another of lower rank," *Psychometrika*, vol. 1, pp. 211-218, 1936.

[19] D. C. Rife and R. R. Boorstyn, "Multiple tone parameter estimation from discrete time observations," *Bell Syst. Tech. J.*

[20] V. F. Pisarenko, "The retrieval of harmonics from a covariance function," *Geophys. J. Roy. Astron. Soc.*, vol. 33, pp. 247-266.

[21] R. Kumaresan and D. W. Tufts, "Improved Pisarenko-type methods in spectral analysis," *IEEE Trans. Acoust., Speech, Signal Processing*, submitted for publication.

The High-Resolution Spectrum Estimator—A Subjective Entity

STEVEN KAY AND CEDRIC DEMEURE

It is shown by example that the attribute of high resolution for a spectrum estimator can be a subjective one. The particular spectrum estimator considered is the MUSIC algorithm, although the observations are valid for any spectrum estimator. The main point is that visual examination of spectra is not a scientifically valid means of assessing resolution.

I. High-Resolution versus Low-Resolution Spectra

Consider the case of two complex sinusoids in white noise

$$x_t = A_1 e^{j(\omega_1 t + \phi_1)} + A_2 e^{j(\omega_2 t + \phi_2)} + w_t, \qquad t = 0, 1, \cdots, N-1 \quad (1)$$

where w_t is complex white Gaussian noise. The usual test case to assess the resolution capability of a spectrum estimator is when the two sinusoidal frequencies are spaced closer than $1/N$, the classical Fourier resolution limit. Hence, let $N = 25$ and $\omega_1 = 2\pi(0.25)$, $\omega_2 = 2\pi(0.27)$ so that the two sinusoids "will not be resolved by a Fourier spectrum estimator." Also, let $A_1 = A_2 = 1$ and $\phi_1 = 0$, $\phi_2 = 0$. The signal-to-noise ratio (SNR) is defined as

$$\text{SNR} = 10 \log_{10} 1/\text{var}(w_t) \quad \text{dB.} \qquad (2)$$

For our example we will let SNR = 30 dB.

For these conditions consider the two spectrum estimates shown in Fig. 1 for one noise realization. From a visual inspection of the spectra one would generally conclude that the spectrum estimator which generated the spectrum in Fig. 1(a) has a higher resolution than the one which generated the spectrum in Fig. 1(b). Other noise realizations will produce similar results. The spectrum of Fig. 1(a) is the result of applying the MUSIC algorithm or [1]

$$P_A(\omega) = \frac{1}{E^H \sum_{i=p+1}^{M} V_i V_i^H E} \qquad (3)$$

where

$$E = \frac{1}{\sqrt{M}} [1 \; e^{j\omega} \; e^{j2\omega} \; \cdots \; e^{j(M-1)\omega}]^T$$

and V_i is the ith eigenvector of $\hat{R}$. Here $\hat{R}$ is an estimate of the $M \times M$ autocorrelation matrix and $\{V_{p+1}, V_{p+2}, \cdots, V_M\}$ correspond to the noise subspace eigenvectors. For our example (Fig. 1(a)) we let $p = 2$, $M = 12$, and we used the forward–backward estimate of the autocorrelation matrix. [2] [3] The MUSIC algorithm is claimed to be a high-resolution spectrum estimator.

The spectrum in Fig. 1(b) is the result of applying a Bartlett (i.e., a Fourier-type) spectrum estimator based on principal components [4], [5]. Specifically, the Bartlett spectrum estimator used is

$$P_B(\omega) = E^H \hat{R}' E \qquad (4)$$

where $\hat{R}'$ is the $M \times M$ forward–backward estimate of the autocorrelation matrix obtained by including only the principal eigenvectors (to effect noise reduction) and setting the corresponding principal eigenvalues equal to one. Hence

$$\hat{R}' = \sum_{i=1}^{p} V_i V_i^H.$$

Manuscript received January 30, 1984.
The authors are with the Electrical Engineering Department, University of Rhode Island, Kingston, RI 02881, USA.

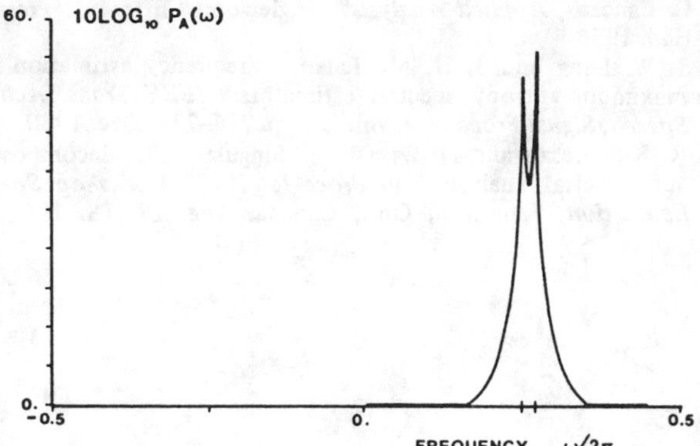

(a)

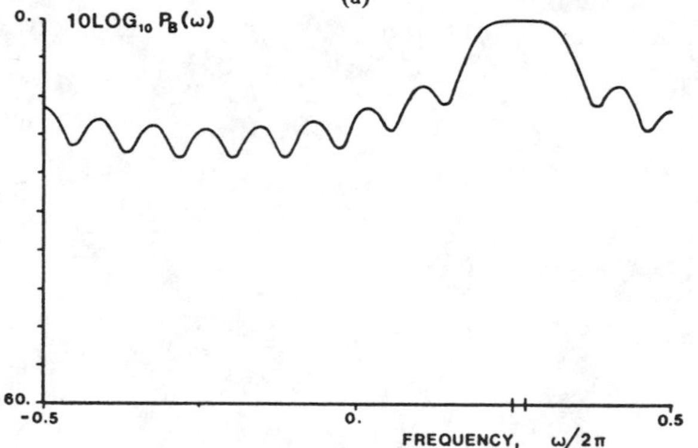

(b)

Fig. 1. Spectrum estimates. (a) MUSIC spectrum estimate. (b) Principal component Bartlett spectrum estimate.

For our example (Fig. 1(b)) $p = 2$ and $M = 12$. It is clear then that the MUSIC algorithm has higher resolution than the Fourier-based spectrum estimator.

II. Relating Spectrum of Fig. 1(a) to Spectrum of Fig. 1(b)

We now show that the spectrum estimators given by (3) and (4) are related. Since

$$I = \sum_{i=1}^{M} V_i V_i^H = \sum_{i=1}^{p} V_i V_i^H + \sum_{i=p+1}^{M} V_i V_i^H$$

where I is the identity matrix, one can rewrite (3) as

$$P_A(\omega) = \frac{1}{E^H \left[I - \sum_{i=1}^{p} V_i V_i^H \right] E} = \frac{1}{1 - E^H \hat{R}' E} = \frac{1}{1 - P_B(\omega)}. \qquad (5)$$

Reprinted from *Proc. IEEE*, vol. 72, pp. 1815–1816, Dec. 1984.

Note that since

$$E^H \hat{R}' E = E^H \sum_{i=1}^{p} V_i V_i^H E$$

$$= \sum_{i=1}^{p} |V_i^H E|^2 \leqslant \sum_{i=1}^{M} |V_i^H E|^2$$

$$= E^H \sum_{i=1}^{M} V_i V_i^H E$$

$$= E^H E = 1$$

we have that

$$0 \leqslant P_B(\omega) \leqslant 1. \tag{6}$$

Hence, the higher resolution visual appearance of the spectrum of Fig. 1(a) over that of the spectrum of Fig. 1(b) is due to the transformation $f(x) = 1/(1 - x)$ over the region $0 \leqslant x \leqslant 1$. It is clear also then that since the transformation is monotonic with x the peaks of the spectrum of Fig. 1(b), which are not plainly visible, are identical to the sharp peaks of the spectrum of Fig. 1(a). For frequency estimation in which one estimates the frequencies by the peaks of the spectrum the two spectrum estimators will produce identical results. Similar observations have been made in [6].

III. Discussion

The example discussed raises several important points. Firstly, visual inspection of spectra is not a scientifically valid means of assessing resolution of a spectrum estimator. In fact the "resolution" of the MUSIC algorithm could be improved by defining a new spectrum estimator

$$P_C(\omega) = \frac{1}{1 - \sqrt{P_B(\omega)}} \tag{7}$$

as an example. Secondly, and most importantly, one must first determine what resolution is. Assume, for example, that resolution relates to the ability of a spectrum estimator to determine whether one or two sinusoidal components, closely spaced in frequency, is present in white noise. Then, the problem of assessing the resolution of a spectrum estimator is naturally addressed within the framework of statistical hypothesis testing. One could, for example, determine the probability of detecting the two sinusoids when they are present versus the probability of deciding there are two sinusoids when only one sinusoid is present, i.e., the probability of false alarm [7].

In summary, it is hoped that this example will serve to illustrate the difficulties arising in assessing the resolution of a spectrum estimator by solely visual means.

References

[1] R. Schmidt, "Multiple emitter location and signal parameter estimation," in *Proc. RADC Spectral Estimation Workshop*, pp. 243–258, Rome, NY, 1979.

[2] A. Nuttall, "Spectral analysis of a univariate process with bad data points, via maximum entropy, and linear predictive techniques," NUSC Tech. Rep. TR 5303, New London, CT, Mar. 26, 1976.

[3] T. T. Ulrych and R. W. Clayton, "Time series modeling and maximum entropy," *Phys. Earth Planetary Interiors*, vol. 12, pp. 188–200, Aug. 1976.

[4] W. S. Liggett, "Passive sonar: Fitting models to multiple time series," in *Signal Processing*, J. W. R. Griffiths *et al.*, Eds. New York: Academic Press, 1973.

[5] N. L. Owsley, "A recent trend in adaptive spatial processing for sensor arrays: Constrained adaptation," in *Signal Processing*, J. W. R. Griffiths *et al.*, Eds. New York: Academic Press, 1973.

[6] D. H. Johnson, "The application of spectral estimation methods to bearing estimation problems," *Proc. IEEE*, vol. 70, no. 9, pp. 1018–1028, Sept. 1982.

[7] Personal communication with D. W. Tufts.

Part III
Nonparametric Methods

SPECTRAL analysis became a practical discipline only in the mid-1960's, when the fast Fourier transform algorithms were developed. However, it was long before that that the need for frequency-domain representation of finite time series was recognized. In 1905, Sir Arthur Schuster described the periodogram, which he introduced in 1898, and pointed out the need for spectral averaging. His paper and the paper by M. S. Bartlett, published in 1948 are reprinted in this section because of their historical interest. Bartlett established a statistical basis for spectral smoothing and introduced the triangular window which now bears his name. A very thorough tutorial review of windows and their use in Fourier transform based spectral analysis is presented in the paper by F. J. Harris. Most of the known windows are included, and the practical guidelines for their usage are given.

The following two papers, by G. C. Carter and A. H. Nuttall, and by A. H. Nuttall, respectively, discuss a generalized framework for the nonparametric spectral analysis. Included in the framework are windowed covariance and weighted overlapped segment averaging methods. Spectral smoothing as well as the analysis of the first- and second-order statistics are also treated. They give us a compact review of the classical spectral analysis.

Procedures based on extrapolation for band-limited signals are presented in Cadzow's paper. The subject received much attention in mid-70's but, since then, the interest gradually waned, because it was learned that the methods are hypersensitive on the assumption of signals being strictly band-limited. They are also sensitive to noise.

As mentioned in the Preface, several papers in the September 1982 special issue of the *Proceedings of the IEEE* should have been included in this part. In particular, we should mention the paper by D. J. Thompson, which analyzes a very attractive nonparametric method, as well as the paper by A. H. Nuttall and G. C. Carter which gives a more detailed treatment of the subject of the fourth and fifth papers in this section.

The Periodogram and its Optical Analogy

By Arthur Schuster, F.R.S.

(Received November 23,—Read December 7, 1905.)

I have recently applied the periodogram method to the investigation of several fluctuating quantities, and the experience thus gained has led me to modify slightly the original definition.* Having always laid stress on the fact that the periodogram supplies by calculation the transformation which the spectroscope instrumentally impresses on a luminous disturbance, I may now enter a little more closely into this optical analogy, and thus lead up to what I hope will be the final definition.

Consider a parallel beam of light falling on a grating, the reflected light being collected at the focus of an observing telescope in the usual way. For simplicity of calculation I assume that the grating considered is of a particular type, which, in a former paper, I have called a simple grating. Such a grating only gives two spectra of the first order.

If $\phi\,(\mathrm{V}t+x)$ be the velocity at any point of the incident beam, the displacement at the focus of the observing telescope is†

$$\frac{hl\cos\beta}{2\pi f\mathrm{NV}\lambda}\,\mathrm{R},$$

where

$$\mathrm{R}=\int_{-\frac{1}{2}\mathrm{N}\lambda}^{+\frac{1}{2}\mathrm{N}\lambda}\cos nx\,\phi\,(\mathrm{V}t+x)\,dx. \qquad (1)$$

In these equations f denotes the focal length of the telescope, h is the length of the lines ruled on the grating, l the width of ruled space measured at right angles to the lines, N gives the number of lines, and β the angle between the direction of the optic axis of the observing telescope and the normal to the grating. For the sake of shortness n is written for $2\pi/\lambda$. The quantity denoted by λ is the wave-length of homogeneous light which would have its first principal maximum at the focus of the telescope. It may be said to be the wave-length towards which the telescope points, and its strict definition is given by the relation

$$l\,(\sin\alpha-\sin\beta)=\mathrm{N}\lambda,$$

where α is the angle of incidence.

In order not to complicate needlessly the calculations, I shall assume that the resolving power is sufficient to ensure that at any point of the spectrum the vibrations are nearly homogeneous; this involves that the average squares of the velocities are sensibly equal to the average squares of the displacements multiplied by $4\pi^2\,\mathrm{V}^2/\lambda^2$. The average square of the velocity at the focus of the telescope is in that case—

$$\frac{h^2l^2\cos^2\beta}{f^2\mathrm{N}^2\lambda^4}\,\mathrm{R}^2.$$

where for R^2 we must put its average value. This expression represents the measure of the intensity at the point considered. Its line and surface integrals may be called the total linear intensity and the total intensity respectively.

In observing a spectrum, we associate with a particular wave-length all the light which lies in a straight line parallel to the rulings of the grating.

* 'Cambridge Phil. Soc. Trans.,' vol. 18, p. 107.

† 'Phil. Mag.,' vol. 37, p. 545 (1894).

Reprinted with permission from *Proc. Roy. Soc. London*, vol. 77, Series A, pp. 136–140, June 1906.

The distribution of light along a vertical line for nearly homogeneous light takes place according to the law $\alpha^{-2}\sin^2\alpha$, where $\alpha = \pi hy/f\lambda$, y being the vertical distance. Multiplying by dy, and integrating from minus to plus infinity, the total intensity in a vertical line is found to be $\lambda f/h$ when the intensity at the central maximum is unity. With the value for the central intensity previously found, we now obtain the total linear intensity associated with λ to be

$$\frac{hl^2\cos^2\beta}{fN^2\lambda^3}R^2. \tag{2}$$

Changing the variable, the expression for R takes the form

$$\int_{x_0}^{x_0+N\lambda} \cos n(Vt-x)\,\phi(x)\,dx,$$

where x_0 is put for $Vt-\frac{1}{2}N\lambda$.

Write

$$A = \int_{x_0}^{x_0+N\lambda} \cos nx\,\phi(x)\,dx; \qquad B = \int_{x_0}^{x_0+N\lambda} \sin nx\,\phi(x)\,dx.$$

The mean value of R^2 is then equal to the mean value of

$$\tfrac{1}{2}(A^2+B^2).$$

A grating such as that to which the above equations apply forms two spectra and absorbs part of the light; we must now estimate what fraction of the incident beam is utilised to form the spectrum under consideration. For this purpose we imagine homogeneous light to fall on the grating, and put $\phi(x) = \cos nx$. The mean value of $\frac{1}{2}(A^2+B^2)$ is then easily found to be $\frac{1}{8}N^2\lambda^2$. By substitution into (2) we find that the total linear intensity in the central line is now

$$\frac{hl^2\cos^2\beta}{8f\lambda}R^2. \tag{3}$$

To either side of the principal maximum the intensity varies according to the law $\alpha^{-2}\sin^2\alpha$, where α is equal to $\pi\xi l\cos\beta/f\lambda$, ξ representing a distance measured at right angles to the spectroscopic line. The total energy measured in the focal plane of the telescope is obtained by multiplying (3) with $\alpha^{-2}\sin^2\alpha d\xi$, and integrating. This gives $\frac{1}{8}hl\cos\beta$.

If the incident light is normal to the grating, its total energy is $\frac{1}{2}hl$, the factor $\frac{1}{2}$ representing the fact that we have taken the average square of the velocity which is half the square of the greatest velocity as the measure of intensity. We conclude that $\frac{1}{4}\cos\beta$ is the fraction of light utilised to form the spectrum.

Taking account in (2) of this, we find that the type of spectroscope considered estimates the intensity of light passing through its central meridian as being

$$\frac{2hl^2\cos\beta}{fN^2\lambda^3}(A^2+B^2),$$

where for A^2 and B^2 their mean values are to be substituted.

To obtain the total light within a small angular distance $d\beta$, we must multiply by $Fd\beta$; as $Nd\lambda = l\cos\beta d\beta$, we find that the total energy within a range $d\lambda$ is

$$\frac{2hl}{N\lambda^3}(A^2+B^2)\,d\lambda.$$

If the total energy of the light incident on the grating is unity, the energy assigned by the grating to a range dn is therefore finally—

$$\frac{A^2+B^2}{\pi N\lambda}dn. \tag{4}$$

In the application of the periodogram it is more convenient to take the time as the independent variable. Defining therefore—

$$A = \int_{t_0}^{t_0+NT} \cos \kappa t\, \phi(t)\, dt, \qquad B = \int_{t_0}^{t_0+NT} \sin \kappa t\, \phi(t)\, dt, \tag{5}$$

(4) becomes equal to $\dfrac{A^2+B^2}{\pi NT}d\kappa$.

Leaving out the constant factor, I now define

$$S = (A^2+B^2)/NT$$

to be the ordinate of the periodogram. The definition differs from the previous one by the factor NT, which occurs in the denominator instead of its square.

The present definition is not only justified by the close optical analogy which has now been formally proved, but also by the resulting convenience. I have previously shown that, in the absence of any homogeneous periodicities, the average of A^2+B^2 increases in proportion to the time interval, NT, which occurs in the limits of the integrals for A and B. It follows that for such variations, the ordinate of the periodogram as at present defined is independent of the time limits chosen. This is an advantage. On the other hand, the former definition gave directly the amplitude of the periodic variation when it was of an absolutely homogeneous character. For such homogeneous variation the present ordinate increases proportionally to the time interval chosen.

The optical analogy explains the reason of this, and gives its justification. When homogeneous light falls on an instrument of definite resolving power the light in the central meridian does not by itself give sufficient indication of the intensity of the incident light. It is only when correction has been made for the lateral spreading that the true intensity can be deduced, the correction depending on the resolving power. It is otherwise when the spectrum is continuous, for in that case the light lost by lateral spreading is replaced by that which properly belongs to the neighbouring wave-lengths. Hence, in this case, the intensity in the central meridian is a true measure of the intensity of the incident light.

It need hardly be pointed out how constant use is made of the fact that increased resolving power (*not* increased dispersion) brings out the homogeneous lines of a spectrum by increasing their intensity beyond that of the continuous background. It is correspondingly one of the principal advantages of the periodogram method that it gives a measure of the resolving power necessary to isolate a true homogeneous period from the irregular fluctuations.

Light is thrown on parts of the previous investigation by a formula given by Lord Rayleigh for the intensity to be assigned to the homogeneous components of a disturbance. If $\phi(x)$ be the velocity at any point of a linear disturbance, so that the total intensity is

$$\int_{-\infty}^{+\infty} \{\phi(x)\}^2 dx,$$

Lord Rayleigh shows that the energy to be assigned to a range dn, where

$n = 2\pi/\lambda$, is

$$(A^2+B^2)\,dn/\pi,$$

in which

$$A = \int_{-\infty}^{+\infty} \cos \kappa v\, \phi(v)\, dv, \qquad B = \int_{-\infty}^{+\infty} \sin \kappa v\, \phi(v)\, dv.$$

The average intensity spread over a certain length L may be estimated by taking v_0 and v_0+L as lower and upper limits of the integrals, and averaging the values obtained by a change of v_0. The energy per unit length would then be found on dividing the expression in (5) by L. We arrive in this manner at equation (4).

I might have confined myself to this simple deduction had I not wished to lay stress on the equations for instrumental resolution. This seemed all the more desirable because for absolutely homogeneous radiation a definition of the periodogram based on the average intensity per unit length would fail. If a simple periodicity exists, its amplitude may easily be derived from the ordinate of the periodogram, for, if S be that ordinate, the amplitude is $2\,(S/NT)^{\frac{1}{2}}$.

In practical applications the function $\phi(t)$ will generally be given for successive intervals (α) of the time. The integrals occurring in (5) are then replaced by summations, unless a harmonic analyser is used. It is most convenient to write in this case

$$A = \sum_{s=0}^{s=(n-1)\alpha} \Phi_s \cos\left(\frac{2\pi}{n} s\right), \qquad B = \sum_{s=0}^{s=(n-1)\alpha} \Phi_s \sin\left(\frac{2\pi}{n} s\right), \tag{6}$$

$$S = (A^2+B^2)\,\alpha^2/NT,$$

where ϕ_s represents the values which $\phi(t)$ takes at the successive times considered.

If we take p to be equal to the total number of separate values of $\phi(t)$ used in the calculations, we may put

$$S = (A^2+B^2)\,\alpha/p. \tag{7}$$

If a harmonic analyser be used, and a b are the two Fourier coefficients,

$$S = \tfrac{1}{4}\,(a^2+b^2)\,NT = \tfrac{1}{4}\,(a^2+b^2)\,\alpha p,$$

where αp represents the complete time interval to which the Fourier analysis has been applied.

Smoothing Periodograms from Time-Series with Continuous Spectra

In his review[1] of M. G. Kendall's brochure[2] on oscillatory time-series, David G. Kendall made the pertinent observation that the smoothing of periodograms obtained from autoregressive or other time-series with continuous spectra is equivalent to considering the first few sample autocorrelations. I had arrived at a similar conclusion, though possibly by an alternative route, having noticed that the averaging of periodograms obtained from contiguous lengths of series is approximately equivalent to a truncation of the correlogram at a point represented by the length of the subseries. From preliminary computation already made on M. G. Kendall's artificial series to test out this smoothing device, it appears promising.

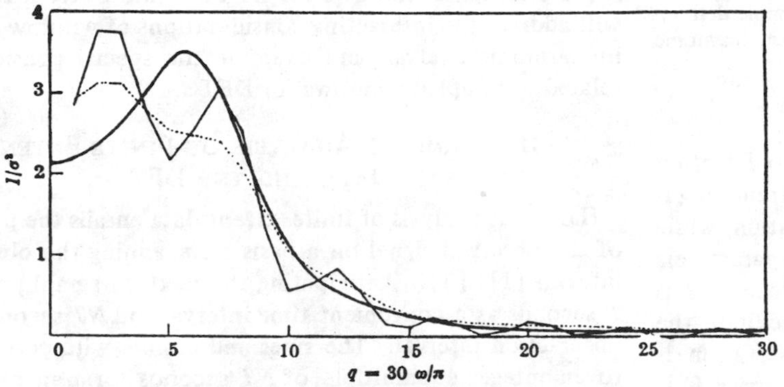

SMOOTHED PERIODOGRAM (I_{16} JAGGED LINE, I'_{16} DASHED LINE) COMPARED WITH THEORETICAL SPECTRUM (SMOOTH CURVE).

In terms of the periodogram intensity $I(\omega)$, standardized to have expectation equal to the variance σ^2 of the series if the latter were entirely random, and related to the correlogram by the identity

$$I(\omega) \equiv \sum_{s=-N+1}^{N-1} C_s \cos \omega s, \qquad (1)$$

where C_s is defined in terms of the observations X_r as

$$C_s = \frac{1}{N} \sum_{r=1}^{N-|s|} X_r X_{r+|s|}, \qquad (2)$$

the (continuous) spectrum of X_r, apart from a factor $1/(\pi\sigma^2)$, is the limit of $E(I)$ as N increases, where E denotes expectation[3]. For M. G. Kendall's series, with autocorrelations ρ_s satisfying $\rho_{s+2} + a\rho_{s+1} + b\rho_s = 0$, $(s \geqslant -1)$, it may be shown that this limit is

$$\frac{\sigma^2(1-b)(1-a^2+b^2+2b)}{(1+b)\{1+a^2+b^2-2b+2a(1+b)\cos\omega+4b\cos^2\omega\}}, \quad (0 < \omega \leqslant \pi). \qquad (3)$$

However, in accordance with general spectral theory and as remarked by the late P. J. Daniell, $I(\omega)$ unless smoothed never tends to this limit whatever N, random fluctuations being of the same order of magnitude as the expected value and only the 'resolving power' increasing.

The smoothing procedure first considered was to calculate

$$I_n(\omega) = \sum_{s=-n+1}^{n-1} C_s \cos \omega s, \qquad (4)$$

where n is a submultiple of N, chosen in practice to be a compromise between the averaging power required and the avoidance of bias in the expected value of $I_n(\omega)$. The standard deviation of fluctuations should be approximately reduced by the factor $\sqrt{(n/N)}$ and the bias should remain small if n is still large enough for ρ_n to be effectively zero. A variant of (4) which should be less subject to systematic error, though also rather less convenient since it implies fixing n in advance, is to replace C_s by

$$C_s' = C_s\left(1 - \frac{|s|}{n}\right) \Big/ \left(1 - \frac{|s|}{N}\right). \qquad (5)$$

In the diagram the expected intensity (divided by σ^2) is shown for Series 1 ($N = 480$), together with the values of $I_n(\omega)$ computed for $q = 30\omega/\pi = 1 \ldots 30$ and $n = 16$. There is a tendency to systematic oscillation in $I_n(\omega)$, but this is largely eliminated in $I_n'(\omega)$ (corresponding to C_s' in (5)), the values for which are also shown. Even $I_n(\omega)$, however, is a very great improvement over the unsmoothed periodogram, which includes a peak of 12 at $q = 3$ (see M. G. Kendall's Fig. 4.1, where it should be noted that the intensity contains an extra factor of approximately $2/N$ and is plotted against the trial period p, which is $N/8q$).

The arithmetical results also suggested the increase in the fluctuations about the mean as n was increased, but the computation (for most of which I am indebted to Mr. D. F. Ferguson) was carried out with the aid of the serial correlations recorded by M. G. Kendall, and hence was approximate and limited to n less than 48. It is hoped to test out and compare $I_n(\omega)$ and $I_n'(\omega)$ more precisely in due course. For Series 3 ($N = 240$) M. G. Kendall found the unsmoothed periodogram less confusing, owing to the closer approach of this series to an undamped harmonic, but the results for the modified periodogram still suggested an improvement even for n as low as 16.

Two further comments are: (i) this method of estimating the true spectrum could, of course, be improved if the correct form of the autoregressive scheme were assumed, its unknown constants estimated and the corresponding spectrum calculated, but a direct and more empirical estimation of the spectrum will often be required; (ii) the most rapid methods of obtaining spectra for series defined for continuous time avoid computation (for example, employ optical or electrical methods), and it may be possible to arrange for the smoothing to be carried out automatically without the above machinery of the truncated correlogram.

M. S. BARTLETT

University of Manchester.

Feb. 13.

[1] *Nature*, 161, 187 (1948).

[2] "Contributions to the Study of Oscillatory Time-Series" (Cambridge, 1946).

[3] See Wold, H., "Analysis of Stationary Time-Series" (Uppsala, 1938).

On the Use of Windows for Harmonic Analysis with the Discrete Fourier Transform

FREDRIC J. HARRIS, MEMBER, IEEE

Abstract—This paper makes available a concise review of data windows and their affect on the detection of harmonic signals in the presence of broad-band noise, and in the presence of nearby strong harmonic interference. We also call attention to a number of common errors in the application of windows when used with the fast Fourier transform. This paper includes a comprehensive catalog of data windows along with their significant performance parameters from which the different windows can be compared. Finally, an example demonstrates the use and value of windows to resolve closely spaced harmonic signals characterized by large differences in amplitude.

I. Introduction

There is much signal processing devoted to detection and estimation. Detection is the task of determining if a specific signal set is present in an observation, while estimation is the task of obtaining the values of the parameters describing the signal. Often the signal is complicated or is corrupted by interfering signals or noise. To facilitate the detection and estimation of signal sets, the observation is decomposed by a basis set which spans the signal space [1]. For many problems of engineering interest, the class of signals being sought are periodic which leads quite naturally to a decomposition by a basis consisting of simple periodic functions, the sines and cosines. The classic Fourier transform is the mechanism by which we are able to perform this decomposition.

By necessity, every observed signal we process must be of finite extent. The extent may be adjustable and selectable, but it must be finite. Processing a finite-duration observation imposes interesting and interacting considerations on the harmonic analysis. These considerations include detectability of tones in the presence of nearby strong tones, resolvability of similar-strength nearby tones, resolvability of shifting tones, and biases in estimating the parameters of any of the aforementioned signals.

For practicality, the data we process are N uniformly spaced samples of the observed signal. For convenience, N is highly composite, and we will assume N is even. The harmonic estimates we obtain through the discrete Fourier transform (DFT) are N uniformly spaced samples of the associated periodic spectra. This approach is elegant and attractive when the processing scheme is cast as a spectral decomposition in an N-dimensional orthogonal vector space [2]. Unfortunately, in many practical situations, to obtain meaningful results this elegance must be compromised. One such compromise consists of applying windows to the sampled data set, or equivalently, smoothing the spectral samples.

Manuscript received September 10, 1976; revised April 11, 1977 and September 1, 1977. This work was supported by Naval Undersea Center (now Naval Ocean Systems Center) Independent Exploratory Development Funds.

The author is with the Naval Ocean Systems Center, San Diego, CA, and the Department of Electrical Engineering, School of Engineering, San Diego State University, San Diego, CA 92182.

The two operations to which we subject the data are sampling and windowing. These operations can be performed in either order. Sampling is well understood, windowing is less so, and sampled windows for DFT's significantly less so! We will address the interacting considerations of window selection in harmonic analysis and examine the special considerations related to sampled windows for DFT's.

II. Harmonic Analysis of Finite-Extent Data and the DFT

Harmonic analysis of finite-extent data entails the projection of the observed signal on a basis set spanning the observation interval [1], [3]. Anticipating the next paragraph, we define T seconds as a convenient time interval and NT seconds as the observation interval. The sines and cosines with periods equal to an integer submultiple of NT seconds form an orthogonal basis set for continuous signals extending over NT seconds. These are defined as

$$\left.\begin{array}{l}\cos\left[\dfrac{2\pi}{NT}kt\right]\\[2ex] \sin\left[\dfrac{2\pi}{NT}kt\right]\end{array}\right\}\quad \begin{array}{l}k=0,1,\cdots,N-1,N,N+1,\cdots\\[2ex] 0\leqslant t<NT.\end{array} \tag{1}$$

We observe that by defining a basis set over an ordered index k, we are defining the spectrum over a line (called the frequency axis) from which we draw the concepts of bandwidth and of frequencies close to and far from a given frequency (which is related to resolution).

For sampled signals, the basis set spanning the interval of NT seconds is identical with the sequences obtained by uniform samples of the corresponding continuous spanning set up to the index $N/2$,

$$\left.\begin{array}{l}\cos\left[\dfrac{2\pi}{NT}knT\right]=\cos\left[\dfrac{2\pi}{N}kn\right]\\[2ex] \sin\left[\dfrac{2\pi}{NT}knT\right]=\sin\left[\dfrac{2\pi}{N}kn\right]\end{array}\right\}\quad \begin{array}{l}k=0,1,\cdots,N/2\\[2ex] n=0,1,\cdots,N-1.\end{array} \tag{2}$$

We note here that the trigonometric functions are unique in that uniformly spaced samples (over an integer number of periods) form orthogonal sequences. Arbitrary orthogonal functions, similarly sampled, do not form orthogonal sequences. We also note that an interval of length NT seconds is not the same as the interval covered by N samples separated by intervals of T seconds. This is easily understood when we

Reprinted from *Proc. IEEE*, vol. 66, pp. 51–83, Jan. 1978.

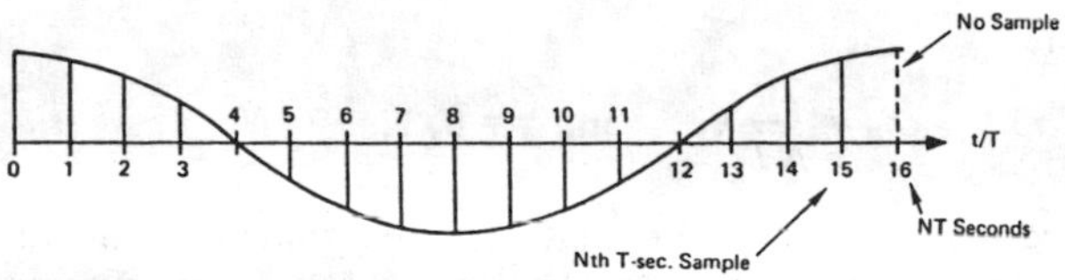

Fig. 1. *N* samples of an even function taken over an *NT* second interval.

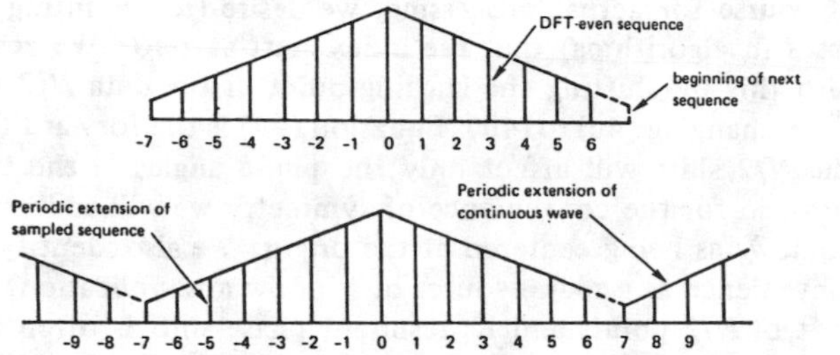

Fig. 2. Even sequence under DFT and periodic extension of sequence under DFT.

realize that the interval over which the samples are taken is closed on the left and is open on the right (i.e., [——)). Fig. 1 demonstrates this by sampling a function which is even about its midpoint and of duration *NT* seconds.

Since the DFT essentially considers sequences to be periodic, we can consider the missing end point to be the beginning of the next period of the periodic extension of this sequence. In fact, under the periodic extension, the next sample (at 16 s in Fig. 1.) is indistinguishable from the sample at zero seconds.

This apparent lack of symmetry due to the missing (but implied) end point is a source of confusion in sampled window design. This can be traced to the early work related to convergence factors for the partial sums of the Fourier series. The partial sums (or the finite Fourier transform) always include an odd number of points and exhibit even symmetry about the origin. Hence much of the literature and many software libraries incorporate windows designed with true even symmetry rather than the implied symmetry with the missing end point!

We must remember for DFT processing of sampled data that even symmetry means that the projection upon the sampled sine sequences is identically zero; it does not mean a matching left and right data point about the midpoint. To distinguish this symmetry from conventional evenness we will refer to it as DFT-even (i.e., a conventional even sequence with the right-end point removed). Another example of DFT-even symmetry is presented in Fig. 2 as samples of a periodically extended triangle wave.

If we evaluate a DFT-even sequence via a finite Fourier transform (by treating the $+N/2$ point as a zero-value point), the resultant continuous periodic function exhibits a non zero imaginary component. The DFT of the same sequence is a set of samples of the finite Fourier transform, yet these samples exhibit an imaginary component equal to zero. Why the disparity? We must remember that the missing end point under the DFT symmetry contributes an imaginary sinusoidal component of period $2\pi/(N/2)$ to the finite transform (corresponding to the odd component at sequence position $N/2$). The sampling positions of the DFT are at the multiples of $2\pi/N$, which, of course, correspond to the zeros of the imaginary sinusoidal component. An example of this fortuitous sampling is shown in Fig. 3. Notice the sequence $f(n)$,

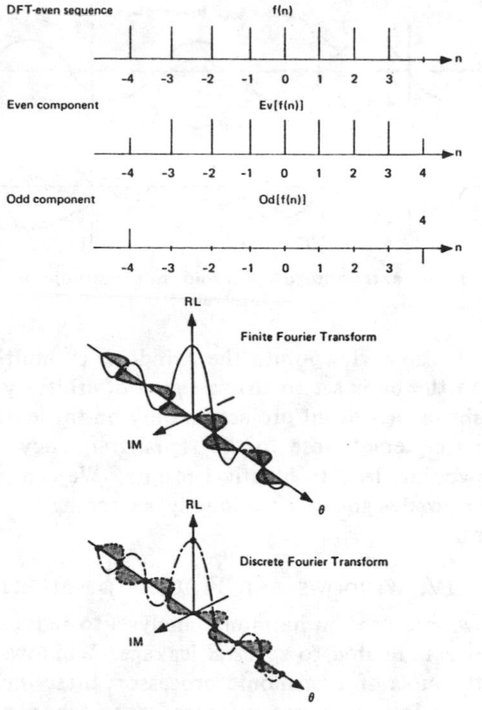

Fig. 3. DFT sampling of finite Fourier transform of a DFT even sequence.

is decomposed into its even and odd parts, with the odd part supplying the imaginary sine component in the finite transform.

III. Spectral Leakage

The selection of a finite-time interval of *NT* seconds and of the orthogonal trigonometric basis (continuous or sampled) over this interval leads to an interesting peculiarity of the spectral expansion. From the continuum of possible frequencies, only those which coincide with the basis will project onto a single basis vector; all other frequencies will exhibit non zero projections on the entire basis set. This is often referred to as spectral leakage and is the result of processing finite-duration records. Although the amount of leakage is influenced by the sampling period, leakage is not caused by the sampling.

An intuitive approach to leakage is the understanding that signals with frequencies other than those of the basis set are not periodic in the observation window. The periodic extension of a signal not commensurate with its natural period exhibits discontinuities at the boundaries of the observation. The discontinuities are responsible for spectral contributions (or leakage) over the entire basis set. The forms of this discontinuity are demonstrated in Fig. 4.

Windows are weighting functions applied to data to reduce the spectral leakage associated with finite observation intervals. From one viewpoint, the window is applied to data (as a multiplicative weighting) to reduce the order of the discontinuity at the boundary of the periodic extension. This is accomplished by matching as many orders of derivative (of the weighted data) as possible at the boundary. The easiest way to achieve this matching is by setting the value of these derivatives to zero or near to zero. Thus windowed data are smoothly brought to zero at the boundaries so that the periodic extension of the data is continuous in many orders of derivative.

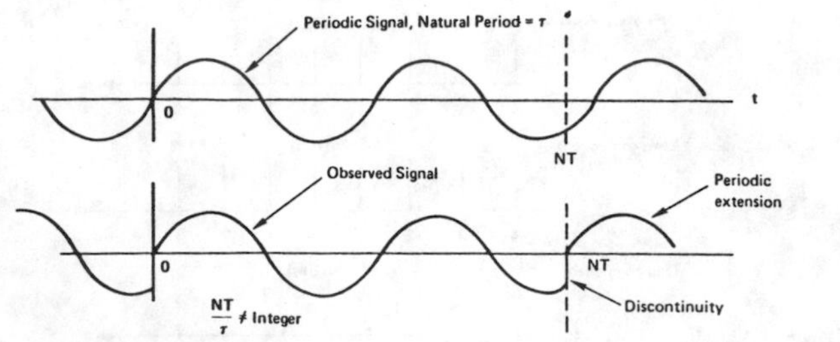

Fig. 4. Periodic extension of sinusoid not periodic in observation interval.

From another viewpoint, the window is multiplicatively applied to the basis set so that a signal of arbitrary frequency will exhibit a significant projection only on those basis vectors having a frequency close to the signal frequency. Of course both viewpoints lead to identical results. We can gain insight into window design by occasionally switching between these viewpoints.

IV. Windows and Figures of Merit

Windows are used in harmonic analysis to reduce the undesirable effects related to spectral leakage. Windows impact on many attributes of a harmonic processor; these include detectability, resolution, dynamic range, confidence, and ease of implementation. We would like to identify the major parameters that will allow performance comparisons between different windows. We can best identify these parameters by examining the effects on harmonic analysis of a window.

An essentially bandlimited signal $f(t)$ with Fourier transform $F(\omega)$ can be described by the uniformly sampled data set $f(nT)$. This data set defines the periodically extended spectrum $F^T(\omega)$ by its Fourier series expansion as identified as

$$F(\omega) = \int_{-\infty}^{+\infty} f(t) \exp(-j\omega t)\, dt \tag{3a}$$

$$F^T(\omega) = \sum_{n=-\infty}^{+\infty} f(nT) \exp(-j\omega nT) \tag{3b}$$

$$f(t) = \int_{-\pi/T}^{+\pi/T} F^T(\omega) \exp(+j\omega t)\, d\omega/2\pi \tag{3c}$$

$$|F(\omega)| = 0, \qquad |\omega| \geqslant \tfrac{1}{2}[2\pi/T]$$

and where

$$F^T(\omega) = F(\omega), \qquad |\omega| \leqslant \tfrac{1}{2}\,[2\pi/T].$$

For (real-world) machine processing, the data must be of finite extent, and the summation of (3b) can only be performed as a finite approximation as indicated as

$$F_a(\omega) = \sum_{n=-N/2}^{+N/2} f(nT) \exp(-j\omega nT)\,, \qquad N \text{ even} \tag{4a}$$

$$F_b(\omega) = \sum_{n=-N/2}^{(N/2)-1} f(nT) \exp(-j\omega nT)\,, \qquad N \text{ even} \tag{4b}$$

$$F_c(\omega_k) = \sum_{n=-N/2}^{(N/2)-1} f(nT) \exp(-j\omega_k nT), \qquad N \text{ even} \tag{4c}$$

$$F_d(\omega_k) = \sum_{n=0}^{N-1} f(nT) \exp(-j\omega_k nT), \qquad N \text{ even} \tag{4d}$$

where

$$\omega_k = \frac{2\pi}{NT}k, \qquad \text{and } k = 0, 1, \cdots, N-1.$$

We recognize (4a) as the finite Fourier transform, a summation addressed for the convenience of its even symmetry. Equation (4b) is the finite Fourier transform with the right-end point deleted, and (4c) is the DFT sampling of (4b). Of course for actual processing, we desire (for counting purposes in algorithms) that the index start at zero. We accomplish this by shifting the starting point of the data $N/2$ positions, changing (4c) to (4d). Equation (4d) is the forward DFT. The $N/2$ shift will affect only the phase angles of the transform, so for the convenience of symmetry we will address the windows as being centered at the origin. We also identify this convenience as a major source of window misapplication. The shift of $N/2$ points and its resultant phase shift is often overlooked or is improperly handled in the definition of the window when used with the DFT. This is particularly so when the windowing is performed as a spectral convolution. See the discussion on the Hanning window under the $\cos^\alpha(X)$ windows.

The question now posed is, to what extent is the finite summation of (4b) a meaningful approximation of the infinite summation of (3b)? In fact, we address the question for a more general case of an arbitrary window applied to the time function (or series) as presented in

$$F_w(\omega) = \sum_{n=-\infty}^{+\infty} w(nT) f(nT) \exp(-j\omega nT) \tag{5}$$

where

$$w(nT) = 0, \qquad |n| > \frac{N}{2}, \qquad N \text{ even}$$

and

$$w(nT) = w(-nT), \qquad n \neq \frac{N}{2},\ w\left(\frac{N}{2}T\right) = 0.$$

Let us now examine the effects of the window on our spectral estimates. Equation (5) shows that the transform $F_w(\omega)$ is the transform of a product. As indicated in the following equation, this is equivalent to the convolution of the two corresponding transforms (see Appendix):

$$F_w(\omega) = \int_{-\infty}^{+\infty} F(x)\, W(\omega - x)\, dx/2\pi \tag{6}$$

or

$$F_w(\omega) = F(\omega) * W(\omega).$$

Equation (6) is the key to the effects of processing finite-extent data. The equation can be interpreted in two equivalent ways, which will be more easily visualized with the aid of an example. The example we choose is the sampled rectangle window; $w(nT) = 1.0$. We know $W(\omega)$ is the Dirichlet kernel [4] presented as

$$W(\omega) = \exp\left(+j\frac{\omega T}{2}\right) \frac{\sin\left[\frac{N}{2}\omega T\right]}{\sin\left[\frac{1}{2}\omega T\right]}. \tag{7}$$

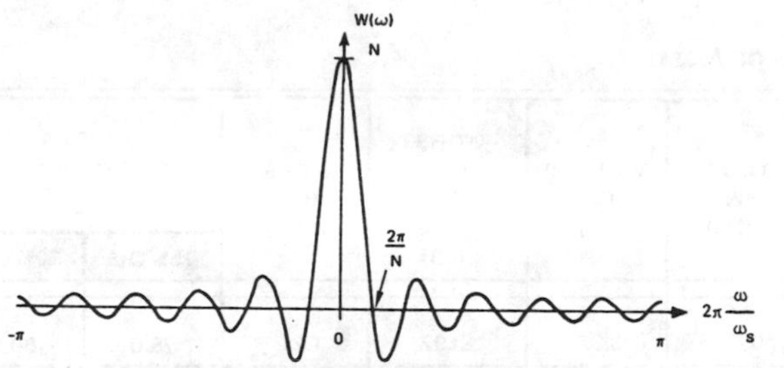

Fig. 5. Dirichlet kernel for N point sequence.

Except for the linear phase shift term (which will change due to the $N/2$ point shift for realizability), a single period of the transform has the form indicated in Fig. 5. The observation concerning (6) is that the value of $F_w(\omega)$ at a particular ω, say $\omega = \omega_0$, is the sum of all of the spectral contributions at each ω weighted by the window centered at ω_0 and measured at ω (see Fig. 6).

A. Equivalent Noise Bandwidth

From Fig. 6, we observe that the amplitude of the harmonic estimate at a given frequency is biased by the accumulated broad-band noise included in the bandwidth of the window. In this sense, the window behaves as a filter, gathering contributions for its estimate over its bandwidth. For the harmonic detection problem, we desire to minimize this accumulated noise signal, and we accomplish this with small-bandwidth windows. A convenient measure of this bandwidth is the equivalent noise bandwidth (ENBW) of the window. This is the width of a rectangle filter with the same peak power gain that would accumulate the same noise power (see Fig. 7).

The accumulated noise power of the window is defined as

$$\text{Noise Power} = N_0 \int_{-\pi/T}^{+\pi/T} |W(\omega)|^2 \, d\omega/2\pi \tag{8}$$

where N_0 is the noise power per unit bandwidth. Parseval's theorem allows (8) to be computed by

$$\text{Noise Power} = \frac{N_0}{T} \sum_n w^2(nT). \tag{9}$$

The peak power gain of the window occurs at $\omega = 0$, the zero frequency power gain, and is defined by

$$\text{Peak } \textit{Signal} \text{ Gain} \doteq W(0) = \sum_n w(nT) \tag{10a}$$

$$\text{Peak } \textit{Power} \text{ Gain} = W^2(0) = \left[\sum_n w(nT)\right]^2 \tag{10b}$$

Thus the ENBW (normalized by N_0/T, the noise power per bin) is given in the following equation and is tabulated for the windows of this report in Table I

$$\text{ENBW} = \frac{\sum_n w^2(nT)}{\left[\sum_n w(nT)\right]^2}. \tag{11}$$

B. Processing Gain

A concept closely allied to ENBW is processing gain (PG) and processing loss (PL) of a windowed transform. We can think of the DFT as a bank of matched filters, where each filter is matched to one of the complex sinusoidal sequences of the basis set [3]. From this perspective, we can examine the PG (sometimes called the coherent gain) of the filter, and we can examine the PL due to the window having reduced the data to zero values near the boundaries. Let the input sampled sequence be defined by (12):

$$f(nT) = A \exp(+j\omega_k nT) + q(nT) \tag{12}$$

where $q(nT)$ is a white-noise sequence with variance σ_q^2. Then the signal component of the windowed spectrum (the matched filter output) is presented in

$$F(\omega_k)\big|_{\text{signal}} = \sum_n w(nT) A \exp(+j\omega_k nT) \exp(-j\omega_k nT)$$

$$= A \sum_n w(nT). \tag{13}$$

We see that the noiseless measurement (the expected value of the noisy measurement) is proportional to the input amplitude A. The proportionality factor is the sum of the window terms, which is in fact the dc signal gain of the window. For a rectangle window this factor is N, the number of terms in the window. For any other window, the gain is reduced due to the window smoothly going to zero near the boundaries. This

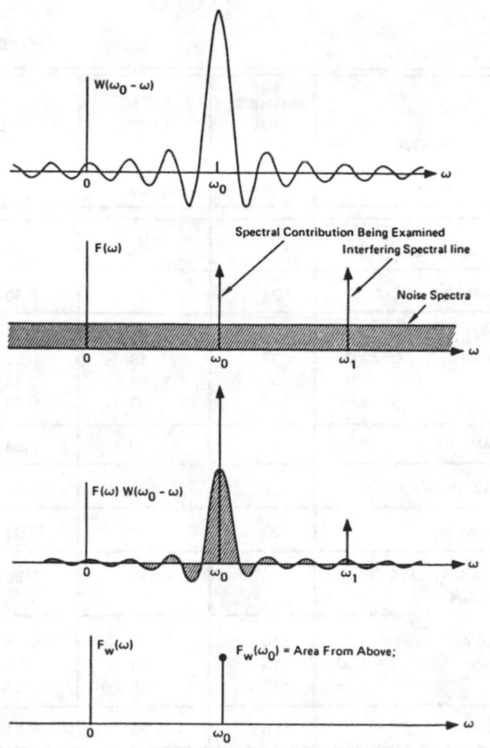

Fig. 6. Graphical interpretation of equation (6). Window visualized as a spectral filter.

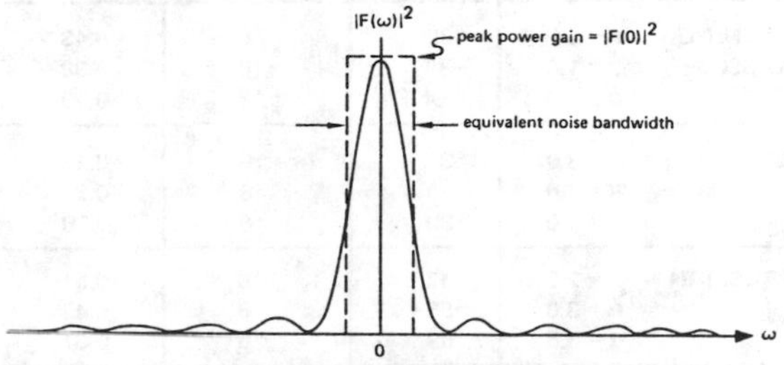

Fig. 7. Equivalent noise bandwidth of window.

TABLE I
WINDOWS AND FIGURES OF MERIT

WINDOW		HIGHEST SIDE-LOBE LEVEL (dB)	SIDE-LOBE FALL-OFF (dB/OCT)	COHERENT GAIN	EQUIV NOISE BW (BINS)	3.0-dB BW (BINS)	SCALLOP LOSS (dB)	WORST CASE PROCESS LOSS (dB)	6.0-dB BW (BINS)	OVERLAP CORRELATION (PCNT) 75% OL	OVERLAP CORRELATION (PCNT) 50% OL
RECTANGLE		−13	−6	1.00	1.00	0.89	3.92	3.92	1.21	75.0	50.0
TRIANGLE		−27	−12	0.50	1.33	1.28	1.82	3.07	1.78	71.9	25.0
$\cos^{\alpha}(X)$ HANNING	$\alpha = 1.0$	−23	−12	0.64	1.23	1.20	2.10	3.01	1.65	75.5	31.8
	$\alpha = 2.0$	−32	−18	0.50	1.50	1.44	1.42	3.18	2.00	65.9	16.7
	$\alpha = 3.0$	−39	−24	0.42	1.73	1.66	1.08	3.47	2.32	56.7	8.5
	$\alpha = 4.0$	−47	−30	0.38	1.94	1.86	0.86	3.75	2.59	48.6	4.3
HAMMING		−43	−6	0.54	1.36	1.30	1.78	3.10	1.81	70.7	23.5
RIESZ		−21	−12	0.67	1.20	1.16	2.22	3.01	1.59	76.5	34.4
RIEMANN		−26	−12	0.59	1.30	1.26	1.89	3.03	1.74	73.4	27.4
DE LA VALLE-POUSSIN		−53	−24	0.38	1.92	1.82	0.90	3.72	2.55	49.3	5.0
TUKEY	$\alpha = 0.25$	−14	−18	0.88	1.10	1.01	2.96	3.39	1.38	74.1	44.4
	$\alpha = 0.50$	−15	−18	0.75	1.22	1.15	2.24	3.11	1.57	72.7	36.4
	$\alpha = 0.75$	−19	−18	0.63	1.36	1.31	1.73	3.07	1.80	70.5	25.1
BOHMAN		−46	−24	0.41	1.79	1.71	1.02	3.54	2.38	54.5	7.4
POISSON	$\alpha = 2.0$	−19	−6	0.44	1.30	1.21	2.09	3.23	1.69	69.9	27.8
	$\alpha = 3.0$	−24	−6	0.32	1.65	1.45	1.46	3.64	2.08	54.8	15.1
	$\alpha = 4.0$	−31	−6	0.25	2.08	1.75	1.03	4.21	2.58	40.4	7.4
HANNING-POISSON	$\alpha = 0.5$	−35	−18	0.43	1.61	1.54	1.26	3.33	2.14	61.3	12.6
	$\alpha = 1.0$	−39	−18	0.38	1.73	1.64	1.11	3.50	2.30	56.0	9.2
	$\alpha = 2.0$	NONE	−18	0.29	2.02	1.87	0.87	3.94	2.65	44.6	4.7
CAUCHY	$\alpha = 3.0$	−31	−6	0.42	1.48	1.34	1.71	3.40	1.90	61.6	20.2
	$\alpha = 4.0$	−35	−6	0.33	1.76	1.50	1.36	3.83	2.20	48.8	13.2
	$\alpha = 5.0$	−30	−6	0.28	2.06	1.68	1.13	4.28	2.53	38.3	9.0
GAUSSIAN	$\alpha = 2.5$	−42	−6	0.51	1.39	1.33	1.69	3.14	1.86	67.7	20.0
	$\alpha = 3.0$	−55	−6	0.43	1.64	1.55	1.25	3.40	2.18	57.5	10.6
	$\alpha = 3.5$	−69	−6	0.37	1.90	1.79	0.94	3.73	2.52	47.2	4.9
DOLPH-CHEBYSHEV	$\alpha = 2.5$	−50	0	0.53	1.39	1.33	1.70	3.12	1.85	69.6	22.3
	$\alpha = 3.0$	−60	0	0.48	1.51	1.44	1.44	3.23	2.01	64.7	16.3
	$\alpha = 3.5$	−70	0	0.45	1.62	1.55	1.25	3.35	2.17	60.2	11.9
	$\alpha = 4.0$	−80	0	0.42	1.73	1.65	1.10	3.48	2.31	55.9	8.7
KAISER-BESSEL	$\alpha = 2.0$	−46	−6	0.49	1.50	1.43	1.46	3.20	1.99	65.7	16.9
	$\alpha = 2.5$	−57	−6	0.44	1.65	1.57	1.20	3.38	2.20	59.5	11.2
	$\alpha = 3.0$	−69	−6	0.40	1.80	1.71	1.02	3.56	2.39	53.9	7.4
	$\alpha = 3.5$	−82	−6	0.37	1.93	1.83	0.89	3.74	2.57	48.8	4.8
BARCILON-TEMES	$\alpha = 3.0$	−53	−6	0.47	1.56	1.49	1.34	3.27	2.07	63.0	14.2
	$\alpha = 3.5$	−58	−6	0.43	1.67	1.59	1.18	3.40	2.23	58.6	10.4
	$\alpha = 4.0$	−68	−6	0.41	1.77	1.69	1.05	3.52	2.36	54.4	7.6
EXACT BLACKMAN		−51	−6	0.46	1.57	1.52	1.33	3.29	2.13	62.7	14.0
BLACKMAN		−58	−18	0.42	1.73	1.68	1.10	3.47	2.35	56.7	9.0
MINIMUM 3-SAMPLE BLACKMAN-HARRIS		−67	−6	0.42	1.71	1.66	1.13	3.45	1.81	57.2	9.6
*MINIMUM 4-SAMPLE BLACKMAN-HARRIS		−92	−6	0.36	2.00	1.90	0.83	3.85	2.72	46.0	3.8
*61 dB 3-SAMPLE BLACKMAN-HARRIS		−61	−6	0.45	1.61	1.56	1.27	3.34	2.19	61.0	12.6
74 dB 4-SAMPLE BLACKMAN-HARRIS		−74	−6	0.40	1.79	1.74	1.03	3.56	2.44	53.9	7.4
4-SAMPLE KAISER-BESSEL	$\alpha = 3.0$	−69	−6	0.40	1.80	1.74	1.02	3.56	2.44	53.9	7.4

*REFERENCE POINTS FOR DATA ON FIGURE 12 – NO FIGURES TO MATCH THESE WINDOWS.

reduction in proportionality factor is important as it represents a known bias on spectral amplitudes. Coherent power gain, the square of coherent gain, is occasionally the parameter listed in the literature. Coherent gain (the summation of (13)) normalized by its maximum value N is listed in Table I.

The incoherent component of the windowed transform is given by

$$F(\omega_k)\big|_{\text{noise}} = \sum_n w(nT)\, q(nT) \exp(-j\omega_k nT) \tag{14a}$$

and the incoherent power (the mean-square value of this component where $E\{\ \}$ is the expectation operator) is given by

$$E\{|F(\omega_k)|_{\text{noise}}|^2\} = \sum_n \sum_m w(nT)\, w(mT)\, E\{q(nT)\, q^*(mT)\} \cdot \exp(-j\omega_k nT) \exp(+j\omega_k mT)$$
$$= \sigma_q^2 \sum_n w^2(nT). \tag{14b}$$

Notice the incoherent power gain is the sum of the squares of the window terms, and the coherent power gain is the square of the sum of the window terms.

Finally, PG, which is defined as the ratio of output signal-to-noise ratio to input signal-to-noise ratio, is given by

$$\text{PG} = \frac{S_o/N_o}{S_i/N_i} = \frac{A^2 \left[\sum\limits_n w(nT)\right]^2 \Big/ \sigma_q^2 \sum\limits_n w^2(nT)}{A^2/\sigma_q^2}$$
$$= \frac{\left[\sum\limits_n w(nT)\right]^2}{\sum\limits_n w^2(nT)}. \tag{15}$$

Notice PG is the reciprocal of the normalized ENBW. Thus large ENBW suggests a reduced processing gain. This is reasonable, since an increased noise bandwidth permits additional noise to contribute to a spectral estimate.

C. Overlap Correlation

When the fast Fourier transform (FFT) is used to process long-time sequences a partition length N is first selected to establish the required spectral resolution of the analysis. Spectral resolution of the FFT is defined in (16) where Δf is the resolution, f_s is the sample frequency selected to satisfy the Nyquist criterion, and β is the coefficient reflecting the bandwidth increase due to the particular window selected. Note that $[f_s/N]$ is the minimum resolution of the FFT which we denote as the FFT bin width. The coefficient β is usually selected to be the ENBW in bins as listed in Table I

$$\Delta f = \beta\left(\frac{f_s}{N}\right). \tag{16}$$

If the window and the FFT are applied to nonoverlapping partitions of the sequence, as shown in Fig. 8, a significant part of the series is ignored due to the window's exhibiting small values near the boundaries. For instance, if the transform is being used to detect short-duration tone-like signals, the non overlapped analysis could miss the event if it occurred near the boundaries. To avoid this loss of data, the transforms are usually applied to the overlapped partition sequences as shown in Fig. 8. The overlap is almost always 50 or 75 percent. This overlap processing of course increases the work load to cover the total sequence length, but the rewards warrant the extra effort.

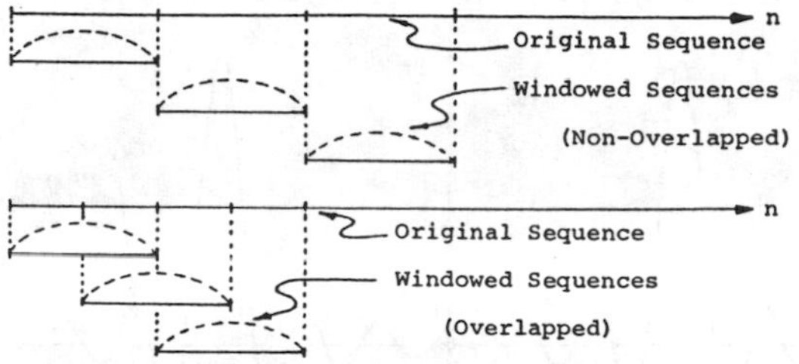

Fig. 8. Partition of sequences for nonoverlapped and for overlapped processing.

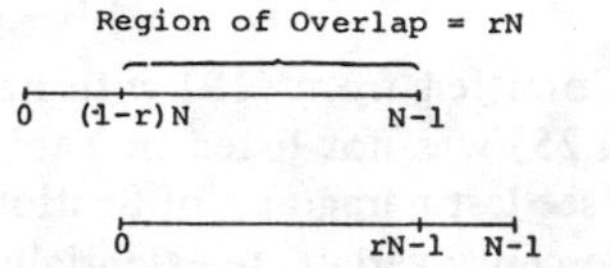

Fig. 9. Relationship between indices on overlapped intervals.

An important question related to overlapped processing is what is the degree of correlation of the random components in successive transforms? This correlation, as a function of fractional overlap r, is defined for a relatively flat noise spectrum over the window bandwidth by (17). Fig. 9 identifies how the indices of (17) relate to the overlap of the intervals. The correlation coefficient

$$c(r) = \frac{\left\{\sum\limits_{n=0}^{rN-1} (W(n)\, W(n + [1 - r]N))\right\}}{\left\{\sum\limits_{n=0}^{N-1} W^2(n)\right\}} \tag{17}$$

is computed and tabulated in Table I. for each of the windows listed for 50- and 75-percent overlap.

Often in a spectral analysis, the squared magnitude of successive transforms are averaged to reduce the variance of the measurements [5]. We know of course that when we average K identically distributed independent measurements, the variance of the average is related to the individual variance of the measurements by

$$\frac{\sigma^2_{\text{Avg.}}}{\sigma^2_{\text{Meas.}}} = \frac{1}{K}. \tag{18}$$

Now we can ask what is the reduction in the variance when we average measurements which are correlated as they are for overlapped transforms? Welch [5] has supplied an answer to this question which we present here, for the special case of 50- and 75-percent overlap

$$\frac{\sigma^2_{\text{Avg.}}}{\sigma^2_{\text{Meas.}}} = \frac{1}{K}[1 + 2c^2(0.5)] - \frac{2}{K^2}[c^2(0.5)],$$

50 percent overlap

$$= \frac{1}{K}[1 + 2c^2(0.75) + 2c^2(0.5) + 2c^2(0.25)]$$
$$- \frac{2}{K^2}[c^2(0.75) + 2c^2(0.5) + 3c^2(0.25)],$$

75 percent overlap. (19)

The negative terms in (19) are the edge effects of the average and can be ignored if the number of terms K is larger than ten. For good windows, $c^2(0.25)$ is small compared to 1.0,

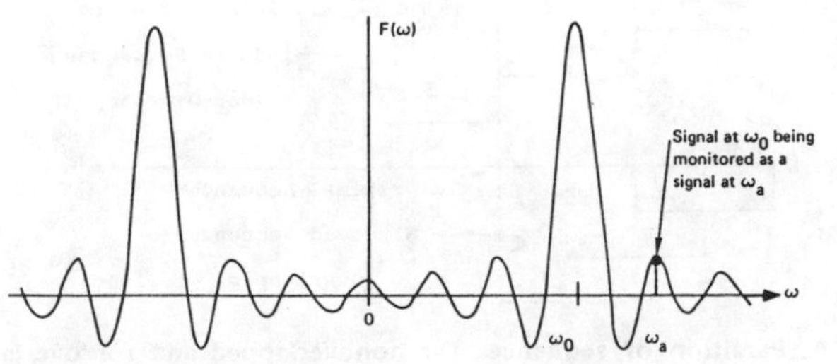

Fig. 10. Spectral leakage effect of window.

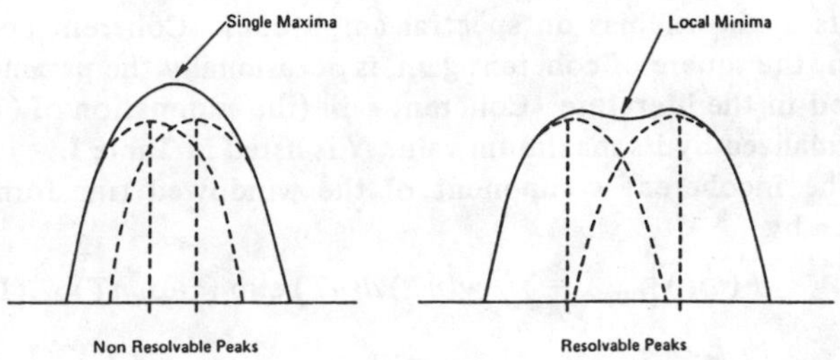

Fig. 11. Spectral resolution of nearby kernels.

and can also be omitted from (19) with negligible error. For this reason, $c(0.25)$ was not listed in Table I. Note, that for good windows (see last paragraph of Section IV-F), transforms taken with 50-percent overlap are essentially independent.

D. Scalloping Loss

An important consideration related to minimum detectable signal is called scalloping loss or picket-fence effect. We have considered the windowed DFT as a bank of matched filters and have examined the processing gain and the reduction of this gain ascribable to the window for tones matched to the basis vectors. The basis vectors are tones with frequencies equal to multiples of f_s/N (with f_s being the sample frequency). These frequencies are sample points from the spectrum, and are normally referred to as DFT output points or as DFT bins. We now address the question, what is the additional loss in processing gain for a tone of frequency midway between two bin frequencies (that is, at frequencies $(k + 1/2) f_s/N$)?

Returning to (13), with ω_k replaced by $\omega_{(k+1/2)}$, we determine the processing gain for this half-bin frequency shift as defined in

$$F(\omega_{(1/2)})\big|_{\text{signal}} = A \sum_n w(nT) \exp(-j\omega_{(1/2)}nT),$$

$$\text{where } \omega_{(1/2)} = \frac{1}{2}\frac{\omega s}{N} = \frac{\pi}{NT}. \quad (20a)$$

We also define the scalloping loss as the ratio of coherent gain for a tone located half a bin from a DFT sample point to the coherent gain for a tone located at a DFT sample point, as indicated in

$$\text{Scalloping Loss} = \frac{\left|\sum_n w(nT) \exp\left(-j\frac{\pi}{N}n\right)\right|}{\sum_n w(nT)} = \frac{\left|W\left(\frac{1}{2}\frac{\omega_s}{N}\right)\right|}{W(0)}. \quad (20b)$$

Scalloping loss represents the maximum reduction in PG due to signal frequency. This loss has been computed for the windows of this report and has been included in Table I.

E. Worst Case Processing Loss

We now make an interesting observation. We define worst case PL as the sum of maximum scalloping loss of a window and of PL due to that window (both in decibel). This number is the reduction of output signal-to-noise ratio as a result of windowing and of worst case frequency location. This of course is related to the minimum detectable tone in broadband noise. It is interesting to note that the worst case loss is always between 3.0 and 4.3 dB. Windows with worst case PL exceeding 3.8 dB are very poor windows and should not be used. Additional comments on poor windows will be found in Section IV-G. We can conclude from the combined loss figures of Table I and from Fig. 12 that for the detection of single tones in broad-band noise, nearly any window (other than the rectangle) is as good as any other. The difference between the various windows is less than 1.0 dB and for good windows is less than 0.7 dB. The detection of tones in the presence of other tones is, however, quite another problem. Here the window does have a marked affect, as will be demonstrated shortly.

F. Spectral Leakage Revisited

Returning to (6) and to Fig. 6, we observe the spectral measurement is affected not only by the broadband noise spectrum, but also by the narrow-band spectrum which falls within the bandwidth of the window. In fact, a given spectral component say at $\omega = \omega_0$ will contribute output (or will be observed) at another frequency, say at $\omega = \omega_a$ according to the gain of the window centered at ω_0 and measured at ω_a. This is the effect normally referred to as spectral leakage and is demonstrated in Fig. 10 with the transform of a finite duration tone of frequency ω_0

This leakage causes a bias in the amplitude and the position of a harmonic estimate. Even for the case of a single real harmonic line (not at a DFT sample point), the leakage from the kernel on the negative-frequency axis biases the kernel on the positive-frequency line. This bias is most severe and most bothersome for the detection of small signals in the presence of nearby large signals. To reduce the effects of this bias, the window should exhibit low-amplitude sidelobes far from the central main lobe, and the transition to the low sidelobes should be very rapid. One indicator of how well a window suppresses leakage is the peak sidelobe level (relative to the main lobe): another is the asymptotic rate of falloff of these sidelobes. These indicators are listed in Table I.

G. Minimum Resolution Bandwidth

Fig. 11 suggests another criterion with which we should be concerned in the window selection process. Since the window imposes an effective bandwidth on the spectral line, we would be interested in the minimum separation between two equal-strength lines such that for arbitrary spectral locations their respective main lobes can be resolved. The classic criterion for this resolution is the width of the window at the half-power points (the 3.0-dB bandwidth). This criterion reflects the fact that two equal-strength main lobes separated in frequency by less than their 3.0-dB bandwidths will exhibit a single spectral peak and will not be resolved as two distinct lines. The problem with this criterion is that it does not work for the coherent addition we find in the DFT. The DFT output points are the coherent addition of the spectral components weighted through the window at a given frequency.

If two kernels are contributing to the coherent summation, the sum at the crossover point (nominally half-way between them) must be smaller than the individual peaks if the two peaks are to be resolved. Thus at the crossover points of the kernels, the gain from each kernel must be less than 0.5, or the crossover points must occur beyond the 6.0-dB points of the windows. Table I lists the 6.0-dB bandwidths of the various windows examined in this report. From the table, we see that the 6.0-dB bandwidth varies from 1.2 bins to 2.6 bins, where a bin is the fundamental frequency resolution ω_s/N. The 3.0-dB bandwidth does have utility as a performance indicator as shown in the next paragraph. Remember however, it is the 6.0-dB bandwidth which defines the resolution of the windowed DFT.

From Table I, we see that the noise bandwidth always exceeds the 3.0-dB bandwidth. The difference between the two, referenced to the 3.0-dB bandwidth, appears to be a sensitive indicator of overall window performance. We have observed that for all the good windows on the table, this indicator was found to be in the range of 4.0 to 5.5 percent. Those windows for which this ratio is outside that range either have a wide main lobe or a high sidelobe structure and, hence, are characterized by high processing loss or by poor two-tone detection capabilities. Those windows for which this ratio is inside the 4.0 to 5.5-percent range are found in the lower left corner of the performance comparison chart (Fig. 12), which is described next.

While Table I does list the common performance parameters of the windows examined in this report, the mass of numbers is not enlightening. We do realize that the sidelobe level (to reduce bias) and the worst case processing loss (to maximize detectability) are probably the most important parameters on the table. Fig. 12 shows the relative position of the windows as a function of these parameters. Windows residing in the lower left corner of the figure are the good-performing windows. They exhibit low-sidelobe levels and low worst case processing loss. We urge the reader to read Sections VI and VII; Fig. 12 presents a lot of information, but not the full story.

V. Classic Windows

We will now catalog some well-known (and some not well-known windows. For each window we will comment on the justification for its use and identify its significant parameters. All the windows will be presented as even (about the origin) sequences with an odd number of points. To convert the window to DFT-even, the right end point will be discarded and the sequence will be shifted so that the left end point coincides with the origin. We will use normalized coordinates with sample period $T = 1.0$, so that ω is periodic in 2π and, hence, will be identified as θ. A DFT bin will be considered to extend between DFT sample points (multiples of $2\pi/N$) and have a width of $2\pi/N$.

A. Rectangle (Dirichlet) Window [6]

The rectangle window is unity over the observation interval, and can be thought of as a gating sequence applied to the data so that they are of finite extent. The window for a finite Fourier transform is defined as

$$w(n) = 1.0, \quad n = -\frac{N}{2}, \cdots, -1, 0, 1, \cdots, \frac{N}{2} \tag{21a}$$

and is shown in Fig. 13. The same window for a DFT is

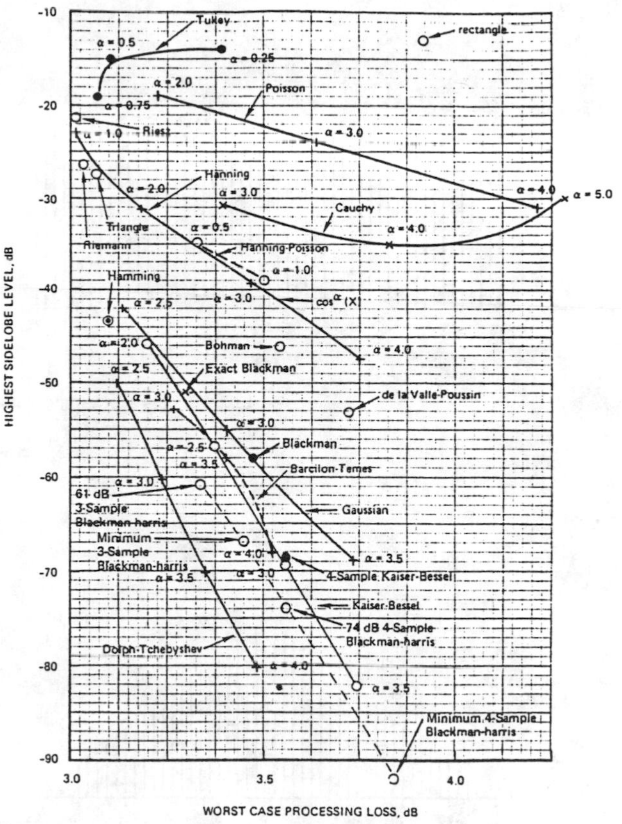

Fig. 12. Comparison of windows: sidelobe levels and worst case processing loss.

defined as

$$w(n) = 1.0, \quad n = 0, 1, \cdots, N-1. \tag{21b}$$

The spectral window for the DFT window sequence is given in

$$W(\theta) = \exp\left(-j\,\frac{N-1}{2}\,\theta\right) \frac{\sin\left[\dfrac{N}{2}\theta\right]}{\sin\left[\dfrac{1}{2}\theta\right]}. \tag{21c}$$

The transform of this window is seen to be the Dirichlet kernel, which exhibits a DFT main-lobe width (between zero crossings) of 2 bins and a first sidelobe level approximately 13 dB down from the main-lobe peak. The sidelobes fall off at 6.0 dB per octave, which is of course the expected rate for a function with a discontinuity. The parameters of the DFT window are listed in Table I.

With the rectangle window now defined, we can answer the question posed earlier: in what sense does the finite sum of (22a) approximate the infinite sum of (22b)?

$$F(\theta) = \sum_{n=-N/2}^{+N/2} f(n) \exp(-jn\theta) \tag{22a}$$

$$F(\theta) = \sum_{n=-\infty}^{+\infty} f(n) \exp(-jn\theta). \tag{22b}$$

We observe the finite sum is the rectangle-windowed version of the infinite sum. We recognize that the infinite sum is the Fourier series expansion of some periodic function for which the $f(n)$'s are the Fourier series coefficients. We also recognize

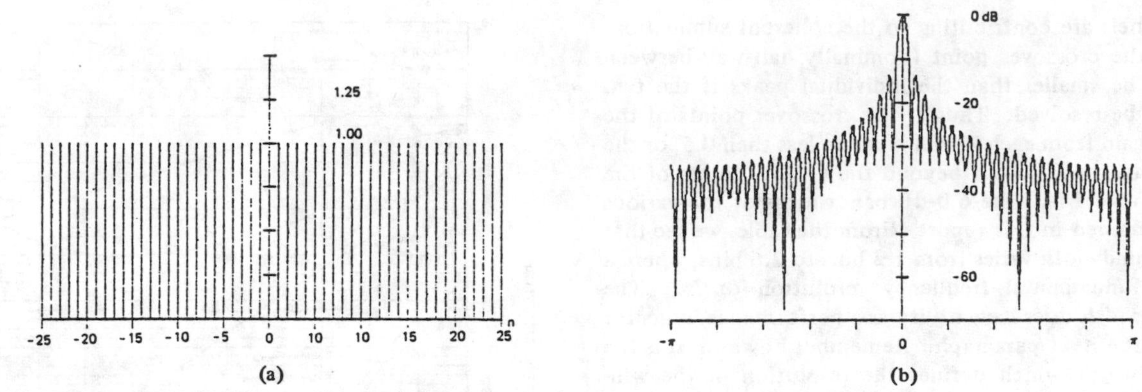

Fig. 13. (a) Rectangle window. (b) Log-magnitude of transform.

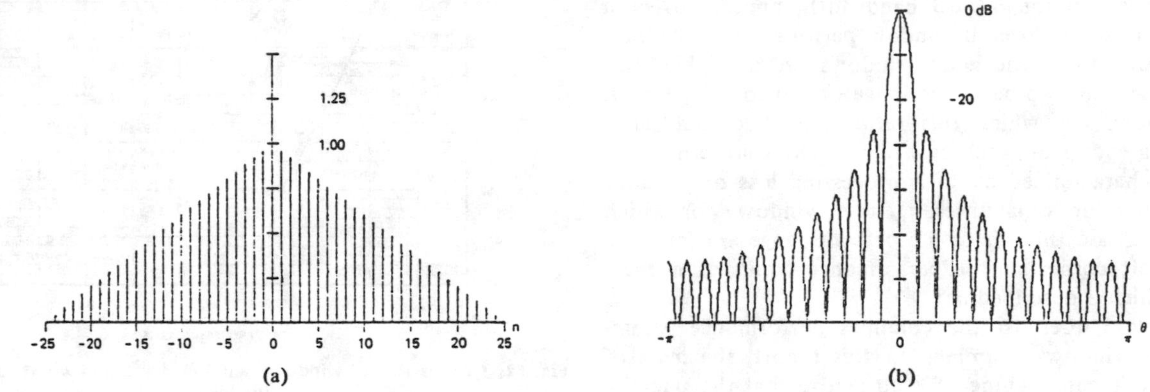

Fig. 14. (a) Triangle window. (b) Log-magnitude of transform.

that the finite sum is simply the partial sum of the series. From this viewpoint we can cast the question in terms of the convergence properties of the partial sums of Fourier series. From this work we know the partial sum is the least mean-square error approximation to the infinite sum.

We observe that mean square convergence is a convenient analytic concept, but it is not attractive for finite estimates or for numerical approximations. Mean-square estimates tend to oscillate about their means, and do not exhibit uniform convergence. (The approximation in a neighborhood of a point of continuity may get worse if more terms are added to the partial sum.) We normally observe this behavior near points of discontinuity as the ringing we call Gibbs phenomenon. It is this oscillatory behavior we are trying to control by the use of other windows.

B. Triangle (Fejer, Bartlet) Window [7]

The triangle window for a finite Fourier transform is defined as

$$W(n) = 1.0 - \frac{|n|}{N/2}, \qquad n = -\frac{N}{2}, \cdots, -1, 0, 1, \cdots, \frac{N}{2} \tag{23a}$$

and is shown in Fig. 14. The same window for a DFT is defined as

$$W(n) = \begin{cases} \dfrac{n}{N/2}, & n = 0, 1, \cdots, \dfrac{N}{2} \\ W(N-n), & n = \dfrac{N}{2}, \cdots, N-1 \end{cases} \tag{23b}$$

and the spectral window corresponding to the DFT sequence is given in

$$W(\theta) = \frac{2}{N} \exp\left[-j\left(\frac{N}{2} - 1\right)\theta\right] \left[\frac{\sin\left(\frac{N}{4}\theta\right)}{\sin\left(\frac{1}{2}\theta\right)}\right]^2 \tag{23c}$$

The transform of this window is seen to be the squared Dirichlet kernel. Its main-lobe width (between zero crossings) is twice that of the rectangle's and the first sidelobe level is approximately 26 dB down from the main-lobe peak, again, twice that of the rectangle's. The sidelobes fall off at -12 dB per octave, reflecting the discontinuity of the window residing in the first derivative (rather than in the function itself). The triangle is the simplest window which exhibits a nonnegative transform. This property can be realized by convolving any window (of half-extent) with itself. The resultant window's transform is the square of the original window's transform.

A window sequence derived by self-convolving a parent window contains approximately twice the number of samples as the parent window, hence corresponds to a trigonometric polynomial (its Z-transform) of approximately twice the order. (Convolving two rectangles each of $N/2$ points will result in a triangle of $N + 1$ points when the zero end points are counted.) The transform of the window will now exhibit twice as many zeros as the parent transform (to account for the increased order of the associated trigonometric polynomial). But how has the transform applied these extra zeros available from the increased order polynomial? The self-

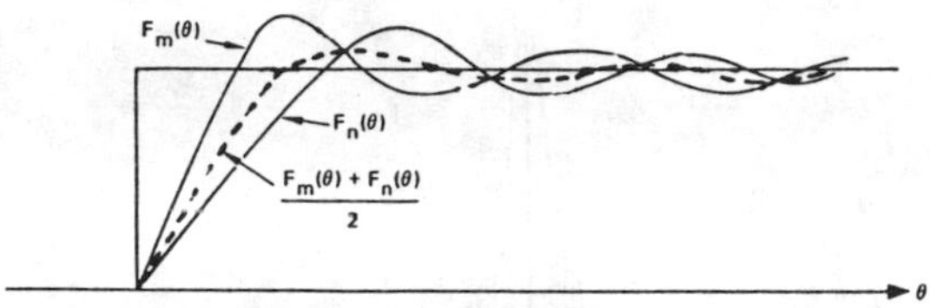

Fig. 15. Two partial sums and their average.

TABLE II
FEJER CONVERGENCE FACTORS AS AN AVERAGE TRANSFORM

$F_0(\theta)$					f_0				
$F_1(\theta)$				f_{-1}	f_0	f_{+1}			
$F_2(\theta)$			f_{-2}	f_{-1}	f_0	f_{+1}	f_{+2}		
$F_3(\theta)$		f_{-3}	f_{-2}	f_{-1}	f_0	f_{+1}	f_{+2}	f_{+3}	
$F^4(\theta)$	$\frac{0}{4}f_{-4}$	$\frac{1}{4}f_{-3}$	$\frac{2}{4}f_{-2}$	$\frac{3}{4}f_{-1}$	$\frac{4}{4}f_0$	$\frac{3}{4}f_{+1}$	$\frac{2}{4}f_{+2}$	$\frac{1}{4}f_{+3}$	$\frac{0}{4}f_{+4}$

convolved window simply places repeated zeros at each location for which the parent transform had a zero. This, of course, not only sets the transform to zero at those points, but also sets the first derivative to zero at those points. If the intent of the increased order of polynomial is to hold down the sidelobe levels, then doubling up on the zeros is a wasteful tactic. The additional zeros might better be placed between the existing zeros (near the local peaks of the sidelobes) to hold down the sidelobes rather than at locations for which the transform is already equal to zero. In fact we will observe in subsequent windows that very few good windows exhibit repeated roots.

Backing up for a moment, it is interesting to examine the triangle window in terms of partial-sum convergence of Fourier series. Fejer observed that the partial sums of Fourier series were poor numerical approximations [8]. Fourier coefficients were easy to generate however, and he questioned if some simple modification of coefficients might lead to a new set with more desirable convergence properties. The oscillation of the partial sum, and the contraction of those oscillations as the order of the partial sum increased, suggested that an average of the partial sums would be a smoother function. Fig. 15 presents an expansion of two partial sums near a discontinuity. Notice the average of the two expansions is smoother than either. Continuing in this line of reasoning, an average expansion $F^N(\theta)$ might be defined by

$$F^N(\theta) = \frac{1}{N}\left[F_{N-1}(\theta) + F_{N-2}(\theta) + \cdots + F_0(\theta)\right] \qquad (24)$$

where $F_M(\theta)$ is the M-term partial sum of the series. This is easily visualized in Table II, which lists the nonzero coefficients of the first four partial sums and their average summation. We see that the Fejer convergence factors applied to the Fourier series coefficients is, in fact, a triangle window. The averaging of partial sums is known as the method of Cesàro summability.

C. $Cos^\alpha(X)$ Windows

This is actually a family of windows dependent upon the parameter α, with α normally being an integer. Attractions of this family include the ease with which the terms can be generated, and the easily identified properties of the transform of the cosine function. These properties are particularly attractive under the DFT. The window for a finite Fourier transform is defined as

$$w(n) = \cos^\alpha\left[\frac{n}{N}\pi\right], \qquad n = -\frac{N}{2}, \cdots, -1, 0, 1, \cdots, \frac{N}{2} \qquad (25a)$$

and for a DFT as

$$w(n) = \sin^\alpha\left[\frac{n}{N}\pi\right], \qquad n = 0, 1, 2, \cdots, N-1. \qquad (25b)$$

Notice the effect due to the change of the origin. The most common values of α are the integers 1 through 4, with 2 being the most well known (as the Hanning window). This window is identified for values of α equal to 1 and 2 in (26a), (26b), (27a), and (27b), (the "a" for the finite transforms, the "b" for the DFT):

$\alpha = 1.0$ (cosine lobe)

$$w(n) = \cos\left[\frac{n}{N}\pi\right], \qquad n = -\frac{N}{2}, \cdots, -1, 0, 1, \cdots, \frac{N}{2} \qquad (26a)$$

$\alpha = 1.0$ (sine lobe)

$$w(n) = \sin\left[\frac{n}{N}\pi\right], \qquad n = 0, 1, 2, \cdots, N-1 \qquad (26b)$$

$\alpha = 2.0$ (cosine squared, raised cosine, Hanning)

$$w(n) = \cos^2\left[\frac{n}{N}\pi\right] = 0.5\left[1.0 + \cos\left[\frac{2n}{N}\pi\right]\right], \qquad n = -\frac{N}{2}, \cdots, -1, 0, 1, \cdots, \frac{N}{2} \qquad (27a)$$

$\alpha = 2.0$ (sine squared, raised cosine, Hanning)

$$w(n) = \sin^2\left[\frac{n}{N}\pi\right] = 0.5\left[1.0 - \cos\left[\frac{2n}{N}\pi\right]\right], \qquad n = 0, 1, 2, \cdots, N-1. \qquad (27b)$$

The windows are shown for α integer values of 1 through 4 in Figs. 16 through 19. Notice as α becomes larger, the windows become smoother and the transform reflects this increased smoothness in decreased sidelobe level and faster falloff of the sidelobes, but with an increased width of the main lobe.

Of particular interest in this family, is the Hann window (after the Austrian meteorologist, Julius Von Hann)[1] [7]. Not only is this window continuous, but so is its first derivative. Since the discontinuity of this window resides in the second derivative, the transform falls off at $1/\omega^3$ or at -18 dB per octave. Let us closely examine the transform of this window. We will gain some interesting insight and learn of a clever application of the window under the DFT.

[1]The correct name of this window is "Hann." The term "Hanning" is used in this report to reflect conventional usage. The derived term "Hann'd" is also widely used.

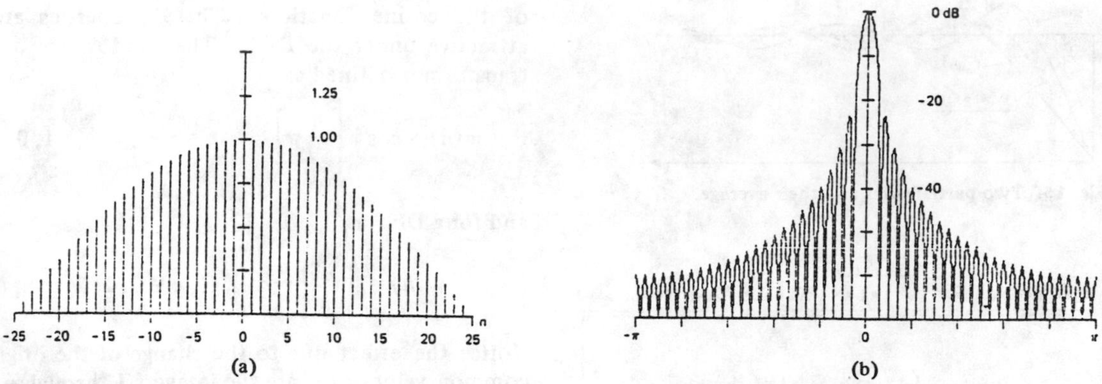

Fig. 16. (a) Cos $(n\pi/N)$ window. (b) Log-magnitude of transform.

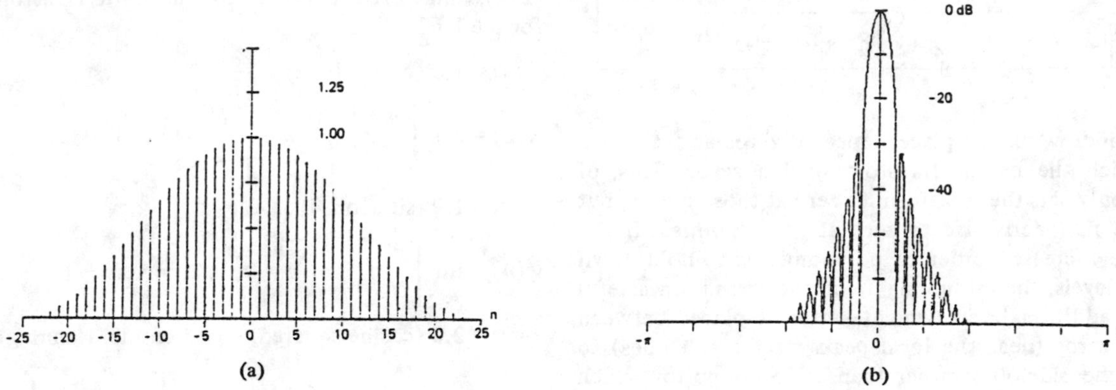

Fig. 17. (a) Cos^2 $(n\pi/N)$ window. (b) Log-magnitude of transform.

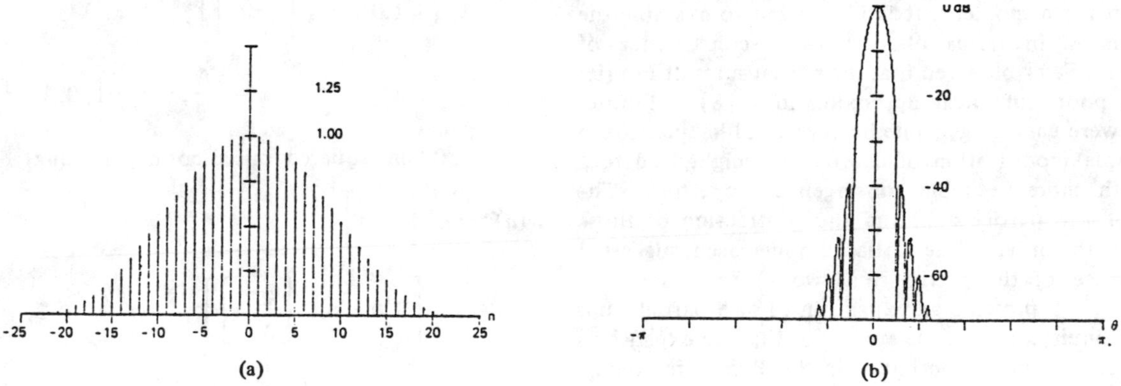

Fig. 18. (a) Cos^3 $(n\pi/N)$ window. (b) Log-magnitude of transform.

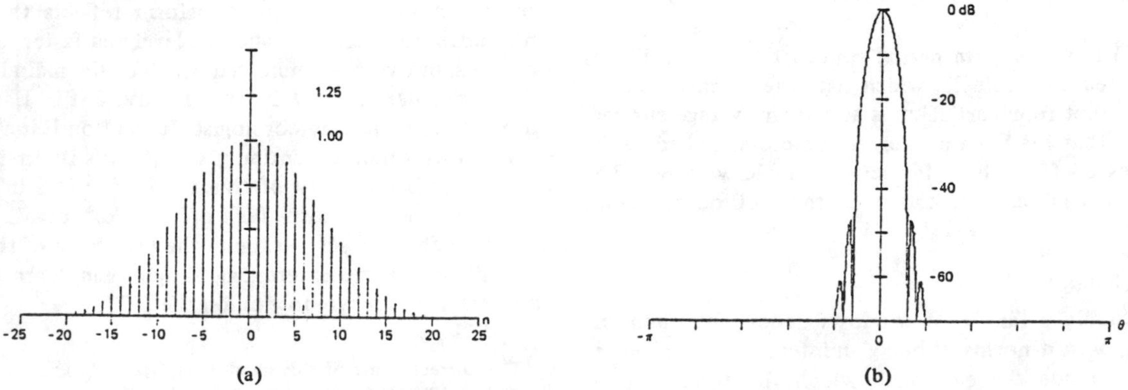

Fig. 19. (a) Cos^4 $(n\pi/N)$ window. (b) Log-magnitude of transform.

The sampled Hanning window can be written as the sum of the sequences indicated in

$$w(n) = 0.5 + 0.5 \cos\left[\frac{2n}{N}\pi\right],$$

$$n = -\frac{N}{2}, \cdots, -1, 0, 1, \cdots, \frac{N}{2} - 1. \quad (28a)$$

Each sequence has the easily recognized DFT indicated in

$$W(\theta) = 0.5\, D(\theta) + 0.25\left[D\left(\theta - \frac{2\pi}{N}\right) + D\left(\theta + \frac{2\pi}{N}\right)\right] \quad (28b)$$

where

$$D(\theta) = \exp\left(+j\frac{\theta}{2}\right)\frac{\sin\left[\frac{N}{2}\theta\right]}{\sin\left[\frac{1}{2}\theta\right]}.$$

We recognize the Dirichlet kernel at the origin as the transform of the constant 0.5 samples and the pair of translated kernels as the transform of the single cycle of cosine samples. Note that the translated kernels are located on the first zeros of the center kernel, and are half the size of the center kernel. Also the sidelobes of the translated kernel are about half the size and are of opposite phase of the sidelobes of the central kernel. The summation of the three kernels' sidelobes being in phase opposition, tends to cancel the sidelobe structure. This cancelling summation is demonstrated in Fig. 20 which depicts the summation of the Dirichlet kernels (without the phase-shift terms).

The partial cancelling of the sidelobe structure suggests a constructive technique to define new windows. The most well-known of these are the Hamming and the Blackman windows which are presented in the next two sections.

For the special case of the DFT, the Hanning window is sampled at multiples of $2\pi/N$, which of course are the locations of the zeros of the central Dirichlet kernel. Thus only three nonzero samples are taken in the sampling process. The positions of these samples are at $-2\pi/N$, 0, and $+2\pi/N$. The value of the samples obtained from (28b) (including the phase factor $\exp(-j(N/2)\theta)$ to account for the $N/2$ shift) are $-\frac{1}{4}$, $+\frac{1}{2}$, $-\frac{1}{4}$, respectively. Note the minus signs. These results from the shift in the origin for the window. Without the shift, the phase term is missing and the coefficients are all positive $\frac{1}{4}$, $\frac{1}{2}$, $\frac{1}{4}$. These are incorrect for DFT processing, but they find their way into much of the literature and practice.

Rather than apply the window as a product in the time domain, we always have the option to apply it as a convolution in the frequency domain. The attraction of the Hanning window for this application is twofold; first, the window spectra is nonzero at only three data points, and second, the sample values are binary fractions, which can be implemented as right shifts. Thus the Hanning-windowed spectral points obtained from the rectangle-windowed spectral points are obtained as indicated in the following equation as two real adds and two binary shifts (to multiply by $\frac{1}{2}$):

$$F(k)\big|_{\text{Hanning}} = \tfrac{1}{2}\left[F(k) - \tfrac{1}{2}\left[F(k-1) + F(k+1)\right]\right]\big|_{\text{Rectangle}}. \quad (29)$$

Thus a Hanning window applied to a real transform of length N can be performed as N real multiplies on the time sequence or as $2N$ real adds and $2N$ binary shifts on the spectral data. One other mildly important consideration, if the window is to be applied to the time data, is that the samples of the window must be stored somewhere, which normally means additional memory or hardware. It so happens that the samples of the cosine for the Hanning window are already stored in the machine as the trig-table for the FFT; thus the Hanning window requires no additional storage.

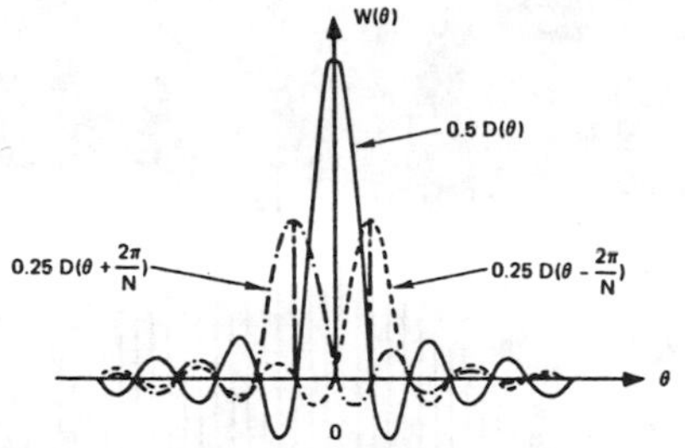

Fig. 20. Transform of Hanning window as a sum of three Dirichlet kernels.

D. Hamming Window [7]

The Hamming window can be thought of as a modified Hanning window. (Note the potential source of confusion in the similarities of the two names.) Referring back to Figs. 17 and 20, we note the inexact cancellation of the sidelobes from the summation of the three kernels. We can construct a window by adjusting the relative size of the kernels as indicated in the following to achieve a more desirable form of cancellation:

$$w(n) = \alpha + (1 - \alpha)\cos\left[\frac{2\pi}{N}n\right]$$

$$W(\theta) = \alpha D(\theta) + 0.5\,(1 - \alpha)\left[D\left(\theta - \frac{2\pi}{N}\right) + D\left(\theta + \frac{2\pi}{N}\right)\right]. \quad (30a)$$

Perfect cancellation of the first sidelobe (at $\theta = 2.5[2\pi/N]$) occurs when $\alpha = 25/46$ ($\alpha \doteq 0.543\,478\,261$). If α is selected as 0.54 (an approximation to 25/46), the new zero occurs at $\theta \doteq 2.6[2\pi/N]$ and a marked improvement in sidelobe level is realized. For this value of α, the window is called the Hamming window and is identified by

$$w(n) = \begin{cases} 0.54 + 0.46\cos\left[\frac{2\pi}{N}n\right], & n = -\frac{N}{2}, \cdots, -1, 0, 1, \cdots, \frac{N}{2} \\ 0.54 - 0.46\cos\left[\frac{2\pi}{N}n\right], & n = 0, 1, 2, \cdots, N-1. \end{cases} \quad (30b)$$

The coefficients of the Hamming window are nearly the set which achieve minimum sidelobe levels. If α is selected to be 0.53856 the sidelobe level is -43 dB and the resultant window is a special case of the Blackman–Harris windows presented in Section V-E. The Hamming window is shown in Fig. 21. Notice the deep attenuation at the missing sidelobe position. Note also that the small discontinuity at the boundary of the window has resulted in a $1/\omega$ (6.0 dB per octave) rate of falloff. The better sidelobe cancellation does result in a much lower initial sidelobe level of -42 dB. Table I lists the param-

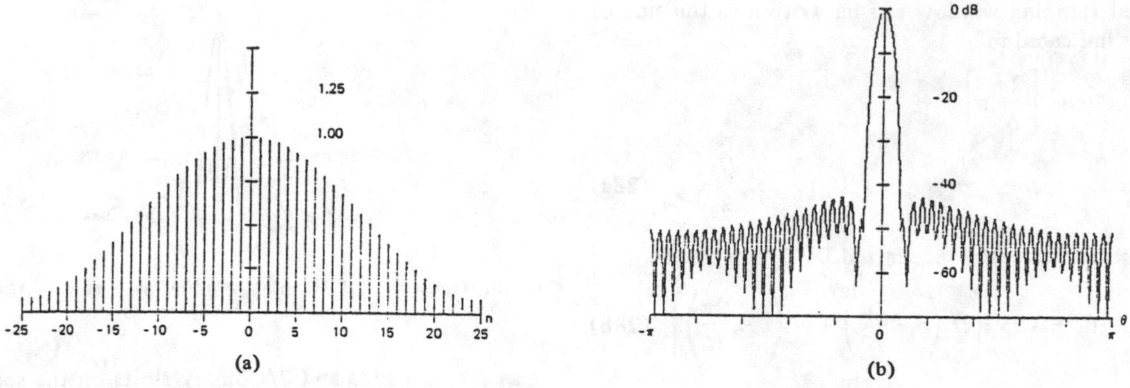

Fig. 21. (a) Hamming window. (b) Log-magnitude of Fourier transform.

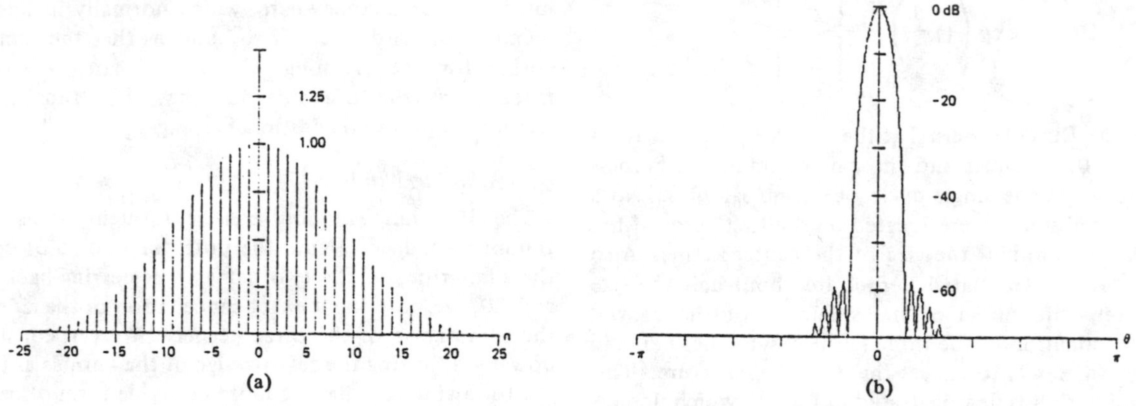

Fig. 22. (a) Blackman window. (b) Log-magnitude of transform.

eters of this window. Also note the loss of binary weighting; hence the need to perform multiplication to apply the weighting factors of the spectral convolution.

E. Blackman Window [7]

The Hamming and Hanning windows are examples of windows constructed as the summation of shifted Dirichlet kernels. This data window is defined for the finite Fourier transform in (31a) and for the DFT in (31b); equation (31c) is the resultant spectral window for the DFT given as a summation of the Dirichlet kernels $D(\theta)$ defined by $W(\theta)$ in (21c);

$$W(n) = \sum_{m=0}^{N/2} a_m \cos\left[\frac{2\pi}{N} mn\right], \qquad n = -\frac{N}{2}, \cdots, -1, 0, 1, \cdots, \frac{N}{2} \tag{31a}$$

$$W(n) = \sum_{m=0}^{N/2} (-1)^m a_m \cos\left[\frac{2\pi}{N} mn\right], \qquad n = 0, 1, \cdots, N-1 \tag{31b}$$

$$W(\theta) = \sum_{m=0}^{N/2} (-1)^m \frac{a_m}{2}\left[D\left(\theta - \frac{2\pi}{N} m\right) + D\left(\theta + \frac{2\pi}{N} m\right)\right]. \tag{31c}$$

Subject to constraint

$$\sum_{m=0}^{N/2} a_m = 1.0.$$

We can see that the Hanning and the Hamming windows are of this form with a_0 and a_1 being nonzero. We see that their spectral windows are summations of three-shifted kernels.

We can construct windows with any K nonzero coefficients and achieve a $(2K-1)$ summation of kernels. We recognize, however, that one way to achieve windows with a narrow main lobe is to restrict K to a small integer. Blackman examined this window for $K = 3$ and found the values of the nonzero coefficients which place zeros at $\theta = 3.5\ (2\pi/N)$ and at $\theta = 4.5\ (2\pi/N)$, the position of the third and the fourth sidelobes, respectively, of the central Dirichlet kernel. These exact values and their two place approximations are

$$a_0 = \frac{7938}{18608} \doteq 0.426\,590\,71 \simeq 0.42$$

$$a_1 = \frac{9240}{18608} \doteq 0.496\,560\,62 \simeq 0.50$$

$$a_2 = \frac{1430}{18608} \doteq 0.076\,848\,67 \simeq 0.08.$$

The window which uses these two place approximations is known as the Blackman window. When we describe this window with the "exact" coefficients we will refer to it as the exact Blackman window. The Blackman window is defined for the finite transform in the following equation and the window is shown in Fig. 22:

$$W(n) = 0.42 + 0.50 \cos\left[\frac{2\pi}{N} n\right] + 0.08 \cos\left[\frac{2\pi}{N} 2n\right], \qquad n = -\frac{N}{2}, \cdots, -1, 0, 1, \cdots, \frac{N}{2}. \tag{32}$$

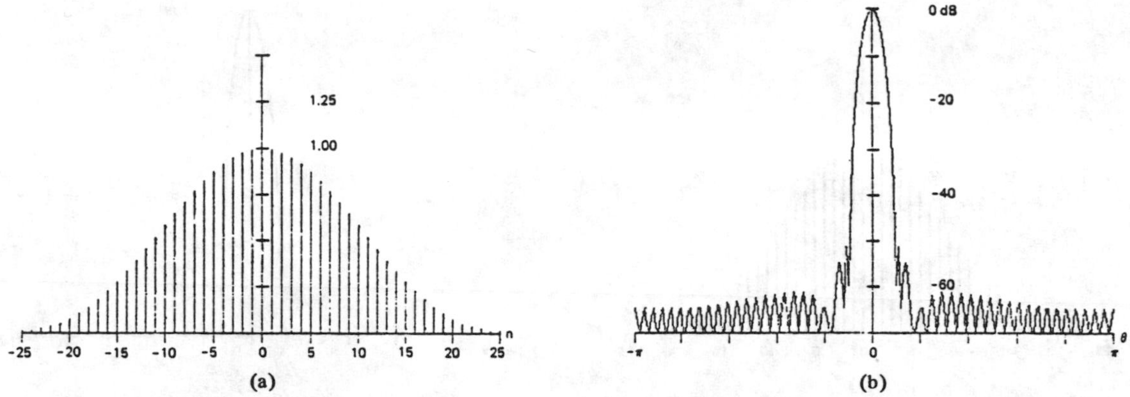

Fig. 23. (a) Exact Blackman window. (b) Log-magnitude of transform.

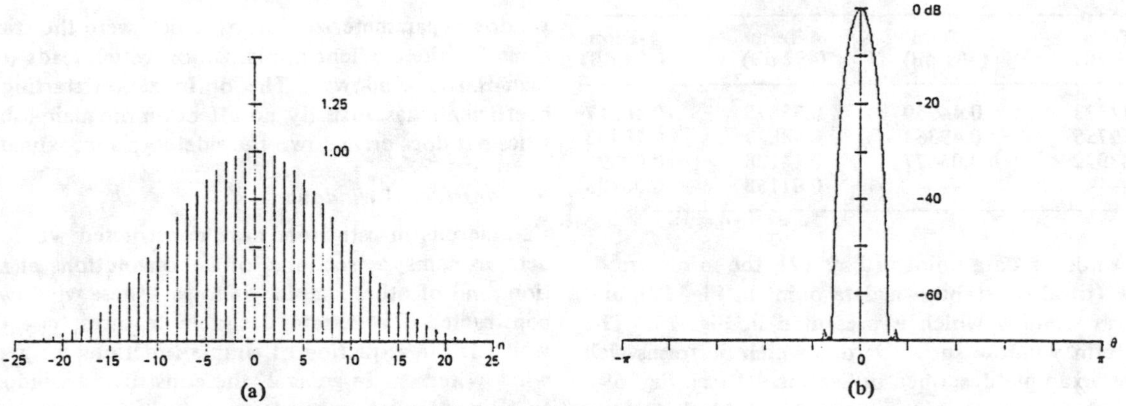

Fig. 24. (a) Minimum 3-term Blackman-Harris window. (b) Log-magnitude of transform.

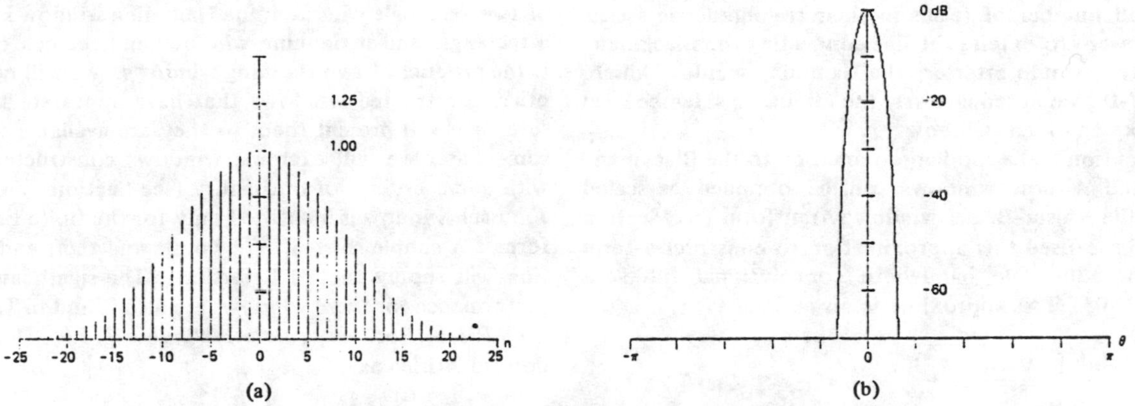

Fig. 25. (a) 4-term Blackman-Harris window. (b) Log-magnitude of transform.

The exact Blackman window is shown in Fig. 23. The sidelobe level is 51 dB down for the exact Blackman window and is 58 dB down for the Blackman window. As an observation, note that the coefficients of the Blackman window sum to zero (0.42 -0.50 +0.08) at the boundaries while the exact coefficients do not. Thus the Blackman window is continuous with a continuous first derivative at the boundary and falls off like $1/\omega^3$ or 18 dB per octave. The exact terms (like the Hamming window) have a discontinuity at the boundary and falls off like $1/\omega$ or 6 dB per octave. Table I lists the parameters of these two windows. Note that for this class of windows, the a_0 coefficient is the coherent gain of the window.

Using a gradient search technique [9], we have found the windows which for 3- and 4-nonzero terms achieve a minimum sidelobe level. We have also constructed families of 3- and 4-term windows in which we trade main-lobe width for sidelobe level. We call this family the Blackman-Harris window. We have found that the minimum 3-term window can achieve a sidelobe level of -67 dB and that the minimum 4-term window can achieve a sidelobe level of -92 dB. These windows are defined for the DFT by

$$w(n) = a_0 - a_1 \cos\left(\frac{2\pi}{N} n\right) + a_2 \cos\left(\frac{2\pi}{N} 2n\right) - a_3 \cos\left(\frac{2\pi}{N} 3n\right),$$
$$n = 0, 1, 2, \cdots, N-1. \quad (33)$$

The listed coefficients correspond to the minimum 3-term window which is presented in Fig. 24, another 3-term window

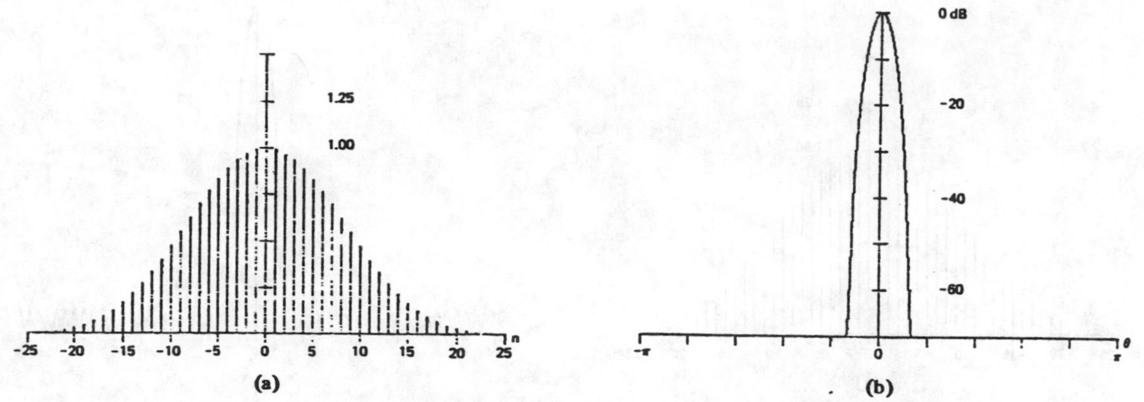

Fig. 26. (a) 4-sample Kaiser–Bessel window. (b) Log-magnitude of transform.

	3-Term (-67 dB)	3-Term (-61 dB)	4-Term (-92 dB)	4-Term (-74 dB)
a_0	0.42323	0.44959	0.35875	0.40217
a_1	0.49755	0.49364	0.48829	0.49703
a_2	0.07922	0.05677	0.14128	0.09392
a_3	---	---	0.01168	0.00183

(to establish another data point in Fig. 12), the minimum 4-term window (to also establish a data point in Fig. 12), and another 4-term window which is presented in Fig. 25. The particular 4-term window shown is one which performs well in a detection example described in Section VI (see Fig. 69). The parameters of these windows are listed in Table I. Note in particular where the Blackman and the Blackman–Harris windows reside in Fig. 12. They are surprisingly good windows for the small number of terms in their trigonometric series. Note, if we were to extend the line connecting the Blackman–Harris family it would intersect the Hamming window which, in Section V-D , we noted is nearly the minimum sidelobe level 2-term Blackman–Harris window.

We also mention that a good approximation to the Blackman–Harris 3- and 4-term windows can be obtained as scaled samples of the Kaiser–Bessel window's transform (see Section V-H). We have used this approximation to construct 4-term windows for adjustable bandwidth convolutional filters as reported in [10]. This approximation is defined as

$$b_m = \frac{\sinh\left[\pi\sqrt{\alpha^2 - m^2}\right]}{\pi\sqrt{\alpha^2 - m^2}}, \quad m \leqslant \alpha, \quad 2 \leqslant \alpha \leqslant 4$$

$$c = b_0 + 2b_1 + 2b_2 + (2b_3)$$

$$a_0 = \frac{b_0}{c} \quad a_m = 2\frac{b_m}{c}, \quad m = 1, 2, (3). \tag{34}$$

The 4 coefficients for this approximation when $\alpha = 3.0$ are $a_0 = 0.40243$, $a_1 = 0.49804$, $a_2 = 0.09831$, and $a_3 = 0.00122$. Notice how close these terms are to the selected 4-term Blackman–Harris (-74 dB) window. The window defined by these coefficients is shown in Fig. 26. Like the prototype from which it came (the Kaiser–Bessel with $\alpha = 3.0$), this window exhibits sidelobes just shy of -70 dB from the main lobe. On the scale shown, the two are indistinguishable. The parameters of this window are also listed in Table I and the window is entered in Fig. 12 as the "4-sample Kaiser–Bessel." It was these 3- and 4-sample Kaiser–Bessel prototype windows (parameterized on α) which were the starting conditions for the gradient minimization which leads to the Blackman–Harris windows. The optimization starting with these coefficients has virtually no effect on the main-lobe characteristics but does drive down the sidelobes approximately 5 dB.

F. Constructed Windows

Numerous investigators have constructed windows as products, as sums, as sections, or as convolutions of simple functions and of other simple windows. These windows have been constructed for certain desirable features, not the least of which is the attraction of simple functions for generating the window terms. In general, the constructed windows tend not to be good windows, and occasionally are very bad windows. We have already examined some simple window constructions. The Fejer (Bartlett) window, for instance, is the convolution of two rectangle windows; the Hamming window is the sum of a rectangle and a Hanning window; and the $\cos^4(X)$ window is the product of two Hanning windows. We will now examine other constructed windows that have appeared in the literature. We will present them so they are available for comparison. Later we will examine windows constructed in accord with some criteria of optimality (see Sections V-G, H, I, and J). Each window is identified only for the finite Fourier transform. A simple shift of $N/2$ points and right end-point deletion will supply the DFT version. The significant figures of performance for these windows are also found in Table I.

1) Riesz (Bochner, Parzen) Window [11]: The Riesz window, identified as

$$w(n) = 1.0 - \left|\frac{n}{N/2}\right|^2, \quad 0 \leqslant |n| \leqslant \frac{N}{2} \tag{35}$$

is the simplest continuous polynomial window. It exhibits a discontinuous first derivative at the boundaries; hence its transform falls off like $1/\omega^2$. The window is shown in Fig. 27. The first sidelobe is -22 dB from the main lobe. This window is similar to the cosine lobe (26) as can be demonstrated by examining its Taylor series expansion.

2) Riemann Window [12]: The Riemann window, defined by

$$w(n) = \frac{\sin\left[\frac{n}{N}2\pi\right]}{\left[\frac{n}{N}2\pi\right]}, \quad 0 \leqslant |n| \leqslant \frac{N}{2} \tag{36}$$

is the central lobe of the SINC kernel. This window is con-

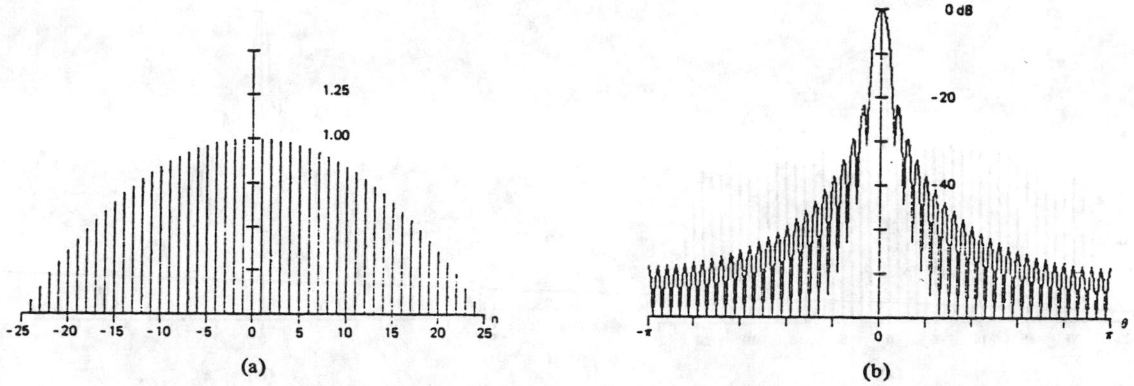

Fig. 27. (a) Riesz window. (b) Log-magnitude of transform.

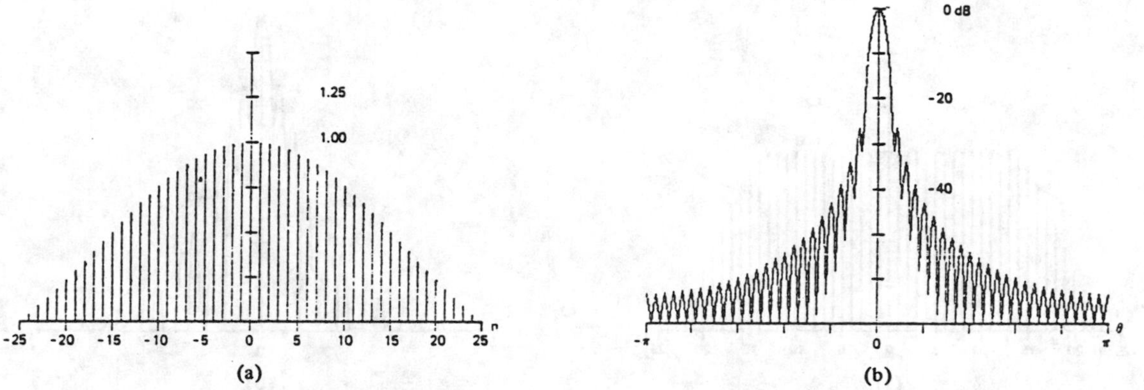

Fig. 28. (a) Riemann window. (b) Log-magnitude of transform.

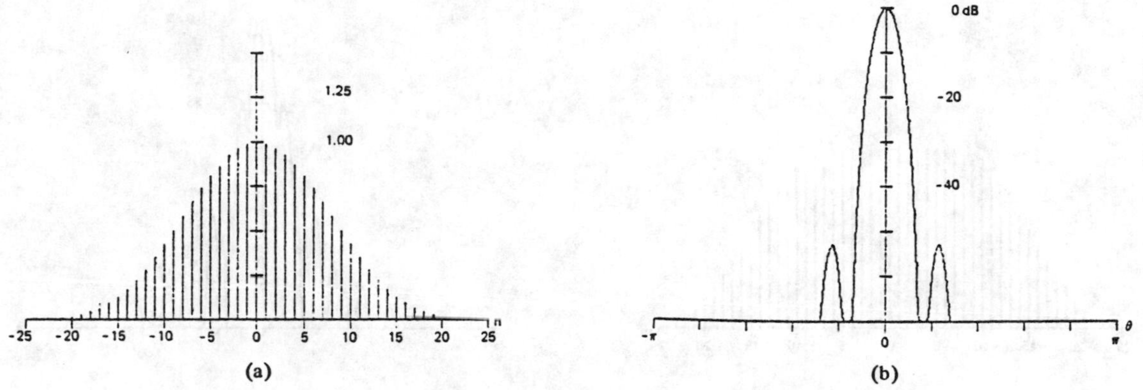

Fig. 29. (a) The de la Vallé-Poussin window. (b) Log-magnitude of transform.

tinuous, with a discontinuous first derivative at the boundary. It is similar to the Riesz and cosine lobe windows. The Riemann window is shown in Fig. 28.

3) de la Vallé-Poussin (Jackson, Parzen) Window [11]: The de la Vallé-Poussin window is a piecewise cubic curve obtained by self-convolving two triangles of half extent or four rectangles of one-fourth extent. It is defined as

$$w(n) = \begin{cases} 1.0 - 6\left[\dfrac{n}{N/2}\right]^2 \left[1.0 - \dfrac{|n|}{N/2}\right], & 0 \leq |n| \leq \dfrac{N}{4} \\[2ex] 2\left[1.0 - \dfrac{|n|}{N/2}\right]^3, & \dfrac{N}{4} \leq |n| \leq \dfrac{N}{2}. \end{cases} \tag{37}$$

The window is continuous up to its third derivative so that its sidelobes fall off like $1/\omega^4$. The window is shown in Fig. 29. Notice the trade off of main-lobe width for sidelobe level. Compare this with the rectangle and the triangle. It is a nonnegative window by virtue of its self-convolution construction.

4) Tukey Window [13]: The Tukey window, often called the cosine-tapered window, is best imagined as a cosine lobe of width $(\alpha/2)N$ convolved with a rectangle window of width $(1.0 - \alpha/2)N$. Of course the resultant transform is the product of the two corresponding transforms. The window represents an attempt to smoothly set the data to zero at the boundaries while not significantly reducing the processing gain of the windowed transform. The window evolves from the rectangle to the Hanning window as the parameter α varies from zero to unity. The family of windows exhibits a confusing array of

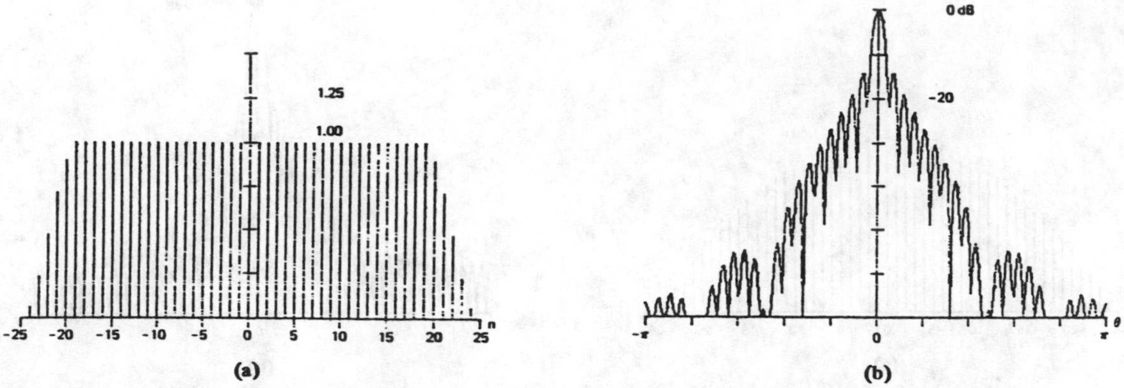

Fig. 30. (a) 25-percent cosine taper (Tukey) window. (b) Log-magnitude of transform.

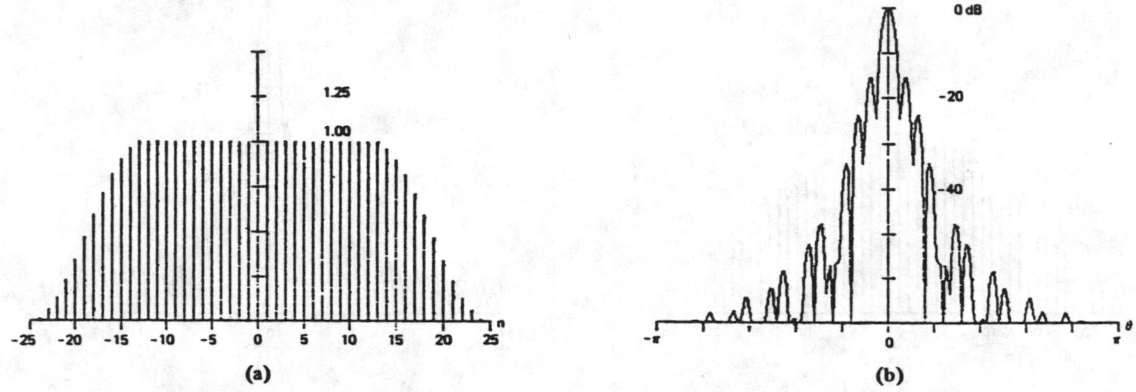

Fig. 31. (a) 50-percent cosine taper (Tukey) window. (b) Log-magnitude of transform.

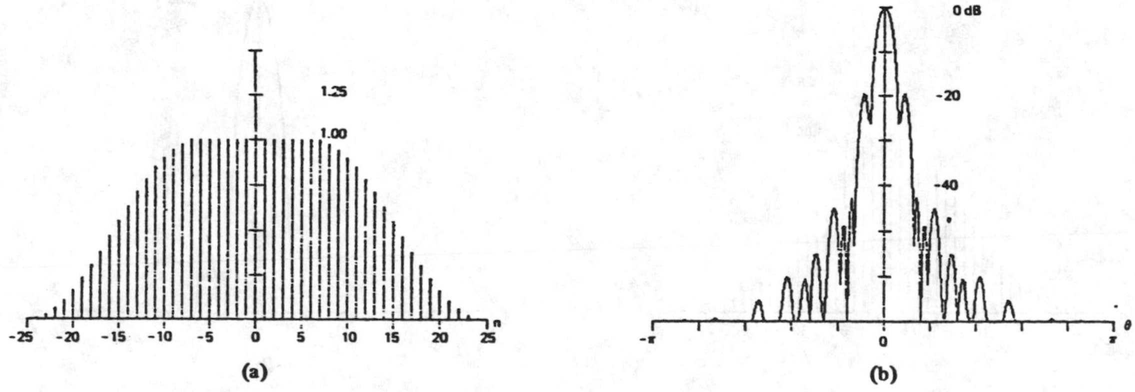

Fig. 32. (a) 75-percent cosine taper (Tukey) window. (b) Log-magnitude of transform.

sidelobe levels arising from the product of the two component transforms. The window is defined by

$$w(n) = \begin{cases} 1.0, & 0 \leqslant |n| \leqslant \alpha\dfrac{N}{2} \\ 0.5\left[1.0 + \cos\left[\pi \dfrac{n - \alpha\dfrac{N}{2}}{2(1-\alpha)\dfrac{N}{2}}\right]\right], & \alpha\dfrac{N}{2} \leqslant |n| \leqslant \dfrac{N}{2} \end{cases} \tag{38}$$

The window is shown in Figs. 30–32 for values of α equal to 0.25, 0.50, and 0.75, respectively.

5) Bohman Window [14]: The Bohman window is obtained by the convolution of two half-duration cosine lobes (26a), thus its transform is the square of the cosine lobe's transform (see Fig. 16). In the time domain the window can be described as a product of a triangle window with a single cycle of a cosine with the same period and, then, a corrective term added to set the first derivative to zero at the boundary. Thus the second derivative is continuous, and the discontinuity resides in the third derivative. The transform falls off like $1/\omega^4$. The window is defined in the following and is shown in Fig. 33:

$$w(n) = \left[1.0 - \frac{|n|}{N/2}\right]\cos\left[\pi\frac{|n|}{N/2}\right] + \frac{1}{\pi}\sin\left[\pi\frac{|n|}{N/2}\right], \quad 0 \leqslant |n| \leqslant \frac{N}{2}. \tag{39}$$

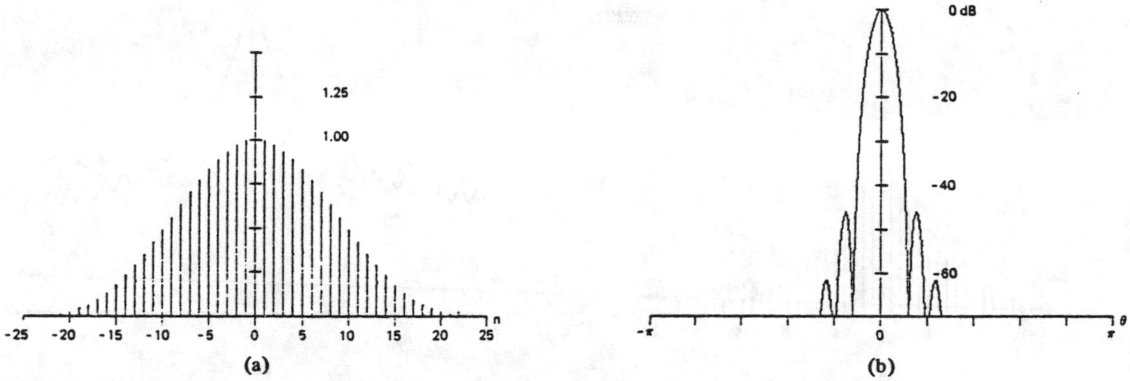

Fig. 33. (a) Bohman window. (b) Log-magnitude of transform.

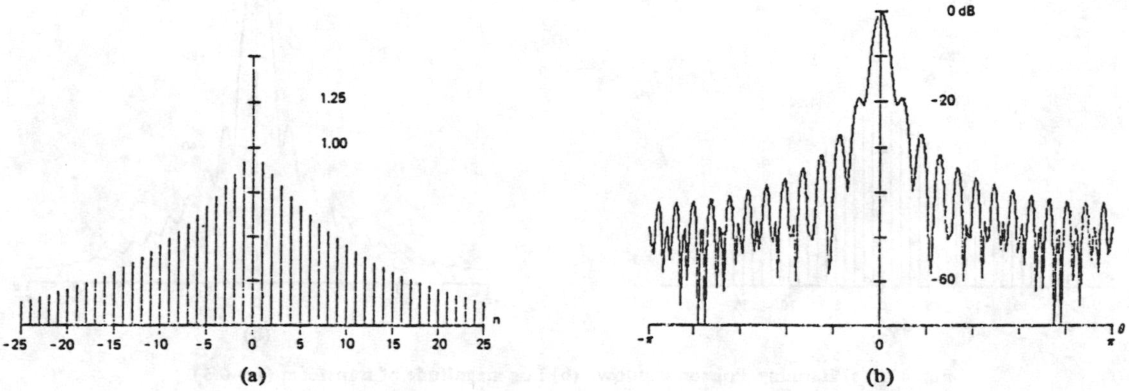

Fig. 34. (a) Poisson window. (b) Log-magnitude of transform (a = 2.0).

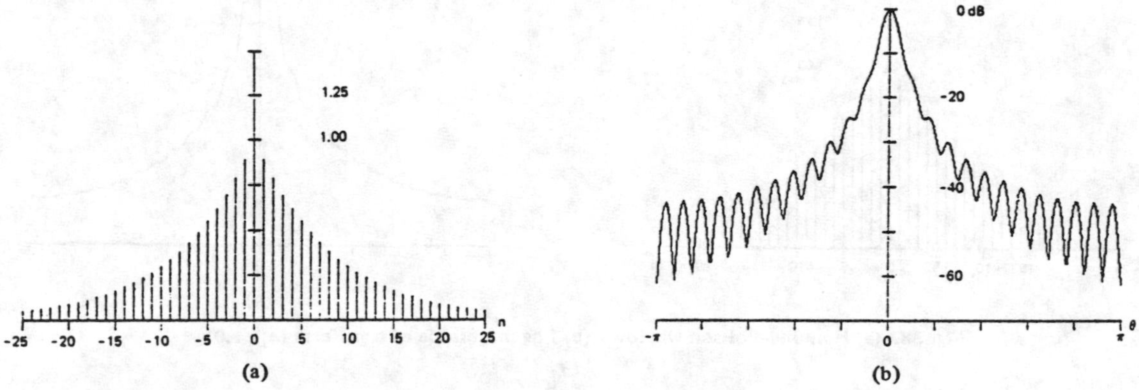

Fig. 35. (a) Poisson window. (b) Log-magnitude of transform (a = 3.0).

6) Poisson Window [12]: The Poisson window is a two-sided exponential defined by

$$w(n) = \exp\left(-\alpha \frac{|n|}{N/2}\right), \qquad 0 \leqslant |n| \leqslant \frac{N}{2}. \tag{40}$$

This is actually a family of windows parameterized on the variable α. Since it exhibits a discontinuity at the boundaries, the transform can fall off no faster than $1/\omega$. The window is shown in Figs. 34–36 for values of α equal to 2.0, 3.0, and 4.0, respectively. Notice as the discontinuity at the boundaries becomes smaller, the sidelobe structure merges into the asymptote. Also note the very wide main lobe; this will be observed in Table I as a large equivalent noise bandwidth and as a large worst case processing loss.

7) Hanning–Poisson Window: The Hanning–Poisson window is constructed as the product of the Hanning and the Poisson windows. The family is defined by

$$w(n) = 0.5\left[1.0 + \cos\left[\pi \frac{n}{N/2}\right]\right] \exp\left(-\alpha \frac{|n|}{N/2}\right), \qquad 0 \leqslant |n| \leqslant \frac{N}{2}. \tag{41}$$

This window is similar to the Poisson window. The rate of sidelobe falloff is determined by the discontinuity in the first

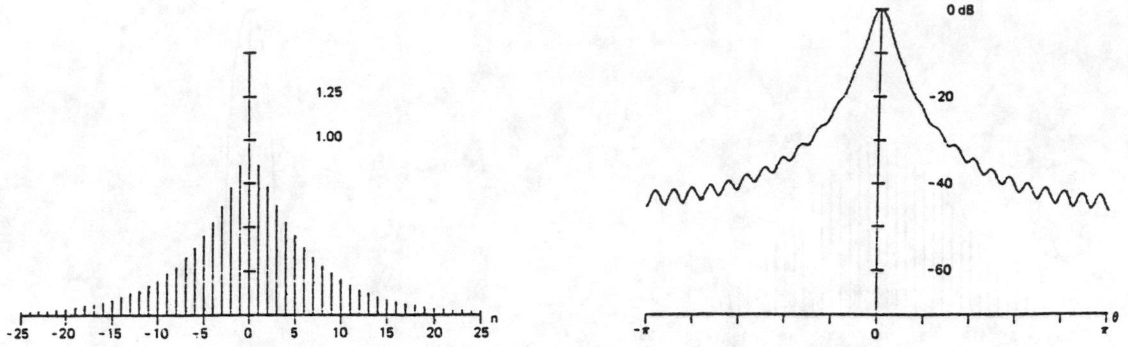

(a) (b)

Fig. 36. (a) Poisson window. (b) Log-magnitude of transform ($a = 4.0$).

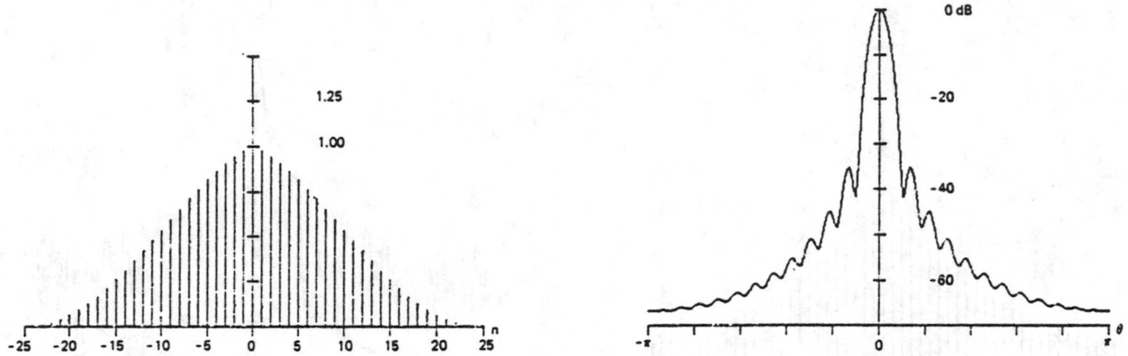

(a) (b)

Fig. 37. (a) Hanning-Poisson window. (b) Log-magnitude of transform ($a = 0.5$).

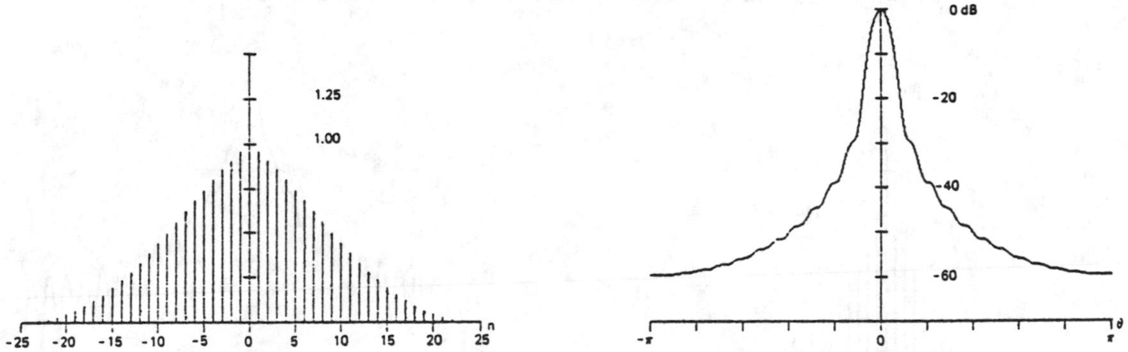

(a) (b)

Fig. 38. (a) Hanning-Poisson window. (b) Log-magnitude of transform ($a = 1.0$).

derivative at the origin and is $1/\omega^2$. Notice as α increases, forcing more of the exponential into the Hanning window, the zeros of the sidelobe structure disappear and the lobes merge into the asymptote. This window is shown in Figs. 37–39 for values of α equal to 0.5, 1.0, and 2.0, respectively. Again note the very large main-lobe width.

8) Cauchy (Abel, Poisson) Window [15]: The Cauchy window is a family parameterized on α and defined by

$$w(n) = \frac{1}{1.0 + \left[\alpha \dfrac{n}{N/2}\right]^2}, \qquad 0 \leqslant |n| \leqslant \frac{N}{2}. \tag{42}$$

The window is shown in Figs. 40–42 for values of α equal to 3.0, 4.0, and 5.0, respectively. Note the transform of the Cauchy window is a two-sided exponential (see Poisson windows), which when presented on a log-magnitude scale is essentially an isosceles triangle. This causes the window to exhibit a very wide main lobe and to have a large ENBW.

G. Gaussian or Weierstrass Window [15]

Windows are smooth positive functions with tall thin (i.e., concentrated) Fourier transforms. From the generalized uncertainty principle, we know we cannot simultaneously concentrate both a signal and its Fourier transform. If our measure of concentration is the mean-square time duration T and the mean-square bandwidth W, we know all functions satisfy the inequality of

$$TW \geqslant \frac{1}{4\pi} \tag{43}$$

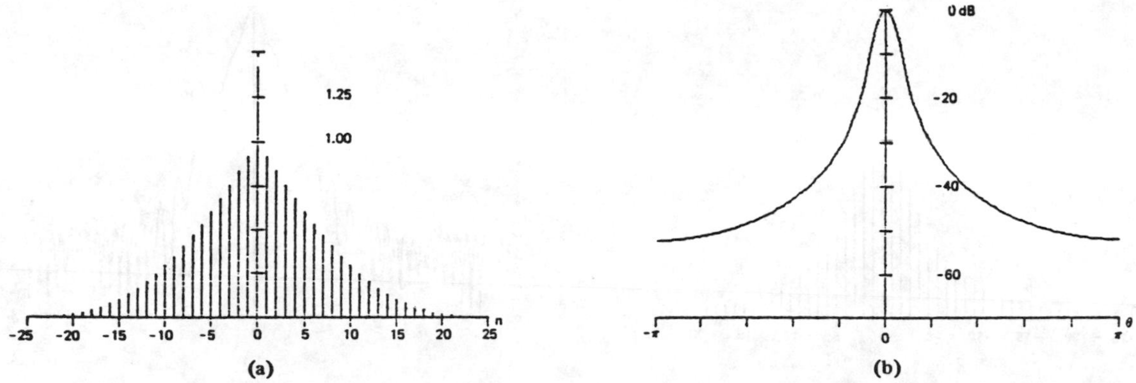

Fig. 39. (a) Hanning-Poisson window. (b) Log-magnitude of transform (a = 2.0).

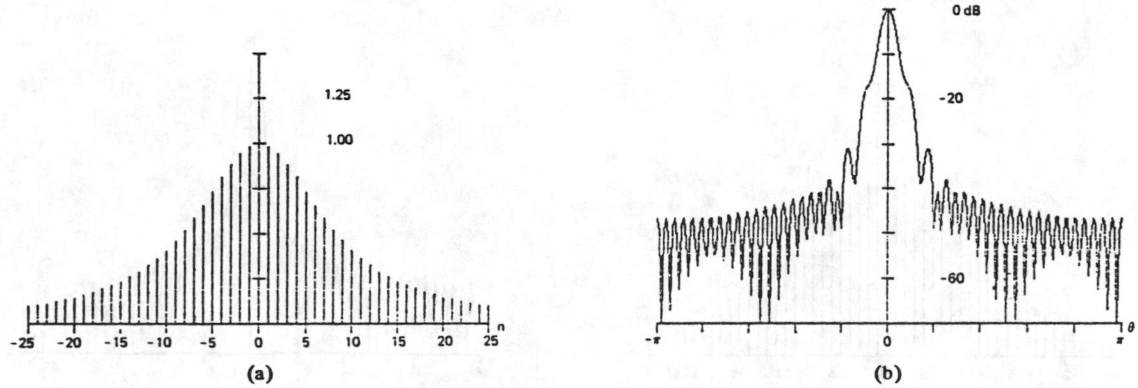

Fig. 40. (a) Cauchy window. (b) Log-magnitude of transform (a = 3.0).

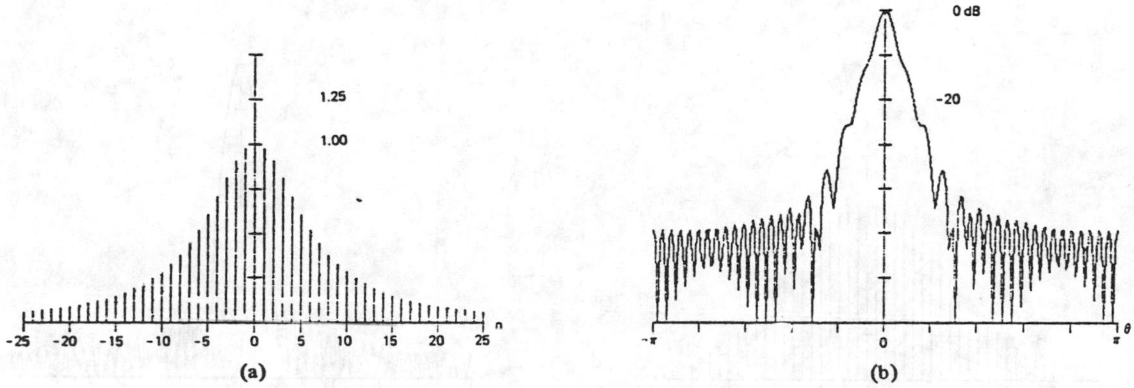

Fig. 41. (a) Cauchy window. (b) Log-magnitude of transform (a = 4.0).

with equality being achieved only for the Gaussian pulse [16]. Thus the Gaussian pulse, characterized by minimum time-bandwidth product, is a reasonable candidate for a window. When we use the Gaussian pulse as a window we have to truncate or discard the tails. By restricting the pulse to be finite length, the window no longer is minimum time-bandwidth. If the truncation point is beyond the three-sigma point, the error should be small, and the window should be a good approximation to minimum time-bandwidth.

The Gaussian window is defined by

$$w(n) = \exp\left[-\frac{1}{2}\left[\alpha\frac{n}{N/2}\right]^2\right] \tag{44a}$$

The transform is the convolution of a Gaussian transform with a Dirichlet kernel as indicated in

$$W(\theta) = \frac{1}{2}\frac{\sqrt{2\pi}}{\alpha}\exp\left[-\frac{1}{2}\left[\frac{1}{\alpha}\theta\right]^2\right] * D(\theta)$$

$$\simeq \frac{N}{2}\frac{\sqrt{2\pi}}{\alpha}\exp\left[-\frac{1}{2}\left[\frac{1}{\alpha}\theta\right]^2\right], \qquad \text{for } \alpha > 2.5, \text{ and } \theta \text{ small.} \tag{44b}$$

This window is parameterized on α, the reciprocal of the standard deviation, a measure of the width of its Fourier transform. Increased α will decrease with the width of the window and reduce the severity of the discontinuity at the boundaries. This will result in an increased width transform

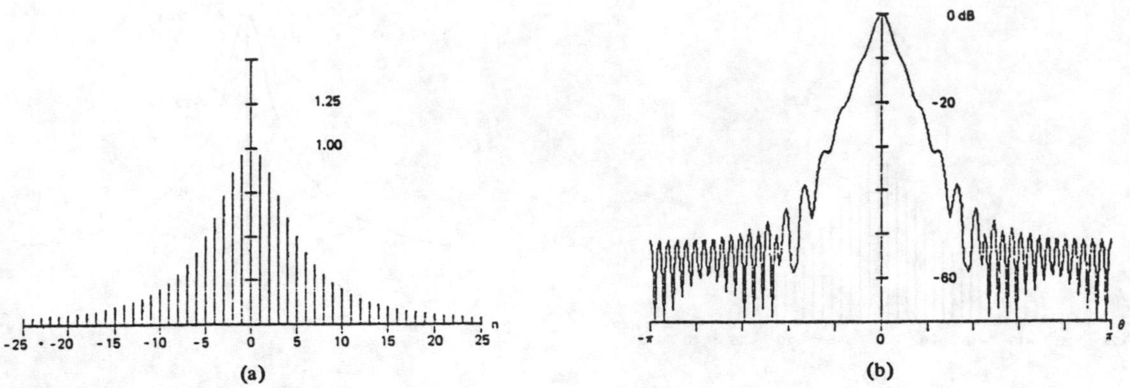

Fig. 42. (a) Cauchy window. (b) Log-magnitude of transform (a = 5.0).

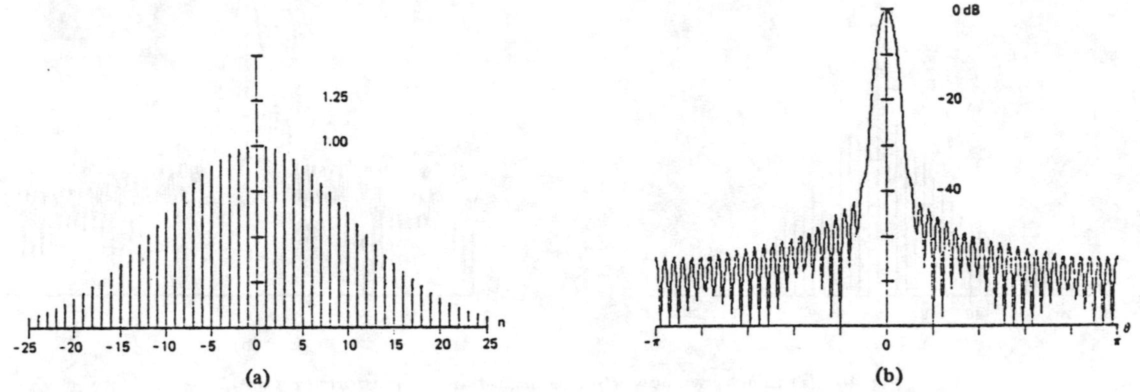

Fig. 43. (a) Gaussian window. (b) Log-magnitude of transform (a = 2.5).

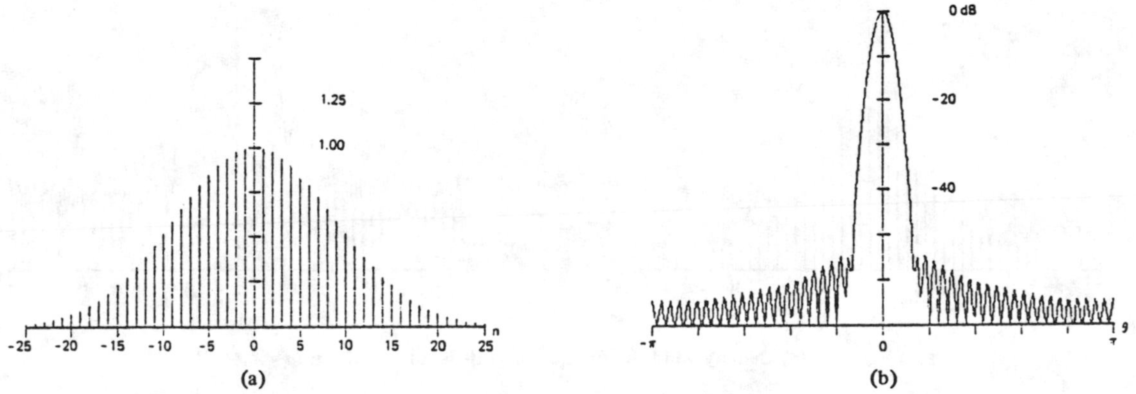

Fig. 44. (a) Gaussian window. (b) Log-magnitude of transform (a = 3.0).

main lobe and decreased sidelobe levels. The window is presented in Figs. 43, 44, and 45 for values of α equal to 2.5, 3.0, and 3.5, respectively. Note the rapid drop-off rate of sidelobe level in the exchange of sidelobe level for main-lobe width. The figures of merit for this window are listed in Table I

H. Dolph–Chebyshev Window [17]

Following the reasoning of the previous section, we seek a window which, for a known finite duration, in some sense exhibits a narrow bandwidth. We now take a lead from the antenna design people who have faced and solved a similar problem. The problem is to illuminate an antenna of finite aperture to achieve a narrow main-lobe beam pattern while simultaneously restricting sidelobe response. (The antenna designer calls his weighting procedure *shading*.) The closed-form solution to the minimum main-lobe width for a given sidelobe level is the Dolph–Chebyshev window (shading). The continuous solution to the problem exhibits impulses at the boundaries which restricts continuous realizations to approximations (the Taylor approximation). The discrete or sampled window is not so restricted, and the solution can be implemented exactly.

The relation $T_n(X) = \cos(n\theta)$ describes a mapping between the nth-order Chebyshev (algebraic) polynomial and the nth-order trigonometric polynomial. The Dolph–Chebyshev

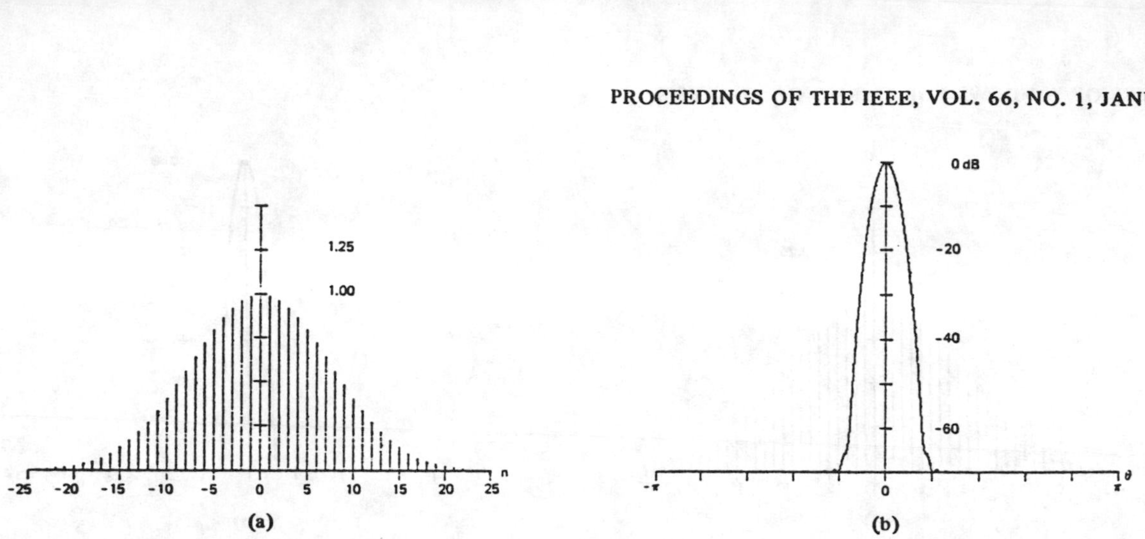

Fig. 45. (a) Gaussian window. (b) Log-magnitude of transform ($a = 3.5$).

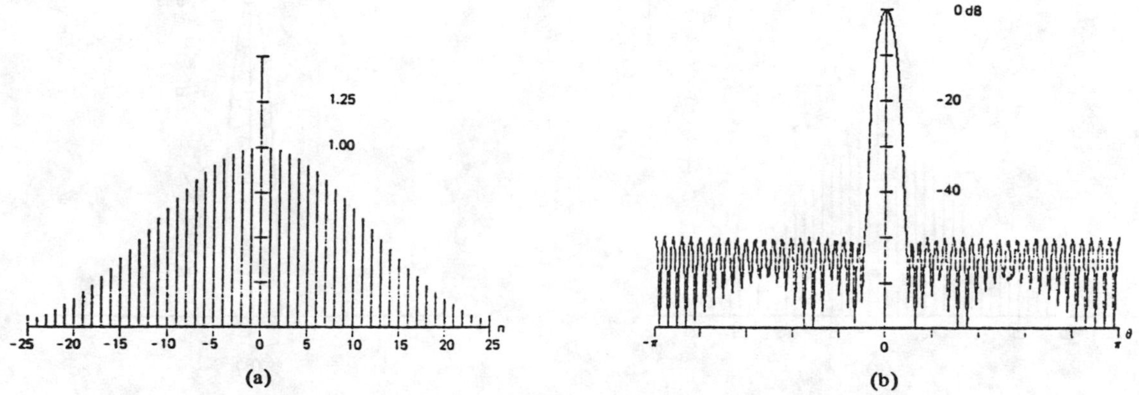

Fig. 46. (a) Dolph–Chebyshev window. (b) Log-magnitude of transform ($a = 2.5$).

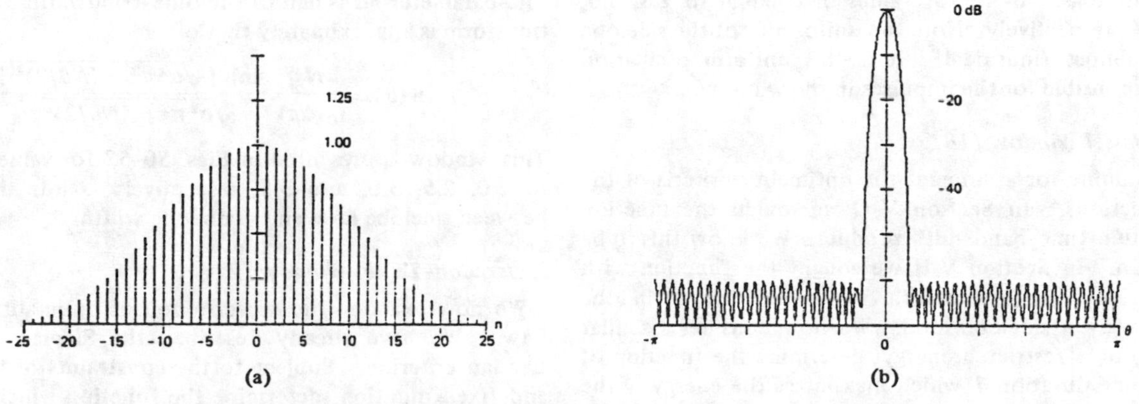

Fig. 47. (a) Dolph–Chebyshev window. (b) Log-magnitude of transform ($a = 3.0$).

window is defined with this mapping in the following equation, in terms of uniformly spaced samples of the window's Fourier transform,

$$W(k) = (-1)^k \frac{\cos\left[N \cos^{-1}\left[\beta \cos\left(\pi \frac{k}{N}\right)\right]\right]}{\cosh\left[N \cosh^{-1}(\beta)\right]}, \qquad 0 \leqslant |k| \leqslant N-1 \quad (45)$$

where

$$\beta = \cosh\left[\frac{1}{N}\cosh^{-1}(10^{\alpha})\right]$$

and

$$\cos^{-1}(X) = \begin{cases} \frac{\pi}{2} - \tan^{-1}\left[X/\sqrt{1.0 - X^2}\right], & |X| \leqslant 1.0 \\ \ln\left[X + \sqrt{X^2 - 1.0}\right], & |X| \geqslant 1.0. \end{cases}$$

To obtain the corresponding window time samples $w(n)$, we simply perform a DFT on the samples $W(k)$ and then scale for unity peak amplitude. The parameter α represents the log of the ratio of main-lobe level to sidelobe level. Thus a value of α equal to 3.0 represents sidelobes 3.0 decades down from the main lobe, or sidelobes 60.0 dB below the main lobe. The $(-1)^k$ alternates the sign of successive transform samples to reflect the shifted origin in the time domain. The window is

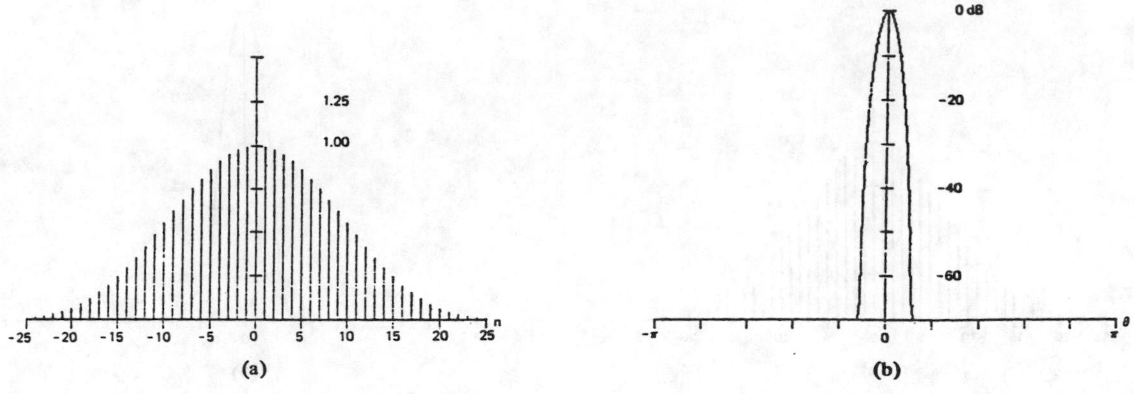

Fig. 48. (a) Dolph–Chebyshev window. (b) Log-magnitude of transform ($a = 3.5$).

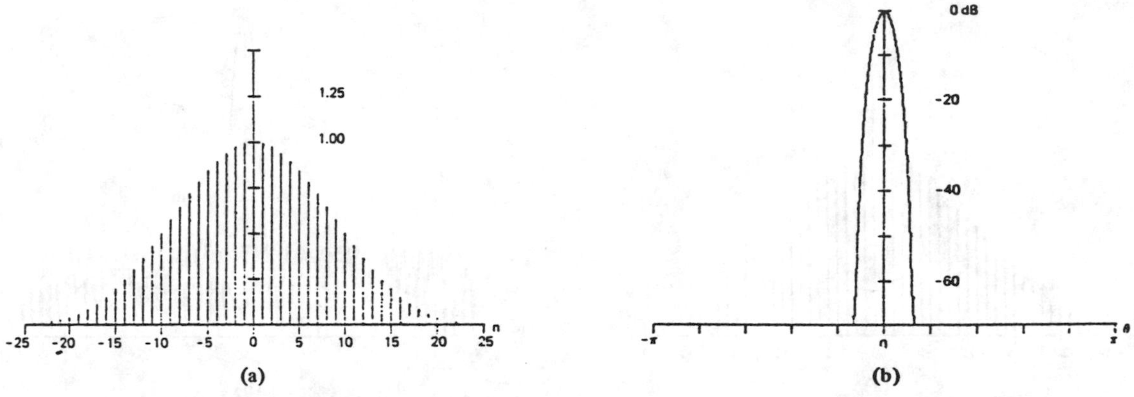

Fig. 49. (a) Dolph–Chebyshev window. (b) Log-magnitude of transform ($a = 4.0$).

presented in Figs. 46–49 for values of α equal to 2.5, 3.0, 3.5, and 4.0, respectively. Note the uniformity of the sidelobe structure; almost sinusoidal! It is this uniform oscillation which is responsible for the impulses in the window.

I. Kaiser-Bessel Window [18]

Let us examine for a moment the optimality criteria of the last two sections. In Section V-G we sought the function with minimum time–bandwidth product. We know this to be the Gaussian. In Section V-H we sougnt the function with restricted time duration, which minimized the main-lobe width for a given sidelobe level. We now consider a similar problem. For a restricted energy, determine the function of restricted time duration T which maximizes the energy in the band of frequencies W. Slepian, Pollak, and Landau [19], [20] have determined this function as a family parameterized over the time–bandwidth product, the prolate-spheroidal wave functions of order zero. Kaiser has discovered a simple approximation to these functions in terms of the zero-order modified Bessel function of the first kind. The Kaiser–Bessel window is defined by

$$w(n) = \frac{I_0\left[\pi\alpha\sqrt{1.0 - \left(\dfrac{n}{N/2}\right)^2}\,\right]}{I_0[\pi\alpha]}, \qquad 0 \leqslant |n| \leqslant \frac{N}{2} \tag{46a}$$

where

$$I_0(X) = \sum_{k=0}^{\infty}\left[\frac{\left(\dfrac{x}{2}\right)^k}{k!}\right]^2$$

The parameter $\pi\alpha$ is half of the time–bandwidth product. The transform is approximately that of

$$W(\theta) \doteq \frac{N}{I_0(\alpha\pi)}\,\frac{\sinh\left[\sqrt{\alpha^2\pi^2 - (N\theta/2)^2}\,\right]}{\sqrt{\alpha^2\pi^2 - (N\theta/2)^2}} \tag{46b}$$

This window is presented in Figs. 50–53 for values of α equal to 2.0, 2.5, 3.0, and 3.5, respectively. Note the trade off between sidelobe level and main-lobe width.

J. Barcilon–Temes Window [21]

We now examine the last criterion of optimality for a window. We have already described the Slepian, Pollak, and Landau criterion. Subject to the constraints of fixed energy and fixed duration, determine the function which maximizes the energy in the band of frequencies W. A related criterion, subject to the constraints of fixed area and fixed duration, is to determine the function which minimizes the energy (or the weighted energy) outside the band of frequencies W. This is a reasonable criterion since we recognize that the transform of a good window should minimize the energy it gathers from frequencies removed from its center frequency. Till now, we have been responding to this goal by maximizing the concentration of the transform at its main lobe.

A closed-form solution of the unweighted minimum-energy criterion has not been found. A solution defined as an expansion of prolate-spheroidal wave functions does exist and it is of the form shown in

$$H\left(\frac{\omega}{W}\right) = \sum_n \frac{\psi_{2n}(\pi\alpha, 0)}{1 - \lambda_{2n}}\,\psi_{2n}\left(\pi\alpha, \frac{\omega}{W}\right) \tag{47}$$

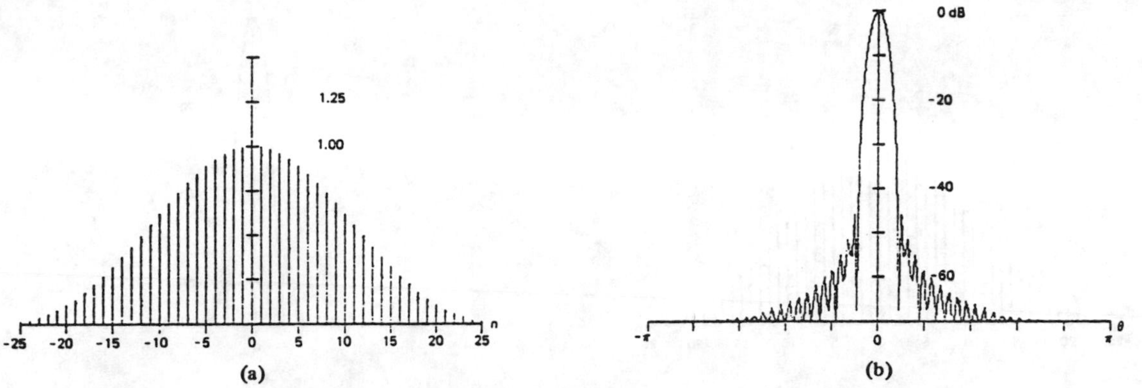

Fig. 50. (a) Kaiser–Bessel window. (b) Log-magnitude of transform ($a = 2.0$).

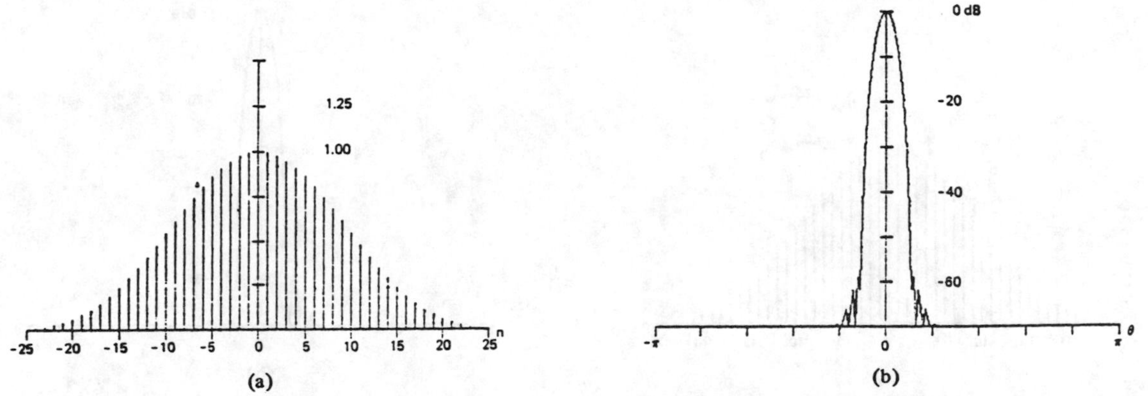

Fig. 51. (a) Kaiser–Bessel window. (b) Log-magnitude of transform ($a = 2.5$).

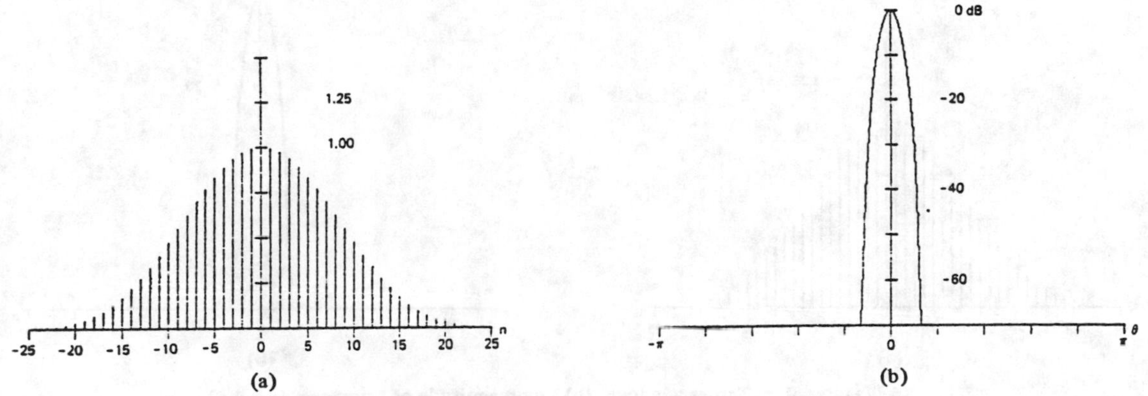

Fig. 52. (a) Kaiser–Bessel window. (b) Log-magnitude of transform ($a = 3.0$).

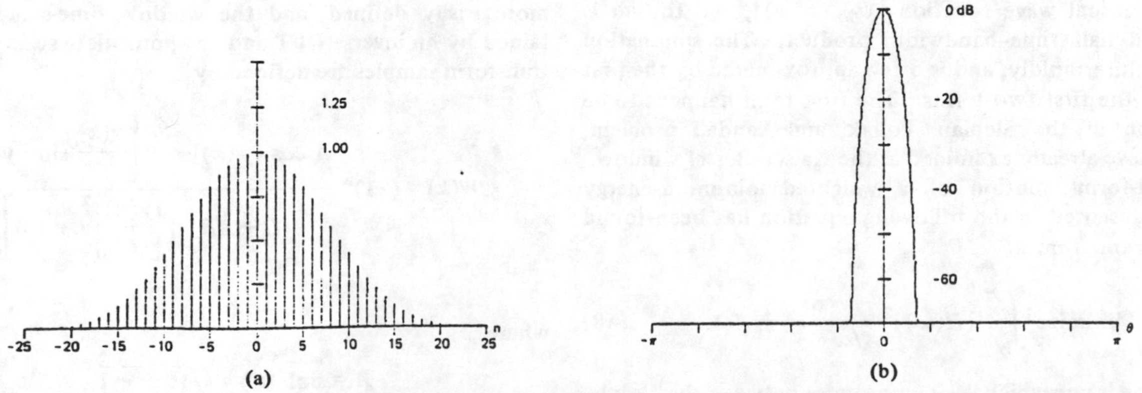

Fig. 53. (a) Kaiser–Bessel window. (b) Log-magnitude of transform ($a = 3.5$).

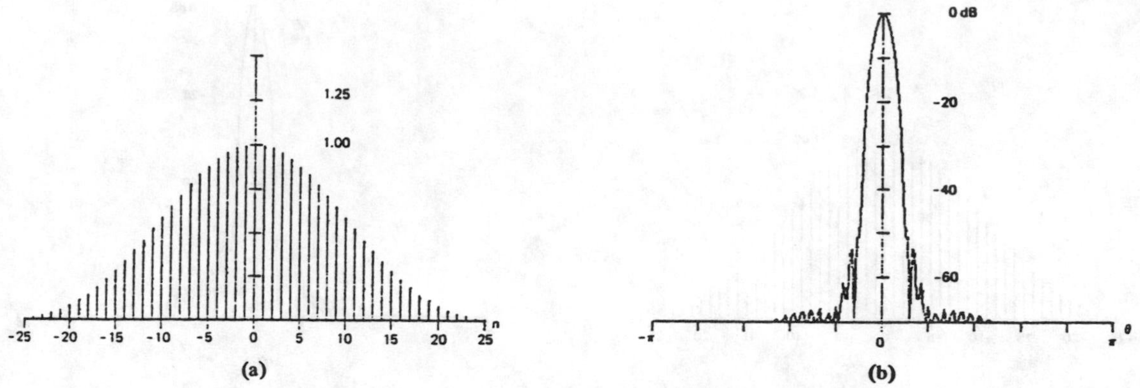

Fig. 54. (a) Barcilon–Temes window. (b) Log-magnitude of transform ($a = 3.0$).

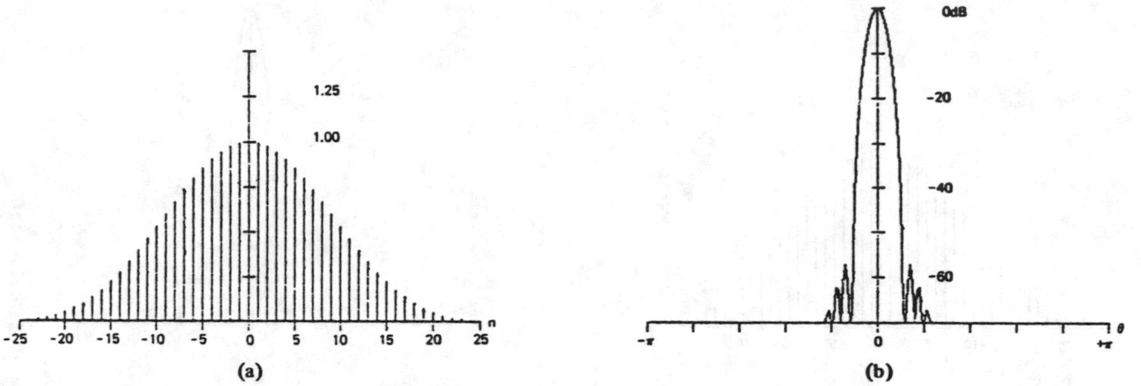

Fig. 55. (a) Barcilon–Temes window. (b) Log-magnitude of transform ($a = 3.5$).

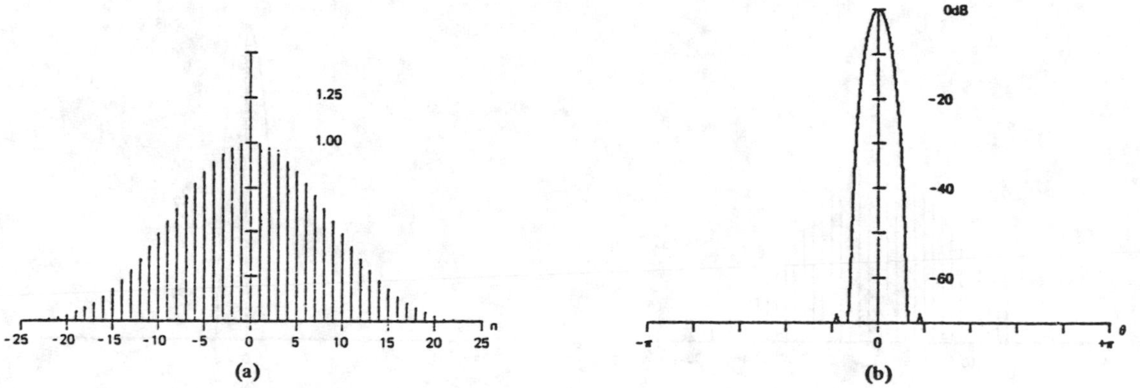

Fig. 56. (a) Barcilon–Temes window. (b) Log-magnitude of transform ($a = 4.0$).

Here the λ_{2n} is the eigenvalue corresponding to the associated prolate-spheroidal wave function $|\psi_{2n}(x, y)|$, and the $\pi\alpha$ is the selected half time-bandwidth product. The summation converges quite rapidly, and is often approximated by the first term or by the first two terms. The first term happens to be the solution of the Slepian, Pollak, and Landau problem, which we have already examined as the Kaiser-Bessel window.

A closed-form solution of a weighted minimum-energy criterion, presented in the following equation has been found by Barcilon and Temes:

$$\text{Minimize} \int_W^\infty |H(\omega)|^2 \frac{\omega}{\sqrt{\omega^2 - W^2}} \, d\omega. \tag{48}$$

This criterion is one which is a compromise between the Dolph–Chebyshev and the Kaiser-Bessel window criteria.

Like the Dolph–Chebyshev window, the Fourier transform is more easily defined, and the window time-samples are obtained by an inverse DFT and an appropriate scale factor. The transform samples are defined by

$$W(k) = (-1)^k \frac{A \cos [y(k)] + B \left[\dfrac{y(k)}{C} \sin [y(k)] \right]}{[C + AB] \left[\left[\dfrac{y(k)}{C} \right]^2 + 1.0 \right]} \tag{49}$$

where

$$A = \sinh (C) = \sqrt{10^{2\alpha} - 1}$$

$$B = \cosh (C) = 10^\alpha$$

$$C = \cosh^{-1}(10^{\alpha})$$

$$\beta = \cosh\left[\frac{1}{N}C\right]$$

$$y(k) = N\cos^{-1}\left[\beta\cos\left(\pi\frac{k}{N}\right)\right].$$

(See also (45).) This window is presented in Figs. 54-56 for values of α equal to 3.0, 3.5, and 4.0, respectively. The main-lobe structure is practically indistinguishable from the Kaiser-Bessel main-lobe. The figures of merit listed on Table I suggest that for the same sidelobe level, this window does indeed reside between the Kaiser-Bessel and the Dolph-Chebyshev windows. It is interesting to examine Fig. 12 and note where this window is located with respect to the Kaiser-Bessel window; striking similarity in performance!

VI. Harmonic Analysis

We now describe a simple experiment which dramatically demonstrates the influence a window exerts on the detection of a weak spectral line in the presence of a strong nearby line. If two spectral lines reside in DFT bins, the rectangle window allows each to be identified with no interaction. To demonstrate this, consider the signal composed of two frequencies 10 f_s/N and 16 f_s/N (corresponding to the tenth and the sixteenth DFT bins) and of amplitudes 1.0 and 0.01 (40.0 dB separation), respectively. The power spectrum of this signal obtained by a DFT is shown in Fig. 57 as a linear interpolation between the DFT output points.

We now modify the signal slightly so that the larger signal resides midway between two DFT bins; in particular, at 10.5 f_s/N. The smaller signal still resides in the sixteenth bin. The power spectrum of this signal is shown in Fig. 58. We note that the sidelobe structure of the larger signal has completely swamped the main lobe of the smaller signal. In fact, we know (see Fig. 13) that the sidelobe amplitude of the rectangle window at 5.5 bins from the center is only 25 dB down from the peak. Thus the second signal (5.5 bins away) could not be detected because it was more than 26 dB down, and hence, hidden by the sidelobe. (The 26 dB comes from the 25-dB sidelobe level minus the 3.9-dB processing loss of the window plus 3.0 dB for a high confidence detection.) We also note the obvious asymmetry around the main lobe centered at 10.5 bins. This is due to the coherent addition of the sidelobe structures of the pair of kernels located at the plus and minus 10.5 bin positions. We are observing the self-leakage between the positive and the negative frequencies. Fig. 59 is the power spectrum of the signal pair, modified so that the large-amplitude signal resides at the 10.25-bin position. Note the change in asymmetry of the main-lobe and the reduction in the sidelobe level. We still can not observe the second signal located at bin position 16.0.

We now apply different windows to the two-tone signal to demonstrate the difference in second-tone detectability. For some of the windows, the poorer resolution occurs when the large signal is at 10.0 bins rather than at 10.5 bins. We will always present the window with the large signal at the location corresponding to worst-case resolution.

The first window we apply is the triangle window (see Fig. 60). The sidelobes have fallen by a factor of two over the rectangle windows' lobes (e.g., the −35-dB level has fallen to −70 dB). The sidelobes of the larger signal have fallen to approximately −43 dB at the second signal so that it is barely detectable. If there were any noise in the signal, the second tone would probably not have been detected.

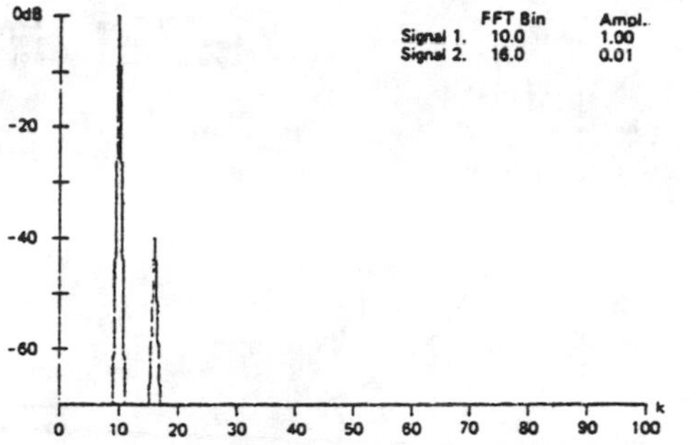

Fig. 57. Rectangle window.

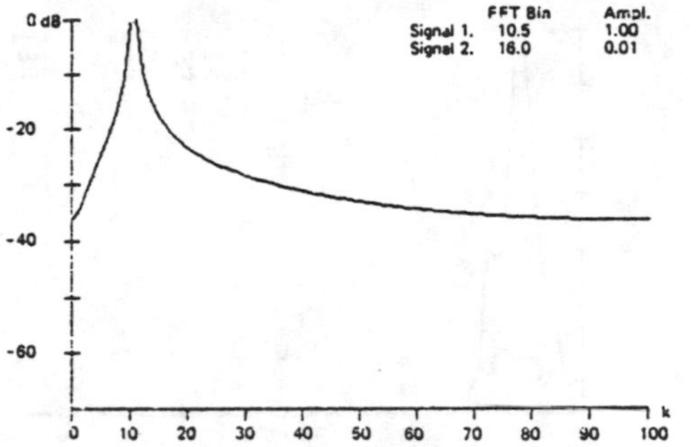

Fig. 58. Rectangle window.

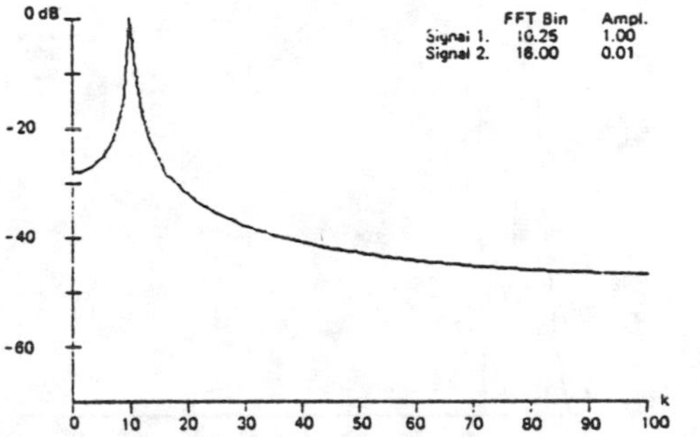

Fig. 59. Rectangle window.

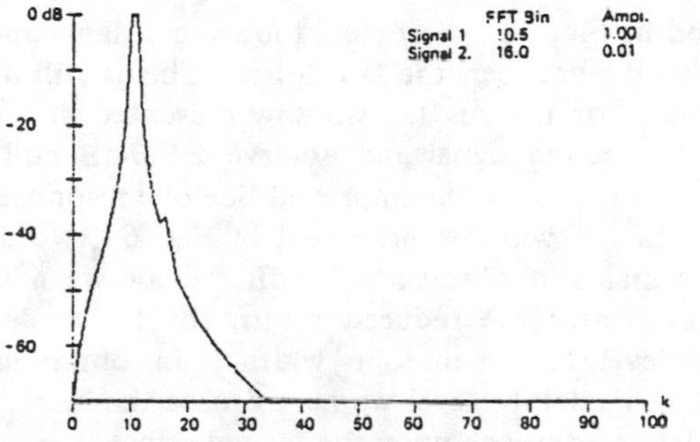

Fig. 60. Triangle window.

The next windows we apply are the $\cos^{\alpha}(x)$ family. For the cosine lobe, $\alpha = 1.0$, shown in Fig. 61 we observe a phase cancellation in the sidelobe of the large signal located at the small signal position. This cannot be considered a detection. We also see the spectral leakage of the main lobe over the frequency axis. Signals below this leakage level would not be detected. With $\alpha = 2.0$ we have the Hanning window, which is

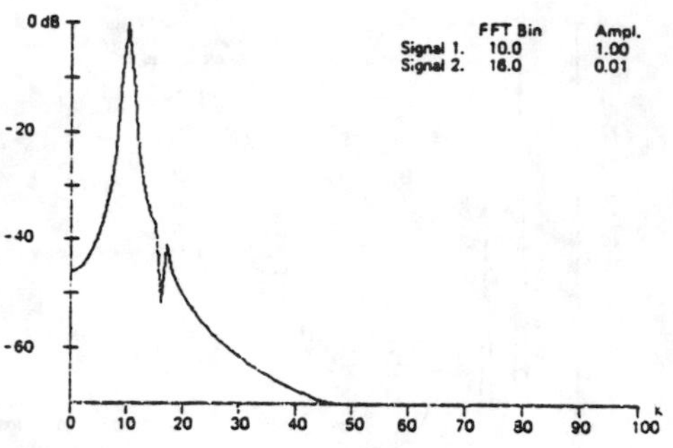

Fig. 61. Cos $(n\pi/N)$ window.

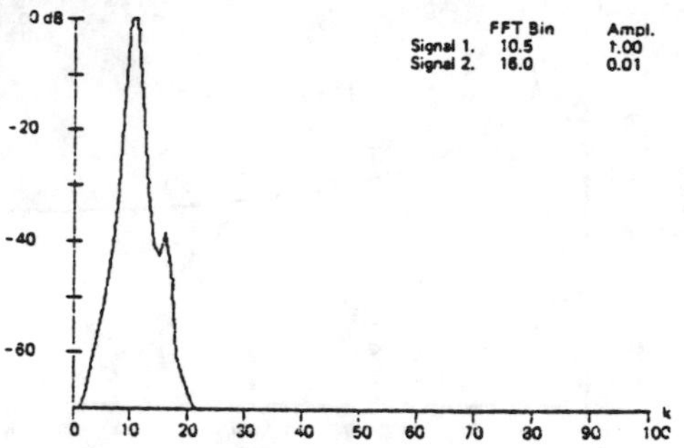

Fig. 62. Cos² $(n\pi/N)$ window.

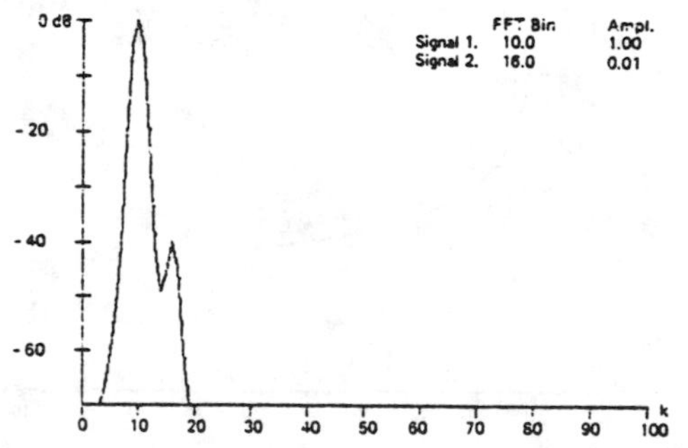

Fig. 63. Cos³ $(n\pi/N)$ window.

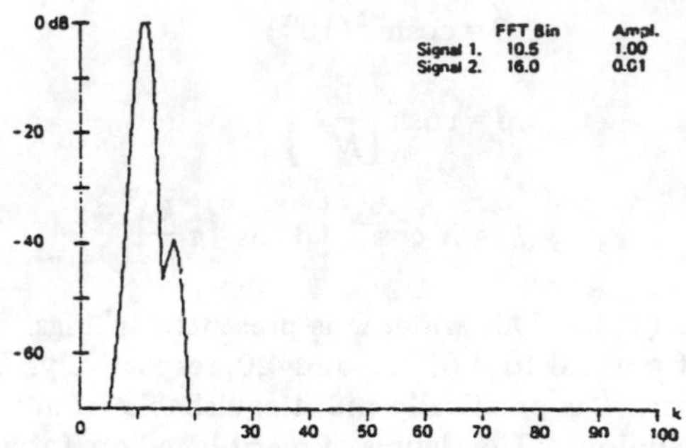

Fig. 64. Cos⁴ $(n\pi/N)$ window.

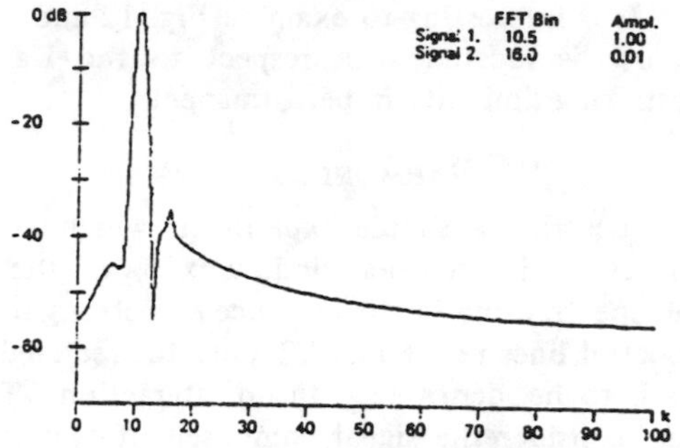

Fig. 65. Hamming window.

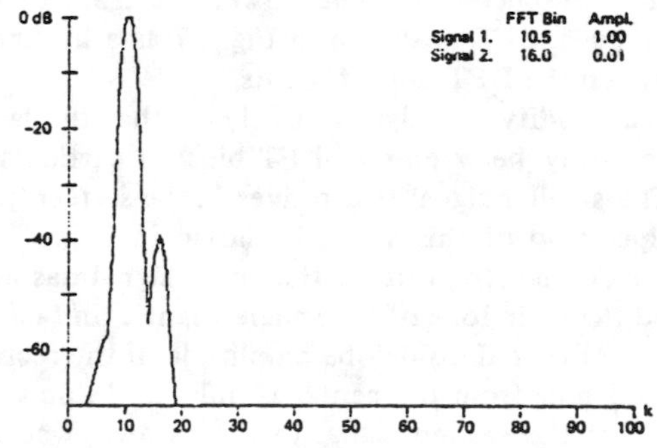

Fig. 66. Blackman window.

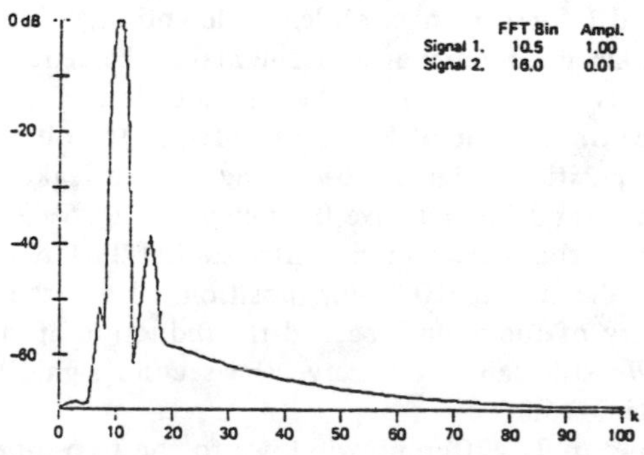

Fig. 67. Exact Blackman window.

presented in Fig. 62. We detect the second signal and observe a 3.0-dB null between the two lobes. This is still a marginal detection. For the $\cos^3(x)$ window presented in Fig. 63, we detect the second signal and observe a 9.0-dB null between the lobes. We also see the improved sidelobe response. Finally for the $\cos^4(x)$ window presented in Fig. 64, we detect the second signal and observe a 7.0-dB null between the lobes. Here we witness the reduced return for the trade between sidelobe level and main-lobe width. In obtaining further reduction in sidelobe level we have caused the increased main-lobe width to encroach upon the second signal.

We next apply the Hamming window and present the result in Fig. 65. Here we observe the second signal some 35 dB down, approximately 3.0 dB over the sidelobe response of the large signal. Here, too, we observe the phase cancellation and the leakage between the positive and the negative frequency components. Signals more than 50 dB down would not be detected in the presence of the larger signal.

The Blackman window is applied next and we see the results in Fig. 66. The presence of the smaller amplitude kernel is now very apparent. There is a 17-dB null between the two signals. The artifact at the base of the large-signal kernel is the sidelobe structure of that kernel. Note the rapid rate of falloff of the sidelobe leakage has confined the artifacts to a small portion of the spectral line.

We next apply the exact Blackman coefficients and witness the results in Fig. 67. Again the second signal is well defined with a 24-dB null between the two kernels. The sidelobe structure of the larger kernel now extends over the entire spectral range. This leakage is not terribly severe as it is nearly

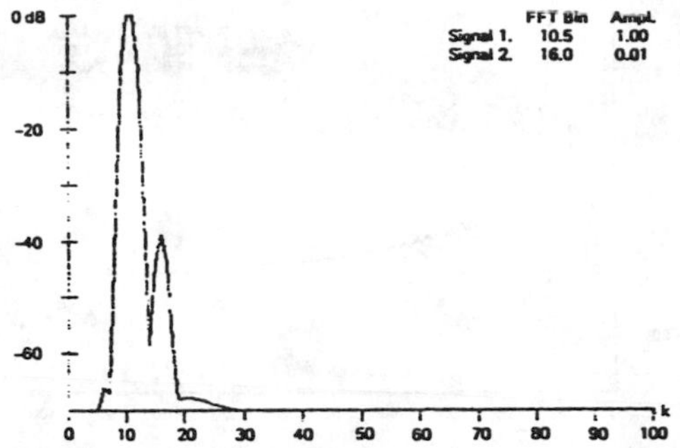

Fig. 68. Minimum 3-term Blackman–Harris window.

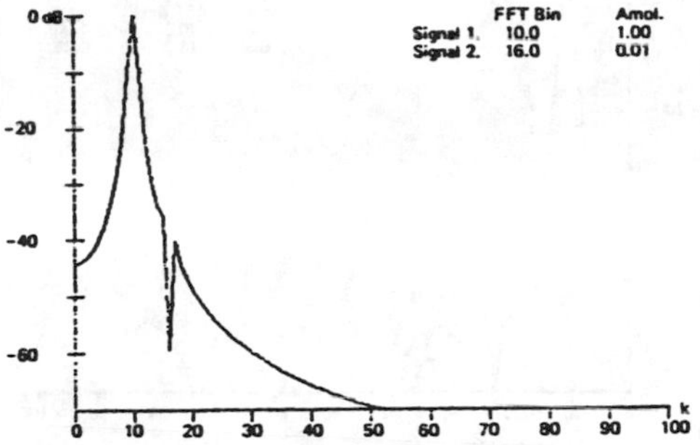

Fig. 71. Riesz window.

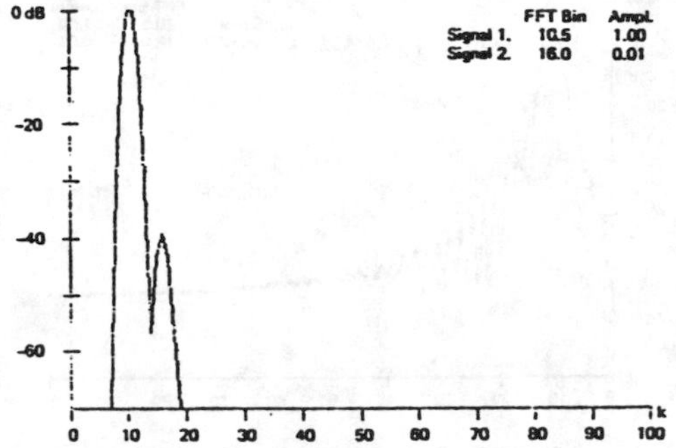

Fig. 69. 4-term Blackman–Harris window.

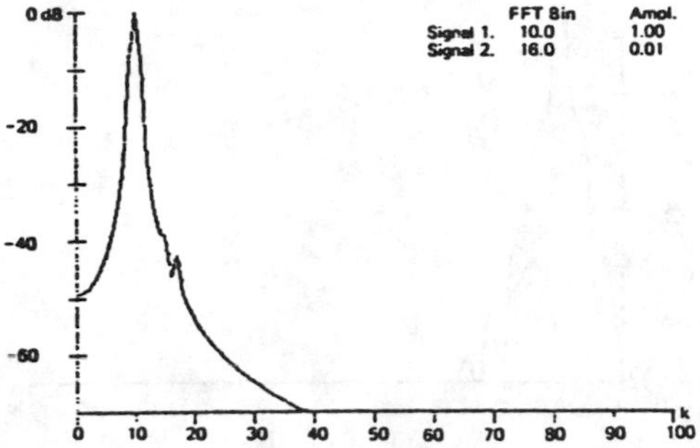

Fig. 72. Riemann window.

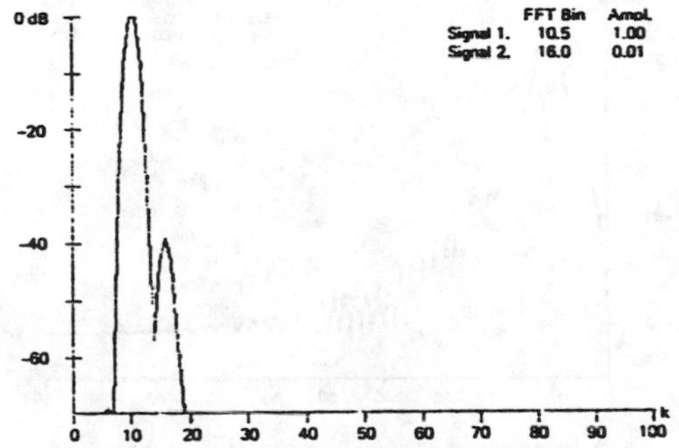

Fig. 70. 4-sample Kaiser–Bessel window.

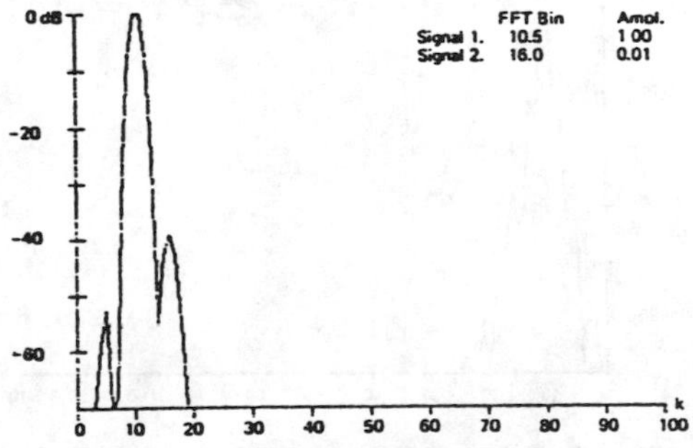

Fig. 73. de la Vallé–Poussin window.

60-dB down relative to the peak. There is another small artifact at 50-dB down on the low frequency side of the large kernel. This is definitely a single sidelobe of the large kernel. This artifact is essentially removed by the minimum 3-term Blackman–Harris window which we see in Fig. 68. The null between the two signal main lobes is slightly smaller, at approximately 20 dB.

Next the 4-term Blackman–Harris window is applied to the signal and we see the results in Fig. 69. The sidelobe structures are more than 70-dB down and as such are not observed on this scale. The two signal lobes are well defined with approximately a 19-dB null between them. Now we apply the 4-sample Kaiser–Bessel window to the signal and see the results in Fig. 70. We have essentially the same performance as with the 4-term Blackman–Harris window. The only observable difference on this scale is the small sidelobe artifact 68 dB down on the low frequency side of the large kernel. This group of Blackman-derived windows perform admirably well for their simplicity.

The Riesz window is the first of our constructed windows and is presented in Fig. 71. We have not detected the second signal but we do observe its affect as a 20.0-dB null due to the phase cancellation of a sidelobe in the large signal's kernel.

The result of a Riemann window is presented in Fig. 72. Here, too, we have no detection of the second signal. We do have a small null due to phase cancellation at the second signal. We also have a large sidelobe response.

The next window, the de la Vallé–Poussin or the self-convolved triangle, is shown in Fig. 73. The second signal is easily found and the power spectrum exhibits a 16.0-dB null. An artifact of the window (its lower sidelobe) shows up, however, at the fifth DFT bin as a signal approximately 53.0 dB down. See Fig. 29.

The result of applying the Tukey family of windows is presented in Figs. 74–76. In Fig. 74 (the 25-percent taper) we see the lack of second-signal detection due to the high sidelobe structure of the dominant rectangle window. In Fig. 75 (the 50-percent taper) we observe a lack of second-signal detection, with the second signal actually filling in one of the nulls of the first signals' kernel. In Fig. 76 (the 76-percent taper) we witness a marginal detection in the still high sidelobes of the larger signal. This is still an unsatisfying window because of the artifacts.

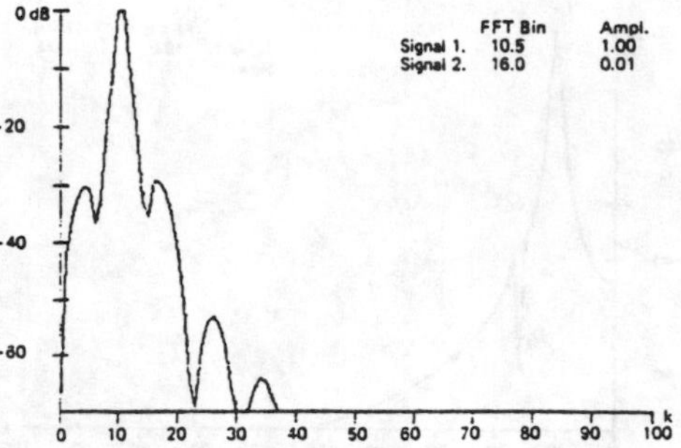

Fig. 74. Tukey (25-percent cosine taper) window.

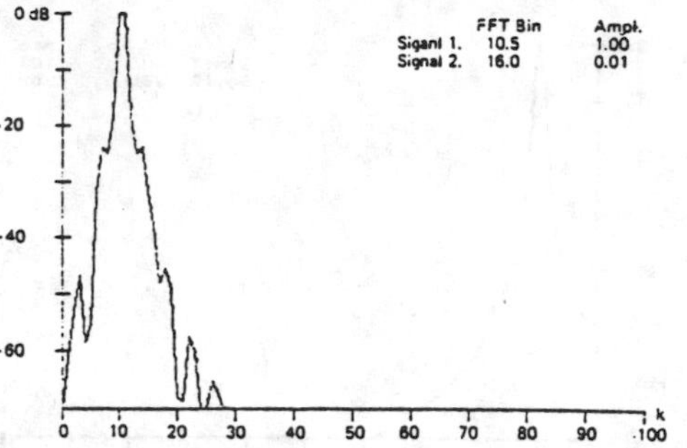

Fig. 75. Tukey (50-percent cosine taper) window.

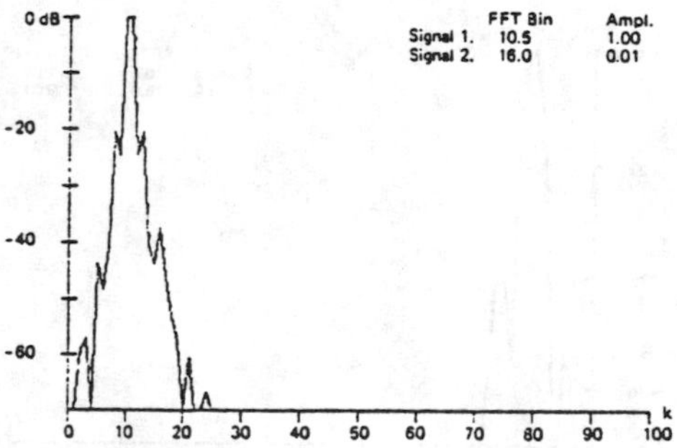

Fig. 76. Tukey (75-percent cosine taper) window.

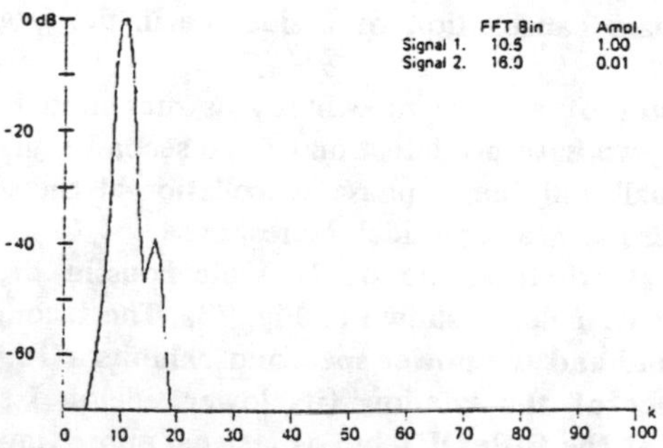

Fig. 77. Bohman window.

The Bohman construction window is applied and presented in Fig. 77. The second signal has been detected and the null between the two lobes is approximately 6.0 dB. This is not bad, but we can still do better. Note where the Bohman window resides in Fig. 12.

The result of applying the Poisson-window family is presented in Figs. 78–80. The second signal is not detected for any of the selected parameter values due to the high-sidelobe

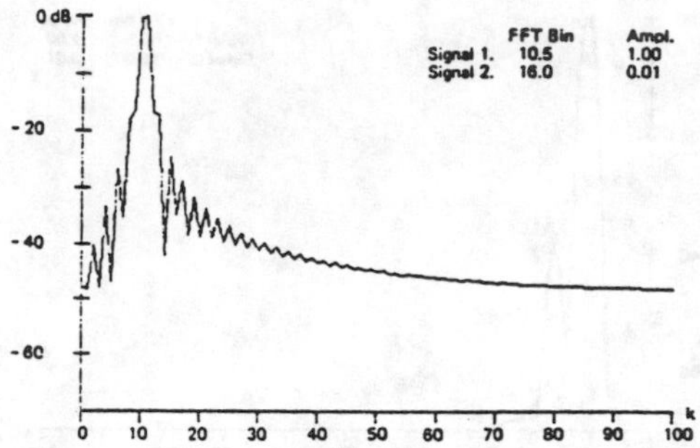

Fig. 78. Poisson window (a = 2.0).

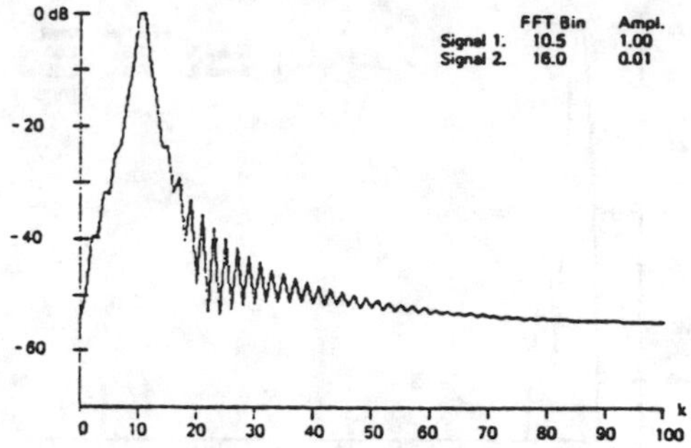

Fig. 79. Poisson window (a = 3.0).

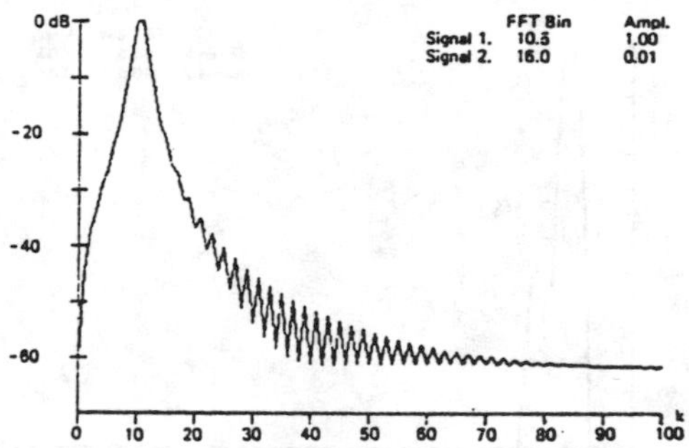

Fig. 80. Poisson window (a = 4.0).

levels of the larger signal. We anticipated this poor performance in Table I by the large difference between the 3.0 dB and the ENBW.

The result of applying the Hanning–Poisson family of windows is presented in Figs. 81–83. Here, too, the second signal is either not detected in the presence of the high-sidelobe structure or the detection is bewildered by the artifacts.

The Cauchy-family windows have been applied and the results are presented in Figs. 84–86. Here too we have a lack of satisfactory detection of the second signal and the poor sidelobe response. This was predicted by the large difference between the 3.0 dB and the equivalent noise bandwidths as listed in Table I.

We now apply the Gaussian family of windows and present the results in Figs. 87–89. The second signal is detected in all three figures. We note as we further depress the sidelobe structure to enhance second-signal detection, the null deepens to approximately 16.0 dB and then becomes poorer as the main-lobe width increases and starts to overlap the lobe of the smaller signal.

The Dolph–Chebyshev family of windows is presented in Figs. 90–94. We observe strong detection of the second signal

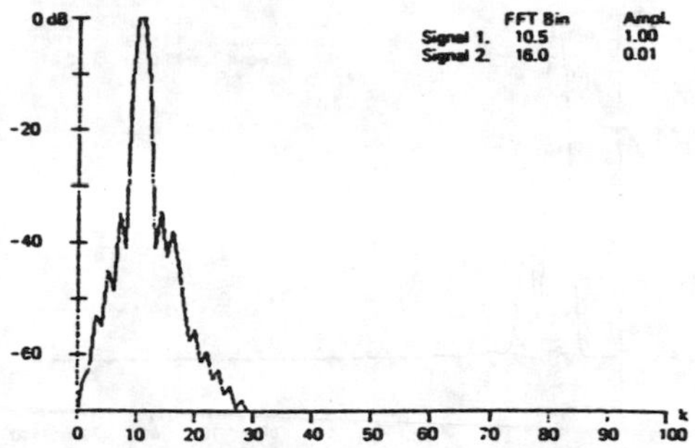

Fig. 81. Hanning–Poisson window ($a = 0.5$).

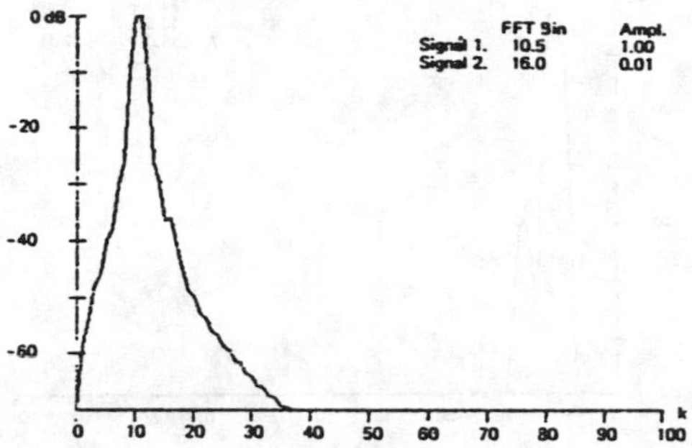

Fig. 82. Hanning–Poisson window ($a = 1.0$).

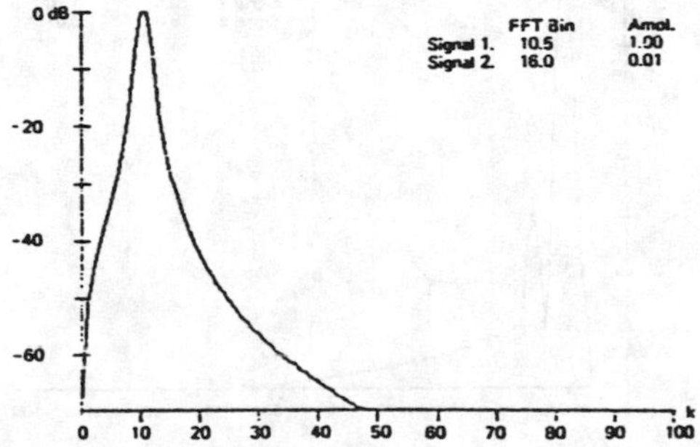

Fig. 83. Hanning–Poisson window ($a = 2.0$).

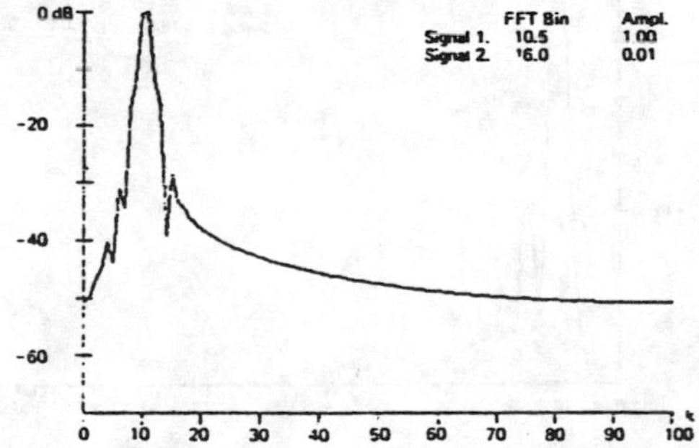

Fig. 84. Cauchy window ($a = 3.0$).

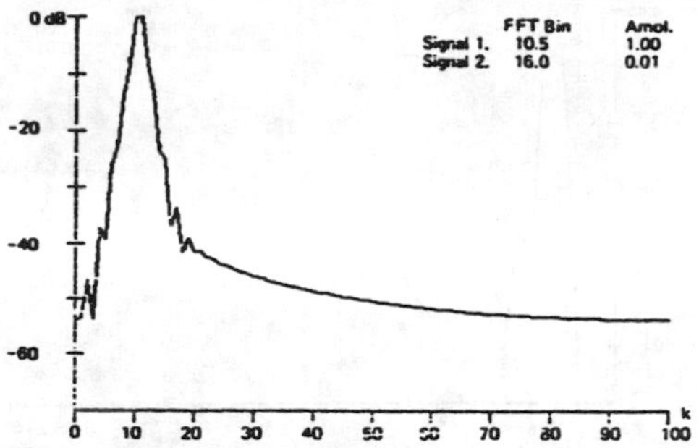

Fig. 85. Cauchy window ($a = 4.0$).

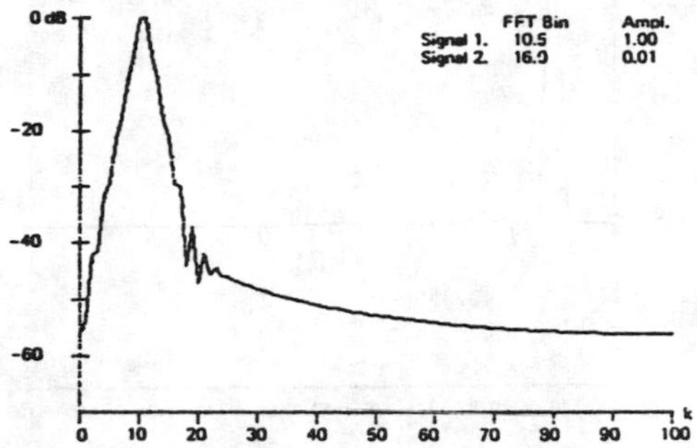

Fig. 86. Cauchy window ($a = 5.0$).

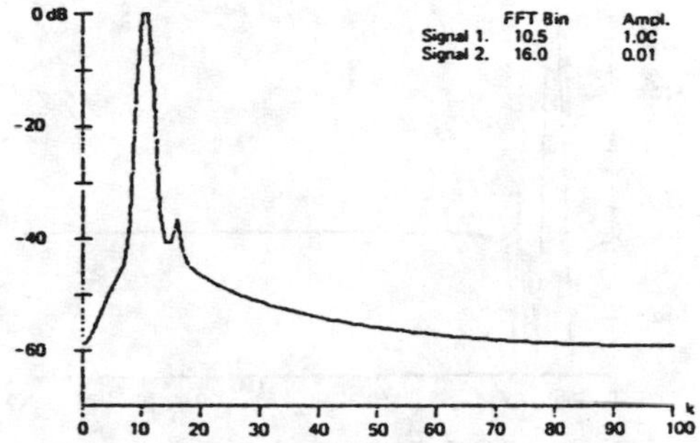

Fig. 87. Gaussian window ($a = 2.5$).

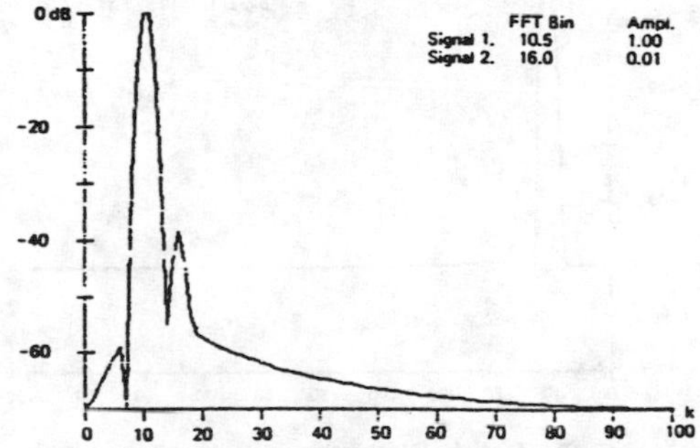

Fig. 88. Gaussian window ($a = 3.0$).

in all cases, but it is distressing to see the uniformly high sidelobe structure. Here, we again see the coherent addition of the sidelobes from the positive and negative frequency kernels. Notice that the smaller signal is not 40-dB down now. What we are seeing is the scalloping loss of the large signals' main-lobe being sampled off of the peak and being referenced as zero dB. Figs. 90 and 91 demonstrate the sensitivity of the sidelobe coherent addition to main-lobe position. In Fig. 90 the larger signal is at bin 10.5; in Fig. 91 it is at bin 10.0. Note the difference in phase cancellation near the base of the large signal. Fig. 93, the 70-dB-sidelobe window, exhibits an 18-dB null between the two main lobes but the sidelobes have added constructively (along with the scalloping loss) to the −62.0-dB level. In Fig. 94, we see the 80-dB sidelobe window exhibited sidelobes below the 70-dB level and still managed to hold the null between the two lobes to approximatley 18.0 dB.

The Kaiser–Bessel family is presented in Figs. 95–98. Here,

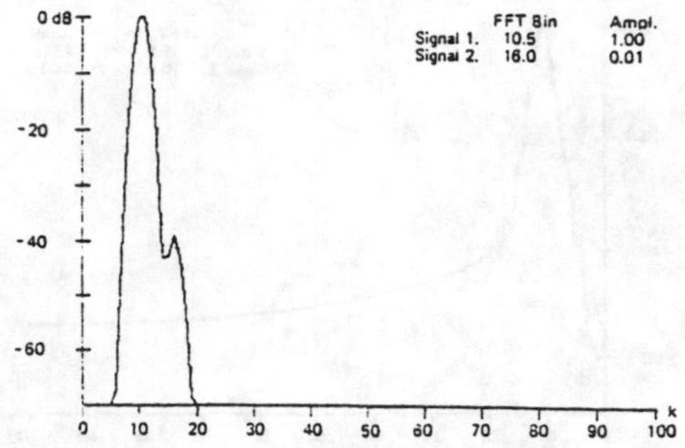

Fig. 89. Gaussian window (a = 3.5).

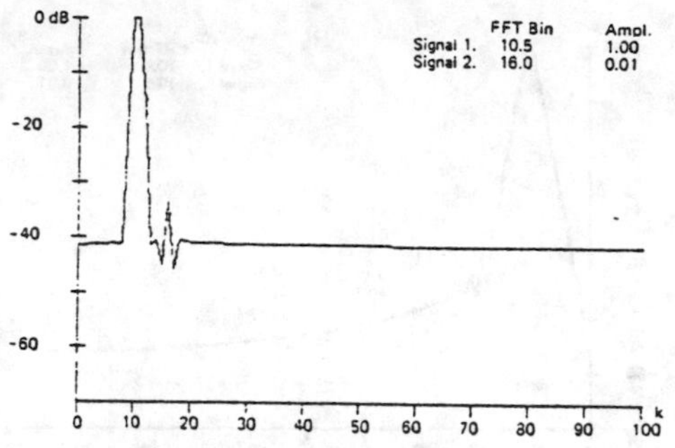

Fig. 90. Dolph–Chebyshev window (a = 2.5).

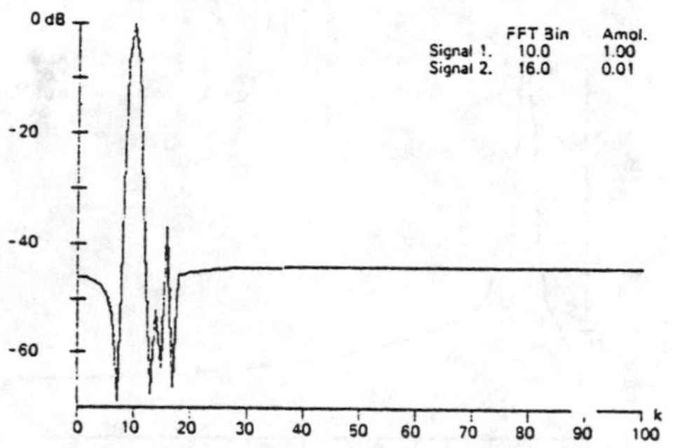

Fig. 91. Dolph–Chebyshev window (a = 2.5).

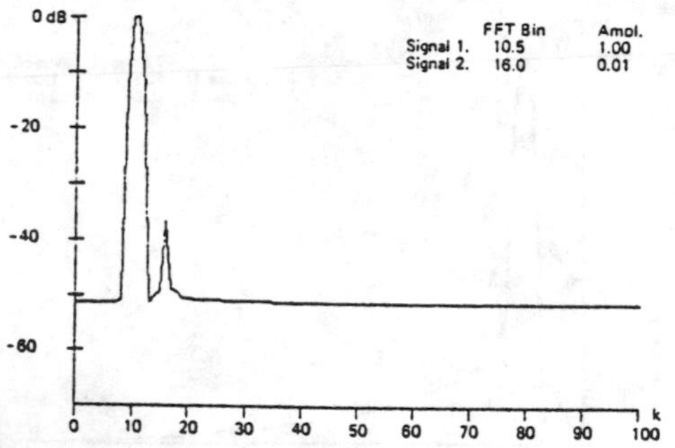

Fig. 92. Dolph–Chebyshev window (a = 3.0).

too, we have strong second-signal detection. Again, we see the effect of trading increased main-lobe width for decreased sidelobe level. The null between the two lobes reaches a maximum of 22.0 dB as the sidelobe structure falls and then becomes poorer with further sidelobe level improvement. Note that this window can maintain a 20.0-dB null between the two signal lobes and still hold the leakage to more than 70 dB down over the entire spectrum.

Figs. 99–101 present the performance of the Barcilon–Temes window. Note the strong detection of the second signal.

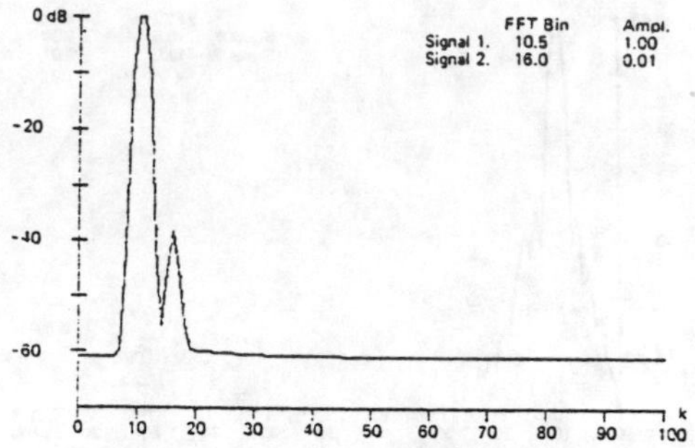

Fig. 93. Dolph–Chebyshev window (a = 3.5).

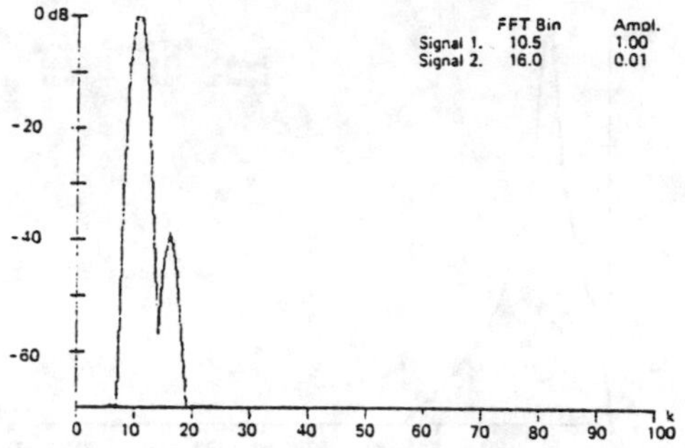

Fig. 94. Dolph–Chebyshev window (a = 4.0).

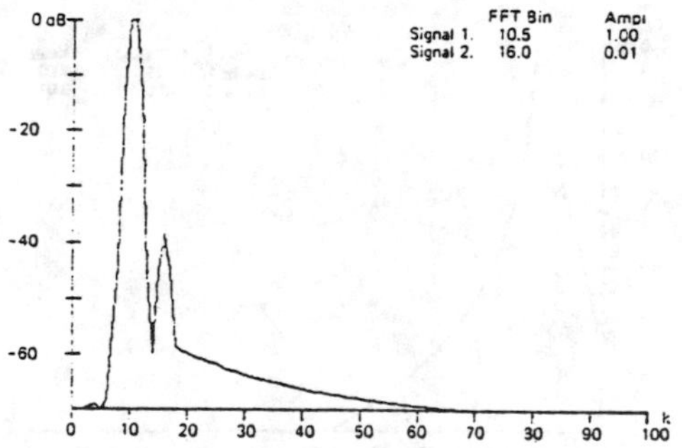

Fig. 95. Kaiser–Bessel window (a = 2.0).

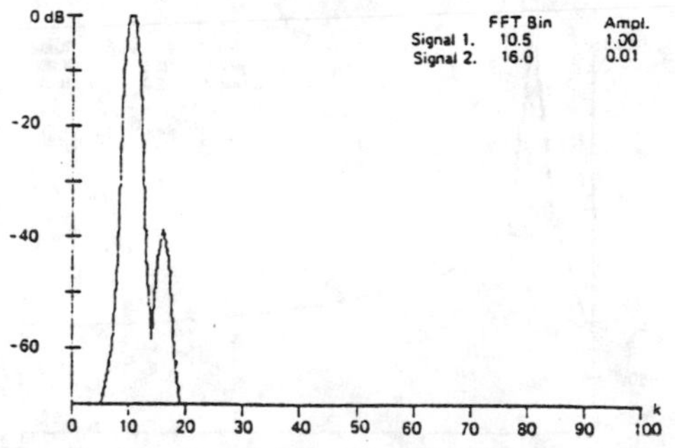

Fig. 96. Kaiser–Bessel window (a = 2.5).

There are slight sidelobe artifacts. The window can maintain a 20.0-dB null between the two signal lobes. The performance of this window is slightly shy of that of the Kaiser–Bessel window, but the two are remarkably similar.

VII. Conclusions

We have examined some classic windows and some windows which satisfy some criteria of optimality. In particular, we have described their effects on the problem of general har-

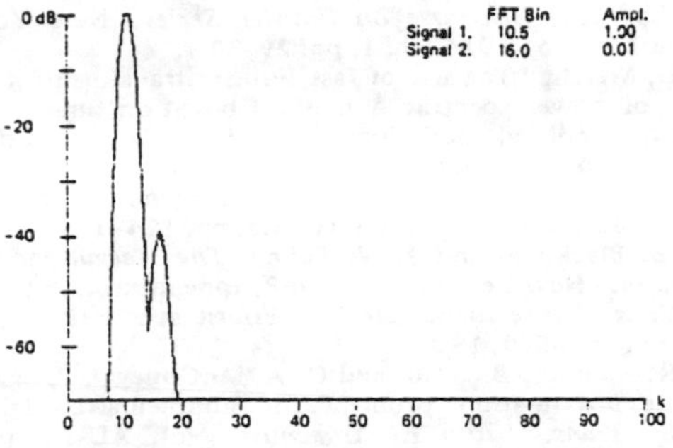

Fig. 97. Kaiser–Bessel window (a = 3.0).

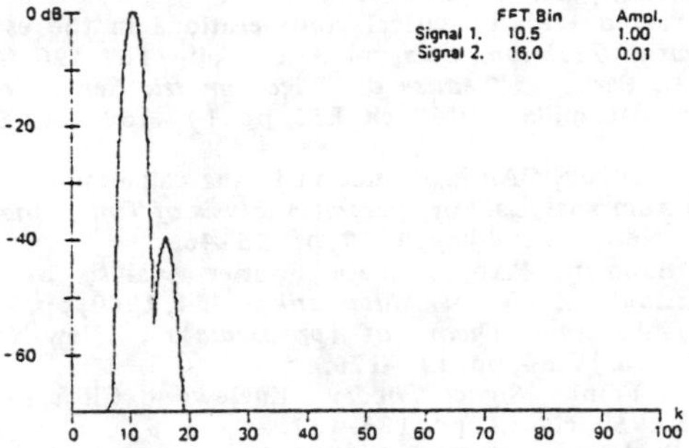

Fig. 98. Kaiser–Bessel window (a = 3.5).

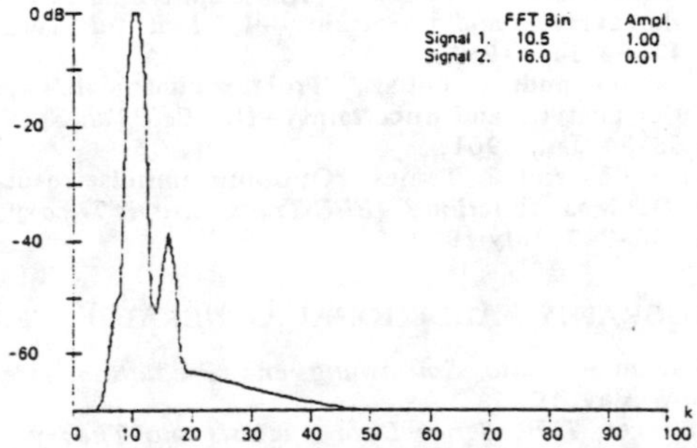

Fig. 99. Barcilon–Temes window (a = 3.0).

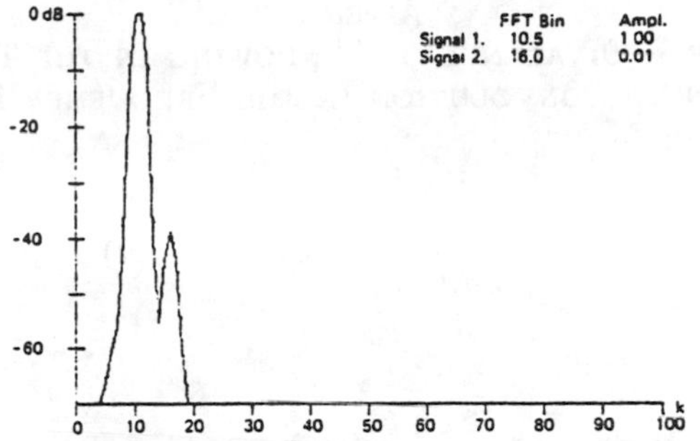

Fig. 100. Barcilon–Temes window (a = 3.5).

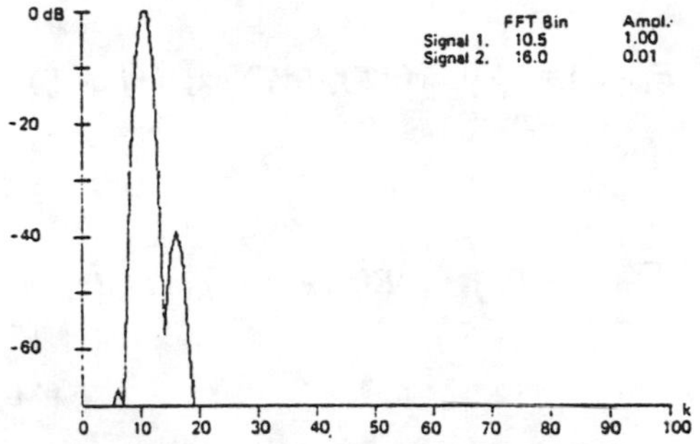

Fig. 101. Barcilon–Temes window (a = 4.0).

monic analysis of tones in broadband noise and of tones in the presence of other tones. We have observed that when the DFT is used as a harmonic energy detector, the worst case processing loss due to the windows appears to be lower bounded by 3.0 dB and (for good windows) upper bounded near 3.75 dB. This suggests that the choice of particular windows has very little effect on worst case performance in DFT energy detection. We have concluded that a good performance indicator for the window is the difference between the equivalent noise bandwidth and the 3.0-dB bandwidth normalized by the 3.0-dB bandwidth. The windows which perform well (as indicated in Fig. 12) exhibit values for this ratio between 4.0 and 5.5 percent. The range of this ratio for the windows listed in Table I is 3.2 to 22.9 percent.

For multiple-tone detection via the DFT, the window employed does have a considerable effect. Maximum dynamic range of multitone detection requires the transform of the window to exhibit a highly concentrated central lobe with very-low sidelobe structure. We have demonstrated that many classic windows satisfy this criterion with varying degrees of success and some not at all. We have demonstrated the optimal windows (Kaiser–Bessel, Dolph–Chebyshev, and Barcilon–Temes) and the Blackman–Harris windows perform best in detection of nearby tones of significantly different amplitudes. Also for the same dynamic range, the three optimal windows and the Blackman–Harris window are roughly equivalent with the Kaiser–Bessel and the Blackman–Harris, demonstrating minor performance advantages over the others. We note that while the Dolph–Chebyshev window appears to be the best window by virtue of its relative position in Fig. 12, the coherent addition of its constant-level sidelobes detracts from its performance in multi tone detection. Also the sidelobe structure of the Dolph–Chebyshev window exhibits extreme sensitivity to coefficient errors. This would affect its performance in machines operating with fixed-point arithmetic. This suggests that the Kaiser–Bessel or the Blackman–Harris window be declared the top performer. My preference is the Kaiser–Bessel window. Among other reasons, the coefficients are easy to generate and the trade-off of sidelobe level as a function of time-bandwidth product is fairly simple. For many applications, the author would recommend the 4-sample Blackman–Harris (or the 4-sample Kaiser–Bessel) window. These have the distinction of being defined by a few easily generated coefficients and of being able to be applied as a spectral convolution after the DFT.

We have called attention to a persistent error in the application of windows when performing convolution in the frequency domain, i.e., the omission of the alternating signs on the window sample spectrum to account for the shifted time origin. We have also identified and clarified a source of confusion concerning the evenness of windows under the DFT.

Finally, we comment that all of the conclusions presented about window performance in spectral analysis are also applicable to shading for array processing of spatial sampled data, including FFT beamforming.

APPENDIX
THE EQUIVALENCE OF WINDOWING IN THE TIME DOMAIN TO CONVOLUTION IN THE FREQUENCY DOMAIN

Let

$$f(t) = \int_{-\infty}^{+\infty} F(\omega) \exp(-j\omega t)\, d\omega/2\pi$$

and

$$W(\omega) = \sum_{n=-N/2}^{+N/2} w(nT) \exp(+j\omega nT).$$

Then

$$F_w(\omega) = \sum_{n=-\infty}^{+\infty} w(nT) f(nT) \exp(+j\omega nT)$$

becomes

$$F_w(\omega) = \sum_{n=-\infty}^{+\infty} w(nT) \int_{-\infty}^{+\infty} F(x) \exp(-jxnT)\, dx/2\pi \cdot \exp(+j\omega nT)$$

$$= \int_{-\infty}^{+\infty} F(x) \sum_{n=-\infty}^{+\infty} w(nT) \exp[+j(\omega - x)nT]\, dx/2\pi$$

$$= \int_{-\infty}^{+\infty} F(x) \sum_{n=-N/2}^{+N/2} w(nT) \exp[+j(\omega - x)nT]\, dx/2\pi$$

$$= \int_{-\infty}^{+\infty} F(x)\, W(\omega - x)\, dx/2\pi$$

or

$$F_w(\omega) = F(\omega) * W(\omega).$$

REFERENCES

[1] C. W. Helstrom, *Statistical Theory of Signal Detection*, 2nd ed. New York: Pergamon Press, 1968, Ch. IV, 4, pp. 124-130.

[2] J. W. Cooley, P. A. Lewis, and P. D. Welch, "The finite Fourier transform," *IEEE Trans. Audio Electroacoust.*, vol. AU-17, pp. 77-85, June 1969.

[3] J. W. Wozencraft and I. M. Jacobs, *Principles of Communication Engineering*. New York: Wiley, 1965, ch. 4.3, pp. 223-228.

[4] C. Lanczos, *Discourse on Fourier Series*. New York: Hafner Publishing Co., 1966, ch. 1, pp. 29-30.

[5] P. D. Welch, "The use of fast Fourier transform for the estimation of power spectra: A method based on time averaging over short, modified periodograms," *IEEE Trans. Audio Electroacoust.*, vol. AU-15, pp. 70-73, June 1967.

[6] J. R. Rice, *The Approximation of Functions*, Vol. I. Reading, MA: Addison-Wesley, 1964, ch. 5.3, pp. 124-131.

[7] R. B. Blackman and J. W. Tukey, *The Measurement of Power Spectra*. New York: Dover, 1958, appendix B.5, pp. 95-100.

[8] L. Fejer, "Untersuchunger uber Fouriersche Reihen," *Mat. Ann.*, 58, pp. 501-569, 1904.

[9] L. R. Rabiner, B. Gold, and C. A. McGonegal, "An approach to the approximation problem for nonrecursive digital filters," *IEEE Trans. Audio Electroacoust.*, vol. AU-18, pp. 83-106, June 1970.

[10] F. J. Harris, "High-resolution spectral analysis with arbitrary spectral centers and adjustable spectral resolutions," *J. Comput. Elec. Eng.*, vol. 3, pp. 171-191, 1976.

[11] E. Parzen, "Mathematical considerations in the estimation of spectra," *Technometrics*, vol. 3, no. 2, pp. 167-190, May 1961.

[12] N. K. Bary, *A Treatise on Trigonometric Series*, Vol. I. New York: Macmillan, 1964, ch. I.53, pp. 149-150, ch. I.68, pp. 189-192.

[13] J. W. Tukey, "An introduction to the calculations of numerical spectrum analysis," in *Spectral Analysis of Time Series*, B. Harris, Ed. New York: Wiley, 1967, pp. 25-46.

[14] H. Bohman, "Approximate Fourier analysis of distribution functions," *Arkiv Foer Matematik*, vol. 4, 1960, pp. 99-157.

[15] N. I. Akhiezer, *Theory of Approximation*. New York: Ungar, 1956, ch. IV.64, pp. 118-120.

[16] L. E. Franks, *Signal Theory*. Englewood Cliffs, NJ: Prentice-Hall, 1969, ch. 6.1, pp. 136-137.

[17] H. D. Helms, "Digital filters with equiripple or minimax responses," *IEEE Trans. Audio Electroacoust.*, vol. AU-19, pp. 87-94, Mar. 1971.

[18] F. F. Kuo and J. F. Kaiser, *System Analysis by Digital Computer*. New York: Wiley, 1966, ch. 7, pp. 232-238.

[19] D. Slepian and H. Pollak, "Prolate-spheroidal wave functions, Fourier analysis and uncertainty—I," *Bell Tel. Syst. J.*, vol. 40, pp. 43-64, Jan. 1961.

[20] H. Landau and H. Pollak, "Prolate-spheroidal wave functions, Fourier analysis and uncertainty—II," *Bell Tel. Syst. J.*, vol. 40, pp. 65-84, Jan. 1961.

[21] V. Barcilon and G. Temes, "Optimum impulse response and the Van Der Maas function," *IEEE Trans. Circuit Theory*, vol. CT-19, pp. 336-342, July 1972.

BIBLIOGRAPHY—ADDITIONAL GENERAL REFERENCES

R. B. Blackman, *Data Smoothing and Prediction*. Reading, MA: Addison-Wesley, 1965.

D. R. Brillinger, *Time Series Data Analysis and Theory*. New York: Holt, Rinehart, and Winston, 1975.

D. Gingras, "Time series windows for improving discrete spectra estimation," Naval Undersea Research and Development Center, Rep. NUC TN-715, Apr. 1972.

F. J. Harris, "Digital signal processing," Class notes, San Diego State Univ., 1971.

G. M. Jenkins, "General considerations in the estimation of spectra," *Technometrics*, vol. 3, no. 2, pp. 133-166, May 1961.

Analysis of a generalised framework for spectral estimation

Part 1: The technique and its mean value

G. Clifford Carter, B.S., M.S., Ph.D., Sen. Mem. I.E.E.E., and Albert H. Nuttall, B.S., M.S., Sc.D.

Indexing term: *Signal processing*

Abstract: The paper presents a generalised framework for spectral estimation. Included in the framework are the classical Blackman-Tukey and widely used weighted overlapped segment averaging methods. An analysis and discussion of the mean value of the spectral estimate is provided here. Part 2 of the paper presents results on the variance of the technique.

1 Introduction

The purpose of this paper is to present a generalised technique for spectral estimation [1–6] useful in sonar applications and an analysis of its mean value. Included in the generalised framework are the classical Blackman-Tukey method and the now widely used weighted overlapped segment averaging (WOSA) method. A companion paper [7] presents the variance of the technique (see also Reference 5).

The navies of the world are interested in sonars with sensors, signal processing and displays that perform three vital functions. Sonar systems must:

(*a*) detect (decide if an acoustic signal source is present or absent)

(*b*) classify (type the signal as much as possible, for example warship or merchant ship)

(*c*) localise the signal source (determine source position, for example range and bearing).

To perform these functions, sonar signal processors must estimate not only autospectra, but also cross-spectra, weighted cross-spectra (such as the complex coherence function) and Fourier transforms of these three spectral functions. For example, the generalised crosscorrelator for time delay estimation is the Fourier transform of the weighted cross-spectrum (see the June 1981 special issue of the *IEEE Transactions on Acoustics, Speech, and Signal Processing)*.

The characteristics of the signal and background noise in the naval application are unique. Many commercially available devices provide virtually all the processing power needed for researchers to process one- and two-channel data. These devices almost all rely on the WOSA fast Fourier transform (FFT) method, discussed by Bartlett [8], Welch [9], Nuttall [10], Carter, Knapp and Nuttall [11], Rabiner and Allen [12] and Bingham, Godfrey and Tukey [13]. Selection of a proper time weighting or taper has been an important part of the WOSA method [10, 14], especially for signals of the type experienced in passive underwater acoustic environments. In particular, large dynamic range, multiple tonals and multiple narrowband and broadband signals and noises are routinely present. Often the signals and noises are of comparable levels and characteristics in the problems of interest. Readers interested in more detail should consult References 15–31, especially the recent work in References 17 and 19.

While we limit our mathematical presentation here to continuous functions of time and frequency, we in fact recognise that today's technology is predominantly digital. However, a discussion of quantisation and time/frequency sampling would only confuse the main issues we are trying to present here; sampling considerations are treated in Reference 5. Included in our generalised framework [1–6] are the now classical Blackman-Tukey (correlation) method [16] and the widely used WOSA method. Not included are techniques that first model the data as white noise exciting a linear filter and then attempt to characterise the spectrum of the output by estimating the filter characteristics. (See References 27 and 28 for a discussion of these methods.)

Spectral analysis techniques have received a great deal of attention in the past, ranging from the original autocorrelation approach of Blackman-Tukey to the more recent WOSA FFT approach. These two apparently different approaches are shown here to be limiting special cases of a generalised framework for spectral analysis.

2 Generalised spectral analysis method

The generalised method consists of several steps, at the end of which the desired spectral estimates are available. We refer to some intermediate results as first- and second-stage spectral/correlation estimates. Greater detail is given in Reference 5.

2.1 First step

First, the observed time-limited data record (of an assumed wide-sense stationary process) is partitioned into P segments. Each segment is multiplied by the same real symmetric (time) weighting function $w_1(t)$. Specifically, the pth weighted segment is

$$y_p(t) = x(t)\, w_1\left(t - \frac{L_1}{2} - pS\right)$$

where t is time, p is an integer, S is the time shift and L_1 is the length of the segment (or piece). Note that, by proper selection of shift S, the segments can be disjoint or overlapped. Two typical time weighting functions are sketched in Fig. 1.

2.2 Second step

The second step is to compute the (discrete) Fourier transform of each segment, compute the magnitude-squared values (see Reference 11 for a slightly more complicated method for the cross-power spectrum), and average over the available segments. In particular, the first-stage spectral estimate is

$$\hat{G}_1(f) = \frac{1}{P} \sum_{p=0}^{P-1} |F[y_p(t)]|^2$$

Paper 2453F, first received 1st April and in revised form 20th August 1982

The authors are with the New London Laboratory, Naval Underwater Systems Center, New London, CT 06320, USA where f is frequency, the circumflex denotes an estimate and F denotes the Fourier transform. When P is large, and smooth time weighting like a cosine (sometimes called 'Hanning' after Julius von Hann) is used, this is the popular WOSA method. With proper overlap, the WOSA method can achieve all the available spectral stability for a fixed frequency resolution [2, 10]. While there are still compelling reasons for the continued use of the WOSA method, we now explore some promising extensions to that method.

2.3 *Third step*

Continuing the development of our generalised method, we have the third step:

$$\hat{R}_1(\tau) = F^{-1}[\hat{G}_1(f)]$$

where τ is the time delay (lag) variable. In this step, the inverse Fourier transform of the first-stage power spectral estimate gives the first-stage correlation function estimate.

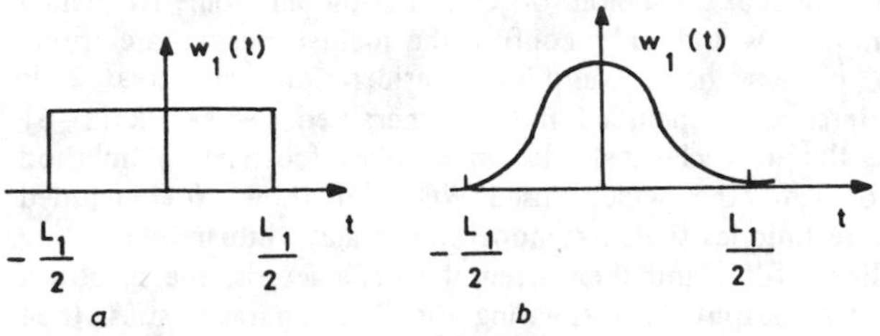

Fig. 1 *Two examples of time weighting $w_1(t)$*

a Rectangular
b Cosine

2.4 *Fourth step*

In the Blackman-Tukey method [16], we could have reached the correlation estimate $\hat{R}_1(\tau)$ without resort to steps one and two. Being familiar with their work (see also Jenkins and Watts [18]), one can entertain the next logical computation, step four, as

$$\hat{R}_2(\tau) = \hat{R}_1(\tau)\, w_2(\tau)$$

i.e. the second-stage correlation estimate $\hat{R}_2(\tau)$ is formed by multiplying the first-stage estimate by a smooth real symmetric lag weighting $w_2(\tau)$.

2.5 *Fifth step*

The fifth step is then to transform back into the frequency domain to yield the second-stage spectral estimate

$$\hat{G}_2(f) = F[\hat{R}_2(\tau)]$$

When $P = 1$ (i.e. one segment) and the rectangular time weighting on the left-hand side of Fig. 1 is selected, the method is equivalent to the Blackman-Tukey method, even though the correlation estimate is arrived at via the frequency domain. When P is large, the major computational costs are in steps one and two. Indeed, steps four and five could be repeated many times with different lag weightings $w_2(\tau)$ if desired.

With any measurement technique, it is important to understand the mean and the variance of the estimates. In this case, studying the mean of the spectral estimate will suggest a method for selecting the time and lag weightings. A derivation and discussion of the variance of the spectral estimates in our generalised method can be found in Reference 5 and the companion paper [7]. Empirical examples of mean squared errors (variance plus squared bias) for four spectral estimation methods that fall within our generalised framework are given in Reference 2.

3 Mean value

The mean value of the first-stage correlation estimate is given by

$$\overline{\hat{R}_1(\tau)} = R(\tau) \int_{-\infty}^{\infty} w_1(t)\, w_1(t-\tau)dt$$

where $R(\tau)$ is the true autocorrelation of process $x(t)$. For the power spectral estimate, the mean of the first-stage estimate is

$$\overline{\hat{G}_1(f)} = G(f) \otimes \{F[w_1(t)]\}^2$$

where $\otimes$ denotes convolution and $G(f)$ is the true power spectrum of process $x(t)$. The Fourier transform of $w_1(t)$ is real since time weighting $w_1(t)$ is symmetric. (In a practical application where digital computers are employed for the data manipulation, this convolution will be digital. Excellent discussions of digital convolution are given in Oppenheim and Shafer [23] and Rabiner and Gold [24].)

The mean value of the second-stage correlation function estimate is given by

$$\overline{\hat{R}_2(\tau)} = \overline{\hat{R}_1(\tau)}\, w_2(\tau) = R(\tau)\, w_e(\tau)$$

where the effective lag weighting is

$$w_e(\tau) = w_2(\tau) \int_{-\infty}^{\infty} w_1(t)\, w_1(t-\tau)dt$$

Without loss of generality, we take $w_e(0) = 1$, which means that the mean second-stage power estimate is unbiased, i.e. $\overline{\hat{R}_2(0)} = R(0)$. The mean value of the second-stage power spectral estimate is given by

$$\overline{\hat{G}_2(f)} = G(f) \otimes W_e(f)$$

where the effective frequency window is

$$W_e(f) = \{F[w_1(t)]\}^2 \otimes F[w_2(\tau)]$$

and where $W_e(f)$ is the Fourier transform of $w_e(\tau)$. Thus, as we expect, both the time weighting and lag weighting affect the mean value of the second-stage spectral estimate $\hat{G}_2(f)$. The mean results above apply to any wide-sense stationary process $x(t)$; the process need not be Gaussian.

4 Selection of time and lag weightings

Now suppose we had a desired frequency window $W_d(f)$ through which we desired to view the true power spectrum $G(f)$. The Fourier transform of $W_d(f)$ is then the desired effective lag weighting

$$w_d(\tau) = w_2(\tau) \int_{-\infty}^{\infty} w_1(t)\, w_1(t-\tau)\, dt$$

i.e. we could realise the desired effective weighting by choosing lag weighting

$$w_2(\tau) = \frac{w_d(\tau)}{\int_{-\infty}^{\infty} w_1(t)\, w_1(t-\tau)\, dt}$$

In so far as the mean of the spectral estimate is concerned, we can select a lag weighting $w_2(\tau)$ that undoes the effect of the time weighting, provided that the length of $w_d(\tau)$ does not exceed that of the correlation of $w_1(t)$.

For example, suppose that, instead of multiplying each time segment of $x(t)$ by a time weighting (a time-consuming process), we simply selected a piece of the original waveform; i.e. we gate it out or equivalently 'multiply' it by the rectangular time weighting shown on the left-hand side of Fig. 1. Now we select $w_2(\tau)$ equal to the desired lag weighting divided by the triangular function shown in Fig. 2, i.e. the auto-correlation of $w_1(t)$.

In particular, suppose the desired lag weighting function is symmetric and monotonically decreasing for lags greater than zero; for example, see Fig. 3. Then dividing the Fig. 3 weighting by the function shown in Fig. 2 results in a function as sketched in Fig. 4, which may no longer be a monotonically decreasing function of lag for positive lags.

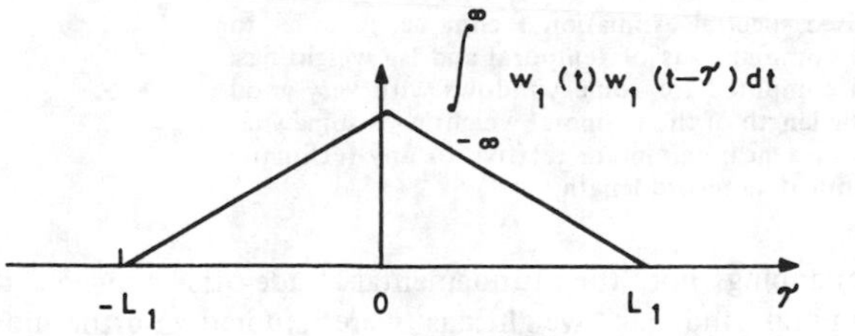

Fig. 2 *Autocorrelation of rectangular $w_1(t)$*

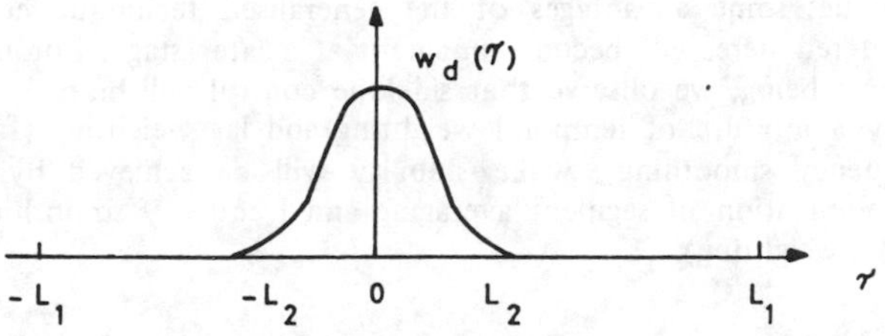

Fig. 3 *Typical desired lag weighting*

Note that the extent L_2 of lag weighting is less than L_1

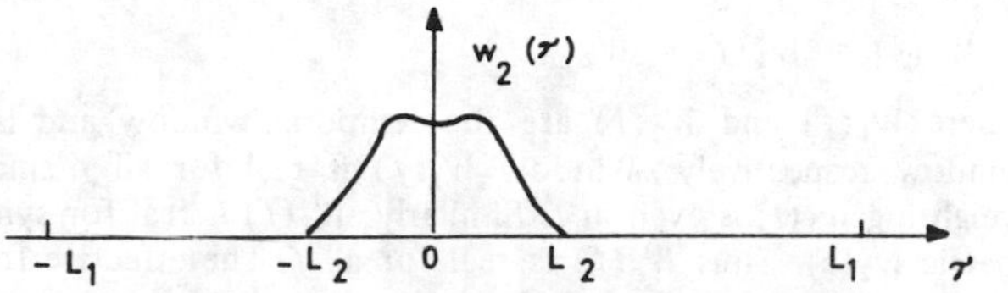

Fig. 4 *Lag reshaping to compensate rectangular time weighting and achieve desired effective weighting (Fig. 3 divided by Fig. 2)*

In so far as the mean of the spectral estimate is concerned, by selection of a proper (although unusual) lag weighting, as in Fig. 4, we can avoid the computational expense of multiplying each data segment by a smooth time weighting function. We also note that some time weightings lend themselves to complex voltage convolution in the frequency domain. Hence the lag weighting, or spectrum convolution, discussed in this paper should not be confused with the separate issue of how best to do time weighting.

These comments apply only to the mean of the spectral estimate. A separate question is whether the variance has been increased by this lag reshaping procedure; the companion paper [7] treats this problem.

5 Conclusions

This paper has presented a generalised framework for power spectral estimation. An outline of the processing steps in the procedure was also presented, as well as a discussion of how two well known past procedures fit into the two-stage method. Mean values of the spectral estimates were given. A new method that avoids time weighting, by modifying a correlation estimate, and that realises equivalent mean behaviour was presented. The method appears to offer reduced computational cost relative to the WOSA method.

6 References

1 NUTTALL, A.H., and CARTER, G.C.: 'A generalized framework for power spectral estimation', *IEEE Trans.*, 1980, **ASSP-28**, pp. 334–335

2 CARTER, G.C., and NUTTALL, A.H.: 'On the weighted overlapped segment-averaging method for power spectral estimation', *Proc. IEEE*, 1980, **68**, pp. 1352–1354

3 CARTER, G.C., and NUTTALL, A.H.: 'A brief summary of a generalized framework for spectral estimation', *Signal Process.*, 1980, **2**, pp. 387–390

4 CARTER, G.C.: 'A generalized framework for coherence estimation'. Proceedings of 1980 conference on information sciences and systems, Princeton, New Jersey, USA, Mar. 1980

5 NUTTALL, A.H.: 'Spectral analysis via quadratic frequency-smoothing of Fourier-transformed, overlapped, weighted data segments'. NUSC technical report 6459, New London, Connecticut, USA, June 1981

6 CARTER, G.C., and NUTTALL, A.H.: 'Analysis of a generalized framework for power spectral estimation'. Proceedings of ASSP Workshop on spectral estimation, McMaster University, Ontario, Canada, Aug. 1981

7 NUTTALL, A.H.: 'Analysis of a generalised framework for spectral estimation. Part 2: Reshaping and variance results', *IEE Proc. F, Commun., Radar & Signal Process.*, 1983, **130**, (3), pp. 242–245

8 BARTLETT, M.S.: 'An introduction to stochastic processes with special reference to methods and applications' (Cambridge University Press, 1953)

9 WELCH, P.D.: 'The use of FFT for the estimation of power spectra: A method based on time averaging over short modified periodograms' *IEEE Trans.*, 1967, **AU-15**, pp. 70–73

10 NUTTALL, A.H.: 'Spectral estimation by means of overlapped FFT processing of windowed data'. NUSC report 4169, New London, Connecticut, USA, Oct. 1971

11 CARTER, G.C., KNAPP, C.H., and NUTTALL, A.H.: 'Estimation of the magnitude-squared coherence function via overlapped fast Fourier transform processing', *IEEE Trans.*, 1973, **AU-21**, pp. 337–344

12 RABINER, L.R., and ALLEN, J.B.: 'On the implementation of a short-time spectral analysis method for system identification', *ibid.*, 1980, **ASSP-28**, pp. 69–78

13 BINGHAM, C., GODFREY, M.D., and TUKEY, J.W.: 'Modern techniques of power spectrum estimation', *ibid.*, 1967, **AU-15**, pp. 56–66

14 BRILLINGER, D.R.: 'The key role of tapering in spectrum estimation', *ibid.*, 1981, **ASSP-29**, pp. 1075–1076

15 PARZEN, E.: 'Mathematical considerations in the estimation of spectra', *Technometrics*, 1961, **3**, pp. 167–190

16 BLACKMAN, R.B., and TUKEY, J.W.: 'The measurement of power spectra from the point of view of communications engineering' (Dover Publications, 1959)

17 KAY, S.M. and MARPLE, S.L. Jr.: 'Spectrum analysis – A modern perspective', *Proc. IEEE*, 1981, **69**, pp. 1380–1419

18 JENKINS, G.M., and WATTS, D.G.: 'Spectral analysis and its applications' (Holden-Day, 1968)

19 KNIGHT, W.C., PRIDHAM, R.G., and KAY, S.M.: 'Digital signal processing for sonar', *Proc. IEEE*, 1981, **69**, pp. 1451–1506

20 BENDAT, J.S., and PIERSOL, A.G.: 'Random data: Analysis and measurement procedures', (Wiley, 1971)

21 KOOPMANS, L.H.: 'The spectral analysis of time series' (Academic Press, 1974)

22 BRILLINGER, D.R.: 'Time series: Data analysis and theory' (Holt, Rinehart & Winston, 1975)

23 OPPENHEIM, A.V., and SCHAFER, R.W.: 'Digital signal processing' (Prentice-Hall, 1975)

24 RABINER, L.R., and GOLD, B.: 'Theory and application of digital signal processing' (Prentice-Hall, 1975)

25 BLOOMFIELD, P.: 'Fourier analysis of time series: An introduction' (Wiley, 1976)

26 OTNES, R.K., and ENOCHSON, L.: 'Applied time series analysis' (Wiley, 1978)

27 CHILDERS, D.G.: 'Modern spectrum analysis' (IEEE Press, 1978)

28 HAYKIN, S. (Ed.): 'Nonlinear methods of spectral analysis' (Springer Verlag, 1979)

29 Digital Signal Processing Committee (Eds.): 'Programs for digital signal processing' (IEEE Press, 1979)

30 BENDAT, J.S., and PIERSOL, A.G.: 'Engineering applications of correlation and spectral analysis' (Wiley, 1980)

31 URICK, R.J.: 'Principles of underwater sound for engineers' (McGraw-Hill, 1967)

Analysis of a generalised framework for spectral estimation

Part 2: Reshaping and variance results

Albert H. Nuttall, B.S., M.S., Sc. D.

Indexing term: *Signal processing*

Abstract: The capabilities of lag reshaping for the generalised spectral estimation technique, in order to realise desirable effective windows, is illustrated for several combinations of temporal and lag weightings. Also the variance of the spectral estimate is presented and computed for some windows with very good sidelobe behaviour. It is found that, with proper overlap, if the length of the temporal weighting is somewhat larger than the length of the lag weighting, the variance is at a near minimum relative to any technique which realises the same frequency resolution with the given finite data record length.

1 Introduction

The generalised framework for spectral estimation was presented in Reference 1, and the mean value of the spectral estimate was given in Reference 2 and Part 1 of the paper [3]. The possibilities of doing lag reshaping so that the effective window of the technique has a desirable frequency response were also outlined. We consider here some practical temporal and lag weightings and give quantitative results on the capabilities of lag reshaping.

The variance of the spectral estimate is also of considerable interest. Unless it can be maintained near the minimum possible for the given data record length and desired frequency resolution, the technique will not be a viable alternative to available methods. The generalised spectral technique has been analysed [2] in terms of the mean and variance of the spectral estimate, thereby revealing the fundamental dependence of its performance on the temporal weighting, lag weighting, amount of overlap, number of data pieces, available data record length and desired frequency resolution. To enable a fair trade-off study and comparison between many different special cases of the technique, it is required that all cases achieve the same specified frequency resolution with the given data record length. Without this reasonable constraint, valid conclusions about relative performance of different cases are tentative at best.

There are two fundamental parameters that critically affect the performance of any spectral estimation technique. They are the available data record length T of the stationary random process under investigation and the effective frequency resolution B_e of the technique under consideration. We would like to be able to attain fine resolution (small B_e) with short data lengths and storage (small T); however, stable results (small fluctuations) are achievable only if the product TB_e is much larger than unity. The problem is how to make optimum use of a given *limited* amount of data, in order to realise a specified desired frequency resolution with maximum stability, and to determine what trade-offs are available regarding windowing and weighting at different stages of the spectral analysis procedure. Trade-offs are presented in Reference 2 for the weighted overlapped segment averaging FFT procedure considered here.

A related procedure to the one presented here has been given in References 4 and 5; however, neither incorporate overlapping nor the fundamental trade-offs between the temporal and lag weightings were studied. Furthermore, the only frequency-smoothing case considered was a rectangular boxcar, which severely limits the potential of the technique; some advantages of the generalised technique considered here will become apparent at a later stage. For the time being, we observe that sidelobe control will be realised by a mixture of temporal weighting and lag weighting (frequency smoothing), while stability will be achieved by a combination of segment averaging and frequency smoothing (lag weighting).

Paper 2454F, first received 1st April and in revised form 20th August 1982

The author is with the New London Laboratory, Naval Underwater Systems Center, New London, CT 06320, USA

2 Examples of effective windows

The effective window $W_e(f)$ is given by Reference 3 and eqn. 51 of Reference 2 as the convolution

$$W_e(f) = W_1^2(f) \oplus W_2(f) \tag{1}$$

where $W_1(f)$ and $W_2(f)$ are the temporal window and lag window respectively. Window $W_1(f)$ is real for all f since weighting $w_1(t)$ is even in t. Similarly, $W_2(f)$ is real for symmetric $w_2(\tau)$. Thus $W_e(f)$ is real for all f. The effective frequency resolution B_e is then defined as

$$B_e = \frac{[\int W_e(f)df]^2}{\int W_e^2(f)df} \tag{2}$$

This bandwidth measure is called the statistical bandwidth of $W_e(f)$ on page 265 of Reference 6. Realisation of a specified B_e will force a relationship between the lengths L_1 and $2L_2$ of the temporal weighting $w_1(t)$ and lag weighting $w_2(\tau)$; this is beyond the scope of this paper, but is discussed in considerable detail in pages 16–22 of Reference 2.

We consider first a rectangular temporal weighting $w_1(t)$, for which the correlation

$$\phi_1(\tau) = \int w_1(t)w_1(t-\tau)dt \tag{3}$$

is triangular in τ. Examples are given in Figs. 1 and 2, where we have plotted

$$dB \equiv 10 \log \frac{W_e(f)}{W_e(0)} \quad \text{against} \quad \frac{f}{B_e} \tag{4}$$

the C_1 window is discussed in Reference 7. The curve corresponding to $L_2/L_1 = 0$ (i.e. $L_1 = \infty$) is that for the lag window alone.

The overriding impression of Figs. 1 and 2 is that the

Reprinted with permission from *Proc. Inst. Elec. Eng.*, vol. 130, Part F, pp. 242–245, Apr. 1983.

effective window has poor sidelobe behaviour and decay unless L_2/L_1 is chosen to be very small; i.e. the poor sidelobe behaviour of the temporal window is difficult to suppress, even by choice of good lag windows. It would be desirable to realise the bottom-most figures in these plots, since these latter curves have good sidelobes and decay; a procedure for accomplishing this goal is presented in the following Section.

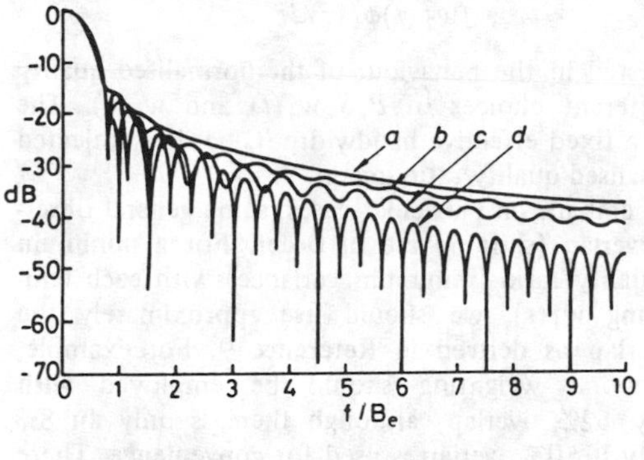

Fig. 1 *Effective window for rectangular temporal weighting and Hanning lag weighting*

a $L_2/L_1 = 1$, 6 dB/octave decay
b $L_2/L_1 = 0.5$, 6 dB/octave decay
c $L_2/L_1 = 0.25$, 6 dB/octave decay
d $L_2/L_1 = 0$, 9 dB/octave decay

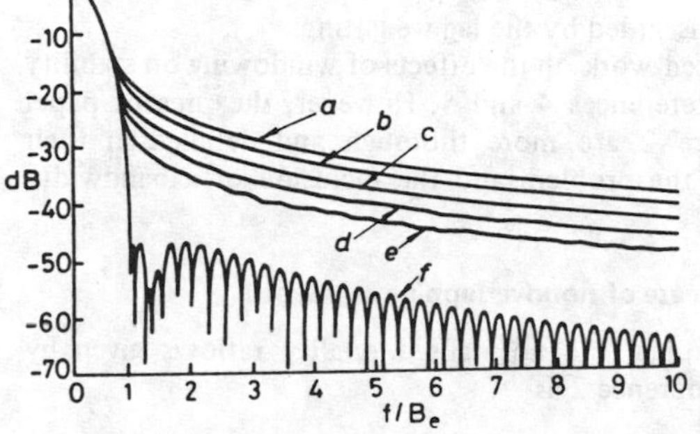

Fig. 2 *Effective window for rectangular temporal weighting and C_1 lag weighting*

a $L_2/L_1 = 1$, 6 dB/octave decay
b $L_2/L_1 = 0.5$, 6 dB/octave decay
c $L_2/L_1 = 0.25$, 6 dB/octave decay
d $L_2/L_1 = 0.125$, 6 dB/octave decay
e $L_2/L_1 = 0.0625$, 6 dB/octave decay
f $L_2/L_1 = 0$, 9 dB/octave decay

The situation is significantly improved when the temporal weighting is tapered. An example for Hanning temporal and lag weightings is given in Fig. 3. The bottom-most curve has an eventual 18 dB/octave decay because $\phi_1(\tau)$ has a discontinuous fifth derivative at $\tau = 0$, which is not compensated by $w_2(\tau)$. [$\phi_1(\tau)$ also has a discontinuous fifth derivative at $\tau = \pm L_1$, but this is converted to a discontinuous seventh derivative for $w_e(\tau)$ by means of $w_2(\tau)$ when $L_2 = L_1$.]

3 Lag reshaping for desired effective windows

The effective weighting of the generalised spectral analysis technique is given by eqn. 48 of Reference 2 as

$$w_e(\tau) \equiv \phi_1(\tau) w_2(\tau) \tag{5}$$

Now suppose that for a given temporal weighting $w_1(t)$, with associated correlation $\phi_1(\tau)$, we choose lag weighting

$$w_2(\tau) = \frac{w_d(\tau)}{\phi_1(\tau)} \quad \text{for } |\tau| \leqslant L_2 \leqslant L_1 \tag{6}$$

where $w_d(\tau)$ is a desired weighting, and $w_d(\tau)$ is zero for $|\tau| > L_2$. Note that $L_2 > L_1$ is disallowed. Then substitution of eqn. 6 in eqn. 5 yields

$$\left.\begin{aligned} w_e(\tau) &= w_d(\tau) \\ W_e(f) &= W_d(f) \end{aligned}\right\} \tag{7}$$

i.e. the effective weighting and window are equal to the desired behaviour. In so far as the mean of the spectral estimate is concerned, we have 'undone' the effects of bad sidelobes in the temporal window by reshaping according to lag weighting $w_2(\tau)$ in eqn. 6. (The effect on the variance of the second-stage spectral estimate $\hat{G}_2(f)$ is considered below.)

To see how much can be accomplished by this approach, some attainable effective windows that can be realised via lag reshaping, for continuous rectangular temporal weighting, are given in Figs. 4 and 5 for the largest possible value of L_2, namely $L_2 = L_1$. Superposed on the window $W_1^2(f)$ for rectangular temporal weighting are the effective windows for candidate lag reshapings, for $L_2 = L_1$. These are the narrowest possible effective windows for a *given* L_1. The first window in Fig. 4 corresponds to an effective Hanning weighting. The peak sidelobe is only reduced from -13.3

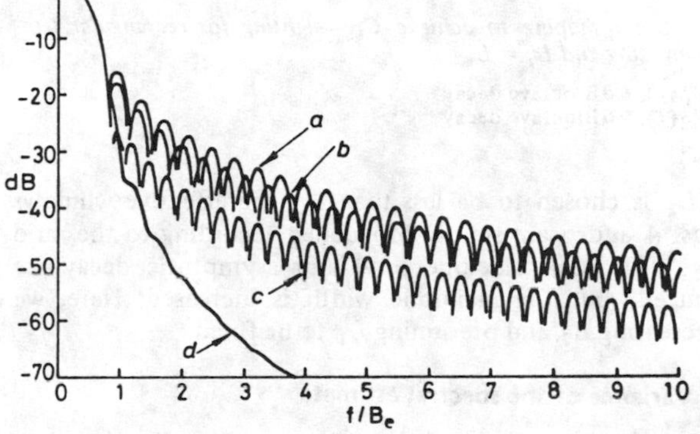

Fig. 3 *Effective window for Hanning temporal weighting and Hanning lag weighting*

a $L_2/L_1 = 0$, 9 dB/octave decay
b $L_2/L_1 = 0.25$, 9 dB/octave decay
c $L_2/L_1 = 0.5$, 9 dB/octave decay
d $L_2/L_1 = 1$, 18 dB/octave decay

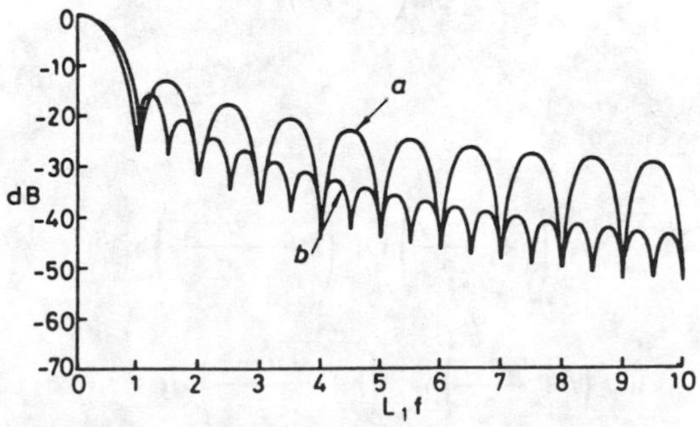

Fig. 4 *Lag reshaping to achieve Hanning weighting for rectangular temporal weighting and $L_2 = L_1$*

a $W_1^2(f)$, 6 dB/octave decay
b $W_d(f)$, 9 dB/octave decay

to -15.7 dB, and the asymptotic decay is improved from 6 to 9 dB/octave. The mainlobe width is only slightly broadened. Much greater improvements in sidelobe behaviour are possible with other lag weightings, as illustrated in Fig. 5. Other illustrations are available in Reference 2.

Figs. 4 and 5 illustrate how advantageous the reshaping technique can be in terms of peak sidelobe and asymptotic decay, although the mainlobe width is significantly increased. In fact, the peak sidelobe at $|f| = 1.5/L_1$ for the rectangular window is really not suppressed so much as smeared out; however, the other peaks of $W_1^2(f)$ for $|f| > 2/L_1$ are indeed significantly reduced. Thus reduction of leakage via lag reshaping is a very effective method, *provided* that we accept the nearest sidelobe of the temporal window; this conclusion is in contrast to page 57 of Reference 8. These general conclusions on lag reshaping hold also for temporal weightings other than rectangular, although the exact degree of improvement will be different.

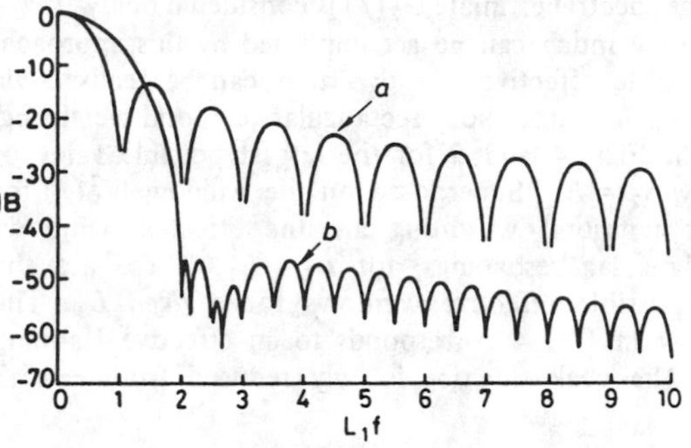

Fig. 5 *Lag reshaping to achieve C_1 weighting for rectangular temporal weighting and $L_2 = L_1$*

a $W_1^2(f)$, 6 dB/octave decay
b $W_d(f)$, 9 dB/octave decay

If L_2 is chosen to be less than L_1, the effective windows in Figs. 4 and 5 are simply broadened according to the ratio L_1/L_2. The peak sidelobe level and asymptotic decay are unchanged, but the mainlobe width is increased. Here, we are decreasing L_2 and presuming L_1 to be fixed.

4 Variance of the spectral estimate

The variance of the second-stage spectral estimate $\hat{G}_2(f)$ is developed in appendix D of Reference 2 for a Gaussian data process with arbitrary spectrum. When the true spectrum G varies slowly in a bandwidth B_e about a frequency f of interest, the variance result simplifies to

$$\text{var}\,\{\hat{G}_2(f)\} = G^2(f)\frac{1}{P}\sum_{p=1-P}^{P-1}\left(1-\frac{|p|}{P}\right)\int w_2^2(\tau)\phi_3(\tau, pS)d\tau \tag{8}$$

where

$$\phi_3(\tau, u) \equiv \int w_1\left(t+\frac{u-\tau}{2}\right) w_1\left(t+\frac{-u-\tau}{2}\right) \times w_1\left(t+\frac{u+\tau}{2}\right) w_1\left(t+\frac{-u+\tau}{2}\right) dt$$

is the third-order correlation of temporal weighting $w_1(t)$.

We adopt as a measure of the stability of the spectral estimator the quality ratio Q, equal to the variance of $\hat{G}_2$ divided by the square of the mean of $\hat{G}_2$. It can be shown (see pages 3–5 of Reference 2) that the minimum possible value of Q is $(TB_e)^{-1}$. Therefore we consider the normalised quality ratio $\text{NQR} = QTB_e$, which can be developed as the form

$$\text{NQR} = T\frac{\frac{1}{P}\sum_{p=1-P}^{P-1}\left(1-\frac{|p|}{P}\right)\int w_2^2(\tau)\phi_3(\tau, pS)d\tau}{\int w_2^2(\tau)\phi_1^2(\tau)d\tau} \tag{9}$$

We are interested in the behaviour of the normalised quality ratio for different choices of $P, S, w_1(t)$ and $w_2(\tau)$. The constraint of a fixed effective bandwidth B_e has been injected into the normalised quality ratio in eqn. 9.

Before we embark on particular cases, some general observations on overlap (shift S) are in order. For a minimum normalised quality ratio (minimum variance) with each temporal weighting $w_1(t)$, we should use approximately the optimum overlap as derived in Reference 9. For example, Hanning temporal weighting should be employed with approximately 62% overlap, although there is only an 8% loss in stability if 50% overlap is used for convenience. There is no point in considering excessive or inadequate overlap, since this leads to excessive computational effort or more variance, respectively. Inadequate temporal overlap cannot be made up, in terms of variance reduction, by any amount of quadratic frequency smoothing. This can be seen by observing that poor first-stage correlation estimates $\hat{R}_1(\tau)$ are merely multiplied by lag weighting $w_2(\tau)$, and are not improved statistically in any way for $|\tau| < L_2$; those estimates for $|\tau| > L_2$ are discarded by the lag weighting.

Some related work on the effects of windowing on stability is given in References 4 and 5. However, the present paper and Reference 2 are more thorough and detailed in their treatment of the problem and the inclusion of a bandwidth constraint.

5 Special case of nonoverlapping segments

For lag reshaping, the normalised quality ratio is given by page 48 of Reference 2 as

$$\text{NQR}_P \text{ (lag reshaping)} = \frac{L_1 \int w_d^2(\tau)\frac{\phi_2(\tau)}{\phi_1^2(\tau)}d\tau}{\int w_d^2(\tau)d\tau}$$

$$\text{for } L_2 \leqslant L_1 \tag{10}$$

where $\phi_2(\tau)$ is the autocorrelation of $w_1^2(t)$. The division by $\phi_1(\tau)$ in reshaping (eqn. 6) increases the variance and the normalised quality ratio above the minimum value of 1.

Eqn. 10 is plotted in Fig. 6 for rectangular temporal weighting and for desired effective weightings of Hanning, C_1, C_3 and C_5. Weightings C_1, C_3 and C_5 are defined in terms of continuous first, third and fifth derivatives in Figs. 10–12 of Reference 7. Note that the abscissa is limited to $L_2/L_1 \leqslant 1$. As expected, the normalised quality ratio tends to 1 as L_2/L_1 tends to zero; i.e. we can do lag reshaping for good sidelobe behaviour and lose little in terms in stability, provided that L_1 is chosen to be sufficiently larger than L_2. As an example, for desired effective weighting C_1, if we take $L_1 = 2L_2$, the increase in variance over the ideal value is only 9%. Thus lag reshaping is an attractive procedure for spectral estimation; recall that L_2 is set by the specified B_e and the shape of $w_d(\tau)$.

6 Summary

The possibilities and performance of a generalised spectral

estimation technique employing temporal and lag weighting have been investigated in terms of the mean and variance of the spectral estimate. The only assumption required about the process under analysis, in so far as the mean of the spectral estimate is concerned, was that it be second-order stationary over the observation interval. We then were able to extract a simple expression for the effective window involving the temporal and lag windows.

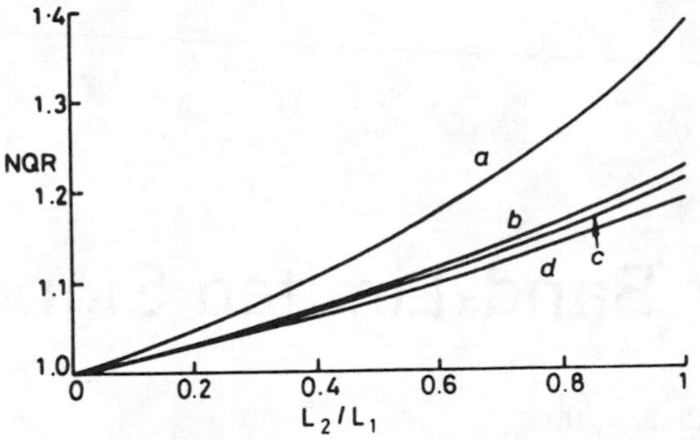

Fig. 6 *Normalised quality ratio for rectangular temporal weighting and no overlap, with reshaping to desired lag weighting* $w_d(\tau)$

a $w_d(\tau)$ = Hanning
b C_1
c C_3
d C_5

The possibility of doing lag reshaping to achieve desirable effective windows was considered in detail and found to be reasonable for a wide variety of windows with good sidelobe behaviour and decay rates. In particular, if rectangular temporal weighting is employed, its inherent poor sidelobe structure can be corrected via proper lag weighting, in so far as the effective window is concerned. Strictly speaking, the closest sidelobe cannot be eliminated; however, all the other sidelobes can be suppressed.

The effect of temporal and lag weighting on the variance of the spectral estimate was evaluated and compared with the ideal value for large B_eT. For rectangular temporal weighting, it was found that small values of L_2/L_1 and no overlap lead to values of the normalised quality ratio virtually equal to the best attainable by any spectral analysis technique. The comparison is made under the constraint that the effective frequency resolution B_e is maintained the same for all techniques under consideration. On the other hand, if Hanning temporal weighting is employed, overlapping must be used for maximum variance reduction, and the length ratio L_2/L_1 ought to be of the order of unity (see pages 55–57 of Reference 2).

Since Hanning temporal weighting requires multiplication of each and every data segment (P pieces) and significant overlap (~ 50%), whereas rectangular temporal weighting requires no multiplication and no overlap, the latter approach is a strong candidate for spectral estimation, particularly since excellent effective windows (low sidelobe and rapid decay) and virtually ideal variance reduction can be achieved by proper lag weighting and choice of ratio L_2/L_1.

7 References

1 NUTTALL, A.H., and CARTER, G.C.: 'A generalized framework for power spectral estimation', *IEEE Trans.*, 1980, **ASSP-28**, pp. 334–335

2 NUTTALL, A.H.: 'Spectral analysis via quadratic frequency-smoothing of Fourier-transformed overlapped weighted data segments'. NUSC technical report 6459, New London, Connecticut, USA, June 1981

3 CARTER, G.C., and NUTTALL, A.H.: 'Analysis of a generalised framework for spectral estimation. Part 1: The technique and its mean value', *IEE Proc. F, Commun., Radar & Signal Process.*, 1983, **130**, (3), pp. 239–241

4 SLOANE, E.A.: 'Comparison of linearly and quadratically modified spectral estimates of Gaussian signals', *IEEE Trans.*, 1969, **AU-17**, pp. 133–137

5 ROBERTSON, G.H.: 'Influence of data window shape on detectability of small CW signals in white noise', *J. Acoust. Soc. Am.*, 1980, **67**, pp. 1274–1276

6 BENDAT, J.S., and PIERSOL, A.G.: 'Measurement and analysis of random data' (J. Wiley, 1966)

7 NUTTALL, A.H.: 'Some windows with very good sidelobe behaviour', *IEEE Trans.*, 1981, **ASSP-29**, pp. 84–91

8 BINGHAM, C., GODFREY, M.D., and TUKEY, J.W.: 'Modern techniques of power spectrum estimation', *ibid.*, 1967, **AU-15**, pp. 56–66

9 NUTTALL, A.H.: 'Spectral estimation by means of overlapped fast Fourier transform processing of windowed data'. NUSC report 4169, New London, Connecticut, USA, Oct. 1971 (See also Supplement: 'Estimation of cross-spectra via overlapped fast Fourier transform processing'. NUSC technical report 4169-S, July 1975)

An Extrapolation Procedure for Band-Limited Signals

JAMES A. CADZOW, SENIOR MEMBER, IEEE

***Abstract*—In this paper, the task of extrapolating a time-truncated version of a band-limited signal shall be considered. It will be shown that the basic extrapolation operation is feasible for only a particular subset of the class of band-limited signals (i.e., the operation is well-posed mathematically). An efficient algorithmic method for achieving the desired extrapolation on this subset is then presented. This algorithm is structured so that all necessary signal manipulations involve signals which are everywhere zero except possibly on a finite "observation time" set. As a consequence, its implementation is straightforward and can be carried out in real time. This is to be contrasted with many existing extrapolation algorithms which theoretically involve operations on signals that are nonzero for almost all values of time. Their numerical implementation thereby necessitates an error producing time-truncation and a resultant deleterious effect on the corresponding extrapolation.**

Using straightforward algebraic operations, a convenient one-step extrapolation procedure is next developed. This is noteworthy in that this procedure thereby enables one to effectively circumvent any potentially slow convergence rate difficulties which typically characterize extrapolation algorithms. The effectiveness of this one-step procedure is demonstrated by means of two examples.

I. INTRODUCTION

A classical tool for characterizing the intrinsic features of a signal x whose domain is the infinite-extent time interval $(-\infty, \infty)$ is that of Fourier analysis. In particular, such an analysis is begun by first evaluating the Fourier transform integral

$$X(\omega) = \int_{-\infty}^{\infty} x(t)\, e^{-j\omega t}\, dt \tag{1}$$

Manuscript received August 25, 1977; revised March 1, 1978 and August 11, 1978. This work was supported in part by the Signal Processing Section, Surveillance Technology Branch, Rome Air Development Center through the Post Doctoral Program under Contract F30602-75-0018.

The author is with the Department of Electrical Engineering, Virginia Polytechnic Institute and State University, Blacksburg, VA 24061.

in which ω is taken to be a real frequency variable with values in $(-\infty, \infty)$. The Fourier spectrum associated with signal x is then defined to be the magnitude quantity $|X(\omega)|$, and its behavior as a function of ω often yields useful information relative to the signal x. Conceptually, the evaluation of this transform integral is straightforward and its utility is well known.

In any real-world application, however, it must be realized that only a finite-time observation of the signal to be transformed is ever available. Namely, the values of $x(t)$ are known only for those t contained in some "observation-time" set Λ which has a nonzero but finite measure. This observation set is a subset of the infinite-extent time interval and will typically be a simple finite interval such as (t_0, t_1). Clearly, since only the partial behavior of signal x as specified by

$$x(t) \quad \text{for } t \in \Lambda \tag{2}$$

is given, one is unable to evaluate the Fourier transform integral (1). In such practical situations, it is then necessary to apply suitably "good" estimates of the Fourier transform based on this partial information. A variety of methods have been presented for achieving satisfactory estimates. In general, these methods assume Λ to be a finite interval and usually fall into one of the following three categories.

1) Periodogram: The known finite-time segment of the signal $x(t)$ is multiplied by a window signal $w(t)$ and the Fourier transform of the finite-time product $w(t)\,x(t)$ is determined (e.g., see [1] and [10]).

2) Predictive Filtering: The coefficients of a fixed-order linear digital filter are determined which will yield optimum one-step prediction of a sampled version of the finite-time segment signal. The resultant filter is then used to estimate the signal's spectrum. This procedure includes the essentially equivalent maximum entropy and autoregressive methods (e.g., see [2] and [12]).

Reprinted from *IEEE Trans. Acoust., Speech, Signal Processing*, vol. ASSP-27, pp. 4–12, Feb. 1979.

3) Extrapolation: In this approach, the finite-time segment of the signal is extended using an appropriate extrapolation rule (e.g., see [6] and [11]).

In this paper, we shall be concerned with developing a systematic procedure for extrapolating a finite-time segment of a band-limited signal. The signal x is said to be band limited if its energy content is finite and its Fourier transform is identically zero outside some frequency set Ω which has nonzero finite measure; that is,

$$X(\omega) \equiv 0 \quad \text{for } \omega \notin \Omega. \tag{3}$$

Typically, the set Ω will be a simple interval of the ω axis (e.g., low-pass signals) or a union of a finite number of intervals (e.g., bandpass signals).

Recently, an algorithmic procedure for extrapolating signals of a low-pass band-limited nature has been presented (see [6] and [11]). It consists essentially of sequentially applying a set of signal operations to generate an approximation sequence with convergence to the desired extrapolation being guaranteed in theory. Unfortunately, when this procedure is implemented numerically, it is necessary to truncate the domain of the approximation signals that are in fact infinite in extent. As such, the numerical algorithm need not converge to the desired solution although published empirical evidence is promising. This algorithm is in fact a special case of the noteworthy general alternating orthogonal projection algorithm of Youla [15].

In what is to follow, we shall treat the more general problem of extrapolating the observed signal (2) which is known to be band limited in the sense of expression (3). An iterative method for achieving the desired extrapolation will be presented which has guaranteed convergence. Furthermore, this algorithm may be numerically approximated without introducing the undesirable domain truncations as previously mentioned, thereby offering a potentially better behaved algorithm. Most importantly, an efficient procedure for obtaining the desired extrapolation analytically in a one-step fashion is then offered. This one-step procedure alleviates the slow convergence rates which characterize extrapolation algorithms such as [6] and [11].

II. Extrapolation Problem

In what is to follow, we shall be interested in the set of signals defined on the infinite-extent time axis which possesses finite energy. The signal x is said to contain finite energy if the integral expression

$$\int_{-\infty}^{\infty} x(t)\, x^*(t)\, dt$$

is finite where $x^*(t)$ denotes the complex conjugate of $x(t)$ and the integration is carried out in the sense of Lebesgue. This set of signals in conjunction with the standard inner product

$$\langle x_1, x_2 \rangle = \int_{-\infty}^{\infty} x_1(t)\, x_2^*(t)\, dt \tag{4}$$

in fact constitutes a Hilbert space which has been classically designated as L_2.

If a signal x is contained in Hilbert space L_2, it is well known that its associated Fourier transform as defined by expression (1) exists. Furthermore, the signal x can be recovered from its Fourier transform by means of the inverse Fourier transform integral

$$x(t) = \frac{1}{2\pi} \int_{-\infty}^{\infty} X(\omega)\, e^{j\omega t}\, d\omega. \tag{5}$$

Much of signal theory is dependent on decomposing a given signal into components whose Fourier transforms occupy different portions of the frequency spectrum. To demonstrate how this decomposition is effected, let us consider the aforementioned band-limited frequency set Ω, which is taken to have a nonzero, but finite measure.[1] This set typically identifies that frequency spectrum component of x which is of interest in a given application. With this in mind, we now decompose the inverse Fourier transform integral as follows:

$$\begin{aligned} x(t) &= \frac{1}{2\pi} \int_{\omega \in \Omega} X(\omega)\, e^{j\omega t}\, d\omega + \frac{1}{2\pi} \int_{\omega \notin \Omega} X(\omega)\, e^{j\omega t}\, d\omega \\ &= x_1(t) + x_2(t) \end{aligned} \tag{6}$$

in which the first integral has been designated as signal $x_1(t)$ and the second as signal $x_2(t)$.

Using this signal decomposition, it is apparent that the Fourier transforms of the component signals $x_1(t)$ and $x_2(t)$ are identically zero for all $\omega \notin \Omega$ and $\omega \in \Omega$, respectively. This in turn implies that the component signals x_1 and x_2 are "orthogonal" in the sense that $\langle x_1, x_2 \rangle = 0$. To establish this property, let us introduce the "ideal band-limited" filter transfer function which corresponds to set Ω as defined by

$$H(\omega) = \begin{cases} 1 & \text{for } \omega \in \Omega \\ 0 & \text{for } \omega \notin \Omega. \end{cases} \tag{7}$$

Upon taking the Fourier transform of relationship (6) and using this definition of $H(\omega)$, it is seen that

$$\begin{aligned} X(\omega) &= X_1(\omega) + X_2(\omega) \\ &= H(\omega)\, X(\omega) + [1 - H(\omega)]\, X(\omega). \end{aligned} \tag{8}$$

Clearly, the transform product $X_1(\omega)\, X_2^*(\omega)$ is identically zero due to the structure of $H(\omega)$. This fact in conjunction with Parseval's identity, which states that

$$\int_{-\infty}^{\infty} X_1(\omega)\, X_2^*(\omega)\, d\omega = 2\pi \int_{-\infty}^{\infty} x_1(t)\, x_2^*(t)\, dt$$

[1] An example that will be subsequently used is the low-pass frequency set as specified by $\Omega = \{\omega: -\sigma < \omega < \sigma\}$ where σ is a positive constant.

or in the equivalent inner product representation

$$\langle X_1, X_2 \rangle = 2\pi \langle x_1, x_2 \rangle \tag{9}$$

establishes the orthogonality of the component signals x_1 and x_2. This orthogonality property is of primary importance in many signal analysis studies.

It has thereby been established that every signal contained in Hilbert space L_2 can be "uniquely" decomposed into the sum of two signals, one of which has a Fourier transform that is identically zero for ω outside a given set Ω and the other which has a Fourier transform which is identically zero for ω inside the set Ω. Moreover, the constituent signals x_1 and x_2 which arise by this process are always orthogonal. With this in mind, it is apparent that the underlying Hilbert space L_2 has been thereby decomposed into the specific direct sum (e.g., see [8])

$$L_2 = B_\Omega \oplus B_\Omega^\perp \tag{10}$$

where $B_\Omega^\perp$ denotes that subspace of L_2 whose elements are orthogonal to the band-limited subspace

$$B_\Omega = \{x \in L_2 : X(\omega) \equiv 0 \quad \text{for } \omega \notin \Omega\}. \tag{11}$$

This direct sum decomposition is noteworthy in that many signal processing applications are characterized by constituent signals which will be contained (or be approximately contained) in a band-limited subspace as given by expression (11). In such situations, the following extrapolation problem is of interest.

Extrapolation Problem: Given the time-truncated version of the signal $x \in B_\Omega$ as specified by

$$x(t) \quad \text{for } t \in \Lambda$$

determine the behavior of $x(t)$ outside the observation time set Λ.

In this problem formulation, the two characterizing sets Λ and Ω are each taken to have a finite, but nonzero measure. A solution to this extrapolation task is found by introducing two projection operators.

III. Fundamental Orthogonal Projection Operators

In order to give the extrapolation problem a desirable algebraic structure, it will be beneficial to consider orthogonal projection operators. The linear operator P is said to be a projection operator on the vector space X if it is idempotent, that is, $Px^2 = Px$ for all $x \in X$. Furthermore, this projection operator is said to be orthogonal if its range and nullspace are orthogonal [i.e., $\langle x, y \rangle = 0$ for all $x \in N(P)$ and $y \in R(P)$]. With this basic definition, we shall introduce two orthogonal projection operators which will play a central role in our extrapolation investigation.

Time-Truncation Operator: T

The fundamental time-truncation operator T is motivated by the aforementioned observation-time set Λ. In particular, if this operator is applied to the pre-image (input) signal x, there is generated the image (output) signal y as denoted by

$$y = Tx \tag{12a}$$

in which the value of y at time t is specified by

$$y(t) = \begin{cases} x(t) & \text{for } t \in \Lambda \\ 0 & \text{for } t \notin \Lambda. \end{cases} \tag{12b}$$

This time truncation operator is clearly linear and its range and nullspace are formally given by

$$R(T) = \{x \in L_2 : x(t) \equiv 0 \quad \text{for } t \notin \Lambda\}$$

and

$$N(T) = \{x \in L_2 : x(t) \equiv 0 \quad \text{for } t \in \Lambda\}.$$

Upon examination of relationship (12), it is clear that $T^2 x = Tx$ for all $x \in L_2$, which establishes that T is a projection operator. Furthermore, since the inner product between elements of subspaces $R(T)$ and $N(T)$ is obviously zero, it follows that these subspaces are orthogonal. We have therefore established that the time-truncation operator is an orthogonal projection operator and, consequently, the underlying signal space L_2 can be decomposed by the direct sum

$$L_2 = R(T) \oplus R(T)^\perp \tag{13}$$

where $R(T)^\perp = N(T)$ denotes the subspace orthogonal to $R(T)$. This direct sum decomposition implies that any signal $x \in L_2$ can be uniquely expressed as $x = x_1 + x_2$ in which the individual components $x_1 \in R(T)$ and $x_2 \in R(T)^\perp$. Moreover, these components are generated by the orthogonal projection operations $x_1 = Tx$ and $x_2 = (I - T)x$, respectively, in which I denotes the identity operator.

Band-Limited Operator: B

The second orthogonal projection operator to be considered is motivated by the Fourier transform decomposition of a general signal in L_2 as called out by relationship (8). In particular, the individual components of the signal $X(\omega)$ as specified by $H(\omega)\,X(\omega)$ and $[1 - H(\omega)]\,X(\omega)$ were found to be contained in closed subspaces B_Ω and $B_\Omega^\perp$, respectively. In the time domain, the component $x_1 \in B_\Omega$ is given by

$$x_1(t) = \int_{-\infty}^{\infty} h(t-\tau)\,x(\tau)\,d\tau \tag{14}$$

where the unit-impulse response characterizing this linear convolution operation is the inverse Fourier transform of the ideal band-limited filter transfer function (7) as formally specified by

$$h(t) = \frac{1}{2\pi} \int_{\omega \in \Omega} e^{j\omega t}\,d\omega. \tag{15}$$

It will be desirable to express relationship (14) in an operator notation in line with our objective of providing an algebraic structure to the extrapolation problem. With this in mind, the following natural correspondence

$$x_1 = Bx \tag{16}$$

is made where B denotes the ideal band-limited filter operation as specified by convolution integral relationship (14). To establish that B is a projection operator, we use the fact that $H^2(\omega) = H(\omega)$ for all ω to conclude that $B^2 x = Bx$ for all

$x \in L_2$. Furthermore, the range and nullspace of this projection operator as given by

$$R(B) = \{x \in L_2: X(\omega) \equiv 0 \quad \text{for} \quad \omega \notin \Omega\}$$

$$N(B) = \{x \in L_2: X(\omega) \equiv 0 \quad \text{for} \quad \omega \in \Omega\}$$

are orthogonal as can be shown by using Parseval's identity (9). Thus, the ideal band-limited operator B is an orthogonal projection operator whose range space is identical to the band-limited subspace B_Ω. This is not coincidental since operator B was purposely defined so as to achieve this objective. It is interesting to note that this ideal band-limited operator provides yet another direct sum decomposition of the underlying signal space as given by

$$L_2 = R(B) \oplus R(B)^\perp$$

where we have $R(B)^\perp = N(B)$. Thus, any signal $x \in L_2$ can therefore be uniquely decomposed as the sum of a signal in $R(B)$ and $R(B)^\perp$. This decomposition is achieved by the basic operation $x = Bx + [I - B]\, x$ in which $Bx \in R(B)$ and $[I - B]\, x \in R(B)^\perp$.

IV. Extrapolation Algorithm

An algorithmic procedure shall now be presented which has proven effective in achieving the operation of signal extrapolation. In particular, given the time-truncated version of a signal x contained in subspace B_Ω as operationally specified by

$$y = Tx \tag{17}$$

an iterative method for generating the signal x given y [i.e., inverting relationship (17)] will be developed. This algorithm's development stems from the basic signal identity

$$x = (I - BT)\, x + By \tag{18}$$

which is seen to hold for all signals x and y related by expression (17).[2] In what is to follow, we shall not initially assume that $x \in B_\Omega$ although this will be an eventual restriction.

The extrapolation problem can then be reformulated as that of finding the "fixed point" x which will satisfy operator relationship (18) given only its time-truncated mapping y (i.e., given Tx). With this in mind, we shall use the classical principle of successive approximations for finding this fixed point. This results in the iterative relationship

$$x_n = (I - BT)\, x_{n-1} + B(Tx) \qquad n = 1, 2, 3, \cdots \tag{19}$$

in which x_n designates the nth approximation to x, x_0 is the initial approximation, and we have suggestively replaced the time-truncated signal y by its equivalent representation Tx. To determine under what conditions this algorithm will generate a sequence $\{x_n\}$ which converges to the desired result x, it will be beneficial to investigate the error signal. At the nth iteration, relationship (19) indicates that this error signal is given by

$$x_n - x = [I - BT]\, (x_{n-1} - x) \qquad n = 1, 2, 3, \cdots. \tag{20}$$

[2]More generally, the signal $Tx - y$ need only lie within the nullspace of operator B. However, due to the nature of operator B and the fact that $y \in R(T)$, this infers that $y = Tx$.

If this error relationship is iterated, the following expression is readily obtained:

$$x_n - x = [I - BT]^n\, (x_0 - x) \qquad n = 1, 2, 3, \cdots \tag{21}$$

from which the following lemma is an obvious consequence.

Lemma 1: The sequence $\{x_n\}$ as generated by algorithm (19) will converge to x if and only if

$$[I - BT]^n\, (x_0 - x) \to \theta \tag{22}$$

where θ denotes the zero signal.

We shall now concern ourselves with establishing a condition on the initial approximation x_0 which will ensure the satisfaction of convergence requirement (22). Specifically, let x_0 be selected in a manner so that the initial error signal $x_0 - x$ is contained in B_Ω. This will be the case if either 1) both x and x_0 are in B_Ω, or 2) the out-of-band components of x and x_0 are equal. Whichever the case may be, if $x_0 - x \in B_\Omega$, it follows from relationship (20) that the entire error signal sequence $\{x_n - x\}$ will also be contained in B_Ω. This furthermore implies that relationship (20) may be equivalently expressed as

$$x_n - x = B\,[I - T]\, (x_{n-1} - x) \qquad \text{if} \quad x_0 - x \in B_\Omega \tag{23}$$

since the generally different operators $I - BT$ and $B\,[I - T]$ are in fact identical on the restricted domain B_Ω.

An examination of relationship (23) reveals that the energy in the nth error signal as measured by the standard inner product $E_n = \langle x_n - x, x_n - x\rangle$ is always less than or equal to the energy in the $(n-1)$st error signal (i.e., $E_n \leqslant E_{n-1}$ for $n = 1, 2, 3, \cdots$). This is a direct consequence of the fact that each of the operators $I - T$ and B composing relationship (23) never increase, and generally decrease the energy of the signal upon which they operate. Thus, when the initial approximation x_0 is selected so that $x_0 - x \in B_\Omega$, algorithm (19) will always generate an approximation signal sequence $\{x_n\}$ for which the approximation error's energy decreases, or at worst stays the same, at each iteration. There is then reason to believe that the error signal sequence $\{x_n - x\}$ will be such that $E_n \to 0$, thereby establishing the convergence of the sequence $\{x_n\}$ to x.

To establish that the approximation sequence $\{x_n\}$ does in fact converge to x whenever $x_0 - x \in B_\Omega$, we next appeal to Von Neumann's alternating projection theorem (see [9, p. 55]). This theorem, when applied to the sequence of composite orthogonal projection operators $I - T, B\,[I - T], [I - T]\, B\,[I - T], B\,[I - T]\, B\,[I - T], \cdots$, yields the following.

Theorem 1: The sequence of approximation error signals $\{x_n - x\}$ converges to zero if and only if $x_0 - x \in B_\Omega$.

This is a consequence of the fact that the sequences $\{[I - BT]^n\, (x_0 - x)\}$ and $\{[B(I - T)^n]\, (x_0 - x)\}$ are identical whenever $x_0 - x \in B_\Omega$. Von Neumann's theorem applied to this latter sequence indicates that it will converge to the signal determined by orthogonally projecting $x_0 - x$ onto the closed subspace $R(B) \cap R(I - T)$. Since a signal cannot be simultaneously band limited and time limited, this closed subspace contains only the zero signal, thereby establishing condition (22). Conversely, if $x_0 - x \notin B_\Omega$, then the orthogonal projection will be nonzero implying that $x_n \not\to x$.

It has been proven, therefore, that algorithm (19) will yield the desired extrapolation (i.e., $x_n \to x$) if and only if $x_0 - x \in$

B_Ω. One should appreciate the fact that x need not be contained in B_Ω for $x_n \to x$. If we somehow knew the out-of-band component of x, then by setting the initial approximation x_0 equal to that component, the algorithm would generate an approximation sequence which would converge to x irrespective of the fact that x is not contained in B_Ω. Unfortunately, the determination of this out-of-band component is generally not possible and we must suitably restrict the class of signals to be extrapolated. More specifically, when the signal class of interest is taken to be that one for which the out-of-band component is always zero (i.e., the subspace B_Ω), then the algorithm will always converge provided $x_0 \in B_\Omega$. This observation will now be formally stated in lemma format.

Lemma 2: Extrapolation algorithm (19) will generate an approximation signal sequence $\{x_n\}$ which converges to the desired extrapolation provided that the signal x and its initial approximation x_0 are each contained in subspace B_Ω.

V. Implementation of Algorithm

The proposed extrapolation algorithm (19) has been found to have guaranteed convergence (i.e., $x_n \to x$) provided that the initial approximation x_0 is such as to ensure that $x - x_0 \in B_\Omega$. In the time domain, this algorithm has the following equivalent integral representation:

$$x_n(t) = x_{n-1}(t) + \int_{\tau \in \Lambda} h(t-\tau)\,[x(\tau) - x_{n-1}(\tau)]\,d\tau \tag{24}$$

where the fixed time t may take on any real value whatsoever. Since our basic objective is that of evaluating $x(t^*)$ for a fixed "extrapolation-time instant" $t^* \notin \Lambda$ (i.e., performing an extrapolation), an efficient procedure for accomplishing this task using algorithm (24) shall now be outlined. This is begun by setting $t = t^*$ in relationship (24) and observing that the nth extrapolation approximation $x_n(t^*)$ is dependent on the previous extrapolation $x_{n-1}(t^*)$ as well as on the behavior of $x_{n-1}(t)$ for all $t \in \Lambda$. Based on this observation, it will be advisable to decompose the algorithm's implementation into a two-step procedure whereby at the nth iteration:

1) The signal $x_n(t)$ for all $t \in \Lambda$ is evaluated using algorithm (24).

2) The nth extrapolation approximation $x_n(t^*)$ is evaluated by using algorithm (24) with t set equal to $t^* \notin \Lambda$.

The desirability of this approach is that the results of step 1) are totally independent of the choice t^*. This has obvious computational savings appeal when we are seeking to extrapolate a signal for more than one value of t^*, since the most arduous task at the nth iteration is step 1).

An evaluation of algorithm (24) will generally entail a numerical approximation of the constituent integral. Any variety of different numerical integration procedures can be employed, such as rectangular or trapezoidal approximation (e.g., see [4]). The explicit numerical scheme that results will be obviously dependent on the observation-time set Λ and the band-limit set Ω. This numerical approximation must be carefully carried out in order to reduce any numerical error build up during the course of the iterations.

It is important to note that the proposed algorithm (24) requires the evaluation of integrals only over the finite domain Λ. This is to be contrasted with other extrapolation algorithms which, in theory, require integral evaluations over the infinite domain $(-\infty, \infty)$ (e.g., see [11]). This practical difficulty is there avoided by truncating the range of integration. This range of integration truncation, however, may have a deleterious effect on the resultant algorithm's behavior.

VI. One-Step Extrapolation Procedure

Although the extrapolation algorithm (19) has guaranteed convergence, its rate of convergence can be exceedingly slow in most practical situations. With this factor in mind, a tractable analytical procedure for obtaining the desired extrapolation in an efficient manner will be now presented. This is begun by first reformulating algorithmic relationship (19) as

$$x_n = x_{n-1} + B[T(x - x_{n-1})]. \tag{25}$$

Furthermore, we shall take the initial approximation x_0 to be zero (i.e., $x_0 = \theta$). If one iterates on relationship (25) commencing with $n = 1$, it is apparent that the general element of the extrapolation sequence is given by

$$x_n = B\left[\sum_{k=0}^{n-1} T(x - x_k)\right] \qquad n = 1, 2, 3, \cdots.$$

Next, we incorporate relationship (21) into this expression and use the fact that $x_0 = \theta$ to yield

$$x_n = B\left[\sum_{k=0}^{n-1} T[I - BT]^k x\right].$$

We now make use of the readily established operator identity $T[I - BT]^k = [I - TB]^k T$ to obtain

$$x_n = B\left[\sum_{k=0}^{n-1} [I - TB]^k Tx\right]. \tag{26}$$

This summation expression can be given a convenient closed form by using the readily established operator identity

$$\sum_{k=0}^{n-1} [I - TB]^k TBz = z - [I - TB]^n z \tag{27}$$

which holds for all signals contained in Hilbert space L_2. Upon comparison of expressions (26) and (27), it is apparent that if a signal $z = R(T)$ can be found which satisfies the operator equation

$$TBz = Tx, \tag{28}$$

then expression (26) simplifies to

$$x_n = Bz - B[I - TB]^n z. \tag{29}$$

If the observed truncated signal Tx was generated by a band-limited signal x (i.e., $x = Bx$), it was shown in Section IV that the proposed algorithm generates an approximation sequence which converges to x. With this in mind, let us now take the limit of expression (29) as n approaches infinity to obtain

$$x = Bz. \tag{30}$$

This result is a consequence of the fact that the sequence $\{B[I - TB]^n z\}$ must converge to an element of $N(T)$ if expression (29) is to converge to a solution of relationship (28) as hypothesized. However, the sequence $\{B[I - TB]^n z\}$ is totally contained in B_Ω and it therefore must converge to the zero signal which is the only signal shared in common by B_Ω and $N(T)$.

In summary, the solution to the extrapolation problem using this one-step approach is contingent on one being able to find a finite energy signal $z \in R(T)$ which will satisfy operator relationship (28). If such a solution exists, then the desired extrapolation is achieved by operating upon z by the ideal band-limited operator B as indicated by expression (30). The most difficult task in this procedure is that of solving operator relationship (28) which in the time domain takes the form

$$\int_{\tau \in \Lambda} h(t - \tau)\, z(\tau)\, d\tau = x(t) \qquad \text{for} \quad t \in \Lambda \tag{31}$$

where $h(t)$ is the ideal band-limited filter's unit-impulse response as given in expression (15). This is recognized as being a Fredholm integral equation of the first kind with kernal $h(t - \tau)$. Once a solution to operator relationship (31) has been obtained, the desired extrapolation is obtained, using extrapolation expression (30). In the time domain, this extrapolation expression takes the form

$$x(t^*) = \int_{\tau \in \Lambda} h(t^* - \tau)\, z(\tau)\, d\tau \tag{32}$$

where $t^* \notin \Lambda$ is the extrapolation time instant.

VII. Class of Well-Behaved Band-Limited Signals

In virtually all cases of practical interest, the ideal band-limited filter transfer function (7) is an even function of ω. This then implies that its associated inverse Fourier transform $h(t)$, which serves as the kernel of integral equation (31), is an even function of t. Since this kernel is symmetric, it is then possible to characterize the solution(s) to integral equation (31) by incorporating the eigenvalue-eigenfunction structure of composite linear operator TB. In particular, the scalar λ and signal $\phi \in R(T)$ are said to be an eigenvalue-eigenfunction pair associated with operator TB if they satisfy the operational expression

$$TB\phi = \lambda\phi$$

or, equivalently, in the time domain

$$\int_{\tau \in \Lambda} h(t - \tau)\, \phi(\tau)\, d\tau = \lambda\phi(t) \qquad \text{for} \quad t \in \Lambda.$$

A number of interesting properties relative to the algebraic structure of these eigenvalue-eigenfunction pairs are well documented (e.g., see [14], pp. 180-181). Of particular interest is a general theorem developed by Picard (see [5]) which states that there will exist a finite energy solution $z \in R(T)$ to integral equation (31) if and only if the observed signal Tx can be expressed in the series expansion

$$Tx = \sum_{n=1}^{\infty} \langle Tx, \phi_n \rangle\, \phi_n \tag{33a}$$

and the coefficients of this expansion are such that

$$\sum_{n=1}^{\infty} [\langle Tx, \phi_n \rangle / \lambda_n]^2 \tag{33b}$$

is finite. In these expressions, λ_n and $\phi_n \in R(T)$ are the totality of eigenvalues and associated orthonormal eigenfunctions, respectively, of operator TB. Furthermore, the solution to integral relationship (31) will be unique if the eigenvector set $\{\phi_n\}$ is complete in $R(T)$. As an example, the set of prolate spheroidal eigenfunctions associated with the class of time-truncated band-limited low-pass signals to be considered in Section IX are complete [14].

It must be noted that there will generally exist a subset of B_Ω for which series expansions of the form (33) do not hold. One is then unable to use the one-step extrapolation procedure for any signal from this poorly behaved subset. More importantly, the basic operation of extrapolation is itself ill-posed on this subset of signals as can be readily established. With this in mind, it is therefore advisable to only seek extrapolations of band-limited signals which are well-behaved in the sense of satisfying conditions (33).

VIII. Numerical Extrapolation

The extrapolated signal as generated by the one-step procedure of the last section entails solving the Fredholm integral equation (31). Assuming that the truncated signal Tx satisfies conditions (33), a finite energy solution $z \in R(T)$ is guaranteed. It should be apparent, however, that an analytical solution is all but impossible except in the most contrived of examples. We then, of necessity, must appeal to numerical techniques to attain an approximate solution. To illustrate the approach which is typically taken, let us assume that the observation-time set is a simple interval which, without loss of generality, will be taken to be

$$\Lambda = \{t\colon -\delta \leqslant t \leqslant \delta\}.$$

With this choice of observation-time set, the integral equation to be solved is given by

$$\int_{-\delta}^{\delta} h(t - \tau)\, z(\tau)\, d\tau = x(t) \qquad -\delta \leqslant t \leqslant \delta \tag{34}$$

whereby given the signal $x(t)$ for $t \in \Lambda$, we are to determine a finite energy solution $z(t)$ for $t \in \Lambda$. The kernel of this integral operation is given by expression (15).

A numerical solution to integral equation (34) generally entails a decomposing of the integration interval $(-\delta, \delta)$ into a finite number of subintervals over which each subintegral is approximated as the area under a rectangle, trapezoid, etc. As an example, suppose the integration interval is divided into $2M + 1$ equal subintervals each of width $\Delta = 2\delta/2M + 1$. If the rectangular integration method is then applied to relationship (34), evaluated at $t = k\Delta$, we have

$$\sum_{i=-M}^{M} \Delta h(k\Delta - i\Delta)\, z(i\Delta) = x(k\Delta) \quad \text{for } k = 0, \pm 1, \pm 2, \cdots, \pm M. \tag{35}$$

The accuracy of this approximation will be dependent on the selection of the positive integer M with larger values of M generally yielding more accurate approximations. This numerical approximation of integral relationship (34) is recognized as being a system of $2M + 1$ linear equations in the $2M + 1$ unknowns $z(i\Delta)$ which will be suggestively expressed in the matrix format

$$H\hat{z} = \hat{x}. \tag{36}$$

The elements of the square matrix H of order $2M + 1$ are sampled versions of the ideal band-limited filter's unit-impulse response (15) as given by

$$h_{ij} = \Delta h(i\Delta - j\Delta) \qquad i, j = 1, 2, \cdots, 2M + 1 \tag{37}$$

while the components of the $2M + 1$ column vectors $\hat{x}$ and $\hat{z}$ are

$$\hat{x}_i = x(-M\Delta - \Delta + i\Delta) \text{ and } \hat{z}_i = z(-M\Delta - \Delta + i\Delta) \quad i = 1, 2, \cdots, 2M + 1. \tag{38}$$

It will be convenient to interpret $\hat{x}$ as being a uniformly sampled version of Tx and $\hat{z}$ as a uniformly sampled approximation of a solution to relationship (34).

One next solves the system of $2M + 1$ linear equations in $2M + 1$ unknowns (36) for the vector $\hat{z}$. Finally, using the standard rectangular integration approximation for extrapolation relationship (32), the following extrapolation algorithm results:

$$x(q\Delta) = \sum_{i=-M}^{M} \Delta h(q\Delta - i\Delta)\, \hat{z}(i\Delta) \tag{39}$$

where q may take on any integer value. It is important to realize that once the solution $\hat{z}$ to expression (36) has been obtained, one need only perform $2M + 1$ multiplications and M additions to obtain each extrapolated point desired using relationship (39). In summary, the procedure to be followed is that of 1) solving the system of equations (36) once for $\hat{z}$, and 2) evaluating the extrapolation relationship (39) for any extrapolation time $q\Delta$ desired.

In using this numerical scheme to implement the one-step procedure, one must recognize some potential difficulties which may arise. For example, if the signal Tx belongs to the class of ill-posed signals mentioned in the previous section, then the linear system of equations (36) will themselves be ill-posed. In such cases, small perturbations of the right side vector $\hat{x}$ will cause large changes in the solution $\hat{z}$. This is a most undesirable characteristic when imprecise measurements of $\hat{x}$ are anticipated. In addition, it has been previously shown [3] that an infinity of different band-limited signals will be contained in B_Ω which will generate the same set of $2M + 1$ samples $\{x(k\Delta)\}$ used in relationship (35). As such, the postulated extrapolation procedure will be effective only if M is selected large enough to significantly reduce the effect of this phenomenon. How large one must make M will be dependent on the structure of the underlying band-limited signal and remains a topic for future investigation.

It is to be noted that Sabri and Steenaart [13] have developed an alternative one-step extrapolation procedure which evolved from the Papoulis extrapolation algorithm [11]. Their procedure requires the inversion of a square matrix whose order $N = 2M + 1 +$ (number of points to be extrapolated). This is to be contrasted with this paper's method which requires only the inversion of a square matrix of order $2M + 1$. Generally, since N is a much larger integer than $2M + 1$, this paper's method is far more efficient than theirs. More importantly, the alternate procedure must truncate signals whose domains are of infinite extent in order to achieve a tractable numerical implementation. Unfortunately, this can have a degrading effect on their resultant extrapolation.

An important and practical case arises when the ideal band-limited filter transfer function $H(\omega)$ is an even function of ω. Namely, it will then follow that the resulting H matrix characterizing relationship (36) is Toeplitz (i.e., $h_{ij} = \Delta h(|i - j|\Delta)$). This is a consequence of the fact that the associated impulse response $h(t) = F^{-1}[H(\omega)]$ is an even function of time. This is noteworthy in that one may then apply the Levinson algorithm (e.g., see [7]) for solving the resultant Toeplitz system of linear equations (36) for $\hat{z}$. This solution procedure is computationally efficient and our experience indicates it to be relatively insensitive to cases where matrix H is almost singular.

IX. Numerical Examples

In this section, we shall demonstrate the effectiveness of the given one-step extrapolation procedure. This will be accomplished by considering the two specific cases of low-pass and bandpass signal extrapolation. These fairly well depict the class of problems normally encountered in applications or the literature.

Extrapolation of Low-Pass Signals

In the low-pass signal extrapolation problem, the characterizing observation-time and band-limit sets are specified by

$$\Lambda = \{t: \ -\delta \leqslant t \leqslant \delta\} \tag{40}$$

$$\Omega = \{\omega: \ -\sigma \leqslant \omega \leqslant \sigma\} \tag{41}$$

where δ and σ are positive constants which identify the observation-time interval and passband, respectively. Clearly, the set Ω as given above is seen to correspond to low-pass signals whose "highest" frequency does not exceed σ rad/s. According to relationship (15), the associated ideal unit-impulse is then given by

$$h(t) = \frac{1}{2\pi} \int_{-\sigma}^{\sigma} e^{j\omega t}\, d\omega = \frac{\sin[\sigma t]}{\pi t}. \tag{42}$$

This ideal signal in turn specifies the elements of the H matrix of relationship (36) used to numerically approximate the extraction operation. In particular, it follows from expression (37) that these elements are given by

$$h_{ij} = \frac{\sin[(i-j)\,\sigma\Delta]}{\pi(i-j)} \tag{43}$$

for $i, j = 1, 2, \cdots, 2M+1$. This matrix is seen to be Toeplitz (i.e., the general element value h_{ij} is a function of $|i-j|$).

The extrapolation procedure for this low-pass example would then proceed as follows: 1) uniformly sample the observed segment of $x(t)$ over the observation interval $(-\delta, \delta)$ with sampling period $\Delta = 2\delta/(2M+1)$ to generate the vector $\hat{x}$ given by (38); 2) determine a solution, $\hat{z}$, of the linear system of equations (36) whereby the elements of matrix H are given by expression (43); 3) the extrapolated signal is then generated using expression (39) for any choice of extrapolation time $q\Delta$. To illustrate this procedure, let us consider the specific low-pass signal

$$x(t) = \frac{\sin(0.94\sigma t)}{0.94\sigma t} \tag{44}$$

whose spectrum is wholly contained within the set (41).[3] It will be assumed that the center 20.68 percent of this signal's main lobe is observed as shown in Fig. 1 and we desire to perform an extrapolation based on this partial information (this corresponds to $\delta = 11\pi/50\sigma$). The one-step extrapolation procedure was then implemented in which M was arbitrarily taken to be 5 which yielded the unformly sampled set

$$x(k\Delta) = \frac{\sin[0.0376\,\pi\,k]}{0.0376\,\pi\,k} \qquad k = 0, \pm 1, \pm 2, \cdots, \pm 5.$$

Applying the aforementioned Levinson algorithm for solving the resultant linear system of eleven equations in eleven unknowns (36), we determine $\hat{z}$. Finally, the extrapolation expression (39) is evaluated at $q = \pm 6, \pm 7, \cdots, \pm 99$ to generate the desired extrapolation. The results are plotted in Fig. 1, in which the solid line designates the untruncated signal (44), the solid dots denote the extrapolated values, and the eleven samples of $x(t)$ within the designated observation window are those used in the extrapolation process. It is apparent that the extrapolation procedure has worked remarkably well in this example (i.e., the extrapolated points agree with the actual values to within the fourth place to the right of the decimal point).

Bandpass Extrapolation

In a variety of applications, the signal under consideration has a spectrum which is known to fall within a given frequency band as specified by the set

$$\Omega = \{\omega\colon\ \sigma_1 \leqslant |\omega| \leqslant \sigma_2\} \tag{45}$$

where σ_1 and σ_2 are positive constants which identify the bandpass interval. The observation-time set for this extrapolation problem will again be taken to be the symmetrical set (40). Using relationship (15), the corresponding ideal unit-

[3] The time-truncated version of this signal is much more difficult to extrapolate than is the low-pass signal $\sin(\sigma t)/\sigma t$ as considered by Papoulis [11].

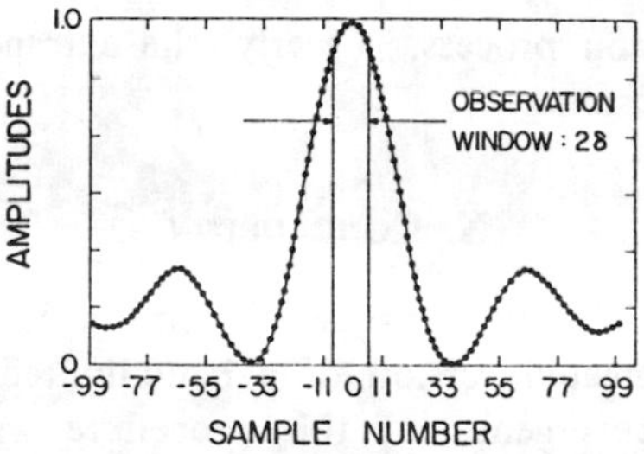

Fig. 1. Extrapolation of a time-truncated version of the low-pass signal $\sin(0.94\sigma t)/0.94\sigma t$ where the dots denote the extrapolated values.

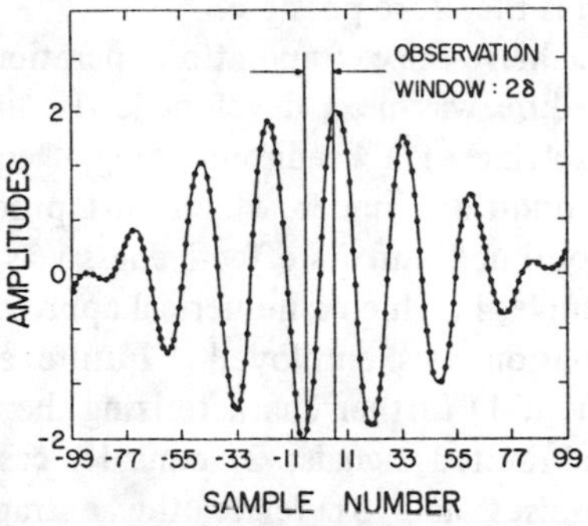

Fig. 2. Extrapolation of a time-truncated version of the bandpass signal $\sin(0.99\sigma t) + \sin(0.85\sigma t)$ where the dots denote the extrapolated values.

impulse response is found to be

$$h(t) = \frac{\sin[\sigma_2 t] - \sin[\sigma_1 t]}{\pi t}. \tag{46}$$

Accordingly, the elements of the associated H matrix are then

$$h_{ij} = \frac{\sin[(i-j)\,\sigma_2\Delta] - \sin[(i-j)\,\sigma_1\Delta]}{\pi(i-j)} \tag{47}$$

for $i, j = 1, 2, \cdots, 2M+1$.

To illustrate the bandpass extrapolation procedure, let us consider the case in which the parameters which identify set Ω are taken to be $\sigma_2 = \sigma$ and $\sigma_1 = 0.8\,\sigma$. Moreover, the signal whose time-truncated version is to be extrapolated is given by

$$x(t) = \sin[0.99\,\sigma t] + \sin[0.85\,\sigma t]. \tag{48}$$

It is clear that this signal is band limited and that its spectrum is wholly contained within the specified set Ω. An observation of this signal over the interval (40) with $\delta = 11\pi/25\sigma$ is made. This corresponds to only 0.374 of one period of the lowest sinusoidal component of $x(t)$ as shown in Fig. 2.

Using relationship (36) with $M = 5$, the Toeplitz system of linear equations (36) is next solved whereby the components $x(k\Delta)$ for $k = 0, \pm 1, \pm 2, \cdots, \pm 5$ are obtained by uniformly sampling signal (48) with $\Delta = 2\delta/11$. Upon using the Levinson algorithm for solving this system of equations for $\hat{z}$, we then incorporate extrapolation expression (39) to generate the extrapolated values $x(q\Delta)$ for $q = \pm 6, \pm 7, \cdots, \pm 99$. The results of this process are shown in Fig. 2 in which the solid line specifies the untruncated signal (48), the solid dots denote the extrapolated values, and the eleven samples of $x(t)$ within the designated observation window are those used in the extrapolation process. Clearly, the extrapolation yields almost exact results.

X. Conclusion

An effective algorithm was developed for the extrapolation of a time-truncated version of a band-limited signal. The guaranteed convergence of this procedure was established using Von Neumann's alternating projection theorem. Its implementation requires the iterative evaluation of an integral defined over finite integration limits so that a numerical approximation is therefore possible.

In order to achieve the extrapolation operation efficiently, a one-step procedure was next developed. Its implementation involved the solving of a Fredholm integral equation of the first kind. Conditions under which this procedure is well posed where outlined and its effectiveness was demonstrated using two examples in which a numerical approximation of the Fredholm equation was employed. Future areas of study would be that of 1) further characterizing the class of "well-behaved" band-limited signals, 2) consider cases of additive out-of-band noise, and 3) relate the extrapolation error to the structure of the numerical integration scheme used in either the algorithm or the one-step procedure.

As a final comment, it should be noted that the concepts herein developed hold for a much more general class of problems. Specifically, let B and T be any pair of orthogonal projection operators on a Hilbert space X. Furthermore, let the transformed image signal $y = Tx$ be given and suppose it is desired to find the pre-image signal x which is hypothesized to be contained in $R(B)$. It is readily shown that algorithm (19) will generate an approximation sequence which converges to x provided that subspace $R(B) \cap N(T)^{\perp}$ contains only the zero element. Furthermore, the one-step procedure is also applicable and requires solving the operator equation $TBz = Tx$ for a $z \in R(T)$ and then setting $x = Bz$.

References

[1] R. B. Blackman and J. W. Tukey, *The Measurement of Power Spectra*. New York: Dover, 1958.

[2] J. P. Burg, "Maximum entropy spectral analysis," Ph.D. dissertation, Stanford Univ., Stanford, CA, May 1975.

[3] J. A. Cadzow, "Improved spectral estimation from incomplete sampled-data observations," presented at the RADC Spectrum Estimation Workshop, Rome, NY, May 24-26, 1977.

[4] L. G. Chambers, *Integral Equations: A Short Course*. London, England: International, 1976.

[5] R. Courant and D. Hilbert, *Methods of Mathematical Physics*. New York: Interscience, 1973.

[6] R. W. Gerchberg, "Super resolution through error energy reduction," *Opt. Acta*, vol. 21, no. 9, pp. 709-720, 1974.

[7] N. Levinson, "The Wiener RMS (root mean square) error criterion in filter design and prediction," *J. Math. Phys.*, vol. 25, pp. 261-278, 1947.

[8] A. W. Naylor and G. R. Sell, *Linear Operator Theory*. New York: Holt, Rhinehart and Winston, 1971.

[9] J. von Neumann, *The Geometry of Orthogonal Spaces*, vol. II. Princeton, NJ: Princeton University Press, 1950.

[10] A. V. Oppenheim and R. W. Schafer, *Digital Signal Processing*. Englewood Cliffs, NJ: Prentice-Hall, 1975.

[11] A. Papoulis, "A new algorithm in spectral analysis and band-limited extrapolation," *IEEE Trans. Circuits Syst.*, vol. CAS-22, pp. 735-742, Sept. 1975.

[12] E. Parzen, "Multiple time series modeling," in *Multivariate Analysis*, vol. 2, P. R. Krishnaiah, Ed. New York: Academic, 1969.

[13] M. S. Sabri and W. Steenaart, "An approach to band-limited signal extrapolation: The extrapolation matrix," *IEEE Trans. Circuits Syst.*, vol. CAS-25, pp. 74-78, Feb. 1978.

[14] H. L. van Trees, *Detection, Estimation, and Modulation Theory–Part I*. New York: Wiley, 1968.

[15] D. C. Youla, "Generalized image restoration by the method of alternating orthogonal projections," presented for publication.

Part IV
Multichannel, Multidimensional, and Spatial Spectral Analysis

IN point-to-point communications, the transmission channel often carries a variety of signals from different sources, and delivers them to a number of users. Thus, signal processing blocks within channels (e.g., in transponder units) have multiple inputs and outputs. Often, an interaction exists among signals, in a sense that there are nonzero power cross-spectra in addition to auto-spectra of the single-channel series. There are also the other examples of multichannel signal processing. Some of the spectral analysis methods are extended to include the multisignal environment. Some of the most important ones are reviewed in the paper by Marple and Nuttall.

Applications in such areas as image processing, geophysical data analysis, biomedical imaging, radar and sonar, require power spectrum estimates which are functions of more than one independent variable. Multidimensional procedures have therefore been developed concurrently with its one-dimensional counterparts. Because of the specific problems encountered in a multidimensional environment, straightforward extensions of one-dimensional procedures are usually not feasible. Some of the most representative methods are given in papers by Jackson and Chien, Lim and Malik, and by Nikias and Raghuveer, which appear as papers 2, 3, and 4, respectively, in this collection.

A very important example of multidimensional spectral analysis is the processing of spatio-temporal signals received by an array of sensors. In their paper, Lang and McClellan develop a mathematical framework for spectral analysis of sensor signals, and treat the question of feasibility of extending the one-dimensional procedures to multidimensional cases. The last two papers deal with the eigensystem (ES) approach to spatio-temporal problem, from somewhat different points of view. Multitude of ES based methods in the recent literature of which the papers by Johnson and DeGraaf, and by Bienvenu and Kopp are typical examples, indicate the trend in array signal processing.

Experimental comparison of three multichannel linear prediction spectral estimators

S. Lawrence Marple Jr. and Albert H. Nuttall

Indexing term: *Signal processing*

Abstract: Single-channel spectral estimators based on linear prediction techniques, such as the maximum-entropy method, have been shown to often provide better spectral stability and resolution than standard FFT procedures for short data sequences. Based on this improved performance, a multitude of multichannel linear prediction techniques have been promoted for processing multichannel data sequences. Three of these are examined in the paper: a multichannel generalisation of the single-channel Burg algorithm by Nuttall, a maximum-entropy type of algorithm by Morf, Vieira, Lee and Kailath, and a multichannel extension of the covariance method of linear prediction implemented by Marple. For purposes of experimental comparison, various two-channel data sets were processed by the three methods to produce the two autospectra, the magnitude-squared coherence and the coherence phase associated with each data set. A possible deleterious effect of signal 'feed-across' between autospectra and in the coherence has been discovered in all three methods. The phenomenon, due to inexact pole-zero cancellation, is especially prominent for short data sequences. Based on the multichannel results given here, the Nuttall method generally produced the best spectral estimates.

1 Introduction

Multichannel digital signal processing is being used in an increasing number of application areas, particularly in the sonar and geoseismic communities. Until recently, most multichannel digital signal processing was based on fast Fourier transform (FFT) methods. The success of unichannel high-resolution spectral estimation techniques, like the autoregressive or so-called maximum-entropy methods, has encouraged researchers to develop multichannel extensions in hope of obtaining performance improvements for multichannel applications comparable with that seen in unichannel applications. The multichannel extensions are all based on linear prediction concepts, since a linear prediction filter whitens an autoregressive process.

Three multichannel linear prediction spectrum analysis algorithms are examined in this paper. They are the multichannel generalisation of the Burg algorithm developed by Nuttall [1–4], a multichannel maximum-entropy spectral estimate by Morf, Vieira, Lee and Kailath [5, 6], and a multichannel generalisation of the covariance method of linear prediction as implemented by Marple [7]. All three algorithms make estimates of the multichannel linear prediction coefficients, from which the multichannel autoregressive autospectra and cross-spectra may be computed. From the cross-spectra, the magnitude-squared coherence and coherence phase may be computed.

There has been no previous attempt in the literature to compare and characterise the various multichannel linear-prediction/autoregressive spectral estimators. An experimental approach is used in this paper to empirically characterise performance with respect to two signal classes. One class is an actual two-channel autoregressive process. The other class is a set of tones (sinusoids) imbedded in a coloured noise process. An analytic approach for calculating multichannel algorithm performance was felt to be mathematically intractable, given the very complex analysis that was required to characterise the single-channel autoregressive spectral estimate [8, 9].

Paper 2461F, first received 1st April 1982 and in revised form 4th January 1983

Dr. Marple was formerly with The Analytic Sciences Corporation, 8301 Greensboro Drive, Suite 1200, McLean, VA 22102, USA, and is now with Schlumberger Well Services-Engineering, PO Box 4594, Houston, TX 77210-4594, USA. Dr. Nuttall is with the Naval Underwater Systems Center, New London Laboratory, New London, CT 06320, USA

Relative to the particular multichannel results reported here, the Nuttall algorithm generally produced the best results. Best in this case means with less frequency estimation bias and variance than the other methods when tested against autoregressive and tonal processes. During the testing, a 'feed-across' effect was discovered that was common to all techniques. Narrowband components that should be present in only one channel's autospectra were found coupled into another channel's autospectral estimate. This is shown to be due to inexact pole-zero cancellation in the autospectral estimate. Conditions that would cause spectral line splitting in the single-channel case were found to also cause splitting in two of the three linear prediction techniques considered here.

2 Summary of techniques

This Section provides an overview of the three multichannel linear prediction methods considered. If we define X_n as the vector of M channel samples at time index n

$$X_n = \begin{bmatrix} x_n(1) \\ \vdots \\ x_n(M) \end{bmatrix}$$

from a stationary zero-mean multichannel process, then the covariance function at time lag k is given by

$$R_k = E[X_{n+k}X_n^H] = R_{-k}^H$$

The symbol H denotes Hermitian transpose, and E denotes statistical expectation. Define the forward linear prediction error for prediction order p as

$$Y_k = \sum_{n=0}^{p} A_n^{(p)} X_{k-n}$$

and the backward linear prediction error as

$$Z_k = \sum_{n=0}^{p} B_n^{(p)} X_{k-p+n}$$

where $\{A_n^{(p)}\}_1^p$ and $\{B_n^{(p)}\}_1^p$ are, respectively, the forward and backward linear prediction coefficient matrices of dimension

Reprinted with permission from *Proc. Inst. Elec. Eng.*, vol. 130, Part F, pp. 218–229, Apr. 1983.

$M \times M$. Note that $A_0^{(p)} = B_0^{(p)} = I$ (the identity matrix) by convention.

Minimisation of the forward and backward linear prediction mean square errors $\mathrm{tr}\,\{E[Y_k Y_k^H]\} = \mathrm{tr}\,\{U_p\}$ and $\mathrm{tr}\,\{E[Z_k Z_k^H]\} = \mathrm{tr}\,\{V_p\}$, respectively, (tr denotes the matrix trace) will yield coefficient matrices that satisfy the block-Toeplitz normal equation

$$\begin{bmatrix} I & A_1^{(p)} & \cdots & A_{p-1}^{(p)} & A_p^{(p)} \\ B_p^{(p)} & B_{p-1}^{(p)} & \cdots & B_1^{(p)} & I \end{bmatrix} \begin{bmatrix} R_0 & R_1^H & \cdots & R_p^H \\ R_1 & R_0 & \cdots & R_{p-1}^H \\ \vdots & \vdots & \ddots & \vdots \\ R_p & R_{p-1} & \cdots & R_0 \end{bmatrix} = \begin{bmatrix} U_p & 0 & \cdots & 0 & 0 \\ 0 & 0 & \cdots & 0 & V_p \end{bmatrix}$$

The solution to this normal equation, when the covariance matrices $\{R_k\}_0^p$ are known, is provided by the multichannel (vector) Levinson-Wiggins-Robinson (LWR) algorithm. Briefly, this algorithm relates the order p solution to the order $p-1$ solution according to the recursions

$$\Delta_p = \sum_{n=0}^{p-1} A_n^{(p-1)} R_{p-n}$$

$$A_p^{(p)} = -\Delta_p (V_{p-1})^{-1}$$

$$B_p^{(p)} = -\Delta_p^H (U_{p-1})^{-1}$$

$$U_p = (I - A_p^{(p)} B_p^{(p)}) U_{p-1}$$

$$V_p = (I - B_p^{(p)} A_p^{(p)}) V_{p-1}$$

$$A_k^{(p)} = A_k^{(p-1)} + A_p^{(p)} B_{p-k}^{(p-1)} \quad \text{for} \quad 1 \leqslant k \leqslant p-1$$

$$B_k^{(p)} = B_k^{(p-1)} + B_p^{(p)} A_{p-k}^{(p-1)} \quad \text{for} \quad 1 \leqslant k \leqslant p-1$$

with initial conditions

$$U_0 = V_0 = R_0$$

$$A_0^{(k)} = B_0^{(k)} = I \quad \text{for} \quad 0 \leqslant k \leqslant p$$

The $M \times M$ matrix Δ_p is often called the reflection coefficient matrix.

Based on the linear prediction coefficients, the multichannel autoregressive power spectral density matrix estimate may be shown to be [1]

$$G_A(z) = \Delta t\,[A(z)]^{-1} U_p\,[A(1/z^*)]^{-H}$$
$$= \Delta t\,[B(z)]^{-1} V_p\,[B(1/z^*)]^{-H}$$
$$= G_B(z)$$

where Δt is the sample interval, $-H$ denotes the Hermitian transpose of the inverse, the asterisk denotes complex conjugate, and

$$A(z) = \sum_{n=0}^{p} A_n^{(p)} z^{-n}$$

$$B(z) = \sum_{n=0}^{p} B_n^{(p)} z^{-n}$$

The substitution $z = \exp(j2\pi f \Delta t)$ is made, and G_A is evaluated as a function of the frequency f. With this substitution having been made, then

$$G(f) = \begin{bmatrix} G_{11}(f) & G_{12}(f) \\ G_{21}(f) & G_{22}(f) \end{bmatrix}$$

is the power spectral density matrix in the two-channel case. Here $G_{11}(f)$ and $G_{22}(f)$ are the autospectra of channel 1 and channel 2, respectively, and the magnitude-squared coherence (MSC) is given by

$$\mathrm{MSC} = |G_{12}(f)|^2 / \{G_{11}(f) G_{22}(f)\}$$

The coherence phase spectrum (CPS) is simply

$$\mathrm{CPS} = \arg\,[G_{12}(f)]$$

3 Unknown covariance: Measured data

The three linear prediction algorithms examined in this paper are concerned with the situation in which there are data samples available, but no covariance values are known at any lags. Assuming N vectors X_n of channel samples have been collected, then the following squared-error and cross-error covariance estimates may be formed:

$$E_p = \frac{1}{N-p} \sum_{k=p+1}^{N} Y_k Y_k^H$$

$$F_p = \frac{1}{N-p} \sum_{k=p+1}^{N} Z_{k-1} Z_{k-1}^H$$

$$G_p = \frac{1}{N-p} \sum_{k=p+1}^{N} Y_k Z_{k-1}^H$$

The Nuttall extension of the Burg algorithm to the multichannel case makes use of the LWR algorithm, with the exception of how the reflection coefficient Δ_p is computed, since covariance values are now unavailable. Using the error covariance estimates, Δ_p is obtained as the solution of the bilinear equation

$$(V_{p-1})^{-1} F_p \Delta_p + \Delta_p (U_{p-1})^{-1} E_p = -2G_p$$

in order to satisfy a weighted arithmetic mean criterion between the forward and backward squared errors E_p and F_p.

The so-called maximum entropy algorithm of Morf *et al.* also uses the LWR algorithm, but the reflection coefficient Δ_p is computed as the weighted geometric mean of the forward and backward squared errors:

$$\Delta_p = [E_p]^{-1/2} G_p\,[F_p^H]^{-1/2}$$

The superscript notation $-1/2$ means the square-root matrix of the inverse, since a normalised form of the LWR algorithm is used by the authors of this linear prediction method.

Instead of forcing the LWR recursions to hold in the given data case, as in the two previous algorithms, one could separately minimise the forward and backward squared errors

$$\mathrm{tr}\left[\sum_{n=p+1}^{N} Y_n Y_n^H\right]$$

$$\mathrm{tr}\left[\sum_{n=p+1}^{N} Z_{n-1} Z_{n-1}^{H}\right]$$

in the least-squares sense. This yields the normal equation

$$\begin{bmatrix} I & A_1^{(p)} \ldots\ldots A_{p-1}^{(p)} & A_p^{(p)} \\ B_p^{(p)} & B_{p-1}^{(p)} \ldots\ldots B_1^{(p)} & I \end{bmatrix} \begin{bmatrix} R_{00}^{(p)} & \ldots & R_{0p}^{(p)} \\ \vdots & \ddots & \vdots \\ R_{p0}^{(p)} & \ldots & R_{pp}^{(p)} \end{bmatrix}$$

$$= \begin{bmatrix} U_p & 0 \ldots\ldots 0 & 0 \\ 0 & 0 \ldots\ldots 0 & V_p \end{bmatrix}$$

where

$$R_{ij}^{(p)} = \sum_{n=p+1}^{N} X_{n-i} X_{n-j}^{H} \qquad \text{for } 0 \leqslant i, j \leqslant p$$

and the matrix of $R^{(p)}$ terms is no longer block-Toeplitz. However, a fast recursive algorithm similar in structure to the LWR algorithm can solve the above least-squares normal equation; see Reference 7 for details. The fast algorithm is a multichannel extension of the single-channel covariance linear prediction algorithm used in speech processing [10]. Since the block-Toeplitz property is lost in the least-squares minimisation of this method, the estimated power spectral densities $G_A(z)$ and $G_B(z)$ do not, in general, yield mathematically identical values, although the plotted spectra are often of comparable shapes.

4 Experimental comparisons

Plots of autospectra, MSC and CPS from various two-channel test cases are used here to examine relative experimental performance differences among the three linear prediction spectrum analysis techniques. The discussion is restricted to real-valued data and filters.

In the first test, a first-order ($p = 1$) two-channel autoregressive process was used to generate data ensembles. Data were generated according to the recursion

$$\left.\begin{aligned} X_n &= A_1 X_{n-1} + W_n \\ A_1 &= \begin{bmatrix} 0.85 & -0.75 \\ 0.65 & 0.55 \end{bmatrix} \end{aligned}\right\} \qquad (1)$$

where W_n is a white process, uncorrelated from channel to channel and of unit variance in both channels. The exact autospectra for each of the two channels are shown in Figs. 1*a* and *b*, while the MSC and phase of the complex coherence (cross-spectrum) are shown in Figs. 1*c* and *d*. There is seen to be a strong narrowband component at approximately 1/8 of the sampling frequency; the peak magnitude of the squared coherence is 0.999013 at this point. 20 sequences, each of 100 sample points, were processed by each of the three algorithms, with the order p restricted to be 1. The 20 spectral plots were overlapped, as shown in Figs. 2–5. All plots are made to the same reference values established in Fig. 1. Of the four sets of plots, clearly the Nuttall linear prediction algorithm has less variance and bias (location of the spectral

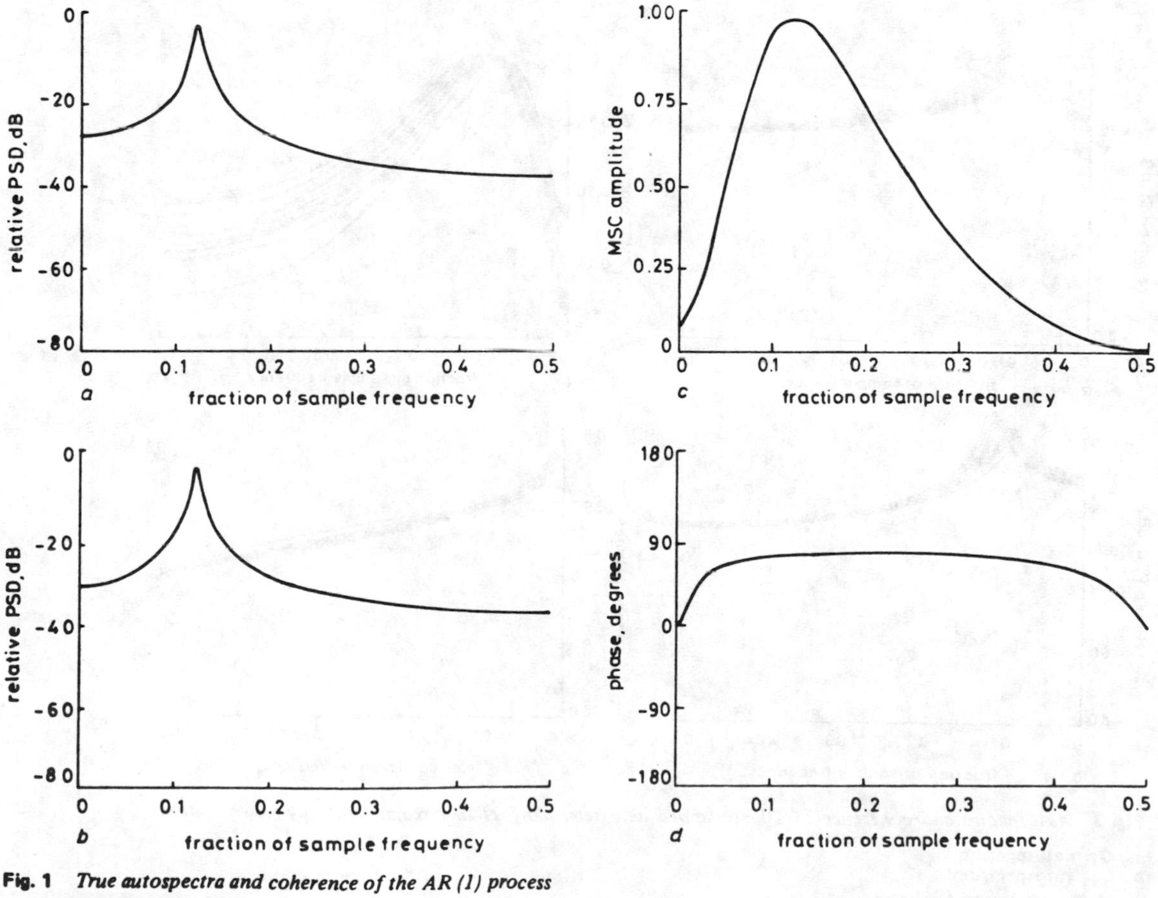

Fig. 1 *True autospectra and coherence of the AR (1) process*

a G_{11} autospectrum
b G_{22} autospectrum
c Magnitude-squared coherence
d Coherence phase

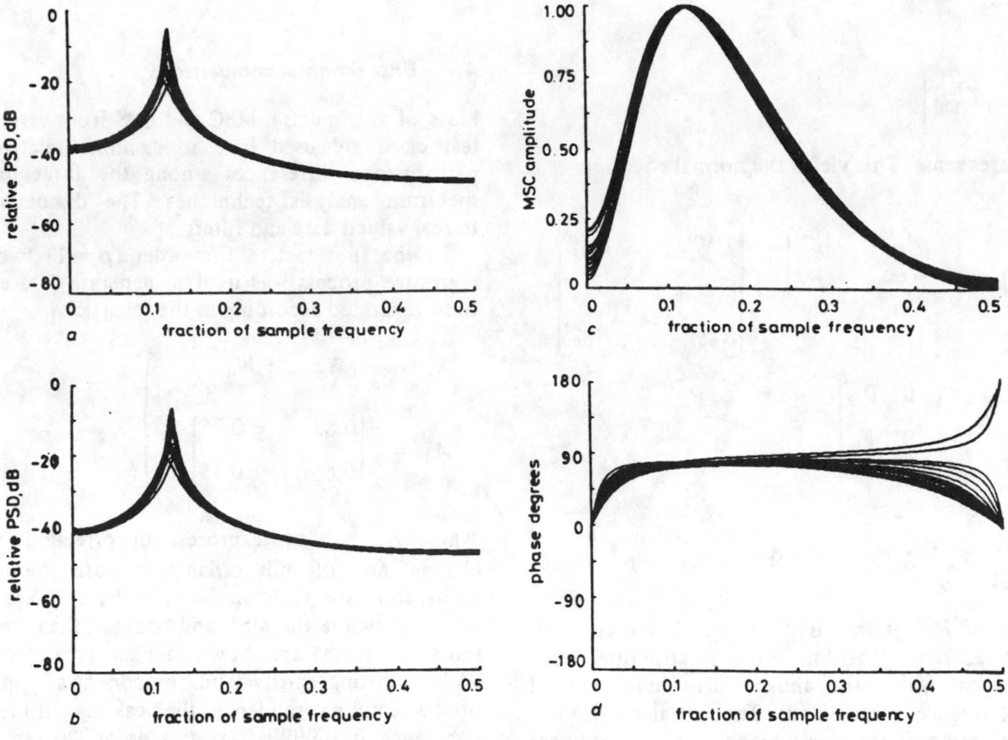

Fig. 2 *Autospectra and coherence of 20 overlapped data sets: Nuttall-Burg algorithm generalisation*

a G_{11} autospectrum
b G_{22} autospectrum
c Magnitude-squared coherence
d Coherence phase

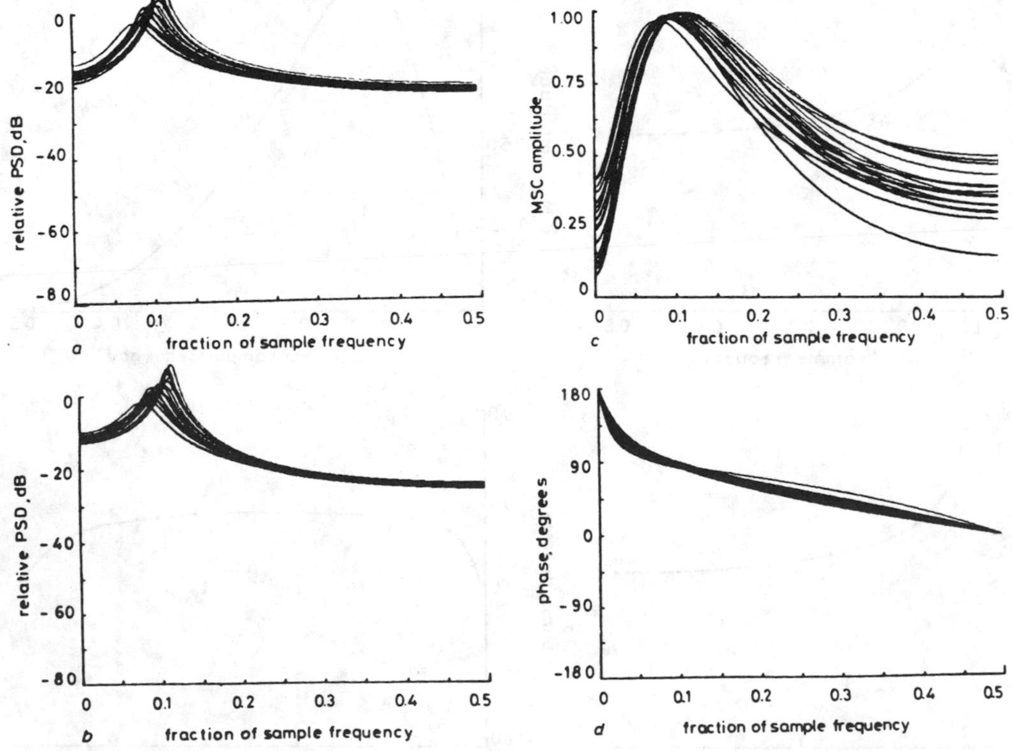

Fig. 3 *Autospectra and coherence of 20 overlapped data sets: Morf et al. maximum-entropy generalisation*

a G_{11} autospectrum
b G_{22} autospectrum
c Magnitude-squared coherence
d Coherence phase

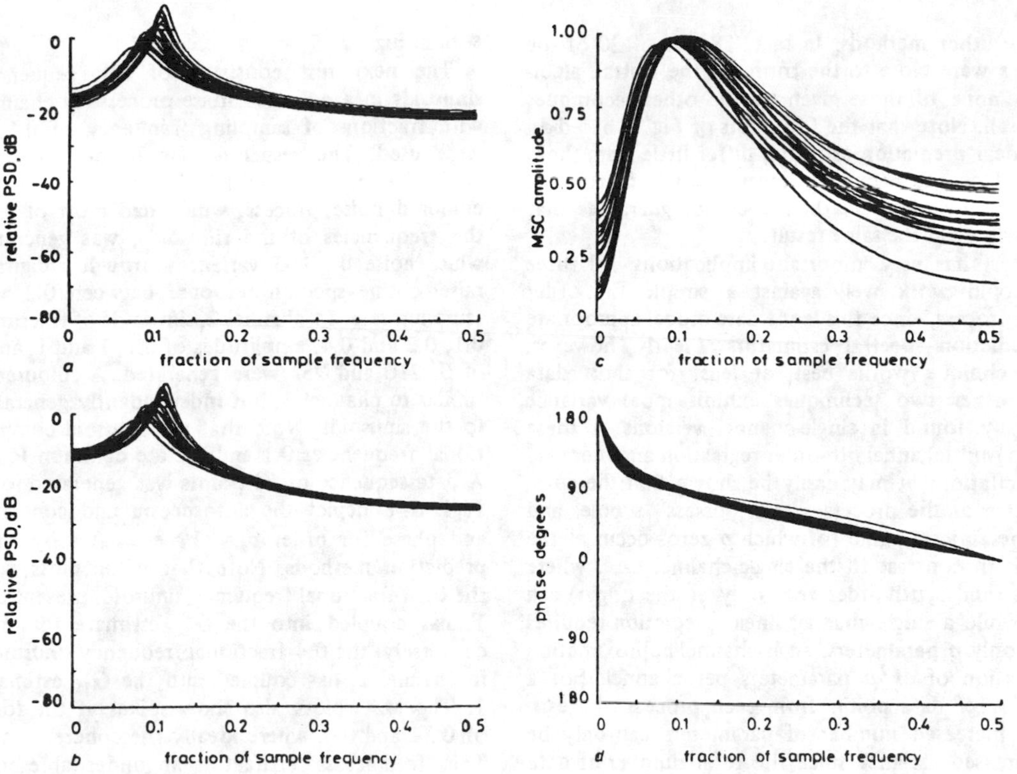

Fig. 4 *Autospectra and coherence of 20 overlapped data sets: Marple covariance generalisation (forward)*

a G_{11} autospectrum
b G_{22} autospectrum
c Magnitude-squared coherence
d Coherence phase

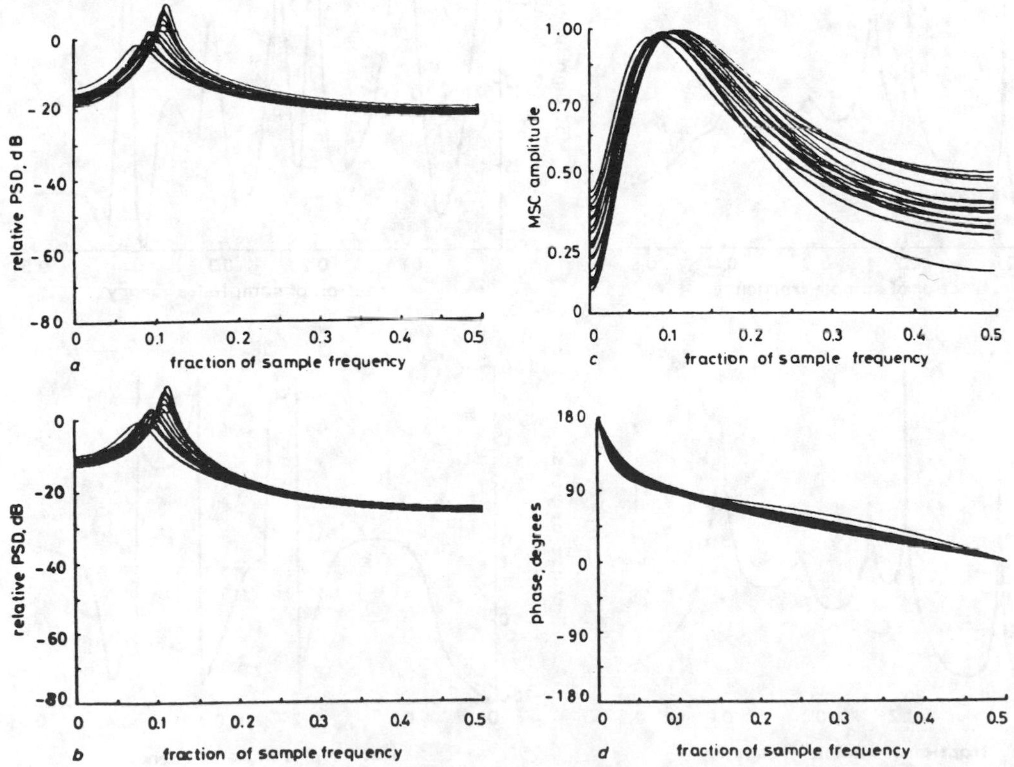

Fig. 5 *Autospectra and coherence of 20 overlapped data sets: Marple covariance generalisation (backward)*

a G_{11} autospectrum
b G_{22} autospectrum
c Magnitude-squared coherence
d Coherence phase

peak) than the other methods. In fact, 18 out of 20 of the phase estimates were close to the truth for the Nuttall algorithm, whereas none of those given by the other techniques was correct at all. Note that the four plots of Fig. 4, based on the forward linear prediction estimate, differ little from those in Fig. 5, based on the backward linear prediction estimate, even though the covariance method does not guarantee that the two cases will yield the same result.

This first test has very important implications. All three techniques should work well against a simple first-order autoregressive process, since this is the data model appropriate for linear prediction spectral estimators. Clearly, however, the Nuttall technique works best, at least for short data records. The other two techniques exhibit more variance than is typically found in single-channel versions of these methods. For a multichannel pth-order regression and independent white excitations, it may easily be shown that the auto- and cross-spectra of the processes each possess $4p$ poles and $3p$ zeros in the finite z-plane (of which p zeros occur at the origin). This is in contrast to the single-channel case, where only $2p$ poles (and a pth-order zero only at the origin) can occur. Thus, while a single-channel linear prediction requires estimation of only p parameters, an M-channel approximation requires estimation of M^2p parameters per channel. For a fixed number N of data points from each process, the estimation of an increased number of parameters can only be done with increased variance since the total number of data points is only MN. This is a partial explanation of the results seen in Figs. 2–5.

The next test consisted of data sequences with three sinusoids in a coloured noise process. In channel 1, sinusoids with fractions of sampling frequency of 0.1, 0.2 and 0.24 were used. The respective amplitudes were 0.1, 1 and 1, while the respective initial phases were 0, 90 and 235°. The coloured noise process, which had most of its power above the frequencies of the sinusoids, was generated by passing white noise of 0.05 variance through a digital filter with a raised-cosine spectral response between 0.2 and 0.5 of the sampling rate. In channel 2, sinusoids of fractional frequencies 0.1, 0.2 and 0.4, amplitudes of 0.1, 1 and 1, and initial phases of 0, 210 and 25° were generated. A coloured noise process similar to channel 1, but independently generated, was added to the sinusoids. Note that the sinusoid components at fractional frequencies 0.1 and 0.2 are common to both channels. A data sequence of 64 points was generated for each channel. Figs. 6–8 depict the autospectra and coherence magnitude and phase for order $p = 12$ estimates by the three linear prediction methods. Note that in all plots, some energy of the 0.24 fractional frequency sinusoid, present only in channel 1, has coupled into the G_{22} estimate for channel 2, and, conversely, the 0.4 fractional frequency sinusoid, present only in channel 2, has coupled into the G_{11} estimate for channel 1. The MSC plots also show spikes at the tone frequencies of 0.24 and 0.4, where ideally the coherence should be zero. This 'feed-across' artifact is an undesirable effect; however, it appears unavoidable whenever processing short data

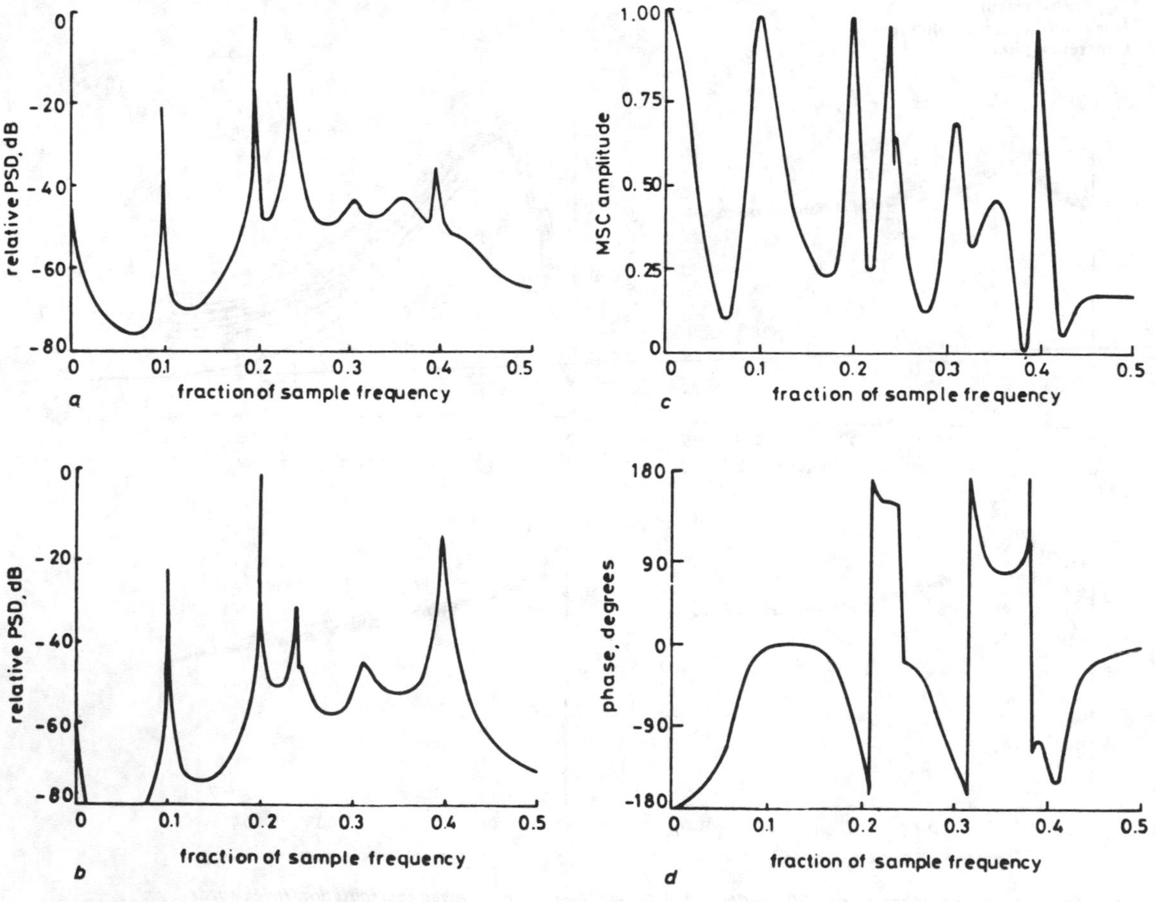

Fig. 6 *Autospectra and coherence of sinusoids in noise (64 samples): Nuttall-Burg algorithm generalisation*

a G_{11} autospectrum
b G_{22} autospectrum
c Magnitude-squared coherence
d Coherence phase

sequences with multichannel algorithms. If the data sequences from each channel are processed separately by the single-channel Burg linear prediction algorithm, the autospectra shown in Fig. 9 are the result. Note that these spectra do not give a false tonal indication. By increasing the data sequence length to 500 samples, there is an improvement in the feed-across effect in the multichannel autospectra in Fig. 10, although the coherence still has strong sharp spikes. Thus it seems that the attractive properties of single-channel linear prediction spectrum analysis relative to periodogram spectrum analysis when processing short data records do not extend to the multichannel case without problems. The feed-across effect leads to false indications of narrowband components, especially when processing short data sequences. A detailed analysis of the cause of the feed-across effect is presented in the following Section.

The final test case examined the spectral line splitting phenomenon that has been reported in the literature for single-channel linear prediction spectrum analysis. A data sequence of 101 samples of a single tone at 0.0725 fraction of sampling frequency, unit amplitude and 45° initial phase, added to the same coloured noise process (variance 10^{-4}) as in the last test, was generated for channel 1. A comparable 101-sample sequence was generated for channel 2, but the tone was placed at a fractional frequency of 0.2175. Both cases are known to cause line splitting in the single-channel Burg algorithm. Figs. 11–13 show the autospectra generated by the three multichannel linear prediction methods for order $p = 15$. Line splitting is present in the Nuttall and Morf methods, but not in the covariance method. Thus line splitting behaviour carries over into some of the multichannel techniques.

5 Properties of linear predictive spectral analysis techniques

The behaviour of the linear predictive techniques in the presence of tonals is of importance to anyone using these spectral analysis procedures. We now present some of the important properties of multichannel linear prediction techniques and illustrate these properties by an example from the Nuttall algorithm for the two-channel autoregressive process in eqn. 1, with a tone added solely to the channel-2 process. The tone strength is − 6.6 dB relative to the sample power in the second component of X_n, and the tone is located at fractional frequency $f_c = 0.3$; this is basically the example considered in Fig. 11 of Reference 4.

The autospectra and coherence estimates for a data run with $N = 1000$ data points and $p = 8$ are given in Fig. 14. There is the desired strong tonal indication at f_c in G_{22}, but, in addition, there is a weak undesired contribution at f_c in G_{11}. Since this tone was never added to process 1, this indication in G_{11} has 'leaked across' in the mathematical data manipulations of the linear predictive algorithm. This leakage is unavoidable and will be present in all multichannel linear predictive procedures; it is due to the fact that we must work with finite data sets, thereby resulting in slightly inaccurate filter coefficients. The effects on the poles and zeros of the autospectral and coherence estimates are discussed below. The MSC estimate in Fig. 14*c* has developed two notches and a sharp spike near frequency f_c, while the phase in Fig. 14*d* has gone through an abrupt 2π change in that neighbourhood.

In order to explain the various behaviours of the spectral

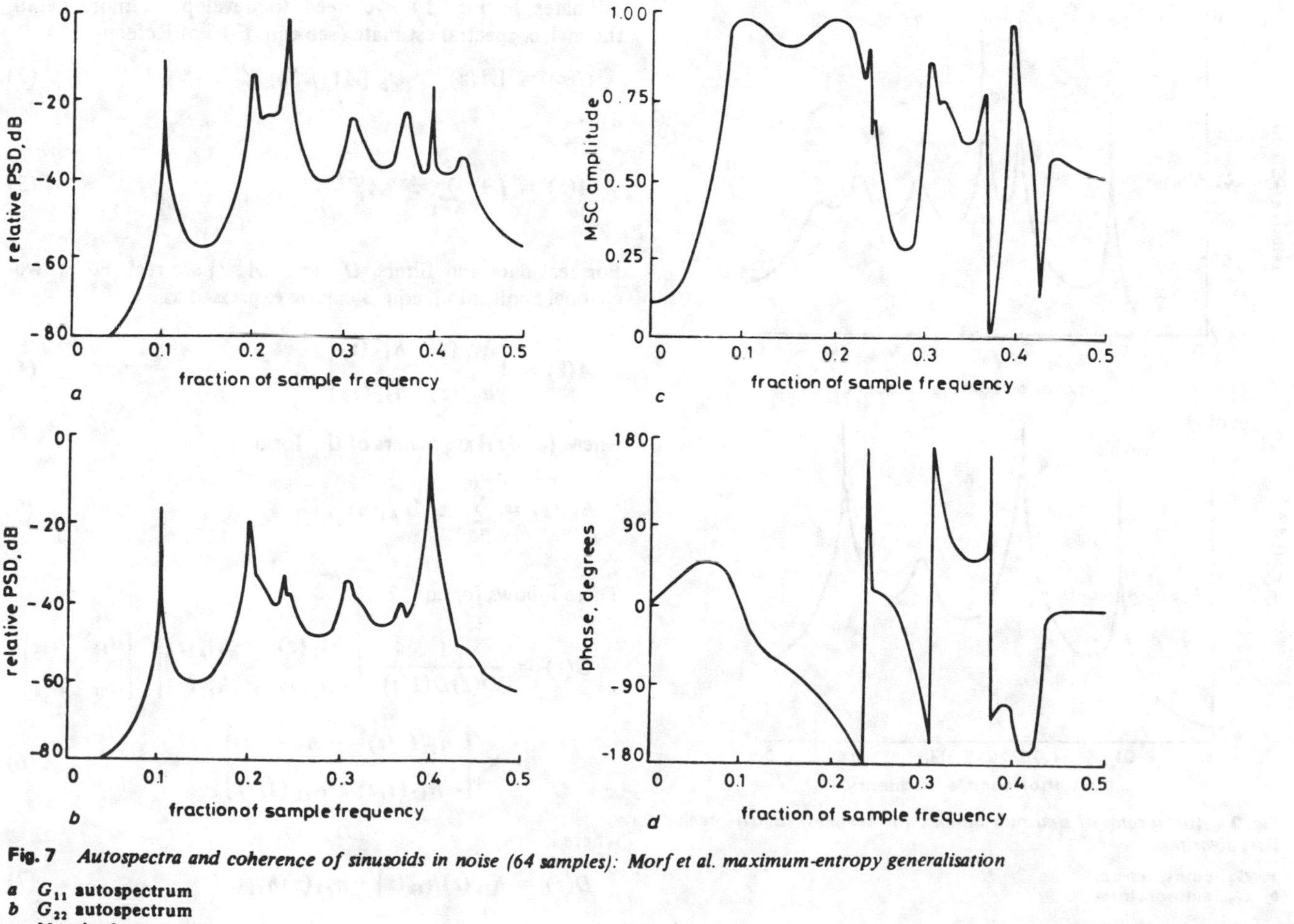

Fig. 7 *Autospectra and coherence of sinusoids in noise (64 samples): Morf et al. maximum-entropy generalisation*

a G_{11} autospectrum
b G_{22} autospectrum
c Magnitude-squared coherence
d Coherence phase

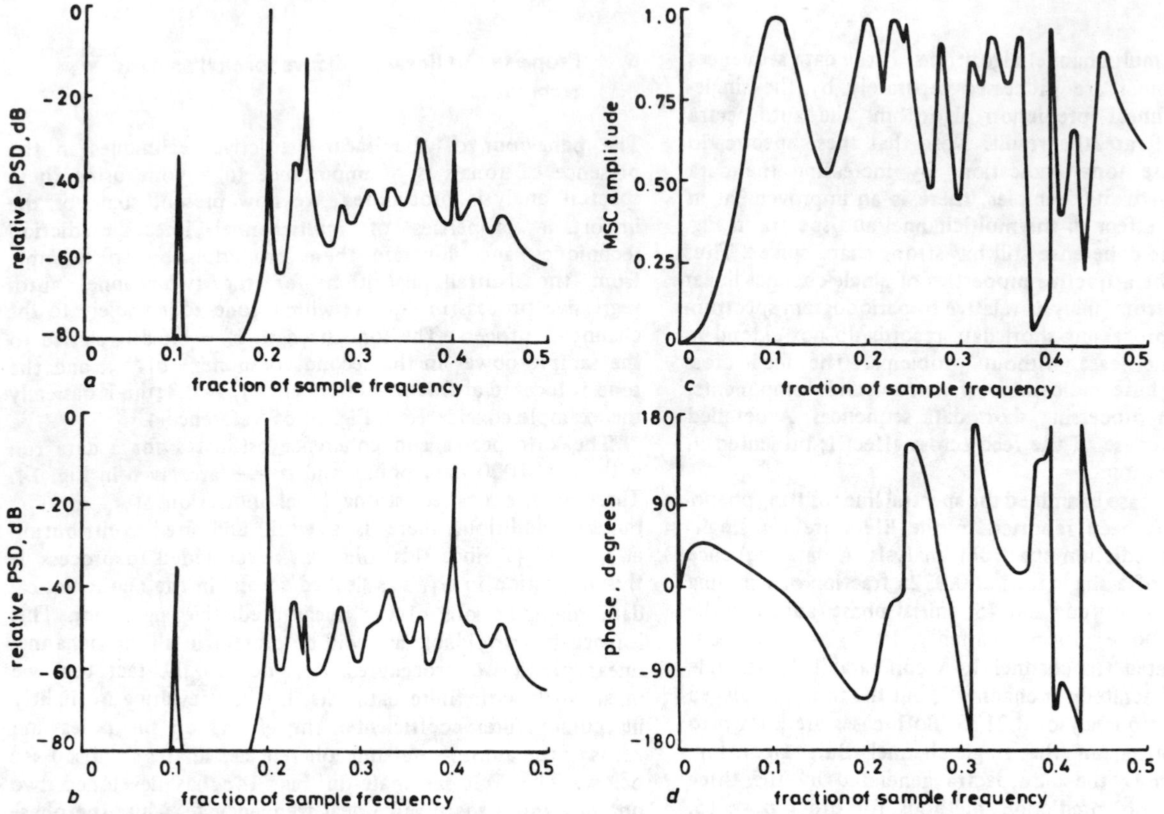

Fig. 8 *Autospectra and coherence of sinusoids in noise (64 samples): Marple covariance generalisation (forward)*

a G_{11} autospectrum
b G_{22} autospectrum
c Magnitude-squared coherence
d Coherence phase

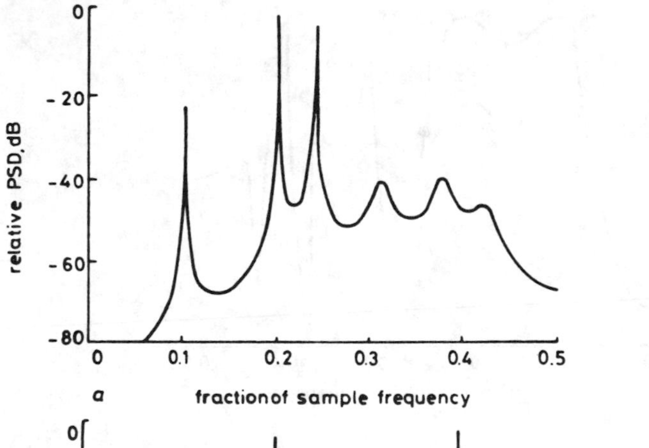

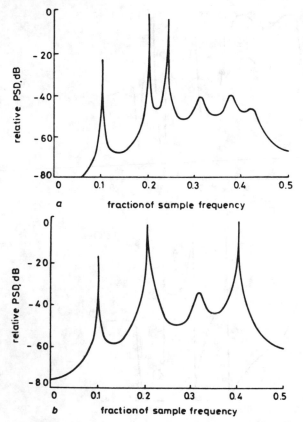

Fig. 9 *Autospectra of sinusoids in noise (64 samples): Single-channel Burg algorithm*

a G_{11} autospectrum
b G_{22} autospectrum

estimates in Fig. 14, we need to develop, in more detail, the matrix spectral estimate (see eqn. F-16 of Reference 1)

$$G(z) = [A(z)]^{-1} U_p [A(1/z^*)]^{-H} \tag{2}$$

where

$$A(z) = I + \sum_{n=1}^{p} z^{-n} A_n^{(p)} \tag{3}$$

For real data and filters, U_p and $\{A_n^{(p)}\}$ are real. For a two-channel application, eqn. 3 can be expressed as

$$A(z) = \begin{bmatrix} h_{11}(z) & h_{12}(z) \\ h_{21}(z) & h_{22}(z) \end{bmatrix} \tag{4}$$

where $\{h_{kl}(z)\}$ are scalars of the form

$$h_{kl}(z) = \sum_{n=0}^{p} z^{-n} a_{kl}(n) \tag{5}$$

There follows for eqn. 2

$$G(z) = \frac{1}{D(z)D(1/z)} \begin{bmatrix} h_{22}(z) & -h_{12}(z) \\ -h_{21}(z) & h_{11}(z) \end{bmatrix} \begin{bmatrix} u_{11} & u_{12} \\ u_{21} & u_{22} \end{bmatrix}$$

$$\times \begin{bmatrix} h_{22}(1/z) & -h_{21}(1/z) \\ -h_{12}(1/z) & h_{11}(1/z) \end{bmatrix} \tag{6}$$

where

$$D(z) = h_{11}(z)h_{22}(z) - h_{12}(z)h_{21}(z) \tag{7}$$

We denote

$$\left.\begin{aligned} h_{kl}(z) &= z^{-p} q_{kl}(z) \\ h_{kl}(1/z) &= r_{kl}(z) \end{aligned}\right\} \tag{8}$$

where $q_{kl}(z)$ and $r_{kl}(z)$ are polynomials of degree p in z with real coefficients. Then

$$\left.\begin{aligned} D(z) &= z^{-2p} Q(z) \\ D(1/z) &= R(z) \end{aligned}\right\} \tag{9}$$

where $Q(z)$ and $R(z)$ are polynomials of degree $2p$ in z. It follows that

$$Q(0) = a_{11}(p)a_{22}(p) - a_{12}(p)a_{21}(p)$$

$$R(0) = 1$$

neither of which is zero; thus $Q(z)$ and $R(z)$ have no zeros at the origin of the z-plane. The spectral estimate can now be expressed as

$$G(z) = \frac{z^p}{Q(z)R(z)} \begin{bmatrix} N_{11}(z) & N_{12}(z) \\ N_{21}(z) & N_{22}(z) \end{bmatrix} \tag{10}$$

where the 2 x 2 matrix in eqn. 10 is given by

$$\begin{bmatrix} q_{22}(z) & -q_{12}(z) \\ -q_{21}(z) & q_{11}(z) \end{bmatrix} \begin{bmatrix} u_{11} & u_{12} \\ u_{21} & u_{22} \end{bmatrix} \begin{bmatrix} r_{22}(z) & -r_{21}(z) \\ -r_{12}(z) & r_{11}(z) \end{bmatrix} \tag{11}$$

All the functions in eqn. 10 and expr. 11 are polynomials in z.

For the real data and filters employed here, the $2p$ zeros of $Q(z)$ occur in conjugate pairs (or in real pairs). Furthermore, they are all inside the unit circle C_1 in the complex z-plane; this property was proved in eqn. 33 of Reference 3. Thus we only need to search the interior of the unit hemisphere in the upper-half z-plane for p zeros of $Q(z)$.

Furthermore, if $Q(z)$ has a zero at z_0, then $R(z)$ has a zero at $1/z_0$; thus $G(z)$ in eqn. 10 has $4p$ poles in the z-plane, $2p$ in the upper halfplane, of which p are inside the unit hemisphere. A typical 4-tuple of poles in $G(z)$ is given by z_0, z_0^*, $1/z_0$, $1/z_0^*$; there are p such 4-tuples. These poles are common to all the auto- and cross-spectral estimates in eqn. 10.

Similarly, a typical term, $N_{kl}(z)$, in the numerator of $G(z)$ in eqn. 10 is a polynomial of degree $2p$ in z, of which p zeros will be located in the upper-half z-plane. Thus eqn. 10 shows that (every term of) $G(z)$ has p zeros at 0, p zeros at ∞ and $2p$ zeros in the finite plane. We need only search for the p zeros of $N_{kl}(z)$ in the upper-half plane, for $k, l = 1, 2$.

In order to explain the behaviour of the coherence estimates given by the linear predictive techniques, we start with the complex coherence at frequency f as given (for $\Delta t = 1$) by

$$C(f) = \frac{N_{12}(\exp(j2\pi f))}{[N_{11}(\exp(j2\pi f))N_{22}(\exp(j2\pi f))]^{1/2}} \tag{12}$$

The squared coherence, generalised to the entire z-plane, is

$$\mathscr{C}^2(z) = \frac{N_{12}^2(z)}{N_{11}(z)N_{22}(z)} \quad \left(\text{not} \quad \frac{N_{12}(z)N_{21}(z)}{N_{11}(z)N_{22}(z)}\right) \tag{13}$$

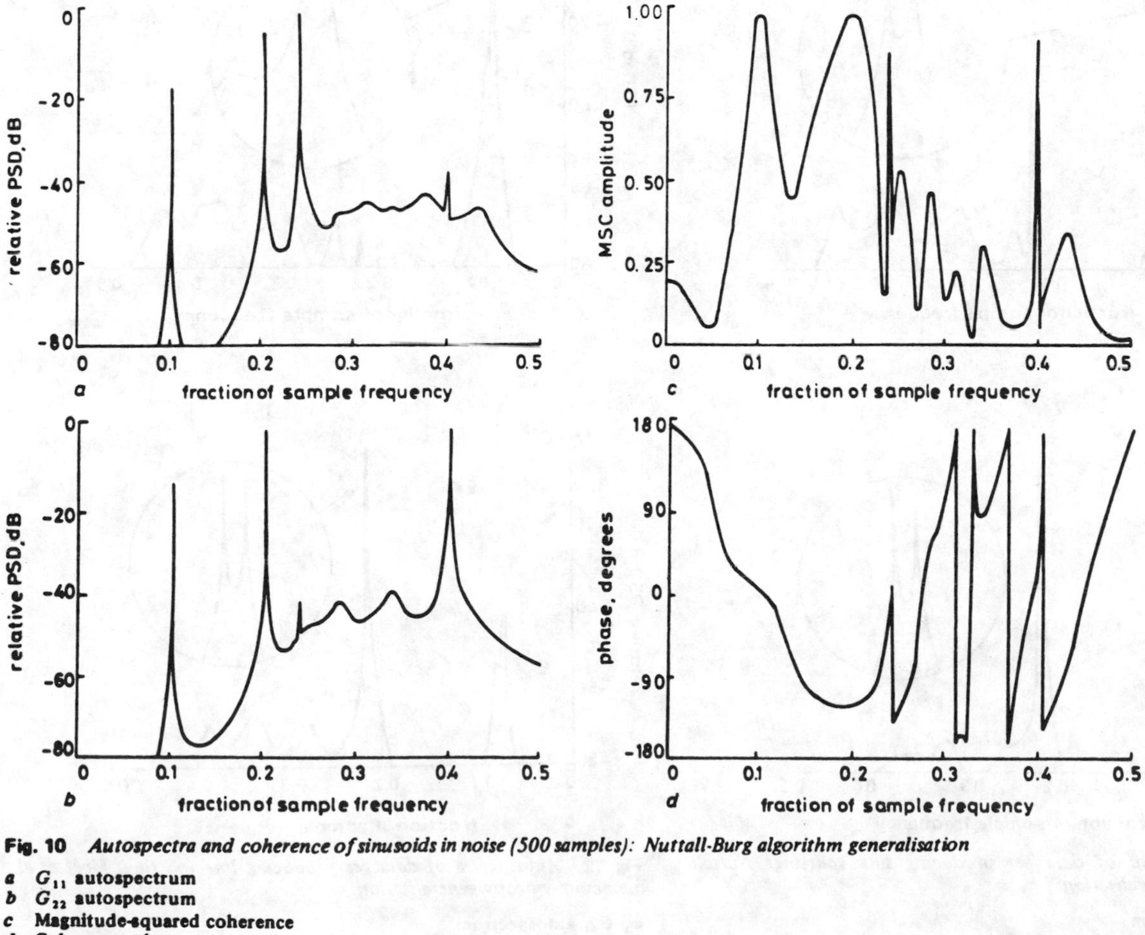

Fig. 10 *Autospectra and coherence of sinusoids in noise (500 samples): Nuttall-Burg algorithm generalisation*

a G_{11} autospectrum
b G_{22} autospectrum
c Magnitude-squared coherence
d Coherence phase

The following are general properties of the $N_{kl}(z)$ polynomials:

(*a*) If $N_{11}(z)$ has a zero at $r_1 \exp(j\theta_1)$, it also has a zero at $(1/r_1)\exp(j\theta_1)$.

(*b*) If $N_{22}(z)$ has a zero at $r_2 \exp(j\theta_2)$, it also has a zero at $(1/r_2)\exp(j\theta_2)$.

(*c*) If $N_{12}(z)$ has a zero at $r_0 \exp(j\theta_0)$, $N_{21}(z)$ has a zero at $(1/r_0)\exp(j\theta_0)$.

In fact, property (*c*) is a special case of the general rule that

$$N_{21}(z) = z^{2p} N_{12}(1/z) \qquad \text{for all } z \tag{14}$$

Now we return to the example of Fig. 14, where a tone was present only in the channel-2 data, at frequency f_c. Since process 1 has no tone at f_c, $N_{11}(z)$ develops *two* zeros near $z_c \equiv \exp(j2\pi f_c) = \exp(j\pi 0.6)$, one inside the unit circle C_1 and one outside C_1, tending to cancel the one zero of $D(z)$ inside C_1 and the one zero of $D(1/z)$ outside C_1, which are near z_c, so that the autospectral estimate $G_{11}(f)$ is well behaved near f_c. However, the cancellation is not perfect, and the small spike at f_c in Fig. 14*a* remains. For this example, the pole location inside C_1 is 0.993591 exp ($j\pi$ 0.600104), whereas the zero location inside C_1 is 0.992697 exp ($j\pi$ 0.600015). Since the pole is closer to C_1, a positive-going perturbation occurs in Fig. 14*a* at f_c.

Similarly, the cross-spectrum of $G(z)$ ideally should have no indication at f_c since there is no tone in process 1. In order to counter the zeros of $D(z)$ and $D(1/z)$ near z_c, $N_{12}(z)$ develops *two* zeros near z_c. For the same example, they are at 0.996958 exp ($j\pi$0.592593) and 0.983192 exp ($j\pi$ 0.615533), both of which happen to be inside C_1. Other examples have shown that these two zeros can both be outside C_1, or one can be inside and one outside C_1.

Finally, since process 2 does have a tone at f_c, $N_{22}(f)$ does *not* develop any zeros near z_c. Thus the zeros of $D(z)$ and $D(1/z)$ dominate near z_c, and the estimate $G_{22}(f)$ in Fig. 14*b* develops a large value near f_c, as desired.

The squared coherence estimate in eqn. 13 is independent of $D(z)$ or $Q(z)$. However, according to the discussion above, it has two double-zeros near z_c, owing to the $N_{12}^2(z)$ term. These four zeros can lie either all inside C_1, all outside C_1, or two inside and two outside C_1. $\mathscr{C}^2(z)$ also has two poles near z_c, owing to the two zeros of $N_{11}(z)$ near z_c: one pole lies inside C_1; the other lies outside C_1. Since there are no poles and zeros near z_c that cancel exactly, some very fine detail can develop in the coherence estimate in the neighbourhood of z_c. Sharp notches and spikes are the rule, not the exception, in the MSC evaluated on C_1 in the neighbourhood of a tonal frequency owing to this imperfect cancellation of poles and zeros. The phase variations can be so rapid that large FFT sizes are required to track it accurately. The two double-zeros of $N_{12}^2(z)$ cause this rapid variation, especially if they are all located on the same side of C_1; the two zeros of $N_{11}(z)$ are always located on opposite sides of C_1 and so do not themselves lead to a very rapid phase change near z_c, although they greatly influence the MSC in that region. In general, the squared coherence $\mathscr{C}^2(z)$ has $2p$ double-zeros and $4p$ poles in the complex z-plane, none restricted to be inside or outside of C_1; however, they all occur in conjugate pairs.

Since order $p = 8$ used in this example is larger than

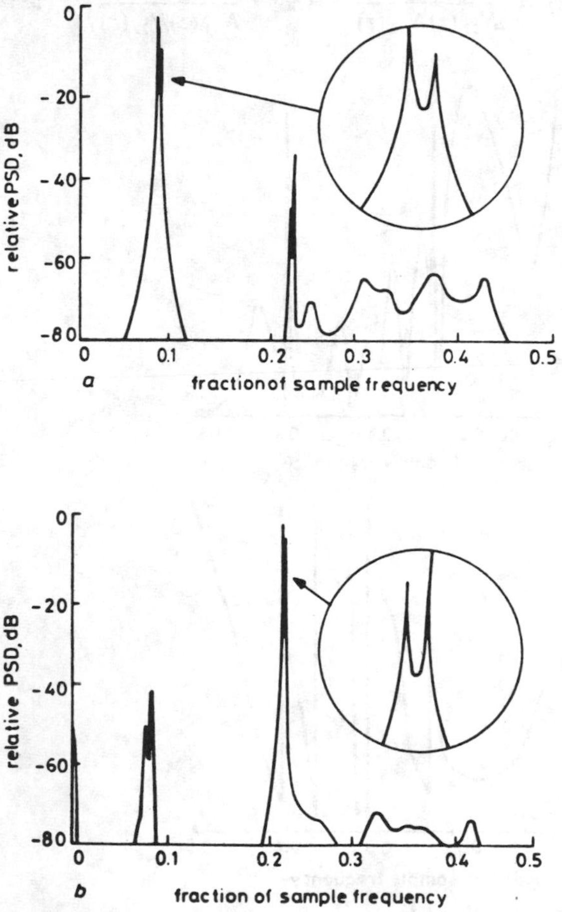

Fig. 11 *Autospectra of data set producing line splitting: Nuttall-Burg algorithm generalisation*

a G_{11} **autospectrum**
b G_{22} **autospectrum**

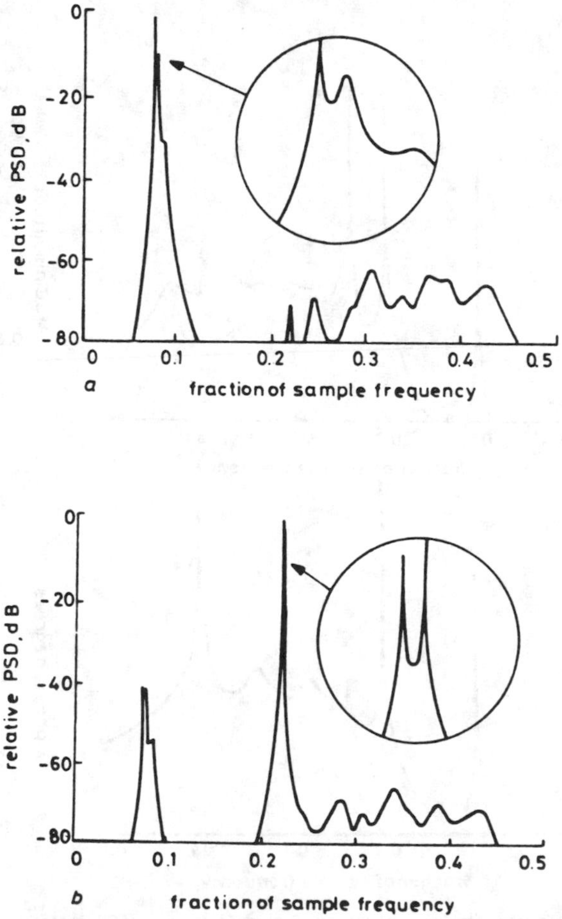

Fig. 12 *Autospectra of data set producing line splitting: Morf et al. maximum-entropy generalisation*

a G_{11} **autospectrum**
b G_{22} **autospectrum**

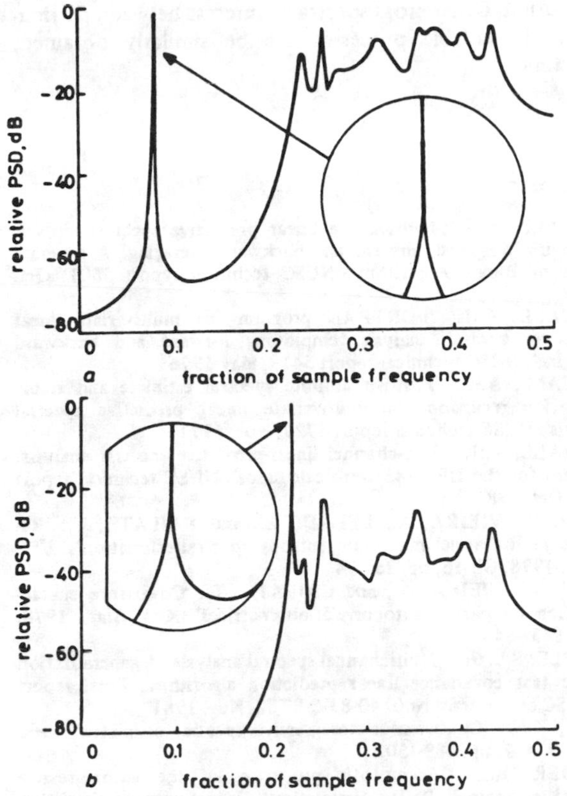

Fig. 13 *Autospectra of data set that produced line splitting in Figs. 11 and 12: Marple covariance algorithm generalisation (forward)*

a G_{11} **autospectrum**
b G_{22} **autospectrum**

necessary to account for a single tone and a narrowband spectral component, the extra poles [zeros of $Q(z)$] are distributed fairly uniformly inside C_1, with radii in the range 0.5 to 0.7. The extra zeros of polynomials $N_{kl}(z)$ can lie anywhere, either inside or outside of C_1.

6 Discussion and conclusions

Experimental results for two-channel linear prediction auto- and cross-spectral estimation have been presented for a first-order autoregressive process and for cases with interfering tones. It has been shown that some misleading estimates may be obtained because of feed-across in the mathematical manipulations of the finite lengths of data from each channel. This feed-across manifests itself as narrow spurious spikes in the spectral and coherence estimates. In order to circumvent this problem, while maintaining the high-resolution properties of linear prediction techniques, the following philosophy for multichannel spectrum analysis is suggested.

Suppose we are given finite data records of three stationary processes $x(t)$, $y(t)$ and $z(t)$, and we wish to estimate all the autospectra and cross-spectra involved. The Blackman and Tukey and weighted-FFT approaches evaluate the autospectrum of each process separately. Thus the spectrum of $x(t)$ is estimated without interference from $y(t)$ and $z(t)$; the availability of the data records for $y(t)$ and $z(t)$ plays no part in the eventual autospectral estimate for $x(t)$. Additionally, the cross-spectral estimate for processes $x(t)$ and $y(t)$ is independent of the available data on the $z(t)$ process. Finally, the coherence estimate between two processes is independent of any additional data records for other (statistically related) processes.

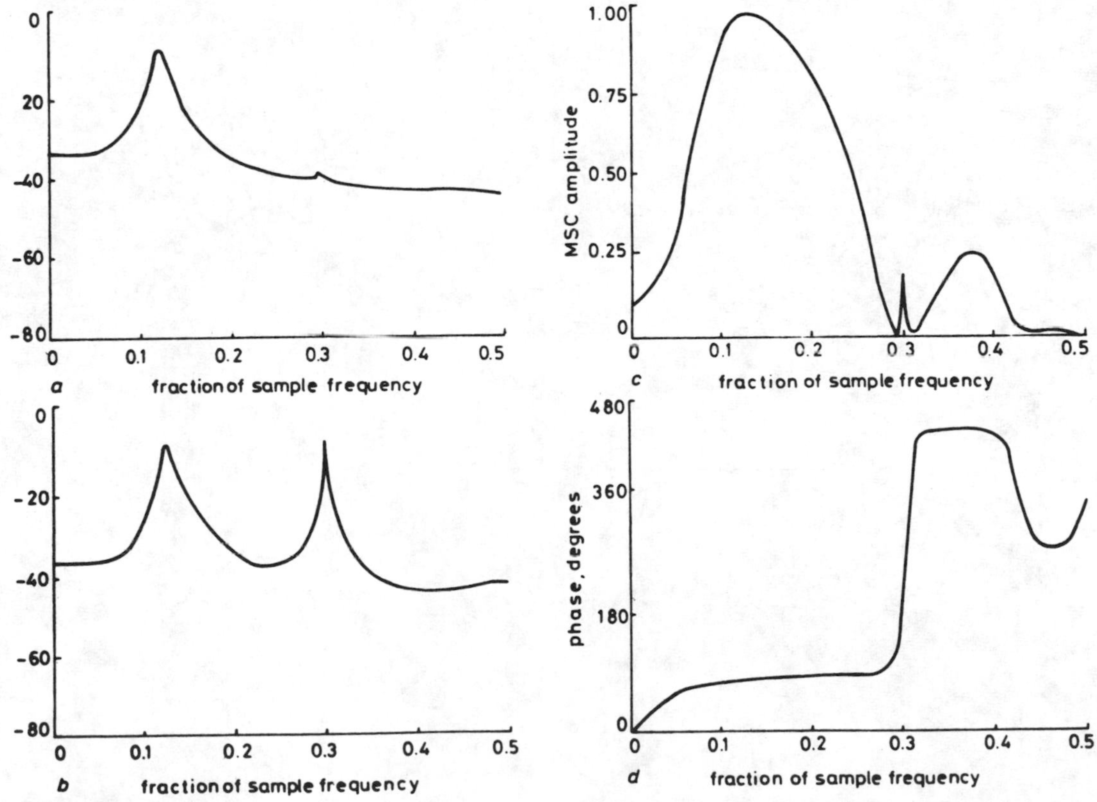

Fig. 14 *Autospectra and coherence of first-order autoregressive process plus single sinusoid (1000 samples, p = 8): Nuttall-Burg algorithm generalisation*

a G_{11} **autospectrum**
b G_{22} **autospectrum**
c **Magnitude-squared coherence**
d **Coherence phase**

On the other hand, the three multichannel linear predictive spectral analysis approaches give autospectral estimates of the $x(t)$ process that are dependent on the available values of $y(t)$ and $z(t)$. Also, the cross-spectral estimate between $x(t)$ and $y(t)$ is dependent on the particular $z(t)$ data available. This procedure can be poor for short data lengths if, for example, $y(t)$ contains a strong tone at f_0 that is not present in $x(t)$ or $z(t)$. Thus estimates of spectra $G_{xx}(f)$, $G_{xy}(f)$ and $G_{xz}(f)$ all contain tonal indications at f_0 that should not be there. These spurious tonal indications are due to feed-across between the available finite data segments of the various processes.

This raises the following questions:

(i) Should the estimate of $G_{xx}(f)$ be determined only from the available $x(t)$ data record?

(ii) Should the estimate of $G_{xy}(f)$ be determined only from the available $x(t)$ and $y(t)$ data records?

(iii) If coherence $C_{xy}(f_0) = 0$, why use $y(t)$ to estimate $G_{xx}(f_0)$?

(iv) If coherence $C_{xy}(f_0) = 1$, why use the completely statistically dependent $y(t)$ data to estimate $G_{xx}(f_0)$?

This philosophy of discarding 'irrelevant' data would be consistent with the Blackman and Tukey and FFT approaches. Carrying this philosophy on, we are led to the following: estimate $G_{xx}(f)$ solely from the $x(t)$ data by some single-channel linear predictive technique. Then estimate cross-spectrum $G_{xy}(f)$ or coherence $C(f)$ directly, by some linear predictive technique whose sole goal is linear prediction of $x(t)$ from $y(t)$ and vice versa, with no interest in or diversion from simultaneous estimation of $G_{xx}(f)$ or $G_{yy}(f)$. By this means, we can concentrate on extracting all the relevant cross-spectral information with maximum stability and resolution. Other cross-spectra of interest between particular pairs of available processes can be similarly obtained, one at a time.

7 References

1 NUTTALL, A.H.: 'Multivariate linear predictive spectral analysis employing weighted forward and backward averaging: A generalization of Burg's Algorithm'. NUSC technical report 5501, Oct. 1976

2 NUTTALL, A.H.: 'FORTRAN program for multivariate linear predictive spectral analysis, employing forward and backward averaging'. NUSC technical report 5419, May 1976

3 NUTTALL, A.H.: 'Positive definite spectral estimate and stable correlation recursion for multivariate linear predictive spectral analysis'. NUSC technical report 5729, Nov. 1977

4 NUTTALL, A.H.: 'Two-channel linear-predictive spectral analysis: Program for the HP 9845 desk calculator'. NUSC technical report 6533, Oct. 1981

5 MORF, M., VIEIRA, A., LEE, D.T.L., and KAILATH, T.: 'Recursive multichannel maximum entropy spectral estimation', *IEEE Trans.*, 1978, GE-16, pp. 85–94

6 MORF, M., VIEIRA, A., and KAILATH, T.: 'Covariance characterization by partial autocorrelation matrices', *Ann. Stat.*, 1978, **6**, pp. 643–648

7 MARPLE, S.L. Jr.: 'Multichannel spectral analysis: A generalization of the fast covariance linear prediction algorithm'. Final report to NUSC on contract N00140-81-C-BT82, Nov. 1981

8 BERK, K.N.: 'Consistent autoregressive spectral estimates', *Ann. Stat.*, 1974, **2**, pp. 489–502

9 KROMER, R.E.: 'Asymptotic properties of the autoregressive spectral estimator'. Ph.D. dissertation, Department of Statistics, Stanford University, California, USA, Dec. 1969

10 MORF, M., DICKINSON, B., KAILATH, T., and VIEIRA, A.: 'Efficient solution of covariance equations for linear prediction', *IEEE Trans.*, 1977, **ASSP-25**, pp. 429–433

FREQUENCY AND BEARING ESTIMATION BY TWO-DIMENSIONAL LINEAR PREDICTION

Leland B. Jackson and H. C. Chien
University of Rhode Island
Kingston, Rhode Island 02881

ABSTRACT

Simultaneous frequency and bearing estimation using 2-D spectral analysis of the space-time data array is investigated. The spectral estimates are generated using 2-D linear prediction. It is shown that single-quadrant prediction can lead to severe asymmetry and bias in the estimated spectra; while a certain combination of the results for two adjacent quadrants yields well-behaved spectral estimates.

INTRODUCTION

Results have recently appeared in the literature on the general properties of two-dimensional linear prediction (1),(2), and related work has been published in specific fields such as radar and seismic data processing (3),(4). We consider the specific problem of simultaneous frequency and bearing (angle of arrival) estimation in passive sonar, although the results should find application in other areas as well. This general problem area and associated DFT methods are described in (5).

We assume a line array of uniformly spaced sensors and treat the sampled outputs from the sensors as a two-dimensional data sequence. One dimension is thus spatial, and the other temporal. These two dimensions are usually processed separately, with spatial filters or "beams" being formed to estimate the bearings of the received signals and temporal filters (spectral analysis) being used to estimate the frequency content of the signals. This implicitly assumes, however, that the two estimation problems are separable, which is only approximately true; and hence improved performance is to be expected if the data are processed as a two-dimensional (2-D) sequence. In particular, 2-D spectra can be formed to exhibit frequency and bearing information simultaneously for detection and identification purposes.

To illustrate these points, we consider the simple case of a monochromatic plane wave with angular frequency ω received by the line array at an azimuthal angle (bearing) θ. The received 2-D signal is thus

$$x(n,m) = e^{j\omega(nT+mD)} + u(n,m) \tag{1}$$

where T is the sampling period, D is the time delay between elements of the array due to the bearing, $u(n,m)$ is the received noise, and we arbitrarily assume unit signal variance. A complex-valued signal is assumed because, as we have previously shown (6), significant advantages including reduced frequency bias and variance result from forming and processing the complex-valued (analytic) version of the received real-valued signal. The corresponding 2-D Discrete Fourier Transform is

$$X(e^{j\omega_1 T}, e^{j\omega_2 d}) = \sum_{n=1}^{N}\sum_{m=1}^{M} x(n,m)\, e^{-j(\omega_1 nT + \omega_2 md)}$$

$$= \sum_{n=1}^{N} e^{jnT(\omega-\omega_1)} \sum_{m=1}^{M} e^{jm(D\omega-d\omega_2)} + U(e^{j\omega_1 T}, e^{j\omega_2 d}) \tag{2}$$

where d is the time delay associated with a bearing of $\theta = \pi/2$ (i.e., $D = d \sin\theta$). Note that the signal portion of the spectrum in (2) is indeed separable, but the noise transform $U(e^{j\omega_1 T}, e^{j\omega_2 d})$ will not be, in general. Hence, degraded performance is to be expected if the data processing methods impose an assumption of separability.

DEVELOPMENT

The extension of 1-D linear prediction to two dimensions is relatively straightforward although certain difficulties arise, as we will note. If the predicted signal $\hat{x}(n,m)$ is assumed to be of the form

$$\hat{x}(n,m) = -\sum_{\substack{i=0 \\ (i,k)\neq(0,0)}}^{I}\sum_{k=0}^{K} a_{ik}\, x(n-i, m-k)$$

and the mean-square prediction error

$$e_o = E\left(|x(n,m) - \hat{x}(n,m)|^2 \right)$$

is minimized, the resulting normal equations for the prediction coefficients a_{ik} are readily shown to be

$$\sum_{i=0}^{I}\sum_{k=p}^{K} a_{ik}\, R(p-i, q-k) = 0 \tag{3}$$

Reprinted from *IEEE Int. Conf. Acoust., Speech, Signal Processing,* 1979, pp. 665–668.

$p = 0,1,\ldots I;\ q = 0,1,\ldots K;\ (p,q) \neq (0,0);$

where $a_{00} = 1$ and $R(p,q) = E\left(x(n,m)\ x^*(n-p,m-q)\right)$. The corresponding minimum mean-square error is

$$e_{min} = \sum_{i=0}^{I} \sum_{k=0}^{K} a_{ik}\ R(i,k)$$

and the corresponding inverse (whitening) 2-D filter is

$$H_1(z_1,z_2) = \sum_{i=0}^{I} \sum_{k=0}^{K} a_{ik}\ z_1^{-i}\ z_2^{-k}\ . \tag{4}$$

The associated 2-D spectral estimate is

$$\hat{S}_1(\omega_1,\omega_2) = \frac{K_o}{\left|H(e^{j\omega_1 T}, e^{j\omega_2 d})\right|^2} \tag{5}$$

where K_o is a normalizing constant.

For the case of the monochromatic plane wave in white noise, the space-time correlation function is given by

$$R(p,q) = e^{j\omega(pT+qD)} + \sigma^2\ \delta(p,q) \tag{6}$$

where σ^2 is the noise power. The corresponding solution to the normal equations is simply

$$a_{ik} = g_{ik}\ e^{j\omega(iT + kD)} \qquad (a_{00} = 1) \tag{7}$$

where the g_{ik} are real-valued constants. In the noise-free case ($\sigma^2 = 0$), the g_{ik} are not uniquely determined; but with noise, the g_{ik} are unique and equal with value

$$g = \frac{-1}{(I+1)\ (K+1)\ -1 + \sigma^2}\ . \tag{8}$$

Hence in the simplest (first-order) case of I=1, K=1, we have $g = -1/(3+\sigma^2)$, and the inverse filter is

$$H_1(z_1,z_2) = 1 + ge^{j\omega T} z_1^{-1} + ge^{j\omega D} z_2^{-1} \tag{9}$$

$$+\ ge^{j\omega(T+D)} z_1^{-1}\ z_2^{-1}\ .$$

Note that $H_1(z_1,z_2)$ is not separable, i.e., it cannot be written in the form $(1 - \sigma_1 z_1^{-1})(1 - \sigma_2 z_2^{-1})$, even in the noise-free case.

The spectral estimate $\hat{S}_1(\omega_1,\omega_2)$ is readily evaluated from $H_1(z_1,z_2)$ as given in (5), and one finds here a single spectral peak at $\omega_1 = \omega$, $\omega_2 = \omega \sin\theta$. Hence, the location of this peak is an asymptotically unbiased estimator of ω (or, given ω, of sin θ), as is true in the corresponding 1-D case (6). However, in other respects, the spectral estimate in (5) is not satisfactory. Figure 1 shows $\hat{S}_1(\omega_1,\omega_2)$ for T = d sin θ, I = K = 2, and $\sigma^2 = 1$, and we note that the spectral peak has an elliptic symmetry, rather than the circular symmetry we might expect. If we change the sign of θ, we get Fig. 2, which is a simple translation of Fig. 1, not a mirror image. Since the temporal spectrum for a signal at bearing θ corresponds to a radial "slice" of $\hat{S}_1(\omega_1,\omega_2)$ for $\omega_2=(\omega_1 d\sin\theta)/T$, this means that the temporal resolution is better for positive θ than for negative θ, and vice versa for the spatial resolution.

The spectral estimate $\hat{S}_1(\omega_1,\omega_2)$ is even worse in the case of multiple sinusoidal components. Fig. 3 shows the case of two components at the same bearing θ with I=K=3 and $\sigma^2=1$, and severe bias is noted (the correct locations for the spectral peaks are marked by x's).

The problem here is related to that encountered in 2-D spectral factorization (7), where it is found that a single-quadrant filter is not sufficient to realize a general 2-D magnitude response. That is, the spectral estimate should incorporate a second inverse filter $H_2(z_1,z_2)$ corresponding to second-(or fourth)-quadrant prediction, i.e.,

$$H_2(z_1,z_2) = \sum_{i=0}^{I} \sum_{k=0}^{K} b_{ik} z_1^{-i}\ z_2^{k}\ . \tag{10}$$

The coefficients b_{ik} are found to be

$$b_{ik} = ge^{j\omega(iT-kD)} \tag{11}$$

with g given, as before, by (8). The corresponding $\hat{S}_2(\omega_1,\omega_2)$ for the case in Fig. 1 is shown in Fig. 4 and is seen to be the mirror image of Fig. 2 about the ω_1 axis.

One might now expect the appropriate spectral estimate to be

$$\hat{S}_3(\omega_1,\omega_2) = \frac{K_o}{|H_1 \cdot H_2|} \tag{12}$$

which is shown in Fig. 5. However, a better choice appears to be

$$\hat{S}_4(\omega_1,\omega_2) = \frac{2K_o}{|H_1|^2 + |H_2|^2} \tag{13}$$

as shown in Fig. 6. Note the approximately circular symmetry of the peak in $\hat{S}_4$. Even more significant are the cases of $\hat{S}_3$ and $\hat{S}_4$ for multiple components at the same bearing θ in Figures 7 and 8, respectively. In particular, the bias in the location of the spectral peaks in $\hat{S}_4$ is imperceptible (although nonzero); while in $\hat{S}_3$, there is significant bias. Fig. 9 shows $\hat{S}_4$ for different θ's.

ACKNOWLEDGMENT

This research was supported by the National Science Foundation under Grant ENG76-00735.

REFERENCES

1. T.L. Marzetta, "A Linear Prediction Approach to Two-Dimensional Spectral Factorization and Spectral Estimation", Ph.D. Thesis, M.I.T., Cambridge, MA., Feb. 1978.

2. J.H. Justice, "A Levinson-Type Algorithm for Two-Dimensional Wiener Filtering using Bivariate Szego Polynomials", Proc. IEEE, Vol. 65, No. 6, pp. 882-886, June 1977.
3. L.E. Brennan and I.S. Reed, "Theory of Adaptive Radar", IEEE Trans., Vol. AES-9, p. 237, 1973.
4. E.A. Robinson, Statistical Communication and Detection with Special Reference to Digital Data Processing of Radar and Seismic Signals, New York: Hafner, 1967.
5. O.S. Halpery and D.G. Childers, "Composite Wavefront Decomposition via Multidimensional Digital Filtering of Array Data", IEEE Trans., Vol. CAS-22, No. 6, pp. 552-563, June 1975.
6. L.B. Jackson, D.W. Tufts, F.K. Soong and R.M. Rao, "Frequency Estimation by Linear Prediction", Proc. IEEE ICASSP, Tulsa, OK., pp. 352-6, 1978.
7. M.P. Ekstrom and J.W. Woods, "Two-Dimensional Spectral Factorization with Applications in Recursive Digital Filtering", IEEE Trans., Vol. ASSP-24, No. 2, pp. 115-128, April 1976.

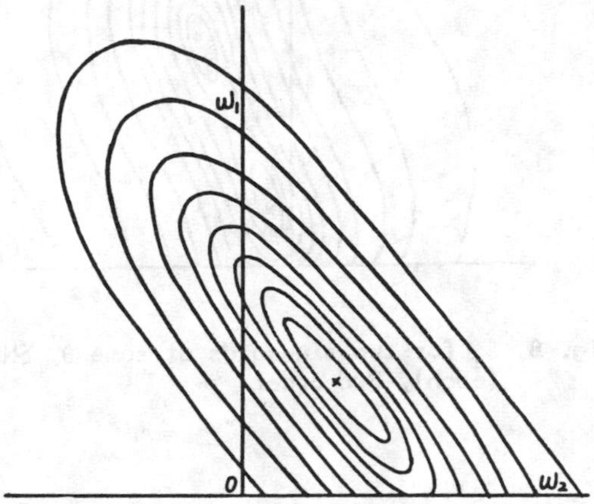

Fig. 1. $\hat{S}_1$ for one sinusoid, $\theta>0$, SNR=0db, 2nd order.

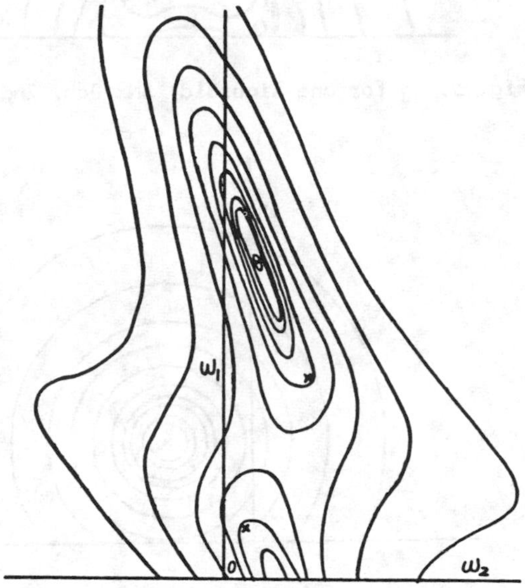

Fig. 3. $\hat{S}_1$ for two sinusoids at same θ, SNR=0db (each), 3rd order, showing large bias.

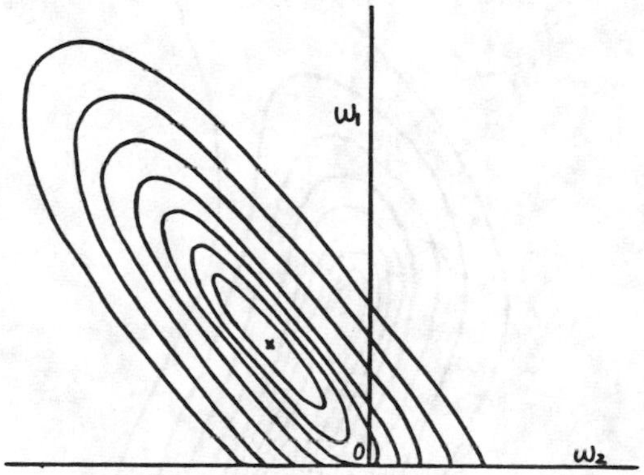

Fig. 2. $\hat{S}_1$ for one sinusoid, $\theta<0$, SNR=0db, 2nd order.

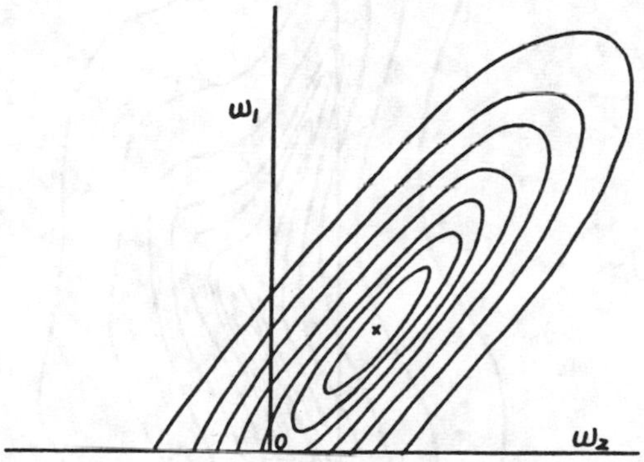

Fig. 4. $\hat{S}_2$ for one sinusoid, $\theta>0$, SNR=0db, 2nd order.

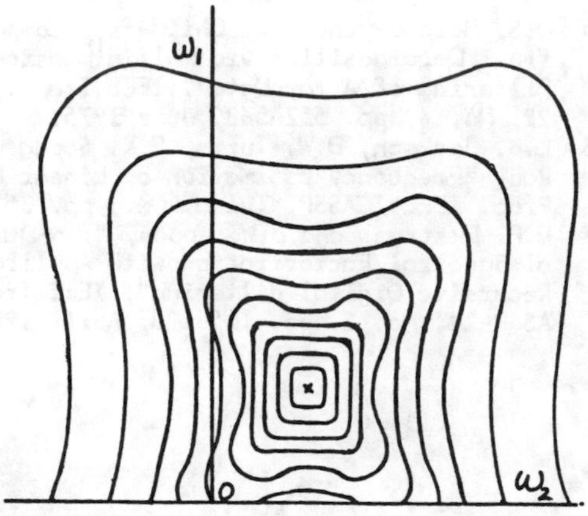

Fig. 5. $\hat{S}_3$ for one sinusoid, SNR=0db, 2nd order.

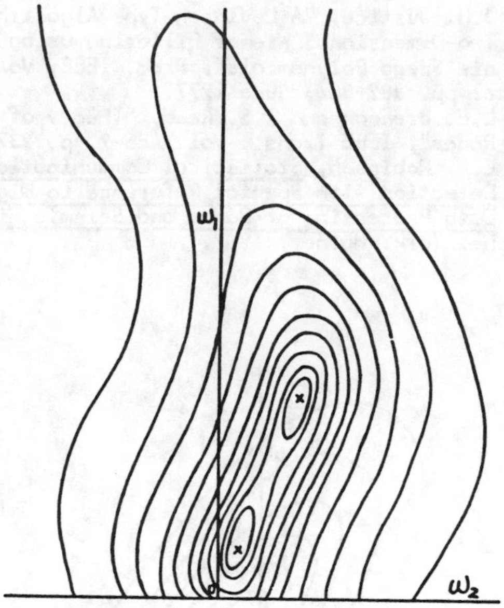

Fig. 8. $\hat{S}_4$ for two sinusoids at same θ, SNR=0db (each), 3rd order.

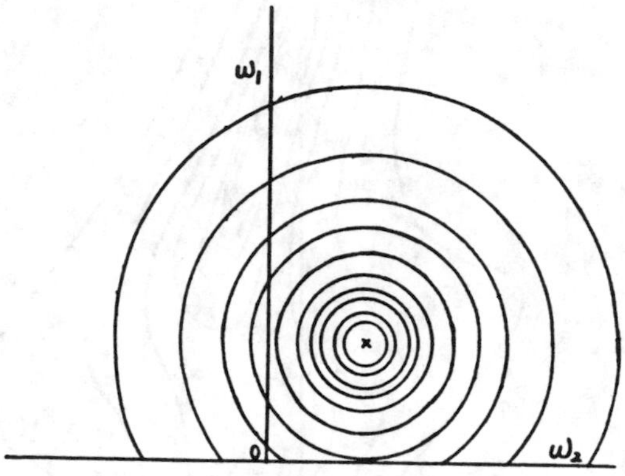

Fig. 6. $\hat{S}_4$ for one sinusoid, SNR=0db, 2nd order.

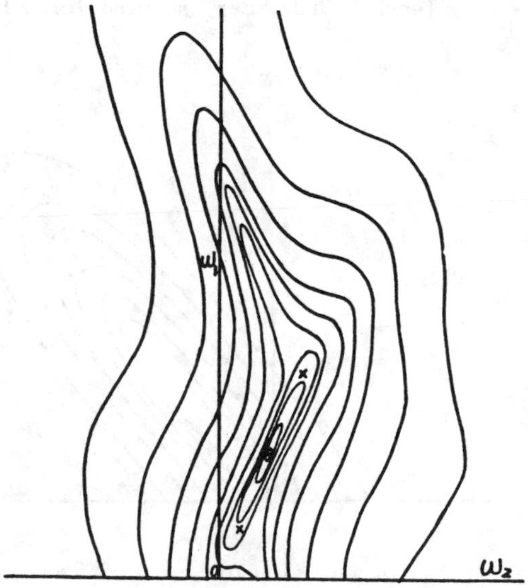

Fig. 7. $\hat{S}_3$ for two sinusoids at same θ, SNR=0db (each), 3rd order, showing only one peak.

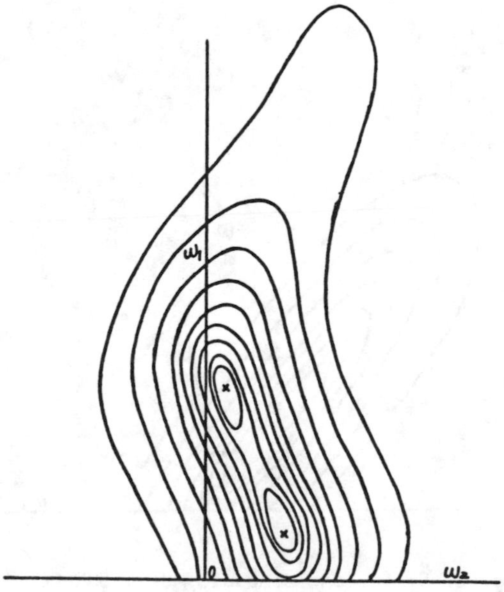

Fig. 9. $\hat{S}_4$ for two sinusoids at different θ, SNR=0db (each), 3rd order.

A New Algorithm for Two-Dimensional Maximum Entropy Power Spectrum Estimation

JAE S. LIM, MEMBER, IEEE, AND NAVEED A. MALIK, STUDENT MEMBER, IEEE

Abstract—**A new iterative algorithm for the maximum entropy power spectrum estimation is presented in this paper. The algorithm, which is applicable to two-dimensional signals as well as one-dimensional signals, utilizes the computational efficiency of the fast Fourier transform (FFT) algorithm and has been empirically observed to solve the maximum entropy power spectrum estimation problem. Examples are shown to illustrate the performance of the new algorithm.**

Manuscript received August 6, 1980; revised January 6, 1981. This work was supported in part by the Defense Advanced Research Projects Agency monitored by ONR under Contract N00014-75-C-0951-NR049-328 at RLE and in part by the Department of the Air Force under Contract F19628-78-C-0002 at Lincoln Laboratory. The views and conclusions contained in this document are those of the contractor and should not be interpreted as necessarily representing the official policies, either expressed or implied, of the United States Government.

The authors are with the Lincoln Laboratory and the Research Laboratory of Electronics, Department of Electrical Engineering and Computer Science, Massachusetts Institute of Technology, Cambridge, MA 02139.

I. Introduction

THE problem of power spectrum estimation (PSE) arises in various fields such as speech processing [1], seismic signal processing [2], image restoration [3], radar [4], sonar [5], radio astronomy, etc., and its applications range from identifying signal source parameters and transmission channel characteristics to removing noise from images [3]. Consequently, this problem has received considerable attention in the literature and a variety of techniques for power spectrum estimation have been developed.

One technique which has been studied extensively due to its high resolution characteristics is the maximum entropy (ME) method. For one-dimensional (1-D) signals, this method is equivalent [7] to autoregressive signal modeling, and thus it leads to a linear problem formulation that is theoretically tractable and computationally attractive [7]. Unlike most

Reprinted from *IEEE Trans. Acoust., Speech, Signal Processing*, vol. ASSP-29, pp. 401–413, June 1981.

other PSE techniques such as the conventional methods [8], [9] and the maximum likelihood method [10], however, the ME method does not extend from 1-D signals to two-dimensional (2-D) signals in a straightforward manner, and the ME method for 2-D signal PSE remains a highly nonlinear problem.

To solve the nonlinear ME PSE problem for 2-D signals, various attempts [2], [11]-[13] have been made in the literature. In all cases, however, the algorithms are computationally unattractive, and there is no guarantee of a solution or only an approximate solution can be obtained. For example, Burg [11] has proposed an iterative solution which requires the inversion of a matrix in each iteration where the dimension of the matrix is of the order of the number of the given autocorrelation points. No experimental results using this technique have yet been reported. As another example, Wernecke and D'Addario [12] have proposed a scheme in which an attempt is made to numerically maximize the entropy. The maximization is done by continuously adjusting the power spectrum (PS) estimate and evaluating the expressions for the entropy and its gradient. The procedure is computationally expensive and is not guaranteed to have a solution. As a third example, Woods [2] expresses the ME PS estimate as a power series in the frequency domain and attempts to approximate the ME PS estimate by truncating the power series expansion. Even though such an approach has some computational advantages relative to others, the method is restricted to the class of signals for which the power series expansion is possible. Furthermore, examples have been found in which the algorithm does not converge to the desired ME PS estimate.

In this paper, we develop a new iterative algorithm which is computationally simple due to its utilization of the fast Fourier transform (FFT) algorithm, and which has been empirically observed to lead to the true ME power spectrum estimate for 2-D signals as well as 1-D signals. In Section II, we briefly review previous results on the ME PSE for 2-D signals. In Section III, we develop a new algorithm for the ME PSE. In Section IV, we illustrate and discuss the performance of this algorithm by way of various examples.

II. Previous Results on Maximum Entropy Power Spectrum Estimation for 2-D Signals

In this section, we briefly review important previous results relevant to this paper on the ME PSE for 2-D signals. In reviewing these results, we use the following notations.

- $x(n_1, n_2)$: A 2-D random signal whose power spectrum we wish to estimate.
- $R_x(n_1, n_2)$: Autocorrelation function of $x(n_1, n_2)$.
- $\hat{R}_x(n_1, n_2)$: An estimate of $R_x(n_1, n_2)$.
- $P_x(\omega_1, \omega_2)$: Power spectrum of $x(n_1, n_2)$.
- $\hat{P}_x(\omega_1, \omega_2)$: An estimate of $P_x(\omega_1, \omega_2)$.
- $\lambda(n_1, n_2)$: Autocorrelation function whose power spectrum is $1/P_x(\omega_1, \omega_2)$.
- A: A set of points (n_1, n_2) for which $R_x(n_1, n_2)$ is known.
- F: Discrete time Fourier transform.
- F^{-1}: Inverse discrete time Fourier transform.

With the above notations, the ME PSE problem can be stated as follows.

Given $R_x(n_1, n_2)$ for $(n_1, n_2) \in A$, determine $\hat{P}_x(\omega_1, \omega_2)$ such that the entropy H given by

$$H = \int_{\omega_1=-\pi}^{\pi} \int_{\omega_2=-\pi}^{\pi} \log \hat{P}_x(\omega_1, \omega_2)\, d\omega_1\, d\omega_2 \tag{1}$$

is maximized and

$$R_x(n_1, n_2) = F^{-1}[\hat{P}_x(\omega_1, \omega_2)] \quad \text{for} \quad (n_1, n_2) \in A. \tag{2}$$

By rewriting $\hat{P}_x(\omega_1, \omega_2)$ in terms of $R_x(n_1, n_2)$ for $(n_1, n_2) \in A$ and $\hat{R}_x(n_1, n_2)$ for $(n_1, n_2) \notin A$ and then setting $dH/d\hat{R}_x(n_1, n_2) = 0$ for $(n_1, n_2) \notin A$, it can be shown that the above problem is equivalent to the following.

Given $R_x(n_1, n_2)$ for $(n_1, n_2) \in A$, determine $\hat{P}_x(\omega_1, \omega_2)$ such that $\hat{P}_x(\omega_1, \omega_2)$ is in the form of

$$\hat{P}_x(\omega_1, \omega_2) = \frac{1}{\sum\sum_{(n_1,n_2) \in A} \lambda(n_1, n_2) \cdot e^{-j\omega_1 n_1} \cdot e^{-j\omega_2 n_2}} \tag{3}$$

and

$$R_x(n_1, n_2) = F^{-1}[\hat{P}_x(\omega_1, \omega_2)] \quad \text{for} \quad (n_1, n_2) \in A. \tag{4}$$

The above problem statement for the ME PSE applies, with appropriate dimensionality changes, to all signals regardless of their dimensionality. The solutions to the problem, however, strongly depend on the signal dimensionality. For 1-D signals, the mean-square error minimization of the prediction filter based on autoregressive signal modeling requires solving a set of linear equations for the filter coefficients, and the power spectrum obtained from the estimated filter coefficients is identical to the ME PS estimate. For 2-D signals, this is no longer the case. Specifically, even though minimizing the mean-square error of the autoregressive filter still requires solving a set of linear equations, the power spectrum obtained from the estimated filter coefficients is not the ME PS estimate. The reason for this can be seen by examining the form of the normal equations for the filter coefficients in the autoregressive signal modeling. The derivation of the general form of the normal equations for 2-D signals is analogous to that for 1-D signals and is given by

$$\sum\sum_{(i,j) \in B} a_{ij} R_x(r - i, s - j) = R_x(r, s) \quad \text{for} \quad (r, s) \in B \tag{5}$$

where a_{ij} represents the autoregressive filter coefficients to be estimated, the set B consists of all points where the filter mask has nonzero values, and the power spectrum obtained from a_{ij} is given by

$$\hat{P}_x(\omega_1, \omega_2) = \frac{1}{\left| \sum\sum_{(k,l) \in B} a_{kl} \cdot e^{-j\omega_1 k} \cdot e^{-j\omega_2 l} \right|^2}. \tag{6}$$

From (5), for any nontrivial choice of B, that is, if B does not consist of a set of collinear points, the size of independent values of $R_x(n_1, n_2)$ required to solve the above set of equations is greater than the size of the filter mask. For example,

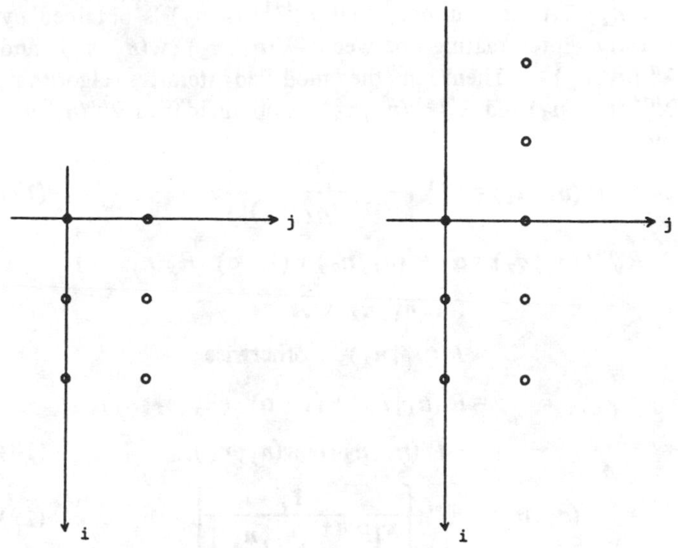

(a) (b)

Fig. 1. (a) First quadrant autoregressive filter mask of size 3 × 2. (b) Independent autocorrelation points required to solve the normal equations for the mask of (a).

consider the filter mask shown in Fig. 1(a) in which the open dots represent the range for which a_{ij} is nonzero. Fig. 1(b) shows the size of independent values of $R_x(n_1, n_2)$ required to solve for a_{ij} in Fig. 1(a) by (5). Clearly, the number of correlation points needed is greater than the number of filter coefficients. Since the estimated power spectrum given by (6) is completely determined by the filter coefficients alone, it does not possess enough degrees of freedom to satisfy (4), which is required for the ME PS estimate. Due to this difficulty, a closed form solution for the 2-D ME PSE problem has not yet been found.

In the absence of a closed form solution, it is important to know the conditions for the existence and uniqueness of the solution. In this regard, Woods [2] has obtained the theoretical result that, if the given $R_x(n_1, n_2)$ for $(n_1, n_2) \in A$ is a part of some positive definite correlation function [meaning that its Fourier transform is positive for all (ω_1, ω_2)], a solution to the ME PSE problem exists and is unique. In general, it is difficult [14] to determine if the given segment of the correlation function is a part of some positive definite correlation function, even though this is generally the case in most practical problems. In this paper, we assume that the given segment of the correlation function indeed forms a part of some positive definite correlation function, so that the solution to the ME PSE problem exists and is unique.

III. A New Iterative Algorithm

In this section, we develop a new iterative algorithm for the ME PS estimates which is applicable to both 1-D and 2-D signals. This algorithm is computationally simple, since it utilizes the computational efficiency of the FFT algorithm.

Suppose we are given $R_x(n_1, n_2)$ for $(n_1, n_2) \in A$ such that $R_x(n_1, n_2)$ is a segment of some positive definite correlation function. To find the unique ME PS estimate $\hat{P}_x(\omega_1, \omega_2)$, we express a power spectrum $P_y(\omega_1, \omega_2)$ as follows.

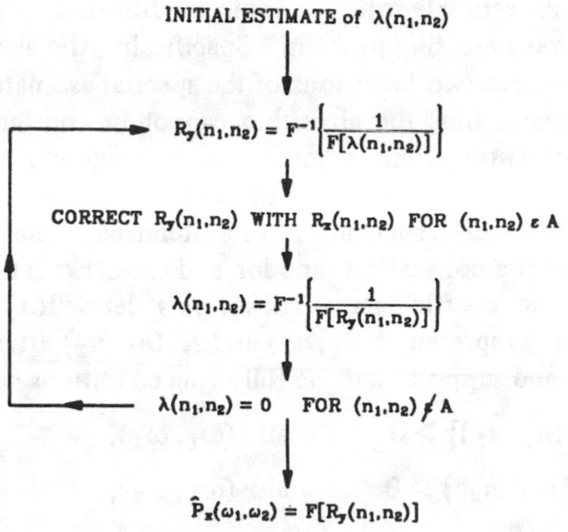

Fig. 2. A new approach to 2-D maximum entropy (ME) power spectrum estimation (PSE).

$$P_y(\omega_1, \omega_2) = F[R_y(n_1, n_2)] = \sum_{n_1=-\infty}^{\infty} \sum_{n_2=-\infty}^{\infty} R_y(n_1, n_2) \cdot e^{-j\omega_1 n_1} \cdot e^{-j\omega_2 n_2} \tag{7}$$

and

$$\frac{1}{P_y(\omega_1, \omega_2)} = F[\lambda(n_1, n_2)] = \sum_{n_1=-\infty}^{\infty} \sum_{n_2=-\infty}^{\infty} \lambda(n_1, n_2) \cdot e^{-j\omega_1 n_1} \cdot e^{-j\omega_2 n_2}. \tag{8}$$

From (7) and (8), it is clear that $R_y(n_1, n_2)$ can be obtained from $\lambda(n_1, n_2)$ and vice versa through Fourier transform operations. Now, from (3) and (4), $P_y(\omega_1, \omega_2)$ is the unique ME PS estimate if and only if $\lambda(n_1, n_2) = 0$ for $(n_1, n_2) \notin A$ and $R_y(n_1, n_2) = R_x(n_1, n_2)$ for $(n_1, n_2) \in A$. Thus, we see that for $P_y(\omega_1, \omega_2)$ to be the desired ME PS estimate, we have a constraint on $R_y(n_1, n_2)$ and a constraint on $\lambda(n_1, n_2)$. Recognizing this, it is straightforward to develop a simple iterative algorithm to find the unique ME PS estimate. Specifically, we go back and forth between $R_y(n_1, n_2)$ (the correlation domain) and $\lambda(n_1, n_2)$ (the coefficient domain) and at each time impose the constraints on $R_y(n_1, n_2)$ and $\lambda(n_1, n_2)$. Thus, starting with some initial estimate for $\lambda(n_1, n_2)$, we obtain an estimate for $R_y(n_1, n_2)$. This estimate is then corrected by the given $R_x(n_1, n_2)$ over the region A and is used to generate a new $\lambda(n_1, n_2)$. The new $\lambda(n_1, n_2)$ is then truncated to the desired limits and this procedure is repeated. The above iterative procedure is illustrated in Fig. 2 and forms the basis for a new iterative algorithm for the ME PSE.

The iterative procedure discussed above is very similar in form to other iterative techniques [15], [16] that have been successfully used in image processing. Even though the conditions under which the algorithm converges are not yet known, if the algorithm converges, the converging solution satisfies both (3) and (4) and consequently is the desired ME PS estimate.

The algorithm in Fig. 2 cannot, in general, be used to obtain the ME PS estimate without some modifications due to the spectral zero crossing problem. Specifically, the algorithm in Fig. 2 requires two inversions of the spectral estimates in each iteration, and thus the algorithm cannot be continued if the power spectrum estimate has a zero crossing at any stage in the iterative procedure. Unfortunately, zero crossings can occur in two different ways in each iteration. One is the correction of the correlation function and the other is the truncation of the coefficients. To see this, let $\lambda^m(n_1, n_2)$ and $R_y^m(n_1, n_2)$ represent $\lambda(n_1, n_2)$ and $R_y(n_1, n_2)$ after the mth iteration and suppose that the following conditions hold:

$$F[\lambda^m(n_1, n_2)] > 0 \quad \text{for all } (\omega_1, \omega_2), \tag{9}$$

$$F[R_y^m(n_1, n_2)] > 0 \quad \text{for all } (\omega_1, \omega_2), \tag{10}$$

and

$$\lambda^m(n_1, n_2) = F^{-1}\left[\frac{1}{F[R_y^m(n_1, n_2)]}\right] \cdot w(n_1, n_2) \tag{11}$$

where $w(n_1, n_2)$ represents a rectangular type window such that

$$w(n_1, n_2) = \begin{cases} 1 & \text{for } (n_1, n_2) \in A \\ 0 & \text{otherwise.} \end{cases} \tag{12}$$

Similarly, let $\lambda^{m+1}(n_1, n_2)$ and $R_y^{m+1}(n_1, n_2)$ represent $\lambda(n_1, n_2)$ and $R_y(n_1, n_2)$ after the $m + 1$th iteration. In the iterative algorithm of Fig. 2, $\lambda^{m+1}(n_1, n_2)$ and $R_y^{m+1}(n_1, n_2)$ are obtained from $\lambda^m(n_1, n_2)$ by

$$R'(n_1, n_2) = F^{-1}\left[\frac{1}{F[\lambda^m(n_1, n_2)]}\right], \tag{13}$$

$$\begin{aligned} R^{m+1}(n_1, n_2) &= R_x(n_1, n_2) \quad \text{for } (n_1, n_2) \in A \\ &\quad R'(n_1, n_2) \quad \text{otherwise} \\ &= R'(n_1, n_2) + (R_x(n_1, n_2) - R'(n_1, n_2)) \\ &\quad \cdot w(n_1, n_2), \end{aligned} \tag{14}$$

$$\lambda'(n_1, n_2) = F^{-1}\left[\frac{1}{F[R_y^{m+1}(n_1, n_2)]}\right], \tag{15}$$

and

$$\begin{aligned} \lambda^{m+1}(n_1, n_2) &= \lambda'(n_1, n_2) \quad \text{for } (n_1, n_2) \in A \\ &\quad 0 \quad \text{otherwise} \\ &= \lambda'(n_1, n_2) \cdot w(n_1, n_2). \end{aligned} \tag{16}$$

From (13)-(16), it is clear that $R'(n_1, n_2)$ is positive definite since $\lambda^m(n_1, n_2)$ is assumed to be positive definite, but $R_y^{m+1}(n_1, n_2)$ may not be positive definite due to the rectangular windowing $w(n_1, n_2)$ in (14). Furthermore, even if $R_y^{m+1}(n_1, n_2)$ were positive definite so that $\lambda'(n_1, n_2)$ is positive definite, $\lambda^{m+1}(n_1, n_2)$ may not be positive definite due to $w(n_1, n_2)$ in (16).

To ensure that the resulting $R_y^{m+1}(n_1, n_2)$ and $\lambda^{m+1}(n_1, n_2)$ are positive definite so that the iterations can be continued, we make modifications to (14) and (16). Specifically, suppose that $R_y^{m+1}(n_1, n_2)$ is obtained by linearly interpolating between $R'(n_1, n_2)$ and the known values $R_x(n_1, n_2)$ for $(n_1, n_2) \in A$, and suppose that $\lambda^{m+1}(n_1, n_2)$ is obtained by linearly interpolating between $\lambda'(n_1, n_2) \cdot w(n_1, n_2)$ and $\lambda^m(n_1, n_2)$. Then, in the modified iterative algorithm, $\lambda^{m+1}(n_1, n_2)$ and $R_y^{m+1}(n_1, n_2)$ are obtained from $\lambda^m(n_1, n_2)$ by

$$R'(n_1, n_2) = F^{-1}\left[\frac{1}{F[\lambda^m(n_1, n_2)]}\right], \tag{17}$$

$$\begin{aligned} R_y^{m+1}(n_1, n_2) &= \alpha \cdot R'(n_1, n_2) + (1 - \alpha) \cdot R_x(n_1, n_2) \\ &\quad \text{for } (n_1, n_2) \in A \\ &\quad R'(n_1, n_2) \quad \text{otherwise} \\ &= R'(n_1, n_2) + (1 - \alpha) \cdot (R_x(n_1, n_2) \\ &\quad - R'(n_1, n_2)) \cdot w(n_1, n_2), \end{aligned} \tag{18}$$

$$\lambda'(n_1, n_2) = F^{-1}\left[\frac{1}{F[R_y^{m+1}(n_1, n_2)]}\right], \tag{19}$$

and

$$\begin{aligned} \lambda^{m+1}(n_1, n_2) &= \beta \cdot \lambda^m(n_1, n_2) + (1 - \beta) \cdot \lambda'(n_1, n_2) \\ &\quad \cdot w(n_1, n_2). \end{aligned} \tag{20}$$

Comparing (14) and (18), (18) reduces to (14) when $\alpha = 0$. With any other choice of α, (18) represents a nonideal correction of $R'(n_1, n_2)$ with the known values $R_x(n_1, n_2)$ for $(n_1, n_2) \in A$, with a larger deviation of α from zero corresponding to a more nonideal correction. However, with proper choice of α, the resulting $R_y^{m+1}(n_1, n_2)$ can be guaranteed to be positive definite. This can be seen by noting that $\lambda^m(n_1, n_2)$ and therefore $R'(n_1, n_2)$ are assumed to be positive definite and by considering α sufficiently close to 1 so that $R_y^{m+1}(n_1, n_2)$ can be made arbitrarily close to $R'(n_1, n_2)$. Similarly, (20) reduces to (16) when $\beta = 0$. With any other choice of β, $\lambda^{m+1}(n_1, n_2)$ now corresponds to an autocorrelation function which is a kind of "parallel resistor average" of $R_y^m(n_1, n_2)$ and $R_y^{m+1}(n_1, n_2)$. With proper choice of β, $\lambda^{m+1}(n_1, n_2)$ can also be guaranteed to be positive definite, which can be seen by noting that $\lambda^m(n_1, n_2)$ is assumed to be positive definite and by considering β sufficiently close to 1 so that $\lambda^{m+1}(n_1, n_2)$ can be made arbitrarily close to $\lambda^m(n_1, n_2)$. Therefore, by properly choosing α and β in the ranges $0 \leqslant \alpha < 1$, $0 \leqslant \beta < 1$, $R_y^{m+1}(n_1, n_2)$ and $\lambda^{m+1}(n_1, n_2)$ can be guaranteed to be positive definite, and thus the spectral zero crossing problem can be avoided so that the iterations can be continued.

From (13)-(16), it is clear that if $\lambda^m(n_1, n_2)$ and $R_y^m(n_1, n_2)$ satisfy (9)-(11), then $\lambda^{m+1}(n_1, n_2)$ and $R_y^{m+1}(n_1, n_2)$ obtained by the modified iterative algorithm also satisfy (9)-(11). With proper choice of α and β, then if $\lambda^0(n_1, n_2)$ and $R_y^0(n_1, n_2)$, the initial estimates of $\lambda(n_1, n_2)$ and $R_y(n_1, n_2)$, satisfy (9)-(11), the iterations specified by (9)-(12) and (17)-(20) form an iterative algorithm. This algorithm is shown in Fig. 3.

In implementing the algorithm in Fig. 3, there are several important issues that need to be discussed. One of them is the determination of the length of the discrete Fourier transform (DFT) and inverse discrete Fourier transform (IDFT) to be used for the Fourier transform and inverse Fourier transform operations. In general, a large DFT length should be used in

POSITIVE DEFINITE INITIAL ESTIMATES
$R_y^0(n_1,n_2)$ AND $\lambda^0(n_1,n_2)$

$R'(n_1,n_2) = F^{-1}\left\{\dfrac{1}{F[\lambda^m(n_1,n_2)]}\right\}$

DESIRED SOLUTION ?　YES

NO

$R_y^{m+1}(n_1,n_2) = R'(n_1,n_2) + (1-\alpha)[R_x(n_1,n_2) - R'(n_1,n_2)]\cdot w(n_1,n_2)$
WHERE $0 \leq \alpha < 1$ IS CHOSEN SUCH THAT $F[R_y^{m+1}(n_1,n_2)] > 0$

$\lambda'(n_1,n_2) = F^{-1}\left\{\dfrac{1}{F[R_y^{m+1}(n_1,n_2)]}\right\}$

$\lambda^{m+1}(n_1,n_2) = \beta\lambda^m(n_1,n_2) + (1-\beta)\lambda'(n_1,n_2)\cdot w(n_1,n_2)$
WHERE $0 \leq \beta < 1$ IS CHOSEN SUCH THAT $F[\lambda^{m+1}(n_1,n_2)]>0$

$\hat{P}_x(\omega_1,\omega_2) = F[R'(n_1,n_2)] = \dfrac{1}{F[\lambda^m(n_1,n_2)]}$

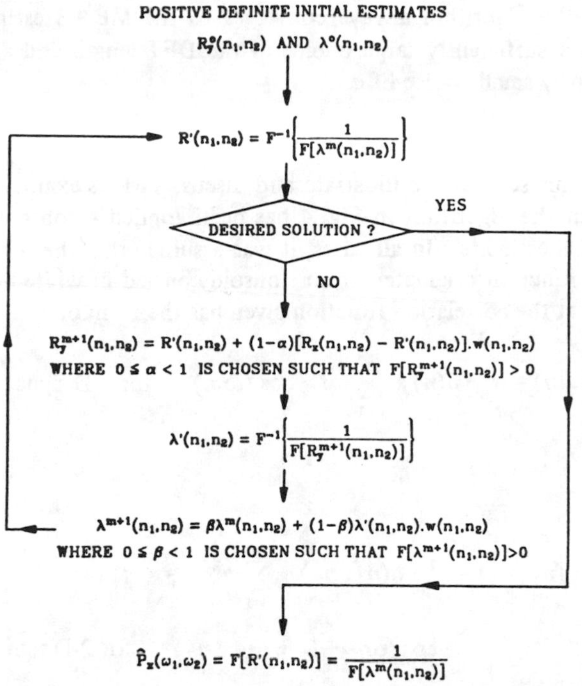

Fig. 3. An iterative algorithm for 2-D ME PSE based on Fig. 2.

the implementation to avoid any aliasing problem. Specifically, the ME method of PSE is essentially an attempt to extrapolate the correlation function beyond the limits of the known segment. Since the DFT is used in the implementation instead of the true Fourier transform, the length of the DFT should be chosen such that the extended correlation function corresponding to the ME PS estimate is essentially zero beyond the DFT limits. Choice of the DFT length and its effect on the system performance will be further discussed in Section IV.

Another issue to be considered in the implementation is determination of $\lambda^0(n_1,n_2)$ and $R_y^0(n_1,n_2)$, the initial estimates of $\lambda(n_1,n_2)$ and $R_y(n_1,n_2)$. Even though various different choices are possible, one choice which is particularly simple and satisfies (9)–(11) is

$$R_y^0(n_1,n_2) = R_x(0,0)\cdot\delta(n_1,n_2) \tag{21}$$

and

$$\lambda^0(n_1,n_2) = \frac{1}{R_x(0,0)}\cdot\delta(n_1,n_2) \tag{22}$$

where it is assumed that the region A includes the point $(n_1,n_2) = (0,0)$ and that $R_x(0,0)>0$. Furthermore, the choice of $\lambda^0(n_1,n_2)$ and $R_y^0(n_1,n_2)$ given by (21) and (22) is reasonable in that $R_y^m(n_1,n_2) = R_x(n_1,n_2)\cdot w(n_1,n_2)$ for $m = 1, \alpha = 0$, and $\beta = 0$ in the iterative algorithm of Fig. 3.

A third issue to be discussed in the implementation is how to choose the values of α and β in each iteration. The choice of α and β is dictated by two considerations. One is the requirement that the resulting $R_y^{m+1}(n_1,n_2)$ and $\lambda^{m+1}(n_1,n_2)$

be positive definite. The second is our desire to choose α and β as close to zero as possible so that the algorithm progresses at the fastest possible rate. In this regard, we have empirically observed that choosing the smallest possible values of α and β consistent with the positive definite requirement on $R_y^{m+1}(n_1,n_2)$ and $\lambda^{m+1}(n_1,n_2)$ can lead to a limit cycle behavior where the algorithm does not converge. A similar behavior has also been observed to occur if α is chosen to decrease in the course of the algorithm. Further, we have observed that the rate of convergence must be slowed down if the error between $R'(n_1,n_2)$ and $R_x(n_1,n_2)$ for $(n_1,n_2)\in A$ increases from one iteration to the next.

One method to choose α and β, which avoids the divergence problem discussed above and is used in this paper, is to begin with $\alpha_0 = \beta_0 = 0$ and change α and β in the following manner;

$$\alpha_{m+1} \leftarrow \max\left[\alpha_m, 1 - k\right.$$
$$\left.\cdot\frac{\displaystyle\min_{(\omega_1,\omega_2)}[F[R'(n_1,n_2)]]}{\left|\displaystyle\min_{(\omega_1,\omega_2)}F[(R_x(n_1,n_2) - R'(n_1,n_2))\cdot w(n_1,n_2)]\right|}\right]$$

$$\tag{23}$$

$$\text{if}\quad\min_{(\omega_1,\omega_2)}[F[(R_x(n_1,n_2) - R'(n_1,n_2))$$
$$\cdot w(n_1,n_2)]] < 0$$

and

$$\alpha_{m+1} \leftarrow \alpha_m \quad\text{otherwise,}$$

$$\beta_{m+1} \leftarrow \left[1 + (1-k)\cdot\left(\frac{1}{\beta_{min}} - 1\right)\right]\cdot\beta_{min}$$

$$\text{if}\quad\min_{(\omega_1,\omega_2)}[F[\lambda'(n_1,n_2)\cdot w(n_1,n_2)]] < 0 \tag{24}$$

and

$$\beta_{m+1} \leftarrow 0 \quad\text{otherwise}$$

where

$$\beta_{min} \triangleq \frac{\left|\displaystyle\min_{(\omega_1,\omega_2)}F[\lambda'(n_1,n_2)\cdot w(n_1,n_2)]\right|}{\displaystyle\min_{(\omega_1,\omega_2)}F[\lambda^m(n_1,n_2)] + \left|\displaystyle\min_{(\omega_1,\omega_2)}F[\lambda'(n_1,n_2)\cdot w(n_1,n_2)]\right|}\cdot$$

In (23) and (24), α_i and β_i represent the values of α and β in the ith iteration, β_{min} represents the minimum value of β that results in a positive semidefinite estimate for the coefficient set, max $[\,,\,]$ represent the maximum of two arguments, $\min_{(\omega_1,\omega_2)}[\,]$ represents the minimum of the argument expression over (ω_1,ω_2), and "k" is the convergence rate parameter with $0 < k \leqslant 1$. The initial value of "k" is chosen to be large (0.5) and then subsequently reduced if the error between $R'(n_1,n_2)$ and $R_x(n_1,n_2)$ for $(n_1,n_2)\in A$ increases. Thus, the algorithm moves towards the desired solution rapidly at first, and if necessary, it is slowed down as convergence is approached. When α and β are chosen according to (23) and (24), it is straightforward to show from (18) and (20) that the resulting $R_y^{m+1}(n_1,n_2)$ and $\lambda^{m+1}(n_1,n_2)$ are guaranteed to be

positive definite. Further, computing α_{m+1} and β_{m+1} by (23) and (24) requires little extra computations since the individual terms in the two equations are computed in the course of the algorithm.

Finally, another important issue to be considered in implementing the algorithm in Fig. 3 is the decision on when the algorithm converged so that the iteration can be stopped. One reasonable approach is to consider that the algorithm has converged when the following condition is satisfied:

$$\frac{\sum\sum_{(n_1,n_2)\in A} (R'(n_1,n_2) - R_x(n_1,n_2))^2}{\sum\sum_{(n_1,n_2)\in A} R_x^2(n_1,n_2)} \leqslant \epsilon. \tag{25}$$

Clearly, if $\epsilon = 0$ with $R'(n_1,n_2)$ computed from $\lambda^m(n_1,n_2)$ using the discrete time Fourier transform rather than the DFT, the resulting solution corresponds to the desired ME PS estimate. However, due to a finite DFT length and finite precision arithmetic used, it may not be possible to reduce the error exactly to zero. On the other hand, the use of a finite DFT length may reduce the error to a very small value without leading to the desired ME PS estimate. This again brings into sharp focus the fact that the DFT length must not be underestimated in implementing the algorithm. However, to avoid unnecessary computations, the algorithm can be started with a reasonable DFT length and a one-time test for the solution made at the end. Specifically, with a reasonable choice of the DFT length, the algorithm is continued until the error ϵ reaches a very small value, typically 10^{-4}. If this error cannot be reached, then the DFT length has to be increased. Then the coefficient set λ obtained as the ME solution is tested for positive definiteness over a much finer grid (a much larger DFT) than that used in the iterations. If the solution is not positive definite, a longer DFT is required. If the coefficient set λ is positive definite, then the error ϵ given by (25) is rechecked by computing $R'(n_1,n_2)$ using a much longer DFT. If the new error is of the same order as that obtained in the iterations, the solution is declared to be good; otherwise, more iterations are required. Since the minimum error that can be achieved with a given DFT length is dictated by the amount of aliasing the autocorrelation function undergoes, if it is necessary to continue the iterations, it is preferable to use a DFT length longer than that used previously. Since the resulting coefficient set λ is zero for $(n_1,n_2)\notin A$ and $R'(n_1,n_2)$ needs to be computed only for $(n_1,n_2)\in A$ for the test, the much longer DFT used in the test can be computed directly rather than using the FFT algorithms without significant additional computations. Thus, the test performed only once in the last stage of the algorithm does not require any significant additional computations and requires no extra storage, even though a much longer DFT is used in the test.

Fig. 4 shows a more detailed flowchart of an algorithm which incorporates the important implementation issues discussed above. It is not theoretically known under what conditions the algorithm in Fig. 4 converges. As will be discussed in the next section, however, we have empirically observed that the algorithm always converges to the ME PS estimate with a sufficiently large choice of the DFT length and a sufficiently small choice of ϵ.

IV. Examples and Discussions

In this section, we illustrate and discuss various examples in which the algorithm in Fig. 4 has been applied to obtain the ME PS estimate. In all cases, it was assumed that the correlation function originated from sinusoids buried in white noise, so that the correlation function given has the form of

$$R_x(n) = \sigma^2 \cdot \delta(n) + \sum_{i=1}^{M} a_i^2 \cdot \cos(\omega_i n) \quad \text{for 1-D signals} \tag{26}$$

and

$$R_x(n_1,n_2) = \sigma^2 \cdot \delta(n_1,n_2) + \sum_{i=1}^{M} a_i^2 \cdot \cos(\omega_{i1}\cdot n_1 + \omega_{i2}\cdot n_2) \quad \text{for 2-D signals} \tag{27}$$

where σ^2 represents the white noise power, M represents the number of sinusoids, a_i^2 represents the power of the ith sinusoid, and ω_i for 1-D signals and ω_{i1} and ω_{i2} for 2-D signals represent the frequencies of the ith sinusoid.

We first consider the case of 1-D signals, which is ideal, due to the existence of a simple closed form solution, in illustrating that the solution obtained from the iterative algorithm in Fig. 4 indeed leads to the ME PS estimate. Fig. 5 shows the results obtained using the parameters in Table I. Fig. 5(a) shows the results obtained from the iterative algorithm as a function of the number of iterations. The converging solution with the choice of ϵ in Table I was obtained after 15 iterations. Fig. 5(b) shows the ME PS estimate obtained from the closed form solution of (6). Fig. 5(c) shows the PS estimate obtained from the iterative algorithm (solid line) and the closed form solution (dotted line). It is clear from Fig. 5(c) that the iterative algorithm leads to the ME PS estimate.

In addition to the above example, a variety of other examples have been considered. In all cases that we have considered so far, the iterative algorithm in Fig. 4 leads to the PS estimates which are visually indistinguishable from the closed form solutions.

We now consider the case of 2-D signals. For 2-D signals, a closed form solution for ME PSE has not yet been found and, consequently, the iterative algorithm developed in this paper has practical significance. Fig. 6 shows the results obtained using the parameters in Table II. Fig. 6(a) shows the periodogram [9] obtained by Fourier transforming $R_x(n_1,n_2)\cdot w(n_1,n_2)$. Fig. 6(b) shows the PS estimate based on autoregressive signal modeling of (6) using a backward L-shaped filter mask [17]. This particular mask was chosen due to its high performance [17] relative to other filter mask shapes, such as symmetric and first quadrant filters. Fig. 6(c) shows

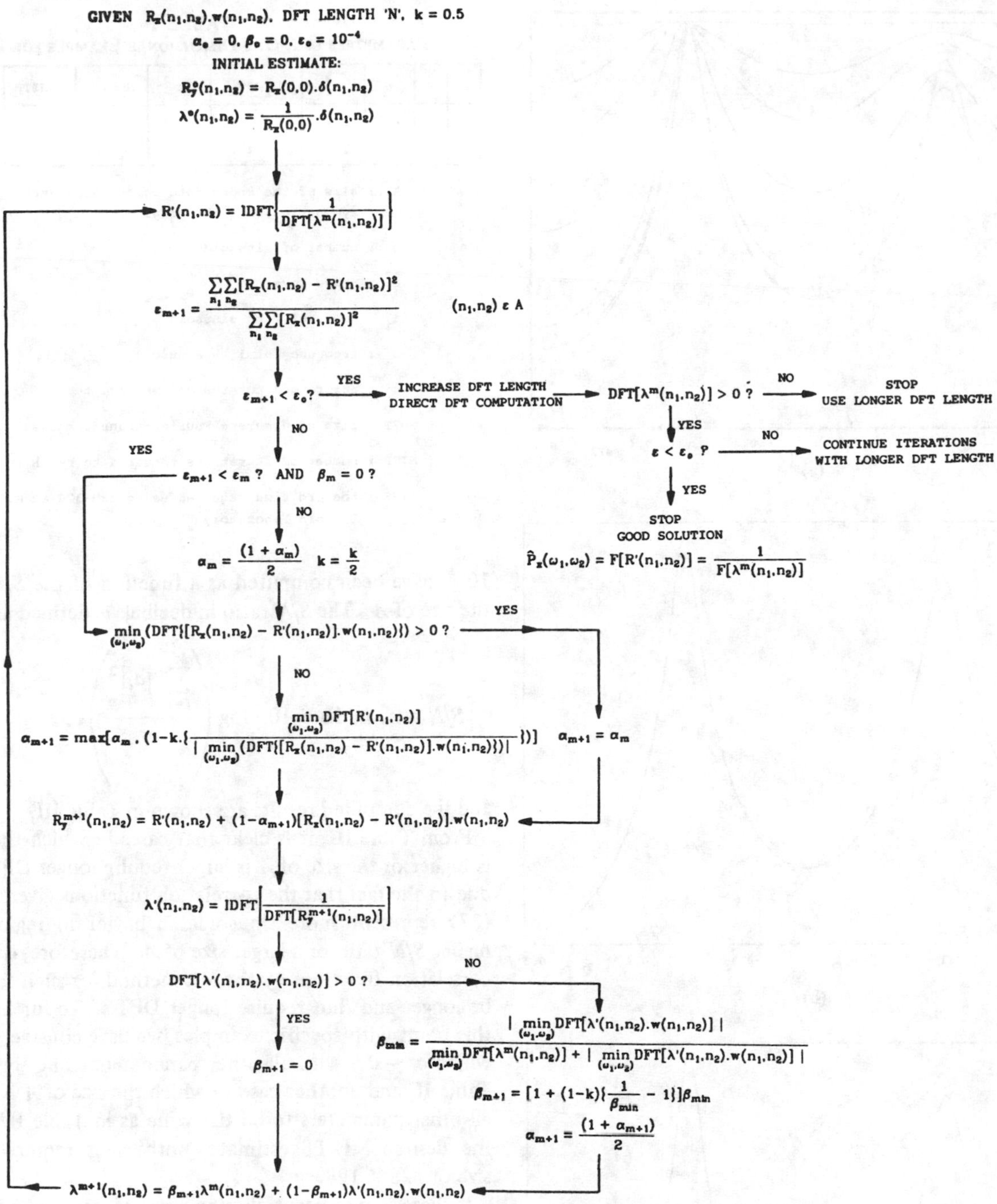

Fig. 4. A detailed flowchart of the iterative algorithm for 2-D ME PSE implemented in this paper.

the PS estimate obtained from the iterative algorithm developed in this paper. To ensure that the PS estimate shown in Fig. 6(c) corresponds to the ME PS estimate, an additional estimate was obtained using a larger DFT size of 128 X 128 points and a much smaller ϵ of 10^{-6}, and was compared to the result in Fig. 6(c). The two estimates were visually indistinguishable. It is clear from Fig. 6 that the two sinusoids are resolved only in Fig. 6(c), implying that the ME method for PSE has the high resolution characteristics for 2-D signals as well as 1-D signals.

In using the iterative algorithm developed in this paper, care should be exercised in properly choosing the DFT length. As has been discussed in Section III, a smaller aliasing error requires a longer DFT. However, once the DFT length is chosen so that the aliasing error is sufficiently small relative to ϵ, a further increase in the DFT length makes little improvement in the spectral estimates, but only increases the computational requirements. To give a rough estimate of the DFT length needed in practice, a number of examples have been considered, and the ranges of DFT lengths required for $\epsilon =$

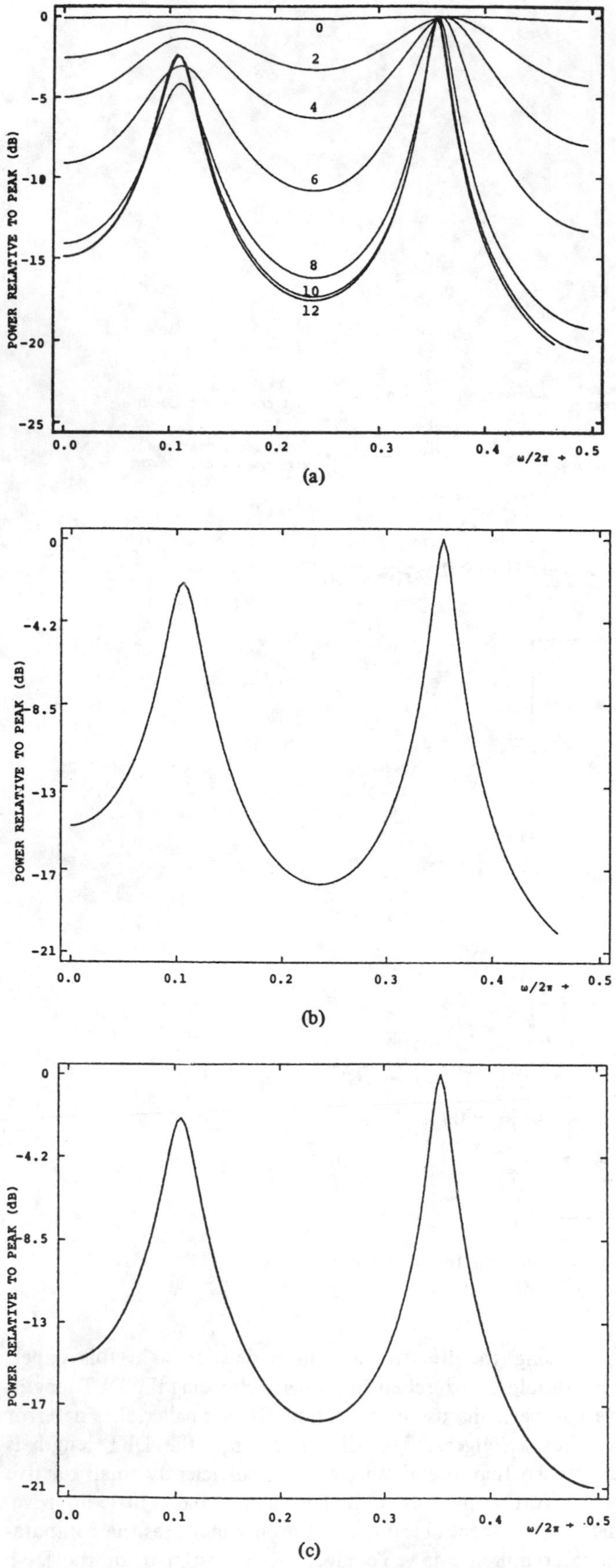

Fig. 5. 1-D PSE for the data in Table I. (a) The PS estimate as a function of the number of iterations (NITR). NITR = 0, 2, 4, 6, 8, 10, 12. (b) Results of ME PSE by the closed form solution of (5) and (6). (c) Result of the iterative method (solid line). Result of the closed form solution (dotted line). The two results are indistinguishable.

TABLE I

PARAMETERS OF A ONE-DIMENSIONAL EXAMPLE FOR FIG. 5

A	M	σ^2	a_i^2	$\omega_i/2\pi$	ϵ	NDFT	NITR	TIME
9	1	1.0	1.0	0.1	10^{-4}	128	15	0.25 secs.
			1.0	0.3456				

A : size of the known auto-correlation array, symmetric about the origin

M : number of sinusoids

σ^2 : noise power

a_i^2 : power of i'th sinusoid

ω_i : frequency of i'th sinusoid

ϵ : error used for the convergence test

NDFT : size of discrete Fourier transform used

NITR : number of iterations required to reach ϵ

TIME : the CPU time required using IBM-370 at M.I.T. Lincoln Laboratory

10^{-4} have been computed as a function of the S/N ratio and the size of A. The S/N ratio in decibels is defined as

$$S/N \text{ ratio in dB} \triangleq 10 \cdot \log \left(\frac{\sum_{i=1}^{M} |a_i|^2}{\sigma^2} \right) \tag{28}$$

and the computed results are shown in Table III.

From Table III, it is clear that cases in which the S/N ratio is higher or the size of A is larger require longer DFT's. This is due to the fact that the correlation functions given in (26) and (27) represent more sinusoidal behavior in region A for a higher S/N ratio or a larger size of A. Therefore, the extended correlation functions by the ME method for such cases tend to be longer and thus require longer DFT's. To further illustrate this point with specific examples, we have considered a case in which $\sigma^2 = 0.5$ with all other parameters to be the same as in Table II, and another case in which the size of A is 7×7 with all other parameters to be the same as in Table II. To obtain the desired ME PS estimate, both cases required the DFT size of 128×128.

If there is significant aliasing error relative to the ϵ used for the convergence test, then the algorithm does not converge. For example, if the DFT size of 32×32 is used for the example of Fig. 6, the algorithm does not converge. In such a case, the DFT length has to be increased to obtain the ME PS estimate. If the DFT length cannot be increased due to computational constraints, ϵ has to be increased to tolerate a larger aliasing error. In such a case, the resulting PS estimate will only be an approximation to the ME PS estimate.

In general, cases in which the S/N ratio is lower or the size of A is smaller require fewer computations to reach the desired ME PS estimate for two reasons. One reason is that such cases require shorter DFT lengths, as is clear from Table IV. Another reason is that a larger noise power (σ^2) contributes a larger positive spectral component in the correlation correction step, and the Fourier transform of a smaller rectangular

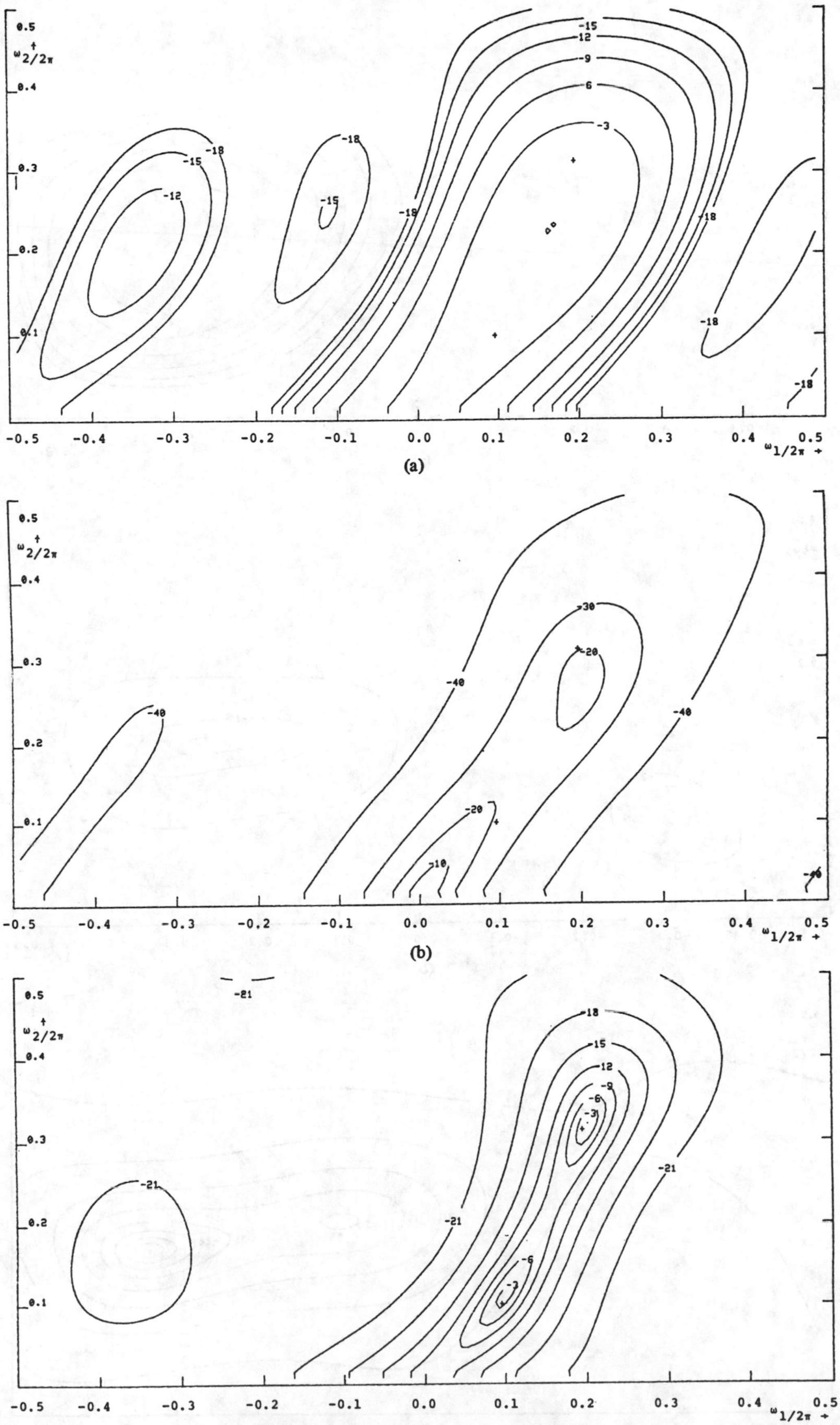

Fig. 6. 2-D PSE for data in Table II. "+" in all figures represents the true peak locations of the signal sinusoids and the numbers on the contours represent "dB" relative to the maximum value of the estimated power spectrum. (a) Results of PSE by periodogram. (b) Result of PSE based on the backward L autoregressive filter. (c) ME PS estimate by the iterative algorithm in Fig. 4.

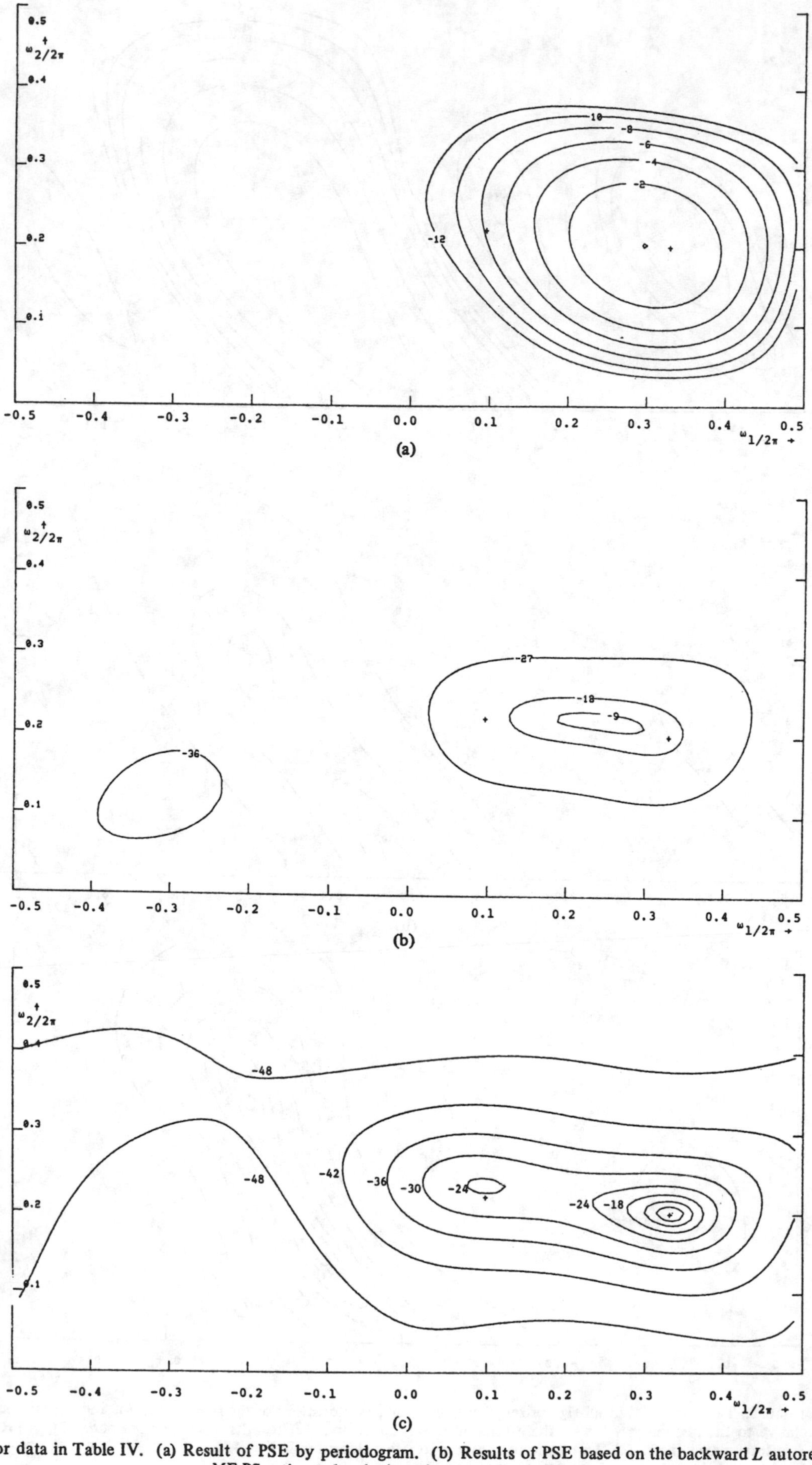

Fig. 7. 2-D PSE for data in Table IV. (a) Result of PSE by periodogram. (b) Results of PSE based on the backward L autoregressive filter. (c) ME PS estimate by the iterative algorithm in Fig. 4.

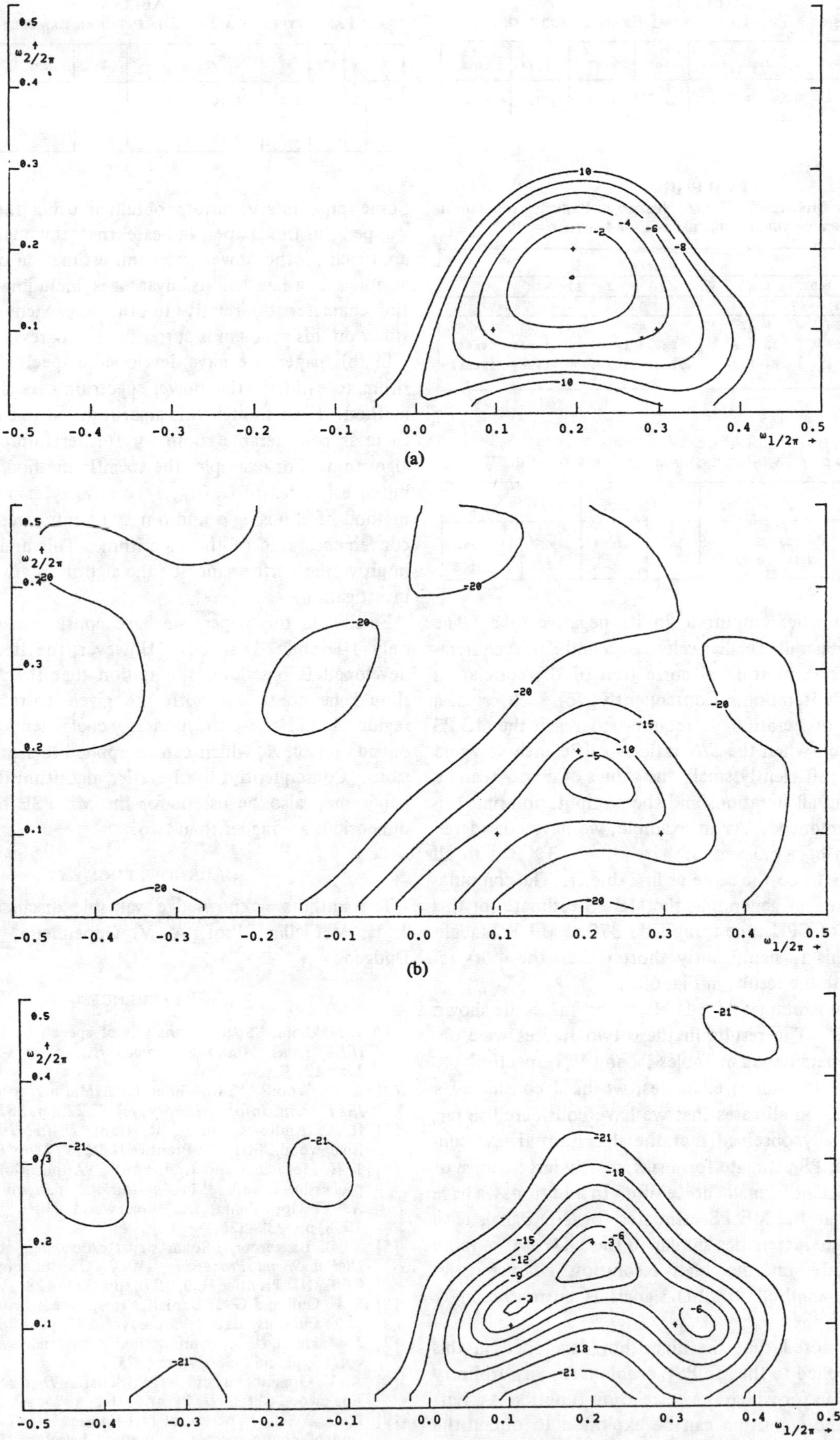

Fig. 8. 2-D PSE for data in Table V. (a) Result of PSE by periodogram. (b) Result of PSE based on the backward *L* autoregressive filter. (c) ME PS estimate by the iterative algorithm in Fig. 4.

TABLE II
PARAMETERS OF A TWO-DIMENSIONAL EXAMPLE FOR FIG. 6

A	M	σ^2	a_i^2	$(\omega_{i1}/2\pi, \omega_{i2}/2\pi)$	ϵ	NDFT	NITR	TIME
5x5	2	1.0	1.0 1.0	0.10 0.10 0.20 0.3125	10^{-4}	64x64	43	4.6 secs.

TABLE III
TYPICAL RANGES OF MINIMUM DFT SIZE NEEDED IN PRACTICE TO ACHIEVE CONVERGENCE OF THE ALGORITHM IN FIG. 4 WITH $\epsilon = 10^{-4}$

S/N ratio	Size of A			
	1-D Signals		2-D Signals	
	5	9	3x3	5x5
-10 dB	8-16	8-32	8x8	8x8 - 16x16
0 dB	16-64	32-128	8x8 - 16x16	16x16- 64x64
+10 dB	64-512	64-1024	32x32-256x256	64x64-512x512

TABLE IV
PARAMETERS OF A TWO-DIMENSIONAL EXAMPLE FOR FIG. 7

A	M	σ^2	a_i^2	$\omega_{i1}/2\pi, \omega_{i2}/2\pi$	ϵ	NDFT	NITR	TIME
5x5	2	2.0	2.0 1.0	0.3333 0.20 0.10 0.22	10^{-4}	64x64	87	9 secs.

window has a smaller amplitude in its negative lobe. The effect of this is generally smaller values of α and β in each iteration and, therefore, more ideal correction of the correlation function in each iteration. Consequently, for such cases, a smaller number of iterations is required to reach the ME PS estimate. In fact, when the S/N ratio is sufficiently low and the size of A is sufficiently small, the values of α and β can be chosen to be 0 in all iterations and the computation time can be significantly reduced. As an example, we have considered a case in which $\sigma^2 = 6.0$ and the size of $A = 3 \times 3$ with all other parameters to be the same as in Table II. The computation time required in generating the ME PS estimate for this case was 0.27 s of CPU time using IBM 370 at M.I.T. Lincoln Laboratory. This is significantly shorter than the 4.6 s required to generate the results in Fig. 6(c).

Two additional examples of 2-D ME PS estimates are shown in Figs. 7 and 8. The results in these two figures were obtained using the parameters in Tables IV and V, respectively.

In addition to the above examples, we have considered a variety of others. In all cases that we have considered so far, we have empirically observed that the algorithm always converges to the ME PS estimate for a sufficiently large choice of DFT and a sufficiently small choice of ϵ. In addition, we have also observed that the ME PS estimates for 2-D signals have high resolution characteristics similar to the 1-D case. A more quantitative study on the high resolution characteristics of the ME PSE method for 2-D signals is currently under investigation.

It should be noted that the algorithm developed in this paper can be applied to the ME PSE problem for any arbitrary shape of the region for which the correlation function is given. This aspect of the algorithm can be exploited to obtain the ME PS estimate of nonuniformly sampled 1-D and 2-D data for which a closed form solution has not yet been found.

TABLE V
PARAMETERS OF A TWO-DIMENSIONAL EXAMPLE FOR FIG. 8

A	M	σ^2	a_i^2	$\omega_{i1}/2\pi, \omega_{i2}/2\pi$	ϵ	NDFT	NITR	TIME
7x7	3	6.0	1.0 1.0 1.0	0.10 0.10 0.30 0.10 0.20 0.20	10^{-4}	32x32	13	1.6 secs

Some preliminary results obtained using the algorithm developed in this paper indicate that the maximum entropy approach to the power spectrum estimation of nonuniformly sampled data has various advantages, including the high resolution characteristics relative to other approaches. More detailed study on this problem is currently in progress.

In this paper, we have developed a specific numerical algorithm to estimate the power spectrum of a signal by the ME method. Even though this algorithm led to successful results, there is considerable room for further improvements of the algorithm. For example, the specific method of choosing the initial estimates of $R_y(n_1, n_2)$ and $\lambda(n_1, n_2)$ and the specific method of choosing α and β may be improved to increase the convergence rate of the algorithm. This and other ways to improve the performance of the algorithm are currently under investigation.

Finally, in this paper, we have considered the ME PSE of only 1-D and 2-D signals. However, the iterative algorithm developed is based on the notion that the ME PS estimate should be consistent with the given correlation points in region A and the corresponding coefficients should be zero outside region A, which can be applied to signals of all dimensions. Consequently, the iterative algorithm developed in this paper may also be useful for the ME PSE of signals whose dimensions are higher than two.

ACKNOWLEDGMENT

The authors acknowledge valuable discussions with Prof. J. H. McClellan, Prof. A. V. Oppenheim, and Dr. D. E. Dudgeon.

REFERENCES

[1] J. Makhoul, "Spectral analysis of speech by linear prediction," *IEEE Trans. Audio Electroacoust.*, vol. AU-21, pp. 140–148, June 1973.

[2] J. W. Woods, "Two-dimensional Markov spectral estimation," *IEEE Trans. Inform. Theory*, vol. IT-22, pp. 552–559, Sept. 1976.

[3] H. C. Andrews and B. R. Hunt, *Digital Image Restoration.* Englewood Cliffs, NJ: Prentice-Hall, 1977, pp. 90–112.

[4] J. H. McClellan and R. J. Purdy, "Applications of digital signal processing to radar," in *Applications of Digital Signal Processing*, A. V. Oppenheim, Ed. Englewood Cliffs, NJ: Prentice-Hall, 1978, pp. 239–326.

[5] A. B. Baggeroer, "Sonar signal processing," in *Applications of Digital Signal Processing*, A. V. Oppenheim, Ed. Englewood Cliffs, NJ: Prentice-Hall, 1978, pp. 331–428.

[6] S. F. Gull and G. J. Daniell, "Image reconstruction from incomplete and noisy data," *Nature*, vol. 272, pp. 686–690, 1978.

[7] J. Makhoul, "Linear prediction: A tutorial review," *Proc. IEEE*, vol. 63, pp. 561–580, Apr. 1975.

[8] A. V. Oppenheim and R. W. Schafer, *Digital Signal Processing.* Englewood Cliffs, NJ: Prentice-Hall, 1975, pp. 532–571.

[9] P. D. Welch, "The use of fast Fourier transform for the estimation of power spectra: A method based on time averaging over short, modified periodograms," *IEEE Trans. Audio Electroacoust.*, vol. AU-15, pp. 70–73, June 1967.

[10] R. T. Lacoss, "Data adaptive spectral analysis methods," *Geophysics*, vol. 36, pp. 661–675, 1971.
[11] J. P. Burg, unpublished notes, 1974. See pp. 20–21 of [17] below.
[12] S. J. Wernecke and L. R. D'Addario, "Maximum entropy image reconstruction," *IEEE Trans. Comput.*, vol. C-26, pp. 351–364, Apr. 1977.
[13] W. I. Newman, "A new method of multidimensional power spectral analysis," *Astron. Astrophys.*, vol. 54, p. 369, 1977.
[14] B. W. Dickinson, "Two-dimensional Markov spectrum estimates need not exist," *IEEE Trans. Inform. Theory*, vol. IT-20, pp. 120–121, Jan. 1980.
[15] R. W. Gerchberg and W. D. Saxton, "A practical algorithm for the determination of phase from image and diffraction plane pictures," *Optik*, vol. 35, pp. 237–246, 1972.
[16] J. R. Fienup, "Reconstruction of an image from the modulus of its Fourier transform," *Opt. Lett.*, vol. 3, pp. 27–29, 1978.
[17] J. V. Pendrell, "The maximum entropy principle in two-dimensional spectral analysis," Ph.D. dissertation, York Univ., Toronto, Ont., Canada, Nov. 1979, pp. 210–219.

MULTI-DIMENSIONAL SPECTRAL ESTIMATION VIA PARAMETRIC MODELS*

Chrysostomos L. Nikias and Mysore R. Raghuveer

Department of Electrical Engineering and Computer Science, U-157
The University of Connecticut
Storrs, Connecticut 06268

ABSTRACT

The approach taken toward estimating the power spectral density function of multi-dimensional data fields is sometimes too general considering that in many areas of application, the PSD is not arbitrary but is low order parametric. However, if the parametric form of such fields is not taken into account, then much is lost in estimating their power spectrum. In this paper a new class of parametric multi-dimensional (m-d) estimation algorithms is introduced based on the minimum variance representations of m-d data fields. These representations are defined in integrated and compact linear predictive forms with their spectrum interpretations being generally ARMA. The selected model parameters for spectral estimation achieve the minimum square error in fitting a recursion among the estimated single-point covariance elements of an m-d data field. Spectra computed from short/long, noisy narrow-band and wide-band data fields are compared with spectra computed by standard techniques and show improvement in resolution of peak detection and in accuracy of spectrum matching.

INTRODUCTION

There is a variety of fields where multi-dimensional power spectrum estimation is employed. Examples of these can be found in sonar (Ref. 1), geophysics (Ref. 2) and biomedicine (Ref. 3-4). In all such applications, the spectrum estimate is used to obtain information about the number, intensity and velocity vector of travelling wave components of the data field. Important requirements to be satisfied by the spectrum estimation method are high resolution, and tolerance toward inhomogeneities in the data field while making use of a small sized data set.

Among the most widely used techniques for m-d spectral estimation are the conventional (autocorrelation and periodogram with their smooth versions). These methods exhibit poor resolution and are unfit for short data. Capon (Ref. 2) proposed the maximum likelihood (ML) method where a linear m-d FIR filter is designed so as to pass undistorted any monochromatic plane wave and suppress noise in a least squares manner. The ML method exhibits higher resolution than the conventional methods.

An active area of research is Maximum Entropy (MEM) spectrum estimation. The maximum entropy spectrum is obtained as the solution to a constrained maximization problem; the constraints are the correlation matching constraints on the inverse fourier transform of the spectrum (Ref. 5). The problem is nonlinear. The solution when it exists is unique because the entropy functional is convex. Various approaches have been reported in the literature that tackle variants of the problem using correlation estimates derived from the data (Ref. 6).

Recently there has been an increased emphasis on methods to construct finite-parameter stochastic models for m-d spectral estimation (Ref. 7). These models do not satisfy the correlation matching property of MEM, but they still provide high resolution spectrum estimates. For the most part, these models require solution of a finite set of linear equations making them very attractive. Jackson and Chien (Ref. 8) and Kumaresan and Tufts (Ref. 9) have proposed methods to obtain parameters of 2-d autoregressive (AR) models for spectrum estimation. Cadzow and Ogino (Ref. 10) proposed a method for generating 2-d quarter-plane causal AR and ARMA spectrum estimation models.

Recognizing that in the case of 2-d data fields, one need not stick strictly to causality except along the time axis, Jain (Ref. 11-12) suggested the use of 2-d minimum variance causal, semicausal and noncausal models for spectral estimation. The 2-d semicausal estimate is shown to be of high resolution and to be accurate in terms of peak location and covariance match when compared to conventional and AR methods (Ref. 11-12). Nikias et al (Ref. 13) while employing similar models estimate the parameters directly from the data field bypassing covariance estimates by using the 2-d covariance least squares (CLS) criterion. This method exhibits high-resolution and has been successfully applied to bipolar cardiac data analysis (Ref. 4).

In this paper, we present the minimum variance representations (MVR) for m-d (m=3,4) data fields in a form that is concise and useful for spectrum estimation. It is shown that there are four possible linear predictions in the 3-d case and five in the 4-d case. The CLS approach for m-d spectrum estimation is outlined. Estimated spectra of short/long and noisy data fields as obtained using this approach are compared with those obtained by standard techniques.

---

*This work was supported in part by the National Institute of Health under Grant HL-29280

Reprinted from *IEEE ASSP Spectrum Estimation Workshop II*, 1983, pp. 213–218.

m-d PARAMETRIC MODELS

In many physical application areas, real random fields are unnecessarily general because their power spectrum is not arbitrary but is a low order parametric rational function. However, if the parametric form of such a random field is not taken into account, then much is lost in estimating the power spectrum. In other words, if the given random field is better described by a finite set of parameters which is much smaller than the total number of observations, then parametric spectrum estimation approach will be much more accurate than any of the conventional approaches. Such random fields are often the result of a physical process whose propagation can be approximated arbitrarily closely by the model of a stochastic partial difference equation (Ref. 7). In this section a class of m-d parametric models is defined in the framework of linear prediction and based on the minimum variance representation of m-d stationary random fields. These parametric models lead to analytically tractable expressions for the power spectral density as a function of the model parameters.

Consider an m-d (m=2,3,4) zero-mean, stationary random field $\{X(\underline{r})\}$, $\underline{r}\varepsilon R^m$ where at each point $\underline{r}=(r_1,r_2,\ldots,r_m)$ there is an associated random variable $X(\underline{r})=X(r_1,r_2,\ldots r_m)$. Thus $\{X(\underline{r})\}$ can be thought as a usual random process where $\underline{r}$ is multidimensional. Let $\hat{X}(\underline{r})$ denote a linear prediction estimate of $X(\underline{r})$, i.e.,

$$\hat{X}(\underline{r}) = \sum_{\underline{\ell}\varepsilon\bar{V}} a(\underline{\ell})\ X(\underline{r}-\underline{\ell}) \qquad (1)$$

where $\bar{V}$ is the prediction space and $a(\underline{\ell})$, $\underline{\ell}=(\ell_1,\ldots,\ell_m)$ are the predictor coefficients. The prediction space $\bar{V}$ is a subspace of R^m and the point $(0,0,0)\notin\bar{V}$.

If $\hat{X}(\underline{r})$ is chosen to be a minimum variance predictor of $X(\underline{r})$, we can define a stochastic representation of the random field $\{X(\underline{r})\}$ as

$$\hat{X}(\underline{r}) = \sum_{\underline{\ell}\varepsilon\bar{V}} \alpha(\underline{\ell})\ X(\underline{r}-\underline{\ell}) + W(\underline{r}) \qquad (2)$$

where $\{W(\underline{r})\}$ is another random field with statistics such that the covariance properties of $\{X(\underline{r})\}$ are satisfied. The set of parameters $\{\alpha(\underline{\ell})\}$, $\underline{\ell}\varepsilon\bar{V}$ is obtained from the minimization of $||\hat{X}(\underline{r})-X(\underline{r})||^2$ and the use of the orthogonality principle. In practice one is interested in finite prediction spaces V, which are subsets of $\bar{V}$. Hence, assuming that the prediction parameters $\alpha(\underline{\ell})$ are nonzero only in a finite space V, the stationary random field may be represented by a constant parameter stochastic difference equation of the form

$$X(\underline{r}) = \sum_{\underline{\ell}\varepsilon V} \alpha(\underline{\ell})\ X(\underline{r}-\underline{\ell}) + W(\underline{r}) \qquad (3)$$

whose power spectral density (PSD) function is given by

$$S(\underline{z}) = S_w(\underline{z})/D(\underline{z})\ D(\underline{z}^{-1}), \qquad (4)$$

where $D(\underline{z}) = 1-\sum_{\underline{\ell}\varepsilon V} \alpha(\underline{\ell})\ \underline{z}^{-\underline{\ell}}$, $S_w(\underline{z})$ is the PSD of $\{W(\underline{r})\}$, $\underline{z} = (z_1, z_2,\ldots,z_m)$, $\underline{z}^{-\underline{\ell}} = z_1^{-\ell_1}\ldots z_m^{-\ell_m}$ and $|\underline{z}| = |z_1| = \ldots = |z_m|$. One can easily show that the noise field $\{W(\underline{r})\}$ is white, i.e., $Sw(\underline{z})=Q$ only in the special case where the prediction space V is causal. In any other case, $S(\underline{z})$ is generally an ARMA type spectral function. Since Eq. 3 defines a transformation from the input $W(\underline{r})$ to the output $X(\underline{r})$, it is required that a bounded input must be transformed to a bounded output, i.e., the parametric model must be stable. In this paper we shall assume that the parameters of Eq. 3 are restricted to stable regions, although to assure that this is the case in practice, convenient stability constraints must be considered.

In the two-dimensional case, A. Jain (Ref. 11-12) first introduced the 2-d minimum variance parametric representations for spectral estimation. He suggested three types of finite prediction regions (V) for the 2-d case, namely the causal, semicausal and noncausal (Ref. 11-12). While their spectrum representation is described by Eq. 4 for $\underline{z}=(z_1,z_2)$, the analytical expression of the noise field PSD function $Sw(z_1,z_2)$ can be found in (Ref. 11). In the 3-d case, four possible geometries of the prediction space V have been defined by the present authors (Ref. 14) in an integrated and compact form useful for spectral estimation: the causal, semicausal I, semicausal II and noncausal. The analytical description of these four geometries is given by

$V=\{(-r\le\ell<0,\ -q\le n\le q,\ -p\le m\le p)U(\ell=0,\ -q\le n<0,\ -p\le m\le p)$
$U(\ell=0,\ n=0,\ -p\le m<0)\}$ causal (5.a)

$\{(-r\le\ell<0,\ -q\le n\le q,\ -p\le m\le p)U(\ell=0,\ -q\le n<0,\ -p\le m\le p)$
$U(\ell=0,\ n=0,\ -p\le m\le p,\ m\ne0)\}$ semicausal I (5.b)

$\{(-r\le\ell<0,\ -q\le n\le q,\ -p\le m\le p)U(\ell=0,\ -q\le n\le q,\ -p\le m\le p,$
$(m,n)\ne(0,0))\}$ semicausal II (5.c)

$\{(-r\le\ell\le r,\ -q\le n\le q,\ -p\le m\le p,\ (m,n,\ell)\ne(0,0,0)\}$
noncausal (5.d)

The spectral density function of the noise random field $\{W_{i,j,k}\}$ is given by

$$S_w(z_1,z_2,z_3)=Q \text{ causal prediction space} \qquad (6.a)$$

$$= Q[1-\sum_{m=-p}^{p} a_{m,o,o}\ z_1^{-m}] \text{ semicausal I} \qquad (6.b)$$

$$= Q[1-\sum_{m=-p}^{p}\sum_{n=-q}^{q} a_{m,n,o}\ z_1^{-m}z_2^{-n}] \text{semicausal II} \qquad (6.c)$$

$$= Q\ D(z_1,z_2,z_3) \text{ non-causal} \qquad (6.d)$$

where $|z_1|=|z_2|=|z_3|=1$ and $a_{o,o,o}=1$. Note that there is symmetry associated with the model parameters of Eq. 6. For example, in the semicausal II case $a_{m,n,o}=a_{-m,-n,o}$. Closer examination of Eqs. (4) and (6) reveals that the SDF of the random field $\{X_{i,j,k}\}$ is of "all pole" nature in the causal and noncausal cases whereas in the two semicausal cases is ARMA type.

The development of the 4-D minimum variance models can be achieved by following steps similar to those taken above. In this case there are five possible geometries for the prediction space $\bar{V}$, viz.

$\bar{V}=\{(t<0)U(z<0,t=0)U(y<0,(z,t)=(0,0))U(x<0,(y,z,t)=$
$(0,0,0))\}$ causal (7.a)

$\{(t\ 0)U(z\ 0,t=0)U(y\ 0,(z,t)=(0,0))U(x\neq 0,(y,z,t)=(0,0,0))\}$ 1st-order non-causal (7.b)

$\{(t<0)U(z<0,t=0)U((x,y)\neq(0,0),(z,t)=(0,0))\}$ 2nd-order non-causal (7.c)

$\{(t<0)U(x,y,z)\neq(0,0,0),t=0))\}$ 3rd-order non-causal (7.d)

$\{(x,y,z,t)\neq(0,0,0,0)\}$ non-causal (7.e)

where (x,y,z,t) is an arbitrary point in R^4. As a closing remark of this section, the parametric models discussed here are limited to cases where the data field or covariance function are known on a regularly spaced grid. Thus in many practical application areas where the array configurations do not consist of uniformly spaced sensors, interpolation algorithms should be used first to generate uniformly spaced data field.

m-d PARAMETRIC SPECTRAL ESTIMATION

Given a finite length record from a zero-mean, m-d random field with unknown covariance sequence, the m-d model parameters for spectral estimation can not be determined exactly. Thus the criterion-of-fit that is employed to generate the model parameters from the data field determines the quality of the spectral estimate. Different criteria of fit will demonstrate different spectrum performance and perhaps only asymptotically will converge to identical estimates.

One approach suggested by A. Jain (Ref. 12) in the 2-d case which can be easily extended to 3- and 4-d cases is to be first compute estimates of the covariance sequence from the available data and then to formulate the "normal equations" of Eq. 3, viz:

$$\hat{R}(\underline{n}) - \sum_{\underline{\ell}\varepsilon V} \alpha(\underline{\ell})\ \hat{R}(\underline{n}-\underline{\ell}) = \hat{Q}\ \delta(\underline{n},\underline{0}) \qquad (8)$$

where $\hat{Q}$ and $\hat{R}(\lambda)$ are estimates of $Q=E\{W^2(\underline{r})\}$ and $R(\lambda)=E\{X(\underline{r}+\underline{\lambda})X(\underline{r})\}$, respectively and $\underline{n}\varepsilon\{(\underline{V})U(\underline{0})\}$. $E\{\cdot\}$ denotes the expectation operation and $\delta(\underline{n},\underline{0})$ is the Kronecker delta function. The parameter $\alpha(\underline{\ell})$ are obtained by matrix inversion. This approach is similar to the Yule-Walker technique of 1-d AR spectral estimation which is known to exhibit limited resolution spectral estimates. On the other hand, the covariance estimates approach achieves very good spectrum quality in the case of wideband random fields.

An alternative mathematical structure for estimating the model parameters from finite length data fields, which is briefly described below, fits m-d minimum variance representations optimally in the sense of minimizing the least-squares error covariance function within the model prediction space and was thus designated as the m-d covariance least-squares (CLS) algorithm. This approach has been successfully used for 1-d and 2-d spectral estimation (Ref. 4,13,15).

Assuming that $\{X(\underline{r})\}$, $\underline{r}\varepsilon R_f^m$ is a finite set of observations of a zero-mean random field described by Eq. 3, its covariance recursion equation with the stationarity assumption relaxed is of the form

$$\hat{\phi}(\underline{r};\underline{k};\underline{\alpha}) = \hat{R}(\underline{r};\underline{k}) - \sum_{\underline{\ell}\varepsilon V} \alpha(\underline{\ell})\ \hat{R}(\underline{\ell};\underline{k}), \qquad (9)$$

where $\hat{R}(\underline{r};\underline{k})$ denotes unbiased estimates of the $E\{X(\underline{r})\ X(\underline{k})\}$ and $\underline{k}\varepsilon V$, $\underline{r}\notin V$. Note that $E\{\hat{\phi}(\underline{r};\underline{k}\underline{\alpha})\}=0$, i.e., the expected curve-fitting error at the true parameter value is zero. By substituting the single-point unbiased estimates $\hat{R}(\underline{r};\underline{k})=X(\underline{r})X(\underline{k})$ into Eq. 9 and employing

$$c(\hat{\underline{\alpha}}) = \sum_{\underline{r}\varepsilon F}\sum_{\underline{k}\varepsilon V} \hat{\phi}^2(\underline{r};\underline{k};\hat{\underline{\alpha}}) \qquad (10)$$

as the criterion-of-fit, one obtains a set of optimum $\hat{\underline{\alpha}}$ that accurately reflects the covariance structure of the data. This is achieved by minimizing $c(\hat{\underline{\alpha}})$ with respect to $\hat{\underline{\alpha}}$ and solving the resulting system of equations by matrix inversion. The set F in Eq. 10 is the subset of R_f which can be predicted without assuming that the unavailable data is zero.

As regards admissiblity of the m-d CLS spectral estimates (positive), the causal model is guaranteed to provide positive values of the spectrum. We have run several m-d (m=2,3) semicausal cases with simulated data field without yet observing inadmissible estimates. In some non-causal cases, such as the one illustrated in Fig. 1 we have observed inadmissible estimates but larger orders had minimized this effect.

Since all methods for power spectrum estimation on m-d finite length random fields involve approximations, one of the most practical ways to demonstrate if a specific approach works is to investigate its numerical and asymptotic statistical behavior on data generated from a model with known parameters. It is shown in (Ref. 16) that the m-d CLS approach provides asymptotically unbiased parameter estimates, a property also shared by the m-d covariance estimates approach. The numerical behavior of the m-d CLS technique as well as that of some comparison algorithms is discussed in the next section.

SIMULATION RESULTS

In this section the CLS, Maximum Likelihood (ML) of Capon, Kumaresan - Tufts, DFT algorithms as well as the parametric approach based on covariance estimates are applied to narrow-band and wide-band data fields and their spectrum performance is evaluated. To illustrate the comparative performance of the 2-d CLS, Kumaresan-Tufts and Jain's approach (based on covariance estimates), the standard problem of resolving with high-resolution two closely-spaced plane waves is considered. The sequence is described by $x(i,j)=\cos\{2\pi f_1(i+j)\}+\cos\{2\pi f_2(i+j)\}+W(i,j)$; $1\leq(i,j)\leq 16$ and its PSD function has two peaks (f_1,f_1) and (f_2,f_2) in the frequency-wave-number plane. Fig. 2 shows the estimated spectra for $f_1=0.0625$ Hz, $f_2=0.12$ Hz and noise variance $Q_w=0.01$. A causal window with 35 parameters was employed for CLS and Jain's approach. While Jain's approach (2.c) fails to resolve the two plane waves, the peaks of the Kumaresan-Tufts spectrum (2.b) and CLS spectrum (2.a) are distinct and at the correct locations but with those of the CLS being much sharper.

The second simulation example was run to test the resolution and dynamic range of the 3-D CLS and ML methods. This is a situation that arises in many areas of application where strong and weak closely-spaced plane waves need to be detected. The example considered will be for a 6x6 planar array with 2 cm spacing between sensors. The data field is generated by $s(t;x,y)=A1\cos[2\pi(f_o t+k_{x1}x+k_{y1}y)]+ +A2\cos[2\pi(f_o t+k_{x2}x+k_{y2}y)]+W(t;x,y)$, where $A1=1.0$, $A2=0.01$, $f_o=12.5$ Hz, time sampling interval is 0.01 s, number of time samples is 16, $k_{x1} = -k_{y1}$, $k_{x2} = -k_{y2}$, $k_{x1} = -0.0625$ c/cm, $k_{x2} = -0.125$ c/cm and the variance of the additive Gaussian noise is 0.01. The estimated 3-D CLS (Semicausal II with 26 parameters) and ML spectra are illustrated in Fig. 3. From this figure, it appears that the 3-D CLS algorithm is superior over the ML algorithm in resolving the two closely spaced plane waves. Although the CLS technique has correctly positioned the peaks with high-resolution, the variation in peak heights does not sensibly follow the true one. This can be justified on the grounds that direction and speed information may well be more important in practice.

In order to test the resolution of 3-D CLS and ML algorithms in the case of multiple wavefront patterns, an example was considered for a 6x6 array with 2 cm spacing between sensors and 16 samples per sensor of sampling interval 0.01 sec. Four equal amplitude plane waves of $f_o=12.5$ Hz and wavenumbers (c/cm) (-.0625, .0625), (-.0625, -.0625), (.0625, -.0625) and (.0625, .0625) are present in this short length data field. The variance of the additive white Gaussian noise is 0.0001. Fig. 4 illustrates the estimated CLS and ML spectra. While both techniques have resolved all four plane waves, the resolution of the CLS spectrum is superior to that of the ML.

The last simulation example was run to test the ability of the CLS criterion to match a wide-band spectral density function by employing data generated from a causal 3-d model with known parameters. The "stationary" random field of size 6x6x32 was generated by the following causal model driven by white Gaussian noise of variance 0.1

$$X(i,j,k) = \sum_{\ell=o}^{1}\sum_{n=o}^{2}\sum_{m=o}^{2} \alpha(m,n,\ell)X(i-m,j-n,k-\ell) +$$

$$+ W(i,j,k), \quad (m,n,\ell) \neq (0,0,0)$$

where the model parameters (total of 17) $\{\alpha(2,2,1), \alpha(1,2,1), \alpha(0,2,1), \alpha(2,1,1), \alpha(1,1,1),\ldots, \alpha(1,0,0)\}$ take values {.1, -.2, .08, -.2, .1, -.2, .08, -.2, .1, -.08, .2, -.1, .2, -.1, .2, -.1, .2}. The true spectrum of the generated 3-D random field is illustrated in Fig. 5(a). Using the correct parameter model of this field, the parameters were estimated by the CLS algorithm and the CLS spectrum is shown in Fig. 5(b). Comparing Fig. 5(b) with Fig. 5(a), it is apparent that they are quite similar. A number of alternative models with various parametric forms and sizes were also fitted to the data. Figures 5(c)-5(e) illustrate the semicausal I, II and noncausal CLS spectra respectively, and by comparing those with Fig. 5(a) it appears that the semicausal II and noncausal spectra are closer to the true one than semicausal I. A causal model with 26 parameters was also fitted to the data by the CLS algorithm and its estimated spectrum (Fig. 5(f)) is quite similar to the true (Fig. 5(a)) and CLS causal with 17 parameters (Fig. 5(b)) spectra. Figures 5(g) and 5(h) illustrate the conventional and ML spectra, respectively, of the same 3-d random field and clearly demonstrate that both techniques fail to match the true wide-band spectrum.

The last example above illustrates the point we made earlier namely that when the parametric nature of the process is ignored, the conventional or ML spectrum estimate may not be reliable. Also in this example, we find that both causal models (17 and 26 parameters) provide sensibly identical estimates that closely match the true spectrum indicating that variations in model order may not seriously affect the nature of the estimated spectrum. With real data one would have to answer questions related to selection of proper prediction models and the right model order. We feel that the choice will be dictated by some prior knowledge of the nature of the process and that further research is needed in this area.

REFERENCES

1. W.C. Knight, R.G. Pridham, and S.M. Kay, "Digital Signal Processing for Sonar", Proc. IEEE, Vol. 69, No. 11, pp. 1451-1506, Nov. 1981.
2. J. Capon, "High-Resolution frequency-wave-number spectrum analysis", Proc. IEEE, Vol. 57, pp. 1408-1418, Aug. 1969.
3. L.J. Pinson and D.G. Childers, "Frequency-Wavenumber spectrum analysis of EEG multi-electrode array data", IEEE Trans. Biom. Eng., pp. 192-207, May 1974.
4. C.L. Nikias, P.D. Scott, and J.H. Siegel, "Computer based 2-D spectral estimation for the detection of prearrhythmic states after hypothermic myocardial preservation", J. Comp. Med. Biolog. In press 1983.
5. T.E. Barnard and J.P. Burg, "Analytical studies of techniques for the computation of high-resolution wavenumber spectra", Rep. no. 9, Texas Instr., Inc. May 1969.
6. J.H. McClellan, "Multidimensional spectral estimation", Proc. 1st IEEE ASSP Workshop on Spectral Estimation, pp. 3.1.1-3.1.10, Hamilton, Ontario, Canada, August 1981.
7. W.E. Larimore, "Statistical inference on stationary random fields", Proc. IEEE, Vol. 65, No. 6, pp. 961-970, June 1977.
8. L.B. Jackson and H.C. Chien, "Frequency and bearing estimation by two-dimensional linear prediction", Proc. ICASSP 79, pp. 665-668, April 1979.
9. R. Kumaresan and D.W. Tufts, "A two-dimensional technique for frequency wavenumber estimation", Proc. IEEE, Vol. 69, pp. 1515-1517, Nov. 1981.
10. J.A. Cadzow and K. Ogino, "Two-dimensional spectral estimation", IEEE Trans. Acous., Speech Signal Proc., Vol. ASSP-29, pp. 396-401, June 1981.
11. A.K. Jain, "Advances in mathematical models for image processing", Proc. IEEE, Vol. 69, No. 5, pp. 502-528, May 1981.
12. A.K. Jain and S. Ranganath, "High-resolution spectrum estimation in two dimensions", Proc. 1st ASSP Workshop on Spectral Estimation, pp. 3.4.1-3.4.5, Hamilton, Ont., Canada, Aug. 1981.
13. C.L. Nikias, P.D. Scott, and J.H. Siegel, "A new robust 2-D spectral estimation method and its application in cardiac data analysis", Proc. ICASSP 82, pp. 729-732, Paris, France, May 1982.
14. C.L. Nikias and M.R. Raghuveer, "A new class of

high-resolution and robust multi-dimensional spectral estimation algorithms", Proc. IEEE Int. Conf. ASSP, pp. 859-862, Boston, April 1983.
15. C.L. Nikias and P.D. Scott, "The Covariance least-squares algorithm for spectral estimation of processes of short length data", IEEE Trans. Geosc. Remote Sens., GE-21, No. 2, April 1983.
16. M.R. Raghuveer and C.L. Nikias, "A new approach to 3-d spectrum estimation based on minimum variance representations of data fields", Rep. no. TR-83-1, Dept. EECS, Univ. of Connecticut, January 1983.

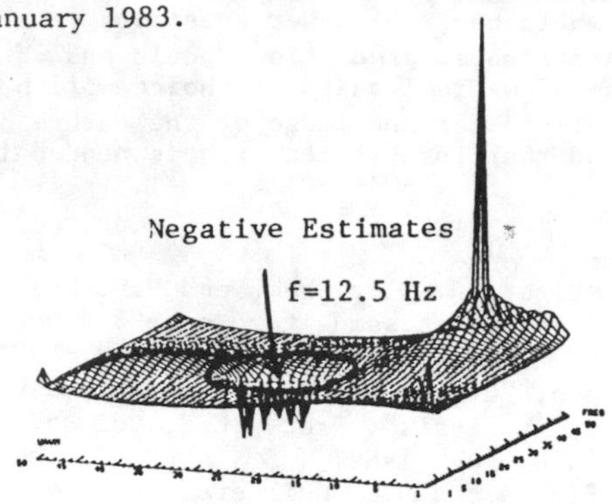

Fig. 1 3-d CLS noncausal spectrum of a single plane wave: Region of negative estimates.

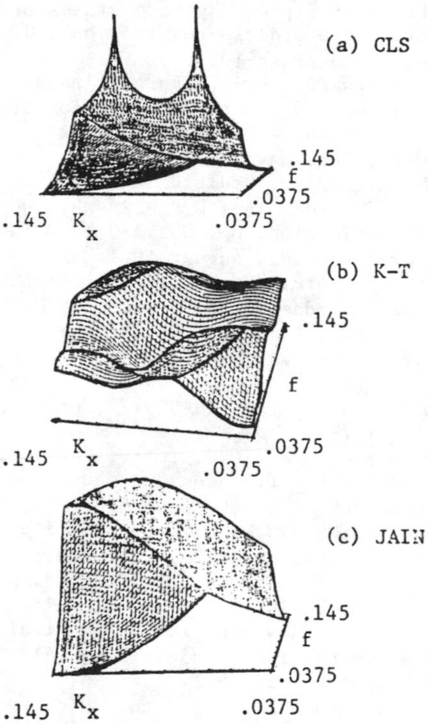

Fig. 2 Two closely-spaced plane waves in a 16x16 data field. (a) causal CLS, (b) Kumaresan-Tufts, (c) causal Jain spectrum (based on covariance estimates).

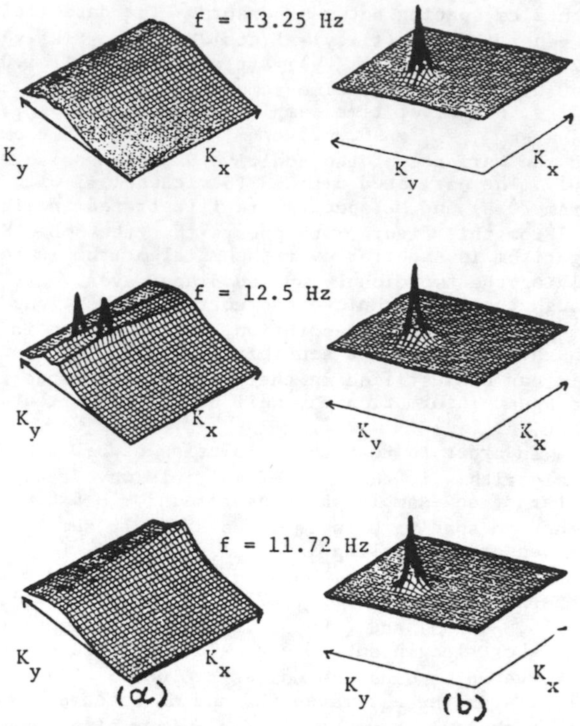

Fig. 3 Resolution and dynamic range 3-d example. (a) semicausal II CLS (b) ML spectrum.

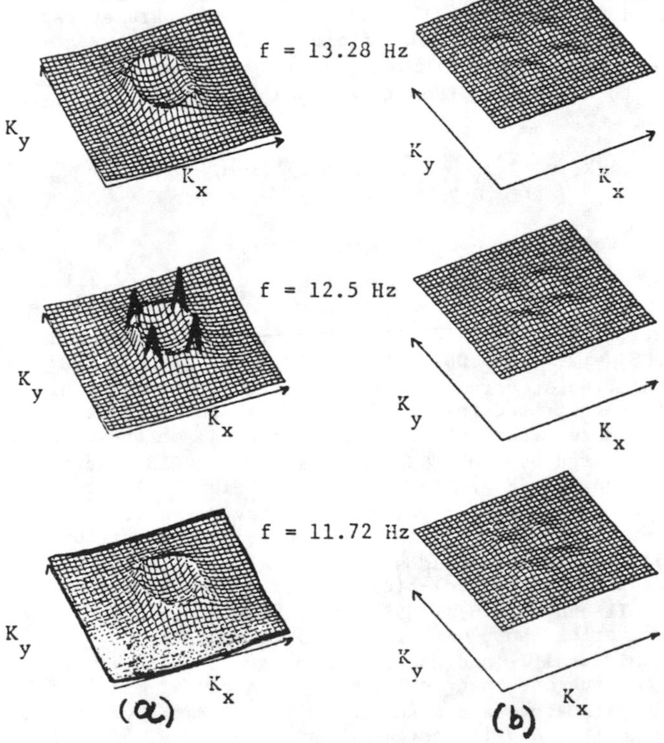

Fig. 4 Resolution and multiple waves (four) 3-d example. (a) semicausal II CLS (b) ML spectrum.

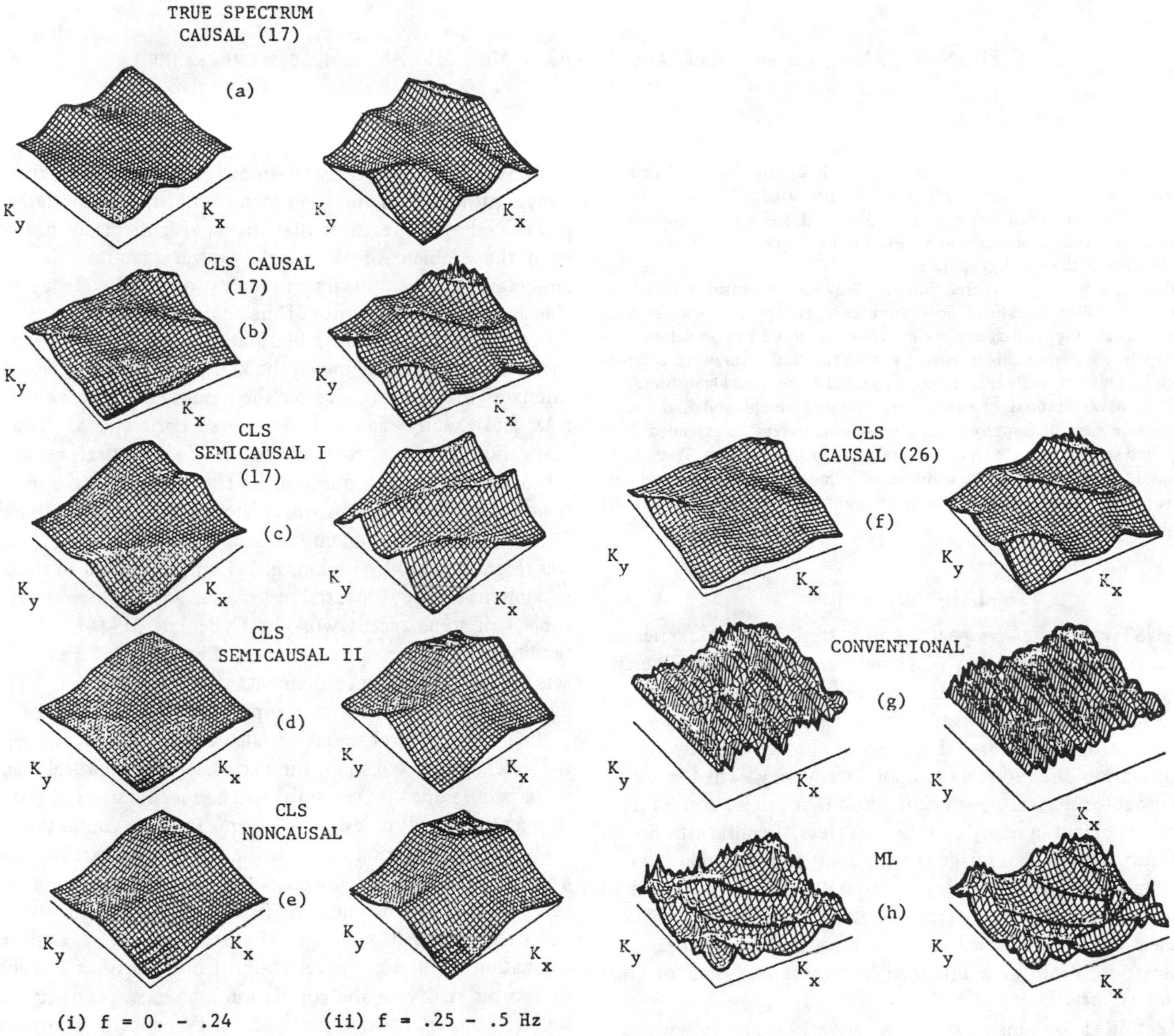

Fig. 5 Spectra of a 6x6x32 causal (17) wideband field: (a) true spectrum, (b) CLS causal (17), (c) CLS semicausal I (17), (d) CLS semicausal II (17), (e) CLS noncausal (17), (f) CLS causal (26), (g) Conventional, (h) ML spectrum. Sliced wave-number spectra at frequencies (i) f = 0. - .24 and (ii) f = .25 - .5 Hz.

Spectral Estimation for Sensor Arrays

STEPHEN W. LANG, MEMBER, IEEE, AND JAMES H. McCLELLAN, SENIOR MEMBER, IEEE

Abstract—The array processing problem is briefly discussed and an abstract spectral estimation problem is formulated. This problem involves the estimation of a multidimensional frequency-wave vector power spectrum from measurements of the correlation function and knowledge of the spectral support.

The investigation of correlation-matching spectral estimates leads to the extendibility question: does there exist any positive spectrum on the spectral support that exactly matches a given set of correlation samples? In answering this question, a mathematical framework is developed in which to analyze and design spectral estimation algorithms.

Pisarenko's method of spectral estimation, which models the spectrum as a sum of impulses plus a noise component, is extended from the time series case to the more general array processing case. Pisarenko's estimate is obtained as the solution of a linear optimization problem, which can be solved using a linear programming algorithm such as the simplex method.

I. INTRODUCTION

JUST as the power spectrum of a stationary time series describes a distribution of power versus frequency, the frequency-wave vector power spectrum of a homogeneous and stationary wavefield describes a distribution of power versus wave vector and temporal frequency, or equivalently, versus propagation direction and temporal frequency. The frequency-wave vector spectrum, or information that can be derived from it, is important in many applications areas. In radio astronomy, the formation of an image is equivalent to the estimation of a power spectrum. The detection and bearing estimation of targets in radar and sonar can be based upon information contained in a power spectral estimate. Hence, power spectral estimation, from data provided by sensor arrays, is of great practical interest.

Section II contains a synopsis of wavefields and sensor arrays and an introduction to the spectral estimation problem. Alternative mathematical representations of power spectra, as measures and as spectral density functions, are discussed. Section II introduces the *coarray*, the set of vector separations and time lags for which correlation samples are available, and the *spectral support*, the region of frequency-wave vector space containing power to which the sensors are sensitive. No particular structure is assumed for either the coarray or the spectral support. Section II concludes with the formulation of an abstract problem: the estimation of a power spectrum given only that it is positive on the spectral support, zero outside, and has certain known correlations for separations in the coarray. Although simpler than many problems encountered in practice, the key features that distinguish the array problem from the problem of time series power spectral estimation are retained: the multidimensionality of the frequency variable and the nonuniformity of the coarray.

Given this problem formulation, it is natural to consider spectral estimates that match the known information: spectral estimates that are positive on the spectral support, zero outside, and exactly match the measured correlations. The investigation of such *correlation-matching* spectral estimates raises two important questions. The first and more fundamental question concerns the existence of any such estimate. This *extendibility* problem has deep historical roots [1], and was recently raised by Dickinson [2] with reference to the 2-D maximum entropy spectral estimation method, and is the subject of some recent work by Cybenko [3]-[4]. The extendibility problem is explored in Section III. Extendible sets of correlation measurements are characterized. Their dependence on the spectral support and the effect of discretizing the spectral support is also considered. In answering the extendibility question, the necessary mathematical framework is developed in which to analyze specific spectral estimation methods and to design algorithms for their computation.

The second question raised is that of uniqueness: is there a unique correlation-matching spectral estimate, and if not, how can a specific one be chosen? In fact, a unique estimate does not exist except in very special cases; the task of a spectral estimation method is the selection of one out of an ensemble of spectra satisfying the correlation matching, positivity, and spectral support constraints. Section IV concerns Pisarenko's method [5], which involves modeling the correlation measurements as a sum of two components. One, a noise component of known spectral shape but unknown amplitude, is made as large as possible without rendering the other component nonextendible. The spectral estimate resulting from Pisarenko's method is shown to solve a linear optimization problem. A solution to this optimization problem will always exist if the correlation measurements are extendible. In fact, Pisarenko's method is shown to be intimately related to the extendibility question and an algorithm for the computation of Pisarenko's estimate will also serve as an extendibility test. It is shown that Pisarenko's estimate is not always unique in the general case, although it is unique in the time series case, where the linear optimization problem reduces to an eigenvalue problem.

II. THE ARRAY PROCESSING PROBLEM

Imagine a multidimensional homogeneous medium supporting a complex-valued wavefield $u(\boldsymbol{x}, t)$ and containing an array of sensors. The wavefield will be assumed to be homogeneous

Manuscript received January 25, 1982; revised September 1, 1982.

This work was supported in part by a Fannie and John Hertz Foundation Fellowship, in part by the National Science Foundation under Grant ECS79-15226, and in part by the Army Research Office under Contract DAAG29-81-K-0073.

S. W. Lang is with Schlumberger-Doll Research, Ridgefield, CT 06877.

J. H. McClellan is with Schlumberger Well Services, Austin Engineering Center, Austin, TX 78759.

Reprinted from *IEEE Trans. Acoust., Speech, Signal Processing,* vol. ASSP-31, pp. 349–358, Apr. 1983.

and stationary so that its second order statistics are described by a correlation function r, or equivalently, by a power spectrum μ [6].

$$r(\boldsymbol{\delta}, \tau) = E[u^*(\boldsymbol{x}, t)\, u(\boldsymbol{x} + \boldsymbol{\delta}, t + \tau)] = \int e^{j(\boldsymbol{k} \cdot \boldsymbol{\delta} + \omega\tau)}\, d\mu. \tag{2.1}$$

The representation of the power spectrum by a positive measure μ[1] provides the flexibility needed to deal with a range of spectral supports in a unified manner and to handle spectra that contain impulses: finite power at a single wave vector.

It is more common, in the engineering literature, to represent a power spectrum by means of a positive spectral density function $S(\boldsymbol{k}, \omega)$. In this representation

$$r(\boldsymbol{\delta}, \tau) = \int e^{j(\boldsymbol{k} \cdot \boldsymbol{\delta} + \omega\tau)}\, S(\boldsymbol{k}, \omega)\, d\nu, \tag{2.2}$$

where ν is some fixed measure which allows (2.2) to be interpreted as a multidimensional surface or volume integral, possibly weighted, over frequency-wave vector space.

Given a power spectral density function $S(\boldsymbol{k}, \omega)$, it is possible to define a corresponding positive measure by requiring the measure of a subset B of frequency-wave vector space to equal the integral of the spectral density function over B:

$$\mu(B) = \int_B d\mu = \int_B S(\boldsymbol{k}, \omega)\, d\nu. \tag{2.3}$$

A simple spectral estimation problem will now be formulated. Particular attention will be paid to modeling features of the data collection process that are common to many array processing problems. These features include measurement of the correlation function at a finite number of nonuniformly distributed points, and constraints on the region of frequency-wave vector space in which power may be present.

The sensors each produce a time function that is the wavefield u sampled at a point in space. The collection of time functions produced by all the sensors, the array output or response, is to be processed so as to provide an estimate of the frequency-wave vector power spectrum. The stochastic character of the wavefield invariably leads to random variations of any spectral estimate based on the array output. To combat this effect, spectral estimates are often based on stable statistics derived from the array output. A common example of such a statistic is a correlation estimate calculated by multiplying the output of one sensor with the time-delayed output of a second sensor, and averaging over time. This process results in an estimate of the correlation function at a temporal lag corresponding to the delay time and a spatial separation that is the vector distance between the two sensors. The averaging process provides statistically stable correlation estimates and results in the statistical stability of a spectral estimate based on these correlation estimates. It is important to note that estimates of the correlations are only available for a finite set of intersensor separations and time delays, the *coarray* [8]. The topic of error in the correlation estimates will not be addressed. Rather, this paper is concerned with the properties of sets of actual correlation samples, and with spectral estimates based on correlation samples.

It is assumed that the spectrum is known to be confined to a bounded region of frequency-wave vector space, the *spectral support*. Outside of this support the spectrum is assumed to be zero. A bounded spectral support can arise naturally in several ways. For example, in a medium that supports scalar waves, known source, medium, and sensor characteristics can be used to construct an appropriate spectral support. The source may have a known temporal bandwidth, or a known finite angular extent. The dispersion relation and attenuation in the medium limits the region of frequency-wave vector space in which power may be present. The sensors may have finite temporal bandwidth and may be directional. All of these effects can be modeled by assuming that no power is present outside of a certain region of frequency-wave vector space. A known spectral support, based on the physics of a particular problem, constitutes important prior information that can be brought to bear on the spectral estimation problem.

In many applications much more data is available in the time dimension than in the space dimension. In these cases it is convenient to separate out the time variable by Fourier analyzing the time series output of each sensor and then doing a separate wave vector spectral estimate for each temporal frequency by using the Fourier coefficients as data for a wave vector spectral estimator. Thus the estimation problem is formulated for complex data even though physical wavefields are real valued. Fortunately, conventional Fourier analysis is often satisfactory when data are abundant, as well as being implicit in the narrow-band character of many sensors. Where limited data in the time dimension make the above approach impractical and wide-band sensor arrays are available, the full problem may be treated by including the temporal variables τ and ω in the vectors $\boldsymbol{\delta}$ and $\boldsymbol{k}$. Thus $\boldsymbol{\delta}$ would describe a separation in both space and time, and $\boldsymbol{k}$ a space-time wave vector. It shall be assumed that one of these two approaches has been taken; hence, the temporal variables τ and ω will be dropped.

A simple example of the spectral estimation model developed above is provided by a sensor array composed of uniformly oriented dish antennas.

Example 2.1: A three dish array. Imagine that an array of dish antennas, shown in Fig. 1, is used to receive a single temporal frequency ω_0, corresponding to a wavelength λ.

A dish antenna of diameter d has a passband that is roughly described by

$$\sqrt{k_2^2 + k_3^2} \leqslant \frac{0.61}{d}.$$

Assuming that the wavefield satisfies the dispersion relation for a homogeneous, nondispersive medium, the support for the

[1] A *positive measure* is a set function that assigns a nonnegative power to each measurable subset of frequency-wave vector space [7]. The power in a subset B of frequency-wave vector space is denoted $\int_B d\mu = \mu(B)$.

Fig. 1. A three dish array.

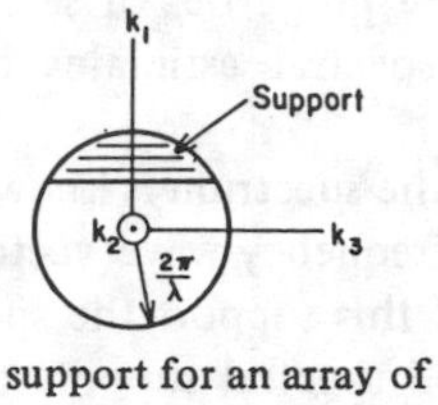

Fig. 2. Spectral support for an array of dish antennas.

spectral estimate should be the polar cap described by the two equations

$$k_1^2 + k_2^2 + k_3^2 = \left[\frac{2\pi}{\lambda}\right]^2$$

$$\sqrt{k_2^2 + k_3^2} \leqslant \frac{0.61}{d}$$

and shown in Fig. 2.

The coarray for this problem is just the set of all 3-dimensional spatial separations between antennas in the array.

III. Extendibility

A simple model of the array processing problem was constructed in the last section: given certain correlation measurements and a spectral support, produce a spectral estimate. It is natural to use the known information about the spectrum to constrain the spectral estimate by requiring that the spectral estimate match the measured correlations, be positive, and be confined to the spectral support. Such spectral estimates are called *correlation-matching* spectral estimates.

The investigation of correlation-matching spectral estimates raises a fundamental existence question. Given a finite collection of measured correlations and a spectral support, does there exist at least one correlation-matching spectral estimate? If such a spectral estimate exists, the measured correlations are said to be *extendible*.[2] After some necessary mathematical definitions, this existence question will be answered by characterizing the set of extendible correlation measurements.

A. Spectral Supports and Coarrays

It is first necessary to define more carefully the terms spectral support and coarray. The spectral support K is assumed to be a compact subset of R^D, i.e., K is closed and bounded. Assuming that K is compact leads to a certain technical advantage: a continuous function on a compact set attains its infimum and supremum. Furthermore, compactness should always hold in a physical problem. As discussed in the previous section, knowledge of source, medium, and sensor characteristics can be used to construct an appropriate spectral support.

[2]The correlation function obtained through the inverse Fourier transform of a correlation-matching spectral estimate is a suitable *extension* of the correlation measurements to all spatial separations.

The coarray Δ will be defined as a finite subset of R^D with the properties

i) $\mathbf{0} \in \Delta$;
ii) if $\boldsymbol{\delta} \in \Delta$, then $-\boldsymbol{\delta} \in \Delta$;
iii) $\{e^{j\boldsymbol{k}\cdot\boldsymbol{\delta}}: \boldsymbol{\delta} \in \Delta\}$ is a set of linearly independent functions on $K \subset R^D$.

Condition i) implies knowledge of $r(\mathbf{0})$, the total power in the spectrum. Condition ii) reflects the fact that the correlation function is always conjugate symmetric; thus if $r(\boldsymbol{\delta})$ is known, so to is $r(-\boldsymbol{\delta})$. Together, conditions i) and ii) imply that Δ is of the form

$$\Delta = \{\mathbf{0}, \pm\boldsymbol{\delta}_1, \cdots, \pm\boldsymbol{\delta}_M\}. \tag{3.1}$$

Condition iii) guarantees that the correlation measurements are independent; each measurement gives new information about the spectrum.

If $D > 1$ then the spectral estimation problem is *multidimensional*. If $D = 1$, $K = [-\pi, \pi]$, and $\Delta = \{0, \pm 1, \cdots, \pm M\}$ then the spectral estimation problem is that of the familiar *time series case* and the extendibility question reduces to the famous trigonometric moment problem [9].

B. Conjugate-Symmetric Functions and Their Vector Representation

The spectral support and the coarray naturally suggest a vector-space setting for the spectral estimation problem, one in which conjugate-symmetric complex-valued functions on Δ will play a central role. A *conjugate-symmetric* function f on Δ is one for which $f(-\boldsymbol{\delta}) = f^*(\boldsymbol{\delta})$ for all $\boldsymbol{\delta} \in \Delta$. Correlation samples, from which spectral estimates are to be made, are such functions. (Because of this symmetry, many of the expressions to follow are real valued even though, for the sake of simplicity, they have been written in a form which suggests that they might be complex valued.) The coarray Δ has $2M + 1$ elements, and so a conjugate-symmetric function on Δ is characterized by $2M + 1$ independent real numbers. Thus a conjugate-symmetric function on Δ may be thought of as a vector in R^{2M+1}.[3] Both the functional notation $f(\boldsymbol{\delta})$ and the vector notation $\boldsymbol{f}$ will be used.

Since $\{e^{j\boldsymbol{k}\cdot\boldsymbol{\delta}}: \boldsymbol{\delta} \in \Delta\}$ is a linearly independent set of functions on K, it follows that each vector $\boldsymbol{p}$ in R^{2M+1} can be uniquely associated with a real-valued Δ-*polynomial* $P(\boldsymbol{k})$ on K through the relation

$$P(\boldsymbol{k}) = \sum_{\boldsymbol{\delta} \in \Delta} p(\boldsymbol{\delta})\, e^{-j\boldsymbol{k}\cdot\boldsymbol{\delta}}. \tag{3.2}$$

The vector $\boldsymbol{p}$ shall be termed *positive* if $P(\boldsymbol{k}) \geqslant 0$ on K. P shall denote the set of those vectors associated with positive Δ-polynomials. From the compactness of K, it can be shown that P is a closed convex cone with vertex at the origin.[4]

[3]A vector space over the real numbers is chosen because it is only multiplication by a *real* number which sends a correlation function into another correlation function.

[4]A set C is a *cone with vertex at the origin* if $\boldsymbol{x} \in C$ implies $\alpha\boldsymbol{x} \in C$ for all $\alpha > 0$ [10]. Cones are important kinds of sets in the spectral estimation problem because it is only multiplication by *positive* real numbers that sends a correlation function into another correlation function, and a Δ-polynomial into another Δ-polynomial.

The inner product between a vector $\boldsymbol{r}$ of correlation samples and a vector $\boldsymbol{p}$ of polynomial coefficients shall be defined as

$$(\boldsymbol{r}, \boldsymbol{p}) = \sum_{\boldsymbol{\delta} \in \Delta} \boldsymbol{r}^*(\boldsymbol{\delta})\, \boldsymbol{p}(\boldsymbol{\delta}).$$

$$= r(\mathbf{0})\, p(\mathbf{0}) + 2 \sum_{i=1}^{M} \cdot [\mathrm{Re}\, r(\boldsymbol{\delta}_i)\, \mathrm{Re}\, p(\boldsymbol{\delta}_i) + \mathrm{Im}\, r(\boldsymbol{\delta}_i)\, \mathrm{Im}\, p(\boldsymbol{\delta}_i)]. \quad (3.3)$$

This inner product gives a new way of writing a Δ-polynomial: $P(\boldsymbol{k}) = (\boldsymbol{\phi_k}, \boldsymbol{p})$, where $\boldsymbol{\phi_k}$ denotes the vector with components $\phi_{\boldsymbol{k}}(\boldsymbol{\delta}) = e^{j\boldsymbol{k}\cdot\boldsymbol{\delta}}$. Also note that if $\boldsymbol{r} = \int_K \boldsymbol{\phi_k}\, d\mu$ then $(\boldsymbol{r}, \boldsymbol{p}) = \int_K P(\boldsymbol{k})\, d\mu$, an expression of Parseval's relation.

C. Characterizations of Extendibility

Let E denote the set of extendible correlation vectors. That is, $\boldsymbol{r} \in E$ if

$$\boldsymbol{r} = \int_K \boldsymbol{\phi_k}\, d\mu \quad (3.4)$$

for some positive measure μ on K. From the properties of the integral, it follows that E is a closed convex cone with vertex at the origin. Furthermore, a section through E at $r(\mathbf{0}) = 1$;

$$E' = \{\boldsymbol{r} \in E:\ r(\mathbf{0}) = 1\} \quad (3.5)$$

is the convex hull of the compact set

$$A = \{\boldsymbol{\phi_k}:\ \boldsymbol{k} \in K\}. \quad (3.6)$$

Thus E is the closed convex cone, with vertex at the origin, generated by A. This characterization of extendible correlations is similar to that given originally by Carathéodory in 1907 for the trigonometric moment problem [1]. It is important in that the set of extendible correlation vectors is described in terms of the simple set A. It also gives a clear geometric picture of extendibility, and will be useful in proofs.

A second characterization of extendibility which is more useful in the development of spectral estimation methods results from expressing E as the intersection of all the closed half-spaces containing it [10]. This characterization involves duality, since half-spaces are defined by linear functionals, i.e., elements of the dual space. A closed half space is defined by a vector $\boldsymbol{q}$ and a real number c as the set

$$\{\boldsymbol{r}:\ (\boldsymbol{r}, \boldsymbol{q}) \geq c\}. \quad (3.7)$$

To determine the particular half-spaces containing E, it is sufficient to consider those correlation vectors that generate E: positive multiples of vectors in the set A. A closed half-space contains E if and only if $(\alpha\boldsymbol{\phi_k}, \boldsymbol{q}) = \alpha Q(\boldsymbol{k}) \geq c$ for every $\boldsymbol{k} \in K$ and every $\alpha \geq 0$. Since α may be made arbitrarily large, it must be true that $Q(\boldsymbol{k}) \geq 0$, i.e., $\boldsymbol{q}$ is a member of the cone P. The smallest half-space containing E for such a $\boldsymbol{q}$ corresponds to choosing $c = 0$. Thus

$$E = \bigcap_{\boldsymbol{p} \in P} \{\boldsymbol{r}:\ (\boldsymbol{r}, \boldsymbol{p}) \geq 0\} \quad (3.8)$$

or, in words, the following.

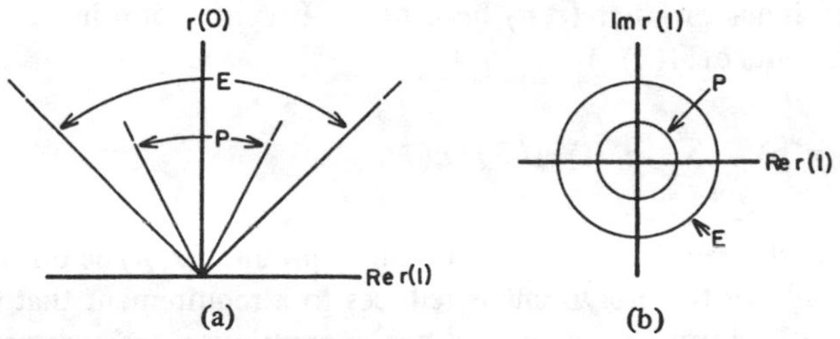

Fig. 3. E and P for $K = [-\pi, \pi]$ and $\Delta = \{0, \pm 1\}$. (a) A section through E and P at $\mathrm{Im}\, r(1) = 0$, and (b) a section through E and P at $r(0) = 1$.

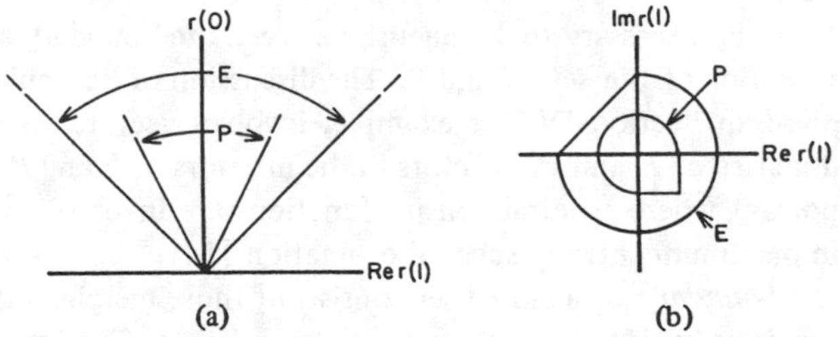

Fig. 4. E and P for $K = [-\pi, \pi/2]$ and $\Delta = \{0, \pm 1\}$. (a) A section through E and P at $\mathrm{Im}\, r(1) = 0$, and (b) a section through E and P at $r(0) = 1$.

The Extension Theorem: The vector $\boldsymbol{r}$ is extendible if and only if $(\boldsymbol{r}, \boldsymbol{p}) \geq 0$ for all positive $\boldsymbol{p}$.

Positive polynomials thus occur naturally in the extendibility problem, since they define the supporting hyperplanes of the set E of extendible correlation vectors. In the language of functional analysis, the extension theorem, which is a form of Farkas' Lemma [11], simply states that E and P are positive conjugate cones [10]. This theorem has the important effect of transferring the simple characterization of P, in terms of positivity, to a characterization of E. Although the incorporation of a spectral support into the problem is new, essentially the same characterization of extendibility was originally used by Calderón and Pepinsky [12], and Rudin [13].

Fig. 4 demonstrates the dependence of E on the spectral support. There are two ways of looking at this dependence. The direct way is to note that E is the convex cone generated by A; because K has been reduced, A has shrunk and E is smaller than in Fig. 3. The indirect way involves constraints; the set K constrains the set P via the positivity condition and the set P constrains the set E via the extendibility theorem. Thus when K shrinks, P grows, and E shrinks.

In the time series case, the extendibility theorem reduces to a test of the positive definiteness of a Toeplitz matrix formed from the correlation samples. Hence, extendibility may be thought of as a general analog of positive-definiteness.

Example 3.1: The Time Series Case; $D = 1$, $K = [-\pi, \pi]$, $\Delta = \{0, \pm 1, \cdots, \pm M\}$. In this case, the extendibility problem reduces to the trigonometric moment problem [9]. Although not generally true, it follows in the time series case, from the fundamental theorem of algebra, that a positive polynomial may be factored as the squared magnitude of an Mth degree trigonometric polynomial

$$P(k) = |A(k)|^2.$$

The inner product $(\boldsymbol{r}, \boldsymbol{p})$ becomes a Toeplitz form in the coefficients of $A(k)$

$$(\boldsymbol{r}, \boldsymbol{p}) = \sum_{i,j=0}^{M} a^*(i)\, r(i-j)\, a(j).$$

Thus the requirement that the inner product $(\boldsymbol{r}, \boldsymbol{p})$ be positive for all positive polynomials reduces to a requirement that the Toeplitz form corresponding to the correlation measurements be positive definite.

D. Boundary and Interior

It will be necessary to distinguish between the boundary and the interior of the sets E and P. The discussion of Pisarenko's method in Section IV, for example, involves vectors on the boundaries of E and P. Vectors in the interiors of E and P are important where spectral density functions are involved, such as in maximum entropy spectral estimation [14].

The *boundary* of a closed set consists of those members that are arbitrarily close to some vector outside the set. The *interior* of a closed set consists of those members that are not on the boundary. The boundary and the interior of a finite dimensional set do not depend upon a particular choice of a vector norm [15]. Moreover, since P and E are convex sets, they have interiors and boundaries which are particularly simple to characterize.

The boundary of P, denoted ∂P, consists of those positive polynomials which are zero for some $\boldsymbol{k} \in K$. The interior of P, denoted P°, consists of those polynomials which are strictly positive on K.

Positive polynomials may be used to define the boundary and the interior of E. The boundary of E, denoted ∂E, consists of those extendible correlation vectors which make a zero inner product with some nonzero positive polynomial. The interior of E, denoted E°, consists of those correlation vectors which make strictly positive inner products with every nonzero positive polynomial.

E. Power Spectral Density Functions

Many spectral estimation methods represent the power spectrum, not as a measure, but as a spectral density function. This leads to a modification of the extendibility problem: given a fixed finite positive measure ν, which defines the integral

$$\boldsymbol{r} = \int_K \boldsymbol{\phi}_{\boldsymbol{k}}\, S(\boldsymbol{k})\, d\nu, \tag{3.9}$$

which correlation vectors $\boldsymbol{r}$ can be derived from some strictly positive function $S(\boldsymbol{k})$? Under one additional constraint on ν, easily satisfied in practice, it can be shown that vectors which can be represented in this fashion are exactly those vectors in the interior of E. Furthermore, it can be shown that any vector in the interior of E can be represented in the form (3.9) for some continuous, strictly positive $S(\boldsymbol{k})$.

An Extension Theorem for Spectral Density Functions: If every neighborhood of every point in K has strictly positive ν-measure, then

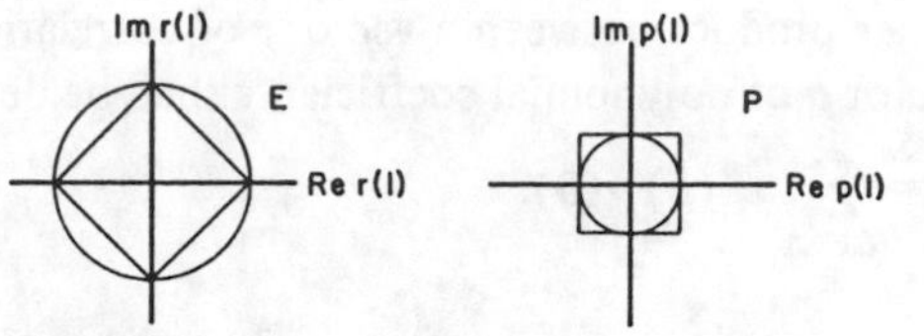

Fig. 5. Approximation of a spectral support by sampling; a section at $r(0) = 1$.

1) if $S(\boldsymbol{k})$ is uniformly bounded away from zero over K, then

$$\boldsymbol{r} = \int_K \boldsymbol{\phi}_{\boldsymbol{k}}\, S(\boldsymbol{k})\, d\nu \in E^\circ;$$

2) if $\boldsymbol{r} \in E^\circ$ then

$$\boldsymbol{r} = \int_K \boldsymbol{\phi}_{\boldsymbol{k}}\, S(\boldsymbol{k})\, d\nu$$

for some continuous, strictly positive function $S(\boldsymbol{k})$.

A proof of this theorem is contained in Appendix A.

F. Discretization of the Spectral Support

Many spectral supports of interest contain an infinite number of points. These spectral supports must often be approximated, in computational algorithms, by a finite number of points. It is important, therefore, to understand the effects of such approximation.

Consider the discrete spectral support

$$K = \{\boldsymbol{k}_i \in R^D\colon\ i = 0, \cdots, N-1\}. \tag{3.10}$$

A measure μ on a discrete support is completely characterized by its value $\mu(\boldsymbol{k}_i)$ at each point. Thus the inverse Fourier integral reduces to a finite sum

$$\int_K \boldsymbol{\phi}_{\boldsymbol{k}}\, d\mu = \sum_{i=0}^{N-1} \boldsymbol{\phi}_{\boldsymbol{k}_i}\, \mu(\boldsymbol{k}_i). \tag{3.11}$$

Similarly, for spectral density functions

$$\int_K \boldsymbol{\phi}_{\boldsymbol{k}}\, S(\boldsymbol{k})\, d\nu = \sum_{i=0}^{N-1} \boldsymbol{\phi}_{\boldsymbol{k}_i}\, S(\boldsymbol{k}_i)\, \nu(\boldsymbol{k}_i). \tag{3.12}$$

The measure ν can be considered to define a quadrature rule for integrals over the spectral support.

From the definitions of extendible correlation vectors and of positive polynomials, it can be seen that if a spectral support is formed by choosing a finite number of points out of some original spectral support, then the new set E is a convex polytope inscribed within the original set E and the new set P is a convex polytope circumscribed about the original set P. Hence, the new E is smaller than the original E and the new P is larger than the original P. A sufficiently dense sampling of the original spectral support will result in polytopes which approximate the original sets to arbitrary precision. For example, Fig. 5 shows the effect of approximating the spectral support $[-\pi, \pi]$ by the four samples $\{0, \pm(\pi/2), \pi\}$ for $\Delta = \{0, \pm 1\}$. The original E and P cones have a circular cross section at

$r(0) = 1$, as shown in Fig. 3. The cones corresponding to the sampled support both have a square cross section. The boundaries of the new and old cones intersect at vectors corresponding to the sample points.

IV. Pisarenko's Method

Pisarenko described a time series spectral estimation method in which the spectrum is modeled as a sum of impulses plus a white noise component [5]. If the white noise component is chosen as large as possible, he showed that the position and amplitudes of the impulses needed to match the measured correlations are uniquely determined. Pisarenko's method will be derived in the more general array setting and for a more general noise component. The relationship of Pisarenko's method to the extendibility question will be demonstrated.

The extended Pisarenko's estimate will be derived as the solution of an optimization problem involving the minimization of a linear functional over a convex region defined by linear constraints. A solution to this optimization problem always exists, but it may not be unique. A dual optimization problem is derived which, in the time series case, leads to the familiar interpretation of Pisarenko's method as the design of a constrained least squares smoothing filter. Again, a solution to this dual problem always exists, but may not be unique.

Algorithms for the computation of Pisarenko's method are discussed. A primal optimization problem is written, for a spectral support composed of a finite number of points, as a standard form linear program. The application of the simplex method to the solution of this primal linear program is discussed. A dual linear program is presented. The possibility of computational algorithms faster than the simplex method is also discussed.

A. Pisarenko's Method for Sensor Arrays

The basis of Pisarenko's method is the unique decomposition (Fig. 6) of a correlation vector $\boldsymbol{r}$ into the sum of a scaled noise correlation vector $\boldsymbol{n}$, in the interior of E, plus a remainder $\boldsymbol{r}'$, on the boundary of E

$$\boldsymbol{r} = \boldsymbol{r}' + \alpha \boldsymbol{n}. \tag{4.1}$$

The assumption that $\boldsymbol{n}$ is in E° implies that such a decomposition of an arbitrary vector $\boldsymbol{r}$ exists and is unique. Consider the one-parameter family of correlation vectors

$$\boldsymbol{r}_c = \boldsymbol{r} - c\boldsymbol{n}. \tag{4.2}$$

For c sufficiently positive, $\boldsymbol{r}_c$ must be nonextendible and, for c sufficiently negative, $\boldsymbol{r}_c$ must be extendible, since the assumption that $\boldsymbol{n} \in E^\circ$ implies that E contains a neighborhood of $\boldsymbol{n}$. The convexity of E implies that there is some greatest number α such that $\boldsymbol{r}' = \boldsymbol{r} - \alpha\boldsymbol{n}$ is extendible. Since there are nonextendible vectors arbitrarily close to $\boldsymbol{r}'$, $\boldsymbol{r}'$ must be on the boundary of E. Furthermore, since $\alpha \geqslant 0$ if and only if $\boldsymbol{r}$ is extendible, this decomposition of $\boldsymbol{r}$ can also be used as an extendibility test.

This unique decomposition of $\boldsymbol{r}$ can be formulated as a primal linear optimization problem over all positive power spectra. Note that $\boldsymbol{r}'$ has at least one positive spectral representation μ' and that, from (4.1) for $\boldsymbol{\delta} = \boldsymbol{0}$

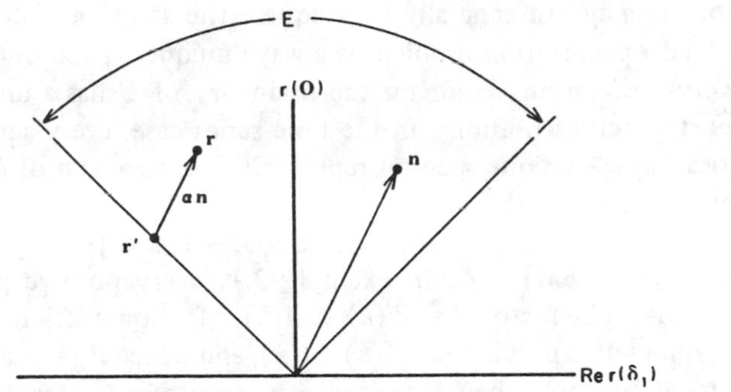

Fig. 6. Decomposition of a vector $\boldsymbol{r}$ into vector $\boldsymbol{r}'$ on the boundary of E plus a multiple of a given vector $\boldsymbol{n}$.

$$\alpha = \left[\frac{r(0)}{n(0)} - \frac{1}{n(0)} \int_K d\mu'\right]. \tag{4.3}$$

The statement that α is the largest number such that the remainder $\boldsymbol{r}' = \boldsymbol{r} - \alpha\boldsymbol{n}$ is extendible leads to the linear optimization problem

$$\max_{\mu \geqslant 0} \left[\frac{r(0)}{n(0)} - \frac{1}{n(0)} \int_K d\mu\right] \tag{4.4a}$$

such that

$$\boldsymbol{r} = \int_K \boldsymbol{\phi_k}\, d\mu + \left[\frac{r(0)}{n(0)} - \frac{1}{n(0)} \int_K d\mu\right] \boldsymbol{n}. \tag{4.4b}$$

The maximum is α and is attained at $\mu = \mu'$.

Since $\boldsymbol{n}$ is extendible, it corresponds to some positive measure μ_n. Hence, (4.1) becomes

$$\boldsymbol{r} = \int_K \boldsymbol{\phi_k} [d\mu' + \alpha d\mu_n]. \tag{4.5}$$

If $\alpha \geqslant 0$ then $\mu' + \alpha\mu_n$ is a positive measure which matches the correlation measurements and which has the largest possible noise component.

Some further information about the remainder $\boldsymbol{r}'$ and its spectral representation can be derived. $\boldsymbol{r}'$ is on the boundary of E; hence, it makes a zero inner product with some nonzero positive polynomial

$$(\boldsymbol{r}', \boldsymbol{p}') = \int_K P'(\boldsymbol{k})\, d\mu' = 0. \tag{4.6}$$

It follows that the support of μ' must be on the zero set of $P'(\boldsymbol{k})$. More precisely, the support of any spectral representation of $\boldsymbol{r}'$ must be on the intersection of the zero sets of all positive polynomials that make a zero inner product with $\boldsymbol{r}'$. This suggests the final step in the derivation of Pisarenko's method; namely the association of the remainder $\boldsymbol{r}'$ with an impulsive spectrum.

The fact that the objective functional of the primal optimization problem is not strictly convex suggests that the solution μ' may not generally be unique. The solution μ' to the primal optimization problem is always unique if, and only if, every correlation vector on the boundary of E has a unique spectral representation. In the time series case, every such $\boldsymbol{r}'$ does have a unique spectral representation, as a sum of M or fewer impulses [5].

Example 4.1: The Time Series Case, $D = 1, K = [-\pi, \pi], \Delta = \{0, \pm 1, \cdots, \pm M\}$. As in example 3.1, every positive polynomial can be factored as $P'(k) = |A(k)|^2$ for some Mth degree trigonometric polynomial $A(k)$. $A(k)$, and hence $P'(k)$, can be zero at no more than M points. The spectrum μ', therefore, must be a sum of impulses at these points. Furthermore, since it is possible to construct a positive polynomial that is zero at $N \leqslant M$ arbitrarily selected points, and nowhere else, it follows that $\boldsymbol{r}'$ has a unique spectral representation as a sum of impulses at the common zeros of all positive polynomials $\boldsymbol{p}'$ such that $(\boldsymbol{r}, \boldsymbol{p}') = 0$.

More generally, the extension theorem combined with Carathéodory's theorem [16] shows that there is at least one spectral representation of $\boldsymbol{r}'$ as a sum of no more than $2M$ impulses.

The Representation Theorem: If $\boldsymbol{r}' \in \partial E$, then there exist $a(i) \geqslant 0$ and $\boldsymbol{k}_i \in K$ such that

$$\boldsymbol{r}' = \sum_{i=1}^{2M} a(i) \boldsymbol{\phi}_{\boldsymbol{k}_i}. \tag{4.7}$$

A proof of the representation theorem can be found in Appendix B. This representation, and thus the solution to the primal optimization problem may not be unique. A further discussion of this uniqueness problem can be found in Appendix C.

If α and the locations of the impulses in a unique solution μ' could be determined for a given $\boldsymbol{r}'$, then the impulse amplitudes could be calculated simply by solving a set of linear equations. A dual optimization problem will now be derived which gives α and $\boldsymbol{p}'$ such that $(\boldsymbol{r}', \boldsymbol{p}') = 0$. Then, if $\boldsymbol{r}'$ has a unique spectral representation, the impulse locations can be determined from the zeros of $P(\boldsymbol{k})$. From the extendibility theorem

$$\begin{aligned}(\boldsymbol{r}', \boldsymbol{p}) &= (\boldsymbol{r} - \alpha \boldsymbol{n}, \boldsymbol{p}) \\ &= (\boldsymbol{n}, \boldsymbol{p}) \, [(\boldsymbol{r}, \boldsymbol{p})/(\boldsymbol{n}, \boldsymbol{p}) - \alpha] \geqslant 0.\end{aligned} \tag{4.8}$$

Since $\boldsymbol{n} \in E^\circ$ and $\boldsymbol{r}' \in \partial E$, it follows that $(\boldsymbol{n}, \boldsymbol{p}) > 0$ and $(\boldsymbol{r}', \boldsymbol{p}) \geqslant 0$ for all $\boldsymbol{p} \in P$. Furthermore, since $(\boldsymbol{r}', \boldsymbol{p}) = 0$ for some $\boldsymbol{p}' \in P$, it follows that

$$\alpha = \min \, (\boldsymbol{r}, \boldsymbol{p}) \tag{4.9a}$$

over the set

$$\{\boldsymbol{p} \in P\colon \; (\boldsymbol{n}, \boldsymbol{p}) = 1\} \tag{4.9b}$$

and that the minimum is attained at $\boldsymbol{p}'$. The solution to this dual problem may not be unique even in the time series case, where it reduces to the eigenvector problem derived by Pisarenko and leads to the interpretation of Pisarenko's method as determining a constrained least squares smoothing filter.

Example 4.2: The Time Series Case, $D = 1, K = [-\pi, \pi]$, $\Delta = \{0, \pm 1, \cdots, \pm M\}$. As in example (3.1)

$$(\boldsymbol{r}, \boldsymbol{p}) = \sum_{i,j=0}^{M} a^*(i) \, r(i-j) \, a(j).$$

Furthermore, if $\boldsymbol{n}$ corresponds to white noise of unit power,

$$(\boldsymbol{n}, \boldsymbol{p}) = p(0) = \sum_{i=0}^{M} |a(i)|^2.$$

Thus the dual optimization problem reduces to finding the eigenvector of the Toeplitz matrix associated with $\boldsymbol{r}$ corresponding to the smallest eigenvalue. If there are several such eigenvectors, the impulses are located at the common zeros of the corresponding polynomials. Any normalized eigenvector corresponding to the minimal eigenvalue gives the coefficients of a smoothing filter, the sum of whose squared magnitudes is constrained to be one, which gives the least output power when fed an input process whose correlations are described by $\boldsymbol{r}$ [17].

B. The Computation of Pisarenko's Estimate

In the design of algorithms to compute Pisarenko's estimate, one may be concerned with a discrete spectral support

$$K = \{\boldsymbol{k}_i \in R^D\colon \; i = 0, \cdots, N-1\}. \tag{4.10}$$

On such a support, the primal problem (4.4) can be rewritten as the standard form *linear program*

$$\min_{\nu \geqslant 0} \sum_{i=0}^{N-1} \nu(\boldsymbol{k}_i) \tag{4.11a}$$

such that, for $\boldsymbol{\delta} \in \Delta$, $\boldsymbol{\delta} \neq 0$,

$$\sum_{i=0}^{N-1} \left[e^{j\boldsymbol{k}_i \cdot \boldsymbol{\delta}} - \frac{n(\boldsymbol{\delta})}{n(\boldsymbol{0})} \right] \nu(\boldsymbol{k}_i) = \frac{r(\boldsymbol{\delta})}{r(\boldsymbol{0})} - \frac{n(\boldsymbol{\delta})}{n(\boldsymbol{0})} \tag{4.11b}$$

with N variables and $2M$ constraints. The minimum is $1 - (n(\boldsymbol{0})/r(\boldsymbol{0})) \, \alpha$ and is attained for $\nu' = (1/r(\boldsymbol{0})) \, \mu'$. The fundamental theorem of linear programming [18] is equivalent to the representation theorem in this case. Given that a solution exists to this linear program, as shown in the previous section, the fundamental theorem guarantees a solution in which no more than $2M$ of the $\nu(\boldsymbol{k}_i)$'s are nonzero, a so-called *basic* solution.

The dual linear program [15]

$$\min_{q} \sum_{\boldsymbol{\delta} \neq 0} \left[\frac{r^*(\boldsymbol{\delta})}{r(\boldsymbol{0})} - \frac{n^*(\boldsymbol{\delta})}{n(\boldsymbol{0})} \right] q(\boldsymbol{\delta}) \tag{4.12a}$$

such that, for $i = 0, \cdots, N-1$

$$\sum_{\boldsymbol{\delta} \neq 0} \left[e^{-j\boldsymbol{k}_i \cdot \boldsymbol{\delta}} - \frac{n^*(\boldsymbol{\delta})}{n(\boldsymbol{0})} \right] q(\boldsymbol{\delta}) \geqslant -1 \tag{4.12b}$$

is equivalent to the dual problem (4.9) for a discrete spectral support, where the constraint

$$(\boldsymbol{n}, \boldsymbol{p}) = 1 \tag{4.13}$$

has been used to eliminate $p(\mathbf{0})$ and where $\boldsymbol{q} = n(\mathbf{0})\,\boldsymbol{p}$. Its minimum is $(n(\mathbf{0})/r(\mathbf{0}))\,\alpha - 1$ and is attained at $\boldsymbol{q}' = n(\mathbf{0})\,\boldsymbol{p}'$.

The primal problem can be solved using the *simplex method* [18]. The application of the simplex method to the primal problem results in essentially the same computational algorithm as the application of the (single) *exchange method* to the dual problem [19]. By incorporating a technique to avoid cycling [20], an algorithm can be obtained which is guaranteed to converge to an optimal solution in a finite number of steps, although implementations have typically been slow.

The problem of Chebyshev approximation is related to the computation of Pisarenko's estimate; it also can be formulated as the minimization of a linear functional over a convex space defined by linear inequality constraints [16]. It also has been solved using the simplex (single exchange) method. However, for the particular problem of the Chebyshev approximation of continuous functions by polynomials in one variable, a computational method exists which is significantly faster than the simplex method, a *multiple-exchange* method due to Remes. Although attempts have been made to extend this method to more general problems [21], the resulting algorithms are not well understood; in particular, there is no proof of convergence.

Finally, the undiscretized optimization problems involved in the computation of Pisarenko's estimate, (4.4) and (4.9), are a form known as *semi-infinite programs*. Both theoretical and computational aspects of such programs are discussed in a collection of papers edited by Hettich [22].

V. Summary

This paper has been concerned with what is probably the simplest interesting problem in array processing; the estimation of a power spectrum with a known support, given certain samples of its correlation function. Although simple, this problem retains several features which are common to many array processing problems: multidimensional spectra, nonuniformly sampled correlation functions, and arbitrary spectral supports.

The investigation of correlation-matching spectral estimates led to the extendibility problem. Two characterizations of extendibility were given. This problem, in the time series case, is known as the trigonometric moment problem and its solution involves consideration of the positive definiteness of the correlation samples. Positive definiteness can therefore be considered as a special case of extendibility.

Building on the theoretical framework developed in solving the extendibility problem, Pisarenko's method was extended from the time series case to the array processing problem. Pisarenko's method was shown to be intimately related to the extendibility problem. The computation of Pisarenko's estimate was shown to involve the solution of a linear optimization problem. The solution of this problem was shown not to be unique in general, although it is unique in the time-series case, where the linear optimization problem reduces to an eigenvalue problem.

Although the spectral estimation problem considered in this paper was developed for array processing, the theoretical framework and the resulting algorithms should be useful in other multidimensional problems, such as image processing.

Appendix A
The Extension Theorem for Spectral Density Functions

This appendix concerns the extension theorem for spectral density functions discussed in Section III-E. It is assumed that every neighborhood of every point in K has strictly positive ν-measure. This condition guarantees that correlation vectors corresponding to impulses in K can be approximated by correlation vectors corresponding to continuous, strictly positive spectral density functions.

An Extension Theorem for Spectral Density Functions: If every neighborhood of every point in K has strictly-positive ν-measure, then

1) if $S(\boldsymbol{k})$ is uniformly bounded away from zero over K, then

$$\boldsymbol{r} = \int_K \boldsymbol{\phi}_{\boldsymbol{k}}\, S(\boldsymbol{k})\, d\nu \in E^\circ,$$

2) if $\boldsymbol{r} \in E^\circ$ then

$$\boldsymbol{r} = \int_K \boldsymbol{\phi}_{\boldsymbol{k}}\, S(\boldsymbol{k})\, d\nu,$$

for some continuous, strictly positive functions $S(\boldsymbol{k})$.

Proof: The first statement may be proved by consideration of the mapping from a bounded function $Q(\boldsymbol{k})$ to a vector $\boldsymbol{r}$ defined by

$$\boldsymbol{r} = \int_K \boldsymbol{\phi}_{\boldsymbol{k}}\, Q(\boldsymbol{k})\, d\nu. \tag{A1}$$

That $S(\boldsymbol{k})$ is uniformly bounded away from zero means that, for some $\epsilon > 0$, $S(\boldsymbol{k}) > \epsilon$ for all $\boldsymbol{k} \in K$. Because the functions $\{e^{j\boldsymbol{k}\cdot\boldsymbol{\delta}}: \boldsymbol{\delta} \in \Delta\}$ are linearly independent functions on K, and since every neighborhood of every point in K contains a set of strictly positive measure, it follows that the image of the set of bounded Δ-polynomials

$$\{\boldsymbol{q} \in R^{2M+1}: |Q(\boldsymbol{k})| < \epsilon\} \tag{A2}$$

under (A1), is a neighborhood of $\mathbf{0}$. Therefore the image of

$$\{S(\boldsymbol{k}) + Q(\boldsymbol{k}): |Q(\boldsymbol{k})| < \epsilon\} \tag{A3}$$

is a subset of E which is a neighborhood of $\boldsymbol{r}$. Hence, $\boldsymbol{r} \in E^\circ$.

The second statement may be proved by considering the set E_B of correlation vectors corresponding to spectral density functions which are integrable, continuous, and strictly positive (hence, bounded away from zero),

$$\{S(\boldsymbol{k}) \in C(K): S(\boldsymbol{k}) > 0\}. \tag{A4}$$

E_B is convex and, from the argument above, it follows that E_B is open. It is easily shown that the vectors $\boldsymbol{\phi}_{\boldsymbol{k}}$, for $\boldsymbol{k} \in K$, are in the closure of E_B. From Carathéodory's theorem [16], it follows that every $\boldsymbol{r} \in E$ can be written as a positive sum of some $2M + 1$ such $\boldsymbol{\phi}_{\boldsymbol{k}_i}$. Since each $\boldsymbol{\phi}_{\boldsymbol{k}_i}$ is in the closure of E_B, it follows that each $\boldsymbol{r} \in E$ is also. Therefore, the closure of

E_B is E. Two open convex sets with the same closure must be identical. Since E is the closure of both E_B and E°, it follows that $E_B = E^\circ$.

Appendix B
The Representation Theorem

The representation theorem of Section IV-A is a simple extension of Carathéodory's theorem [16] for correlation vectors on the boundary of E, making use of the extension theorem. It is the generalization of the "A theorem of C." Carathéodory [9, ch. 4] to multiple dimensions. In view of the derivation of Pisarenko's method, in Section IV, as a linear program, the representation theorem may also be viewed as a form of the fundamental theorem of linear programming [18].

The Representation Theorem: If $\boldsymbol{r}'$ is on the boundary of E, then for some $2M$ nonnegative $a(i)$ and some $\boldsymbol{k}_i \in K$:

$$\boldsymbol{r}' = \sum_{i=1}^{2M} a(i)\boldsymbol{\phi}_{\boldsymbol{k}_i}. \tag{B1}$$

Proof: Consider the compact convex set $E' = \{\boldsymbol{r} \in E: \boldsymbol{r}(\boldsymbol{0}) = 1\} \subset R^{2M}$, which is the convex hull of $A = \{\boldsymbol{\phi}_{\boldsymbol{k}}: \boldsymbol{k} \in K\}$. From Carathéodory's theorem, any element in E' can be expressed as a convex combination of $2M + 1$ elements of A

$$\boldsymbol{r} = \sum_{i=1}^{2M+1} \theta_i \boldsymbol{\phi}_{\boldsymbol{k}_i} \tag{B2}$$

with $\theta_i \geq 0$, $\Sigma_{i=1}^{2M+1} \theta_i = 1$, and $\boldsymbol{k}_i \in K$. If one of the θ_i is zero, the proof is complete. Otherwise, since $\boldsymbol{r}$ is on the boundary of E', there is some nonzero $\boldsymbol{p} \in P$ such that

$$0 = (\boldsymbol{r}, \boldsymbol{p}) = \sum_{i=1}^{2M+1} \theta_i (\boldsymbol{\phi}_{\boldsymbol{k}_i}, \boldsymbol{p}). \tag{B3}$$

Thus, for each i, $(\boldsymbol{\phi}_{\boldsymbol{k}_i}, \boldsymbol{p}) = 0$. The $\boldsymbol{\phi}_{\boldsymbol{k}_i}$'s must be linearly dependent, so there are some $\beta_i \in R$, not all zero, such that $\Sigma_{i=1}^{2M+1} \beta_i \boldsymbol{\phi}_{\boldsymbol{k}_i} = 0$. Let λ be the number with the smallest magnitude such that $\theta_i + \lambda\beta_i = 0$ for some i. Then

$$\boldsymbol{r} = \sum_{i=1}^{2M+1} (\theta_i + \lambda\beta_i)\boldsymbol{\phi}_{\boldsymbol{k}_i}. \tag{B4}$$

One of the coefficients is zero, reducing this to a sum over only $2M$ terms. Recognizing that any element of E is a scaled version of an element of E' completes the proof.

Note that, in the times series case, $\boldsymbol{r}'$ could be expressed as a sum of no more than M complex exponentials while the above theorem only guarantees a representation in terms of $2M$ exponentials. This is not a deficiency in the proof, but a genuine feature of the problem, as the following one-dimensional example shows.

Example B.1: $D = 1$, $K = [-\pi, \pi/2]$, $\Delta = \{0, \pm 1\}$. Suppose that $\boldsymbol{r}$ is on the straight portion of the boundary of E, as indicated in Fig. 7. Clearly, $\boldsymbol{r}$ has a unique representation as a convex sum of members of A in terms of the *two* correlation vectors corresponding to $k = \pi/2$ and $k = -\pi$,

$$\boldsymbol{r}(\delta) = \tfrac{1}{2} e^{j(\pi/2)\delta} + \tfrac{1}{2} e^{-j\pi\delta}, \quad \delta \in \Delta.$$

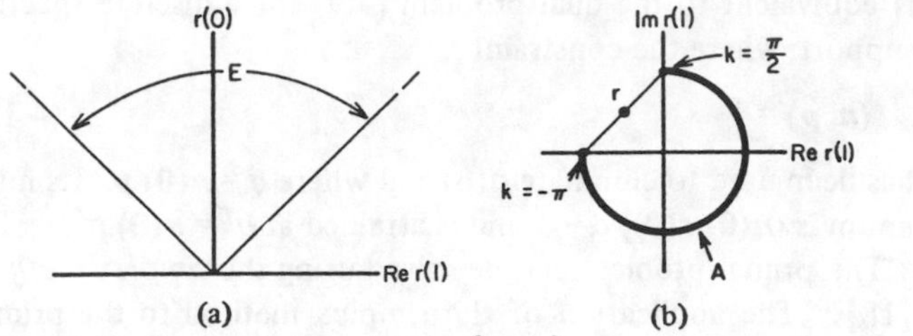

Fig. 7. E for $K = [-\pi, \pi/2]$ and $\Delta = \{0, \pm 1\}$. (a) A section through E at $\operatorname{Im} r(1) = 0$, and (b) a section through E at $r(0) = 1$.

Appendix C
The Uniqueness of Pisarenko's Estimate

As discussed in Section IV-A, Pisarenko's estimate is unique if one and only one spectrum can be associated with each correlation vector on the boundary of E. Trivial uniqueness problems result if two distinct $\boldsymbol{k}$'s in K give rise to the same $\boldsymbol{\phi}_{\boldsymbol{k}}$'s. More generally, consider the set of correlation vectors corresponding to the zero set of some nonzero positive polynomial $\boldsymbol{p}$

$$Z = \{\boldsymbol{\phi}_{\boldsymbol{k}}: (\boldsymbol{\phi}_{\boldsymbol{k}}, \boldsymbol{p}) = P(\boldsymbol{k}) = 0, \quad \boldsymbol{k} \in K\}. \tag{C1}$$

Any vector $\boldsymbol{r}' \in E$ which makes a zero inner product with $\boldsymbol{p}$ can be expressed as a sum of positive multiples of vectors from the set Z. It follows that, if this set is linearly independent the representation is unique. Conversely, if this set is linearly dependent, then an $\boldsymbol{r}'$ on the boundary of E can be constructed which has more than one spectral representation. If the set is linearly dependent then there is a finite collection of nonzero real numbers $c(i)$ and $\boldsymbol{\phi}_{\boldsymbol{k}_i} \in Z$ such that

$$\sum_i c(i)\boldsymbol{\phi}_{\boldsymbol{k}_i} = 0. \tag{C2}$$

Because $\boldsymbol{\phi}_{\boldsymbol{k}_i}(\boldsymbol{0}) = 1$ for all i, there must be at least one $c(i)$ which is strictly positive and one which is strictly negative. Thus

$$\boldsymbol{r}' = \sum_{c(i)>0} c(i)\boldsymbol{\phi}_{\boldsymbol{k}_i} = \sum_{c(i)<0} [-c(i)]\,\boldsymbol{\phi}_{\boldsymbol{k}_i} \tag{C3}$$

is a nonzero correlation vector on the boundary of E with at least two spectral representations.

Therefore, Pisarenko's estimate is unique if and only if the set of correlation vectors corresponding to the zero of each nonzero positive polynomial is linearly independent. In particular, for Pisarenko's estimate to be unique, no nonzero positive polynomial can have more than $2M$ zeros. This condition is similar to, though not quite as strict as the Haar condition [23], which involves all polynomials, not just positive ones.

The factorability of polynomials, in the time series case, leads to a strong result. In the time series case, a nonzero positive polynomial can have no more than M zeros. Furthermore, a nonzero positive polynomial can be constructed that is zero at M or fewer arbitrary locations, and nowhere else. This implies (Example 4.1) that a correlation vector in ∂E has a unique spectral representation and that this spectrum is composed of M or fewer impulses. Furthermore, it implies that any spectrum composed of M or fewer impulses has a correlation vector in ∂E.

However, a simple example shows that Pisarenko's estimate

is not guaranteed to be unique in most multidimensional situations. Consider the nonzero positive polynomial

$$P(\boldsymbol{k}) = 1 - \cos(\boldsymbol{k} \cdot \boldsymbol{\delta}) \geqslant 0 \qquad \text{(C4)}$$

for some nonzero $\delta \in \Delta$. The zero set of $P(k)$ includes the portion of the hyperplane

$$\{\boldsymbol{k}\colon \boldsymbol{k} \cdot \delta = 0\} \qquad \text{(C5)}$$

that is in K. Many spectral supports, of practical interest, intersect this hyperplane at an infinite number of points, implying the existence of some correlation vector on the boundary of E with a nonunique spectral representation. This nonuniqueness problem is similar to the nonuniqueness problem in multi-dimensional Chebyshev approximation [24].

Acknowledgment

The work of B. Dickinson was instrumental in starting the authors on this course of research, and they are grateful to G. Cybenko for references to the semi-infinite programming literature.

References

[1] J. Stewart, "Positive definite functions and generalizations, an historical survey," *Rocky Mountain J. Math.*, vol. 6, pp. 409-434, Summer 1976.

[2] B. W. Dickinson, "Two-dimensional Markov spectrum estimates need not exist," *IEEE Trans. Inform. Theory*, vol. IT-26, pp. 120-121, Jan. 1980.

[3] G. Cybenko, "Moment problems and low rank Toeplitz approximations," presented at the Int. Symp. on Rational Approximation Syst., Katholieke Universiteit Leuven, Leuven, Holland, Aug. 31-Sept. 1, 1981 (to appear in *Circuits, Syst., Signal Processing*).

[4] —, "Affine minimax problems and semi-infinite programming," *Math. Programming*, to be published.

[5] V. F. Pisarenko, "The retrieval of harmonics from a covariance function," *Geophys. J. R. Astr. Soc.*, vol. 33, pp. 347-366, 1973.

[6] A. B. Baggeroer, "Space/time random processes and optimum array processing," Naval Undersea Center, San Diego, CA., Rep. NUC TP 506, Apr. 1976.

[7] H. L. Royden, *Real Analysis*. New York: Macmillan, 1968.

[8] R. A. Haubrich, "Array design," *Bull. Seismological Soc. Amer.*, vol. 58, pp. 977-991, June 1968.

[9] U. Grenander and G. Szegö, *Toeplitz Forms and Their Applications*. Berkeley and Los Angeles: Univ. of California Press, 1958.

[10] D. G. Luenberger, *Optimization by Vector Space Methods*. New York: Wiley, 1969.

[11] M. Avriel, *Nonlinear Programming*. Englewood Cliffs, NJ: Prentice-Hall, 1976.

[12] A. Calderón and R. Pepinsky, "On the phases of Fourier coefficients for positive real functions," *Computing Methods and the Phase Problem in X-Ray Crystal Analysis*, The X-Ray Crystal Analysis Laboratory, Dep. Physics, Pennsylvania State College, pp. 339-348, 1952.

[13] W. Rudin, "The extension problem for positive-definite functions," *Ill. J. Math.*, vol. 7, pp. 532-539, 1963.

[14] S. W. Lang and J. H. McClellan, "Multidimensional MEM spectral estimation," *IEEE Trans. Acoust., Speech, Signal Processing*, vol. ASSP-30, pp. 880-887, Dec. 1982.

[15] K. Hoffman, *Analysis in Euclidean Space*. Englewood Cliffs, NJ: Prentice-Hall, 1975.

[16] E. W. Cheney, *Introduction to Approximation Theory*. New York: McGraw-Hill, 1966.

[17] T. M. Sullivan, O. L. Frost, and J. R. Treichler, "High resolution signal estimation," Argosystems, Inc., Sunnyvale, CA, June 1978.

[18] D. G. Luenberger, *Introduction to Linear and Nonlinear Programming*. Reading, MA: Addison-Wesley, 1973.

[19] E. Stiefel, "Note on Jordan elimination, linear programming, and Tchebycheff approximation," *Numerische Mathematik*, vol. 2, pp. 1-17, 1960.

[20] C. H. Papadimitriou and K. Steiglitz, *Combinatorial Optimization*. Englewood, NJ: Prentice-Hall, 1982.

[21] D. B. Harris and R. M. Mersereau, "A comparison of algorithms for minimax design of two-dimensional linear phase FIR digital filters," *IEEE Trans. Acoust., Speech, Signal Processing*, vol. ASSP-25, pp. 492-500, Dec. 1977.

[22] R. Hettich, Ed., *Semi-Infinite Programming*. Berlin: Springer-Verlag, 1979.

[23] J. R. Rice, *The Approximation of Functions, Vol. 1–Linear Theory*. Reading, MA: Addison-Wesley, 1964.

[24] —, *The Approximation of Functions, vol. 2–Nonlinear and Multivariate Theory*. Reading, MA: Addison-Wesley, 1969.

[25] S. W. Lang and J. H. McClellan, "Spectral estimation for sensor arrays," in *Proc. 1st Acoust., Speech, Signal Processing Workshop Spectral Estimation*, Hamilton, Ont., Canada, Aug. 17-18, 1981, pp. 3.2.1-3.2.7.

[26] S. W. Lang, "Spectral estimation for sensor arrays," Ph.D. dissertation, Massachusetts Institute of Technology, Cambridge, MA, Aug. 1981.

Improving the Resolution of Bearing in Passive Sonar Arrays by Eigenvalue Analysis

DON H. JOHNSON, MEMBER, IEEE, AND STUART R. DEGRAAF

Abstract—A method of improving the bearing-resolving capabilities of a passive array is discussed. This method is an adaptive beamforming method, having many similarities to the minimum energy approach. The evaluation of energy in each steered beam is preceded by an eigenvalue–eigenvector analysis of the empirical correlation matrix. Modification of the computations according to the eigenvalue structure results in improved resolution of the bearing of acoustic sources. The increase in resolution is related to the time-bandwidth product of the computation of the correlation matrix. However, this increased resolution is obtained at the expense of array gain.

I. INTRODUCTION

An array of acoustic sensors is placed in a known spatial pattern to record the acoustic environment. Measurement of the acoustic field by an array offers two basic improvements over the signal processing capabilities of a single sensor. The first is the determination of the bearing of the acoustic source(s). Bearing cannot be obtained with a single sensor, whereas an array offers some bearing-resolving capability. This capability is usually measured by the just detectable separation of two equistrength sources for a given signal-to-noise ratio at one sensor in the array [6]. The determination of the bearing of a remote source of acoustic energy remains one of the fundamental problems of passive sonar systems. The waveforms recorded by each sensor can be acquired with as much fidelity as desired. In contrast, the size and geometry of the array are usually limited by physical considerations; these limitations restrict the spatial resolution of source bearing. The second improvement is the increase of signal-to-noise ratio. If the noise field is uncorrelated at each sensor location with respect to all other locations, the signal-to-noise ratio in the array output is increased by a factor equal to the number of sensors comprising the array. This factor decreases when the noise field is correlated at the sensor locations. The measure of improvement in the signal-to-noise ratio is array gain: the ratio of the signal-to-noise ratio at the array output to that obtained with a single sensor.

Adaptive beamforming methods (ABF) are known to have superior bearing-resolution capabilities when compared to conventional beamformers (in a theoretical sense) [6]. Specifically, the minimum energy method has been analyzed extensively in this regard. However, there is no theoretical basis indicating that this method has the best possible bearing-resolution properties. On the other hand, the array gain provided by this method is optimum: no other beamforming technique can yield a larger increase in signal-to-noise ratio. This paper is concerned with a new ABF scheme which is similar in many respects to the minimum energy method. It can demonstrate greatly increased bearing-resolution properties, but at the expense of array gain. The scheme is based on an eigenvector-eigenvalue decomposition of the empirical correlation matrix, which is then truncated so as to retain only those terms which best contribute to increased bearing resolution. This approach is similar to those described by Schmidt [13] in his MUSIC system, by Owsley [12] in his modal decomposition approach, and by Bienvenu [2], [3]. Analytic results are presented here which contrast the bearing-resolving properties of these various eigenvector methods and the minimum energy method.

II. PRELIMINARIES

Let $x_m(t)$ denote the outputs taken from an array of sensors having a known geometry. *Beamforming* consists of computing the quantity

$$y(t) = \sum_{m=0}^{M-1} a_m x_m(t - \tau_m) \tag{1}$$

where a_m is the amplitude weighting (shading) applied to the mth sensor output, τ_m is the delay applied to the mth sensor output, and M is the number of sensors in the array. The parameters $\{a_m\}$ and $\{\tau_m\}$ of a beam are chosen according to some desired criterion (e.g., steering the beam in a particular direction, minimizing sidelobe height, etc.).

Beamforming can also be viewed as a type of multidimensional spectral analysis [1], [9]. Evaluating the Fourier transform of the beam $y(t)$, we have

$$Y(f) = \sum_{m=0}^{M-1} a_m e^{-j2\pi f\tau_m} X_m(f). \tag{2}$$

Therefore, at each temporal frequency f, the Fourier transform of the beam can be written as the dot product of two vectors

$$Y(f) = \boldsymbol{A}'\boldsymbol{X} \tag{3}$$

where $\boldsymbol{A}$ denotes the steering vector consisting of the elements

$$A_m = a_m e^{+j2\pi f\tau_m}$$

Manuscript received August 14, 1981; revised November 17, 1981. This work was supported in part by an American Society for Engineering Education Fellowship and by the Office of Naval Research under Contract N00014-81-K-0565. Portions of this paper were presented at the Acoustics, Speech, and Signal Processing Spectral Estimation Workshop, McMaster University, Hamilton, Ont., Canada, August 1981.

The authors are with the Department of Electrical Engineering, Rice University, Houston, TX 77001.

Reprinted from *IEEE Trans. Acoust., Speech, Signal Processing*, vol. ASSP-30, pp. 638–647, Aug. 1982.

and X denotes the vector comprised of the Fourier transforms of the sensor outputs. Here, A' denotes the conjugate transpose of A. Assume that we have a linear array of equally spaced sensors; in this instance, the delays τ_m will be of the form $\tau_m = mT$. Equation (2) becomes

$$Y(f,T) = \sum_{m=0}^{M-1} a_m e^{-j2\pi mfT} X_m(f). \quad (4)$$

Consequently, $Y(f, T)$ is the Fourier transform of the sequence $a_m X_m(f)$. For a linear array, computing a transform along the index m is identical to computing the transform in space across the array. $Y(f, T)$ is, therefore, the spatial transform of $a_m X_m(f)$ evaluated at the spatial frequency $k = fT$. The result of applying a particular shading a_m is to convolve the true (infinite aperture) spatial spectrum with the Fourier transform of a_m, $m = 0, \cdots, M-1$, thereby smearing the true spectrum and limiting the resolution that can be obtained.

III. High-Resolution Techniques

One can obtain a set of weights (or, equivalently, a steering vector) to achieve better resolution by adapting them to the particular noise field and signal field impinging on the array. In these adaptive beamforming schemes, the steering vector is the solution to an optimization problem [4], [11]: find the steering vector which minimizes the energy in the beam subject to the constraint $A'Z = 1$ where Z is the constraint vector. The energy contained in a beam can be expressed by the quadratic form $A'\mathrm{R}A$ where R denotes the empirical spatial correlation matrix of the Fourier transforms of the sensor outputs. The correlation matrix R is usually estimated from the sensor outputs by a variation of the Bartlett procedure. The output of each sensor is sectioned and windowed. The Fourier transform of each section is evaluated and the vector $X_i(f_0)$ of Fourier transform values at the frequency f_0 across the array for the ith section is formed. Assuming that K sections are available, R is computed according to

$$\mathrm{R} = \frac{1}{K}\sum_{i=0}^{K-1} X_i(f_0)X_i'(f_0). \quad (5)$$

Usually K is taken to be the number of statistically independent terms used in the empirical computation of the correlation matrix R; K is frequently referred to as the time-bandwidth product.

The solution to this optimization problem is

$$A = \frac{\mathrm{R}^{-1}Z}{Z'\mathrm{R}^{-1}Z}. \quad (6)$$

The resulting value of the energy in the beam is

$$A'\mathrm{R}A = (Z'\mathrm{R}^{-1}Z)^{-1}. \quad (7)$$

In the so-called high-resolution [5] or *minimum energy* scheme, the constraint vector Z is a plane-wave direction-of look vector W, each element of which is given by $W_m = e^{j2\pi mk}$ where k corresponds to a specific spatial frequency. As $k = fT$ and f is known, each value of k corresponds to a specific per-channel delay. Defining θ to the bearing of the source relative to array-broadside, $T = (d/f\lambda)\sin\theta$. Consequently, spatial frequency k is related to bearing θ as $k = (d/\lambda)\sin\theta$. The constraint $A'W = 1$ fixes the gain of the steering vector in the direction-of-look W to be unity. Forcing the energy to be minimum thereby reduces the contributions from plane waves arriving from other directions and from the noise field. The energy in the beam corresponding to the spatial frequency k is expressed by

$$S_{ME}(k) = (W'\mathrm{R}^{-1}W)^{-1}. \quad (8)$$

The bearing of the target(s) is determined by finding the spatial frequency(s) at which the quantity in (8) achieves maxima. The maximum likelihood spectral estimate is closely related to the minimum energy estimate, but differs from it in an important way. The maximum likelihood estimate uses the noise-only correlation matrix in its evaluation of the beam energy:

$$S_{ML}(k) = (W'\mathrm{R}_n^{-1}W)^{-1}. \quad (9)$$

Here $\mathrm{R}_n = E[NN']$, the theoretical correlation matrix of the noise component of X. Note specifically that the matrix R_n is assumed to be a known quantity and is not computed from empirical data. The maximum likelihood method yields the optimum array gain [6], [7]. The maximum likelihood and minimum energy methods yield the same array gain when the beam is steered toward the acoustic source [6].

IV. Improving Resolution

From the viewpoint presented here, there is great flexibility in choosing the constraint vector Z. One wonders if there is a particular choice for the constraint vector which can maximize the spatial resolution of source bearing. For example, choose a constraint vector of the form

$$Z = \mathrm{C}W \quad (10)$$

where C is a matrix to be described. The energy in the beam when steered toward the source would then be

$$S(k) = (W'\mathrm{C}'\mathrm{R}^{-1}\mathrm{C}W)^{-1}. \quad (11)$$

Suppose the vector W corresponded to an actual plane-wave source. If C were a matrix having the property that this choice of W lay in the null space of the matrix ($\mathrm{C}W = \mathbf{0}$), the energy in the beam when steered in this direction would be infinite. If the direction vector corresponding to the plane-wave source were the *only* direction vector that lay in the null space of C, one would therefore obtain a marked indication of the bearing of the source.

While it is theoretically possible to have a perfect indication of source bearing by this approach, the difficulty lies in finding the matrix C. This matrix must have the property that direction vectors lying in its null space correspond only to plane waves eminating from actual sources. Assuming that the bearing of the sources is not known, construction of the matrix C would seem impossible. However, one can construct a matrix having a null space consisting of vectors which closely resemble direction vectors of source plane waves. The key idea of this procedure is to carefully analyze the eigenvectors and eigenvalues of the correlation matrix.

A. The Eigenvector Method

The eigenvectors of the matrix R are defined by the property

$$\mathrm{R}V_i = \lambda_i V_i \quad i = 1, \cdots, M \tag{12}$$

where λ_i is the eigenvalue associated with the eigenvector V_i. As correlation matrices are conjugate-symmetric (Hermitian), the eigenvectors form an orthonormal set. Assume that the sensor outputs contain one signal and noise uncorrelated with the signal ($X = S + N$). The result of computing R according to (5) for sufficiently large K is

$$\mathrm{R} = \sigma_n^2 \mathrm{Q} + \sigma_s^2 SS'. \tag{13}$$

Q is the noise correlation matrix normalized to have a trace equal to the dimension M of the matrix R. S is the direction vector of a plane wave source and has a squared norm equal to M. The cross terms involving signal and noise are assumed to be negligible. If the noise correlation matrix equals the identity matrix (i.e., only sensor noise is assumed to be present), the eigenvector corresponding to the largest eigenvalue (hereby referred to as the "largest eigenvector") is equal to the vector S with eigenvalue equal to $\sigma_n^2 + M\sigma_s^2$. The remaining $M - 1$ eigenvectors of R consist of those vectors orthogonal to S and each has eigenvalue σ_n^2. If p *incoherent* linearly independent signals S_i, $i = 1, \cdots, p$, are present so that the correlation matrix is of the form

$$\mathrm{R} = \sigma_n^2 \mathrm{I} + \sum_{i=1}^{p} \sigma_{s,i}^2 S_i S_i', \tag{14}$$

the p largest eigenvectors correspond to the signal terms and the $M - p$ smallest are orthogonal to all of the signal direction vectors. Note that the largest eigenvectors are not necessarily equal to the signal direction vectors in this case; these eigenvectors comprise an orthonormal basis for the vector space containing the signal vectors. Consequently, one cannot always inspect the largest eigenvectors and determine the signal vectors directly.

Define C_{EV} to be the sum of the outer products of the $M - p$ smallest eigenvectors:[1]

$$\mathrm{C}_{EV} = \sum_{i=1}^{M-p} V_i V_i'. \tag{15}$$

As the p largest eigenvectors are orthogonal to each of the $M - p$ smallest, $\mathrm{C}_{EV} V_i = \mathbf{0}$, $i = M - p + 1, \cdots, M$. As the p largest eigenvectors span the space containing the signal vectors, each signal lies in the null space of C_{EV} and $\mathrm{C}_{EV} S_i = \mathbf{0}$. In this manner, perfect resolution of the bearing of multiple sources can be obtained from an eigenvector analysis of the correlation matrix R.

Note that in computing the beam energy (11), the matrix C_{EV} need never be computed. The correlation matrix R can be expressed in terms of its eigenvectors as

$$\mathrm{R} = \sum_{i=1}^{M} \lambda_i V_i V_i' \tag{16}$$

[1]By convention, the smallest eigenvector is denoted by the subscript 1, the next smallest by 2, etc.

and the inverse of R as

$$\mathrm{R}^{-1} = \sum_{i=1}^{M} \frac{1}{\lambda_i} V_i V_i'. \tag{17}$$

Because of the orthogonality property of the eigenvectors of a correlation matrix, the quantity $\mathrm{C}_{EV}' \mathrm{R}^{-1} \mathrm{C}_{EV}$ appearing in (11) becomes

$$\mathrm{C}_{EV}' \mathrm{R}^{-1} \mathrm{C}_{EV} = \sum_{i=1}^{M-p} \frac{1}{\lambda_i} V_i V_i'. \tag{18}$$

It is this matrix which is computed in the evaluation of the quadratic form of (11). .

$$S_{EV}(k) = (W' \mathrm{C}_{EV}' \mathrm{R}^{-1} \mathrm{C}_{EV} W)^{-1} = \left[\sum_{i=1}^{M-p} \frac{1}{\lambda_i} |W' V_i|^2 \right]^{-1}. \tag{19}$$

The following sequence of computations constitute the *eigenvector method.*

1) Compute the correlation matrix R.

2) Decompose the matrix R into its eigenvectors and eigenvalues.

3) Determine the number p of sources present in the acoustic field.

4) Compute the energy in the beams corresponding to all possible bearings (19).

5) The major peaks in this spectrum correspond to acoustic sources.

The methods of Owsley [12], Bienvenu [2], [3], and Schmidt [13] are similar, but differ somewhat from the eigenvector method just described. The expression for R^{-1} is truncated as in (5); however, the small eigenvalues are set to the same value (taken here to be unity). Instead of evaluating the quadratic form of (11), the spectral estimate of the MUSIC method can be expressed by

$$S_{\mathrm{MUSIC}}(k) = (W' \mathrm{C}_{EV}' \mathrm{C}_{EV} W)^{-1} = \left[\sum_{i=1}^{M-p} |W' V_i|^2 \right]^{-1}. \tag{20}$$

This estimate can also be evaluated in the same manner as (11) if the matrix C is redefined to be

$$\mathrm{C}_{\mathrm{MUSIC}} = \sum_{i=1}^{M-p} \sqrt{\lambda_i}\, V_i V_i'.$$

A steering vector can be defined (6) when the latter approach is used to express the MUSIC spectral estimate. Under the conditions just described, this spectral estimate also has the capability of resolving source bearing perfectly.

The presumption of the preceding analysis has been that the noise correlation matrix Q is equal to the identity matrix. It is this key assumption which leads to perfect bearing resolution. Imperfection in bearing resolution occurs when this presumption is false. The noise field may contain more than just sensor noise: for example, isotropic noise may also be present. In addition, a finite amount of averaging is used to compute the correlation matrix R. Even if the noise field were spatially white, an empirical noise correlation matrix would not be an identity matrix. Either of these situations can reduce the resolution of the eigenvector and MUSIC methods.

The performances of the eigenvector method and of the MUSIC method under these more realistic conditions are analyzed mathematically in the Appendix. From this analysis, an approximate lower limit on the energy in the beam when steered on-target using the eigenvector method is found to be

$$S(k) = \frac{\sigma_s^2}{\frac{\sigma_n^2}{\sigma_s^2}\frac{\gamma^2}{M} + \frac{M}{K}}. \tag{21}$$

A similar result is obtained for the MUSIC method. The quantity γ is a measure of the spread of the eigenvalues α_i of Q. For the eigenvector method, this quantity is given by

$$\gamma_{EV}^2 = \frac{1}{M}\sum_{i=1}^{M}\left(\alpha_i - \frac{1}{M}\sum_m \alpha_m^2\right)^2 \tag{22}$$

and for the MUSIC method by

$$\gamma_{\text{MUSIC}}^2 = \frac{1}{M}\sum_{i=1}^{M}\alpha_i\left(\alpha_i - \frac{1}{M}\sum_m \alpha_m^2\right)^2. \tag{23}$$

γ_{MUSIC}^2 tends to be smaller than γ_{EV}^2. In either case, the quantity γ is an implicit function of the time-bandwidth product K. Generally speaking, γ will decrease with increasing K, tending toward the spread of the eigenvalues of the spatial correlation matrix of the underlying noise process. If the theoretical noise correlation matrix equals the identity matrix (i.e., spatially white noise), the spread of its eigenvalues is zero. Otherwise, the spread is nonzero.

The first term in the denominator of (21) determines how large the energy peak will be if an infinite time-bandwidth product were available. The second term describes how the size of the energy peak depends on K. The larger of these will dominate the expression in (21). When steered off-target, the energy in the beam produced by the eigenvector method coincides with that produced with the minimum energy approach. The ratio ν of the on-target to off-target beam energy can be used to assess the size of the peak in the beam energy as the direction vector $\boldsymbol{W}$ is scanned through all possible bearings. When only sensor noise is present, this quantity is given in the minimum energy method by [10]

$$\nu_{ME} = 1 + \frac{M\sigma_s^2}{\sigma_n^2}. \tag{24}$$

In the eigenvector method and the MUSIC method, this quantity is given by

$$\nu = \frac{1}{\frac{\sigma_n^2}{\sigma_s^2}\frac{\gamma^2}{M^2} + \frac{1}{K}} \cdot \frac{\sigma_s^2}{\sigma_n^2}. \tag{25}$$

From simulation studies, this quantity can be somewhat larger (a few decibels) in the MUSIC method than in the eigenvector method. Comparing (24) and (25), one concludes that when the eigenvector method is used, the array appears to consist of a number of elements $\tilde{M}$ given by

$$\tilde{M} = \frac{1}{\frac{\sigma_n^2}{\sigma_s^2}\frac{\gamma^2}{M^2} + \frac{1}{K}}. \tag{26}$$

In most instances, this quantity is larger than the actual number of sensors M.

B. Simulation Results

Computer simulations were used to evaluate the eigenvector and MUSIC methods and to compare them with the minimum energy method. In these studies, sequences of the data vector $\boldsymbol{X}_i = \boldsymbol{S} + \boldsymbol{N}_i$ were produced. The parameters of the signal vector $\boldsymbol{S}$ were defined as described in Section II. The noise vector $\boldsymbol{N}_i$ consisted of identically distributed complex Gaussian noise components. The covariance matrix of this random vector could be specified by the user. Each noise vector was generated to be statistically independent of all other noise vectors. The empirical correlation matrix R was then computed as in (5), and its eigenvectors and eigenvalues were computed according to a *QL* algorithm [8].

With one exception, each step of the procedure outlined above for the eigenvector method was followed. As the number p of sources present in the acoustic field was known by the user, it was supplied by him. In a physical situation, this parameter would not be so readily known. However, the purpose of the simulations was to determine the validity of the theory and to test how well the methods could perform. The effects of inaccurate choices for p are described in a later section.

A comparison of the beam energies produced by the minimum energy method and by the eigenvector method is shown in Fig. 1(a) when one source is present in the acoustic field. A similar comparison is found in Fig. 1(b) for the MUSIC method. Note that the height of the main peak relative to the background noise level varies with K, the time-bandwidth product, in both methods, whereas it does not in the minimum energy method. In this example, sensor noise (i.e., spatially white) is present. For the array length used, the second term in the denominator of (21) was larger than the first. Simulated and theoretical values of ν_{EV} are compared in Table I. The theoretical prediction of the value of this quantity is close to that obtained from the simulations.

The spectra obtained from the eigenvector method and the MUSIC method differ only slightly in these examples. The latter method tends to produce much flatter off-target spectra. The small eigenvalues of the correlation matrix R tend to represent the noise field [see the discussion following (13)], and the equalization of these values will "whiten" the background noise, thereby resulting in a flat spectrum. This effect is further illustrated by considering a case where isotropic noise is present in the acoustic field (Fig. 2).

C. Resolution of Multiple Targets

Cox [6] derives analytic expressions for the limits to which the minimum energy method applied to a linear array can resolve two equistrength incoherent acoustic sources. There, "resolution" is defined as the minimum bearing separation at broadside at which two targets can be distinguished by evaluating beam energy. Here, a slight dip in the beam energy is re-

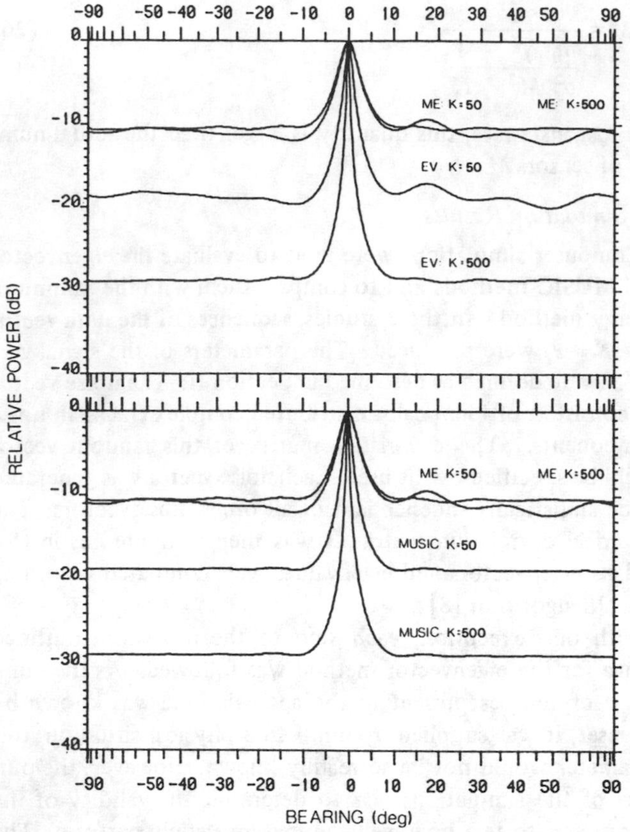

Fig. 1. Energy in beams formed by the minimum energy method (*ME*) and by the eigenvector method (*EV*) are plotted against bearing for a linear array. Energy is expressed in decibels relative to the peak value. Bearing is expressed in degrees with zero corresponding to broadside. The ten sensors are equally spaced and separated by half a wavelength. The source was assumed to be narrow-band, with all of its source energy concentrated in one temporal-frequency analysis bin. In each subfigure, sensor noise and one source located at 0° are present in the sound field; the sensor signal-to-noise ratio is 0 dB. The results obtained when two time-bandwidth products (*K*) are used are shown in each subfigure for each method.

TABLE I
EMPIRICAL AND THEORETICAL PEAK-TO-BACKGROUND RATIOS

K	$\widetilde{M}$	$\hat{\nu}_{ME}$(dB)	$\hat{\nu}_{EV}$(dB)	ν_{EV}(dB)
20	18	10	13	13
50	46	11	18	17
100	90	12	22	20
200	183	11	27	23
500	461	11	29	27
1000	940	11	32	30
2000	1784	11	34	33
5000	4597	11	39	37

The results of computer simulations and theoretical predictions of the value of ν are shown. In the simulations, an equally spaced linear array containing ten sensors was assumed to be present in an acoustic field containing sensor noise and a plane-wave source. The sensor signal-to-noise ratio was 0 dB. Values of γ were computed in separate simulations from noise-only correlation matrices having the same time-bandwidth product. The source was assumed to be narrow-band, with all of its source energy concentrated in one temporal-frequency analysis bin. The sensor spacing is one-half wavelength. Plot of beam energy versus bearing were obtained and empirical values of ν estimated. The quantities $\hat{\nu}_{ME}$ and $\hat{\nu}_{EV}$ correspond to these empirical values.

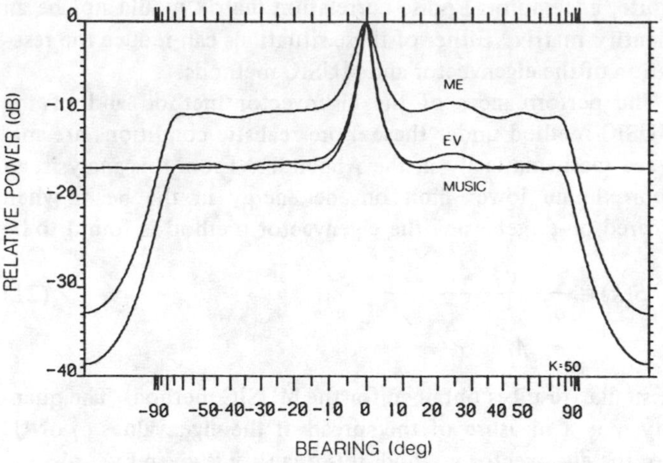

Fig. 2. Beam energy is plotted against bearing when one source (located at 0°) and isotropic noise are present in the sound field. The array configuration is similar to that described in Fig. 1, the only exception being that the sensor spacing is three-eighths of a wavelength. The sensor signal-to-noise ratio is 0 dB. The results of applying the minimum energy (*ME*), eigenvector (*EV*), and MUSIC methods are shown. The time-bandwidth product here is 50; the theoretical values of ν corresponding to this situation are ν_{EV} = 15.7 dB and ν_{MUSIC} = 16.4 dB.

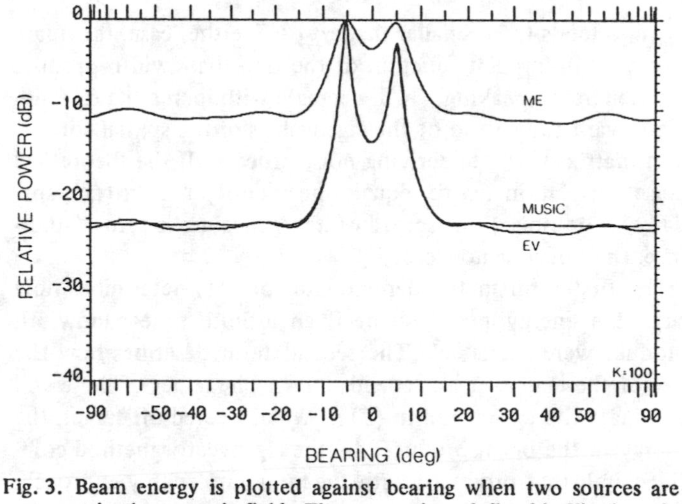

Fig. 3. Beam energy is plotted against bearing when two sources are present in the acoustic field. The conventions defined in Fig. 1 apply to this plot. Here, the sensor signal-to-noise ratio of each source is 0 dB and the sources are located at -5° and +5°.

quired when the array is steered between the sources. The critical factors determining the resolution of beams formed by the minimum energy approach are aperture [defined as the spatial extent of the array relative to a wavelength $-(D/\lambda)$], the number of elements in the array (M), and the sensor signal-to-noise ratio (σ_s^2/σ_n^2). Cox's results are summarized in [6, Fig. 5]. An approximation to those results is

$$\frac{M\sigma_s^2}{\sigma_n^2} \propto \left(\theta\,\frac{D}{\lambda}\right)^4 \tag{27}$$

where θ is bearing separation of the two targets.

When the eigenvector method is applied in situations such as these, targets are more easily resolved, and furthermore, the resolution capabilities of the array are increased. Figs. 3 and 4

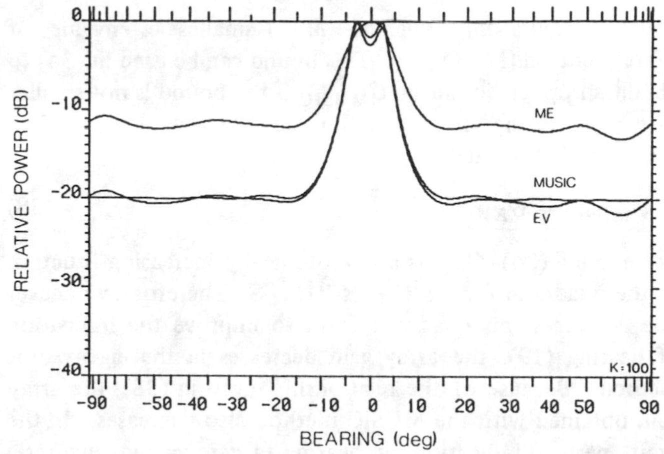

Fig. 4. Beam energy is plotted against bearing when two sources are present in the acoustic field. The conventions defined in Fig. 1 apply to this plot. Here, the sensor signal-to-noise ratio of each source is 0 dB and the sources are located at -3° and +3°.

TABLE II
EMPIRICAL AND THEORETICAL RESOLUTION LIMITS

K	θ_{ME}	$\hat{\theta}_{EV}$	θ_{EV}
20	8	7.5	7.5
50	8	6	5.8
100	8	6	5.4
200	8	4.5	4.6
500	8	3.5	3.7
1000	8	3.5	3.0
2000	8	3	2.6
5000	8	2.5	2.1

The results of computer simulations and theoretical predictions of resolution limits are summarized. The array configuration used in Table I was used here. The sources were symmetrically located about broadside (0 degrees). Plots of beam energy versus bearing were obtained, and the separation between the sources was reduced until they could just be resolved. The measurements of separation were made in half-degree increments. The angular quantities θ are indicated in degrees.

display typical examples of these cases. A theoretical prediction of the degree to which resolution is increased can be obtained from (27). If one substitutes $\tilde{M}$ evaluated by (26) for M, the value of θ thus obtained is the resolution limit of the eigenvector method and the MUSIC method. A comparison of the resolution obtained from some of the simulations with that predicted by the theory is shown in Table II. The degree of agreement between theory and simulation results implied by Table II is valid for all of the simulations.

D. Effect of an Improper Choice for p

The theoretical and simulation results presented thus far are valid only when the parameter p equals the actual number of acoustic sources. In practice, this quantity may not be known, and one questions the sensitivity of the eigenvector and MUSIC methods to an incorrect choice of p. This sensitivity was studied through the simulations; no analytic results were obtained on this issue.

An incorrect choice for p has different effects on the eigenvector and MUSIC methods. In both methods, choosing p too small does *not* result in a beamformer having superior bearing-resolution properties to the minimum energy method. In the eigenvector method, the spectra tend to resemble those obtained with the minimum energy method. In particular, by assuming that no sources are present (setting $p = 0$ or the matrix $C_{EV} = I$), the eigenvector method is exactly the minimum energy method. In contrast, the MUSIC method tends to produce a number of spectral peaks equal to p. For example, if a value of zero is chosen for p, a uniformly flat spectrum results. The peaks that result from nonzero choices tend to correspond to the bearings of acoustic sources; however, which sources are thus located is not easily predicted. When p is chosen too large, the bearing-resolving capabilities of either method are not greatly reduced. The spectra produced by the eigenvector method tend not to vary from that obtained with the proper value of p. The MUSIC method tends to produce spurious peaks that do not correspond to physical sources. The effects are illustrated in Fig. 5.

V. RESOLUTION AND ARRAY GAIN

While these approaches increase the resolution of bearing, this increase in resolution is accompanied by a decrease in the array gain. To show this, assume that the correlation matrix R is of the form given in (13). The signal-to-noise ratio at each sensor is therefore σ_s^2/σ_n^2. The signal-to-noise ratio in the beam output is the quantity

$$\frac{\sigma_s^2 A'SS'A}{\sigma_n^2 A'QA}. \tag{28}$$

The array gain G is the ratio of these signal-to-noise ratios:

$$G = \frac{A'SS'A}{A'QA}. \tag{29}$$

Recalling that the steering vector in this case is given by

$$A = \frac{R^{-1}CW}{W'C'R^{-1}CW} \tag{30}$$

and setting the direction vector to correspond to the source ($W = S$), the array gain becomes

$$G = \frac{S'C'R^{-1}SS'R^{-1}CS}{S'C'R^{-1}QR^{-1}CS}. \tag{31}$$

As the matrix Q is given by

$$Q = \frac{1}{\sigma_n^2}(R - \sigma_s^2 SS'), \tag{32}$$

the denominator of (31) becomes

$$S'C'R^{-1}QR^{-1}CS = \frac{1}{\sigma_n^2}(S'C'R^{-1}CS - \sigma_s^2 S'C'R^{-1}S \cdot S'R^{-1}CS) \tag{33}$$

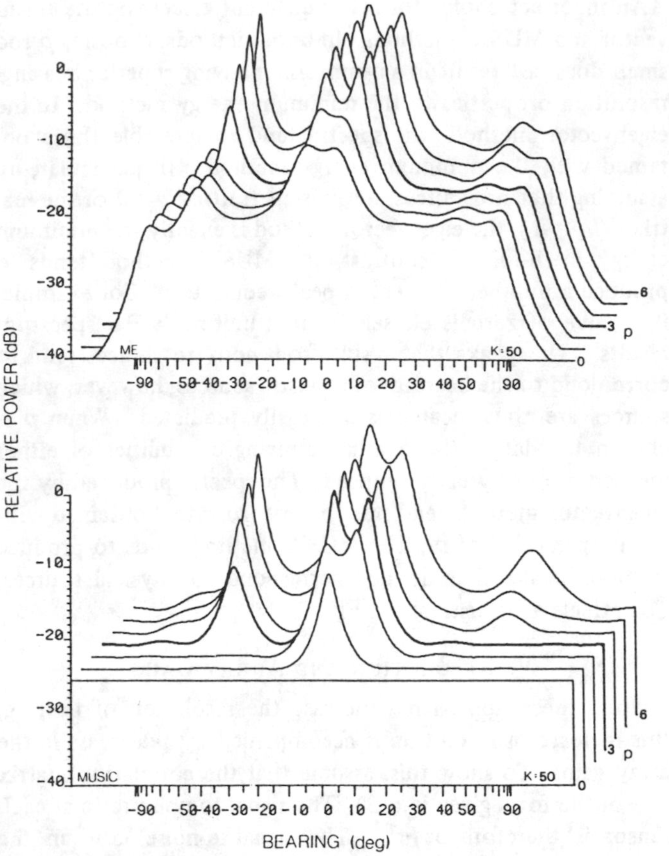

Fig. 5. Beam energy is plotted against bearing with the number p of terms truncated from the eigenvector expansion of R^{-1} as a parameter. The parametric beam energy functions in each panel is plotted with the same vertical scaling. A linear array of ten equally spaced sensors (spacing equal to three-eighths of a wavelength) is present in an acoustic field. Three incoherent sources are present in the acoustic field: two have unity amplitude and are located at bearings +5° and -5°, while the third has an amplitude of one-half and a bearing of -40°. Isotropic noise is also present in the acoustic field; the sensor signal-to-noise ratio (relative to the larger signals) is 0 dB. The time-bandwidth product in both panels is 50. The upper panel displays the result of applying the eigenvector method, and the bottom panel illustrates the result of applying the MUSIC method for the same set of data. Note that the proper value of p for these data is $p = 3$.

so that we obtain

$$G = \frac{\sigma_n^2}{\dfrac{S'C'R^{-1}CS}{S'C'R^{-1}S \cdot S'R^{-1}CS} - \sigma_s^2}. \tag{34}$$

For the eigenvector method,

$$C'_{EV}R^{-1} = R^{-1}C_{EV} = C'_{EV}R^{-1}C_{EV} \tag{35}$$

and the array gain becomes

$$G_{EV} = \frac{\sigma_n^2 S'C'_{EV}R^{-1}C_{EV}S}{1 - \sigma_s^2 S'C'_{EV}R^{-1}C_{EV}S}. \tag{36}$$

In the MUSIC method, the ratio appearing in the denominator of (34) can be bounded using the Schwarz inequality:

$$\frac{S'C'_{MUSIC}R^{-1}C_{MUSIC}S}{S'C'_{MUSIC}R^{-1}S \cdot S'R^{-1}C_{MUSIC}S} \geqslant \frac{1}{S'C'_{EV}R^{-1}C_{EV}S}. \tag{37}$$

Equality occurs only when the $M - 1$ smallest eigenvalues of R are identical (i.e., Q = I). This bound can be used in (34) to obtain an upper bound on G_{MUSIC} if the bound is not smaller than σ_s^2. As the expression thus obtained equals G_{EV} (36), this condition is satisfied. Consequently,

$$G_{MUSIC} \leqslant G_{EV}. \tag{38}$$

Considering (36), G_{EV} is a monotonically increasing function of the quadratic form $S'C'_{EV}R^{-1}C_{EV}S$. Therefore, whenever one decreases this quadratic form to improve the indication of bearing (19), the array gain decreases in the eigenvector method. Because of the relationship given in (38), the array gain obtained with the MUSIC method also decreases. In the limit, perfect indication of bearing (a zero-valued quadratic form) corresponds to zero array gain with either method.

VI. Conclusions

The eigenvector method can enhance the bearing-resolving capabilities of an array. Here, the eigenvectors and eigenvalues of the correlation matrix must be found and the weighted sum of Fourier transforms of the eigenvectors computed. In the minimum energy method, the inverse of the correlation matrix must be found and the quadratic form of (8) computed. Roughly speaking, the computational complexities involved in the use of the eigenvector method are not excessive when compared to those required in the minimum energy method.

The degree to which the resolving power is increased is related to the quantity $\tilde{M}$. Because of (27), this increase is proportional to $(\tilde{M}/M)^{1/4}$. Consequently, to increase the bearing resolution by a factor of two requires the virtual number of sensors $\tilde{M}$ to be 16 times the actual number. Referring to Table I, such large virtual array lengths can be obtained only when large time-bandwidth products are possible. Under these circumstances, enhanced bearing resolution is possible. For a given time-bandwidth product, the smaller the number of elements in the array, the greater the increase in bearing resolution.

The eigenvector method and the MUSIC method produce quite similar results. They differ in at least two respects, however. The first is that a nonzero value of p, the number of assumed sources in the acoustic field, must be chosen in the MUSIC method. If the value of p is not close to the actual value, the spectra thus obtained can differ from that obtained with the proper value: spurious peaks appear and/or peaks can be missed. In contrast, the eigenvector method is less sensitive to the choice of p. Second, the shape of the spectrum of the background "noise" is drastically altered in the MUSIC method. For example, the variations due to low-level sources or to the physical noise spectrum are lost (see Fig. 2, for example). This portion of the spectrum can also vary as p is changed; this effect is much less pronounced in the eigenvector method.

The increase in bearing resolution is obtained at the expense of array gain. Consequently, if more than bearing information is required, other techniques should probably be used to obtain them. One can conceive of the eigenvector method being used to acquire source bearing, and this information being used to steer a beam with the minimum energy method so as to analyze the waveform produced by the source. Note that this two-step procedure need only be sequential in a conceptual manner.

Because of the close relationship between the two methods, obtaining the steering vector for the minimum energy beamformer means including more terms in the eigenvector decomposition of R.

The decrease of array gain with increasing resolution raises many theoretical issues. The minimum energy method is known to yield the optimal value of array gain. Consequently, any method which has greater resolution capabilities cannot also increase array gain. Can array gain be maintained while increasing resolution or is increased resoltuion always obtained at the expense of array gain? The present method has the latter property. A theoretical understanding of the limits to which array gain and resolution can be traded against each other would be of interest.

The main issue not addressed in this study is the determination of the number of sources, p. From the analysis presented in Section IV, the number of sources corresponds to the number of dominant eigenvalues in the matrix R. Determining p in this way can be difficult when small time-bandwidth products are available and isotropic noise is present. Reasonably accurate methods of determining p from the eigenvalues of R are not known at this time.

Appendix

Let the vector X of sensor outputs be of the form $X = \sigma_s S + \sigma_n N$ where S denotes the source direction vector as before and N denotes additive noise. The correlation matrix R is computed empirically according to (5) to yield

$$\mathrm{R} = \sigma_n^2 Q + \sigma_n \sigma_s S\tilde{N}' + \sigma_n \sigma_s \tilde{N} S' + \sigma_s^2 SS' \qquad \text{(A1)}$$

where $\mathrm{Q} = (1/K)\, \Sigma_{i=1}^{K}\, N_i N_i'$, a statistical estimate of the noise correlation matrix, and $\tilde{N} = (1/K)\, \Sigma_{i=1}^{K}\, N_i$, an estimate of the average noise component. The vectors N_i are assumed to be statistically independent random vectors; each component of N_i has zero mean and unity variance. Consequently, the components of the vector $\tilde{N}$ have zero mean and energy $1/K$. Define the matrix P to be the noise-related terms in (A1):

$$\mathrm{P} = \mathrm{Q} + \frac{\sigma_s}{\sigma_n} S\tilde{N}' + \frac{\sigma_s}{\sigma_n} \tilde{N} S'. \qquad \text{(A2)}$$

Consequently, the expression for R in (A1) can be written more simply as

$$\mathrm{R} = \sigma_n^2 \mathrm{P} + \sigma_s^2 SS'. \qquad \text{(A3)}$$

The estimate of the energy in the beam pointed in the W direction is given by (11). Following Cox [6], this expression can be written as

$$S(k) = \frac{\sigma_n^2}{W'\mathrm{C}'\mathrm{P}^{-1}\mathrm{C}W} \left[\frac{1 + \left(\dfrac{S}{N}\right)_{\max}}{1 + \left(\dfrac{S}{N}\right)_{\max} \sin^2 (\mathrm{C}W, S; \mathrm{P}^{-1})} \right] \qquad \text{(A4)}$$

where $(S/N)_{\max} = S'\mathrm{P}^{-1} S(\sigma_s^2/\sigma_n^2)$ is the signal-to-noise ratio of the beam output obtained with the optimally chosen steering vector and $\sin^2 (\mathrm{C}W, S; \mathrm{P}^{-1})$ is the sine squared of the angle between the vectors $\mathrm{C}W$ and S with respect to the matrix P^{-1}. The matrix C_{EV} is given by (15) when p, the number of signals in the acoustic field, is assumed to be one. The critical aspect of (A4) is the quantity $\mathrm{C}_{EV}W$. This quantity can be viewed as the projection of the vector W onto the set of eigenvectors orthogonal to the largest eigenvector of R. As indicated earlier, these eigenvectors are approximately orthogonal to the signal vector. To a good approximation, the vector $\mathrm{C}_{EV}W$ is orthogonal to S, thereby implying for all W that $\sin^2 (\mathrm{C}_{EV}W, S; \mathrm{P}^{-1}) = 1$. Expression (A4) therefore becomes

$$S(k) = \frac{\sigma_n^2}{W'\mathrm{C}_{EV}'\mathrm{P}^{-1}\mathrm{C}_{EV}W}. \qquad \text{(A5)}$$

When W does not correspond to the signal direction vector S, the matrix C_{EV} has little effect on the vector W. In this case, one obtains

$$S(k) = \frac{\sigma_n^2}{W'\mathrm{P}^{-1}W},$$

the result obtained with the minimum energy estimate. Consequently, one should expect the eigenvector procedure and the minimum energy procedure to yield the same numerical values when steered off-target. As the beam is steered toward the source, the two estimates will begin to differ as the matrix C_{EV} begins to affect the vector W.

In the succeeding analysis, the inverse of the matrix P is assumed to be approximately equal to the inverse of Q in the computation of the quadratic form appearing in (A5). Inspecting (A2), this approximation will be less accurate as the signal-to-noise ratio (σ_s/σ_n) increases and as the amount of averaging (K) decreases. The energy estimate can be written approximately as

$$S(k) \approx \frac{\sigma_n^2}{W'\mathrm{C}_{EV}'\mathrm{Q}^{-1}\mathrm{C}_{EV}W}. \qquad \text{(A6)}$$

The quadratic form in (A6) can be interpreted as the squared length of the vector $\mathrm{C}_{EV}W$ with respect to the norm induced by the matrix Q^{-1}. We therefore seek an expression for this quantity when $W = S$.

Assume that V_M, the largest eigenvector of R, is given by $V_M = S + \epsilon$ where is a vector orthogonal to S. Consider the vector diagram shown in Fig. 6. The vector a is defined to be orthogonal to the eigenvector $S + \epsilon$. What is sought in (A6) is the square of the length L of the vector S projected onto a. As shown below, the length of ϵ is small compared to S; therefore, the quantity L will be approximately equal to the length of ϵ. To a good approximation, the energy in the beam steered on-target is given by

$$S(k) = \frac{\sigma_n^2}{\|\epsilon\|_{\mathrm{Q}^{-1}}^2} \qquad \text{(A7)}$$

where $\|\epsilon\|_{\mathrm{Q}^{-1}}^2$ denotes the squared norm of ϵ with respect to Q^{-1}:

$$\|\epsilon\|_{\mathrm{Q}^{-1}}^2 = \epsilon'\mathrm{Q}^{-1}\epsilon.$$

The definition of an eigenvector implies that this vector must satisfy

$$\mathrm{R}(S + \epsilon) = \lambda_M (S + \epsilon). \qquad \text{(A8)}$$

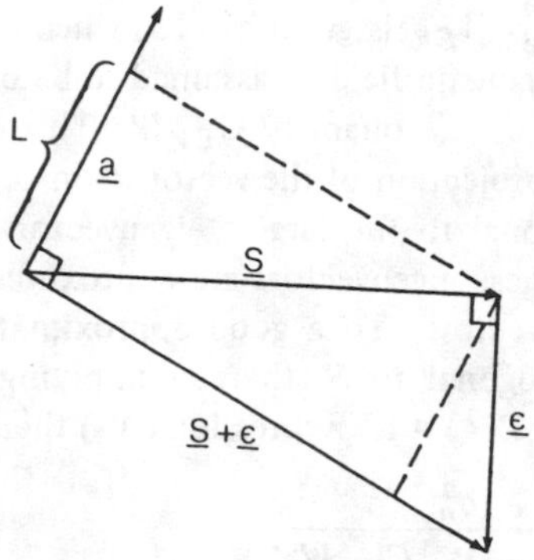

Fig. 6. Relationship of the signal vector and the largest eigenvector.

As the vectors S and ϵ are assumed to be orthogonal, the eigenvalue λ_M can be found through the relationship

$$S'\mathrm{R}(S+\epsilon) = \lambda_M S'(S+\epsilon) = M\lambda_M.$$

Using (A3) for R, we have

$$\lambda_M = \frac{\sigma_n^2 S'\mathrm{Q}S}{M} + \sigma_s\sigma_n(S'\tilde{N} + \tilde{N}'S + \tilde{N}'\epsilon) + M\sigma_s^2. \tag{A9}$$

An expression for the vector ϵ is obtained by evaluating the quantity $\mathrm{R}(S+\epsilon) - \lambda_M S$. After some manipulation and assuming that the length of ϵ is small compared to the length of S, we have

$$\epsilon = (\lambda_M \mathrm{I} - \sigma_n^2 \mathrm{Q})^{-1}\left(\sigma_n^2 \mathrm{Q}S - \sigma_n^2 \frac{S'\mathrm{Q}S}{M} S + M\sigma_s\sigma_n\tilde{N} - \sigma_s\sigma_n S'\tilde{N}S\right) \tag{A10}$$

This expression for the veetor ϵ consists of a matrix $(\lambda_M \mathrm{I} - \sigma_n^2 \mathrm{Q})^{-1}$ times the sum of two terms. The first term, denoted by B_1, is comprised only of the signal-related terms and the matrix Q:

$$B_1 = \sigma_n^2 \mathrm{Q}S - \sigma_n^2 \frac{S'\mathrm{Q}S}{M} S. \tag{A11}$$

The second term contains the terms depending on the average noise vector $\tilde{N}$:

$$B_2 = M\sigma_s\sigma_n\tilde{N} - \sigma_s\sigma_n(S'\tilde{N})S. \tag{A12}$$

If one assumes the noise vector $\tilde{N}$ to be zero (implying infinite statistical averaging), the vector ϵ is given by the quantity $\epsilon = (\lambda_M \mathrm{I} - \sigma_n^2\mathrm{Q})^{-1}B_1$. Note that if $\mathrm{Q} = \mathrm{I}$, the quantity $B_1 = 0$ which implies $\epsilon = 0$. Therefore, the signal vector S corresponds to the largest eigenvector of R; this result is consistent with the analysis described while leading to (5). The term B_2 expresses the effect on the eigenvector of the statistical averaging process. Note that if noise field can be described as containing only sensor noise ($\mathrm{Q} = \mathrm{I}$), the expression for ϵ is dominated by B_2. The vector B_2 can be interpreted as the noise vector which results when all of its components in the direction of the signal vector S are eliminated (A12). The squared magnitude of this vector therefore depends on the "angle" between the vectors $\tilde{N}$ and S. This angle will be a random quantity when the computation of R is completed. This vector will be largest when $\tilde{N}$ and S are assumed to be orthogonal. In this case, $S'\tilde{N} = 0$, and the expression for B_2 becomes

$$B_2 = M\sigma_s\sigma_n\tilde{N}. \tag{A13}$$

Define $\tilde{B}_1$ ($\tilde{B}_2$) to be the product of $(\lambda_M \mathrm{I} - \sigma_n^2\mathrm{Q})^{-1}$ and B_1 (B_2). The vector ϵ is therefore written as

$$\epsilon = \tilde{B}_1 + \tilde{B}_2. \tag{A14}$$

An expression for the norm of ϵ can now be obtained. Let U_i denote an eigenvector of the matrix Q and α_i the associated eigenvalue. As these eigenvectors are orthonormal, any vector can be expressed as a linear combination of them. As U_i is also an eigenvector of $(\lambda_M \mathrm{I} - \sigma_n^2\mathrm{Q})^{-1}$, $\tilde{B}_1$ and $\tilde{B}_2$ can be written as

$$\tilde{B}_1 = \sigma_n^2 \sum_i \frac{\left(\alpha_i - \frac{1}{M}\sum_m \alpha_m |S'U_m|^2\right) U_i'S}{\lambda_M - \alpha_i\sigma_n^2} U_i \tag{A15a}$$

$$\tilde{B}_2 = M\sigma_s\sigma_n \sum_i \frac{U_i'\tilde{N}}{\lambda_M - \alpha_i\sigma_n^2} U_i. \tag{A15b}$$

The quantity $\lambda_M - \alpha_i\sigma_n^2$ can be simplified. Using (A9) and assuming that both ϵ and $\tilde{N}$ are small compared to S, we have

$$\lambda_M - \alpha_i\sigma_n^2 = M\sigma_s^2 + \left(\frac{S'\mathrm{Q}S}{M} - \alpha_i\right)\sigma_n^2.$$

The quantities within the parentheses are comparable, and further assuming that $M\sigma_s^2 > \sigma_n^2$, we have

$$\lambda_M - \alpha_i\sigma_n^2 \approx M\sigma_s^2. \tag{A16}$$

The norm of ϵ depends upon the angle between these two vectors. Based on statistical arguments, a zero-mean vector obtained from averaging ($\tilde{B}_2$ in this case) will be nearly orthogonal to any fixed vector. Consequently,

$$\|\epsilon\|^2_{\mathrm{Q}^{-1}} = \|\tilde{B}_1\|^2_{\mathrm{Q}^{-1}} + \|\tilde{B}_2\|^2_{\mathrm{Q}^{-1}}. \tag{A17}$$

Now

$$\|\tilde{B}_1\|^2_{\mathrm{Q}^{-1}} = \sigma_n^4 \sum_i \frac{\left(\alpha_i - \frac{1}{M}\sum_m \alpha_m |S'U_m|^2\right)^2}{\alpha_i(\lambda_M - \alpha_i\sigma_n^2)^2} |S'U_i|^2.$$

To evaluate this expression, the relationship between the signal direction vector S and the eigenvectors of Q must be specified. If S were proportional to an eigenvector of Q, the quantity $\|\tilde{B}_1\|^2_{\mathrm{Q}^{-1}}$ would be zero. As the norm of $\tilde{B}_1$ will appear in the denominator of the expression for beam energy (A7), one can obtain a lower bound on the energy in the beam by assuming the largest possible value for its length. To approximate the maximal length of $\tilde{B}_1$, assume that S does not prefer *any* of the eigenvector directions of Q. A reasonable mathematical description of this situation is that $|S'U_i|^2 = \alpha_i$. In this instance, we have, using (A16), that

$$\|\tilde{B}_1\|^2_{\mathrm{Q}^{-1}} = \frac{\sigma_n^4}{\sigma_s^4}\frac{1}{M^2}\sum_i \left(\alpha_i - \frac{1}{M}\sum_m \alpha_m^2\right)^2. \tag{A18}$$

The quantity in the summation depends only on the eigenvalues of the matrix Q. Define the quantity γ_{EV}^2 to be

$$\gamma_{EV}^2 = \frac{1}{M}\sum_i \left(\alpha_i - \frac{1}{M}\sum_m \alpha_m^2\right)^2. \tag{A19}$$

Then (A18) becomes

$$\|\tilde{B}_1\|^2_{Q^{-1}} = \frac{\sigma_n^4}{\sigma_s^4} \cdot \frac{\gamma_{EV}^2}{M}. \tag{A20}$$

As the term $\tilde{B}_2$ is a random quantity, its squared norm is defined to be

$$\|\tilde{B}_2\|^2_{Q^{-1}} = E\left[\frac{\sigma_n^2}{\sigma_s^2} \sum_i \frac{|\tilde{N}'U_i|^2}{\alpha_i}\right]. \tag{A21}$$

where $E[\cdot]$ denotes expected value. To a good approximation, U_i is an eigenvector of the correlation matrix associated with the random vector $\tilde{N}$. Consequently,

$$E[|\tilde{N}'U_i|^2] = \frac{\alpha_i}{K}$$

so that

$$\|\tilde{B}_2\|^2_{Q^{-1}} = \frac{\sigma_n^2}{\sigma_s^2} \cdot \frac{M}{K}. \tag{A22}$$

Substituting equations (A18), (A22), and (A17) into (A7), we have finally

$$S_{EV}(k) = \frac{\sigma_s^2}{\frac{\sigma_n^2}{\sigma_s^2}\frac{\gamma_{EV}^2}{M} + \frac{M}{K}} \tag{A23}$$

as an expression for the energy in the beam when steered toward the source.

The analysis for the MUSIC method differs only in detail from that just described. In this method, the spectral estimate is given by

$$S_{\text{MUSIC}}(k) = (W'C'_{EV}C_{EV}W)^{-1}. \tag{A24}$$

Off-target, $C_{EV}W$ approximately equals W, implying that $S_{\text{MUSIC}}(k) = M^{-1}$. When steered on-target, the expression for the MUSIC estimate differs little from that given in (A7). The significant difference is that the norm of ϵ is computed with respect to the identity matrix instead of Q^{-1}. The quantity of interest is therefore

$$\|\epsilon\|^2 = \|\tilde{B}_1\|^2 + \|\tilde{B}_2\|^2.$$

The norm of $\tilde{B}_2$ with respect to I equals that computed with respect to Q^{-1}; the expression for the norm of $\tilde{B}_1$ is

$$\|\tilde{B}_1\|^2 = \frac{\sigma_n^4}{\sigma_s^4}\frac{1}{M^2}\sum_i \alpha_i\left(\alpha_i - \frac{1}{M}\sum_m \alpha_m^2\right)^2.$$

Therefore, the on-target beam energy in the MUSIC method is given by

$$S_{\text{MUSIC}}(k) = \frac{\sigma_s^2/\sigma_n^2}{\frac{\sigma_n^2}{\sigma_s^2}\frac{\gamma_{\text{MUSIC}}^2}{M} + \frac{M}{K}} \tag{A25}$$

where γ_{MUSIC}^2 is defined to be

$$\gamma_{\text{MUSIC}}^2 = \frac{1}{M}\sum^M \alpha_i\left(\alpha_i - \frac{1}{M}\sum \alpha_m^2\right)^2.$$

Acknowledgment

The authors are indebted to D. J. Edelblute of the Naval Ocean Systems Center for many fruitful discussions concerning this research.

References

[1] A. B. Baggeroer, "Sonar signal processing," in *Applications of Digital Signal Processing*, A. V. Oppenheim, Ed. Englewood Cliffs, NJ: Prentice-Hall, 1978, pp. 331–437.

[2] G. Bienvenu and L. Kopp, "Adaptivity to background noise spatial coherence for high resolution passive methods," in *Proc. IEEE ICASSP*, Denver, CO, 1980, pp. 307–310.

[3] —, "Source power estimation method associated with high resolution bearing estimation," in *Proc. IEEE ICASSP*, Atlanta, GA, 1981, pp. 153–156.

[4] J. A. Cadzow and T. P. Bronez, "An algebraic approach to superresolution adaptive array processing," in *Proc. IEEE ICASSP*, Atlanta, GA, 1981, pp. 302–305.

[5] J. Capon, "High-resolution frequency-wavenumber spectrum analysis," *Proc. IEEE*, vol. 57, pp. 1408–1418, Aug. 1969.

[6] H. Cox, "Resolving power and sensitivity to mismatch of optimum array processors," *J. Acoust. Soc. Amer.*, vol. 54, no. 3, pp. 771–785, 1973.

[7] D. J. Edelbute, J. M. Fisk, and G. L. Kinnison, "Criteria for optimum-signal-detection theory for arrays," *J. Acoust. Soc. Amer.*, vol. 41, no. 1, pp. 199–205, 1967.

[8] B. S. Garbow, J. M. Boyle, J. J. Dongarra, and C. B. Moler, "Matrix eigensystem routines–EISPACK guide extension," in *Lecture Notes in Computer Science*, vol. 51, G. Goos and J. Hartmanis, Eds. New York: Springer-Verlag, 1977.

[9] M. J. Hinich, "Frequency-wavenumber array processing," *J. Acoust. Soc. Amer.*, vol. 69, no. 3, pp. 732–737, 1981.

[10] D. H. Johnson, "On improving the resolution of bearing in passive sonar arrays," Naval Ocean Syst. Cen., San Diego, CA, Tech. Note 914, 1980.

[11] R. T. Lacoss, "Data adaptive spectral analysis methods," *Geophysics*, vol. 36, no. 4, pp. 661–675, 1971.

[12] N. L. Owsley, "Modal decomposition of data adaptive spectral estimates," presented at the Yale Univ. Workshop on Appl. of Adaptive Syst. Theory, New Haven, CT, 1981.

[13] R. Schmidt, "Multiple emitter location and signal parameter estimation," in *Proc. RADC Spectral Estimation Workshop*, Rome, NY, 1979, pp. 243–258.

Optimality of High Resolution Array Processing Using the Eigensystem Approach

GEORGES BIENVENU, MEMBER, IEEE, AND LAURENT KOPP

Abstract—In the classical approach to underwater passive listening, the medium is sampled in a convenient number of "look-directions" from which the signals are estimated in order to build an image of the noise field. In contrast, a modern trend is to consider the noise field as a global entity depending on few parameters to be estimated simultaneously. In a Gaussian context, it is worthwhile to consider the application of likelihood methods in order to derive a detection test for the number of sources and estimators for their locations and spectral levels. This paper aims to compute such estimators when the wavefront shapes are not assumed known *a priori*. This justifies results previously found using the asymptotical properties of the eigenvalue-eigenvector decomposition of the estimated spectral density matrix of the sensor signals: they have led to a variety of "high resolution" array processing methods. More specifically, a covariance matrix test for equality of the smallest eigenvalues is presented for source detection. For source localization, a "best fit" method and a test of orthogonality between the "smallest" eigenvectors and the "source" vectors are discussed.

I. INTRODUCTION

IN the classical approach to underwater passive listening, the medium is sampled in a convenient number of "look directions" and one attempts to estimate the signal coming from each of them.

For each given look direction, the optimal estimator has been derived (Gaussian context) and this has led to "adaptive" array processors. The kind of optimality achieved by this approach might be called "local optimality," but it is not likely to best answer the basic questions: how many sources are present and what are their locations and spectral levels? Modeling the noise field as an entity which depends on a few parameters to be estimated simultaneously appears to be the suitable approach in order to achieve "global optimality." This modern trend [1] has led to several array processing methods which are characterized by a significant improvement in resolution over conventional methods.

The reasons for the improvement in resolution achieved by these "high resolution" methods, besides the approach itself, lie essentially in a more accurate modeling of the noise field features, and specifically a more accurate modeling of the background noise [2]. Such a requirement may seem to bring severe restrictions to the applicability of these methods, but in fact, similar assumptions are needed in the conventional approach.

The conventional processors provide noise field power estimates as a function of location, generally bearing angle. These estimates need further processing to provide answers concerning the number of sources, their locations and spectral levels. In order to make valid comparisons between the high resolution approach and the conventional approach, this last "deconvolution" step should be described. In practice, this operation is performed either by a human operator or by a computer. In either case, some specific assumptions will be made about the background noise. The way these assumptions are handled in the high resolution approach is probably the most suitable.

This paper is concerned with several questions about the optimality of these "high resolution" methods. So far, these methods have been justified only on the basis of algebraic properties of the spectral density matrix of the sensor signals "heuristically" extended to the finite estimation time situation. Some of the results presented in this paper have already been anticipated. For instance, in the study of the detection of one source in the presence of incoherent background noise, estimating the wavefront shape by the eigenvector of the estimated spectral density matrix related to the largest eigenvalue has been suggested [3].

Here, we shall be concerned with the problem of detecting an *a priori* unknown number of sources with unknown wavefront shapes.

Let us first recall how the high resolution methods have been justified.

Let $\vec{x}(t)$ be the signal vector for which the kth component is the signal received on the kth of K sensors in an array. Its correlation matrix (stationarity being assumed) is

$$R(\tau) = E[\vec{x}(t)\,\vec{x}^+(t+\tau)]\,. \tag{1}$$

$E\{\cdot\}$ is the statistical expectation and $\vec{x}^+$ is the complex conjugate and transpose of $\vec{x}$.

In a linear medium $\vec{x}(t)$ may be written

$$\vec{x}(t) = \vec{b}(t) + \sum_{i=1}^{p} \vec{s}_i(t)$$

where $\vec{s}_i(t)$ is the signal due to the ith source alone and $\vec{b}(t)$ is the background noise contribution.

Assuming mutual decorrelation between the various contributions, the correlation matrix (1) becomes

$$R(\tau) = E[\vec{b}(t)\,\vec{b}^+(t+\tau)] + \sum_{i=1}^{p} E[\vec{s}_i(t)\,\vec{s}_i^+(t+\tau)]\,.$$

Manuscript received February 26, 1982; revised February 2, 1983. This work was supported by Direction des Recherches et Moyens d'Essais, Paris, France.

The authors are with Thomson-CSF ASM Division, 06801 Cagnes-Sur-Mer, France.

Reprinted from *IEEE Trans. Acoust., Speech, Signal Processing*, vol. ASSP-31, pp. 1235–1247, Oct. 1983.

The sources are assumed to be point-like, and to have perfect spatial coherence. The wavefront shape is a generally assumed known function of the source position (an essential assumption for beamforming which will be dropped in later sections). The so-called spectral density matrix is the Fourier transform of the correlation matrix and, with the above assumptions, may be written as

$$\Gamma(f) = \Gamma_b(f) + \sum_{i=1}^{p} \Gamma_i(f)$$

where $\Gamma_i(f)$ is the spectral density matrix of the ith source alone, and $\Gamma_b(f)$ is the spectral density matrix of the background noise. These matrices take the form

$$\Gamma_i(f) = \gamma_i^2(f)\, \vec{d}_i(f)\, \vec{d}_i^{\,+}(f)$$

wherein $\gamma_i^2(f)$ is the power spectral density of the ith source signal, and $\vec{d}_i(f)$ is the ith source "position vector" which describes the propagation "transfer function" between the ith source and the various sensors. This vector actually represents the wavefront shape as sampled by the array geometry and is a function of the source position, including direction and range. In the general case, these p vectors are linearly independent.

$\Gamma_i(f)$ has a dyadic form and is of rank one which is a characteristic feature of a perfectly coherent source.

To measure the power spectral densities a convention has to be taken. For instance, the first sensor may be taken as a spatial reference but any other choice is suitable.

By measuring similarly the power due to the background noise, its coherence matrix is defined as

$$\Gamma_b = \sigma^2(f)\, J(f)$$

where $J(f)$ is the "background noise spatial coherence matrix," a full rank matrix, and $\sigma^2(f)$ is the power spectral density of the background noise.

The total noise field spectral density matrix takes the final form

$$\Gamma(f) = \sum_{i=1}^{p} \gamma_i^2(f)\, \vec{d}_i(f)\, \vec{d}_i^{\,+}(f) + \sigma^2(f)\, J(f). \tag{2}$$

In conventional beamforming or adaptive array processing, the only assumptions needed concern the wavefront shapes represented by $\vec{d}_i(f)$. However, as mentioned previously, it will be necessary to assume something about $J(f)$ for further processing.

The additional assumptions needed for high resolution processing are that $J(f)$ is known and the noise field solvable $(p < K)$.

Generally, the background noise is assumed to be incoherent, that is, independent between sensors such that $J(f) = I$, the identity matrix. This is not restrictive since it has been shown [5] that if $J(f)$ is known, it is possible to "spatially whiten" the noise by using a matrix $C(f)$ such that

$$C(f)\, J(f)\, C^+(f) = I.$$

Thus, if the input signal vector is linearly transformed by $C(f)$, the spectral density matrix becomes

$$\begin{aligned}\Gamma_c(f) &= C(f)\, \Gamma(f)\, C^+(f)\\ &= \sigma^2(f)\, I + \sum_{i=1}^{p} \gamma_i^2(f)\, \vec{d}_{c_i}(f)\, \vec{d}_{c_i}^{\,+}(f)\\ &= \sigma^2(f)\, I + \Gamma_{cs}(f)\end{aligned}$$

with

$$\vec{d}_{c_i}(f) = C(f)\, \vec{d}_i(f).$$

The "source" term in the covariance matrix is now equal to

$$\Gamma_{cs}(f) = \sum_{i=1}^{p} \gamma_i^2(f)\, \vec{d}_{c_i}(f)\, \vec{d}_{c_i}^{\,+}(f).$$

High resolution methods are based on the eigensystem decomposition of $\Gamma_c(f)$.

For eigenvector $\vec{v}(f)$ related to an eigenvalue $\lambda(f)$ we have either

$$\Gamma_c(f)\, \vec{v}(f) = \lambda(f)\, \vec{v}(f)$$

or

$$\Gamma_{cs}(f)\, \vec{v}(f) + \sigma^2(f)\, \vec{v}(f) = \lambda(f)\, \vec{v}(f).$$

Now $\Gamma_{cs}(f)$ and $\Gamma_c(f)$ have the same eigenvectors, the set of eigenvalues of $\Gamma_{cs}(f)$ being translated by $\sigma^2(f)$ to provide those of $\Gamma_c(f)$.

It can be shown that $\Gamma_{cs}(f)$ is a rank p matrix with p nonzero eigenvalues where the p eigenvectors $\vec{v}_k(f)$ $(k \in [1, p])$ are the eigenvectors of the matrix $\Gamma_{cs}(f)$ corresponding to p nonzero eigenvalues $\lambda_{csk}(f)$ of the signal only spectral density matrix $\Gamma_{cs}(f)$ shifted by $\sigma^2(f)$

$$\lambda_{ck}(f) = \lambda_{csk}(f) + \sigma^2(f).$$

These p eigenvectors are an orthogonal basis of the p-dimensional source subspace spanned by the p position vectors $\vec{d}_{ci}(f)$, and the other eigenvalues of $\Gamma_{cs}(f)$ are zero. Thus, there results

$$\sum_{i=1}^{p} \gamma_i^2(f)\, \vec{d}_{c_i}(f)\, \vec{d}_{c_i}^{\,+}(f) = \sum_{k=1}^{p} \lambda_{csk}(f)\, \vec{v}_k(f)\, \vec{v}_k^{\,+}(f). \tag{3}$$

The remaining $K - p$ eigenvectors $\vec{v}_l(f)$ are orthogonal to the previous ones ($\Gamma_{cs}(f)$ Hermitian) and therefore to each position vector

$$\vec{v}_l^{\,+}(f)\, \vec{d}_{c_i}(f) = 0 \qquad \forall l \in [p+1, K], \ \ \forall i \in [1, p]. \tag{4}$$

These vectors constitute a basis for the $(K - p)$-dimensional noise subspace orthogonal to the source subspace.

The corresponding eigenvalues are all equal to $\sigma^2(f)$ and, therefore, are smaller than those corresponding to the source subspace. From the above results the following are deduced:

a) The number q of equal and minimum eigenvalues provides the number of sources by $p = K - q$.

b) Two procedures may be applied to achieve source localization:

i) Either utilizing the source subspace and (3) ("model fitting" methods [1] and processing derived from conventional minimum variance and maximum entropy methods [6] have been proposed), or

ii) utilizing the noise subspace [2], [4], and (4). Let $\vec{d}_c(f, \vec{\theta})$ be the model for a source position vector, according to the hypothesis derived for the relationship between the wavefront shape and the source position $\vec{\theta}$, and transformed by $C(f)$. Because the eigenvectors of the noise subspace are orthogonal to each source position vector $\vec{d}_{ci}(f)$, the values of $\vec{\theta}$ for which the function

$$G(f,\vec{\theta}) = \sum_{l=p+1}^{K} |\vec{d}_c^{\,+}(f,\vec{\theta})\,\vec{v}_l(f)|^2 \tag{5}$$

is equal to zero give the positions of the sources.

In practice, the limited observation time provides only an estimation $\hat{\Gamma}(f)$ of $\Gamma(f)$. This affects the previous results because the set of minimum eigenvalues of $\hat{\Gamma}_c(f)$ will not all be equal but will suffer some dispersion which depends on the observation time.

The question of optimality of these high resolution methods remains open and it is the intent herein to show the context in which they are optimal.

In addition to knowledge of background noise spatial coherence, an important requirement is a detection test for the number of sources. This information is important in itself, and is necessary to derive the source parameter estimates. The intent of this paper is to obtain an optimal detection test, based on generalized likelihood ratio philosophy. The wavefront shapes are not assumed known in this likelihood ratio approach; rather an attempt is made to estimate them.

II. Problem Framework

The main objective is to derive a detection test for the number of sources p, when neither the background noise spectral density $\sigma^2(f)$, the source spectral densities $\gamma_i^2(f)$, nor the source position vectors $\vec{d}_i(f)$ are known. By $\vec{d}_i(f)$ unknown, it is meant that each component of $\vec{d}_i(f)$ is unknown. We suppose only that the sources have perfect spatial coherence. As an important byproduct we shall compute estimates for these parameters which will allow us to appreciate optimality of high resolution methods.

As indicated above, the generalized likelihood ratio test procedure is used [7]. Accordingly, if $\vec{X}$ is the observation vector which depends on a set of unknown parameters $\vec{\beta}$, the generalized likelihood ratio test is given by

$$\text{Max}_{\vec{\beta}_1} p_{H_1}(\vec{X}|\vec{\beta}_1)/\text{Max}_{\vec{\beta}_0} p_{H_0}(\vec{X}|\vec{\beta}_0)$$

where $p_{Hi}(\vec{X}/\vec{\beta}_i)$ is the conditional probability density of the observation $\vec{X}$ when $\vec{\beta} = \vec{\beta}_i$ in the ith hypothesis. This ratio is equivalently written

$$p_{H_1}(\vec{X}|\hat{\vec{\beta}}_1)/p_{H_0}(\vec{X}|\hat{\vec{\beta}}_0)$$

where $\hat{\vec{\beta}}_i$ is the maximum likelihood estimate of $\vec{\beta}_i$.

We then need to derive the optimal estimators for $\sigma(f)$, $\gamma_i(f)$, and $\vec{d}_i(f)$ for every fixed number of sources.

The background noise spatial coherence $J(f)$ is assumed known so that it may be taken as incoherent without loss of generality

$$J(f) = I.$$

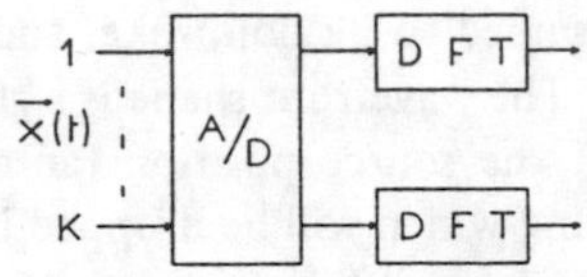

Fig. 1. The observation space is defined by the complex vector Fourier transform (DFT) of the sensor signals.

In this section, the computation of the likelihood function is detailed. An accurate description of the observation space, the statistic of the observations in this space, and the relations between the parameters and this statistic are specified. In the following, the frequency f will be omitted from the equations.

A. The Observation Space

The observation vector space is spanned by the set of N output samples of the system illustrated in Fig. 1. The K sensors signals $\vec{x}(t)$ are digitized (A/D) and Fourier transformed (DFT) at frequency f into the complex N-vector of DFT coefficients

$$\vec{X}_i = \sum_{k=1}^{L} \vec{x}[k\Delta + (i-1)L\Delta] \exp(-2i\pi f k\Delta)$$

where Δ is the sampling period and $1/L\Delta$ is the spectral resolution. An observation $\vec{X}$ will be defined as the set $(\vec{X}_1, \cdots, \vec{X}_N)$ of N of the complex vectors $\vec{X}_i$.

B. Statistical Properties of the Observation

The time sample vector $\vec{x}(t)$ is assumed to be stationary, zero-mean, and Gaussian. Thus, neglecting the digitization errors (which introduce nonlinearities), each vector $\vec{X}_i$ will then be zero-mean and Gaussian (complex). It will be further assumed that successive samples are not correlated. The observation is then distributed according to the density

$$p(\vec{X}) = (\pi)^{-KN} |A|^{-N} \exp\left(-\sum_{i=1}^{N} \vec{X}_i^{+} A^{-1} \vec{X}_i\right) \tag{7}$$

where A is the covariance matrix

$$A = E[\vec{X}_i \vec{X}_i^{+}] \qquad \forall i \in [1, N]. \tag{8}$$

$|A|$ is the matrix A determinant and A^{-1} its inverse.

C. Computation of the Likelihood Function

Using (6) and (8), the covariance matrix A can be written

$$A = \sum_{k,l=1}^{L} \exp[-2i\pi f(k-l)\Delta]\, E[\vec{x}(k\Delta)\vec{x}^{+}(l\Delta)].$$

From (1), it is deduced

$$A = \sum_{k,l=1}^{L} \exp[-2i\pi f(k-l)\Delta]\, R[(k-l)\Delta].$$

Thus, A is equal to

$$A = \sum_{k=-(L-1)}^{(L-1)} (L-|k|) \exp[-2i\pi f k\Delta]\, R(k\Delta).$$

In terms of the spectral density matrix $\Gamma(\nu)$, we have

$$A = \int_{-\infty}^{\infty} d\nu\, \Gamma(\nu) \left\{ \sum_{k=-(L-1)}^{(L-1)} (L - |k|) \cdot \exp\left[2i\pi(\nu - f)k\Delta\right] \right\}$$

which may also be written as

$$A = \int_{-\infty}^{\infty} d\nu\, \Gamma(\nu) \left[\frac{\sin\left[L\pi(\nu - f)\Delta\right]}{\sin\left[\pi(\nu - f)\Delta\right]}\right]^2.$$

Except for a multiplying factor which is equal to one when using a suitable DFT normalization, A is considered an approximation of $\Gamma(f)$.

The quality of the approximation $A = \Gamma(f)$ is controlled by the spectral resolution $(1/L\Delta)$. That is, for the errors made by identifying A and $\Gamma(f)$ to be negligible, $\Gamma(f)$ should not vary too much in a resolution interval. The smoothness of $\Gamma(f)$ is controlled by two factors: the noise field spectral structure and the noise field spatial structure.

Considering a typical (k, l) term of the spectral density matrix when there is one source present gives

$$\{\Gamma(f)\}_{kl} = \gamma^2(f) \exp(-2i\pi f \tau_{kl})$$

where τ_{kl} is the travel time of the wavefront from sensor k to sensor l. The source is remote enough to justify neglecting the attenuation between two sensors.

The smoothness conditions are then

$$\tau_c \ll L\Delta$$

$$\tau_{kl} \ll L\Delta$$

where τ_c is the correlation time of the source signal. If D is the array aperture size and C the speed of the wavefront, we should have

$$L\Delta \gg D/C.$$

However, several other factors have to be considered in choosing L. For example, the stationarity of the data must be such that N is large enough to reduce the variance of the observation sufficiently.

Assuming these conditions are fulfilled, the likelihood function is given by (7) where A is replaced by Γ given by (2).

$$p(\vec{x}) = (\pi)^{-KN} |\Gamma|^{-N} \exp\left[-N \operatorname{Tr}(\hat{\Gamma}\Gamma^{-1})\right]$$

where

$$\hat{\Gamma} = \frac{1}{N} \sum_{i=1}^{N} \vec{X}_i \vec{X}_i^{+}$$

$$\Gamma = \sigma^2 J + \sum_{i=1}^{p} \gamma_i^2 \vec{d}_i \vec{d}_i^{+}$$

and $\operatorname{Tr}(A)$ is a notation for the trace of the matrix A.

III. Noise Field Parameter Likelihood Estimate Equations

The likelihood function must be maximized with respect to the parameter vector $\vec{\beta}$, or equivalently, its logarithm

$$L = -KN \ln(\pi) - N \ln |\Gamma| - N \operatorname{Tr}(\hat{\Gamma}\Gamma^{-1}) \tag{9}$$

must be maximized in an identical manner.

In the development which follows an optimal (maximum likelihood) estimate of a parameter β is written $\hat{\beta}$.

A. Source Power Equations

To derive the condition on the kth source spectral level γ_k^2 the following decomposition of Γ is suitable:

$$\Gamma = \Gamma_k + \gamma_k^2 \vec{d}_k \vec{d}_k^{+}.$$

Γ_k depends neither on γ_k nor on $\vec{d}_k$; then

$$\Gamma^{-1} = \Gamma_k^{-1} - \left[\gamma_k^2/(1 + \gamma_k^2 \vec{d}_k^{+} \Gamma_k^{-1} \vec{d}_k)\right] \Gamma_k^{-1} \vec{d}_k \vec{d}_k^{+} \Gamma_k^{-1}$$

(from Woodbury's identity [8, Appendix E]) (Appendix A)

$$|\Gamma| = |\Gamma_k| (1 + \gamma_k^2 \vec{d}_k^{+} \Gamma_k^{-1} \vec{d}_k).$$

Then

$$L = -KN \ln(\pi) - N \ln(|\Gamma_k|) - N \operatorname{Tr}(\hat{\Gamma}\Gamma_k^{-1}) + L_k$$

with

$$L_k = -N \ln(\delta_k) + N(\gamma_k^2/\delta_k) \vec{d}_k^{+} \Gamma_k^{-1} \hat{\Gamma} \Gamma_k^{-1} \vec{d}_k$$

and

$$\delta_k = 1 + \gamma_k^2 \vec{d}_k^{+} \Gamma^{-1} \vec{d}_k.$$

The condition on γ_k is

$$\left.\frac{\partial L}{\partial \gamma_k}\right|_{\gamma_k = \hat{\gamma}_k, \vec{d}_k = \hat{\vec{d}}_k, \sigma = \hat{\sigma}} = 0.$$

The resulting equation (Appendix B) is

$$\hat{\vec{d}}_k^{+} \tilde{\Gamma}^{-1} \hat{\Gamma} \tilde{\Gamma}^{-1} \hat{\vec{d}}_k = \hat{\vec{d}}_k^{+} \tilde{\Gamma}^{-1} \hat{\vec{d}}_k, \quad k \in [1, p]. \tag{10}$$

The notation $\tilde{\Gamma}$ means that Γ, considered as a function of the parameters, is taken at the optimal point

$$\tilde{\Gamma} = \hat{\sigma}^2 J + \sum_{i=1}^{p} \hat{\gamma}_i^2 \hat{\vec{d}}_i \hat{\vec{d}}_i^{+}. \tag{11}$$

This notation will be used extensively in subsequent developments, even when it is not possible to express the quantity of interest as an explicit function of the optimum parameters.

B. Wavefront Shape Equations

To derive the condition on $\vec{d}_k$, the gradient of the log-likelihood L relative to $\vec{d}_k$ is needed. Using $\vec{\nabla}_{\vec{d}}(\cdot)$ as a notation for the gradient relative to $\vec{d}$, we have

$$\vec{\nabla}_{\vec{d}_k}(L) = \vec{\nabla}_{\vec{d}_k}(L_k).$$

As mentioned previously, a convention has to be made which defines the spectral power.

Consider the total power received by the array coming from a source

$$\gamma^2 \vec{d}^+ \vec{d}.$$

This quantity may be measured, but to decide which is γ^2 and which is $\vec{d}^+ \vec{d}$ is a matter of convention. We retain

$$\vec{d}^+ \vec{d} = K. \tag{12}$$

Any other convention is suitable. For instance, [11] uses the convention $\gamma^2 = 1$, which leads to equivalent results. Equation (12) makes the computation easier. The optimal estimation of $\vec{d}_k$ is then constrained optimization problem

$$\begin{cases} \text{maximize } L_k \text{ with respect to } \vec{d}_k \\ \text{with } \vec{d}_k^+ \vec{d}_k = K. \end{cases}$$

This may be turned into an unconstrained optimization problem using Lagrange's multiplier technique

$$\vec{\nabla}_{\vec{d}_k} [L_k - \mu_k(\vec{d}_k^+ \vec{d}_k - K)] \Big|_{\vec{d}_k = \hat{\vec{d}}_k, \gamma_k = \hat{\gamma}_k, \sigma = \hat{\sigma}} = \vec{0}$$

which may be written at the optimum point as

$$\vec{\nabla}_{\vec{d}_k} (L_k) = 2\mu_k \vec{d}_k$$

where μ_k is a Lagrange's multiplier.

The computation of the gradient (Appendix C) leads to the conditions

$$\begin{cases} \mu_k = 0 \\ \hat{\Gamma}\tilde{\Gamma}^{-1} \vec{d}_k = \vec{d}_k, \quad k \in [1, p]. \end{cases} \tag{13}$$

Equation (10) is easily deduced from (13) (Appendix C).

Notice that the various vectors $\vec{d}_k$ are not parameterized (as they would be if the propagation model was known). Each one of their components is free to take any value provided the constraint (12) is respected. Later on (see Section VI) it will be seen that, except for $p = 1$, the solution of (13) is not unique; at best, we can localize $\vec{d}_k$ in a linear subspace.

Notice also that, in general, Γ is not equal to $\tilde{\Gamma}$. $\hat{\Gamma}$ is the maximum likelihood estimate of the observation covariance matrix [see (8)] and $\tilde{\Gamma}$ is a notation [see (11)]. However, if $p = K - 1$, it will be seen that $\hat{\Gamma} = \tilde{\Gamma}$, so that (13) provides no information (see Section VI).

C. Background Noise Level Equation

Similarly, the maximum likelihood estimate of σ requires the solution of

$$\frac{\partial L}{\partial \sigma} \Big|_{\sigma = \hat{\sigma}, \gamma_k = \hat{\gamma}_k, \vec{d}_k = \hat{\vec{d}}_k} = 0.$$

The resulting equation (Appendix D) is

$$\mathrm{Tr}\,(\hat{\Gamma}\tilde{\Gamma}^{-1}) = K. \tag{14}$$

IV. Basic Properties of the Noise Field Parameter Estimates

It will now be shown, from the previous set of equations, that it is possible to derive a relationship between the noise field maximum likelihood parameter estimates and the eigensystem of the spectral density matrix estimate $\hat{\Gamma}$.

This relationship is exactly the same as that between the actual noise field parameters and the eigensystem of Γ which have been referred to in the Introduction and are the basis of high resolution methods. That important property justifies the utilization of an estimate of the spectral density matrix instead of the spectral density matrix itself as stated by the existing asymptotical theory.

A. Property 1

The maximum likelihood estimate $\hat{\Gamma}$ of the spectral density matrix of the received signals Γ has the same eigenvectors as the matrix $\tilde{\Gamma}$ formed with the maximum likelihood estimates of the noise field parameters (11).

Moreover, p of these eigenvectors are in the p-dimensional linear subspace $\mathcal{E}$ spanned by the maximum likelihood estimates of the p position vectors $(\hat{\vec{d}}_1, \cdots, \hat{\vec{d}}_p)$. The $(K - p)$ remaining eigenvectors are orthogonal to $(\hat{\vec{d}}_1, \cdots, \hat{\vec{d}}_p)$.

Proof: First notice that $\mathcal{E}$ is also spanned by

$$(\tilde{\Gamma}^{-1} \hat{\vec{d}}_1, \cdots, \tilde{\Gamma}^{-1} \hat{\vec{d}}_p).$$

Indeed, from the generalized Woodbury's identity (Appendix E), $\tilde{\Gamma}^{-1}$ can be written

$$\tilde{\Gamma}^{-1} = \frac{I}{\hat{\sigma}^2} - \sum_{i,j=1}^{p} \tilde{\alpha}_{ij} \hat{\vec{d}}_i \hat{\vec{d}}_j^+.$$

The actual expression for $\tilde{\alpha}_{ij}$ is not needed explicitly. Post-multiplying $\tilde{\Gamma}^{-1}$ by $\hat{\vec{d}}_k$ gives the vector $\tilde{\vec{w}}_k$:

$$\tilde{\vec{w}}_k = \tilde{\Gamma}^{-1} \hat{\vec{d}}_k = \frac{1}{\hat{\sigma}^2} \hat{\vec{d}}_k - \sum_{i,j=1}^{p} \tilde{\alpha}_{ij} (\hat{\vec{d}}_j^+ \hat{\vec{d}}_k) \hat{\vec{d}}_i.$$

Thus, $\tilde{\vec{w}}_k$ is a linear combination of the source position vector estimates and lies in $\mathcal{E}$. Therefore, the p vectors $\tilde{\vec{w}}_k$ also span $\mathcal{E}$. An eigenvector $\tilde{\vec{v}}$ of $\tilde{\Gamma}$ and its related eigenvalue $\tilde{\lambda}$ are such that

$$\tilde{\Gamma}\tilde{\vec{v}} = \tilde{\lambda}\tilde{\vec{v}}.$$

From (11)

$$\hat{\sigma}^2 \tilde{\vec{v}} = \sum_{i=1}^{p} \hat{\gamma}_i^2 (\hat{\vec{d}}_i^+ \tilde{\vec{v}}) \hat{\vec{d}}_i = \tilde{\lambda}\tilde{\vec{v}}. \tag{15}$$

If $\tilde{\vec{v}}$ belongs to $\mathcal{E}^\perp$ the linear subspace orthogonal to $\mathcal{E}$, the various scalar products in (15) are zero so that

$$\hat{\sigma}^2 \tilde{\vec{v}} = \tilde{\lambda}\tilde{\vec{v}}.$$

Therefore, any vector in $\mathcal{E}^\perp$ is an eigenvector of $\tilde{\Gamma}$, and the related eigenvalue is equal to $\hat{\sigma}^2$. Since we may find $(K - p)$ linearly independent vectors in $\mathcal{E}^\perp$, there are $(K - p)$ eigenvectors of $\tilde{\Gamma}$ orthogonal to $\mathcal{E}$, with the same eigenvalue $\hat{\sigma}^2$. Since $\tilde{\Gamma}$ is Hermitian, its eigenvectors can be chosen orthogonal so that the remaining p eigenvectors are orthogonal to $\mathcal{E}^\perp$ and are then lying in $\mathcal{E}$. Let $\tilde{\vec{v}}_k$ be such an eigenvector. Since the vectors $\tilde{\vec{w}}_l$ span $\mathcal{E}$, $\tilde{\vec{v}}_k$ can be written

$$\tilde{\vec{v}}_k = \sum_{l=1}^{p} x_l^k \tilde{\vec{w}}_l$$

in terms of its components x_l^k.

We may now write

$$\hat{\Gamma}\tilde{\vec{v}} = \hat{\Gamma}\left(\sum_{l=1}^{p} x_l^k \tilde{\vec{w}}_l\right) = \sum_{l=1}^{p} x_l^k \hat{\Gamma}\tilde{\vec{w}}_l. \quad (16)$$

Equation (13),

$$\hat{\Gamma}\tilde{\Gamma}^{-1}\hat{\vec{d}}_l = \hat{\vec{d}}_l,$$

written in terms of $\tilde{\vec{w}}_l$ gives

$$\hat{\Gamma}\tilde{\vec{w}}_l = \tilde{\Gamma}\tilde{\vec{w}}_l. \quad (17)$$

Using (17), (16) becomes

$$\hat{\Gamma}\tilde{\vec{v}}_k = \sum_{l=1}^{p} x_l^k \tilde{\Gamma}\tilde{\vec{w}}_l$$

$$= \tilde{\Gamma}\left(\sum_{l=1}^{p} x_l^k \tilde{\vec{w}}_l\right)$$

$$= \tilde{\Gamma}\tilde{\vec{v}}_k = \tilde{\lambda}_k \tilde{\vec{v}}_k \quad (18)$$

showing that $\tilde{\vec{v}}_k$ is an eigenvector of $\hat{\Gamma}$ with the same eigenvalue. Therefore, the p eigenvectors of Γ lying in $\mathcal{E}$ are also eigenvectors of $\hat{\Gamma}$. Because $\hat{\Gamma}$ is Hermitian, its eigenvectors are orthogonal, thus its $(K-p)$ remaining eigenvectors are in $\mathcal{E}^{\perp}$. In other words, they are orthogonal to the maximum likelihood estimates of the source position vectors.

Moreover, we shall show that the p eigenvalues $\tilde{\lambda}_k$ of $\tilde{\Gamma}$ corresponding to the eigenvectors lying in $\mathcal{E}$ are the largest eigenvalues of $\tilde{\Gamma}$.

Taking (15) (in which $\tilde{\vec{v}} = \tilde{\vec{v}}_k$) and left-multiplying it by $\tilde{\vec{v}}_k^{+}$ gives

$$\hat{\sigma}^2 \tilde{\vec{v}}_k^{+}\tilde{\vec{v}}_k + \sum_{l=1}^{p} \hat{\gamma}_l^2 \,|\hat{\vec{d}}_l^{+}\tilde{\vec{v}}_k|^2 = \tilde{\lambda}_k \tilde{\vec{v}}_k^{+}\tilde{\vec{v}}_k$$

so that (the eigenvectors are normalized)

$$\tilde{\lambda}_k = \hat{\sigma}^2 + \sum_{i=1}^{p} \hat{\gamma}_i^2 \,|\hat{\vec{d}}_i^{+}\tilde{\vec{v}}_k|^2.$$

As $\tilde{\vec{v}}_k$ is lying in $\mathcal{E}$ the dot products $(\hat{\vec{d}}_i^{+}\tilde{\vec{v}}_k)$ are not all equal to zero, so that

$$\tilde{\lambda}_k > \hat{\sigma}^2, \quad k \in [1, p].$$

As shown above, the $(K-p)$ eigenvalues of $\tilde{\Gamma}$ related to the eigenvectors lying in $\mathcal{E}^{\perp}$ are all equal to $\hat{\sigma}^2$.

Finally, the eigenvalues of $\tilde{\Gamma}$ are split in two sets, namely the first $(K-p)$ all equal to $\hat{\sigma}^2$, the eigenvectors of which are in $\mathcal{E}^{\perp}$, and the last p, all strictly larger than $\hat{\sigma}^2$, the eigenvectors of which are in $\mathcal{E}$.

B. *Property 2*

The p eigenvectors of $\hat{\Gamma}$ lying in the space $\mathcal{E}$ spanned by the p maximum likelihood estimates of the sources positions vectors $(\hat{\vec{d}}_1, \cdots, \hat{\vec{d}}_p)$ are related to the p largest eigenvalues of $\hat{\Gamma}$. Thus the $(K-p)$ eigenvectors of $\hat{\Gamma}$ related to the $(K-p)$ smallest eigenvalues are orthogonal to the maximum likelihood estimates of the source position vectors. The maximum likelihood estimate of the background noise spectral density $\hat{\sigma}^2$ is given by

$$\hat{\sigma}^2 = \frac{1}{K-p} \sum_{i=p+1}^{K} \hat{\lambda}_i$$

which is the average of the $(K-p)$ smallest eigenvalues. Finally, the maximum likelihood estimates of the noise field parameters are such that

$$\sum_{i=1}^{p} \hat{\gamma}_i^2 \hat{\vec{d}}_i \hat{\vec{d}}_i^{+} = \sum_{i=1}^{p} (\hat{\lambda}_i - \hat{\sigma}^2)\, \hat{\vec{\omega}}_i \hat{\vec{\omega}}_i^{+}$$

where $\hat{\lambda}_i$ and $\hat{\vec{\omega}}_i$ $(i \in [1, p])$ are the p largest eigenvalues and related eigenvectors of $\hat{\Gamma}$. This corresponds to the asymptotical (3).

Proof: It has been shown in the preceding section that the eigenvectors of $\tilde{\Gamma}$ are eigenvectors of $\hat{\Gamma}$. It has also been shown that the p largest eigenvalues of $\tilde{\Gamma}$ are related to eigenvectors lying in $\mathcal{E}$ and that p eigenvalues of $\hat{\Gamma}$ are equal to those, but it is not clear that they are the largest.

Let us rank the eigenvalues of $\hat{\Gamma}$ in decreasing order:

$$\hat{\lambda}_1 \geqslant \hat{\lambda}_2 \geqslant \cdots \geqslant \hat{\lambda}_K > 0.$$

From (18), if $\tilde{\vec{v}}_k$ is an eigenvector of $\tilde{\Gamma}$ lying in $\mathcal{E}$,

$$\tilde{\Gamma}\tilde{\vec{v}}_k = \tilde{\lambda}_k \tilde{\vec{v}}_k.$$

Then

$$\hat{\Gamma}\tilde{\vec{v}}_k = \tilde{\lambda}_k \tilde{\vec{v}}_k$$

so that there exists an index $e(k)$ such that

$$\hat{\lambda}_{e(k)} = \tilde{\lambda}_k$$

and

$$\hat{\vec{\omega}}_{e(k)} = \tilde{\vec{v}}_k. \quad (19)$$

However, all that is known is that $e(k) \geqslant k$ and that $e(1) < e(2) < \cdots < e(p)$. We are going to show that $e(k) = k$.

Let $E(p)$ be the set of indexes $e(k)$, $k \in [1, p]$, so that the corresponding eigenvectors $\hat{\vec{\omega}}_{e(k)}$ are in $\mathcal{E}$.

$$E(p) = [e(1), \cdots, e(p)].$$

The problem is to find this set.

As the maximum likelihood estimates of the noisefield parameters are such that the likelihood function or equivalently its logarithm is maximum, the set $E(p)$ should be such that L is a maximum. From (9) we have

$$\tilde{L} = -KN \ln(\pi) - N \ln\ (|\tilde{\Gamma}|) - N\,\mathrm{Tr}\,(\hat{\Gamma}\tilde{\Gamma}^{-1})$$

and using (14), we deduce

$$\tilde{L} = -KN \ln(\pi) - N \ln(|\tilde{\Gamma}|) - KN. \tag{20}$$

Thus, we see that $|\tilde{\Gamma}|$ should be minimum.

$|\tilde{\Gamma}|$ can be written in terms of the eigenvalues of $\tilde{\Gamma}$:

$$|\tilde{\Gamma}| = (\hat{\sigma}^2)^{K-p} \prod_{i \in E(p)} \hat{\lambda}_i. \tag{21}$$

It is shown in Appendix F that

$$\hat{\sigma}^2 = \frac{1}{K-p} \sum_{i \notin E(p)} \hat{\lambda}_i. \tag{22}$$

Thus, (21) becomes

$$|\tilde{\Gamma}| = F[p, E(p)]$$

with

$$F[p, E(p)] = \left[\sum_{i \notin E(p)} \hat{\lambda}_i/(K-p)\right]^{K-p} \prod_{i \in E(p)} \hat{\lambda}_i \tag{23}$$

such that $\tilde{L}$ will be maximized by minimizing $F[p, E(p)]$.

The eigenvalues $\tilde{\lambda}_i$ of $\tilde{\Gamma}$ related to eigenvectors lying in $\mathcal{E}$ are such that

$$\tilde{\lambda}_i > \hat{\sigma}^2.$$

Using (19) we have

$$\hat{\lambda}_i > \hat{\sigma}^2 \qquad \forall\, i \in E(p).$$

In other words, from (22)

$$\hat{\lambda}_i > \sum_{j \notin E(p)} \hat{\lambda}_j/(K-p) \qquad \forall\, i \in E(p) \tag{24}$$

where $e(p)$ is the largest index of $E(p)$. Thus, $\hat{\lambda}_{e(p)}$ is the smallest eigenvalue of the set $\{\lambda_{e(i)}; i \in [1, p]\}$. Let $E(p)$ be written

$$E(p) = [E(p-1), e(p)]$$

with

$$E(p-1) = [e(1), \cdots, e(p-1)]$$

$F[p, E(p)]$ can be written

$$F[p, E(p)] = \left(\prod_{j \in E(p-1)} \hat{\lambda}_j\right) \hat{\lambda}_{e(p)} \cdot \left(a - \sum_{j \in E(p-1)} \hat{\lambda}_j - \hat{\lambda}_{e(p)}\right)^{K-p} (k-p)^{-(k-p)}$$

where a is the trace of the matrix $\hat{\Gamma}$. Introducing $F[p-1, E(p-1)]$ in $F[p, E(p)]$, we get

$$F[p, E(p)] = F[p-1, E(p-1)] \frac{\hat{\lambda}_{e(p)}}{a - \sum_{j \in E(p-1)} \hat{\lambda}_j} \cdot \left(1 - \frac{\hat{\lambda}_{e(p)}}{a - \sum_{j \in E(p)} \hat{\lambda}_j}\right)^{K-p} \frac{(K-p+1)^{K-p+1}}{(K-p)^{K-p}}$$

which can be rewritten as

$$F[p, E(p)] = F[p-1, E(p-1)] \frac{(K-p+1)^{K-p+1}}{(K-p)^{K-p}} \cdot \Psi\left(\frac{\hat{\lambda}_{e(p)}}{a - \sum_{j \in E(p-1)} \hat{\lambda}_j}, K-p\right)$$

where $\Psi(x, K-p) = x(1-x)^{K-p}; x \in (0, 1]$.

It is shown in Appendix G that Ψ is decreasing for

$$x > 1/(K-p+1).$$

However, because

$$a = \mathrm{Tr}(\hat{\Gamma}) = \sum_{j \in E(p-1)} \hat{\lambda}_j + \hat{\lambda}_{e(p)} + \sum_{j \notin E(p)} \hat{\lambda}_j$$

and [from (24)]

$$\hat{\lambda}_{e(p)} > \sum_{j \notin E(p)} \hat{\lambda}_j/(K-p),$$

we have

$$a - \sum_{j \in E(p-1)} \hat{\lambda}_j < (K-p+1)\, \hat{\lambda}_{e(p)}.$$

This results in

$$\hat{\lambda}_{e(p)} \Big/ \left[a - \sum_{j \in E(p-1)} \hat{\lambda}_j\right] > 1/(K-p+1).$$

This implies that L is maximum only if $E(p)$ is such that $\hat{\lambda}_{e(p)}$ is larger than any eigenvalue $\hat{\lambda}_j; j \notin E(p)$. Indeed, if we could find $q \notin E(p)$ such that $\hat{\lambda}_q > \hat{\lambda}_{e(p)}$ by permuting $e(p)$ and q, $E(p-1)$ would not be changed. Moreover, Ψ would be decreased as would be $F(p, E(p))$. This means that the $(K-p)$ eigenvectors $\hat{\lambda}_j; j \notin E(p)$ are the smallest ones, and so $e(k) = k, k \in [1, p]$.

$$E(p) = [1, \cdots, p].$$

Consequently, the $(K-p)$ eigenvectors of $\hat{\Gamma}$ related to the $(K-p)$ smallest eigenvalues are orthogonal to $(\hat{\vec{d}}_1, \cdots, \hat{\vec{d}}_p)$. Using (22), we obtain

$$\hat{\sigma}^2 = \frac{1}{K-p} \sum_{i=p+1}^{K} \hat{\lambda}_i \tag{25}$$

where now $(\hat{\lambda}_{p+1}, \cdots, \hat{\lambda}_K)$ are the smallest eigenvalues of $\hat{\Gamma}$.

From (F1) we deduce

$$\sum_{i=1}^{p} \hat{\gamma}_i^2 \hat{\vec{d}}_i \hat{\vec{d}}_i^+ = \sum_{i=1}^{p} (\hat{\lambda}_i - \hat{\sigma}^2)\, \hat{\vec{\omega}}_i \hat{\vec{\omega}}_i^+. \tag{26}$$

V. Detection Test for the Number of Sources

To find the number of sources p is a multiple testing problem; p has to be chosen between 0 and $(K-1)$ where $(K-1)$ is the maximum allowed number of sources in the model. The method used is the generalized likelihood ratio test, as introduced in Section II.

Symbolizing the set of unknown parameters in the p source assumption by $\vec{\beta}_p$, we have

$$\vec{\beta}_p = \{\sigma; (\gamma_k, \vec{d}_k), k \in [1, p]\}.$$

To test a "p-sources" assumption against a "q-sources" assumption we use

$$\mathrm{Max}_{\vec{\beta}_p}\, p(\vec{X} | p, \vec{\beta}_p) / \mathrm{Max}_{\vec{\beta}_q}\, p(\vec{X} | q, \vec{\beta}_q) \underset{H_q}{\overset{H_p}{\gtrless}} \eta_{pq}(f_a).$$

In other words,

$$p(\vec{X} | p, \hat{\vec{\beta}}_p) / p(\vec{X} | q, \hat{\vec{\beta}}_q) \underset{H_q}{\overset{H_p}{\gtrless}} \eta_{pq}(f_a)$$

where f_a is the false alarm probability and $\eta_{p,q}(f_a)$ the corresponding threshold fixed by

$$\Pr\{p(\vec{X} | p, \hat{\vec{\beta}}_p) / p(\vec{X} | q, \hat{\vec{\beta}}_q) \geqslant \eta_{pq}(f_a) | H_q\} = f_a.$$

This method would be suitable if we are sure to be in either of these situations, but in general this is not the case.

The various quantities $p(\vec{X} | p, \hat{\vec{\beta}}_p)$ have been computed in the previous section. Using (9), (14), and (23) gives

$$\tilde{L}(p) = p(\vec{X} | p, \hat{\vec{\beta}}_p) = (\pi e)^{-KN} [F(p)]^{-N}. \tag{27}$$

$F(p)$ is given by (23) where $E(p)$ is now known:

$$F(p) = \left[\sum_{i=p+1}^{K} \hat{\lambda}_i / (K - p)\right]^{K-p} \prod_{i=1}^{p} \hat{\lambda}_i. \tag{28}$$

To define a detection test for multiple hypothesis is possible from the theoretical point of view, when we are able to attach a cost to any decision H_p (given that H_q is true) and when the *a priori* probabilities of the various assumptions are known. However, in practice, it is almost impossible to evaluate the cost of H_p given H_q when the *a priori* probabilities are unknown. A suitable criterion would be a generalization of the Neyman-Pearson approach to multiple testing. However, few results are available in this field and even the concept of false alarm is not quite clear in this case. Despite the problem of choosing a reasonable criterion, the procedure always leads to a comparison of likelihood ratios (or a linear combination of likelihood ratios) to thresholds. Therefore, these are the important quantities to compute. Thus, we want to evaluate the form

$$[L(p)/L(q)] = [F(q)/F(p)]^N. \tag{29}$$

Finally, the detection tests have the following important properties: dependence only on the eigenvalues of the estimated spectral density matrix (whatever the criterion), and invariance with respect to the total received power for the noisefield. This latter property follows because if the signals received by the sensors are all multiplied by G, then the estimated spectral density is multiplied by G^2, as are its eigenvalues. Thus, (29) is unchanged. These tests will only be sensitive to the distribution of the total received power between the various eigenvalues. The set of eigenvalues is a "mirror" of the power distribution between the sources and the background noise.

An interesting test which makes use of a false alarm criterion can be established. A sequence of binary (composite) tests is performed for each value of p to decide between the following two hypotheses. The first hypothesis is the most likely of the $(K - p - 1)$ assumptions to have more than p sources.

The likelihood of this situation is

$$\mathrm{Max}_{p<q\leqslant K-1}\, \hat{L}(q) \qquad (\text{situation } \bar{H}_p).$$

The second hypothesis is the most likely of the p assumptions to have a maximum of p sources. Its likelihood is

$$\mathrm{Max}_{q\leqslant p}\, \hat{L}(q) \qquad (\text{situation } H_p).$$

The test, being the generalized likelihood ratio test, is given by

$$\frac{\mathrm{Max}_{p<q\leqslant K-1}\, \hat{L}(q)}{\mathrm{Max}_{q\leqslant p}\, \hat{L}(q)} = \Lambda(p) \underset{H_p}{\overset{\bar{H}_p}{\gtrless}} \eta_p[f_a(p)]$$

where the threshold is defined by

$$\Pr\{\Lambda(p) > \eta_p[f_a(p)] \,|\, H_0 (\text{no source})\} = f_a(p). \tag{30}$$

$f_a(p)$ may depend on p or not ("false alarm" policy). $\hat{p}$ is the value of p such that

$$\Lambda(\hat{p}) \geqslant \eta_{\hat{p}}[f_a(p)]$$

and

$$\Lambda(\hat{p} + 1) < \eta_{\hat{p}}[f_a(p)].$$

Due to the monotonic form of $\hat{L}(q)$ as a function of q (Appendix G), we have

$$\Lambda(p) = \frac{\hat{L}(K-1)}{\hat{L}(p)} = \left\{\left[\sum_{i=p+1}^{K} \hat{\lambda}_i \Big/ (K - p)\right]^{K-p} \Big/ \prod_{i=p+1}^{K} \hat{\lambda}_i\right\}^N. \tag{31}$$

Starting sequentially from $p = 0$, the likelihood of having one more source than in the preceding step is tested. The decision is controlled by the probability of detecting one more source in background noise only.

This is actually a test which concerns the repartition of the smallest eigenvalues of a covariance matrix. As the test function (31) at step p depends only on the $(K - p)$ smallest eigenvalues, the actual hypothesis tested is that the $(K - p)$ smallest eigenvalues are likely to be distributed as the eigenvalues of a "noise-only" situation. This, together with the way the thresholds are adjusted, is justified only if the distribution of the $(K - p)$ smallest eigenvalues does not depend on the p largest eigenvalues (when there are actually p sources). To a first order approximation, this seems to be justified [10].

The above test can be written equivalently as

$$T(p) = (K - p) \ln \left[\sum_{i=p+1}^{K} \hat{\lambda}_i \Big/ (K - p)\right] - \sum_{i=p+1}^{K} \ln(\hat{\lambda}_i).$$

This test, known in the time series analysis literature as a "covariance matrix test for the equality of the $(K - p)$ smallest eigenvalues," has been proposed in the underwater passive listening domain by Ligget [1] on a "heuristic" ground. We can see now in which context it can be justified.

Establishing the thresholds defined by (30) is difficult. $\hat{\Gamma}$ is distributed according to the complex Wishart distribution (if $N > K - 1$) and the theoretical distribution of the eigenvalues of $\hat{\Gamma}$ has been worked out [9] so that it is theoretically possible to solve the problem. Unfortunately, the expressions are quite untractable for practical use.

An asymptotical (approximate) theory has been developed in [10]. This approach is only valid for large N with well separated largest eigenvalues. This is not the case for low signal-to-noise ratios or closely spaced sources.

As has been pointed out above, the likelihood ratios are invariant with the total power. Thus, the decision thresholds can always be determined empirically for a given false alarm probability.

VI. Comments

Estimators for the wavefront $\hat{\vec{d}}_k$ and sources spectral levels $\hat{\gamma}_k^2$ have not been found explicitly and we shall now show that the problem is ambiguous by nature. Fortunately (but quite naturally), this ambiguity does not affect the derivation of a detection test.

It has been shown that

$$\sum_{i=1}^{p} \hat{\gamma}_i^2 \hat{\vec{d}}_i \hat{\vec{d}}_i^+ = \sum_{i=1}^{p} (\hat{\lambda}_i - \hat{\sigma}^2) \hat{\vec{\omega}}_i \hat{\vec{\omega}}_i^+ \tag{26}$$

with

$$\hat{\sigma}^2 = \frac{1}{K - p} \sum_{i=p+1}^{K} \hat{\lambda}_i \tag{25}$$

where $\hat{\vec{\omega}}_i$ is the eigenvector of $\hat{\Gamma}$ related to the eigenvalue $\hat{\lambda}_i$ (ith largest).

Equations (25) and (26) summarize the conditions which must be checked by the various estimators in order to fulfill the initial requirements.

It is known [1], [11], [12] that (26) cannot be solved completely without *a priori* information about the wavefront shapes. But the problem would be of a different nature if such information were available, since it should have been used in the model to derive the optimal conditions [13].

The solution of (26) is given by

$$\hat{\vec{d}}_k = \hat{\gamma}_k^{-1} V \Lambda^{1/2} \vec{u}_k, \qquad k \in [1, p] \tag{32}$$

where Λ is the diagonal $(p \times p)$ matrix

$$\Lambda = \text{diag}\ \{\hat{\lambda}_i - \hat{\sigma}^2\}_{i=1,p}$$

and V the $(K \times p)$ matrix of eigenvectors

$$V = \{\hat{\vec{\omega}}_i\}_{i=1,p}.$$

The p unknown vectors $(\vec{u}_k)$ $k = 1, p$ are p-component vectors which form an orthonormal system. The derivation of this result may be found in [14]. A better proof is given in Appendix H.

Provided the stipulated conditions are satisfied, any choice on $\vec{u}_k$ is suitable to optimize the likelihood. We say the problem is ambiguous.

An equation may be found relating the maximum likelihood estimates of γ_k and $\vec{d}_k$ [14].

$$\hat{\gamma}_k^2 = (\hat{\vec{d}}_k^+ V \Lambda^{-1} V^+ \hat{\vec{d}}_k)^{-1}$$

which is independent of $\vec{u}_k$ $(k = 1, p)$.

However, there is an important exception to this ambiguity situation, namely when only one source is present in the medium. In this case, the only vector $\vec{u}_k$ is reduced to a scalar which may be taken equal to one.

The optimal estimate of the only position vector $\vec{d}_1$ is then the eigenvector of $\hat{\Gamma}$ related to the largest eigenvalue. We have been able to estimate this shape with no *a priori* assumptions on the wavefront shape. This may be useful in studies of propagation models.

In the peculiar case $p = K - 1$, (25) and (26), which summarize the problem conditions, take a very compact form. From (25)

$$\hat{\sigma}^2 = \hat{\lambda}_K.$$

From (26)

$$\sum_{i=1}^{K-1} \hat{\gamma}_i^2 \hat{\vec{d}}_i \hat{\vec{d}}_i^+ = \sum_{i=1}^{K-1} (\hat{\lambda}_i - \hat{\lambda}_K) \hat{\vec{\omega}}_i \hat{\vec{\omega}}_i^+$$

then

$$\sum_{i=1}^{K-1} \hat{\gamma}_i^2 \hat{\vec{d}}_i \hat{\vec{d}}_i^+ + \hat{\sigma}^2 I$$

$$= \sum_{i=1}^{K-1} \hat{\lambda}_i \hat{\vec{\omega}}_i \hat{\vec{\omega}}_i^+ + \hat{\lambda}_K \left[I - \sum_{i=1}^{K-1} \hat{\vec{\omega}}_i \hat{\vec{\omega}}_i^+ \right]$$

yielding to

$$\tilde{\Gamma} = \hat{\Gamma}.$$

VII. Conclusion

The general conclusion of the present study is that with *a priori* hypotheses made on the noise field, the optimum processing obtained by using a generalized maximum likelihood strategy leads to the utilization of the eigenvalues and eigenvectors of the spectral density matrix estimate. More specifically, two important results have been obtained.

First, a justification has been developed for the optimality of high resolution methods. Previously, only heuristic arguments in the asymptotical case have been proposed, stating that the eigensystem of the spectral density matrix is algebraically related to the noise field parameters. It has been proved that the same relationship exists between the maximum likelihood estimates of the noise field parameters and the eigensystem of the maximum likelihood estimate of the spectral density matrix.

Secondly, detection tests for the number of sources are proposed which are independent of any assumption about the wavefront shapes. They depend only on the eigenvalues of the spectral density matrix estimate. These tests are useful for detection in complex propagation conditions as in coherent multipath situations. The determination of the number of sources is compulsory for high resolution methods to be applied. A covariance matrix test suggested by Ligget [1] for equality of the eigenvalues of the spectral density matrix has been derived exactly and the conditions in which it is justified have been made clear.

Numerous issues remain to be considered. The criterion which has been obtained for detection seems reasonable particularly in view of preliminary performance results [15]. However, at this stage, the criterion is still somewhat arbitrary due to a conceptual difficulty encountered in any multiple hypothesis testing situation. Specifically, it is necessary to define practical decision costs. A feasible way to set the decision thresholds would also need further investigation. This requires the development of good quality approximations of the eigenvalue statistics. It should be kept in mind that a small loss [15] of detection performance over conventional methods when working on undistorted wavefronts is an unescapable price of robustness. Moreover, the value of these methods is in the high spatial resolution exhibited compared to conventional methods.

Appendix A
A Matrix Determinant Identity

It is desired to prove that

$$|\Gamma| = |\Gamma_k| (1 + \gamma_k^2 \vec{d}_k^+ \Gamma_k^{-1} \vec{d}_k).$$

The matrix Γ_k is a Hermitian, positive matrix, which is therefore reducible under the Cholesky rule to

$$\Gamma_k = CC^+.$$

Now

$$\Gamma = \Gamma_k + \gamma_k^2 \vec{d}_k \vec{d}_k^+$$
$$= C(I + \gamma_k^2 C^{-1} \vec{d}_k \vec{d}_k^+ C^{-1+}) C^+$$

so that

$$|\Gamma| = |C| \, |I + \gamma_k^2 \vec{v}\vec{v}^+| \, |C^+|$$
$$= |CC^+| \, |I + \gamma_k^2 \vec{v}\vec{v}^+|$$

with

$$\vec{v} = C^{-1} \vec{d}_k.$$

The matrix $I + \gamma_k^2 \vec{v}\vec{v}^+$ has $(K - 1)$ eigenvalues equal to 1; the last, largest eigenvalue is $(1 + \gamma_k^2 \vec{v}^+ \vec{v})$ since

$$(I + \gamma_k^2 \vec{v}\vec{v}^+) \vec{v} = \vec{v} + \gamma_k^2 (\vec{v}^+ \vec{v}) \vec{v}$$
$$= (1 + \gamma_k^2 \vec{v}^+ \vec{v}) \vec{v}.$$

The determinant of any matrix is the product of its eigenvalues, so that

$$|I + \gamma_k^2 \vec{v}\vec{v}^+| = 1 + \gamma_k^2 \vec{v}^+ \vec{v}$$
$$= 1 + \gamma_k^2 (C^{-1} \vec{d}_k)^+ (C^{-1} \vec{d}_k)$$
$$= 1 + \gamma_k^2 \vec{d}_k^+ C^{-1+} C^{-1} \vec{d}_k$$

since

$$\Gamma_k = CC^+$$
$$\Gamma_k^{-1} = (C^+)^{-1} C$$

and

$$|I + \gamma_k^2 \vec{v}\vec{v}^+| = 1 + \gamma_k^2 \vec{d}_k^+ \Gamma_k^{-1} \vec{d}_k.$$

Finally,

$$|\Gamma| = |\Gamma_k| (1 + \gamma_k^2 \vec{d}_k^+ \Gamma_k^{-1} \vec{d}_k).$$

Appendix B
Source Power Equations

The maximum likelihood estimates of $(\gamma_k, \sigma, \vec{d}_k)$ are such that

$$\left. \frac{\partial L_k}{\partial \gamma_k} \right|_{\gamma_k = \hat{\gamma}_k, \sigma = \hat{\sigma}, \vec{d}_k = \hat{\vec{d}}_k} = 0.$$

The quantities $\Gamma, \Gamma_k, \delta_k$ are functions of the noisefield parameters. In order to simplify the notation, the values of these functions at the optimal point are simply written $\tilde{\Gamma}$, $\tilde{\Gamma}_k$, and $\tilde{\delta}_k$, respectively. This general notation with the "wiggle" is used throughout the main text. Thus,

$$\frac{\partial L_k}{\partial \gamma_k} = -\frac{2N\gamma_k}{\delta_k} (\vec{d}_k^+ \Gamma_k^{-1} \vec{d}_k) + \frac{2N\gamma_k}{\delta_k} \vec{d}_k^+ \Gamma_k^{-1} \hat{\Gamma} \Gamma_k^{-1} \vec{d}_k$$
$$- \frac{2N\gamma_k^3}{(\delta_k)^2} (\vec{d}_k^+ \Gamma_k^{-1} \hat{\Gamma} \Gamma_k^{-1} \vec{d}_k) \vec{d}_k^+ \Gamma_k^{-1} \vec{d}_k$$

can be written, after multiplying by $\delta_k / 2\gamma_k$, at the optimal point as

$$-N(\hat{\vec{d}}_k^+ \tilde{\Gamma}^{-1} \hat{\vec{d}}_k) + N \hat{\vec{d}}_k^+ \tilde{\Gamma}_k^{-1} \hat{\Gamma} \tilde{\Gamma}_k^{-1} \hat{\vec{d}}_k \left(1 - \frac{\hat{\gamma}_k^2 \hat{\vec{d}}_k^+ \tilde{\Gamma}_k^{-1} \hat{\vec{d}}_k}{\tilde{\delta}_k}\right) = 0.$$

From the expression for $\tilde{\delta}_k$ there results

$$\hat{\vec{d}}_k^+ \tilde{\Gamma}_k^{-1} \hat{\Gamma} \tilde{\Gamma}_k^{-1} \hat{\vec{d}}_k = \tilde{\delta}_k \hat{\vec{d}}_k^+ \tilde{\Gamma}_k^{-1} \hat{\vec{d}}_k. \tag{B1}$$

But, from Woodbury's identity

$$\tilde{\Gamma}^{-1} = \tilde{\Gamma}_k^{-1} - \frac{\hat{\gamma}_k^2}{\tilde{\delta}_k} \tilde{\Gamma}_k^{-1} \hat{\vec{d}}_k \hat{\vec{d}}_k^+ \tilde{\Gamma}_k^{-1}$$

so that

$$\tilde{\Gamma}^{-1} \hat{\vec{d}}_k = \tilde{\Gamma}_k^{-1} \hat{\vec{d}}_k \left(1 - \frac{\hat{\gamma}_k^2 \hat{\vec{d}}_k^+ \tilde{\Gamma}_k^{-1} \hat{\vec{d}}_k}{\tilde{\delta}_k}\right) = \frac{\tilde{\Gamma}_k^{-1} \hat{\vec{d}}_k}{\tilde{\delta}_k} \tag{B2}$$

with the final result

$$\hat{\vec{d}}_k^+ \tilde{\Gamma}^{-1} \hat{\Gamma} \tilde{\Gamma}^{-1} \hat{\vec{d}}_k = \hat{\vec{d}}_k^+ \tilde{\Gamma}^{-1} \hat{\vec{d}}_k.$$

Appendix C
Wavefront Shape Equations

The maximum likelihood estimates of the noise field parameters are such that

$$\vec{\nabla}_{\hat{\vec{d}}_k} (L_k) \Big|_{\vec{d}_k = \hat{\vec{d}}_k, \gamma_k = \hat{\gamma}_k, \sigma = \hat{\sigma}} = \tilde{\mu}_k \hat{\vec{d}}_k.$$

The gradient of L_k relative to $\vec{d}_k$ is

$$\vec{\nabla}_{\vec{d}_k} (L_k) = -\frac{2N\gamma_k^2}{\delta_k} \Gamma_k^{-1} \vec{d}_k - \frac{2N\gamma_k^2}{\delta_k^2} (\vec{d}_k^+ \Gamma_k^{-1} \hat{\Gamma} \Gamma_k^{-1} \vec{d}_k)$$
$$\cdot \gamma_k^2 \Gamma_k^{-1} \vec{d}_k + \frac{2N\gamma_k^2}{\delta_k} \Gamma_k^{-1} \hat{\Gamma} \Gamma_k^{-1} \vec{d}_k.$$

At the optimal point, using (B1) for the second term yields

$$\vec{\nabla}_{\hat{\vec{d}}_k}(L_k)\Big|_{\vec{d}_k=\hat{\vec{d}}_k,\gamma_k=\hat{\gamma}_k,\sigma=\hat{\sigma}}$$

$$= -2\,\frac{\gamma_k^2}{\delta_k}\,(N\tilde{\delta}_k\tilde{\Gamma}_k^{-1}\hat{\vec{d}}_k - N\tilde{\Gamma}_k^{-1}\,\hat{\Gamma}\tilde{\Gamma}_k^{-1}\hat{\vec{d}}_k) = \tilde{\mu}_k\hat{\vec{d}}_k \qquad \text{(C1)}$$

and using (B2) yields

$$-2\hat{\gamma}_k^2\tilde{\Gamma}_k^{-1}N(\hat{\vec{d}}_k - \hat{\Gamma}\tilde{\Gamma}^{-1}\hat{\vec{d}}_k) = \tilde{\mu}_k\hat{\vec{d}}_k. \qquad \text{(C2)}$$

Multiplying (C2) on the left by $\hat{\vec{d}}_k^{+}$ and using (10) gives

$$\tilde{\mu}_k = 0$$

so that

$$\hat{\Gamma}\tilde{\Gamma}^{-1}\hat{\vec{d}}_k = \hat{\vec{d}}_k. \qquad (13)$$

Equation (13) implies (10) after a left multiplication of (13) by $(\hat{\vec{d}}_k^{+}\tilde{\Gamma}^{-1})$.

Appendix D
Noise Power Equation

The maximum likelihood estimates of the noise field parameters are such that

$$\frac{\partial L}{\partial\sigma}\Big|_{\sigma=\hat{\sigma},\gamma_k=\hat{\gamma}_k,\vec{d}_k=\hat{\vec{d}}_k} = 0$$

$$\frac{\partial L}{\partial\sigma} = -N\,\frac{\partial}{\partial\sigma}\,[\ln(|\Gamma|)] - N\,\mathrm{Tr}\left(\hat{\Gamma}\,\frac{\partial\Gamma^{-1}}{\partial\sigma}\right).$$

Since

$$|\Gamma| = (\sigma^2)^{K-p}(\sigma^2+\lambda_1)\cdots(\sigma^2+\lambda_p)$$

where $\lambda_1,\cdots,\lambda_p$ do not depend on σ, then

$$\frac{\partial}{\partial\sigma}\,[\ln(|\Gamma|)] = \frac{2(K-p)\,\sigma}{\sigma^2} + \sum_{i=1}^{p}\frac{2\sigma}{\lambda_i+\sigma^2} = 2\sigma\,\mathrm{Tr}\,(\Gamma^{-1}).$$

Moreover, since

$$\Gamma\Gamma^{-1} = I$$

$$\Gamma\,\frac{\partial\Gamma^{-1}}{\partial\sigma} + \frac{\partial\Gamma}{\partial\sigma}\,\Gamma^{-1} = 0$$

and

$$\frac{\partial\Gamma}{\partial\sigma} = 2\sigma I,$$

$$\frac{\partial\Gamma^{-1}}{\partial\sigma} = -\Gamma^{-1}(2\sigma I)\,\Gamma^{-1} = -2\sigma\Gamma^{-2}.$$

Finally,

$$\frac{\partial L}{\partial\sigma} = -2N\sigma\,\mathrm{Tr}\,(\Gamma^{-1}) + 2N\sigma\,\mathrm{Tr}\,(\hat{\Gamma}\Gamma^{-2}) \qquad \text{(D1)}$$

computing

$$\hat{\Gamma}\Gamma^{-2} = \hat{\Gamma}\Gamma^{-1}(\Gamma^{-1})$$

with

$$\Gamma = \sigma^2 I + \sum_{i=1}^{p}\gamma_i^2\,\vec{d}_i\vec{d}_i^{+}.$$

To compute Γ^{-1} Woodbury's identity may be applied (Appendix E). This takes the form

$$\Gamma^{-1} = \frac{I}{\sigma^2} - \sum_{i,j=1}^{p}\alpha_{ij}\,\vec{d}_i\vec{d}_j^{+}. \qquad \text{(D2)}$$

An explicit expression of α_{ij} is not needed. Now

$$\hat{\Gamma}\Gamma^{-2} = \hat{\Gamma}\Gamma^{-1}(I/\sigma^2 - \sum_{i,j=1}^{p}\alpha_{ij}\vec{d}_i\vec{d}_j^{+})$$

$$= \frac{\hat{\Gamma}\Gamma^{-1}}{\sigma^2} - \sum_{i,j=1}^{p}\alpha_{ij}\,\hat{\Gamma}\Gamma^{-1}\,\vec{d}_i\vec{d}_j.$$

At the optimal point, using (13),

$$\hat{\Gamma}\tilde{\Gamma}^{-1}\hat{\vec{d}}_i = \hat{\vec{d}}_i \qquad (13)$$

$$\hat{\Gamma}\tilde{\Gamma}^{-2} = \frac{\hat{\Gamma}\tilde{\Gamma}^{-1}}{\hat{\sigma}^2} - \sum_{i,j=1}^{p}\tilde{\alpha}_{ij}\,\hat{\vec{d}}_i\hat{\vec{d}}_j^{+} = \frac{\hat{\Gamma}\tilde{\Gamma}^{-2}}{\hat{\sigma}^2} + \left(\tilde{\Gamma}^{-1} - \frac{I}{\hat{\sigma}^2}\right)$$

[from (D2)].

Finally,

$$\mathrm{Tr}\,(\hat{\Gamma}\tilde{\Gamma}^{-2}) = \frac{\mathrm{Tr}\,(\hat{\Gamma}\tilde{\Gamma}^{-1})}{\hat{\sigma}^2} + \mathrm{Tr}\,(\tilde{\Gamma}^{-1}) - \frac{K}{\hat{\sigma}^2}$$

and, from (D1), the equation $\partial L/\partial\sigma = 0$ leads to

$$\mathrm{Tr}\,(\hat{\Gamma}\tilde{\Gamma}^{-1}) = K. \qquad (14)$$

Appendix E
Woodbury's Identity [8]

The inversion identity is

$$(A+UV^{+})^{-1} = A^{-1} - A^{-1}\,U(I_p + V^{+}A^{-1}U)^{-1}\,V^{+}A^{-1},$$

an expression which may be easily checked for:

$A = K \times K$ matrix

U, V: any $K \times p$ matrix

I_p: $p \times p$ identity matrix.

This expression is particularly useful when $p = 1$ (U, V are vectors) wherein

$$(A + \vec{u}\vec{v}^{+})^{-1} = A^{-1} - \frac{A^{-1}\vec{u}\vec{v}^{+}A^{-1}}{1 + \vec{v}^{+}A^{-1}\vec{u}}.$$

If

$$A + UV^{+} = \sigma^2 I + \mathfrak{D}\,\gamma\,\mathfrak{D}^{+}$$

with

$\mathfrak{D} = (\vec{d}_1,\cdots,\vec{d}_p)$ a $(K \times p)$ matrix

$\gamma = \mathrm{diag}\,(\gamma_1^2,\cdots,\gamma_p^2)$ a $(p \times p)$ diagonal matrix defining $U = \mathfrak{D}\gamma$ and $V = \mathfrak{D}$

$$(A + UV^+)^{-1} = (\sigma^2 I + \mathfrak{D}\gamma\mathfrak{D}^+)^{-1} = \left(\sigma^2 I + \sum_{i=1}^{p} \gamma_i^2 \vec{d}_i \vec{d}_i^+\right)^{-1}$$

$$(A + UV^+)^{-1} = \frac{I}{\sigma^2} - \frac{1}{\sigma^2} \mathfrak{D}\gamma(\sigma^2 I_p + \mathfrak{D}^+\mathfrak{D}\gamma)^{-1}\mathfrak{D}^+.$$

Defining

$$\alpha = \frac{\gamma}{\sigma^2}(\sigma^2 I_p + \mathfrak{D}^+\mathfrak{D}\gamma)^{-1},$$

it may be written

$$(\sigma^2 I + \mathfrak{D}\gamma\mathfrak{D}^+)^{-1} = \frac{I}{\sigma^2} - \mathfrak{D}\gamma\mathfrak{D}^+$$

or, in terms of the coefficients of the matrix α

$$(\sigma^2 I + \mathfrak{D}\gamma\mathfrak{D}^+)^{-1} = \frac{I}{\sigma^2} - \sum_{i,j=1}^{p} \alpha_{ij} \vec{d}_i \vec{d}_j^+$$

$$\left(\sigma^2 I + \sum_{i=1}^{p} \gamma_i^2 \vec{d}_i \vec{d}_i^+\right)^{-1} = \frac{I}{\sigma^2} - \sum_{i,j=1}^{p} \alpha_{ij} \vec{d}_i \vec{d}_j^+.$$

This expression is used in the main text.

Appendix F
Matrix in Terms of Its Eigensystem

The matrix $\tilde{\Gamma}$ (11) can be written in term of its eigensystem as

$$\tilde{\Gamma} = \sum_{i=1}^{p} \tilde{\lambda}_i \tilde{v}_i \tilde{v}_i^+ + \hat{\sigma}^2 \sum_{i=p+1}^{K} \tilde{v}_i \tilde{v}_i^+.$$

Since the eigenvectors form an orthonormal base

$$\sum_{i=1}^{K} \tilde{v}_i \tilde{v}_i^+ = I,$$

so

$$\sum_{i=p+1}^{K} \tilde{v}_i \tilde{v}_i^+ = I - \sum_{i=1}^{p} \tilde{v}_i \tilde{v}_i^+$$

and

$$\tilde{\Gamma} = \sum_{i=1}^{p} (\tilde{\lambda}_i - \hat{\sigma}^2) \tilde{v}_i \tilde{v}_i^+ + \hat{\sigma}^2 I.$$

Therefore,

$$\sum_{i=1}^{p} \hat{\gamma}_i^2 \hat{d}_i \hat{d}_i^+ = \sum_{i=1}^{p} (\tilde{\lambda}_i - \hat{\sigma}^2) \tilde{v}_i \tilde{v}_i^+ \quad \text{(F1)}$$

by using the relations

$$\tilde{\lambda}_i = \hat{\lambda}_{e(i)}$$

$$\tilde{v}_i = \hat{\omega}_{e(i)}.$$

From (F1) we get

$$\sum_{i=1}^{p} \hat{\gamma}_i^2 \hat{d}_i \hat{d}_i^+ = \sum_{i \in E(p)} (\hat{\lambda}_i - \hat{\sigma}^2) \hat{\omega}_i \hat{\omega}_i^+.$$

The inverse of $\tilde{\Gamma}$ can be written as

$$\tilde{\Gamma}^{-1} = \sum_{i=1}^{K} \tilde{\lambda}_i^{-1} \tilde{v}_i \tilde{v}_i^+$$

$$= \sum_{i=1}^{p} \tilde{\lambda}_i^{-1} \tilde{v}_i \tilde{v}_i^+ + \hat{\sigma}^{-2}\left(I - \sum_{i=1}^{p} \tilde{v}_i \tilde{v}_i^+\right)$$

$$= \sum_{i=1}^{p} (\tilde{\lambda}_i^{-1} - \hat{\sigma}^{-2}) \tilde{v}_i \tilde{v}_i^+ + \frac{I}{\hat{\sigma}^2}.$$

Therefore,

$$\hat{\Gamma}\tilde{\Gamma}^{-1} = \sum_{i=1}^{p} (\tilde{\lambda}_i^{-1} - \hat{\sigma}^{-2}) \hat{\Gamma}\tilde{v}_i \tilde{v}_i^+ + \frac{\hat{\Gamma}}{\hat{\sigma}^2}$$

so that

$$\mathrm{Tr}(\hat{\Gamma}\tilde{\Gamma}^{-1}) = \sum_{i=1}^{p} (\tilde{\lambda}_i^{-1} - \hat{\sigma}^{-2}) \tilde{v}_i^+ \hat{\Gamma}\tilde{v}_i + \frac{\mathrm{Tr}(\hat{\Gamma})}{\hat{\sigma}^2} \quad \text{(F2)}$$

$$\tilde{v}_i^+ \hat{\Gamma} \tilde{v}_i = \hat{\lambda}_{e(i)} = \tilde{\lambda}_i$$

and

$$\mathrm{Tr}(\hat{\Gamma}) = \sum_{i=1}^{K} \hat{\lambda}_i = \sum_{i \in E(p)} \hat{\lambda}_i + \sum_{i \notin E(p)} \hat{\lambda}_i.$$

Equation (F2) becomes

$$\mathrm{Tr}(\hat{\Gamma}\tilde{\Gamma}^{-1}) = \sum_{i \in E(p)} (\hat{\lambda}_i^{-1} - \hat{\sigma}^{-2}) \hat{\lambda}_i + \sum_{i=1}^{K} \hat{\lambda}_i / \hat{\sigma}^2$$

$$= p + \frac{1}{\hat{\sigma}^2} \sum_{i \notin E(p)} \hat{\lambda}_i. \quad \text{(F3)}$$

From (14)

$$\mathrm{Tr}(\hat{\Gamma}\tilde{\Gamma}^{-1}) = K \quad \text{(14)}$$

which combined with (F3) leads to

$$\hat{\sigma}^2 = \frac{1}{K - p} \sum_{i \notin E(p)} \hat{\lambda}_i.$$

Appendix G
Monotonicity of the Generalized Likelihood

The function $\Psi(x, K - p) = x(1 - x)^{K-p}$ is defined on the domain $x \in [0, 1]$. Its maximum value is reached when $x = x_0$ for which

$$\Psi'(x_0) = (1 - x_0)^{K-p-1} [1 - (K - p + 1) x_0] = 0.$$

Then with

$$x_0 = (K - p + 1)^{-1}$$

and since $(1 - x_0)^{K-p-1} > 0$, it follows that ψ is decreasing for $x > x_0$.

The maximum value of Ψ is then

$$\Psi(x_0) = (K - p)^{(K-p)}/(K - p + 1)^{(K-p+1)}$$

so that

$$\Psi(x) \leq \frac{(K-p)^{K-p}}{(K-p+1)^{K-p+1}} \qquad \forall x \in [0, 1[.$$

The function $F(p) = F[p, E(p)]$ with $E(p) = (1, \cdots, p)$ introduced in the main text was such that

$$F[p, E(p)] = F[p-1, E(p-1)] \left[\frac{(K-p+1)^{K-p+1}}{(K-p)^{K-p}}\right] \cdot \Psi\left[\frac{\hat{\lambda}_{e(p)}}{a - \sum_{i \in E(p-1)} \hat{\lambda}_i}, (K-p)\right].$$

In other words

$$F(p) = F(p-1)\,\frac{\Psi(x)}{\Psi(x_0)}$$

with

$$x = \frac{\hat{\lambda}_{e(p)}}{a - \sum_{i \in E(p-1)} \hat{\lambda}_i} \in [0, 1[.$$

But $\Psi(x) \leq \Psi(x_0)$, then $F(p) \leq F(p-1)$, so that since the likelihood is

$$\hat{L}(p) = (\pi e)^{-KN}\,[F(p)]^{-N},$$

there results

$$\hat{L}(p) \geq \hat{L}(q) \qquad \text{if } p > q.$$

Appendix H
Proof of (27) by the SVD Theorem

Let us define the K rows, p columns matrix D

$$D = (\hat{\vec{d}}_1, \cdots, \hat{\vec{d}}_p)$$

and the diagonal $(p \times p)$ matrix

$$\gamma = \operatorname{diag}\,(\hat{\gamma}_i^2)_{i=1,p}.$$

We may then define the $(K \times p)$, rank p matrix

$$A = D\gamma^{1/2}$$

and apply the SVD (singular value decomposition) theorem to A, which says that we may write

$$A = U\Delta W^+ \tag{H1}$$

where U is $a (K \times K)$ unitary matrix, W is $a (p \times p)$ unitary matrix, and Δ is $a (K \times p)$ "diagonal" matrix

$$(\Delta)_{kl} = 0 \qquad \forall k \neq l.$$

Equation (26) may be written as

$$AA^+ = V\Lambda V^+ \tag{H2}$$

where V is a $(K \times p)$ matrix

$$V = (\hat{\vec{\omega}}_1, \cdots, \hat{\vec{\omega}}_p)$$

and Λ is a $(p \times p)$ diagonal matrix

$$\Lambda = \operatorname{diag}\,(\hat{\lambda}_i - \hat{\sigma}^2)_{i=1,p}.$$

Comparing (H_1) to (H_2),

$$AA^+ = U\Delta\Delta^+U^+ = V\Lambda V^+$$

yields to

$$U\Delta = V\Lambda^{1/2}.$$

So that we get finally

$$D = V\Lambda^{1/2}W^+\gamma^{-1/2}$$

which can be written column by column as (32). Notice that there are no conditions on W except that it be unitary. This is because there is nothing like condition (H2) concerning the product A^+A.

Previous derivation of (32) (as in [14]) may now be seen as proof of the SVD theorem.

Acknowledgment

The authors wish to thank the reviewers for their fruitful help in reaching the final form of this paper.

References

[1] W. S. Ligget, "Passive sonar: Fitting models to multiple time series," in *Proc. NATO ASI Signal Processing*, Loughborough, England, 1972, pp. 327-345.

[2] G. Bienvenu, "Influence of the spatial coherence of the background noise on high resolution passive methods," in *Proc. ICASSP '79*, Washington, DC, Apr. 2-4, 1979, pp. 306-309.

[3] N. L. Owsley, "An adaptive search and track array (ASTA)," NUSL Tech. Memo. 2242-166-69, July 7, 1969.

[4] R. O. Schmidt, "Multiple emitter location and signal parameters estimation," in *Proc. RADC Spectrum Estimation Workshop*, Oct. 1979.

[5] G. Bienvenu and L. Kopp, "Adaptivity to background noise spatial coherence for high resolution passive methods," in *Proc. ICASSP '79*, Washington, DC, Apr. 2-4, 1979, pp. 306-309.

[6] N. L. Owsley and J. F. Law, "Dominant mode power spectrum estimation," in *Proc. ICASSP '82*, Paris, France, May 3-5, 1982, pp. 775-778.

[7] H. L. Van Trees, *Detection, Estimation and Modulation Theory, Part I.* New York: Wiley, 1968.

[8] A. S. Householder, *The Theory of Matrices in Numerical Analysis.* New York: Blaisdell, 1964.

[9] A. T. James, "Distributions of matrix variates and latent roots derived from normal samples," *Ann. Math. Statist.*, vol. 35, pp. 475-501, 1964.

[10] —, "Test of equality of latent roots of the covariance matrix," in *Multivariate Analysis*, P. R. Krishnaiah, Ed. New York: Academic, 1969, pp. 205-208.

[11] H. Mermoz, "Imagerie, corrélation et modèles," *Ann. Télécommun.*, vol. 31, pp. 17-36, Jan.-Feb. 1976.

[12] G. Bienvenu and L. Kopp, "Adaptive high resolution passive methods," in *Proc. EUSIPCO '80*, Lausanne, Switzerland, Sept. 16-19, 1980, pp. 715-721.

[13] F. C. Schweppe, "Sensor-array data processing for multiple-signal sources," *IEEE Trans. Inform. Theory*, vol. IT-14, pp. 294-305, Mar. 1968.

[14] G. Bienvenu and L. Kopp, "Source power estimation method associated with high resolution bearing estimation," in *Proc. ICASSP '81*, Atlanta, GA, Mar. 30-Apr. 2, 1981, pp. 153-156.

[15] L. Kopp, G. Bienvenu, and M. Aiach, "New approach to source detection in passive listening," in *Proc. ICASSP '82*, Paris, France, May 3-5, 1982, pp. 779-782.

Part V
Algorithms and Adaptive Techniques

CONSTANT demand for reduced computation time in performing data analysis is leading to the development of fast algorithms which, in turn lead to modifications of the original techniques. Of course, reasons other than computational efficiency may motivate the development of new algorithms. These are, among others, statistical stability, reduction of sensitivity to roundoff noise, etc. In this section we have included some of the efficient algorithms. Three of them are concerned with the autoregressive (AR) analysis, reflecting its relative significance. These are papers by S. L. Marple, Jr., C. L. Nikias and P. D. Scott, and by S. M. Kay. The paper by S. Levy *et al.* introduces a linear programming approach to spectrum estimation. Paper by Cadzow *et al.*, which presents a singular value decomposition approach to signal modeling is included. This approach is the basis for a number of recently published algorithms.

Adaptive techniques are used to process data samples one at a time, in an attempt to perform a real-time computation. A typical example of the need for such processing is tracking of signals. Another aspect of adaptivity is that the certain nonstationary signal environments can be handled. The last three papers in this section present some of the adaptive techniques. W. F. Gabriel's paper summarizes some of the most frequently used methods. V. U. Reddy *et al.* treat the Pisarenko algorithm, while the paper by Bowyer *et al.* shows an example of adaptive AR procedure. The latter is also a fine example of the application of spectral estimation in radar clutter filtering.

A New Autoregressive Spectrum Analysis Algorithm

LARRY MARPLE

Abstract—A new recursive algorithm for autoregressive (AR) spectral estimation is introduced, based on the least squares solution for the AR parameters using forward and backward linear prediction. The algorithm has computational complexity proportional to the process order squared, comparable to that of the popular Burg algorithm. The computational efficiency is obtained by exploiting the structure of the least squares normal matrix equation, which may be decomposed into products of Toeplitz matrices. AR spectra generated by the new algorithm have improved performance over AR spectra generated by the Burg algorithm. These improvements include less bias in the frequency estimate of spectral components, reduced variance in frequency estimates over an ensemble of spectra, and absence of observed spectral line splitting.

I. Introduction

Autoregressive (AR) spectrum analysis, sometimes termed maximum entropy spectrum analysis (MESA), has become a popular alternative to the periodogram as an estimate of the power spectral density (PSD) for a sampled process. For signal-to-noise ratios (SNR's) greater than 0 dB, the AR PSD estimate has better frequency resolution than that of the conventional periodogram estimate [1]. Autoregressive spectral estimates also do not have the distortion produced by sidelobe leakage effects that are inherent in the periodogram approach to spectrum analysis. For short data records, the AR method yields reasonable spectral estimates. These are three of several attractive features of AR spectral estimation that have created interest in this technique.

The method used to estimate the autoregressive model parameters is the key to the performance of the AR technique. If $M + 1$ lags of the autocorrelation function for a process are known, the M autoregressive parameters are obtained by solving the Yule-Walker normal equations using the Levinson recursion algorithm [2]. The algorithm requires a number of computational operations proportional to M^2.

A host of techniques are available for estimating the AR parameters from data samples. The most obvious approach is to first make estimates of the autocorrelation lags with the available data, and then to apply the usual Levinson recursion using the estimated lags. This approach is rarely used since better resolution may be obtained with direct AR parameter estimation methods. If unbiased autocorrelation estimates are used, one may also run into numerical ill-conditioning during the solution of the normal equations. Biased autocorrelation estimates reduce the risk of ill-conditioning, but at the expense of a degradation of the AR spectral resolution and a shifting of spectral peaks from their true locations [1]. The shift effect is termed a frequency estimation bias. A third reason that has made this a seldom-used technique is the problem of spectral line splitting. Spectral line splitting is the occurrence of two or more closely-spaced peaks in an AR spectral estimate where only one spectral peak should be present. The reasons for spectral line splitting in the Yule-Walker technique have been documented by Kay and Marple [3].

The most popular approach for AR parameter estimation is the Burg algorithm [4], [5]. This algorithm utilizes a constrained least squares estimation procedure to obtain the M estimated autoregressive parameters from N data samples. The constraint requires the AR parameter estimates to satisfy the Levinson recursion. The Burg algorithm requires computational operations proportional to the product NM.

AR spectral estimates based on the Burg algorithm suffer from two of the same problems observed in Yule-Walker estimates of the AR spectrum. The problem of spectral line splitting in AR spectra produced by the Burg algorithm was first documented by Fougere *et al.* [6]. They noted that spectral line splitting was most likely to occur when 1) the SNR is high, 2) the initial phase of sinusoidal components is some odd multiple of 45°, 3) the time duration of the data sequence is such that sinusoidal components have an odd number of quarter cycles, and 4) the number of AR parameters estimated is a large percentage of the number of data values used for the estimation. Many spurious spectral peaks often accompany spectra that exhibit line splitting.

The connection between line splitting and the number of AR parameters estimated (model order) highlights a problem area common to all methods of AR spectrum analysis—the selection of an AR model order. Akaike [7] has provided the FPE and AIC criteria for AR order determination that are perhaps the most well-known criteria. However, this author's experience has shown that most order selection rules, including Akaike's, are not always effective in preventing the line splitting phenomenon.

A second major problem area with the Burg algorithm, as with the Yule-Walker case, is the bias in the positioning of spectral peaks with respect to the true frequency location of those peaks. Swingler [8] has shown that this bias can pull a peak away from the true frequency by as much as 16 percent of a resolution cell (a resolution cell = $1/(N\Delta t)$ Hz, where Δt is the sample interval).

In order to alleviate the spectral line splitting problem, Fougere [9] devised a rather complicated gradient descent

Manuscript received August 14, 1979; revised March 5, 1980.
The author is with the Washington Systems Engineering Division, The Analytic Sciences Corporation, McLean, VA 22102.

Reprinted from *IEEE Trans. Acoust., Speech, Signal Processing*, vol. ASSP-28, pp. 441–454, Aug. 1980.

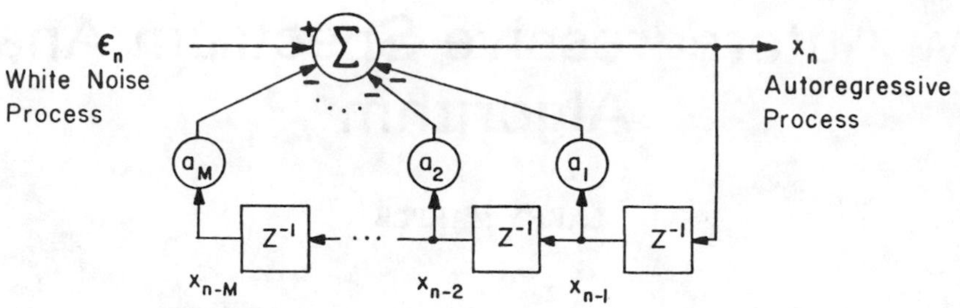

Fig. 1. Autoregressive process block diagram.

algorithm for AR parameter estimation. The algorithm has been shown to work for selected one and two sinusoid examples, but it is an iterative procedure that requires a much higher computational effort than the popular Burg algorithm.

This paper presents a new algorithm for AR parameter estimation that yields AR spectra with no apparent line splitting and reduced spectral peak frequency estimation biases. A set of sensitive stopping rules for order selection has been found for the algorithm. The method is based on an unconstrained least squares estimation of the AR parameters proposed independently by Ulrych and Clayton [10], who termed it the least squares (LS) AR spectral estimate, and Nuttall [11], who termed it the forward and backward prediction method. In their experiments with the LS estimate, Ulrych and Clayton observed, for processes with one and two sinusoids in noise, that LS-generated spectral estimates had less variation of the spectral peaks from their actual frequencies (as a function of initial phase) than Burg algorithm spectra. Nuttall compared the LS spectral estimate, for a large ensemble of sampled autoregressive processes, to other AR spectral estimates, including the Burg estimate, and found the LS estimate to be as good as, and often better than, the other estimators. In fact, among all AR estimation techniques examined by Nuttall, the LS method exhibited the least variation in frequency over an ensemble of spectra.

A straightforward matrix inversion solution of the linear simultaneous equations for this LS method of AR parameter estimation has been the usual computational approach. This requires a number of computational operations proportional to NM^2, making it computationally unattractive relative to the more efficient Burg algorithm. This paper presents an algorithm for the solution of the LS linear equations with a computational complexity proportional to NM, making it comparable to that of the Burg algorithm.

II. Yule-Walker Estimate of the AR Spectrum

The use of an AR spectral estimate tacitly assumes that the underlying model for the sampled process is an autoregression, as shown in Fig. 1. The observed complex sequence $x_1, \cdots, x_N$ is the output from an Mth order AR model driven by a white noise process ϵ_n. Mathematically, the current output sample x_n is a weighted sum of M past output samples plus a noise term

$$x_n = -\sum_{m=1}^{M} a_{M,m} x_{n-m} + \epsilon_n \tag{1}$$

where $a_{M,m}$ is AR parameter m of the Mth order AR process. In terms of vectors,

$$\epsilon_n = X_{M,n}^T A_M \tag{2}$$

where $A_M^T = [1, a_{M,1}, \cdots, a_{M,M}]$, $X_{M,n}^T = [x_n, x_{n-1}, \cdots, x_{n-M}]$, and T denotes vector transpose. If each side of (2) is premultiplied by the vector $X_{M,n}^*$, where "*" denotes complex conjugate, and the expected value taken, then

$$\Phi_M A_M = P_M \tag{3}$$

where

$$\Phi_M = E[X_{M,n}^* X_{M,n}^T] = \begin{bmatrix} \phi_0 & \phi_1 & \cdots & \phi_M \\ \phi_1^* & \phi_0 & \cdots & \phi_{M-1} \\ \vdots & \vdots & & \vdots \\ \phi_M^* & \phi_{M-1}^* & \cdots & \phi_0 \end{bmatrix}$$

$= (M+1) \times (M+1)$ Toeplitz autocorrelation matrix

$\phi_i = E[x_j x_{j+i}^*]$ = autocorrelation function at lag i

($\phi_i^* = \phi_{-i}$ for a stationary process)

$P_M = [p_M, 0, \cdots, 0]^T$

$p_M = E[\epsilon_n \epsilon_n^*]$ = white noise power spectral density.

By definition, the white noise ϵ_n is uncorrelated with all x_m for $m < n$; thus $E[\epsilon_n x_n^*] = E[\epsilon_n \epsilon_n^*]$. The set of linear equations represented by (3) is often called the Yule-Walker normal equations.

The output power spectral density $S_x(f)$ is related to the input noise power spectral density $S_\epsilon(f) = p_M$ by

$$S_x(f) = p_m \Bigg/ \left| 1 + \sum_{m=1}^{M} a_{M,m} \exp(-j2\pi f m \Delta t) \right|^2 \tag{4}$$

where Δt is the autocorrelation lag interval in seconds. Expression (4) is popularly termed the autoregressive, or maximum entropy, spectral estimate. Given the $M+1$ autocorrelation lags $\phi_0, \cdots, \phi_M$, the solution for the AR parameter vector A_M in (3) is provided by the recursive Levinson algorithm, in which the AR parameters for order M may be obtained recursively from the AR parameters previously determined for order $M-1$

$$a_{M,k} = a_{M-1,k} + a_{M,M} a_{M-1,M-k}^* \quad \text{for } k = 1, \cdots, M-1 \tag{5}$$

where $a_{M,M}$ is given by

$$a_{M,M} = -\sum_{k=0}^{M-1} a_{M-1,k}\phi^*_{M-k} \Big/ p_{M-1}. \tag{6}$$

Recursion (5) is obtained by taking advantage of the positive-definite Toeplitz structure of the autocorrelation matrix. The recursion for the noise spectral density is given by

$$p_M = p_{M-1}[1 - |a_{M,M}|^2]. \tag{7}$$

Relationship (5) and the fact that $|a_{M,M}| \leq 1$ are sufficient to guarantee that the roots of the polynomial

$$1 + \sum_{m=1}^{M} a_{M,m} z^{-m} = 0$$

are within the unit Z-plane circle, i.e., the modulus of each pole of (4) is less than unity. This guarantees that a synthesis of an AR filter using the AR parameters computed by the Levinson recursion will be a stable, minimum-phase filter.

If only N data samples, rather than actual autocorrelation function lags, are available, then biased lag estimates

$$\hat{\phi}_m = \frac{1}{N}\sum_{n=0}^{N-m} x_n x^*_{n+m} \qquad \text{for} \quad m = 0, \cdots, M \tag{9}$$

are typically computed when selecting the Yule-Walker approach for AR parameter estimation. Use of (9) guarantees a positive-definite autocorrelation matrix. This is termed the autocorrelation method of linear prediction analysis [14].

III. Burg Algorithm Estimate of the AR Spectrum

A popular alternative approach for AR parameter estimation with data samples was introduced by Burg in 1968. The Burg algorithm may be viewed as a constrained least squares estimation procedure using the sum of the forward and backward linear prediction error energies. Assuming an all-pole stationary stochastic process, the forward linear prediction error $f_{M,k}$ is given by

$$f_{M,k} = x_{k+M} + \sum_{i=1}^{M} a_{M,i}x_{k+M-i} = \sum_{i=0}^{M} a_{M,i}x_{k+M-i} \tag{10}$$

for $1 \leq k \leq N - M$ and the backward linear prediction error $b_{M,k}$ is given by

$$b_{M,k} = \sum_{i=0}^{M} a^*_{M,i}x_{k+i}, \tag{11}$$

also for $1 \leq k \leq N - M$. Note that complex-valued data is assumed and $a_{n,0}$ is defined as unity. Since stationarity was assumed, the backward AR coefficients are simply the conjugates of the forward AR coefficients.

To obtain estimates of the AR parameters, Burg minimized the sum of the backward and forward prediction error energies

$$e_M = \sum_{k=1}^{N-m} |f_{M,k}|^2 + \sum_{k=1}^{N-m} |b_{M,k}|^2 \tag{12}$$

subject to the constraint that the AR parameters satisfy the Levinson recursion, (5), for all orders from 1 to M. This constraint was motivated by Burg's desire to guarantee a stable AR filter (poles within the unit circle). A close evaluation of (5) and (12) will show that the prediction error energy e_M at order M is a function of $a_{M,M}$, which is unknown, and the AR parameters $a_{M-1,1}$ through $a_{M-1,M-1}$, which will always be known from order $(M - 1)$ of the recursion. Thus (12) is a function of the single parameter $a_{M,M}$, often termed a reflection coefficient for its physical meaning in seismological applications.

Substitution of (5) into (10) and (11) yields the following recursive relationships between the forward and backward prediction errors:

$$f_{M,k} = f_{M-1,k+1} + a_{M,M}b_{M-1,k} \tag{13}$$

$$b_{M,k} = b_{M-1,k} + a^*_{M,M}f_{M-1,k+1} \tag{14}$$

for $1 \leq k \leq N - M$. Substituting (13) and (14) into (12) and minimizing with respect to $a_{M,M}$ (set derivative to zero), one obtains

$$a_{M,M} = -\frac{2\sum_{k=1}^{N-M} b^*_{M-1,k}f_{M-1,k+1}}{\text{DEN}_M} \tag{15}$$

where

$$\text{DEN}_M = \sum_{k=1}^{N-M} [|b_{M-1,k}|^2 + |f_{M-1,k+1}|^2]. \tag{16}$$

It is possible to show using (15) that $|a_{M,M}| \leq 1$, a necessary condition for a minimum-phase linear prediction filter. Anderson [12] discovered the following recursive relationship for the denominator term DEN_M:

$$\text{DEN}_M = [1 - |a_{M-1,M-1}|^2]\,\text{DEN}_{M-1} - |b_{M-1,N-M+1}|^2 - |f_{M-1,1}|^2, \tag{17}$$

that reduces the computational effort for $a_{M,M}$.

Since the Levinson recursion is maintained in the Burg algorithm, then

$$e_M = e_{M-1}[1 - |a_{M,M}|^2]. \tag{18}$$

Fig. 2 is a flowchart of the Burg algorithm. A computational complexity analysis of the Burg algorithm indicates that $3NM - M^2 - 2N - M$ complex adds, $3NM - M^2 - N + 3M$ complex multiplications, and M real divisions are required. The algorithm requires storage of $3N + M + 2$ complex words.

IV. The Marple Least Squares Algorithm Estimate of the AR Spectrum

In this section a recursive algorithm that provides the exact least squares solution, without matrix inversion, for the AR parameter estimates is outlined. The Levinson recursion constraint is removed. The algorithm will be shown to be as computationally efficient as the Burg algorithm for most practical situations.

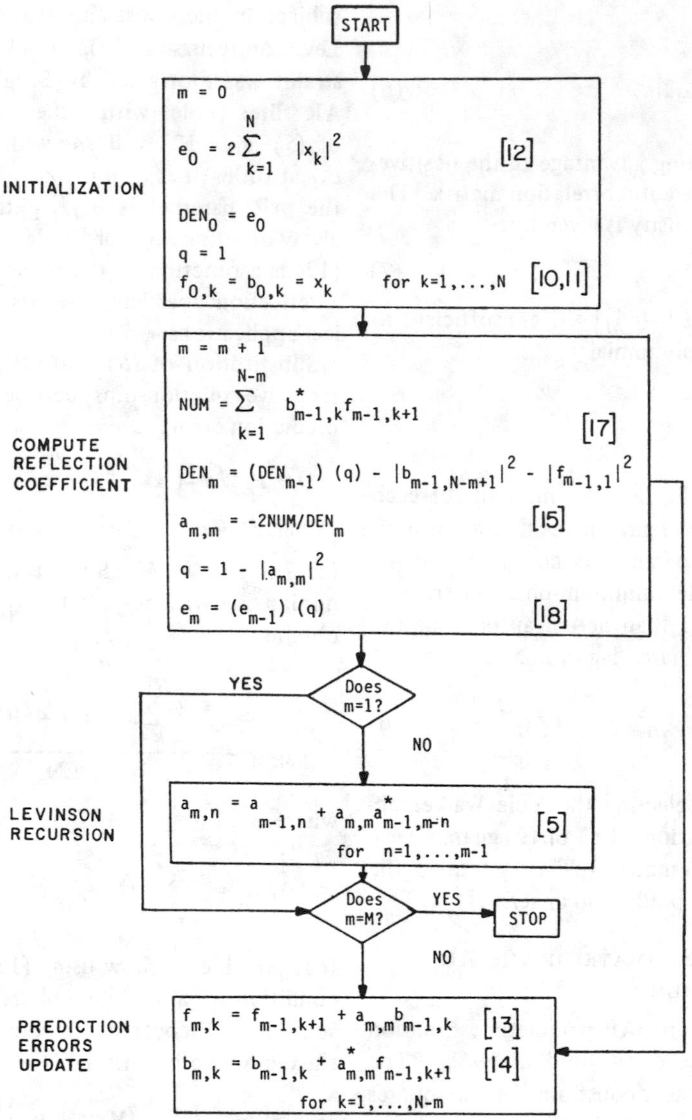

Fig. 2. Burg algorithm flowchart. Numbers in brackets refer to equations in text.

If (10) and (11) are substituted into (12), the least squares minimum for e_M is found by setting the derivatives of e_M with respect to the AR parameters $a_{M,1}$ through $a_{M,M}$ to zero,

$$\frac{\partial e_M}{\partial a_{M,i}} = 2 \sum_{j=0}^{M} a_{M,j} r_M(i,j) = 0 \tag{19}$$

for $i = 1, \cdots, M$ ($a_{M,o} = 1$ by definition), where

$$r_M(i,j) = \sum_{k=1}^{N-M} (x_{k+M-j} x^*_{k+M-i} + x^*_{k+j} x_{k+i}) \tag{20}$$

for $0 \leq i, j \leq M$. Using the M normal equations (19) and definitions (10) and (11), the minimum prediction error energy is found to be

$$e_M = \sum_{j=0}^{M} a_{M,j} r_M(0,j). \tag{21}$$

Expressions (19) and (21) can be combined into a single $(M+1) \times (M+1)$ matrix expression

$$R_M A_M = E_M \tag{22}$$

where

$$A_M = \begin{bmatrix} 1 \\ a_{M,1} \\ \vdots \\ a_{M,M} \end{bmatrix}, \quad E_M = \begin{bmatrix} e_M \\ 0 \\ \vdots \\ 0 \end{bmatrix},$$

$$R_M = \begin{bmatrix} r_M(0,0) & \cdots & r_M(0,M) \\ \vdots & & \vdots \\ r_M(M,0) & \cdots & r_M(M,M) \end{bmatrix}. \tag{23}$$

Ulrych and Clayton [10] and Nuttall [11] independently proposed this least squares approach for AR parameter estimation. Both found the LS estimate by computation of the $r_M(i,j)$ terms directly and then by solution of (22) for vector A_M using matrix inversion. This requires on the order of M^3 computational operations. The Burg algorithm requires on the order of M^2 operations. This gives it a computational advantage over the LS estimate if (22) is solved by general matrix inversion techniques.

Expression (22), though, has a structure that can be exploited to produce an algorithm requiring a number of operations proportional to M^2 rather than M^3. The remainder of this section will establish this LS algorithm. The approach presented here was motivated by a similar algorithm developed by Morf *et al.* [13]. Examination of (20) shows that R_M has both Hermitian symmetry $[r_M(i,j) = r_M^*(j,i)]$ and Hermitian persymmetry $[r_M(i,j) = r_M^*(M-i, M-j)]$. It does *not* have the Toeplitz symmetry $[r_M(i,j) = r_M(i-j)]$, as (3) has, that would permit solution with the Levinson recursion procedure. However, upon closer examination, R_M has a special structure composed of Toeplitz matrices

$$R_M = (T_M)^H T_M + (T_M^\nu)^H T_M^\nu. \tag{24}$$

Matrix T_M is an $(N-M) \times (M+1)$ Toeplitz matrix of data samples,

$$T_M = \begin{bmatrix} x_{M+1} & x_M & \cdots & x_1 \\ x_{M+2} & x_{M+1} & \cdots & x_2 \\ \vdots & \vdots & & \vdots \\ x_N & x_{N-1} & \cdots & x_{N-M} \end{bmatrix}, \tag{25}$$

with T_M^ν denoting the conjugated and reversed matrix

$$T_M^\nu = \begin{bmatrix} x_1^* & \cdots & x_{M+1}^* \\ \vdots & & \vdots \\ x_{N-M}^* & \cdots & x_N^* \end{bmatrix} \tag{26}$$

and H denoting the complex conjugate transpose operation. Thus, R_M has a structure composed of the sum of two Toeplitz data matrix products. It is this underlying structure that allows a recursive algorithm of order M^2 operations to be generated.

To exploit the special structure, it is first necessary to introduce two additional prediction error energy terms

$$e_M' = \sum_{k=1}^{N-M-1} \left[|f_{M,k+1}|^2 + |b_{M,k}|^2\right] \tag{27}$$

and

$$e_M'' = \sum_{k=1}^{N-M-1} \left[|f_{M,k}|^2 + |b_{M,k+1}|^2\right]. \tag{28}$$

The terms e_M' and e_M'' are time-index-shifted variants of the definition (12) for e_M, in which one error term index has been altered by one. Minimizing e_M' and e_M'' in a manner similar to that used for e_M, the following expressions, comparable to (22) and (23), were obtained

$$R_M' A_M' = E_M' \tag{29}$$

$$R_M'' A_M'' = E_M'' \tag{30}$$

where

$$A_M' = \begin{bmatrix} 1 \\ a_{M,1}' \\ \vdots \\ a_{M,M}' \end{bmatrix}, \quad E_M' = \begin{bmatrix} e_M' \\ 0 \\ \vdots \\ 0 \end{bmatrix}, \quad A_M'' = \begin{bmatrix} 1 \\ a_{M,1}'' \\ \vdots \\ a_{M,M}'' \end{bmatrix}, \quad E_M'' = \begin{bmatrix} e_M'' \\ 0 \\ \vdots \\ 0 \end{bmatrix}. \tag{31}$$

The elements of R_M' and R_M'' are

$$r_M'(i,j) = \sum_{k=1}^{N-M-1} \left[x_{k+M+1-j} x_{k+M+1-i}^* + x_{k+j}^* x_{k+i}\right] \tag{32}$$

and

$$r_M''(i,j) = \sum_{k=1}^{N-M-1} \left[x_{k+M-j} x_{k+M-i}^* + x_{k+1+j}^* x_{k+1+i}\right] \tag{33}$$

for $0 \leqslant i, j \leqslant M$. From these, it is easy to see the persymmetry relationship $r_M'(i,j) = [r_M''(M-i, M-j)]^*$.

The following relationships exist among the correlation matrices R_M, R_M', and R_M'':

$$R_M' = R_M - \begin{pmatrix} x_{M+1}^* \\ \vdots \\ x_1^* \end{pmatrix} (x_{M+1}, \cdots, x_1) - \begin{pmatrix} x_{N-M} \\ \vdots \\ x_N \end{pmatrix} (x_{N-M}^*, \cdots, x_N^*) \tag{34}$$

$$R_M'' = R_M - \begin{pmatrix} x_N^* \\ \vdots \\ x_{N-M}^* \end{pmatrix} (x_N, \cdots, x_{N-M}) - \begin{pmatrix} x_1 \\ \vdots \\ x_{M+1} \end{pmatrix} (x_1^*, \cdots, x_{M+1}^*) \tag{35}$$

$$R_{M+1} = \left[\begin{array}{ccc|c} & R_M' & & r_{M+1}(0, M+1) \\ \hline & & & \vdots \\ r_{M+1}(M+1, 0) & \cdots & & r_{M+1}(M+1, M+1) \end{array}\right] \tag{36}$$

$$R_{M+1} = \left[\begin{array}{c|ccc} r_{M+1}(0,0) & \cdots & & r_{M+1}(0, M+1) \\ \hline \vdots & & & \\ r_{m+1}(M+1, 0) & & R_M'' & \end{array}\right]. \tag{37}$$

The auxiliary $(M+1)$ element column vectors

$$C_M = \begin{bmatrix} c_{M,0} \\ \vdots \\ c_{M,M} \end{bmatrix}, \quad C_M'' = \begin{bmatrix} c_{M,0}'' \\ \vdots \\ c_{M,M}'' \end{bmatrix}, \quad D_M = \begin{bmatrix} d_{M,0} \\ \vdots \\ d_{M,M} \end{bmatrix},$$

$$D_M'' = \begin{bmatrix} d_{M,0}'' \\ \vdots \\ d_{M,M}'' \end{bmatrix}$$

will be required for the algorithm. They are defined by the following matrix-vector products:

$$R_M C_M = \begin{pmatrix} x_{M+1}^* \\ \vdots \\ x_1^* \end{pmatrix} \tag{38}$$

$$R_M'' C_M'' = \begin{pmatrix} x_{M+1}^* \\ \vdots \\ x_1^* \end{pmatrix} \tag{39}$$

$$R_M D_M = \begin{pmatrix} x_{N-M} \\ \vdots \\ x_N \end{pmatrix} \tag{40}$$

$$R_M'' D_M'' = \begin{pmatrix} x_{N-M} \\ \vdots \\ x_N \end{pmatrix}. \tag{41}$$

The notation A_M^I denotes the column vector formed from A_M by reversing the element order and conjugating

$$A_M^I = \begin{pmatrix} a_{M,M}^* \\ \vdots \\ a_{M,1}^* \\ 1 \end{pmatrix}. \tag{42}$$

The vectors E_M^I, C_M^I, and D_M^I are defined accordingly. The following relationships will then hold as a result of the Hermetian persymmetry of R_M

$$R_M A_M^I = E_M^I \tag{43}$$

$$R_M C_M^I = \begin{pmatrix} x_1 \\ \vdots \\ x_{M+1} \end{pmatrix} \tag{44}$$

$$R_M D_M^I = \begin{pmatrix} x_N^* \\ \vdots \\ x_{N-M}^* \end{pmatrix}. \tag{45}$$

From (10) and (11), two forward and backward prediction error scalar terms to be used in the algorithm may be expressed in vector notation as

$$f_{M,1} = (x_{M+1}, \cdots, x_1) A_M \tag{46}$$

and

$$b_{M,N-M} = (x_{N-M}, \cdots, x_N) A_M^*. \tag{47}$$

Other scalar definitions required by the algorithm are

$$g_M = (x_{M+1}, \cdots, x_1) C_M \tag{48}$$

$$h_M = (x_{N-M}^*, \cdots, x_N^*) C_M \tag{49}$$

$$s_M = (x_N, \cdots, x_{N-M}) C_M \tag{50}$$

$$u_M = (x_N, \cdots, x_{N-M}) D_M \tag{51}$$

$$v_M = (x_1^*, \cdots, x_{M+1}^*) C_M \tag{52}$$

$$w_M = (x_{N-M}^*, \cdots, x_N^*) D_M. \tag{53}$$

Several useful scalar identities can be derived at this point. The vector product $A_M^H R_M C_M$ represents a complex-valued scalar. From the identity

$$A_M^H R_M C_M = (C_M^H R_M^H A_M)^*, \tag{54}$$

one may obtain

$$A_M^H \begin{pmatrix} x_{M+1}^* \\ \vdots \\ x_1^* \end{pmatrix} = \left[C_M^H \begin{pmatrix} e_M \\ 0 \\ \vdots \\ 0 \end{pmatrix} \right]^* \tag{55}$$

by substituting (22) and (38) for $R_M^H A_M$ and $R_M C_M$, and noting that $R_M^H = R_M$ since there is Hermitian symmetry. Expression (55) is further reduced to

$$c_{M,0} = f_{M,1}^* / e_M \tag{56}$$

by substituting definition (46). Similarly, the identities

$$A_M^H R_M D_M = (D_M^H R_M^H A_M)^* \tag{57}$$

$$D_M^H R_M C_M = (C_M^H R_M^H D_M)^* \tag{58}$$

$$D_M^H R_M C_M^I = (C_M^{IH} R_M^H D_M)^* \tag{59}$$

$$C_M^H R_M C_M = (C_M^H R_M^H C_M)^* \tag{60}$$

$$D_M^H R_M D_M = (D_M^H R_M^H D_M)^* \tag{61}$$

yield, respectively,

$$d_{M,0} = b_{M,N-M} / e_M \tag{62}$$

$$h_M^* = (x_{M+1}, \cdots, x_1) D_M \tag{63}$$

$$s_M = (x_1^*, \cdots, x_{M+1}^*) D_M \tag{64}$$

$$g_M^* = g_M \tag{65}$$

$$w_M^* = w_M. \tag{66}$$

Thus, g_M and w_M are real-valued scalars. The relationship for the time-shift update (a term used by Morf *et al.* [13] to describe the change in index of definitions e_M' and e_M'') for A_M' is

$$A_M' = \alpha_M (A_M + \beta_1 C_M + \gamma_1 D_M). \tag{67}$$

Premultiplying (67) by R_M' and substituting relationship (34) for R_M' in the multiplication on the right-hand side of (67) yields

$$\begin{pmatrix} e_M' \\ 0 \\ \vdots \\ 0 \end{pmatrix} = \alpha_M \left[\begin{pmatrix} e_M \\ 0 \\ \vdots \\ 0 \end{pmatrix} + [-f_{M,1} + (1 - g_M)\beta_1 - \gamma_1 h_M^*] \begin{pmatrix} x_{M+1}^* \\ x_M^* \\ \vdots \\ x_1^* \end{pmatrix} + \begin{pmatrix} x_{N-M} \\ x_{N-M+1} \\ \vdots \\ x_N \end{pmatrix} [-b_{M,N-M}^* - \beta_1 h_M + (1 - w_M)\gamma_1] \right]. \tag{68}$$

The relationship

$$e_M' = \alpha_M e_M \tag{69}$$

must hold, which means

$$(1 - g_M)\beta_1 - h_M^* \gamma_1 - f_{M,1} = 0$$
$$-h_M \beta_1 + (1 - w_M)\gamma_1 - b_{M,N-M}^* = 0. \tag{70}$$

Solving the set of linear equations (70) for α_M, β_1, and γ_1,

$$\beta_1 = [f_{M,1}(1 - w_M) + b_{M,N-M}^* h_M^*] / \text{DEN}_M \tag{71}$$
$$\gamma_1 = [b_{M,N-M}^*(1 - g_M) + f_{M,1} h_M] / \text{DEN}_M \tag{72}$$

where

$$\text{DEN}_M = (1 - g_M)(1 - w_M) - |h_M|^2. \tag{73}$$

Noting that the first elements of both A_M and A_M' must be unity, α_M may be determined from the relationship

$$1 = \alpha_M [1 + \beta_1 c_{M,0} + \gamma_1 d_{M,0}]. \tag{74}$$

Solving for α_M, after substitution of (56), (62), (71), and (72), yields

$$\alpha_M = \left[1 + \frac{|f_{M,1}|^2(1 - w_M) + |b_{M,N-M}|^2(1 - g_M) + 2\,\text{Real}\,(f_{M,1} h_M b_{M,N-M})}{e_M\,\text{DEN}_M}\right]^{-1}. \tag{75}$$

The relationship for the time-shift update of C_M'' is

$$C_M'' = C_M + \beta_2 C_M^I + \gamma_2 D_M^I. \tag{76}$$

Premultiplying (76) by R_M'' and substituting time-shift relationship (35) for R_M'' in the product on the right-hand side of (76) yields

$$\begin{pmatrix} x_{M+1}^* \\ \vdots \\ x_1^* \end{pmatrix} = \begin{pmatrix} x_{M+1}^* \\ \vdots \\ x_1^* \end{pmatrix} + \underbrace{[-s_M - \beta_2 h_M^* + \gamma_2(1 - w_M)]}_{\text{zero}} \begin{pmatrix} x_N^* \\ \vdots \\ x_{N-M}^* \end{pmatrix}$$
$$+ \underbrace{[-v_M + \beta_2(1 - g_M) - \gamma_2 h_M]}_{\text{zero}} \begin{pmatrix} x_1 \\ \vdots \\ x_{M+1} \end{pmatrix}. \tag{77}$$

The indicated terms must be zero for equality to hold. The solution for β_2 and γ_2 is then

$$\beta_2 = [s_M h_M + v_M(1 - w_M)] / \text{DEN}_M \tag{78}$$
$$\gamma_2 = [v_M h_M^* + s_M(1 - g_M)] / \text{DEN}_M. \tag{79}$$

Finally, the time-shift update of D_M'' has the form

$$D_M'' = D_M + \beta_3 C_M^I + \gamma_3 D_M^I. \tag{80}$$

Premultiplying (80) by R_M'' and substituting time-shift relationship (35) for R_M'' in the product of the right-hand side of expression (80), then

$$\begin{pmatrix} x_{N-M} \\ \vdots \\ x_N \end{pmatrix} = \begin{pmatrix} x_{N-M} \\ \vdots \\ x_N \end{pmatrix} + \underbrace{[-u_M - \beta_3 h_M^* + \gamma_3(1 - w_M)]}_{\text{zero}} \begin{pmatrix} x_N^* \\ \vdots \\ x_{N-M}^* \end{pmatrix}$$
$$+ \underbrace{[-s_M + \beta_3(1 - g_M) - \gamma_3 h_M]}_{\text{zero}} \begin{pmatrix} x_1 \\ \vdots \\ x_{M+1} \end{pmatrix}. \tag{81}$$

Again, the indicated terms must be zero for equality to hold. The solution for β_3 and γ_3 is

$$\beta_3 = [u_M h_M + s_M(1 - w_M)] / \text{DEN}_M \tag{82}$$
$$\gamma_3 = [s_M h_M^* + u_M(1 - g_M)] / \text{DEN}_M. \tag{83}$$

The relationship for the order update of A_M was found to be of the form

$$A_{M+1} = \begin{bmatrix} A_M' \\ 0 \end{bmatrix} + \alpha_2 \begin{bmatrix} 0 \\ (A_M')^I \end{bmatrix}. \tag{84}$$

Multiplying (84) by R_{M+1} and substituting either order relationship (36) or order relationship (37), as appropriate, for R_{M+1} in each of the two product terms on the right-hand side of (84) yields the expression

$$\begin{pmatrix} e_{M+1} \\ 0 \\ \vdots \\ 0 \\ 0 \end{pmatrix} = \begin{pmatrix} e_M' \\ 0 \\ \vdots \\ 0 \\ \Delta_{M+1} \end{pmatrix} + \alpha_2 \begin{pmatrix} \Delta_{M+1}^* \\ 0 \\ \vdots \\ 0 \\ e_M' \end{pmatrix} \tag{85}$$

where Δ_{M+1} is defined as

$$\Delta_{M+1} = [r_{M+1}(M+1, 0), \cdots, r_{M+1}(M+1, M)]\, A_M'. \tag{86}$$

For equality to hold in (85), then

$$\Delta_{M+1} + \alpha_2 e_M' = 0 \tag{87}$$

and

$$e_M' + \alpha_2 \Delta_{M+1}^* = e_{M+1}. \tag{88}$$

The solution is

$$\alpha_2 = a_{M+1,M+1} = -\Delta_{M+1}/e_M' \tag{89}$$

and

$$e_{M+1} = e_M'[1 - |a_{M+1,M+1}|^2]. \tag{90}$$

Note that $|a_{M+1,M+1}|^2 < 1$ since e_{M+1} and e_M' must be positive by definition.

Based on (84) and (89), the AR parameters satisfy the recursive relationship

$$a_{M+1,i} = a_{M,i}' + a_{M+1,M+1}(a_{M,M+1-i}')^* \tag{91}$$

for $i = 1, \cdots, M$. Equation (91) is of the same form as the Levinson recursion (5); however, the recursion here is a function of the time-shifted AR parameter $a_{M,i}'$ rather than a function of $a_{M,i}$.

The computation of Δ_{M+1} can be simplified by computing the r_{M+1} terms recursively,

$$r_{M+1}(M+1, i) = r_M(M, i-1) - x_{N-i+1} x_{N-M}^* - x_{M+1} x_i^* \tag{92}$$

for $i = 1, \cdots, M$. Equation (92) may be derived directly by inspection of definition (20). The order update relationship for C_{M+1} is given by

$$C_{M+1} = \begin{pmatrix} 0 \\ C''_M \end{pmatrix} + \alpha_3 A_{M+1}. \tag{93}$$

Since the first coefficient of A_{M+1} is unity, then

$$\alpha_3 = c_{M+1,0} = f^*_{M+1,1}/e_{M+1} \tag{94}$$

using (56). Similarly, the order update relationship for D_{M+1} has the form

$$D_{M+1} = \begin{pmatrix} 0 \\ D''_M \end{pmatrix} + \alpha_4 A_{M+1} \tag{95}$$

and the solution is

$$\alpha_4 = d_{M+1,0} = b_{M+1,N-M-1}/e_{M+1} \tag{96}$$

using (62). Recursive order update relationships for the real-valued scalars g_M and w_M are easily determined. From the definition of g_M and using (76) and (93),

$$g_{M+1} = [x_{M+2}, \cdots, x_1] C_{M+1} = [x_{M+2}, \cdots, x_1] \left[\begin{pmatrix} 0 \\ C_M + \beta_2 C^I_M + \gamma_2 D^I_M \end{pmatrix} + \alpha_3 A_{M+1}\right]. \tag{97}$$

A bit of algebra yields

$$g_{M+1} = g_M + \frac{|f_{M+1,1}|^2}{e_{M+1}} + \frac{|v_M|^2(1 - w_M) + |s_M|^2(1 - g_M) + 2 \text{ Real } (s_M h_M v^*_M)}{\text{DEN}_M}. \tag{98}$$

A similar development for w_M based on definition (53) and the use of (80) and (95) yields the following recursive relationship:

$$w_{M+1} = w_M + \frac{|b_{M+1,N-M-1}|^2}{e_{M+1}} + \frac{|s_M|^2(1 - w_M) + |u_M|^2(1 - g_M) + 2 \text{ Real } (u_M h_M s^*_M)}{\text{DEN}_M}. \tag{99}$$

To complete the algorithm, the following initial conditions for zero order and first order are required:

$$e_0 = 2 \sum_{k=1}^{N} |x_k|^2 = \text{twice total signal energy}$$

$$r_{1,0} = 2 \sum_{k=1}^{N-1} (x_{k+1} x^*_k)$$

$$f_{0,1} = x_1$$

$$b_{0,N} = x_N$$

$$g_0 = |x_1|^2/e_0$$

$$w_0 = |x_N|^2/e_0$$

$$h_0 = (x_1 x_N)^*/e_0$$

$$s_0 = x^*_1 x_N/e_0$$

$$u_0 = x^2_N/e_0$$

$$v_0 = (x^*_1)^2/e_0$$

$$\text{DEN}_0 = 1 - g_0 - w_0$$

$$e'_0 = e_0 \text{DEN}_0 = e_0 - |x_1|^2 - |x_N|^2$$

$$c''_{0,0} = x^*_1/e'_0$$

$$d''_{0,0} = x_N/e'_0$$

$$a_{1,1} = -r_{1,0}/e'_0$$

$$e_1 = e'_0[1 - |a_{1,1}|^2]. \tag{100}$$

Fig. 3 is a flowchart for the LS algorithm. Numbers in parentheses on the flowchart refer to the appropriate equations in the text. The LS algorithm requires $NM + 8M^2 + N + 7M - 8$ complex additions, $NM + 9M^2 + 2N + 25M - 3$ complex multiplications, and $5M + 3$ real divisions. The LS algorithm needs $N + 4M + 15$ complex-valued computer memory locations. As a typical case, consider $N = 100$ samples from an hypothesized AR process of order $M = 30$. The total number of multiplications, adds, divisions, and storage locations for the Burg algorithm are 8181, 8059, 30, and 335, respectively. For the LS algorithm the numbers are 12047, 10402, 153, and 235, which is quite comparable to that for the Burg algorithm. In fact, less storage is required for the LS algorithm.

Appendix B contains a Fortran subroutine for computation of the AR parameters via the LS algorithm. The computer version of the algorithm contains several simple checks for both numerical ill-conditioning and order selection indication.

The key performance indicators for the algorithm are the error energy e_M and the divisor term DEN_M. Changes in these have been found empirically to be sensitive indicators of reasonable order selection when using the LS algorithm. The Akaike FPE and AIC order selection criteria, although useful in most cases for the Burg algorithm, were not found to be as useful for the LS algorithm as were the two terms e_M and DEN_M. The FPE and AIC criteria tended to select orders that were too low.

V. Performance of the LS Algorithm

Experiments with the LS approach to AR spectral estimation have always yielded spectra with comparable performance to spectra generated with the Burg algorithm, and with superior performance in many cases. AR spectra computed with the LS algorithm have been found to have less frequency bias, less phase dependence, and somewhat sharper resolution than AR spectra computed with the Burg algorithm. No evidence of spectral line splitting, observed in the Burg algorithm, has been seen with the new technique. The LS and Burg algorithm both have computational complexities proportional to the square of the hypothesized order of the AR estimate, so there is no computational advantage in favor of the Burg algorithm.

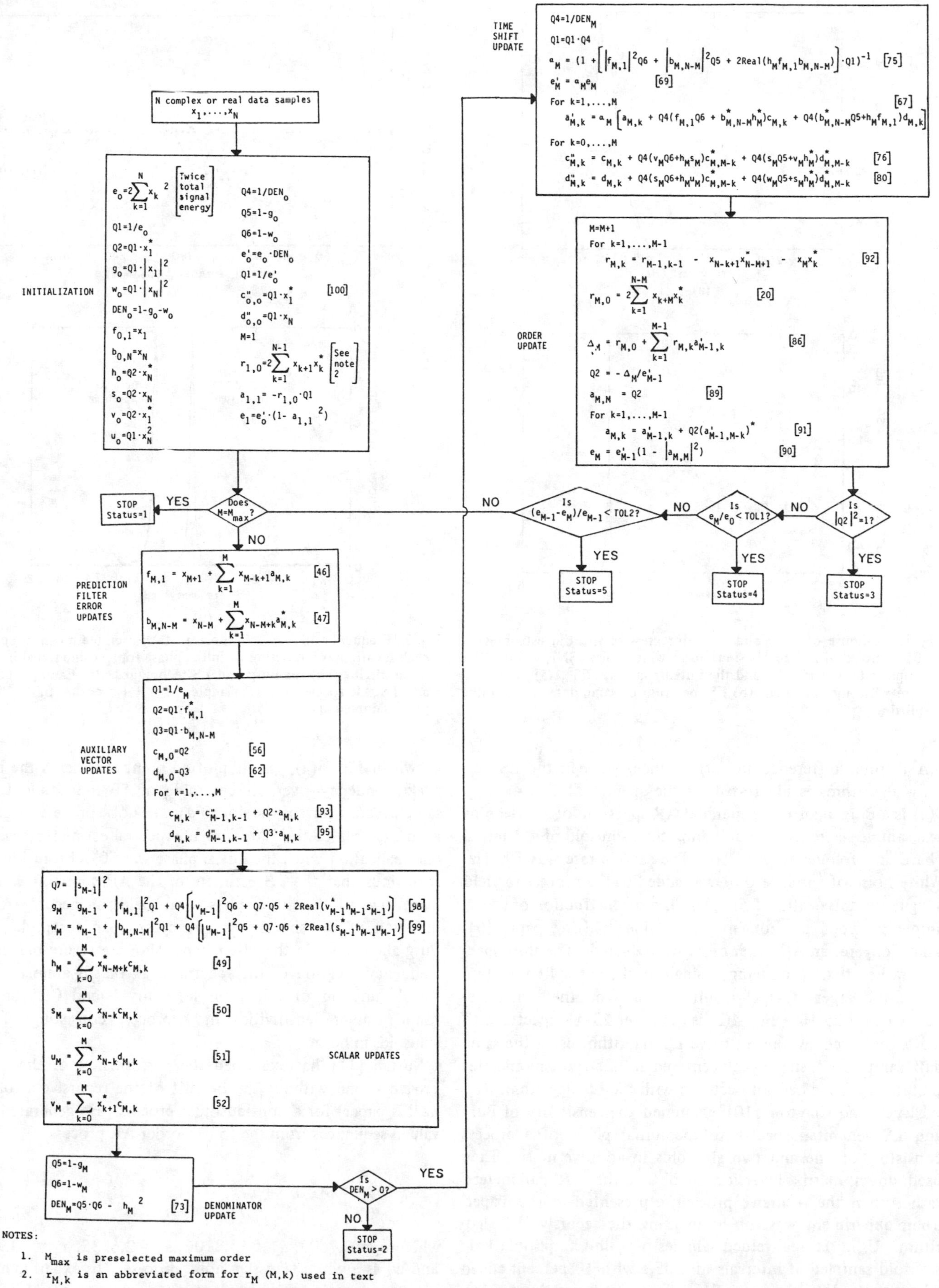

Fig. 3. Least squares algorithm flowchart. Numbers in brackets refer to equations in text.

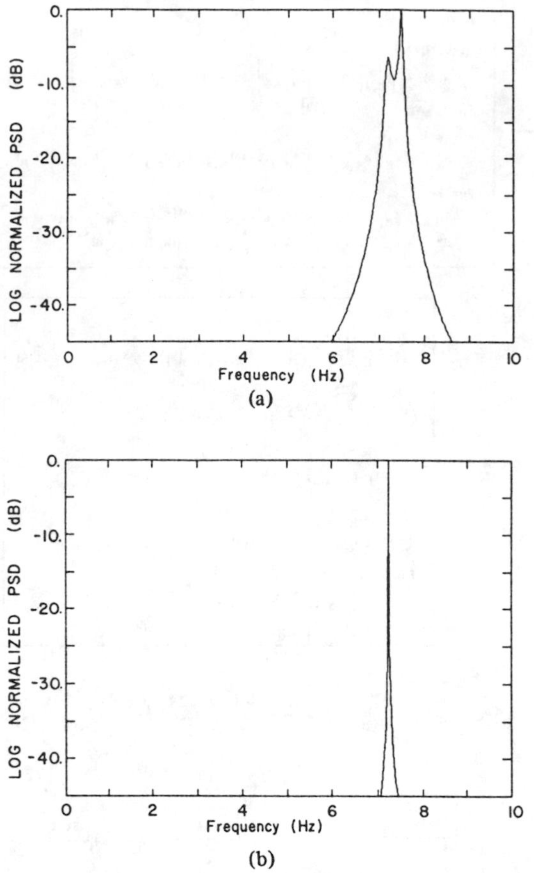

Fig. 4. Response of Burg and LS autoregressive spectral estimates to 101 samples of a 7.25 Hz sinusoid in white noise (SNR = 50). The sample rate was 100 Hz and the initial phase was 50°. (a) Burg spectrum with line splitting. (b) LS spectrum of same data with no line splitting.

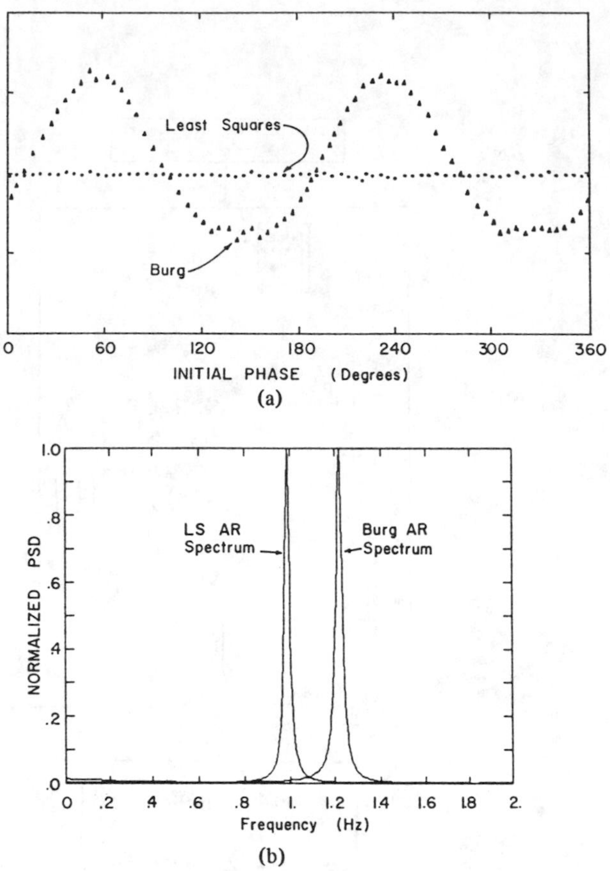

Fig. 5. Frequency bias in AR spectra. (a) Ensemble mean of spectral peak location as a function of initial phase for a cosine signal in white noise (SNR = 10) for Burg and LS techniques. (b) Example of Burg and LS AR spectra for 15 samples of a 1 Hz cosine of 50° initial phase sampled at 20 Hz rate.

A distinct difference in performance between the LS and Burg algorithms is illustrated by the spectra of Fig. 4. Fig. 4(a) is a Burg algorithm generated AR spectrum of a 101 point real sample sequence of a unit amplitude sinusoid of 45° initial phase and frequency 7.25 Hz. The sample rate was 100 Hz. White noise of variance 0.01 was added to the process to yield a signal-to-noise ratio of 50. This is a reconstruction of an example provided by Fougere in his line splitting paper [6]. Like Fougere, an AR order of 25 was selected. The Burg spectra, rather than producing a single peak, has split into two peaks, the larger of which is shifted away from the desired frequency of 7.25 Hz. Fig. 4(b) is the order 25 AR spectral estimate generated by the recursive LS algorithm using the same data samples. A single peak centered at 7.25 Hz demonstrates that line splitting has not occurred with the LS algorithm.

Ulrych and Clayton [10] examined the sensitivity of Burg and LS generated spectra to the initial phase of a process consisting of one and two sinusoids in additive noise. They used direct matrix inversion to obtain the AR parameters, rather than the recursive procedure presented in this paper. Their experiment was run again using the recursive LS algorithm. Using 15 real-valued samples of a unit amplitude 1 Hz sinusoid sampled at intervals of 0.05 s with 10 percent added white noise, the mean variation of the spectral estimate peak as a function of sinusoid initial phase was determined, as shown in Fig. 5(a). Each plotted point represents the mean peak frequency over an ensemble of 50 independent data sample realizations of the sinusoid in the noise process. Fig. 5(b) is an example of the Burg and LS spectral estimates based on one realization when the initial phase was 50°. From Fig. 5, it is obvious that the LS estimate of the AR spectrum is fairly insensitive to the initial phase and yields an accurate determination of the sinusoid frequency over an ensemble. The Burg algorithm, on the other hand, shows a severe bias in the frequency location of the spectral peak that is almost a sinusoidal function of initial phase. Ulrych and Clayton also found similar sensitivities in a process consisting of two sinusoids in noise.

Nuttall [11] has examined the performance of the LS approach (again without the benefit of the recursive algorithm in this paper) for a nonsinusoidal process. He generated real-valued sequences from the fourth-order AR process

$$x_k = \sum_{n=1}^{4} a_n x_{k-n} + w_k, \tag{101}$$

where $a_1 = 2.7607$, $a_2 = -3.8106$, $a_3 = 2.6535$, $a_4 = -0.9238$, and w_k is a white Gaussian noise process. His experiment was performed again, this time with the recursive LS algorithm. One hundred Burg and LS algorithm AR spectra were gener-

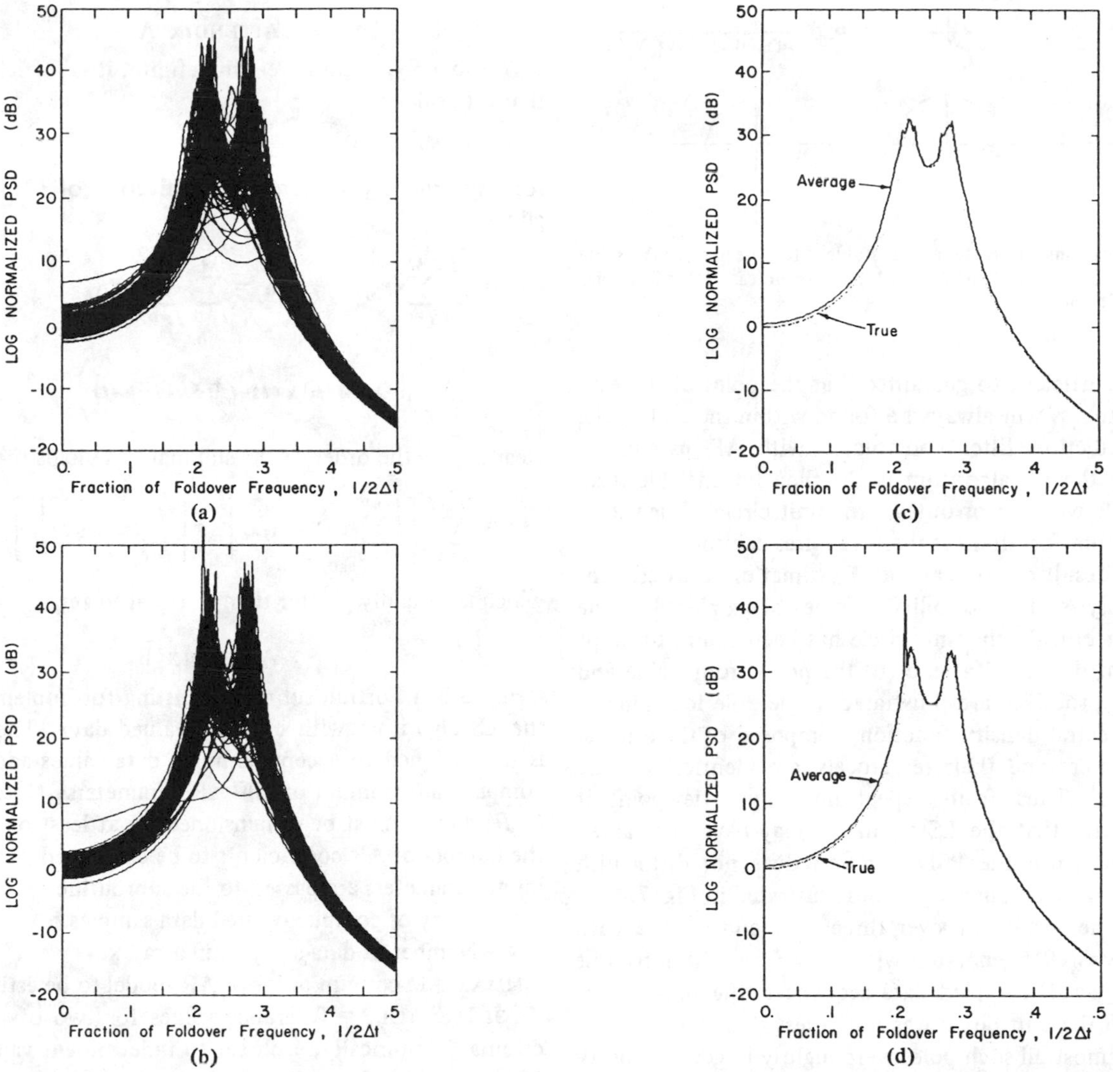

Fig. 6. Response of Burg and LS AR spectra to a fourth-order AR process. (a) Overlapped ensemble of 100 Burg spectra from independent 40 sample realizations. (b) Overlapped ensemble of 100 LS spectra based on same realizations used for Burg ensemble. (c) Average of 100 Burg spectra compared to true AR(4) spectrum. (d) Average of 100 LS spectra compared to true spectrum.

ated from independent 40 sample sequences of the AR (4) process in steady state. Fig. 6(a) and (b) shows the 100 overlapped estimated spectra generated by the two algorithms, while Fig. 6(c) and (d) indicates the average of all 100 spectra for each algorithm compared to the true AR spectrum. The model order was preselected at the correct value $M = 4$. The results here are comparable to that of Nuttall's. One observation that can be made is that the LS technique tends to have slightly less variability in the skirt, but more spiky estimates near the peaks of the spectrum, than seen in Burg algorithm spectra. That is, the LS algorithm produces AR spectra with slightly less frequency variability, but more power spectral density variability. The greater PSD variability can be attributed to the fact that, unlike the Burg algorithm, the LS algorithm does not restrict the poles from moving close to the unit circle. Since the area under the spectral density curve, (rather than the peak height) is proportional to power, the variability in PSD amplitude is not generally of much concern. Rather, obtaining unbiased accurate estimates of the spectral peak frequencies is more important for most applications.

The least squares AR spectrum resolution (ability to resolve two close spectral peaks) is never worse than the resolution of AR spectra generated by the Burg algorithm. At higher SNR's, the resolution has been observed to be almost always better than the Burg resolution since the LS algorithm can place the poles closer to the unit circle to produce spectra with narrower skirts.

Much of the performance improvement obtained with the LS algorithm can be attributed to the removal of the constraint that the estimation procedure always place poles inside the unit circle. The matrix R_M of (23) is always positive-semidefinite, as proved in Appendix A. If R_M is nonsingular, then it is positive-definite. The latter is most common since recursive determination for the AR parameters would otherwise not be possible. An indication of the singularity of, or the ill-conditioning in, R_M is provided by the prediction error energy e_M and the term DEN_M of (73). If these become zero or very close to zero, then R_M is either singular or is at least ill-conditioned for further recursion. The positive-definite, Hermitian symmetric, and Hermitian persymmetric proper-

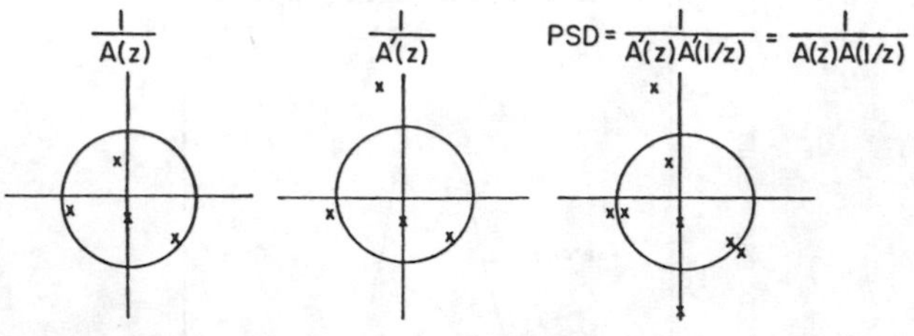

Fig. 7. Pole locations of stable and unstable AR processes. (a) Stable AR process poles. (b) Unstable AR process poles. (c) AR power spectral density poles.

ties are not sufficient to guarantee that the poles of the estimated AR process will always be found within the unit circle. A linear prediction filter constructed with AR parameters generated by the LS algorithm might then be unstable if an estimated pole were on or outside the unit circle. The use of the LS algorithm for filter synthesis requires caution.

If the end result of AR parameter estimation is strictly for spectrum analysis, the possibility of poles being placed by the LS algorithm outside the unit circle has been found to be of little concern. In Fig. 7(a) and (b) the poles from stable and unstable AR processes are illustrated. The pole locations of the power spectral density function, composed of the original AR process poles and their reciprocals, are identical for the two processes. Thus, from a spectrum analysis viewpoint, it does not matter that the LS algorithm may have "unstable" pole locations since the PSD function does not distinguish between the stable and unstable cases illustrated in Fig. 7.

Based on the results of over three thousand spectra with high and low SNR's generated with the LS algorithm by this author from actual sampled data sequences, the presence of poles outside the unit circle was noted less than 1 percent of the time. Almost all such poles were slightly larger than unity in magnitude and usually corresponded to the highest SNR component in the spectrum. Nuttall [11] in his computer simulations of a narrow-band process with the LS algorithm found no pole estimates outside the unit circle.

No cases of spectral line splitting have been observed using the order selection criteria given in Appendix B. In practice, then, the LS algorithm appears to yield AR parameters that produce stable spectra, even with pole estimates that occassionally fall outside the unit circle.

VI. Summary

A new recursive algorithm that provides AR parameters for an AR spectral estimate based on forward and backward linear prediction has been introduced. It has the same order of computational complexity as the popular Burg algorithm. Examples have been provided to illustrate the improved performance of spectra generated with the LS algorithm when compared to spectra generated with the Burg algorithm. Improvements include reduced sensitivity to initial phase, reduced bias in the frequency estimate, less frequency variability over an ensemble of spectra generated from the same process, and absence of spectral line splitting. All these factors suggest that the LS algorithm is an attractive alternative to the Burg algorithm for AR spectral estimation.

Appendix A

To show R_M is positive-semidefinite, it is sufficient to show that the scalar

$$s = \alpha^H R_M \alpha \geqslant 0 \tag{A1}$$

for any arbitrary vector α. Substitution of (21) for $r_M(i,j)$ gives

$$s = \sum_{i=0}^{M}\sum_{j=0}^{M} r_M(i,j)\, \alpha_i^* \alpha_j = \sum_{i=0}^{M}\sum_{j=0}^{M} \alpha_i^* \alpha_j$$

$$\left(\sum_{k=1}^{N-M} [x_{k+M-j}x^*_{k+M-i} + x^*_{k+j}x_{k+i}]\right). \tag{A2}$$

Rearranging the order of the summations yields the result

$$s = \sum_{k=1}^{N-M}\left[\left|\sum_{i=0}^{M} \alpha_i x_{k+M-i}\right|^2 + \left|\sum_{j=0}^{M} \alpha_j^* x_{k+j}\right|^2\right], \tag{A3}$$

which is certainly greater than or equal to zero.

Appendix B

Fig. 8 is a Fortran subroutine listing for implementation of the LS algorithm with complex-valued data. The subroutine is dimensioned to accept complex data values and is fixed to compute a maximum of 100 AR parameters. Note that arrays C, R, and D must be dimensioned by at least one more than the number of AR coefficients to be computed. The following input parameters are passed to the subroutine

X = Array of complex-valued data samples.

N = Number of data samples in array X.

MMAX = Maximum order of AR model to be estimated.

TOL1 & TOL2 = Tolerance values for two of the stopping criteria. Empirically, each set to independent values between 10^{-2} and 10^{-3}, depending upon the data.

The following output parameters are passed from the subroutine

M = Order of AR model computed when a stopping criterion was satisfied. Note: M $\leqslant$ MMAX.

A = Array of complex-valued AR parameters.

E = Prediction error energy at order M.

EO = Twice the total signal energy in the data samples.

STATUS = Integer indicating stopping criteria that terminated the recursion at order M.

Five values of STATUS are possible. STATUS = 1 is the normal program exit when the maximum order is reached, M = MMAX. STATUS = 4 indicates the program terminated when $e_M/e_0 <$ TOL1, that is, the residual prediction error energy is a small fraction of the total signal energy. STATUS = 5 indicates the program terminated when $(e_{M-1} - e_M)/e_{M-1} <$ TOL2, that is, the residual prediction error energy at order M has changed by only a small fraction from the previous order $M - 1$. This is the stopping criteria encountered most frequently. STATUS = 3 occurs when $|a_{M,M}| > 1$, indicating numerical ill-conditioning or possibly a singular matrix, since $0 < |a_{M,M}| \leqslant 1$. STATUS = 2 indicates the algorithm terminated when $\text{DEN}_M \leqslant 0$. This is also an indicator of numerical ill-conditioning within the algorithm since DEN_M must be positive-valued.

```
      SUBROUTINE LSTSQS(N, M, MMAX, X, A, E, TOL1, TOL2, STATUS, EO)
C
C     N = NUMBER OF DATA SAMPLES (Input Parameter)
C     M = MODEL ORDER AT TIME OF EXIT FROM SUBROUTINE (Output Parameter)
C     MMAX = MAXIMUM MODEL ORDER DESIRED (Input Parameter)
C            (M=MMAX for normal subroutine exit)
C     X = ARRAY OF N COMPLEX-VALUED DATA SAMPLES (Input Parameters)
C     A = ARRAY OF M COMPLEX-VALUED AUTOREGRESSIVE COEFFICIENTS (Output)
C     E = PREDICTION ERROR ENERGY (FORWARD AND BACKWARD) FOR
C         ORDER M (Output Parameter)
C     TOL1 = TOLERANCE FACTOR THAT STOPS RECURSION AT ORDER M
C            WHEN E(M)/E(O) < TOL1
C     TOL2 = TOLERANCE FACTOR THAT STOPS RECURSION AT ORDER M
C            WHEN (E(M)-E(M-1))/E(M-1) < TOL2
C     STATUS = INTEGER INDICATOR OF WHICH OF FIVE CONDITIONS EXISTED
C              AT TIME OF SUBROUTINE EXIT (See Flowchart)
C     EO = TOTAL SIGNAL ENERGY
      COMPLEX X(1), A(1), C(101), D(101), R(101), F, B, H, S, U, V
      COMPLEX SAVE1, SAVE2, SAVE3, SAVE4, Q2, Q3, DELTA, C1, C2, C3, C4, C5, C6
      INTEGER STATUS
C
C     INITIALIZATION
C
      EO=0.
      DO 10 K=1, N
10      EO=EO+REAL(X(K))**2+AIMAG(X(K))**2
      EO=2. *EO
      Q1=1. /EO
      Q2=CMPLX(Q1*REAL(X(1)), -Q1*AIMAG(X(1)))
      G=Q1*(REAL(X(1))**2+AIMAG(X(1))**2)
      W=Q1*(REAL(X(N))**2+AIMAG(X(N))**2)
      DEN=1. -G-W
      Q4=1. /DEN
      Q5=1. -G
      Q6=1. -W
      F=X(1)
      B=X(N)
      H=Q2*CONJG(X(N))
      S=Q2*X(N)
      U=Q1*X(N)*X(N)
      V=Q2*CONJG(X(1))
      E=EO*DEN
      Q1=1. /E
      C(1)=Q1*CONJG(X(1))
      D(1)=Q1*X(N)
      M=1
      SAVE1=(0. , 0. )
      N1=N+1
      NM=N-1
      DO 20 K=1, NM
20      SAVE1=SAVE1+X(K+1)*CONJG(X(K))
      R(1)=2. *SAVE1
      A(1)=CMPLX(-Q1*REAL(R(1)), -Q1*AIMAG(R(1)))
      E=E*(1. -REAL(A(1))**2-AIMAG(A(1))**2)
30    IF (M .LT. MMAX) GO TO 40
      STATUS=1
      RETURN
C
C     PREDICTION FILTER ERROR UPDATE
C
40    EOLD=E
      M1=M+1
      F=X(M1)
      B=X(NM)
      DO 50 K=1, M
        F=F+X(M1-K)*A(K)
50      B=B+X(NM+K)*CONJG(A(K))
C
C     AUXILIARY VECTORS ORDER UPDATE
C
      Q1=1. /E
      Q2=CMPLX(Q1*REAL(F), -Q1*AIMAG(F))
      Q3=CMPLX(Q1*REAL(B), Q1*AIMAG(B))
      DO 60 K=M, 1, -1
        K1=K+1
        C(K1)=C(K)+Q2*A(K)
60      D(K1)=D(K)+Q3*A(K)
      C(1)=Q2
      D(1)=Q3
C
C     SCALAR ORDER UPDATE
C
      Q7=REAL(S)**2+AIMAG(S)**2
      Y1=REAL(F)**2+AIMAG(F)**2
      Y3=REAL(B)**2+AIMAG(B)**2
      Y2=REAL(V)**2+AIMAG(V)**2
      Y4=REAL(U)**2+AIMAG(U)**2
      G=G+Y1*Q1+Q4*(Y2*Q6+Q7*Q5+2. *REAL(CONJG(V)*H*S))
      W=W+Y3*Q1+Q4*(Y4*Q5+Q7*Q6+2. *REAL(CONJG(S)*H*U))
      H=(0. , 0. )
      S=(0. , 0 )
      U=(0. , 0. )
      V=(0. , 0. )
      DO 70 K=0, M
        K1=K+1
        NK=N-K
        H=H+CONJG(X(NM+K))*C(K1)
        S=S+X(NK)*C(K1)
        U=U+X(NK)*D(K1)
70      V=V+CONJG(X(K1))*C(K1)
C
C     DENOMINATOR UPDATE
C
      Q5=1. -G
      Q6=1. -W
      DEN=Q5*Q6-REAL(H)**2-AIMAG(H)**2
      IF (DEN  GT  0. ) GO TO 80
      STATUS=2
      RETURN
C
C     TIME SHIFT VARIABLES UPDATE
C
```

Fig. 8. Fortran subroutine of the Marple least squares algorithm.

```
80    Q4=1. /DEN
      Q1=Q1*Q4
      ALPHA=1. /(1. +(Y1*Q6+Y3*Q5+2. *REAL(H*F*B))*Q1)
      E=ALPHA*E
      C1=Q4*(F*Q6+CONJG(B*H))
      C2=Q4*(CONJG(B)*Q5+H*F)
      C3=Q4*(V*Q6+H*S)
      C4=Q4*(S*Q5+V*CONJG(H))
      C5=Q4*(S*Q6+H*U)
      C6=Q4*(U*Q5+S*CONJG(H))
      DO 90 K=1, M
        K1=K+1
90      A(K)=ALPHA*(A(K)+C1*C(K1)+C2*D(K1))
      M2=M/2+1
      DO 100 K=1, M2
        MK=M+2-K
        SAVE1=CONJG(C(K))
        SAVE2=CONJG(D(K))
        SAVE3=CONJG(C(MK))
        SAVE4=CONJG(D(MK))
        C(K)=C(K)+C3*SAVE3+C4*SAVE4
        D(K)=D(K)+C5*SAVE3+C6*SAVE4
        IF (MK .EQ. K) GO TO 100
        C(MK)=C(MK)+C3*SAVE1+C4*SAVE2
        D(MK)=D(MK)+C5*SAVE1+C6*SAVE2
100     CONTINUE
C
C     ORDER UPDATE
C
      M=M+1
      NM=N-M
      M1=M-1
      DELTA=(0. , 0. )
      C1=CONJG(X(N1-M))
      C2=X(M)
      DO 110 K=M1, 1, -1
        R(K+1)=R(K)-X(N1-K)*C1-CONJG(X(K))*C2
110     DELTA=DELTA+R(K+1)*A(K)
      SAVE1=(0. , 0. )
      DO 120 K=1, NM
120     SAVE1=SAVE1+X(K+M)*CONJG(X(K))
      R(1)=2. *SAVE1
      DELTA=DELTA+R(1)
      Q2=CMPLX(-REAL(DELTA)/E, -AIMAG(DELTA)/E)
      A(M)=Q2
      M2=M/2
      DO 130 K=1, M2
        MK=M-K
        SAVE1=CONJG(A(K))
        A(K)=A(K)+Q2*CONJG(A(MK))
        IF (K .EQ. MK) GO TO 130
        A(MK)=A(MK)+Q2*SAVE1
130     CONTINUE
      Y1=REAL(Q2)**2+AIMAG(Q2)**2
      E=E*(1. -Y1)
      IF (Y1 .LT. 1. ) GO TO 140
      STATUS=3
      RETURN
140   IF (E .GE. EO*TOL1) GO TO 150
      STATUS=4
      RETURN
150   IF ((EOLD-E) .GE. EOLD*TOL2) GO TO 30
      STATUS=5
      RETURN
      END
```

Fig. 8. (Continued.)

References

[1] S. L. Marple, Jr., "Frequency resolution of high resolution spectrum analysis techniques," in *Proc. 1st RADC Spectrum Estimation Workshop*, 1978, pp. 19-35.

[2] T. J. Ulrych and T. N. Bishop, "Maximum entropy spectral analysis and autoregressive decomposition," *Rev. Geophys.*, vol. 13, pp. 183-200, 1975.

[3] S. M. Kay and S. L. Marple, Jr., "Sources of and remedies for spectral line splitting in autoregressive spectrum analysis," in *Proc. IEEE Int. Conf. Acoust., Speech, Signal Processing*, 1979, pp. 151-154.

[4] N. O. Anderson, "On the calculation of filter coefficients for maximum entropy analysis," *Geophys.*, vol. 39, pp. 69-72, Feb. 1974.

[5] J. P. Burg, "Maximum entropy spectral analysis," Ph.D. dissertation, Dep. Geophys., Stanford Univ., Stanford, CA, May 1975.

[6] P. F. Fougere, E. J. Zawalick, and H. R. Radoski, "Spontaneous line splitting in maximum entropy power spectrum analysis," *Phys. Earth and Plan. Inter.*, vol. 12, pp. 201-207, Aug. 1976.

[7] A. Akaike, "Statistical predictor identification," *Ann. Inst. Stat. Math.*, vol. 22, pp. 203-217, 1970.

[8] D. N. Swingler, "A comparison between Burg's maximum entropy method and a nonrecursive technique for the spectral analysis of deterministic signals," *J. Geophys. Res.*, vol. 84, pp. 679-685, Feb. 10, 1979.

[9] P. F. Fougere, "A solution to the problem of spontaneous line splitting in maximum entropy power spectrum analysis," *J. Geophys. Res.*, vol. 82, pp. 1051-1054, Mar. 1, 1977.

[10] T. J. Ulrych and R. W. Clayton, "Time series modeling and maxi-

mum entropy," *Phys. Earth and Plan. Int.*, vol. 12, pp. 188–200, Aug. 1976.

[11] A. H. Nuttal, "Spectral analysis of a univariate process with bad data points, via maximum entropy and linear predictive techniques," Naval Underwater Systems Center, New London, CT, Tech. Rep. 5303, Mar. 26, 1976.

[12] N. O. Anderson, "Comments on the performance of maximum entropy algorithms," *Proc. IEEE*, vol. 66, pp. 1581–1582, Nov. 1978.

[13] M. Morf, B. Dickinson, T. Kailath, and A. Vieira, "Efficient solution of covariance equations for linear prediction," *IEEE Trans. Acoust., Speech, Signal Processing*, vol. ASSP-25, pp. 429–433, Oct. 1977.

[14] J. Makhoul, "Linear prediction: A tutorial review," *Proc. IEEE*, vol. 63, pp. 561–580, Apr. 1975.

The Covariance Least-Squares Algorithm for Spectral Estimation of Processes of Short Data Length

CHRYSOSTOMOS L. NIKIAS, MEMBER, IEEE, AND PETER D. SCOTT, MEMBER, IEEE

***Abstract*—A new method for generating the autoregressive (AR) process parameters for spectral estimation is introduced. The method fits AR models to the data optimally in the sense of minimizing the sum of squares of the error covariance function within the model prediction region, and is thus designated as the Covariance Least-Squares (CLS) algorithm. This minimization is shown to be identical with minimizing the weighted average one-step, linear prediction errors with adaptive weights corresponding to the energy of the data within the prediction region. The CLS algorithm is compared to the Least-Squares (LS) algorithm [1], [2] by simulation and asymptotic properties. It is shown that the CLS method combines all the desirable properties of the comparison algorithm with improved robustness in the presence of nonstationarity, namely, additive transients and envelope modulation. It is also shown that the CLS algorithm provides asymptotically unbiased AR parameters, a property also shared by the comparison LS algorithm.**

I. Introduction

POWER spectral estimation is a methodology which has been found useful in many areas of physical applications where harmonic phenomena are to be detected and analyzed, such as geoscience [3], [4], [5], sonar [6], antennas [7], and biomedical applications [8]. In using the power spectral density function to characterize a process, a special attention is often required to design a reliable estimator based on finite length data records.

The first class of estimators to be used extensively was the conventional (Fourier-type) [9], [10] which is now known to yield poor resolution when applied to short length data records. A seminal step in the development of spectral estimation methods was the approach taken by Burg and which resulted in the widely used Maximum Entropy (ME) method [11]. Burg showed how a high-resolution spectrum can be estimated from *precise* knowledge of the first $M+1$ covariance lags of a stationary Gaussian random process. The principal motivation for the use of the ME method is the apparently higher resolution when compared with either convention methods [9], [10] or the Maximum likelihood (ML) [12] method of spectral analysis. Lacoss [13] has compared the ME with the ML and conventional methods and demonstrated the superiority of the ME spectrum in terms of spectral resolution.

Manuscript received April 26, 1982; revised August 10, 1982. This work was partially supported by the National Institute of Health under Grants HL 15676 and HL 29280. Portions of this paper were presented at the ASSP Spectral Estimation Workshop held at McMaster University, Hamilton, Ontario, Canada, August 1981.

C. L. Nikias is with the Department of Electrical Engineering and Computer Science, The University of Connecticut, Storrs, CT 06268.

P. D. Scott is with the Department of Electrical and Computer Engineering, State University of New York at Buffalo, Amherst, NY 14260.

Furthermore, Burg [14] very elegantly showed the superior resolving power of ME compared with the ML approach. In an important short communication, van den Bos [15] pointed out that ME method of spectral analysis is equivalent to matching the parameters of an Mth-order autoregressive (AR) model to the $M+1$ given covariances, thus linking autoregressive modelling techniques long used in time series analysis to ME spectral analysis.

Since van den Bos' observation, there has been increased emphasis in construction of AR models of time series for spectral estimation. Research has been basically directed towards finding new criteria-of-fit (either constrained or unconstrained) to estimate the AR model parameters. This is due to the fact that in practice the covariance lags of a finite length random process can not be determined exactly and thus the criterion-of-fit used to generate the AR model parameters (either directly from data samples or through covariance estimates) determines the performance of the spectral estimator. It is important to note that, in general, AR spectral estimation *is not* maximum entropy.

One of the earliest methods in all poles or AR spectral estimation is the Yule-Walker (YW) or autocorrelation technique [16]. This method substitutes biased estimates of the autocorrelation lags generated from the available data into the Yule-Walker normal equations. Biased autocorrelation estimates guarantee a positive-definite autocorrelation matrix, but at the expense of a considerable loss of spectral resolution, frequency bias [17] and line splitting [18]. Spectral line-splitting is the resolution of two or more closely spectral peaks when only one is truly present.

Burg [19] introduced a new method of estimating the AR parameters from data samples by minimizing the sum of the average forward and backward squared one-step linear prediction errors. The AR parameters so estimated are further constrained to satisfy the Levinson recursion [20]. Since the introduction of Burg's algorithm [20], a variety of authors have demonstrated its strong points as well as the limitations. On the one hand, the Burg technique provides dramatically increased resolution over conventional methods and the YW method, especially for short length data records [21], [22], [23]. On the other hand, AR spectra generated by the Burg algorithm suffer from line splitting and frequency bias. The line splitting problem was originally reported by Chen and Stegen [24] and later documented more thoroughly by Fougere *et al.* [25]. Frequency bias in processing sinusoidal signals has been experimentally investigated by Chen and Stegen [24] for the noisy case and theoretically analyzed by Swingler

Reprinted from *IEEE Trans. Geosci. Remote Sensing*, vol. GE-21, pp. 180–190, Apr. 1983.

[26] for the noiseless case. Another limitation of the Burg algorithm is the fact that it usually fails to provide acceptable spectral estimates when the assumed stationarity of the data is violated, i.e., first-order exponential decays [27] and sinusoid modulated by envelope function [28].

The present authors [28], [29] have shown that the introduction of energy-weights into Burg's linear prediction filter associated the AR spectral estimation demonstrates improved performance over Burg's uniform weights. These improvements include less frequency bias in processing sinusoidal signals, less tendency to line-splitting and robustness in noise presence and envelope function modulation. The Energy-Weighted (EW) algorithm also employs the Levinson recursion and the lattice formulation of an AR process [30].

The Burg method assumes that the covariance matrices implicitly involved in processing the algorithm are Toeplitz, thus permitting the use of the Levinson recursion. This assumption is responsible for the frequency bias incurred on processing sinusoids because the covariance matrices involved in the Burg algorithm are, in fact, symmetrical but not Toeplitz [2]. A common problem exists in the EW algorithm since the Levinson recursion is also employed, but as the results in [28], [29] demonstrate, the bias is considerably reduced.

Ulrych and Clayton [2], and independently Nuttal [1] have shown considerable improvement in the Burg results by removing the Levinson constraint. Their method, termed the Least-Squares (LS) algorithm, is based on an unconstrained least-squares minimization of the average forward and backward squared one-step linear prediction errors. Marple [31] has recently reported that AR spectra computed with the LS algorithm have been found to have less frequency bias and somewhat sharper resolution than Burg-generated spectra. Further, no evidence of spectral line splitting has been observed with the LS technique [18], [31]. A detailed review of AR spectral estimation techniques can be found in the paper by Kay and Marple [32]. In a similar manner, however, removal of the Levinson constraint in the EW method yields further improvement over the EW results provided in [28], [29], though at the expense of greater complexity. The method, the Covariance Least-Squares (CLS) algorithm, is presented in this paper and is shown to demonstrate improved robustness over the LS algorithm in the presence of decaying transients and envelope modulation, while retaining the superior performance of the LS method in processing AR processes and pure sinusoidal signals.

II. Covariance Least-Squares Estimates

An Mth-order zero mean stationary AR process is characterized by

$$X_n + \sum_{m=1}^{M} a_{M,m} X_{n-m} = W_n \tag{1}$$

where $a_{M,m}$ is the mth AR model parameter and W_n is a zero-mean white random process $E\{W_i W_j\} = Q\delta_{i,j}$. $E\{\cdot\}$ denotes the expectation operator and $\delta_{i,j}$ is the Kronecker delta function. By definition, $E\{W_n X_i\} = 0$ for $i < n$. Since the AR process $\{X_n\}$ is stationary, its covariance elements are shift-invariant and the process has power spectrum

$$S(f) = Q/\left|1 + \sum_{m=1}^{M} a_{M,m} \exp(-i2\pi f m T)\right|^2 \tag{2}$$

where T is the sampling period in seconds and frequency f is measured in Hertz. The spectral estimation algorithm described in this section proceeds by determining an optimal set of parameters $\{a_{M,m}\}$, $m = 1, \cdots, M$ in a well-defined sense and taking $S(f)/Q$, $S(f)$ in (2) as the normalized spectral estimate.

Given a finite length data record $\{X_n\}, n = 1, \cdots, N$ an AR model order M is selected and the model parameters are determined which minimize the sum of the forward and backward covariance recursion errors

$$C_M \triangleq C_M^f + C_M^b \tag{3}$$

$$C_M^f = \sum_{k=1}^{N-M} \sum_{i=k}^{k+M-1} \left(R(k+M, i) + \sum_{m=1}^{M} a_{M,m} R(k+M-m, i)\right)^2 \tag{4a}$$

$$C_M^b = \sum_{k=1}^{N-M} \sum_{i=k}^{k+M-1} \left(R(k, i) + \sum_{m=1}^{M} a_{M,m} R(k+m, i)\right)^2 \tag{4b}$$

and $R(\lambda, \mu) = E\{X_\lambda X_\mu\}$. Since the statistics of the process are unknown, their estimates are generated by the unbiased single-point estimates of $R(\lambda, \mu)$, namely $\hat{R}(\lambda, \mu) = X_\lambda X_\mu$. Thus substituting the unbiased single-point estimates of the corresponding true covariance elements into (4a) and (4b), after some algebra, one obtains

$$\hat{C}_M^f = \sum_{k=1}^{N-M} \left(X_{k+M} + \sum_{i=1}^{M} \alpha_{M,i} X_{k+M-i}\right)^2 E_{M,k}^f \tag{5a}$$

$$E_{M,k}^f = \sum_{i=1}^{M} X_{k+M-i}^2 \tag{5b}$$

$$\hat{C}_M^b = \sum_{k=1}^{N-M} \left(X_k + \sum_{i=1}^{M} \alpha_{M,i} X_{k+i}\right)^2 E_{M,k}^b \tag{6a}$$

$$E_{M,k}^b = \sum_{i=1}^{M} X_{k+i}^2. \tag{6b}$$

The factors of (5b) and (6b) in each term of (5a) and (6a), respectively, may be interpreted as adaptive weights which scale the one-step prediction errors by the energy of the associated data in each translated prediction window of size M thus emphasizing the high "local SNR" data points. For a justification of this criterion-of-fit based on the covariance recursion equation of an Mth order AR process, the reader is directed to [Appendix A [29]].

Unconstrained minimization of $\hat{C}_M$ leads to the following linear system of equations

$$R_M A_M = E_M \tag{7}$$

where

$$R_M = \{r_M(i,j)\} = \sum_{k=1}^{N-M} (E^f_{M,k} X_{k+M-i} X_{k+M-j} + E^b_{M,k} X_{k+i} X_{k+j})$$

$$i = 0, 1, \cdots, M \quad \text{and} \quad j = 0, 1, \cdots, M$$

$$A_M = [1, \alpha_{M,1}, \cdots, \alpha_{M,M}]^T$$

$$E_M = [C^*_M, 0, \cdots, 0]^T$$

"T" denotes the transpose operation, $E^f_{M,k}$, $E^b_{M,k}$ are given by (5b) and (6b), respectively, and C^*_M is defined as

$$C^*_M \triangleq \sum_{j=0}^{M} \alpha_{M,j} r_M(0,j), \qquad \alpha_{M,0} = 1.$$

The R_M matrix of (7) is always positive-semidefinite as proved in Appendix I. It is also symmetric but it is neither Toeplitz ($r_M(i,j) \neq r_M(i-j)$) nor persymmetric ($r_M(i,j) \neq (r_M(M-i, M-j))$.

The AR parameters for spectral estimation are found by computation of the $r_M(i,j)$ terms directly from the data and then by solution of (7) for vector A_M using matrix inversion. Thus the required computational operations are on the order of M^3. To reduce the algorithm's computational effort the identity $E^f_{M,k+1} = E^b_{M,k}$ is employed.

Since the exact statistics for the estimates of the AR parameters of the AR spectrum are not known for finite data records, the asymptotic CLS case (as the number of data points $N \to \infty$, AR order M fixed) is also examined. It is assumed that the process is a stationary pure Mth-order AR and that W_k (driving noise) and hence X_k is Gaussian. It is shown in Appendix II that the CLS estimates $\underline{\alpha} = [\alpha_{M,1}, \cdots, \alpha_{M,M}]^T$ are asymptotically unbiased, i.e.

$$\lim_{N \to \infty} E\{\alpha\} = \alpha$$

where $\alpha = [a_{M,1}, \cdots, a_{M,M}]^T$ is the true parameter vector. The asymptotic ($N \to \infty$) error covariance matrix of the CLS estimate α is also derived in Appendix II but as is illustrated in the Section III does not always yield useful estimates for short data records.

III. Results

In this section the Least-Squares (LS) and the Covariance Least-Squares (CLS) algorithms are applied to short and long sinusoidal and nonsinusoidal data. The results here serve to illustrate the comparative performance of the algorithms. Specifically, the algorithms are compared in terms of robustness in the presence of exponential decays and envelope function modulation, resolution, frequency bias, phase dependence, spectral variance, and spectral line splitting. Additionally, in some cases, results are given for the Burg [20] and Energy-Weighted (EW) [29] algorithms. In the following, f is the frequency (in Hertz) of sinusoidal signal, ϕ the initial phase (degrees), f_s the sampling rate (in Hertz), Q the noise variance, N the total number of samples, and M is the AR model order used.

A. Absence of Line Splitting

Kay *et al.* [18] and Marple [31] have reported that the LS-generated spectra have no tendency to line splitting. According to Fougere *et al.* [25], the worst Burg linesplitting problems occur for long portions of sinusoidal data an odd number of quarter cycles with initial phase 45°. Reconstructing an example provided by Fougere a unit amplitude sine wave in additive Gaussian white noise ($f = 26.25$, $f_s = 100$, $\phi = 45°$, $N = 101$, $M = 25$, $Q = 10^{-5}$) was employed. Fig. 1 illustrates the Burg and EW computed spectra where the Burg spectrum exhibits line splitting and the EW sidelobe of amplitude -27 dB. On the other hand, Fig. 2 shows the LS and CLS generated spectra and clearly demonstrates that line splitting has not occurred in either spectral estimate. (Running a considerable amount of simulations, line splitting tendency of LS and CLS schemes has not been observed.)

B. Effect of Initial Phase and Number of Data Samples

Ulrych and Clayton [2] and Marple [31] examined the sensitivity of LS generated spectra to the initial phase and number of data samples of a process consisting of sinusoids in additive white noise and found it to be fairly insensitive. Their experiments were performed again using the CLS algorithm and the derived results were found to be entirely consistent with those of the LS algorithm. In the case of the noiseless sinusoidal data, it is shown in Appendix III that both LS and CLS algorithms do not exhibit frequency bias due to the initial phase and number of samples (i.e., yield to the "correct" second-order AR model). Note that the Burg and EW estimated spectra exhibit bias in the frequency location, with Burg's bias being much more severe [28], [29].

C. Ten Closely-Spaced Spectral Peaks of Sinusoidal Process

One of the most attractive features of the LS algorithm is its well-documented ability to resolve closely-spaced peaks [2], [27], [31]. However, it would be very interesting to investigate the ability of the CLS algorithm to resolve closely-spaced spectral peaks of a process consisting of ten sinusoids in additive white noise. By using $N = 32$ samples ($f_s = 64$) of a process defined mathematically as

$$X_k = \sum_{i=1}^{10} A_i(\cos(\omega_i k)) + W_k$$

$$k = 1, 2, \cdots, 32$$

where

$$A_i = 0.1i$$

$$\omega_i = 2\pi(10 + 1.5(i-1))/f_s$$

$$Q = 10^{-4}$$

its spectrum was computed by the CLS algorithm, and shown in Fig. 3(a). The LS spectrum was computed and found to be sensibly identical to that of Fig. 3(a). This is an example used by Swingler [27] to examine the performance of the LS algorithm. Following Swingler, the AR model order was set $M = 20$. From Fig. 3(a), it appears that the CLS algorithm has

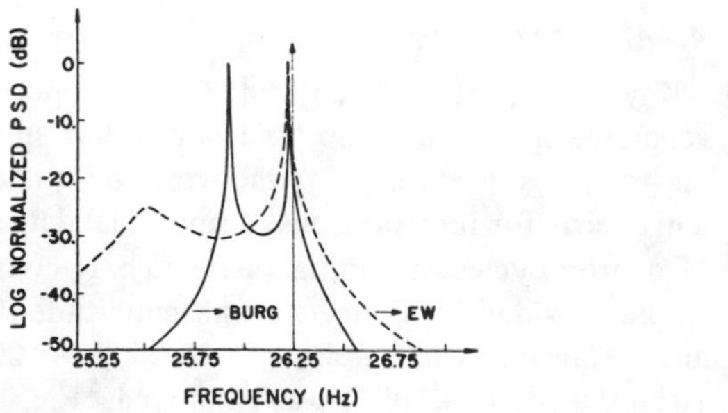

Fig. 1. Line splitting for Burg spectrum. Spectra for 101 samples, sampled at 100-Hz rate, of a unit amplitude sine wave with frequency 26.25 Hz in white noise of variance 10^{-5} (SNR = 220). The initial phase is 45^0 and the AR model order is 25.

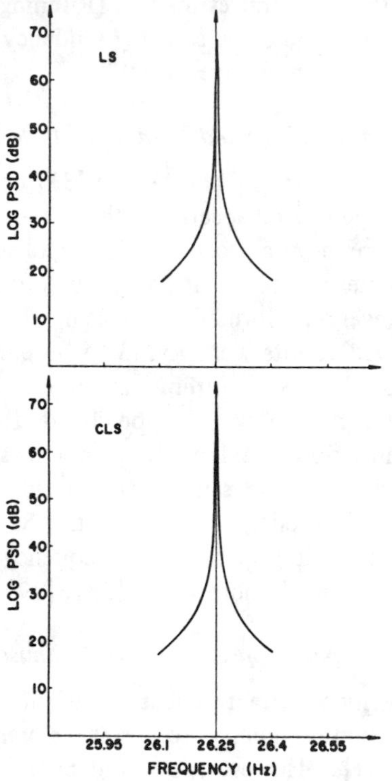

Fig. 2. LS and CLS spectra of the same process realization of Fig. 1. LS and CLS spectra with no line splitting.

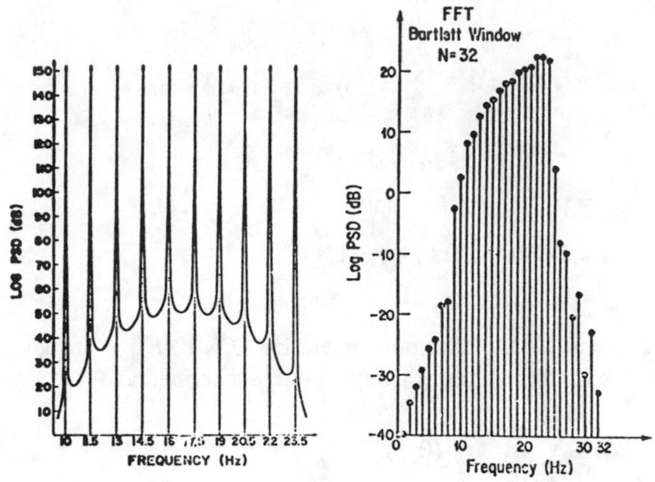

(a) (b)

Fig. 3. Spectra of 32 samples consisting of 10 sinusoids in additive noise (see text). (a) CLS spectrum. (b) FFT spectrum.

correctly positioned the peaks with high-resolution, though the variation in peak heights doesn't sensibly follow the true one. This can be justified on the grounds that frequency information may well be the more important in practice. Fig. 3(b) illustrates the computed spectrum by the conventional FFT algorithm and demonstrates the failure of the conventional spectral estimator to resolve the closely-spaced peaks.

D. Two Closely-Spaced Spectral Peaks of Nonsinusoidal Process

Nuttal [1] and Marple [31] have examined the performance of the LS algorithm for nonsinusoidal sequences generated from a harmonic fourth-order AR process with parameters $a_{4,1} = -2.7607$, $a_{4,2} = 3.8106$, $a_{4,3} = -2.6535$, $a_{4,4} = 0.9238$, driven by white noise Gaussian process. Their experiment was performed again with the LS and CLS algorithms. Specifically, LS and CLS AR spectra were generated from independent 40 samples sequences of the AR process driven by a white Gaussian noise process of variance $Q = 1.0$. The model order was set at the correct value $M = 4$. Fig. 4 shows the estimated 100 overlapped spectra. The results here reveal that both methods yielded sensibly identical estimates.

With "clean" data, there is little difference in performance between the CLS and LS algorithms. The main differences arise in the presence of significant nonstationarity.

E. First-Order Stable Exponential Decay

Swingler [27] has shown that the nonstationarity introduced by a significant transient in a stable first-order AR process leads to unacceptable Burg estimates. The latter limitation also exists in the LS algorithm as it provides the same first-order AR model. In spite of the fact that the CLS algorithm is bidirectional, it is shown that it tolerates exponential decays well.

Taking the deterministic output of an undriven first-order stable AR model

$$X_k = \alpha_{11}^k, \quad k = 1, \cdots, N \tag{8}$$

and acquiring a first-order ($M = 1$) AR model $\{1, -\alpha_{11}\}$, the sequence of (8) was operated on by the LS and CLS schemes. The spectral estimate obtained from the LS algorithm $\{1, -2\alpha_{11}/(1 + \alpha_{11}^2)\}$ has little apparent relevance to the sequence, whereas the spectral estimate derived from the CLS scheme $\{1, -\alpha_{11}(1 + \alpha_{11}^2)/(1 + \alpha_{11}^4)\}$ is indeed very near to the true one. Fig. 5 compares the ratio of the estimated $\hat{\alpha}_{11}$ over the true α_{11} for different values of α_{11} and clearly demonstrates that the CLS estimates are uniformly closer to the true values. The LS algorithm yields to a pole in the model that is too close to the unit circle. Further, as the true paramter α_{11} decreases and approaches zero, the percentage error in the CLS model approaches zero, whereas the LS error approaches 100 percent. If we now proceed to second-order model for the sequence of (8), both schemes estimate two real poles, α_{11} and $1/\alpha_{11}$.

The dynamic range (DR) of the spectrum may be defined as the ratio of the maximum to the minimum values of log $S(f)$, where $S(f)$ is given by (2). For the first-order AR

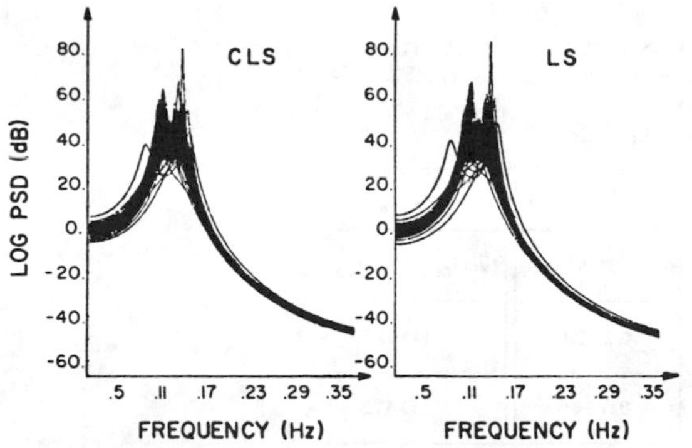

(a) (b)

Fig. 4. Response of LS and CLS spectra to a fourth-order AR process {-2.7606, 3.8106, -2.6536, 0.9238}. (a) Overlapped ensemble of 100 CLS spectra from independent 40 sample realizations. (b) Overlapped ensemble of 100 LS spectra based on the same realizations used in (a).

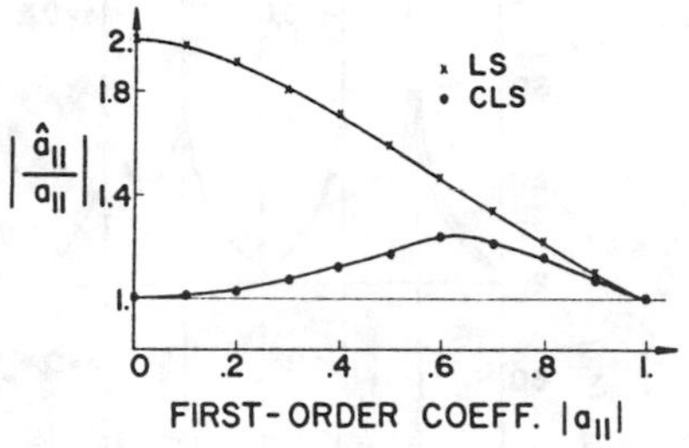

Fig. 5. LS and CLS estimates of first-order AR coefficients from undriven first-order AR process.

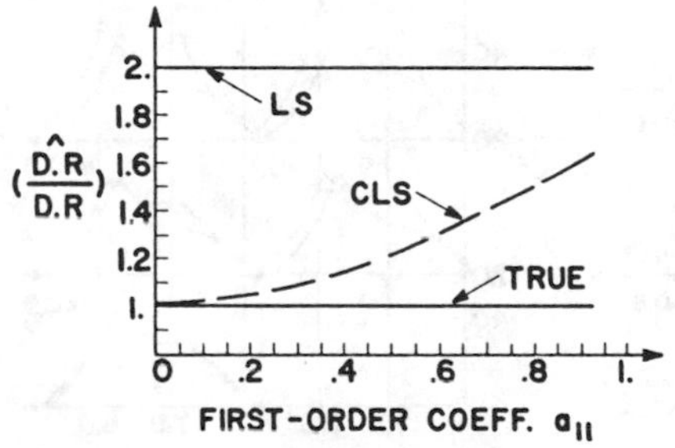

Fig. 6. LS and CLS estimates of the dynamic range (DR) of the spectrum of a first-order AR model.

model, with a pole at α_{11}, this is found to be

$$\mathrm{DR} = \log \{(1 + |\alpha_{11}|)/(1 - |\alpha_{11}|)\}.$$

For the estimate models, $|\alpha_{11}|$ is replaced by the magnitude of the parameter estimate, and the ratio of the estimated dynamic range $\widehat{\mathrm{DR}}$ to the true DR with respect α_{11} using the LS and CLS schemes is shown in Fig. 6. From this figure, it appears that in the model obtained with the LS algorithm, the dynamic range becomes twice the number of decibels of the correct result, while that of CLS is uniformly closer to the true values with the percentage error approaching zero as the parameter α_{11} value approaches zero.

The asymptotic ($N \to \infty$) covariance matrix of the CLS parameter estimates has been derived in Appendix II and is given by (B18). As an example to demonstrate that this result is only valid asymptotically, consider the first-order AR process

$$X_{k+1} + aX_k = W_{k+1} \tag{9}$$

driven by Gaussian white noise process W_{k+1} of variance Q. Then, from (B18) we have the asymptotic variance of the CLS estimate α

$$\operatorname{var}(\alpha)_{\mathrm{CLS}} = 1.66 \frac{Q}{NR(0)}.$$

The asymptotic variance of the LS estimate α is easily estimated.

$$\operatorname{var}(\alpha)_{\mathrm{LS}} = \frac{R}{NR(0)}$$

and thus the CLS asymptotic variance is 1.66 times as large as LS asymptotic variance. This result does not hold for $N = 500$ and the process given by (9). This may be seen by generating CLS and LS parameter estimates from independent 500 stationary samples sequences of the AR process of (9) driven by a white Gaussian noise process of variance $Q = 1$. The model order is set at the correct value $M = 1$. Table I shows the sample mean and variance of 150 CLS and LS estimated parameters. From this table, it is apparent that when the AR parameter of (9) is $a = 0.2$, the ratio of the CLS to LS sample variances is 1.15025, and in the cases where $a = 0.5$ and $a = 0.8$ the ratios are 1.17373 and 1.30176, respectively. Note that the good performance of the LS algorithm in this case is overshadowed by its inability to tolerate first-order exponential decays as demonstrated in Fig. 5.

F. Effect of Additive Transients in Spectral Resolution and Bias

In the field of geophysics (i.e., reflection seismographs [4], annual geophysical time series [33], biomedical applications, etc.), one is often faced with data sets which are nonstationary or have a short time duration. Most of these processes are well modelled by sinusoids corrupted by additive exponential decays, or by sinusoids of fixed frequency whose amplitude is modulated by an envelope function. Therefore, the ability of the LS and CLS algorithms to tolerate additive exponential decay-type nonstationarity in sinusoidal data was tested. By using a process consisting of two closely-spaced cosine waves in additive exponential decay of the form

$$\begin{aligned} X_k = {} & \cos(2\pi f_1 (k - 1)/f_s + \phi) \\ & + \cos(2\pi f_2 (k - 1)/f_s) + \alpha^{(k-1)} \\ & k = 1, 2, \cdots, N \end{aligned} \tag{10}$$

where $f_s = 32, f_1 = 8, \phi = 135°, f_2 = 7.5, \alpha = 0.1, 0.2, \cdots, 0.9$, and acquiring a fifth-order AR model ($M = 5$), its spectra were computed by the LS and CLS algorithms, and shown in Fig. 7(a) ($N = 32$), (b) ($N = 64$), and (c) ($N = 100$). From this figure, it is apparent that the LS scheme fails to resolve the two closely spaced spectral peaks in most of the cases whereas the CLS scheme appears to be more robust, degrading much more slowly as the parameter value (α) increases. Note also that in all the cases where both algorithms resolved the closely spaced spectral peaks the CLS resolution is superior and the frequency

TABLE I
SAMPLE MEAN AND VARIANCE OF 150 CLS AND LS ESTIMATED PARAMETERS FROM INDEPENDENT 500 STATIONARY SAMPLES SEQUENCES OF THE AR PROCESS OF (9)
(Three Cases: a = 0.2, 0.5, 0.8)

AR PARAM.	SAMPLE MEAN		SAMPLE VARIANCE $(\hat{Var}) \times 10^{-2}$		RATIO
a	CLS	LS	CLS	LS	$(\hat{Var})_{CLS}/(\hat{Var})_{LS}$
0.2	0.19950	0.20057	0.25324	0.22016	1.15025
0.5	0.49427	0.49764	0.20970	0.17866	1.17373
0.8	0.79001	0.79448	0.12523	0.09620	1.30176

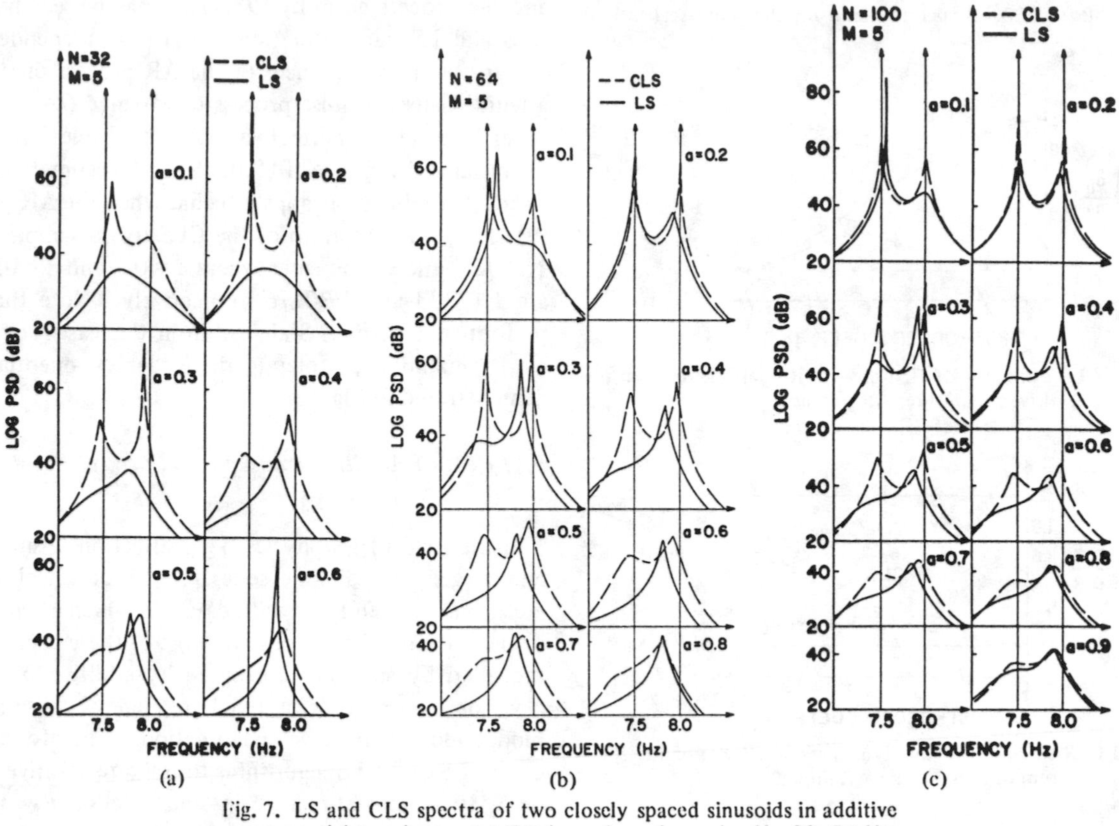

Fig. 7. LS and CLS spectra of two closely spaced sinusoids in additive exponential transients. (a) Total number of samples N = 32; (b) N = 64; and (c) N = 100.

TABLE II
TRUE AND ESTIMATED POLES OF THE PROCESS OF (10) WITH EXPONENTIAL DECAY PARAMETERS VALUE a = 0.2 AND TOTAL NUMBER OF SAMPLES N = 32.

POLES	TRUE	CLS	LS
$P_{1,2}$	0.098017±j0.99518	0.097373±j0.995346	0.09240±j0.950325
$P_{3,4}$	0.000000±j1.0000	0.009770±j1.007137	0.0290±j1.05155
P_5	0.2	0.244	0.38733

bias is less. As an example, Table II illustrates the true and estimated poles of the process of (10) with parameter value $\alpha = 0.2$ and total number of samples $N = 32$. These results demonstrate that the CLS algorithm provides estimates very close to the true ones, while the LS algorithm shows a severe bias in the poles estimates.

G. Effect of Envelope Modulation

The performance of LS and CLS generated spectra for sinusoidal signals modulated by envelope function has also been investigated. By using a noiseless cosine wave ($f = 1$, $f_s = 20$, $N = 41$, $M = 2$) modulated by an envelope function of the form $R(t)$ = at exp $(-at)$, with $\alpha = \frac{1}{3}\pi$, the location of the peak of the estimated spectrum and the resolution (peakiness) index (RI) as functions of initial phase ϕ were determined by the LS and CLS schemes, and shown in Fig. 8. The resolution index is defined as RI $\triangleq 1/(1 - R^2)$, where R is the magnitude of the estimated pair of conjugate poles. From this figure it

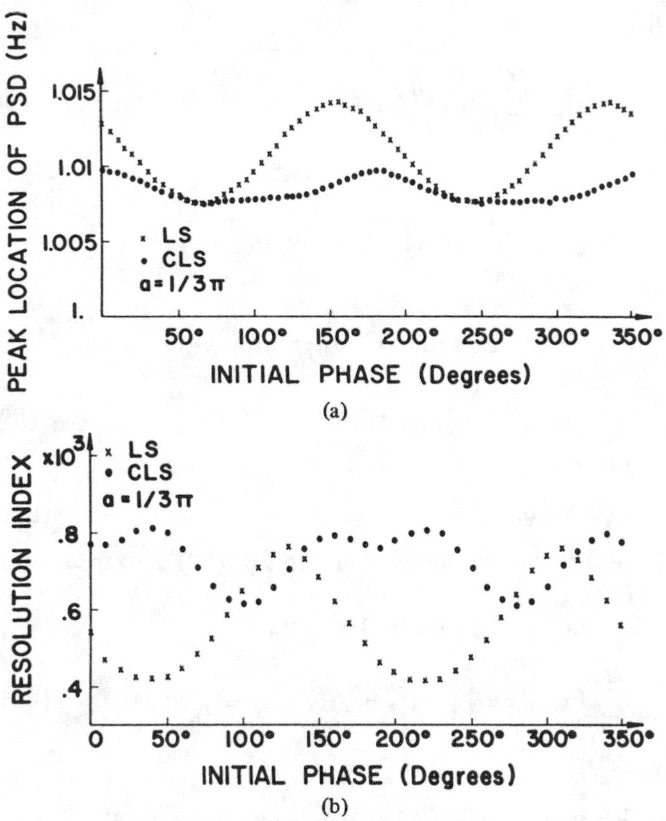

Fig. 8. Effect of envelope function modulation on the locations of spectral peaks and resolutions of LS and CLS generated spectra. Unit amplitude cosine wave ($f = 1, f_s = 20, N = 41, M = 2$) modulated by an envelope function at exp $(-at)$: (a) The spectral peak location as a function of initial phase, (b) the resolution as a function of initial phase.

appears that the CLS algorithm is much less sensitive to initial phase and achieves higher spectral resolution than LS.

H. Extended Performance Index

The performance index shown by (3) may be extended to include all the information available in the process, though at the expense of greater complexity. Initial results have shown that the use of a more general index of the form.

$$C_M = \sum_{m=0}^{N-M-1} \sum_{k=m+1}^{N-M} W(m) \{F^2(k,M)\, E^f(k,m,M) + B^2(k,M)\, E^b(k,m,M)\}$$

where

$$F(k,M) = \sum_{i=0}^{M} \alpha_{M,i} X_{k+M-i}$$

$$B(k,M) = \sum_{i=0}^{M} \alpha_{m,i} X_{k+i}, \qquad \alpha_{M,0} = 1$$

$$E^f(k,m,M) = \sum_{n=0}^{M-1} X_{k+n-m}^2$$

$$E^b(k,m,M) = \sum_{n=0}^{M-1} X_{k+1+n-m}^2$$

provides a satisfactory set of AR parameters for spectral estimation. The sequence $\{W(m)\}$ is selected to be nonnegative and nonincreasing in order to compensate the decreasing summand terms with respect to M.

IV. Conclusion

A fundamental limitation in the use of modern spectral estimation techniques based on AR representations is their intolerance of nonstationarities: additive transients, buildup and decay of process power, envelope function modulation, etc. In many practical applications where stationarity or short-term stationarity cannot be guaranteed a priori this lack of robustness precludes the choice of AR networks for harmonic analysis.

Here a new method for estimating the AR parameters for spectral estimation has been introduced (the CLS algorithm), which exhibits considerably improved performance under these conditions. It has been demonstrated that the CLS algorithm is markedly more robust in the presence of transients and envelope modulation than the comparison LS autoregressive whose derivation requires stationarity. Often, it is the case that robustness in algorithms is traded off for performance in the "ideal" case, here corresponding to stationary or pure sinusoidal data. The simulations and asymptotic calculations show that not to be the case here. The CLS algorithm shows equal or improved performance relative to resolution, bias, spectral variance and line splitting when applied to stationary or sinusoidal data. It is anticipated that development of robust algorithms such as the CLS will expand the practical value of AR spectral methods.

Appendix I

It is shown here t ıat the matrix R_M of (7) is positive-semidefinite. By definition, a symmetric matrix R_M is said to be positive-semidefinite if and only if the scalar

$$v = A^T R_M A \geqslant 0 \tag{A1}$$

for all A in R^n, and the equality holds for some nonzero A in R^n. Since the R_M matrix of (7) is symmetric, substituting for $r_M(i,j)$ in (A1), one obtains

$$v = \sum_{i=0}^{M} \sum_{j=0}^{M} r_M(i,j)\, \alpha_i \alpha_j \tag{A2}$$

or, after some algebra

$$v = \sum_{k=1}^{N-M} \sum_{i=0}^{M} \sum_{j=0}^{M} \alpha_i \alpha_j (E_{M,k}^f X_{k+M-j} X_{k+M-i} + E_{M,k}^b X_{k+j} X_{k+i}) \tag{A3}$$

where

$$E_{M,k}^f = \sum_{\lambda=1}^{M} X_{k+M-\lambda}^2, \qquad E_{M,k}^b = \sum_{\lambda=1}^{M} X_{k+\lambda}^2. \tag{A4}$$

Rearranging the order of the summations, (A3) takes the form

$$v = \sum_{k=1}^{N-M} \left\{ E_{M,k}^f \left(\sum_{i=0}^{M} \alpha_i X_{k+M-i} \right)^2 + E_{M,k}^b \left(\sum_{j=0}^{M} \alpha_j X_{k+j} \right)^2 \right\}. \tag{A5}$$

Since $E^f_{M,k} \geqslant 0$ and $E^b_{M,k} \geqslant 0$ for all k (see (A4)), then certainly $v \geqslant 0$. Q.E.D.

Appendix II

It is shown here that the CLS algorithm provides asymtotically unbiased AR model parameters. The derivation which follows is similar to that adopted by Whitman [34] and Kay [35] for the Yule-Walker estimates.

It is assumed that $\{X_n\}$ is a zero mean Gaussian stationary AR process of order M. Furthermore, it is also assumed that the process is ergodic, i.e., time averages equal ensemble averages. Then, we have

$$\sum_{m=0}^{M} a_{M,m} X_{k+M+m} = W_{k+M} \tag{B1}$$

where $a_{M,0} = 1$ and $W \sim N(0, Q)$, i.i.d. The covariance recursion of (B1) is given by

$$\sum_{m=0}^{M} a_{M,m} R(j-m) = 0, \quad j = k, k+1, \cdots, k+M-1 \tag{B2}$$

where $R(j-m) = E\{X_j X_m\}$. On the other hand, given $\{X_n\}$, $n = 1, \cdots, N$, the CLS algorithm yields the following system of equations

$$\sum_{\tau=0}^{M} \alpha_{M,\tau} r_{\lambda,\tau} = 0, \quad \lambda = 1, 2, \cdots, M, \tag{B3}$$

where $\alpha_{M,0} = 1$,

$$r_{\lambda,\tau} = \frac{1}{N-M} \sum_{k=1}^{N-M} E^f_{M,k} X_{k+M-\lambda} X_{k+M-\tau} + E^b_{M,k} X_{k+\lambda} X_{k+\tau}, \tag{B4}$$

where $E^f_{M,k}$, $E^b_{M,k}$ are given by (5) and (6), respectively. The set of AR parameters $\boldsymbol{\alpha} = [\alpha_{M,1}, \cdots, \alpha_{M,M}]^T$ is the CLS estimate of the true parameter set $\boldsymbol{a} = [a_{M,1}, \cdots, a_{M,M}]^T$. Equation (B4) can also be written in the form

$$r_{\lambda,\tau} = \frac{1}{N-M} \sum_{k=1}^{N-M} \sum_{i=1}^{M} (X^2_{k+M-i} X_{k+M-\lambda} X_{k+M-\tau} + X^2_{k+i} X_{k+\lambda} X_{k+\tau}). \tag{B5}$$

Since $\{X_n\}$ is a zero mean Gaussian random process, then [36]

$$E\{X_i X_j X_k X_l\} = E\{X_i X_j\} \cdot E\{X_k X_l\} + E\{X_i X_k\} \cdot E\{X_j X_l\} + E\{X_i X_l\} \cdot E\{X_j X_k\}.$$

So

$$E\{X^2_m X_k X_l\} = R(0)\, R(k-1) + 2R(k-m)\, R(l-m). \tag{B6}$$

Employing (B5) and (B6) we obtain after some algebra

$$E\{r_{\lambda,\tau}\} = \sum_{i=1}^{M} 2[R(0)\, R(\tau-\lambda) + 2R(\lambda-i)\, R(\tau-i)]. \tag{B7}$$

Now let

$$u_\lambda = \sum_{\tau=0}^{M} a_{M,\tau} r_{\lambda,\tau}, \quad \lambda = 1, \cdots, M$$

then

$$E\{u_\lambda\} = \sum_{\tau=0}^{M} a_{M,\tau} E\{r_{\lambda,\tau}\}. \tag{B8}$$

Substitution of (B7) into (B8) yields

$$E\{u_\lambda\} = 2 \sum_{i=1}^{M} R(0) \left\{ \sum_{\tau=0}^{M} a_{M,{}^{0}\tau} R(\lambda-\tau) + 2R(\lambda-i) \sum_{\tau=0}^{M} a_{M,{}^{0}\tau} R(i-\tau) \right\}$$

$$\lambda = 1, 2, \cdots, M, \tag{B9}$$

and therefore

$$E\{u_\lambda\} = 0. \tag{B10}$$

Note that the two summation terms of (B9) are zero because of (B2).

Combining (B8) with (B3), we obtain

$$\sum_{\tau=0}^{M} (\alpha_{M,\tau} - a_{M,\tau}) r_{\lambda,\tau} = -u_\lambda, \quad \lambda = 1, \cdots, M. \tag{B11}$$

Let

$$\boldsymbol{e} = \boldsymbol{\alpha} - \boldsymbol{a} = [(\alpha_{M,1} - a_{M,1}), \cdots, (\alpha_{M,M} - a_{M,M})]^T$$

$$\boldsymbol{u} = [u_1, \cdots, u_M]^T \quad \text{and}$$

$$\boldsymbol{r} = \{r_{\lambda,\tau}\}, \tau = 1, \cdots, M, \quad \lambda = 1, \cdots, M.$$

So, $dim(\boldsymbol{e}) = M \times 1$, $dim(\boldsymbol{u}) = M \times 1$ and $dim(\boldsymbol{r}) = (M \times M)$. Since, $\alpha_{M,0} = a_{M,0} = 1$ then

$$\boldsymbol{re} = -\boldsymbol{u} \tag{B12}$$

or

$$\underline{e} = -\boldsymbol{r}^{-1} \boldsymbol{u}. \tag{B13}$$

But

$$M \lim_{N \to \infty} \boldsymbol{r} = E\{\boldsymbol{r}\} = \boldsymbol{R}_r$$

so

$$M \lim_{N \to \infty} \boldsymbol{e} = -\boldsymbol{R}_r^{-1} \boldsymbol{u}.$$

Thus, in the limit $E\{\boldsymbol{e}\} = -\boldsymbol{R}_r^{-1}$. $E\{\boldsymbol{u}\} = 0$ and $\boldsymbol{\alpha}$ is *asymptotically unbiased*, i.e., $\lim E\{\boldsymbol{\alpha}\} = \boldsymbol{a}$.

$N \to \infty$

Since we have already shown that the CLS estimates are asymptotically unbiased, it will also be very interesting to compute the covariance of $\boldsymbol{\alpha}, \boldsymbol{\phi}_\alpha$ as $N \to \infty$, where

$$\boldsymbol{\phi}_\alpha = (\boldsymbol{\alpha} - E\{\boldsymbol{\alpha}\})\, (\boldsymbol{\alpha} - E\{\boldsymbol{\alpha}\})^T.$$

Without loss of generality, take only the forward case. Then

$$u_\lambda = \sum_{\tau=0}^{M} a_{M,\tau} \left\{ \frac{1}{N-M} \sum_{k=1}^{N-M} \sum_{i=1}^{M} X^2_{k+M-i} X_{k+M-\lambda} X_{k+M-\tau} \right\},$$

$$\lambda = 1, 2, \cdots, M$$

or

$$u_\lambda = \frac{1}{N-M} \sum_{k=1}^{N-M} \sum_{i=1}^{M} X_{k+M-i}^2 X_{k+M-\lambda} \sum_{\tau=0}^{M} a_{M,\tau} X_{k+M-\tau}. \tag{B14}$$

Combining (B14) with (B1), one obtains

$$u_\lambda = \frac{1}{N-M} \sum_{k=1}^{N-M} \sum_{i=1}^{M} X_{k+M-i}^2 X_{k+M-\lambda} W_{k+M}. \tag{B15}$$

The elements of the covariance matrix $\boldsymbol{\phi}_u$ of $\boldsymbol{u} = [u_1, \cdots, u_M]^T$ are

$$\begin{aligned} \mathrm{Cov}(u_l, u_m) = \frac{1}{(N-M)^2} \cdot \sum_{k=1}^{N-M} \sum_{j=1}^{N-M} \sum_{\rho=1}^{M} \sum_{\mu=1}^{M} & \cdot E\{X_{k+M-\rho}^2 X_{j+M-\mu}^2 X_{k+M-l} X_{j+M-m} \\ & \cdot W_{k+M} W_{j+M}\} \end{aligned} \tag{B16}$$

Since $W_k \sim N(0, Q)$, then X_k is Gaussian and (W_k, X_k) are jointly Gaussian. So, the problem one faces is the determination of [36]

$$m_{221111} = \partial^8 \phi(0, 0, \cdots, 0)/\partial X_1^2 \partial X_2^2 \partial X_3 \partial X_4 \partial X_5 \partial X_6$$

where

$$\phi(\omega_1, \cdots, \omega_6) = \exp - \tfrac{1}{2} \boldsymbol{\omega}^T \boldsymbol{C} \boldsymbol{\omega}$$

$$\boldsymbol{\omega} = [\omega_1, \cdots, \omega_6]^T$$

$$\boldsymbol{C} = \{c_{i,j}\} = E\{X_i X_j\}, \quad i, j = 1, \cdots, 6$$

$$c_{11} = c_{22} = c_{33} = c_{44} = R(0),$$

$$c_{55} = c_{66} = Q, \quad c_{12} = R(j-k+\rho-\mu)$$

$$c_{13} = R(\rho - l), \quad c_{14} = R(j-k+\rho-m),$$

$$c_{15} = 0, \quad c_{16} = E\{X_{k+M-\rho} W_{j+M}\}$$

$$c_{23} = R(k-j+\mu-l), \quad c_{24} = R(\mu-m),$$

$$c_{25} = E\{X_{j+M-\mu} W_{k+M}\}, \quad c_{26} = 0$$

$$c_{34} = R(j-k+l-m), \quad c_{35} = 0,$$

$$c_{36} = E\{X_{k+M-l} W_{j+M}\}, \quad c_{45} = E\{X_{j+M-m} W_{k+M}\}$$

$$c_{46} = 0, \quad c_{56} = Q\delta(k-j).$$

Note that the products $c_{16}c_{25} = 0, c_{25}c_{36} = 0, c_{16}c_{45} = 0$, and $c_{36}c_{45} = 0$ because one of the two terms is always zero.

Since

$$\exp(-X) = 1 - X + \frac{X^2}{2!} - \frac{X^3}{3!} + \frac{X^4}{4!} - \frac{X^5}{5!} + \cdots,$$

we seek the coefficient of $\omega_1^2 \omega_2^2 \omega_3 \omega_4 \omega_5 \omega_6$ which is inherent in $X^4/4!$ term.

After much nontrivial algebra, (B16) becomes

$$\begin{aligned} \boldsymbol{\phi}_u = \mathrm{Cov}(u_l, u_m) = \frac{Q}{N-M} \sum_{\rho=1}^{M} \sum_{\mu=1}^{M} & R^2(0) R(l-m) \\ & + 2R(0) R(\mu-l) R(\mu-m) \\ & + 2R(0) R(\rho-l) R(\rho-m) + 2R^2(\rho-\mu) R(l-m) \\ & + 4R(\rho-\mu) R(\rho-l) R(\mu-m) \\ & + 4R(\rho-\mu) R(\rho-m) R(\mu-l). \end{aligned} \tag{B17}$$

Also

$$\boldsymbol{\phi}_e = E\{ee^T\} = \boldsymbol{\phi}_\alpha = R_r^{-1} \boldsymbol{\phi}_u R_r^{-T}. \tag{B18}$$

Note that the result of (B18) is only valid asymptotically.

Appendix III

It is shown that the LS and CLS estimated AR model of the noise free sinusoidal signal

$$X_k = \cos(\theta k + \phi), \quad k = 0, 1, \cdots, N \tag{C1}$$

is the "correct" one with parameters $(1, -2\cos\theta, 1)$.

The solution to the system of (7) for $M = 2$ yields parameter estimates

$$\alpha_{21} = \Delta_{12}/\Delta_2, \quad \alpha_{22} = \Delta_{22}/\Delta_2 \tag{C2}$$

where

$$\Delta_{12} = r_2(1,2) r_2(2,0) - r_2(2,2) r_2(1,0) \tag{C3}$$

$$\Delta_{22} = r_2(1,2) r_2(1,0) - r_2(1,1) r_2(2,0) \tag{C4}$$

$$\Delta_2 = r_2(1,1) r_2(2,2) - r_2^2(1,2) \tag{C5}$$

with

$$r_2(1,1) = \sum_{k=0}^{N-2} (E_{2,k}^f + E_{2,k}^b) X_{k+1}^2 \tag{C6}$$

$$r_2(2,2) = \sum_{k=0}^{N-2} (E_{2,k}^f X_k^2 + E_{2,k}^b X_{k+2}^2) \tag{C7}$$

$$r_2(1,2) = \sum_{k=0}^{N-2} (E_{2,k}^f X_k X_{k+1} + E_{2,k}^b X_{k+1} X_{k+2}) \tag{C8}$$

$$r_2(1,0) = \sum_{k=0}^{N-2} (E_{2,k}^f X_{k+2} X_{k+1} + E_{2,k}^b X_k X_{k+1}) \tag{C9}$$

$$r_2(2,0) = \sum_{k=0}^{N-2} (E_{2,k}^f + E_{2,k}^b) X_k X_{k+2}. \tag{C10}$$

A. LS Estimates

In this case,

$$E_{2,k}^f = E_{2,k}^b = 1, \quad \text{for every} \quad k. \tag{C11}$$

Substituting (C4) and (C11) into (C2)-(C10), and employing the identity

$$\sum_{k=0}^{N-\lambda} \cos(2\theta k + \gamma) = \cos[\gamma - \theta + (N+1-\lambda)\theta] \cdot \sin(N+1-\lambda)\theta / \sin\theta, \tag{C12}$$

one easily obtains

$$r_2(1,1) = 1 + C_1 S_1 \tag{C13}$$

$$r_2(2,2) = 1 + \cos 2\theta C_1 S_1 \tag{C14}$$

$$r_2(1,2) = r_2(1,0) = \cos\theta \, r_2(1,1) \tag{C15}$$

$$r_2(2,0) = \cos 2\theta + C_1 S_1 \tag{C16}$$

where

$$C_1 = \cos(2\phi + N\theta), \quad S_1 = \sin(N-1)\theta/(N-1)\sin\theta. \tag{C17}$$

By substituting (C13)-(C16) into the system of (C3)-(C5) and then into (C2), the following LS parameters are obtained LS:

$$\alpha_{21} = -2\cos\theta X/X = -2\cos\theta$$

$$\alpha_{22} = X/X = 1, \qquad \text{Q.E.D.}$$

where

$$X = 1 + C_1 S_1.$$

B. CLS Estimates

In this case,

$$E^f_{2,k} = X_k^2 + X_{k+1}^2, \quad E^b_{2,k} = X_{k+1}^2 + X_{k+2}^2. \tag{C18}$$

The substitution of (C1) into the energy-weights of (C18) yields

$$E^f_{2,k} = 1 + \cos\theta\cos(2\theta k + \theta + 2\phi) \tag{C19}$$

$$E^b_{2,k} = 1 + \cos\theta\cos(2\theta k + 3\theta + 2\phi). \tag{C20}$$

Substituting (C1), (C19), and (C20) into (C6)-(C10), and employing the identity of (C12), one obtains

$$r_{\bar{2}}(1,1) = \left(1 + \frac{1}{2}\cos^2\theta\right) C_1 S_1 + \frac{1}{2}\cos^2\theta C_2 S_2 \tag{C21}$$

$$r_2(2,2) = \left(1 + \frac{1}{2}\cos^2\theta\right) + (\cos 2\theta + \cos^2\theta) C_1 S_1 + \frac{1}{2}\cos 3\theta\cos\theta C_2 S_2 \tag{C22}$$

$$r_2(1,2) = \frac{3}{2}\cos\theta + (\cos\theta + \cos^3\theta) C_1 S_1 + \frac{1}{2}\cos\theta\cos 2\theta\, C_2 S_2 \tag{C23}$$

$$r_2(1,0) = \left(1 + \frac{1}{2}\cos 2\theta\right)\cos\theta + (\cos\theta + \cos^3\theta) C_1 S_1 + \frac{1}{2}\cos\theta C_2 S_2 \tag{C24}$$

$$r_2(2,0) = \cos 2\theta + \frac{1}{2}\cos^2\theta + (1 + \cos 2\theta\cos^2\theta) C_1 S_1 + \frac{1}{2}\cos^2\theta C_2 S_2 \tag{C25}$$

where C_1, S_1 are defined by (C17) and

$$C_2 = \cos(4\theta + 2N\theta), \quad S_2 = \sin 2(N-1)\theta/(N-1)\sin 2\theta \tag{C26}$$

By substituting (C21)-(C25) into (C3)-(C5) and then into (C2), the following CLS parameter estimates are obtained CLS:

$$\alpha_{21} = -2\cos\theta Y/Y = -2\cos\theta$$

$$\alpha_{22} = Y/Y = 1$$

where

$$Y = 1 + \frac{1}{4}\cos^4\theta - \frac{5}{4}\cos^2\theta + (\cos^2\theta - \cos^4\theta) C_1 S_1 - \frac{1}{8}\sin^2 2\theta\cos 2\theta\, C_2 S_2 + 0(N-1)^{-2}. \qquad \text{Q.E.D.}$$

REFERENCES

[1] A. H. Nuttal, "Spectral analysis of a univariate process with bad data points, via maximum entropy and linear predictive techniques," Naval Underwater Systems Center, New London, CT, Tech. Rep. 5303, Mar. 16, 1976.

[2] T. J. Ulrych and R. W. Clayton, "Time series modeling and maximum entropy," *Phys. of the Earth and Planet Inter.*, vol. 12, pp. 188-200, 1976.

[3] H. R. Radoski, E. J. Zawalick, and P. F. Fougere, "The superiority of maximum entropy spectrum techniques applied to geomagnetic micropulsations," *Phys. of the Earth and Planetary Inter.*, vol. 12, pp. 208-216, Aug. 1976.

[4] L. J. Griffiths and R. Prieto-Diaz, "Spectral analysis of natural seismic events using autoregressive techniques," *IEEE Trans. Geosci. Electron.*, vol. GE-15, no. 1, pp. 13-24, Jan. 1977.

[5] T. E. Landers and R. T. Lacoss, "Some geophysical applications of autoregressive spectral estimates," *IEEE Trans. Geosci. Electron.*, vol. GE-15, no. 1, pp. 26-32, Jan. 1977.

[6] W. C. Knight, R. G. Proidham, and S. M. Kay, "Digital signal processing for sonar," *Proc. IEEE*, vol. 69, no. 11, pp. 1451-1506, Nov. 1981.

[7] W. F. Gabriel, "Spectral analysis and adaptive array superresolution techniques," *Proc. IEEE*, vol. 68, pp. 654-666, June 1980.

[8] P. D. Scott, "Spectral array processing for electrocardiographic torso wave estimation," in *Proc. 9th Ann. Pittsburgh Conf. on Modelling and Simulation*, vol. 9, p. 991, 1978.

[9] R. B. Blackman and J. W. Tukey, *The Measurement of Power Spectra*. New York: Dover Publications, Inc., 1958.

[10] G. M. Jenkins and D. G. Watts, *Spectral Analysis and Its Applications*. San Francisco, CA: Hoden Bay, Inc., 1968.

[11] J. P. Burg, "Maximum entropy spectral analysis," in *Proc. 37th Meet. of the Society of Exploration Geophysicists*, 1967.

[12] J. Capon, "High-resolution frequency-wavenumber spectrum analysis," *Proc. IEEE*, vol. 57, pp. 1408-1418, 1969.

[13] R. T. Lacoss, "Data adaptive spectral analysis methods," *Geophys.*, vol. 36, pp. 661-675, Aug. 1971.

[14] J. Burg, "The relationship between maximum entropy spectra and maximum likelihood spectra," *Geophys.*, vol. 37, pp. 375-376, Apr. 1972.

[15] A. van den Bos, "Alternative interpretation of maximum entropy spectral analysis," *IEEE Trans. Inform. Theory*, vol. IT-17, pp. 493-494, 1971.

[16] J. Makhoul, "Linear prediction: A tutorial review," *Proc. IEEE*, vol. 63, pp. 561-580, Apr. 1975.

[17] L. Marple, Jr., "Frequency resolution of high resolution spectrum analysis techniques," in *Proc. 1st RADC Spectrum Estimation Workshop*, pp. 19-35, 1975.

[18] S. M. Kay and S. L. Marple, "Sources of and remedies of spectral line-splitting in AR spectrum analysis," in *Proc. IEEE Int. Conf. ASSP*, pp. 151-154, 1979.

[19] J. P. Burg, "A new analysis technique for time series data," presented at *NATO Advanced Study Institute of Signal Processing with Emphasis on Underwater Acoustics*, Aug. 1968.

[20] N. Anderson, "On calculation of filter coefficients for ME spectral analysis," *Geophys.*, vol. 39, pp. 69-72, Feb. 1974.

[21] T. J. Ulrych and T. N. Bishop, "Maximum entropy spectral analysis and autoregressive decomposition," *Rev. Geophys. and Space Phys.*, vol. 13, pp. 183-209, Feb. 1975.

[22] T. J. Ulrych, "Maximum entropy power spectrum of truncated sinusoids," *J. Geophys. Res.*, vol. 77, no. 8, pp. 1396-1400, Mar. 1972.
[23] O. L. Frost, "Power spectrum estimation," in *Proc. NATO Adv. Study Inst. Signal Processing*, pp. 701-736, Sept. 1976.
[24] W. Y. Chen and G. R. Stegen, "Experiments with maximum entropy power spectra of sinusoids," *J. Geophys. Res.*, vol. 79, no. 20, pp. 3019-3022, July 1974.
[25] P. F. Fougere, E. J. Zawalick, H. R. Radoski, "Spontaneous line splitting in maximum entropy power spectrum analysis," *Phys. of the Earth and Planetary Inter.*, vol. 12, pp. 201-207, 1976.
[26] D. N. Swingler, "Frequency errors in MEM processing," *IEEE Trans. Acoust., Speech, Signal Processing*, vol. ASSP-28, no. 2, pp. 257-259, Apr. 1980.
[27] D. N. Swingler, "A comparison between Burg's maximum entropy method and nonrecursive technique for the spectral analysis of deterministic signals," *J. Geophys. Res.*, vol. 84, no. B2, pp. 679-685, Feb. 1979.
[28] C. L. Nikias and P. D. Scott, "Improved spectral resolution by energy-weighted prediction method," *Proc. IEEE Int. Conf. ASSP*, vol. 2, pp. 496-499, Mar. 1981.
[29] P. D. Scott and C. L. Nikias, "Energy-weighted linear predictive spectral estimation: A new method combining robustness and high resolution," *IEEE Trans. Acoust., Speech, Signal Processing*, vol. ASSP-30, no. 2, Apr. 1982.
[30] F. Itakura and S. Saito, "Digital filtering technique for speech analysis and synthesis," presented at the *7th Int. Conf. Acoustics* (Budapest, Hungary) paper 25-C-1, 1971.
[31] L. Marple, "A new autoregressive spectrum analysis algorithm," *IEEE Trans. Acoust., Speech, and Signal Processing*, ASSP-26, no. 4, Aug. 1980.
[32] S. H. Kay and L. Marple, "Spectrum analysis: A modern perspective," *Proc. IEEE*, vol. 69, no. 11, pp. 1380-1418, Nov. 1981.
[33] K. W. Hipel, "Geophysical model discrimination using the Akaike information criterion," *IEEE Trans. Automat. Contr.*, vol. AC-26, no. 2, pp. 358-378, Apr. 1981.
[34] E. C. Whitman, "The Spectral analysis of discrete time series in terms of linear regressive models," *NOLTR*, Tech. Rep., pp. 70-109, June 23, 1970.
[35] S. M. Kay, "Noise compensation for autoregressive spectral estimates," *IEEE Trans. Acoust., Speech, Signal Processing*, vol. ASSP-28, no. 3, June 1980.
[36] A. Papoulis, *Probability, Random Variables, and Stochastic Processes.* New York: McGraw-Hill, 1965.

Recursive Maximum Likelihood Estimation of Autoregressive Processes

STEVEN M. KAY, MEMBER, IEEE

Abstract—A new method of autoregressive parameter estimation is presented. The technique is a closer approximation to the true maximum likelihood estimator than that obtained using linear prediction techniques. The advantage of the new algorithm is mainly for short data records and/or sharply peaked spectra. Simulation results indicate that the parameter bias as well as the variance is reduced over the Yule-Walker and the forward-backward approaches of linear prediction. Also, spectral estimates exhibit more resolution and less spurious peaks. A stable all-pole filter estimate is guaranteed. The algorithm operates in a recursive model order fashion, which allows one to successively fit higher order models to the data.

I. INTRODUCTION

ALL presently used autoregressive (AR) parameter estimators are approximations to the maximum likelihood estimator (MLE) [1], [2]. The use of the MLE for parameter estimation can be justified for large data records on the basis of its asymptotic unbiased and minimum variance properties [1]. For shorter data records, it is well known that the MLE of the AR parameters does not attain the Cramer-Rao lower bound, i.e., it is not efficient [1]. Hence, the use of the MLE in this case can only be justified in retrospect on the basis of the estimation accuracy afforded in relation to other possible estimators. Accepting the desirability of determining the MLE for the AR parameters, the actual computation of the MLE proves to be a difficult nonlinear maximization problem. Hence, approximations are generally made in order to obtain an AR estimator which is computationally efficient. To understand the limitation which these approximations impose upon the resultant estimator, consider a zero-mean Gaussian AR process X_t of order p. If

$$X = [X_1 \; X_2 \cdots X_N]^T$$

is observed, then the probability density function of X is

$$p(X|a, \sigma_\epsilon^2) = \frac{1}{(2\pi)^{N/2}|R|^{1/2}} \exp\left(-\frac{1}{2}X^T R^{-1} X\right) \tag{1}$$

where $a = [a_1 a_2 \cdots a_p]^T$ is the AR filter parameter vector, σ_ϵ^2 is the excitation white noise variance, and $R = E(XX^T)$ is the covariance matrix. The true MLE of a, σ_ϵ^2 is found by maximizing $p(X|a, \sigma_\epsilon^2)$ or, equivalently, $\ln p(X|a, \sigma_\epsilon^2)$ with respect to a, σ_ϵ^2. To separate a from σ_ϵ^2 in (1), one notes that the autocorrelation function $R(k) = E(X_t X_{t+k})$ can be written as [1]

$$R(k) = \sigma_\epsilon^2 R_f(k). \tag{2}$$

$R_f(k)$ is the filter autocorrelation function which is defined as

$$R_f(k) = \mathcal{Z}^{-1}\left[\frac{1}{A(z)\,A(z^{-1})}\right]$$

where $\mathcal{Z}^{-1}$ denotes the inverse Z transform and

$$A(z) = 1 + \sum_{k=1}^{p} a_k Z^{-k}.$$

Then using (2) in (1),

$$p(X|a, \sigma_\epsilon^2) = \frac{1}{(2\pi)^{N/2}\sigma_\epsilon^N |R_f|^{1/2}} \exp\left(-\frac{1}{2\sigma_\epsilon^2}X^T R_f^{-1} X\right) \tag{3}$$

where $[R_f]_{ij} = R_f(i-j)$ for $i, j = 1, 2, \cdots, N$ and R_f depends only upon the filter parameters.

If we first maximize $\ln p$ with respect to σ_ϵ^2, we obtain [1]

$$\frac{\partial \ln p}{\partial \sigma_\epsilon^2} = -\frac{N}{2}\frac{1}{\sigma_\epsilon^2} + \frac{1}{2\sigma_\epsilon^4} X^T R_f^{-1} X = 0$$

$$\Rightarrow \sigma_\epsilon^2 = \frac{1}{N} X^T R_f^{-1} X. \tag{4}$$

Note that σ_ϵ^2 as given by (4) would be the MLE if R_f were replaced by its MLE in terms of the MLE of the AR filter parameters.

Substituting (4) into (3), we now need to maximize

$$\frac{1}{(2\pi)^{N/2}\left(\frac{1}{N}X^T R_f^{-1} X\right)^{N/2} |R_f|^{1/2}} e^{-N/2}$$

over a or, equivalently, maximize

$$\frac{|R_f^{-1}|^{1/N}}{\frac{1}{N} X^T R_f^{-1} X}. \tag{5}$$

The usual approach is to ignore the $|R_f^{-1}|^{1/N}$ factor and minimize some approximation to $(1/N)\, X^T R_f^{-1} X$ [2]. For large N, this is a valid approximation since [3]

$$\lim_{N \to \infty} |R_f^{-1}|^{1/N} = 1 \tag{6}$$

so that for large N, $|R_f^{-1}|^{1/N}$ does not depend upon a.

However, for small N, and in particular for peaky spectra, this approximation is poor as we now show by example [4].

Manuscript received November 17, 1981; revised July 29, 1982. This work was supported in part by the Office of Naval Research under ONR Grant N00014-81-K-0144.

The author is with the Department of Electrical Engineering, University of Rhode Island, Kingston, RI 02881.

Reprinted from *IEEE Trans. Acoust., Speech, Signal Processing*, vol. ASSP-31, pp. 56–65, Feb. 1983.

Consider an AR process of order 1. Then

$$R_f^{-1} = \begin{bmatrix} 1 & a_1 & & 0\cdots 0 \\ a_1 & 1+a_1^2 & a_1 & 0\cdots 0 \\ 0 & a_1 & 1+a_1^2 & a_1 \cdots 0 \\ & & \vdots & \\ 0 & \cdots & 0 \quad a_1 & 1+a_1^2 \quad a_1 \\ 0 & \cdots & 0 \quad 0 \quad a_1 & 1 \end{bmatrix} \tag{7}$$

It is easily verified that [1] [see also (14)]

$$|R_f^{-1}| = 1 - a_1^2 \tag{8}$$

so that

$$|R_f^{-1}|^{1/N} = (1 - a_1^2)^{1/N}. \tag{9}$$

Clearly, as $N \to \infty$, (9) approaches 1 regardless of the value of a_1. For finite data records, N will need to be larger as $|a_1| \to 1$ for this approximation to be valid. Note that as $|a_1| \to 1$, the power spectral density becomes more sharply peaked.

If one attempts to maximize (5) directly, it is soon discovered that a nonlinear optimization routine must be employed. As an example, again consider the AR(1) process. Then, using (7) and (9) in (5), it is found that one must maximize

$$\frac{(1-a_1^2)^{1/N}}{\frac{1}{N}(S_{00} + 2a_1 S_{01} + a_1^2 S_{11})}$$

where

$$S_{00} = \sum_{t=1}^{N} X_t^2$$

$$S_{01} = \sum_{t=1}^{N-1} X_t X_{t+1}$$

$$S_{11} = \sum_{t=1}^{N-2} X_{t+1}^2.$$

The approach described in Section II is to maximize (5) recursively. The procedure is exactly analogous to Burg's technique for estimation of the reflection coefficients [5]. The fundamental difference is that here we maximize the likelihood function for each model order, while Burg minimizes the sum of the forward and backward prediction error powers. The result is that the estimation algorithm does not require a nonlinear optimization, but instead, one needs to solve for the roots of p cubic equations. In Section III, the character of the roots of the cubic equation is examined, while in Section IV, the performance of the new algorithm is described. Finally, a summary and some conclusions are given in Section V.

II. Derivation of Algorithm

From (5), we must maximize

$$l' = \frac{|R_f^{-1}|^{1/N}}{\frac{1}{N} X^T R_f^{-1} X}$$

or, equivalently, $l = N \ln l'$ where

$$l = -N \ln \frac{1}{N} X^T R_f^{-1} X + \ln |R_f^{-1}|. \tag{10}$$

To maximize l recursively, we assume that X_t is an AR$(n-1)$ process and that l has been maximized with respect to $\boldsymbol{a}_{n-1} = [a_1^{(n-1)}\ a_2^{(n-1)} \cdots a_{n-1}^{(n-1)}]^T$ where $a_i^{(j)}$ is the ith AR parameter for the jth-order model. We then fix $\boldsymbol{a}_{n-1}$ at the values determined from the previous maximization and express $\boldsymbol{a}_n$ in terms of $\boldsymbol{a}_{n-1}$ by the Levinson-Durbin algorithm, i.e. [5],

$$a_i^{(n)} = \begin{cases} a_i^{(n-1)} + K_n a_{n-i}^{(n-1)} & i = 1, 2, \cdots, n-1 \\ K_n & i = n \end{cases} \tag{11}$$

Thus, l_n (where the subscript n now denotes the assumed model order) depends only on K_n, the nth reflection coefficient. Next we maximize with respect to K_n and use (11) to generate $\boldsymbol{a}_n$. In this fashion, we recursively maximize the likelihood function, and in the process estimate the AR parameters for model orders $n = 1, 2, \cdots, p$. Now,

$$l_n = -N \ln \frac{1}{N} X^T R_{f_n}^{-1} X + \ln |R_{f_n}^{-1}|. \tag{12}$$

To determine $|R_{f_n}^{-1}|$, we note that for an AR(n) process [5],

$$B^T R B = P \tag{13}$$

where

$$B = \begin{bmatrix} 1 & a_1^{(1)} & a_2^{(2)} \cdots a_n^{(n)} & 0 & 0 \cdots 0 \\ 0 & 1 & a_1^{(2)} \cdots a_{n-1}^{(n)} & a_n^{(n)} & 0 \cdots 0 \\ & & \ddots & & \\ 0 & 0 & 0 \quad \cdots & & 1 \end{bmatrix} \quad (N \times N)$$

$$P = \text{diag}\,(P_0, P_1, \cdots, P_{n-1}, P_n, P_n, \cdots, P_n) \quad (N \times N)$$

and P_i is the prediction error power for the ith-order predictor and $P_n = \sigma_\epsilon^2$. Since $|B| = 1$, from (13),

$$|R| = |P| = \prod_{i=0}^{n-1} P_i \prod_{i=n}^{N-1} P_n$$

$$|R_{f_n}|^{-1} = |\sigma_\epsilon^2 R^{-1}| = \sigma_\epsilon^{2N}/|R|$$

$$= \frac{1}{\prod_{i=0}^{n-1} P_i/P_n \prod_{i=n}^{N-1} P_n/P_n} = \prod_{i=0}^{n-1} P_n/P_i,$$

and finally, using the fact that $P_i = R(0) \prod_{j=1}^{i} (1 - K_j^2)$,

$$|R_{f_n}^{-1}| = \prod_{i=1}^{n} (1 - K_i^2)^i$$

$$= (1 - K_n^2)^n \prod_{i=1}^{n-1} (1 - K_i^2)^i$$

$$= (1 - K_n^2)^n |R_{f_{n-1}}^{-1}| \tag{14}$$

which is the desired recursive relationship.

To determine a recursive relationship for $X^T R_{f_n}^{-1} X$, we first note that [1]

$$\mathcal{E}_n = X^T R_{f_n}^{-1} X = a_n'^T S_n a_n' \tag{15}$$

where

$$a_n' = [1\ \ a_1^{(n)}\ \ a_2^{(n)} \cdots a_n^{(n)}]^T$$

$$[S_n]_{ij} = S_{ij} = \sum_{t=1}^{N-i-j} X_{t+i}X_{t+j}$$

$$i = 0, 1, \cdots, n;\quad j = 0, 1, \cdots, n.$$

S_n is a symmetric matrix, although it is not positive definite (see the Appendix for a counterexample). However, $a_n'^T S_n a_n'$ will be positive if a_n' is a minimum phase sequence. This is because then R_{f_n} is a valid autocorrelation matrix, and hence R_{f_n} and also $R_{f_n}^{-1}$ will be positive definite.

We now use (11) in vector form and partition a_n' and S_n as follows:

$$\mathcal{E}_n = \begin{bmatrix} 1 \\ (I_{n-1} + K_n J_{n-1})\, a_{n-1} \\ K_n \end{bmatrix}^T \cdot \begin{bmatrix} \underset{1\times 1}{S_{00}} & \underset{1\times(n-1)}{q_{n-1}^T} & \underset{1\times 1}{S_{0n}} \\ \underset{(n-1)\times 1}{q_{n-1}} & \underset{(n-1)\times(n-1)}{S_{n-2}} & \underset{(n-1)\times 1}{p_{n-1}} \\ \underset{1\times 1}{S_{n0}} & \underset{1\times(n-1)}{p_{n-1}^T} & \underset{1\times 1}{S_{nn}} \end{bmatrix} \cdot \begin{bmatrix} 1 \\ (I_{n-1} + K_n J_{n-1})\, a_{n-1} \\ K_n \end{bmatrix}$$

where I_{n-1} is the identity matrix of dimension $(n-1)\times(n-1)$ and

$$J_{n-1} = \begin{bmatrix} 0 & & & 1 \\ & & \cdot & \\ & 1 & & \\ 1 & & & 0 \end{bmatrix}$$

with dimension $(n-1)\times(n-1)$.

But a_n' may be further decomposed as

$$a_n' = \begin{bmatrix} a_{n-1}' \\ 0 \end{bmatrix} + K_n \begin{bmatrix} 0 \\ J_n a_{n-1}' \end{bmatrix} = \begin{bmatrix} a_{n-1}' \\ 0 \end{bmatrix} + K_n \begin{bmatrix} 0 \\ b_{n-1}' \end{bmatrix}$$

where $b_{n-1}' = J_n a_{n-1}'$.

After some algebra, one can show that

$$\mathcal{E}_n = \mathcal{E}_{n-1} + 2C_nK_n + d_nK_n^2$$

where

$$C_n = a_{n-1}'^T \begin{bmatrix} q_{n-1}^T & S_{0n} \\ S_{n-2} & p_{n-1} \end{bmatrix} b_{n-1}'$$

$$d_n = b_{n-1}'^T \begin{bmatrix} S_{n-2} & p_{n-1} \\ p_{n-1}^T & S_{nn} \end{bmatrix} b_{n-1}'$$

$$b_{n-1}' = [a_{n-1}^{(n-1)}\ \ a_{n-2}^{(n-1)} \cdots a_1^{(n-1)}\ \ 1]^T. \tag{16}$$

Note that $\mathcal{E}_{n-1}$, C_n, d_n do not depend upon K_n. Also $\mathcal{E}_n/N$ and d_n/N can be interpreted as the estimates of the forward and backward prediction error powers, respectively, while C_n/N is an estimate of the covariance between the forward and backward prediction errors. Thus, using (15) and (16) in (12),

$$l_n = -N \ln \frac{1}{N}(\mathcal{E}_{n-1} + 2C_nK_n + d_nK_n^2) + \ln |R_{f_{n-1}}^{-1}| + n \ln (1 - K_n^2). \tag{17}$$

Since it is assumed that $K_1, K_2, \cdots, K_{n-1}$ have already been estimated, $\mathcal{E}_{n-1}$, C_n, d_n, and $|R_{f_{n-1}}^{-1}|$ can be computed and so are treated as constants. To maximize l_n with respect to K_n, we set $\partial l_n/\partial K_n = 0$;

$$\Rightarrow K_n^3 + \frac{(N-2n)\,C_n}{(N-n)\,d_n} K_n^2 - \frac{n\,\mathcal{E}_{n-1} + Nd_n}{(N-n)\,d_n} K_n - \frac{NC_n}{(N-n)\,d_n} = 0. \tag{18}$$

The solution of this cubic produces the estimate of K_n.

Once K_n has been determined from (18), the estimate of $\sigma_\epsilon^{2(n)}$, the excitation white noise variance estimate for the nth-order AR model, may be obtained from (4), (15), and (16) as

$$\sigma_\epsilon^{2(n)} = 1/N\, \mathcal{E}_n = 1/N\,(\mathcal{E}_{n-1} + 2C_nK_n + d_nK_n^2). \tag{19}$$

It is shown in Section III that for data sets not consisting of n or fewer sinusoids, $\sigma_\epsilon^{2(n)}$ is positive and decreases or stays the same at each step of the recursion. The entire algorithm is summarized in Fig. 1.

It should be noted that to ease the computational burden of determining C_n and d_n, one can use [6]

$$[S_n]_{ij} = [S_{n-1}]_{ij} \qquad i,j = 0, 1, \cdots, n-1$$

$$[S_n]_{ij} = [S_{n-1}]_{i-1,j-1} - X_iX_j - X_{N+1-i}X_{N+1-j} \qquad i = n,\ j = 1, 2, \cdots, n$$

$$[S_n]_{ij} = [S_n]_{ji} \qquad i = 1, 2, \cdots, n,\ j = n. \tag{20}$$

III. Character of Cubic Roots

In order to estimate K_n, (18) must be solved. Since the equation is a cubic, we have either three real roots or one real root and two complex conjugate roots. It is now proven that the solution of (18) will produce at least one real root within the interval $(-1, 1)$.

Proof: From (17), it is seen that we are maximizing

$$f(K_n) = \frac{(1 - K_n^2)^{n/2}}{\left[\frac{1}{N}(\mathcal{E}_{n-1} + 2C_nK_n + d_nK_n^2)\right]^{N/2}} \tag{21}$$

with respect to K_n. (Recall that $|R_{f_{n-1}}^{-1}|$ is a constant.)

We first prove that if X_t, $t = 1, 2, \cdots, N$ does not consist solely of n or fewer sinusoids, the denominator of (21) is positive for $|K_n| \leq 1$. Recall that

$$\mathcal{E}_n = \mathcal{E}_{n-1} + 2C_nK_n + d_nK_n^2 = X^T R_{f_n}^{-1} X. \tag{22}$$

If $|K_n| < 1$ and assuming that $|K_i| < 1$ for $i = 1, 2, \cdots,$

$$\mathcal{E}_0 = S_{00}$$

for $n = 1$: $C_n = S_{01}$, $d_1 = S_{11}$

to find K_1 solve

$$K_1^3 + \frac{N-2}{N-1}\frac{C_1}{d_1}K_1^2 - \frac{{}_0 + Nd_1}{(N-1)\,d_1} - \frac{N}{N-1}\frac{C_1}{d_1} = 0$$

and choose root within [−1, 1].

$$a_1^{(1)} = K_1$$

$$\mathcal{E}_1 = S_{00} + 2K_1S_{01} + K_1^2S_{11}$$

$$\sigma_\epsilon^{2(1)} = \frac{1}{N}\mathcal{E}_1$$

For $n = 2, 3, \cdots, p$:

$$C_n = a_{n-1}'^{T}\begin{bmatrix} q_{n-1}^T & S_{0n} \\ S_{n-2} & p_{n-1} \end{bmatrix} b_{n-1}'$$

$$d_n = b_{n-1}'^{T}\begin{bmatrix} S_{n-2} & p_{n-1} \\ p_{n-1}^T & S_{nn} \end{bmatrix} b_{n-1}'^{T}.$$

to find K_n solve (for $d_n > 0$)*

$$K_n^3 + \frac{N-2n}{N-n}\frac{C_n}{d_n}K_n^2 - \frac{n \quad {}_{n-1} + Nd_n}{(N-n)\,d_n} - \frac{N}{N-n}\frac{C_n}{d_n} = 0$$

and choose root within [−1, 1].

$$a_i^{(n)} = \begin{cases} a_i^{(n-1)} + K_n a_{n-i}^{(n-1)}, & i = 1, 2, \cdots, n-1 \\ K_n, & i = n \end{cases}$$

$$\mathcal{E}_n = \mathcal{E}_{n-1} + 2C_nK_n + d_nK_n^2$$

$$\sigma_\epsilon^{2(n)} = \frac{1}{N}\mathcal{E}_n$$

where

$$a_{n-1}' = [1 \;\; a_1^{(n-1)} \;\; a_2^{(n-1)} \cdots a_{n-1}^{(n-1)}]^T$$

$$b_{n-1}' = [a_{n-1}^{(n-1)} \;\; a_{n-2}^{(n-1)} \cdots a_1^{(n-1)} \;\; 1]^T$$

$$[S_n]_{ij} = S_{ij} = \sum_{t=1}^{N-i-j} x_{t+i}x_{t+j} \qquad \begin{matrix} i = 0, 1, \cdots, n \\ j = 0, 1, \cdots, n \end{matrix}$$

$$S_n = \begin{bmatrix} \underset{1\times 1}{S_{00}} & \underset{1\times(n-1)}{q_{n-1}^T} & \underset{1\times 1}{S_{0n}} \\ \underset{(n-1)\times 1}{q_{n-1}} & \underset{(n-1)\times(n-1)}{S_{n-2}} & \underset{(n-1)\times 1}{p_{n-1}} \\ \underset{1\times 1}{S_{n0}} & \underset{1\times(n-1)}{p_{n-1}^T} & \underset{1\times 1}{S_{nn}} \end{bmatrix}$$

*For $d_n < 0$ see Section III for discussion of roots.

Fig. 1. Summary of recursive MLE algorithm.

$n - 1$, then $\mathcal{E}_n > 0$ since R_{f_n} and hence $R_{f_n}^{-1}$ is positive definite. If $|K_n| = 1$, $R_{f_n}^{-1}$ is no longer positive definite (see (9) for an example). However, $\mathcal{E}_n > 0$ for nonsinusoidal data sets. To show this, use (2) and (13) in (22):

$$\mathcal{E}_n = X^TB\sigma_\epsilon^2P^{-1}B^TX.$$

But

$$\sigma_\epsilon^2 = P_n = R(0)\prod_{i=1}^{n}(1 - K_i^2)$$

$$\Rightarrow \sigma_\epsilon^2P^{-1} = \text{diag}\left(\frac{P_n}{P_0}, \frac{P_n}{P_1}, \cdots, \frac{P_n}{P_{n-1}}, 1, 1, \cdots, 1\right)$$

$$= \text{diag}\left(\prod_{i=1}^{n}(1-K_i)^2, \prod_{i=2}^{n}\cdot(1-K_i^2), \cdots, 1-K_n^2, 1, 1, \cdots, 1\right).$$

For $|K_n| = 1$ and $K_i < 1$, $i = 1, 2, \cdots, n-1$,

$$\sigma_\epsilon^2P^{-1} = \text{diag}\,(\underbrace{0, 0, \cdots, 0}_{n \text{ terms}}, \underbrace{1, 1, \cdots, 1}_{N-n \text{ terms}}).$$

Thus,

$$\mathcal{E}_n = \sum_{i=n+1}^{N} g_i^2$$

where g_i is the ith component of B^TX:

$$g_i = \sum_{k=1}^{N}[B^T]_{ik}X_k.$$

But from (13), for $i = n+1, \cdots, N$,

$$g_i = \sum_{k=0}^{n} a_k^{(n)}X_{i-k}$$

where $a_0^{(n)} = 1$.

Thus,

$$\mathcal{E}_n = \sum_{i=n+1}^{N}\left(\sum_{k=0}^{n} a_k^{(n)}X_{i-k}\right)^2.$$

Clearly, $\mathcal{E}_n = 0$ if and only if $\sum_{k=0}^{n} a_k^{(n)}X_{i-k} = 0$ for $i = n+1, \cdots, N$. These conditions could be met if $X_1, X_2, \cdots, X_N$ consisted of n or fewer exponentials. However, since $K_n = \pm 1$, from (11) we have that

$$a_k^{(n)} = a_k^{(n-1)} + K_na_{n-k}^{(n-1)} = a_k^{(n-1)} \pm a_{n-k}^{(n-1)}$$
$$= \pm a_{n-k}^{(n)}$$

is a symmetric or antisymmetric sequence. Thus, $\mathcal{E}_n = 0$ if and only if

$$\sum_{k=0}^{n} a_k^{(n)}X_{i-k} = \pm\sum_{k=0}^{n} a_{n-k}^{(n)}X_{i-k}$$
$$= \pm\sum_{k=0}^{n} a_k^{(n)}X_{i-n+k} = 0 \qquad i = n+1, \cdots, N.$$

Hence, $X_1, X_2, \cdots, X_n$ must also consist of n or fewer exponentials when reversed in time. We conclude that for $\mathcal{E}_n = 0$, a necessary condition is that the data must consist of n or fewer pure sinusoids. The condition is not sufficient, however, in that for $\mathcal{E}_n = 0$, the $a_k^{(n-1)}$ sequence must be chosen appropriately. We henceforth neglect the possibility that the data set is purely harmonic since it occurs with low probability. We can now conclude from (21) that $f(1) = f(-1) = 0$ since the denominator of (21) is positive. Furthermore, for $|K_n| < 1$, $f(K_n) > 0$ and $f(K_n)$ is continuous and differentiable. Therefore, by Rolle's theorem [7], there is at least one maximum within $(-1, 1)$. To complete the proof, we must verify that $|K_i| < 1, i = 1, 2, \cdots, n-1$. Since the above results are valid for any n, this condition results from induction.

To actually determine the roots, let

$$p = \frac{N - 2n}{N - n} \frac{C_n}{d_n}$$

$$q = -\frac{n\mathcal{E}_{n-1} + Nd_n}{(N-n)\, d_n}$$

$$r = -\frac{N}{N-n} \frac{C_n}{d_n}$$

$$a = \tfrac{1}{3}(3q - p^2)$$

$$b = \tfrac{1}{27}(2p^3 - 9pq + 27r).$$

If $b^2/4 + a^3/27 > 0$, then there is one real root given by

$$K_n = A + B - p/3$$

where

$$A = \left[-\frac{b}{2} + \sqrt{\frac{b^2}{4} + \frac{a^3}{27}}\right]^{1/3}$$

$$B = \left[-\frac{b}{2} - \sqrt{\frac{b^2}{4} + \frac{a^3}{27}}\right]^{1/3}$$

If $b^2/4 + a^3/27 \leqslant 0$, there will be three real roots which can be found as follows:

$$K_{n_1} = 2\sqrt{-a/3}\cos\theta - p/3$$

$$K_{n_2} = 2\sqrt{-a/3}\cos(\theta + 2\pi/3) - p/3$$

$$K_{n_3} = 2\sqrt{-a/3}\cos(\theta + 4\pi/3) - p/3$$

where $\cos 3\theta = 3b/(2a\sqrt{-a/2})$.

In this case, the root which maximizes (21) is chosen if more than one root falls within $(-1, 1)$.

In the majority of cases, there will be three real roots with only one falling inside $(-1, 1)$. This is assured if $d_n > 0$. We now prove some properties of the roots of (18).

1) If $d_n > 0$, then there will be three real roots, $K_{n_1}, K_{n_2}, K_{n_3}$ where

$$-1 < K_{n_1} < 1$$

$$K_{n_2} > 1$$

$$K_{n_3} < 1. \tag{23}$$

Proof: Before proceeding, we replace $f(K_n)$ by $g(K_n)$ where

$$g(K_n) = \frac{(|1 - K_n^2|)^{n/2}}{[|1/N(\mathcal{E}_{n-1} + 2C_nK_n + d_nK_n^2)|]^{N/2}} \quad \text{for } |K_n| > 1$$

$$= f(K_n) \quad \text{for } |K_n| \leqslant 1.$$

This is implicit in (17) in order to make ln meaningful for $|K_n| > 1$. We examine two cases. In the first case, if $\mathcal{E}_n = \mathcal{E}_{n-1} + 2C_nK_n + d_nK_n^2$ does not have any real roots for $K_n > 1$, $K_n < -1$, then $g(K_n)$ is continuous outside the interval $(-1, 1)$. Now as $|K_n| \to \infty$,

$$g(K_n) \to \frac{|K_n|^n}{\left(\frac{d_n}{N}\right)^{N/2} |K_n|^N} = \left(\frac{N}{d_n}\right)^{N/2} |K_n|^{n-N} \to 0$$

for $N > n$. Also, $g(1) = g(-1) = 0$. Therefore, by a slight generalization of Rolle's theorem to account for a nonfinite internal, there must be at least one root of (18) in the interval $(-\infty, -1)$ and at least one real root of (18) within the interval $(1, \infty)$. Since there is at least one real root within $(-1, 1)$ and there are at most three roots, then (23) must follow. Case 2 occurs if $\mathcal{E}_n$ has real roots outside the interval $(-1, 1)$. Note that the preceding argument is invalid since $g(K_n)$ is discontinuous outside the $[-1, 1]$ interval. The roots of $\mathcal{E}_n$ cannot occur within $[-1, 1]$ since there $\mathcal{E}_n > 0$. The roots of $\mathcal{E}_n$ are the solutions of

$$\mathcal{E}_n = \mathcal{E}_{n-1} + 2C_nK_n + d_nK_n^2 = 0$$

or

$$K_n^2 + \frac{2C_n}{d_n} K_n + \frac{\mathcal{E}_{n-1}}{d_n} = 0.$$

The real roots, K_{n_1}, K_{n_2}, must both be positive or both be negative since

$$K_{n_1}K_{n_2} = \frac{\mathcal{E}_{n-1}}{d_n} > 0.$$

Note that $d_n > 0$ by assumption and $\mathcal{E}_{n-1} > 0$. If, for instance, the roots of $\mathcal{E}_n$ are both positive, then there must be at least one real maximum of $g(K_n)$ within $(-\infty, -1)$ by a generalization of Rolle's theorem. Also, $1/g(K_n) = 0$ at $K_n = K_{n_1}$, K_{n_2} and is continuous and differentiable within (K_{n_1}, K_{n_2}). Thus, there is at least one maximum of $1/g(K_n)$ within this interval, and hence $g(K_n)$ has at least one minimum within $(1, \infty)$. By the same argument as before, then, (23) is proven.

2) If $d_n = 0$, there will be one real root within $(-1, 1)$ and one root outside this interval. Note that in this case, (18) should be replaced by

$$K_n^2 - \frac{n}{N - 2n} \frac{\mathcal{E}_{n-1}}{C_n} K_n - \frac{N}{N - 2n} = 0.$$

Proof: As before, we are guaranteed at least one real root within $(-1, 1)$. But

$$K_{n_1}K_{n_2} = -\frac{N}{N - 2n} < -1 \quad \text{for } N > 2n.$$

Thus, the roots are of opposite sign, and if $|K_{n_1}| < 1$, then $|K_{n_2}| > 1$ and vice versa.

3) If $d_n < 0$, then there will be at least one real root within $(-1, 1)$. If there is more than one real root within this interval, then the one that maximizes (21) should be chosen. Simulation results indicate that if $d_n < 0$, there is always one real root within $(-1, 1)$ and two complex roots, although a proof is not available. To see how d_n can be negative, consider the following example. We compute d_2, which is the coefficient of K_2^2 in $\mathcal{E}_2$ where $\mathcal{E}_2 = X^T R_{f_2}^{-1} X$. It can be shown that [9]

$$R_{f_2}^{-1} = \begin{bmatrix} 1 & a_1 & a_2 & 0 & 0 & 0 \cdots 0 \\ a_2 & 1 + a_1^2 & a_1(1 + a_2) & a_2 & 0 & 0 \cdots 0 \\ a_2 & a_1(1 + a_2) & 1 + a_1^2 + a_2^2 & a_1(1 + a_2) & a_2 & 0 \cdots 0 \\ 0 & a_2 & a_1(1 + a_2) & 1 + a_1^2 + a_2^2 & a_1(1 + a_2) & 0 \cdots 0 \\ & & & \vdots & & \\ 0 & \cdots & a_2 & a_1(1 + a_2) & 1 + a_1^2 + a_2^2 & a_1(1 + a_2) \quad a_2 \\ 0 & \cdots & 0 & a_2 & a_1(1 + a_2) & 1 + a_1^2 \quad a_1 \\ 0 & \cdots & 0 & 0 & a_2 & a_1 \quad 1 \end{bmatrix}$$

From (11), we have

$$a_1 = K_1(1 + K_2)$$

$$a_2 = K_2$$

so that the contribution to d_2 of each term is

$$1 + a_1^2 = 1 + K_1^2(1 + K_2)^2 \Rightarrow K_1^2$$

$$a_1(1 + a_2) = K_1(1 + K_2)(1 + K_2) \Rightarrow K_1$$

$$1 + a_1^2 + a_2^2 = 1 + K_1^2(1 + K_2)^2 + K_2^2 \Rightarrow K_1^2 + 1.$$

Thus, d_2 is given by

$$d_2 = X^T \begin{bmatrix} 0 & 0 & 0 & 0 & 0 & 0 & 0 & 0 \\ 0 & K_1^2 & K_1 & 0 & 0 & 0 & 0 & 0 \\ 0 & K_1 & 1 + K_1^2 & K_1 \cdots 0 & 0 & 0 & 0 \\ & & & \vdots & & & & \\ 0 & 0 & 0 & 0 \cdots K_1 & 1 + K_1^2 & K_1 & 0 \\ 0 & 0 & 0 & 0 \cdots 0 & K_1 & K_1^2 & 0 \\ 0 & 0 & 0 & 0 \cdots 0 & 0 & 0 & 0 \end{bmatrix} X$$

$$= \sum_{t=3}^{N-1} (X_t + K_1 X_{t-1})^2 - X_{N-1}^2(1 - K_1^2).$$

If, for instance, $X_t = (-K_1)^t$, $t = 2, \cdots, N - 1$,

$$d_2 = -K_1^{2(N-1)}(1 - K_1^2) < 0 \quad \text{for } |K_1| < 1.$$

4) If $N >> n$, then the roots will be at $-1, +1$, and $-C_n/d_n$.

Proof: From (18), we see that for $N >> n$, the cubic equation becomes

$$K_n^3 + \frac{C_n}{d_n} K_n^2 - K_n - \frac{C_n}{d_n} = 0$$

whose roots are $-1, +1, -C_n/d_n$.

We now show that the estimate of the excitation noise variance satisfies

$$0 < \sigma_\epsilon^{2(n)} \leqslant \sigma_\epsilon^{2(n-1)}.$$

Proof: To show that $\sigma_\epsilon^{2(n)} > 0$, we note that

$$\sigma_\epsilon^{2(n)} = 1/N \, \mathcal{E}_n.$$

As shown previously, $\mathcal{E}_n > 0$ if X_t does not consist solely of sinusoids. We note that to estimate K_n, we have maximized

$$f(K_n) = \frac{(1 - K_n^2)^{n/2}}{[1/N(\mathcal{E}_{n-1} + 2C_n K_n + d_n K_n^2)]^{N/2}}$$

over the interval $[-1, 1]$ to produce $\hat{K}_n$. Thus,

$$\frac{(1 - \hat{K}_n^2)^{n/2}}{[1/N(\mathcal{E}_{n-1} + 2C_n \hat{K}_n + d_n \hat{K}_n^2)]^{N/2}} > \frac{(1 - K_n^2)^{n/2}}{[(1/N(\mathcal{E}_{n-1} + 2C_n K_n + d_n K_n^2)]^{N/2}} \tag{24}$$

for all $K_n \neq \hat{K}_n$. Assuming that $\hat{K}_n \neq 0$,

$$\frac{1}{\left[\frac{1}{N}\mathcal{E}_n\right]^{N/2}} > \frac{(1 - \hat{K}_n^2)^{n/2}}{\left[\frac{1}{N}\mathcal{E}_n\right]^{N/2}} > \frac{1}{\left[\frac{1}{N}\mathcal{E}_{n-1}\right]^{N/2}}$$

where the last inequality follows from (24) with $K_n = 0 \neq \hat{K}_n$. Therefore, $\mathcal{E}_n < \mathcal{E}_{n-1}$, and hence $\sigma_\epsilon^{2(n)} < \sigma_\epsilon^{2(n-1)}$. Also, from (16), $\mathcal{E}_n = \mathcal{E}_{n-1}$, and hence $\sigma_\epsilon^{2(n)} = \sigma_\epsilon^{2(n-1)}$ if and only if $\hat{K}_n = 0$.

Finally, it should be mentioned that if X_t is an AR(1) process, then the solution of (18) produces the exact MLE [10]. In the next section, we present some simulation results concerning the performance of the algorithm, which we will denote as the recursive MLE (RMLE).

IV. Performance of Algorithm

A. Parameter Estimation

To determine the performance of the RMLE algorithm, we first examine the means and variances of the estimated AR parameters and compare these to the Yule-Walker (YW) and forward-backward (FB) techniques of linear prediction [2]. The FB method is reported to be the most accurate of the linear prediction techniques, at least for spectral estimation

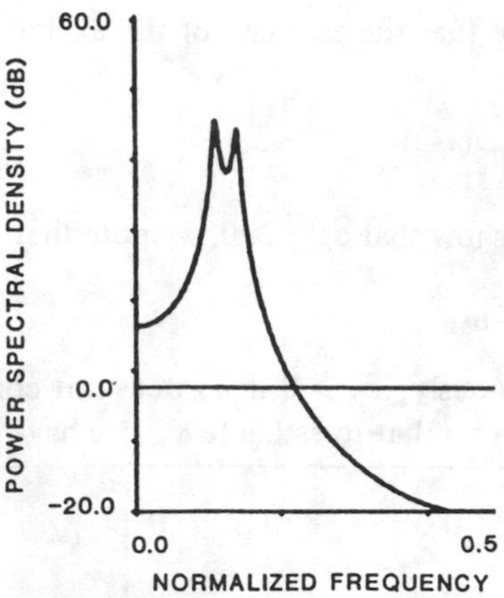

Fig. 2. True power spectral density for AR(4) process.

[11]-[13]. The YW technique, in which the biased autocorrelation estimator is employed, is popular in speech processing (known as the autocorrelation method of linear prediction [14]). The process chosen for the simulation is an AR(4) process with coefficients $a_1 = -2.76070$, $a_2 = 3.81060$, $a_3 = -2.65350$, $a_4 = 0.92380$, $\sigma_\epsilon^2 = 1$, and power spectral density as given in Fig. 2. The sharpness of this power spectral density indicates that the approximation $|R_f^{-1}|^{1/N} \approx 1$ will be a poor one, especially for short data records. The results of a simulation, using 1000 realizations and $p = 4$, is shown in Table I for the data record lengths $N = 11, 50$. It is seen that the YW approach produces heavily biased and highly variable estimates. This result is to be expected from Nuttall's work [11]. The FB approach is much more accurate than the YW approach. For the short data record, however, the RMLE algorithm has less bias and about half the variance of the FB approach. For the longer data record, the two techniques are comparable. It should be noted that the FB technique does not guarantee a stable AR filter [11], while the RMLE does.

TABLE I
COMPARISON OF AR PARAMETER ESTIMATION METHODS

$N = 11$ Data Points	a_1	a_2	a_3	a_4
True AR Parameters	-2.76070	3.81060	-2.65350	0.92380
Mean-YW	-1.01527	0.61949	0.04466	0.05735
Mean-FB	-2.59333	3.42185	-2.27303	0.77262
Mean-RMLE	-2.62975	3.51086	-2.37156	0.81412
Variance-YW	3.13780	10.29627	7.33884	0.75840
Variance-FB	0.11598	0.52574	0.49303	0.08462
Variance-RMLE	0.10960	0.33371	0.26227	0.04148

$N = 50$ Data Points	a_1	a_2	a_3	a_4
True AR Parameters	-2.76070	3.81060	-2.65350	0.92380
Mean-YW	-1.30552	0.92378	-0.05153	0.06719
Mean-FB	-2.72368	3.71024	-2.54984	0.87639
Mean-RMLE	-2.71515	3.69876	-2.54345	0.87765
Variance-YW	2.22544	8.56804	6.90835	0.74526
Variance-FB	0.00758	0.04138	0.04219	0.00858
Variance-RMLE	0.01076	0.04485	0.04069	0.00741

B. Spectral Estimation

Spectral estimates were obtained for the RMLE and the FB techniques for the same AR(4) process. We arbitrarily let the estimate of σ_ϵ^2 be the true value $\sigma_\epsilon^2 = 1$ so that the comparison is between the ability of the algorithms to estimate a_1, a_2, a_3, a_4. Typical spectral estimates are shown in Figs. 3 and 4 for both a short and longer data record, $N = 11, 50$, and for $p = 4$. The true power spectral density is shown at the top. The same set of data was used to generate the side-by-side comparison. These spectral estimates are typical and were not chosen for aesthetic or other reasons. General conclusions are difficult. However, it would appear from Figs. 3 and 4 and from numerous other simulations that the following holds.

1) For short data records, the RMLE more often resolves the peaks than the FB method. In one experiment, for instance, which consisted of 50 realizations, for $N = 11$, $p = 4$ it was found that the RMLE resolved the two peaks 39 times, while the FB did so only 31 times. The RMLE, however, tends to be less accurate in determining the peak locations. In another experiment consisting of 1000 realizations for $N = 11$, $p = 4$, the peak locations of the spectral estimate were computed and analyzed for statistical accuracy. If only one peak was found, then both peak locations were set equal to the observed peak location. The results are shown in Table II. It is seen that the overall bias is about the same. However, the RMLE, in this case, has about twice the variance.

2) For longer data records, the RMLE and FB techniques are comparable, although the latter seems to lead to overly sharp peaks. This is probably why some researchers prefer the FB for sinusoidal data.

Finally, a simulation was run to determine the susceptibility of the RMLE and FB to spurious peaks. For $N = 11$ data points and $p = 6$ (one spurious peak is possible), 25 realizations are overlayed in Fig. 5. It is seen that the RMLE exhibits less spurious peaks than the FB.

V. SUMMARY AND CONCLUSIONS

A new method for estimating the parameters of an AR process has been described. The estimator, termed the recursive maximum likelihood estimator (RMLE), is a closer approximation to the true MLE than that obtained by linear prediction methods. As such an improvement in statistical accuracy is obtained for short data records and/or sharply peaked spectra. The RMLE operates in a recursive fashion in that it allows one to successively fit higher order AR models to the data. In addition, the estimated all-pole filter is guaranteed to be stable.

The improvement in statistical accuracy has been demonstrated with respect to estimation of the AR parameters. If one wishes to ultimately estimate the power spectral density, however, then a fidelity criterion must first be specified before a reasonable evaluation of the RMLE can be made. The RMLE does well with respect to resolvability of closely spaced peaks and the absence of spurious peaks. Its performance is not as good as the forward-backward method for spectral peak location estimation. Further work, however, is required for a definitive evaluation of the relative merits of RMLE for spectral estimation.

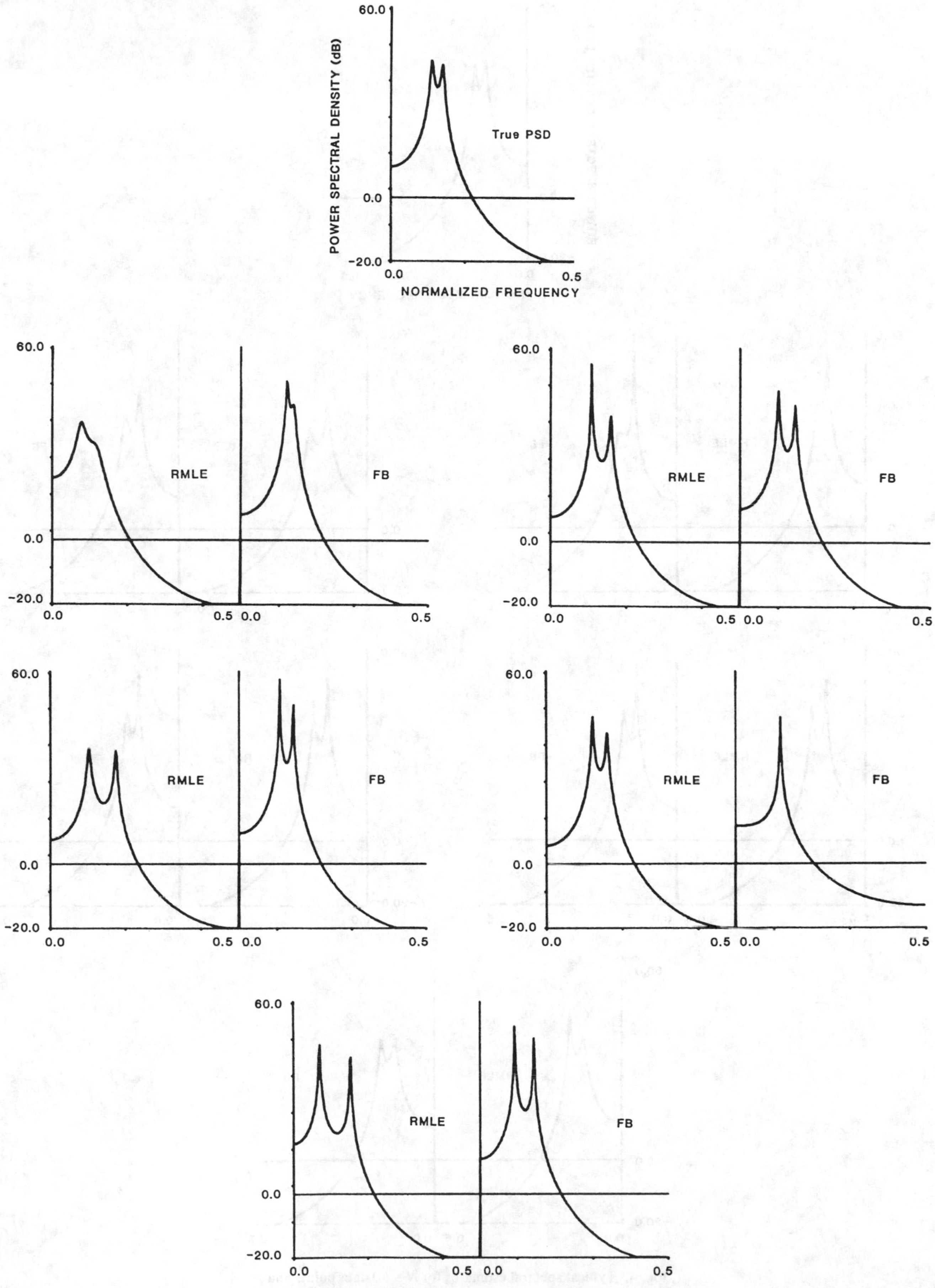

Fig. 3. Typical spectral estimates for $N = 11$ data points and $p = 4$.

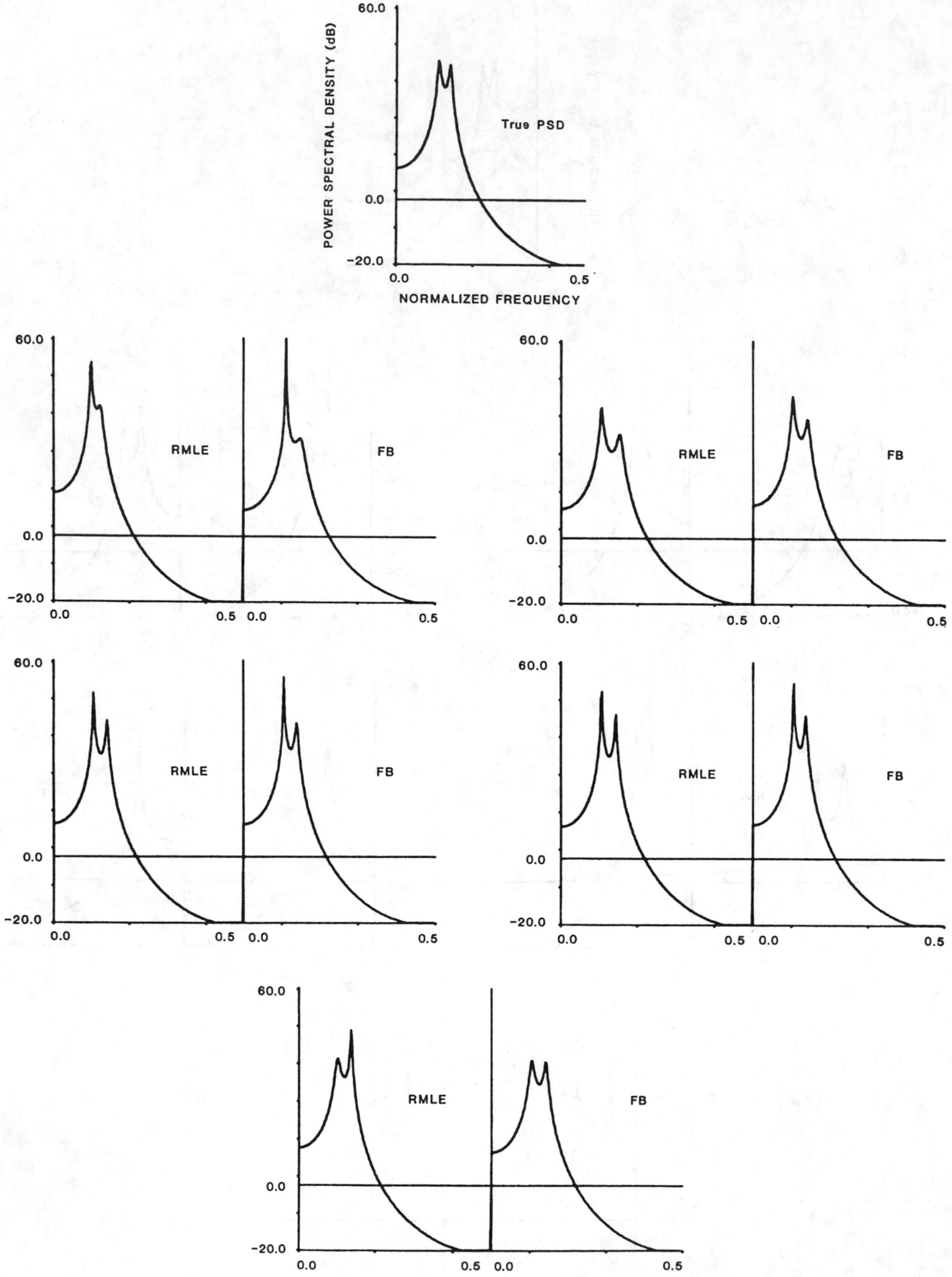

Fig. 4. Typical spectral estimates for $N = 50$ data points and $p = 4$.

TABLE II
COMPARISON OF AR SPECTRAL PEAK ESTIMATION METHODS

True Peak Locations	0.109375	0.138672
Mean–FB	0.107936	0.14281
Mean–RMLE	0.114066	0.137434
Variance–FB	0.279×10^{-3}	0.372×10^{-3}
Variance–RMLE	0.559×10^{-3}	0.535×10^{-3}

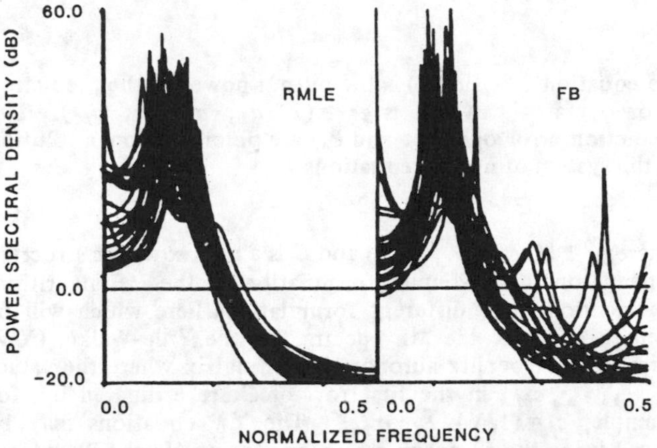

Fig. 5. Ensemble of spectral estimates for $N = 11$ data points and $p = 6$.

APPENDIX

COUNTEREXAMPLE FOR POSITIVE DEFINITENESS OF S_n

To show that S_n is not positive definite, consider $n = 2$, $V = 3$. Then

$$S_2 = \begin{bmatrix} S_{00} & S_{01} \\ S_{10} & S_{11} \end{bmatrix} = \begin{bmatrix} X_1^2 + X_2^2 + X_3^2 & X_1X_2 + X_2X_3 \\ X_1X_2 + X_2X_3 & X_1^2 \end{bmatrix}.$$

If $X_1 = 0$, $|S_2| = -X_2^2X_3^2 < 0$.

ACKNOWLEDGMENT

The author would like to thank N. Juddell of the University of Rhode Island, Kingston, for the many valuable discussions pertaining to the work reported herein.

REFERENCES

[1] G. E. P. Box and G. M. Jenkins, *Time Series Analysis–Forecasting Control*. San Francisco, CA: Holden-Day, 1970, ch. 7.
[2] S. Kay and S. L. Marple, Jr., "Spectrum analysis–A modern perspective," *Proc. IEEE*, Nov. 1981.
[3] U. Grenander and G. Szego, *Toeplitz Forms and Their Applications*. Berkeley, CA: Univ. of California Press, 1958.
[4] N. Juddell, "Maximum likelihood estimation of autoregressive and autoregressive-moving average processes," M.S. thesis, Univ. Rhode Island, Kingston, in preparation.
[5] J. P. Burg, "Maximum entropy spectral analysis," Ph.D. dissertation, Stanford Univ., Stanford, CA, 1975.
[6] B. Dickinson, "Two recursive estimates of autoregressive models based on maximum likelihood," *J. Statist. Computation Simulation*, vol. 7, pp. 85-92, 1978.
[7] E. Wilson, *Advanced Calculus*. New York: Dover, 1958.
[8] S. Selby, *Standard Mathematical Tables*. Cleveland, OH: Chemical Rubber Co., 1972, pp. 103-105.
[9] J. Wise, "The autocorrelation function and the spectral density function," *Biometrika*, no. 42, pp. 151-159, 1955.
[10] T. W. Anderson, *The Statistical Analysis of Time Series*. New York: Wiley, 1971, pp. 353-355.
[11] A. H. Nuttall, "Spectral analysis of a univariate process with bad data points, via maximum entropy and linear predictive techniques," in *NUSC Sci. and Eng. Studies, Spectral Estimation*, NUSC, New London, CT.
[12] T. J. Ulrych and R. W. Clayton, "Time series modeling and maximum entropy," *Phys. Earth and Planetary Interiors*, vol. 12, pp. 188-200, Aug. 1976.
[13] D. N. Swingler, "A comparison between Burg's maximum entropy method and a nonrecursive technique for the spectral analysis of deterministic signals," *J. Geophys. Res.*, vol 84, pp. 674-685, Feb. 1974.
[14] J. Makhoul, "Linear prediction: A tutorial review," *Proc. IEEE*, pp. 561-580, Apr. 1975.

A Linear Programming Approach to the Estimation of the Power Spectra of Harmonic Processes

SHLOMO LEVY, COLIN WALKER, TAD J. ULRYCH, AND PETER K. FULLAGAR

Abstract—A new approach to the estimation of high resolution power spectra of harmonic processes is presented. This approach uses linear programming (LIP) to exploit the spiky character of the L_1 norm. Results for the LIP power spectral estimate are contrasted with those using various formulations of the maximum entropy algorithm.

I. Introduction

In recent years, a very powerful technique for determining high resolution power spectra has been extensively discussed in the literature. This technique, the maximum entropy (ME) method, exhibits high resolution properties because it does not entail the truncation of the autocorrelation function. Another, and complementary, point of view is to consider the ME estimate as the power spectral density of the autoregressive process which best describes the data.

In this note, we wish to present a rather different approach to the problem of determining the power spectrum of a noisy harmonic process. This formulation uses linear programming (LIP) to find the solution, with minimum L_1 norm, to a set of underdetermined linear equations. We show, in the ensuing discussion, that this approach introduces the constraint of "spikiness" into the computation of the power spectrum, a constraint which is appropriate in the frequency analysis of harmonic processes.

In order to illustrate the performance of our LIP technique, we will contrast the results of this method with those of various formulations of the ME approach.

Theoretical Considerations

The Maximum Entropy Spectrum

We will first of all take a brief look at the salient features of the ME method. Details may be found in [1] and [2].

The ME spectral density estimator is given by

$$S_{\mathrm{ME}}(f) = \frac{P_M \Delta t}{\left|1 - \sum_{n=1}^{M} a_n \exp(-2\pi i f n \Delta t)\right|^2} \tag{1}$$

where $\boldsymbol{a}^T = (a_1, a_2, \cdots, a_M)$ is the Mth-order prediction filter, or, alternatively, $\boldsymbol{a}^T$ is the vector of autoregressive parameters which model the input data x_t as

$$x_t = a_1 x_{t-1} + a_2 x_{t-2} + \cdots + a_M x_{t-M} + e_t. \tag{2}$$

In (2) we have assumed that $\Delta t = 1$ for convenience in writing the equation. P_M in (1) is the output power of the prediction error operator $\boldsymbol{g}$ which is $\boldsymbol{g}^T = (1, -a_1, -a_2, \cdots, -a_M)$. The prediction error operator and P_M are obtained from a solution of the system of normal equations

$$C\boldsymbol{g} = \boldsymbol{p} \tag{3}$$

where $\boldsymbol{p}^T = (P_M, 0, 0, \cdots, 0)$ and C is a matrix whose structure depends on the particular formulation of the ME algorithm. We mention three different formulations here which will be used to compute the ME spectra. In the Yule-Walker (YW) form, C is a Toeplitz autocorrelation matrix where the values $\{c_0, c_1, \cdots, c_M\}$ in the first row of C are estimated by, for example, $c_j = (1/N) \sum_i x_i x_{i+j}$. The YW equations may be solved using the Levinson recursion scheme. In the Burg form of the normal equations, C is also a Toeplitz autocorrelation matrix, but the c_j's are determined during the recursive solution of (3). Finally, in the least squares (LS) approach proposed in [3], C is a symmetric (but non-Toeplitz) matrix. The LS form has been shown to have resolution properties which, for short data sets, are superior to those of the YW and Burg estimates.

One other aspect of the ME estimator which is important to mention is the actual determination of the power. As we have mentioned, $S_{\mathrm{ME}}(f)$, given in (1), is a spectral density estimator [4], and consequently, the power may be determined by integration. Although this procedure is straightforward when spectral peaks are well resolved, difficulty arises when peaks are only partly resolved. An interesting approach to this problem is discussed in [5] where it is suggested that the power of the spectral component with frequency f be computed by means of the residue of the function $S_{\mathrm{ME}}(z)/z$ evaluated at $z = z_k = \exp(2\pi i f_k)$. As we will see, the LIP approach, presented below, provides an estimate of the power directly.

The "Spikiness" of the L_1 Norm

In many practical problems, a time series is comprised of contributions from a small number of harmonics, and consequently, the power spectrum consists of a limited number of spikes. Mathematically, the problem of determining a complete power spectrum from a short data segment is underdetermined, and therefore admits an infinite number of solutions in general. If the form of the unknown solution can be restricted on physical grounds, it may be possible to determine the most physically appropriate solution by maximizing or minimizing a particular norm. In the case when a "spiky" solution is sought, the solutions with minimum L_1 norm or minimum L_2 norm might be considered appropriate choices. We have found, however, that the minimization of the L_1 norm using linear programming has yielded superior results. This enhanced "spikiness" or simple structure of the solution with minimum L_1 norm, in contrast to the more distributed character of the solution with minimum L_2 norm, will now be demonstrated with a simple numerical example in which we

Manuscript received August 12, 1980; revised February 17, 1981.

S. Levy, C. Walker, and T. J. Ulrych are with the Department of Geophysics and Astronomy, University of British Columbia, Vancouver, B.C., Canada V6T 1W5.

P. K. Fullagar was with the Department of Geophysics and Astronomy, University of British Columbia, Vancouver, B.C., Canada V6T 1W5. He is now with W.M.C., Belmont, West Australia 6104.

Reprinted from *IEEE Trans. Acoust., Speech, Signal Processing*, vol. ASSP-30, pp. 675–679, Aug. 1982.

determine three unknowns from two equations. In other words, we consider an underdetermined problem, precisely the kind of problem with which we will be faced in the determination of the power spectrum.

Consider two equations of the form

$$a_{11}x_1 + a_{12}x_2 + a_{13}x_3 = d_1, \qquad (4)$$
$$a_{21}x_1 + a_{22}x_2 + a_{23}x_3 = d_2$$

where $x^T = (x_1, x_2, x_3)$ is the vector of unknowns and the matrix $A = \{a_{mn}\}$ and the vector $d^T = (d_1, d_2)$ are known.

First of all, we consider the least squares (or L_2 norm) determination of x. In this approach, $\Sigma_{i=1}^3 x_i^2$ is minimized subject to exactly satisfying the constraint equations (4). This solution corresponds to the one which satisfies a criterion of minimum energy. It is also commonly known as the smallest solution. If we call the L_2 solution ${}^2s^T = {}^2s_1, {}^2s_2, {}^2s_3)$ we obtain the set of equations

$$\frac{\partial}{\partial^2 s_j}\left[\sum_{i=1}^{3} {}^2s_i^2 - 2\sum_{k=1}^{2} \lambda_k(a_{k1}{}^2s_1 + a_{k2}{}^2s_2 + a_{k3}{}^2s_3 - d_k)\right] = 0$$

where the λ's are Lagrange multipliers.

Performing the differentiation, we obtain

$$A^T\lambda = {}^2s \qquad (5)$$

where $\lambda^T = (\lambda_1, \lambda_2)$.

Since (4) may be written as $A\,{}^2s = d$ for the least squares solution, (4) and (5) may be combined to give

$${}^2s = A^T(AA^T)^{-1}\, d. \qquad (6)$$

As a simple example we consider a true spiky vector $x^T = (0, 1, 0)$ with

$$A = \begin{pmatrix} 1 & 2 & 3 \\ 3 & 1 & -1 \end{pmatrix}.$$

The data vector is, therefore, $d^T = (2, 1)$. Substituting these values into (6), we obtain

$${}^2s = \frac{1}{150}\begin{pmatrix} 1 & 3 \\ 2 & 1 \\ 3 & -1 \end{pmatrix}\begin{pmatrix} 11 & -2 \\ -2 & 14 \end{pmatrix}\begin{pmatrix} 2 \\ 1 \end{pmatrix} = \begin{pmatrix} \frac{1}{3} \\ \frac{1}{3} \\ \frac{1}{3} \end{pmatrix}.$$

Thus, the least squares solution for this particular example is absolutely flat.

Let us now find the solution of the above example with minimum L_1 norm using linear programming. In the Simplex algorithm, all the unknowns are positive, so we designate each unknown ${}^1s_i$ as ${}^1s_i = a_i - b_i$ where $a_i \geqslant 0$ and $b_i \geqslant 0$. This allows the solution vector to have both positive and negative elements. The solution with minimum L_1 norm is obtained using the Simplex algorithm to minimize $\Sigma_{i=1}^3 a_i + b_i = \Sigma_{i=1}^3 |a_i - b_i|$. The result obtained is, in this case, ${}^1s^T = (0, 1, 0)$ and is, in fact, the "true" solution x. The LIP approach returns the true spiky solution, in contrast to the flat L_2 solution, because at most two of the six variables $\{a_i, b_i : i = 1, 2, 3\}$ can be nonzero when only two constraint equations are specified. Thus, it is a combination of the properties of the Simplex algorithm with those of the L_1 norm which provide the LIP solution with the required spiky structure. Another example of the use of linear programming to delineate spiky structure is in the recent deconvolution study presented in [6].

It is possible to demonstrate the distributed character of the least squares solution more generally in the case where the coefficient matrix is composed of orthonormal basis vectors. Suppose the columns of the $N \times N$ matrix U are comprised of a set $\{u_j; j = 1, N\}$ of orthonormal vectors. Any N vector P can be expressed as a linear combination of these basis vectors, i.e.,

$$P = \sum_{j=1}^{N} r_j u_j$$

where $r_j = P^T \cdot u_j$ is the jth expansion coefficient. If only M expansion coefficients are known ($M < N$), then the smallest approximation to P is given by

$$P_L = \sum_{j=1}^{M} r_j u_j = P - \sum_{j=M+1}^{N} r_j u_j. \qquad (7)$$

If P represents the power spectrum of a harmonic process, then it is clear from (7) that its spikiness will be seriously compromised by the superposition of a linear combination of smoothly varying basis functions, such as cosines. Hence, the least squares solution P_L will not, in general, retain the spiky character of P.

Details of the LIP Solution

The LIP solution is developed in two parts. In the first part, we determine the frequencies and power of the spectral components. In the second part, Fourier coefficients for these frequencies are determined to enable the extrapolation of the data outside the known interval. This procedure has important implications for data processing [7].

The relationship between the autocorrelation function $r(\tau)$ and the power spectrum $P(f)$ is

$$r(\tau) = P(0) + 2\sum_{n=1}^{\infty} P(n)\cos(2\pi f n\tau).$$

For a sampling interval $\Delta t = 1$ and N_f points in the discrete Fourier transform, we may write

$$r_m = P(0) + 2\sum_{n=1}^{N_f-1} P(n) \cdot \cos[mn\pi/(N_f - 1)] \qquad m = 0, 1, \cdots, M$$

where M is the length of the autocorrelation function.

Since the autocorrelation function is known imperfectly, the above equation should not be satisfied exactly. Consequently, denoting by e_m the error associated with each estimated r_m, we obtain the following system of constraint equations:

$$r_m + e_m = P(0) + 2\sum_{n=1}^{N_f-1} P(n) \cdot \cos[mn\pi/(N_f - 1)] \qquad m = 1, 2, \cdots, M. \qquad (8)$$

We note that the addition of e_m is, in effect, equivalent to the addition of $2M$ variables to the problem because each e_m is expressed as a difference of positive values.

In light of our discussion above, we choose as our objective function the minimization of $\Sigma_{n=0}^{N_f-1} P(n)$ in order to obtain a spectrum consisting of isolated spikes.

In addition to the constraint equations in (8), a constraint on the sum of the e_m's is introduced. Specifically, we constrain $\sum_{M-1}^{M} |e_m| = \phi$. (Otherwise, our solution will result in all the $P(n)$'s equal to zero and e_m's equal to r_m.) As we do not know the distribution of the e_m's, ϕ is determined empirically and is set to a percentage of $r_0 M$. It is important to note from this latter constraint that we do not insist on the smallest error. The Simplex algorithm is free to distribute the errors in a manner most conducive to the minimization of the objective function.

In order to keep the number of variables for a specified frequency increment to a minimum, we have found it advantageous to modify the LIP equations by the inclusion of a spectral bandpass. This bandpass is chosen with reference to the width of the peak in the periodogram. If the bandpass is defined to be $H(n)$ where

$$H(n) = 1 \qquad N_l < n < N_h$$

$$H(n) = 0 \qquad \text{otherwise},$$

then the objective function becomes

$$\sum_{n=0}^{N_f - 1} H(n)\,P(n) = \sum_{n=N_l}^{N_h} P(n)$$

and the constraint equations become

$$r_m + e_m = P(0) + 2 \sum_{n=N_l}^{N_n} P(n) \cdot \cos\,[mn\pi/(N_f - 1)] \qquad m = 1, 2, \cdots, M.$$

The second part of the LIP solution follows using the frequencies determined from the spectral indentification. Assuming the frequencies to be f_k, $k = 1, 2, \cdots, K$, we model the data x_t as the sum of K harmonics with additive white noise n_t assumed to be distributed as $N(0, \sigma_n^2)$:

$$x_t = \sum_{k=1}^{K} [b_k \sin(2\pi f_k t) + c_k \cdot \cos(2\pi f_k t)] + n_t \qquad t = 1, 2, \cdots, N. \tag{9}$$

A solution for the Fourier coefficients b_k and c_k, subject to the constraints given by (9), is obtained by minimizing the objective function

$$\sum_{k=1}^{K} (|b_k| + |c_k|)$$

with an additional constraint

$$\sum_{t=1}^{N} |n_t| \leqslant \frac{2}{\pi} N\sigma_n. \tag{10}$$

(It is shown in [6] that for n_t distributed as $N(0, \sigma_n^2)$, $E[|n_t|] = \sqrt{2/\pi}\,\sigma_n$.)

The determination of the b_k's and c_k's can, of course, be accomplished in the usual least squares manner since the spikiness of the L_1 norm is not an issue here, but we have chosen linear programming to preserve a uniformity of approach.

Numerical Example

We consider as a numerical example a data set, 30 points long, composed of two equiamplitude harmonics with frequencies of 0.1984 and 0.2222 Hz. The noise level is 5 dB. The frequencies were chosen so that the frequency separation of 0.0238 Hz is 30 percent less than that required for the often quoted resolution limit of the reciprocal of the record length (0.0333 Hz in this case).

The first part of our LIP approach is illustrated in Fig. 1. Fig. 1(a) is the unsmoothed periodogram and demonstrates the difficulty in resolving the two harmonics. Choosing the bandpass for the function $H(n)$ to be 0.15-0.25 Hz [in accordance with Fig. 1(a)] and transforming to determine the smoothed autocorrelation function with $M = 19$ lags, we obtain as the solution to the LIP problem the power spectrum depicted in Fig. 1(b). The parameter ϕ was set to $0.05\, r_0 M$. Four harmonics were obtained with values $f_1 = 0.196$ Hz, $P_1 = 0.295$, $f_2 = 0.198$ Hz, $P_2 = 0.204$, $f_3 = 0.218$ Hz, $P_3 = 0.347$, $f_4 = 0.220$ Hz, and $P_4 = 0.216$. Since Δf was chosen to be 0.002 Hz, we conclude that there are probably only two frequencies, f'_1 and f'_2, which together with the appropriate powers are approximately computed to be $f'_1 = 0.197$ Hz, $P'_1 = 0.50$ and $f'_2 = 0.219$ Hz, $P'_2 = 0.56$. In the above, we have used the following weighted linear interpolations, e.g.,

$$f'_n = (f_{2n-1} P_{2n-1} + f_{2n} P_{2n})/(P_{2n-1} + P_{2n}) \quad \text{and } P'_n = P_{2n-1} + P_{2n}.$$

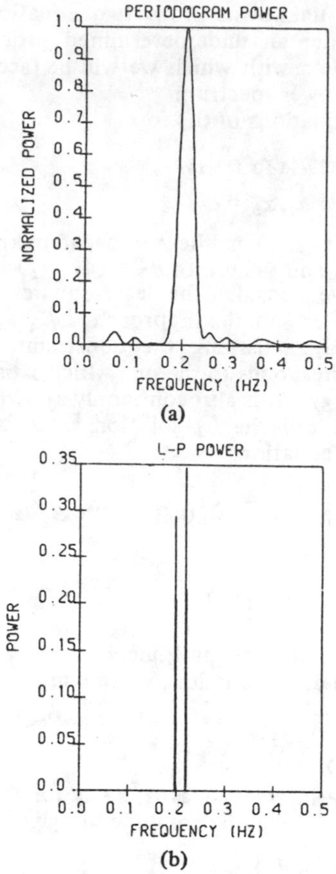

Fig. 1. (a) Periodogram of the 30-point input signal described in the text. (b) Power spectrum determined using the linear programming approach.

Results from various maximum entropy estimators are illustrated in Fig. 2. The best result is obtained from the YW estimate with the autocorrelations obtained from the bandpassed periodogram. Both the LS and the Burg estimators exhibit the well-known sensitivity of the ME method to filter length. Fig. 2 illustrates the results for an AR model of order 10, the extraneous spikes being characteristic of the ME method when long filter lengths are used with a noisy signal. Unfortunately, in this case, lower order models do not provide the required resolution.

The second part of the LIP solution is illustrated in Fig. 3. The noise variance was estimated to be $\sigma_n^2 = 0.19$ from the periodogram. The actual noiseless input signal and the version reconstructed from the calculated Fourier coefficients are shown in Fig. 3(a). Fig. 3(b) is a comparison of a longer 90-point noiseless input signal and the corresponding reconstructed

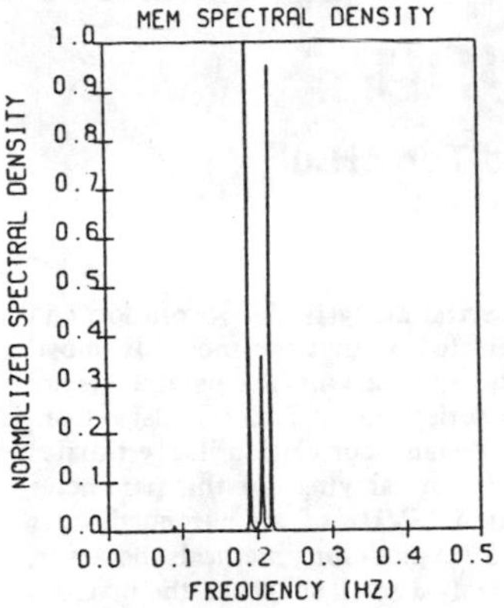

(a)

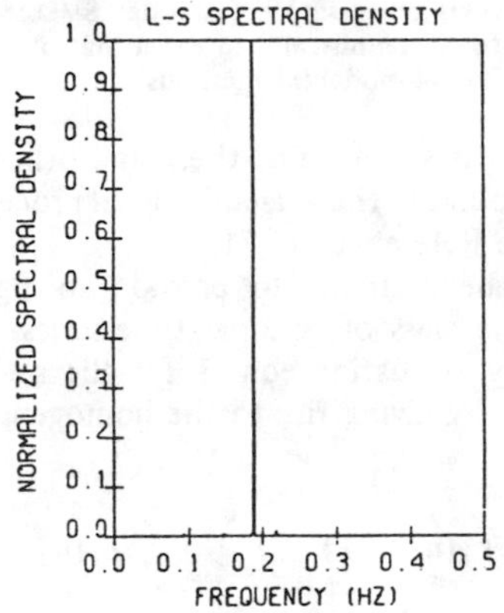

(b)

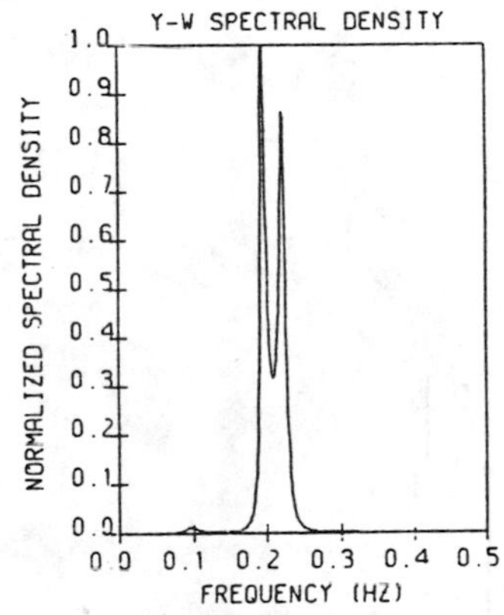

(c)

Fig. 2. (a) ME spectral density computed using the Burg algorithm. (b) ME spectral density computed using the LS algorithm. (c) ME spectral density computed using the YW algorithm with autocorrelation obtained from the bandpassed periodogram.

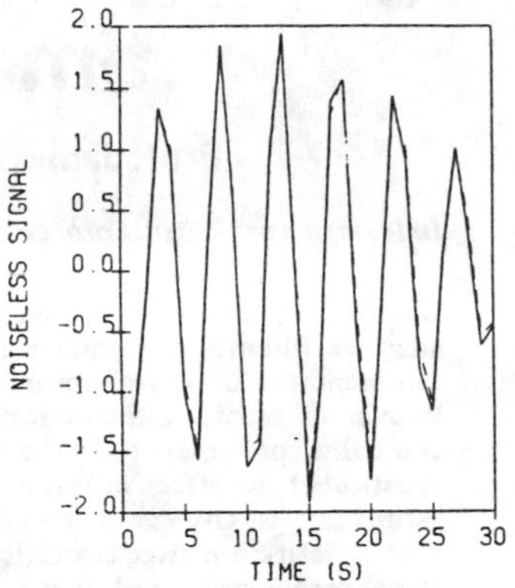

(a)

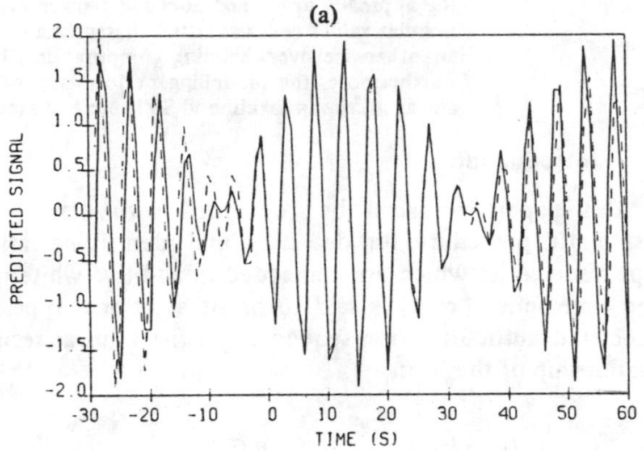

(b)

Fig. 3. (a) A comparison of the actual noiseless input signal (full curve) and the signal reconstructed from the computed Fourier coefficients (broken curve). (b) A similar comparison to that in Fig. 3(a) except that a 90-point version is shown illustrating backward and forward extrapolation. The input signal is contained between 0 and 29 s.

signal. In spite of some amplitude and phase discrepancy, the correspondence is generally quite good. The reconstructed signal shown in Fig. 3(b) can now be used in the same manner as the forward and backward predicted signal in [7].

Conclusions

We have presented in this note a new approach to the determination of the power spectrum of a harmonic process. The philosophy behind this method is quite different from the one which underlies the ME method. Whereas the ME development looks for an extension of the measured or estimated autocorrelation, which is most uncommitted with respect to the data outside the known interval, the LIP approach constrains the power spectrum to have a spiky structure. It appears that this constraint allows a well-resolved spectral estimate to be made even in cases where other high resolution estimators encounter problems.

It is hard to match the ME formulation in its elegance and simplicity, and the LIP method presented here is not meant as a substitute. Where the greatest possible resolution is a prime requirement however, as, for example, in the determination of the splitting parameters associated with the free oscillation of the earth [8], the LIP spectral estimator should prove to be a very useful one.

References

[1] S. Haykin and S. Kesler, "Prediction-error filtering and maximum-entropy spectral estimation," *Topics in Applied Physics*, vol. 34, *Nonlinear Methods of Spectral Analysis*. 1979, ch. 2, pp. 9-72.

[2] T. J. Ulrych and M. Ooe, "Autoregressive and mixed autoregressive-moving average models and spectra," *Topics in Applied Physics*, vol. 34, *Nonlinear Methods of Spectral Analysis*. 1979, ch. 3, pp. 73-125.

[3] T. J. Ulrych and R. W. Clayton, "Time series modelling and maximum entropy," *Phys. Earth Planet. Int.*, vol. 12, pp. 188-200, 1976.

[4] R. T. Lacoss, "Data adaptive spectral analysis methods," *Geophysics*, vol. 36, pp. 661-675, 1971.

[5] S. J. Johnsen and N. Andersen, "On power estimation in maximum entropy spectral analysis," *Geophysics*, vol. 43, pp. 681-690, 1978.

[6] S. Levy and P. K. Fullagar, "Reconstruction of a sparse spike train from a portion of its spectrum, and application to high resolution deconvolution," *Geophysics*, 1981, accepted for publication.

[7] T. J. Ulrych, D. E. Smylie, O. G. Jensen, and G. K. C. Clarke, "Predictive filtering and smoothing of short records using maximum entropy," *J. Geophys. Res.*, vol. 78, pp. 4959-4964, 1973.

[8] B. F. Chao and F. Gilbert, "Autoregressive estimation of complex eigenfrequencies in low frequency seismic spectra," *Geophys. J. R. Astr. Soc.*, vol. 63, pp. 641-657, 1980.

Singular-value decomposition approach to time series modelling

Prof. James A. Cadzow, Behshad Baseghi and Tony Hsu

Indexing term: *Signal processing*

Abstract: In various signal processing applications, as exemplified by spectral analysis, deconvolution and adaptive filtering, the parameters of a linear recursive model are to be selected so that the model is 'most' representative of a given set of time series observations. For many of these applications, the parameters are known to satisfy a theoretical recursive relationship involving the time series' autocorrelation lags. Conceptually, one may then use this recursive relationship, with appropriate autocorrelation lag estimates substituted, to effect estimates for the operator's parameters. A procedure for carrying out this parameter estimation is given which makes use of the singular-value decomposition (SVD) of an extended-order autocorrelation matrix associated with the given time series. Unlike other SVD modelling methods, however, the approach developed does not require a full-order SVD determination. Only a small subset of the matrix's singular values and associated characteristic vectors need be computed. This feature can significantly alleviate an otherwise overwhelming computational burden that is necessitated when generating a full-order SVD. Furthermore, the modelling performance of this new method has been found empirically to excel that of a near maximum-likelihood SVD method as well as several other more traditional modelling methods.

1 Introduction

In this paper, we shall be concerned with developing a systematic procedure for detecting and identifying rational-type time series which are embedded in additive white noise. The time series $\{x(n)\}$ is said to be of a rational type if its associated autocorrelation sequence satisfies a linear recursive relationship of the form

$$\sum_{k=0}^{p} a_k r_x(n-k) = 0 \qquad \text{for } n \in I \tag{1}$$

in which I is an appropriate integer set associated with the application at hand. Unless otherwise indicated, the time series will be taken to be a complex-valued wide-sense stationary random process with autocorrelation sequence

$$r_x(n) = E\{x(n+k)\bar{x}(k)\} \tag{2}$$

in which the symbols E and the overbar denote the operations of statistical expectation and complex conjugation, respectively. To gain an appreciation for the wide range of relevant signal processing applications which are describable by a homogeneous relationship of the form of eqn. 1, the following abbreviated listing is offered:

(*a*) In modelling a wide-sense stationary time series by an autoregressive moving-average model of order (p, q) [i.e. an ARMA (p, q) model], it can be shown that $I = [q+1, \infty)$ specifies the integer index set.†

(*b*) When the time series is composed of a sum of sinusoidal components of different frequencies that have initial random phase angles uniformly distributed on $(-\pi, \pi)$, the integer index set is $I = (-\infty, \infty)$ in the noise-free case.

(*c*) In transient response analyses associated with linear recursive operations, a homogeneous relationship of the form of eqn. 1 will hold for suitably chosen I.

(*d*) In narrowband interference rejection applications where the a_k coefficients are required to be symmetric so as to achieve zero phase filtering, the relationship is also applicable.

A more detailed description of these and other related applications can be found in the available signal processing literature (for example see References 1–7).

It will now be desirable to provide an algebraic characterisation for the class of rational time series. This is readily accomplished by evaluating eqn. 1 for the set of indices $n_1 \leqslant n \leqslant n_2$, thereby giving rise to the homogeneous system of linear equations

$$\begin{bmatrix} r_x(n_1) & r_x(n_1-1) & \cdots & r_x(n_1-p) \\ r_x(n_1+1) & r(n_1) & \cdots & r_x(n_1-p+1) \\ \vdots & \vdots & & \vdots \\ r_x(n_2) & r_x(n_2-1) & \cdots & r_x(n_2-p) \end{bmatrix} \times \begin{bmatrix} a_0 \\ a_1 \\ \vdots \\ a_p \end{bmatrix} = \begin{bmatrix} 0 \\ 0 \\ \vdots \\ 0 \end{bmatrix} \tag{3a}$$

or, more compactly,

$$R\boldsymbol{a} = \boldsymbol{0} \tag{3b}$$

It is here tacitly assumed that the integers n_1 and n_2 are selected so that $[n_1, n_2] \in I$. In eqns. 3, R is the $(n_2 - n_1 + 1) \times (p+1)$ autocorrelation matrix, $\boldsymbol{a}$ is the $(p+1) \times 1$ model parameter vector and $\boldsymbol{0}$ is the $(n_2 - n_1 + 1) \times 1$ zero vector. Owing to the manner in which the autocorrelation lags are interrelated through eqn. 1, it follows that the rank of the autocorrelation matrix is specified by

$$\text{rank}\,[R] = \min(p, n_2 - n_1 + 1)$$

In most practical applications, the integers n_1 and n_2 will be selected so that $n_2 - n_1 > p$. For such a selection, it

†The symbol $[n_1, n_2]$ denotes the set of integers satisfying $n_1 \leqslant n \leqslant n_2$ while $[n_1, \infty)$ specifies the set of integers satisfying $n \geqslant n_1$.

Paper 2431 F, first received 1st April and in revised form 20th December 1982

The authors are with the Department of Electrical & Computer Engineering, Arizona State University, Tempe, AZ 85287, USA follows that 'any' vector lying in the one-dimensional null space of the matrix R will provide a solution to eqns. 3. Thus, in the ideal case in which exact autocorrelation lag information is available, the required a_k model parameters may be readily obtained by solving the consistent system of linear homogeneous equations given in eqn. 3. When only time series observations are available for the modelling, however, it will be necessary to suitably adapt this solution procedure. An approach for effecting this adaptation will now be given.

2 Modelling approach

We shall now describe a typical application scenario involving the fitting of a linear model to a given data set. Specifically, it will be assumed that there is available a set of time series observations whose underlying autocorrelation sequence is postulated as satisfying a homogeneous relationship of the form of eqn. 1. Without loss of generality, this observation set will be taken to be composed of the N contiguous measurements

$$x(1), x(2), \ldots, x(N) \tag{4}$$

The task at hand is to use these measurements in some systematic manner so as to generate estimates for the order parameter p and the recursive parameters a_k identifying the model given in eqn. 1. Typically, the recursive parameters are normalised so that $a_0 = 1$. For reasons which will be shortly made apparent, however, this normalisation will not be imposed here.

To effect the required parameter estimation in accordance with eqns. 3, a straightforward procedure is suggested. Namely, one first generates estimates for the underlying autocorrelation lags from the observation set given in expr. 4. Although a variety of procedures are available for this purpose, empirical evidence indicates that the biased estimates as given by

$$\hat{r}_x(n) = \frac{1}{N}\sum_{k=1}^{N-n} x(n+k)\bar{x}(k) \qquad 0 \leqslant n \leqslant N-1 \tag{5}$$

typically provide excellent modelling results. Appropriate negative autocorrelation lag estimates may be readily generated from these estimates by appealing to the complex-conjugate symmetric property $r_x(-n) = \bar{r}_x(n)$ which characterises wide-sense stationary processes. These computed autocorrelation lag estimates are next substituted into eqns. 3, and the resultant system of 'estimated' homogeneous linear equations are then solved to obtain the model's parameters. Although this approach has much intuitive appeal, it possesses the following two serious shortcomings:

(i) In most applications, the order of the underlying model, p, is not known *a priori*.

(ii) Owing to errors in the autocorrelation lag estimates, the autocorrelation matrix estimate will have full rank, thereby indicating that the resultant 'estimated' homogeneous equations $Ra = \mathbf{0}$ will generally only have the uninteresting zero-vector solution (i.e. $a = \mathbf{0}$).

To overcome these deficiencies, it will be necessary to appropriately adapt this modelling procedure in a manner now to be outlined.

2.1 Extended-order model

Since the order parameter is not normally known *a priori*, we shall take the conservative approach by overestimating the model's order. In particular, let us consider a recursive model of the form of eqn. 1 with order p_e (i.e. p is replaced by p_e) in which p_e is selected to be much larger than the expected value of the underlying unknown order p of the time series being analysed. This extended-order model would be characterised by

$$\sum_{k=0}^{p_e} a_k r(n-k) = 0 \tag{6}$$

Using appropriate lag estimates (for example the biased estimates given in eqn. 5) as entries, this extended-order model is next evaluated for $n_1 \leqslant n \leqslant n_2$ to yield the estimated homogeneous relationship

$$\hat{R}_e a = \mathbf{0} \tag{7}$$

in which a is the $(p_e + 1) \times 1$ extended-order model parameter vector and R_e is the $(n_2 - n_1 + 1) \times (p_e + 1)$ extended-order autocorrelation matrix estimate given by

$$\hat{R}_e = \begin{bmatrix} \hat{r}_x(n_1) & \hat{r}_x(n_1-1) & \cdot \quad \cdot & \hat{r}_x(n_1-p_e) \\ \hat{r}_x(n_1+1) & \hat{r}_x(n_1) & \cdot \quad \cdot & \hat{r}_x(n_1-p_e+1) \\ \vdots & & & \vdots \\ \hat{r}_x(n_2) & \hat{r}_x(n_2-1) & & \hat{r}_x(n_2-p_e) \end{bmatrix} \tag{8}$$

It should be noted that, by taking this extended-order model approach, the resultant modelling as represented by eqn. 7 will be less sensitive to the statistical quirks of the data being analysed than would be the case for the minimal-order choice p.

If the autocorrelation lag estimates used in the matrix $\hat{R}_e$ were exact, it would follow that the rank of $\hat{R}_e$ would be p. This of course assumes that these autocorrelation lags entries would satisfy a pth-order recursive relationship of the form of eqn. 1 and that $p_e \geqslant p$ and $n_2 - n_1 > p$. In the more realistic case in which the autocorrelation lag estimates are in error, however, it follows that the rank of the extended autocorrelation matrix estimate of eqn. 8 will always be full [i.e. min $(p_e + 1, n_2 - n_1 + 1)$]. This in turn implies that there will not exist a nontrivial parameter vector solution to the homogeneous relationship given in eqn. 7. Nonetheless, it is still desirable to select a nontrivial parameter vector a which will cause this homogeneous relationship to be best satisfied in some sense. With this objective in mind, we shall seek a parameter vector which will minimise the l_2 norm of the equation error vector $\hat{R}_e a$, in which the mth component of a shall be constrained to be one. This constrained minimisation problem is formally specified by

$$\min_{a_m = 1} \; a^* \hat{R}_e^* \hat{R}_e a \tag{9}$$

Empirical evidence has demonstrated that the parameter vector solution to this problem will typically provide excellent modelling performance. In this expression, the asterisk denotes the operation of complex-conjugate transposition, and m is a fixed integer in $[0, p_e]$. The selection of the integer m is critical, and fundamentally different modelling behaviour will arise as a function of m. In the standard approach to spectral estimation, m is taken to be zero [1]. We shall say more about this important issue in later Sections.

Using straightforward methods, the parameter vector $\mathbf{a}$ which minimises the functional given in expr. 9 is found to be

$$\mathbf{a} = \lambda\,[\hat{R}_e^*\,\hat{R}_e]^{-1}\,\mathbf{e}_{m+1} \tag{10}$$

where $\mathbf{e}_{m+1}$ is the $(p_e+1)\times 1$ standard basis vector whose components are all zero except for its $(m+1)$th, which is one, and λ is a normalising scalar selected so that the $(m+1)$th component of $\mathbf{a}$ is one as required. The main computational requirement in generating this parameter vector estimate is seen to be the inversion of the $(p_e+1)\times(p_e+1)$ matrix $\hat{R}_e^*\hat{R}_e$. Although the parameter vector estimate given in eqn. 10 typically provides satisfactorily good performance, it is possible to obtain a further performance improvement by taking full advantage of the algebraic properties associated with the theoretical extended-order autocorrelation matrix. The following Section will provide the analytical tools for this purpose.

3 SVD representation algorithm

The extended-order autocorrelation matrix of eqn. 8 has a theoretical rank of p when its entries corresponded to the actual autocorrelation lags interrelated according to the pth-order recursion of eqn. 1. As suggested earlier, when lag estimates are incorporated, the actual rank of R_e will be full or near full. Nonetheless, the 'effective rank' of R_e will still tend to be near p for even moderately low signal/noise ratios. The ambiguous terminology 'effective rank of a matrix' can be quantified by appealing to the concept of singular-value decomposition (SVD). In particular, the SVD representation for the extended-order autocorrelation matrix given in eqn. 8 will be of the form [8]

$$\hat{R}_e = \sum_{k=1}^{h} \sigma_k\,\mathbf{u}_k\,\mathbf{v}_k^* \tag{11a}$$

in which

$$h = \min\,(p_e+1,\,n_2-n_1+1) \tag{11b}$$

The terms composing this SVD representation may be obtained by solving the following two eigenvalue-eigenvector problems:

$$\left.\begin{aligned} \hat{R}_e\hat{R}_e^*\,\mathbf{u}_k &= \sigma_k^2\,\mathbf{u}_k \\ \hat{R}_e^*\hat{R}_e\,\mathbf{v}_k &= \sigma_k^2\,\mathbf{v}_k \end{aligned}\right\} \quad 1 \leqslant k \leqslant h \tag{12}$$

From these Hermitian eigenvector relationships, it is apparent that the normalised $(n_2-n_1+1)\times 1$ and $(p_e+1)\times 1$ characteristic vector sets $\{\mathbf{u}_k\}$ and $\{\mathbf{v}_k\}$, respectively, are each orthonormal. Moreover, the non-negative 'singular' values σ_k are here ordered in the conventional nonincreasing-size manner (i.e. $\sigma_k \geqslant \sigma_{k+1}$).

To obtain the effective rank of the matrix $\hat{R}_e$ using this SVD representation, let us now consider the related problem of finding that $(n_2-n_1+1)\times(p_e+1)$ matrix of rank p which will best approximate R_e in the least-squares sense. It is readily shown that the best rank p approximation is obtained by simply truncating the SVD representation to its largest p singular-value components [8], i.e.

$$\hat{R}_e^{(p)} = \sum_{k=1}^{p} \sigma_k\,\mathbf{u}_k\,\mathbf{v}_k^* \tag{13}$$

where it is assumed that $p \leqslant h$. The degree to which this rank p matrix approximates R_e is conveniently given by the norm ratio

$$\frac{\|\hat{R}_e^{(p)}\|}{\|\hat{R}_e\|} = \sqrt{\frac{\sigma_1^2+\sigma_2^2+\ldots+\sigma_p^2}{\sigma_1^2+\sigma_2^2+\ldots+\sigma_h^2}} \qquad 1 \leqslant p \leqslant h \tag{14}$$

in which the matrix norm symbol $\|\cdot\|$ designates the standard l_2 matrix norm. This norm ratio is seen to equal its largest value of one at $p = h$, where $\hat{R}_e^{(h)} = \hat{R}_e$. The extended-order autocorrelation matrix estimate $\hat{R}_e$ is said to have a low effective rank if this ratio is near one for small values of p relative to h. Otherwise, $\hat{R}_e$ is said to have a high effective rank.

Upon examination of the above norm ratio as a function of p, one is typically able to form a good estimate for the order integer p characterising the underlying recursive relationship given in eqn. 1. Namely, the p largest singular values will tend to dominate the remaining $h-p$ smaller sized singular values. Once this order selection has been obtained, the parameter vector $\mathbf{a}$ is next estimated by considering the associated reduced-order homogeneous relationship

$$\hat{R}_e^{(p)}\,\mathbf{a} = \mathbf{0} \tag{15}$$

Unlike the original expression given in eqn. 7, this system of equations is guaranteed to be consistent provided that $p < h$. From this expression, the following fundamental result applies.

Lemma 1

The set of model parametric vectors which satisfy the homogeneous relationship given in eqn. 15 lie in the $(h-p)$-dimensional null space associated with the reduced-order matrix $\hat{R}_e^{(p)}$.

The implications of this lemma are profound since it implies that there will exist an infinite number of candidates for the model's parameter vector. Clearly, from this infinity of choices, we wish to select one which will yield an acceptably good modelling representation. With this objective in mind, extensive empirical evidence has suggested that the solution to the following problem will yield such a choice:‡

$$\min_{\mathbf{a}\in A}\ \mathbf{a}^*\mathbf{a} \tag{16a}$$

where

$$A = \{\mathbf{a}: \hat{R}_e^{(p)}\mathbf{a} = \mathbf{0} \text{ and } a_m = 1\} \tag{16b}$$

Namely, we seek that $(p_e+1)\times 1$ parameter vector of minimum l_2 norm which lies in the null space of $\hat{R}_e^{(p)}$ and has its $(m+1)$th component constrained to be one. Using standard procedures, it is found that the solution to this constrained minimisation problem is given by the $(p_e+1)\times 1$ parameter vector

$$\mathbf{a} = \frac{\left[\mathbf{e}_m - \sum_{k=1}^{p} \bar{v}_k(m)\,\mathbf{v}_k\right]}{\left[1 - \sum_{k=1}^{p} |v_k(m)|^2\right]} \tag{17a}$$

$$= \frac{\left[\sum_{k=p+1}^{h} \bar{v}_k(m)\,\mathbf{v}_k\right]}{\left[\sum_{k=p+1}^{h} |v_k(m)|^2\right]} \tag{17b}$$

where the $(p_e+1)\times 1$ vectors $\mathbf{v}_k$ are those used in the SVD

‡ A more general modelling procedure is to be found in the Appendix.

representation in eqn. 11. Although the two parameter vector relationships specified by eqn. 17 are mathematically equivalent, one may be preferable to the other when computational considerations are taken into account. Namely, the first representation, eqn. 17*a* is seen to entail knowledge of the p 'largest' $\mathbf{v}_k$ characteristic vectors, while the latter expression, eqn. 17*b*, requires knowledge of the $h - p$ 'smallest' $\mathbf{v}_k$ characteristic vectors. Typically, p is much smaller than h so that one need only compute a relatively small subset of the full set of eigenvectors $\mathbf{v}_k$ in order to generate the desired parameter vector. An appealing method for taking advantage of this feature is outlined in Reference 2.

In summary, the modelling approach taken in this paper for representing rational-type data entails the four-step procedure listed in Table 1. How one is to interpret step (iv) may be dependent on the particular application at hand, as will be demonstrated in Sections 4 and 5. An important consideration in using this approach is the selection of the extended-order parameter p_e and the reduced-order parameter p. *Truly exceptional* modelling performance can be achieved by a proper choice of these parameters. This selection process may require empirical experimentation in some applications.

Table 1: Basic steps of an SVD modelling approach

Step	
(i)	Generate autocorrelation lag estimates from the given time series observations.
(ii)	To reduce the model's sensitivity to autocorrelation lag errors, one forms the extended-order autocorrelation matrix estimate given in eqn. 8.
(iii)	The effective rank p of this extended-order autocorrelation matrix estimate is next obtained by generating the matrix's SVD and evaluating its associated singular-value behaviour using the norm ratio eqn. 14.
(iv)	The required parameter vector is then generated by solving the associated reduced pth-ordered system of equations.

4 Sinusoids in white noise example

In order to illustrate how one might apply the SVD representation approach as developed in the preceding Section, let us consider the important practical problem of detecting the presence of sinusoids in additive white noise. The time series under consideration is then taken to be specified by

$$x(n) = \sum_{k=1}^{p} A_k \exp\,[j(\omega_k n + \phi_k)] + w(n) \tag{18}$$

in which $\{w(n)\}$ is a zero-mean white-noise process whose statistically independent real and imaginary components each have a variance of $0.5\,\sigma^2$, and the p complex sinusoids have unknown real amplitudes A_k, frequencies ω_k, and independent random variable phases ϕ_k which are uniformly distributed on $[-\pi, \pi]$. Under these assumptions, it is readily shown that the autocorrelation sequence associated with the time series given by eqn. 18 is given by

$$r_x(n) = \sum_{k=1}^{p} A_k^2 \exp\,(j\omega_k n) + \sigma^2\,\delta(n) \tag{19}$$

in which $\delta(n)$ designates the Kronecker delta sequence. Appealing to linear systems theory, it can be established that the autocorrelation sequence given in eqn. 19 will satisfy the homogeneous linear recursion expression

$$\sum_{k=0}^{p} a_k r_x(n-k) = 0 \qquad \text{for } |n| \geqslant p+1 \tag{20}$$

in which the required a_k coefficients are theoretically linked to the sinusoidal frequency parameters according to

$$\sum_{k=0}^{p} a_k z^{-k} = a_0 \prod_{k=1}^{p} (1 - \exp\,(j\omega_k) z^{-1}) \tag{21}$$

Thus there is seen to be a one-to-one relationship between the recursive a_k parameters and the sinusoidal frequencies (provided that the ω_k are restricted to the Nyquist interval). If one is able to effect evaluations of the a_k parameters, it follows that the corresponding values for the sinusoidal frequencies may then be obtained through the root relationship given in eqn. 21 and vice versa.

In the ideal case in which exact autocorrelation lag information is available, the procedure for obtaining the required a_k parameters is conceptually straightforward. This simply entails evaluating relationship given in eqn. 20 over any set of p or more indices satisfying $n \geqslant p + 1$. The resultant consistent system of homogeneous equations is then solved to yield the required unique (to within a scalar factor) parameter vector. It is, of course, assumed here that the autocorrelation lags which are used do in fact satisfy a recursive relationship of the form of eqn. 20. Finally, the polynomial given in eqn. 21 associated with this parameter vector is factored to determine the sinusoidal frequencies ω_k.

In the more realistic case in which autocorrelation lag estimates (for example the biased lag estimates given in eqn. 5) are used in this solution process, however, the errors inherent in the lag entries will give rise to inaccurate estimates for the a_k parameters. As suggested earlier, the deleterious effects caused by the lag errors can be significantly reduced by considering an extended-order model such as

$$\sum_{k=0}^{p_e} a_k r_x(n-k) = 0 \qquad \text{for } |n| \geqslant p_e + 1 \tag{22}$$

in which p_e is normally taken to be significantly larger than p. To generate an estimate for the $p_e + 1$ recursive a_k parameters, one might then evaluate eqn. 22 for $p_e + 1 \leqslant n \leqslant 2p_e + 1$ to yield

$$\begin{bmatrix} \hat{r}_x(p_e+1) & \hat{r}_x(p_e) & \cdots & \hat{r}_x(1) \\ \hat{r}_x(p_e+2) & \hat{r}_x(p_e+1) & \cdots & \hat{r}_x(2) \\ \vdots & \vdots & & \vdots \\ \hat{r}_x(2p_e+1) & \hat{r}_x(2p_e) & \cdots & \hat{r}_x(p_e+1) \end{bmatrix} \begin{bmatrix} a_0 \\ a_1 \\ \vdots \\ a_{p_e} \end{bmatrix} = \begin{bmatrix} 0 \\ 0 \\ \vdots \\ 0 \end{bmatrix} \tag{23a}$$

or equivalently

$$\hat{R}_e \mathbf{a} = \mathbf{0} \tag{23b}$$

in which autocorrelation lag estimates have been used as matrix entries. Since the rank of the $(p_e + 1) \times (p_e + 1)$ extended-order autocorrelation matrix estimate is generally full (i.e. $p_e + 1$), it will be then necessary to appeal to a solution procedure such as represented by eqn. 10 or eqns. 17 in order to obtain a suitable nontrivial parameter vector estimate. In using this extended-order approach, the resultant parameter estimates will be generally found to be less sensitive to the errors contained in the autocorrelation lag estimates than for the minimal ideal order choice of p. Moreover, the spectral estimate which is generated from the extended-order estimate typically will have a much more sharply defined peak behaviour at the p frequencies ω_k than will the minimum-order choice, and the resolution capabilities will also be better.

4.1 Improved method

Although the above approach for generating spectral estimates of the sinusoids in white noise will generally provide satisfactory results, the method now to be outlined can yield a significant performance improvement. It is predicated on first observing that in the noise-free case (i.e. $\sigma^2 = 0$), the autocorrelation lags given in eqn. 19 will satisfy the homogeneous relationship

$$\sum_{k=0}^{p} a_k r_x(n-k) = 0 \tag{24}$$

for *all* integer choices of n. The ability to use the specific indices $0 \leqslant n \leqslant p$ for the noise-free case (in contrast to the noisy case) will have profound computational and algebraic consequences. To illustrate one such effect, let us now evaluate the theoretical relationship given in eqn. 24 for the indices $0 \leqslant n \leqslant p$ to obtain

$$\begin{bmatrix} r_x(0) & r_x(-1) & \dots & r_x(-p) \\ r_x(1) & r_x(0) & \dots & r_x(-p+1) \\ \vdots & & & \vdots \\ r_x(p) & r_x(p-1) & \dots & r_x(0) \end{bmatrix} \begin{bmatrix} a_0 \\ a_1 \\ \vdots \\ a_p \end{bmatrix} = \begin{bmatrix} 0 \\ 0 \\ \vdots \\ 0 \end{bmatrix} \tag{25}$$

The matrix involved in this system of homogeneous equations is seen to have a complex-conjugate symmetric Toeplitz structure. This structure is of particular interest since it may be effectively used in algebraically characterising the required parameter vector solution.

As suggested earlier, when lag estimates are used in eqn. 25, the resultant system of homogeneous equations will generally not have a nontrivial parameter vector solution. To overcome the deleterious effects caused by the errors inherent in the lag estimates, it will then be desirable to consider the extended-order version of eqn. 25 in which p is replaced by p_e and p_e is taken to be much larger than p. The corresponding system of extended-order homogeneous relationships with autocorrelation lag entries then becomes

$$\begin{bmatrix} \hat{r}_x(0) & \hat{r}_x(-1) & \dots & \hat{r}_x(-p_e) \\ \hat{r}_x(1) & \hat{r}_x(0) & \dots & \hat{r}_x(1-p_e) \\ \vdots & & & \vdots \\ \hat{r}_x(p_e) & \hat{r}_x(p_e-1) & \dots & \hat{r}_x(0) \end{bmatrix} \begin{bmatrix} a_0 \\ a_1 \\ \vdots \\ a_{p_e} \end{bmatrix} = \begin{bmatrix} 0 \\ 0 \\ \vdots \\ 0 \end{bmatrix} \tag{26a}$$

or

$$\hat{\boldsymbol{R}}_e \boldsymbol{a} = \boldsymbol{0} \tag{26b}$$

The $(p_e+1) \times (p_e+1)$ extended-order autocorrelation matrix estimate $\hat{R}_e$ is again seen to have a complex-conjugate symmetric Toeplitz structure. Although the theoretical rank of matrix R_e is p, its estimate $\hat{R}_e$ will almost always have a full rank (i.e. $p_e + 1$) because of errors inherent in the autocorrelation lag estimate process.

Owing to the full-rank nature of matrix R_e, it will be advisable to take the approach described in the preceding Section so as to obtain a suitably good parameter vector estimate. Specifically, an SVD representation for R_e is first made so as to determine its effective rank p. The required choice for the order parameter p will be based on an examination of the singular values associated with the matrix R_e. In most situations, it will be found that there will exist p readily identifiable dominant singular values and consequently $p_e - p + 1$ relatively insignificant singular values. Once the determination of p has been made, the optimum $(p_e+1) \times 1$ parameter vector $\boldsymbol{a}$ is next obtained using eqns. 17, and this set of parameters in turn ultimately gives rise to the factored polynomial

$$A(z) = \sum_{k=0}^{p_e} a_k z^{-k} = a_0 \prod_{k=1}^{p_e} (1 - p_k z^{-1}) \tag{27}$$

The presence of sinusoids in the time series will be indicated whenever any of the roots p_k are sufficiently close to the unit circle (i.e. $|p_k| \simeq 1$), as suggested by the theoretical expression given in eqn. 21. For the time series given in eqn. 18, it will generally be found that p of the p_e roots will tend to be located close to the values $\exp(j\omega_k)$ for $1 \leqslant k \leqslant p$, while the remaining $p_e - p$ roots will tend to be sufficiently removed from the unit circle. A particularly convenient method for displaying the sinusoidal content of a given time series is then obtained by making a plot of $1/(A(\exp(j\omega))$ against ω. It will be found that this spectral function will produce sharply defined peaks for values of ω in which $\exp(j\omega)$ is close to roots located near the unit circle and will otherwise be more smoothly behaved. The location and the sharpness of these peaks then provides a mechanism for detecting sinusoidal components.

In using eqns. 17 for estimating the model parameter vector $\boldsymbol{a}$, the selection of the parameter $m \in [0, p_e]$ is of fundamental importance. Typically, this parameter is selected to be $m = 0$, as exemplified in linear prediction methods and other procedures [1]. For the sinusoids in white noise time series given in eqn. 18, however, other choices may be more preferable. This will be a consequence of the complex-conjugate symmetric Toeplitz structure of the matrix R_e. Specifically, because of this structure, it has been shown in Reference 2 that the SVD characteristic vectors $\boldsymbol{v}_k$ which appear in the parameter vector expression given in eqns. 17 will be either complex-conjugate symmetric or complex-conjugate skew-symmetric; i.e.

$$\bar{\boldsymbol{v}}_k = J\boldsymbol{v}_k \quad \text{or} \quad \bar{\boldsymbol{v}}_k = -J\boldsymbol{v}_k \tag{28}$$

where J is the $[(p_e+1) \times (p_e+1)]$-order reversal matrix whose components are all zero except for ones which appear along its main antidiagonal. This characteristic will now be used to obtain a parameteric vector $\boldsymbol{a}$ which itself will be complex-conjugate symmetric.

It will now be assumed that we have chosen the extended-order parameter p_e to be even. Furthermore, let us take the integer m used in the model parameter vector expression given in eqns. 17 to be equal to $0.5p_e$ (i.e. the integer midway between 0 and p_e). For this selection of m, it follows that the $\boldsymbol{v}_k(m)$ as required in eqns. 17 will be zero for all complex-conjugate skew-symmetric vectors $\boldsymbol{v}_k$ therein appearing (a requirement so that $\bar{\boldsymbol{v}}_k = -J\boldsymbol{v}_k$). Thus the parameter vector given in eqns. 17 is simply equal to a linear combination of complex-conjugate symmetric vectors, which implies that $\boldsymbol{a}$ is itself similarly characterised; i.e.

$$\bar{\boldsymbol{a}} = J\boldsymbol{a} \tag{29}$$

Since $\boldsymbol{a}$ is complex-conjugate symmetric, it then follows that the roots of the associated polynomial $A(z)$ in eqn. 27 will appear in conjugate reciprocal pairs, i.e.

$$\text{if} \quad A(p_k) = 0 \quad \text{then} \quad A(\bar{p}_k^{-1}) = 0 \tag{30}$$

A little thought will convince one of the desirability of this property when seeking to detect the presence of sinusoids. Specifically, a sinusoid will have its root in $A(z)$ located on the unit circle [i.e. $p_k = \exp(j\omega_k)$] and the complex conjugate reciprocal of this root will be equal to itself [i.e. $\bar{p}_k^{-1} = \exp(j\omega_k)$]. Empirical evidence has demonstrated that there will be a marked tendency to position a pole *exactly* on the unit circle whenever a sinusoidal component is present in the signal. For the situation in which there are p sinusoids as specified in eqn. 18, the autoregressive parameter vector will typically produce p roots on the unit circle at or near the desired locations $\exp(j\omega_k)$ for $1 \leqslant k \leqslant p$, and the remaining $p_e - p$ roots will tend to be evenly located on two concentric circles located inside and outside the unit circle such that the reciprocal relationship given in eqn. 30 is satisfied.

Although the approach just described was developed for the noise-free case in which $w(n) = 0$, its utilisation for situations in which additive noise is present also gives rise to satisfactory spectral estimation performance. The reason for this is readily explained by noting that the additive white noise will only directly effect the zero-lag term $r_x(0)$ in eqns. 23. Thus only the $p_e + 1$ diagonal terms in the extended autocorrelation matrix given in eqns. 23 will be primarily affected by the presence of additive white noise. It can then be argued that the SVD representation of R_e for a sufficiently large selection of the extended-order parameter p_e will be typically only moderately influenced by the additive white noise since only the diagonal elements of R_e are explicitly influenced. This behaviour has been empirically observed for even low-signal/noise (i.e. $A_k^2/2\sigma^2$) environments, as the numerical example given in Section 6 will demonstrate.

5 Autoregressive modelling

It is possible to straightforwardly adapt the philosophy behind the SVD approach taken in Section 3 to evolve ARMA model representations for satationary time series. To demonstrate this, let us now consider the task of estimating the parameters of a hypothesised AR(p) model as specified by

$$x(n) + \sum_{k=1}^{p} a_k x(n-k) = b_0 \epsilon(n) \tag{31}$$

with this estimation being based on a finite set of time series observations. In this expression, the excitation $\{\epsilon(n)\}$ is composed of a set of zero-mean unit-variance uncorrelated random variables (i.e. white noise). It is well known that the autocorrelation sequence of this AR(p) model satisfies the homogeneous relationship [1]

$$r_x(n) + \sum_{k=1}^{p} a_k r_x(n-k) = 0 \qquad \text{for } n \geqslant 1 \tag{32}$$

If the underlying time series is an AR(p) process, then the required a_k parameters of the model given in eqn. 31 may be conceptually obtained by solving the consistent system of eqn. 32 for any set of p or more indices satisfying $n \geqslant 1$.

When one only has available time series observations to effect the modelling, it will then be necessary to compute autocorrelation lag estimates from these observations and use these estimates in eqn. 32. Owing to errors inherent in the lag estimate procedure, however, the resultant system of homogeneous equations given in eqn. 32 with lag estimates will not, in general, be consistent. To overcome the deleterious effects due to these errors, it will again be beneficial to consider an extended order AR model, which will give rise to the following relationships in which lag estimates are incorporated:

$$\hat{r}_x(n) + \sum_{k=1}^{p_e} a_k \hat{r}_x(n-k) = 0 \qquad \text{for } n \geqslant 1 \tag{33}$$

The extended-order parameter p_e is here selected to satisfy $p_e > p$. An evaluation of these equations for the indices $1 \leqslant n \leqslant p_e$ is seen to yield the system of equations

$$\begin{bmatrix} \hat{r}_x(0) & \hat{r}_x(-1) & \dots & \hat{r}_x(1-p_e) \\ \hat{r}_x(1) & \hat{r}_x(0) & & \hat{r}_x(2-p_e) \\ \cdot & \cdot & & \cdot \\ \cdot & \cdot & & \cdot \\ \cdot & \cdot & & \cdot \\ \hat{r}_x(p_e-1) & \hat{r}_x(p_e-2) & \dots & \hat{r}_x(0) \end{bmatrix} \begin{bmatrix} a_1 \\ a_2 \\ \cdot \\ \cdot \\ \cdot \\ a_{p_e} \end{bmatrix} = - \begin{bmatrix} \hat{r}_x(1) \\ \hat{r}_x(2) \\ \cdot \\ \cdot \\ \cdot \\ \hat{r}_x(p_e) \end{bmatrix} \tag{34a}$$

or in the more compact matrix form

$$\hat{R}_e a = -r \tag{34b}$$

We have here elected to separate out the vector r so as to cause the matrix R_e to have the desired complex-conjugate symmetric Toeplitz structure. This will enable us to take advantage of the algebraic and computational properties associated with such matrices. Whatever the case, because of the aforementioned lag estimate errors, the parameter vector solution to this system of generally consistent equations as specified by

$$a = -\hat{R}_e^{-1} r \tag{35}$$

will not, in general, adequately represent the dynamics of the underlying theoretical process given in eqn. 33. By choosing p_e to be much larger than p, however, there will be a tendency to reduce the sensitivity to lag estimate errors, and thereby to produce a more representative model.

Although the overextended parameter vector selection given in eqn. 35 will typically yield satisfactory modelling, a significant improvement in modelling performance can still be achieved by a suitable modification of this approach. This improvement takes advantage of the fact that the order of the underlying AR process is p. This in turn implies that the theoretical rank of the $p_e \times p_e$ matrix R_e (i.e. with exact autocorrelation lag entries) appearing in eqns. 34 is also p. Thus, by obtaining the best rank p approximation to the matrix estimate R_e through an SVD representation, the following system of equations will more properly reflect the dynamics of the underlying process:

$$\hat{R}_e^{(p)} a = -r \tag{36}$$

The parameter vector which provides the least-squares solution to this generally inconsistent system of equations is next found by using the expression

$$a = -[\hat{R}_e^{(p)}]^{\#} r$$
$$= -\sum_{k=1}^{p} \sigma_k^{-1} (v_k^* r) v_k \tag{37}$$

where $[R_e^{(p)}]^{\#}$ denotes the pseudoinverse of matrix $R_e^{(P)}$, and the pairs (σ_k, v_k) for $1 \leqslant k \leqslant p$ correspond to the p 'largest' singular-value-characteristic-vector pairs associated with the SVD of the matrix R_e.[§] On examination of eqn. 37,

§ The complex-conjugate symmetric Toeplitz matrix R_e will have an SVD of the form $\sum_{k=1}^{p_e} \sigma_k v_k v_k^*$, and its reduced pth-order pseudo-inverse will then be given by $[R_e^{(p)}]^{\#} = \sum_{k=1}^{p} \sigma_k^{-1} v_k v_k^*$, as shown in Reference 2.

it is important to note that *only* the p 'largest' singular-value-characteristic pairs (σ_k, v_k) need be calculated when determining the optimum parameter vector. This observation can result in a significant computational saving whenever $p << p_e$, since one may suitably adapt the method of exhaustion to this situation [2].

In most practical applications, one does not know the value of the order parameter p *a priori*. Fortunately, it is possible to effect an estimate for p by examining the singular-value behaviour of the matrix R_e. Since the theoretical rank of $\hat{R}_e$ is p, it follows that the effective rank of the $p_e \times p_e$ matrix estimate $\hat{R}_e$ will be close to p. This implies that the p largest singular values of $\hat{R}_e$ will tend to dominate its remaining $p_e - p$ singular values. The degree to which this is so is measured by the norm ratio

$$\frac{\sum_{k=1}^{p} \sigma_k^2}{\sum_{k=1}^{p_e} \sigma_k^2} = \frac{\sum_{k=1}^{p} \sigma_k^2}{\|\hat{R}_e\|^2} \qquad (38)$$

where $\|\hat{R}_e\|$ designates the standard l_2 norm of the matrix $\hat{R}_e$. This ratio is bounded above by one (achieved at $p = p_e$), and it provides a measure for the effective rank of $\hat{R}_e$. The matrix $\hat{R}_e$ is said to have an effective rank of p if this ratio is satisfactorily close to one. Thus, when p is not known *a priori*, one simply applies the adapted method of exhaustion [2] to the matrix $\hat{R}_e$ to compute the pair (σ_k, v_k) as k is incrementally increased from one. Once the right-hand side of eqn. 38 is determined to be suitably close to one, that value of k is taken to be the required value of p.

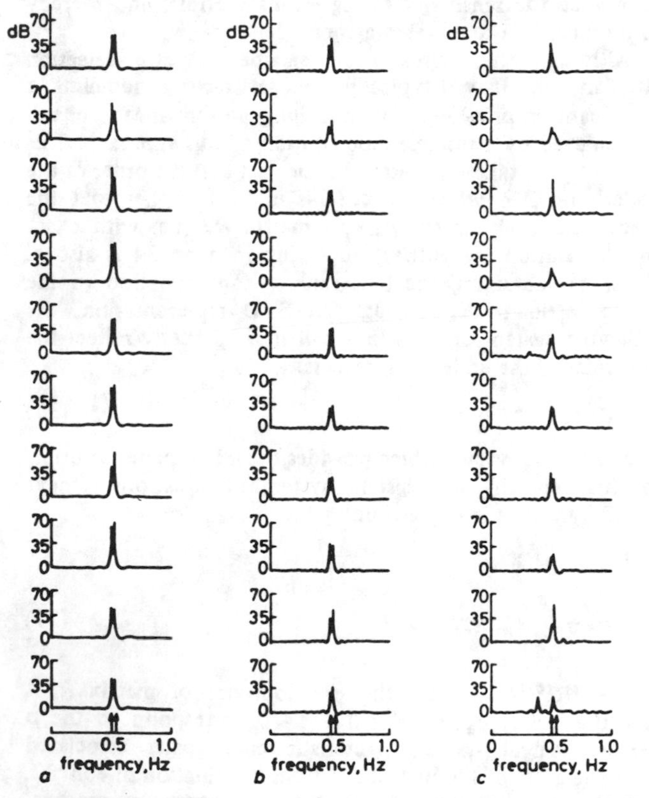

Fig. 1 *Spectral estimates corresponding to ten statistically independent trials of the time series given in eqn. 39 at a signal/noise ratio of 10 dB*

a Paper's method: $p_e = 20, p = 2, m = 10$
b Paper's method: $p_e = 20, p = 2, m = 0$
c Tufts-Kumaresan method: $L = 18, M = 2$

Table 2: Norm ratio for this paper's method and the Tufts-Kumaresan method time series given in eqn. 39

p	1	2	3	4	5
Norm ratio (paper's method)	0.982	0.998	0.999	0.999	0.999
Norm ratio (Tufts-Kumaresan method)	0.980	0.983	0.986	0.989	0.992

6 Numerical example

To illustrate the effectiveness of the method presented in Section 4, we shall consider examples which involve both real- and complex-valued sinusoidal data. The resultant spectral estimates are compared with those obtained using the Tufts-Kumaresan method [9] for the complex-valued data example, a procedure claimed to perform in a near maximum-likelihood fashion. The Tufts-Kumaresan method serves as an excellent bench mark in that it has been found to provide exceptional spectral estimates for the sinusoids in white noise case.

Example 1
In the first example, we shall use the numerical example considered by Tufts-Kumaresan [9], i.e.

$$x(n) = \exp\{j(2\pi f_1 n + \pi/4)\} + \exp(j2\pi f_2 n) + w(n) \qquad 1 \leqslant n \leqslant 25 \qquad (39)$$

where $f_1 = 0.52$ Hz, $f_2 = 0.5$ Hz and the $w(n)$ are independent complex Gaussian random variables with variance σ^2 for each of the uncorrelated real and imaginary components. The variance σ^2 is selected to be equal to 0.05 so as to give a signal/noise ratio [as measured by $10\log(1/2\sigma^2)$] of 10 dB. Ten statistically independent trials of the time series given in eqn. 39 were made next with each trial being of length 25. Finally, the individual spectral estimates and superimposed plots of the roots of $A(z)$ for the ten trials obtained using this paper's method, as represented by eqns. 17 with $p_e = 20$, $p = 2$ and $m = 10$, and $p_e = 20$, $p = 2$ and $m = 0$, were obtained and are shown displayed in Figs. 1*a* and *b* and 2*a* and *b*, respectively. From the plots, it is apparent that a spectral resolution of the two complex sinusoids was achieved in an impressive manner for each of the ten trials. When the Tufts-Kumaresan method was applied to the same data for a parameter selection of $L = 18$ (their declared optimum selection) and $M = 2$ (the Tufts-Kumaresan parameter M corresponds to our p), it produced the individual spectral estimates and superimposed root plots shown in Figs. 1*c* and 2*c*, respectively. The Tufts-Kumaresan estimate was able to produce a distinctive resolution of the two complex sinusoids in only five of the ten trials. Furthermore, an examination of the associated root plots reveals that this paper's method produced an almost exact overlay in position location (a desirable feature) for each of the ten statistically independent trials, while the Tufts-Kumaresan method provided a somewhat more scattered root location behaviour.

To illustrate the effectiveness of using the singular values of the SVD of matrix R_e for model order determination, the norm ratio given in eqn. 14 for a typical trial is shown in Table 2. From this Table, it is apparent that both this paper's method and the Tufts-Kumaresan method provide convincing evidence for selecting a low-order effective rank. This paper's method, however, produced a better order behaviour in that its norm ratio at the correct order $p = 2$ was much closer to one than the Tufts-Kumaresan norm ratio (i.e. 0.99 against 0.983).

Using the same time series description as in eqn. 39, the additive noise variance was next increased to 0.5 to give a

0 dB signal/noise ratio. Applying this paper's method with $p_e = 20$, $p = 2$ and $m = 10$, and $p_e = 20$, $p = 2$ and $m = 0$, and the Tufts-Kumaresan method with $L = 18$ and $M = 2$ to the ten trials, each of length 25, produced the ten individual spectral plots shown in Fig. 3 and the superimposed root plots shown in Figs. 2*d–f*. The Tufts-Kumaresan method did not resolve the two sinusoids in any of the ten trails, while this paper's method produced a resolution for each of the ten trials for $m = 10$, and a resolution in seven out of the ten trials for $m = 0$. Moreover, although somewhat deteriorated because of the low signal/noise ratio, the root pattern behaviour continued to display a desirable uniformity.

Example 2

In the second example, the case of real sinusoids in white noise as specified by

$$x(n) = \cos(2\pi f_1 n + \pi/4) + \cos(2\pi f_2 n) + w(n) \qquad 1 \leqslant n \leqslant 16 \tag{40}$$

was treated, in which $f_1 = 0.2$ and $f_2 = 0.25$, where the $w(n)$ are independent real Gaussian random variables with variance σ^2. The variance was taken to be 0.5, thereby giving a signal/noise ratio of 0 dB. Ten statistically independent trials of the time series given in eqn. 40 were then made, with each trial being of the relatively *short* length 16. The spectral estimates obtained using the method given in Section 4 with parameter selections $p_e = 14$, $p = 4$, $m = 7$ are shown in Fig. 4*a*. A resolution of the two real-valued sinusoids was achieved in each of the ten trials. The Tufts-Kumaresan method was next applied to the same data and produced the spectral estimates shown in Fig. 4*b* for the parameter selection $L = 12$ (their optimal selection) and $M = 4$. In only five of the ten trials did the Tufts-Kumaresan method yield a resolution of the two sinusoids, and these resolutions were not of a sharply defined nature. Next, the periodogram method was applied to the ten sets of data length 16. To avoid loss of resolution due to sampling, each of these data sets was padded with 496 zeros. The 512-point FFT spectral plots for the ten padded data sets are shown in Fig. 4*c*. A resolution of the two sinusoids was not achieved in a sharply defined manner for any of the ten FFT spectral estimates. Finally, the forward data version of the Burg algorithm was applied to the same ten sets of data length 16 with an order choice of eight. The ten resultant spectral plots are shown in Fig. 4*d*, in which a frequency resolution was achieved in only four estimates.

7 Conclusion

An SVD approach for estimating the model parameters characterising important classes of time series has been presented. This method entails forming an extended-order autocorrelation matrix estimate and then determining the 'best' rank p approximation of that matrix from its SVD representation. Finally, the model's parameters are obtained by finding the least-squares solution to a system of equations associated with the reduced-order autocorrelation matrix estimate.

One of the more important applications of this approach is

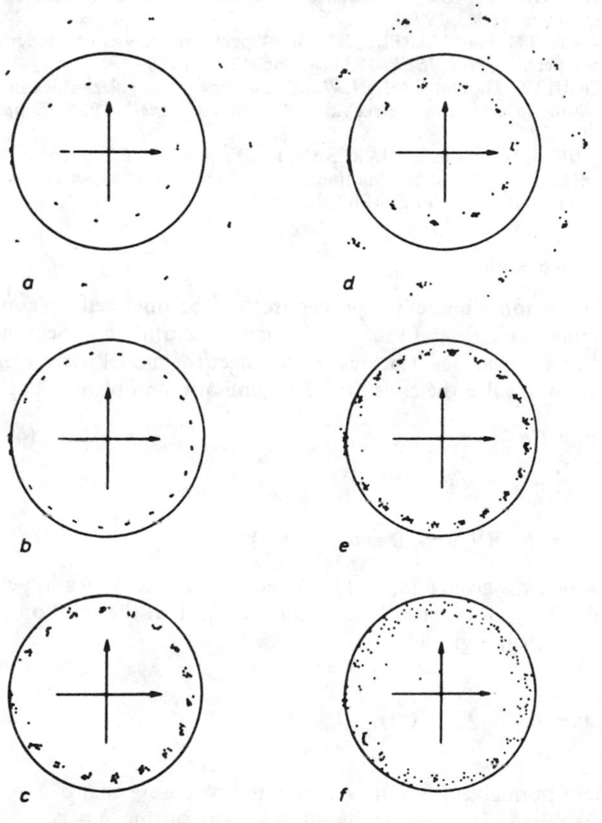

Fig. 2 *Superimposed roots corresponding to ten statistically independent trials of the time series given in eqn. 39 at signal/noise ratios of 10 and 0 dB*

a 10 dB, paper's method $p_e = 20, p = 2, m = 10$
b 10 dB, paper's method $p_e = 20, p = 2, m = 0$
c 10 dB, Tufts-Kumaresan method $L = 18, M = 2$
d 0 dB, paper's method $p_e = 20, p = 2, m = 10$
e 0 dB, paper's method $p_e = 20, p = 2, m = 0$
f 0 dB, Tufts-Kumaresan method $L = 18, M = 2$

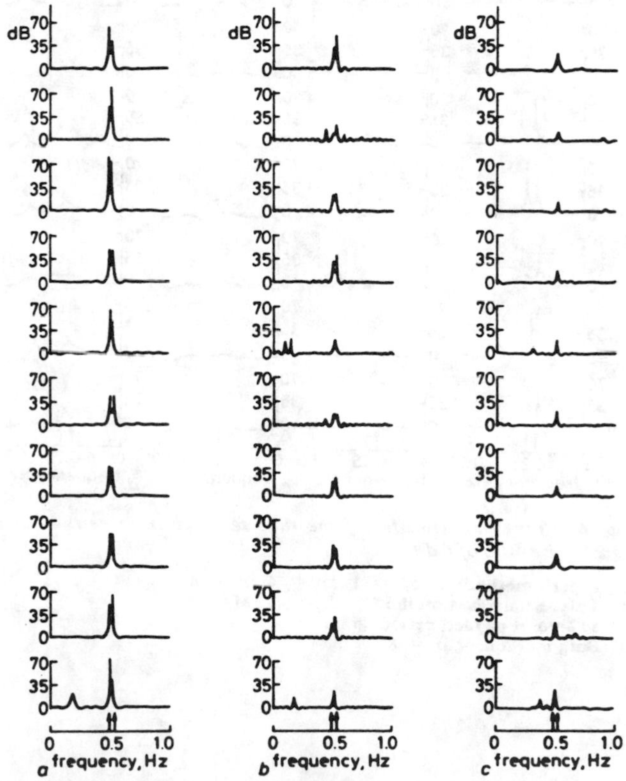

Fig. 3 *Spectral estimates corresponding to ten statistically independent trials of the time series given in eqn. 39 at a signal/noise ratio of 0 dB*

a Paper's method: $p_e = 20, p = 2, m = 10$
b Paper's method: $p_e = 20, p = 2, m = 0$
c Tufts-Kumaresan method: $L = 18, M = 2$

that of detecting the presence of sinusoids in noise-contaminated data. In this case, the solution of the reduced-order system of equations is sought in which the $(m+1)$th component of the parameter vector is constrained to be one. By appropriately selecting the integer m, the resultant spectral estimate may be made to take on different characteristics. For example, the selection of $m = 0$ will tend to cause the nonsinusoidal roots of $A(z)$ to lie within the unit circle. On the other hand, a selection of $m = 0.5 p_e$ will tend to cause the nonsinusoidal roots to be equally distributed on concentric circles located inside and outside the unit circle. Moreover, this selection will create a strong tendency to locate 'sinusoidal' roots exactly on the unit circle owing to the complex-conjugate symmetrical property possessed by the resultant parameter vector. Clearly, this is a desirable feature to have when seeking to detect the presence of sinusoidal components in additive noise. Moreover, the associated polynomial $A(z)$ will have a linear phase characterisation, which can be of importance in such applications as interference rejection filtering. It is further noted that the nonsinusoidal roots tend to lie further from the unit circle than in the $m = 0$ case (a desirable quality). In a similar fashion, it can be empirically demonstrated that a choice of $m = p_e$ will tend to result in all the nonsinusoidal roots being located outside the unit circle.

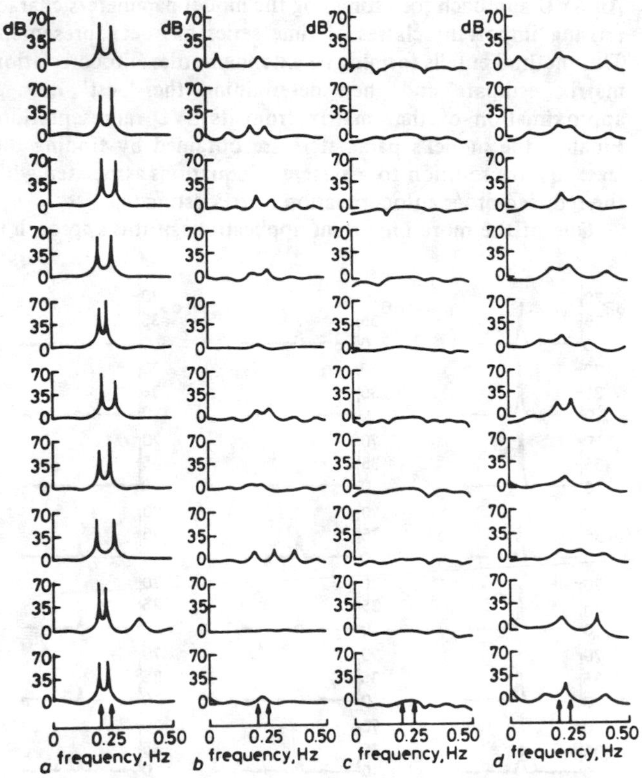

Fig. 4 *Ten trial estimates of the time series given in eqn. 40 at a signal/noise ratio of 0 dB*

a Paper's method: $p_e = 14, p = 4, m = 8$
b Tufts-Kumaresan method: $L = 12, M = 4$
c 512-point padded periodogram
d Burg method: $p = 8$

Although we have here concentrated our attention on the sinusoids in white noise case, the spectral modelling procedure presented herein is applicable to a much broader class of problems. For example, it has produced excellent performance when generating AR and ARMA models, in detecting rational type signals and in performing deconvolution operations.

8 Acknowledgment

The research reported here was sponsored by the Statistics and Probability Program of the US Office of Naval Research under ONR contract N00014-82-K-0257.

9 References

1 CADZOW, J.A.: 'Spectral estimation: An overdetermined rational model equation approach', *Proc. IEEE*, 1982, **70**, pp. 907–939.
2 CADZOW, J.A.: 'SVD representation of an important class of matrices', *IEEE Trans.*, 1983, **ASSP-31**
3 CHILDERS D.G. (Ed.): 'Modern spectral analysis' (IEEE Press, 1978)
4 HAYKIN, S., and CADZOW, J.A., (Eds.): 'Special issue on Spectral estimation', *Proc. IEEE*, 1982, **70**, (9)
5 HAYKIN, S., and CADZOW, J.A., (Co-Chairmen): First ASSP workshop on Spectral Estimation. Sponsored by ASSP Society of the IEEE, Vols. I and II, Aug. 1981
6 HAYKIN, S. (Ed.): 'Nonlinear methods of spectral analysis' (Springer-Verlag, 1979)
7 KAY, S.M., and MARPLE, S.L. Jr.: 'Spectrum analysis–A modern perspective', *Proc. IEEE*, 1981, **69**, pp. 1380–1419
8 GOULUB, G., and KAHAN, W.: 'Calculating the singular values and pseudo-inverse of a matrix', *SIAM J. Numer. Anal.*, 1965, **2**, pp. 205–224
9 TUFTS, D.W., and KUMARESAN, R.: 'Estimation of frequencies of multiple sinusoids: Making linear prediction perform like maximum likelihood', *Proc. IEEE*, 1982, **70**, pp. 975–989

10 Appendix

Other rational modelling procedures can be obtained by considering more general functionals than those utilised in Section 3. In particular, let it be desired to select the model parameter vector $\boldsymbol{a}$ to solve the constrained minimisation problem

$$\min_{\boldsymbol{a} \in A} \boldsymbol{a}^* \boldsymbol{a} \tag{41}$$

where

$$A = \{\boldsymbol{a}: R^{(p)} \boldsymbol{a} = \boldsymbol{0} \text{ and } \boldsymbol{a}^* \boldsymbol{h} = 1\}$$

in which $\boldsymbol{h}$ is given a $(p_e + 1) \times 1$ vector characterising a hyperplane in $l_{p_e} + 1$ space. It is readily shown that the solution to this problem is given by

$$\boldsymbol{a} = \alpha \boldsymbol{h} - \sum_{k=1}^{p} \bar{v}_k(m)\, v_k \tag{42}$$

where normalising scalar α is selected to ensure that $\boldsymbol{a}^* \boldsymbol{h} = 1$ as required. It is to be noted that, on setting $\boldsymbol{h} = \boldsymbol{e}_m$, we obtain the solution procedure found in Section 3.

nts.) They are a prediction filter in the sense
vergence, the weighted auxiliary elements are
signal data sequence arriving at the phase center
n element.

ce of the adaptive sidelobe canceller to our
n filter is that their spatial filter pattern analysis
ed and can be applied directly to achieve a
anding of the superresolution performance
ther point is that real-time operation is readily
ost of the current adaptive algorithms, pro-
number of snapshots is sufficient to reach
whitening ϵ. Convergence may require as
napshots or as many as several thousand,
he particular algorithm and the parameters
bution. Several examples will be discussed.

SPATIAL FILTER PATTERNS

function for the array of Fig. 3 is simply
after convergence, which is determined
ent weights. All examples contained in
ted from simulated signal data sequences
he nth "snapshot" signal sample at the
of independent Gaussian receiver noise,
ource voltages,

$$\exp [j(ku_i + \phi_{in})], \quad 1 \leqslant k \leqslant K \qquad (12)$$

$$2\pi \left(\frac{d}{\lambda}\right) \sin \theta_i$$

med near $\lambda/2$

f ith source
1

rce, nth sample
l of K elements
vith total of N snapshots.

simultaneous sampling of the
ents. Note that snapshots are
ther because of the random

ample is computed from the
Weiner matrix inverse algo-
le covariance matrix [16],

(13)

0, 0, 1] (14)

(15)

(16)

om because of a concur-
ces sampled at random

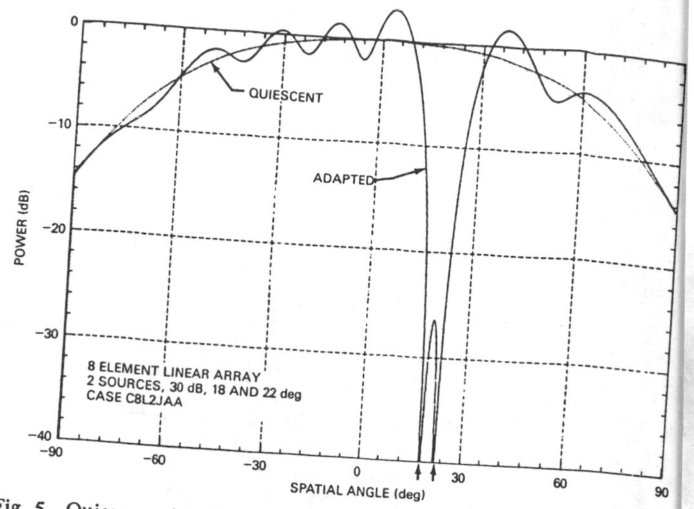

Fig. 5. Quiescent (single element) and adapted patterns for two-source case, covariance matrix inverse algorithm, 1024 data snapshots.

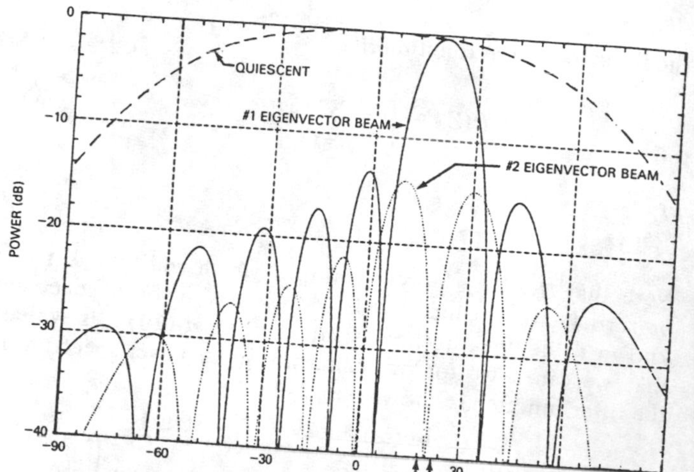

Fig. 6. Eigenvector component beam patterns for the two-source case of Fig. 5.

where E_n is the nth snapshot signal sample vector whose element components are given by (12), M_n is the nth snapshot contribution to the sample covariance matrix, M is the sample covariance matrix averaged over N snapshots, S^* is the quiescent weight steering vector, μ is a scalar quantity, and W_0 is the optimum weight vector. The sample covariance matrix can never exactly represent the true covariance of an ensemble mean, but it usually comes close enough for reasonable values of N such as $N > 2K$ [17]. Note that the steering vector S^* injects a zero weight on every element except for the end element, thus causing the quiescent pattern of the array to be that of the single end element. Fig. 5 illustrates a typical quiescent (single element) pattern and an adapted pattern obtained from an 8 element linear array with two far-field incoherent 30-dB sources located at 18° and 22°. The adapted pattern weights were computed per (13) from the inverse of the sample covariance matrix averaged over 1024 simulated snapshots. Note that the two pattern nulls (zeros) align perfectly with the locations of the two sources. Of course, the array signals in this simulation were corrupted only by receiver noise (no element errors are included) and an average over 1024 snapshots results in an overall high SNR of about 69 dB (source + elements + snapshots). Another important point t

Spectral Analysis and Adaptive Array Superresolution Techniques

WILLIAM F. GABRIEL, SENIOR MEMBER, IEEE

Abstract—Recent nonlinear "superresolution" techniques reported in the field of spectral analysis are of great interest in other fields as well, including radio-frequency (RF) adaptive array antenna systems. This paper is primarily a "cross-fertilization" treatise which takes the two most popular nonlinear techniques, the Burg maximum entropy method and the maximum likelihood method, and relates them to their similar nonlinear adaptive array antenna counterparts, which consist of the generic sidelobe canceller and directional gain constraint techniques. The comparison analysis permits an examination of their principles of operation from the antenna spatial pattern viewpoint, and helps to qualify their actual superresolution performance.

A summary of the resolution performance of several adaptive algorithms against multiple-incoherent sources is provided, including a universal graph of signal-to-noise ratio (SNR) versus source separation in beamwidths for the case of two equal-strength sources. Also, a significant dividend in the easy resolution of unequal-strength sources is reported. The superresolution of coherent spatial sources or radar targets is more difficult for these techniques, but successful results have been obtained whenever sufficient relative motion or "Doppler cycles" are available. Two alternate adaptive spatial spectrum estimators are suggested, consisting of a circular array predicting to its center point, and a new "thermal noise" algorithm.

Manuscript received August 2, 1979; revised February 12, 1980.
This contribution is derived largely from [1], and also, a subsequent report [2].
The author is with the Radar Division, Naval Research Laboratory, Washington, DC 20375.

I. INTRODUCTION

NONLINEAR spectral analysis techniques are currently a subject area of intense interest in the fields of spectrum analysis, geophysics, underwater acoustics, and radio-frequency (RF) array antennas. The major reason for the

Reprinted from *Proc. IEEE*, vol. 68, pp. 654–666, June 1980.

interest is reported "superresolution" capabilities beyond the conventional periodogram or the Blackman–Tukey windowed Fourier transform [3]. Two methods, in particular, which have demonstrated a considerable increase in resolution are the maximum entropy spectral analysis (MESA) technique introduced by J. P. Burg [4], [5], and the maximum-likelihood method (MLM) demonstrated by J. Capon [6]–[8]. Since these techniques are most significant when processing short data sets, it is natural to consider their use for RF array antennas with a modest number of elements [9], [10].

Adaptive processing techniques have been associated with these spectral estimation methods to some extent [11]–[13], but the literature indicates that cross fertilization has been rather sparse. This situation is surprising, because both MESA and MLM bear a very close relationship to nonlinear adaptive array processing techniques. It is the intent of this paper to increase the cross fertilization by relating the MESA and MLM methods to their similar adaptive array antenna counterparts. The comparison analysis contained in Sections II, III, and IV, permits an examination of their principles of operation from the antenna array spatial pattern viewpoint, and helps to qualify their superresolution performance behavior. The paper is written in a tutorial style with simplified examples, a minimum of mathematics, and adequate references for obtaining further details.

Some contribution to the state of the art may be found in Sections V–VIII. Section V briefly describes phase-center signal prediction utilizing a circular array antenna, and a new adaptive "thermal noise" algorithm. Section VI demonstrates that two sources within a beamwidth may be resolved in real time even with the simple Howells–Applebaum adaptive algorithm. This algorithm requires no data storage and may be implemented in either analogue or digital form. Section VII summarizes the resolution performance of several adaptive algorithms against multiple incoherent sources. Fig. 14 is of particular interest because it plots signal-to-noise ratio (SNR) versus source separation in beamwidths for two incoherent sources of equal strength, and it is applicable to linear arrays of any number of elements. The relatively easy resolution of sources of unequal strength is also discussed. Section VIII then turns to the more difficult superresolution of coherent sources or radar targets, and describes how successful results can be obtained if sufficient relative motion or "Doppler cycles" are available.

In addition to discussing the similarities between spectral analysis and adaptive array techniques, some attention is devoted to the significant differences in Section IX. The two greatest differences are the problem of processing two-dimensional data; i.e., the array signals are correlated in both space and time, and the impractical nature of single snapshot operation for RF arrays.

II. The Burg MESA Linear Prediction Filter

The Burg MESA method has been shown by Van den Bos [14] to be equivalent to least mean square (LMS) error linear prediction. It runs a K-point linear prediction filter across a data sequence of N samples, where N should be at least twice the value of K. Referring to the discrete filter diagram illustrated in Fig. 1, an optimum K point prediction filter predicts the nth value of the sequence from K past values,

$$\hat{X}_n = \sum_{k=1}^{K} A_k X_{n-k} \tag{1}$$

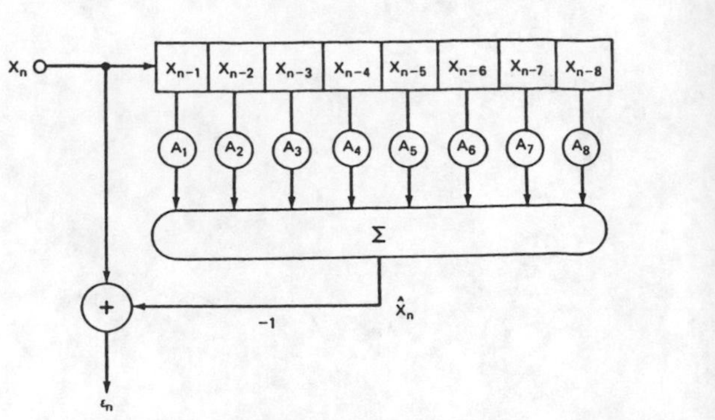

Fig. 1. Maximum entropy filter with one-step linear prediction estimator.

where $\hat{X}_n$ is the predicted sample, the A_k are optimum weighting coefficients, and the K past samples of X_{n-k} are presumed known. Define the difference between this predicted value and the current true value of X_n as the error ϵ_n, which is to be LMS minimized,

$$\epsilon_n = X_n - \hat{X}_n. \tag{2}$$

We minimize the total squared error E over the complete data sequence of N samples

$$E = \sum_{n=K}^{N-1} \epsilon_n^2 \tag{3}$$

and

$$\frac{\partial E}{\partial A_i} = 0, \qquad 1 \leq i \leq K \tag{4}$$

thus obtaining a set of K equations in K unknowns, i.e., the A_k filter weights,

$$\sum_{k=1}^{K} A_k \phi_{ki} = -\phi_{oi}, \qquad 1 \leq k \leq K, \quad 1 \leq i \leq K \tag{5}$$

where

$$\phi_{ki} = \sum_{n=K}^{N-1} X_{n-k} X_{n-i}. \tag{6}$$

There are several different techniques, including Burg's, for manipulating this set of equations to solve for the optimum A_k filter weights, and [3] is recommended if the reader is interested in pursuing the details further. When this error has been minimized, its power spectrum will be equivalent to "white" noise. Thus the uncertainty in ϵ_n has been maximized, hence we have a maximum entropy filter.

Upon substituting (1) into (2) it is seen that we have the form of a discreet convolution,

$$\epsilon_n = -\sum_{k=0}^{K} A_k X_{n-k} \tag{7}$$

where $A_0 = -1$. The associated Z-transforms may be written,

$$\mathcal{E}(Z) = \left[1 - \sum_{k=1}^{K} A_k Z^{-k}\right] X(Z) \tag{8}$$

where the expression within the brackets may be defined as

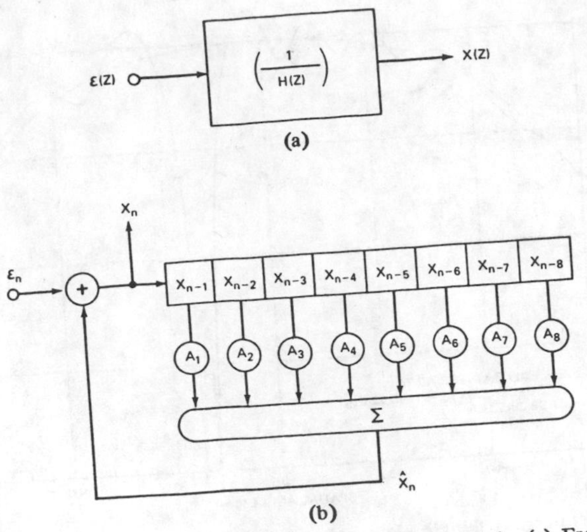

Fig. 2. Discrete all-pole linear prediction filter model. (a) Frequency domain. (b) Time domain.

the filter transform function $H(Z)$

$$H(Z) = \left[1 - \sum_{k=1}^{K} A_k Z^{-k}\right] \tag{9}$$

or

$$\mathcal{E}(Z) = H(Z)\, X(Z). \tag{10}$$

Note that $H(Z)$ is a polynomial in Z which will have K roots or zero factors. Since "white" noise has a power spectrum known to be equal to a constant, then from (10) it is eviden that we can solve for our unknown input power spectrum the filter function is known, i.e., from $Z = e^{j\omega}$

$$|X(\omega)|^2 = \frac{|\mathcal{E}(\omega)|^2}{|H(\omega)|^2} = \frac{(\text{constant})}{\left|\prod_{k=1}^{K} (1 - D_k e^{-j\omega})\right|^2}$$

where the peaks (poles) of the unknown power spec occur at the zeros of the filter function. This per model the input sequence with the powerful, discret linear prediction filter illustrated in Fig. 2, both quency domain and in the time domain. The pre is driven by white noise.

If one considers the action of such a filter corrupted with a small amount of noise, it is filter can synchronize with even a short sectio time waveform of the sinusoids and, once then proceed to "predict" many addition waveform with little error.

Another aspect of this filter is that i filter, so-called because it estimates th directly from the reciprocal of the filt Note in (10) that the error spectrum multiplication of the unknown spectr tion, and that no convolution of the ventional windowed Fourier transf hand, the unknown spectrum is e spectrum with the window filter convolution usually smears or peaked spectra. Reference [3] is rec readers who desire additional information on sp tion filters.

all array that, afte predicting of the main The perti linear predict is well-develo better underst behavior. A fu achieved via m vided that the convergence in little as a few depending upon of the source distr

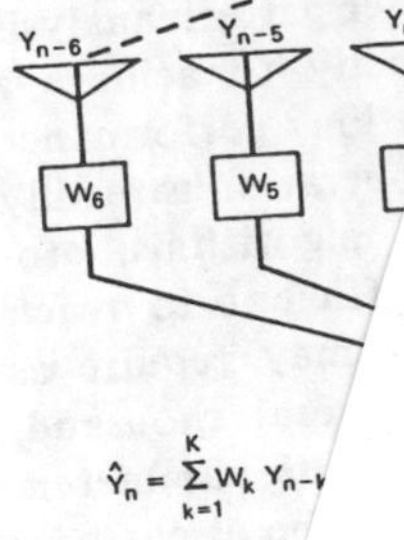

Fig. 3.

INTERFE SOURC

AU AF

III.

The spatial filter the adapted patter by the adaptive ele this paper were comp derived as follows: kth element consists η_{kn}, plus I incoherent

$$E_{kn} = \eta_{kn} + \sum_{i=1}^{I} J_i$$

where

and $u_i =$

d element spacing, assu
λ wavelength
θ_i spacial location angle
J_i amplitude of ith source
ϕ_{in} random phase of ith so
k element index, with tot
n snapshot sample index,

A "snapshot" is defined as one aperture signals at all array elem not time coherent with one and phase variable ϕ_{in}.

The first spatial filter pattern ex optimum weights of the familiar rithm, i.e., the inverse of the sam [17],

$$W_0 = \mu M^{-1} S^*$$

$$S^{*t} = [0, 0, 0, 0,$$

$$M = \frac{1}{N} \sum_{n=1}^{N} M_n$$

$$M_n = [E_n^* \cdot E_n^t]$$

[1] J_i is a constant amplitude rather than ran rent measurement program involving CW sou times.

note is that nulls in such an adapted pattern may be located arbitrarily close together in terms of beamwidth, without violating any physical principle, such as the Rayleigh criterion on the diffraction limit of our aperture. Yet, because the nulls have served to locate two sources within a beamwidth, one may describe this as a "superresolution" pattern.

It is readily shown that this adapted pattern is obtained by subtracting the summed array output pattern from the element (mainbeam) pattern and, furthermore, that the summed array pattern consists of properly weighted "eigenvector beams" [18]. Written in terms of the eigenvector weights, we can express the optimum weights in the form,

$$W_0 = S^* - \sum_{i=1}^{K} \left(\frac{B_i - B_0}{B_i + B_0}\right) \hat{W}_{qi} e_i$$

$$\hat{W}_{qi} = (e_i^{*t} S^*) \tag{17}$$

where e_i is the ith eigenvector of the covariance matrix, B_i is the ith eigenvalue, and B_0 is the smallest eigenvalue corresponding to receiver noise power. Note that only the significant eigenvectors corresponding to $B_i > B_0$ need be considered here. An adaptive array forms one such eigenvector beam for each degree of freedom consumed in nulling out the spacial source distribution. Fig. 6 illustrates the two eigenvector beams required for this two-source example. It should be emphasized here that *the true resolution and signal gain of the array is reflected in these eigenvector beams.* They demonstrate the importance of having as wide an aperture as possible, because the superresolution capability in the adapted pattern is a percentage of the true resolution of these beams. Also, since the superresolution nulls are formed via the subtraction of these beams of conventional width, it follows that the nulls will be rather delicate and very sensitive to system imperfections and signal fluctuations.

The desired "spacial spectrum pattern" is then obtained from (11) as simply the inverse of the adapted pattern. Fig. 7 illustrates this inverse for the two-source example, in comparison with the output of a conventional beam scanned through the two sources. Several comments are in order concerning such inverse patterns:

1) They are not true antenna patterns, because there is no combination of the element weights that could produce such a peaked spacial pattern. They are simply a function computed from the reciprocal of a true antenna pattern.

2) Linear superposition does not hold in either the inverse or the original adapted pattern, because of the nonlinear processing involved in the inverse of the sample covariance matrix (or the equivalent).

3) The heights of the peaks do not correspond with the relative strengths of the sources, because the depths of the adapted pattern nulls do not. In general, the adaptive null depth will be proportional to the square of the SNR of a source [18], but even this relationship fails when there are multiple sources closely spaced.

4) There is no real-signal output port associated with such a pattern, because it is not a true antenna pattern. An output could be simulated, of course, by implementing the equivalent all-pole filter of Fig. 2 and driving it with white noise.

5) They do emphasize the locations of the zeros (nulls) of the adaptive array filter polynomial.

6) They are inherently capable of superresolution.

7) They achieve good "contrast" with the spatial background "passband ripple" of the filter because of the aforementioned proportionality to the square of source strengths.

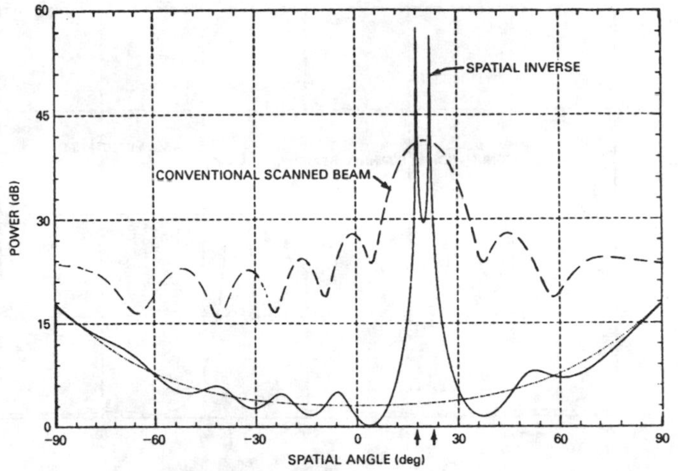

Fig. 7. Spacial spectrum inverse pattern for the two-source case of Fig. 5, and comparison with output of conventional scanned beam.

8) Spatial information is gained beyond that obtained from a conventional array beam which is scanned through the sources, because the array degrees of freedom are utilized in a more effective, data adaptive manner.

IV. MLM and Adaptive Directional Constraints

The maximum likelihood spectral estimate is defined as a filter designed to pass the power in a narrow band about the signal frequency of interest, and to minimize or reject all other frequency components in an optimal manner [6], [7]. This is identical to the use of a zeroth-order mainbeam directional gain constraint in adaptive arrays [19], [20], where the "spatial spectrum" would be estimated by the output residual power P_0 from the optimized adapted array weights,

$$P_0 = W_0^{*t} M W_0 \tag{18}$$

where

W_0 optimized weights $= \mu M^{-1} S^*$
M covariance matrix estimate
S^* mainbeam direction steering vector
μ scalar quantity.

Under the zeroth-order gain constraint, we require $S^t W_0 = 1$, whereupon μ becomes

$$\mu = (S^t M^{-1} S^*)^{-1}. \tag{19}$$

Substituting μ and W_0 into (18) then results in,

$$P_0 = \frac{1}{S^t M^{-1} S^*}. \tag{20}$$

Upon sweeping the steering vector S^* for a given covariance matrix inverse, P_0 will estimate the spatial spectrum. Interestingly, this result is very similar to the adapted angular response (AAR) technique described in [21], which consists of equation (18) normalized to the output thermal noise. Also, the denominator of (20) is identical (within a constant) to the mainbeam signal output voltage from an unconstrained optimized adapted array.

Referring back to Section III and the eigenvector beams shown in Fig. 6, one may describe the principle of operation in terms of subtracting the eigenvector beams from the quiescent uniform-illumination steering vector "mainbeam," for

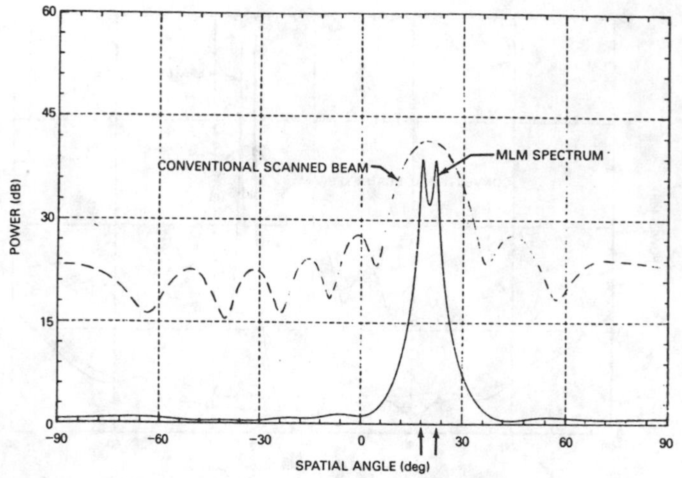

Fig. 8. MLM spatial spectrum plotted from residual power of adaptive zeroth-order mainbeam constraint for the two-source case of Fig. 5.

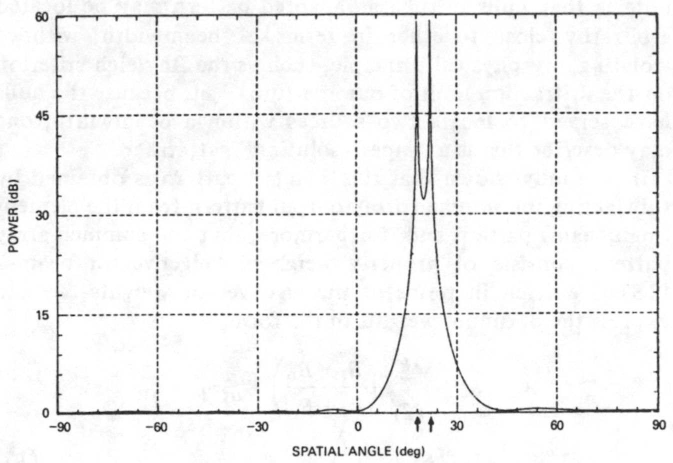

Fig. 9. "Thermal noise" algorithm spatial spectrum plotted for the two-source case of Fig. 5.

each position of the mainbeam. Thus we have a continuously changing adapted pattern as the mainbeam is scanned, subject to the above gain constraint in the steering direction.

Fig. 8 illustrates the output spectrum plotted from P_0 for the two-source case utilized for Figs. 5, 6, and 7. Note that in comparison with Fig. 7, this MLM spectrum has peaks which are about 18 dB lower and thus of less resolution capability. However, the two peaks have located the sources correctly and, in addition, the peak values reflect the true power levels of the sources. This is in agreement with the observations of Lacoss [7] and others. Although this technique has less resolution than the previous one and requires more computation in plotting the output spectrum, it does offer several rather significant advantages:

1) The output power is directly referenced to receiver noise power, thus permitting calibration and measurement of relative source strength.

2) If the sources can be resolved, then a psuedolinear superposition holds at the peaks, and they should reflect the true relative source strengths.

3) The output of this filter is a real signal, and if the filter passband is steered to a particular source, one can monitor that source at full array gain while rejecting all other sources.

4) The residual background spatial ripple is very low and well behaved.

5) It is not necessary to have the elements equally spaced or spaced near $\lambda/2$. Thus one should take advantage of this property to spread them out for a wider aperture and substantially increase the resolution for a given number of elements. This is done in the field of geophysics [6].

V. Alternate Adaptive Processing for Spatial Spectra

A. Phase Center Prediction

The adaptive array processing described in Section III did not utilize the configuration of Fig. 3 in the true sense of a K-point linear prediction filter which runs across a larger aperture of data samples. The K elements involved were the total aperture, and a series of N snapshots of data were utilized to estimate or "predict" the signal data at the phase center of the unweighted "mainbeam" element. Phase center prediction of this type is very flexible in that the location of the phase center of the "mainbeam" is rather arbitrary, and it is not necessary for the elements to be equally spaced. For example, the "mainbeam" may be a weighted summation of some or all of the elements in the array.

Carrying phase center prediction to its logical conclusion from the standpoint of estimating spacial spectra, it would appear that an ideal element configuration is a circular aperture array with an omnimode "mainbeam," i.e., the array would be predicting to the exact center of the circle. An example of such an array is contained in [18], and it could readily utilize the processing described herein.

B. The "Thermal Noise" Algorithm

In discussing the MLM spectral estimate in Section IV and its identical relationship to a zeroth-order mainbeam directional constraint, it was mentioned that the denominator of (20) was equal (within a constant) to the mainbeam signal output voltage from an unconstrained, optimized, adapted array. If we define this residual output signal voltage as R_0, then

$$R_0 = (S^t W_0) = \mu(S^t M^{-1} S^*) \tag{21}$$

and we have the reciprocal relationship to (20) except for μ.

An interesting feature about R_0 is that it approaches zero whenever the steering vector sweeps through the position of a source, and the reason it approaches zero is that the optimum weight vector W_0 approaches zero. Therefore, it appears prudent to formulate the dot product of W_0 with its own conjugate, and define this as the adapted thermal noise power output N_0,

$$N_0 = \lambda W_0^{*t} W_0 \tag{22}$$

where λ is a quiescent noise power level constant.

The reciprocal of N_0 then estimates the spacial spectrum, and we may refer to this as the "thermal noise" algorithm for spectrum estimation. Fig. 9 illustrates the application of this algorithm to the same two-source case as was utilized for Figs. 5–8. Comparing it against the equivalent MESA and MLM techniques, note that it exhibits resolution peaks fully equal to MESA, and yet retains the very low residual background ripple of MLM referenced to quiescent receiver noise power level. Like MESA, it cannot measure relative source strength, but it should be proportional to the square of source strengths whenever it can resolve the sources. Thus the "ther-

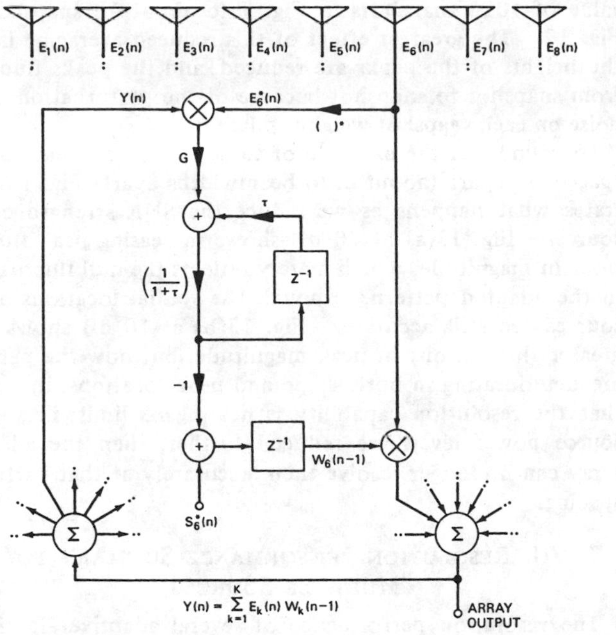

Fig. 10. The Howells-Applebaum algorithm in a recursive digital form.

mal noise" algorithm appears to possess an interesting combination of the characteristics of both MESA and MLM.

VI. Real-Time Filter Operation Using Howells-Applebaum Algorithm

To get a feel for real-time operation performance with weight update averaging, simulations were run in which an eight element array had its weights computed from the Howells-Applebaum algorithm [16]. This well-known algorithm is simple, robust, requires no data storage, and may be implemented in either analog or digital form. It is shown in digital form in Fig. 10. The associated recursive relationship for the kth weight may be written,

$$(1+\tau)\, W_k(n) = \tau W_k(n-1) + S_k^*(n) - \left(\frac{1}{B_0}\right) E_k^*(n)\, Y(n) \tag{23}$$

where

$$Y(n) = \sum_{k=1}^{K} E_k(n)\, W_k(n-1) \tag{24}$$

$Y(n)$ is the current array output, $E_k(n)$ is the current snapshot signal sample at the kth element (similar to (12)), $W_k(n-1)$ is the previous value of the kth weight, $S_k^*(n)$ is the injected kth beamsteering weight, and B_0 is a constant equal to the mean receiver noise power.

The digital integration loop shown in Fig. 10 is designed to simulate a simple low-pass RC filter with a time constant of τ, but we choose to make τ dynamic in order to get faster convergence for most situations. Thus let τ become $\tau(n)$,

$$\tau(n) = \tau_0 + TP_r(n) \tag{25}$$

where

T high-power fast time constant
τ_0 quiescent conditions slow time constant

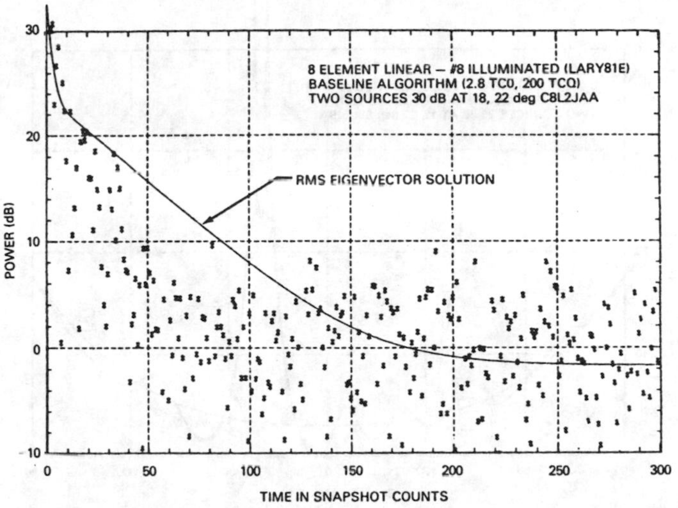

Fig. 11. Time sequence snapshot output of array in decibel above receiver noise power, Howells-Applebaum recursive algorithm with dynamic time constant.

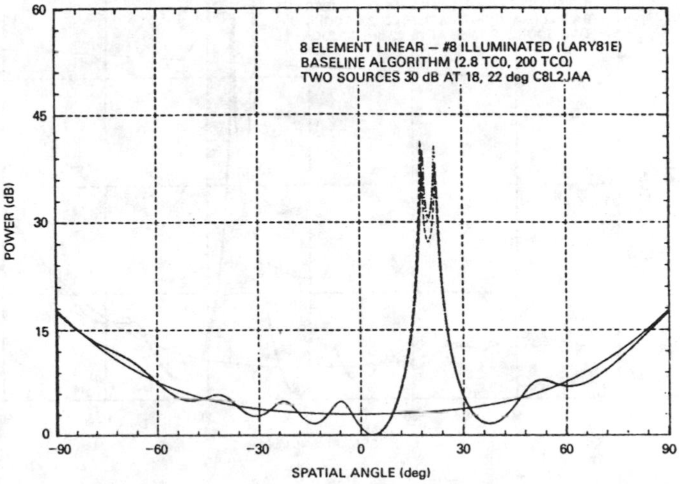

Fig. 12. Typical spatial spectrum snapshot plots after convergence, snapshot weight sets numbers 100, 200, and 300. Two 30-dB sources located at 18° and 22°.

and

$$P_r(n) \text{ snapshot power ratio} = \frac{1}{B_0} \sum_{k=1}^{K} E_k(n)\, E_k^*(n).$$

This formulation permits us to satisfy the 10 percent bandwidth criterion at high power levels to avoid noisy weights [18] by choosing the value of $T = 3.2$, and yet the quiescent condition time constant need be no worse than $\tau_0 = 200$. The larger value for τ_0 is necessary in order to have a relatively stable quiescent pattern. Actual weight update averaging is performed in accordance with the reciprocal of the closed-loop bandwidth α

$$\frac{1}{\alpha} = \frac{\tau}{1 + P_r(n)} = \frac{\tau_0 + TP_r(n)}{1 + P_r(n)} \tag{26}$$

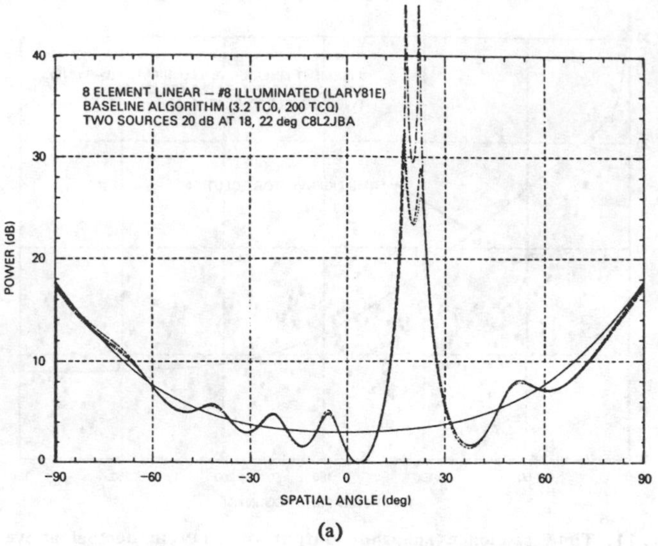

(a)

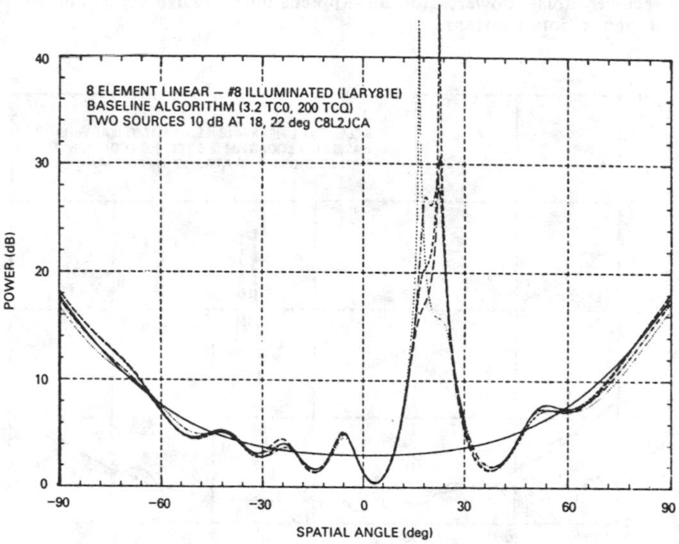

(b)

Fig. 13. Typical spatial spectrum snapshot plots after convergence. Two sources located at 18° and 22°. (a) 20-dB SNR sources. (b) 10-dB SNR sources.

where we approach the value $\tau_0/2$ under quiescent conditions when $P_r(n) \approx 1$, and we approach the value of T when $P_r(n) \gg 1$. The increase in system SNR due to this averaging is approximately equal to (π/α), i.e., the ratio of receiver channel bandwidth to the (two-sided) RC filter bandwidth.

A typical time sequence plot of snapshot output power in dB above receiver noise level is illustrated in Fig. 11 for the case of two 30-dB sources located at 18° and 22°. Note that the system converges and nulls out the two strong sources within about 60 snapshots after processing commences. The principal point demonstrated here is that convergence occurs reasonably fast in terms of snapshot counts, even for this simple LMS adaptive algorithm and, after convergence, any of the snapshot weight sets can be utilized to compute the spatial spectrum. Fig. 12 shows the spatial spectrum plots associated with snapshots numbers 100, 200, and 300. Comparing Fig. 12 against Fig. 7, note that the pattern has changed very little, in spite of the fact that the integration or averaging has been reduced by two orders of magnitude, i.e., from a value of 1024 snapshots in Fig. 7 to about 10 snapshots in Fig. 12. The greatest effect of this reduced averaging is that the heights of the peaks are reduced, and the peaks fluctuate from snapshot to snapshot because of the perturbation of the noise on each snapshot weight update.

To round out the example of these two incoherent sources spaced 4° apart (about 0.26 beamwidths apart) Fig. 13 illustrates what happens as we reduce the SNR strength of the sources. Fig. 13(a) at 20 dB shows increasing peak fluctuations in magnitude, which merely reflects the null fluctuations in the adapted pattern, although the spatial locations of the sources are still accurate. Fig. 13(b) at 10 dB shows even greater fluctuations in peak magnitude, but now the patterns are deteriorating in both shape and peak locations, indicating that the resolution capability is nearing its limit, i.e., if the source power levels are reduced further, then the adaptive array can no longer resolve them accurately at that particular spacing.

VII. Resolution Performance Summary for Multiple Sources

The resolution performance of several adaptive algorithms against multiple sources has been tested with the simulated signal data samples described in Section III. These algorithms consisted of the ones described in previous sections; namely, the simple Howells–Applebaum, "thermal noise," constrained-optimized beam scan, and the Weiner matrix inverse; plus the orthonormal lattice filter algorithm described by Alam [22]. The latter two algorithms have very fast convergence in terms of snapshot counts, and are "medium" in their processing computer burden.

A very useful initial test is the resolution of two incoherent sources of equal strength in which we vary the parameters: K is the number of array elements; P_s is the SNR of each source at the elements; θ_i is the spatial location of ith source; and N is the number of snapshots averaged. Fig. 14 summarizes the results of many trial runs in the form of a "lower limit" plot of array output SNR $(K \cdot P_s)$ versus source separation in beamwidths. The plot is universal in nature and can be utilized for any number of array elements in a linear array configuration. The 3-dB beamwidth $\Delta\theta_a$ of the linear array sampling aperture was computed from the expression [23],

$$\Delta\theta_a = \sin^{-1}\left(\frac{\lambda}{D}\right) \approx \sin^{-1}\left[\frac{2}{K\cos(\theta_1+\theta_2)/2)}\right] \tag{27}$$

where D is the effective aperture width. A point of interest here is that D cannot be considered to be increased by N, because the snapshots are not time coherent with one another. The plot itself is similar to an ensemble average limit and tells us that the two sources at a given separation cannot be resolved unless the array output SNR exceeds the plot value, even though N may have a large value. The importance of N relates to averaging the interelement signal correlations which determines the variance of fluctuations in the adaptive weights, and thus the variance of fluctuations of the resolution peaks. A criterion for choosing the value of N must compromise between location accuracy and speed of response. During the course of the simulation data runs, it was found that a convenient formula for choosing the minimum value of N was,

$$N \geqslant \frac{300 + 10\,KP_s}{1 + KP_s}. \tag{28}$$

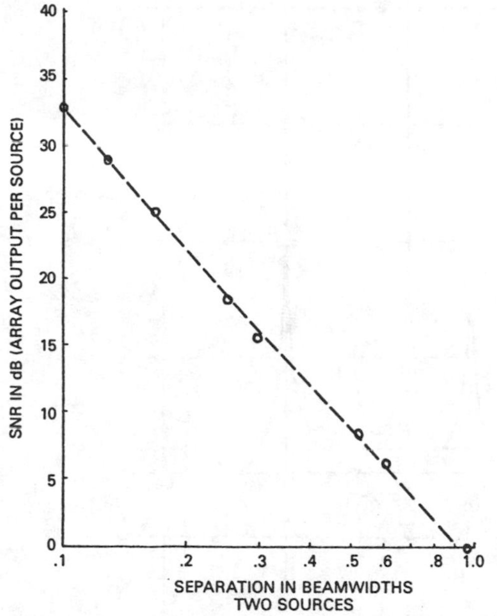

Fig. 14. Universal approximate resolution limit for two equal-strength incoherent sources. Simulation conditions: narrow band, no array errors, λ/2 element spacing, linear array, Gaussian receiver noise.

Note that the total SNR for adaptive weight determination will be the product $(K \cdot P_s \cdot N)$.

The reader is cautioned that the plot in Fig. 14 represents a "best estimate" of resolution deterioration from the trial data runs and does not represent the application of a rigorous resolution criterion. Also, it does not apply to the constrained-optimized beam scan algorithm, which requires approximately 12 dB more SNR for the same resolution. On the positive side, however, it does tell us that we can resolve two incoherent sources at arbitrarily small spacings, provided we have sufficient array output SNR and, also, provided that the data is accurate. (Recall that no element errors were included here.)

If there are more than two sources within a beamwidth, then difficulties mount rapidly and the filter null points may not accurately represent the spatial locations of the sources. For example, the simple Howells–Applebaum algorithm was tested against three incoherent sources of 30-dB strength spaced 4° apart, and it could not resolve all three even after 2048 snapshots because it never converged sufficiently (the eigenvalue spread was too great). In order to separate the three sources in a reasonable number of snapshots, it is necessary to use an adaptive algorithm of faster convergence such as the Weiner matrix inverse or the orthonormal lattice filter mentioned above.

Fig. 15 illustrates resolution of the three sources via the latter algorithm, but it should be noted that even with this fast algorithm, resolution was not reliable until after about 200 snapshots, and the locations of the peaks fluctuates considerably. This was a good illustration of the "delicacy" of null formation for the case of closely spaced multiple sources. Another difficulty is that the spatial background "passband ripple" fluctuations usually increase in magnitude as you increase the number of sources within a beamwidth, thus increasing the risk of "false sources."

Lest the impression be left that more than two sources is always bad news, the final resolution example has been deliberately chosen to consist of four incoherent sources, with

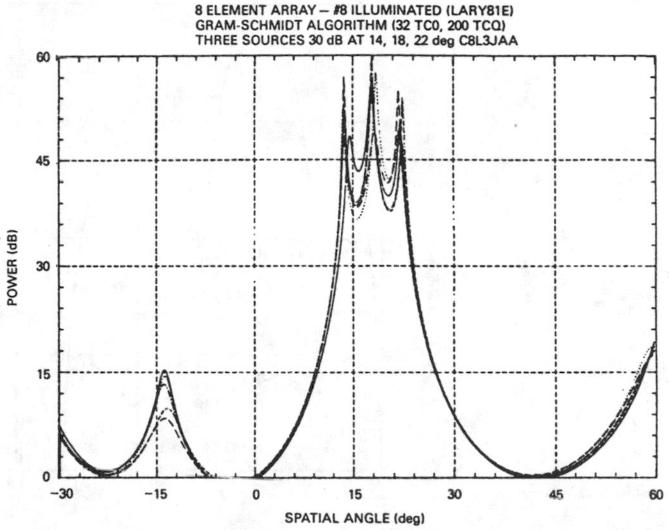

Fig. 15. Typical spatial spectrum snapshot plots after convergence, Gram–Schmidt algorithm, three 30-dB sources located at 14°, 18°, and 22°.

the additional complication that the sources are of unequal strengths:

$$\begin{aligned} P_{s1} &= 0 \text{ dB} & \theta_1 &= -15^\circ \\ P_{s2} &= 30 \text{ dB} & \theta_2 &= 0^\circ \\ P_{s3} &= 10 \text{ dB} & \theta_3 &= +15^\circ \\ P_{s4} &= 20 \text{ dB} & \theta_4 &= +30^\circ. \end{aligned}$$

Fig. 16 illustrates the easy adaptive resolution of all four sources by an 8 element array which has a uniform illumination beamwidth equal to their separations. The secret to success here is to avoid crowding all four sources within a beamwidth. Notice in Fig. 16(a) that, even though the sources are separated by a full beamwidth, a conventional beam scan cannot resolve them because of its sidelobes [23]. Tapering the conventional beam illumination to reduce its sidelobes would only make matters worse because the new beamwidth then becomes greater than the source separations. Fig. 16(b) is the adapted pattern inverse obtained via the Weiner matrix inverse algorithm for $N = 1024$ snapshots, and it exhibits excellent resolution but inaccurate relative source strengths. Fig. 16(c) is the constrained-optimized beam scan of Section IV (equivalent to MLM), and here we obtain the double advantage of resolving the sources and also accurately indicating their relative strengths. Fig. 16(d) illustrates several time-sequence patterns obtained via the orthogonal lattice filter algorithm, wherein the solution firms up after about 50 snapshots.

VIII. Superresolution of Coherent Sources/Targets

The resolution of coherent sources or equivalent radar targets is more difficult for these nonlinear adaptive processing techniques, because correlated signals produce fields that are nonstationary in the space dimension [24]–[26]. However, successful results have been obtained whenever sufficient relative motion of "Doppler cycles" are available. To address the coherent problem, consider the aperture "modulation" function produced by two equal-strength coherent sources located 8° apart in the far field, as illustrated in Fig. 17. This aperture modulation function is produced by the interaction of the two coherent wavefronts along the line containing

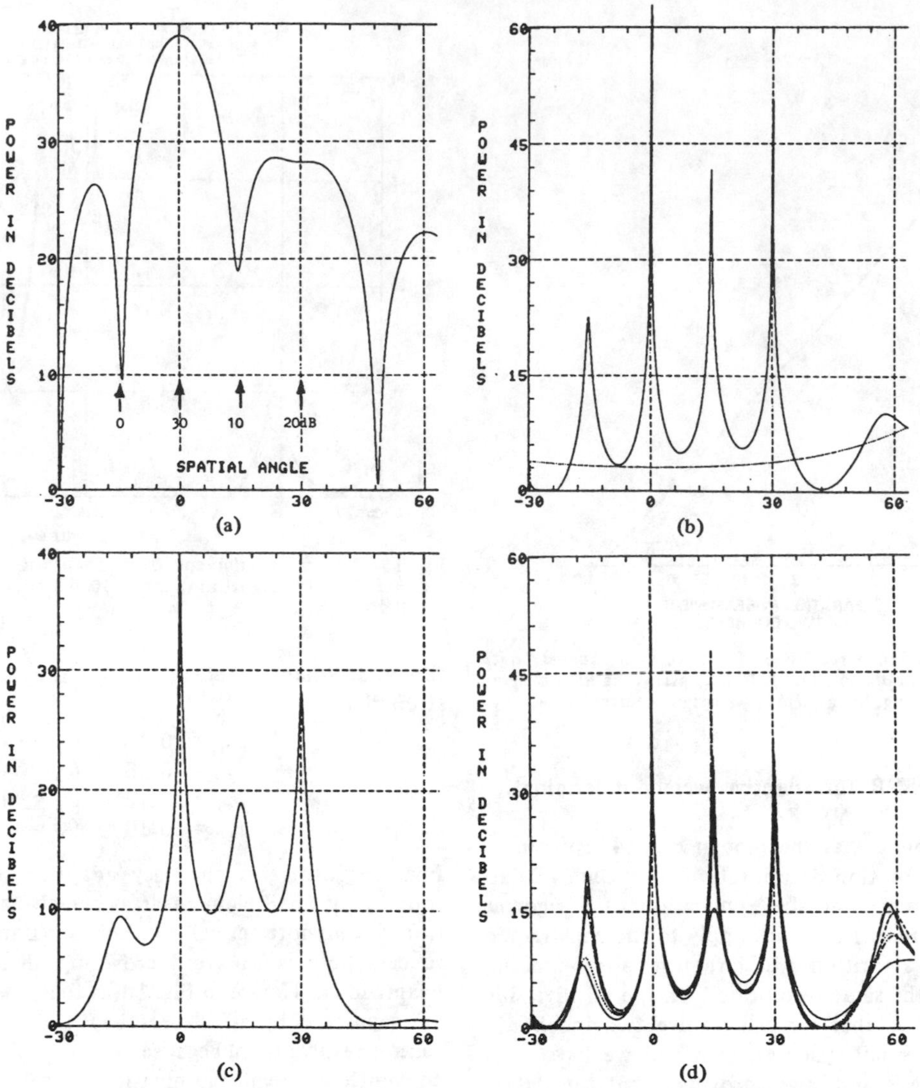

Fig. 16. Spatial spectrum estimates for four incoherent sources of 0 dB, 30 dB, 10 dB, and 20 dB strength located at −15°, 0°, +15°, and +30°; 8 element linear array; (a), (b), and (c) computed from sample covariance matrix of 1024 snapshots. (a) conventional beam scan. (b) adapted pattern inverse. (c) constrained-optimized beam scan. (d) orthogonal lattice filter solutions, snapshots nos. 25, 40, 55, 70, 85, and 100.

the aperture sampling elements. If the coherent sources maintain their fixed-phase relationship and if the 8-element sampling aperture does not move, then the following two difficulties arise with the resulting sample covariance matrix, which prevents resolution via the techniques described above:

1) *It is not Toeplitz*, i.e., the values along diagonals differ in accordance with the fixed element sampling of the modulation function amplitude.

2) *It has only one unique eigenvalue*, i.e., the two sources are blended into a single eigenvector instead of two.

Some examples for this fixed-phase case are discussed by White [26].

One solution to these difficulties is to have sideways movement of the 8 element sampling aperture such that the aperture line remains parallel for all snapshot positions. Note that whenever we shift the position sideways between snapshots, we are changing the received phase difference between the two coherent sources because the distances from the sources to the aperture phase center have changed. If we denote the incremental phase change per snapshot as $\Delta\phi_n$, then the coherent phase difference for the nth snapshot referenced to the aperture phase center may be written,

$$\phi_{cn} = \phi_{co} + \sum_{n=1}^{n} \Delta\phi_n \tag{29}$$

where ϕ_{co} is the starting phase difference. Thus, as we move sideways, ϕ_{cn} is changing with time and *each element is sampling the aperture modulation function at different points as it moves along*. After sampling one or more complete "cycles" of the modulation function, then similar interelement-spacing covariance terms become approximately equal, the sample covariance matrix approaches Toeplitz condition, and the resolution performance of the various algorithms becomes similar to their performance against incoherent sources.

Some further thought on the time rate of change of ϕ_{cn} in (29) leads quickly to the conclusion that we have formed a coherent "Doppler shift" between the two sources via move-

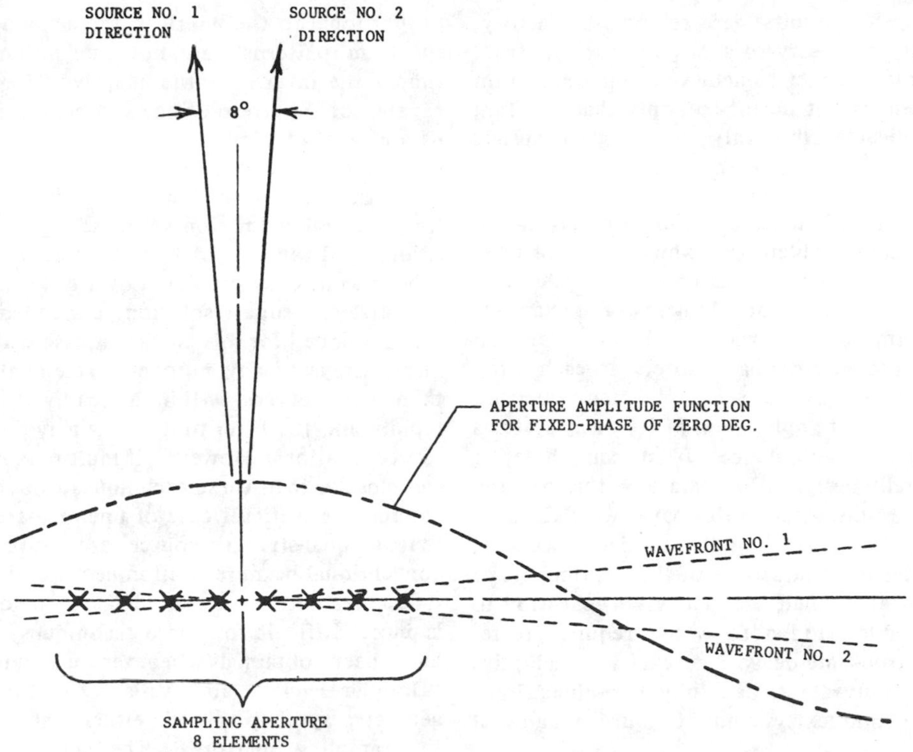

Fig. 17. Aperture space modulation function produced by two equal-strength coherent sources located 8° apart in far field, linear array sampling aperture of 8 elements with half-wavelength spacing.

ment of the phase center of our sampling aperture. Furthermore, it is obvious that the identical result could be produced by keeping the position of the sampling aperture fixed, but let the corresponding coherent Doppler shift exist between the two sources. Note that for Fig. 17, this would cause the aperture modulation function to move sideways with respect to the fixed sampling aperture. Thus a Doppler shift between coherent sources/targets gives us a second possibility for resolving them within a beamwidth. As indicated above, it is desirable to have enough snapshots to sample one or more complete "Doppler cycles" in order that the covariance matrix may approach Toeplitz condition.

A. Superresolution from a Single Snapshot

The single snapshot case is the most difficult one to solve satisfactorily. To begin with, it is a fixed-phase coherent case even when the sources are nominally incoherent, simply because only one snapshot is available. For the techniques described herein, a solution is sometimes possible via synthetic motion of a smaller sampling subaperture along the single snapshot data sample. If L represents the number of elements in the sampling subaperture, and K is the total number of elements, then we can form a reduced $L \times L$ sample covariance matrix averaged over $(K - L + 1)$ "samplings." Such synthetic movement of a subaperture is very similar to the action which occurs in the MESA technique of Burg [3]–[5], where he moves an L-point linear prediction filter across a larger data sequence of K samples. The Burg MESA technique has been utilized by several investigators for this single snapshot case [10], [24], [25]. Both techniques require equal element spacing, and they both attempt to resolve the sources under the condition that there is less than one complete modulation cycle across the aperture (an inherent implication in superresolution from a single snapshot). Successful results for the sampling subaperture technique are usually obtained when the $L \times L$ sample covariance matrix approaches Toeplitz condition. However, the reader is cautioned that the single snapshot case is very sensitive to the particular portion of the modulation cycle which is sampled, both its fractional length and phase position, in addition to the number of sampling elements, element spacing, and element SNR. False peaks and/or location errors are to be expected in experimenting with these parameters, and additional research is needed. The papers by Evans [24], [25] and by White [26] are recommended for additional details and pertinent examples.

IX. Significant Differences with Spectral Analysis

Although the similarities are extensive enough to create a favorable climate for technique interchanges, there are also some significant differences which arise from the very nature of the applications and their data. For example, assume that we utilize the Burg MESA technique to run a L-point prediction filter across an RF array aperture of K elements, where L must be smaller than K by at least 50 percent in order to obtain reasonable averaged correllation estimates for determination of the L filter weights. This type of spectral analysis processing has a great advantage in being able to operate with a single snapshot of element data, but it is almost "unthinkable" from an RF array point of view because it is wasting expensive aperture elements. It is far more preferable to operate on the basis of many snapshots of data from the smallest number of elements possible. In fact, a recent study of the Burg MESA technique by King [10], as applied to an RF spatial array, found that single snapshot results were seldom satisfactory, and that it was usually necessary to average the results from ten or twenty snapshots in order to achieve a stable spectrum estimate. This comment is not meant to imply that the Burg technique is not applicable, but only that single snapshot operation is not very practical for RF arrays.

A related difference is the fact that RF array element signal samples are correlated in both space and time, thus giving rise to a two-dimensional data problem [9] which does not exist in spectrum analysis. To overcome this problem, it is usually necessary to perform filtering in both domains. For example, one may handle the time domain via tapped delay lines, an FFT operation, or actual narrow-band filters at each array element. Note that in each case, we imply many aperture snapshots. Fortunately, most applications of RF array systems are such as to produce an abundance of data snapshots, so that processing is naturally designed to operate on this premise. The simulation examples discussed in this paper were all based upon narrow-band filters preceeding the spatial processing, because it is much easier to demonstrate the basic principles of spatial superresolution at a single frequency. Application to RF signals of appreciable bandwidth would require greater sophistication such as true-time-delay matrix network adaptive processing [27] or its equivalent, i.e., fully two-dimensional processing wherein the complexity, computer burden, and cost all increase substantially.

A difference of less significance is the manner in which correlation matrix coefficients are estimated. In spectral analysis, one generally deals with a single data sequence in X_n of M samples, from which is computed the autocorrelation matrix coefficients of the form,

$$\phi_T = \frac{1}{N} \sum_{n=0}^{M-T-1} X_n X_{n+T}^*, \qquad 0 \leqslant T \leqslant L \tag{30}$$

where the lag T is restricted to some maximum value L which is a fraction of M. In adaptive array processing, on the other hand, one generally deals with N "snapshots" of data sampled from K elements, from which is computed sample covariance matrix coefficients of the form,

$$\phi_{k1} = \frac{1}{N} \sum_{n=1}^{N} X_{kn}^* X_{1n}, \qquad 1 \leqslant k \leqslant K \quad 1 \leqslant 1 \leqslant K. \tag{31}$$

The application of (30) and (31) to incoherent source cases with many snapshots results in little difference. However, when applied to coherent source cases as illustrated by Fig. 17, note that these coefficients may differ and thereby result in somewhat different spectrum estimates.

X. Conclusions

The Burg MESA and MLM nonlinear spectral analysis techniques have been related to their similar nonlinear adaptive array antenna counterparts, which consist of the "sidelobe canceller" and directional gain constraint techniques. The comparison study was conducted in the interests of achieving some cross fertilization by examining their principles of operation from the antenna array spatial pattern viewpoint, and the analysis helps to qualify their superresolution performance behavior. It was shown that the superresolution derives from the subtraction of eigenvector beams which embody the true conventional resolution and signal gain of the array, such that physical principles are not really violated. Resolution is still proportional to the width of the array aperture. The "spatial spectrum patterns" are not true antenna patterns, but are simply the inverse of the adaptive filter patterns. They are capable of superresolution, and spatial information is gained beyond that obtained from a conventional array beam which is scanned through the sources.

The adaptive array counterpart is naturally suited to real-time spectral estimation via most of the current adaptive algorithms, and the case of two incoherent sources located within a beamwidth was simulated over a SNR range of 0 to 40 dB. A universal superresolution performance curve, Fig. 14, was developed for this particular case which can be utilized for linear arrays of any number of elements. If there are more than two sources within a beamwidth, difficulties mount rapidly and the filter null points may not accurately represent source locations. However, if multiple sources are not crowded too closely, then these techniques can easily resolve them, including the difficult case of unequal-strength sources. Recall that unequal-strength sources are a difficult problem area for conventional beam resolution because of their sidelobes [23].

Superresolution of coherent spatial sources or radar targets is more difficult for these techniques, but successful results have been obtained whenever sufficient relative motion or "Doppler cycles" are available. "Doppler cycles" can be generated synthetically via either real or electronic motion of the sampling aperture. The worst superresolution results occur when only a single snapshot of data is available. This case is seldom resolved satisfactorily and requires additional research.

In addition to the direct adaptive counterparts, two alternate adaptive spatial spectrum estimators were suggested. One is a circular array aperture arrangement which predicts to the center of the circle, and the other is a new adaptive "thermal noise" algorithm which appears to possess an interesting combination of both MESA and MLM pattern characteristics.

There are some significant differences between spectral analysis techniques and adaptive array techniques which relate to the nature of their applications and the two-dimensional data problem. However, it appears that there is much to be gained through careful analysis of the others' techniques. For example, in addition to the obvious applications in target detection, direction finding location systems, and source classification, spectral analysis techniques should be of benefit in problem areas involving aperture data interpolation and extrapolation [24].

References

[1] "Antenna spatial pattern viewpoint of MEM, MLM, and adaptive array resolution," in *Proc. RADC Spectrum Estimation Workshop* (Rome Air Development Center) Griffiss AFB, NY, Oct. 1979.

[2] "Nonlinear spectral analysis and adaptive array superresolution techniques," Naval Res. Lab. Rep. 8345, Feb. 1980.

[3] D. G. Childers, Ed., *Modern Spectrum Analysis*. New York: IEEE Press, 1978. (Note: This book contains complete copies of the following references [4]-[9], [11], [13], [14].)

[4] J. P. Burg, "Maximum entropy spectral analysis," in *Proc. 37th Meeting Society Exploration Geophysicists*, 1967.

[5] ——, "A new analysis technique for time series data," presented at the NATO Advanced Study Inst. Signal Processing with Emphasis on Underwater Acoustics, Enschede, The Netherlands, 1968.

[6] J. Capon, "High-resolution frequency-wavenumber spectrum analysis," *Proc. IEEE*, vol. 57, pp. 1408-1418, Aug. 1969.

[7] R. T. Lacoss, "Data adaptive spectral analysis methods," *Geophysics*, vol. 36, pp. 661-675, Aug. 1971.

[8] J. P. Burg, "The relationship between maximum entropy spectra and maximum likelihood spectra," *Geophysics*, vol. 37, pp. 375-376, Apr. 1972.

[9] R. N. McDonough, "Maximum-entropy spatial processing of

array data," *Geophysics*, vol. 39, pp. 843–851, Dec. 1974.
[10] W. R. King, "Maximum entropy spectral analysis in the spatial domain," Naval Res. Lab. Rep. 8298, Mar. 1979.
[11] L. J. Griffiths, "Rapid measurement of digital instantaneous frequency," *IEEE Trans. Acoust., Speech, Signal Processing*, vol. ASSP-23, pp. 207–222, Apr. 1975.
[12] D. R. Morgan and S. E. Craig, "Real-time adaptive linear prediction using the least mean square gradient algorithm," *IEEE Trans. Acoust. Speech, Signal Processing*, vol. ASSP-24, pp. 494–507, Dec. 1976.
[13] M. A. Alam, "Adaptive spectral estimation," in *Proc. 1977 Joint Automatic Control Conf.*, June 1977.
[14] A. Van den Bos, "Alternative interpretation of maximum entropy spectral analysis," *IEEE Trans. Inform. Theory*, vol. IT-17, pp 493–494, July 1971.
[15] P. W. Howells, "Explorations in fixed and adaptive resolution at GE and SURC," *IEEE Trans. Antennas Propagat.*, vol. AP-24, pp. 575–584, Sept. 1976.
[16] S. P. Applebaum, "Adaptive arrays," *IEEE Trans. Antennas Propagat.*, vol. AP-24, pp. 585–598, Sept. 1976.
[17] I. S. Reed, J. O. Mallett, and L. E. Brennan, "Rapid convergence rate in adaptive arrays," *IEEE Trans. Aerosp. Electron. Syst.*, vol. AES-10, pp. 853–863, Nov. 1974.
[18] W. F. Gabriel, "Adaptive arrays—An introduction," *Proc. IEEE*, vol. 64, pp. 239–272, Feb. 1976.
[19] O. L. Frost, "An algorithm for linearly constrained adaptive array processing," *Proc. IEEE*, vol. 60, pp. 926–935, Aug. 1972.
[20] S. P. Applebaum and D. J. Chapman, "Adaptive arrays with main-beam constraints," *IEEE Trans. Antennas Propagat.*, vol. AP-24, pp. 650–662, Sept. 1976.
[21] G. V. Borgiotti and L. J. Kaplan, "Superresolution of uncorrelated interference sources by using adaptive array techniques," *IEEE Trans. Antennas Propagat.*, vol. AP-27, pp. 842–845, Nov. 1979.
[22] M. A. Alam, "Orthonormal lattice filter—A multistage, multichannel estimation technique," *Geophysics*, vol. 43, pp. 1368–1383, Dec. 1978.
[23] B. D. Steinberg, *Principles of Aperture & Array System Design* New York: Wiley, 1976, reference ch. 10.
[24] *Proc. RADC Spectrum Estimation Workshop* (Rome Air Development Center) Griffiss AFB, NY, Oct. 1979.
[25] J. E. Evans, "Aperture sampling techniques for precision direction finding," *IEEE Trans. Aerosp. Electron. Syst.*, vol. AES-15, pp. 891–895, Nov. 1979. See also pp. 899–903.
[26] W. D. White, "Angular spectra in radar applications," *IEEE Trans. Aerosp. Electron. Syst.*, vol. AES-15, pp. 895–899, Nov. 1979.
[27] L. E. Brennan, J. D. Mallett, and I. S. Reed, "Adaptive arrays in airborne MTI radar," *IEEE Trans. Antennas Propagat.*, vol. AP-24, pp. 607–615, Sept. 1976.

Least Squares Type Algorithm for Adaptive Implementation of Pisarenko's Harmonic Retrieval Method

V. U. REDDY, MEMBER, IEEE, B. EGARDT, MEMBER, IEEE, AND T. KAILATH, FELLOW, IEEE

Abstract—Pisarenko's harmonic retrieval method involves determining the minimum eigenvalue and the corresponding eigenvector of the covariance matrix of the observed random process. Recently, Thompson [9] suggested a constrained gradient search procedure for obtaining an adaptive version of Pisarenko's method, and his simulations have verified that the frequency estimates provided by his procedure were unbiased. However, the main cost of this technique was that the initial convergence rate could be very slow for certain poor initial conditions.

Restating the constrained minimization as an unconstrained nonlinear problem, we derived an alternative Gauss–Newton type recursive algorithm, which also used the second derivative matrix (or Hessian); this algorithm may also be viewed as an approximate least squares algorithm.

Simulations have been performed to compare this algorithm to (a slight variant of) Thompson's original algorithm. The most important conclusions are that the least squares type algorithm has faster convergence in the beginning, while its convergence rate close to the true parameters depends on the signal-to-noise ratio of the input signal. The approximate least squares algorithm resolves the sinusoids much faster than the gradient version.

I. Introduction

The development of adaptive techniques for estimating the parameters of sinusoidal signals in additive noise is important in many applications. The so-called adaptive line enhancer (ALE) proposed by Widrow *et al.* [1] has been a popular solution. The ALE is a tapped delay line filter of some fixed length L whose tap gains are recursively adjusted by using the Widrow–Hoff algorithm [2] (a kind of stochastic approximation or stochastic gradient technique) so that they converge to the solution of the normal equations for the one-step minimum mean-square error prediction problem. Another popular solution uses a so-called lattice filter whose parameters are adjusted by using a technique, due to Burg [3], based on minimizing the sum of certain forward and backward one-step prediction residuals. Griffiths [4] merged the above approaches by proposing a lattice filter whose coefficients were adapted by use of the Widrow–Hoff LMS algorithm, leading to what is often called a gradient lattice filter.

The three approaches of [1], [3], and [4] all yield spectral estimates with fairly sharp peaks (with Burg's method doing better with short data lengths), but the estimates of the sinusoidal frequencies invariably appear to be biased when the sinusoids are observed in the presence of additive white noise.

In an attempt to improve the above methods with respect to bias, Nuttall [5] and, independently, Ulrych and Clayton [6] proposed the least squares fitting of an autoregressive (AR) model, based on a criterion involving both forward and backward prediction errors but, unlike Burg's method, without using a lattice filter model. Marple [7] derived a recursive form of the above method, and has shown by extensive simulations that the bias in the spectral estimates can be reduced significantly compared to the Burg technique. It should be noted, though, that this is true only for short data lengths, since all the above techniques give identical results for large data lengths.

Possibilities for improvement of the above approaches become apparent when one considers that they were all designed

Manuscript received June 7, 1981; revised December 7, 1981. This work was supported in part by the Joint Services Electronics Program under Contract DAAG29-79-0047, by the National Science Foundation under Grant ENG78-10003, by the U.S. Army Research under Contract DAAG29-79-C-0215, and by the Air Force Office of Scientific Research, Air Force Systems Command, under Contract AF49-620-79-0058.

V. U. Reddy was with the Information Systems Laboratory, Stanford University, Stanford, CA 94305. He is now with the Department of ECE, College of Engineering, Osmania University, Hyderabad, India.

B. Egardt was with the Information Systems Laboratory, Stanford University, Stanford, CA 94305. He is now with the Department of Research and Innovation, ASEA, Inc., Västerås, Sweden.

T. Kailath is with the Information Systems Laboratory, Stanford University, Stanford, CA 94305.

Reprinted from *IEEE Trans. Acoust., Speech, Signal Processing*, vol. ASSP-30, pp. 399–405, June 1982.

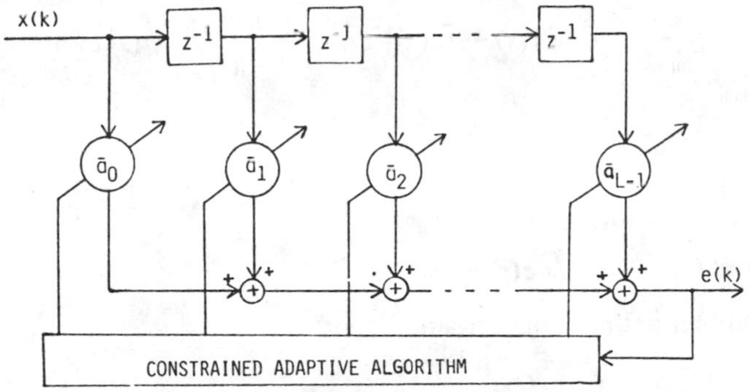

Fig. 1. Constrained AR spectral estimator.

to converge to the optimum linear least squares solution for the prediction of any random process, and do not specifically exploit the fact that the signals are sinusoidal. Pisarenko [8] was perhaps the first to attempt to do this in his so-called "harmonic retrieval" method, which involves determining the minimum eigenvalue and the corresponding eigenvector of the covariance matrix of the observed random process. Thompson [9] noted that this eigenvalue–eigenvector computation was equivalent to a certain constrained gradient-search procedure, and he exploited this fact to obtain an adaptive version of Pisarenko's method. Thompson's simulations verified that the frequency estimates provided by his procedure were unbiased, and Larimore and Calvert [10] have described some studies on the costs accompanying this nice unbiasedness property—in particular, that the initial convergence rate may be very slow for certain poor "initial conditions."

One of the goals of this paper is to consider a way of providing faster initial convergence by using a different (so-called Gauss-Newton type) algorithm for the constrained minimization problem. This is done in Section II. In Section III we compare this method with Thompson's method (actually, a very similar, but slightly improved version of his method) in a number of different examples. The new method does have faster initial convergence, but the rate of convergence close to the true parameters depends on the signal-to-noise ratio of the input signal.

II. The Constrained Adaptive Algorithms

The adaptive filter with constrained filter coefficients as suggested by Thompson is shown in Fig. 1. The observed process, consisting of a sum of sinusoids and white noise, is denoted by a time series $x(k)$. The filter output $e(k)$ can be expressed as the inner product[1]

$$e(k) = \bar{A}^T X(k) \tag{1}$$

where

$$\bar{A} = [\bar{a}_0, \cdots, \bar{a}_{L-1}]^T \tag{2a}$$

$$X(k) = [x(k), x(k-1), \cdots, x(k-L+1)]^T \tag{2b}$$

and

$$\bar{A} = A/\|A\|$$

[1] The superscript T denotes transpose.

is the constrained unit-norm weight vector. The key observation of Thompson was that minimizing the variance of $e(k)$ with respect to $\bar{A}$ gives the desired eigenvector in Pisarenko's method as the resulting weight vector.

Two algorithms to adaptively adjust the parameter vector A will be given below. We first present a gradient-type least mean square (LMS) scheme that is similar to Thompson's, and then derive an approximate least squares (or Gauss–Newton) algorithm.

A. LMS Algorithm

The adaptation criterion for the filter of Fig. 1 is the minimization of

$$J = \tfrac{1}{2} E\{e^2(k)\} \tag{3}$$

with respect to the vector A where $E\{\cdot\}$ denotes expectation. The gradient vector ∇J is

$$\nabla J = E[e(k)\, \nabla e(k)] \tag{4}$$

where ∇J and $\nabla e(k)$ represent the gradient of J and $e(k)$, respectively, with respect to the weight vector A. Expressing $e(k)$ in terms of the unnormalized weight vector A,

$$e(k) = A^T X(k)/\|A\| = A^T X(k)/(A^T A)^{1/2}$$

$$\nabla e(k) = \frac{1}{(A^T A)} \left\{ (A^T A)^{1/2} X(k) - \frac{A^T X(k) A}{(A^T A)^{1/2}} \right\}. \tag{5}$$

Substituting (5) into (4) and simplifying, we obtain

$$\nabla J = \frac{1}{(A^T A)^{1/2}} E \left[e(k)\, X(k) - e^2(k) \frac{A}{(A^T A)^{1/2}} \right]$$

which gives the instantaneous gradient estimate of J as

$$\hat{\nabla}_k J = \frac{1}{\|\hat{A}(k-1)\|} \left[e(k)\, X(k) - e^2(k) \frac{\hat{A}(k-1)}{\|\hat{A}(k-1)\|} \right]. \tag{6}$$

Here $\hat{\nabla}_k J$ represents the estimate of ∇J at the kth sample instant. The time update for the ith coefficient is then given by

$$\hat{a}_i(k) = \hat{a}_i(k-1) - \frac{\mu'}{\|\hat{A}(k-1)\|} \cdot \left[e(k)\, x(k-i) - e^2(k) \frac{\hat{a}_i(k-1)}{\|\hat{A}(k-1)\|} \right] \tag{7}$$

where μ' is a positive scalar constant.

Dividing both sides of (7) by $\|\hat{A}(k)\|$ and simplifying, the time update for the normalized coefficient vector can be expressed as

$$\hat{\bar{A}}(k) = \alpha(k) \{ \hat{\bar{A}}(k-1) - \mu [X(k) - \hat{\bar{A}}(k-1) e(k)]\, e(k) \} \tag{8}$$

where

$$\alpha(k) = \|\hat{A}(k-1)\| / \|\hat{A}(k)\|$$

and μ is a new positive scalar constant. Equation (8) gives the normalized version of the weight vector of (7). The update of (8) is performed in two steps. The expression contained inside the curly brackets is computed first, and the resulting coefficient vector is then normalized to unit norm. Equations (7) or (8) thus describe the constrained LMS algorithm.

The alternate forms of (8) are

$$\hat{A}(k) = \alpha(k)\{\hat{A}(k-1)[1+\mu e^2(k)] - \mu X(k)e(k)\} \tag{9a}$$

$$= \alpha'(k)\left\{\hat{A}(k-1) - \frac{\mu}{[1+\mu e^2(k)]} X(k)e(k)\right\} \tag{9b}$$

where

$$\alpha'(k) = \alpha(k) \cdot [1+\mu e^2(k)].$$

We should state that there is a slight difference between (9b) and the algorithm of Thompson (cf. [9, eq. (26)]. It is interesting to note further that the weight vector update portion of (9a) is similar to the one in the γ-LMS algorithm, which is described by Treichler [11] and attributed to Frost [12]. The $\hat{\sigma}^2$ in the γ-LMS algorithm (the estimate of noise power) is replaced by its instantaneous estimate in (9a).

Thus, the above algorithm differs slightly from Thompson's. We remark here that our algorithm has been derived from a precise formulation of the constrained minimization problem. The implication is that we use the gradient of the criterion *with* constraint, namely, $((X(k) - e(k)\hat{A}(k-1))e(k))$, whereas Thompson's algorithm involves the gradient of the unconstrained criterion, i.e., $X(k-1)e(k)$. Nevertheless, the two algorithms have been found to behave almost the same and the asymptotic properties can be shown to be identical.

In particular, it is straightforward to calculate the stationary points of the above algorithm. It is seen from (6) that these are given by the equation

$$E\{(X(k) - e(k)\hat{A})\, e(k)\} = 0$$

which, using (1), can be simplified to give

$$E\{X(k)\, X^T(k)\}\, \hat{A} = \hat{A}^T E\{X(k)\, X^T(k)\}\, \hat{A} \cdot \hat{A}.$$

Clearly, every eigenvector of the covariance matrix satisfies this equation. By a somewhat more involved argument (as in [10]), it can be shown that only the eigenvector corresponding to the minimum eigenvalue gives a stable stationary point.

B. Approximate Deterministic Least Squares Algorithm

To derive a deterministic least squares type algorithm, we choose the adaptation criterion for the filter of Fig. 1 as the minimization of [compared to (3)]

$$V = \tfrac{1}{2}\sum_{s=0}^{t} e^2(s) \tag{10}$$

with respect to the unit-norm vector $\bar{A}$. Because of the constraint on the weight vector, the minimization of (10) is a nonlinear problem, and an exact least squares solution does not appear to exist. We therefore derive an approximate solution using a Gauss-Newton type algorithm.

Recalling from the previous section that

$$e(s) = \frac{A^T X(s)}{\|A\|}, \tag{11}$$

we obtain

$$\nabla V = \frac{1}{\|A\|}\sum_{s=0}^{t} [X(s) - \bar{A}e(s)]\, e(s) = \frac{1}{\|A\|}\sum_{s=0}^{t} \psi(s)\, e(s) \tag{12}$$

where ∇V represents the gradient of V with respect to the weight vector A and

$$\psi(s) = X(s) - \bar{A}e(s).$$

Differentiating V once again, we get

$$\nabla^2 V = \frac{1}{\|A\|^2}\sum_{s=0}^{t} \{[\psi(s) - e(s)\bar{A}]\,[\psi(s) - e(s)\bar{A}]^T - e^2(s)\cdot I\} \tag{13}$$

where I denotes the identity matrix. To examine whether the eigenvector corresponding to the minimum eigenvalue is the only stable stationary point of the criterion (10), consider the following.

At a stationary point, $E\{\psi(s)e(s)\} = 0$, giving

$$E[X(s)X(s)^T]\,\bar{A} = E[e^2(s)] \cdot \bar{A}, \tag{14}$$

which shows that $E[e^2(s)]$ is an eigenvalue of $E[X(s)X(s)^T]$. Further, at a stationary point,

$$\lim_{t\to\infty} \frac{1}{t}\nabla^2 V = \frac{1}{\|A\|^2} E[X(t)\, X(t)^T - e^2(t)\cdot I]$$

$$= \frac{1}{\|A\|^2}\{R_{ss} + \sigma_n^2 \cdot I - E[e^2(t)]\cdot I\} \tag{15}$$

where R_{ss} is the covariance matrix of the sinusoids and σ_n^2 is the variance of the observation noise. From (14) and (15), the only eigenvalue giving a positive semidefinite matrix $E[X(t)X(t)^T - e^2(t)\cdot I]$ is the minimum one, i.e., σ_n^2. Therefore, the only stable stationary point of the criterion is the desired one.

In order to obtain a recursive algorithm for the minimization problem of (10), we need to get a recursion directly in terms of the inverse of the second derivative matrix. This, however, is not possible with the exact version of (13) because of the presence of the term $-e^2(s)\cdot I$ which is a full rank matrix. We therefore make an approximation for the Hessian $\nabla^2 V$. A natural approximation that can be seen from (13) is

$$\nabla^2 V \approx \frac{1}{\|A\|^2}\sum_{s=0}^{t} \psi(s)\,\psi^T(s). \tag{16}$$

From (12) and (16), we obtain the following approximate version of the Gauss-Newton algorithm:

$$\hat{A}(k) = \alpha(k)[\hat{A}(k-1) - P(k)\psi(k)e(k)] \tag{17a}$$

$$P(k) = \left[\sum_{s=0}^{k} \psi(s)\psi^T(s)\right]^{-1} \tag{17b}$$

$$\psi(k) = X(k) - \hat{A}(k-1)e(k) \tag{17c}$$

$$\alpha(k) = \|\hat{A}(k-1)\| / \|\hat{A}(k)\|. \tag{17d}$$

Using the matrix-inversion lemma, a recursion for $P(k)$ can be obtained and the resulting algorithm is then given by

$$\hat{\bar{A}}(k) = \alpha(k)[\hat{\bar{A}}(k-1) - P(k)\psi(k)e(k)] \tag{18a}$$

$$P(k) = P(k-1) - \frac{P(k-1)\psi(k)\psi^T(k)P(k-1)}{1 + \psi^T(k)P(k-1)\psi(k)} \tag{18b}$$

$$\psi(k) = X(k) - \hat{\bar{A}}(k-1)e(k). \tag{18c}$$

$\alpha(k)$ is a scalar constant whose value is chosen such that the updated weight vector has unit norm. Because of its similarity to the recursive least squares algorithm, we refer to (18) as the least squares type algorithm.

To see the effect of the approximation, we consider the following. Recall that the true Hessian is more important close to the true parameters than elsewhere. Consequently, the convergence rate in the neighborhood of the true minimum depends on how close the approximate Hessian is to the true Hessian. Before we examine this, we may remark that by using the ODE technique of Ljung [13], it can be shown that the only possible convergence point of the algorithm (17) is the true parameter vector; we shall not show the details here.

Using (14) and (16), the limiting value of the approximate Hessian can be shown to be

$$\frac{1}{\|A\|^2} E[\psi(t)\,\psi^T(t)]$$

$$= \frac{1}{\|A\|^2} \{R_{ss} + \sigma_n^2 \cdot I - E[e^2(t)]\,\bar{A}\bar{A}^T\}. \tag{19}$$

Since $E[e^2(t)] = \sigma_n^2$ at the true minimum, (15) and (19) show that when the noise variance is small compared to the signal power, i.e., when the signal-to-noise ratio (SNR) is high, the approximation in (16) is good close to the true minimum. Consequently, the convergence rate of the algorithm [(17) or (18)] in the neighborhood of the true parameters is much less affected by the approximation in the case of high SNR. In the next section, we shall illustrate the effect of SNR on the convergence rate close to the true parameters with some numerical examples.

When the statistics of the observed process vary slowly, an exponential weighting is applied to the data so as to track the slowly varying parameters of the process. Weighting of the data with a sliding exponential window is equivalent to minimizing

$$V = \frac{1}{2} \sum_{s=0}^{t} \lambda^{t-s} e^2(s), \qquad \lambda \leq 1 \tag{20}$$

where λ is a so-called forgetting factor. The effect of λ in (20) reflects itself in the recursion of (18b) as

$$P(k) = \frac{1}{\lambda}\left[P(k-1) - \frac{P(k-1)\psi(k)\psi^T(k)P(k-1)}{\lambda + \psi^T(k)P(k-1)\psi(k)}\right]. \tag{21}$$

We comment here that the least squares type algorithm (18) requires more computations than the LMS algorithm (8). But this is not likely to be of great concern with today's computing power costs. Instead, what is probably more important is the number of samples required to get a good spectral estimate.

III. Simulation Results

In this section, we present simulation results obtained with the two different algorithms of the previous section. In each case, we choose $\bar{A} = [1, 0, \cdots, 0]^T$ as the initial value of the weight vector. The experimental data are the sum of sequences of signal and white noise samples where the signal may be one or two sinusoids. The algorithm is run for a predetermined number of iterations, and from the final value of the weight vector the AR spectrum $|1/\bar{A}(f)|^2$ is computed, where

$$\bar{A}(f) = \bar{A}(z)\big|_{z=e^{j2\pi f}}$$

and

$$\bar{A}(z) = \bar{a}_0 + \bar{a}_1 z^{-1} + \cdots + \bar{a}_{L-1} z^{-(L-1)}.$$

The resulting spectral estimate is normalized with respect to its peak value.

The convergence constant μ in the LMS algorithm is computed as

$$\mu = \beta/\hat{\sigma}_x^2(t)$$

where

$$\hat{\sigma}_x^2(t) = (1-\gamma)\hat{\sigma}_x^2(t-1) + \gamma \sum_{i=0}^{L-1} x^2(t-i).$$

The values of β, γ, and $\hat{\sigma}_x^2(0)$ are selected after many trial experiments. The combination $\beta = 0.01$, $\gamma = 0.01$, and $\hat{\sigma}_x^2(0) = 0.01$ gave the best performance for the examples with single sinusoid, while the combination $\beta = 0.001$, $\gamma = 0$, and $\hat{\sigma}_x^2(0) = 1$ gave the best results for the examples with two sinusoids. We, however, remark that we do not claim that our search for the best parameters was an exhaustive one.

In the case of the least squares type algorithm, we used a time-varying forgetting factor $\lambda(t)$ where

$$\lambda(t) = \lambda_0 \lambda(t-1) + 1 - \lambda_0.$$

In all the examples below, $\lambda_0 = 0.99$, $\lambda(0) = 0.95$, and $P(0) = 100I$, where I is the identity matrix.

In the figures, N denotes the number of data samples used in the simulation and L represents the length of the weight vector. The true frequencies of the sinusoids are indicated by vertical lines.

Example 1

In this example, we considered a single sinusoid of normalized frequency 0.150 at a signal-to-noise ratio (SNR) of 0 dB. From the plots of Fig. 2, it is clear that the least squares type algorithm (18) converged to the desired stationary point much faster than the LMS algorithm. The two spectral estimates of Fig. 2(b) indicate that the algorithm of (18) converges to the neighborhood of the true parameters very fast, and thereafter the convergence appears to be slow. Both algorithms gave unbiased estimates of the frequency.

Example 2

Two equal power sinusoids of normalized frequencies 0.15 and 0.20 at 0 dB SNR and $L = 7$ are considered in this example. The plots of Fig. 3 show that the initial convergence of the least squares version is faster than that of the LMS algorithm. The approximate least squares algorithm resolved the two frequencies with 500 data samples. However, both algorithms performed similarly close to the desired stationary point. The number of parameters used here exceeds the num-

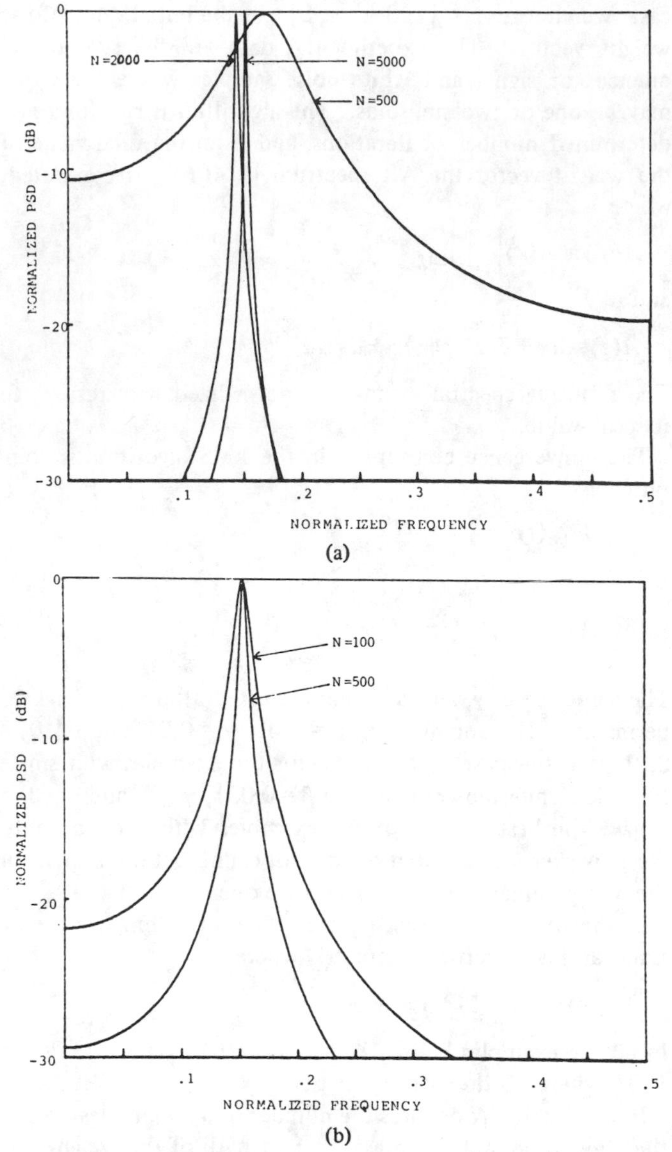

Fig. 2. Spectral estimates for Example 1 (L = 3, SNR = 0 dB). (a) LMS algorithm. (b) Least squares type algorithm.

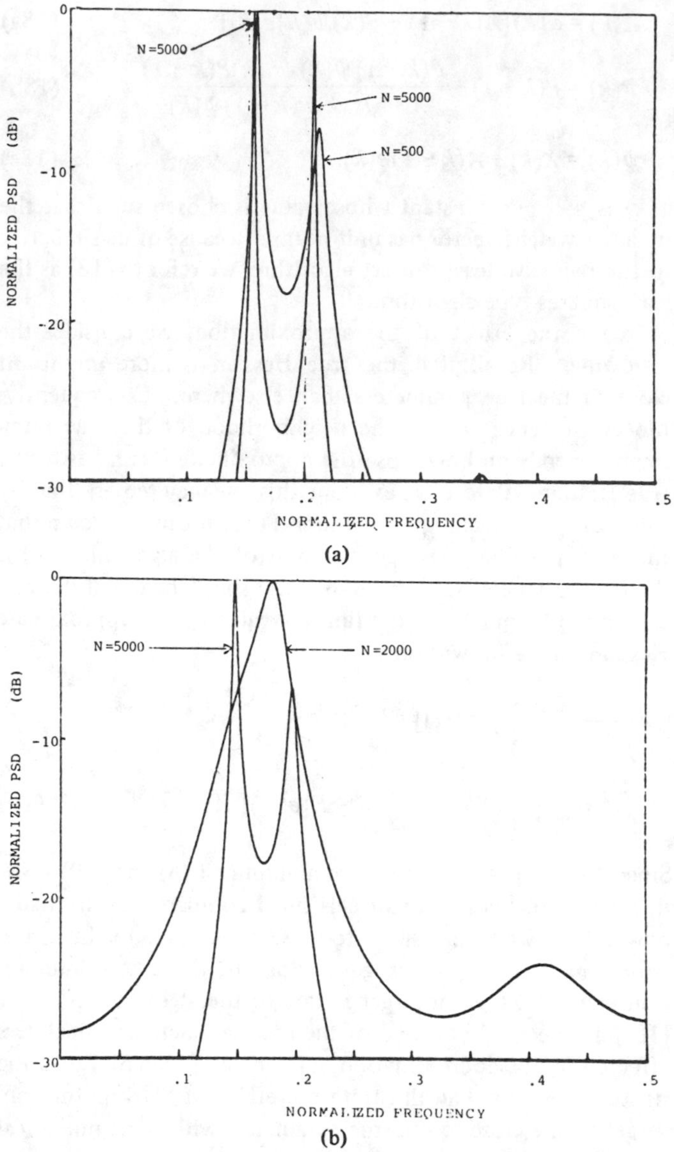

Fig. 3. Spectral estimates for Example 2 (L = 7, SNR = 0 dB). (a) **Least squares type algorithm.** (b) **LMS** algorithm.

ber needed to identify the two sinusoids. The excess zeros seem to have converged to locations well away from the unit circle and hence did not indicate the presence of any other sinusoids. This phenomenon was also observed by Thompson [9].

Example 3

In the previous section, it was seen that when SNR of the input signal is high, the approximate Hessian is close to the true Hessian in the neighborhood of the desired stationary point. To see how the least squares type algorithm performs under this condition, we repeated Example 1 with 10 dB SNR. The plots of Fig. 4 show the faster convergence of the least squares version compared to the LMS algorithm, both in the beginning and also close to the desired stationary point.

Example 4

To see how the algorithms perform at high SNR, but with the minimum number of AR parameters and, also, to see how the least squares version performs in the neighborhood of the true parameters, Example 2 was repeated with 10 dB SNR and L = 5. Taking L = 5 corresponds to choosing a minimum number of AR parameters because this gives a fourth degree $\bar{A}(z)$ polynomial which has four roots allowing one pair for each sinusoid. The plots of Fig. 5 show that the LMS algorithm required as many as 3000 data samples to resolve the sinusoids, while the least squares type algorithm converged to the neighborhood of the desired stationary point with 100 data samples.

Comparing spectral estimates of Fig. 5(b) with those of Fig. 3(b), it is seen that the convergence of the least squares version in the neighborhood of the desired stationary point has improved significantly with high SNR. This property of the algorithm shows that the approximation of (16) does not introduce any noticeable degradation in convergence, close to the true parameters, with high SNR.

IV. Conclusions

This paper was motivated by an adaptive algorithm proposed by Thompson [9] to give an on-line implementation of Pisarenko's harmonic retrieval method [8]. It has been shown

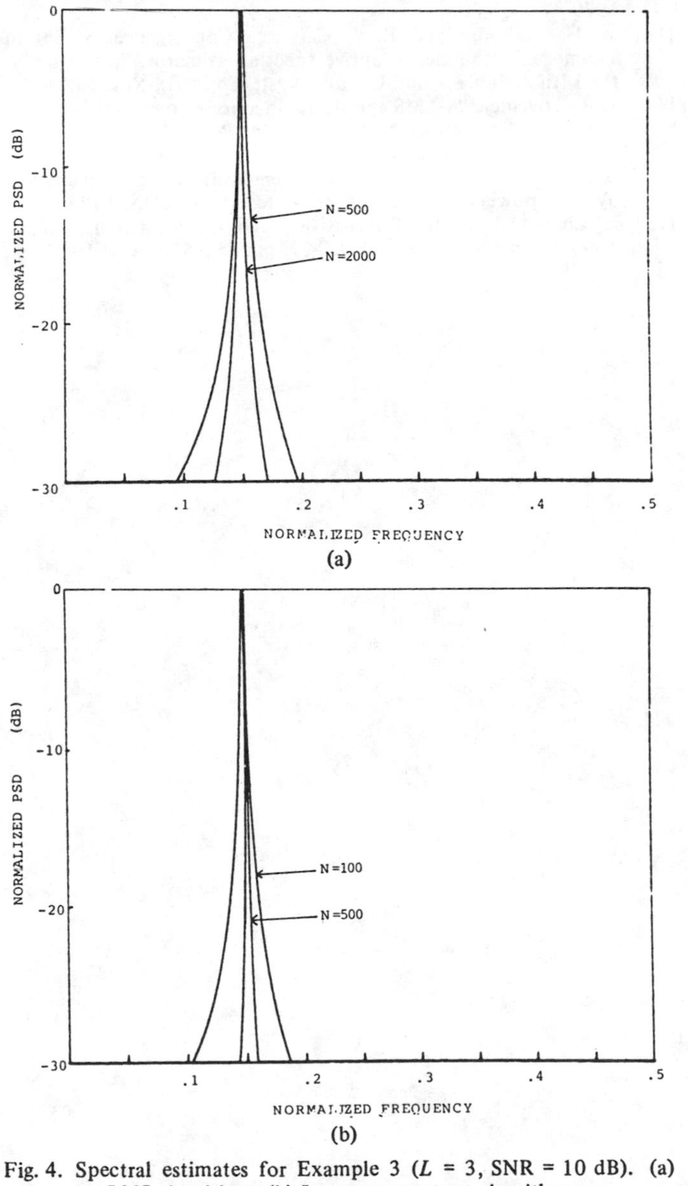

Fig. 4. Spectral estimates for Example 3 (L = 3, SNR = 10 dB). (a) LMS algorithm. (b) Least squares type algorithm.

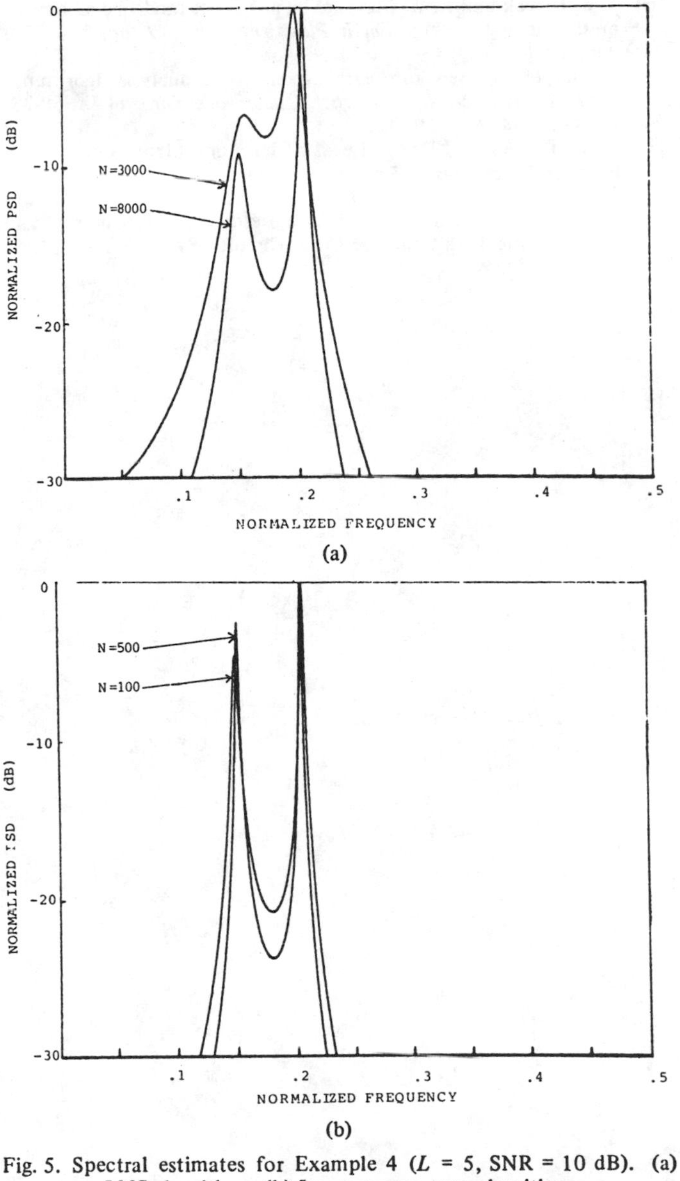

Fig. 5. Spectral estimates for Example 4 (L = 5, SNR = 10 dB). (a) LMS algorithm. (b) Least squares type algorithm.

that Thompson's interpretation of Pisarenko's scheme as a constrained minimization problem can be restated as an unconstrained nonlinear problem. Based on this formulation, we derived two adaptive algorithms to recursively update the filter parameters. The first one is a gradient-type algorithm and is similar to Thompson's. The second one is derived as a Gauss-Newton algorithm, making use also of the second derivative matrix (the Hessian) of the criterion. The latter algorithm can also be viewed as an approximate least squares algorithm.

Some numerical examples illustrated the properties of the two schemes. The most important conclusions are as follows. 1) The initial convergence of the least squares type algorithm is faster than the LMS algorithm for both low and high SNR input signals. 2) The convergence rate of the least squares version close to the desired stationary point is a function of the input SNR. This property of the algorithm is a consequence of the approximation used for the Hessian in the derivation. 3) The removal of the bias in the frequency estimates is slower than the resolution between different frequencies.

The above observations are not conclusive. The basic objective of the paper is to point out alternatives in the adaptive implementation of Pisarenko's scheme. Further experience with simulations and studies of convergence behavior are needed to fully understand the properties of the least squares type algorithm.

References

[1] B. Widrow *et al.*, "Adaptive noise cancelling principles and applications," *Proc. IEEE*, vol. 63, pp. 1692–1716, Dec. 1975.

[2] B. Widrow and M. E. Hoff, "Adaptive switching circuits," in *IRE 1960 Wescon Conv. Rec.*, part 4, pp. 96–104.

[3] J. P. Burg, "A new analysis technique for time series data," NATO Adv. Study Inst. Signal Processing with Emphasis on Underwater Acoust., The Netherlands, Aug. 1968.

[4] L. J. Griffiths, "A continuously-adaptive filter implemented as a lattice structure," in *Proc. IEEE Int. Conf. Acoust., Speech, Signal Processing*, Hartford, CT, May 1977, pp. 683–686.

[5] A. H. Nuttall, "Spectral analysis of a univariate process with bad data points, via maximum entropy and linear predictive techniques," Naval Underwater Syst. Cent., New London, CT, Tech. Rep. 5303, Mar. 1976.

[6] T. J. Ulrych and R. W. Clayton, "Time series modeling and maximum entropy," *Phys. Earth Plan. Int.*, vol. 12, pp. 188–200, Aug. 1976.
[7] L. Marple, "A new autoregressive spectrum analysis algorithm," *IEEE Trans. Acoust., Speech, Signal Processing*, vol. ASSP-28, pp. 441–454, Aug. 1980.
[8] V. F. Pisarenko, "The retrieval of harmonics from a covariance function," *Geophys. J. Roy. Astron. Soc.*, pp. 347–366, 1973.
[9] P. Thompson, "An adaptive spectral analysis technique for unbiased frequency estimation in the presence of white noise," presented at the 13th Asilomar Conf. Circuits, Syst., Comput., Nov. 1979.
[10] M. G. Larimore and R. J. Calvert, "Convergence studies of Thompson's unbiased adaptive spectral estimator," presented at the 14th Asilomar Conf. Circuits, Syst., Comput., Nov. 1980.
[11] J. R. Treichler, "γ-LMS and its use in a noise-compensating adaptive spectral analysis technique," in *Proc. IEEE Int. Conf. Acoust., Speech, Signal Processing*, 1979, pp. 933–936.
[12] O. L. Frost, III, "Resolution improvement in AR spectral analysis by noise power cancellation," presented at EASCON, 1977.
[13] L. Ljung, "Analysis of recursive stochastic algorithms," *IEEE Trans. Automat. Contr.*, vol. AC-22, pp. 551–575, Aug. 1977.

Adaptive Clutter Filtering Using Autogressive Spectral Estimation

D.E. BOWYER, Member, IEEE
System's Control, Inc.
P.K. RAJASEKARAN
Texas Instruments
W.W. GEBHART, Member, IEEE
General Research Corp.

Abstract

A new adaptive filter to reject clutter is derived using autoregressive spectral analysis techniques. The adaptive filter performs open-loop processing, resulting in a shorter transient response, and is therefore suitable for radar waveforms containing only a small number of samples. A number of examples including application to ballistic missile defense are presented to demonstrate the performance capabilities of the new adaptive filter.

Manuscript received July 21, 1978; revised December 5, 1978.

The Authors were formerly with Teledyne Brown Engineering.

Authors' address: D.E. Bowyer, Systems Control, Inc., 357 North Eglin Parkway, Ft. Walton Beach, FL 37548.

0018-9251/79/0700-0538 $00.75 ©1979 IEEE

I. Introduction

This paper presents a new method of adaptively rejecting radar clutter based on the use of autoregressive (AR) estimates of the clutter spectrum. This technique involves implementing a whitening filter and a filter matched to the distorted received signal as shown in Fig. 1. These filter implementations are changed in real time based on a parametric estimate of the clutter spectrum under the assumption that the clutter can be modeled as an AR random process. The functional form of the adaptive filter implementation is shown in Fig. 2.

Previous work in the area of adaptive clutter filters [1] utilized the tapped delay line/feedback-loop approach shown in Fig. 3. This implementation was based on the adaptive array algorithms of Applebaum [2] and Widrow [3] and has the disadvantage of long convergence times (or a long transient response). For short radar dwell times or short bursts of pulses (20-30 pulses), the transient response of the feedback-loop adaptive filter can be on the order of the dwell time or burst length. If the clutter range is unknown or the clutter is extended in range for more than a pulse repetition interval (PRI) (as for a high-PRF waveform), then the clutter transients can obscure any target returns. The AR adaptive filter avoids the transient response problem by processing the signal through tapped delay line filters with no feedback-loops, as shown in Section II, and by using the AR spectral estimator which converges very quickly to the correct estimate of the clutter spectrum (about four pulses for a four-parameter estimator).

In addition to minimizing any transient response in the adaptive filter, the AR spectral estimation provides rapid spectral estimates for cases in which the clutter Doppler frequency changes significantly over short range intervals. This situation occurs in the case of ballistic missile defense (BMD) tank break-up clutter, and airborne chaff, where significant wind shear effects are present. Another application of the AR adaptive filter is in airborne mutiple target indicator (AMTI) systems in which a highly maneuvering aircraft or missile causes a rapid change in the ground clutter spectrum. This implementation could replace the clutter tracking and notch filter circuitry presently used in AMTI system.

The theoretical development of the AR adaptive filter is presented in Section II. The important point is that the assumption of an all-pole AR model for the clutter allows an all-zero (tapped delay line) implementation for the whitening filter and modified matched filter.

Section III describes the results obtained from applying the AR adaptive filter to a variety of situations. The most important example is the application to recorded BMD, tank break-up clutter data. Although this clutter may not be truly an AR process, the 10-20 dB of improvement afforded by the AR adaptive filter compared to the matched filter justifies the use of this model. Good results were obtained using two and four complex parameters in the AR spectral estimates. It can be shown that

Reprinted from *IEEE Trans. Aerosp. Electron. Syst.*, vol. AES-15, pp. 538–546, July 1979.

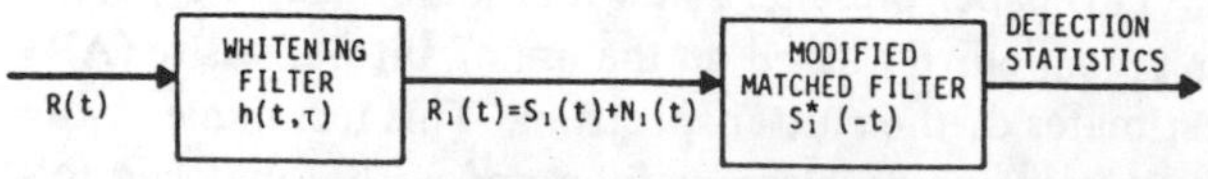

Fig. 1. Optimum receiver for clutter environment.

Fig. 2. Conceptual structure of the adaptive filter.

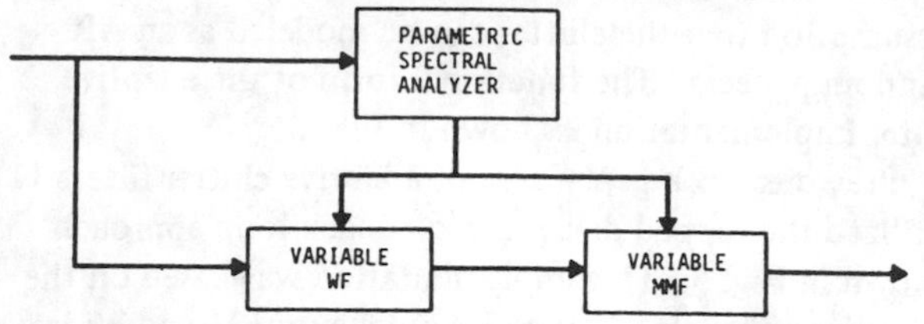

Fig. 3. Conventional adaptive filter.

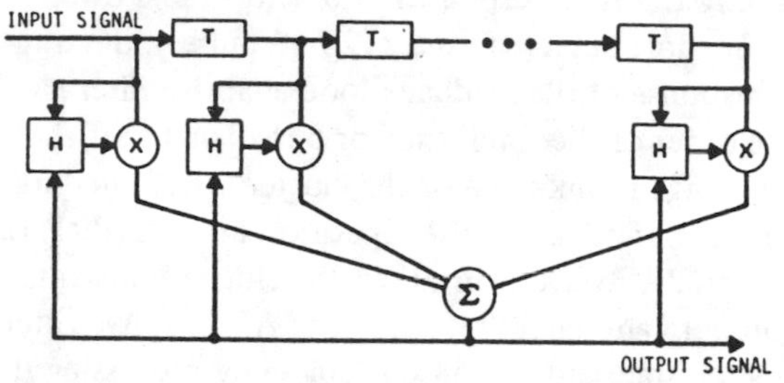

each complex parameter implemented allows the estimation of a distinct mode of the clutter spectrum. In addition to the BMD clutter results, the performance of the AR adaptive filter is quantitatively compared to the true optimum filter (maximum signal-to-interference ratio filter) for a Markov source clutter input.

Section IV describes the implementation aspect of the AR spectral esitmator. In particular, the performance of a digital implementation of a two-parameter estimator for 8, 10, and 12 bits of accuracy is compared to the full-precision floating-point estimator performance. The results indicate that a real-time digital-hardware implementation is feasible. Some key points are discussed in Section V and the article concludes with a summary.

II. Technical Approach

We consider the detection of a target signal $S(t)$ in clutter $C(t)$ and white noise $N(t)$ as a preliminary step in deriving the adaptive filter structure. This can be posed as a hypothesis testing problem. Under hypothesis H_1 the target is present and the return signal $R(t)$ can be written as

$$H_1 : R(t) = S(t) + C(t) + N(t).$$

Under hypothesis H_0 the target is not present and the return signal can be written as

$$H_0 : R(t) = C(t) + N(t).$$

The objective is to be able to detect the target in a strong clutter environment, and minimize the false alarms due to clutter.

The first assumption in solving the problem is that clutter is a correlated noise process. In accordance with this assumption, let the autocorrelation function of $C(t)$ be

$$\phi_c(t_1, t_2) = E[C(t_1)\, C(t_2)]$$

where E is the expected value operator. Let the thermal noise be a stationary process with a power spectral density of $N_0/2$. In this case, the optimum solution [4] can be obtained as shown in Fig. 1.

The whitening filter (WF) decorrelates the interference $C(t) + N(t)$ to a thermal noise like signal $N_1(t)$. In the process of doing so, the signal $S(t)$, if present, is modified to $S_1(t)$. The output of the WF is given by

$$H_1 : R_1(t) = S_1(t) + N_1(t)$$

$$H_0 : R_1(t) = N_1(t).$$

This corresponds to the classical problem of detecting the known signal $S_1(t)$ in the white noise $N_1(t)$. The solution consists of passing $R_1(t)$ through a filter matched to the signal $S_1(t)$. This filter is called the *modified matched filter* (MMF). In other words the problem of detecting a signal in colored noise has been converted to a standard form. The mathematical details may be found in [4].

The impulse response of the whitening filter is given by the solution of an integral equation [4]

$$\int_{T_i}^{T_f} \phi_I(x, z) \left[\int_{T_i}^{T_f} h(u, v)\, h(u, x)\, du\right] dx = \delta(z - v), \quad T_i \leqslant z, v \leqslant T_f \tag{1}$$

where $\phi_I(x, z) = \phi_c(x, z) + N_0/2\, \delta(x - z)$ is the autocorrelation of the interference. Note that in the integral equation the quantity inside the parentheses is the inverse kernel of the correlation function of the interference. Naturally, the whitening filter impulse response $h(t, \tau)$ is the solution to the integral equation with this inverse kernel. The above integral equation can be derived by assuming that the clutter process is nonstationary. This leads to a time-varying whitening filter. In practice, clutter is not stationary, and the interval of observation is only finite. Therefore, the true optimum solution may be expected to involve a time-varying whitening filter and MMF. However, if a short-time clutter spectrum is defined, it can be used to obtain the whitening filter. As the clutter spectrum varies with time, the impulse responses of the WF and MMF will vary, insuring real-time adaptation to the changing environment.

In general the clutter correlation function $\phi_c(t_1, t_2)$ is not known and must be estimated. Or, equivalently, the short term spectrum $\Phi_c(f)$ must be estimated, as in this study. Over a short period of time the clutter is modeled as a stationary AR process; the parameters of the AR process are estimated; and, in turn, these estimates are used automatically to design the whitening filter and MMF. Because the estimates depend on the data, they change with time resulting in a time-varying receiver structure.

Specifically, the clutter samples are modeled as an M-parameter AR process. That is,

$$C(k) = \sum_{i=1}^{M} \alpha_i C(k-i) + e(k) \tag{2}$$

where the α_i are some constants, $e(k)$ is a white noise process with unit variance, and $C(k-i)$ is the clutter sample lagging by i samples from the present sample. It is simple to show in this case that the power spectrum of the clutter process is given by

$$\Phi_c(f) = \left|1/[1 - \sum_{i=1}^{M} \alpha_i \exp(-j2\pi i f T)]\right|^2 \tag{3}$$

where T is the sampling interval.

Let $\hat{\alpha}_i$ be the estimate of α_i based on the observation of the clutter samples. Then, an estimate of the clutter spectrum is given by

$$\hat{\Phi}_c(f) = \left|1/[1 - \sum_{i=1}^{M} \hat{\alpha}_i \exp(-j2\pi i f T)]\right|^2 . \tag{4}$$

In this case the whitening filter (assuming a large clutter to noise ratio) is given by

$$H_w(f) = 1 - \sum_{i=1}^{M} \hat{\alpha}_i \exp(-j2\pi i f T). \tag{5}$$

Note that the above whitening filter can be realized as a tapped delay line with tap weights given by 1 and $-\hat{\alpha}_i$, and has a simple structure which can be easily changed as the $\hat{\alpha}$ vary. This was one of the motivating factors in modeling clutter as an AR process. Other factors include the noted success with which AR models have been applied to speech signals, which are also nonstationary processes.

Many different algorithms exist to estimate the parameters α_i, and invariably involve time averages of correlation functions and matrix inversions [5-7]. In this paper a Kalman filtering technique is employed to estimate the WF parameters. The advantages are that no matrix inversion is involved in this case and the estimates are obtained sequentially in real time.

To utilize Kalman filtering techniques, the spectral parameter estimation problem must be first converted to the state variable form. Accordingly, the AR process parameters $\alpha_1, \alpha_2, \ldots, \alpha_M$ are defined as state variables and modeled as constants

$$\alpha(k+1) = \alpha(k) \quad M \times 1 \text{ vector}. \tag{6}$$

That is, the state transition matrix is an identity matrix. The process noise term, usually present in Kalman filtering, is absent in this case. The observation is a scalar variable $C(k+1)$ given by a *time-varying* linear combination of the state variables $\alpha(k)$ plus the white noise term $e(k+1)$. This can be written compactly in a vector notation as

$$C(k+1) = \mathbf{H}(k+1)\,\dot{\alpha}(k+1) + e(k+1) \tag{7}$$

where

$$\mathbf{H}(K+1) = [C(k), C(k-1), \ldots, C(k-M+1)]. \tag{8}$$

Let the variance of the white noise component be denoted by G^2. By applying a standard discrete Kalman filtering algorithm [8] to the message and observation models given by (6) and (7), the following equations are obtained:

$$\hat{\alpha}(k+1) = \hat{\alpha}(k) + \mathbf{K}(k+1)[C(k+1) - \mathbf{H}(k+1)\,\hat{\alpha}(k)] \tag{9}$$

$$V_\alpha(k+1) = [I - \mathbf{K}(k+1)\,\mathbf{H}(k+1)]\; V_\alpha(k)$$

$$\text{– covariance update equation} \tag{10}$$

$$\mathbf{K}(k+1) = \{1/[G^2 + \mathbf{H}(k+1)\, V_\alpha(k)\, \mathbf{H}^*(k+1)]\}$$

$$\cdot\, [V_\alpha(k)\, \mathbf{H}^*(k+1)] \text{ – gain vector computation.} \tag{11}$$

In the above equations the asterisk indicates conjugate transpose. Note the $\mathbf{H}(k+1)\,\hat{\alpha}(k)$ is the one step prediction of the present clutter observation sample $C(k+1)$. The term $[C(k+1) - \mathbf{H}(k+1)\,\hat{\alpha}(k)]$ can be denoted by $\nu(k+1)$ and is known as residual or innovation process. It is basically a sequence of "new information" and is ideally a white process [9]. Note that the algorithm does not need to invert any matrix. This is because the observation $C(k+1)$ is a scalar quantity.

The effect of using M parameters for the AR process is to interpolate the clutter spectrum with the frequency response of M poles in the complex frequency plane. The number chosen for M is generally a compromise between good spectral estimation and desirability of short transient response. It is also worthwhile noting that the AR process model can represent discrete target interference as well. Therefore, the method can also be used for reducing interference from other targets in a multiple target environment.

Certain drawbacks exist in the Kalman filter algorithm for clutter spectral estimation. If the clutter samples are present only up to a certain time (say N samples) beyond which only noise is present, the estimates do not drop off to a small value and are preserved with a memory typical of recursive techniques. In order to reduce this memory a number of solutions have been proposed. One of the more attractive schemes is to have an exponentially fading memory for the Kalman spectral estimator. This is implemented by artificially increasing the covariance $V_\alpha(k)$ at each stage by a certain percentage [10]. This method has been fairly effective in the present study. Another method, called the limited-memory filter [11], is a little more complicated, and periodically erases the memory of the estimator completely. In many cases, it is not clear how this periodicity can be chosen a priori. On the other hand, the exponentially fading memory length can be chosen such that the covariance is not changed by more than 20 percent each time. This increase reflects the uncertainty in the estimates, and is a divergence prevention technique. In such a case, the results are only near optimum.

III. Performance Results

In this section we present a number of examples illustrating the studies conducted with the proposed adaptive filter. In subsection A we derive the structure of the adaptive filter for the case of an N-pulse weighted burst waveform. This structure is used in the examples that follow. In subsection B we compare the signal-to-clutter (S/C) and signal-to-interference (S/I) ratios theoretically obtainable by an optimum interference rejection filter (the interference statistics exactly known) with that achieved by the adaptive filter for the case of a Markov random process as an interference source.

In subsection C we demonstrate the ability of the adaptive filter to place nulls in its frequency response corresponding to the interference frequencies. Subsection D presents the application of the adaptive filter to suppress the tank break-up clutter of a reentry booster in a BMD application.

A. Adaptive Filter Structure for an N-Pulse Burst

Let the transmit weights of an N-pulse burst waveform be $\{a_1, a_2, ..., a_N\}$, which may be complex in general. Then the sampled return from a desired target of unit strength, zero range, and Doppler velocity ω_d rad/s will be

$$r(k) = a_k \exp(j\omega_d kT), \quad k = 1, 2, ..., N. \tag{12}$$

The matched filter to process this return will be a tapped delay line with weights $\{b_k\}$ where

$$b_k = a^*_{N-k+1} \exp(-j\omega_d kT), \quad k = 1, 2, ..., N. \tag{13}$$

Without loss of generality we can set $\omega_d = 0$ in all our discussions, in which case the esponential in (13) can be set to unity. Let the spectral parameter estimates corresponding to the interference be $\hat{\alpha}_1, \hat{\alpha}_2, ..., \hat{\alpha}_M$ at any given time sample. Then the impulse response of the whitening filter is given by $\{1, -\hat{\alpha}_1, -\hat{\alpha}_2, ..., -\hat{\alpha}_M\}$ and can be implemented as a tapped delay line. The return signal from the desired target is passed through the whitening filter along with the interference. In this process the target signal is modified and corresponds to the sequence $\{c_k\}$, $k = 1, 2, ..., M + N - 1$ given by convolving $\{a_k\}$ with $\{1, -\hat{\alpha}_1, ..., -\hat{\alpha}_M\}$. In order to maximize the signal-to-noise ratio at the output of the processor, the filter following the whitening filter must be matched to this modified signal return $\{c_k\}$. Thus the MMF has an impulse response $\{d_k\}$, and is given by

$$d_k = c^*_{M+N-k}, \quad k = 1, 2, ..., M + N - 1. \tag{14}$$

The MMF can be implemented as a tapped delay line with weights d_k. The output of the MMF can now be processed further for detection processing.

The receiver structure for processing the N-pulse burst data is shown in Fig. 4. This structure is basically used in our subsequent examples and discussions. Note that in practice we can combine the WF and MMF to implement a single tapped delay line filter with weights g_k, $k = 1, 2, ..., N + 2M - 2$, where $\{g_k\}$ is the result of convolving $\{1, -\hat{\alpha}_1, -\hat{\alpha}_2, ..., -\hat{\alpha}_M\}$ with the sequence $\{d_k\}$.

B. S/C and S/I Performance of the Adaptive Filter

Consider that the clutter is a stationary, first-order Gaussian Markov process generated by $C_{k+1} = \Theta C_k + w_k$, where C_k is the interference clutter sample, $|\Theta| < 1$, and w_k is a white Gaussian sequence. Assuming zero mean for the processes involved, the covariance matrix of the clutter for N samples is given by the $N \times N$ Toeplitz matrix

$$\phi_c = \begin{bmatrix} 1 & \Theta & \Theta^2 & \cdots & \Theta^{N-1} \\ \Theta & 1 & \Theta & \cdots & \Theta^{N-2} \\ \cdot & \cdot & \cdot & \cdot & \cdot \\ \Theta^{N-1} & \Theta^{N-2} & \cdot & \cdot & 1 \end{bmatrix}$$

Let the N-sample signal return be denoted by the vector $\mathbf{S}$. Let the return vector $\mathbf{R}$ denote the sum of signal return, clutter, and thermal noise $\mathbf{R} = \mathbf{S} + \mathbf{C} + \mathbf{V}$, where $\mathbf{C}$ is an N vector of clutter samples and $\mathbf{V}$ is an N vector of thermal noise samples. Let the vector $\mathbf{W}_{opt}$ denote the weights of the optimum tapped delay line filter. Then the S/C and S/I ratios are biven by

$$S/C = |\mathbf{W}^*_{opt}\mathbf{S}|^2 / |\mathbf{W}^*_{opt}\phi_c\mathbf{W}_{opt}|$$

and

$$S/I = |\mathbf{W}^*_{opt}\mathbf{S}|^2 / |\mathbf{W}^*_{opt}\phi_I\mathbf{W}_{opt}|$$

where ϕ_I refers to the interference covariance matrix, the sum of the clutter and thermal noise covariance matrices ϕ_c and ϕ_v. For the case of the adaptive filter, we can obtain S/C and S/I with $\mathbf{W}_{opt}$ replaced by the appropriate adaptive filter weights $\mathbf{W}_{AF}$.

The performance of the autoregressive adaptive filter (ARAF) was evaluated for a clutter source which consisted of the Markov process just described, with different values of the correlation coefficient θ. This performance was

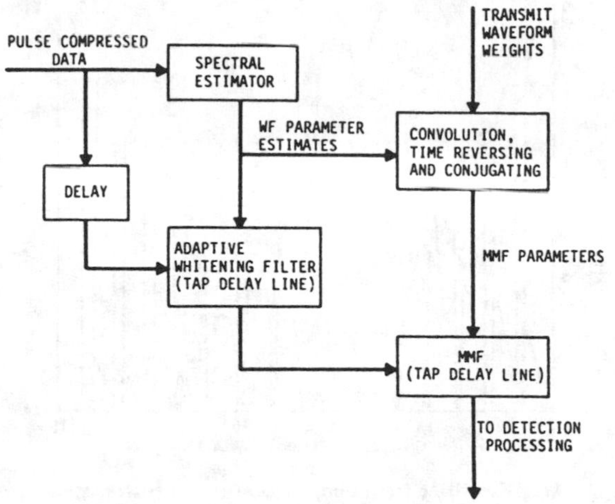

Fig. 4. Adaptive filter structure to process coherent burst of *N* weighted pulses.

Fig. 5. S/C performance of ARAF and LMSAF relative to optimum filter. (A) $\theta = 0.7$ and (B) $\theta = 0.9$.

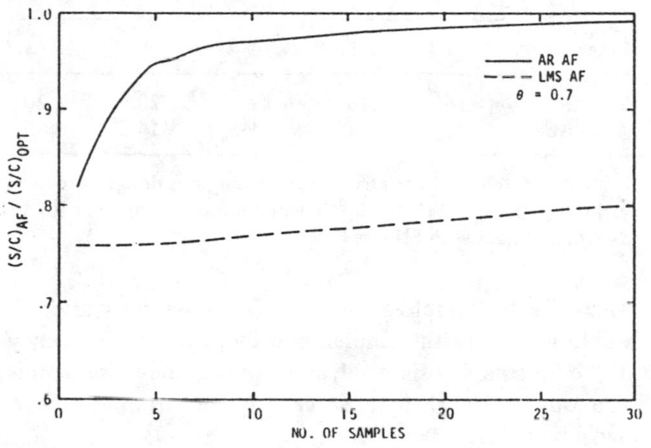

(A)

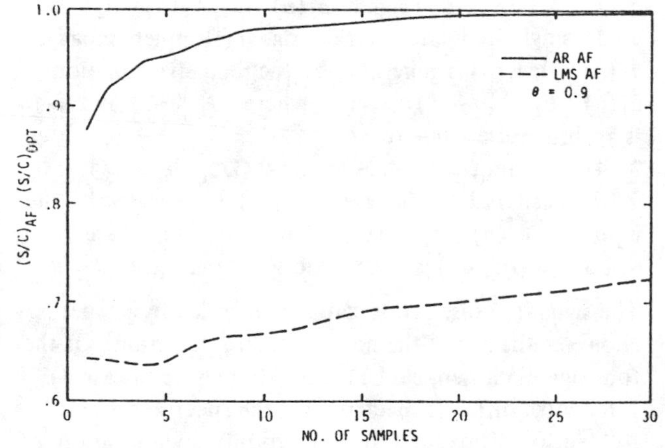

(B)

evaluated in terms of S/C and S/I at each iteration of the adaptive process, and is presented in the curves of Figs. 5 and 6 as a ratio to the optimum S/I and S/C. The assumed signal, in this case, is a 10-pulse Taylor-weighted burst. Each point on these curves is an average of 100 Monte Carlo runs of the adaptive process. Figs. 5 and 6 also show the performance of the LMS adaptive filter (LMSAF)

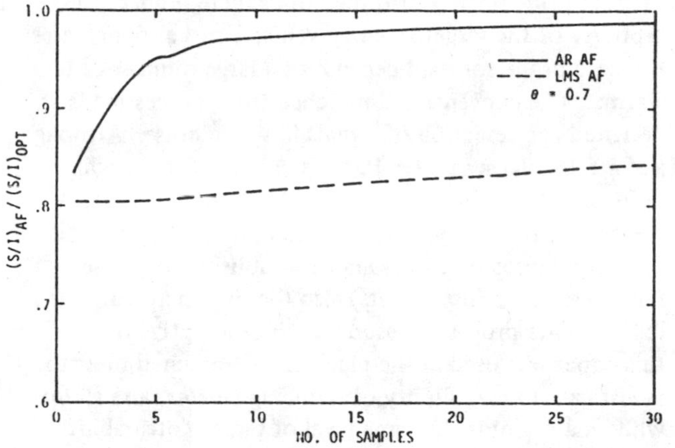

(A)

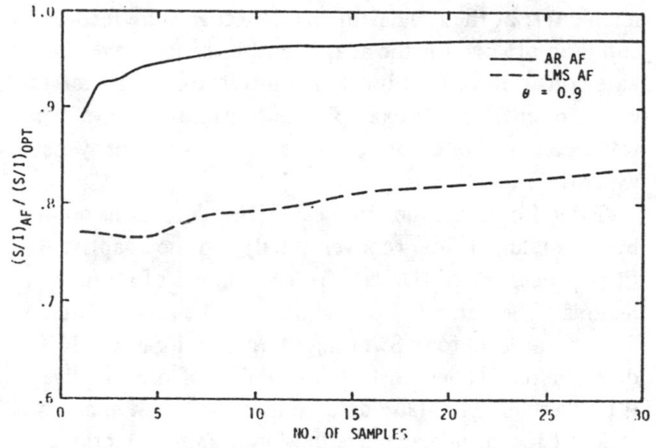

(B)

Fig. 6. S/I performance of ARAF and LMSAF relative to optimum filter. (A) $\theta = 0.7$ and (B) $\theta = 0.9$.

of Fig. 3 for the same input. From these curves the advantage of the much faster convergence of the ARAF is obvious.

C. Adaptive Filter Frequency Response

The adaptive filter of this paper may be considered as derived from a frequency domain viewpoint. The filter automatically adjusts its weight to place a null in its frequency response corresponding to the significant frequency components of the clutter spectrum. We demonstrate the ability of the adaptive filter in this section.

Clutter data was simulated by means of two closely spaced discrete frequency scatterers with a transmit waveform of 16 uniformly weighted coherent pulses. The clutter-to-noise ratio was 37 dB. Fig. 7 presents the adaptive filter frequency response at the end of 15 adaptive samples. The clutter spectrum is also shown in the figure of comparison. Note that the null in the adaptive filter is down −100 dB from the zero frequency gain. The number of spectral parameters used in this case was four.

D. BMD Applications

In a BMD environment the clutter due to tank break-up

(TBU) during reentry into the atmosphere reduces the visibility of the target (reentry vehicle) and also increases the data processor load because of a large number of false alarms. The conventional matched filter processing is designed for detection of signals in white noise. Although a low false alarm rate ($\cong 10^{-6}$) is possible for the white noise situation, the false alarms (clutter detections) increase significantly in a TBU clutter environment. This in turn requires that tracking be established on these false detections (nontargets) also thereby increasing the load on data processor resources. The adaptive filter of this paper was used in the place of the matched filter to investigate the possibility of reducing clutter false alarms while maintaining the same level of target detectability.

As a specific example of the performance of the adaptive filter on real radar data, radar returns from a ballistic missile test at the Kwajalein Missile Range were recorded and used off-line for the adaptive filter processing. The data consisted of tank break-up clutter; that is, no targets were present. In this example the transmitted waveform was a coherent burst of 16 pulses with 50-dB Chebyshev weighting.

Table I presents the number of false detections reported by the matched filter (conventional) and the adaptive filter processing of the data from a number of altitude regions. The detection probability was maintained at 99 percent level for a Swerling I target with SNR = 34 dB. The results demonstrate the ability of the adaptive filter to reduce the false detections by as much as an order of magnitude. This reduction in false detections is obtained as a consequence of the clutter suppression properties of the adaptive filter.

Many sets of these TBU data have been successfully processed with the adaptive filter demonstrating its clutter suppression capability in real world applications.

IV. Implementation Study

The implementation of the adaptive filter structure of Fig. 3 depends mainly on the feasibility of realizing the spectral estimator with digital hardware. For practical application in a real-time processor, fixed point arithmetic with a small number of bits (yet performing adeqately) is necessary. This section presents the results of a study of the second-order spectral estimator with finite-bit fixed-point arithmetic for acceptable spectral estimation performance.

Because of the intractable nature of the analytical approach to determining the performance of a finite-bit processor, a computer simulation of the spectral estimator with fixed-point finite-bit operations was carried out.

In particular, the spectral estimator was analyzed for 8-, 10-, and 12-bit internal accuracy implementations with an 8-bit complex signal input from a simulated analog-to-digital converter. The general model of the generated input signal can be written as

$$X(nT) = a_1(n)\exp(jw_1 nT) + a_2(n)\exp(jw_1 nT) + V(nT)$$

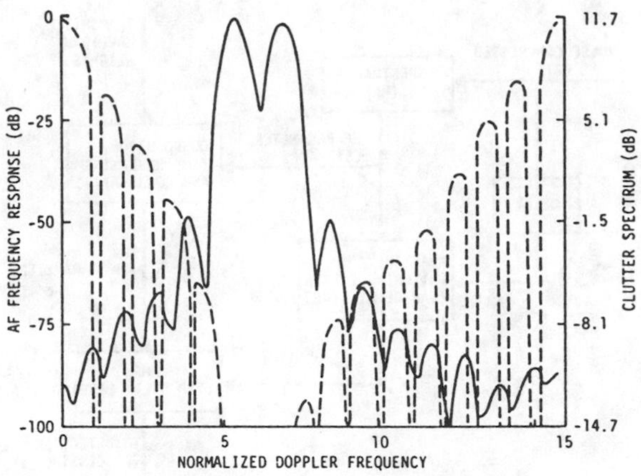

Fig. 7. Adaptive filter frequency response and clutter spectrum.

TABLE I

Filter	Number of False Detections in TBU Environment — Altitude Region 1	2	3	4	5
Matched	16	16	22	20	20
Adaptive	0	1	4	14	2

Note: Matched filter detection processing was designed to yield $P_D = 0.99$, and $P_F \cong 10^{-6}$ in white noise environment for a Swerling I target with SNR = 34 dB.

where T is the sampling interval, $a_i(n)$ is the time-varying amplitude of the ith complex sinusoid, w_i is the frequency of the ith complex sinusoid, and $V(nT)$ is the noise sample. The four test signals that can be generated from the above signal model are

1) single-frequency signal: $a_1(n) = a_1$, $a_2(n) = 0$;
2) two-frequency signal: $a_1(n) = a_1$, $a_2(n) = a_2$;
3) single-frequency Markov signal (Doppler spread): $a_2(n) = 0$; $a_1(n)$ is generated by the recursive equation: $a_1(n) = \theta_1 a_1(n-1) + w_1(n)$, where $|\theta_1| < 1$ and $w_1(n)$ is a white noise sequence;
4) two-frequency Markov signal (Doppler spread): $a_1(n)$ is generated as in case 3); $a_2(n)$ is generated by the equation: $a_2(n) = \theta_2 a_2(n-1) + w_2(n-1)$, where $|\theta_2| < 1$ and $w_2(n-1)$ is a white noise sequence.

The signal-to-noise ratios can be controlled by proper choice of the a_i and the noise standard deviation. Of the four signals chosen, case 1) is the simplest, and case 4) is the most difficult spectrum for the spectral estimator to identify. Thus the test signals typify a wide variety of signals that may be encountered in practice, and serve to investigate the performance of the different spectral estimator implementations.

The performance of the floating-point implementation was used as a comparison yardstick in evaluating the fixed-point implementations. Care was taken to maintain the same SNR and average signal power for any fixed-point implementation, and the floating-point implementation with which it is compared. This was necessary because

TABLE II

Signal Description	Spectral Estimator Performance Relative to Full-Precision Floating-Point Realization		
	8 Bit	10 Bit	12 Bit
One line frequency	Acceptable. Spectral peaks lower by about 10 dB.	Comparable to floating point. ≅2-3 dB peak difference.	Almost as good as floating point.
One frequency Markov	Acceptable	Comparable to floating point. ≅2-3 dB peak difference.	Almost as good as floating point.
Two line frequencies	Acceptable, but spectral peaks lower by about 10 dB.	Comparable to floating point. ≅2-3 dB peak difference.	Almost as good as floating point.
Two frequencies Markov	Unacceptable	Comparable to floating point. ≆2-3 dB peak difference.	Almost as good as floating point.

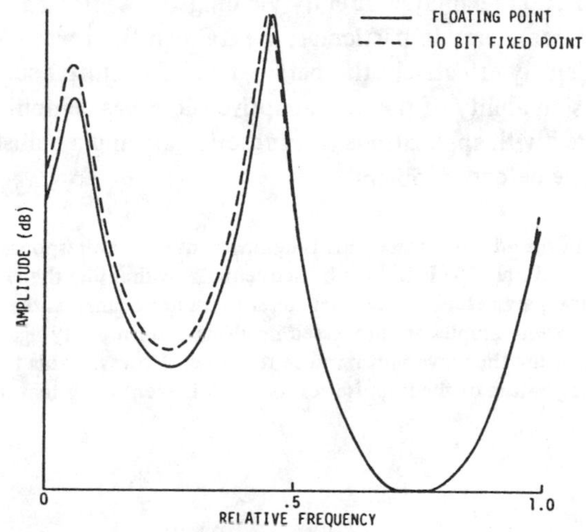

Fig. 8. Comparison of 10-bit and full-precision floating-point spectral estimator performances for Markov inputs with center frequencies at 1/16 and 27/64 of the Nyquist frequency.

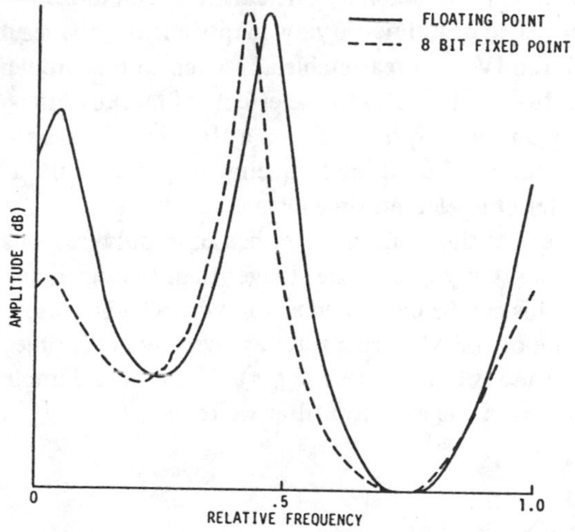

Fig. 9. Comparison of 8-bit and full-precision floating-point spectral estimator performances for Markov inputs with center frequencies at 1/16 and 27/64 of the Nyquist frequency.

the signal power generally tends to influence the spectral amplitudes.

The fixed-point implementation used a finite number of bits for the arithmetic operations and storage registers (accumulators). In the case of multiplication of two n-bit numbers, the $2n$-bit output was truncated to retain the most significant n bits. In the case of addition of two n-bit numbers, no truncation is required. However, an overflow could occur resulting in an $n + 1$ bit output. If this happened, the adder output was saturated at its full scale value of n bits.

Some difficulty was experienced whenever the input signal was very large. This caused the premature convergence to zero value of the elements of the covariance terms, and resulted in neglecting further inputs in updating the α estimates. To overcome this difficulty, the lower bounds of diagonal elements of the covariance matrix V_α were limited to a small value. This is similar to plant noise tuning in standard Kalman tracking filter algorithms. Necessarily suboptimal, this technique has indeed proved to be useful.

Table II presents the results of the study for the four typical test signals and 8-, 10-, and 12-bit spectral estimators. Typically a signal-to-noise ratio on the order of 20 dB was used. Figs. 8 and 9 show the spectral estimates obtained for the case of two Markov sources (Doppler spread) centered at 1/16 and 27/64 of the Nyquist frequency with the floating-point 8-bit and 10-bit fixed-point implementations. These figures along with Table I bring out the inadequacy of the 8-bit implementation and indicate that 10-bit implementation must be the minimal design in digital hardware configurations.

This study demonstrated the feasibility of implementing the second-order spectral estimator with fixed-point finite precision arithemtic. Similar results may be expected with a fourth-order spectral estimator.

V. Discussion

The adaptive filter presented in this paper is an open-loop technique unlike the sidelobe canceller [2] or Widrow's adaptive filter [3]. This generally reduces the transient response duration. The convergence of the adaptive filter to the optimum filter depends on the convergence of the spectral estimator. In all our studies, this convergence has not been a problem. In particular, in BMD applications where only 16 samples from a range cell are available for processing, the conventional feedback techniques appear to be inadequate, while the new adaptive filter has performed satisfactorily. This seems to be primarily due to the rapid identification of clutter spectral parameters by the Kalman filter technique.

Although it is generally difficult to characterize the performance of a time-varying adaptive filter, the results of Section IV are a reasonable approach to this problem. Subsection IV B analyzed the effects of Markov interference on the adaptive filter on a Monte Carlo basis. Subsection IV C used the frequency response of the adaptive filter at a selected time instant.

Note that the adaptive filter has an impulse response duration (at any given time) longer than the signal duration. This can be explained as follows. The impulse response of the whitering filter as given by (1) is time varying and is indicated by $h(t, \tau)$. In the actual implementation of the adaptive filter we replace $h(t, \tau)$ by a "time-invariant"[1] impulse response $h(t - \tau)$ where $h(t - \tau)$ is the solution of (1) with the assumptions of stationarity and infinite interval of observation. Such a replacement results in the modified signal $S_1(t)$ that is longer in duration than the original signal $S(t)$, This in turn results in increasing the impulse response duration of the MMF thereby causing the increased impulse response duration of the adaptive filter. This difficulty can be overcome by appropriately truncating the impulse response of the filter to the signal duration. For example, in the adaptive filter structure of Section III A, the truncated impulse response could correspond to the central N samples out of the complete impulse response of $N + 2M - 2$ samples.

In spite of the abovementioned shortcomings, the new adaptive filter has performed satisfactorily in practical applications, such as in BMD clutter suppression.

VI. Summary

In this paper we presented a new open-loop approach to adaptive filtering. By modeling the interference as an autoregressive process we could use a parametric spectral estimation technique directly yielding the whitening filter parameters. In particular, the Kalman filter was used to identify effectively the parameters. The clutter suppression ability of the new adaptive filter was demonstrated with applications to synthetic data and a ballistic missile defense problem.

[1] In the adaptive filter, this time-invariant impulse response is estimated only. This is done by sequentially estimating the clutter spectral parameters. These parameter estimates change as the observation samples are processed resulting in a time-varying estimate of the time-invariant impulse response. However, this time-varying nature of the impulse response is different from that implied by (1).

References

[1] L.J. Spafford, "Optimum radar signal processing in clutter," *IEEE Trans. Inform. Theory*, vol. IT-14, pp. 734-743, 1968.

[2] S.O. Applebaum, "Adaptive arrays," *IEEE Trans. Antennas Propagat.*, vol. AP-24, pp. 585-589, 1976.

[3] B. Widrow, "Adaptive filters," in *Aspects of Network and System Theory*, Ed. R.E. Kalman and N. DeClaris. New York: Holt, Reinhart and Winston, 1971, ch. 5.

[4] H.L. Van Trees, *Detection, Estimation and Modulation: Part I.* New York: Wiley, 1968, ch. 4.

[5] N. Levinson, "The wiener RMS error criterion in filter design and prediction," *J. Math. Phys.*, vol. 25, pp. 261-278, 1974.

[6] M.D. Srinath and M.M. Viswanathan, "Sequential algorithm for identification of parameters of an autoregressive process, *IEEE Trans. Automat. Contr.*, vol. AC-20, pp. 542-546, 1975.

[7] J. Makhoul, "Linear prediction: A tutorial review, *Proc. IEEE*, vol. 63, pp. 561-580, 1975.

[8] A.P. Sage and J.L. Melsa, *Estimation Theory with Applications to Communications and Control.* New York: McGraw-Hill, 1971, ch. 7.

[9] T. Kailath, "An innovation approach to least square estimation – Part I: Linear filtering in additive white noise," *IEEE Trans. Automat. Contr.*, vol. AC-13, pp. 646-654, 1968.

[10] T.J. Zarn and J. Zaborsky, "A practical nondiverging filter," *AIAA J.*, vol. 8, pp. 1127-1133, 1970.

[11] A.H. Jazwinsky, "Limited memory optimal filtering," *IEEE Trans. Automat. Contr.*, vol. AC-13, pp. 558-563, 1968.

Part VI
Statistics and Detection

THE exact statistics of the parametric spectral estimators are difficult to determine. The equations involved are often nonlinear, and methods are usually designed for processing relatively short data records. As an alternative to exact statistical analysis, the investigations were directed toward the determination of the asymptotic distributions, and bounds for the estimators. In some cases the statistical performance was demonstrated by examples generated by Monte Carlo simulations. The first three papers deal with the statistics of the autoregressive (AR) estimators. Kay and Makhoul derive the asymptotic statistics of the reflection coefficient estimates. The paper by Sakai investigates the statistics of the AR method by using the periodogram technique devised by himself. Empirical analysis of the AR spectral estimator's probability density function, using generalized Burg algorithm, is presented in paper by Kesler.

Papers 4 and 5 by Bruzzone and Kaveh, and by Friedlander and Porat, respectively, treat the statistics of various ARMA methods and develop the lower bounds for their efficiencies. (Incidentally, both papers appeared in the same issue of the *IEEE Trans. Acoust., Speech, and Signal Processing.*)

Detection of signals is an important issue related to the statistics of the estimator. The final paper in this selection, authored by S. M. Kay, deals with the properties of the AR estimator as a detector.

On the Statistics of the Estimated Reflection Coefficients of an Autoregressive Process

STEVEN KAY, SENIOR MEMBER, IEEE, AND JOHN MAKHOUL, FELLOW, IEEE

Abstract—The exact statistics of the estimated reflection coefficients for an autoregressive process are difficult to determine. However, since almost all the common methods for estimating the reflection coefficients are maximum likelihood estimates for large data records, the asymptotic distribution of the estimates is multivariate Gaussian with a covariance matrix given by the Cramer-Rao bound. A recursive means of computing the covariance matrix bound is described. Simulation results show that the asymptotic expressions are accurate for large data records. However, for relatively short data records, the asymptotic expressions are accurate only for spectra with a small dynamic range.

I. Introduction

INTENSE research activity has been focused on the use of linear prediction or autoregressive (AR) time series modeling in diverse fields, such as speech [1], radar [2], sonar [3], image [4], and biological signal processing [5]. Most of the methods for estimating the AR parameters endeavor to yield a minimum-phase (stable) all-pole model. If the signal under consideration is windowed (autocorrelation method), the resulting all-pole model is guaranteed to be stable. If no windowing is performed (covariance method), then the stability of the resulting model is not guaranteed. To ensure stability under no windowing, a number of methods have been proposed that estimate the reflection coefficients (or partial correlation coefficients) directly. Filter stability is guaranteed by ensuring that the reflection coefficients are less than one in magnitude. Among these methods are those of Itakura [6] and Burg [7], which have been shown to be special cases of a class of *lattice* methods [17] for all-pole modeling. Another method is to set the reflection coefficients equal to the negative of the partial correlation coefficients [8].

Using a classic result by Mann and Wald [9] on the covariance of AR parameter estimates, we derive the asymptotic distribution of reflection coefficient estimates. The basic result was first derived by Barndorf-Nielsen and Schou [10] for the partial correlation coefficients.[1] We derive a recursive procedure for computing the covariance of the reflection coefficient estimates and show that the asymptotic result applies to all the well-known parameter estimation methods. Then, we present the results of simulations that show the basic experimental validity of the derivations.

In this paper, the properties of maximum likelihood estimation are used to determine approximate statistics of the estimated reflection coefficients. Section II contains a summary of the necessary theorems of maximum likelihood estimation. Section III shows that all the common methods of estimating the reflection coefficients are asymptotically equivalent. Section IV derives the asymptotic distribution of reflection coefficients and presents a recursive method of determining the Cramer-Rao bound or the asymptotic covariance matrix. Finally, in Section V, we compare the theoretical results to those obtained via computer simulations.

II. Background on Maximum Likelihood Estimation

Below, we summarize some relevant results from maximum likelihood estimation which will be useful in the sections to follow.

We first consider the Cramer-Rao bound for estimates of vector parameters. Let $\boldsymbol{\theta} = [\theta_1, \theta_2, \cdots, \theta_p]^T$ be a set of parameters to be estimated. It is assumed that the underlying probability density function (PDF) is $p(\boldsymbol{x}|\boldsymbol{\theta})$ where $\boldsymbol{x} = [x_0, x_1, \cdots, x_{N-1}]^T$ is a sample of size N. Let $\hat{\boldsymbol{\theta}}$ be an unbiased estimator, i.e., $E(\hat{\boldsymbol{\theta}}) = \boldsymbol{\theta}$. Then the covariance matrix of $\hat{\boldsymbol{\theta}}$ given by $C_{\hat{\theta}}$ satisfies the Cramer-Rao bound [11], [12]:

$$C_{\hat{\theta}} = E((\hat{\boldsymbol{\theta}} - \boldsymbol{\theta})(\hat{\boldsymbol{\theta}} - \boldsymbol{\theta})^T) \geqslant \boldsymbol{I}(\boldsymbol{\theta})^{-1} \tag{1}$$

where $\boldsymbol{A} \geqslant \boldsymbol{B}$ means that $\boldsymbol{A} - \boldsymbol{B}$ is a positive semidefinite matrix. $\boldsymbol{I}(\boldsymbol{\theta})$ is the Fischer information matrix whose (i, j) element is

$$[\boldsymbol{I}(\boldsymbol{\theta})]_{ij} = E\left[\frac{\partial \ln p(\boldsymbol{x}|\boldsymbol{\theta})}{\partial \theta_i} \frac{\partial \ln p(\boldsymbol{x}|\boldsymbol{\theta})}{\partial \theta_j}\right]. \tag{2}$$

It should be noted that (1) implies

$$\text{var}(\hat{\theta}_i) \geqslant [\boldsymbol{I}(\boldsymbol{\theta})^{-1}]_{ii} \qquad i = 1, 2, \cdots, p. \tag{3}$$

The equality sign in (1) holds if and only if

$$\nabla_\theta \ln p(\boldsymbol{x}|\boldsymbol{\theta}) = \boldsymbol{A}(\boldsymbol{\theta})(\hat{\boldsymbol{\theta}} - \boldsymbol{\theta}) \tag{4}$$

where ∇_θ is the gradient with respect to $\boldsymbol{\theta}$. $\boldsymbol{A}(\boldsymbol{\theta})$ is a matrix which may or may not depend upon $\boldsymbol{\theta}$. Further, note that if

Manuscript received October 28, 1981; revised February 9, 1983 and May 31, 1983. This work was supported in part by the Office of Naval Research under Grant N00014-81-K-0144, by the National Science Foundation under Grant Eng. 79-00337, and by the Advanced Research Projects Agency under Contract F19628-80-C-0165 monitored by RADC.

S. Kay is with the Department of Electrical Engineering, University of Rhode Island, Kingston, RI 02881.

J. Makhoul is with Bolt Beranek and Newman Inc., Cambridge, MA 02238.

[1] As defined in this paper, reflection coefficients have the same asymptotic distribution as the negative of the partial correlation coefficients.

Reprinted from *IEEE Trans. Acoust., Speech, Signal Processing*, vol. ASSP-31, pp. 1447–1455, Dec. 1983.

(4) is satisfied, the unbiased estimate attains the minimum variance of [13]

$$C_{\hat{\theta}} = A(\boldsymbol{\theta})^{-1}. \tag{5}$$

These properties of the Cramer-Rao bound will be essential for our later arguments.

Now let $\hat{\boldsymbol{\theta}}_{ML}$ be the maximum likelihood estimator (MLE) of $\boldsymbol{\theta}$, i.e., $\hat{\boldsymbol{\theta}}_{ML}$ is found from

$$\max_{\boldsymbol{\theta}} p(\boldsymbol{x}|\boldsymbol{\theta}) = p(\boldsymbol{x}|\hat{\boldsymbol{\theta}}_{ML}). \tag{6}$$

The MLE has the following properties [11]:

1) $E(\hat{\boldsymbol{\theta}}_{ML}) \to \boldsymbol{\theta}$ as $N \to \infty$. (7)

2) $C_{\hat{\theta}_{ML}} \to I(\boldsymbol{\theta})^{-1} = A(\boldsymbol{\theta})^{-1}$ as $N \to \infty$ (8)

3) The MLE PDF is multivariate Gaussian with mean $\boldsymbol{\theta}$ and covariance $A(\boldsymbol{\theta})^{-1}$ as $N \to \infty$.

4) $\hat{\boldsymbol{\theta}}_{ML} \to \boldsymbol{\theta}$ as $N \to \infty$. (9)

The condition $N \to \infty$ is termed the asymptotic case. Thus, for sufficiently large data records, $\hat{\boldsymbol{\theta}}_{ML}$ is Gaussian distributed with a mean $\boldsymbol{6}$ and a covariance matrix given by the Cramer-Rao bound. Also, $\hat{\boldsymbol{\theta}}_{ML}$ is a consistent estimator. What follows is another useful property [13].

5) If $\boldsymbol{\theta}' = g(\boldsymbol{\theta})$ is a one-to-one transformation from $\boldsymbol{\theta}$ to $\boldsymbol{\theta}'$, then $\hat{\boldsymbol{\theta}}'_{ML} = g(\hat{\boldsymbol{\theta}}_{ML})$ is an MLE of $\boldsymbol{\theta}'$. Furthermore, the asymptotic covariance matrix of $\hat{\boldsymbol{\theta}}'_{ML}$ is

$$C_{\hat{\theta}'_{ML}} = L\, C_{\hat{\theta}_{ML}} L^T \tag{10}$$

where

$$[L]_{ij} = \frac{\partial g_i(\boldsymbol{\theta})}{\partial \theta_j}.$$

L is just a linearization of the function $g(\boldsymbol{\theta})$ about the true parameters, $\boldsymbol{\theta}$.

III. Equivalence of Asymptotic Distribution of Reflection Coefficient Estimates

There are numerous methods for estimating the reflection coefficients of an AR process. It will be shown in this section that all the popular methods yield the same numerical estimates for large enough data records. One popular approach to estimating the AR parameters $\{a(1), a(2), \cdots, a(p)\}$ is to solve the normal equations

$$\begin{bmatrix} \hat{R}(0) & \hat{R}(1) & \cdots & \hat{R}(p-1) \\ \hat{R}(1) & \hat{R}(0) & \cdots & \hat{R}(p-2) \\ \vdots & \vdots & & \vdots \\ \hat{R}(p-1) & \hat{R}(p-2) & \cdots & \hat{R}(0) \end{bmatrix} \begin{bmatrix} \hat{a}(1) \\ \hat{a}(2) \\ \vdots \\ \hat{a}(p) \end{bmatrix} = - \begin{bmatrix} \hat{R}(1) \\ \hat{R}(2) \\ \vdots \\ \hat{R}(p) \end{bmatrix} \tag{11}$$

or in matrix notation

$$\hat{R}\hat{a} = -\hat{r} \tag{12}$$

where

$$\hat{R}(k) = \frac{1}{N}\sum_{t=k}^{N-1} x_t x_{t-k}, \quad k \geq 0 \tag{13}$$

and it is assumed that $\{x_0, x_1, \cdots, x_{N-1}\}$ are the observed data. This method is termed the autocorrelation method or the Yule-Walker approach [1]. $\hat{R}(k)$ is just an estimate of the statistical autocorrelation function $R(k) = E(x_t x_{t-k})$. (In this paper we assume that the process x_t is stationary and ergodic. Therefore, estimating the ensemble average by a time average is justified.)

The Yule-Walker equations given by (11) are usually solved using the Levinson-Durbin algorithm [1], which recursively solves (11) as follows:

$$a_n(0) = 1, \quad n = 0, 1, \cdots, p$$
$$\sigma_0^2 = \hat{R}(0).$$

For $n = 1, 2, \cdots, p$

$$\hat{K}_n = a_n(n) = -\frac{\sum_{i=0}^{n-1} a_{n-1}(i)\hat{R}(n-i)}{\sigma_{n-1}^2} \tag{14}$$
$$a_n(i) = a_{n-1}(i) + \hat{K}_n a_{n-1}(n-i), \quad i = 1, 2, \cdots, n-1$$
$$\sigma_n^2 = (1 - \hat{K}_n^2)\sigma_{n-1}^2.$$

The solution of (11) is given by $\hat{a}(i) = a_p(i)$ for $i = 1, 2, \cdots, p$. The elements of the sequence defined above $\{\hat{K}_1, \hat{K}_2, \cdots, \hat{K}_p\}$ are termed the reflection coefficient estimates. Clearly, if $\hat{R}(k)$, as defined in (11), is equal to the true autocorrelation function, then the $\hat{K}_n$, as generated by the Levinson-Durbin algorithm, would be the true reflection coefficients.

The recursion to compute $a_n(i)$ in (14) may also be written in matrix form, which will be useful in Section IV. Let $\boldsymbol{a}_n = [a_n(1)\, a_n(2) \cdots a_n(p)]^T$ and

$$\boldsymbol{J} = \begin{bmatrix} \boldsymbol{0} & & & 1 \\ & & \cdot^{\cdot^{\cdot}} & \\ & 1 & & \\ 1 & & & \boldsymbol{0} \end{bmatrix} \tag{15}$$

which is sometimes termed the reverse operator, since $\boldsymbol{JM}$ reverses the rows of $\boldsymbol{M}$ and $\boldsymbol{MJ}$ reverses the columns. Then,

$$\boldsymbol{a}_n = \begin{bmatrix} (\boldsymbol{I} + \hat{K}_n \boldsymbol{J})\boldsymbol{a}_{n-1} \\ \hat{K}_n \end{bmatrix} \tag{16}$$

and it is assumed that $\hat{K}_n$ has been computed.

In addition to the autocorrelation method, there are numerous other approaches to estimating the AR parameters. Two other popular methods are the covariance method [1] and the forward-backward method [14], [15]. In the covariance and forward-backward methods, the AR parameter estimates are given by the solution of (12) with

$$[\hat{R}]_{ij} = \frac{1}{N-p} \sum_{t=p}^{N-1} x_{t-i} x_{t-j}$$

$$[\hat{r}]_i = \frac{1}{N-p} \sum_{t=p}^{N-1} x_t x_{t-i} \tag{17}$$

for the covariance method, and

$$[\hat{R}]_{ij} = \frac{1}{2(N-p)} \left[\sum_{t=p}^{N-1} x_{t-i} x_{t-j} + x_{t-p+i} x_{t-p+j} \right]$$

$$[\hat{r}]_i = \frac{1}{2(N-p)} \left[\sum_{t=p}^{N-1} x_{t-i} x_t + x_{t-p+i} x_{t-p} \right] \tag{18}$$

for the forward-backward method.

In either case, the Levinson-Durbin algorithm cannot be used to solve (12). Hence, to obtain the reflection coefficient estimates, one first finds the AR parameter estimates by solving (12) and then uses the step-down procedure

$$a_p(i) = \hat{a}(i), \quad i = 1, 2, \cdots, p$$

$$\hat{K}_p = a_p(p)$$

for $n = p, p-1, \cdots, 2$

$$a_{n-1}(i) = \frac{a_n(i) - \hat{K}_n a_n(n-i)}{1 - \hat{K}_n^2}, \quad i = 1, 2, \cdots, n-1$$

$$\hat{K}_{n-1} = a_{n-1}(n-1). \tag{19}$$

Equation (19) is nothing more than the recursion as given by (16), written so as to obtain a_{n-1} from a_n. It should be noted that, of the three methods cited above, only the autocorrelation or Yule-Walker approach is guaranteed to yield $|\hat{K}_n| \leq 1$, $n = 1, 2, \cdots, p$.

The three approaches just described all have the same asymptotic statistics for the estimated reflection coefficients. This is because for $N >> p$, the entries for $\hat{R}$ and $\hat{r}$ as given by (17) and (18) will approach in numerical value those as given by (11). The AR parameter estimates will approach each other, and hence, so will the reflection coefficient estimates. Then, the asymptotic statistics of the reflection coefficient estimates must be identical. This is because each realization of the random process will produce the same reflection coefficient estimate for each estimation procedure.

A second class of reflection coefficient estimators comprise the lattice methods [17]. These algorithms operate directly on the data to produce reflection coefficient estimates. As an example, we shall deal here with the Burg estimate [17]. The Burg estimate of the reflection coefficients is

$$f_0(t) = b_0(t) = x_t, \quad t = 0, 1, \cdots, N-1$$

for $n = 1, 2, \cdots, p$

$$\check{K}_n = -\frac{\sum_{t=n}^{N-1} f_{n-1}(t) b_{n-1}(t-1)}{\frac{1}{2} \sum_{t=n}^{N-1} [f_{n-1}^2(t) + b_{n-1}^2(t-1)]}$$

$$f_n(t) = f_{n-1}(t) + \check{K}_n b_{n-1}(t-1)$$

$$b_n(t) = b_{n-1}(t-1) + \check{K}_n f_{n-1}(t). \tag{20}$$

The carat above K_n is used to distinguish this estimator from the previous ones. Note that in this case, $|\check{K}_n| \leq 1$ [7].

To show that the Burg reflection coefficient estimator also has the same asymptotic distribution, we proceed by induction. First, we show that $\check{K}_1 = \hat{K}_1$, where the equality is in an asymptotic sense. From (20), we have

$$\check{K}_1 = -\frac{\sum_{t=1}^{N-1} x_t x_{t-1}}{\frac{1}{2} \sum_{t=1}^{N-1} (x_t^2 + x_{t-1}^2)}. \tag{21}$$

For very large N, the numerator approaches $N\hat{R}(1)$ and the denominator $\hat{R}(0)$. Therefore, asymptotically,

$$\check{K}_1 = -\frac{\hat{R}(1)}{\hat{R}(0)} = \hat{K}_1. \tag{22}$$

Now assume $\check{K}_i = \hat{K}_i, i = 1, 2, \cdots, n-1$. It may be shown that

$$f_{n-1}(t) = \sum_{k=0}^{n-1} a_{n-1}(k) x_{t-k}, \quad n-1 \leq t \leq N-1$$

$$b_{n-1}(t) = \sum_{k=0}^{n-1} a_{n-1}(k) x_{t-n+1+k}, \quad n-1 \leq t \leq N-1 \tag{23}$$

where $a_n(0) = 1$ and $a_{n-1}(k)$ for $k = 1, 2, \cdots, n-1$ obey (11) and are found recursively from (14) with $\check{K}_i$ replacing $\hat{K}_i$. Since it has been assumed that $\check{K}_i = \hat{K}_i$ for $i = 1, 2, \cdots, n-1$, the $a_{n-1}(k)$ coefficients are identical to those generated by the Levinson-Durbin algorithm for the autocorrelation method. Inserting (23) into (20) and rearranging the summations yields

$$\check{K}_n = -\frac{\sum_{k=0}^{n-1} \sum_{i=0}^{n-1} a_{n-1}(k) a_{n-1}(i) \left[\frac{1}{N} \sum_{t=n}^{N-1} x_{t-k} x_{t-n+i} \right]}{\sum_{k=0}^{n-1} \sum_{i=0}^{n-1} a_{n-1}(k) a_{n-1}(i) \frac{1}{2N} \left[\sum_{t=n}^{N-1} x_{t-k} x_{t-i} + \sum_{t=n}^{N-1} x_{t-n+k} x_{t-n+i} \right]}. \tag{24}$$

But for $N >> p$, we have

$$\check{K}_n \to -\frac{\sum_{k=0}^{n-1} a_{n-1}(k) \sum_{i=0}^{n-1} a_{n-1}(i) \hat{R}(n-k-i)}{\sum_{k=0}^{n-1} a_{n-1}(k) \sum_{i=0}^{n-1} a_{n-1}(i) \hat{R}(k-i)} \tag{25}$$

and from (11), we have

$$\sum_{i=0}^{n-1} a_{n-1}(i)\hat{R}(k-i) = 0, \qquad k = 1, 2, \cdots, n-1$$
$$= \sigma_{n-1}^2, \qquad k = 0. \tag{26}$$

Hence, (25) reduces to

$$\check{K}_n \to -\frac{\sum_{i=0}^{n-1} a_{n-1}(i)\hat{R}(n-i)}{\sigma_{n-1}^2} = \hat{K}_n \tag{27}$$

as may be seen from (14). By induction then, the reflection coefficient estimates for large N have the identical numerical values for the autocorrelation, covariance, forward-backward, and Burg methods. The analysis given for the Burg method is easily extendible to a larger class of lattice methods [17].

IV. Statistics of Reflection Coefficient Estimators

A. Probability Density Function

In the previous section, it was shown that all the common methods of estimating the reflection coefficients produce the same estimate (i.e., in numerical value) for $N >> p$. Hence, we will now restrict the discussion to the Yule-Walker or autocorrelation method. It has been shown that, asymptotically, the Yule-Walker method produces an MLE of the AR parameters [18]. The asymptotic PDF for the $\hat{a}$ parameters is Gaussian and is given by

$$p(\hat{a}) = (2\pi)^{-p/2}|C_{\hat{a}}|^{-1/2} \exp -\tfrac{1}{2}(\hat{a}-a)^T C_{\hat{a}}^{-1}(\hat{a}-a). \tag{28}$$

Now, since the transformation from the AR parameters to the reflection coefficients is one-to-one [see (10)], the reflection coefficient estimator is also an MLE. [See property 5) in Section II.] Hence, the asymptotic PDF of $\hat{K} = [\hat{K}_1 \hat{K}_2 \cdots \hat{K}_p]^T$ is Gaussian

$$p(\hat{K}) = (2\pi)^{-p/2}|C_{\hat{K}}|^{-1/2} \exp -\tfrac{1}{2}(\hat{K}-K)^T C_{\hat{K}}^{-1}(\hat{K}-K) \tag{29}$$

where K is the true vector of reflection coefficients and $C_{\hat{K}}$ is the covariance matrix. From property 5), we also have

$$C_{\hat{K}} = A\, C_{\hat{a}}\, A^T \tag{30}$$

where

$$[A]_{ij} = \frac{\partial K_i(a)}{\partial a_j}.$$

Equation (30) can be heuristically derived as follows. Let $\hat{K} = \Phi(\hat{a})$ where Φ is a nonlinear transformation from $\mathbb{R}^p$ to $\mathbb{R}^p$. For large data records, $\hat{a}$ will only exhibit values close to a by property 4). Hence, $\hat{K}$ will be close to $\Phi(a) = K = E(K)$. We can linearize $\Phi(\hat{a})$ about $a = E(\hat{a})$.

$$\hat{K} - E(\hat{K}) = \left.\frac{\partial \Phi(\alpha)}{\partial \alpha}\right|_{\alpha = a} (\hat{a} - E(\hat{a})). \tag{31}$$

Letting

$$\left.\frac{\partial \Phi(\alpha)}{\partial \alpha}\right|_{\alpha = a} = A$$

$$\hat{K} - E(\hat{K}) = A(\hat{a} - E(\hat{a})) \tag{32}$$

from which (30) follows.

Now, assuming x_t is an AR(p) process, it has been shown that asymptotically [9]

$$C_{\hat{a}} = \frac{\sigma_\epsilon^2}{N} R^{-1} \tag{33}$$

where σ_ϵ^2 is the variance of the white noise driving process and $[R]_{ij} = R_x(i-j)$. From (30)

$$C_{\hat{K}} = \frac{\sigma_\epsilon^2}{N} A R^{-1} A^T. \tag{34}$$

However, by the Levinson-Durbin algorithm (16), one can show that [16]

$$B^T R B = E \tag{35}$$

where

$$B = \begin{bmatrix} 1 & a_1(1) & a_2(2) & \cdots & a_{p-1}(p-1) \\ 0 & 1 & a_2(1) & \cdots & a_{p-1}(p-2) \\ \vdots & \vdots & \vdots & & \vdots \\ 0 & 0 & 0 & \cdots & 1 \end{bmatrix} \tag{36}$$

and

$$E = \text{diag}(\sigma_0^2, \sigma_1^2, \cdots, \sigma_{p-1}^2).$$

Therefore,

$$C_{\hat{K}} = \frac{\sigma_\epsilon^2}{N} A B E^{-1} B^T A^T. \tag{37}$$

Below, we derive a recursive means of computing $C_{\hat{K}}$ and $|C_{\hat{K}}|$ based upon the Levinson-Durbin algorithm.

B. Recursive Computation of $C_{\hat{K}}$ and $|C_{\hat{K}}|$

Let B_n, E_n denote the $n \times n$ matrices given by (35). Also, let

$$A_n = \begin{bmatrix} \frac{\partial K_1}{\partial a_n(1)} & \frac{\partial K_1}{\partial a_n(2)} & \cdots & \frac{\partial K_1}{\partial a_n(n)} \\ \frac{\partial K_2}{\partial a_n(1)} & \frac{\partial K_2}{\partial a_n(2)} & \cdots & \frac{\partial K_2}{\partial a_n(n)} \\ \vdots & \vdots & & \vdots \\ \frac{\partial K_n}{\partial a_n(1)} & \frac{\partial K_n}{\partial a_n(2)} & \cdots & \frac{\partial K_n}{\partial a_n(n)} \end{bmatrix} \tag{38}$$

Then, define

$$C_{\hat{K}}(n) = \frac{\sigma_n^2}{N} A_n B_n E_n^{-1} B_n^T A_n^T \tag{39}$$

so that $\sigma_n^2 = \sigma_\epsilon^2$ for $n = p$ and $C_{\hat{K}} = C_{\hat{K}}(p)$. We intend to compute $C_{\hat{K}}(n)$ given $C_{\hat{K}}(n-1)$. Thus, we will sucessively find the covariance matrices for an AR(1) process, AR(2) process, $\cdots$, AR(p) process with reflection coefficients $\{K_1\}$; $\{K_1, K_2\}$; $\cdots$; $\{K_1, K_2, \cdots, K_p\}$ and white noise variances $\sigma_1^2, \sigma_2^2, \cdots, \sigma_p^2 = \sigma_\epsilon^2$. The details of the derivation are given in Appendix A. The results are given below.

1) For $n = 1$

$$a_1 = a_1(1) = K_1$$

$$A_1 = B_1 = E_1 = \sigma_0^2 = 1$$

$$\sigma_1^2 = (1 - K_1^2)$$

$$C_{\hat{K}}(1) = \frac{1}{N}(1 - K_1^2). \tag{40}$$

2) $n \leftarrow n + 1$

$$\sigma_n^2 = (1 - K_n^2)\,\sigma_{n-1}^2$$

$$D_{n-1} = \frac{1}{1 - K_n^2} A_{n-1}(I - K_n J) B_{n-1}$$

$$C_{\hat{K}}(n) = \frac{1}{N}\begin{bmatrix} \sigma_n^2 D_{n-1} E_{n-1}^{-1} D_{n-1}^T & 0 \\ \mathbf{0}^T & 1 - K_n^2 \end{bmatrix}. \tag{41}$$

3) If $n = p$, let $C_{\hat{K}} = C_{\hat{K}}(n)$ and exit.

4)

$$E_n^{-1} = \begin{bmatrix} E_{n-1}^{-1} & \mathbf{0} \\ \mathbf{0}^T & 1/\sigma_{n-1}^2 \end{bmatrix}$$

$$a_{n-1} = \begin{bmatrix} (I + K_{n-1} J) a_{n-2} \\ K_{n-1} \end{bmatrix}$$

$$B_n = \begin{bmatrix} B_{n-1} & J a_{n-1} \\ \mathbf{0}^T & 1 \end{bmatrix}$$

$$A_n = \begin{bmatrix} \frac{1}{1 - K_n^2} A_{n-1}(I - K_n J) & -\frac{1}{1 - K_n^2} A_{n-1}(I - K_n J) J a_{n-1} \\ \mathbf{0}^T & 1 \end{bmatrix}. \tag{42}$$

5) Go to 2.

Note that $\mathbf{0}^T$ is $1 \times (n - 1)$.

The following observations can be made [10].

1) From the form of $C_{\hat{K}}$, one recognizes that for an AR(p) process, $\hat{K}_p$ is uncorrelated with $\hat{K}_1, \hat{K}_2, \cdots, \hat{K}_{p-1}$. Since $\hat{K}$ is Gaussian, $\hat{K}_p$ is independent of $\hat{K}_1, \hat{K}_2, \cdots, \hat{K}_{p-1}$. Also,

$$\operatorname{var}(\hat{K}_p) = (1 - K_p^2)/N. \tag{43}$$

2) For an AR(p) process, $\hat{K}_{p+1}, \hat{K}_{p+2}, \cdots$ are independent and zero mean with

$$\operatorname{var}(\hat{K}_n) = 1/N \qquad n = p+1, p+2, \cdots. \tag{44}$$

We now derive this property from (41). For an AR(p) process, $K_n = 0$ and $\sigma_n^2 = \sigma_p^2 = \sigma_\epsilon^2$ for $n > p$. Hence,

$$C_{\hat{K}}(p+1) = \frac{1}{N}\begin{bmatrix} \sigma_p^2 D_p E_p^{-1} D_p^T & \mathbf{0} \\ \mathbf{0}^T & 1 \end{bmatrix}$$

$$= \frac{1}{N}\begin{bmatrix} \sigma_\epsilon^2 A_p B_p E_p^{-1} B_p^T A_p^T & \mathbf{0} \\ \mathbf{0}^T & 1 \end{bmatrix} \tag{45}$$

so that from (37)

$$C_{\hat{K}}(p+1) = \begin{bmatrix} C_{\hat{K}} & \mathbf{0} \\ \mathbf{0}^T & \frac{1}{N} \end{bmatrix} \tag{46}$$

Proceeding in a similar fashion,

$$C_{\hat{K}}(p+2) = \begin{bmatrix} C_{\hat{K}} & \mathbf{0} & \mathbf{0} \\ \mathbf{0}^T & \frac{1}{N} & 0 \\ \mathbf{0}^T & 0 & \frac{1}{N} \end{bmatrix} \tag{47}$$

etc., from which observation 2 follows.

The determinant of C_K can also be obtained from (41) (see Appendix B). The result is

$$|C_{\hat{K}}| = \frac{1}{N^p} \prod_{i \text{ odd}} (1 - K_i^2) \prod_{j \text{ even}} (1 - K_j)^2 \qquad \text{for } 1 \leqslant i, j \leqslant p. \tag{48}$$

Alternately, a recursive expression may be found as

$$|C_{\hat{K}}(n)| = \frac{1}{N}(1 - K_n^2)\, g_{n-1}\, |C_{\hat{K}}(n-1)| \tag{49}$$

where

$$g_{n-1} = 1, \qquad n \text{ odd}$$
$$= \frac{1 - K_n}{1 + K_n}, \qquad n \text{ even}. \tag{50}$$

C. Covariance Expressions for Low Order

From (41), one can show that the asymptotic covariance of the reflection coefficient estimates for $p = 1, 2, 3$ are given as follows.

$p = 1$:

$$C_{\hat{K}} = \frac{1}{N}(1 - K_1^2) \tag{51}$$

$p = 2$:

$$C_{\hat{K}} = \frac{1}{N}\begin{bmatrix} \frac{(1 - K_1^2)(1 - K_2)}{1 + K_2} & 0 \\ 0 & 1 - K_2^2 \end{bmatrix} \tag{52}$$

$p = 3$:

$$C_{\hat{K}} = \frac{1}{N}\begin{bmatrix} \frac{(1 - K_1^2)(1 - K_2^2)(1 + 2K_1K_3 + K_3^2)}{(1 - K_3^2)(1 + K_2)^2} & \frac{-2K_3(1 - K_1^2)(1 - K_2)}{1 - K_3^2} & 0 \\ \frac{-2K_3(1 - K_1^2)(1 - K_2)}{1 - K_3^2} & \frac{(1 - K_2^2)(1 - 2K_1K_3 + K_3^2)}{1 - K_3^2} & 0 \\ 0 & 0 & (1 - K_3^2) \end{bmatrix}. \tag{53}$$

Note that in general, the $\hat{K}_i$'s are correlated, although $\hat{K}_p$ is uncorrelated with $\{\hat{K}_1, \hat{K}_2, \cdots, \hat{K}_{p-1}\}$ for an AR(p) process.

V. Simulation Results

To determine the possible applicability of the asymptotic expression for $C_{\hat{K}}$ and the unbiasedness of $\hat{K}$ for finite data records, we ran simulations for $p = 3$ utilizing a Gaussian process.

A. Data Computation

The computer program required the following as input: the order p of the process, a set of p reflection coefficients K, the number of data records M, and the number of samples N in each data record. The set of reflection coefficients specified an all-pole filter, which was used to obtain the data. Each data record was generated as the output of the all-pole filter when excited by a zero-mean white Gaussian random source. The estimated reflection coefficients $\hat{K}$ from each data record were computed from (11) and (14), where the autocorrelation coefficients were estimated from (13). The estimate of the true mean was then computed as

$$\overline{K} = \widehat{E(\hat{K})} = \frac{1}{M} \sum_{m=1}^{M} \hat{K}_m \tag{54}$$

where $\hat{K}_m$ is the estimate of K for the mth data record, and M is the total number of records. Similarly, the estimate of the true covariance matrix was computed from

$$\hat{C}_{\hat{K}} = \frac{1}{M} \sum_{m=1}^{M} (\hat{K}_m - \overline{K})(\hat{K}_m - \overline{K})^T. \tag{55}$$

The number of data record M should, of course, be as large as possible, in order to obtain stable estimates of the mean and covariance for any given record length N. We have found that a value of $M = 1000$ gave reasonably stable results without requiring an excessive amount of computing. All the results given below, therefore, were obtained with $M = 1000$.

B. Bias Results

To study the effect of the data record length N on the bias, we computed the following measure as an indicator of the bias magnitude:

$$B = \sqrt{\frac{1}{p} \sum_{i=1}^{p} (\overline{L}_i - L_i)^2} \tag{56}$$

where

$$L = \frac{1}{2} \ln \frac{1+K}{1-K}. \tag{57}$$

The transformation in (57) is linear with K for $|K| \ll 1$ and approaches infinity as K approaches 1. The reason for using this measure is to emphasize the fact that a change in K near $|K| = 1$ has a much larger effect on the spectrum than the same change near $K = 0$. (Also, previous experience has shown a positive linear correlation between the measure B and the mean-squared value between the two spectra. For values of $B < 0.1$, an approximate value of mean-squared spectral difference in dB can be obtained by multiplying B by 25 [19].)

Table I gives the parameters of four AR(3) processes that were used in our simulations. Shown in Table I are the reflection coefficients, the normalized prediction error given by

$$V_p = \frac{\sigma_p^2}{R(0)} = \prod_{i=1}^{p} (1 - K_i^2) \tag{58}$$

TABLE I
PARAMETER VALUES FOR FOUR AR(3) PROCESSES USED IN THE SIMULATION. SHOWN FOR EACH PROCESS ARE THE THREE REFLECTION COEFFICIENTS, THE NORMALIZED PREDICTION ERROR V, AND THE DYNAMIC RANGE OF THE CORRESPONDING ALL-POLE SPECTRUM IN dB.

PROCESS	K_1	K_2	K_3	V	D(dB)
A	0	0	0	1.0	0
B	0.57	0.57	0.57	0.3077	17
C	0.83	0.83	0.83	0.0301	34
D	0.92	0.92	0.92	0.0036	42

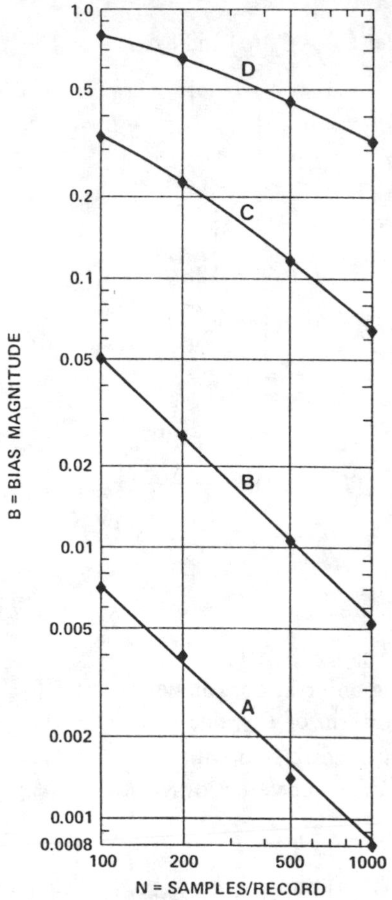

Fig. 1. Bias magnitude as defined by (56) versus the number of samples per data record. A, B, C, and D are the four processes given in Table I.

and the spectral dynamic range D of the corresponding all-pole spectrum. The relation between V and D has been discussed elsewhere [1], [16]; D generally increases as V decreases.

Fig. 1 shows log-log plots of B in (56) as a function of N for the four processes. N was varied between 100 and 1000 samples per data record. The results show that the estimates are clearly biased for finite data records but that the bias decreases as N increases. As N goes to infinity, the bias would be expected to go to zero. We also see from Fig. 1 that, for a given N, the bias generally increases with increased spectral dynamic range D or with decreased normalized prediction error V. For values of $B < 0.1$ and some given process (process A or B in Fig. 1), the bias decreases as $1/N$.

As far as we know, there are no known theoretical results that explain the particular experimental results for finite data records shown in Fig. 1. This would be a good area for further research. Also, the results of Fig. 1 were obtained using one particular method of estimating the K's. It would be interesting to repeat the same computations for a number of other well-known estimates.

C. Covariance Estimates

The asymptotic covariance of reflection coefficient estimates is given by (53) for $p = 3$. We note that the covariance decreases proportionally to $1/N$ as N increases. Also, the actual values of the covariance elements depend on the specific values of K_i. In particular, it is clear from (53) that the sign of K_2 can affect the covariance values substantially, especially the variance of the K_1 estimate, which goes to infinity as K_2 approaches -1. Also, the covariances between the K_1 and K_2 estimates become unbounded as the magnitude of K_3 approaches 1. We ran a number of simulations which verified the correctness of the general expressions for $C_{\hat{K}}$.

Table II gives the results for two examples, with the only difference between the two examples being the sign of K_2. Each example gives the true K vector, the estimate of the mean with $N = 1000$, the asymptotic covariance expressed as $NC_{\hat{K}}$, and the estimate of the covariance also multiplied by N. One can see the general agreement between the theoretical and experimental results.

As with the bias results of the previous section, the experimental results differ more from the theoretical expressions as N decreases. The same is true, generally, as the normalized prediction error V decreases (or the spectral dynamic range increases).

VI. Conclusion

The asymptotic statistics (as $N \to \infty$) of the estimated reflection coefficients are Gaussian, with a mean equal to the true reflection coefficients and a covariance matrix given by the Cramer-Rao bound. We have shown that all the common methods of estimating the reflection coefficients have the same asymptotic PDF. Also, we presented an efficient recursive procedure for estimating the covariance matrix. Using this procedure, the asymptotic covariance matrix has been compared to the true covariance matrix, which was found by employing a computer simulation. The simulation results indicate that the asymptotic covariance matrix is applicable for finite data records if the normalized prediction error variance V_p is close to 1 or the spectral dynamic range is small.

Appendix A
Recursive Computation of Covariance Matrix

From (39), $C_{\hat{K}}(n) = \sigma_n^2/N\, A_n B_n E_n^{-1} B_n^T A_n^T$. We first find A_n given A_{n-1}, where A_n was defined by (38). Let $L_{n-1} = A_{n-1}^{-1}$. Note that A_{n-1}^{-1} exists, since the transformation from $\boldsymbol{a}$ to K is 1-1 and onto [10]. But

$$[L_{n-1}]_{ij} = \frac{\partial a_{n-1}(i)}{\partial K_j}.$$

From (14), we have that

$$a_n(i) = a_{n-1}(i) + K_n a_{n-1}(n-i) \qquad i = 1, 2, \cdots, n$$

where $a_{n-1}(0) = 1$ and $a_{n-1}(n) = 0$.

$$[L_n]_{ij} = \frac{\partial a_n(i)}{\partial K_j} = \frac{\partial a_{n-1}(i)}{\partial K_j} + K_n \frac{\partial a_{n-1}(n-i)}{\partial K_j}$$

$$+ \frac{\partial K_n}{\partial K_j} a_{n-1}(n-i) \qquad i, j = 1, 2, \cdots, n \tag{A1}$$

$$[L_n]_{ij} = \frac{\partial a_{n-1}(i)}{\partial K_j} + K_n \frac{\partial a_{n-1}(n-i)}{\partial K_j} \qquad i, j = 1, 2, \cdots, n-1$$

$$= [L_{n-1}]_{ij} + K_n [L_{n-1}]_{n-i,j} \tag{A2}$$

since

$$\frac{\partial K_n}{\partial K_j} = 0 \qquad \text{for } j = 1, 2, \cdots, n-1.$$

For $i = n$, $j = 1, 2, \cdots, n-1$, from (A1)

$$[L_n]_{ij} = 0 \quad \text{since } a_{n-1}(n) = 0 \text{ and } a_{n-1}(0) = 1.$$

For $j = n$, $i = 1, 2, \cdots, n-1$, from (A1), $[L_n]_{ij} = a_{n-1}(n-i)$, since $a_{n-1}(i)$ does not depend upon K_n. Finally, for $i = j = n$

$$[L_n]_{ij} = 1.$$

Therefore, we can partition L_n as

$$L_n = \begin{bmatrix} (I + K_n J) L_{n-1} & J a_{n-1} \\ 0^T & 1 \end{bmatrix} \tag{A3}$$

To find $A_n = L_n^{-1}$, we use the inversion formula for a partitioned matrix

$$M = \begin{bmatrix} B & C \\ D & E \end{bmatrix}$$

$$M^{-1} = \begin{bmatrix} (B - CE^{-1}D)^{-1} & -B^{-1}C(E - DB^{-1}C)^{-1} \\ -E^{-1}D(B - CE^{-1}D)^{-1} & (E - DB^{-1}C)^{-1} \end{bmatrix}. \tag{A4}$$

It is assumed that E^{-1}, B^{-1} exist. But $E = 1$, which clearly has an inverse, and $B = (I + K_n J) L_{n-1}$, which is invertible if $(I + K_n J)$ is invertible. The latter will have an inverse if $|K_n| \neq 1$, since

$$(I + K_n J)^{-1} = \frac{I - K_n J}{1 - K_n^2}.$$

Then, from (A3) and (A4), and noting that $L_{n-1}^{-1} = A_{n-1}$,

TABLE II
Theoretical and Experimental Values of the Reflection Coefficients and Their Covariances for Two AR(3) Processes. The Two Processes Differ by the Sign of K_2

	THEORETICAL VALUES			EXPERIMENTAL VALUES		
$K_1\ K_2\ K_3$	0.570	0.570	0.570	0.569	0.568	0.564
$NC_{\hat{K}}$	0.541	−0.490	0.000	0.523	−0.479	−0.006
	−0.490	−0.675	0.000	−0.479	0.689	0.028
	0.000	0.000	0.675	−0.006	0.028	0.665

	THEORETICAL VALUES			EXPERIMENTAL VALUES		
$K_1\ K_2\ K_3$	0.570	−0.570	0.570	0.557	−0.563	0.563
$NC_{\hat{K}}$	7.210	−1.790	0.000	6.988	−1.858	−0.151
	−1.790	0.675	0.000	−1.858	0.716	0.016
	0.000	0.000	0.675	−0.151	0.016	0.692

$$A_n = L_n^{-1} = \begin{bmatrix} \frac{1}{1-K_n^2} A_{n-1}(I - K_n J) & -\frac{1}{1-K_n^2} A_{n-1}(I-K_nJ)Ja_{n-1} \\ 0^T & 1 \end{bmatrix}. \tag{A5}$$

Now, from (35) and (36) we have

$$B_n = \begin{bmatrix} B_{n-1} & Ja_{n-1} \\ 0^T & 1 \end{bmatrix} \qquad E_n^{-1} = \begin{bmatrix} E_{n-1}^{-1} & 0 \\ 0^T & 1/\sigma_{n-1}^2 \end{bmatrix}$$

and from (16)

$$a_{n-1} = \begin{bmatrix} (I + K_{n-1}J)a_{n-2} \\ K_{n-1} \end{bmatrix}$$

so that

$$A_nB_n = \begin{bmatrix} \frac{1}{1-K_n^2} A_{n-1}(I-K_nJ)B_{n-1} & 0 \\ 0^T & 1 \end{bmatrix}.$$

Let $D_{n-1} = A_{n-1}(I - K_nJ)B_{n-1}/(1-K_n^2)$

$$C_{\hat{K}}(n) = \frac{\sigma_n^2}{N} \begin{bmatrix} D_{n-1} & 0 \\ 0^T & 1 \end{bmatrix} \begin{bmatrix} E_{n-1}^{-1} & 0 \\ 0^T & 1/\sigma_{n-1}^2 \end{bmatrix} \begin{bmatrix} D_{n-1}^T & 0^T \\ 0 & 1 \end{bmatrix}$$

$$= \frac{\sigma_n^2}{N} \begin{bmatrix} D_{n-1}E_{n-1}^{-1}D_{n-1}^T & 0 \\ 0^T & 1/\sigma_{n-1}^2 \end{bmatrix} \tag{A6}$$

or finally

$$C_{\hat{K}}(n) = \frac{1}{N} \begin{bmatrix} \sigma_n^2 D_{n-1}E_{n-1}^{-1}D_{n-1}^T & 0 \\ 0^T & 1-K_n^2 \end{bmatrix}$$

which is (41). To begin the recursion, note that

$$A_1 = \frac{\partial K_1}{\partial a_1(1)} = 1$$

$$B_1 = 1$$

$$E_1 = \sigma_0^2 = R_x(0).$$

$R_x(0)$ can be set equal to 1, since from (34), one can show that $C_{\hat{K}}$ is independent of $R_x(0)$.

Appendix B

Recursive Computation of Covariance Matrix Determinant

From (41)

$$C_{\hat{K}}(n) = \frac{1}{N} \begin{bmatrix} \sigma_n^2 D_{n-1}E_{n-1}^{-1}D_{n-1}^T & 0 \\ 0^T & 1-K_n^2 \end{bmatrix}$$

$$|C_{\hat{K}}(n)| = \frac{\sigma_n^{2(n-1)}}{N^n}(1-K_n^2)\frac{|D_{n-1}|^2}{|E_{n-1}|}. \tag{B1}$$

But

$$|E_{n-1}| = \sigma_{n-2}^2|E_{n-2}|$$

$$|D_{n-1}| = \left|\frac{1}{1-K_n^2}A_{n-1}(I_{n-1} - K_nJ_{n-1})B_{n-1}\right|$$

$$= \left(\frac{1}{1-K_n^2}\right)^{n-1} |A_{n-1}|\,|B_{n-1}|\,|I_{n-1} - K_nJ_{n-1}|.$$

From (36), $|B_{n-1}| = |B_{n-2}| = 1$, and from (A5)

$$|A_{n-1}| = \left|\frac{1}{1-K_{n-1}^2}A_{n-2}(I_{n-2} - K_{n-1}J_{n-2})\right|$$

$$= \left(\frac{1}{1-K_{n-1}^2}\right)^{n-2} |I_{n-2} - K_{n-1}J_{n-2}|\,|A_{n-2}|$$

$$\Rightarrow |D_{n-1}| = \left(\frac{1}{1-K_n^2}\right)^{n-1}\left(\frac{1}{1-K_{n-1}^2}\right)^{n-2} \cdot |I_{n-1} - K_nJ_{n-1}|\,|I_{n-2} - K_{n-1}J_{n-2}| \cdot |A_{n-2}|\,|B_{n-2}|$$

$$= \left(\frac{1}{1-K_n^2}\right)^{n-1} |I_{n-1} - K_nJ_{n-1}|\,|D_{n-2}|. \tag{B2}$$

From (B1) and (B2)

$$|C_{\hat{K}}(n)| = \frac{\sigma_n^{2(n-1)}}{N^n}(1-K_n^2)\left(\frac{1}{1-K_n^2}\right)^{2n-2} \cdot |I_{n-1} - K_nJ_{n-1}|^2\,|D_{n-1}|^2/(\sigma_{n-2}^2|E_{n-2}|)$$

$$= \frac{\sigma_n^{2(n-1)}(1-K_n^2)}{\sigma_{n-2}^2N^n}\left(\frac{1}{1-K_n^2}\right)^{2n-2} \cdot |I_{n-1} - K_nJ_{n-1}|^2 \cdot \frac{|C_{\hat{K}}(n-1)|}{\frac{\sigma_{n-1}^{2(n-2)}}{N^{n-1}}(1-K_{n-1}^2)}$$

$$= \frac{1}{N}(1-K_n^2)^{-n+2}\,|I_{n-1} - K_nJ_{n-1}|^2 \cdot |C_{\hat{K}}(n-1)|$$

where use is made of the relation [see (14)]

$$\sigma_n^2 = (1-K_n^2)\,\sigma_{n-1}^2$$

but it has been shown that [10]

$$|I_{n-1} - K_nJ_{n-1}| = (1-K_n^2)^{n-1} \qquad n \text{ odd}$$
$$= (1-K_n^2)^{n-2}(1-K_n)^2 \qquad n \text{ even}.$$

Thus,

$$|C_{\hat{K}}(n)| = \frac{1}{N}(1-K_n^2)\,|C_{\hat{K}}(n-1)| \qquad n \text{ odd}$$

$$= \frac{1}{N}(1-K_n)^2\,|C_{\hat{K}}(n-1)| \qquad n \text{ even} \tag{B3}$$

or, by letting

$$g_{n-1} = 1 \qquad n \text{ odd}$$

$$= \frac{1-K_n}{1+K_n} \qquad n \text{ even}$$

$$|C_{\hat{K}}(n)| = \frac{1}{N}(1-K_n^2)\,g_{n-1}\,|C_{\hat{K}}(n-1)|$$

which is (49). Also, (48) easily follows from (B3).

Acknowledgment

The authors thank the reviewers for their constructive suggestions to improve the presentation of this paper.

References

[1] J. Makhoul, "Linear prediction: A tutorial review," *Proc. IEEE*, vol. 63, pp. 561–580, Apr. 1975.

[2] D. E. Bowyer *et al.*, "Adaptive clutter filtering using autoregressive spectral estimation," *IEEE Trans. Aerosp. Electron. Syst.*, vol. AES-15, pp. 538–546, July 1979.

[3] S. M. Kay, "Autoregressive spectral analysis of narrowband processes in white noise with application to sonar signals," Ph.D. dissertation, Georgia Inst. Technol., Atlanta, Mar. 1980.

[4] M. Hsu, "Maximum entropy principle and its application to spectral analysis and image reconstruction," Ph.D. dissertation, Ohio State Univ., Columbus, 1975.

[5] W. Gersch, "Spectral analysis of EEG's by autoregressive decomposition of time series," *Math. Biosci.*, vol. 7, pp. 205–222, 1970.

[6] F. Itakura and S. Saito, "Digital filtering techniques for speech analysis and synthesis," presented at the 7th Int. Conf. Acoust., Budapest, Hungary, 1971.

[7] J. P. Burg, "Maximum entropy spectral analysis," Ph.D. dissertation, Stanford Univ., Stanford, CA, May 1975.

[8] B. S. Atal and M. R. Schroeder, "Predictive coding of speech signals and subjective error criteria," *IEEE Trans. Acoust., Speech, Signal Processing*, vol. ASSP-27, pp. 247–254, June 1979.

[9] H. B. Mann and A. Wald, "On the statistical treatment of stochastic difference equations," *Econometrics*, vol. 11, pp. 173–220, July 1943.

[10] O. Barndorf-Nielsen and G. Schou, "On the parametrization of autoregressive models by partial autocorrelations," *J. Multivariate Anal.*, vol. 3, pp. 408–419, 1973.

[11] M. Kendall and A. Stuart, *The Advanced Theory of Statistics.* New York: Macmillan, 1979.

[12] H. L. Van Trees, *Detection Estimation and Modulation Theory*, Part I. New York: Wiley, 1968.

[13] S. Zacks, *The Theory of Statistical Inference.* New York: Wiley, 1971.

[14] A. H. Nuttall, "Spectral analysis of a univariate process with bad data points, via maximum entropy and linear predictive techniques," Naval Underwater Syst. Center, New London, CT, Tech. Rep. 5303, Mar. 16, 1976.

[15] T. J. Ulrych and R. W. Clayton, "Time series modeling and maximum entropy," *Phys. Earth, Planetary Interiors*, vol. 12, pp. 188–200, Aug. 1976.

[16] J. Markel and A. Gray, *Linear Prediction of Speech.* New York: Springer-Verlag, 1976.

[17] J. Makhoul, "Stable and efficient lattice methods of linear prediction," *IEEE Trans. Acoust., Speech, Signal Processing*, vol. ASSP-25, pp. 423–428, Oct. 1977.

[18] G. M. Jenkins and D. G. Watts, *Spectral Analysis and its Applications.* San Francisco, CA: Holden-Day, 1968.

[19] R. Viswanathan and J. Makhoul, "Quantization properties of transmission parameters in linear predictive systems," *IEEE Trans. Acoust., Speech, Signal Processing*, vol. ASSP-23, pp. 309–320, June 1975.

Statistical Properties of AR Spectral Analysis

HIDEAKI SAKAI, MEMBER, IEEE

***Abstract*—This paper investigates several statistical properties of the autoregressive (AR) spectral analysis method by using the periodogram technique recently devised by the author. When the data are made up of several sinusoids contaminated by stationary noise, the asymptotic variances of the AR spectral estimator are given. It is shown numerically that the behavior of the variances is similar to Kromer and Berk's earlier result for stationary processes.**

As for frequency measurement accuracies, the statistical fluctuation of a peak frequency is analyzed under the assumption that the deviation from the true peak frequency is small. It is shown numerically that the resulting variance is inversely proportional to the data length and the square of the signal-to-noise ratio (SNR).

I. Introduction

In recent years, the autoregressive (AR) spectral analysis method proposed by Burg [1] and Parzen [2] has received much attention and has been frequently used in many fields of science. This is because the AR method has been recognized to have much greater resolution than the traditional methods, such as the lag window technique [3]. As for the theoretical aspects of this method, Lacoss [4] studied the case where time series are made up of several sinusoids and white noise. But the statistical properties of this method have not been well studied. Recently, Baggeroer [5] introduced a particular assumption about the data structure which enables us to apply the theory of multivariate analysis to this AR (MEM) method and derived many interesting statistical properties. In a sense, his result can be considered as a generalization of Akaike's earlier work [6] where the data are assumed to be generated by a pure finite-order autoregressive process.

In this paper, by using the periodogram technique recently devised by the author (Sakai *et al.* [7]), the statistical analysis of the AR method applied to the data consisting of several sinusoids plus stationary noise is performed with emphasis on its asymptotics in a different way from [5]. At first, we shall show that the AR spectral estimator (ARSPE) is asymptotically equivalent to a smoothed periodogram with a data-dependent "spectral window." From this expression, the vari-

Manuscript received May 22, 1978; revised November 6, 1978.

The author was with the Department of Information Science and Systems Engineering, University of Tokushima, Tokushima, Japan. He is now with the Department of Applied Mathematics and Physics, Kyoto University, Kyoto, Japan.

Reprinted from *IEEE Trans. Acoust., Speech, Signal Processing*, vol. ASSP-27, pp. 402–409, Aug. 1979.

ances of the estimator can be determined and their behavior turns out to be similar to Kromer and Berk's earlier results for stationary processes [8], [9]. Next, we consider the frequency measurement aspect of the AR method. Since it is reasonable to expect that the AR spectrum has several sharp peaks at the corresponding sinusoidal frequencies with sufficiently high order, the frequency at which the sample AR spectrum gives a sharp peak can be used as a frequency estimate. Under the assumption that the estimation error is small, the asymptotic variance of that estimator is derived and compared with that of the conventional Fourier method.

II. An Alternative Expression for the ARSPE

We assume that the time series under consideration consist of p sinusoids and a zero-mean stationary Gaussian noise n_t, that is, $\{x_t\}$ is expressed as

$$x_t = \sum_{i=1}^{p} [A_i e^{j\omega_i t} + A_i^* e^{-j\omega_i t}] + n_t \tag{1}$$

where "*" denotes the complex conjugate. Hence, the autocovariance function r_k is

$$r_k = \sum_{i=1}^{p} 2|A_i|^2 \cos(\omega_i k) + q_k \tag{2}$$

where $q_k = E[n_t n_{t+k}]$. In (2) r_k is defined by the limit of $\hat{r}_k$ below as $N \to \infty$, and q_k is written as above because of the ergodicity of $\{n_t\}$. As an alternative definition, we may assume that A_i is expressed by $B_i e^{j\phi_i}$ where ϕ_i is a random phase angle distributed uniformly over $[0, 2\pi]$. Then r_k can be defined as the ensemble average $E[x_t x_{t+k}]$. However, these two definitions lead the same autocovariance function and, as is easily seen later, the same statistical properties of the AR method, so that we assume the first definition to be the following. Given a set of data $\{x_1, x_2, \cdots, x_N\}$, it is well known that the AR estimate is calculated by the following steps. First, estimate r_k by

$$\hat{r}_k = \frac{1}{N} \sum_{t=1}^{N-|k|} x_t x_{t+|k|}. \tag{3}$$

Then, substituting these $\hat{r}_k$ into the Yule-Walker equations

$$Ra = r \tag{4}$$

where $(R)_{i,k} = r_{i-k}$, $(r)_i = r_i$, and $a^T = (a_1, a_2, \cdots, a_m)$, we obtain

$$\hat{R}\hat{a} = \hat{r} \tag{5}$$

where $\hat{R}$, $\hat{r}$, and $\hat{a}$ are defined similarly. From the solution of (5), we calculate the estimate of the "residual"

$$\sigma^2 = r_0 - a_1 r_1 - \cdots - a_m r_m \tag{6}$$

by

$$\hat{\sigma}^2 = \hat{r}_0 - \hat{a}_1 \hat{r}_1 - \cdots - \hat{a}_m \hat{r}_m. \tag{7}$$

Finally, we form the estimate of the AR spectrum

$$f(s) = \frac{\sigma^2}{2\pi B(s) B(-s)} \tag{8}$$

by

$$\hat{f}(s) = \frac{\hat{\sigma}^2}{2\pi \hat{B}(s) \hat{B}(-s)} \tag{9}$$

where

$$B(s) = \sum_{i=0}^{m} (-a_i) e^{-jis}, \hat{B}(s) = \sum_{i=0}^{m} (-\hat{a}_i) e^{-jis} \tag{10}$$

with $a_0 = \hat{a}_0 = -1$. For later convenience, unlike the electrical engineering convention, we divide the usual spectrum by 2π.

Although there are possible choices for $\hat{r}_k$, we prefer (3) because of its relevance to the periodogram. In what follows it will be shown that if the data length is sufficiently long, in terms of the periodogram

$$I_N(s) = \frac{1}{2\pi N} \left| \sum_{t=1}^{N} x_t e^{-jts} \right|^2, \tag{11}$$

the ARSPE is expressed as

$$\hat{f}(s) \simeq \int_{-\pi}^{\pi} G(s, t) I_N(t)\, dt \tag{12}$$

where "$\simeq$" indicates that both sides are asymptotically equivalent. Thus, the ARSPE can be viewed as a smoothed periodogram with a data-dependent "spectral window" $G(s, t)$. Obviously, this indicates the data adaptivity of the AR method which was mentioned but not explicitly shown by Lacoss [4].

The derivation of (12) can be performed in a similar way as in [7], but we include it here solely for completeness. Since $\hat{r}_k$ and $\hat{a}$ are consistent estimators, the errors $\Delta a = \hat{a} - a$, $\Delta r = \hat{r} - r$, and $\Delta R = \hat{R} - R$ can be assumed to be small. By substituting these into (5), neglecting the second-order terms concerning errors, and noting (4), we have

$$R \Delta a \simeq \Delta r - \Delta R\, a = \hat{r} - \hat{R} a. \tag{13}$$

On the other hand, it is well known that

$$\hat{r}_k = \int_{-\pi}^{\pi} I_N(s) e^{jks}\, ds. \tag{14}$$

Substituting this into (13) gives

$$(R\Delta a)_k \simeq \int_{-\pi}^{\pi} B(s) I_N(s) e^{jks}\, ds. \tag{15}$$

In a similar way, from (6), (7), (14), and (15), $\Delta\sigma^2 = \hat{\sigma}^2 - \sigma^2$ is written as

$$\begin{aligned} \Delta\sigma^2 &\simeq -\Delta a^T r + \hat{r}_0 - r_0 - a^T(\hat{r} - r) \\ &= -(R\Delta a)^T R^{-1} r + \int_{-\pi}^{\pi} B(s) I_N(s)\, ds - \sigma^2 \\ &= \int_{-\pi}^{\pi} B(s) B(-s) I_N(s)\, ds - \sigma^2. \end{aligned} \tag{16}$$

Also, approximating the estimation error $\Delta f(s) = \hat{f}(s) - f(s)$ as a linear combination of $\Delta\sigma^2$ and Δa, we obtain

$$\Delta f(s) \simeq f(s)\left[\frac{\Delta\sigma^2}{\sigma^2} + H^T(s)\,\Delta a\right] \tag{17}$$

where $H^T(s) = (h_1(s), \cdots, h_m(s))$ with $h_k(s) = 2\,\mathrm{Re}\,[e^{-jks}/B(s)]$. By defining $R^{-1}H(s) = M(s) = [M_1(s), \cdots, M_m(s)]^T$ and substituting (15) and (16) into (17), we have

$$\Delta f(s) \simeq \int_{-\pi}^{\pi} f(s)\,B(t)\,[K(s,t) + \sigma^{-2}B(-t)]\,I_N(t)\,dt - f(s) \tag{18}$$

with

$$K(s,t) = \sum_{k=1}^{m} M_k(s)\,e^{jkt}. \tag{19}$$

Consequently, we have shown (12) with the smoothing kernel $G(s,t)$ given by

$$G(s,t) = f(s)\,B(t)\,[K(s,t) + \sigma^{-2}B(-t)] \triangleq f(s)\,g(s,t). \tag{20}$$

III. Statistical Properties of the ARSPE

In this section we first derive the asymptotic covariance between $\hat{f}(\mu)$ and $\hat{f}(\nu)$. From (18), it follows that

$$E\left[\frac{\Delta f(\mu)}{f(\mu)}\cdot\frac{\Delta f(\nu)}{f(\nu)}\right] \simeq \int_{-\pi}^{\pi}\int_{-\pi}^{\pi} g(\mu,s)\,g(\nu,t) \cdot \mathrm{cov}\,[I_N(s), I_N(t)]\,ds\,dt + \left\{\int_{-\pi}^{\pi} g(\mu,s)\,E[I_N(s)]\,ds - 1\right\} \cdot \left\{\int_{-\pi}^{\pi} g(\nu,t)\,E[I_N(t)]\,dt - 1\right\}. \tag{21}$$

From Appendix I, the second term of (21) is of order N^{-2} and can be neglected. Applying the result of Appendix II, the remaining term becomes

$$\frac{2\pi}{N}\left\{\sum_{k=1}^{p} 2|A_k|^2[g(\mu,\omega_k)\,g(\nu,-\omega_k) + g(\mu,-\omega_k)\,g(\nu,\omega_k) + g(\mu,\omega_k)\,g(\nu,\omega_k) + g(\mu,-\omega_k)\,g(\nu,-\omega_k)]\;Q(\omega_k) + \int_{-\pi}^{\pi}[g(\mu,s)\,g(\nu,s) + g(\mu,s)\,g(\nu,-s)]\;Q(s)^2\,ds\right\}. \tag{22}$$

At present, the expression (22) does not permit clear-cut analytic interpretations, so that we must resort to numerical calculations to examine its behavior. This will be done in the next section.

Next, we consider the statistical fluctuation of the estimator $\hat{\omega}$ for the frequency ω at which $f(s)$ has a peak value. Since $f(s)$ and $\hat{f}(s)$ take extreme values at ω and $\hat{\omega}$, respectively, it follows that

$$\left.\frac{d}{ds}B(s)\,B(-s)\right|_{s=\omega} = \left.\frac{d}{ds}\hat{B}(s)\,\hat{B}(-s)\right|_{s=\hat{\omega}} = 0$$

or

$$B'(\omega)\,B(-\omega) + B(\omega)\,B'(-\omega) = 0,$$
$$\hat{B}'(\hat{\omega})\,\hat{B}(-\hat{\omega}) + \hat{B}(\hat{\omega})\,\hat{B}'(-\hat{\omega}) = 0 \tag{23}$$

where "$'$" denotes the differential operation with respect to s. As above, the error $\Delta\omega = \hat{\omega} - \omega$ can be assumed to be small for sufficiently large N, so that the following approximations are valid:

$$\hat{B}'(\hat{\omega}) \simeq \hat{B}'(\omega) + \hat{B}''(\omega)\,\Delta\omega,$$
$$\hat{B}(-\hat{\omega}) \simeq \hat{B}(-\omega) + \hat{B}'(-\omega)\,\Delta\omega.$$

Also, from (10)

$$\hat{B}(s) = B(s) - \Delta a^T\,V(s)$$

with $V^T(s) = [e^{-js}, \cdots, e^{-jms}]$. Substituting these into (23) and neglecting a higher order term like $\Delta a^T \Delta\omega$, we have

$$\Delta\omega \simeq \Delta a^T\,\frac{V(-\omega)\,B'(\omega) + V(\omega)\,B'(-\omega) + V'(\omega)\,B(\omega) + V'(-\omega)\,B(\omega)}{2B'(-\omega)\,B'(\omega) + B''(\omega)\,B(-\omega) + B''(-\omega)\,B(\omega)} \triangleq \Delta a^T\,C(\omega). \tag{24}$$

Defining $R^{-1}C(\omega) = [z_1(\omega), \cdots, z_m(\omega)]$ and using (15), give

$$\Delta\omega \simeq \int_{-\pi}^{\pi} B(s)\,Z(\omega,s)\,I_N(s)\,ds$$

with

$$Z(\omega,s) = \sum_{k=1}^{m} z_k(\omega)\,e^{jks}.$$

Hence, the variance $E[(\Delta\omega)^2]$ is calculated by applying the formula (A-27) in Appendix II with

$$F(s,t) = B(s)\,Z(\omega,s)\,B(t)\,Z(\omega,t). \tag{25}$$

It should be noted that the variance is of order N^{-1}.

IV. Numerical Examples

Although we have derived the fundamental formulas for evaluating the statistical properties, they do not allow simple analytical interpretations. Thus, numerical studies were carried out for the case where the data are made up of a single sinusoid and white noise, that is,

$$x_t = A\,e^{j\omega t} + A^*\,e^{-j\omega t} + n_t.$$

It is obvious from the above development that (22) and $N \cdot E[(\Delta\omega)^2]$ depend explicitly on the SNR $P = 2|A|^2/q_0$.

TABLE I
ASYMPTOTIC AND EMPIRICAL VARIANCES OF THE ARSPE FOR A SINUSOID PLUS WHITE NOISE

$S = \pi k/20$	Asymptotic variances		Empirical variances
	Total	Four terms	
k = 0	62.452	0.043	67.875
1	23.932	0.006	24.191
2	29.643	0.034	35.241
3	33.021	0.302	40.570
4	18.616	0.048	17.825
→ 5	41.006	29.324	45.406
6	21.211	0.003	23.183
7	30.238	0.547	30.770
8	32.042	0.177	32.026
9	30.983	0.200	31.729
10	30.151	0.043	31.948
11	30.397	0.089	37.778
⋮	⋮	⋮	⋮
17	33.267	0.017	30.031
18	29.407	0.001	31.636
19	25.062	0.023	23.515
20	61.885	0.043	58.476

Thus, without loss of generality, we assume that $2|A|^2 = P$ and $q_k = \delta_{k,0}(Q(s) = 1/2\pi)$. The first column of Table I shows the numerical values of $N \cdot E[(\Delta f(\nu)/f(\nu))^2]$ for $\omega = \pi/4$, $m = 15$, and $P = -10$ dB. The second column shows the contributions due to the first four terms of (22), i.e., the summation part. It can be seen that except at $\nu = \omega$, these are negligibly small. To see the validity of the asymptotic expression (22) in the third column, we also present the simulational results which were obtained by averaging the squares of the estimation error $\Delta f(s)$ over 100 different data sets each of $N = 1000$ length. The agreements are fairly good. From this table it can be inferred that, except at frequencies near ω, the asymptotic variances are roughly approximated as

$$N \cdot E\left[\left(\frac{\Delta f(\nu)}{f(\nu)}\right)^2\right] \simeq \begin{cases} 2m & \text{for } \nu \neq 0, \pi \\ 4m & \text{for } \nu = 0, \pi \end{cases}. \tag{26}$$

Although we do not present detailed numerical tables here, the relation (26) holds for other cases such as multiple sinusoids plus nonwhite noise. However, the analytic proof of (26) is not known at present. It should be pointed out that Kromer [8] and Berk [9] have obtained the same result with (26) for the AR spectral estimate applied to a fairly wide class of stationary processes. Thus, our result can be considered as a generalization of [8], [9].

Now let us examine the variance of the main peak frequency fluctuation $\Delta\omega$. Here $\Delta\omega$ is interpreted as the difference between the main peak frequencies of $f(s)$ and $\hat{f}(s)$. To see the explicit dependence of $N \cdot E[(\Delta\omega)^2]$ on P, we briefly analyze the case where $P \ll 1$. From (2) and (4), a is of order P. Thus, $B(s) = 1 - 0(P)$, $B'(s) = 0(P)$, and $B''(s) = 0(P)$ where

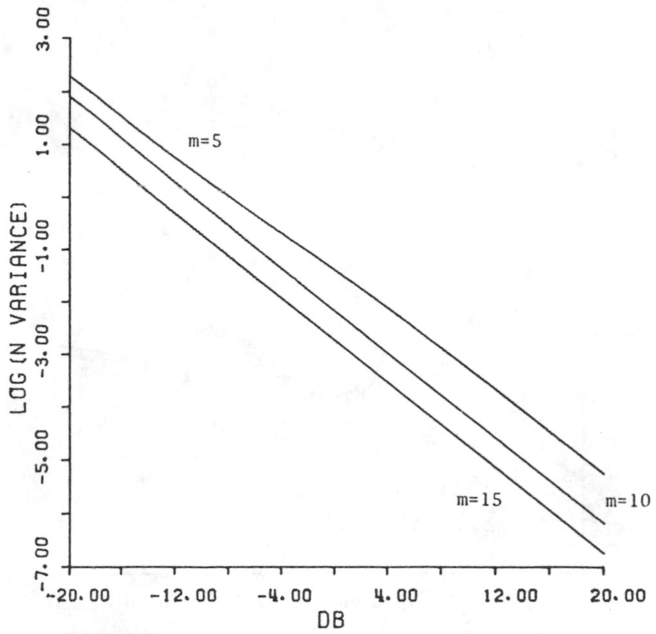

Fig. 1. The asymptotic variances of the AR frequency estimator with various autoregression orders.

$0(P)$ means that this is of order P. But from (24), $C(\omega)$ becomes $0(P^{-1})$, and from (25), $F(s,t) = 0(P^{-2})$. Consequently, we obtain

$$E[(\Delta\omega)^2] \simeq \alpha \frac{P^{-2}}{N} \quad \text{for } P \ll 1 \tag{27}$$

where α is a constant depending on ω and m. Since it is unknown whether (27) holds for intermediate and large P, the numerical calculations were performed. Fig. 1 shows the values of $N \cdot E[(\Delta\omega)^2]$ with $\omega = \pi/4$ and $m = 5, 10, 15$ versus P. The vertical axis is logarithmically scaled. Since the graphs for $m = 10$, 15 are almost straight lines and their inclinations are almost -0.2, it can be deduced that (27) approximately holds at all the range of P. The next thing that can be seen from Fig. 1 is the dependency on the length of autoregression, i.e., m. The gains obtained by increasing m by 5 are about 7 to 8 dB. At first sight, this observation seems to be against our intuition since, in many statistical problems, increasing the number of parameters causes larger statistical variations in final estimators. To explain the above fact, the shapes of $f(s)$ for $m = 5$, 15 are plotted in Figs. 2 and 3 where solid lines indicate $f(s)$ and square marks show the sample spectra $\hat{f}(s)$.

It is intuitively clear that the more the shape is sharp, or in other words, the "bandwidth" of $f(s)$ is narrow, the more the estimation of ω is easy. In this case, the ultimate performance is more affected by the narrower bandwidth than the larger statistical variation.

If the condition $Pm \gg 1$ is met, as shown in Lacoss [4], $f(s)$ has a mean peak at $s = \omega$. Hence, the above-mentioned $\hat{\omega}$ becomes an unbiased estimator for ω. From Figs. 2 and 3, this unbiasedness holds for $P = -10$ dB, $m = 15$, but not for $P = -10$ dB, $m = 5$.

Lastly, we briefly discuss the implication of the assumption that $\Delta\omega$ is small. From Figs. 2 and 3, we see that for

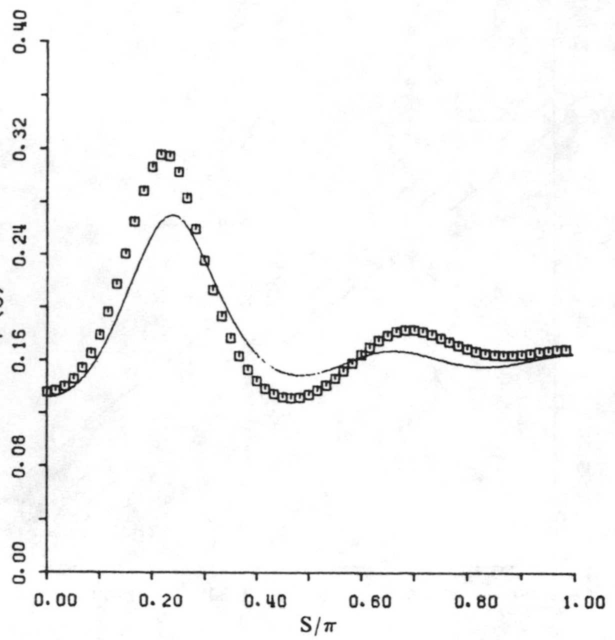

Fig. 2. True and sample AR spectra for $m = 5$, $P = -10$ dB, and $N = 1000$.

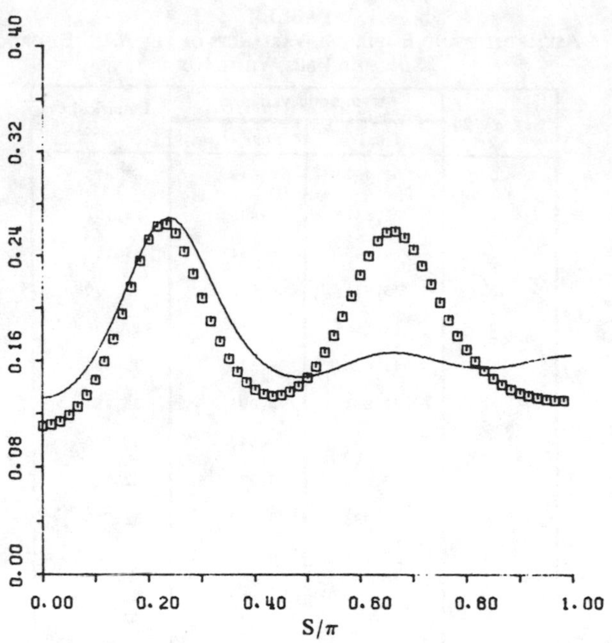

Fig. 4. An ill-behaved sample AR spectrum for $m = 5$, $p = -10$ dB, and $N = 200$.

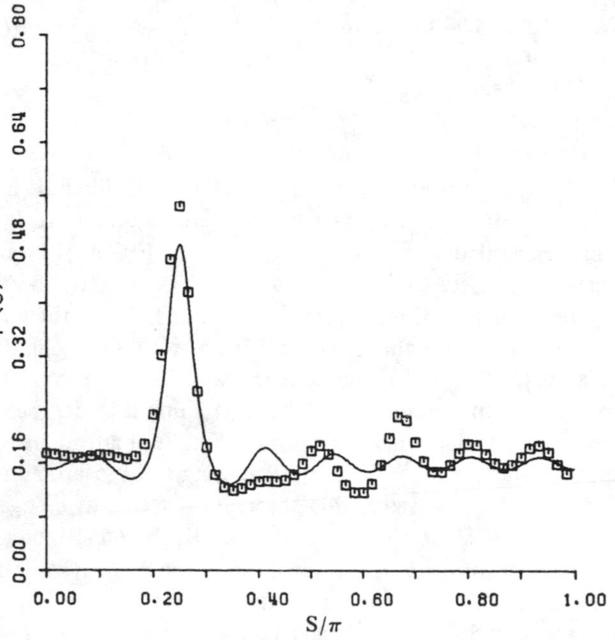

Fig. 3. True and sample AR spectra for $m = 15$, $P = -10$ dB, and $N = 1000$.

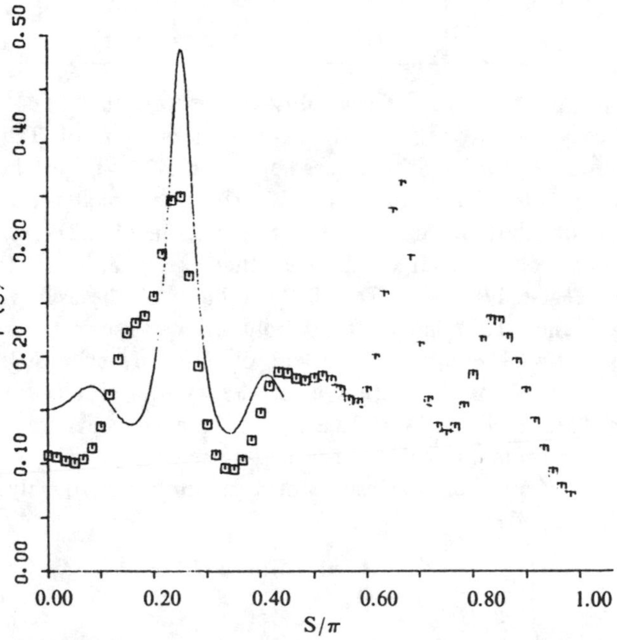

Fig. 5. An ill-behaved sample AR spectrum for $m = 15$, $P = -10$ dB, and $N = 200$.

$N = 1000$, the sample spectra are close to the true spectra $f(s)$, so the assumption is valid. However, when N is reduced to 200, the shapes of $\hat{f}(s)$ are quite different from $f(s)$, and the spurious peaks have emerged. (See Figs. 4 and 5). Hence, taking these spurious peak frequencies as $\hat{\omega}$ causes large estimation error $\Delta\omega$, thereby breaking down our "local" analysis. This phenomenon can be considered a kind of "threshold effect" inherent in any nonlinear processing of data. Thus, (27) is valid only for $N \gtrsim N_{\text{thr}}(P)$ where $N_{\text{thr}}(P)$ is a function of P and specifies the threshold. However, its determination seems to be quite difficult by our asymptotic analysis.

V. Discussion and Conclusion

As another frequency estimation method, Fourier analysis has been widely used. The procedure is simply finding the frequency at which the periodogram (11) takes an extreme value. If the quadrature component of (1) is available, according to Rife and Boorstyn [10], this method coincides with the maximum likelihood (ML) method. Although, as stated above, our data structure is different from [10] and the direct comparison of two methods is not feasible, here we quote their results which may contain the properties of the Fourier method for our situation.

It was shown that if the SNR is sufficiently high, or above the threshold, the Cramer-Rao bound [10, eq. (17)] is almost met by the ML frequency estimator; that is, the variance is inversely proportional to the SNR P and the cube of the data length N. It is interesting to note that the variance of the AR and the Fourier methods have different functional dependencies on P and N.

In conclusion, this paper presents two new points. The first is that Berk's result may be extended to the data containing several sinusoids. The second is that the variance of the frequency estimation is proportional to $N^{-1} \cdot P^{-2}$. However, these points are inferred numerically and must be justified analytically in the future.

Appendix I

In this Appendix we shall show

$$\int_{-\pi}^{\pi} g(\nu, s)\, E[I_N(s)]\; ds = 1 + 0(N^{-1}). \tag{A-1}$$

Noting $E[n_t] = 0$, it obviously follows from (7) that

$$E[I_N(s)] = \frac{1}{2\pi N}\left\{\left|\sum_{t=1}^{N}\sum_{i=1}^{p}(A_i e^{j\omega_i t} + A_i^* e^{-j\omega_i t})\, e^{-jts}\right|^2 + E\left|\sum_{t=1}^{N} n_t e^{-jts}\right|^2\right\}. \tag{A-2}$$

As is well known, the second term of (A-2) is written as

$$\frac{1}{2\pi}\sum_{k=-N+1}^{N-1} q_k\left(1 - \frac{|k|}{N}\right) e^{-jks} = Q(s) + 0(N^{-1}) \tag{A-3}$$

where $Q(s)$ is the spectrum of $\{n_t\}$ and is defined by $\Sigma_{k=-\infty}^{\infty} q_k \cdot \exp(-jks)/2\pi$. Introducing the Dirichlet kernel

$$D_N(s) \triangleq \sum_{t=1}^{N} e^{-jts}, \tag{A-4}$$

the first term of (A-2) becomes

$$\frac{1}{2\pi N}\left|\sum_{i=1}^{p}(A_i D_N(s - \omega_i) + A_i^* D_N(s + \omega_i))\right|^2. \tag{A-5}$$

From the theory of the Fejér kernel, we have

$$\frac{1}{2\pi N}|D_N(s \pm \omega_i)|^2 \to \delta(s \pm \omega_i) \tag{A-6}$$

as $N \to \infty$ where $\delta(\cdot)$ is Dirac's delta function. Also, using the fact

$$D_N(s) = \begin{cases} N & \text{for } s = 0, \pm 2\pi, \pm 4\pi, \cdots \\ 0(1) & \text{otherwise} \end{cases} \tag{A-7}$$

and $\omega_k - \omega_i \neq 2\pi n$ for any i, k and integer n, we obtain

$$\frac{1}{2\pi N} D_N(s - \omega_i)\, D_N(-s - \omega_k) \to 0. \tag{A-8}$$

Thus, apart from the $0(N^{-1})$ term, $E[I_N(s)]$ is given by

$$\sum_{i=1}^{p}|A_i|^2(\delta(s + \omega_i) + \delta(s - \omega_i)) + Q(s) \triangleq W(s) \tag{A-9}$$

where $W(s)$ is the spectrum of $\{x_t\}$. Apparently, this result is consistent with (2) since

$$\int_{-\pi}^{\pi} W(s)\, e^{jks}\, ds = r_k. \tag{A-10}$$

Hence, for proving (A-1) it is sufficient to show

$$\int_{-\pi}^{\pi} g(\nu, s)\, W(s)\, ds = 1. \tag{A-11}$$

Substituting (10), (19), and (20) into the left-hand side of (A-11) and noting (A-10), we have

$$\sum_{i=0}^{m}\sum_{k=1}^{m}(-a_i) M_k(\nu)\, r_{k-i} + \sigma^{-2}\sum_{i=0}^{m}\sum_{k=0}^{m}(-a_i)(-a_k)\, r_{k-i}. \tag{A-12}$$

The first term is written as $M^T(\nu)\, r + (-a^T) R \cdot M(\nu)$. But this is zero since $M^T(\nu) = H^T(\nu) R^{-1}$ and $R^{-1} r = a$. Similarly, the second term is 1 from (4) and (6). This completes the proof of (A-1).

Appendix II

In this Appendix we shall evaluate the asymptotic value of the integral

$$\int_{-\pi}^{\pi}\int_{-\pi}^{\pi} F(s, t)\, \mathrm{cov}\,[I_N(s), I_N(t)]\; ds\, dt \tag{A-13}$$

for a sufficiently smooth periodic real function $F(s, t)$. We first seek the expression for $\mathrm{cov}\,[I_N(s), I_N(t)]$. For notational simplicity, we put

$$X_1 \triangleq \frac{1}{\sqrt{2\pi N}}\sum_{k=1}^{p}[A_k D_N(s - \omega_k) + A_k^* D_N(s + \omega_k)] \tag{A-14}$$

$$Y_1 \triangleq \frac{1}{\sqrt{2\pi N}}\sum_{i=1}^{N} n_i e^{-jis} \tag{A-15}$$

and X_2, Y_2 are defined by replacing s by t in (A-14), (A-15), respectively. Thus, $I_N(s) = |X_1|^2 + 2\,\mathrm{Re}(X_1^* Y_1) + |Y_1|^2$, $I_N(t) = |X_2|^2 + 2\,\mathrm{Re}(X_2^* Y_2) + |Y_2|^2$. From the assumption that $\{n_t\}$ is Gaussian, Y_1 and Y_2 are zero-mean complex Gaussian random variables. Hence, means of their first and third products are zero, i.e., $E[X_1^* Y_1] = E[X_2^* Y_2] = E[(2 \cdot \mathrm{Re}\, X_1^* Y_1) \cdot |Y_2|^2] = E[(2 \cdot \mathrm{Re}\, X_2^* Y_2)|Y_1|^2] = 0$. Consequently, we have

$$\mathrm{cov}\,[I_N(s), I_N(t)] = E[(2 \cdot \mathrm{Re}\, X_1^* Y_1)(2 \cdot \mathrm{Re}\, X_2^* Y_2)] + \mathrm{cov}\,[|Y_1|^2, |Y_2|^2]. \tag{A-16}$$

In (A-16) the first term represents the cross effect due to the signal and the noise and is rewritten as

$$X_1^* X_2^* E[Y_1 Y_2] + X_1 X_2 E[Y_1^* Y_2^*] + X_1^* X_2 E[Y_1 Y_2^*] + X_1 X_2^* E[Y_1^* Y_2]. \tag{A-17}$$

On the other hand, it follows from [11, p. 93] that

$$E[Y_1 Y_2] = \frac{1}{N} Q(s)\, D_N(s+t) + 0(N^{-1}) \tag{A-18}$$

where $D_N(\cdot)$ is defined by (A-4). As is easily seen from the ensuing analysis, the $0(N^{-1})$ term in (A-18) does not affect the final result, so that from now on, we discard this term. Thus, (A-17) becomes

$$\frac{1}{N} Q(s)\, [X_1^* X_2^* D_N(s+t) + X_1 X_2 D_N(-s-t) + X_1^* X_2 D_N(s-t) + X_1 X_2^* D_N(-s+t)]. \tag{A-20}$$

The second term of (A-16), representing the pure effect due to the noise, is asymptotically equal to

$$\frac{1}{N^2} [|D_N(s+t)|^2 + |D_N(s-t)|^2]\, Q(s)\, Q(t). \tag{A-20}$$

(See [12, pp. 238–250] or Sakai *et al.* [7].) Using the approximation (A-6) in the integral (A-13), the contribution due to (A-20) is given by

$$\frac{2\pi}{N} \int_{-\pi}^{\pi} [F(s,s) + F(s,-s)]\, Q(s)^2\, ds. \tag{A-21}$$

Next we examine the contribution due to (A-19). It is sufficient to consider the first and third term of (A-19). Upon substituting (A-14) into these terms and performing the integration for (A-13), we encounter the following type of integrals

$$\frac{1}{2\pi N^2} \int_{-\pi}^{\pi} \int_{-\pi}^{\pi} D_N(\mu - s)\, D_N(\nu \mp t)\, D_N(s \pm t) \cdot F(s,t)\, Q(s)\, ds\, dt. \tag{A-22}$$

From the assumption that $F(s,t)$ is periodic $F(s,t)\, Q(s)$ can be expressed as the following Fourier series:

$$F(s,t)\, Q(s) = \frac{1}{2\pi} \sum_{m=-\infty}^{\infty} \sum_{n=-\infty}^{\infty} f_{m,n}\, e^{-j(sm+tn)}. \tag{A-23}$$

Substituting (A-23) and (A-4) into (A-22), we have

$$\frac{1}{(2\pi N)^2} \int_{-\pi}^{\pi} \int_{-\pi}^{\pi} \sum_{k=1}^{N} e^{-j(\mu - s)k} \sum_{k'=1}^{N} e^{-j(\nu \mp t)k'} \sum_{k''=1}^{N} e^{-j(s \pm t)k''} \sum_{m,n} f_{m,n}\, e^{-j(sm+tn)}\, ds\, dt$$

$$= \frac{1}{(2\pi N)^2} \sum_{k,k',k'',m,n} f_{m,n}\, e^{-j(\mu k + \nu k')} \cdot \int_{-\pi}^{\pi} e^{-j(-k+k''+m)s}\, ds \int_{-\pi}^{\pi} e^{-j(\mp k' \pm k'' + n)t}\, dt$$

$$= \frac{1}{N^2} \sum_{k,k',k'',m,n} f_{m,n}\, e^{-j(\mu k + \nu k')} \cdot \delta_{-k+k''+m,0}\, \delta_{\mp k' \pm k''+n,0}$$

$$= \frac{1}{N^2} \sum_{m,n} f_{m,n}\, e^{-j(\mu m \pm \nu n)} \sum_{k'' \in E_{m,n}} e^{-j(\mu+\nu)k''} \tag{A-24}$$

$$= \frac{2\pi}{N^2} F(\mu, \pm\nu)\, Q(\mu)\, D_N(\mu+\nu) + 0(N^{-2}). \tag{A-25}$$

In (A-24) $E_{m,n}$ denotes the common set of $[1-m, N-m]$, $[1 \pm n, N \pm n]$, and $[1, N]$ and differs from $[1, N]$. This difference causes the extra term in (A-25). But this term is of order N^{-2} because of the smoothness of $F(s,t)\, Q(s)$.

Consequently, it follows by (A-7) that the asymptotic value of (A-22) is

$$\begin{cases} \frac{2\pi}{N} F(\mu, \mp\mu)\, Q(\mu) & \text{if } \mu + \nu = 0 \\ 0(N^{-2}) & \text{otherwise} \end{cases} \tag{A-26}$$

Applying this result to calculate the contributions of the first and third term of (A-19), we obtain

$$\frac{2\pi}{N} \sum_{k=1}^{p} |A_k|^2 \{F(\omega_k, -\omega_k) + F(-\omega_k, \omega_k)\}\, Q(\omega_k)$$

corresponding to the first term, and

$$\frac{2\pi}{N} \sum_{k=1}^{p} |A_k|^2 \{F(\omega_k, \omega_k) + F(-\omega_k, -\omega_k)\}\, Q(\omega_k)$$

corresponding to the third term, respectively. Hence, the asymptotic value of (A-13) is

$$\frac{2\pi}{N} \left\{ \sum_{k=1}^{p} 2|A_k|^2 [F(\omega_k, -\omega_k) + F(-\omega_k, \omega_k) + F(\omega_k, \omega_k) + F(-\omega_k, -\omega_k)]\, Q(\omega_k) + \int_{-\pi}^{\pi} [F(s,s) + F(s,-s)]\, Q(s)^2\, ds \right\}. \tag{A-27}$$

Using (A-9), (A-27) is more compactly written as

$$\frac{2\pi}{N} \int_{-\pi}^{\pi} [F(s,s) + F(s,-s)]\, Q(s)\, [2W(s) - Q(s)]\, ds. \tag{A-28}$$

Acknowledgment

The author wishes to express his sincere thanks to H. Tokumaru (Kyoto University, Japan) and T. Soeda (University of Tokushima, Japan) for their constant encouragement and useful advice, and K. Ohnishi for his assistance in programming. Thanks are also due to the reviewers for their helpful comments.

References

[1] J. P. Burg, "Maximum entropy power spectral analysis," presented at the 37th Annu. Int. SEG Meeting, Oklahoma City, OK, Oct. 31, 1967.

[2] E. Parzen, "Multiple time series modeling," in *Multivariate*

Analysis II, P. R. Krishnaiah, Ed. New York: Academic, 1970, pp. 389–410.

[3] R. B. Blackman and J. W. Tukey, *Measurement of Power Spectra from the Point of View of Communication Engineering.* New York: Dover, 1959.

[4] R. T. Lacoss, "Data adaptive spectral analysis method," *Geophysics*, vol. 36, pp. 661–675, Aug. 1971.

[5] A. B. Baggeroer, "Confidence intervals for regression (MEM) spectral estimates," *IEEE Trans. Inform. Theory*, vol. IT-22, pp. 534–545, Sept. 1976.

[6] H. Akaike, "Power spectrum estimation through autoregressive model fitting," *Ann. Inst. Statist. Math.*, vol. 21, pp. 407–419, 1969.

[7] H. Sakai, T. Soeda, and H. Tokumaru, "On the relation between fitting autoregression and periodogram with applications," *Ann. Statist.*, vol. 7, pp. 96–97, Jan. 1979.

[8] K. N. Berk, "Consistent autoregressive spectral estimate," *Ann. Statist.*, vol. 2, pp. 489–502, May 1974.

[9] R. E. Kromer, "Asymptotic properties of the autoregressive spectral estimator," Ph.D. dissertation, Dep. Statistics, Stanford Univ., Stanford, CA, 1969.

[10] D. C. Rife and R. R. Boorstyn, "Single tone parameter estimation from discrete-time observations," *IEEE Trans. Inform. Theory*, vol. IT-20, pp. 591–598, Sept. 1974.

[11] D. R. Brillinger, *Time Series: Data Analysis and Theory.* New York: Holt, Rinehart, and Winston, Inc., 1975.

[12] G. M. Jenkins and D. G. Watts, *Spectral Analysis and Its Applications.* San Francisco: Holden-Day, 1968.

GENERALIZED BURG ALGORITHM FOR BEAMFORMING IN CORRELATED MULTIPATH FIELD

S.B. Kesler

Communications Research Laboratory
McMaster University
Hamilton, Ontario, Canada L8S 4L7

Array beamforming employing nonlinear methods of spectral estimation presents a difficult problem when the spatial field is inhomogeneous, e.g., when correlated multipath is present, and hence, the corresponding cross-spectral (CS) matrix is non-Toeplitz.

To overcome this difficulty in the maximum entropy (ME) beamforming, we present a generalization of the Burg algorithm for dealing with a non-Toeplitz CS matrix. The stability and minimum phase properties of the original Burg algorithm are retained.

The use of the generalized Burg algorithm for elevation angle estimation is illustrated for the case of a passive sonar field. Statistical properties of the estimator are analyzed using (1) uncorrelated Gaussian noise only, and (2) the direct and specular multipath components of the target signal, embedded in uncorrelated noise and in directional ambient noise.

INTRODUCTION

Use of the maximum entropy method (MEM) [1] has been extended to the spatial processing of line array data [2,3]. It has been shown that this method is capable of resolving two independent signals originating from closely spaced sources. This is true for both the known autocorrelation (KA) algorithm and the Burg algorithm, which uses raw data to obtain the array beam pattern. However, when the incoming signals are correlated, as is often the case in the presence of multipath, we find that the maximum entropy (ME) beam pattern estimate is highly sensitive to the magnitude and phase of the inter-signal correlation, and the resulting patterns are sometimes unreliable. This phenomenon is true even when the signals are well separated spatially. The problem arises because the spatial signal field, in the presence of correlated multipath, is inhomogeneous; hence, the corresponding cross-spectral (CS) matrix for the line array is non-Toeplitz.

In this paper we describe a generalized Burg algorithm (GBA) for array beamforming for the case when the CS matrix of the spatial field is non-Toeplitz. The procedure yields a stable value for the pertinent reflection coefficient.

GENERALIZED BURG'S ALGORITHM

Consider a line array of N uniformly spaced omnidirectional sensors. We assume that the sensor signals are Fourier analyzed via the discrete Fourier transform (DFT). We denote by $x(n,f,i)$ the complex coefficient for the n-th sensor, at frequency f, and for the i-th DFT. We assume that the cross spectrum between sensors n and m, given by

$$\begin{aligned} R(n,m,f,i,j) &= E[x(n,f,i)\, x^*(m,f,j)] \\ &= R(n,m,f)\, \delta(i-j) \end{aligned} \tag{1}$$

where $E[\cdot]$, the expectation in the temporal sense, is known for $n,m = 1,2, \ldots, N$. We now derive the GBA for the beam pattern estimate when the (n,m)-th element of the N by N CS matrix is given by Eq. (1). For convenience, we shall drop dependence on the operating frequency f and on the DFT number i from here on.

The ME beam pattern as a function of the elevation angle, θ, measured with respect to the vertical line array axis, is given by [4]

$$S(\theta) = \frac{P_M d/\lambda}{\left| \sum_{m=0}^{M} a_M(m) \exp[-jm\psi(\theta)] \right|^2} \tag{2}$$

where λ is the wavelength, d is the intersensor spacing, and $\psi(\theta) = 2\pi(d/\lambda)\cos\theta$. Parameters $a_M(m)$, $m = 0, 1, \ldots, M$ ($a_M(0) = 1$) and P_M are the M-th order prediction error filter (PEF) coefficients and output power, respectively. The choice of the order M depends on the application. If we are interested only in field mapping, by which we assume that the cross spectral properties of the field are known over the array aperture, we ultimately choose the maximum PEF order, $M=N-1$ [4]. In the detection problem, where the cross spectrum must be estimated from finite data, a lower order filter may be necessary to produce reasonable stability in the ME field estimate [5].

The PEF coefficients are determined in the following recursive manner. For a PEF of order M-1, the forward prediction error at sensor n and the backward prediction error at sensor n-1, are defined by the convolution sums, respectively, [4,5]

$$f_{M-1}(n) = \underline{a}^T \underline{X}_n = \underline{X}_n^T \underline{a} \tag{3}$$

Reprinted from *IEEE Int. Conf. Acoust., Speech, Signal Processing*, vol. 3, 1982, pp. 1481–1484.

$$b_{M-1}(n-1) = \underline{a}^H \underline{\tilde{X}}_{n-1} = \underline{\tilde{X}}^T_{n-1}\, \underline{a}^* = \underline{X}^T_{n-1}\, \underline{a}^* \quad (4)$$

where the vectors are all of length M, as shown by

$$\underline{a} = [1, a_{M-1}(1), \ldots, a_{M-1}(M-1)]^T \quad (5)$$

$$\underline{X}_n = [X(n), X(n-1), \ldots, X(n-M+1)]^T, \; n \geq M+1 \quad (6)$$

$$\underline{X}_{n-1} = [X(n-1), X(n-2), \ldots, X(n-M)]^T, \; n \geq M+1 \quad (7)$$

In Eqs. (3) through (7) the asterisk signifies complex conjugation, the superscripts T and H signify transposition and Hermitian transposition, respectively, and the tilde over a vector signifies the fact that the components of that vector appear in reversed order. Substituting Eqs. (3) and (4) into the Burg formula for the M-th order filter reflection coefficient $a_M(M)$ [3,4]

$$a_M(M) = \frac{-2 \sum_{n=M+1}^{N} E[f_{M-1}(n) b^*_{M-1}(n-1)]}{\sum_{n=M+1}^{N} E[f_{M-1}(n) f^*_{M-1}(n) + b_{M-1}(n-1) b^*_{M-1}(n-1)]} \quad (8)$$

and using Eq. (1), we obtain [4,5]

$$a_M(M) = \frac{-2\underline{a}^T \{ \sum_{n=M+1}^{N} \underline{\hat{R}}(n,n-1) \} \underline{\tilde{a}}}{\underline{a}^H \{ \sum_{n=M+1}^{N} [\underline{\hat{R}}(n,n) + \underline{\hat{R}}(n-M,n-M)] \} \underline{a}} \quad (9)$$

The matrices in Eq. (9) are all of dimension M by M and are given by

$$\underline{\hat{R}}(n,n-1) = \begin{bmatrix} R(n,n-1) & \cdots & R(n,n-M) \\ \vdots & & \vdots \\ R(n-M+1,n-1) & \cdots & R(n-M+1,n-M) \end{bmatrix} \quad (10)$$

$$\underline{\hat{R}}(n,n) = \begin{bmatrix} R(n,n) & \cdots & R(n-M+1,n) \\ \vdots & & \vdots \\ R(n,n-M+1) & \cdots & R(n-M+1,n-M+1) \end{bmatrix} \quad (11)$$

$$\underline{\hat{R}}(n-M,n-M) = \begin{bmatrix} R(n-M,n-M) & \cdots & R(n-M,n-1) \\ \vdots & & \vdots \\ R(n-1,n-M) & \cdots & R(n-1,n-1) \end{bmatrix} \quad (12)$$

These matrices are obviously non-Toeplitz; moreover, $\underline{R}(n,n-1)$ is not even Hermitian.

Having computed the relfection coefficient $a_M(M)$, we may next compute the remaining PEF coefficients and the power P_M by using the Levinson recursion in the usual way [1,3]. The recursion starts with

$$P_o = \frac{1}{N} \operatorname{tr} \underline{\hat{R}}(n,n) = \frac{1}{N} \sum_{n=1}^{N} R(n,n) \quad (13)$$

The GBA, Eq. (9) is the space-time or wavenumber-frequency extension of the covariance-lattice method derived by Makhoul [6] for the case of a real-valued time series.

BEAMFORMING IN CORRELATED MULTIPATH FIELD

To illustrate the application of the GBA to beamforming, we consider an acoustic field consisting of a direct signal and a multipath component embedded in a spatially uncorrelated noise field and a directional ambient noise field. Such conditions are often encountered in an underwater acoustic environment [4], radar [7], and point-to-point communications. This is the same field model as the one used previously in the application of the GBA in field mapping [4,5]. Here we present some of the statistical properties of the GBA, which are needed in establishing its characteristics as a signal source locator in a passive sonar receiver.

The cross-spectral (CS) components of the K plane wave signals, the directional ambient noise, and the spatially uncorrelated noise are given, respectively, by [4,5]

$$R_s(n,m) = E[X_s(n) X^*_s(m)]$$

$$= \sum_{p=1}^{K} \sum_{q=1}^{K} (S_p S_q)^{1/2} |\alpha_{p,q}| \exp[j(\vec{k}_p \vec{d}_n - \vec{k}_q \vec{d}_m + \phi_{p,q})] \quad (14)$$

$$R_a(n,m) = P_a \exp[-\frac{(V_{n,m}\,\sigma)^2}{2} + jV_{n,m} \cos\theta_a] \quad (15)$$

$$R_u(n,m) = P_u\, \delta_{n,m} \quad (16)$$

where $\vec{d}_n$ is the radius vector of the n-th sensor relative to some arbitrary origin, $V_{n,m}$ is proportional to the distance between sensors n and m, $\vec{k}_p$ is the wavenumber vector of the p-th wave, θ_a and σ are the parameters of the ambient noise distribution, and S_p, P_a, and P_u are the p-th signal power, ambient noise power and the uncorrelated noise power, respectively. In Eq. (14), $|\alpha_{p,q}|$ and $\phi_{p,q}$ are the modulus and the argument of the inter-signal correlation. When the p-th and q-th waves are uncorrelated, $|\alpha_{p,q}|=0$; otherwise $|\alpha_{p,q}| \leq 1$.

The total CS matrix of the resulting spatial field is obtained by adding the three contributions described by Eqs. (14)-(16), as shown by

$$R_t(n,m) = R_s(n,m) + R_a(n,m) + R_u(n,m) \quad (17)$$

where n, m = 1,2, ..., N, i.e., the number of array sensors.

ACOUSTIC FIELD SIMULATION

In the previous work [4,5], we demonstrated some properties of the GBA as applied to mapping of a spatial field involving correlated multipath. The components of the model of Eq. (17) were computed analytically, using Eqs. (14)-(16). To analyze the statistical properties of the GBA we generate the signal in the following way. The computed "model" CS matrix of Eq. (17) is factored into a product of lower and upper triangular N by N matrices, such that

$$\underline{R}_t = \underline{B}\ \underline{B}^H \quad (18)$$

Such factorization is possible since $\underline{R}_t(n,m)$ is Hermitian positive semi-definite.

We now generate a vector, $\hat{\underline{Y}}$, of N independent normally distributed complex variates by using a Gaussian random number generator. This vector is transformed to a vector $\underline{X}$ of N correlated normally distributed complex variates, according to the linear operation

$$\hat{\underline{X}} = \underline{B}\ \hat{\underline{Y}} \quad (19)$$

The correlated variates, $\hat{\underline{X}}$, now possess spatial correlations described by matrix $\underline{R}_t$. Note that the normality of the distribution is preserved under the linear transformation.

The sample CS matrix $\hat{\underline{R}}_t$ is now obtained as

$$\hat{\underline{R}}_t = \hat{\underline{X}}\ \hat{\underline{X}}^H \quad (20)$$

Generation of vector $\hat{\underline{Y}}$ and the steps involving Eqs. (19) and (20) are repeated K times and the sample matrix $\underline{R}_t$ is averaged. To avoid the possibility of $\underline{R}_t$ becoming singular, K should be greater than N, the order of the CS matrix. The sample CS matrix obtained in the manner described above, is processed by the GBA to obtain the estimate of the array beampattern.

RESULTS

Since uncorrelated Gaussian noise is always present at the receiver input, it is important to know what the resulting statistics are after the noise is processed by the ME processor employing the GBA. To determine the statistics, a model CS matrix is constructed which contains the uncorrelated noise component (Eq. (16)), only, and the test is performed by generating L independent sets of averaged sample CS matrices. Each of these L sets is then input into the GBA and the beam pattern is calculated at one designated value of the elevation angle, θ. The histogram of the magnitude of the pattern at θ is then formed, in order to determine its probability density.

Figure 1 shows the normalized histograms of the output of the ME beamformer at broadside ($\theta = 90^o$) when uncorrelated Gaussian noise is at its input, for several values of the PEF order M, and L = 20000. The array consists of N = 10 sensors. These are, as expected, typical of all the histograms obtained at other elevation angles, since the input samples have a flat angular spectrum. These histograms represent the general shape of the first order probability density function (pdf) of the quantities being observed. However, the histograms do not tell us what the pdf's are mathematically. One way to identify the pdf's is to make a direct comparison of the histograms with the theoretical pdf. A large number of pdf's can be completely characterized by the first and second order statistics. Since they are readily available in the sample mean and sample variance of the observed quantities, these pdf curves can be easily generated and superimposed on the experimental curves.

The relatively long tails in the histograms in Fig. 1 suggest that the lognormal pdf is somewhat more likely to fit the experimental data than some other pdf's with a similar shape, e.g., Rayleigh and chi-square pdf's. Since a lognormal variate requries a rather large dynamic range, we first transform the output samples into a logarithmic scale by performing

$$Z = 10 \log X \quad (21)$$

where X is the output sample at one particular elevation angle, θ.

Figure 2 shows the normalized histograms of the output samples at $\theta = 90^o$ after the transformation (21). They resemble the bell shaped Gaussian pdf curves, and the theoretical pdf's with sample means and sample variances obtained from the experimental data closely fit the histograms. As an illustration of the fit, the Gaussian pdf is superimposed on the histogram for the PEF order M=1. The agreement between the experimental curve and the corresponding Gaussian pdf is also very good for M=2 and 4. For the case when M=8, the shape of the histogram is somewhat distorted, and there is a spike at very small magnitudes, which is also apparent in Fig. 1. This is due to the fact that the ME estimates become statistically unreliable when the PEF order M approaches the number of array sensors, N.

The exact explanation of the resulting Gaussian distribution for the log-transformed ME output is not trivial and still awaits study. The conjecture is that, since both the numerator and the denominator of Eq. (2) are results of recursions involving log products, taking the logarithm of products would be equivalent to summing the logarithms of individual components. Thus, by arguments of central limit theorem, the resulting distribution would tend to be Gaussian.

Next, we examined the histograms of the beam pattern when the signal and multipath components are present, as well as some amount of ambient noise. Curves in Fig. 3 show the histograms of the estimate of the direct signal arrival angle $\theta_1 = 60^o$. The direct signal, ambient noise and uncorrelated noise all have equal total powers while the multipath component, coming from $\theta_2 = 120^o$, is 3 dB weaker than the direct one. We see from Fig. 3 that the histograms differ from those in Fig. 1. Moreover, the histograms of the log-transformed pattern magnitude do not fit Gaussian model as in the case of uncorrelated noise. This is also true for all the elevation angles in the vicinity of the signal peak. Only at angles where the uncorrelated noise is a dominant component, the output magnitude tends to be lognormally distributed.

In summary, we observe that the GBA response to uncorrelated Gaussian noise exhibits statistical characteristics quite similar to the response of the original Burg algorithm [8]. It is therefore believed that the detection capabilities of the GBA are comparable to those of the original Burg algorithm.

CONCLUSION

In this paper we have analyzed some statistical properties of the generalized Burg algorithm for beamforming in an inhomogeneous spatial signal field. It was shown that the second order statistics of the algorithm are comparable to that of the original Burg algorithm. This property, along with the absolute stability and minimum-phase properties, makes the generalized algorithm a useful extension of the original algorithm when the underlying cross-spectral matrix is non-Toeplitz.

ACKNOWLEDGEMENTS

The author wishes to thank his colleagues S. Haykin, J. Kesler, B.W. Currie, and R.S. Walker for useful discussions on this research.

REFERENCES

[1] J.P. Burg, "Maximum Entropy Spectral Analysis", presented at 37th Meet. Soc. Explor. Geophysicists, Oklahoma City, OK, October 31, 1967.

[2] W.D. White, "Angular Spectra in Radar Application", IEEE Aerosp. Electron. Systems, vol. AES-15, no. 6, pp. 895-899.

[3] J.E. Evans, "Aperture Sampling Techniques for Precision Direction Finding", ibid, pp. 891-895.

[4] S.B. Kesler, S. Haykin, and R.S. Walker, "Maximum-Entropy Field-Mapping in the Presence of Correlated Multipath", Proc. IEEE ICASSP '81, pp. 157-161, Atlanta, GA, March 30-31, April 1, 1981.

[5] S.B. Kesler and S. Haykin, "Conventional, MLM, and MEM Mapping of Inhomogeneous Spatial Field", Proc. 1-st ASSP Workshop on Spectral Estimation, pp. 7.3.1-7.3.7, McMaster Univ., Hamilton, Ont., August 17-18, 1981.

[6] J. Makhoul, "Stable and Efficient Lattice Methods for Linear Prediction", IEEE Trans. Acoust., Speech, Signal Processing, vol. ASSP-25, no. 5, pp. 423-428, October 1977.

[7] J. Kesler and S. Haykin, "A New Adaptive Antenna for Elevation Angle Estimation in the Presence of Multipath", Proc. AP-S International Symposium, vol. I, pp. 130-133, Université Laval, Québec, PQ, June 2-6, 1980.

[8] S. Haykin and H.C. Chan, "Computer Simulation Study of a Radar Doppler Processor Using the Maximum-Entropy Method", IEE Proc., vol. 127, Pt. F, No. 6, pp. 464-470, December 1980.

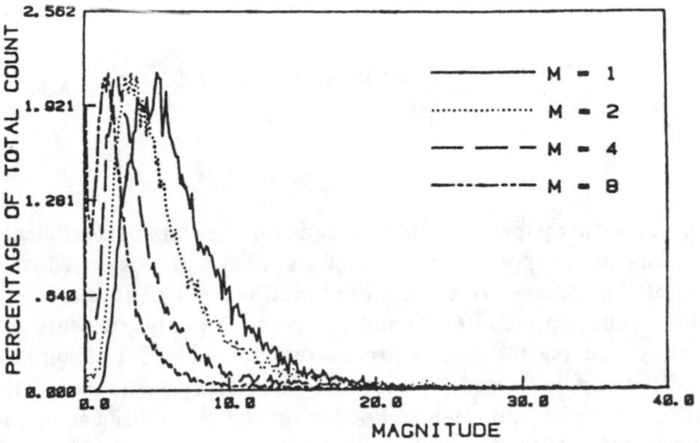

Fig. 1 Histograms of the MEM output with Gaussian noise at its input.

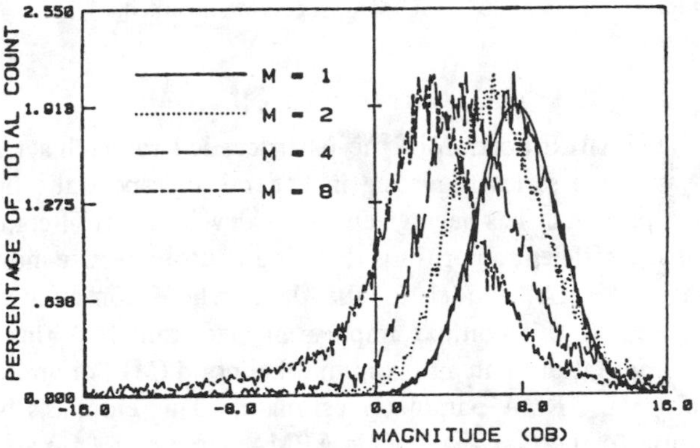

Fig. 2 Histograms of the log-transformed output

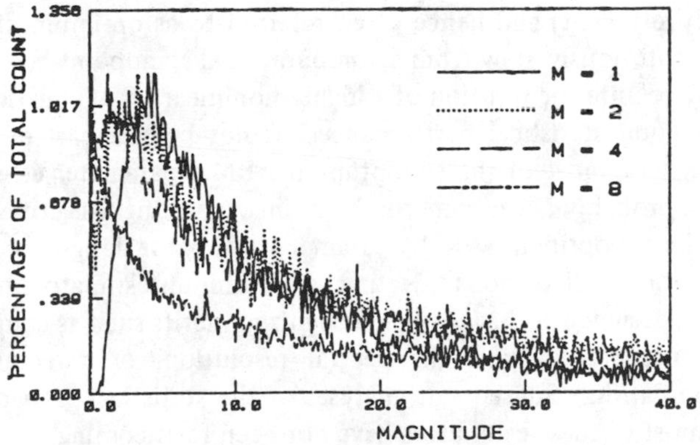

Fig. 3 Histograms of the ME estimate of the direct signal arrival angle

Information Tradeoffs in Using the Sample Autocorrelation Function in ARMA Parameter Estimation

STEPHEN P. BRUZZONE, MEMBER, IEEE, AND M. KAVEH, MEMBER, IEEE

Abstract—This paper considers bounds on the statistical efficiency of estimators of the poles and zeros of an ARMA process based on estimates of the process autocorrelation function (ACF). Special attention is paid to autoregressive (AR) and AR plus white noise processes. It is seen that reducing the ARMA process data to a given set of consecutive lags of the popular lagged-product ACF estimates prior to parameter estimation increases Cramér-Rao bounds on the generalized error covariance. A parametric study of the bound deterioration for some illustrative signal and noise situations reveals some empirical strategies for choosing ACF estimate lags to preserve statistical information. Analysis is based on the relative information index (RII) [2], and derivations of the large sample Fisher's information matrix for the raw data and for the lagged-product ACF estimate of an ARMA process are included.

I. Introduction

A SIZABLE portion of the past decade's research activity in time series modeling in general and spectral estimation in particular has been given to the development of computationally efficient algorithms that fit an autoregressive-moving average (ARMA) model to the data, where computational efficiency in this context implies an algorithm requiring far less computation than a maximum likelihood (ML) or approximately ML ARMA parameter estimate. This emphasis is in response to the fact that, while ARMA parameter estimates in the above ML class are considered the benchmark for statistical performance (asymptotically they are consistent and statistically efficient) and hence often referred to as optimum, they are notoriously slow from a computational standpoint because they require the solution of a highly nonlinear set of equations. Yet good statistical performance has not been a part of the design of most of the suboptimum ARMA parameter estimators, probably because methods for incorporating this criterion in the suboptimum ARMA parameter estimator design process are not at all obvious. Rather, suboptimum estimators have been designed to have other recognized merits such as computational efficiency, high spectral resolution, or a recursive formulation. Subsequent analyses of the statistical properties of most of these estimators have not been forthcoming, for the expressions involved become unduly complicated. As a result, various suboptimum methods have been tested and pruned on subjectively chosen ARMA processes or sinusoids. This has led to a state of confusion as regards any ranking of the suboptimum methods in terms of statistical performance.

Manuscript received February 2, 1983; revised September 11, 1983 and March 2, 1984. This work was supported in part by the U.S. Air Force Office of Scientific Research under Grant AFOSR-78-3628, the National Science Foundation under Grant ECS-8105962, and a University of Minnesota Dissertation Fellowship.

S. P. Bruzzone is with ARGOSystems, Inc., Sunnyvale, CA 94086.

M. Kaveh is with the Department of Electrical Engineering, University of Minnesota, Minneapolis, MN 55455.

One approach to the estimator design problem that favors computationally efficient formulations is to base the estimator on a data reduction (statistic) of much lower dimension than the raw data. This has been a popular approach in suboptimum ARMA parameter estimator design, where the particular data reduction is the lagged-product estimate of the autocorrelation function (ACF), or more briefly the sample ACF. While there appears to be tacit agreement that the sample ACF is a good starting point for the design of a suboptimum ARMA parameter estimator, no statistical arguments to support this approach have been brought out. Nor has the question of which particular lags of the sample ACF to use vis-à-vis intended application been addressed in any depth. This paper examines these issues.

In this paper, a scalar measure of the fractional amount of information for ARMA parameter estimation that is retained when using the sample ACF in lieu of the unreduced data is exhibited for some popular signal classes as a function of the set of sample ACF lags used. This measure, the relative information index (RII) [2], [3], is a ratio that compares the Cramér-Rao bound generalized variance of ARMA parameter estimates from the raw data to that of ARMA parameter estimates from the relevant set of lags of the sample ACF. Complex pole pairs and zero pairs are assumed, as this allows the representation of a wide variety of spectra of practical interest, such as the high resolution case. Results are parameterized in terms of pole and zero magnitudes and angles for two reasons. First, it is more straightforward to derive the bounds on the statistical variability of estimates of the poles and zeros of the process than to do so for estimates of the ARMA parameters. Second, the pole-zero formulation lends itself more directly to the spectral estimation problem. This is because properties of the estimators can be obtained directly as a function of pole angle and magnitude (resonance frequency and bandwidth) and zero angle and magnitude (antiresonance frequency and bandwidth).

A partial list of ARMA parameter estimators that are based on the sample ACF includes the methods of Walker [4], Hsia and Landgrebe [5], Graupe, Krause, and Moore [6], the MLS

Reprinted from *IEEE Trans. Acoust., Speech, Signal Processing*, vol. ASSP-32, pp. 701–715, Aug. 1984.

method of Sakai and Arase [7], Kaveh [8], Kinkel *et al.* [9], Bruzzone and Kaveh [10], Kay [11], Cadzow [12], [13], and Beex and Scharf [28]. The estimation equations are linear or quadratic in most of these, and therefore of lower computational complexity than those of the optimum methods. All of these methods can be considered suboptimum in that Arato [14] has shown that a finite set of lags of the sample ACF is not a sufficient statistic for ARMA parameter estimation. (In fact, Dickinson [23] shows that, among stationary Gaussian processes with rational PSD, only AR processes admit sufficient statistics of dimension less than the number of data points.) For brevity, members of this class are referred to as ACF class estimators.

Looking more closely at the ACF class, it can be seen that [5], [8]-[10], [12], [13], and [28] first compute an estimate of the autoregressive (AR) parameters, and then use these to estimate some function of the moving average (MA) parameters, or the weights of the modal decomposition of the ACF in [28]. The estimator in [6] also estimates AR and MA parameters separately, in this case by estimating MA parameters first, and that in [11] estimates only the AR parameters of an AR process in white noise, and requires prior knowledge of the noise variance to do this. As a result of this nonsimultaneous estimation of the parameters, each of these estimators is suboptimum even with respect to the sample ACF upon which it is based, i.e., it does not utilize all of the information in the lags of the sample ACF. This leaves the possibility of attaining Cramér-Rao bounds only for the method of Walker and that of Sakai and Arase. Walker's method is asymptotically efficient but efficiency obtains only in the limit as the number of sample ACF lags goes to infinity. That of Sakai and Arase has not undergone a detailed statistical analysis.

In the ARMA spectral estimation problem, there is a tendency among some of the nonsimultaneous fitting algorithms to emphasize estimation of the AR parameters and to gloss over the matter of accurately estimating the contribution of zeros to the spectral estimate. This is a reflection of two facts: first, that the poles are usually far more influential than zeros in determining the shape of the PSD in most practical applications and second, that poles are often of more interest in cases where the zeros are merely nuisance parameters introduced by the presence of noise on the observations. To treat these cases, this paper uses a modification of the relative information index, the partial relative information index (PRII) [2], that allows consideration of a subspace of the parameter space.

Some statistical analyses of suboptimum ARMA estimators have been published. Gersch [15] was one of the first to introduce a computationally efficient estimator for ARMA processes to the engineering literature. By solving a set of linear equations, the extended Yule-Walker equations, he provided estimates of just the AR parameters of an ARMA process. He showed that these estimates are asymptotically unbiased and consistent, and computed their asymptotic error covariance matrix. Although the efficiency of these estimates relative to ML has not been published explicitly, Walker [4] mentions that asymptotic efficiencies will not all become high unless MA parameters are all small. More recently, Kay [11] evaluates the asymptotic error covariance of the AR parameter estimates from the extended Yule-Walker equations when the observed time series is strictly AR. All of these results are of interest in that the extended Yule-Walker equations are the backbone of many of the ARMA estimators in the ACF class that estimate the AR parameters first, including [5], [8], [9], [13], [28]. Sakai and Tokumaru [1] derive the asymptotic variances of the spectral estimate produced by [8] and [9], but the resultant expressions are very complicated and therefore can be appreciated only by computing the variances for specific numerical examples. Finally, an asymptotic evaluation of Fisher's information matrix, with respect to poles and zeros, is provided for the raw data by Box and Jenkins [16]. Their result requires the restrictive assumption that poles and zeros are real, however. Aström [29] also derives Fisher's information matrix for the raw data, this with respect to the ARMA parameters. This result is expressed in terms of contour integrals, however, whose exact evaluation is possible only for simple models of low orders. Gersch *et al.* provide a statistical estimate of the raw data information matrix as a byproduct of a reparameterized ARMA process estimator in [31]. In this paper we extend the result of Box and Jenkins to complex pole and zero pairs, providing closed form expressions for the raw data asymptotic information matrix for arbitrary model orders. A new result, closed form expressions for the asymptotic information matrix for the sample ACF of an ARMA process, is also derived herein. These allow exact, speedy calculation of performance bounds without the uncertainties involved in numerical integration and other approximation techniques.

An alternative to using the sample ACF for ARMA parameter estimation is well treated in recent works of Friedlander [26], [27]. These explore recursive and lattice form implementation of ARMA parameter estimators. The goal is to avoid information loss by operating on the unreduced data, while achieving computational efficiency via the estimator's numerical structure. Conclusions as to the statistical efficiency of these methods await the forthcoming statistical analyses for finite data records, indicated in [26].

II. Assumptions and Definitions

We observe N samples of the real, stationary time series x_t, generated by the ARMA (L, M) system

$$x_t - \sum_{i=1}^{L} a_i x_{t-i} = \epsilon_t - \sum_{i=1}^{M} b_i \epsilon_{t-i} \tag{1}$$

where the process ϵ_t is a zero-mean stationary uncorrelated Gaussian sequence of variance σ_ϵ^2, a_i are the autoregressive parameters, and b_i are the moving average parameters. Introducing the notation

$$A_i = \begin{cases} 1, & i = 0 \\ -a_i, & i = 1, \cdots, L \\ 0, & i = L+1, \cdots \end{cases}$$

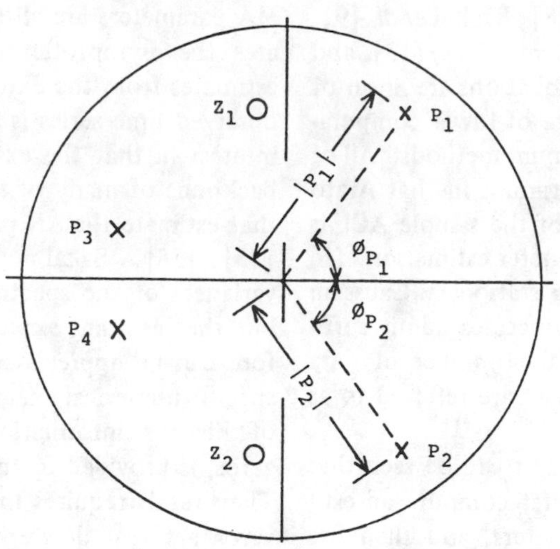

Fig. 1. Locations of poles and zeros for $L = 4, M = 2$.

$$B_i = \begin{cases} 1, & i = 0 \\ -b_i, & i = 1, \cdots, M \\ 0, & i = M+1, \cdots \end{cases} \tag{2}$$

we observe that x_t is generated as the output of a system with transfer function $H(z)$ and input ϵ_t, where

$$H(z) = \frac{B(z)}{A(z)} = \frac{\prod_{i=1}^{M} (1 - Z_i z^{-1})}{\prod_{i=1}^{L} (1 - P_i z^{-1})} \tag{3}$$

and

$$A(z) = A_0 + A_1 z^{-1} + \cdots + A_L z^{-L} \tag{4a}$$

$$B(z) = B_0 + B_1 z^{-1} + \cdots + B_M z^{-M} \tag{4b}$$

and where $P_i, i = 1, \cdots, L$ and $Z_i, i = 1, \cdots, M$ are the system poles and zeros, respectively. It is assumed that there are no repeated poles or zeros, and further that there are no pole-zero cancellations. Poles and zeros are assumed always to occur in complex conjugate pairs, with

$$P_{i+1} = P_i^* = |P_i| e^{-j\phi_{P_i}}, \quad 0 < \phi_{P_i} < \pi, \quad i = 1, 3, \cdots, L-1 \tag{5a}$$

$$Z_{i+1} = Z_i^* = |Z_i| e^{-j\phi_{Z_i}}, \quad 0 < \phi_{Z_i} < \pi, \quad i = 1, 3, \cdots, M-1 \tag{5b}$$

where the asterisk denotes complex conjugation. Roots with odd subscripts are assumed to lie in the upper half of the unit circle in the z plane, as in the example of Fig. 1. These roots are referred to as the principal poles and zeros of the system, in that they uniquely specify the system. Note that stationarity of x_t requires that $|P_i| < 1, i = 1, 3, \cdots, L-1$ and invertibility of x_t requires $|Z_i| < 1, i = 1, 3, \cdots, M-1$ [16, p. 74]. We assume that these conditions are satisfied. Note also that the restriction to root pairs implies that we consider only even values for L and M.

We define

$$\begin{aligned} \boldsymbol{\theta}^T &= [\theta_1, \theta_2, \theta_3, \theta_4, \cdots, \theta_{L-1}, \theta_L, \theta_{L+1}, \theta_{L+2}, \theta_{L+3}, \\ &\quad \theta_{L+4}, \cdots, \theta_{L+M-1}, \theta_{L+M}] \\ &= [|P_1|, \phi_{P_1}, |P_3|, \phi_{P_3}, \cdots, |P_{L-1}|, \phi_{P_{L-1}}, |Z_1|, \\ &\quad \phi_{Z_1}, |Z_3|, \phi_{Z_3}, \cdots, |Z_{M-1}|, \phi_{Z_{M-1}}] \end{aligned} \tag{6}$$

as the vector of parameters for the ARMA system generating x_t. Note that θ_i refers to magnitudes for i odd and angles for i even, these pertaining to poles for $i \leq L$ and zeros for $i > L$. This suggests a useful collection of subsets for the domain of $i, \Theta = \{1, \cdots, L+M\}$:

$$\begin{aligned} \Pi &= \{1, 2, \cdots, L\} \\ \Pi_M &= \{1, 3, \cdots, L-1\} \\ \Pi_A &= \{2, 4, \cdots, L\} \\ \zeta &= \{L+1, L+2, \cdots, L+M\} \\ \zeta_M &= \{L+1, L+3, \cdots, L+M-1\} \\ \zeta_A &= \{L+2, L+4, \cdots, L+M\}. \end{aligned} \tag{7}$$

Thus, θ_i for $i \in \Pi$ refers to poles, with $i \in \Pi_M$ referring to pole magnitudes and $i \in \Pi_A$ referring to pole angles, and similarly for zeros. Except for cases wherein equations are most efficiently expressed in terms of P_i and Z_i directly as in (8), the θ_i notation is used.

Conversion from poles to AR parameters is accomplished by equating coefficients in the denominator of (3), giving

$$a_1 = \sum_{i=1}^{L} P_i$$

$$a_2 = -\sum_{i=2}^{L} \sum_{j=1}^{i-1} P_i P_j$$

$$a_3 = \sum_{i=3}^{L} \sum_{j=1}^{i-1} \sum_{k=1}^{j-1} P_i P_j P_k$$

$$a_4 = -\sum_{i=4}^{L} \sum_{j=1}^{i-1} \sum_{k=1}^{j-1} \sum_{l=1}^{k-1} P_i P_j P_k P_l$$

$$\vdots$$

$$a_L = (-1)^{L-1} \prod_{i=1}^{L} P_i. \tag{8a}$$

Analogously, MA parameters and zeros are related, from the numerator of (3), by

$$b_1 = \sum_{i=1}^{M} Z_i$$

$$\vdots$$

$$b_M = (-1)^{M-1} \prod_{i=1}^{M} Z_i. \tag{8b}$$

The autocorrelation function of a stationary time series is defined by

$$\gamma(i) = \gamma(-i) = E x_t x_{t+i}, \quad i \in J \tag{9}$$

where E denotes statistical expectation, J is the set of integers, and $\gamma(i)$ is called the ith lag autocorrelation. Many methods of spectral estimation utilize one of two popular estimates of the ACF, both of which are dubbed the sample ACF, and neither of which is optimal in any sense. The biased sample ACF is

$$c(i) = c(-i) = \frac{1}{N} \sum_{t=1}^{N-i} x_t x_{t+i}, \quad i = 0, 1, \cdots, K \tag{10a}$$

and the unbiased sample ACF is

$$c'(i) = c'(-i) = \frac{1}{N-i} \sum_{t=1}^{N-i} x_t x_{t+i}, \quad i = 0, 1, \cdots, K. \tag{10b}$$

This investigation considers cases where $K << N$ and $N \to \infty$, so although the sample ACF is subsequently denoted $c(i)$, there is essentially no difference between this and $c'(i)$. Note that these estimates are statistically equivalent, i.e., they contain the same information about the time series, for any value of N, since the lags are scaled by known functions and thus one can be obtained from the other from knowledge of the scaling functions. By the same reasoning, the application of any known taper does not alter the information in the sample ACF. We also mention that the lags $c(0), \cdots, c(L)$ are "asymptotically sufficient" for estimating the parameters of an AR (L) process. This follows from the fact [17] that these lags give asymptotically ML estimates when used to construct the Yule-Walker equations.

The relative information index (RII) is defined [2], [3]

$$R = \frac{|I_c(\boldsymbol{\theta})|}{|I_x(\boldsymbol{\theta})|} \tag{11}$$

where $I_c(\boldsymbol{\theta})$ is Fisher's information matrix for the particular data reduction of interest and $I_x(\boldsymbol{\theta})$ is Fisher's information matrix for the raw data, whose probability density function is parameterized by $\boldsymbol{\theta}$. The properties of R are

P1: $0 \leq R \leq 1$

P2: $R = 1$ if and only if the data reduction $\boldsymbol{C}$ is sufficient for $\boldsymbol{\theta}$.

The proofs are found in [2] or [3]. The RII is proportional to the amount of information about $\boldsymbol{\theta}$ that is retained in $\boldsymbol{C}$ relative to that originally in the raw data $\boldsymbol{X}$. Hence, it is desirable to use statistics that have a value of R near unity for parameter estimation. Rejecting statistics having a value of R that falls below a given threshold is equivalent to rejecting those statistics that lead to an unacceptably large increase in Cramér-Rao bounds for parameter estimation, as compared to such bounds for estimators based on the raw data. Also, the RII is invariant under reparameterization of the estimation problem. Specifically, the value of R is unchanged if we consider estimation of any vector of differentiable functions of the original parameter vector such that the dimension of the new vector equals that of the original parameter vector. This invariance property is well suited to the ARMA spectral estimation problem, in which reparameterizations are common, e.g., some methods estimate ARMA parameters, some estimate poles and zeros, and yet others estimate quadratic functions of the ARMA parameters.

An extension of the RII, the partial relative information index (PRII) is also derived in [2] and [3] to compare Cramér-Rao bounds of statistic and raw data for estimating a subset of the full parameter set. Denoted in this paper by R_1, the PRII is shown in the above references to have the same properties as the RII.

III. Derivations of Fisher's Information Matrices

A. Asymptotic Distribution of the Sample ACF of an ARMA Process

Define the sample ACF vector

$$C_{[k_1, k_2]}^T = [c(k_1), c(k_1+1), \cdots, c(k_2)], \quad k_1 < k_2 << N \tag{12}$$

where $c(i)$ is given in (10a), and the corresponding vector of true autocorrelations

$$\boldsymbol{\Gamma}_{[k_1, k_2]}^T = [\gamma(k_1), \gamma(k_1+1), \cdots, \gamma(k_2)] \tag{13}$$

where $\gamma(i)$ is given in (9). Walker [18] has established that $\sqrt{N}(C_{[k_1,k_2]} - \Gamma_{[k_1,k_2]})$ is asymptotically multivariate normal with mean zero and finite covariance matrix. The asymptotic covariances are given by Bartlett [19] as

$$\lim_{N \to \infty} N \operatorname{cov}[c(k), c(l)] = \sum_{i=-\infty}^{\infty} [\gamma(i)\gamma(i+l-k) + \gamma(i+k)\gamma(i-l)], \tag{14}$$

so we have

$$C_{[k_1, k_2]} \xrightarrow{L} \text{MVN}\left[\boldsymbol{\Gamma}_{[k_1, k_2]}, \frac{1}{N}\Lambda(k_1, k_2)\right] \tag{15}$$

where $\xrightarrow{L}$ denotes convergence in law, MVN $(\boldsymbol{\alpha}, \boldsymbol{\beta})$ denotes a multivariate normal probability density function (pdf) of mean vector $\boldsymbol{\alpha}$ and covariance matrix $\boldsymbol{\beta}$, and $\Lambda(k_1, k_2)$ is given by

$$\Lambda(k_1, k_2)_{k,l} = \sum_{i=-\infty}^{\infty} [\gamma(i)\gamma(i+l-k) + \gamma(i+k-1+k_1)\gamma(i-l+1-k_1)], \quad k, l = 1, 2, \cdots, k_2 - k_1 + 1. \tag{16}$$

B. Fisher's Asymptotic Information Matrix for the Sample ACF of an ARMA Process

Fisher's information matrix for $C_{[k_1, k_2]}$ is defined

$$I_C(\boldsymbol{\theta}) = -E\nabla_\theta [\nabla_\theta \ln f(C_{[k_1,k_2]}|\boldsymbol{\theta})]^T \tag{17}$$

where $\boldsymbol{\theta}$ is the vector of parameters. The log of the asymptotic multivariate normal density $f_\infty(C_{[k_1,k_2]}|\boldsymbol{\theta})$ (15) for this problem is

$$\ln f_\infty(C_{[k_1,k_2]}|\boldsymbol{\theta}) = -\frac{k_2-k_1}{2}\ln 2\pi + \frac{k_2-k_1}{2}\ln N - \frac{1}{2}\ln|\Lambda(k_1,k_2)| - \frac{N}{2}(C_{[k_1,k_2]} - \Gamma_{[k_1,k_2]})^T \Lambda^{-1}(k_1,k_2)(C_{[k_1,k_2]} - \Gamma_{[k_1,k_2]}). \tag{18}$$

Differentiation of the first two terms of (18) gives zero, and the third term is negligible for large N, so the (m, n)th element of the asymptotic information matrix is (dropping subscripts)

$$\lim_{N\to\infty} I_C(\boldsymbol{\theta})_{m,n} = \frac{N}{2}\{E[C^T(\Lambda^{-1})''C] - (\Gamma^T\Lambda^{-1})''E(C) - E(C^T)(\Lambda^{-1}\Gamma)'' + (\Gamma^T\Lambda^{-1}\Gamma)''\}$$
$$= \frac{N}{2}[\Gamma^T(\Lambda^{-1})''\Gamma - 2(\Gamma^T\Lambda^{-1})''\Gamma + (\Gamma^T\Lambda^{-1}\Gamma)''] \tag{19}$$

where, for any matrix A, we denote

$$A'_m = \frac{\partial A}{\partial\theta_m}$$
$$A'' = \frac{\partial^2 A}{\partial\theta_m\partial\theta_n}. \tag{20}$$

Using the chain rule to carry out the differentiation and collecting terms yields

$$I_C(\boldsymbol{\theta})_{m,n} \simeq N(\Gamma'_m)^T\Lambda^{-1}\Gamma'_n. \tag{21}$$

The approximation can be made arbitrarily precise by large enough choice of N. In the remainder of this section the terms of (21) are evaluated in terms of $\boldsymbol{\theta}$. The definitions of Γ in (13) and Λ in (16) indicate that this requires the autocorrelation function of x_t and its first partials.

C. The Autocorrelation Function of an ARMA Process

In all that follows we assume $L \geqslant 2$, i.e., there is at least one pole pair in the system generating x_t. The first several lags of the ACF are obtained in [30] as

$$\begin{bmatrix} 1 & -a_1 & -a_2 & \cdots & -a_{L-1} & -a_L & 0 & 0 & \cdots & 0 \\ -a_1 & 1-a_2 & -a_3 & \cdots & -a_L & 0 & 0 & 0 & \cdots & 0 \\ -a_2 & -a_1-a_3 & 1-a_4 & \cdots & 0 & 0 & 0 & 0 & \cdots & 0 \\ \vdots & \vdots & \vdots & & \vdots & \vdots & \vdots & \vdots & & \vdots \\ -a_L & -a_{L-1} & -a_{L-2} & \cdots & -a_1 & 1 & 0 & 0 & \cdots & 0 \\ 0 & -a_L & -a_{L-1} & \cdots & -a_2 & -a_1 & 1 & 0 & \cdots & 0 \\ 0 & 0 & -a_L & \cdots & -a_3 & -a_2 & -a_1 & 1 & \cdots & 0 \\ \vdots & \vdots & \vdots & & \vdots & \vdots & \vdots & \vdots & & \vdots \\ 0 & 0 & 0 & \cdots & 0 & 0 & 0 & 0 & \cdots & 1 \end{bmatrix} \begin{bmatrix} \gamma(0) \\ \gamma(1) \\ \gamma(2) \\ \vdots \\ \gamma(K) \end{bmatrix} = \begin{bmatrix} D_0 \\ D_1 \\ D_2 \\ \vdots \\ D_K \end{bmatrix}$$

or, more compactly,

$$\boldsymbol{\psi}_K \Gamma_{[0,K]} = D_{[0,K]} \tag{22}$$

where

$$K = \begin{cases} L, & L \geqslant M \geqslant 0, L \geqslant 2 \\ M, & M > L \geqslant 2, \end{cases} \tag{23}$$

$$D_i = \sigma_\epsilon^2 \sum_{j=i+1}^{M} B_j \sum_{l=1}^{j-i} |\Omega_{j-i-l}|(a_l - b_l) + \sigma_\epsilon^2 B_i \tag{24}$$

and where

$$\Omega_i = \begin{cases} \begin{bmatrix} a_1 & a_2 & \cdots & a_i \\ -1 & a_1 & \cdots & a_{i-1} \\ 0 & -1 & \cdots & a_{i-2} \\ \vdots & \vdots & & \vdots \\ 0 & 0 & \cdots & a_1 \end{bmatrix}, & i \geqslant 1 \\ 1, & i = 0. \end{cases} \tag{25}$$

We adopt the convention throughout this paper that $\Sigma_i^j = 0$ for $j < i$. Beyond lag M the ACF obeys the autoregressive recursion [16, p. 75]

$$\gamma(k) = \sum_{i=1}^{L} a_i\gamma(k-i), \qquad k = M+1, M+2, \cdots. \tag{26}$$

Although (26) can be used to extend $\gamma(k)$ arbitrarily far, infinite sums arising from (16) are simplified later by expressing the ACF in terms of its poles

$$\gamma(k) = \sum_{i=1}^{L} \lambda_i P_i^{k-M}, \quad k = M, M+1, \cdots \tag{27}$$

where the P_i are the system poles, and the λ_i are obtained using the extended autocorrelation function (26) to solve (27) for $k = M, M+1, \cdots, M+L-1$, i.e.,

$$\begin{bmatrix} 1 & 1 & \cdots & 1 \\ P_1 & P_2 & \cdots & P_L \\ \vdots & \vdots & & \vdots \\ P_1^{L-1} & P_2^{L-1} & \cdots & P_L^{L-1} \end{bmatrix} \begin{bmatrix} \lambda_1 \\ \lambda_2 \\ \vdots \\ \lambda_L \end{bmatrix} = \begin{bmatrix} \gamma(M) \\ \gamma(M+1) \\ \vdots \\ \gamma(M+L-1) \end{bmatrix},$$

henceforth written

$$\boldsymbol{Q}\boldsymbol{\lambda}_{[1,L]} = \boldsymbol{\Gamma}_{[M,M+L-1]}. \tag{28}$$

Note that the λ_i occur in complex conjugate pairs, with $\lambda_2 = \lambda_1^*$, $\lambda_4 = \lambda_3^*$, etc.

D. First Derivatives of the ACF of an ARMA Process

We begin by differentiating the a_i given in (8a) and the b_i given in (8b). To this end, define the functions $W_i(P_k)$, $i = 0, 1, \cdots, L$ and $k = 1, 3, \cdots, L-1$ as

$$W_0(P_k) = 0,$$

$$W_1(P_k) = 1,$$

$$W_2(P_k) = \sum_{\substack{j=1 \\ j \neq k, k+1}}^{L} P_j,$$

$$W_3(P_k) = \sum_{\substack{j=1 \\ j \neq k, k+1}}^{L} \sum_{\substack{l=j+1 \\ l \neq k, k+1}}^{L} P_j P_l,$$

$$W_4(P_k) = \sum_{\substack{j=1 \\ j \neq k, k+1}}^{L} \sum_{\substack{l=j+1 \\ l \neq k, k+1}}^{L} \sum_{\substack{m=l+1 \\ m \neq k, k+1}}^{L} P_j P_l P_m,$$

$$\vdots$$

$$W_{L-1}(P_k) = \prod_{\substack{j=1 \\ j \neq k, k+1}}^{L} P_j,$$

$$W_L(P_k) = 0. \tag{29}$$

Writing out the expressions for a_i in (8a) term by term and differentiating, it is seen that, for $i = 1, \cdots, L$,

$$\frac{\partial a_i}{\partial \theta_m} = \begin{cases} 2(-1)^{i+1} [W_i(P_m) \cos \theta_{m+1} + W_{i-1}(P_m) \theta_m], & m \in \Pi_M \\ 2(-1)^i W_i(P_{m-1}) \theta_{m-1} \sin \theta_m, & m \in \Pi_A \\ 0, & m \in \zeta. \end{cases} \tag{30}$$

The equations for the b_i are analogous to those for the a_i, with Z_j replacing P_j and M replacing L in (29), so we have for $i = 1, \cdots, M$,

$$\frac{\partial b_i}{\partial \theta_m} = \begin{cases} 0, & m \in \Pi \\ 2(-1)^{i+1} [W_i(Z_{m-L}) \cos \theta_{m+1} + W_{i-1}(Z_{m-L}) \theta_m], & m \in \zeta_M \\ 2(-1)^i W_i(Z_{m-L-1}) \theta_{m-1} \sin \theta_m, & m \in \zeta_A. \end{cases} \tag{31}$$

Now a sequence of steps analogous to that for obtaining $\gamma(k)$ is followed. Equation (22) is differentiated, giving

$$\frac{\partial}{\partial \theta_m} \boldsymbol{\Gamma}_{[0,K]} = \begin{cases} -\boldsymbol{\psi}_K^{-1} \left(\frac{\partial}{\partial \theta_m} \boldsymbol{\psi}_K \right) \boldsymbol{\Gamma}_{[0,K]} + \boldsymbol{\psi}_K^{-1} \left(\frac{\partial}{\partial \theta_m} \boldsymbol{D}_{[0,K]} \right), & m \in \Pi \\ \boldsymbol{\psi}_K^{-1} \left(\frac{\partial}{\partial \theta_m} \boldsymbol{D}_{[0,K]} \right), & m \in \zeta \end{cases} \tag{32}$$

where

$$K = \begin{cases} L, & L \geq M \geq 0, L \geq 2 \\ M, & M > L \geq 2. \end{cases} \tag{33}$$

Also, $\partial/\partial\theta_m \, \boldsymbol{D}_{[0,K]}$ is evaluated by differentiating (24), giving

$$\frac{\partial D_i}{\partial \theta_m} = \begin{cases} \sigma_\epsilon^2 \sum_{j=i+1}^{M} \sum_{l=1}^{j-i} \left[(a_l - b_l) B_j \left(\frac{\partial}{\partial \theta_m} |\boldsymbol{\Omega}_{j-i-l}| \right) + B_j \frac{\partial a_l}{\partial \theta_m} |\boldsymbol{\Omega}_{j-i-l}| \right], & m \in \Pi \\ \sigma_\epsilon^2 \sum_{j=i+1}^{M} \sum_{l=1}^{j-i} |\boldsymbol{\Omega}_{j-i-l}| \left[b_j \frac{\partial b_l}{\partial \theta_m} - (a_l - b_l) \frac{\partial b_j}{\partial \theta_m} \right] + \sigma_\epsilon^2 \frac{\partial B_i}{\partial \theta_m}, & m \in \zeta, \end{cases} \tag{34}$$

where [20, p. 42]

$$\frac{\partial}{\partial \theta_m} |\boldsymbol{\Omega}_k| = \sum_{i=1}^{k} |\boldsymbol{H}_i| \tag{35}$$

and where $\boldsymbol{H}_i$ is the matrix $\boldsymbol{\Omega}_k$ with elements of the ith row replaced by their derivatives, e.g.,

$$\boldsymbol{H}_2 = \begin{bmatrix} a_1 & a_2 & \cdots & a_k \\ 0 & \frac{\partial a_1}{\partial \theta_m} & \cdots & \frac{\partial a_{k-1}}{\partial \theta_m} \\ \vdots & \vdots & & \vdots \\ 0 & 0 & \cdots & a_1 \end{bmatrix} \tag{36}$$

Next, the differentiated autocorrelation function is extended by differentiating (26), giving, for $k = M+1, M+2, \cdots$,

$$\frac{\partial}{\partial \theta_m} \gamma(k) = \begin{cases} \sum_{i=1}^{L} \left[\frac{\partial a_i}{\partial \theta_m} \gamma(k-i) + a_i \frac{\partial \gamma(k-i)}{\partial \theta_m} \right], & m \in \Pi \\ \sum_{i=1}^{L} a_i \frac{\partial \gamma(k-i)}{\partial \theta_m}, & m \in \zeta. \end{cases} \tag{37}$$

E. The Asymptotic Covariance Matrix of the Sample ACF of an ARMA Process

In this subsection, closed form expressions are developed for the infinite sums in (16), from which the asymptotic covariance matrix of the sample ACF follows in closed form. First we show that

$$\sum_{i=-\infty}^{\infty} \gamma(i)\gamma(i+j) = S(j), \quad j=0,1,\cdots, \tag{38}$$

i.e., the series is convergent and expressible in closed form. Expressing (38) in terms of one-sided infinite sums, we have

$$\sum_{i=-\infty}^{\infty} \gamma(i)\gamma(i+j) = \begin{cases} 2\sum_{i=0}^{\infty}\gamma^2(i) - \gamma^2(0), & j=0 \\ 2\left[\sum_{i=0}^{\infty}\gamma(i)\gamma(i+j) + \sum_{i=1}^{(j-1)/2}\gamma(i)\gamma(j-i)\right], & j \text{ odd} \\ 2\left[\sum_{i=0}^{\infty}\gamma(i)\gamma(i+j) + \sum_{i=1}^{(j/2)-1}\gamma(i)\gamma(j-i)\right] + \gamma^2\left(\frac{j}{2}\right), & j \text{ even.} \end{cases} \tag{39}$$

But

$$\sum_{i=0}^{\infty}\gamma(i)\gamma(i+j) = \sum_{i=0}^{M-1}\gamma(i)\gamma(i+j) + \sum_{i=M}^{\infty}\gamma(i)\gamma(i+j). \tag{40}$$

Using (27), we get

$$\sum_{i=M}^{\infty}\gamma(i)\gamma(i+j) = \sum_{i=M}^{\infty}\sum_{r=1}^{L}\sum_{s=1}^{L}\lambda_r\lambda_s P_r^{-M}P_s^{j-M}(P_rP_s)^i. \tag{41}$$

But the assumption that poles are inside the unit circle gives the convergent series

$$\sum_{i=0}^{\infty}(P_rP_s)^i = \frac{1}{1-P_rP_s}$$

so summations in (41) can be interchanged, and

$$\sum_{i=M}^{\infty}\gamma(i)\gamma(i+j) = \sum_{r=1}^{L}\sum_{s=1}^{L}\lambda_r\lambda_s P_r^{-M}P_s^{j-M}\sum_{i=M}^{\infty}(P_rP_s)^i$$

$$= \sum_{r=1}^{L}\sum_{s=1}^{L}\lambda_r\lambda_s\frac{P_s^j}{1-P_rP_s}. \tag{42}$$

Thus, we have the closed form

$$\sum_{i=-\infty}^{\infty}\gamma(i)\gamma(i+j) = S(j)$$

$$= \begin{cases} 2\left[\sum_{i=0}^{M-1}\gamma^2(i) + \sum_{r=1}^{L}\sum_{s=1}^{L}\frac{\lambda_r\lambda_s}{1-P_rP_s}\right] - \gamma^2(0), & j=0 \\ 2\left[\sum_{i=0}^{M-1}\gamma(i)\gamma(i+j) + \sum_{i=1}^{(j-1)/2}\gamma(i)\gamma(j-i) + \sum_{r=1}^{L}\sum_{s=1}^{L}\lambda_r\lambda_s\frac{P_s^j}{1-P_rP_s}\right], & j \text{ odd} \\ 2\left[\sum_{i=0}^{M-1}\gamma(i)\gamma(i+j) + \sum_{i=1}^{(j/2)-1}\gamma(i)\gamma(j-i) + \sum_{r=1}^{L}\sum_{s=1}^{L}\lambda_r\lambda_s\frac{P_s^j}{1-P_rP_s}\right] + \gamma^2\left(\frac{j}{2}\right), & j \text{ even.} \end{cases} \tag{43}$$

Recognizing that

$$\sum_{i=-\infty}^{\infty}\gamma(i)\gamma(i+l-k) = S(l-k) \tag{44a}$$

and

$$\sum_{i=-\infty}^{\infty}\gamma(i+k+k_1-1)\gamma(i-l-k_1+1) = S(l+k+2k_1-2) \tag{44b}$$

we see that the series (16) converges to the sum of these two limits, so

$$\Lambda(k_1,k_2)_{k,l} = S(l-k) + S(l+k+2k_1-2),$$
$$k,l = 1,2,\cdots,k_2-k_1+1. \tag{45}$$

F. Asymptotic Evaluation of Fisher's Information Matrix for the Data of an ARMA Process

The system (3) generating x_t can be expressed in terms of the complex conjugate pole and zero pairs as

$$\epsilon_t = \frac{\prod_{j\in\Pi_M}(1-2\theta_j z^{-1}\cos\theta_{j+1}+\theta_j^2 z^{-2})}{\prod_{i\in\xi_M}(1-2\theta_i z^{-1}\cos\theta_{i+1}+\theta_i^2 z^{-2})}\, x_t. \tag{46}$$

As established by Box and Jenkins [16, p. 270], the information matrix for the data is, for large N,

$$I_X(\boldsymbol{\theta}) = \frac{1}{4\sigma_\epsilon^4}E\left[\left(\nabla_\theta\sum_{t=1}^{N}\epsilon_t^2\right)\left(\nabla_\theta^T\sum_{s=1}^{N}\epsilon_s^2\right)\right]$$

$$= \frac{1}{\sigma_\epsilon^4}E\left[\sum_{t=1}^{N}\epsilon_t(\nabla_\theta\epsilon_t)\sum_{s=1}^{N}\epsilon_s(\nabla_\theta^T\epsilon_s)\right] \tag{47}$$

Gradients are obtained by differentiating (46), yielding

$$\frac{\partial\epsilon_t}{\partial\theta_m} = H_m(z)\,\epsilon_t$$

$$=\begin{cases} J(m)\dfrac{2z^{-1}(\cos\theta_{m+1}-\theta_m z^{-1})}{1-2\theta_m z^{-1}\cos\theta_{m+1}+\theta_m^2 z^{-2}}\epsilon_t, \\ \quad m\in\Pi_M\cup\zeta_M \\ J(m)\dfrac{-2z^{-1}\theta_{m-1}\sin\theta_m}{1-2\theta_{m-1}z^{-1}\cos\theta_m+\theta_{m-1}^2 z^{-2}}\epsilon_t, \\ \quad m\in\Pi_A\cup\zeta_A \end{cases} \tag{48}$$

where

$$J(m)=\begin{cases}1, & m\in\zeta\\ -1, & m\in\Pi.\end{cases} \tag{49}$$

Rewriting (48),

$$d_{m,t}=\frac{\partial\epsilon_t}{\partial\theta_m}, \tag{50}$$

we see that $d_{m,t}$ is the output of the resonant system $H_m(z)$ whose poles consist of the pole or zero pair to which θ_m belongs, driven by the white sequence ϵ_t. The z^{-1} factor in the numerator of each $H_m(z)$ indicates a unit delay, so

$$H_m(z)=h_{m,1}z^{-1}+h_{m,2}z^{-2}+\cdots \tag{51}$$

and hence

$$d_{m,t}=h_{m,1}\epsilon_{t-1}+h_{m,2}\epsilon_{t-2}+\cdots. \tag{52}$$

Incorporating (52) into (47),

$$I_X(\boldsymbol{\theta})_{m,n}=\frac{1}{\sigma_\epsilon^4}E\left[\sum_{t=1}^{N}\epsilon_t\sum_{u=1}^{\infty}h_{m,u}\epsilon_{t-u}\cdot\sum_{s=1}^{N}\epsilon_s\sum_{v=1}^{\infty}h_{n,v}\epsilon_{s-v}\right]$$

$$=\frac{1}{\sigma_\epsilon^4}\left[\sum_{u=1}^{\infty}h_{m,u}h_{n,u}\sum_{t=1}^{N}E(\epsilon_t^2\epsilon_{t-u}^2)\right]$$

$$=N\sum_{u=1}^{\infty}h_{m,u}h_{n,u} \tag{53}$$

so the (m,n)th element of the asymptotic information matrix is N times the inner product of the impulse responses of the systems generating $d_{m,t}$ and $d_{n,t}$. To evaluate these inner products, we first determine the $h_{m,i}$ in (52). With the aid of the expansion [21, p. 40]

$$\frac{1}{1-2P\cos x+P^2}=\frac{1}{\sin x}\sum_{i=1}^{\infty}P^{i-1}\sin ix \tag{54}$$

we rewrite

$$\frac{1}{1-2\theta_j z^{-1}\cos\theta_{j+1}+\theta_j^2 z^{-2}}=\sum_{i=1}^{\infty}\rho_i(j)z^{-(i-1)} \tag{55}$$

where

$$\rho_i(j)=\frac{1}{\theta_j\sin\theta_{j+1}}\theta_j^i\sin i\theta_{j+1}. \tag{56}$$

Substitution of (55) and (56) into (48) gives

$$H_m(z)=\begin{cases} 2J(m)z^{-1}\Big\{\rho_1(m)\cos\theta_{m+1} \\ \quad +\sum_{i=2}^{\infty}[\rho_i(m)\cos\theta_{m+1} \\ \quad -\rho_{i-1}(m)\theta_m]z^{-(i-1)}\Big\}, \quad m\in\Pi_M\cup\zeta_M \\ -2J(m)z^{-1}\theta_{m-1}\sin\theta_m\sum_{i=1}^{\infty}\rho_i(m-1)z^{-(i-1)}, \\ \quad m\in\Pi_A\cup\zeta_A. \end{cases} \tag{57}$$

Equating coefficients of z^{-i} in (51) with those in (57) indicates that (53) involves sums of the form

$$\sum_{i=1}^{\infty}\rho_{i+u}(j)\rho_{i+v}(k),$$

so define

$$F_{u,v}(j,k)=\sum_{i=1}^{\infty}\rho_{i+u}(j)\rho_{i+v}(k) \tag{58}$$

and use the identities [21, p. 40]

$$\sum_{i=1}^{\infty}P^i\sin ix=\frac{P\sin x}{1-2P\cos x+P^2}$$

$$\sum_{i=1}^{\infty}P^i\cos ix=\frac{P\cos x-P^2}{1-2P\cos x+P^2}$$

to get

$$F_{u,v}(j,k)=\frac{\theta_j^u\theta_k^v}{2\sin\theta_{j+1}\sin\theta_{k+1}}\left[\frac{\cos[(u+1)\theta_{j+1}-(v+1)\theta_{k+1}]-\theta_j\theta_k\cos(u\theta_{j+1}-v\theta_{k+1})}{1-2\theta_j\theta_k\cos(\theta_{j+1}-\theta_{k+1})+(\theta_j\theta_k)^2}-\frac{\cos[(u+1)\theta_{j+1}+(v+1)\theta_{k+1}]-\theta_j\theta_k\cos(u\theta_{j+1}+v\theta_{k+1})}{1-2\theta_j\theta_k\cos(\theta_{j+1}+\theta_{k+1})+(\theta_j\theta_k)^2}\right]. \tag{59}$$

Now we reduce (53) using the $h_{m,u}$ and $h_{n,u}$ obtained by equating coefficients of (51) and (57) to get

$$I_X(\boldsymbol{\theta})_{m,n}\simeq\begin{cases} 4NJ(m)J(n)[(\theta_m\theta_n+\cos\theta_{m+1} \\ \quad\cdot\cos\theta_{n+1})F_{0,0}(m,n) \\ \quad-\theta_m\cos\theta_{n+1}F_{0,1}(m,n) \\ \quad-\theta_n\cos\theta_{m+1}F_{1,0}(m,n)], \\ \quad m,n\in\Pi_M\cup\zeta_M \\ -4NJ(m)J(n)\theta_{n-1}\sin\theta_n[F_{0,0}(m,n-1) \\ \quad\cdot\cos\theta_{m+1}-F_{0,1}(m,n-1)\theta_m], \\ \quad m\in\Pi_M\cup\zeta_{M'}, n\in\Pi_A\cup\zeta_A \\ 4NJ(m)J(n)\theta_{m-1}\theta_{n-1}\sin\theta_m \\ \quad\cdot\sin\theta_n F_{0,0}(m-1,n-1), \\ \quad m,n\in\Pi_A\cup\zeta_{A'}. \end{cases} \tag{60}$$

Once again, the approximation can be made arbitrarily precise by large enough choice of N.

IV. Relative Information Index Calculations for Some ARMA Processes

The classes of time series considered in this section are the AR(2) process, the AR(2) process in additive white Gaussian noise, the sum of two AR(2) processes, and an ARMA process having a dominant zero and therefore a PSD with a deep trough or antiresonance. These are chosen because they are simpler processes exhibiting the features of more general processes having a multipeak and/or multitrough PSD. As a guide in interpreting the plots, recall that values of the RII or PRII near unity imply little information loss for the set of $c(i)$ used, and small values indicate substantial increase in Cramér-Rao bounds of parameter estimators based on the relevant set of $c(i)$.

A. AR(2) Process

Fig. 2 details R versus pole magnitude $|P|$ for pole angle $\phi_p = 0.25\ \pi$ and several choices of k_1 and k_2 for the sample ACF vector $C_{[k_1, k_2]}$. Notice that the curve for $k_1 = 0, k_2 = 2$ is unity, irrespective of pole position. This illustrates the consistency of the RII with the fact that $C_{[0, L]}$ is asymptotically sufficient for the parameters of an AR(L) process. The other curves depict the effect of dropping early lags of the sample ACF while still using the same number of lags. Clearly, the early lags, particularly $c(0)$ and $c(1)$, do much to stabilize AR estimators, especially for wide-band ($|P| \leqslant 0.8$, say) processes.

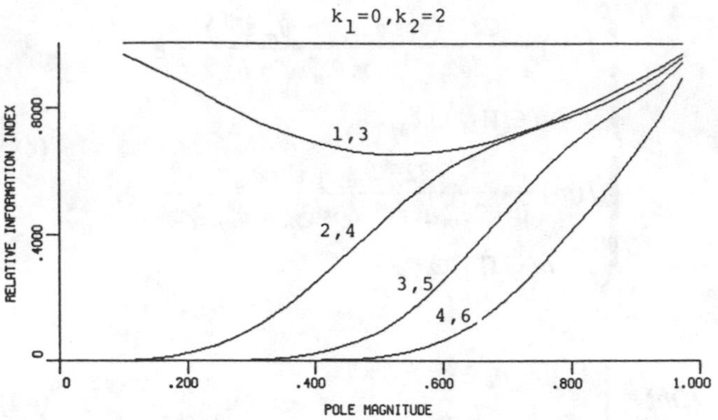

Fig. 2. AR(2) process, $\phi_P = 0.25\ \pi$, using $C_{[k_1, k_2]}$.

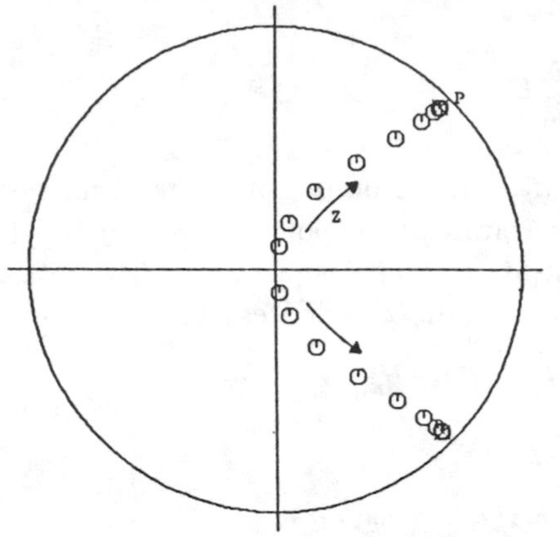

Fig. 3. Migration of zero pair with decreasing SNR for AR(2) process with $|P| = 0.95$, $\phi_P = 0.25\ \pi$ in white noise. Pole pair denoted by "X," zero pair by "⊙."

B. AR(2) Process in White Noise

The AR(2) process in an additive white noise environment is one of the most basic time series encountered in practice. For example, a process having a single peak PSD in a white noise background can often be adequately modeled this way. In studying the information in $C_{[k_1, k_2]}$, we assume first that $k_1 = 0$.

The sum of an AR(L) process and white noise is an ARMA (L, L) process, with zeros determined by pole position and signal-to-noise ratio (SNR). The SNR is defined

$$\text{SNR} = E(x^2)/\sigma_n^2$$

where $E(x^2)$ is the AR(2) process variance and σ_n^2 the noise variance. Before inspecting the plots in detail, we note that Figs. 4 and 5 reveal a nondecreasing behavior of R for $C_{[0, k_2]}$ as SNR or k_2 is increased or $|P|$ is decreased, irrespective in each case of the fixed values of the other independent variables. If no noise were present, Fig. 5 would be identically one. Although not shown, poles-only R_1 is similar. The effect of the noise is to introduce a zero pair starting at the origin for SNR = $+\infty$ dB and traveling outward toward the poles, cancelling them in the limit as SNR $\rightarrow -\infty$ dB. See for example Fig. 3. A derivation of the zero pair corresponding to an AR(2) process in white noise is given in [3] and [25]. It is this zero that is responsible for the suboptimality of $C_{[0, k_2]}$, $k_2 \geqslant L$, for noisy AR(L) processes, with R decreasing as $|Z|$, the zero magnitude, approaches one. More will be said on this in a subsequent subsection. Fig. 4 depicts plots of R versus SNR for different values of $|P|$, with $\phi_P = 0.25\ \pi$ and $k_2 = 4$. Note that this figure shows SNR thresholds, below which R drops rapidly, especially for narrow-band processes. This behooves us to consider the information in the higher lags of $c(i)$ for such processes. Fig. 5 shows R versus ϕ_P for $|P| = 0.95$, SNR = 5 dB and various values of k_2. There is mirror symmetry about $\phi_P = 0.5\ \pi$, so the range $0.5\ \pi \leqslant \phi_P \leqslant \pi$ is not shown. For a given ϕ_P, R is seen to monotonically increase as k_2 increases. This confirms that, for narrow-band processes, there is substantial information in the higher lags. Fig. 5 also shows a threshold ϕ_P, depending on SNR, $|P|$, and k_2, below which R drops rapidly. An interpretation of this phenomenon is to consider the values of ϕ_P near 0 or π radians as a case where the two complex poles are near each other (high resolution case) and interact strongly through the noise-generated zero. Fig. 6 shows the close proximity of the introduced zero to the unit circle as ϕ_P nears the ends of its range. Thus, we expect estimates of spectra requiring high resolution to be "noisier" than low resolution cases even when the same $c(i)$ are used optimally in both cases.

Table I summarizes suggested values of k_2 for different values of $|P|$ and SNR. These are the minimum k_2 required to achieve $R \geqslant 0.5$ at $\phi_P = 0.05\ \pi$. This angle is chosen because R tends to increase with ϕ_P, so an acceptable value of R at $\phi_P = 0.05\ \pi$ should be sufficient over most of the range of ϕ_P.

We now show that estimators which do not use the first few lags of the sample ACF usually have much smaller R and hence much larger Cramér-Rao bounds than those using the same number of lags but including the early ones. This behavior was observed in the earlier study of an AR(2) process in a noiseless environment. Fig. 7 considers using five lags of the sample

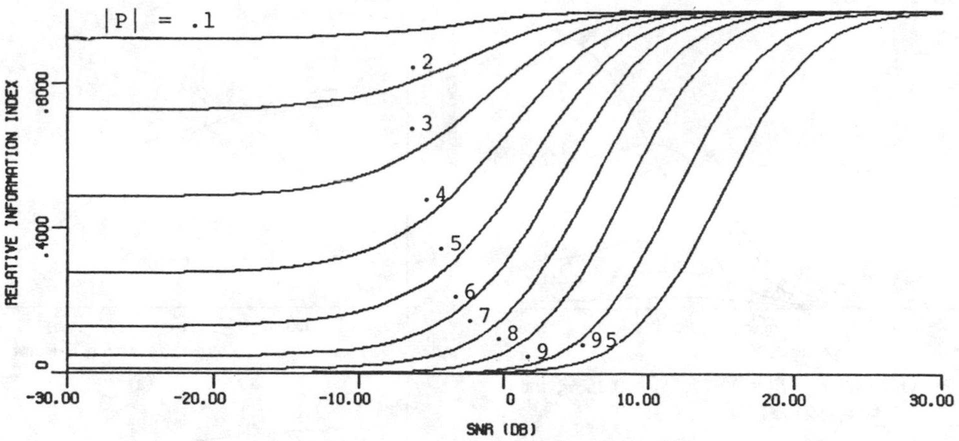

Fig. 4. AR(2) process in white noise, $\phi_P = 0.25\,\pi$, using $C_{[0,4]}$.

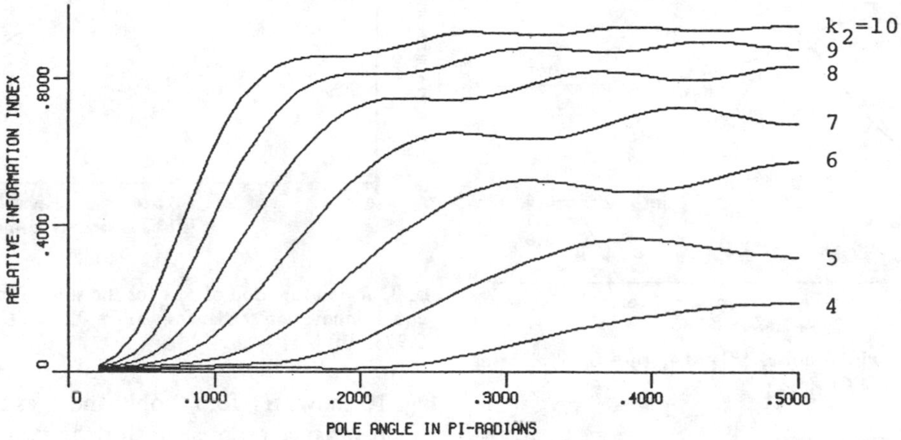

Fig. 5. AR(2) process in white noise, $|P| = 0.95$, SNR = 5 dB, using $C_{[0,k_2]}$.

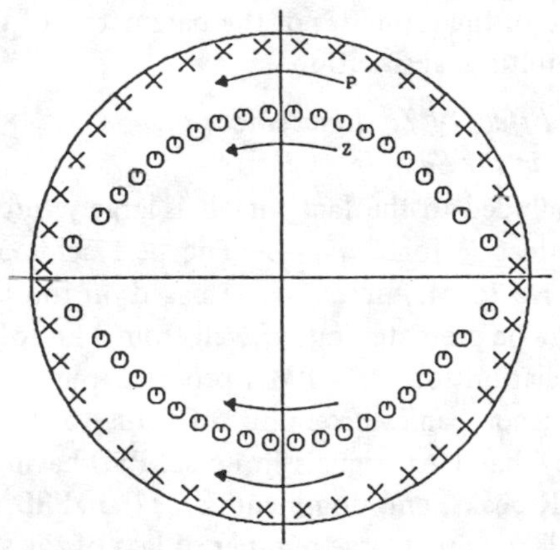

Fig. 6. Migration of zero pair for AR(2) process in white noise with $|P| = 0.95$, SNR = 5 dB, and equal increments of ϕ_P in the range $(0, \pi)$. Pole pair denoted by "X," zero pair by "⊙."

ACF to estimate parameters of a narrow-band AR(2) process $|P| = 0.975$, $\phi_P = 0.25\,\pi$ in white noise for SNR in the range -30 dB to 30 dB. In this case deterioration in R is not present at all SNR's as we move out to higher lags of the sample ACF. For a 10 dB SNR, for example, R is larger for $C_{[1,5]}$ and $C_{[2,6]}$ than for $C_{[0,4]}$. Above about 15 dB, R deteriorates monotonically as k_1 and k_2 of $C_{[k_1,k_2]}$ are incremented

TABLE I

MINIMUM k_2 REQUIRED TO ACHIEVE $R \geqslant 0.5$ AT $\phi_P = 0.05\pi$ FOR $C_{[0,k_2]}$

$\|P\|$	SNR (dB) -5	0	5	10	15	20	25	30
0.975	28	21	16	12	9	7	6	5
0.950	23	17	13	10	8	6	5	4
0.925	19	15	12	9	7	6	5	4
0.900	17	13	10	8	6	5	4	4
0.850	14	11	9	7	6	5	4	4
0.800	12	9	7	6	5	4	4	4
0.750	10	8	7	5	4	4	4	4
0.700	10	8	6	5	4	4	4	4
0.650	9	7	6	5	4	4	4	4
0.600	8	6	5	4	4	4	4	4

equally. Fig. 8 considers five lags of the sample ACF again, this time for the wide-band process $|P| = 0.4$, $\phi_P = 0.25\,\pi$ in white noise. Over a useful range of SNR (SNR > 5 dB), $C_{[0,4]}$ has a larger R than does $C_{[1,5]}$, although both have $R >$ 0.5. Going to $C_{[2,6]}$, $C_{[3,7]}$, and $C_{[4,8]}$, we observe more dramatic drops in R than occurred for the narrow-band AR process, and we find it necessary to use a logarithmic plot to discern these curves from zero. Thus, information about the poles and zeros of an AR(2) process in white noise appears to

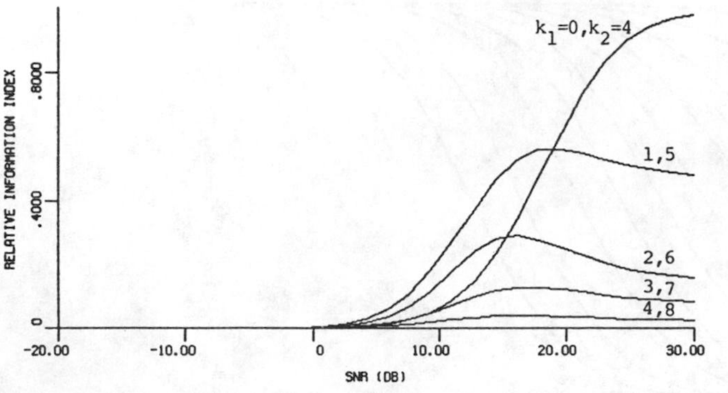

Fig. 7. AR(2) process in white noise, $|P| = 0.975$, $\phi_P = 0.25\ \pi$, using $C_{[k_1, k_2]}$.

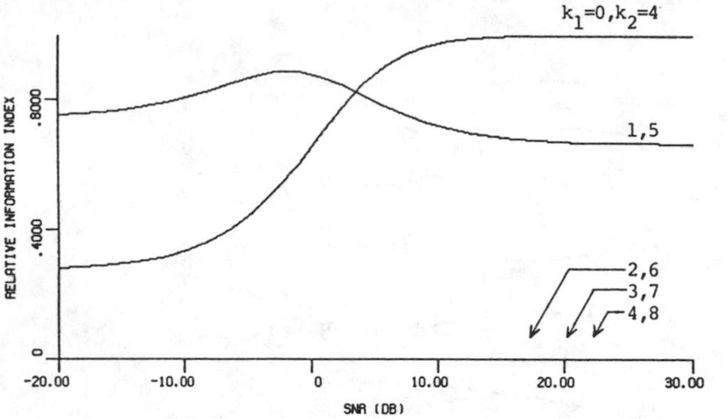

Fig. 8. AR(2) process in white noise, $|P| = 0.4$, $\phi_P = 0.25\ \pi$, using $C_{[k_1, k_2]}$.

be concentrated into the first few lags of the sample ACF if the AR process is wide-band. This is consistent with our findings for an AR(2) process in a noiseless environment.

C. Sum of Two AR(2) Processes

Further results on the dependence of R on the separation of interacting poles are given in this subsection, where by "interacting" we refer here to the introduction of zeros caused by summing independent AR processes. For the case of two AR(2) processes, the result is an ARMA(4, 2) process, and it is once again the introduction of zeros that reduces R for $C_{[0, k_2]}$ below unity for moderate values of k_2.

Fig. 9(a) and (b) show R for $C_{[0,8]}$ as one pole pair, $P1$, is held constant and the other, $P2$, is swept over a range of angle ϕ_{P2}. They clearly show again a loss of information in the sample ACF when high resolution spectra are involved. For example, Fig. 9(a) corresponds to $|P1| = 0.975$, $\phi_{P1} = 0.5\ \pi$, and $|P2| = 0.975$ or $|P2| = 0.6$. As $\phi_{P2} \to 0.5\ \pi$, separation of the two narrow-band spectral peaks ($|P2| = 0.975$) diminishes and so does R. As in the case of an AR(2) signal in white noise, it is the close proximity of the introduced zero to the unit circle that causes a large drop in R. For $|P2| = 0.6$, the zero lies approximately "midway" between the poles and hence is not as close to the unit circle, ergo the higher values of R. The curve in Fig. 9(b) corresponding to $|P1| = 0.6$, $|P2| = 0.6$ shows further improvement in R because $|Z|$ is always less than 0.6.

We now consider the relative information in $C_{[0,8]}$ for two closely spaced AR(2) signals with different innovation variances.

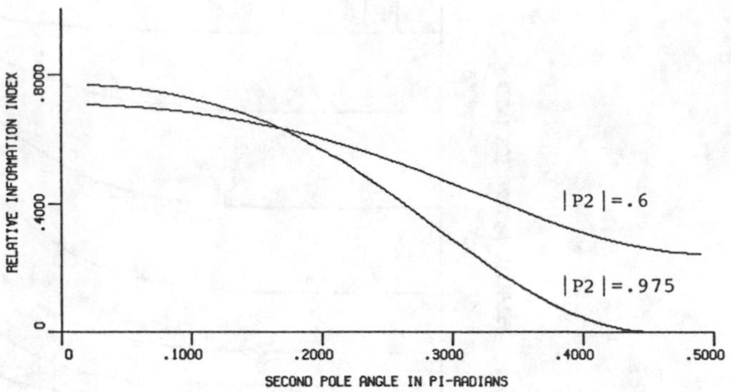

(a)

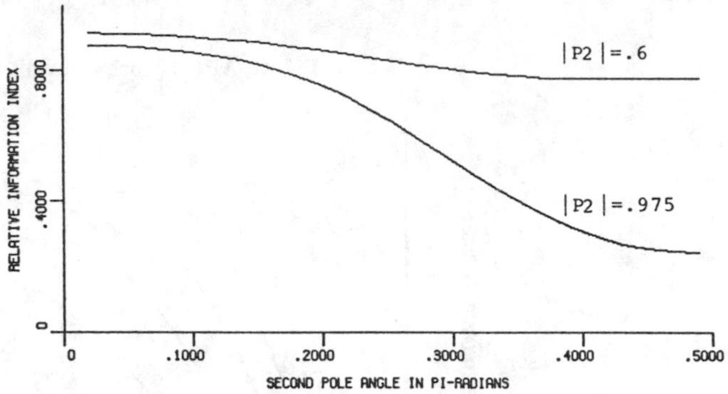

(b)

Fig. 9. R as a function of ϕ_{P2} for the sum of two AR(2) processes with equal innovation variances, $\phi_{P1} = 0.5\ \pi$, $C_{[0,8]}$ used. (a) $|P1| = 0.975$. (b) $|P1| = 0.6$.

Fig. 10 shows R_1 for $P2$ only and R as a function of the innovation variance ratio σ_2^2/σ_1^2 for the two signals. We observe a monotonic growth in R_1 as the relative strength of $P2$ is increased, whereas R achieves its maximum at equal innovation variances. This behavior is as expected and shows the superior performance of the estimates of the parameters of the stronger signal in a multisignal situation.

D. General Effect of Zero Placement on Information in the Sample ACF

We have alluded to the fact that is is largely zero placement that determines R for $C_{[0, k_2]}$. The best achievable performance relative to ML of estimates based on the sample ACF can therefore be predicted rather well from a knowledge of the zero constellation for the ARMA process. Simply stated, deterioration accompanies proximity of zeros to the unit circle. Thus, narrow-band AR signals in noise, PSD having closely interfering AR peaks, and dominant zeros (i.e., PSD with antiresonances) all require a large number of lags of the sample ACF to even allow the possibility that the performance of a parameter estimator based on the sample ACF rivals that of ML class ARMA estimators.

To investigate further, consider using the sample ACF $C_{[0, k_2]}$ to estimate parameters of an ARMA(L, M) process, $L \leqslant k_2$, where all M zeros are on the origin, a special case in which no information loss accompanies the presence of zeros. Then the process is strictly AR(L), and the RII for $C_{[0, k_2]}$ is identically one, irrespective of pole positions. It turns out that this independence of R for $C_{[0, k_2]}$, $k_2 \geqslant L$, to pole position holds approximately for any ARMA(L, M) process whose

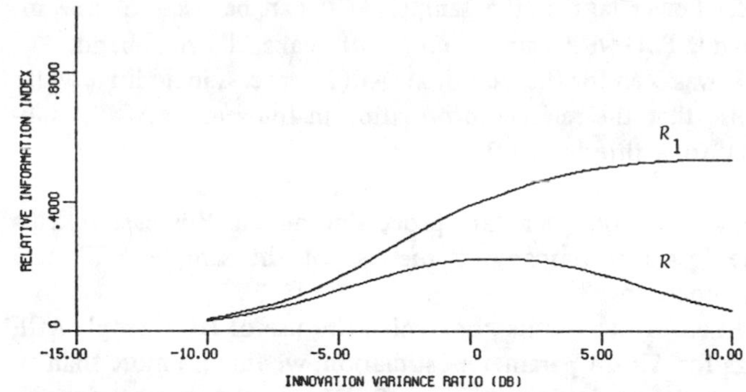

Fig. 10. R and R_1 for $P2$ only as a function of σ_2^2/σ_1^2, $|P1| = 0.95$, $\phi_{P1} = 0.2\ \pi$, $\phi_{P2} = 0.3\ \pi$, $C_{[0,12]}$ used.

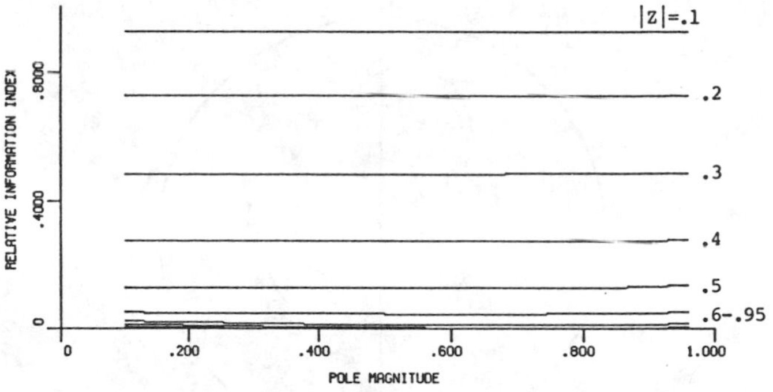

Fig. 11. ARMA(2, 2) process with $\phi_P = 0.252\ \pi$, $\phi_Z = 0.25\ \pi$, and $|P|$ and $|Z|$ indicated, using $C_{[0,4]}$.

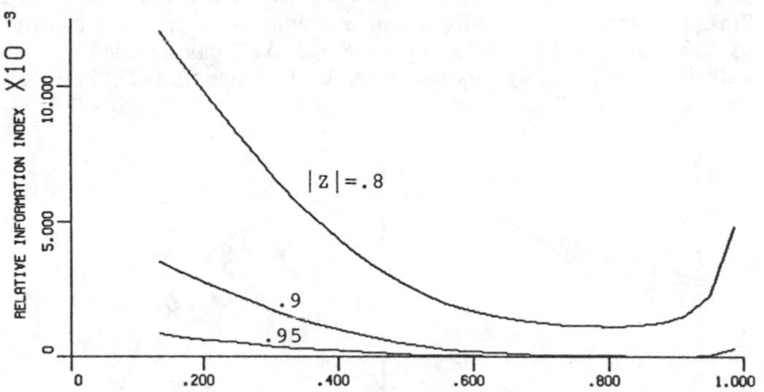

Fig. 12. Expanded scale to view curves of Fig. 11 near zero.

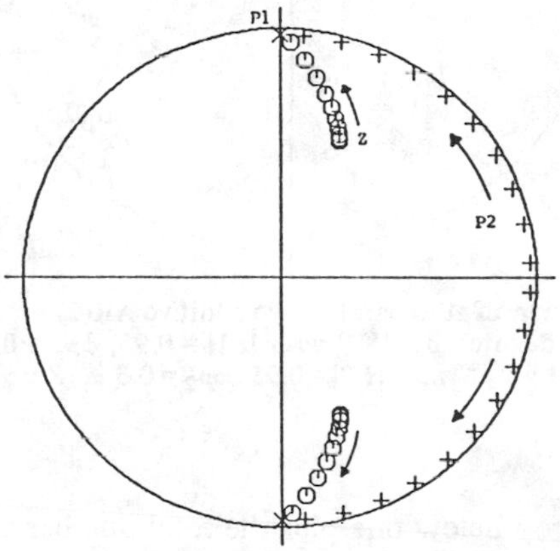

Fig. 13. Migration of zero pair for two additive narrow-band AR(2) processes of Fig. 9(a). $P1$ pair denoted by "X," with $|P1| = 0.975$, $\phi_{P1} = 0.5\ \pi$, and $P2$ pair denoted by "+," with $|P2| = 0.975$ and ϕ_{P2} incrementing in the range $(0, \pi/2)$. Zero pair denoted by "$\odot$."

zeros are not too near the unit circle. In these cases, zero location alone almost entirely determines R.

Turning now to ARMA(2, 2) processes, Fig. 11 depicts the monotonic decrease in R as the zero pair approaches the unit circle. Each curve corresponds to a zero fixed at $\phi_z = 0.25\ \pi$ and magnitude indicated, and is a plot of R versus $|P|$ for $\phi_P = 0.252\ \pi$ and $k_2 = 4$. Throughout this section $k_2 = 4$ is used, but the same behavior holds for arbitrary k_2. Two points are of interest: first, for $|Z|$ less than about 0.6, the curves are essentially constant and hence determined by the zeros only. Second, there are no sudden changes in R as $|P|$ approaches $|Z|$, causing near cancellation of poles and zeros because of their very similar angles. Thus, pole-zero interactions do not influence the information in the sample ACF $C_{[0,k_2]}$ relative to that in the data for small $|Z|$. We mention that the absolute information in $C_{[0,k_2]}$ and the data are profoundly influenced by pole-zero near-cancellation, in that Fisher's information matrix for each becomes nearly singular [3], indicating that both have very large Cramér–Rao bound generalized variances. Fig. 12 contains curves of R for $|Z| \geqslant 0.8$, which were difficult to discern in Fig. 11 because of the relatively low values of R. The emerging influence of the poles on R as zeros approach the unit circle is evident. More importantly, the deterioration in performance of ARMA(2, 2) estimators based on $C_{[0,k_2]}$ when zeros are near the unit circle is clearly severe. This situation arises 1) when the system generating the data has a dominant zero pair and therefore a PSD with a trough or "antiresonance" or 2) when a narrow-band AR(2) process is observed at low SNR in additive white noise.

The ARMA(4, 2) process is subject to the same basic behavior for R as was observed for the ARMA(2, 2) process. Once again, the zero pair plays the major role in determining R when it is sufficiently far from the unit circle. Near cancellation of either pole pair by the zero pair has little effect on R so that pole-zero interactions once again do not influence the information in $C_{[0,k_2]}$ relative to that in the data for small $|Z|$. Fig. 13 shows zero migration corresponding to the case of the two narrow-band additive AR(2) processes in Fig. 9(a) and Fig. 14 shows zero migration for the case in Fig. 9(a) where one of the processes is wide-band. Derivations of these zero locations are found in [3]. The relationship between R and zero position is obvious. The migration of zeros for the study in Fig. 10 of the effect of varying relative powers of two additive AR(2) processes is interesting to observe. Fig. 15 shows that, as one pair is given more power, the weaker one is approached by the zero pair and will ultimately be cancelled by it as the power ratio becomes extreme. What will remain in the PSD is a single peak.

V. Conclusion

The relative information index has been evaluated for several ARMA signal classes for various choices of consecutive lags of the sample ACF. The number of lags considered has not been

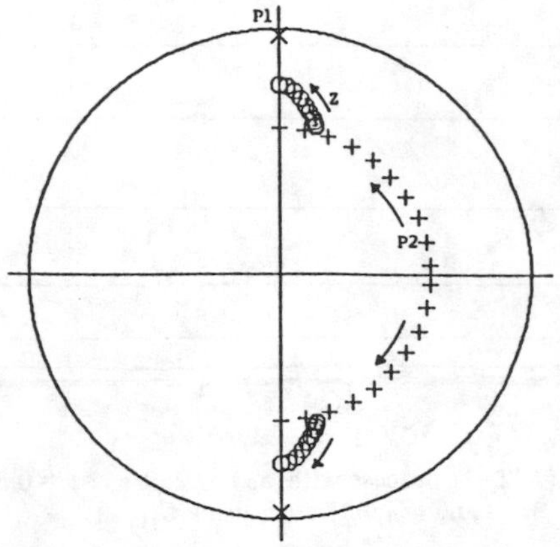

Fig. 14. Migration of zero pair for two additive AR(2) processes of Fig. 9(a), one process narrow-band and one wide-band. $P1$ pair denoted by "X," with $|P1| = 0.975$, $\phi_{P1} = 0.5\,\pi$, and $P2$ pair denoted by "+," with $|P2| = 0.6$ and ϕ_{P2} incrementing in the range $(0, \pi/2)$. Zero pair denoted by "$\odot$."

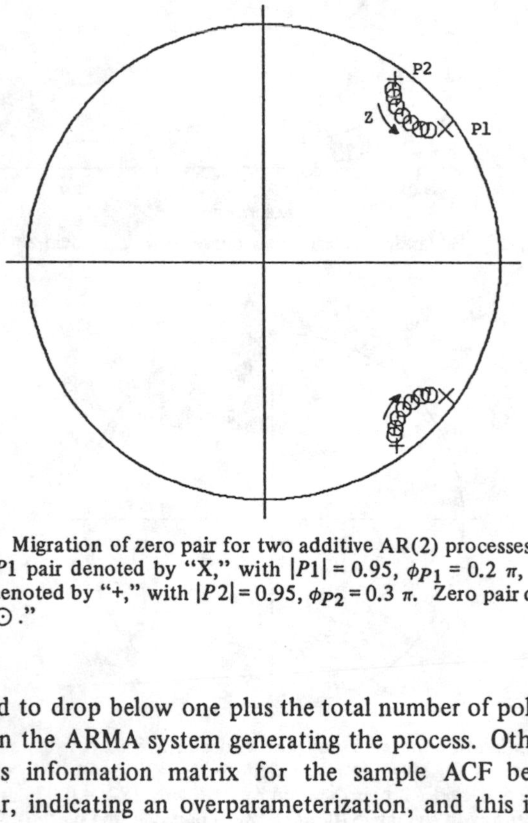

Fig. 15. Migration of zero pair for two additive AR(2) processes of Fig. 10. $P1$ pair denoted by "X," with $|P1| = 0.95$, $\phi_{P1} = 0.2\,\pi$, and $P2$ pair denoted by "+," with $|P2| = 0.95$, $\phi_{P2} = 0.3\,\pi$. Zero pair denoted by "$\odot$."

allowed to drop below one plus the total number of poles and zeros in the ARMA system generating the process. Otherwise Fisher's information matrix for the sample ACF becomes singular, indicating an overparameterization, and this in turn forces the relative information index to zero. The results of this investigation suggest the following rough guidelines for the design of ARMA parameter estimators based on the sample ACF if it is desired to obtain near optimum (maximum likelihood) performance.

1) Include early lags of the sample ACF, especially if wide-band peaks are present in the PSD of the observed time series.

2) Use many lags of the sample ACF if SNR is low and a single narrow-band peak is present in the PSD.

3) Use many lags of the sample ACF, regardless of SNR, if several peaks are present in the PSD and some of them are narrow-band. This recommendation is even more important if some narrow-band peaks are closely adjacent in frequency.

4) Use many lags of the sample ACF if the PSD is known to have antiresonances, indicating dominant zeros.

5) Fewer lags of the sample ACF can be used in cases involving PSD with a small number of peaks, all broad-band.

It was seen for the case of an AR(2) process in additive white noise that the relative information in the sample ACF is subject to a threshold effect, dropping rapidly when SNR dips below what may be called a usable range. This behavior is typical of nonlinear data processing and in this case reflects the losses accompanying the use of the sample ACF data preprocessing.

In all of our results concerning the use of later sample ACF lags for ARMA parameter estimation, we find no more than an approximate doubling of Cramér-Rao bounds when using the set of lags associated with the extended Yule-Walker equations instead of the same number of lags starting with the zeroth one. In some instances involving ARMA(2, 2) and ARMA(4, 2) processes, exclusion of the zeroth lag gives a lower CR bound, although more practical cases (e.g., AR(2) processes in white noise with SNR $>$ 0 dB) tend to favor inclusion of the zeroth lag. This indicates that the algebraic instability of the extended Yule-Walker equations [11] is not necessarily a harbinger of statistical stability problems of similar magnitude. Thus, potentially adequate estimators of ARMA process poles may reside in methods that skirt ill-conditioning in the extended Yule-Walker or related equations, as in [13] and [24]. Exclusion of both the zeroth and the first lags of the sample ACF leads to severe information loss in most cases we have considered [3].

Consistent with our findings, sample ACF based spectral estimators best suited to narrow-band ARMA processes or sinusoids (e.g., [12], [13]) involve a large number of lags of the sample ACF. This is also borne out in [22], where it is shown that the variance of spectral estimates for narrow-band AR(2) processes in white noise is decreased by extending the set of extended Yule-Walker equations and solving the overdetermined set by least squares, whereas an increase in variance accompanies doing the same for a wide-band AR(2) process in white noise. This suggests that, for narrow-band processes, any statistical inefficiencies in extending the set of linear equations are outweighed by the information gain accompanying inclusion of later lags of the sample ACF. The efficiency with which overdetermined schemes use the information in these lags is yet to be quantified, however.

Some care must be exercised in assessing the implications of the results of this study, which pertain to ARMA parameter estimates, on the ARMA spectral estimate, in that the two are not linked in a straightforward manner. Generally, use of an ARMA model or some reparameterization thereof as an intermediate step in constructing a spectral estimate will tie the quality of the spectral estimate to that of the parameter estimates. But the link is tightest to the estimation accuracy of the dominant poles and zeros of the process, i.e., poor estimates of poles and zeros of near-zero magnitude impact the spectral estimate minimally. For example, model order overdetermination is often used to improve the resolution of an ARMA spectral estimator while increasing the variance of the spectral estimate only slightly. The spurious poles and zeros introduced by this practice are usually small in magnitude and highly unstable in angle, yet they seem to provide the additional degrees of freedom necessary to fine-tune the spectral estimate.

Thus, overparameterization may be poor practice in ARMA parameter estimation, but this does not necessarily carry over to spectral estimation. The philosophy of this paper is that poor estimates of poles and zeros of an ARMA process of known model orders from a particular statistic implies poor spectral estimates from that statistic. This is a special case of the generalized Cramér-Rao inequality, in that the spectrum is a function of the parameter vector (see [32]).

One may raise the objection at this point that perhaps some of the bound deterioration indicated by the RII is a result of the ACF-induced ambiguity on each pole and zero pair as to whether it is located inside the unit circle or at its reflection point outside the unit circle, an ambiguity that does not affect the spectral estimate and hence should be transparent to the RII. The reply is that the RII does not reflect this sort of information loss, as the Fisher information measures the local properties of a statistic, not its modal or ambiguity properties. Our assumption that poles and zeros lie inside the unit circle is merely a simplifying convention restricting the analysis to the familiar stable, causal minimum phase system realization. Within this constraint, incidentally, the mapping from ACF to ARMA model is unique [16, p. 195].

By analyzing information in the sample ACF, we are approaching the estimator design problem from the front end, i.e., the statistic selection stage, with the goal of rejecting sample ACF vectors that lose too much information for a given ARMA process and thereby increase the Cramér-Rao bounds by an unacceptably large amount. Although it has been necessary to do an asymptotic analysis, the results presented here hold approximately for moderate data records as evidenced by their agreement with empirical behavior reported in [22], and they reflect the tradeoffs involved in using the sample ACF in ARMA parameter estimation. We do not address the design of an estimator whose performance approaches the bound. It is our hope that these results will help to confine the search to sample ACF vectors having the most promise.

References

[1] H. Sakai and H. Tokumaru, "Statistical analysis of a spectral estimator for ARMA processes," *IEEE Trans. Automat. Contr.*, vol. AC-25, pp. 122-124, Feb. 1980.

[2] S. P. Bruzzone and M. Kaveh, "A criterion for selecting information-preserving data reductions for use in the design of multiple parameter estimators," *IEEE Trans. Inform. Theory*, vol. IT-29, pp. 466-470, May 1983.

[3] S. P. Bruzzone, "Information considerations in the design of autocorrelation-based ARMA parameter estimators," Ph.D. dissertation, Univ. Minnesota, Minneapolis, June 1982.

[4] A. M. Walker, "Large-sample estimation of parameters for autoregressive processes with moving-average residuals," *Biometrika*, vol. 49, pp. 117-131, 1962.

[5] T. C. Hsia and D. A. Landgrebe, "On a method for estimating power spectra," *IEEE Trans. Instrument. Meas.*, vol. IM-16, Sept. 1967.

[6] D. Graupe, D. J. Krause, and J. B. Moore, "Identification of autoregressive moving-average parameters of time series," *IEEE Trans. Automat. Contr.*, pp. 104-107, Feb. 1975.

[7] H. Sakai and M. Arase, "Recursive parameter estimation of an autoregressive process disturbed by white noise," *Int. J. Contr.*, vol. 30, pp. 949-966, 1979.

[8] M. Kaveh, "High resolution spectral estimation for noisy signals," *IEEE Trans. Acoust., Speech, Signal Processing*, vol. ASSP-27, pp. 286-287, June 1979.

[9] J. F. Kinkel, J. Perl, L. L. Scharf, and A. R. Stubberud, "A note on covariance-invariant digital filter design and autoregressive moving average spectrum analysis," *IEEE Trans. Acoust., Speech, Signal Processing*, vol. ASSP-27, pp. 200-202, Apr. 1979.

[10] S. Bruzzone and M. Kaveh, "On some suboptimum ARMA spectral estimators," *IEEE Trans. Acoust., Speech, Signal Processing*, vol. ASSP-28, pp. 753-755, Dec. 1980.

[11] S. M. Kay, "Noise compensation for autoregressive spectral estimates," *IEEE Trans. Acoust., Speech, Signal Processing*, vol. ASSP-28, pp. 292-303, June 1980.

[12] J. A. Cadzow, "ARMA spectral estimation, a model equation error procedure," in *Proc. 1980 Int. Conf. Acoust., Speech, Signal Processing*, Denver, CO, Apr. 14-16, 1980.

[13] —, "Spectral estimation: An overdetermined rational model equation approach," *Proc. IEEE* (Special Issue on Spectral Estimation), vol. 70, pp. 907-939, Sept. 1982.

[14] M. Arato, "On the sufficient statistics for stationary Gaussian random processes," *Theo. Prob. Appl.*, vol. 6, pp. 199-201, 1961.

[15] W. Gersch, "Estimation of the autoregressive parameters of a mixed autoregressive moving average time series," *IEEE Trans. Automat. Contr.*, pp. 583-588, Oct. 1970.

[16] G. E. P. Box and G. M. Jenkins, *Time Series Analysis: Forecasting and Control.* San Francisco, CA: Holden-Day, 1970.

[17] E. J. Hannan, "The estimation of mixed moving average autoregressive systems," *Biometrika*, vol. 56, no. 3, pp. 579-593, 1969.

[18] A. M. Walker, "The asymptotic distribution of serial correlation coefficients for autoregressive processes with dependent residuals," *Proc. Cambridge Philos. Soc.*, vol. 50, pp. 60-64, 1954.

[19] M. S. Bartlett, "On the theoretical specification and sampling properties of autocorrelated time series," *J. Royal Stat. Soc.*, vols. 8-9, pp. 27-41, 1946.

[20] C. R. Rao, *Linear Statistical Inference and its Applications.* New York: Wiley, 1965.

[21] I. S. Gradshteyn and I. M. Ryzhik, *Table of Integrals, Series and Products*, A. Jeffrey, Ed. New York: Academic, 1965.

[22] M. Kaveh and S. P. Bruzzone, "A comparative overview of ARMA spectral estimation," in *Proc. 1st ASSP Workshop on Spectral Estimation*, Hamilton, Ont., Canada, Aug. 17-18, 1981.

[23] B. W. Dickinson, "Structure of stationary finite observation records of discrete-time stochastic linear systems," *IEEE Trans. Automat. Contr.*, to be published.

[24] Y. T. Chan and R. P. Langford, "Spectral estimation via the high-order Yule-Walker equations," *IEEE Trans. Acoust., Speech, Signal Processing*, vol. ASSP-30, pp. 689-698, Oct. 1982.

[25] M. Kaveh and S. P. Bruzzone, "Statistical efficiency of correlation based methods for ARMA spectral estimation," *IEE Proc.* (Part F—Special Issue on Spectral Estimation), Apr. 1983.

[26] B. Friedlander, "A recursive maximum likelihood algorithm for ARMA spectral estimation," *IEEE Trans. Inform. Theory*, vol. IT-28, pp. 639-646, July 1982.

[27] —, "Recursive lattice forms for spectral estimation," *IEEE Trans. Acoust., Speech, Signal Processing*, vol. ASSP-30, pp. 920-930, Dec. 1982.

[28] A. A. Beex and L. L. Scharf, "Covariance sequence approximation for parametric spectrum modeling," *IEEE Trans. Acoust., Speech, Signal Processing*, vol. ASSP-29, pp. 1042-1052, Oct. 1981.

[29] K. J. Aström, "On the achievable accuracy in identification problems," in *Proc. 1967 IFAC Symp. Identification*, Prague, Czechoslovakia.

[30] J. P. Dugré, A.A.L. Beex, and L. L. Scharf, "Generating covariance sequences and the calculation of quantization and rounding error variances in digital filters," *IEEE Trans. Acoust., Speech, Signal Processing*, vol. ASSP-28, pp. 102-104, Feb. 1980.

[31] W. Gersch, N. N. Nielsen, and H. Akaike, "Maximum likelihood estimation of structural parameters from random vibration data," *J. Sound and Vibration*, vol. 31, no. 3, pp. 295-308, 1973.

[32] B. Friedlander and B. Porat, "A general lower bound for parametric spectrum estimation," *IEEE Trans. Acoust., Speech, Signal Processing*, to be published.

A General Lower Bound for Parametric Spectrum Estimation

BENJAMIN FRIEDLANDER, SENIOR MEMBER, IEEE, AND BOAZ PORAT, MEMBER, IEEE

Abstract—**The paper presents a lower bound on the variance of parametric spectral estimators. The bound is potentially useful for evaluating the performance of existing and newly proposed spectrum estimation techniques. Explicit formulas are given for ARMA and AR-plus-noise models. The behavior of the bound is illustrated by several examples.**

I. Introduction

PARAMETRIC modeling of stationary random processes is a common practice in many engineering and statistics applications. Especially popular are the rational models of the autoregressive (AR) and autoregressive moving-average (ARMA) type [1]. Among nonrational models we mention the class of processes with Gaussian spectra, used in some radar and sonar applications [2]. Parametric modeling of random processes can be accomplished by well established parameter estimation techniques, such as maximum likelihood and least squares. By comparison, nonparametric analysis typically relies on heuristic techniques, such as windowing, etc.

The interest in parametric spectrum estimation has been steadily growing in recent years. Estimation techniques proposed in the statistics and signal processing literature are probably in the hundreds (see, e.g., references in the recent survey [3]). Unfortunately, many of these works (especially in the engineering literature) are not accompanied by analysis, and important questions such as the consistency and efficiency of the estimator are often ignored. The performance of the proposed techniques is often demonstrated by isolated examples, usually generated via Monte Carlo simulations. In many cases, no comparison is made to theoretical lower bounds, making it difficult to evaluate the performance of the proposed algorithm.

The aim of this paper is to provide a general lower bound on the performance of any parametric spectral estimation technique. The main novelty is that the bound is expressed in terms of the points of the spectrum, and not just in terms of the parameters. The bound is useful as a reference for evaluating the performance of existing and newly proposed spectrum estimation methods, at least in the asymptotic case (i.e., for a large amount of data).

Explicit expressions are provided for ARMA and AR-plus-noise models. Examples are then given for several cases previously tested by various authors.

Finally, we note that asymptotic bounds on parametric spectral estimates were presented in [11], [13] for the modified Yule-Walker estimator. In [11] the asymptotic variance is derived using the periodogram technique described in [12]. A very different derivation, based on some properties of the instrumental variable method, is presented in [13]. Since the modified Yule-Walker estimator is generally not efficient, these bounds will be larger than the bounds presented here.

Manuscript received August 2, 1983; revised January 23, 1984. This work was supported by the Office of Naval Research under Contract N00014-82-C-0476.

B. Friedlander is with Systems Control Technology, Inc., Palo Alto, CA 94304.

B. Porat is with the Department of Electrical Engineering, Technion, Haifa, Israel.

II. The Bound

Let $\{y_t\}$ be a zero-mean Gaussian stationary time series, and assume that measurements are available in the range $1 \leqslant t \leqslant N$. Denote the power spectral density of the time series $S(\omega, \theta)$, where ω is the angular frequency $(-\pi \leqslant \omega \leqslant \pi)$ and θ is a parameter vector of dimension M. The functional dependence of S on ω and θ is assumed to be known.

A commonly used spectral analysis procedure involves estimating the unknown parameter vector θ and substituting the estimate $\hat{\theta}$ in the spectral function $S(\omega, \theta)$. In other words,

$$\hat{S}(\omega, \theta) \triangleq S(\omega, \hat{\theta}) = \text{spectral estimate}. \tag{1}$$

If $\hat{\theta}$ is a maximum likelihood estimate (MLE), then $\hat{S}(\omega, \theta)$ is also an MLE by the invariance principle [4, p. 223]. The MLE is known to be asymptotically unbiased and efficient. The variance of the estimation error of an MLE approaches the Cramer-Rao lower bound (CRLB) as the number of data points tends to infinity.

Next consider the case where $\hat{\theta}$ is any asymptotically unbiased estimator of θ (e.g., the modified Yule-Walker method), i.e., $E\{\hat{\theta}\} \xrightarrow[N \to \infty]{} \theta$. Then the spectral estimate $\hat{S}(\omega, \theta)$ will again be asymptotically unbiased. To see this we expand $\hat{S}(\omega, \theta)$ in a Taylor series around the true parameter value θ:

$$\hat{S}(\omega, \theta) \triangleq S(\omega, \hat{\theta}) = S(\omega, \theta) + \frac{\partial S(\omega, \theta)}{\partial \theta} (\hat{\theta} - \theta) + (\hat{\theta} - \theta)^T \frac{\partial^2 S(\omega, \theta)}{\partial \theta^2} (\hat{\theta} - \theta) + \cdots. \tag{2}$$

Taking expected values of both sides we get

$$E\{\hat{S}(\omega, \theta)\} = S(\omega, \theta) + \text{terms that tend to zero as } N \to \infty \tag{3}$$

since $E\{(\hat{\theta} - \theta)\} \xrightarrow[N \to \infty]{} 0$ by assumption and since for any rea-

Reprinted from *IEEE Trans. Acoust., Speech, Signal Processing*, vol. ASSP-32, pp. 728–733, Aug. 1984.

sonable estimator $E\{(\hat{\theta}-\theta)(\hat{\theta}-\theta)^T\}$ goes to zero as $1/N$. The higher order terms in the Taylor series will tend to zero even faster.

The CRLB provides an asymptotic bound for asymptotically unbiased estimators. In fact, if $\hat{\theta}$ is an asymptotically efficient and unbiased estimator (with the bias decreasing faster than $1/\sqrt{N}$), then the variance of $\hat{S}(\omega, \theta)$ will approach the CRLB as the number of data points tends to infinity. If $\hat{\theta}$ is not efficient then var $\{\hat{S}(\omega, \theta)\}$ will be strictly greater than the bound as $N \to \infty$. These facts are summarized in the following result.

Let $I_N(\theta)$ be the Fisher information matrix associated with estimating θ from $\{y_1, y_2, \cdots, y_N\}$. Also, denote by $D(\omega, \theta)$ the vector of partial derivatives

$$D(\omega,\theta)=\left[\frac{\partial S(\omega,\theta)}{\partial\theta_1}\ \frac{\partial S(\omega,\theta)}{\partial\theta_2}\cdots\frac{\partial S(\omega,\theta)}{\partial\theta_M}\right]^T. \tag{4}$$

Denote by $\hat{S}(\omega, \theta)$ an estimate of $S(\omega, \theta)$, obtained by substituting an asymptotically unbiased estimate of θ in the function $S(\omega, \theta)$. Then the asymptotic variance of the spectral estimator $\hat{S}(\omega, \hat{\theta})$ is bounded from below by

$$\operatorname{var}\{\hat{S}(\omega,\theta)\} \geqslant D^T(\omega,\theta) I_N^{-1}(\theta) D(\omega,\theta) \tag{5}$$

provided that the Fisher information matrix $I_N(\theta)$ is nonsingular. This result is just a special case of a more general theorem (the generalized CRLB), presented and proven in [4, p. 194]. The Fisher information matrix $I_N(\theta)$ is given by Whittle's formula [5]:

$$\{I_N(\theta)\}_{ij}=\frac{N}{4\pi}\int_{-\pi}^{\pi}\frac{1}{S^2(\omega,\theta)}\frac{\partial S(\omega,\theta)}{\partial\theta_i}\frac{\partial S(\omega,\theta)}{\partial\theta_j}\,d\omega. \tag{6}$$

Substituting (6) and (4) in (5) we get

$$\operatorname{var}\{\hat{S}(\omega,\theta)\} \geqslant \frac{4\pi}{N}D^T(\omega,\theta)\cdot\left\{\int_{-\pi}^{\pi}\frac{1}{S^2(\omega,\theta)}D(\omega,\theta)D^T(\omega,\theta)\,d\omega\right\}^{-1}\cdot D(\omega,\theta). \tag{7}$$

In many cases, spectra are depicted on a logarithmic scale. It is therefore useful to write down a bound for the log-spectrum. Note that

$$\frac{\partial}{\partial\theta_i}\log S(\omega,\theta)=\frac{1}{S(\omega,\theta)}\frac{\partial S(\omega,\theta)}{\partial\theta_i}. \tag{8}$$

Hence,

$$\operatorname{var}\{\log S(\omega,\hat{\theta})\} \geqslant \frac{4\pi}{N}\frac{1}{S^2(\omega,\theta)}D^T(\omega,\theta)\cdot\left\{\int_{-\pi}^{\pi}\frac{1}{S^2(\omega,\theta)}D(\omega,\theta)D^T(\omega,\theta)\,d\omega\right\}^{-1}D(\omega,\theta). \tag{9}$$

A simple numerical integration routine is used to evaluate the bound in (7) or (9).

Sometimes it is useful to summarize the performance of a spectral estimator by a scalar measure. Consider for example the following functional of $\log\hat{S}(\omega, \theta)$:

$$L \triangleq E_\theta\left\{\frac{1}{2\pi}\int_{-\pi}^{\pi}[\log\hat{S}(\omega,\theta)-\log S(\omega,\theta)]^2\,d\omega\right\} \tag{10}$$

In other words, L is the integrated mean square estimation error for the log-spectrum. Next we compute a lower bound for L:

$$\begin{aligned}
L &= \frac{1}{2\pi}\int_{-\pi}^{\pi}\operatorname{var}\{\log\hat{S}(\omega,\theta)\}\,d\omega\\
&\geqslant \frac{2}{N}\frac{1}{2\pi}\int_{-\pi}^{\pi}\frac{1}{S^2(\omega,\theta)}D^T(\omega,\theta)I_N^{-1}(\theta)D(\omega,\theta)\,d\omega\\
&= \frac{2}{N}\cdot\frac{1}{2\pi}\int_{-\pi}^{\pi}\operatorname{tr}\left\{\frac{1}{S^2(\omega,\theta)}\cdot I_N^{-1}(\theta)D(\omega,\theta)D^T(\omega,\theta)\right\}d\omega\\
&= \frac{2}{N}\operatorname{tr}\left\{I_N^{-1}(\theta)\frac{1}{2\pi}\int_{-\pi}^{\pi}\frac{1}{S^2(\omega,\theta)}\cdot D(\omega,\theta)D^T(\omega,\theta)\,d\omega\right\}\\
&= \frac{2}{N}\operatorname{tr}\{I_N^{-1}(\theta)I_N(\theta)\}=\frac{2M}{N}.
\end{aligned} \tag{11}$$

Thus, $2M/N$ is a universal bound for the scalar measure L. This bound is independent of the specific parametric model and of the underlying spectrum. It depends only on the number of parameters and on the number of data points. This bound clearly illustrates the undesirable aspects of overparameterization.

A similar result is derived in [10] where it is shown that

$$E\left\{\frac{1}{2\pi}\int_{-\pi}^{\pi}\left[\frac{\hat{S}(\omega,\theta)-S(\omega,\theta)}{S(\omega,\theta)}\right]^2d\omega\right\}\geqslant\frac{2M}{N}. \tag{12}$$

III. Some Special Cases

In this section we give explicit expressions for three common rational models, as follows.

AR-Plus-Noise

Many practical problems involve noisy measurements of some signal. If the signal is modeled as an AR process, the measurement will be an AR-plus-noise process (which can be shown to be a special ARMA (p, p) process).

Such processes have spectra given by

$$S(\omega,\theta)=\frac{\sigma_u^2}{A(e^{j\omega})A(e^{-j\omega})}+\sigma_v^2 \tag{13}$$

where

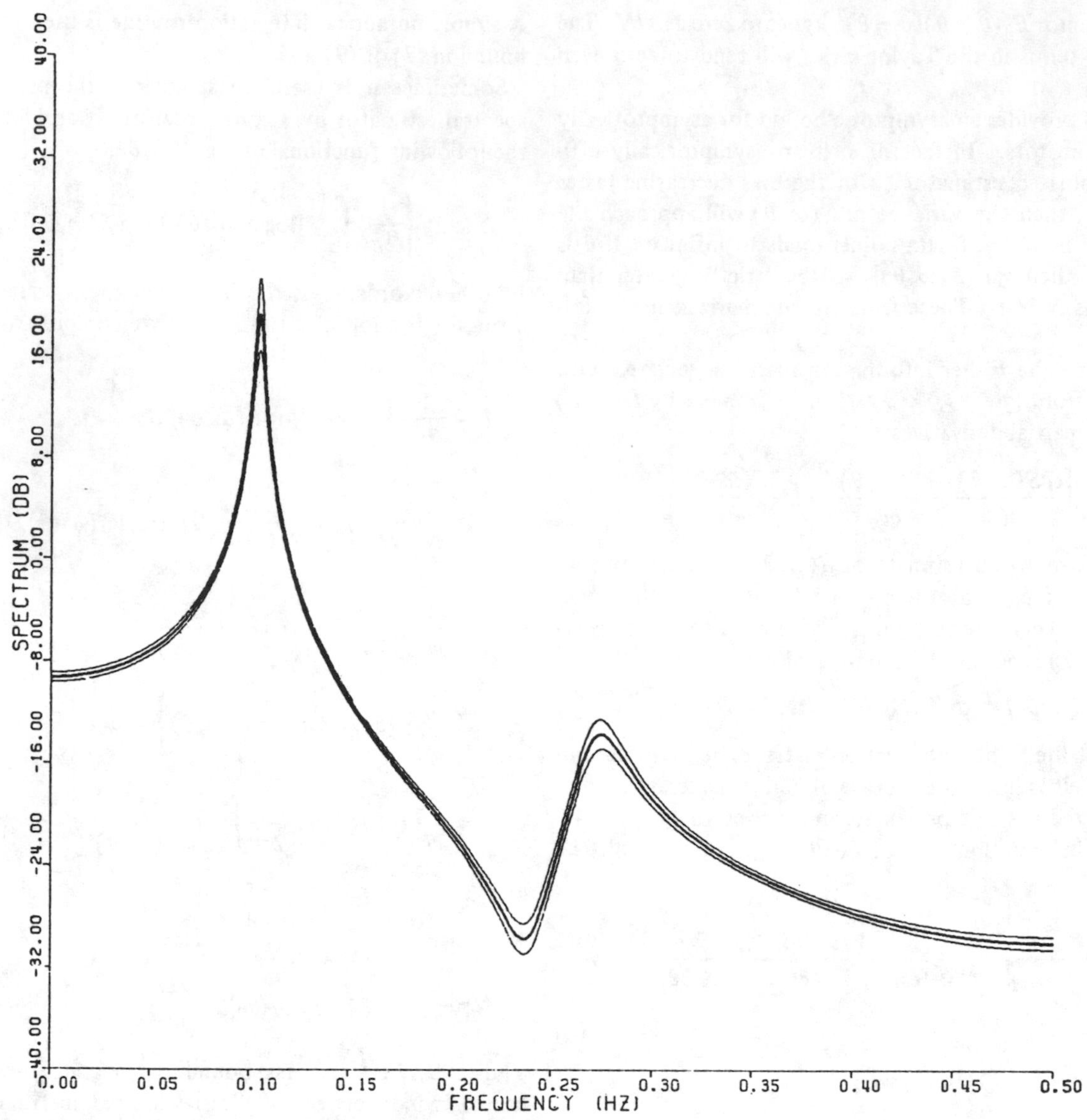

Fig. 1. Spectrum and bounds for Example 1.

$$A(z^{-1}) = 1 + a_1 z^{-1} + \cdots + a_p z^{-p}$$

is a stable polynomial, σ_u^2 is the input noise variance, and σ_v^2 is the measurement noise variance. Pure AR processes can be treated as a special case of this model, with $\sigma_v^2 = 0$. The parameter vector is

$$\theta = [a_1, a_2, \cdots, a_p, \sigma_u^2, \sigma_v^2]^T$$

and the partial derivatives are given by

$$\frac{\partial S(\omega, \theta)}{\partial a_k} = -\frac{2\sigma_u^2}{A(e^{j\omega})A(e^{-j\omega})} \operatorname{Re}\left\{\frac{e^{jk\omega}}{A(e^{j\omega})}\right\} \tag{14a}$$

$$\frac{\partial S(\omega, \theta)}{\partial \sigma_u^2} = \frac{1}{A(e^{j\omega})A(e^{-j\omega})} \tag{14b}$$

$$\frac{\partial S(\omega, \theta)}{\partial \sigma_v^2} = 1. \tag{14c}$$

ARMA Processes (Standard Model)

The spectrum of a ARMA (p, q) process is given by

$$S(\omega, \theta) = \frac{B(e^{j\omega})B(e^{-j\omega})}{A(e^{j\omega})A(e^{-j\omega})} \tag{15}$$

where

$$A(z) = 1 + a_1 z^{-1} + \cdots + a_p z^{-p}$$

and

$$B(z) = b_0 + b_1 z^{-1} + \cdots + b_q z^{-q}.$$

The polynomial $A(z)$ is required to be stable. MA processes can be treated as a special case, with $p = 0$. The parameter vector is in this case

$$\theta = [a_1, a_2, \cdots, a_p, b_0, b_1, \cdots, b_q]^T \tag{16}$$

and the partial derivatives are given by

$$\frac{\partial S(\omega, \theta)}{\partial a_k} = -\frac{2B(e^{j\omega})B(e^{-j\omega})}{A(e^{j\omega})A(e^{-j\omega})} \operatorname{Re}\left\{\frac{e^{jk\omega}}{A(e^{j\omega})}\right\} \tag{17a}$$

$$\frac{\partial S(\omega, \theta)}{\partial b_k} = \frac{2B(e^{j\omega})B(e^{-j\omega})}{A(e^{j\omega})A(e^{-j\omega})} \operatorname{Re}\left\{\frac{e^{jk\omega}}{B(e^{j\omega})}\right\}. \tag{17b}$$

ARMA Process (Modified Model)

An alternative expression for the spectrum of an ARMA process is given by the so-called additive decomposition

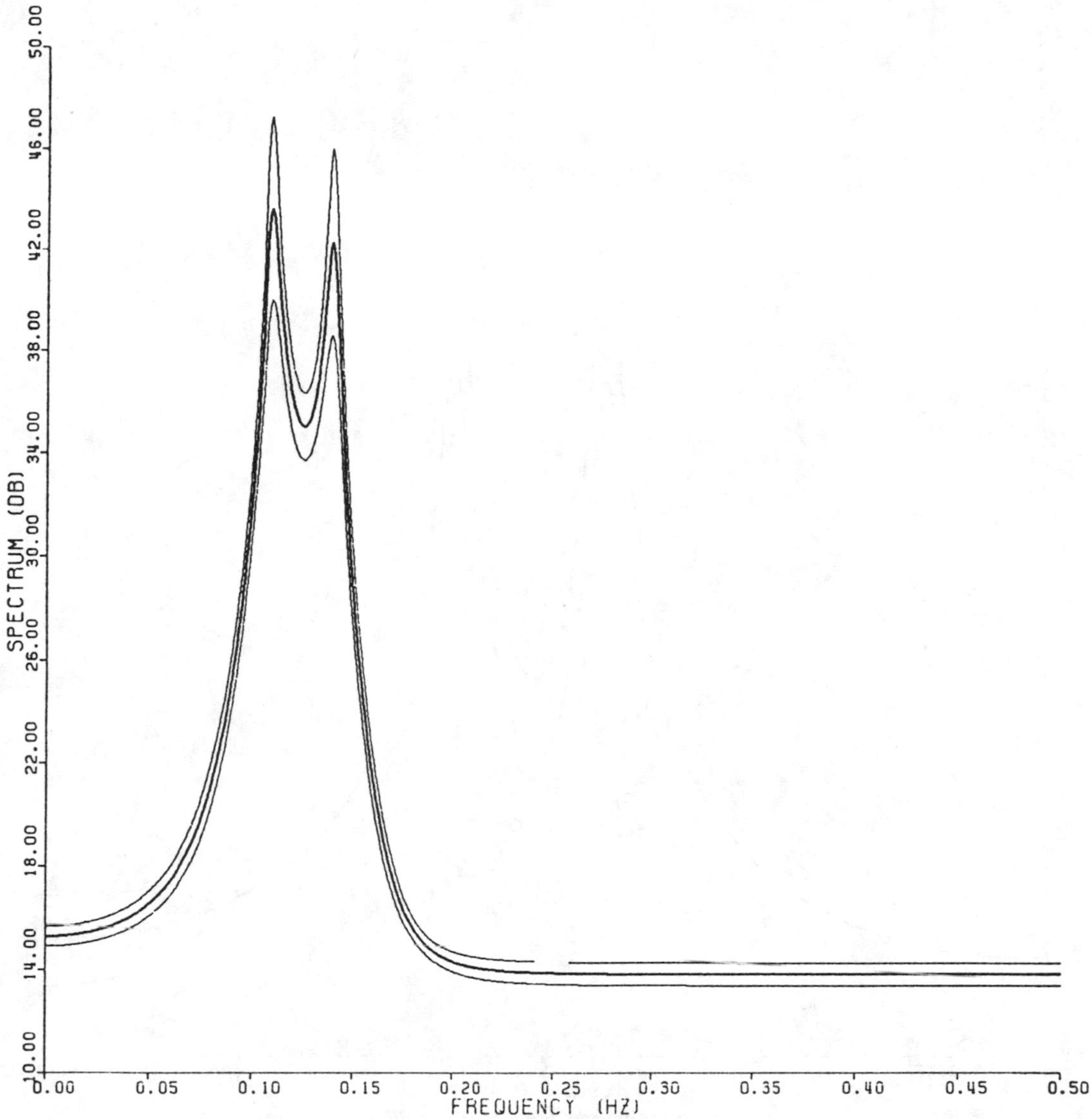

Fig. 2. Spectrum and bounds for Example 2.

$$S(\omega, \theta) = \frac{C(e^{j\omega})}{A(e^{j\omega})} + \frac{C(e^{-j\omega})}{A(e^{-j\omega})} = 2 \text{ Re}\left\{\frac{C(e^{j\omega})}{A(e^{j\omega})}\right\} \tag{18}$$

where

$$A(z) = 1 + a_1 z^{-1} + \cdots + a_p z^{-p}$$

and

$$C(z) = c_0 + c_1 z^{-1} + \cdots + c_r z^{-r}.$$

$A(z)$ is required to be stable and $C(z)/A(z)$ is required to be positive-real on the unit circle. Typically, we have $r = p$.

The parameter vector is in this case

$$\theta = [a_1, a_2, \cdots, a_p, c_0, c_1, \cdots, c_r] \tag{19}$$

and the partial derivatives are given by

$$\frac{\partial S(\omega, \theta)}{\partial a_k} = -2 \text{ Re}\left\{\frac{e^{jk\omega} C(e^{j\omega})}{A^2(e^{j\omega})}\right\} \tag{20a}$$

$$\frac{\partial S(\omega, \theta)}{\partial c_k} = 2 \text{ Re}\left\{\frac{e^{jk\omega}}{A(e^{j\omega})}\right\}. \tag{20b}$$

These partial derivatives define the entries of $D(\omega, \theta)$, needed for computation of the lower bound in (7) or (9).

IV. Numerical Examples

In order to illustrate the behavior of these bounds, we chose three examples from the recent signal processing literature. The first example was used by Gersch and Sharpe [6] and later by Akaike [7]. The process is ARMA (4, 3) with

$$A(z) = 1 - 1.3136z^{-1} + 1.4401z^{-2} - 1.0919z^{-3} + 0.83527z^{-4} \tag{21a}$$

$$B(z) = 0.13137 + 0.023543z^{-1} + 0.10775z^{-2} + 0.03516z^{-3}. \tag{21b}$$

Fig. 1 shows the spectrum (dB scale) and the $\pm 1\sigma$ bounds corresponding to $N = 800$ data points.

The second example is an AR-plus-noise process, where

$$A(z) = 1 - 2.7607z^{-1} + 3.8106z^{-2} - 2.6535z^{-3} + 0.9238z^{-4} \tag{22a}$$

$$\sigma_u^2 = 1; \quad \sigma_v^2 = 24.09 \text{ (SNR} = 15 \text{ dB)}. \tag{22b}$$

This example was used by Kay [8]. Fig. 2 shows the spectrum and the $\pm 1\sigma$ bounds corresponding to $N = 300$ data points. It

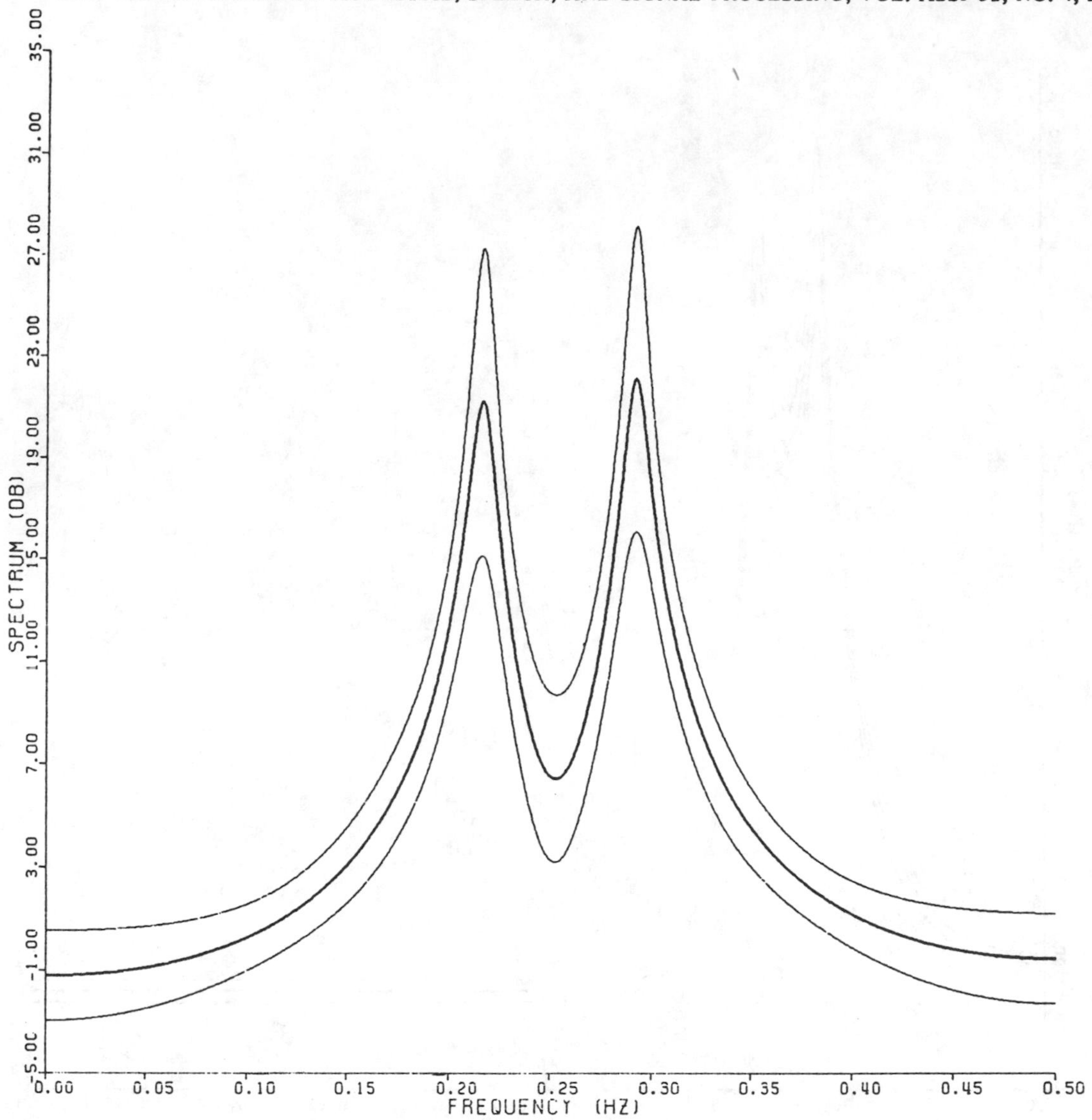

Fig. 3. Spectrum and bounds for Example 3.

is of interest to compare the Monte Carlo results given in [8] to the bounds shown in Fig. 2.

The third example uses the ARMA (4, 4) process with

$$A(z) = 1 + 0.1z^{-1} + 1.66z^{-2} + 0.093z^{-3} + 0.8649z^{-4} \quad (23a)$$

$$B(z) = 1.6814 + 0.038z^{-1} + 1.3745z^{-2} + 0.0109z^{-3} + 0.1284z^{-4}. \quad (23b)$$

This example is essentially the one used by Cadzow in [9, eq. 46] where it was modeled as a sum of two AR processes in white noise. Fig. 3 shows the spectrum and the $\pm 1\sigma$ bounds for $N = 64$ data points. Again, it is of interest to compare the Monte Carlo results shown in [9] to the bounds.

V. Conclusions

We presented a lower bound on the variance of a general parametric spectral estimator. The bound should be of use to researchers who wish to evaluate the performance of new or existing spectral estimation techniques. It should be emphasized that the bound is not necessarily tight for any specific application. However, at least for large-sample problems and asymptotically unbiased estimates, the bound represents a reasonable indicator of best possible performance. Extensions to bounds for multiple time series are relatively straightforward, and are therefore omitted.

References

[1] T. W. Anderson, *The Statistical Analysis of Time Series.* New York: Wiley, 1971.

[2] H. L. Trees, *Detection, Estimation and Modulation Theory, Part III.* New York: Wiley, 1971.

[3] S. M. Kay and S. L. Marple, "Spectrum analysis–A modern perspective," *Proc. IEEE*, vol. 69, pp. 1380–1419, Nov. 1981.

[4] S. Zacks, *The Theory of Statistical Inference.* New York: Wiley, 1971.

[5] P. Whittle, "The analysis of multiple stationary time series," *J. Royal Statist. Soc.*, vol. 15, pp. 125–139, 1953.

[6] W. Gersch and D. R. Sharpe, "Estimation of power spectra with finite-order autoregressive models," *IEEE Trans. Automat. Contr.*, vol. AC-18, pp. 367–369, Aug. 1973.

[7] H. Akaike, "A new look at the statistical model identification," *IEEE Trans. Automat. Contr.*, vol. AC-19, pp. 716–723, Dec. 1974.

[8] S. M. Kay, "Noise compensation for autoregressive spectral estimates," *IEEE Trans. Acoust., Speech, Signal Processing*, vol. ASSP-28, pp. 292–303, June 1980.

[9] J. A. Cadzow and R. L. Moses, "An adaptive ARMA spectral estimator: Part I," in *Proc. 1st ASSP Workshop on Signal Processing*, Hamilton, Ont., Canada, Aug. 1981.

[10] W. E. Larimore, "A survey of some recent developments in system parameter estimation," in *Proc. 6th IFAC Symp. Identifica-*

tion and Syst. Parameter Estimation, Arlington, VA, June 7-11, 1982, pp. 979-984.

[11] H. Sakai and H. Tokumaru, "Statistical analysis of a spectral estimator for ARMA processes," *IEEE Trans. Automat. Contr.*, vol. AC-25, pp. 122-124, Feb. 1980.

[12] H. Sakai, "Statistical properties of AR spectral analysis," *IEEE Trans. Acoust., Speech, Signal Processing*, vol. ASSP-27, pp. 402-409, Aug. 1979.

[13] B. Friedlander, "The asymptotic performance of the modified Yule-Walker estimator," in *Proc. ASSP Spectrum Estimation Workshop II*, Tampa, FL, Nov. 10-11, 1983, pp. 22-26.

Robust Detection by Autoregressive Spectrum Analysis

STEVEN M. KAY, MEMBER, IEEE

Abstract—The problem of detecting a signal with an unknown Doppler shift and random phase in white noise is essentially a problem in spectral analysis. This paper examines the merits of a detector based upon the autoregressive spectral estimator. Some advantages of the autoregressive detector are that the detection performance is independent of Doppler shift and phase and the false alarm rate is independent of noise level. Also, the performance does not depend upon the exact signal form but only upon its autocorrelation function, leading to a robust detector. For the first order autoregressive model investigated, the computational and storage requirements of the autoregressive detector are less than that for a conventional bank of matched filters detector. It is shown by example that when the actual received signal departs appreciably from the signal assumed in a conventional detector, i.e., a bank of matched filters, the AR detection performance exceeds that of the conventional detector.

Manuscript received July 8, 1980; revised April 7, 1981 and September 30, 1981. This work was supported in part by the National Science Foundation under Grant ENG79-00337.

The author was with the Submarine Signal Division, Raytheon Company, Portsmouth, RI 02871. He is now with the Department of Electrical Engineering, University of Rhode Island, Kingston, RI 02881.

I. Introduction

It is well known that the optimum detector for a known complex signal with random phase in complex white Gaussian noise is a quadrature matched filter [1]. In the sonar or radar environment the signal received from a target has associated with it a Doppler shift, which depends upon the target's radial velocity [2]. Since the Doppler shift is generally unknown, the conventional detector employs a bank of matched filters with each filter matched to the same signal but a different assumed Doppler shift [2]. Large detection losses can be incurred with this type of detector if the received signal does not match one of the assumed replicas. This situation arises frequently in practice. For instance, in passive sonar the frequency of a sinusoid emitted by a radiating target can change in time due to target motion. Since this frequency modulation (FM) is unknown, one matches to sinusoids of different frequencies. If the sinusoid frequency moves through several Doppler bins a detection loss is incurred. This case is examined

Reprinted from *IEEE Trans. Acoust., Speech, Signal Processing,* vol. ASSP-30, pp. 256–269, Apr. 1982.

in this paper. In radar detection losses are incurred when the bank of matched filters consists of filters which are identical except for a Doppler frequency shift. This arrangement allows multiplexing of one filter. However, when the compression/expansion of the received signal complex envelope due to Doppler is significant, a detection loss occurs due to a signal mismatch [3]. Another problem with the bank of matched filters detector is that the noise power is needed to set the detection threshold. Since the noise power is generally unknown, an estimate must be used to set the detection threshold [4]. Thus, the performance will vary with the quality of the noise background estimate, which negates the possibility of a truly constant false alarm rate (CFAR) receiver.

A bank of matched filters detector can be interpreted as a spectrum analyzer [5]. When the signal is a sinusoid, the spectrum analyzer is the usual periodogram. Within the last few years, considerable interest has been generated in a high resolution spectral estimator, termed the autoregressive [6] (or maximum entropy [7] or linear prediction [8]) spectral estimator. The performance of the autoregressive (AR) spectral estimator for the detection of a signal of possibly unknown form in white noise of unknown variance is examined in this paper. In particular, it is shown that the use of a modified AR spectral estimator results in a CFAR receiver, and that the detection performance is independent of Doppler frequency shift and phase, and is relatively invariant to signal perturbations as long as the signal autocorrelation function is relatively unperturbed. When the actual received signal departs appreciably from the signal assumed in the bank of matched filters detector the modified AR detector performance exceeds that of the conventional detector. Also, the computational and storage requirements for the modified AR detector with a model order of one are considerably less than those of conventional detectors.

Finally, it should be mentioned that the test statistic to be analyzed in this paper is very similar to one proposed by Lank [10]. The main difference lies in a normalizing factor which is employed by the AR detector.

The paper is organized as follows. Section II defines the modified AR detector and describes its properties. Section III contains a derivation of its detection performance which is used to compare the AR detector to a conventional detector in Section IV. Section V compares the computational and storage requirements of the AR detector to those of the periodogram. Finally, Section VI summarizes the results and offers conclusions.

II. Definition and Properties of AR Detector

It is assumed that the received waveform $\tilde{X}_t$ is a discrete complex time series and that N samples $\{\tilde{X}_0, \tilde{X}_1, \cdots, \tilde{X}_{N-1}\}$, are available to make a decision. Thus,

$$\tilde{X}_t = \begin{cases} \tilde{S}_t e^{j(\omega_d t+\varphi)} + \tilde{w}_t: & H_1 \\ \tilde{w}_t: & H_0 \end{cases} \tag{1}$$

$\tilde{w}_t$ is complex white Gaussian noise, i.e., $\tilde{w}_t = w_{1_t} - jw_{2_t}$ where w_{1_t} and w_{2_t} are each Gaussian white noise processes with each having an unknown variance σ_w^2 and are independent of each other [2]. ω_d is a constant but unknown Doppler shift, φ is a random phase angle uniformly distributed on $[-\pi, \pi)$. $\tilde{S}_t$ is a complex signal of possibly unknown form. Note that $\tilde{X}_t$ could represent the complex envelope of a real bandpass process.

The AR detector is based upon a modified AR power spectral density (PSD) estimator [9]

$$\hat{P}_{\tilde{x}}(\omega) = \frac{1}{|1 + \hat{a}_1 e^{-j\omega} + \cdots + \hat{a}_p e^{-j\omega p}|^2} \tag{2a}$$

where $\{\hat{a}_1, \hat{a}_2, \cdots, \hat{a}_p\}$ are found from

$$\begin{bmatrix} \hat{R}_{\tilde{x}}(0) & \hat{R}_{\tilde{x}}(-1) & \cdots & \hat{R}_{\tilde{x}}(-p+1) \\ \hat{R}_{\tilde{x}}(1) & \hat{R}_{\tilde{x}}(0) & \cdots & \hat{R}_{\tilde{x}}(-p+2) \\ & & \vdots & \\ \hat{R}_{\tilde{x}}(p-1) & \hat{R}_{\tilde{x}}(p-2) & \cdots & \hat{R}_{\tilde{x}}(0) \end{bmatrix} \begin{bmatrix} \hat{a}_1 \\ \hat{a}_2 \\ \vdots \\ \hat{a}_p \end{bmatrix} = - \begin{bmatrix} \hat{R}_{\tilde{x}}(1) \\ \hat{R}_{\tilde{x}}(2) \\ \vdots \\ \hat{R}_{\tilde{x}}(p) \end{bmatrix} \tag{2b}$$

and

$$\hat{R}_{\tilde{x}}(k) = \frac{1}{N} \sum_{t=0}^{N-k-1} \tilde{X}_t^* \tilde{X}_{t+k} \qquad k \geqslant 0$$
$$= \hat{R}_{\tilde{x}}^*(-k) \qquad k < 0. \tag{2c}$$

The modified AR spectral estimator is just the usual AR spectral estimator with the prediction error power estimate set to unity. The AR detector compares the maximum of $\hat{P}_{\tilde{x}}(\omega)$ to a threshold, i.e.,

$$\max_{|\omega| \leqslant \pi} \hat{P}_{\tilde{x}}(\omega) > \eta \quad \text{decide } H_1$$
$$< \eta \quad \text{decide } H_0. \tag{3}$$

We now prove some important properties of the AR detector. The performance of the AR detector is *independent* of:

1) the noise level σ_w^2 under H_0, i.e., it is a constant false alarm rate (CFAR) detector;

2) the Doppler shift ω_d;

3) the random phase angle φ;

4) the signal form $\tilde{S}_t$ as long as the signal autocorrelation function is unchanged, N is large, and $N \gg p$.

Properties 1)–4) are now proven.

Property 1)—Under H_0 the performance of (3) is independent of σ_w^2, i.e., the detector is CFAR.

Proof: Divide both sides of (2b) by $\hat{R}_{\tilde{x}}(0)$. Then, it is seen that $\hat{P}_{\tilde{x}}(\omega)$ depends only upon $\{\hat{r}_{\tilde{x}}(1), \hat{r}_{\tilde{x}}(2), \cdots, \hat{r}_{\tilde{x}}(p)\}$ where

$$\hat{r}_{\tilde{x}}(k) = \hat{R}_{\tilde{x}}(k)/\hat{R}_{\tilde{x}}(0).$$

Under H_0 we have

$$\hat{r}_{\tilde{x}}(k) = \frac{\sum_{t=0}^{N-k-1} \tilde{w}_t^* \tilde{w}_{t+k}}{\sum_{t=0}^{N-1} |\tilde{w}_t|^2} \qquad k \geqslant 0.$$

If we let $\tilde{w}_t = \sqrt{2\sigma_w^2}\,\tilde{Z}_t$ then $\tilde{Z}_t$ is a complex white Gaussian noise process with zero mean and unit variance, and

$$\hat{r}_{\tilde{x}}(k) = \frac{\sum_{t=0}^{N-k-1} \tilde{Z}_t^* \tilde{Z}_{t+k}}{\sum_{t=0}^{N-1} |\tilde{Z}_t|^2}$$

which does not depend upon σ_w^2. Thus, the AR detector is self-normalizing and is a CFAR receiver.

Properties 2) and 3)—The AR detector performance is independent of the random phase angle φ and the Doppler shift ω_d.

Proof: Under H_1 we have that

$$\begin{aligned}\hat{R}_{\tilde{x}}(k) &= \frac{1}{N}\sum_{t=0}^{N-k-1} [\tilde{S}_t^* e^{-j(\omega_d t+\varphi)} + \tilde{w}_t^*] \\ &\quad \cdot [\tilde{S}_{t+k} e^{+j(\omega_d(t+k)+\varphi)} + \tilde{w}_{t+k}] \qquad \text{for } k \geqslant 0 \\ &= \frac{1}{N}\sum_{t=0}^{N-k-1} e^{-j(\omega_d t+\varphi)} [\tilde{S}_t^* + \tilde{w}_t^* e^{j(\omega_d t+\varphi)}] \\ &\quad \cdot e^{j(\omega_d(t+k)+\varphi)} [\tilde{S}_{t+k} + \tilde{w}_{t+k} e^{-j(\omega_d(t+k)+\varphi)}] \\ &= \frac{1}{N} e^{j\omega_d k} \sum_{t=0}^{N-k-1} (\tilde{S}_t + \tilde{u}_t)^* (\tilde{S}_{t+k} + \tilde{u}_{t+k})\end{aligned}$$

where $\tilde{u}_t = \tilde{w}_t e^{-j(\omega_d t+\varphi)}$.

But $\tilde{u}_t$ has the same statistical properties as $\tilde{w}_t$ [10], and hence, $\tilde{u}_t$ is a complex white Gaussian process. Intuitively, this is because $\tilde{w}_t$ has a uniformly distributed phase for all t. Thus, the addition of an independent, uniformly distributed random variable (on even a nonrandom constant phase) results in a uniformly distributed random variable over $[-\pi, \pi)$. Also, the $e^{j\omega_d t}$ factor serves only to shift the power spectral density of $\tilde{w}_t$ by ω_d rad/s. Since the PSD of $\tilde{w}_t$ is white, however, the PSD of $\tilde{u}_t$ is also white. For $k < 0$ it is similarly shown using (2c) that

$$\hat{R}_{\tilde{x}}(k) = \frac{1}{N} e^{j\omega_d k} \sum_{t=0}^{N+k-1} [\tilde{S}_t + \tilde{u}_t]\,[\tilde{S}_{t-k} + \tilde{u}_{t-k}]^*.$$

Thus, $\hat{R}_{\tilde{x}}(k)$ and, hence, $\hat{P}_{\tilde{x}}(\omega)$ is independent of phase. To complete the proof for the independence of Doppler shift note that the autocorrelation estimate may be written as

$$\hat{R}_{\tilde{x}}(k) = e^{j\omega_d k}\tilde{f}(k)$$

where $\tilde{f}(k)$ does not depend upon ω_d. Substituting into (2b):

$$\hat{R}_{\tilde{x}}(k) = -\sum_{l=1}^{p} \hat{a}_l \hat{R}_{\tilde{x}}(k-l) \qquad k = 1, 2, \cdots, p$$

$$e^{j\omega_d k}\tilde{f}(k) = -\sum_{l=1}^{p} \hat{a}_l e^{j\omega_d (k-l)} \tilde{f}(k-l)$$

or the solution for $\hat{a}_l$ is

$$\hat{a}_l = \hat{c}_l e^{j\omega_d l} \tag{4}$$

where $\hat{c}_l$ is the solution for the AR parameters for $\omega_d = 0$ and $\hat{c}_l$ does not depend upon ω_d.

By substituting (4) into (2a) we see that the only effect of ω_d upon $\hat{P}_{\tilde{x}}(\omega)$ is to shift it by ω_d to yield $\hat{P}_{\tilde{x}}(\omega - \omega_d)$. Since the spectral maximum is unchanged, the detection performance is independent of ω_d.

Property 4)—For large data records, N large, and a small model order p relative to the data record length, i.e., $N \gg p$, the AR detector performance depends only upon the signal autocorrelation function for $k = -2p, -2p+1, \cdots, 0, 1, \cdots, 2p$ which is defined as

$$\begin{aligned}R_{\tilde{S}}(k) &= \frac{1}{N}\sum_{t=0}^{N-k-1} \tilde{S}_t^* \tilde{S}_{t+k} \qquad k \geqslant 0 \\ &= R_{\tilde{S}}^*(-k) \qquad\qquad k < 0.\end{aligned}$$

Proof: The detector performance depends upon the joint probability density function (PDF) of $\{\alpha_0, \alpha_1, \beta_1, \cdots, \alpha_p, \beta_p\}$ where $R_{\tilde{x}}(k) = \alpha_k + j\beta_k$. (Note that for $k < 0$ $\hat{R}_{\tilde{x}}(k) = \hat{R}_{\tilde{x}}(-k)^*$. Also, $\beta_0 \equiv 0$.)

For large N these variables can be shown to be jointly Gaussian distributed by a central limit theorem. The first and second order moments are now found. Since, as will be shown, these moments only depend upon the signal autocorrelation function, the PDF depends only upon the signal autocorrelation function and hence, so does the AR detector performance.

Define the circular autocorrelation function estimate as

$$\hat{R}_{\tilde{x}}^c(k) = \frac{1}{N}\sum_{t=0}^{N-1} \tilde{X}_t^* \tilde{X}_{t+k}$$

where the unobserved data points $\{\tilde{X}_N, \tilde{X}_{N+1}, \cdots, \tilde{X}_{N+k-1}\}$ are defined as $\tilde{X}_N = X_0$, $\tilde{X}_{N+1} = \tilde{X}_1$, etc. Thus, $\tilde{X}_t$ can be viewed as a periodic sequence with period N.

For $N \gg p$

$$\hat{R}_{\tilde{x}}^c(k) \approx \hat{R}_{\tilde{x}}(k)$$

since the added terms in the sum are small compared to the rest of the sum. Let $X(i)$ be the discrete Fourier transform (DFT) of the $\tilde{X}_t$ sequence. Then

$$X(i) = \sum_{t=0}^{N-1} \tilde{X}_t e^{-j(2\pi/N)ti} \qquad i = 0, 1, \cdots, N-1$$

It can be shown that [11]

$$\hat{R}_{\tilde{x}}^c(k) = \frac{1}{N^2}\sum_{i=0}^{N-1} |X(i)|^2 e^{j(2\pi/N)ki}$$

Thus,

$$\hat{R}_{\tilde{x}}(k) \approx \frac{1}{N}\sum_{i=0}^{N-1} V_i e^{j(2\pi/N)ki}$$

where

$$V_i = \frac{1}{N}|X(i)|^2.$$

Let

$$\hat{R}_{\tilde{x}}(k) = \alpha_k + j\beta_k$$

Then

$$\alpha_k = \frac{1}{N}\sum_{i=0}^{N-1} V_i \cos\frac{2\pi}{N}ki \qquad k = 0, 1, \cdots, p$$

$$\beta_k = \frac{1}{N}\sum_{i=0}^{N-1} V_i \sin\frac{2\pi}{N}ki \qquad k = 1, 2, \cdots, p.$$

Let

$$\boldsymbol{C}_k = \left[1 \ \cos\frac{2\pi}{N}k \ \cos\frac{2\pi}{N}2k \ \cdots \ \cos\frac{2\pi}{N}(N-1)k\right]^T \qquad k = 0, 1, \cdots, p$$

$$\boldsymbol{S}_k = \left[0 \ \sin\frac{2\pi}{N}k \ \sin\frac{2\pi}{N}2k \ \cdots \ \sin\frac{2\pi}{N}(N-1)k\right]^T \qquad k = 1, 2, \cdots, p$$

$$\boldsymbol{V} = [V_0 \ V_1 \ \cdots \ V_{N-1}]^T.$$

By the orthogonality properties of the DFT, [11] we have

$$\boldsymbol{C}_k^T\boldsymbol{S}_j = 0$$

$$\boldsymbol{C}_k^T\boldsymbol{C}_j = \frac{N}{2}\delta_{kj} + \frac{N}{2}\delta_{k0}\delta_{j0}$$

$$\boldsymbol{S}_k^T\boldsymbol{S}_j = \frac{N}{2}\delta_{kj}.$$

It is shown in the Appendix that if $S(i)$ denotes the DFT of $\tilde{S}_0, \tilde{S}_1, \cdots, \tilde{S}_{N-1}$

$$E(V_i) = 2 + \frac{1}{N}|S(i)|^2$$

$$\text{Var}(V_i) = 4 + \frac{4}{N}|S(i)|^2$$

for all i and the V_i's are independent. Letting $\boldsymbol{K}_v$ be the covariance matrix of $\boldsymbol{v}$,

$$\boldsymbol{K}_v = 4\boldsymbol{I} + \frac{4}{N}\text{diag}(|S(0)|^2, |S(1)|^2, \cdots, |S(N-1)|^2).$$

Finally, we define the circular signal autocorrelation function as

$$R_{\tilde{S}}^c(k) = \frac{1}{N}\sum_{t=0}^{N-1} \tilde{S}_t^*\tilde{S}_{t+k}$$

where as before $\tilde{S}_N = \tilde{S}_0$, $\tilde{S}_{N+1} = \tilde{S}_1$, etc. Considering $\tilde{S}_t$ to be a periodic sequence with period N, $R_{\tilde{S}}^c(k)$ is defined for all k. For

$$N \gg |k|, R_{\tilde{S}}(k) \approx R_{\tilde{S}}^c(k).$$

Now

$$E(\alpha_k) = \frac{1}{N}\sum_{i=0}^{N-1} E(V_i)\cos\frac{2\pi}{N}ki$$

$$= 2\delta_{k0} + \frac{1}{N^2}\sum_{i=0}^{N-1}|S(i)|^2\cos\frac{2\pi}{N}ki$$

$$= 2\delta_{k0} + \text{Re}\,[R_{\tilde{S}}^c(k)] \approx 2\delta_{k0} + \text{Re}\,[R_{\tilde{S}}(k)]$$

where δ_{ij} is the Kronecker delta function. Similarly,

$$E(\beta_k) \approx \text{Im}\,[R_{\tilde{S}}(k)]$$

$$\text{Cov}(\alpha_k, \alpha_j) = \frac{1}{N^2}\text{Cov}(\boldsymbol{C}_k^T\boldsymbol{V}, \boldsymbol{C}_j^T\boldsymbol{v}) = \frac{1}{N^2}\boldsymbol{C}_k^T\boldsymbol{K}_v\boldsymbol{C}_j$$

$$= \frac{1}{N^2}\left[2N\delta_{jk} + 2N\delta_{j0}\delta_{k0} + \frac{4}{N}\sum_{i=0}^{N-1} \cdot\cos\frac{2\pi}{N}ki\cos\frac{2\pi}{N}ji|S(i)|^2\right]$$

$$= \frac{1}{N^2}\left[2N\delta_{jk} + 2N\delta_{j0}\delta_{k0} + \frac{2}{N}\sum_{i=0}^{N-1} \cdot\left(\cos\frac{2\pi}{N}(k+j)i + \cos\frac{2\pi}{N}(k-j)i\right)|S(i)|^2\right]$$

$$= \frac{1}{N^2}[2N\delta_{jk} + 2N\delta_{j0}\delta_{k0} + 2N(\text{Re}\,(R_{\tilde{S}}^c(k+j)) + \text{Re}\,(R_{\tilde{S}}^c(k-j)))]$$

$$\approx \frac{2}{N}\delta_{jk} + \frac{2}{N}\delta_{j0}\delta_{k0} + \frac{2}{N}[\text{Re}\,(R_{\tilde{S}}(k+j)) + \text{Re}\,(R_{\tilde{S}}(k-j))].$$

Similarly,

$$\text{Cov}(\beta_k, \beta_j) \approx \frac{2}{N}\delta_{jk} + \frac{2}{N}[\text{Re}\,(R_{\tilde{S}}(k-j)) - \text{Re}\,(R_{\tilde{S}}(k+j))]$$

$$\text{Cov}(\alpha_k, \beta_j) \approx \frac{2}{N}[\text{Im}\,(R_{\tilde{S}}(j+k)) + \text{Im}\,(R_{\tilde{S}}(j-k))].$$

These expressions hold for $k = 0, 1, \cdots, p$; $j = 1, 2, \cdots, p$ and, as claimed, depend only upon the signal autocorrelation function.

III. Analysis of Detection Performance

The evaluation of the detection performance of the AR detector as given by (2) and (3) is in general a difficult task. In this section, we derive the detection performance for the case $p = 1$. Since the optimal choice for p is probably not $p = 1$, the results to be presented can be interpreted as a lower bound on the AR detector performance.

For $p = 1$ we have from (2a)

$$\hat{P}_{\tilde{x}}(\omega) = \frac{1}{|1 + \hat{a}_1 e^{-j\omega}|^2}.$$

If we let $\hat{a}_1 = -re^{j\theta}$ where $-\hat{a}_1$ is the estimated pole position, then

$$\hat{P}_{\tilde{x}}(\omega) = \frac{1}{|1 - re^{-j(\omega-\theta)}|^2}$$

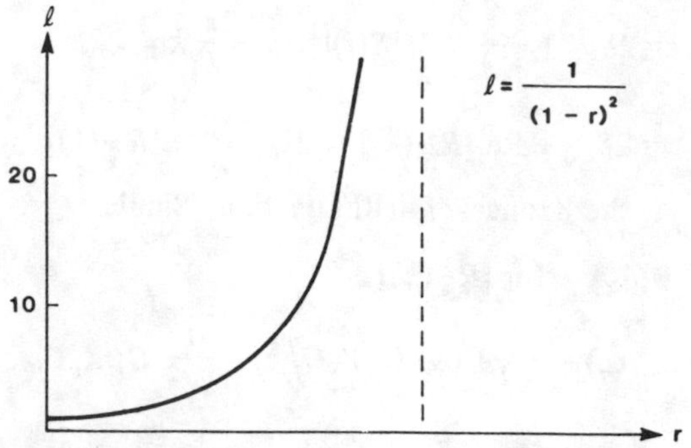

Fig. 1. Plot of maximum of modified AR power spectral density ($p = 1$) versus estimated pole radius.

and we see that the maximum of $\hat{P}_{\tilde{x}}(\omega)$ occurs at $\omega = \theta$. Denoting the maximum by l,

$$l = \max_{|\omega| \leq \pi} \hat{P}_{\tilde{x}}(\omega) = \frac{1}{(1-r)^2}.$$

For the positive semidefinite autocorrelation estimator given by (2c) it can be shown that all the estimated poles lie on or within the unit circle [7]. Hence,

$$0 \leq r \leq 1$$

under both hypotheses. For this range of r, l is a monotonically increasing function of r as shown in Fig. 1. An equivalent detection statistic is either r or r^2. The latter is chosen, with the result that the AR detector is

$$r^2 = |\hat{a}_1|^2 \begin{cases} > \gamma & \text{decide } H_1 \\ < \gamma & \text{decide } H_0. \end{cases}$$

But from (2b) we have

$$\hat{a}_1 = -\hat{R}_{\tilde{x}}(1)/\hat{R}_{\tilde{x}}(0) = -\frac{\sum_{t=0}^{N-2} \tilde{X}_t^* \tilde{X}_{t+1}}{\sum_{t=0}^{N-1} |\tilde{X}_t|^2}$$

so that

$$r^2 = \left| \frac{\sum_{t=0}^{N-2} \tilde{X}_t^* \tilde{X}_{t+1}}{\sum_{t=0}^{N-1} |\tilde{X}_t|^2} \right|^2 \tag{5}$$

The probability density functions (PDF's) for r^2 are derived in the Appendix. The results are based upon a central limit theorem so that they would only be expected to be valid for large N. By extensive simulation, it has been found that the PDF's are valid for $N \geq 50$. They can be expected to yield accurate probability of false alarm P_F and probability of detection P_D for most cases of interest. The PDF's are given as

$$p_{r^2/H_0}(x) = \frac{\sqrt{2}}{4} e^{-N/2} \left[\frac{1}{(x+\frac{1}{2})^{3/2}} + \frac{N/2}{(x+\frac{1}{2})^{5/2}} \right] \cdot e^{(N/4)/[x+(1/2)]} \qquad 0 \leq x \leq 1 \tag{6}$$

$$p_{r^2/H_1}(x) = \frac{d}{dx} F_{r^2/H_1}(x) \qquad 0 \leq x \leq 1 \tag{7}$$

where $F_{r^2/H_1}(x)$ is the cumulative distribution function (CDF) and

$$F_{r^2/H_1}(x) = \frac{1}{2} - \frac{1}{\pi} \int_0^\infty \frac{\mathrm{Im}\,[\varphi(\omega)]}{\omega} d\omega$$

$$\varphi(\omega) = \frac{1}{|\boldsymbol{I} - 2j\omega \boldsymbol{K}\boldsymbol{A}|^{1/2}} e^{j\omega \boldsymbol{u}^T \boldsymbol{A}(\boldsymbol{I} - 2j\omega \boldsymbol{K}\boldsymbol{A})^{-1}\boldsymbol{u}}$$

$$\boldsymbol{u} = [2\sqrt{N\zeta}\, \mathrm{Re}\,(\rho_{\tilde{S}}(1)) \quad 2\sqrt{N\zeta}\, \mathrm{Im}\,(\rho_{\tilde{S}}(1)) \quad 2\sqrt{N} + 2\sqrt{N\zeta}]^T$$

$$\boldsymbol{K} = \begin{bmatrix} 2 + 4\zeta + 4\zeta\, \mathrm{Re}\,(\rho_{\tilde{S}}(2)) & 4\zeta\, \mathrm{Im}\,(\rho_{\tilde{S}}(2)) & 8\zeta\, \mathrm{Re}\,(\rho_{\tilde{S}}(1)) \\ 4\zeta\, \mathrm{Im}\,(\rho_{\tilde{S}}(2)) & 2 + 4\zeta - 4\zeta\, \mathrm{Re}\,(\rho_{\tilde{S}}(2)) & 8\zeta\, \mathrm{Im}\,(\rho_{\tilde{S}}(1)) \\ 8\zeta\, \mathrm{Re}\,(\rho_{\tilde{S}}(1)) & 8\zeta\, \mathrm{Im}\,(\rho_{\tilde{S}}(1)) & 4 + 8\zeta \end{bmatrix}$$

$$\zeta = \frac{E}{2\sigma_w^2 N}$$

$$E = \sum_{t=0}^{N-1} |\tilde{S}_t|^2$$

$$\boldsymbol{A} = \begin{bmatrix} 1 & 0 & 0 \\ 0 & 1 & 0 \\ 0 & 0 & -x \end{bmatrix}$$

$$\rho_{\tilde{S}}(k) = \frac{1}{E} \sum_{t=0}^{N-k-1} \tilde{S}_t^* \tilde{S}_{t+k}.$$

Note that E represents the signal energy and hence, ζ is the signal-to-noise ratio. $\rho_{\tilde{S}}(k)$ is the normalized signal autocorrelation function. To evaluate $\varphi(\omega)$ requires the square root of a complex number. The choice of the correct square root is discussed in the Appendix.

Some observations which can be made are as follows.

1) The PDF under H_0 does not depend upon the noise power $2\sigma_w^2$.

2) The PDF under H_1 does not depend upon the signal form but only upon its autocorrelation function.

3) The PDF under H_1 does not only depend upon the signal energy to noise power, $E/2\sigma_w^2 = N\zeta$ as it does in a matched filter detector. In fact, it appears from $\boldsymbol{u}$ to depend more upon $\sqrt{N}\zeta$. Thus, the performance relative to a matched filter for a signal of known form would be expected to degrade as N increases. We shall see that this is indeed the case.

As an example of the good agreement obtained between the theoretical PDF's and histograms consider the signal to be a sinusoid so that

$$\tilde{X}_t = \begin{cases} Ae^{j((\omega_0+\omega_d)t+\varphi)} + \tilde{w}_t: & H_1 \\ \tilde{w}_t: & H_0. \end{cases}$$

ω_0 was chosen to be $\pi/4$ rad/s although from property 2) the choice is arbitrary.

With $\tilde{S}_t = Ae^{j\omega_0 t}$ it is easily shown that

$$\rho_{\tilde{S}}(k) = \frac{N - |k|}{N}\, e^{j\omega_0 k}.$$

Finally, define the SNR as

$$\text{SNR} = 10 \log_{10} \zeta \text{ dB}$$

For $N = 50$, the theoretical PDF's and histograms are shown in Figs. 2 and 3. To generate the histograms, 1000 realizations of $\tilde{X}_t$ were generated. Very good agreement is noted.

We now determine the probability of false alarm P_F and the probability of detection P_D.

$$P_F = \int_\gamma^1 p_{r^2/H_0}(x)\, dx.$$

Since

$$p_{r^2/H_0}(x) \approx 0$$

for

$$N \geqslant 50, x > 1,$$

$$P_F = \int_\gamma^\infty p_{r^2/H_0}(x)\, dx$$

$$= -\frac{1}{\sqrt{2}}\, e^{-N/2}\, \frac{e^{(N/4)/(x+1/2)}}{\sqrt{x+\frac{1}{2}}}\Bigg|_\gamma^\infty$$

$$P_F = \frac{e^{-N\gamma/1+2\gamma}}{\sqrt{1+2\gamma}}. \tag{8}$$

P_D is simply found as

$$P_D = 1 - F_{r^2/H_1}(\gamma). \tag{9}$$

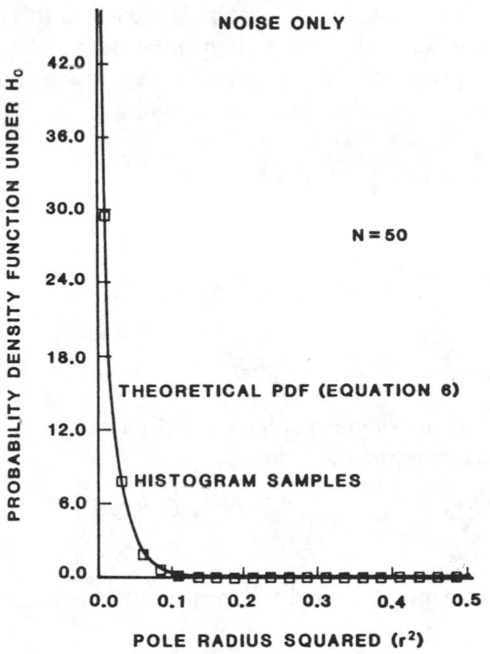

Fig. 2. Histogram and theoretical PDF of AR detection statistic for noise only.

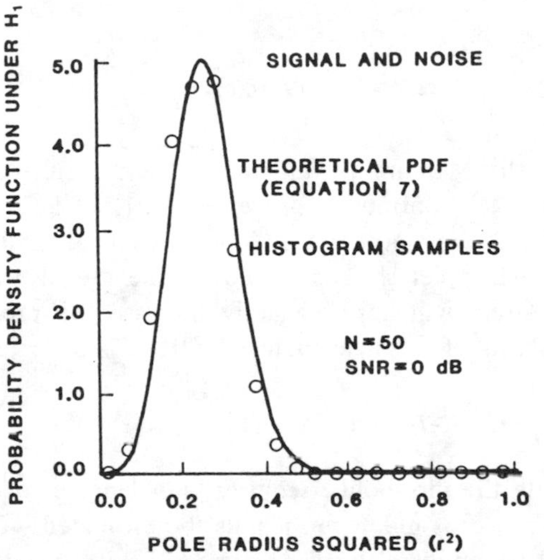

Fig. 3. Histogram and theoretical PDF of AR detection statistic for sinusoid in noise.

Since γ is usually chosen to maintain a given false alarm rate, it is necessary to solve (8) for γ. The nonlinear equation is easily solved by the fixed point iteration [12].

$$\gamma_{n+1} = -\frac{(1 + 2\gamma_n)}{N} \ln\,[P_F\sqrt{1 + 2\gamma_n}]. \tag{10}$$

Convergence has been observed to be quite rapid. Using (8) and (9) some examples of the performance of the AR detector are given and compared to conventional matched filter detectors in the next section.

IV. Examples of Detection Performance

We will consider first the problem of the detection of a signal of *known* form perturbed by an unknown Dopper shift and a random phase shift in white Gaussian noise of unknown variance as given by (1). The usual approach to this problem is to process $\tilde{X}_t$ with a bank of quadrature matched filters, with each matched filter matched to an assumed Doppler shift. Thus, it is standard practice to compute

$$\max_k |\tilde{l}_k|^2 = \max_k \left| \sum_{t=0}^{N-1} \tilde{X}_t \tilde{S}_t^* e^{-j\omega_k t} \right|^2 \begin{matrix} > t_1 & \text{decide } H_1 \\ < t_1 & \text{decide } H_0 \end{matrix} \tag{11}$$

where

$$\omega_k = \frac{2\pi}{N} k \qquad k = 0, 1, \cdots, N-1$$

which is efficiently implemented via an FFT [5].

If ω_d is known, then one can use

$$|\tilde{l}|^2 = \left| \sum_{t=0}^{N-1} \tilde{X}_t \tilde{S}_t^* e^{-j\omega_d t} \right|^2 \begin{matrix} > t_2 & \text{decide } H_1 \\ < t_2 & \text{decide } H_0. \end{matrix}$$

It can be shown that P_F' and P_D' for the latter detector are

$$P_F' = e^{-t_2/2\sigma_w^2 E} \tag{12}$$

$$P_D' = Q(\sqrt{E/\sigma_w^2}, \sqrt{t_2/\sigma_w^2 E}) = Q(\sqrt{E/\sigma_w^2}, \sqrt{2 \ln 1/P_F'}) \tag{13}$$

where

$$Q(\alpha, \beta) = \int_\beta^\infty z e^{-(z^2+\alpha^2/2)} I_0(\alpha z)\, dz.$$

$Q(\alpha, \beta)$ is the Marcum Q function. These results are well known for real continuous processes [13]. The derivation for complex discrete processes follows in a similar fashion.

When the Doppler shift is unknown, the use of a bank of matched filters will approximately increase the probability of false alarm of a single matched filter P_F' by a factor of N [14]. Thus,

$$P_F = NP_F' = Ne^{-t/2\sigma_w^2 E}.$$

To compute the threshold t requires knowledge of σ_w^2. Since the noise power is unknown, it must be estimated. The estimate may be obtained by averaging $|\tilde{l}_k|^2$ over adjacent DFT bins. Assuming those bins do not contain any signal and for $N \geqslant 50$, the detectability loss can be shown to be negligible [4]. Hence, we will assume σ_w^2 to be known. A final correction which needs to be made is for scalloping loss [15], due to the mismatch between the actual signal Doppler, ω_d and the assumed Doppler, ω_k, and windowing loss, due to an intentional windowing of the received data. The scalloping loss is from (11)

$$-10 \log_{10} \frac{\left| \sum_{t=0}^{N-1} |\tilde{S}_t|^2 e^{j(\omega_d - \omega_n)t} \right|^2}{\left(\sum_{t=0}^{N-1} |\tilde{S}_t|^2 \right)^2} \text{ dB}$$

where $|\tilde{l}_k|^2$ is maximized for $k = n$. For the signal we will consider subsequently $|\tilde{S}_t| = 1$. Then, the scalloping loss is at worst 3.9 dB [15]. It is customary to window the received waveform prior to FFT processing, thereby introducing a windowing loss. For a wide class of windows, the total loss due to the worst case scalloping loss and windowing loss is about 3 dB [15]. With this final correction we have

$$P_D = Q(\sqrt{E/2\sigma_w^2}, \sqrt{2 \ln N/P_F})$$

or finally

$$P_D = Q(\sqrt{N\zeta}, \sqrt{2 \ln N/P_F}). \tag{14}$$

The first case to be compared is that of a sinusoidal receive signal, i.e.,

$$\begin{aligned} \tilde{X}_t &= e^{j((\omega_0 + \omega_d)t + \varphi)} + \tilde{w}_t: \quad && H_1 \\ &= \tilde{w}_t: && H_0 \end{aligned}$$

Here, $\tilde{S}_t = e^{j\omega_0 t}$ in (1). The normalized signal autocorrelation function is

$$\rho_{\tilde{S}}(k) = \frac{N - |k|}{N} e^{j\omega_0 k}$$

where $\omega_0 = \pi/4$ is chosen, although arbitrary. Also, we define the SNR at the output of a DFT bin as

$$\begin{aligned} \text{SNR (output)} &= \text{SNR (input)} + 10 \log_{10} N - 3 \text{ dB} \\ &= 10 \log_{10} \zeta + 10 \log_{10} N - 3 \text{ dB} \end{aligned}$$

where

$$\zeta = 1/2\sigma_w^2.$$

The 3 dB correction is that necessary to account for windowing and scalloping loss. The results are shown in Figs. 4 and 5 for $P_F = 10^{-4}$. The bank of matched filters detector is termed the periodogram.

The difference in SNR indicated in Figs. 3 and 4 is the decrease in input SNR that the periodogram could incur before its performance becomes poorer than that of the AR detector. Note that the relative performance of the AR detector with respect to the periodogram decreases with an increasing data record. This result appears to be due to the fact that the AR detector performance depends upon $\sqrt{N}\zeta$ [see (7)] while that of the periodogram depends upon $N\zeta$ (see 14). It would appear that for N small, i.e., a high input SNR, the performance of the AR detector would approach and possibly exceed that of the periodogram. The validity of this conjecture, however, cannot be checked using (6) and (7), since the analysis is valid only for $N \geqslant 50$.

It frequently occurs in practice that the frequency of the sinusoid is not constant with time. This is due to a target's Doppler which varies with speed and bearing. The spectral line then "slews" through several DFT bins. This is the case when the bin width has been chosen to be narrow to accommodate a sinusoid with a nearly constant frequency. In this case the periodogram suffers a decorrelation loss, i.e., the signal output of the DFT bin decreases. As an example, consider the received signal $\tilde{S}_t$ to be

$$\tilde{S}_t = e^{j\pi m t^2}$$

where $m = W/N$ is the sweep rate in (cycles/sample)/sample and W is the range of swept frequencies in the record length time. This signal is an FM slide or linear FM waveform and is characteristic of the received waveform for a narrow-band

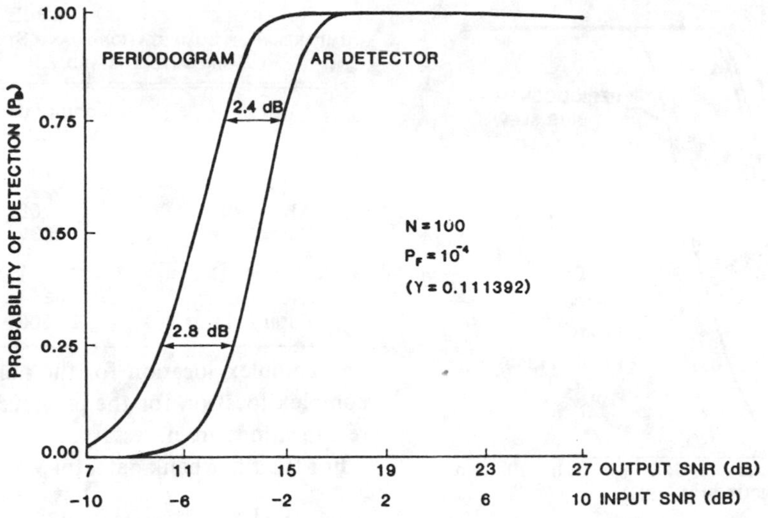

Fig. 4. Comparison of the performance of the periodogram and AR detectors for sinusoidal signal.

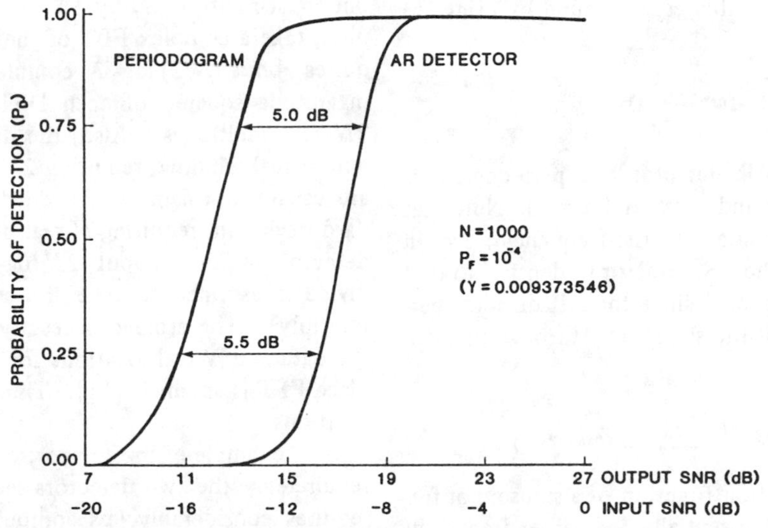

Fig. 5. Comparison of the performance of the periodogram and AR detectors for sinusoidal signal.

emitter as it moves through the closest point of approach. For a periodogram detector the signal output of each DFT bin, from $f = f_d$ to $f = f_d + W$, is decreased over the sinusoid case ($m = 0$) by approximately 10 $\log_{10} \alpha$ dB where α is the number of bins traversed by the FM slide or $\alpha = W/(1/N) = mN^2$. This is because the spectrum of $\tilde{S}_t$ is approximately [16]

$$\left|\sum_{t=0}^{N-1} \tilde{S}_t e^{-j\omega t}\right|^2 = \begin{cases} N/W & 2\pi f_d \leqslant \omega \leqslant 2\pi(f_d + W) \\ 0 & \text{otherwise.} \end{cases}$$

The loss is then

$$10 \log_{10} \frac{N^2}{N/w} = 10 \log_{10} \alpha \text{ dB}. \tag{15}$$

However, one must also account for the fact that the signal appears in α bins and so can be detected in any of these bins. Then

$$P_D = 1 - P_r$$

$$\cdot \{\text{no threshold exceeded in any bin containing signal}\}.$$

Since

$$\tilde{n}_k = \sum_{t=0}^{N-1} \tilde{w}_t e^{-j\omega_k t} \qquad k = 0, 1, \cdots, N-1$$

are uncorrelated and, hence, independent complex Gaussian random variables (see the Appendix)

$$P_D = 1 - P_{M,B}^{\alpha}$$

where $P_{M,B}$ is the probability that the threshold is not exceeded (i.e., a miss) in any bin containing the signal.

$$P_D = 1 - (1 - P_{D,B})^{\alpha}$$

or finally using (14) we have

$$P_D = 1 - [1 - Q(\sqrt{2N\zeta'}, \sqrt{2 \ln N/P_F})]^{\alpha}$$

where

$$10 \log_{10} \zeta' = 10 \log_{10} \zeta - 10 \log_{10} \alpha.$$

Note that the 3 dB loss due to scalloping is not included since

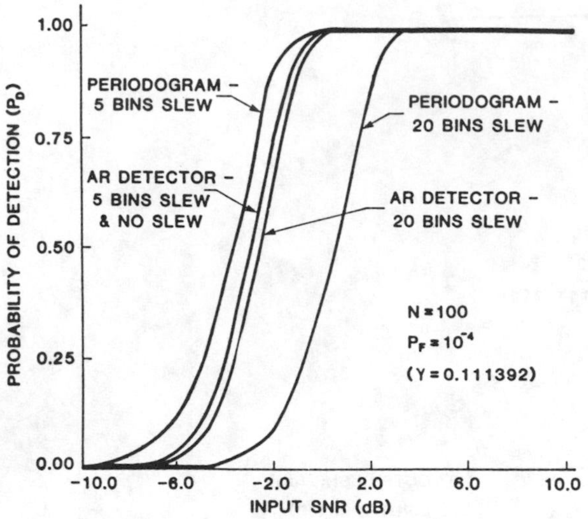

Fig. 6. Comparison of the performance of the periodogram and AR detectors for linear FM signal.

the use of (15) has accounted for the signal mismatch. The AR detector performance for this case is found by using (8) and (9) with

$$\rho_{\tilde{S}}(k) = \frac{\sin[\pi mk(N-k)]}{N \sin \pi mk} e^{j\pi mk(N-1)} \qquad k \geqslant 0.$$

The performance for the AR detector and periodogram is shown in Fig. 6 for $N = 100$ and $\alpha = 5$ and $\alpha = 20$. Note that the AR detector is nearly invariant to the frequency slew. In fact the curve for $\alpha = 5$ for the AR detector is identical to that of Fig. 4. This is not unexpected since the AR detector performance only depends upon the signal only through its autocorrelation function for large N.

For

$$m = \alpha/N^2 \text{ small}, \qquad \rho_{\tilde{S}}(k) \approx \frac{N - |k|}{N} e^{j\pi mk(N-1)}$$

which is just the autocorrelation function of a sinusoid of frequency, $m(N-1)/2$ which is essentially the center frequency of the FM slide. Since the AR detector is independent of the sinusoidal frequency, the AR detector performance for small m is independent of the sweep rate and is the same as the performance for a pure sinusoid. In contrast to this behavior the periodogram detection performance degrades rapidly with increasing sweep rates.

V. Computational and Storage Requirements for AR Detector

We now compare the computational and storage requirements for the AR detector with $p = 1$ and for a bank of matched filters implemented via an FFT

To compute

$$\frac{\left|\sum_{t=0}^{N-2} \tilde{X}_t^* \tilde{X}_{t+1}\right|^2}{\left(\sum_{t=0}^{N-1} |\tilde{X}_t|^2\right)^2}$$

requires about N complex multiplies for the numerator and about $2N$ real multiplies for the denominator. Thus, the total computation is about $6N$ real multiplies. The necessary storage is one real location for the denominator partial sum, one complex location for the numerator partial sum and one complex location for the previous complex sample. Thus, five real locations are necessary.

TABLE I
Comparison of Computational and Storage Requirements for Filter Bank Detector and AR Detector with $p = 1$

	Computation (Real Multiplies)	Storage (Real Locations)
$N = 128$		
AR	768	5
Filter Bank	2432	384
$N = 1024$		
AR	6144	5
Filter Bank	25 600	3072

For the filter bank detector

$$|\tilde{I}_k|^2 = \left|\sum_{t=0}^{N-1} \tilde{X}_t \tilde{S}_t^* e^{-j\omega_k t}\right|^2 \qquad k = 0, 1, \cdots, N-1$$

one must multiply $\tilde{X}_t$ by $\tilde{S}_t^*$ which requires N complex multiplies, take a complex FFT of the resulting sequence which requires about $(N/2)\log_2 N$ complex multiplies [11], take the magnitude squared of each DFT coefficient which requires $2N$ real multiplies. Also, the data $\tilde{X}_t$ is usually windowed with a real window, requiring $2N$ real multiplies and the $|\tilde{I}_k|^2$ are usually normalized or divided by an estimate of the standard deviation, requiring N real divides. Thus, the filter bank detector requires about $2N\log_2 N + 5N$ real multiplies (a divide is assumed to take the same computation time as a multiply). The storage necessary is N complex locations for the data and N real locations for the window weights. An in place FFT is assumed [11]. Thus, the storage required is $3N$ locations.

As an example of the comparison in computation and storage required by the two detectors see Table I. The AR detector requires considerably less computation and storage than the filter bank detector

VI. Summary and Conclusions

For the detection of a signal of unknown form in white Gaussian noise of unknown level the AR detector described here has several important properties. Its performance is independent of the frequency and phase of the signal as given in (1). Thus, one does not need to search over a set of Doppler bins for a maximum as is normally done in a bank of matched filters receiver. This property leads to a substantial savings in computation and storage for the case when the model order p is chosen to be one. Furthermore, the detection performance only depends upon the signal through its autocorrelation function so that complete knowledge of the signal is not required as in a matched filter. Finally, the probability of false alarm is independent of the noise variance, making the AR detector a constant false alarm rate (CFAR) detector. This property is important in sonar and radar applications.

Since the AR detector performance is invariant to signal perturbations as long as the signal autocorrelation function is preserved, the detector is very robust. For the case of a sinusoid of unknown but constant frequency in white noise the AR detector was shown to be inferior to the standard periodogram or bank of matched filters detector by a few dB. As the data record length decreases the AR detector performance will approach that of the periodogram. However, when the sinusoidal frequency is changing in time, in particular, when the signal is a linear FM waveform, the AR detector performance exceeds that of the periodogram by several dB. The linear FM waveform would be encountered in the sonar or radar environment when an emitter moved through the closest point of approach (CPA). It is conjectured that the AR detector performance will be relatively invariant to any frequency modulation of the signal as long as the modulation index does not exceed some value, i.e., as long as the signal autocorrelation function is relatively unchanged. Intuitively, the AR detector performs better than a matched filter when the signal departs from some assumed form since the matched filter correlates against the wrong replica while the AR detector computes autocorrelations, which are correlations between successive samples. If the correlation between successive samples is large, the AR detector performs well. Also, it should be noted that the correlation between successive samples is a correlation between the received waveform and a noisy and time shifted replica of the signal. Thus, the AR detector performs poorer than a bank of matched filters when the signal form is known since the former employs noisy replicas.

Future work should be directed at extending the results reported herein for the AR detector for a model order of one to the general model order case. For a given data record length there is probably some optimum model order to use for maximum detectability. Only a complete statistical analysis, however, will allow one to determine that optimum model order. This task, although possible, appears to be mathematically intractable.

In conclusion, the AR detector offers the potential for improved detection performance in situations where the signal form or noise level is unknown.

Appendix

Autoregressive Detector Performance Derivation

Assume

$$\tilde{X}_t = \begin{cases} \tilde{S}_t + \tilde{w}_t: & H_1 \\ \tilde{w}_t: & H_0 \end{cases}$$

for $t = 0, 1, \cdots, N-1$ and where $\sum_{t=0}^{N-1} |\tilde{S}_t|^2 = E$, $\mathrm{Var}(\tilde{w}_t) = 2$. Without loss of generality we let $\omega_d = 0$, $\phi = 0$, $\sigma_w^2 = 1$ as per properties 1), 2), and 3). To find the performance of the detection statistic

$$r^2 = \frac{\left|\sum_{t=0}^{N-2} \tilde{X}_t^* \tilde{X}_{t+1}\right|^2}{\left(\sum_{t=0}^{N-1} |\tilde{X}_t|^2\right)^2}$$

we proceed as follows.

Consider the circular correlation coefficient

$$r_c^2 = \left| \frac{\sum_{t=0}^{N-2} \tilde{X}_t^* \tilde{X}_{t+1} + \tilde{X}_{N-1} \tilde{X}_0}{\sum_{t=0}^{N-1} |\tilde{X}_t|^2} \right|^2$$

For large N, $r_c^2 \approx r^2$. Let $\boldsymbol{J}$ denote the $N \times N$ permutation matrix

$$\boldsymbol{J} = \begin{bmatrix} 0 & 1 & 0 & 0 & \cdots & 0 \\ 0 & 0 & 1 & 0 & \cdots & 0 \\ 0 & 0 & 0 & 1 & \cdots & 0 \\ & & & & \vdots & \\ 0 & 0 & 0 & 0 & \cdots & 1 \\ 1 & 0 & 0 & 0 & \cdots & 0 \end{bmatrix}$$

Then

$$r_c^2 = \left| \frac{\tilde{\boldsymbol{X}}^H \boldsymbol{J} \tilde{\boldsymbol{X}}}{\tilde{\boldsymbol{X}}^H \tilde{\boldsymbol{X}}} \right|^2$$

where $\tilde{\boldsymbol{X}} = [\tilde{X}_0 \tilde{X}_1 \cdots \tilde{X}_{N-1}]^T$ and H denotes conjugate transpose. Let $\tilde{\boldsymbol{X}}' = \boldsymbol{U}^H \tilde{\boldsymbol{X}}$ where $\boldsymbol{U}$ is the modal matrix of $\boldsymbol{J}$, i.e., the matrix of eigenvectors. It can be shown that [17]

$$[\boldsymbol{U}]_{mn} = \frac{1}{\sqrt{N}} e^{j(2\pi/N)mn} \qquad \begin{matrix} m = 0, 1, \cdots, N-1 \\ n = 0, 1, \cdots, N-1 \end{matrix}$$

and that the eigenvalues of $\boldsymbol{J}$ are

$$\lambda_k = e^{j(2\pi/N)k} \qquad k = 0, 1, \cdots, N-1$$

Furthermore, $\boldsymbol{U}$ is a unitary transformation so that $\boldsymbol{U}^{-1} = \boldsymbol{U}^H$. Letting $\boldsymbol{\Lambda} = \mathrm{diag}(\lambda_0, \lambda_1, \cdots, \lambda_{N-1})$ and noting that $\boldsymbol{U}^H \boldsymbol{J} \boldsymbol{U} = \boldsymbol{\Lambda}$

$$r_c^2 = \left| \frac{(\boldsymbol{U}\tilde{\boldsymbol{X}}')^H \boldsymbol{J} (\boldsymbol{U}\tilde{\boldsymbol{X}}')}{(\boldsymbol{U}\tilde{\boldsymbol{X}}')^H (\boldsymbol{U}\tilde{\boldsymbol{X}}')} \right|^2$$

$$= \left| \frac{\tilde{\boldsymbol{X}}'^H \boldsymbol{\Lambda} \tilde{\boldsymbol{X}}'}{\tilde{\boldsymbol{X}}'^H \tilde{\boldsymbol{X}}} \right|^2 = \left| \frac{\sum_{i=0}^{N-1} \lambda_i |\tilde{X}_i'|^2}{\sum_{i=0}^{N-1} |\tilde{X}_i'|^2} \right|^2$$

Let

$$V_i = |\tilde{X}_i'|^2$$

$$r_c^2 = \frac{\left(\sum_{i=0}^{N-1} V_i \cos \frac{2\pi}{N} i\right)^2 + \left(\sum_{i=0}^{N-1} V_i \sin \frac{2\pi}{N} i\right)^2}{\left(\sum_{i=0}^{N-1} V_i\right)^2}.$$

Let

$$q_1 = \sum_{i=0}^{N-1} V_i \cos \frac{2\pi}{N} i$$

$$q_2 = \sum_{i=0}^{N-1} V_i \sin \frac{2\pi}{N} i$$

$$q_3 = \sum_{i=0}^{N-1} V_i.$$

It is now assumed that N is large enough to invoke the central limit theorem. Then, $\{q_1, q_2, q_3\}$ are jointly Gaussian random variables. Simulations show this approximation to be sufficiently accurate for $N \geqslant 50$. We now proceed to consider the noise only case (H_0)

Let $\tilde{w} = [\tilde{w}_0 \tilde{w}_1 \cdots \tilde{w}_{N-1}]^T$.

Then, $\tilde{w}' = U^H \tilde{w}$. But $\tilde{w}'$ has the same statistical properties as $\tilde{w}$ since

$$E(\tilde{w}') = E(U^H \tilde{w}) = 0$$

$$E(\tilde{w}'\tilde{w}'^H) = U^H E(\tilde{w}\tilde{w}^H) U = 2I$$

and $\tilde{w}'$ is a complex Gaussian vector. Thus, $V_i = |\tilde{X}_i'|^2 = |\tilde{w}_i'|^2$ are $X^2(2)$ random variables which are independent of each other and $E(V_i) = 2$, $\text{Var}(V_i) = 4$ for all i. To find the first and second order moments of $\{q_1, q_2, q_3\}$

$$E(q_1) = E\left(\sum_{i=0}^{N-1} V_i \cos \frac{2\pi}{N} i\right) = 2 \sum_{i=0}^{N-1} \cos \frac{2\pi}{N} i = 0$$

$$E(q_2) = E\left(\sum_{i=0}^{N-1} V_i \sin \frac{2\pi}{N} i\right) = 2 \sum_{i=0}^{N-1} \sin \frac{2\pi}{N} i = 0$$

since

$$\sum_{i=0}^{N-1} e^{j(2\pi/N)i} = 0$$

$$E(q_3) = E\left(\sum_{i=0}^{N-1} V_i\right) = 2N.$$

Define

$$C = \left[1 \quad \cos \frac{2\pi}{N} \quad \cdots \quad \cos \frac{2\pi}{N}(N-1)\right]^T$$

$$S = \left[0 \quad \sin \frac{2\pi}{N} \quad \cdots \quad \sin \frac{2\pi}{N}(N-1)\right]^T$$

$$\mathbf{1} = [1 \quad 1 \quad \cdots \quad 1]^T.$$

Also, we will need $K_v = E[(V - E(V))(V - E(V))^T] = 4I$ where $V = [V_0 V_1 \cdots V_{N-1}]^T$

$$\text{Var}(q_1) = \text{Var}(C^T V) = C^T K_v C = 2N$$

$$\text{Var}(q_2) = \text{Var}(S^T U) = S^T K_v S = 2N$$

since

$$\sum_{i=0}^{N-1} \cos^2 \frac{2\pi}{N} i = \sum_{i=0}^{N-1} \sin^2 \frac{2\pi}{N} i = \frac{N}{2}$$

$$\text{Var}(q_3) = \text{Var}(\mathbf{1}^T V) = \mathbf{1}^T K_v \mathbf{1} = 4N.$$

Finally, q_1, q_2, q_3 are uncorrelated and hence independent since

$$\text{Cov}(q_1, q_2) = C^T K_v S = 4C^T S = 0$$

$$\text{Cov}(q_1, q_3) = C^T K_v \mathbf{1} = 4C^T \mathbf{1} = 0$$

$$\text{Cov}(q_2, q_3) = S^T K_v \mathbf{1} = 4S^T \mathbf{1} = 0.$$

Hence,

$$\left.\begin{aligned} q_1 &\sim N(0, 2N) \\ q_2 &\sim N(0, 2N) \\ q_3 &\sim N(2N, 4N) \end{aligned}\right\} \text{independent}$$

$$r_c^2 = \frac{(q_1/\sqrt{2N})^2 + (q_2/\sqrt{2N})^2}{(q_3/\sqrt{2N})^2} = \frac{X}{Y}$$

X is distributed as a $X^2(2)$ random variable while Y is a noncentral X^2 random variable with PDF [1]

$$p_Y(y) = \begin{cases} \dfrac{1}{\sqrt{4\pi y}} e^{-(1/2)((y+2N)/2)} \cosh\left(\dfrac{\sqrt{2Ny}}{2}\right) & y \geqslant 0 \\ 0 & y < 0. \end{cases}$$

Since X and Y are independent, the quotient formula for $r_c^2 = Z$ yields [18]

$$p_Z(z) = \int_{-\infty}^{\infty} y p_{XY}(zy, y)\, dy$$

$$= \int_0^{\infty} y p_X(zy) p_Y(y)\, dy$$

$$= \int_0^{\infty} \frac{1}{2} y e^{-(1/2)zy} \frac{1}{\sqrt{4\pi y}} e^{-(y+2N)/4} \cdot \cosh(\sqrt{Ny/2})\, dy$$

$$= \frac{1}{4\sqrt{\pi}} e^{-N/2} \int_0^{\infty} \sqrt{y}\, e^{-y((1/2)z+1/4)} \cdot \cosh(\sqrt{Ny/2})\, dy.$$

It can be shown that

$$\int_0^{\infty} \sqrt{y}\, e^{-ay} \cosh(b\sqrt{y})\, dy = \sqrt{\pi/a}\left(\frac{1}{2a} + \frac{b^2}{4a^2}\right) e^{b^2/4a} \qquad \text{for } a > 0, b > 0.$$

Thus, after some simplification, we have

$$p_{r^2/H_0}(x) = \frac{\sqrt{2}}{4} e^{-N/2} \left[\frac{1}{(x+\frac{1}{2})^{3/2}} + \frac{N/2}{(x+\frac{1}{2})^{5/2}}\right] \cdot e^{(N/4)/(x+1/2)} \qquad x \geqslant 0.$$

Note that $p_{r^2/H_0}(x)$ will be nonzero for $x > 1$ even though r^2 must be less than one. This is due to the Gaussian approximation for $\{q_1, q_2, q_3\}$. However, for large N, $p_{r^2/H_0}(x) \approx 0$ for $x > 1$.

To determine the PDF when a signal is present, it is still assumed $\{q_1, q_2, q_3\}$ are jointly Gaussian by the central limit theorem.

Now

$$\tilde{X}' = U^H(\tilde{m} + \tilde{w})$$

where

$$\tilde{m} = [\tilde{S}_0 \tilde{S}_1 \cdots \tilde{S}_{N-1}]^T$$

$$\tilde{X}' = \tilde{m}' + \tilde{w}'$$

where

$$\tilde{m}' = U^H \tilde{m}$$

$\tilde{w}'$ is again a complex Gaussian vector of independent random variables with

$$E(\tilde{w}') = 0$$

$$K'_{\tilde{w}} = 2I.$$

Now, $V_i = |\tilde{m}'_i + \tilde{w}'_i|^2$ is a noncentral X^2 random variable with 2 degrees of freedom. It can be shown that [1]

$$E(V_i) = 2 + |\tilde{m}'_i|^2$$

$$\mathrm{Var}\,(V_i) = 4 + 4|\tilde{m}'_i|^2$$

and again the V_i's are independent random variables. It should be noted that the $\tilde{m}'_i$ sequence is the DFT of $(1/\sqrt{N})\tilde{S}_t$, $t = 0, 1, \cdots, N-1$ since

$$\tilde{m}'_k = \sum_{l=0}^{N-1} [U^H]_{kl} \tilde{m}_l$$

$$= \sum_{l=0}^{N-1} \frac{1}{\sqrt{N}} e^{-j(2\pi/N)kl} \tilde{S}_l = \frac{1}{\sqrt{N}} S(k)$$

where $S(k)$ is the DFT coefficient sequence.

To find the moments of $\{q_1, q_2, q_3\}$

$$E(q_1) = \sum_{i=0}^{N-1} E(V_i) \cos \frac{2\pi}{N} i$$

$$= \sum_{i=0}^{N-1} |\tilde{m}_i|^2 \cos \frac{2\pi}{N} i.$$

Similarly,

$$E(q_2) = \sum_{i=0}^{N-1} |\tilde{m}'_i|^2 \sin \frac{2\pi}{N} i$$

$$E(q_3) = 2N + \sum_{i=0}^{N-1} |\tilde{m}'_i|^2.$$

If we define the circular signal autocorrelation function as

$$R^c_{\tilde{S}}(k) = \frac{1}{N} \sum_{t=0}^{N-1} \tilde{S}^*_t \tilde{S}_{t+k}$$

where $\tilde{S}_N = \tilde{S}_0$, $\tilde{S}_{N+1} = \tilde{S}_1$, etc. it is easily shown that [11]

$$R^c_{\tilde{S}}(k) = \frac{1}{N^2} \sum_{l=0}^{N-1} |S(l)|^2 e^{j(2\pi/N)kl}$$

$$= \frac{1}{N} \sum_{l=0}^{N-1} |\tilde{m}'_l|^2 e^{j(2\pi/N)kl}.$$

Since

$$E = NR^c_{\tilde{S}}(0)$$

letting $\rho^c_{\tilde{S}}(k) = R^c_{\tilde{S}}(k)/R^c_{\tilde{S}}(0)$

$$E(q_1) = E\,\mathrm{Re}\,(\rho^c_{\tilde{S}}(1))$$

$$E(q_2) = E\,\mathrm{Im}\,(\rho^c_{\tilde{S}}(1))$$

$$E(q_3) = 2N + E.$$

Similarly, it may be shown

$$\mathrm{Var}\,(q_1) = 2N + 2E + 2E\,\mathrm{Re}\,(\rho^c_{\tilde{S}}(2))$$

$$\mathrm{Var}\,(q_2) = 2N + 2E - 2E\,\mathrm{Re}\,(\rho^c_{\tilde{S}}(2))$$

$$\mathrm{Var}\,(q_3) = 4N + 4E$$

$$\mathrm{Cov}\,(q_1, q_2) = 2E\,\mathrm{Im}\,(\rho^c_{\tilde{S}}(2))$$

$$\mathrm{Cov}\,(q_1, q_3) = 4E\,\mathrm{Re}\,(\rho^c_{\tilde{S}}(1))$$

$$\mathrm{Cov}\,(q_2, q_3) = 4E\,\mathrm{Im}\,(\rho^c_{\tilde{S}}(1)).$$

If we let

$$X_i = q_i/\sqrt{N} \qquad i = 1, 2, 3$$

$$\zeta = E/2N$$

where ζ is the signal-to-noise ratio

$$r_c^2 = \frac{X_1^2 + X_2^2}{X_3^2}.$$

Then X_1, X_2, X_3 are jointly Gaussian and

$$E(X_1) = 2\sqrt{N}\,\zeta\,\mathrm{Re}\,(\rho^c_{\tilde{S}}(1))$$

$$E(X_2) = 2\sqrt{N}\,\zeta\,\mathrm{Im}\,(\rho^c_{\tilde{S}}(1))$$

$$E(X_3) = 2\sqrt{N} + 2\sqrt{N}\,\zeta$$

$$\mathrm{Var}\,(X_1) = 2 + 4\zeta + 4\zeta\,\mathrm{Re}\,(\rho^c_{\tilde{S}}(2))$$

$$\mathrm{Var}\,(X_2) = 2 + 4\zeta - 4\zeta\,\mathrm{Re}\,(\rho^c_{\tilde{S}}(2))$$

$$\mathrm{Var}\,(X_3) = 4 + 8\zeta$$

$$\mathrm{Cov}\,(X_1, X_2) = 4\zeta\,\mathrm{Im}\,(\rho^c_{\tilde{S}}(2))$$

$$\mathrm{Cov}\,(X_1, X_3) = 8\zeta\,\mathrm{Re}\,(\rho^c_{\tilde{S}}(1))$$

$$\mathrm{Cov}\,(X_2, X_3) = 8\zeta\,\mathrm{Im}\,(\rho^c_{\tilde{S}}(1)).$$

The PDF of r^2 for large N is approximately equal to that of r_c^2. Let $F_{r^2/H_1}(x)$ be the CDF of r^2. Then

$$F_{r^2/H_1}(x) = P_r\{r^2 \leqslant x\}$$

$$= P_r\left\{\frac{X_1^2 + X_2^2}{X_3^2} \leqslant x\right\}$$

$$= P_r\{X_1^2 + X_2^2 - x\,X_3^2 \leqslant 0\}$$

$$= F_Y(0)$$

where

$$Y = X_1^2 + X_2^2 - X\,X_3^2.$$

But $F_Y(y)$ can be found from the characteristic function, $\varphi_Y(\omega)$ as [19]

$$F_Y(y) = \frac{1}{2} - \frac{1}{\pi} \int_0^{\infty} \cdot \frac{\text{Im}\,[\varphi_Y(\omega)] \cos \omega y - \text{Re}\,[\varphi_Y(\omega)] \sin \omega y}{\omega} \, d\omega$$

so that

$$F_{r^2/H_1}(x) = \frac{1}{2} - \frac{1}{\pi} \int_0^{\infty} \frac{\text{Im}\,[\varphi_Y(\omega)]}{\omega} \, d\omega.$$

To find $\varphi_Y(\omega)$ we let

$$Y = X^T A X$$

where

$$X = [X_1 X_2 X_3]^T$$

$$A = \begin{bmatrix} 1 & 0 & 0 \\ 0 & 1 & 0 \\ 0 & 0 & -x \end{bmatrix}.$$

The characteristic function of a quadratic form, where X is a Gaussian vector with mean vector u and covariance matrix, K is [20]

$$\varphi_Y(\omega) = \frac{1}{|I - 2j\omega K A|^{1/2}} e^{j\omega u^T A (I - 2j\omega K A)^{-1} u}.$$

Thus, finally, if we replace the circular correlation coefficients by the serial correlation coefficients, we can compute the PDF of r^2 under H_1 as

$$p_{r^2/H_1}(x) = \begin{cases} \dfrac{d}{dx} F_{r^2/H_1}(x) & 0 \leqslant x \leqslant 1 \\ 0 & \text{elsewhere} \end{cases}$$

where

$$F_{r^2/H_1}(x) = \frac{1}{2} - \frac{1}{\pi} \int_0^{\infty} \frac{\text{Im}\,[\varphi_Y(\omega)]}{\omega} \, d\omega$$

$$\varphi_Y(\omega) = \frac{1}{|I - 2j\omega K A|^{1/2}} e^{j\omega u^T A (I - 2j\omega K A)^{-1} u}$$

$$u = [2\sqrt{N}\zeta \,\text{Re}\,(\rho_{\tilde{S}}(1)) \quad 2\sqrt{N}\zeta \,\text{Im}\,(\rho_{\tilde{S}}(1)) \quad 2\sqrt{N} + 2\sqrt{N}\zeta]^T$$

$$K = \begin{bmatrix} 2 + 4\zeta + 4\zeta \,\text{Re}\,(\rho_{\tilde{S}}(2)) & 4\zeta \,\text{Im}\,(\rho_{\tilde{S}}(2)) & 8\zeta \,\text{Re}\,(\rho_{\tilde{S}}(1)) \\ 4\zeta \,\text{Im}\,(\rho_{\tilde{S}}(2)) & 2 + 4\zeta - 4\zeta \,\text{Re}\,(\rho_{\tilde{S}}(2)) & 8\zeta \,\text{Im}\,(\rho_{\tilde{S}}(1)) \\ 8\zeta \,\text{Re}\,(\rho_{\tilde{S}}(1)) & 8\zeta \,\text{Im}\,(\rho_{\tilde{S}}(1)) & 4 + 8\zeta \end{bmatrix}$$

$$\rho_{\tilde{S}}(k) = \frac{1}{E} \sum_{t=0}^{N-k-1} \tilde{S}_t^* \tilde{S}_{t+k}, \quad E = \sum_{t=0}^{N-1} |\tilde{S}_t|^2, \quad \zeta = E/2N$$

$$A = \begin{bmatrix} 1 & 0 & 0 \\ 0 & 1 & 0 \\ 0 & 0 & -x \end{bmatrix}.$$

It should be noted that to evaluate $\varphi_Y(\omega)$ requires the square root of a complex number. To determine the appropriate sign to use it is necessary to preserve the analyticity of $\varphi_Y(\omega)$. Since from the definition of the characteristic function $\varphi_Y(0) = +1$ for small ω one should choose the positive sign. As ω increases, the sign is chosen to preserve the continuity of $\varphi_Y(\omega)$.

Acknowledgment

The author wishes to express his sincere appreciation to Prof. L. Jackson and Prof. D. Tufts of the University of Rhode Island, Kingston, RI, W. Knight of Raytheon Co., Portsmouth, RI, and A. Nuttall and C. Carter of the Naval Underwater Systems Center, New London, CT for their useful discussions of the ideas presented in this paper, and to K. Roberts of Raytheon Co., Portsmouth, RI, for writing the computer programs necessary for this study.

References

[1] A. D. Whalen, *Detection of Signals in Noise.* New York: Academic, 1971.

[2] H. L. Van Trees, *Detection, Estimation, and Modulation Theory,* Part 3. New York: Wiley, 1971.

[3] A. W. Rihaczek, *Principles of High Resolution Radar.* New York: McGraw-Hill, 1969.

[4] C. N. Pryor, "Effect of background estimation on the sensitivity of sonar and radar receivers," NOLTR 73-53, Apr. 1974.

[5] J. R. Williams and G. R. Ricker, "Signal detectability performance of optimum Fourier receivers," *IEEE Trans. Audio Electroacoust.*, Oct. 1972.

[6] E. C. Whitman, "The spectral analysis of discrete time series in terms of linear regressive models," NOLTR 70-109, June 23, 1970.

[7] J. P. Burg, "Maximum entropy spectral analysis," Ph.D. dissertation, Stanford Univ., 1975.

[8] J. Makhoul, "Linear prediction: A tutorial review," *Proc. IEEE,* vol. 63, pp. 561-580, Apr. 1975.

[9] L. J. Griffiths, "Rapid measurement of digital instantaneous frequency," *IEEE Trans. Acoust., Speech, Signal Processing,* vol. ASSP-23, Apr. 1975.

[10] G. W. Lank *et al.*, "A semicoherent detection and Doppler estimation statistic," *IEEE Trans. Aerosp. Electron. Syst.*, vol. AES-9, Mar. 1973.

[11] A. V. Oppenheim and R. W. Schafer, *Digital Signal Processing.* Englewood Cliffs, NJ: Prentice-Hall, 1975.

[12] S. D. Conte and C. deBoor, *Elementary Numerical Analyses.* New York: McGraw-Hill, 1972.
[13] H. L. Van Trees, *Detection, Estimation, and Modulation Theory, Part I.* New York: Wiley, 1968.
[14] C. W. Helstrom, *Statistical Theory of Signal Detection.* Elmsford, NY: Pergamon, 1968.
[15] F. J. Harris, "On the use of windows for harmonic analysis with the discrete Fourier transform," *Proc. IEEE*, Jan. 1978.
[16] T. H. Glisson, C. I. Black, and A. P. Sage, "On sonar signal analysis," *IEEE Trans. Aerosp. Electron. Syst.*, vol. AES-6, Jan. 1970.
[17] M. Marcus, "Basic theorems in matrix theory," Nat. Bur. Stand., Applied Math. Series, Jan. 22, 1960.
[18] A. Papoulis, *Probability, Random Variables and Stochastic Processes.* New York: McGraw-Hill, 1965.
[19] A. Nuttall, "Numerical evaluation of cumulative probability distribution functions directly from characteristic functions," NUSC Rep. 1032, Aug. 11, 1969.
[20] J. E. Mazo and J. Salz, "Probability of error for quadratic detectors," *Bell Syst. Tech. J.*, vol. 44, 1965.

Bibliography

NO references published before 1979 are included in the bibliography. An extensive listing of earlier publications is given in the predecessor volume by D. G. Childers. Since the large number of listed references deals with more than one aspect of spectrum estimation, it was decided to stack them all together, rather than to divide them into a number of various subject categories. Entries are listed in alphabetical order of the first author's last name.

T. J. Abatzoglou, "A fast maximum likelihood algorithm for frequency estimation of a sinusoid based on Newton's method," *IEEE Trans. Acoust., Speech, Signal Processing*, vol. ASSP-33, no. 1, pp. 77–89, Feb. 1985.

G. D. Achilles, "Spectral estimation," in *Proc. 2nd European Signal Processing Conf.*, EUROSIPCO-83, (Erlangen, Germany, Sept. 12–16, 1983), pp. 447-454.

N. Ahmed and D. H. Youn, "On a realization and related algorithm for adaptive prediction," *IEEE Trans. Acoust., Speech, Signal Processing*, vol. ASSP-28, no. 5, pp. 493–497, Oct. 1980.

O. Alkin and R. Houts, "Comparing the radar clutter-suppression performances of lattice prediction error filters using three variations of Burg's algorithm," in *Proc. IEEE Int. Conf. on ASSP*, (Tampa, FL, Mar. 26–29, 1985), pp. 1372–1375.

M. Amin and L. J. Griffiths, "Time-varying spectral estimation using symmetric smoothing," in *Proc. IEEE Int. Conf. on ASSP*, (Boston, MA, Apr. 14–16, 1983), pp. 9–12.

A. Arcese, "On the method of maximum entropy spectrum estimation," *IEEE Trans. Inform. Theory*, vol. IT-29, no. 1, pp. 161–164, Jan. 1983.

T. E. Barnard, "Two maximum entropy beamforming algorithms for equally spaced line arrays," *IEEE Trans. Acoust., Speech, Signal Processing*, vol. ASSP-30, no. 2, pp. 175–189, Apr. 1982.

N. Beamish and M. B. Preistley, "A study of autoregressive and window spectral estimation," *Appl. Stat.* (GB), vol. 30, no. 1, pp. 41–58, 1981.

N. J. Bershad and P. L. Feintuch, "Correlation function for the weight sequence of the frequency domain adaptive filter," *IEEE Trans. Acoust., Speech, Signal Processing*, vol. ASSP-30, no. 5, pp. 801–804, Oct. 1982.

M. Bertran-Salvans, "A generalized window approach to spectral estimation," *IEEE Trans. Acoust., Speech, Signal Processing*, vol. ASSP-32, no. 1, pp. 7–19, Feb. 1984.

R. Bicocchi, T. Bucciarelli, S. Cacopardi, P. T. Melacci, and G. Picardi, "Autoregressive spectral estimation: A tool to cancel clutter echoes in modern radars," in *Proc. ASSP Spectrum Estimation Workshop II*, (Tampa, FL, Nov. 10–11, 1983), pp. 316–319.

G. Bienvenu, "Influence of the spatial coherence of background noise on high resolution passive methods," in *Proc. IEEE Int. Conf. on ASSP*, (Washington, DC, Apr. 2–4, 1979), pp. 306–309.

G. Bienvenu and L. Kopp, "Adaptivity to background noise spatial coherence for high resolution passive methods," in *Proc. IEEE Int. Conf. on ASSP*, (Denver, CO, Apr. 9–11, 1980), pp. 307–310.

——, "Optimality of high resolution array processing using the eigensystem approach," *IEEE Trans. Acoust., Speech, Signal Processing*, vol. ASSP-31, no. 5, pp. 1235–1248, Oct. 1983.

J. F. Bohme, "On the sensitivity of orthogonal beamforming," in *Proc. IEEE Int. Conf. on ASSP*, (Paris, France, May 3–5, 1982), pp. 787–790.

D. Bordelon, "Complementarity of the Reddi method of source direction estimation with those of Pisarenko and Cantoni and Godara," Vol. I, *J. Acoust. Soc. Amer.*, vol. 69, pp. 1355–1359, May 1981.

D. E. Bowyer, P. K. Rajasekaran, and W. W. Gebhart, "Adaptive clutter filtering using autoregressive spectral estimation," *IEEE Trans. Aerosp. Electron. Syst.*, vol. AES-15, vol. 4, pp. 538–546, July 1979.

Y. Bresler and A. Macovski, "Exact maximum likelihood estimation of superimposed exponential signals in noise," in *Proc. IEEE Int. Conf. on ASSP*, (Tampa, FL, Mar. 26–29, 1985), pp. 1824–1827.

D. R. Brillinger, "A maximum likelihood approach to frequency-wavenumber analysis," *IEEE Trans. Acoust., Speech, Signal Processing*, vol. ASSP-33, no. 5, pp. 1076–1085, Oct. 1985.

T. P. Bronez and J. A. Cadzow, "An algebraic approach to superresolution adaptive array processing," *IEEE Aerosp. Electron. Syst.*, vol. AES-19, no. 1, pp. 123–133, Jan. 1983.

S. P. Bruzzone and M. Kaveh, "Information tradeoffs in using the sample autocorrelation function in ARMA parameter estimation," *IEEE Trans. Acoust., Speech, Signal Processing*, vol. ASSP-32, no. 4, pp. 701–715, Aug. 1984.

K-J. Bry and J. Le Roux, "Comparison of some algorithms for identifying autoregressive signals in the presence of observation noise," in *Proc. IEEE Int. Conf. on ASSP*, (Paris, France, May 3–5, 1982), pp. 224–227.

J. P. Burg, D. G. Luenberger, and D. L. Wenger, "Estimation of structured covariance matrices," *Proc. IEEE*, vol. 70, no. 9, pp. 963–974, Sept. 1982.

J. Casar-Corredera, J. Alcazar-Fernandez, and L. Hernandez-Gomez, "On 2-D Prony methods," in *Proc. IEEE Int. Conf. on ASSP*, (Tampa, FL, Mar. 26–29, 1985), pp. 796–799.

J. A. Cadzow, "High-performance spectral estimation—A new ARMA method," *IEEE Trans. Acoust., Speech, Signal Processing*, vol. ASSP-28, no. 5, pp. 524–529, Oct. 1980.

——, "An effective ARMA modeling method," in *Proc. 2nd Int. Symp. on Computer Aided Seismic Analysis and Discrimination*, North Dartmouth, MA, Aug. 19–21, 1981), pp. 66–75.

——, "ARMA time series modeling: An effective method," *IEEE Trans. Aerosp. Electron. Syst.*, vol. AES-19, no. 1, pp. 49–58, Jan. 1983.

J. A. Cadzow and K. Ogino, "Adaptive ARMA spectral estimation," in *Proc. IEEE Int. Conf. on ASSP*, (Atlanta, GA, Mar. 30–Apr. 1, 1981), pp. 475–479.

G. D. Cain, R. C. S. Morling, and J. Skorkowski, "Performance of adaptive whitening using a CID filter," in *Proc. Digital Processing of Signals in Communications*, (Loughborough, Leics., England, Apr. 7–10, 1981).

A. Cantoni and L. Godara, "Resolving the directions of sources in a correlated signal field incident on an array," *J. Acoust. Soc. Amer.*, vol. 67, no. 4, pp. 1247–1255, Apr. 1980.

G. Carayannis, D. G. Manolakis, and N. Kalouptsidis, "A fast sequential algorithm for least-squares filtering and prediction," *IEEE Trans. Acoust., Speech, Signal Processing*, vol. ASSP-31, no. 6, pp. 1394–1402, Dec. 1983.

Y. T. Chan and J. C. Wood, "A new order determination technique for ARMA processes," *IEEE Trans. Acoust., Speech, Signal Processing*, vol. ASSP-32, no. 3, pp. 517–521, June 1984.

R. Chellappa and G. Sharma, "Two-dimensional spectral estimation using spatial autoregressive models," in *Proc. IEEE Int. Conf. on ASSP*, (Boston, MA, Apr. 14–16, 1983), pp. 855–858.

H. Chen, T. Sarkar, S. Dianat, and J. Brule, "Adaptive spectral estimation by the conjugate gradient method," in *Proc. IEEE Int. Conf. on ASSP*, (Tampa, FL, Mar. 26–29, 1985), pp. 81–84.

Y. Y. Chen, "Stable time- and order-recursive algorithm for the adaptive lattice filter," in *Proc. IEEE Int. Conf. on ASSP*, (San Diego, CA, Mar. 19–21, 1984), pp. 3.8.1–3.8.4.

J. P. Costas, "Residual signal analysis—A search and destroy approach to spectral analysis," in *Proc. First ASSP Workshop on Spectral Estimation*, (Hamilton, ON, Canada, Aug. 17–18, 1981), vol. 2, pp. 6.5.1–6.5.8.

R. L. Cupo, "The iterative NCDE algorithm for ARMA system identification and spectral estimation," *IEEE Trans. Acoust., Speech, Signal Processing*, vol. ASSP-33, no. 4, pp. 1021–1024, Aug. 1985.

H. Dante, "Spectrum estimation of time series with missing data," in *Proc. IEEE Int. Conf. on ASSP*, (Tampa, FL, Mar. 26–29, 1985), pp. 89–92.

R. J. P. De Figueiredo, "Fast algorithms for spectral estimation and spline interpolation of non-bandlimited random signals," in *Proc. ASSP Spectrum Estimation Workshop II*, (Tampa, FL, Nov. 10–11, 1983), pp. 168–171.

S. DeGraaf and D. H. Johnson, "Capability of array processing algorithms to estimate source bearings," *IEEE Trans. Acoust., Speech, Signal Processing*, vol. ASSP-31, no. 2, pp. 344–347, Apr. 1983.

A. Di, "Multiple source location—A matrix decomposition approach," *IEEE Trans. Acoust., Speech, Signal Processing*, vol. ASSP-33, no. 5, pp. 1086–1091, Oct. 1985.

F. U. Dowla and J. S. Lim, "Relationship between maximum likelihood method and autoregressive modeling in multidimensional power spectrum estimation," *IEEE Trans. Acoust., Speech, Signal Processing*, vol. ASSP-32, no. 5, pp. 1083–1087, Oct. 1984.

——, "Resolution property of the improved maximum likelihood method," in *Proc. IEEE Int. Conf. on ASSP*, (Tampa, FL, Mar. 26–29 1985), pp. 820–823.

T. S. Durrani and K. C. Sharman, "Extraction of an eigenvector oriented 'spectrum' for the MESA coefficients," *IEEE Trans. Acoust., Speech, Signal Processing*, vol. ASSP-30, no. 4, pp. 649–651, Aug. 1982.

——, "Eigenfilter approaches to adaptive array processing," *IEE Proc. pt. F* (GB), vol. 130, no. 1, pp. 22–28, Feb. 1983.

R. Dwyer, "Two-dimensional arrays with nonlinear elements," in *Proc. IEEE Int. Conf. on ASSP*, (Tampa, FL, Mar. 26–29, 1985), pp. 1285–1288.

T. Dyson and S. Rao, "Some detection and resolution properties of maximum entropy spectrum analysis," *Signal Process*, (Netherlands, July 1980), vol. 2, no. 3, pp. 261–270.

A. Efron and D. Tufts, "Estimation of frequencies of multiple two-dimensional sinusoids: Improved methods of linear prediction," in *Proc. IEEE Int. Conf. on ASSP*, (Tampa, FL, Mar. 26–29, 1985), pp. 1777–1779.

H. El-Sherief, "Adaptive least-squares for parametric spectral estimation and its application to pulse estimation and deconvolution of seismic data," in *Proc. IEEE Int. Conf. on ASSP*, (San Diego, CA, Mar. 19–21, 1984), pp. 5.4.1–5.5.4.

Y. El-Cherif and S. K. Sarna, "Parametric spectral estimation of gastrointestinal signals," in *Proc. IEEE 1981 Frontiers in Engineering in Health Care Conf.*, (Houston, TX, Sept. 19–21, 1981), pp. 132–136.

M. Er and A. Cantoni, "A new set of linear constraints for broadband time domain element space processors," in *Proc. IEEE Int. Conf. on ASSP*, (Tampa, FL, Mar. 26–29, 1985), pp. 1800–1803.

B. Erickson, G. Johnson, and D. E. Ohlms, "Detection of a sinusoidal signal in the presence of directional interference," in *Proc. IEEE Int. Conf. on ASSP*, (Tampa, FL, Mar. 26–29, 1985), pp. 1277–1280.

C. Esmersoy, "Signal power upper bound in parameter estimation," *IEEE Trans. Acoust., Speech, Signal Processing*, vol. ASSP-33, no. 1, pp. 315–316, Feb. 1985.

J. E. Evans, "Aperture sampling techniques for precision direction finding," *IEEE Trans. Aerosp. Electron. Syst.*, vol. AES-15, no. 6, pp. 891–895, Nov. 1979.

——, "Comments on 'angular spectra in radar applications'," *IEEE Trans. Aerosp. Electron. Syst.*, vol. AES-15, no. 6, pp. 898–903, Nov. 1979.

H. Fan, E. I. El-Masry and W. K. Jenkins, "Resolution enhancement of digital beamformers," *IEEE Trans. Acoust., Speech, Signal Processing*, vol. ASSP-32, no. 5, pp. 1041–1052, Oct. 1984.

D. R. Farrier, "Jaynes' principle and maximum entropy spectral estimation," *IEEE Trans. Acoust., Speech, Signal Processing*, vol. ASSP-32, no. 6, pp. 1176–1183, Dec. 1984.

D. R. Farrier and D. J. Jeffries, "Bearing estimation in the presence of unknown correlated noise," in *Proc. IEEE Int. Conf. on ASSP*, (Tampa, FL, Mar. 26–29, 1985), pp. 1788–1791.

A. Figueiras-Vidal, J. Casar-Coredera, R. Garcia-Gomez, and J. Paez Borrallo, "L_1-norm vs. L_2-norm minimization in parametric spectral analysis: A general discussion," in *Proc. IEEE Int. Conf. on ASSP*, (Tampa, FL, Mar. 26–29, 1985), pp. 304–307.

P. Fougere, "Spectral model order determination via significant reflection coefficients," in *Proc. IEEE Int. Conf. on ASSP*, (Tampa, FL, Mar. 26–29, 1985), pp. 1345–1347.

B. Friedlander, "Recursive lattice forms for spectral estimation and adaptive control," in *Proc. 19th IEEE Conference on Decision and Control Including the Symposium on Adaptive Processes*," (Albuquerque, NM, Dec. 10–12, 1980), pp. 466–471.

——, "Instrumental variable methods for ARMA spectral estimation," in *Proc. IEEE Int. Conf. on ASSP*, (Paris, France, May 3–5, 1982), pp. 248–251.

——, "A recursive maximum likelihood algorithm for ARMA spectral estimation," *IEEE Trans. Inf. Theory*, vol. IT-28, no. 4, pp. 639–646, Jul. 1982.

——, "Recursive lattice forms for spectral estimation," *IEEE Trans. Acoust., Speech, Signal Processing*, vol. ASSP-30, no. 6, pp. 920–930, Dec. 1982.

——, "A recursive maximum likelihood algorithm for ARMA line enhancement," *IEEE Trans. Acoust., Speech, Signal Processing*, vol. ASSP-30, no. 4, pp. 651–657, Aug. 1982.

——, "Instrumental variable methods for ARMA spectral estimation," *IEEE Trans. Acoust., Speech, Signal Processing*, vol. ASSP-31, no. 2, pp. 404–415, Apr. 1983.

——, "On the computation of the Cramer–Rao bound for ARMA parameter estimation," *IEEE Trans. Acoust., Speech, Signal Processing*, vol. ASSP-32, no. 4, pp. 721–727, Aug. 1984.

B. Friedlander and B. Porat, "The modified Yule–Walker method of ARMA spectral estimation," *IEEE Trans. Aerosp. Electron. Syst.*, vol. AES-20, no. 2, pp. 158–173, Mar. 1984.

——, "A spectral matching technique for ARMA parameter estimation," *IEEE Trans. Acoust., Speech, Signal Processing*, vol. ASSP-32, no. 2, pp. 338–343, Apr. 1984.

——, "A general lower bound for parametric spectrum estimation," *IEEE Trans. Acoust., Speech, Signal Processing*, vol. ASSP-32, no. 4, pp. 728–733, Aug. 1984.

——, "Bounds for ARMA spectral analysis based on sample covariances," in *Proc. IEEE Int. Conf. on ASSP*, (Tampa, FL, Mar. 26–29, 1985), pp. 616–619.

B. Friedlander and K. C. Sharman, "Performance evaluation of the modified Yule–Walker estimator," *IEEE Trans. Acoust., Speech, Signal Processing*, vol. ASSP-33, no. 3, pp. 719–725, June 1985.

W. F. Gabriel, "Spectral analysis and adaptive superresolution techniques," *Proc. IEEE*, vol. 68, no. 6, pp. 654–666, June 1980.

——, "Adaptive superresolution of coherent RF spatial sources," in *Proc. 1st IEEE ASSP Workshop on Spectral Analysis*, (Hamilton, ON, Canada, Aug. 17–18, 1981), vol. 2, pp. 5.1.1.–5.1.7.

——, "A 'superresolution' target-tracking concept," in *1983 International Symposium Digest, Antennas and Propagation*, (Houston, TX, May 23–26, 1983), pp. 195–198.

R. G. Gan, K. F. Eman, and S. M. Wu, "An extended FFT algorithm for ARMA spectral estimation," *IEEE Trans. Acoust., Speech, Signal Processing*, vol. ASSP-32, no. 1, pp. 168–170, Feb. 1984.

C. Gibson and S. Haykin, "A comparison of algorithms for the calculation of adaptive lattice filters," in *Proc. IEEE Int. Conf. on ASSP*, (Denver, CO, Apr. 9–11, 1980), pp. 978–983.

——, "Nonstationary learning characteristics of adaptive lattice filters," in *Proc. IEEE Int. Conf. on ASSP*, (Paris, France, May 3–5, 1982), pp. 671–675.

C. J. Gibson, S. Haykin, and S. B. Kesler, "Maximum entropy (adaptive) filtering applied to radar clutter," in *Proc. IEEE Int. Conf. on ASSP*, (Washington, DC, Apr. 2–4, 1979), pp. 166–169.

J. D. Gibson, "Backward adaptive prediction as spectral analysis within a closed loop," *IEEE Trans. Acoust., Speech, Signal Processing*, vol. ASSP-33, no. 5, pp. 1166–1174, Oct. 1985.

D. F. Gingras, "Spectral estimation statistics for noise corrupted autoregressive series—First-order case," in *Proc. IEEE Int. Conf. on ASSP*, (Tampa, FL, Mar. 26–29, 1985), pp. 93–96.

——, "Asymptotic properties of high-order Yule–Walker estimates of the AR parameters of an ARMA time series," *IEEE Trans. Acoust., Speech, Signal Processing*, vol. ASSP-33, no. 5, pp. 1095–1101, Oct. 1985.

B. Golubev and S. Rogers, "Non-recursive frequency estimation for closely-spaced sinusoids," in *Proc. IEEE Int. Conf. on ASSP*, (Tampa, FL, Mar. 26–29, 1985), pp. 316–319.

C. E. Goutis, R. M. Leahy, and P. G. Cassidy, "Spectra using data distribution and covariance modelling," in *Proc. IEEE Int. Conf. on ASSP*, (San Diego, CA, Mar. 19–21, 1984), pp. 31.1.1–31.1.4.

Y. K. Goutsias, "Two dimensional spectral estimation of sinusoidal signals," in *Proc. of MELECON '83 Mediterranean Electrotechnical Conference*, (Athens, Greece, May 24–26, 1983), pp. C2.06.1–C2.06.2.

Y. Grenier, "Time varying lattices and autoregressive models: Parameter estimation," in *Proc. IEEE Int. Conf. on ASSP*, (Paris, France, May 3–5, 1982), pp. 1337–1340.

——, "Time-dependent ARMA modeling of nonstationary signals," *IEEE Trans. Acoust., Speech, Signal Processing*, vol. ASSP-31, no. 4, pp. 899–911, Aug. 1983.

D. W. Griffin and J. S. Lim, "Signal estimation from modified short-time Fourier transform," *IEEE Trans. Acoust., Speech, Signal Processing*, vol. ASSP-32, no. 2, pp. 236–243, Apr. 1984.

L. J. Griffiths, "Adaptive structure for multiple-input noise cancelling applications," in *Proc. IEEE Int. Conf. on ASSP*, (Washington, DC, Apr. 2–4, 1979), pp. 925–928.

——, "A comparison of lattice-based adaptive algorithms," in *Proc. 1980 IEEE International Symposium on Circuits and Systems*, (Houston, TX, Apr. 28–30, 1980), pp. 742–743.

L. J. Griffiths and K. B. McGregor, "Spectral estimation using an adaptive oscillator," *IEE Proc., pt. F*, (GB), vol. 130, no. 3, pp. 246–249, Apr. 1983.

J. K. Hammond, "Parametric methods in signal analysis (with particular reference to autoregressive spectral estimation)," *Agard Report No. 700, Modern Data Analysis Techniques in Noise and Variation Problems* (Agard-R-700), (Rhode-St.-Genese, Belgium, Dec. 7–11, 1981), pp. 117–123.

J. C. Hassab, "Passive bearing estimation of broad-band source," *IEEE Trans. Acoust., Speech, Signal Processing*, vol. ASSP-32, no. 2, pp. 426–431, Apr. 1984.

R. M. Hawkes and C. Hlava, "Adaptive spectral analysis," in *19th International Electronics Convention and Exhibition Digest of Papers*, (IREECON International Sydney '83, Sydney, Australia, Sept. 5–9, 1983), pp. 517–519.

S. Haykin (Ed.), *Nonlinear Methods of Spectral Analysis*, 2nd ed. Berlin, Germany: Springer-Verlag, 1983.

S. Haykin, B. W. Currie, and S. B. Kesler, "Maximum-entropy spectral analysis of radar clutter," *Proc. IEEE*, vol. 70, no. 9, pp. 953–962, Sept. 1982.

S. Haykin and J. Kesler, "Adaptive canceller for evaluation angle estimation in the presence of multipath," *IEE Proc., pt. F*, (GB), vol. 130, no. 4, pp. 303–308, June 1983.

S. Haykin, J. Kesler, and J. Litva, "Evaluation of angle of arrival estimators using real multipath data," in *Proc. IEEE Int. Conf. on ASSP*, (Boston, MA, Apr. 14–16, 1983), pp. 695–698.

S. Haykin and J. P. Reilly, "Maximum likelihood receiver for low-angle tracking radar, part I: The symmetric case," *Proc. IEE*, vol. 129, Part F, pp. 261–272, 1982.

S. Haykin, J. P. Reilly, and D. P. Taylor, "New realization of maximum likelihood receiver of low-angle tracking radar," *Electron Lett.*, vol. 16, pp. 288–289, 1980.

B. Helme and C. L. Nikias, "A high-resolution modified Burg algorithm for spectral estimation," in *Proc. 1983 Int. Electr, Electron. Conf. Proc.*, (Toronto, ON, Canada, Sept. 26–28, 1983), pp. 348–351.

——, "A high-resolution modified algorithm for spectral estimation," in *Proc. IEEE Int. Conf. on ASSP*, (San Diego, CA, Mar. 19–21, 1984), pp. 13.1.1–13.1.4.

B. I. Helme and C. L. Nikias, "Improved spectrum performance via a data-adaptive weighed Burg technique," *IEEE Trans. Acoust., Speech, Signal Processing*, vol. ASSP-33, no. 4, pp. 903–910, Aug. 1985.

R. W. Herring, "The cause of line splitting in Burg maximum-entropy spectral analysis," *IEEE Trans. Acoust., Speech, Signal Processing*, vol. ASSP-28, no. 6, pp. 692–701, Dec. 1980.

W. S. Hodgkiss, "Adaptive array processing: Time versus frequency," in *Proc. IEEE Int. Conf. on ASSP*, (Washington, DC, Apr. 2–4, 1979) pp. 282–285.

S. Holm, "Spectral moments matching in the maximum entropy spectral analysis method," *IEEE Trans. Inform. Theory*, vol. IT-29, no. 2, pp. 311–313, Mar. 1983.

——, "Phase errors in the cross spectrum estimate due to misalignment," in *Proc. IEEE Int. Conf. on ASSP*, (Tampa, FL, Mar. 26–29, 1985), pp. 812–815.

S. Holm and J. M. Hovem, "Estimation of scalar ocean wave spectra by the maximum entropy method," *IEEE J. Ocean Eng.*, vol. OE-4, no. 3, pp. 76–83, July 1979.

M. L. Honig, "Convergence models for joint process estimators and least squares algorithms," *IEEE Trans. Acoust., Speech, Signal Processing*, vol. ASSP-31, no. 2, pp. 415–425, Apr. 1983.

——, "Recursive fixed-order covariance least-squares algorithms," *Bell Syst. Tech. J.*, vol. 62, no. 10, pt. 1, pp. 2961–2992, Dec. 1983.

M. L. Honig and D. G. Messerschmitt, "Convergence properties of adaptive digital lattice filter," *IEEE Trans. Circuits and Systems*, vol. CAS-28, pp. 482–293, June 1981.

Y. Hu, "Adaptive methods for real time Pisarenko spectrum estimate," in *Proc. IEEE Int. Conf. on ASSP*, (Tampa, FL, Mar. 26–29, 1985), pp. 105–108.

S. Ihara, "Maximum entropy spectral analysis and ARMA processes," *IEEE Trans. Inform. Theory*, vol. IT-30, no. 2, pp. 377–380, Mar. 1980.

Y. Inouye, "Maximum entropy spectral estimation for regular time series of degenerate rank," *IEEE Trans. Acoust., Speech, Signal Processing*, vol. ASSP-32, no. 4, pp. 733–740, Aug. 1984.

N. Iwama, A. Inoue, T. Tsukishima, M. Sato, K. Kawahata, and K. Sakai, "Least-squares autoregressive (maximum entropy) spectral estimation for Fourier spectroscopy and its application to the electron cyclotron emission from plasma," *J. Appl. Phys.*, vol. 52, no. 9, pp. 5466–5475, Sept. 1981.

——, "Effects of the phase error on the autoregressive spectral estimate in Fourier transform spectroscopy," *J. Appl. Phys.*, vol. 53, no. 1, pp. 754–755, Jan. 1982.

D. Izraelevitz and J. S. Lim, "Properties of the overdetermined normal equation method for spectral estimation when applied to sinusoids in noise," *IEEE Trans. Acoust., Speech, Signal Processing*, vol. ASSP-33, no. 2, pp. 406-412, Apr. 1985.

P. L. Jackson and R. A. Shuchman, "High-resolution spectral estimation of synthetic aperture radar ocean wave imagery," *J. Geophys. Res.*, vol. 88, no. C4, pp. 2593-2600, Mar. 1983.

L. Jackson, "Approximate factorization of unfactorable spectral models," in *Proc. IEEE Int. Conf. on ASSP*, (Tampa, FL, Mar. 26-29, 1985), pp. 324-326.

A. K. Jain and S. Ranganath, "Application of two dimensional estimation in image restoration," in *Proc. IEEE Int. Conf. on ASSP*, (Tampa, FL, Mar. 26-29, 1985), pp. 1113-1116.

——, "Extrapolation algorithms for discrete signals with application in spectral estimation," *IEEE Trans. Acoust., Speech, Signal Processing*, vol. ASSP-29, no. 4, pp. 830-845, Aug. 1981.

B. H. Jansen, J. R. Bourne, and J. W. Ward, "Autoregressive estimation of short segment spectra for computerized EEG analysis," *IEEE Trans. Biomed. Eng.*, vol. BME-28, no. 9, pp. 630-638, Sept. 1981.

D. Johnson, "Improving the resolution of bearing in passive sonar arrays by eigenvalue analysis," in *Proc. First IEEE ASSP Workshop on Spectral Analysis*, (Hamilton, ON, Canada, Aug. 17-18, 1981), vol. 2, pp. 5.6.1-5.6.9.

D. H. Johnson, "The application of spectral estimation methods to bearing estimation problems," *Proc. IEEE*, vol. 70, pp. 1018-1028, Sept. 1982.

D. Johnson, "Properties of eigenanalysis methods for bearing estimation algorithms," in *Proc. Int. Conf. on ASSP*, (Tampa, FL, Mar. 26-29, 1985), pp. 552-555.

D. H. Johnson and S. R. DeGraaf, "Improving the resolution of bearing in passive sonar arrays by eigenvalue analysis," *IEEE Trans. Acoust., Speech, Signal Processing*, vol. ASSP-30, no. 4, pp. 638-647, Aug. 1982.

R. W. Johnson, J. E. Shore and J. P. Burg, "Multisignal minimum cross-entropy spectrum analysis with weighted initial estimates," *IEEE Trans. Acoust., Speech, Signal Processing*, vol. ASSP-32, no. 3, pp. 531-539, June 1984.

L. S. Joyce, "A separable 2-D autoregressive spectral estimation algorithm," in *Proc. IEEE Int. Conf. on ASSP*, (Washington, DC, Apr. 2-4, 1979), pp. 677-680.

D. Katz, M. Landrum, and L. H. Schick, "Stacking of noisy seismic traces via maximum entropy with a correlation coefficient constraint," vol. ASSP-33, no. 5, pp. 1331-1333, Oct. 1985.

M. Kaveh, "High resolution spectral estimation for noisy signals," *IEEE Trans. Acoust., Speech, Signal Processing*, vol. ASSP-27, no. 3, pp. 286-288, June 1979.

M. Kaveh and S. P. Bruzzone, "Computationally efficient ARMA spectral estimation," in *Proc. of the 18th IEEE Conf. on Decision and Control* (including the Symposium on Adaptive Processes), (Fort Lauderdale, FL, Dec. 12-14, 1979), Part II, pp. 938-939.

——, "Statistical efficiency of correlation-based methods for ARMA spectral estimation," *IEE Proc. Pt. F*, (GB), vol. 130, no. 3, pp. 211-217, Apr. 1983.

M. Kaveh and G. A. Lippert, "On optimum tapered Burg algorithm for linear prediction and spectral analysis," *IEEE Trans. Acoust., Speech, Signal Processing*, vol. ASSP-31, no. 2, pp. 438-444, Apr. 1983.

S. M. Kay, "Fourier-autoregressive spectral estimation," in *Proc. IEEE Int. Conf. on ASSP*, (Washington, DC, Apr. 2-4, 1979), pp. 162-165.

——, "Robust detection by autoregressive spectrum analysis," *IEEE Trans. Acoust., Speech, Signal Processing*, vol. ASSP-30, no. 2, pp. 256-269, Apr. 1982.

——, "Recursive maximum likelihood estimation of autoregressive processes," *IEEE Trans. Acoust., Speech, Signal Processing*, vol. ASSP-31, no. 1, pp. 56-65, Feb. 1983.

——, "Accurate frequency estimation at low signal-to-noise ratio," *IEEE Trans. Acoust., Speech, Signal Processing*, vol. ASSP-32, no. 3, pp. 540-547, June 1984.

S. M. Kay and J. Makhoul, "On the statistics of the estimated reflection coefficients of an autoregressive process," *IEEE Trans. Acoust., Speech, Signal Processing*, vol. ASSP-31, no. 6, pp. 1447-1455, Dec. 1983.

S. Kay and L. Marple, "Sources of and remedies for spectral line splitting in autoregressive spectrum analysis," in *Proc. IEEE Int. Conf. on ASSP*, (Washington, DC, Apr. 2-4, 1979), pp. 151-154.

S. M. Kay and S. L. Marple, Jr., "Spectrum analysis—A modern perspective," *Proc. IEEE*, vol. 69, no. 11, pp. 1380-1419, Nov. 1982.

J. Kesler and S. Haykin, "A new adaptive antenna for elevation angle estimation in the presence of multipath," in *1980 Int. Symp. Digest, Antennas and Propagation*, (McMaster Univ., Hamilton, ON, Canada, June 2-6, 1980), pp. 130-133.

S. B. Kesler, "Generalized Burg algorithm for beamforming in correlated multipath field," in *Proc. IEEE Int. Conf. on ASSP*, (Paris, France, May 3-5, 1982), vol. 3, pp. 1481-1484.

——, "Autoregressive detection by generalized Burg algorithm," in *Proc. IEEE Int. Conf. on ASSP*, (Boston, MA, Apr. 14-16, 1983), pp. 571-574.

S. B. Kesler, S. Boodaghians, and J. Kesler, "Resolution of incoherent and coherent sources by autoregressive beamforming," in *Proc. IEEE Int. Conf. on ASSP*, (San Diego, CA, Mar. 19-21, 1984), pp. 33.6.1-33.6.4.

S. B. Kesler, S. Haykin, and R. S. Walker, "Maximum entropy field-mapping in the presence of correlated multipath," in *Proc. IEEE Int. Conf. on ASSP*, (Atlanta, GA, Mar. 30-Apr. 1, 1981), pp. 157-161.

J. Ketchum and D. Herrick, "Signal detection using autoregressive parameters," in *Proc. IEEE Int. Conf. on ASSP*, (Tampa, FL, Mar. 26-29, 1985), pp. 331-334.

J. F. Kinkel, J. Perl, L. L. Scharf, and A. R. Stubberud, "A note on covariance—invariant digital filter design and autoregressive moving average spectrum analysis," *IEEE Trans. Acoust., Speech, Signal Processing*, vol. ASSP-27, no. 2, pp. 200-202, Apr. 1979.

D. J. Klich, A. H. Nayran, and S. R. Parker, "Two-dimensional spectral estimation with autoregressive lattice parameters," in *Proc. IEEE Int. Conf. on ASSP*, (San Diego, CA, Mar. 19-21, 1984), pp. 411-414.

T. Kobayashi and S. Imai, "Spectral analysis using generalized cepstrum," *IEEE Trans. Acoust., Speech, Signal Processing*, vol. ASSP-32, no. 5, pp. 1087-1089, Oct. 1984.

I. S. Konvalinka, "Iterative nonparametric spectrum estimation," *IEEE Trans. Acoust., Speech, Signal Processing*, vol. ASSP-32, no. 1, pp. 59-69, Feb. 1984.

R. Kumaresan and A. Shaw, "High resolution bearing estimation without eigen deconvolution," in *Proc. IEEE Int. Conf. on ASSP*, (Tampa, FL, Mar. 26-29, 1985), pp. 576-579.

R. Kumaresan and D. Tufts, "Estimating the angles of arrival of multiple plane waves," *IEEE Trans. Aerosp. Electron. Syst.*, vol. AES-19, no. 1, pp. 134-139, Jan. 1983.

S-Y. Kung and D. Bhaskar Rao, "New unbiased methods for narrowband spectral estimation theory," in *Proc. of the IFAC Symposium*, (New Delhi, India, Jan. 5-7, 1982), pp. 235-240.

S. Y. Kung and Y. H. Hu, "A highly concurrent algorithm and pipelined architecture for solving Toeplitz systems," *IEEE Trans. Acoust., Speech, Signal Processing*, vol. ASSP-31, no. 1, pp. 66-75, Feb. 1983.

——, "Highly concurrent Toeplitz eigen-system solver for high-resolution spectral estimation," in *Proc. IEEE, Int. Conf. on ASSP*, (Boston, MA, Apr. 14-16, 1983), pp. 1422-1425.

J. L. Lacoume, M. Gharbi, C. Latombe, and J. L. Nicolas, "Close frequency resolution by maximum entropy spectral estimation," *IEEE Trans. Acoust., Speech, Signal Processing*, vol. ASSP-32, no. 5, pp. 977-984, Oct. 1984.

J. L. Lacoume, C. Latombe, N. Martin, M. Gharbi, and R. Lidin, "AR and ARMA models in spectral and multispectral analysis," *ASSP*

Spectrum Estimation Workshop II, (Tampa, FL, Nov. 10–11, 1983), pp. 240–244.

M. A. Lagunas-Hernández, "Use of the most significant autocorrelation lags in iterative ME spectral estimation," *IEEE Trans. Acoust., Speech, Signal Processing*, vol. ASSP-32, no. 2, pp. 445–448, Apr. 1984.

M. A. Lagunas-Hernández and A. Gasull-Llampalas, "An improved maximum likelihood method for power spectral density estimation," *IEEE Trans. Acoust., Speech, Signal Processing*, vol. ASSP-32, no. 1, pp. 170–173, Feb. 1984.

M. A. Lagunas-Hernandez, M. E. Santamaría-Perez, and A. R. Figueiras-Vidal, "ARMA model maximum entropy power spectral estimation," *IEEE Trans. Acoust., Speech, Signal Processing*, vol. ASSP-32, no. 5, pp. 984–990, Oct. 1984.

M. Lagunas, M. E. Santamaría, A. Gasull, and A. Moreno, "Cross-spectral ML estimate," in *Proc. IEEE Int. Conf. on ASSP*, (Tampa, FL, Mar. 26–29, 1985), pp. 77–80.

S. W. Lang and J. H. McClellan, "Multidimensional MEM spectral estimation," *IEEE Trans. Acoust., Speech, Signal Processing*, vol. ASSP-30, no. 6, pp. 880–887, Dec. 1982.

——, "Spectral estimation for sensor arrays," *IEEE Trans. Acoust., Speech, Signal Processing*, vol. ASSP-31, no. 2, pp. 349–358, Apr. 1983.

M. G. Larimore, "Adaptation convergence of spectral estimation based on Pisarenko harmonic retrieval," *IEEE Trans. Acoust., Speech, Signal Processing*, vol. ASSP-31, no. 4, pp. 955–962, Aug. 1983.

D. T. L. Lee, M. Morf, and B. Friedlander, "Recursive least-squares ladder estimation algorithms," *IEEE Trans. Acoust., Speech, Signal Processing*, vol. ASSP-29, no. 3, pp. 627–641, June 1981.

T-S. Lee, "Identification and spectral estimation of noisy multivariate autoregressive processes," in *Proc. IEEE Int. Conf. on ASSP*, (Atlanta, GA, Mar. 30–Apr. 1, 1981), pp. 503–507.

T-S. Lee, "Large sample identification and spectral estimation of noisy multivariate autoregressive processes," *IEEE Trans. Acoust., Speech, Signal Processing*, vol. ASSP-31, no. 1, pp. 76–82, Feb. 1983.

S. H. Leung and C. W. Barnes, "State-space realizations of fractional-step delay digital filters with applications to array beamforming," *IEEE Trans. Acoust., Speech, Signal Processing*, vol. ASSP-32, no. 2, pp. 371–380, Apr. 1984.

S. Leung and W. Horng, "Superresolution autoregressive spectral estimation technique using multiple step prediction," in *Proc. IEEE Int. Conf. on ASSP*, (Tampa, FL, Mar. 26–29, 1985) pp. 101–104.

S. Levy, C. Walker, T. J. Ulrych, and P. K. Fullagar, "A linear programming approach to the estimation of the power spectra of harmonic processes," *IEEE Trans. Acoust., Speech, Signal Processing*, vol. ASSP-30, no. 4, pp. 675–679, Aug. 1982.

P. Lohnberg and G. H. J. Wisselink," Generalized least-squares iterative inverse filtering parameter estimation for ARMA pulse response and output disturbance," in *Proc. 1982 American Control Conf.*, (Arlington, VA, June 14–16, 1982), pp. 148–152.

——, "Iterative least-squares parameter estimation for ARMA pulse response and output disturbance," *IEEE Trans. Automat. Control*, vol. AC-27, no. 6, pp. 1252–1255, Dec. 1982.

E. Lunde, "Normal mode propagation and high resolution methods," in *Proc. IEEE Int. Conf. on ASSP*, (Tampa, FL, Mar. 26–29, 1985), pp. 568–571.

B. Madan and S. Parker, "Adaptive beam forming in correlated interference environment," in *Proc. IEEE Int. Conf. on ASSP*, (Tampa, FL, Mar. 26–29, 1985), pp. 1792–1795.

P. R. Mahapatra and D. S. Zrnic, "Practical algorithms for mean velocity estimation in pulse Doppler weather radars using a small number of samples," *IEEE Trans. Geosci. Remote Sensing*, vol. GE-21, no. 4, pp. 491–501, Oct. 1983.

N. A. Malik and J. S. Lim, "Properties of two-dimensional maximum entropy power spectral estimates," *IEEE Trans. Acoust., Speech, Signal Processing*, vol. ASSP-30, no. 5, pp. 788–798, Oct. 1982.

S. L. Marple, Jr. "Spectral line analysis by Pisarenko and Prony methods," in *Proc. IEEE Int. Conf. on ASSP*, (Washington, DC, Apr. 2–4, 1979), pp. 159–161.

——, "Efficient least squares FIR system identification," *IEEE Trans. Acoust., Speech, Signal Processing*, vol. ASSP-29, no. 1, pp. 62–73, Feb. 1981.

——, "Fast algorithms for linear prediction and system identification filters with linear phase," *IEEE Trans. Acoust., Speech, Signal Processing*, vol. ASSP-30, no. 6, pp. 942–953, Dec. 1982.

——, "Frequency resolution of Fourier and maximum entropy spectral estimates," *Geophysics*, vol. 47, pp. 1303–1307, 1982.

——, "Fast algorithms for linear prediction and system identification filters with linear phase," in *Proc. IEEE Int. Conf. on ASSP*, (Paris, France, May 3–5, 1982), pp. 178–181.

S. L. Marple, Jr., and L. R. Rabiner, "Performance of a fast algorithm for FIR system identification using least-squares analysis," *Bell Syst. Tech. J.*, vol. 62, pp. 717–742, Mar. 1983.

V. J. Matthews and D-H. Youn, "Analysis of the short-time unbiased spectrum estimation algorithm," *IEEE Trans. Acoust., Speech, Signal Processing*, vol. ASSP-33, no. 1, pp. 136–142, Feb. 1985.

R. J. McAulay, "Maximum likelihood spectral estimation and its application to narrow-band speech," *IEEE Trans. Acoust., Speech, Signal Processing*, vol. ASSP-32, no. 2, pp. 243–251, Apr. 1984.

J. H. McClellan, "Multidimensional spectral estimation," *Proc. IEEE*, vol. 70, no. 9, pp. 1029–1039, Sept. 1982.

J. H. McClellan and S. W. Lang, "Duality for multidimensional MEM spectral analysis," *IEE Proc.*, pt. F, (GB), vol. 130, no. 3, pp. 230–235, Apr. 1983.

C. D. McGillem, J. I. Aunon, and D. G. Childers, "Signal processing in evoked potential research: Applications of filtering and pattern recognition," *CRC Critical Reviews of Bioengineering*, vol. 6, pp. 225–265, Oct. 1981.

R. S. Medaugh and L. J. Griffiths, "A comparison of two fast linear predictors," in *Proc. IEEE Int. Conf. on ASSP*, (Atlanta, GA, Mar. 30–Apr. 1, 1981), pp. 293–296.

——, "Further results of a least squares and gradient adaptive lattice algorithm comparison," in *Proc. IEEE, Int. Conf. on ASSP*, (Paris, France, May 3–5, 1982), pp. 1412–1415.

K. Michael and L. Sibul, "The VLSI implementation of the adaptive lattice filter," in *Proc. 1984 IEEE Int. Symposium on Circuits and Syst.*, (Montreal, PQ, Canada, May 7–10, 1984), vol. 2, pp. 772–775.

K. Minami, S. Kawata, and S. Minami, "Superresolution of Fourier transform spectra by autoregressive model fitting with singular value decomposition," *Appl. Opt.*, vol. 24, no. 2, pp. 162–167, Jan. 1985.

R. A. Monzingo and T. W. Miller, *Introduction to Adaptive Arrays*. New York: Wiley-Interscience, 1980.

R. A. Mucci, "A comparison of efficient beamforming algorithms," *IEEE Trans. Acoust., Speech, Signal Processing*, vol. ASSP-32, no. 3, pp. 548–558, June 1984.

B. R. Musicus, "Fast MLM power spectrum estimation from uniformly spaced correlations," *IEEE Trans. Acoust., Speech, Signal Processing*, vol. ASSP-33, no. 5, pp. 1333–1335, Oct. 1985.

P. S. Naidu and B. Paramasivaiah, "Estimation of sinusoids from incomplete time series," *IEEE Trans. Acoust., Speech, Signal Processing*, vol. ASSP-32, no. 3, pp. 559–562, June 1984.

S. H. Nawab, F. U. Dowla, and R. T. Lacoss, "Direction determination of wideband signals," *IEEE Trans. Acoust., Speech, Signal Processing*, vol. ASSP-33, no. 5, pp. 1114–1122, Oct. 1985.

J. G. Negi and V. P. Dimri, "On a generalization of maximum entropy method for processing of nonstationary multichannel complex valued data," *IEEE Trans. Geosci. Remote Sensing*, vol. GE-22, no. 5, pp. 461–466, Sept. 1984.

H. J. Newton, "On some numerical properties of ARMA parameter estimation procedures," in *Proc. Computer Sci. and Stat: Proc. of the 13th Symp. on the Interface*, (Pittsburgh, PA, Mar. 12–13, 1981), pp. 172–177.

C. L. Nikias and M. R. Raghuveer, "A new class of high-resolution and robust multi-dimensional spectral estimation algorithms," in *Proc. IEEE Int. Conf. on ASSP*, (Boston, MA, Apr. 14–16, 1983), pp. 859-862.

——, "Multi-dimensional spectral estimation via parametric models," in *Proc. ASSP Spectrum Estimation Workshop II*, (Tampa, FL, Nov. 10–11, 1983), pp. 213-218.

C. L. Nikias and P. D. Scott, "Improved spectral resolution by energy-weighted prediction method," in *Proc. IEEE Int. Conf. on ASSP*, (Atlanta, GA, Mar. 30–Apr. 1, 1981), pp. 496–499.

——, "The covariance least-squares algorithm for spectral estimation of processes of short data length," *IEEE Trans. Geosci. and Remote Sensing*, vol. GE-21, no. 2, pp. 180–190, Apr. 1983.

C. L. Nikias, P. D. Scott, and J. H. Siegel, "A new robust 2-D spectral estimation method and its application in cardiac data analysis," in *Proc. IEEE Int. Conf. on ASSP*, (Paris, France, May 3–5, 1982), pp. 729–732.

C. Nikias and A. Venetsanopoulos, "Sufficient condition for extendibility and two-dimensional power spectrum estimation," in *Proc. IEEE Int. Conf. on ASSP*, (Tampa, FL, Mar. 26–29, 1985), pp. 792-795.

T. Ning and C. Nikias, "The optimum approach to multichannel AR spectrum estimation," in *Proc. IEEE Int. Conf. on ASSP*, (Tampa, FL, Mar. 26–29, 1985), pp. 800-803.

R. Nitzberg, "Implementation of an adaptive processor by modern spectral estimation techniques," in *Proc. IEEE 1981 Int. Conf. on Communication*, (Denver, CO, June 14–18, 1981), pp. 10.3.2-10.3.4.

N. Ogino, "A fast algorithm for unmodified ARMA spectrum estimation," in *Proc. IEEE Int. Conf. on ASSP*, (Boston, MA, Apr. 14–16, 1983), pp. 1106–1109.

S. J. Orfanidis and L. M. Vail, "Zero-tracking adaptation algorithms," in *Proc. ASSP Spectrum Estimation Workshop II*, (Tampa, FL, Nov. 10–11, 1983), pp. 209-212.

G. Orlandi, G. Martinelli, and P. Burrascano, "Explicit formulas for super-resolution," in *Proc. IEEE Int. Conf. on ASSP*, (Tampa, FL, Mar. 26–29, 1985), pp. 1356–1359.

M. D. Ortigueira and J. M. Tribolet, "On the double Levinson recursion formulation of ARMA spectral estimation," in *Proc. IEEE Int. Conf. on ASSP*, (Boston, MA, Apr. 14–16, 1983), pp. 1076-1079.

N. L. Owsley, "Modal decomposition of data adaptive spectral estimates," in *Proc. Yale University Workshop on Applications of Adaptive Syst. Theory*, K. S. Narendra, Ed., May 1981.

——, "Wavefront-array shape matching by adaptive focusing," in *Proc. Time Delay Estimation Appl. Conf.* (Naval Post Graduate School), Monteray, CA, May 1979).

N. L. Owsley and J. F. Law, "Dominant mode power spectrum estimation," in *Proc. IEEE Int. Conf. on ASSP*, (Paris, France, May 3-5, 1982), pp. 775-778.

A. Papoulis, "Maximum entropy and spectral estimation: A review," *IEEE Trans. Acoust., Speech, Signal Processing*, vol. ASSP-29, no. 6, pp. 1176-1186, Dec. 1981.

A. Papoulis and C. Chamzas, "Detection of hidden periodicities by adaptive extrapolation," *IEEE Trans. Acoust., Speech, Signal Processing*, vol. ASSP-27, no. 5, pp. 492–500, Oct. 1979.

S. R. Parker, "Modeling of two-dimensional fields with autoregressive lattice parameters," in *Proc. 1984 Amer. Control Conf.*, (San Diego, CA, June 6–8, 1984), pp. 216–221.

S. R. Parker and A. H. Kayran, "Lattice parameter autoregressive modeling of two-dimensional fields," in *Proc. ASSP Spectrum Estimation Workshop II*, (Tampa, FL, Nov. 10–11, 1983), pp. 219–223.

E. Parzen, "Modern empirical statistical spectral analysis," in *Proc. of the NATO Advanced Study Inst.*, (Underwater Acoustics and Signal Processing Conf., Copenhagen, Denmark, Aug. 18–29, 1980), pp. 471–497.

——, "Time series long memory identification and quantile spectral analysis," in *Proc. ASSP Spectrum Estimation Workshop II*, (Tampa, FL, Nov. 10–11, 1983), pp. 1–6.

A. Paulraj and T. Kailath, "On beamforming in presence of multipath," in *Proc. IEEE Int. Conf. on ASSP*, (Tampa, FL, Mar. 26–29, 1985), pp. 564–567.

——, "Direction of arrival estimation by eigenstructure methods with unknown sensor gain and phase," in *Proc. IEEE Int. Conf. on ASSP*, (Tampa, FL, Mar. 26–29, 1985), pp. 640–643.

E. Paulus, "A fast convolution procedure for discrete short-time spectral analysis with frequency dependent resolution," *IEEE Trans. Acoust., Speech, Signal Processing*, vol. ASSP-32, no. 5, pp. 1100-1104, Oct. 1984.

U. Pillai, F. Haber, and Y. Bar-Ness, "A new approach to array geometry for improved spatial spectrum estimation," in *Proc. IEEE Int. Conf. on ASSP*, (Tampa, FL, Mar. 26–29, 1985), pp. 1816-1819.

B. Porat, "ARMA spectral estimation based on partial autocorrelations," *Circuits Syst. and Signal Process*, vol. 2, no. 3, pp. 341–360, 1983.

B. Porat and B. Friedlander, "Estimation of spatial and spectral parameters of multiple sources," *IEEE Trans. Inform. Theory*, vol. IT-29, no. 3, pp. 412–425, May 1983.

——, "ARMA spectral estimation of time series with missing observations," *IEEE Trans. Inf. Theory*, vol. IT-30, no. 6, pp. 823-831, Nov. 1984.

R. D. Preuss and R. Yarlagadda, "Autoregressive spectral estimation in noise in the context of speech analysis," in *Proc. ASSP Spectrum Estimation Workshop II*, (Tampa, FL, Nov. 10–11, 1983), pp. 75–79.

——, "Autoregressive spectral estimation in noise with application to speech analysis," in *Proc. IEEE Int. Conf. on ASSP*, (San Diego, CA, Mar. 19–21, 1984), pp. 6.6.1–6.6.4.

R. G. Pridham and R. A. Mucci, "Digital interpolation beamforming for low pass and bandpass signals," *Proc. IEEE*, vol. 67, no. 6, pp. 904-919, June 1979.

M. P. Quirk and B. Liu, "On the resolution of autoregressive spectral estimation," in *Proc. IEEE Int. Conf. on ASSP*, (Boston, MA, Apr. 14-16, 1983), pp. 1095–1098.

——, "Improved resolution for autoregressive spectral estimation by decimation," *IEEE Trans. Acoust., Speech, Signal Processing*, vol. ASSP-31, no. 3, pp. 630–637, June 1983.

M. R. Raghuveer and C. L. Nikias, "Bispectrum estimation for short length data," in *Proc. IEEE Int. Conf. on ASSP*, (Tampa, FL, Mar. 26-29, 1985), pp. 1352–1355.

——, "Bispectrum estimation: A parametric approach," *IEEE Trans. Acoust., Speech, Signal Processing*, vol. ASSP-33, no. 5, pp. 1213-1230, Oct. 1985.

B. Rao, "Perturbation analysis of a SVD based method for the harmonic retrieval problem," in *Proc. IEEE Int. Conf. on ASSP*, (Tampa, FL, Mar. 26–29, 1985), pp. 624–627.

S. S. Reddi, "Multiple source location—A digital approach," *IEEE Trans. Aerosp. Electron. Syst.*, vol. AES-15, pp. 95–105, Jan. 1979.

V. U. Reddy, B. Egart, and T. Kailath, "Least squares type algorithm for adaptive implementation of Pisarenko's harmonic retrieval method," *IEEE Trans. Acoust., Speech, Signal Processing*, vol. ASSP-30, no. 3, pp. 399–405, June 1982.

J. P. Reilly and S. Haykin, "Maximum-likelihood receiver for low-angle tracking radar, part 2: The nonsymmetric case," *Proc. IEE*, Pt. F, vol. 129, pp. 331–340, 1982.

J. Rissanen, "Universal coding, information, prediction, and estimation," *IEEE Trans. Inform Theory*, vol. IT-30, no. 4, pp. 629–636, July 1984.

E. A. Robinson, "A historical perspective of spectrum estimation," *Proc. IEEE*, vol. 70, no. 9, pp. 885-907, Sept. 1982.

S. E. Roucos and D. G. Childers, "A two-dimensional maximum entropy spectral estimator," *IEEE Trans. Inform. Theory*, vol. IT-26, no. 5, pp. 554–560, Sept. 1980.

H. Sakai, "Statistical properties of AR spectral analysis," *IEEE Trans. Acoust., Speech, Signal Processing*, vol. ASSP-27, no. 4, pp. 402-409, Aug. 1979.

——, "Statistical analysis of Pisarenko's method for sinusoidal frequency estimation," *IEEE Trans. Acoust., Speech, Signal Processing*, vol. ASSP-32, no. 1, pp. 95-101, Feb. 1984.

A. Sano and K. Hashimoto, "Adaptive recursive scheme for spectral analysis of sinusoids in signals with unknown colored spectrum," in *Proc. IEEE Int. Conf. on ASSP*, (Tampa, FL, Mar. 26-29, 1985), pp. 109-112.

C. S. Sarna and H. Stark, "Pattern recognition of waveforms using modern spectral estimation techniques and its application to earthquake/explosion data," in *Proc. 5th Int. Conf. on Pattern Recognition*, (Miami Beach, FL, Dec. 1-4, 1980), pp. 8-10.

R. O. Schmidt, "Multiple emitter location and signal parameter estimation," in *Proc. 1979 RADC Spectrum Estimation Workshop*, (Rome, NY, Oct. 3-5, 1979), pp. 243-258.

J. P. Schott and J. H. McClellan, "Maximum entropy power spectrum estimation with uncertainty in correlation measurements," *IEEE Trans. Acoust., Speech, Signal Processing*, vol. ASSP-32, no. 2, pp. 410-418, Apr. 1984.

P. M. Schultheiss, "Locating a passive source with array measurements: A summary of results," in *Proc. IEEE Int. Conf. on ASSP*, (Washington, DC, Apr. 2-4, 1979), pp. 967-970.

P. D. Scott and C. L. Nikias, "Forward covariance least-squares algorithms: A new method in AR spectral estimation," *Electron. Lett.*, (GB), vol. 17, no. 3, pp. 111-112, Feb. 5, 1981.

——, "Energy-weighted linear predictive estimation: A new method combining robustness and high resolution," *IEEE Trans. Acoust., Speech, Signal Processing*, vol. ASSP-30, no. 2, pp. 287-293, Apr. 1982.

——, "High-resolution frequency estimation via a weighted forward and backward autoregressive modelling," in *Proc. IEEE Int. Conf. on ASSP*, (Boston, MA, Apr. 14-16, 1983), pp. 1072-1075.

T. J. Shan, M. Wax and T. Kailath, "On spatial smoothing for direction-of-arrival estimation of coherent signals," *IEEE Trans. Acoust., Speech, Signal Processing*, vol. ASSP-33, no. 4, pp. 806-811, Aug. 1985.

P. L. Sharma and C. S. Chen, "Auto-regressive spectral estimation of noisy sinusoids," in *Proc. IEEE Int. Conf. on ASSP*, (Paris, France, May 3-5, 1982), pp. 1038-1041.

K. C. Sharman and T. S. Durrani, "A triangular adaptive lattice filter for spatial filtering," in *Proc. IEEE Int. Conf. on ASSP*, (Boston, MA, Apr. 14-16, 1983), pp. 348-351.

——, "Resolving power of signal subspace methods for finite data lengths," in *Proc. IEEE Int. Conf. on ASSP*, (Tampa, FL, Mar. 26-29, 1985), pp. 1501-1504.

J. E. Shore, "Minimum cross-entropy spectral analysis," *IEEE Trans. Acoust., Speech, Signal Processing*, vol. ASSP-29, no. 2, pp. 230-237, Apr. 1981.

L. Sibul and S. Burke, "Error analysis of eigenvector preprocessors used in adaptive beamforming," in *Proc. IEEE Int. Conf. on ASSP*, (Tampa, FL, Mar. 26-29, 1985), pp. 1808-1811.

F. K. Soong and A. A. Peterson, "Fast spectral estimation of speech signal in analytic form," in *Proc. IEEE Int. Conf. on ASSP*, (Denver, CO, Apr. 9-11, 1980), pp. 158-161.

T. Srinivasan, D. C. Swanson, and F. W. Symons, "ARMA model order/data length tradeoff for specified frequency resolution," in *Proc. IEEE Int. Conf. on ASSP*, (San Diego, CA, Mar. 19-21, 1984), pp. 38.2.1-38.2.4.

H. Stark and C. S. Sarna, "Pattern recognition of waveforms using modern spectral estimation techniques and its application to earthquake/explosion data," in *Proc. IEEE Int. Conf. on ASSP*, (Atlanta, GA, Mar. 30-Apr. 1, 1981), pp. 500-502.

A. Steinhardt and C. Bretherton, "Thresholds in frequency estimation," in *Proc. IEEE Int. Conf. on ASSP*, (Tampa, FL, Mar. 26-29, 1985), pp. 1273-1274.

A. Steinhardt, K. Goodrich, and R. Roberts, "Spectral estimation via minimum energy correlation extension," in *Proc. IEEE Int. Conf. on ASSP*, (Tampa, FL, Mar. 26-29, 1985), pp. 97-100.

K. Stewart, T. Durrani, and J. B. Abbiss, "The effects of bandwich mis-estimation in bandlimited signal extrapolation," in *Proc. IEEE Int. Conf. on ASSP*, (Tampa, FL, Mar. 26-29, 1985), pp. 1497-1500.

R. Sudhakar, R. C. Agarwal, and S. C. Dutta Roy, "Frequency estimation based on iterated autocorrelation function," *IEEE Trans. Acoust., Speech, Signal Processing*, vol. ASSP-33, no. 1, pp. 70-76, Feb. 1985.

S. Sugimoto and A. K. Jain, "Identification of a large vector AR model and its application to image processing," in *Proc. 1982 Sixth IFAC Symposium*, (Washington, DC, June 7-11, 1982), pp. 1589-1594.

D. N. Swingler, "On the optimum tapered Burg algorithm," *IEEE Trans. Acoust., Speech, Signal Processing*, vol. ASSP-32, no. 1, pp. 185-186, Feb. 1984.

C. W. Therrien, "Relations between 2-D and multichannel linear prediction," *IEEE Trans. Acoust., Speech, Signal Processing*, vol. ASSP-29, no. 3, pp. 454-456, June 1981.

D. J. Thompson, "Spectrum estimation and harmonic analysis," *Proc. IEEE*, vol. 70, no. 9, pp. 1055-1096, Sept. 1982.

T. Thorvaldsen, "Maximum entropy spectral analysis in antenna spatial filtering," *IEEE Trans. Antennas Propag.*, vol. AP-28, no. 4, pp. 552-560, July 1980.

J. R. Treichler, "Transient and convergent behavior of the adaptive line enhancer," *IEEE Trans. Acoust., Speech, Signal Processing*, vol. ASSP-27, no. 1, pp. 53-62, Feb. 1979.

——, "Gamma—LMS and its use in noise-compensating adaptive spectral analysis," in *Proc. IEEE Int. Conf. on ASSP*, (Washington, DC, Apr. 2-4, 1979), pp. 933-936.

D. W. Tufts and R. Kumaresan, "Singular value decomposition and improved frequency estimation using linear prediction," *IEEE Trans. Acoust., Speech, Signal Processing*, vol. ASSP-30, no. 4, pp. 671-675, Aug. 1982.

——, "Estimation of frequencies of multiple sinusoids: Making linear prediction perform like maximum likelihood," *Proc. IEEE*, vol. 70, no. 9, pp. 975-989, Sept. 1982.

D. Tufts and C. Melissinos, "Simple, effective computation of principal eigenvectors and their eigenvalues and application to high resolution estimation of frequencies," in *Proc. IEEE Int. Conf. on ASSP*, (Tampa, FL, Mar. 26-29, 1985), pp. 320-323.

J. K. Tugnait, "Spectral estimation for noisy signals observed through a linear system," in *Proc. 22nd IEEE Conf. on Decision and Control*, (San Antonio, TX, Dec. 14-16, 1983), pp. 1331-1336.

——, "ARMA spectral estimation for noisy signals observed through a linear system," *IEEE Trans. Acoust., Speech, Signal Processing*, vol. ASSP-33, no. 1, pp. 160-163, Feb. 1985.

J. M. Turner, "Use of digital lattice structure in estimation and filtering signal processing: Theories and applications," in *Proc. First European Signal Processing Conference* (EUSIPCO-80), (Lausanne, Switzerland, Sept. 16-18, 1980), pp. 33-41.

——, "Application of recursive exact least square ladder estimation algorithm for speech recognition," in *Proc. IEEE Int. Conf. on ASSP*, (Paris, France, May 3-5, 1982), pp. 543-545.

P. A. Tyraskis and O. G. Jensen, "Multichannel autoregressive data models," *IEEE Trans. Geosci. Remote Sensing*, vol. GE-21, no. 4, pp. 454-467, Oct. 1983.

M. A. Tzannes, D. Politis, and N. S. Tzannes, "A general method of minimum cross-entropy spectral estimation," vol. ASSP-33, no. 3, pp. 748-752, June 1985.

E. Vertatschitsch, S. Haykin, and K. M. Wong, "Optimum nonredundant arrays in beamforming," in *Proc. 1983 Int. Electrical, Electron. Conf.*, (Toronto, ON, Canada, Sept. 26-28, 1983), pp. 144-147.

A. M. Vural, "Effects of perturbations on the performance of optimum/adaptive arrays," *IEEE Trans. Aerosp. Electron. Syst.*, vol. AES-15, no. 1, pp. 76-87, Jan. 1979.

G. Wakefield and M. Kaveh, "Frequency wavenumber spectral estimation of non-planar random fields," in *Proc. IEEE Int. Conf. on ASSP*, (Tampa, FL, Mar. 26-29, 1985), pp. 808-811.

R. Walker, "Bearing accuracy and resolution bounds of high-resolution beamformers," in *Proc. IEEE Int. Conf. on ASSP*, (Tampa, FL, Mar. 26–29, 1985), pp. 1784–1787.

H. Wang and M. Kaveh, "Sensitivity and performance analysis of coherent signal subspace processing for multiple wideband sources," in *Proc. IEEE Int. Conf. on ASSP*, (Tampa, FL, Mar. 26–29, 1985), pp. 636–639.

——, "Coherent signal-subspace processing for the detection and estimation of angles of arrival of multiple wide-band sources," *IEEE Trans. Acoust., Speech, Signal Processing*, vol. ASSP-33, no. 4, pp. 823–831, Aug. 1985.

M. Wax and T. Kailath, "Extending the threshold of the eigenstructure methods," in *Proc. IEEE Int. Conf. on ASSP*, (Tampa, FL, Mar. 26–29, 1985), pp. 556–559.

——, "Decentralized processing in sensor arrays," *IEEE Trans. Acoust., Speech, Signal Processing*, vol. ASSP-33, no. 5, pp. 1123–1129, Oct. 1985.

M. Wax, T. J. Shan and T. Kailath, "Covariance eigenstructure approach to two-dimensional harmonic retrieval," in *Proc. IEEE Int. Conf.*, (Boston, MA, Apr. 14–16, 1983), pp. 891–894.

——, "Spatio-temporal spectral analysis by eigenstructure methods," *IEEE Trans. Acoust., Speech, Signal Processing*, vol. ASSP-32, no. 4, pp. 817–827, Aug. 1984.

W. D. White, "Angular spectra in radar applications," *IEEE Trans. Aerosp. Electon. Syst.*, vol. AES-15, no. 6, pp. 895–898, Nov. 1979.

——, "Author's reply (To angular spectra in radar resolution)," *IEEE Trans. Aerosp. Electron. Syst.*, vol. AES-15, no. 6, p. 904, Nov. 1979.

J. W. Woods, "Multidimensional digital signal processing," in *Proc. of the 26th Midwest Symposium on Circuits and Syst.*, (Puebla. Mexico, Aug. 15–16, 1983), pp. 627–628.

X. C. Xiao, "An adaptive orthogonal maximum likelihood algorithm for parameter estimation," *Proc. IEEE Int. Conf. on ASSP*, (Tampa, FL, Mar. 26–29, 1985), pp. 308–311.

S. Yuen and H. Subbarum, "A new super-resolution spectral estimation technique using staggered PRF's," in *Proc. IEEE Int. Conf. on ASSP*, (Tampa, FL, Mar. 26–29, 1985), pp. 1360–1363.

W. Zimmer, "Multiple beamformer performance analysis of the coherent mode," in *Proc. IEEE Int. Conf. on ASSP*, (Tampa, FL, Mar. 26–29, 1985), pp. 1780–1783.

L. Zou and B. Liu, "Improvement of resolution of computation in 2D spectral estimation using decimation," *Proc. IEEE Int. Conf. on ASSP*, (San Diego, CA, Mar. 19–21, 1984), pp. 4.7.1–4.7.4.

Author Index

Subject Index

Editor's Biography

Stanislav B. Kesler (S'75–M'77–SM'83) was born in Valjevo, Yugoslavia, on July 3, 1942. He received Dipl. Eng. and M. Eng., both in electrical engineering, from the University of Belgrade, Yugoslavia, in 1965 and 1973, respectively, and Ph.D. degree in electrical engineering from McMaster University, Hamilton, Canada, in 1978.

From 1966 to 1973 he was with the University of Belgrade. From 1977 to 1982 he was a Research Engineer with the Communications Research Laboratory, McMaster University. In 1983 he joined the faculty of Drexel University, Philadelphia, PA, where he is an Associate Professor. His research interests include signal processing, spectral estimation, data communications, stochastic processes, signal detection and estimation, adaptive systems, and radar–sonar systems.

Dr. Kesler is a member of Sigma Xi.

Date Due

TWO PLAYS FOR PURITANS

THE DEVIL'S DISCIPLE

CÆSAR AND CLEOPATRA

GEORGE BERNARD SHAW

TWO PLAYS FOR PURITANS

drawings
by
George Him

THE HERITAGE PRESS
NEW YORK

Acknowledgment is made to

the Public Trustee and

the Society of Authors, London.

Contents

PREFACE

WHY FOR PURITANS?

SINCE I gave my Plays, Pleasant and Unpleasant, to the world two years ago, many things have happened to me. I had then just entered on the fourth year of my activity as a critic of the London theatres. They very nearly killed me. I had survived seven years of London's music, four or five years of London's pictures, and about as much of its current literature, wrestling critically with them with all my force and skill. After that, the criticism of the theatre came to me as a huge relief in point of bodily exertion. The difference between the leisure of a Persian cat and the labor of a cockney cab horse is not greater than the difference between the official weekly or fortnightly playgoings of the theatre critic and the restless daily rushing to and fro of the music critic, from the stroke of three in the afternoon, when the concerts begin, to the stroke of twelve at night, when the opera ends. The pictures were nearly as bad. An Alpinist once, noticing the massive soles of my boots, asked me whether I climbed mountains. No, I replied: these boots are for the hard floors of the London galleries. Yet I once dealt with music and pictures together in the spare time of an active young revolutionist, and wrote plays and books and other toilsome things into

the bargain. But the theatre struck me down like the veriest weakling. I sank under it like a baby fed on starch. My very bones began to perish, so that I had to get them planed and gouged by accomplished surgeons. I fell from heights and broke my limbs in pieces. The doctors said: This man has not eaten meat for twenty years: he must eat it or die. I said: This man has been going to the London theatres for three years; and the soul of him has become inane and is feeding unnaturally on his body. And I was right. I did not change my diet; but I had myself carried up into a mountain where there was no theatre; and there I began to revive. Too weak to work, I wrote books and plays: hence the second and third plays in this volume.* And now I am stronger than I have been at any moment since my feet first carried me as a critic across the fatal threshold of a London playhouse.

Why was this? What is the matter with the theatre, that a strong man can die of it? Well, the answer will make a long story; but it must be told. And, to begin, why have I just called the theatre a playhouse? The well-fed Englishman, though he lives and dies a schoolboy, cannot play. He cannot even play cricket or football: he has to work at them: that is why he beats the foreigner who plays at them. To him playing means playing the fool. He can hunt and shoot and travel and fight: he can, when special holiday festivity is suggested to him, eat and drink, dice and drab, smoke and lounge. But play he cannot. The moment you make his theatre a place of amusement instead of a place of edification, you make it, not a real playhouse, but a place of excitement for the sportsman and the sensualist.

However, this well-fed grown-up-schoolboy Englishman counts for little in the modern metropolitan audience. In the long lines of waiting playgoers lining the pavements outside our fashionable theatres every evening, the men are only the currants in the dumpling. Women are in the majority; and women and men alike belong to that least robust of all our social classes, the class which earns from eighteen to thirty shillings a week in sedentary employment, and lives in a dull lodging or with its intolerably prosaic families. These people preserve the innocence of the theatre: they have neither the philosopher's impatience to get to realities (reality being the one thing they want to es-

* *Captain Brassbound's Conversion* was the third play in the original volume for which Shaw wrote this Preface.

cape from), nor the longing of the sportsman for violent action, nor the fullfed, experienced, disillusioned sensuality of the rich man, whether he be gentleman or sporting publican. They read a good deal, and are at home in the fool's paradise of popular romance. They love the pretty man and the pretty woman, and will have both of them fashionably dressed and exquisitely idle, posing against backgrounds of drawingroom and dainty garden; in love, but sentimentally, romantically; always ladylike and gentlemanlike. Jejunely insipid, all this, to the stalls, which are paid for (when they *are* paid for) by people who have their own dresses and drawingrooms, and know them to be a mere masquerade behind which there is nothing romantic, and little that is interesting to most of the masqueraders except the clandestine play of natural licentiousness.

The stalls cannot be fully understood without taking into account the absence of the rich evangelical English merchant and his family, and the presence of the rich Jewish merchant and *his* family. I can see no validity whatever in the view that the influence of the rich Jews on the theatre is any worse than the influence of the rich of any other race. Other qualities being equal, men become rich in commerce in proportion to the intensity and exclusiveness of their desire for money. It may be a misfortune that the purchasing power of men who value money above art, philosophy, and the welfare of the whole community, should enable them to influence the theatre (and everything else in the market); but there is no reason to suppose that their influence is any nobler when they imagine themselves Christians than when they know themselves Jews. All that can fairly be said of the Jewish influence on the theatre is that it is exotic, and is not only a customer's influence but a financier's influence: so much so, that the way is smoothest for those plays and those performers that appeal specially to the Jewish taste. English influence on the theatre, as far as the stalls are concerned, does not exist, because the rich purchasing-powerful Englishman prefers politics and church-going: his soul is too stubborn to be purged by an avowed make-believe. When he wants sensuality he practises it: he does not play with voluptuous or romantic ideas. From the play of ideas – and the drama can never be anything more – he demands edification, and will not pay for anything else in that arena. Consequently

the box office will never become an English influence until the theatre turns from the drama of romance and sensuality to the drama of edification.

Turning from the stalls to the whole auditorium, consider what is implied by the fact that the prices (all much too high, by the way) range from half a guinea to a shilling, the ages from eighteen to eighty, whilst every age, and nearly every price, represents a different taste. Is it not clear that this diversity in the audience makes it impossible to gratify every one of its units by the same luxury, since in that domain of infinite caprice, one man's meat is another man's poison, one age's longing another age's loathing? And yet that is just what the theatres kept trying to do almost all the time I was doomed to attend them. On the other hand, to interest people of divers ages, classes and temperaments by some generally momentous subject of thought, as the politicians and preachers do, would seem the most obvious course in the world. And yet the theatres avoided that as a ruinous eccentricity. Their wiseacres persisted in assuming that all men have the same tastes, fancies, and qualities of passion; that no two have the same interests; and that most playgoers have no interests at all. This being precisely contrary to the obvious facts, it followed that the majority of the plays produced were failures, recognizable as such before the end of the first act by the very wiseacres aforementioned, who, quite incapable of understanding the lesson, would thereupon set to work to obtain and produce a play applying their theory still more strictly, with proportionately more disastrous results. The sums of money I saw thus transferred from the pockets of theatrical speculators and syndicates to those of wigmakers, costumiers, scene painters, carpenters, doorkeepers, actors, theatre landlords, and all the other people for whose exclusive benefit most London theatres seem to exist, would have kept a theatre devoted exclusively to the highest drama open all the year round. If the Browning and Shelley Societies were fools, as the wiseacres said they were, for producing Strafford, Colombe's Birthday, and The Cenci; if the Independent Theatre, the New Century Theatre, and the Stage Society are impracticable faddists for producing the plays of Ibsen and Maeterlinck, then what epithet is contemptuous enough for the people who produce the would-be popular plays?

The actor-managers were far more successful, because

they produced plays that at least pleased themselves, whereas the others, with a false theory of how to please everybody, produced plays that pleased nobody. But their occasional personal successes in voluptuous plays, and, in any case, their careful concealment of failure, confirmed the prevalent error, which was only exposed fully when the plays had to stand or fall openly by their own merits. Even Shakespear was played with his brains cut out. In 1896, when Sir Henry Irving was disabled by an accident at a moment when Miss Ellen Terry was too ill to appear, the theatre had to be closed after a brief attempt to rely on the attraction of a Shakespearean play performed by the stock company. This may have been Shakespear's fault: indeed Sir Henry later on complained that he had lost a princely sum by Shakespear. But Shakespear's reply to this, if he were able to make it, would be that the princely sum was spent, not on his dramatic poetry, but on a gorgeous stage ritualism superimposed on reckless mutilations of his text, the whole being addressed to a public as to which nothing is certain except that its natural bias is towards reverence for Shakespear and dislike and distrust of ritualism. No doubt the Lyceum ritual appealed to a far more cultivated sensuousness and imaginativeness than the musical farces in which our stage Abbots of Misrule pontificated (with the same financially disastrous result); but in both there was the same intentional brainlessness, founded on the same theory that the public did not want brains, did not want to think, did not want anything but pleasure at the theatre. Unfortunately, this theory happens to be true of a certain section of the public. This section, being courted by the theatres, went to them and drove the other people out. It then discovered, as any expert could have foreseen, that the theatre cannot compete in mere pleasuremongering either with the other arts or with matter-of-fact gallantry. Stage pictures are the worst pictures, stage music the worst music, stage scenery the worst scenery within reach of the Londoner. The leading lady or gentleman may be as tempting to the admirer in the pit as the dishes in a cookshop window are to the penniless tramp on the pavement; but people do not, I presume, go to the theatre to be merely tantalized.

The breakdown on the last point was conclusive. For when the managers tried to put their principle of pleasing everybody into practice, Necessity, ever ironical towards

Folly, had driven them to seek a universal pleasure to appeal to. And since many have no ear for music or eye for color, the search for universality inevitably flung the managers back on the instinct of sex as the avenue to all hearts. Of course the appeal was a vapid failure. Speaking for my own sex, I can say that the leading lady was not to everybody's taste: her pretty face often became ugly when she tried to make it expressive; her voice lost its charm (if it ever had any) when she had nothing sincere to say; and the stalls, from racial prejudice, were apt to insist on more Rebecca and less Rowena than the pit cared for. It may seem strange, even monstrous, that a man should feel a constant attachment to the hideous witches in Macbeth, and yet yawn at the prospect of spending another evening in the contemplation of a beauteous young leading lady with voluptuous contours and longlashed eyes, painted and dressed to perfection in the latest fashions. But that is just what happened to me in the theatre.

I did not find that matters were improved by the lady pretending to be "a woman with a past," violently oversexed, or the play being called a problem play, even when the manager, and sometimes, I suspect, the very author, firmly believed the word problem to be the latest euphemism for what Justice Shallow called a bona roba, and certainly would not either of them have staked a farthing on the interest of a genuine problem. In fact these so-called problem plays invariably depended for their dramatic interest on foregone conclusions of the most heartwearying conventionality concerning sexual morality. The authors had no problematic views: all they wanted was to capture some of the fascination of Ibsen. It seemed to them that most of Ibsen's heroines were naughty ladies. And they tried to produce Ibsen plays by making their heroines naughty. But they took great care to make them pretty and expensively dressed. Thus the pseudo-Ibsen play was nothing but the ordinary sensuous ritual of the stage become as frankly pornographic as good manners allowed.

I found that the whole business of stage sensuousness, whether as Lyceum Shakespear, musical farce, or sham Ibsen, finally disgusted me, not because I was Pharisaical, or intolerantly refined, but because I was bored; and boredom is a condition which makes men as susceptible to disgust and irritation as headache makes them to noise and glare. Being a man, I have my share of the masculine silli-

ness and vulgarity on the subject of sex which so astonishes women, to whom sex is a serious matter. I am not an Archbishop, and do not pretend to pass my life on one plane or in one mood, and that the highest: on the contrary, I am, I protest, as accessible to the humors of The Rogue's Comedy or The Rake's Progress as to the pious decencies of The Sign of the Cross. Thus Falstaff, coarser than any of the men in our loosest plays, does not bore me: Doll Tearsheet, more abandoned than any of the women, does not shock me. I think that Romeo and Juliet would be a poorer play if it were robbed of the solitary fragment it has preserved for us of the conversation of the husband of Juliet's nurse. No: my disgust was not mere thinskinned prudery. When my moral sense revolted, as it often did to the very fibres, it was invariably at the nauseous compliances of the theatre with conventional virtue. If I despised the musical farces, it was because they never had the courage of their vices. With all their labored efforts to keep up an understanding of furtive naughtiness between the low comedian on the stage and the drunken undergraduate in the stalls, they insisted all the time on their virtue and patriotism and loyalty as pitifully as a poor girl of the pavement will pretend to be a clergyman's daughter. True, I may have been offended when a manager, catering for me with coarse frankness as a slave dealer caters for a Pasha, invited me to forget the common bond of humanity between me and his company by demanding nothing from them but a gloatably voluptuous appearance. But this extreme is never reached at our better theatres. The shop assistants, the typists, the clerks, who, as I have said, preserve the innocence of the theatre, would not dare to let themselves be pleased by it. Even if they did, they would not get it from the managers, who, when they are brought to the only logical conclusion from their principle of making the theatre a temple of pleasure, indignantly refuse to change the dramatic profession for Mrs Warren's. For that is what all this demand for pleasure at the theatre finally comes to; and the answer to it is, not that people ought not to desire sensuous pleasure (they cannot help it) but that the theatre cannot give it to them, even to the extent permitted by the honor and conscience of the best managers, because a theatre is so far from being a pleasant or even a comfortable place that only by making us forget ourselves can it prevent us from realizing its inconveniences. A play

rules in the theatre: fitly enough too, because on the stage pretence is all that can exist. Life has its realities behind its shows: the theatre has nothing but its shows. But can the theatre make a show of lovers' endearments? A thousand times no: perish the thought of such unladylike, ungentlemanlike exhibitions. You can have fights, rescues, conflagrations, trials-at-law, avalanches, murders and executions all directly simulated on the stage if you will. But any such realistic treatment of the incidents of sex is quite out of the question. The singer, the dramatic dancer, the exquisite declaimer of impassioned poesy, the rare artist who, bringing something of the art of all three to the ordinary work of the theatre, can enthral an audience by the expression of dramatic feeling alone, may take love for a theme on the stage; but the prosaic walking gentleman of our fashionable theatres, realistically simulating the incidents of life, cannot touch it without indecorum.

Can any dilemma be more complete? Love is assumed to be the only theme that touches all your audience infallibly, young and old, rich and poor. And yet love is the one subject that the drawingroom drama dare not present.

Out of this dilemma, which is a very old one, has come the romantic play: that is, the play in which love is carefully kept off the stage, whilst it is alleged as the motive of all the actions presented to the audience. The result is, to me at least, an intolerable perversion of human conduct. There are two classes of stories that seem to me to be not only fundamentally false but sordidly base. One is the pseudo-religious story, in which the hero or heroine does good on strictly commercial grounds, reluctantly exercising a little virtue on earth in consideration of receiving in return an exorbitant payment in heaven: much as if an odalisque were to allow a cadi to whip her for a couple of millions in gold. The other is the romance in which the hero, also rigidly commercial, will do nothing except for the sake of the heroine. Surely this is as depressing as it is unreal. Compare with it the treatment of love, frankly indecent according to our notions, in oriental fiction. In The Arabian Nights we have a series of stories, some of them very good ones, in which no sort of decorum is observed. The result is that they are infinitely more instructive and enjoyable than our romances, because love is treated in them as naturally as any other passion. There is no cast

iron convention as to its effects; no false association of general depravity of character with its corporealities or of general elevation with its sentimentalities; no pretence that a man or woman cannot be courageous and kind and friendly unless infatuatedly in love with somebody (is no poet manly enough to sing The Old Maids of England?): rather, indeed, an insistence on the blinding and narrowing power of lovesickness to make princely heroes unhappy and unfortunate. These tales expose, further, the delusion that the interest of this most capricious, most transient, most easily baffled of all instincts, is inexhaustible, and that the field of the English romancer has been cruelly narrowed by the restrictions under which he is permitted to deal with it. The Arabian storyteller, relieved of all such restrictions, heaps character on character, adventure on adventure, marvel on marvel; whilst the English novelist, like the starving tramp who can think of nothing but his hunger, seems to be unable to escape from the obsession of sex, and will rewrite the very gospels because the originals are not written in the sensuously ecstatic style. At the instance of Martin Luther we long ago gave up imposing celibacy on our priests; but we still impose it on our art, with the very undesirable and unexpected result that no editor, publisher, or manager, will now accept a story or produce a play without "love interest" in it. Take, for a recent example, Mr H. G. Wells's War of the Worlds, a tale of the invasion of the earth by the inhabitants of the planet Mars: a capital story, not to be laid down until finished. Love interest is impossible on its scientific plane: nothing could be more impertinent and irritating. Yet Mr Wells has had to pretend that the hero is in love with a young lady manufactured for the purpose, and to imply that it is on her account alone that he feels concerned about the apparently inevitable destruction of the human race by the Martians. Another example. An American novelist, recently deceased, made a hit some years ago by compiling a Bostonian Utopia from the prospectuses of the little bands of devout Communists who have from time to time, since the days of Fourier and Owen, tried to establish millennial colonies outside our commercial civilization. Even in this economic Utopia we find the inevitable love affair. The hero, waking up in a distant future from a miraculous sleep, meets a Boston young lady, provided expressly for him to fall in love with. Women have by that time given up

wearing skirts; but she, to spare his delicacy, gets one out of a museum of antiquities to wear in his presence until he is hardened to the customs of the new age. When I came to that touching incident, I became as Paolo and Francesca: "in that book I read no more." I will not multiply examples: if such unendurable follies occur in the sort of story made by working out a meteorologic or economic hypothesis, the extent to which it is carried in sentimental romances needs no expatiation.

The worst of it is that since man's intellectual consciousness of himself is derived from the descriptions of him in books, a persistent misrepresentation of humanity in literature gets finally accepted and acted upon. If every mirror reflected our noses twice their natural size, we should live and die in the faith that we were all Punches; and we should scout a true mirror as the work of a fool, madman, or jester. Nay, I believe we should, by Lamarckian adaptation, enlarge our noses to the admired size; for I have noticed that when a certain type of feature appears in painting and is admired as beautiful, it presently becomes common in nature; so that the Beatrices and Francescas in the picture galleries of one generation, to whom minor poets address verses entitled To My Lady, come to life as the parlormaids and waitresses of the next. If the conventions of romance are only insisted on long enough and uniformly enough (a condition guaranteed by the uniformity of human folly and vanity), then, for the huge School Board-taught masses who read romance and nothing else, these conventions will become the laws of personal honor. Jealousy, which is either an egotistical meanness or a specific mania, will become obligatory; and ruin, ostracism, breaking up of homes, duelling, murder, suicide and infanticide will be produced (often have been produced, in fact) by incidents which, if left to the operation of natural and right feeling, would produce nothing worse than an hour's soon-forgotten fuss. Men will be slain needlessly on the field of battle because officers conceive it to be their first duty to make romantic exhibitions of conspicuous gallantry. The squire who has never spared an hour from the hunting field to do a little public work on a parish council will be cheered as a patriot because he is willing to kill and get killed for the sake of conferring himself as an institution on other countries. In the courts cases will be argued, not on juridical but on romantic principles; and vindictive

damages and vindictive sentences, with the acceptance of nonsensical, and the repudiation or suppression of sensible testimony, will destroy the very sense of law. Kaisers, generals, judges, and prime ministers will set the example of playing to the gallery. Finally the people, now that their Board School literacy enables every penman to play on their romantic illusions, will be led by the nose far more completely than they ever were by playing on their former ignorance and superstition. Nay, why should I say will be? they *are*. Ten years of cheap reading have changed the English from the most stolid nation in Europe to the most theatrical and hysterical.

Is it clear now, why the theatre was insufferable to me; why it left its black mark on my bones as it has left its black mark on the character of the nation; why I call the Puritans to rescue it again as they rescued it before when its foolish pursuit of pleasure sunk it in "profaneness and immorality"? I have, I think, always been a Puritan in my attitude towards Art. I am as fond of fine music and handsome building as Milton was, or Cromwell, or Bunyan; but if I found that they were becoming the instruments of a systematic idolatry of sensuousness, I would hold it good statesmanship to blow every cathedral in the world to pieces with dynamite, organ and all, without the least heed to the screams of the art critics and cultured voluptuaries. And when I see that the nineteenth century has crowned the idolatry of Art with the deification of Love, so that every poet is supposed to have pierced to the holy of holies when he has announced that Love is the Supreme, or the Enough, or the All, I feel that Art was safer in the hands of the most fanatical of Cromwell's major generals than it will be if ever it gets into mine. The pleasures of the senses I can sympathize with and share; but the substitution of sensuous ecstasy for intellectual activity and honesty is the very devil. It has already brought us to Flogging Bills in Parliament, and, by reaction, to androgynous heroes on the stage; and if the infection spreads until the democratic attitude becomes thoroughly Romanticist, the country will become unbearable for all realists, Philistine or Platonic. When it comes to that, the brute force of the strong-minded Bismarckian man of action, impatient of humbug, will combine with the subtlety and spiritual energy of the man of thought whom shams cannot illude or interest. That combination will be on one side; and Romanticism

will be on the other. In which event, so much the worse for Romanticism, which will come down even if it has to drag Democracy down with it. For all institutions have in the long run to live by the nature of things, and not by imagination.

ON DIABOLONIAN ETHICS

THERE is a foolish opinion prevalent that an author should allow his works to speak for themselves, and that he who appends and prefixes explanations to them is likely to be as bad an artist as the painter cited by Cervantes, who wrote under his picture This is a Cock, lest there should be any mistake about it. The pat retort to this thoughtless comparison is that the painter invariably does so label his picture. What is a Royal Academy catalogue but a series of statements that This is The Vale of Rest, This is The School of Athens, This is Chill October, This is The Prince of Wales, and so on? The reason most dramatists do not publish their plays with prefaces is that they cannot write them, the business of intellectually conscious philosopher and skilled critic being no part of the playwright's craft. Naturally, making a virtue of their incapacity, they either repudiate prefaces as shameful, or else, with a modest air, request some popular critic to supply one, as much as to say, Were I to tell the truth about myself I must needs seem vainglorious: were I to tell less than the truth I should do myself an injustice and deceive my readers. As to the critic thus called in from the outside, what can he do but imply that his friend's transcendant ability as a dramatist is surpassed only by his beautiful nature as a man? Now what I say is, why should I get another man to praise me when I can praise myself? I have no disabilities to plead: produce me your best critic, and I will criticize his head off. As to philosophy, I taught my critics the little they know in my Quintessence of Ibsenism; and now they turn their guns—the guns I loaded for them—on me, and proclaim that I write as if mankind had intellect without will, or heart, as they call it. Ingrates: who was it that directed your attention to the distinction between Will and Intellect? Not Schopenhauer, I think, but Shaw.

Again, they tell me that So-and-So, who does not write prefaces, is no charlatan. Well, I am. I first caught the ear of the British public on a cart in Hyde Park, to the blaring of brass bands, and this not at all as a reluctant sacrifice of

my instinct of privacy to political necessity, but because, like all dramatists and mimes of genuine vocation, I am a natural-born mountebank. I am well aware that the ordinary British citizen requires a profession of shame from all mountebanks by way of homage to the sanctity of the ignoble private life to which he is condemned by his incapacity for public life. Thus Shakespear, after proclaiming that Not marble nor the gilded monuments of Princes should outlive his powerful rhyme, would apologize, in the approved taste, for making himself a motley to the view; and the British citizen has ever since quoted the apology and ignored the fanfare. When an actress writes her memoirs, she impresses on you in every chapter how cruelly it tried her feelings to exhibit her person to the public gaze; but she does not forget to decorate the book with a dozen portraits of herself. I really cannot respond to this demand for mock-modesty. I am ashamed neither of my work nor of the way it is done. I like explaining its merits to the huge majority who dont know good work from bad. It does them good; and it does me good, curing me of nervousness, laziness, and snobbishness. I write prefaces as Dryden did, and treatises as Wagner, because I *can;* and I would give half a dozen of Shakespear's plays for one of the prefaces he ought to have written. I leave the delicacies of retirement to those who are gentlemen first and literary workmen afterwards. The cart and trumpet for me.

This is all very well; but the trumpet is an instrument that grows on one; and sometimes my blasts have been so strident that even those who are most annoyed by them have mistaken the novelty of my shamelessness for novelty in my plays and opinions. Take, for instance, the first play in this volume, entitled The Devil's Disciple. It does not contain a single even passably novel incident. Every old patron of the Adelphi pit would, were he not beglamored in a way presently to be explained, recognize the reading of the will, the oppressed orphan finding a protector, the arrest, the heroic sacrifice, the court martial, the scaffold, the reprieve at the last moment, as he recognizes beefsteak pudding on the bill of fare at his restaurant. Yet when the play was produced in 1897 in New York by Mr Richard Mansfield, with a success that proves either that the melodrama was built on very safe old lines, or that the American public is composed exclusively of men of genius, the critics, though one said one thing and another another as

to the play's merits, yet all agreed that it was novel – *original,* as they put it – to the verge of audacious eccentricity.

Now this, if it applies to the incidents, plot, construction, and general professional and technical qualities of the play, is nonsense; for the truth is, I am in these matters a very old-fashioned playwright. When a good deal of the same talk, both hostile and friendly, was provoked by my last volume of plays, Mr Robert Buchanan, a dramatist who knows what I know and remembers what I remember of the history of the stage, pointed out that the stage tricks by which I gave the younger generation of playgoers an exquisite sense of quaint unexpectedness, had done duty years ago in Cool as a Cucumber, Used Up, and many forgotten farces and comedies of the Byron-Robertson school, in which the imperturbably impudent comedian, afterwards shelved by the reaction to brainless sentimentality, was a stock figure. It is always so more or less: the novelties of one generation are only the resuscitated fashions of the generation before last.

But the stage tricks of The Devil's Disciple are not, like some of those of Arms and the Man, the forgotten ones of the sixties, but the hackneyed ones of our own time. Why, then, were they not recognized? Partly, no doubt, because of my trumpet and cartwheel declamation. The critics were the victims of the long course of hypnotic suggestion by which G.B.S. the journalist manufactured an unconventional reputation for Bernard Shaw the author. In England as elsewhere the spontaneous recognition of really original work begins with a mere handful of people, and propagates itself so slowly that it has become a commonplace to say that genius, demanding bread, is given a stone after its possessor's death. The remedy for this is sedulous advertisement. Accordingly, I have advertised myself so well that I find myself, whilst still in middle life, almost as legendary a person as the Flying Dutchman. Critics, like other people, see what they look for, not what is actually before them. In my plays they look for my legendary qualities, and find originality and brilliancy in my most hackneyed claptraps. Were I to republish Buckstone's Wreck Ashore as my latest comedy, it would be hailed as a masterpiece of perverse paradox and scintillating satire. Not, of course, by the really able critics – for example, you, my friend, now reading this sentence. The illusion that makes *you* think me so original is far subtler than that. The Devil's

Disciple has, in truth, a genuine novelty in it. Only, that novelty is not any invention of my own, but simply the novelty of the advanced thought of my day. As such, it will assuredly lose its gloss with the lapse of time, and leave The Devil's Disciple exposed as the threadbare popular melodrama it technically is.

Let me explain (for, as Mr A. B. Walkley has pointed out in his disquisitions on Frames of Mind, I am nothing if not explanatory). Dick Dudgeon, the devil's disciple, is a Puritan of the Puritans. He is brought up in a household where the Puritan religion has died, and become, in its corruption, an excuse for his mother's master passion of hatred in all its phases of cruelty and envy. This corruption has already been dramatized for us by Charles Dickens in his picture of the Clennam household in Little Dorrit: Mrs Dudgeon being a replica of Mrs Clennam with certain circumstantial variations, and perhaps a touch of the same author's Mrs Gargery in Great Expectations. In such a home the young Puritan finds himself starved of religion, which is the most clamorous need of his nature. With all his mother's indomitable selffulness, but with Pity instead of Hatred as his master passion, he pities the devil; takes his side; and champions him, like a true Covenanter, against the world. He thus becomes, like all genuinely religious men, a reprobate and an outcast. Once this is understood, the play becomes straightforwardly simple.

The Diabolonian position is new to the London playgoer of today, but not to lovers of serious literature. From Prometheus to the Wagnerian Siegfried, some enemy of the gods, unterrified champion of those oppressed by them, has always towered among the heroes of the loftiest poetry. Our newest idol, the Superman, celebrating the death of godhead, may be younger than the hills; but he is as old as the shepherds. Two and a half centuries ago our greatest English dramatizer of life, John Bunyan, ended one of his stories with the remark that there is a way to hell even from the gates of heaven, and so led us to the equally true proposition that there is a way to heaven even from the gates of hell. A century ago William Blake was, like Dick Dudgeon, an avowed Diabolonian: he called his angels devils and his devils angels. His devil is a Redeemer. Let those who have praised my originality in conceiving Dick Dudgeon's strange religion read Blake's Marriage of Heaven and Hell; and I shall be fortunate if they do not

rail at me for a plagiarist. But they need not go back to Blake and Bunyan. Have they not heard the recent fuss about Nietzsche and his Good and Evil Turned Inside Out? Mr Robert Buchanan has actually written a long poem of which the Devil is the merciful hero, which poem was in my hands before a word of The Devil's Disciple was written. There never was a play more certain to be written than The Devil's Disciple at the end of the nineteenth century. The age was visibly pregnant with it.

I grieve to have to add that my old friends and colleagues the London critics for the most part showed no sort of connoisseurship either in Puritanism or in Diabolonianism when the play was performed for a few weeks at a suburban theatre (Kennington) in October 1899 by Mr Murray Carson. They took Mrs Dudgeon at her own valuation as a religious woman because she was detestably disagreeable. And they took Dick as a blackguard, on her authority, because he was neither detestable nor disagreeable. But they presently found themselves in a dilemma. Why should a blackguard save another man's life, and that man no friend of his, at the risk of his own? Clearly, said the critics, because he is redeemed by love. All wicked heroes are, on the stage: that is the romantic metaphysic. Unfortunately for this explanation (which I do not profess to understand) it turned out in the third act that Dick was a Puritan in this respect also: a man impassioned only for saving grace, and not to be led or turned by wife or mother, Church or State, pride of life or lust of the flesh. In the lovely home of the courageous, affectionate, practical minister who marries a pretty wife twenty years younger than himself, and turns soldier in an instant to save the man who has saved him, Dick looks round and understands the charm and the peace and the sanctity, but knows that such material comforts are not for him. When the woman nursed in that atmosphere falls in love with him and concludes (like the critics, who somehow always agree with my sentimental heroines) that he risked his life for her sake, he tells her the obvious truth that he would have done as much for any stranger – that the law of his own nature, and no interest nor lust whatsoever, forbad him to cry out that the hangman's noose should be taken off his neck only to be put on another man's.

But then, said the critics, where is the motive? *Why* did Dick save Anderson? On the stage, it appears, people do

things for reasons. Off the stage they dont: that is why your penny-in-the-slot heroes, who only work when you drop a motive into them, are so oppressively automatic and uninteresting. The saving of life at the risk of the saver's own is not a common thing; but modern populations are so vast that even the most uncommon things are recorded once a week or oftener. Not one of my critics but has seen a hundred times in his paper how some policeman or fireman or nursemaid has received a medal, or the compliments of a magistrate, or perhaps a public funeral, for risking his or her life to save another's. Has he ever seen it added that the saved was the husband of the woman the saver loved, or was that woman herself, or was even known to the saver as much as by sight? Never. When we want to read of the deeds that are done for love, whither do we turn? To the murder column; and there we are rarely disappointed.

Need I repeat that the theatre critic's professional routine so discourages any association between real life and the stage, that he soon loses the natural habit of referring to the one to explain the other? The critic who discovered a romantic motive for Dick's sacrifice was no mere literary dreamer, but a clever barrister. He pointed out that Dick Dudgeon clearly did adore Mrs Anderson; that it was for her sake that he offered his life to save her beloved husband; and that his explicit denial of his passion was the splendid mendacity of a gentleman whose respect for a married woman, and duty to her absent husband, sealed his passion-palpitating lips. From the moment that this fatally plausible explanation was launched, my play became my critic's play, not mine. Thenceforth Dick Dudgeon every night confirmed the critic by stealing behind Judith, and mutely attesting his passion by surreptitiously imprinting a heartbroken kiss on a stray lock of her hair whilst he uttered the barren denial. As for me, I was just then wandering about the streets of Constantinople, unaware of all these doings. When I returned all was over. My personal relations with the critic and the actor forbad me to curse them. I had not even the chance of publicly forgiving them. They meant well by me; but if they ever write a play, may I be there to explain!

As to the other plays in this volume, the application of my title is less obvious, since neither Julius Cæsar, Cleopatra nor Lady Cicely Waynflete* have any external political connexion with Puritanism. The very name of Cleopatra suggests at once a tragedy of Circe, with the horrible difference that whereas the ancient myth rightly represents Circe as turning heroes into hogs, the modern romantic convention would represent her as turning hogs into heroes. Shakespear's Antony and Cleopatra must needs be as intolerable to the true Puritan as it is vaguely distressing to the ordinary healthy citizen, because, after giving a faithful picture of the soldier broken down by debauchery, and the typical wanton in whose arms such men perish, Shakespear finally strains all his huge command of rhetoric and stage pathos to give a theatrical sublimity to the wretched end of the business, and to persuade foolish spectators that the world was well lost by the twain. Such falsehood is not to be borne except by the real Cleopatras and Antonys (they are to be found in every public house) who would no doubt be glad enough to be transfigured by some poet as immortal lovers. Woe to the poet who stoops to such folly! The lot of the man who sees life truly and thinks about it romantically is Despair. How well we know the cries of that despair! Vanity of vanities, all is vanity! moans the Preacher, when life has at last taught him that Nature will not dance to his moralist-made tunes. Thackeray, scores of centuries later, is still baying the moon in the same terms. Out, out, brief candle! cries Shakespear, in his tragedy of the modern literary man as murderer and witch consulter. Surely the time is past for patience with writers who, having to choose between giving up life in despair and discarding the trumpery moral kitchen scales in which they try to weigh the universe, superstitiously stick to the scales, and spend the rest of the lives they pretend to despise in breaking men's spirits. But even in pessimism there is a choice between intellectual honesty and dishonesty. Hogarth drew the rake and the harlot without glorifying their end. Swift, accepting our system of morals and religion, delivered the inevitable verdict of that system on us through the mouth of the king of Brobdingnag, and described man as the Yahoo, shocking his superior the horse by his every action. Strindberg, the only living gen-

* Lady Cicely is an important character in *Captain Brassbound's Conversion.*

uine Shakespearean dramatist, shows that the female Yahoo, measured by romantic standards, is viler than her male dupe and slave. I respect these resolute tragicomedians: they are logical and faithful: they force you to face the fact that you must either accept their conclusions as valid (in which case it is cowardly to continue living) or admit that your way of judging conduct is absurd. But when your Shakespears and Thackerays huddle up the matter at the end by killing somebody and covering your eyes with the undertaker's handkerchief, duly onioned with some pathetic phrase, as The flight of angels sing thee to thy rest, or Adsum, or the like, I have no respect for them at all: such maudlin tricks may impose on tea-drunkards, not on me.

Besides, I have a technical objection to making sexual infatuation a tragic theme. Experience proves that it is only effective in the comic spirit. We can bear to see Mrs Quickly pawning her plate for love of Falstaff, but not Antony running away from the battle of Actium for love of Cleopatra. Let realism have its demonstration, comedy its criticism, or even bawdry its horselaugh at the expense of sexual infatuation, if it must; but to ask us to subject our souls to its ruinous glamor, to worship it, deify it, and imply that it alone makes our life worth living, is nothing but folly gone mad erotically – a thing compared to which Falstaff's unbeglamored drinking and drabbing is respectable and rightminded. Whoever, then, expects to find Cleopatra a Circe and Cæsar a hog in these pages, had better lay down my book and be spared a disappointment.

In Cæsar, I have used another character with which Shakespear has been beforehand. But Shakespear, who knew human weakness so well, never knew human strength of the Cæsarian type. His Cæsar is an admitted failure: his Lear is a masterpiece. The tragedy of disillusion and doubt, of the agonized struggle for a foothold on the quicksand made by an acute observation striving to verify its vain attribution of morality and respectability to Nature, of the faithless will and the keen eyes that the faithless will is too weak to blind: all this will give you a Hamlet or a Macbeth, and win you great applause from literary gentlemen; but it will not give you a Julius Cæsar. Cæsar was not in Shakespear, nor in the epoch, now fast waning, which he inaugurated. It cost Shakespear no pang to write Cæsar down for the merely technical purpose of writing Brutus

up. And what a Brutus! A perfect Girondin, mirrored in Shakespear's art two hundred years before the real thing came to maturity and talked and stalked and had its head duly cut off by the coarser Antonys and Octaviuses of its time, who at least knew the difference between life and rhetoric.

It will be said that these remarks can bear no other construction than an offer of my Cæsar to the public as an improvement on Shakespear's. And in fact, that is their precise purport. But here let me give a friendly warning to those scribes who have so often exclaimed against my criticisms of Shakespear as blasphemies against a hitherto unquestioned Perfection and Infallibility. Such criticisms are no more new than the creed of my Diabolonian Puritan or my revival of the humors of Cool as a Cucumber. Too much surprise at them betrays an acquaintance with Shakespear criticism so limited as not to include even the prefaces of Dr Johnson and the utterances of Napoleon. I have merely repeated in the dialect of my own time and in the light of its philosophy what they said in the dialect and light of theirs. Do not be misled by the Shakespear fanciers who, ever since his own time, have delighted in his plays just as they might have delighted in a particular breed of pigeons if they had never learnt to read. His genuine critics, from Ben Jonson to Mr Frank Harris, have always kept as far on this side idolatry as I.

As to our ordinary uncritical citizens, they have been slowly trudging forward these three centuries to the point which Shakespear reached at a bound in Elizabeth's time. Today most of them have arrived there or thereabouts, with the result that his plays are at last beginning to be performed as he wrote them; and the long line of disgraceful farces, melodramas, and stage pageants which actor-managers, from Garrick and Cibber to our own contemporaries, have hacked out of his plays as peasants have hacked huts out of the Coliseum, are beginning to vanish from the stage. It is a significant fact that the mutilators of Shakespear, who never could be persuaded that Shakespear knew his business better than they, have ever been the most fanatical of his worshippers. The late Augustin Daly thought no price too extravagant for an addition to his collection of Shakespear relics; but in arranging Shakespear's plays for the stage, he proceeded on the assumption that Shakespear was a botcher and he an artist. I am far too

good a Shakespearean ever to forgive Sir Henry Irving for producing a version of King Lear so mutilated that the numerous critics who had never read the play could not follow the story of Gloster. Both these idolaters of the Bard must have thought Mr Forbes Robertson mad because he restored Fortinbras to the stage and played as much of Hamlet as there was time for instead of as little. And the instant success of the experiment probably altered their minds no further than to make them think the public mad. Mr Benson actually gives the play complete at two sittings, causing the aforesaid numerous critics to remark with naïve surprise that Polonius is a complete and interesting character. It was the age of gross ignorance of Shakespear and incapacity for his works that produced the indiscriminate eulogies with which we are familiar. It was the revival of genuine criticism of those works that coincided with the movement for giving genuine instead of spurious and silly representations of his plays. So much for Bardolatry!

It does not follow, however, that the right to criticize Shakespear involves the power of writing better plays. And in fact – do not be surprised at my modesty – I do not profess to write better plays. The writing of practicable stage plays does not present an infinite scope to human talent; and the dramatists who magnify its difficulties are humbugs. The summit of their art has been attained again and again. No man will ever write a better tragedy than Lear, a better comedy than Le Festin de Pierre or Peer Gynt, a better opera than Don Giovanni, a better music drama than The Niblung's Ring, or, for the matter of that, better fashionable plays and melodramas than are now being turned out by writers whom nobody dreams of mocking with the word immortal. It is the philosophy, the outlook on life, that changes, not the craft of the playwright. A generation that is thoroughly moralized and patriotized, that conceives virtuous indignation as spiritually nutritious, that murders the murderer and robs the thief, that grovels before all sorts of ideals, social, military, ecclesiastical, royal and divine, may be, from my point of view, steeped in error; but it need not want for as good plays as the hand of man can produce. Only, those plays will be neither written nor relished by men in whose philosophy guilt and innocence, and consequently revenge and idolatry, have no meaning. Such men must rewrite all the old plays in terms of their own philosophy; and that is why, as Mr Stuart-

Glennie has pointed out, there can be no new drama without a new philosophy. To which I may add that there can be no Shakespear or Goethe without one either, nor two Shakespears in one philosophic epoch, since, as I have said, the first great comer in that epoch reaps the whole harvest and reduces those who come after to the rank of mere gleaners, or, worse than that, fools who go laboriously through all the motions of the reaper and binder in an empty field. What is the use of writing plays or painting frescoes if you have nothing more to say or show than was said and shown by Shakespear, Michael Angelo, and Raphael? If these had not seen things differently, for better or worse, from the dramatic poets of the Townley mysteries, or from Giotto, they could not have produced their works: no, not though their skill of pen and hand had been double what it was. After them there was no need (and *need* alone nerves men to face the persecution in the teeth of which new art is brought to birth) to redo the already done, until in due time, when their philosophy wore itself out, a new race of nineteenth century poets and critics, from Byron to William Morris, began, first to speak coldly of Shakespear and Raphael, and then to rediscover, in the medieval art which these Renascence masters had superseded, certain forgotten elements which were germinating again for the new harvest. What is more, they began to discover that the technical skill of the masters was by no means superlative. Indeed, I defy anyone to prove that the great epoch makers in fine art have owed their position to their technical skill. It is true that when we search for examples of a prodigious command of language and of graphic line, we can think of nobody better than Shakespear and Michael Angelo. But both of them laid their arts waste for centuries by leading later artists to seek greatness in copying their technique. The technique was acquired, refined on, and surpassed over and over again; but the supremacy of the two great exemplars remained undisputed. As a matter of easily observable fact, every generation produces men of extraordinary special faculty, artistic, mathematical and linguistic, who for lack of new ideas, or indeed of any ideas worth mentioning, achieve no distinction outside music halls and class rooms, although they can do things easily that the great epoch makers did clumsily or not at all. The contempt of the academic pedant for the original artist is often founded on a genuine superiority of

technical knowledge and aptitude: he is sometimes a better anatomical draughtsman than Raphael, a better hand at triple counterpoint than Beethoven, a better versifier than Byron. Nay, this is true not merely of pedants, but of men who have produced works of art of some note. If technical facility were the secret of greatness in art, Mr Swinburne would be greater than Browning and Byron rolled into one, Stevenson greater than Scott or Dickens, Mendelssohn than Wagner, Maclise than Madox Brown. Besides, new ideas make their technique as water makes its channel; and the technician without ideas is as useless as the canal constructor without water, though he may do very skilfully what the Mississippi does very rudely. To clinch the argument, you have only to observe that the epoch maker himself has generally begun working professionally before his new ideas have mastered him sufficiently to insist on constant expression by his art. In such cases you are compelled to admit that if he had by chance died earlier, his greatness would have remained unachieved, although his technical qualifications would have been well enough established. The early imitative works of great men are usually conspicuously inferior to the best works of their forerunners. Imagine Wagner dying after composing Rienzi, or Shelly after Zastrozzi! Would any competent critic then have rated Wagner's technical aptitude as high as Rossini's, Spontini's, or Meyerbeer's; or Shelley's as high as Moore's? Turn the problem another way: does anyone suppose that if Shakespear had conceived Goethe's or Ibsen's ideas, he would have expressed them any worse than Goethe or Ibsen? Human faculty being what it is, is it likely that in our time any advance, except in external conditions, will take place in the arts of expression sufficient to enable an author, without making himself ridiculous, to undertake to say what he has to say better than Homer or Shakespear? But the humblest author, and much more a rather arrogant one like myself, may profess to have something to say by this time that neither Homer nor Shakespear said. And the playgoer may reasonably ask to have historical events and persons presented to him in the light of his own time, even though Homer and Shakespear have already shown them in the light of their time. For example, Homer presented Achilles and Ajax as heroes to the world in the Iliads. In due time came Shakespear, who said, virtually: I really cannot accept this spoiled child and this

brawny fool as great men merely because Homer flattered them in playing to the Greek gallery. Consequently we have, in Troilus and Cressida, the verdict of Shakespear's epoch (our own) on the pair. This did not in the least involve any pretence on Shakespear's part to be a greater poet than Homer.

When Shakespear in turn came to deal with Henry V and Julius Cæsar, he did so according to his own essentially knightly conception of a great statesman-commander. But in the XIX century comes the German historian Mommsen, who also takes Cæsar for his hero, and explains the immense difference in scope between the perfect knight Vercingetorix and his great conqueror Julius Cæsar. In this country, Carlyle, with his vein of peasant inspiration, apprehended the sort of greatness that places the true hero of history so far beyond the mere *preux chevalier*, whose fanatical personal honor, gallantry and self-sacrifice, are founded on a passion for death born of inability to bear the weight of a life that will not grant ideal conditions to the liver. This one ray of perception became Carlyle's whole stock-in-trade; and it sufficed to make a literary master of him. In due time, when Mommsen is an old man, and Carlyle dead, come I, and dramatize the by-this-time familiar distinction in Arms and the Man, with its comedic conflict between the knightly Bulgarian and the Mommsenite Swiss captain. Whereupon a great many playgoers who have not yet read Shakespear, much less Mommsen and Carlyle, raise a shriek of concern for their knightly ideal as if nobody had ever questioned its sufficiency since the middle ages. Let them thank me for educating them so far. And let them allow me to set forth Cæsar in the same modern light, taking the same liberty with Shakespear as he with Homer, and with no thought of pretending to express the Mommsenite view of Cæsar any better than Shakespear expressed a view which was not even Plutarchian, and must, I fear, be referred to the tradition in stage conquerors established by Marlowe's Tamerlane as much as to even the chivalrous conception of heroism dramatized in Henry V.

For my own part, I can avouch that such powers of invention, humor and stage ingenuity as I have been able to exercise in Plays, Pleasant and Unpleasant, and in these Plays for Puritans, availed me not at all until I saw the old facts in a new light. Technically, I do not find myself able

to proceed otherwise than as former playwrights have done. True, my plays have the latest mechanical improvements: the action is not carried on by impossible soliloquies and asides; and my people get on and off the stage without requiring four doors to a room which in real life would have only one. But my stories are the old stories; my characters are the familiar harlequin and columbine, clown and pantaloon (note the harlequin's leap in the third act of Cæsar and Cleopatra); my stage tricks and suspenses and thrills and jests are the ones in vogue when I was a boy, by which time my grandfather was tired of them. To the young people who make their acquaintance for the first time in my plays, they may be as novel as Cyrano's nose to those who have never seen Punch; whilst to older playgoers the unexpectedness of my attempt to substitute natural history for conventional ethics and romantic logic may so transfigure the eternal stage puppets and their inevitable dilemmas as to make their identification impossible for the moment. If so, so much the better for me: I shall perhaps enjoy a few years of immortality. But the whirligig of time will soon bring my audiences to my own point of view; and then the next Shakespear that comes along will turn these petty tentatives of mine into masterpieces final for their epoch. By that time my twentieth century characteristics will pass unnoticed as a matter of course, whilst the eighteenth century artificiality that marks the work of every literary Irishman of my generation will seem antiquated and silly. It is a dangerous thing to be hailed at once, as a few rash admirers have hailed me, as above all things original: what the world calls originality is only an unaccustomed method of tickling it. Meyerbeer seemed prodigiously original to the Parisians when he first burst on them. Today, he is only the crow who followed Beethoven's plough. I am a crow who have followed many ploughs. No doubt I seem prodigiously clever to those who have never hopped, hungry and curious, across the fields of philosophy, politics and art. Karl Marx said of Stuart Mill that his eminence was due to the flatness of the surrounding country. In these days of Board Schools, universal reading, cheap newspapers, and the inevitable ensuing demand for notabilities of all sorts, literary, military, political and fashionable, to write paragraphs about, that sort of eminence is within the reach of very moderate ability. Reputations are cheap nowadays. Even were they

dear, it would still be impossible for any public-spirited citizen of the world to hope that his reputation might endure; for this would be to hope that the flood of general enlightenment may never rise above his miserable high-water-mark. I hate to think that Shakespear has lasted 300 years, though he got no further than Koheleth the Preacher, who died many centuries before him; or that Plato, more than 2000 years old, is still ahead of our voters. We must hurry on: we must get rid of reputations: they are weeds in the soil of ignorance. Cultivate that soil, and they will flower more beautifully, but only as annuals. If this preface will at all help to get rid of mine, the writing of it will have been well worth the pains.

G.B.S.

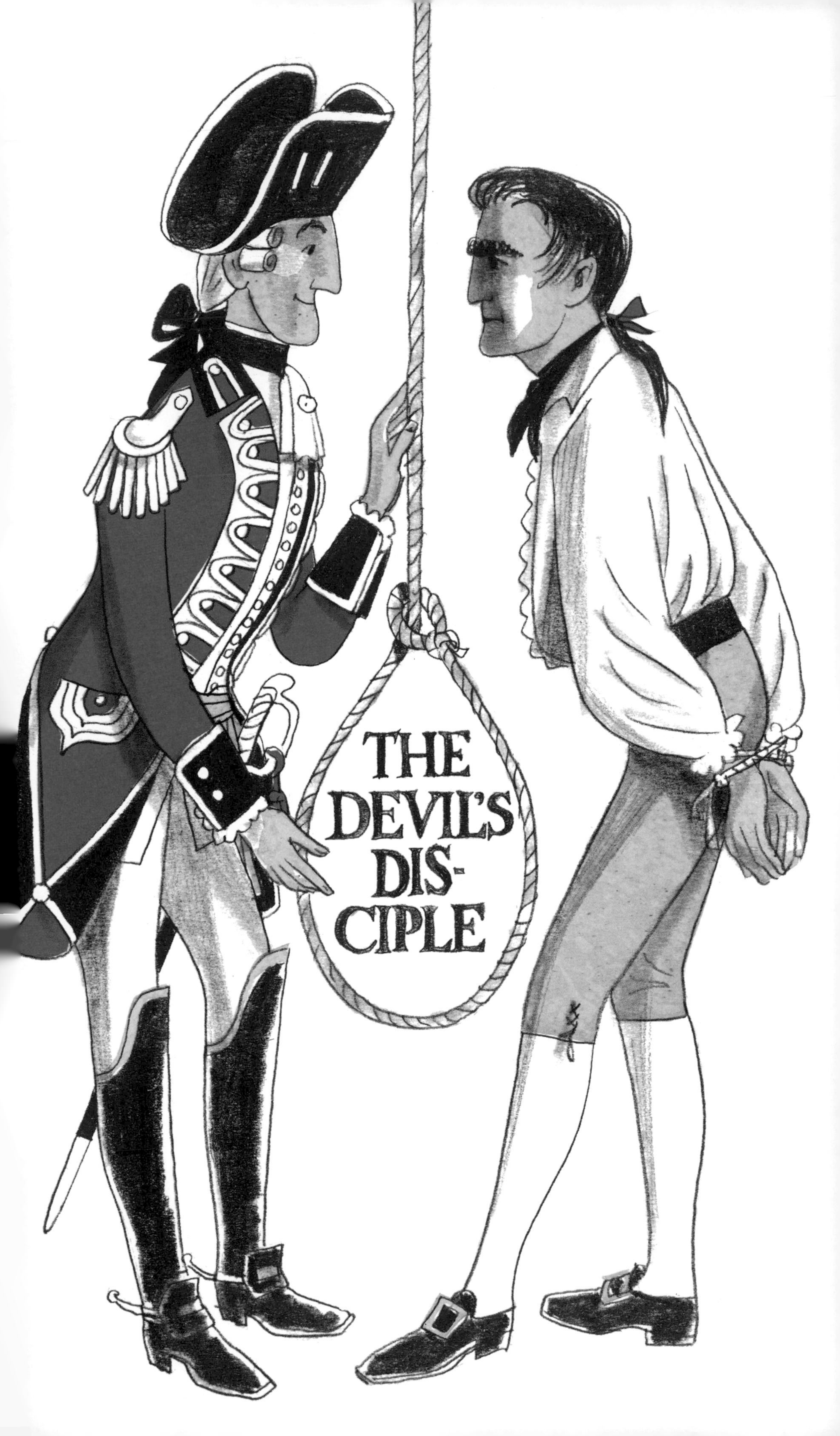
THE
DEVIL'S
DIS-
CIPLE

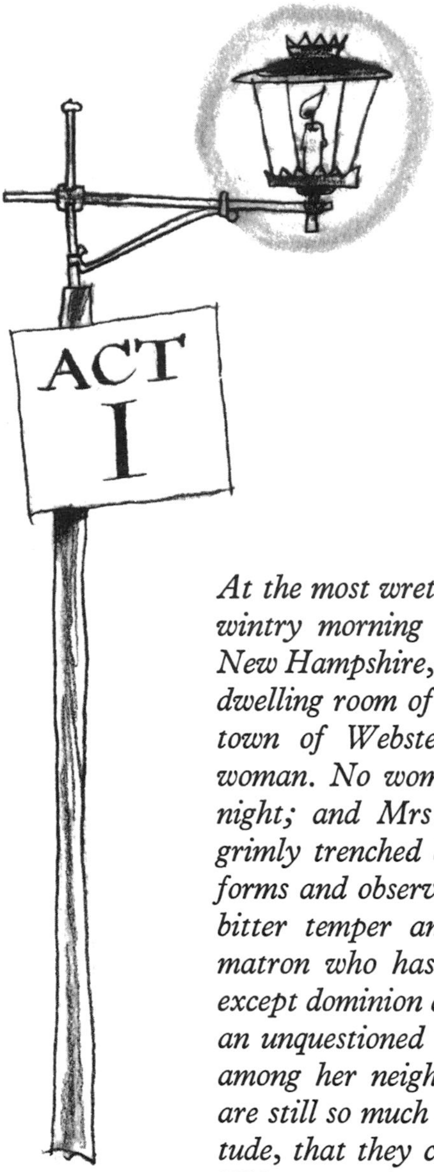

At the most wretched hour between a black night and a wintry morning in the year 1777, Mrs Dudgeon, of New Hampshire, is sitting up in the kitchen and general dwelling room of her farm house on the outskirts of the town of Websterbridge. She is not a prepossessing woman. No woman looks her best after sitting up all night; and Mrs Dudgeon's face, even at its best, is grimly trenched by the channels into which the barren forms and observances of a dead Puritanism can pen a bitter temper and a fierce pride. She is an elderly matron who has worked hard and got nothing by it except dominion and detestation in her sordid home, and an unquestioned reputation for piety and respectability among her neighbors, to whom drink and debauchery are still so much more tempting than religion and rectitude, that they conceive goodness simply as self-denial. This conception is easily extended to others-denial, and finally generalized as covering anything disagreeable. So Mrs Dudgeon, being exceedingly disagreeable, is held to be exceedingly good. Short of flat felony, she enjoys complete license except for amiable weaknesses of any sort, and is consequently, without knowing it, the most licentious woman in the parish on the strength of never having broken the seventh commandment or missed a Sunday at the Presbyterian church.

The year 1777 is the one in which the passions roused by the breaking-off of the American colonies from England, more by their own weight than their own will, boiled up to shooting point, the shooting being idealized to the English mind as suppression of rebellion

and maintenance of British dominion, and to the American as defence of liberty, resistance to tyranny, and self-sacrifice on the altar of the Rights of Man. Into the merits of these idealizations it is not here necessary to inquire: suffice it to say, without prejudice, that they have convinced both Americans and English that the most highminded course for them to pursue is to kill as many of one another as possible, and that military operations to that end are in full swing, morally supported by confident requests from the clergy of both sides for the blessing of God on their arms.

Under such circumstances many other women besides this disagreeable Mrs Dudgeon find themselves sitting up all night waiting for news. Like her, too, they fall asleep towards morning at the risk of nodding themselves into the kitchen fire. Mrs Dudgeon sleeps with a shawl over her head, and her feet on a broad fender of iron laths, the step of the domestic altar of the fireplace, with its huge hobs and boiler, and its hinged arm above the smoky mantel-shelf for roasting. The plain kitchen table is opposite the fire, at her elbow, with a candle on it in a tin sconce. Her chair, like all the others in the room, is uncushioned and unpainted; but as it has a round railed back and a seat conventionally moulded to the sitter's curves, it is comparatively a chair of state. The room has three doors, one on the same side as the fireplace, near the corner, leading to the best bedroom; one, at the opposite end of the opposite wall, leading to the scullery and washhouse; and the house-door, with its latch, heavy lock, and clumsy wooden bar, in the front wall, between the window in its middle and the corner next the bedroom door. Between the door and the window a rack of pegs suggests to the deductive observer that the men of the house are all away, as there are no hats or coats on them. On the other side of the window the clock hangs on a nail, with its white wooden dial, black iron weights, and brass pendulum. Between the clock and the corner, a big cupboard, locked, stands on a dwarf dresser full of common crockery.

On the side opposite the fireplace, between the door and the corner, a shamelessly ugly black horsehair sofa stands against the wall. An inspection of its stridulous surface shews that Mrs Dudgeon is not alone. A girl of

sixteen or seventeen has fallen asleep on it. She is a wild, timid looking creature with black hair and tanned skin. Her frock, a scanty garment, is rent, weather-stained, berrystained, and by no means scrupulously clean. It hangs on her with a freedom which, taken with her brown legs and bare feet, suggests no great stock of underclothing.

Suddenly there comes a tapping at the door, not loud enough to wake the sleepers. Then knocking, which disturbs Mrs Dudgeon a little. Finally the latch is tried, whereupon she springs up at once.

MRS DUDGEON [*threateningly*] Well, why dont you open the door? [*She sees that the girl is asleep, and immediately raises a clamor of heartfelt vexation*]. Well, dear, dear me! Now this is – [*shaking her*] wake up, wake up: do you hear?

THE GIRL [*sitting up*] What is it?

MRS DUDGEON Wake up; and be ashamed of yourself, you unfeeling sinful girl, falling asleep like that, and your father hardly cold in his grave.

THE GIRL [*half asleep still*] I didnt mean to. I dropped off –

MRS DUDGEON [*cutting her short*] Oh yes, youve plenty of excuses, I daresay. Dropped off! [*Fiercely, as the knocking recommences*] Why dont you get up and let your uncle in? after me waiting up all night for him! [*She pushes her rudely off the sofa*]. There: I'll open the door: much good you are to wait up. Go and mend that fire a bit.

The girl, cowed and wretched, goes to the fire and puts a log on. Mrs Dudgeon unbars the door and opens it, letting into the stuffy kitchen a little of the freshness and a great deal of the chill of the dawn, also her second son Christy, a fattish, stupid, fair-haired, roundfaced man of about 22, muffled in a plaid shawl and grey overcoat. He hurries, shivering, to the fire, leaving Mrs Dudgeon to shut the door.

CHRISTY [*at the fire*] F – f – f! but it *is* cold. [*Seeing the girl, and staring lumpishly at her*] Why, who are you?

THE GIRL [*shyly*] Essie.

MRS DUDGEON Oh, you may well ask. [*To Essie*] Go to your room,

child, and lie down, since you havnt feeling enough to keep you awake. Your history isnt fit for your own ears to hear.

ESSIE I –

MRS DUDGEON [*peremptorily*] Dont answer me, Miss; but shew your obedience by doing what I tell you. [*Essie, almost in tears, crosses the room to the door near the sofa*]. And dont forget your prayers. [*Essie goes out*]. She'd have gone to bed last night just as if nothing had happened if I'd let her.

CHRISTY [*phlegmatically*] Well, she cant be expected to feel Uncle Peter's death like one of the family.

MRS DUDGEON What are you talking about, child? Isnt she his daughter – the punishment of his wickedness and shame? [*She assaults her chair by sitting down*].

CHRISTY [*staring*] Uncle Peter's daughter!

MRS DUDGEON Why else should she be here? D'ye think Ive not had enough trouble and care put upon me bringing up my own girls, let alone you and your good-for-nothing brother, without having your uncle's bastards –

CHRISTY [*interrupting her with an apprehensive glance at the door by which Essie went out*] Sh! She may hear you.

MRS DUDGEON [*raising her voice*] Let her hear me. People who fear God dont fear to give the devil's work its right name. [*Christy, soullessly indifferent to the strife of Good and Evil, stares at the fire, warming himself*]. Well, how long are you going to stare there like a stuck pig? What news have you for me?

CHRISTY [*taking off his hat and shawl and going to the rack to hang them up*] The minister is to break the news to you. He'll be here presently.

MRS DUDGEON Break what news?

CHRISTY [*standing on tiptoe, from boyish habit, to hang his hat up, though he is quite tall enough to reach the peg, and speaking with callous placidity, considering the nature of the announcement*] Father's dead too.

MRS DUDGEON [*stupent*] Your *father*!

CHRISTY [*sulkily, coming back to the fire and warming himself again, attending much more to the fire than to his mother*] Well, it's not my fault. When we got to

Nevinstown we found him ill in bed. He didnt know us at first. The minister sat up with him and sent me away. He died in the night.

MRS DUDGEON [*bursting into dry angry tears*] Well, I do think this is hard on me – very hard on me. His brother, that was a disgrace to us all his life, gets hanged on the public gallows as a rebel; and your father, instead of staying at home where his duty was, with his own family, goes after him and dies, leaving everything on my shoulders. After sending this girl to me to take care of, too! [*She plucks her shawl vexedly over her ears*]. It's sinful, so it is: downright sinful.

CHRISTY [*with a slow, bovine cheerfulness, after a pause*] I think it's going to be a fine morning, after all.

MRS DUDGEON [*railing at him*] A fine morning! And your father newly dead! Wheres your feelings, child?

CHRISTY [*obstinately*] Well, I didnt mean any harm. I suppose a man may make a remark about the weather even if his father's dead.

MRS DUDGEON [*bitterly*] A nice comfort my children are to me! One son a fool, and the other a lost sinner thats left his home to live with smugglers and gypsies and villains, the scum of the earth!

Someone knocks.

CHRISTY [*without moving*] That's the minister.

MRS DUDGEON [*sharply*] Well, arnt you going to let Mr Anderson in?

Christy goes sheepishly to the door. Mrs Dudgeon buries her face in her hands, as it is her duty as a widow to be overcome with grief. Christy opens the door, and admits the minister, Anthony Anderson, a shrewd, genial, ready Presbyterian divine of about 50, with something of the authority of his profession in his bearing. But it is an altogether secular authority, sweetened by a conciliatory, sensible manner not at all suggestive of a quite thoroughgoing other-worldliness. He is a strong, healthy man too, with a thick sanguine neck; and his keen, cheerful mouth cuts into somewhat fleshy corners. No doubt an excellent parson, but still a man capable of making the most of this world, and perhaps a little apologetically conscious of getting on better with it than a sound Presbyterian ought.

ANDERSON [*to Christy, at the door, looking at Mrs Dudgeon whilst he takes off his cloak*] Have you told her?

CHRISTY She made me. [*He shuts the door; yawns; and loafs across to the sofa, where he sits down and presently drops off to sleep*].

Anderson looks compassionately at Mrs Dudgeon. Then he hangs his cloak and hat on the rack. Mrs Dudgeon dries her eyes and looks up at him.

ANDERSON Sister: the Lord has laid his hand very heavily upon you.

MRS DUDGEON [*with intensely recalcitrant resignation*] It's His will, I suppose; and I must bow to it. But I do think it hard. What call had Timothy to go to Springtown, and remind everybody that he belonged to a man that was being hanged? – and [*spitefully*] that deserved it, if ever a man did.

ANDERSON [*gently*] They were brothers, Mrs Dudgeon.

MRS DUDGEON Timothy never acknowledged him as his brother after we were married: he had too much respect for me to insult me with such a brother. Would such a selfish wretch as Peter have come thirty miles to see Timothy hanged, do you think? Not thirty yards, not he. However, I must bear my cross as best I may: least said is soonest mended.

ANDERSON [*very grave, coming down to the fire to stand with his back to it*] Your eldest son was present at the execution, Mrs Dudgeon.

MRS DUDGEON [*disagreeably surprised*] Richard?

ANDERSON [*nodding*] Yes.

MRS DUDGEON [*vindictively*] Let it be a warning to him. He may end that way himself, the wicked, dissolute, godless – [*she suddenly stops; her voice fails; and she asks, with evident dread*] Did Timothy see him?

ANDERSON Yes.

MRS DUDGEON [*holding her breath*] Well?

ANDERSON He only saw him in the crowd: they did not speak. [*Mrs Dudgeon, greatly relieved, exhales the pent up breath and sits at her ease again*]. Your husband was greatly touched and impressed by his brother's awful death. [*Mrs Dudgeon sneers. Anderson breaks off*

to demand with some indignation] Well, wasnt it only natural, Mrs Dudgeon? He softened towards his prodigal son in that moment. He sent for him to come to see him.

MRS DUDGEON [*her alarm renewed*] Sent for Richard!

ANDERSON Yes; but Richard would not come. He sent his father a message; but I'm sorry to say it was a wicked message – an awful message.

MRS DUDGEON What was it?

ANDERSON That he would stand by his wicked uncle, and stand against his good parents, in this world and the next.

MRS DUDGEON [*implacably*] He will be punished for it. He will be punished for it – in both worlds.

ANDERSON That is not in our hands, Mrs Dudgeon.

MRS DUDGEON Did I say it was, Mr Anderson? We are told that the wicked shall be punished. Why should we do our duty and keep God's law if there is to be no difference made between us and those who follow their own likings and dislikings, and make a jest of us and of their Maker's word?

ANDERSON Well, Richard's earthly father has been merciful to him; and his heavenly judge is the father of us all.

MRS DUDGEON [*forgetting herself*] Richard's earthly father was a softheaded –

ANDERSON [*shocked*] Oh!

MRS DUDGEON [*with a touch of shame*] Well, I am Richard's mother. If I am against him who has any right to be for him? [*Trying to conciliate him*] Wont you sit down, Mr Anderson? I should have asked you before; but I'm so troubled.

ANDERSON Thank you. [*He takes a chair from beside the fireplace, and turns it so that he can sit comfortably at the fire. When he is seated he adds, in the tone of a man who knows that he is opening a difficult subject*] Has Christy told you about the new will?

MRS DUDGEON [*all her fears returning*] The new will! Did Timothy – ? [*She breaks off, gasping, unable to complete the question*].

ANDERSON Yes. In his last hours he changed his mind.

MRS DUDGEON [*white with intense rage*] And you let him rob me?

ANDERSON I had no power to prevent him giving what was his to his own son.

MRS DUDGEON He had nothing of his own. His money was the money I brought him as my marriage portion. It was for me to deal with my own money and my own son. He dare not have done it if I had been with him; and well he knew it. That was why he stole away like a thief to take advantage of the law to rob me by making a new will behind my back. The more shame on you, Mr Anderson, – you, a minister of the gospel – to act as his accomplice in such a crime.

ANDERSON [*rising*] I will take no offence at what you say in the first bitterness of your grief.

MRS DUDGEON [*contemptuously*] Grief!

ANDERSON Well, of your disappointment, if you can find it in your heart to think that the better word.

MRS DUDGEON My heart! *My* heart! And since when, pray, have *you* begun to hold up our hearts as trustworthy guides for us?

ANDERSON [*rather guiltily*] I – er –

MRS DUDGEON [*vehemently*] Dont lie, Mr Anderson. We are told that the heart of man is deceitful above all things, and desperately wicked. My heart belonged, not to Timothy, but to that poor wretched brother of his that has just ended his days with a rope round his neck – aye, to Peter Dudgeon. You know it: old Eli Hawkins, the man to whose pulpit you succeeded, though you are not worthy to loose his shoe latchet, told it you when he gave over our souls into your charge. He warned me and strengthened me against my heart, and made me marry a Godfearing man – as he thought. What else but that discipline has made me the woman I am? And you, you, who followed your heart in your marriage, you talk to me of what I find in my heart. Go home to your pretty wife, man; and leave me to my prayers. [*She turns from him and leans with her elbows on the table, brooding over her wrongs and taking no further notice of him*].

ANDERSON [*willing enough to escape*] The Lord forbid that I should come between you and the source of all comfort! [*He goes to the rack for his coat and hat*].

MRS DUDGEON [*without looking at him*] The Lord will know what to forbid and what to allow without your help.

ANDERSON And whom to forgive, I hope – Eli Hawkins and myself, if we have ever set up our preaching against His law. [*He fastens his cloak, and is now ready to go*]. Just one word – on necessary business, Mrs Dudgeon. There is the reading of the will to be gone through; and Richard has a right to be present. He is in the town; but he has the grace to say that he does not want to force himself in here.

MRS DUDGEON He *shall* come here. Does he expect us to leave his father's house for his convenience? Let them all come, and come quickly, and go quickly. They shall not make the will an excuse to shirk half their day's work. I shall be ready, never fear.

ANDERSON [*coming back a step or two*] Mrs Dudgeon: I used to have some little influence with you. When did I lose it?

MRS DUDGEON [*still without turning to him*] When you married for love. Now youre answered.

ANDERSON Yes: I am answered. [*He goes out, musing*].

MRS DUDGEON [*to herself, thinking of her husband*] Thief! Thief!! [*She shakes herself angrily out of her chair; throws back the shawl from her head; and sets to work to prepare the room for the reading of the will, beginning by replacing Anderson's chair against the wall, and pushing back her own to the window. Then she calls, in her hard, driving, wrathful way*] Christy. [*No answer: he is fast asleep*]. Christy. [*She shakes him roughly*]. Get up out of that; and be ashamed of yourself – sleeping, and your father dead! [*She returns to the table; puts the candle on the mantelshelf; and takes from the table drawer a red table cloth which she spreads*].

CHRISTY [*rising reluctantly*] Well, do you suppose we are never going to sleep until we are out of mourning?

MRS DUDGEON I want none of your sulks. Here: help me to set this table. [*They place the table in the middle of the room, with Christy's end towards the fireplace and Mrs Dudgeon's towards the sofa. Christy drops the table as soon as possible, and goes to the fire, leaving his mother to make the final adjustments of its position*]. We shall

have the minister back here with the lawyer and all the family to read the will before you have done toasting yourself. Go and wake that girl; and then light the stove in the shed: you cant have your breakfast here. And mind you wash yourself, and make yourself fit to receive the company. [*She punctuates these orders by going to the cupboard; unlocking it; and producing a decanter of wine, which has no doubt stood there untouched since the last state occasion in the family, and some glasses, which she sets on the table. Also two green ware plates, on one of which she puts a barnbrack with a knife beside it. On the other she shakes some biscuits out of a tin, putting back one or two, and counting the rest*]. Now mind: there are ten biscuits there: let there be ten there when I come back after dressing myself. And keep your fingers off the raisins in that cake. And tell Essie the same. I suppose I can trust you to bring in the case of stuffed birds without breaking the glass? [*She replaces the tin in the cupboard, which she locks, pocketing the key carefully*].

CHRISTY [*lingering at the fire*] Youd better put the inkstand instead, for the lawyer.

RS DUDGEON Thats no answer to make to me, sir. Go and do as youre told. [*Christy turns sullenly to obey*]. Stop: take down that shutter before you go, and let the daylight in: you cant expect me to do all the heavy work of the house with a great heavy lout like you idling about.

Christy takes the window bar out of its clamps, and puts it aside; then opens the shutter, shewing the grey morning. Mrs Dudgeon takes the sconce from the mantelshelf; blows out the candle; extinguishes the snuff by pinching it with her fingers, first licking them for the purpose; and replaces the sconce on the shelf.

CHRISTY [*looking through the window*] Here's the minister's wife.

MRS DUDGEON [*displeased*] What! Is she coming here?

CHRISTY Yes.

MRS DUDGEON What does she want troubling me at this hour, before I'm properly dressed to receive people?

CHRISTY Youd better ask her.

MRS DUDGEON [*threateningly*] *Youd* better keep a civil tongue in your head. [*He goes sulkily towards the door. She comes after him, plying him with instructions*]. Tell that girl to come to me as soon as she's had her breakfast. And tell her to make herself fit to be seen before the people. [*Christy goes out and slams the door in her face*]. Nice manners, that! [*Someone knocks at the house door: she turns and cries inhospitably*] Come in. [*Judith Anderson, the minister's wife, comes in. Judith is more than twenty years younger than her husband, though she will never be as young as he in vitality. She is pretty and proper and ladylike, and has been admired and petted into an opinion of herself sufficiently favorable to give her a self-assurance which serves her instead of strength. She has a pretty taste in dress, and in her face the pretty lines of a sentimental character formed by dreams. Even her little self-complacency is pretty, like a child's vanity. Rather a pathetic creature to any sympathetic observer who knows how rough a place the world is. One feels, on the whole, that Anderson might have chosen worse, and that she, needing protection, could not have chosen better*]. Oh, it's you, is it, Mrs Anderson?

JUDITH [*very politely – almost patronizingly*] Yes. Can I do anything for you, Mrs Dudgeon? Can I help to get the place ready before they come to read the will?

MRS DUDGEON [*stiffly*] Thank you, Mrs Anderson, my house is always ready for anyone to come into.

JUDITH [*with complacent amiability*] Yes, indeed it is. Perhaps you had rather I did not intrude on you just now.

MRS DUDGEON Oh, one more or less will make no difference this morning, Mrs Anderson. Now that youre here, youd better stay. If you wouldnt mind shutting the door! [*Judith smiles, implying "How stupid of me!" and shuts it with an exasperating air of doing something pretty and becoming*]. Thats better. I must go and tidy myself a bit. I suppose you dont mind stopping here to receive anyone that comes until I'm ready.

JUDITH [*graciously giving her leave*] Oh yes, certainly. Leave them to me, Mrs Dudgeon; and take your time. [*She hangs her cloak and bonnet on the rack*].

MRS DUDGEON [*half sneering*] I thought that would be more in your way than getting the house ready. [*Essie comes back*]. Oh, here *you* are! [*Severely*] Come here: let me see you. [*Essie timidly goes to her. Mrs Dudgeon takes her roughly by the arm and pulls her round to inspect the results of her attempt to clean and tidy herself – results which shew little practice and less conviction*]. Mm! Thats what you call doing your hair properly, I suppose. It's easy to see what you are, and how you were brought up. [*She throws her arm away, and goes on, peremptorily*] Now you listen to me and do as youre told. You sit down there in the corner by the fire; and when the company comes dont dare to speak until youre spoken to. [*Essie creeps away to the fireplace*]. Your father's people had better see you and know youre there: theyre as much bound to keep you from starvation as I am. At any rate they might help. But let me have no chattering and making free with them, as if you were their equal. Do you hear?

ESSIE Yes.

MRS DUDGEON Well, then go and do as youre told. [*Essie sits down miserably on the corner of the fender furthest from the door*]. Never mind her, Mrs Anderson: you know who she is and what she is. If she gives you any trouble, just tell me; and I'll settle accounts with her. [*Mrs Dudgeon goes into the bedroom, shutting the door sharply behind her as if even it had to be made to do its duty with a ruthless hand*].

JUDITH [*patronizing Essie, and arranging the cake and wine*

on the table more becomingly] You must not mind if your aunt is strict with you. She is a very good woman, and desires your good too.

ESSIE [*in listless misery*] Yes.

JUDITH [*annoyed with Essie for her failure to be consoled and edified, and to appreciate the kindly condescension of the remark*] You are not going to be sullen, I hope, Essie.

ESSIE No.

JUDITH Thats a good girl! [*She places a couple of chairs at the table with their backs to the window, with a pleasant sense of being a more thoughtful housekeeper than Mrs Dudgeon*]. Do you know any of your father's relatives?

ESSIE No. They wouldnt have anything to do with him: they were too religious. Father used to talk about Dick Dudgeon; but I never saw him.

JUDITH [*ostentatiously shocked*] Dick Dudgeon! Essie: do you wish to be a really respectable and grateful girl, and to make a place for yourself here by steady good conduct?

ESSIE [*very half-heartedly*] Yes.

JUDITH Then you must never mention the name of Richard Dudgeon – never even think about him. He is a bad man.

ESSIE What has he done?

JUDITH You must not ask questions about him, Essie. You are too young to know what it is to be a bad man. But he is a smuggler; and he lives with gypsies; and he has no love for his mother and his family; and he

wrestles and plays games on Sunday instead of going to church. Never let him into your presence, if you can help it, Essie; and try to keep yourself and all womanhood unspotted by contact with such men.

ESSIE Yes.

JUDITH [*again displeased*] I am afraid you say Yes and No without thinking very deeply.

ESSIE Yes. At least I mean –

JUDITH [*severely*] What do you mean?

ESSIE [*almost crying*] Only – my father was a smuggler; and – [*Someone knocks*].

JUDITH They are beginning to come. Now remember your aunt's directions, Essie; and be a good girl. [*Christy comes back with the stand of stuffed birds under a glass case, and an inkstand, which he places on the table*]. Good morning, Mr Dudgeon. Will you open the door, please: the people have come.

CHRISTY Good morning. [*He opens the house door*].

The morning is now fairly bright and warm; and Anderson, who is the first to enter, has left his cloak at home. He is accompanied by Lawyer Hawkins, a brisk, middleaged man in brown riding gaiters and yellow breeches, looking as much squire as solicitor. He and Anderson are allowed precedence as representing the learned professions. After them comes the family, headed by the senior uncle, William Dudgeon, a large, shapeless man, bottle-nosed and evidently no ascetic at table. His clothes are not the clothes, nor his anxious wife the wife, of a prosperous man. The junior uncle, Titus Dudgeon, is a wiry little terrier of a man, with an immense and visibly purseproud wife, both free from the cares of the William household.

Hawkins at once goes briskly to the table and takes the chair nearest the sofa, Christy having left the inkstand there. He puts his hat on the floor beside him, and produces the will. Uncle William comes to the fire and stands on the hearth warming his coat tails, leaving Mrs William derelict near the door. Uncle Titus, who is the lady's man of the family, rescues her by giving her his disengaged arm and bringing her to the sofa, where he sits down warmly between his own lady and his brother's.

Anderson hangs up his hat and waits for a word with Judith.

JUDITH She will be here in a moment. Ask them to wait. [*She taps at the bedroom door. Receiving an answer from within, she opens it and passes through*].

ANDERSON [*taking his place at the table at the opposite end to Hawkins*] Our poor afflicted sister will be with us in a moment. Are we all here?

CHRISTY [*at the house door, which he has just shut*] All except Dick.

The callousness with which Christy names the reprobate jars on the moral sense of the family. Uncle William shakes his head slowly and repeatedly. Mrs Titus catches her breath convulsively through her nose. Her husband speaks.

UNCLE TITUS Well, I hope he will have the grace not to come. I *hope* so.

The Dudgeons all murmur assent, except Christy, who goes to the window and posts himself there, looking out. Hawkins smiles secretively as if he knew something that would change their tune if they knew it. Anderson is uneasy: the love of solemn family councils, especially funereal ones, is not in his nature. Judith appears at the bedroom door.

JUDITH [*with gentle impressiveness*] Friends, Mrs Dudgeon. [*She takes the chair from beside the fireplace; and places it for Mrs Dudgeon, who comes from the bedroom in black, with a clean handkerchief to her eyes. All rise, except Essie. Mrs Titus and Mrs William produce equally clean handkerchiefs and weep. It is an affecting moment*].

UNCLE WILLIAM Would it comfort you, sister, if we were to offer up a prayer?

UNCLE TITUS Or sing a hymn?

ANDERSON [*rather hastily*] I have been with our sister this morning already, friends. In our hearts we ask a blessing.

ALL [*except Essie*] Amen.

They all sit down, except Judith, who stands behind Mrs Dudgeon's chair.

JUDITH [*to Essie*] Essie: did you say Amen?

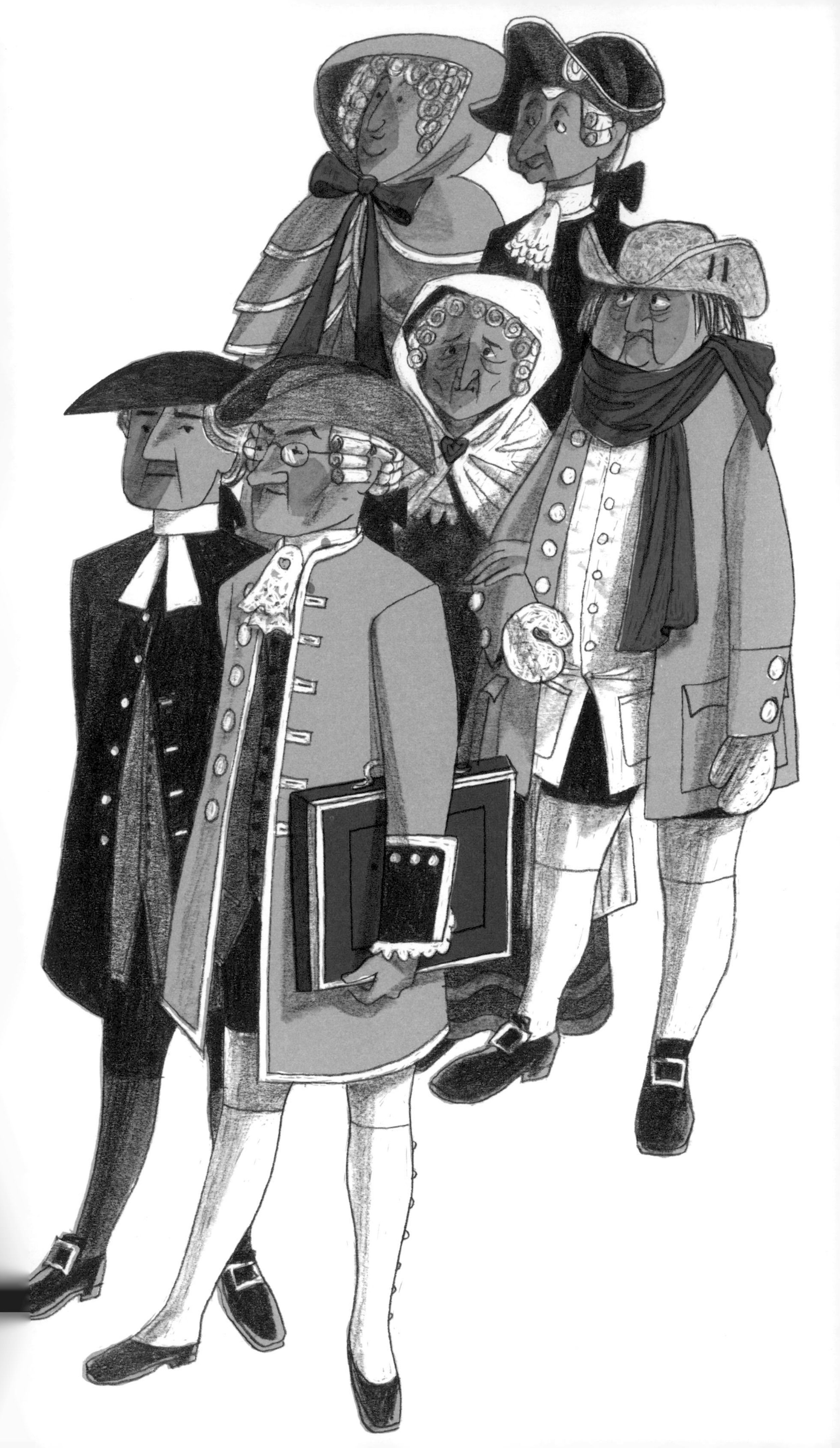

ESSIE [*scaredly*] No.

JUDITH Then say it, like a good girl.

ESSIE Amen.

UNCLE WILLIAM [*encouragingly*] Thats right: thats right. We know who you are; but we are willing to be kind to you if you are a good girl and deserve it. We are all equal before the Throne.

This republican sentiment does not please the women, who are convinced that the Throne is precisely the place where their superiority, often questioned in this world, will be recognized and rewarded.

CHRISTY [*at the window*] Here's Dick.

Anderson and Hawkins look round sociably. Essie, with a gleam of interest breaking through her misery, looks up. Christy grins and gapes expectantly at the door. The rest are petrified with the intensity of their sense of Virtue menaced with outrage by the approach of flaunting Vice. The reprobate appears in the doorway, graced beyond his alleged merits by the morning sunlight. He is certainly the best looking member of the family; but his expression is reckless and sardonic, his manner

defiant and satirical, his dress picturesquely careless. Only, his forehead and mouth betray an extraordinary steadfastness; and his eyes are the eyes of a fanatic.

RICHARD [*on the threshold, taking off his hat*] Ladies and gentlemen: your servant, your very humble servant. [*With this comprehensive insult, he throws his hat to Christy with a suddenness that makes him jump like a negligent wicket keeper, and comes into the middle of the room, where he turns and deliberately surveys the company*]. How happy you all look! how glad to see me! [*He turns towards Mrs Dudgeon's chair; and his lip rolls up horribly from his dog tooth as he meets her look of undisguised hatred*]. Well, mother: keeping up appearances as usual? thats right, thats right. [*Judith pointedly moves away from his neighborhood to the other side of the kitchen, holding her skirt instinctively as if to save it from contamination. Uncle Titus promptly marks his approval of her action by rising from the sofa, and placing a chair for her to sit down upon*]. What! Uncle William! I havnt seen you since you gave up drinking. [*Poor Uncle William, shamed, would protest; but Richard claps him heartily on his shoulder, adding*] you have given it up, havnt you? [*releasing him with a playful push*] of course you have: quite right too: you overdid it. [*He turns away from Uncle William and makes for the sofa*]. And now, where is that upright horsedealer Uncle Titus? Uncle Titus: come forth. [*He comes upon him holding the chair as Judith sits down*]. As usual, looking after the ladies!

UNCLE TITUS [*indignantly*] Be ashamed of yourself, sir –

RICHARD [*interrupting him and shaking his hand in spite of him*] I am: I am; but I am proud of my uncle – proud of all my relatives – [*again surveying them*] who could look at them and not be proud and joyful? [*Uncle Titus, overborne, resumes his seat on the sofa. Richard turns to the table*]. Ah, Mr Anderson, still at the good work, still shepherding them. Keep them up to the mark, minister, keep them up to the mark. Come! [*with a spring he seats himself on the table and takes up the decanter*] clink a glass with me, Pastor, for the sake of old times.

ANDERSON You know, I think, Mr Dudgeon, that I do not drink before dinner.

RICHARD You will, some day, Pastor: Uncle William used to drink before breakfast. Come: it will give your sermons unction. [*He smells the wine and makes a wry face*]. But do not begin on my mother's company sherry. I stole some when I was six years old; and I have been a temperate man ever since. [*He puts the decanter down and changes the subject*]. So I hear you are married, Pastor, and that your wife has a most ungodly allowance of good looks.

ANDERSON [*quietly indicating Judith*] Sir: you are in the presence of my wife. [*Judith rises and stands with stony propriety*].

RICHARD [*quickly slipping down from the table with instinctive good manners*] Your servant, madam: no offence. [*He looks at her earnestly*]. You deserve your reputation; but I'm sorry to see by your expression that youre a good woman. [*She looks shocked, and sits down amid a murmur of indignant sympathy from his relatives. Anderson, sensible enough to know that these demonstrations can only gratify and encourage a man who is deliberately trying to provoke them, remains perfectly goodhumored*]. All the same, Pastor, I respect you more than I did before. By the way, did I hear, or did I not, that our late lamented Uncle Peter, though unmarried, was a father?

UNCLE TITUS He had only one irregular child, sir.

RICHARD *Only* one! He thinks one a mere trifle! I blush for you, Uncle Titus.

ANDERSON Mr Dudgeon: you are in the presence of your mother and her grief.

RICHARD It touches me profoundly, Pastor. By the way, what has become of the irregular child?

ANDERSON [*pointing to Essie*] There, sir, listening to you.

RICHARD [*shocked into sincerity*] What! Why the devil didnt you tell me that before? Children suffer enough in this house without – [*He hurries remorsefully to Essie*]. Come, little cousin! never mind me: it was not meant to hurt you. [*She looks up gratefully at him. Her tear-stained face affects him violently; and he bursts out, in a*

transport of wrath] Who has been making her cry? Who has been ill-treating her? By God –

MRS DUDGEON [*rising and confronting him*] Silence your blasphemous tongue. I will bear no more of this. Leave my house.

RICHARD How do you know it's your house until the will is read? [*They look at one another for a moment with intense hatred; and then she sinks, checkmated, into her chair. Richard goes boldly up past Anderson to the window, where he takes the railed chair in his hand*]. Ladies and gentlemen: as the eldest son of my late father, and the unworthy head of this household, I bid you welcome. By your leave, Minister Anderson: by your leave, Lawyer Hawkins. The head of the table for the head of the family. [*He places the chair at the table between the minister and the attorney; sits down between them; and addresses the assembly with a presidential air*]. We meet on a melancholy occasion: a father dead! an uncle actually hanged, and probably damned. [*He shakes his head deploringly. The relatives freeze with horror*]. *Thats* right: pull your longest faces [*his voice suddenly sweetens gravely as his glance lights on Essie*] provided only there is hope in the eyes of the child. [*Briskly*] Now then, Lawyer Hawkins: business, business. Get on with the will, man.

TITUS Do not let yourself be ordered or hurried, Mr Hawkins.

HAWKINS [*very politely and willingly*] Mr Dudgeon means no offence, I feel sure. I will not keep you one second, Mr Dudgeon. Just while I get my glasses – [*he fumbles for them. The Dudgeons look at one another with misgiving*].

RICHARD Aha! They notice your civility, Mr Hawkins. They are prepared for the worst. A glass of wine to clear your voice before you begin. [*He pours out one for him and hands it; then pours one for himself*].

HAWKINS Thank you, Mr Dudgeon. Your good health, sir.

RICHARD Yours, sir. [*With the glass half way to his lips, he checks himself, giving a dubious glance at the wine, and adds, with quaint intensity*] Will anyone oblige me with a glass of water?

Essie, who has been hanging on his every word and move-

ment, rises stealthily and slips out behind Mrs Dudgeon through the bedroom door, returning presently with a jug and going out of the house as quietly as possible.

HAWKINS The will is not exactly in proper legal phraseology.

RICHARD No: my father died without the consolations of the law.

HAWKINS Good again, Mr Dudgeon, good again. [*Preparing to read*] Are you ready, sir?

RICHARD Ready, aye ready. For what we are about to receive, may the Lord make us truly thankful. Go ahead.

HAWKINS [*reading*] "This is the last will and testament of me Timothy Dudgeon on my deathbed at Nevinstown on the road from Springtown to Websterbridge on this twenty-fourth day of September, one thousand seven hundred and seventy seven. I hereby revoke all former wills made by me and declare that I am of sound mind and know well what I am doing and that

this is my real will according to my own wish and affections."

RICHARD [*glancing at his mother*] Aha!

HAWKINS [*shaking his head*] Bad phraseology, sir, wrong phraseology. "I give and bequeath a hundred pounds to my younger son Christopher Dudgeon, fifty pounds to be paid to him on the day of his marriage to Sarah Wilkins if she will have him, and ten pounds on the birth of each of his children up to the number of five."

RICHARD How if she wont have him?

CHRISTY She will if I have fifty pounds.

RICHARD Good, my brother. Proceed.

HAWKINS "I give and bequeath to my wife Annie Dudgeon, born Annie Primrose" – you see he did not know the law, Mr Dudgeon: your mother was not born Annie: she was christened so – "an annuity of fifty-two pounds a year for life [*Mrs Dudgeon, with all eyes on her, holds herself convulsively rigid*] to be paid out of the interest on her own money" – *there's* a way to put it, Mr Dudgeon! Her own money!

MRS DUDGEON A very good way to put God's truth. It was every penny my own. Fifty-two pounds a year!

HAWKINS "And I recommend her for her goodness and piety to the forgiving care of her children, having stood between them and her as far as I could to the best of my ability."

MRS DUDGEON And this is my reward! [*Raging inwardly*] You know what I think, Mr Anderson: you know the word I gave to it.

ANDERSON It cannot be helped, Mrs Dudgeon. We must take what comes to us. [*To Hawkins*] Go on, sir.

HAWKINS "I give and bequeath my house at Websterbridge with the land belonging to it and all the rest of my property soever to my eldest son and heir, Richard Dudgeon."

RICHARD Oho! The fatted calf, Minister, the fatted calf.

HAWKINS "On these conditions – "

RICHARD The devil! Are there conditions?

HAWKINS "To wit: first, that he shall not let my brother Peter's natural child starve or be driven by want to an evil life."

RICHARD [*emphatically, striking his fist on the table*] Agreed.

Mrs Dudgeon, turning to look malignantly at Essie, misses her and looks quickly round to see where she has moved to; then, seeing that she has left the room without leave, closes her lips vengefully.

HAWKINS "Second, that he shall be a good friend to my old horse Jim" – [*again shaking his head*] he should have written James, sir.

RICHARD James shall live in clover. Go on.

HAWKINS – "and keep my deaf farm labourer Prodger Feston in his service."

RICHARD Prodger Feston shall get drunk every Saturday.

HAWKINS "Third, that he make Christy a present on his marriage out of the ornaments in the best room."

RICHARD [*holding up the stuffed birds*] Here you are, Christy.

CHRISTY [*disappointed*] I'd rather have the china peacocks.

RICHARD You shall have both. [*Christy is greatly pleased*]. Go on.

HAWKINS "Fourthly and lastly, that he try to live at peace with his mother as far as she will consent to it."

RICHARD [*dubiously*] Hm! Anything more, Mr Hawkins?

HAWKINS [*solemnly*] "Finally I give and bequeath my soul into my Maker's hands, humbly asking forgiveness for all my sins and mistakes, and hoping that He will so guide my son that it may not be said that I have done wrong in trusting to him rather than to others in the perplexity of my last hour in this strange place."

ANDERSON Amen.

THE UNCLES AND AUNTS Amen.

RICHARD My mother does not say Amen.

MRS DUDGEON [*rising, unable to give up her property without a struggle*] Mr Hawkins: is that a proper will? Remember, I have his rightful, legal will, drawn up by yourself, leaving all to me.

HAWKINS This is a very wrongly and irregularly worded will, Mrs Dudgeon; though [*turning politely to Richard*] it contains in my judgment an excellent disposal of his property.

ANDERSON [*interposing before Mrs Dudgeon can retort*] That is not what you are asked, Mr Hawkins. Is it a legal will?

HAWKINS The courts will sustain it against the other.

ANDERSON But why, if the other is more lawfully worded?

HAWKINS Because, sir, the courts will sustain the claim of a man – and that man the eldest son – against any woman, if they can. I warned you, Mrs Dudgeon, when you got me to draw that other will, that it was not a wise will, and that though you might make him sign it, he would never be easy until he revoked it. But you wouldnt take advice; and now Mr Richard is cock of the walk. [*He takes his hat from the floor; rises; and begins pocketing his papers and spectacles*].

This is the signal for the breaking-up of the party. Anderson takes his hat from the rack and joins Uncle William at the fire. Titus fetches Judith her things from the rack. The three on the sofa rise and chat with Hawkins. Mrs Dudgeon, now an intruder in her own house, stands inert, crushed by the weight of the law on women, accepting it, as she has been trained to accept all monstrous calamities, as proofs of the greatness of the power that inflicts them, and of her own wormlike insignificance. For at this time, remember, Mary Wollstonecraft is as yet only a girl of eighteen, and her Vindication of the Rights of Women is still fourteen years off. Mrs Dudgeon is rescued from her apathy by Essie, who comes back with the jug full of water. She is taking it to Richard when Mrs Dudgeon stops her.

MRS DUDGEON [*threatening her*] Where have you been? [*Essie, appalled, tries to answer, but cannot*]. How dare you go out by yourself after the orders I gave you?

ESSIE He asked for a drink – [*she stops, her tongue cleaving to her palate with terror*].

JUDITH [*with gentler severity*] *Who* asked for a drink? [*Essie, speechless, points to Richard*].

RICHARD What! I!

JUDITH [*shocked*] Oh Essie, Essie!

RICHARD I believe I did. [*He takes a glass and holds it to Essie to be filled. Her hand shakes*]. What! afraid of me?

ESSIE [*quickly*] No. I – [*She pours out the water*].

RICHARD [*tasting it*] Ah, youve been up the street to the market gate spring to get that. [*He takes a draught*]. Delicious! Thank you. [*Unfortunately, at this moment he chances to catch sight of Judith's face, which expresses the most prudish disapproval of his evident attraction for Essie, who is devouring him with her grateful eyes. His mocking expression returns instantly. He puts down the glass; deliberately winds his arm round Essie's shoulders; and brings her into the middle of the company. Mrs Dudgeon being in Essie's way as they come past the table, he says*] By your leave, mother [*and compels her to make way for them*]. What do they call you? Bessie?

ESSIE Essie.

RICHARD Essie, to be sure. Are you a good girl, Essie?

ESSIE [*greatly disappointed that he, of all people, should begin at her in this way*] Yes. [*She looks doubtfully at Judith*]. I think so. I mean I – I hope so.

RICHARD Essie: did you ever hear of a person called the devil?

ANDERSON [*revolted*] Shame on you, sir, with a mere child –

RICHARD By your leave, Minister: I do not interfere with your sermons: do not you interrupt mine. [*To Essie*] Do you know what they call me, Essie?

ESSIE Dick.

RICHARD [*amused: patting her on the shoulder*] Yes, Dick; but something else too. They call me the Devil's Disciple.

ESSIE Why do you let them?

RICHARD [*seriously*] Because it's true. I was brought up in the other service; but I knew from the first that the Devil was my natural master and captain and friend. I saw that he was in the right, and that the world cringed to his conqueror only through fear. I prayed secretly to him; and he comforted me, and saved me from having my spirit broken in this house of children's tears. I promised him my soul, and swore an oath that I would stand up for him in this world and stand by him in the next. [*Solemnly*] That promise and that oath made a man of me. From this day this house is

his home; and no child shall cry in it: this hearth is his altar; and no soul shall ever cower over it in the dark evenings and be afraid. Now [*turning forcibly on the rest*] which of you good men will take this child and rescue her from the house of the devil?

JUDITH [*coming to Essie and throwing a protecting arm about her*] I will. You should be burnt alive.

ESSIE But I dont want to. [*She shrinks back, leaving Richard and Judith face to face*].

RICHARD [*to Judith*] Actually doesnt want to, most virtuous lady!

UNCLE TITUS Have a care, Richard Dudgeon. The law –

RICHARD [*turning threateningly on him*] Have a care, you. In an hour from this there will be no law here but martial law. I passed the soldiers within six miles on my way here: before noon Major Swindon's gallows for rebels will be up in the market place.

ANDERSON [*calmly*] What have we to fear from that, sir?

RICHARD More than you think. He hanged the wrong man at Springtown: he thought Uncle Peter was respectable, because the Dudgeons had a good name. But his next example will be the best man in the town to whom he can bring home a rebellious word. Well, we're all rebels; and you know it.

ALL THE MEN [*except Anderson*] No, no, no!

RICHARD Yes, you are. You havnt damned King George up hill and down dale as I have; but youve prayed for his defeat; and you, Anthony Anderson, have conducted the service, and sold your family bible to buy a pair of pistols. They maynt hang me, perhaps; because the moral effect of the Devil's Disciple dancing on nothing wouldnt help them. But a minister! [*Judith, dismayed, clings to Anderson*] or a lawyer! [*Hawkins smiles like a man able to take care of himself*] or an upright horsedealer! [*Uncle Titus snarls at him in rage and terror*] or a reformed drunkard! [*Uncle William, utterly unnerved, moans and wobbles with fear*] eh? Would that shew that King George meant business – ha?

ANDERSON [*perfectly self-possessed*] Come, my dear: he is only

trying to frighten you. There is no danger. [*He takes her out of the house. The rest crowd to the door to follow him, except Essie, who remains near Richard*].

RICHARD [*boisterously derisive*] Now then: how many of you will stay with me; run up the American flag on the devil's house; and make a fight for freedom? [*They scramble out, Christy among them, hustling one another in their haste*]. Ha ha! Long live the devil! [*To Mrs Dudgeon, who is following them*] What, mother! Are you off too?

MRS DUDGEON [*deadly pale, with her hand on her heart as if she had received a deathblow*] My curse on you! My dying curse! [*She goes out*].

RICHARD [*calling after her*] It will bring me luck. Ha ha ha!

ESSIE [*anxiously*] Maynt I stay?

RICHARD [*turning to her*] What! Have they forgotten to save your soul in their anxiety about their own bodies? Oh yes: you may stay. [*He turns excitedly away again and shakes his fist after them. His left fist, also clenched, hangs down. Essie seizes it and kisses it, her tears falling on it. He starts and looks at it*]. Tears! The devil's baptism! [*She falls on her knees, sobbing. He stoops goodnaturedly to raise her, saying*] Oh yes, you may cry that way, Essie, if you like.

Minister Anderson's house is in the main street of Websterbridge, not far from the town hall. To the eye of the eighteenth century New Englander, it is much grander than the plain farmhouse of the Dudgeons; but it is so plain itself that a modern house agent would let both at about the same rent. The chief dwelling room has the same sort of kitchen fireplace, with boiler, toaster hanging on the bars, movable iron griddle socketed to the hob, hook above for roasting, and broad fender, on which stand a kettle and a plate of buttered toast. The door, between the fireplace and the corner, has neither panels, fingerplates nor handles: it is made of plain boards, and fastens with a latch. The table is a kitchen table, with a treacle colored cover of American cloth, chapped at the corners by draping. The tea service on it consists of two thick cups and saucers of the plainest ware, with milk jug and bowl to match, each large enough to contain nearly a quart, on a black japanned tray, and, in the middle of the table, a wooden trencher with a big loaf upon it, and a square half

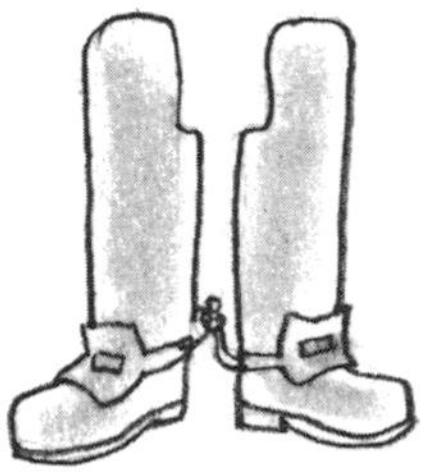

pound block of butter in a crock. The big oak press facing the fire from the opposite side of the room, is for use and storage, not for ornament; and the minister's house coat hangs on a peg from its door, shewing that he is out; for when he is in, it is his best coat that hangs there. His big riding boots stand beside the press, evidently in their usual place, and rather proud of themselves. In fact, the evolution of the minister's kitchen, dining room and drawing room into three separate apartments has not yet taken place; and so, from the point of view of our pampered period, he is no better off than the Dudgeons.

But there is a difference, for all that. To begin with, Mrs Anderson is a pleasanter person to live with than Mrs Dudgeon. To which Mrs Dudgeon would at once reply, with reason, that Mrs Anderson has no children to look after; no poultry, pigs nor cattle; a steady and sufficient income not directly dependent on harvests and prices at fairs; an affectionate husband who is a tower of strength to her: in short, that life is as easy at the minister's house as it is hard at the farm. This is true; but to explain a fact is not to alter it; and however little credit Mrs Anderson may deserve for making her home happier, she has certainly succeeded in doing it. The outward and visible signs of her superior social pretensions are, a drugget on the floor, a plaster ceiling between the timbers, and chairs which, though not upholstered, are stained and polished. The fine arts are represented by a mezzotint portrait of some Presbyterian divine, a copperplate of Raphael's St Paul preaching at Athens, a rococo presentation clock on the mantelshelf, flanked by a couple of miniatures, a pair of crockery dogs with baskets in their mouths, and, at the corners, two large cowrie shells. A pretty feature of the room is the low wide latticed window, nearly its whole width, with little red curtains running on a rod half way up it to serve as a blind. There is no sofa; but one of the seats, standing near the press, has a railed back and is long enough to accommodate two people easily. On the whole, it is rather the sort of room that the nineteenth century has ended in struggling to get back to under the leadership of Mr Philip Webb and his disciples in domestic architecture, though no genteel clergyman would have tolerated it fifty years ago.

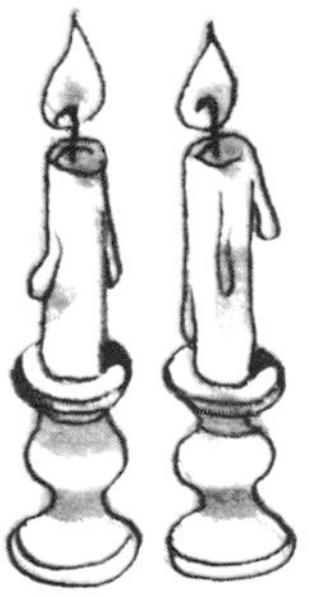

The evening has closed in; and the room is dark except for the cosy firelight and the dim oil lamps seen through the window in the wet street, where there is a quiet, steady, warm, windless downpour of rain. As the town clock strikes the quarter, Judith comes in with a couple of candles in earthenware candlesticks, and sets them on the table. Her self-conscious airs of the morning are gone; she is anxious and frightened. She goes to the window and peers into the street. The first thing she sees there is her husband, hurrying home through the rain. She gives a little gasp of relief, not very far removed from a sob, and turns to the door. Anderson comes in, wrapped in a very wet cloak.

JUDITH [*running to him*] Oh, here you are at last, at last! [*She attempts to embrace him*].

ANDERSON [*keeping her off*] Take care, my love: I'm wet. Wait till I get my cloak off. [*He places a chair with its back to the fire; hangs his cloak on it to dry; shakes the rain from his hat and puts it on the fender; and at last turns with his hands outstretched to Judith*]. Now! [*She flies into his arms*]. I am not late, am I? The town clock struck the quarter as I came in at the front door. And the town clock is always fast.

JUDITH I'm sure it's slow this evening. I'm so glad youre back.

ANDERSON [*taking her more closely in his arms*] *Anxious*, my dear?

JUDITH A little.

ANDERSON Why, youve been crying.

JUDITH Only a little. Never mind: it's all over now. [*A bugle call is heard in the distance. She starts in terror and retreats to the long seat, listening*]. Whats that?

ANDERSON [*following her tenderly to the seat and making her sit down with him*] Only King George, my dear. He's returning to barracks, or having his roll called, or getting ready for tea, or booting or saddling or something. Soldiers dont ring the bell or call over the banisters when they want anything: they send a boy out with a bugle to disturb the whole town.

JUDITH Do you think there is really any danger?

ANDERSON Not the least in the world.

JUDITH You say that to comfort me, not because you believe it.

ANDERSON My dear: in this world there is always danger for those who are afraid of it. There's a danger that the house will catch fire in the night; but we shant sleep any the less soundly for that.

JUDITH Yes, I know what you always say; and youre quite right. Oh, quite right: I know it. But – I suppose I'm not brave: thats all. My heart shrinks every time I think of the soldiers.

ANDERSON Never mind that, dear: bravery is none the worse for costing a little pain.

JUDITH Yes, I suppose so. [*Embracing him again*] Oh how brave you are, my dear! [*With tears in her eyes*] Well, I'll be brave too: you shant be ashamed of your wife.

ANDERSON Thats right. Now you make me happy. Well. well! [*He rises and goes cheerily to the fire to dry his shoes*]. I called on Richard Dudgeon on my way back; but he wasnt in.

JUDITH [*rising in consternation*] You called on that man!

ANDERSON [*reassuring her*] Oh, nothing happened, dearie. He was out.

JUDITH [*almost in tears, as if the visit were a personal humiliation to her*] But why did you go there?

ANDERSON [*gravely*] Well, it is all the talk that Major Swindon is going to do what he did in Springtown – make an example of some notorious rebel, as he calls us. He pounced on Peter Dudgeon as the worst character there; and it is the general belief that he will pounce on Richard as the worst here.

JUDITH But Richard said –

ANDERSON [*goodhumoredly cutting her short*] Pooh! Richard said! He said what he thought would frighten you and frighten me, my dear. He said what perhaps (God forgive him!) he would like to believe. It's a terrible thing to think of what death must mean for a man like that. I felt that I must warn him. I left a message for him.

JUDITH [*querulously*] What message?

ANDERSON Only that I should be glad to see him for a moment

on a matter of importance to himself, and that if he would look in here when he was passing he would be welcome.

JUDITH [*aghast*] You asked that man to come here!

ANDERSON I did.

JUDITH [*sinking on the seat and clasping her hands*] I hope he wont come! Oh, I pray that he may not come!

ANDERSON Why? Dont you want him to be warned?

JUDITH He must know his danger. Oh, Tony, is it wrong to hate a blasphemer and a villain? I do hate him. I cant get him out of my mind: I know he will bring harm with him. He insulted you: he insulted me: he insulted his mother.

ANDERSON [*quaintly*] Well, dear, let's forgive him; and then it wont matter.

JUDITH Oh, I know it's wrong to hate anybody; but –

ANDERSON [*going over to her with humorous tenderness*] Come, dear, youre not so wicked as you think. The worst sin towards our fellow creatures is not to hate them, but to be indifferent to them: thats the essence of inhumanity. After all, my dear, if you watch people carefully, youll be surprised to find how like hate is to love. [*She starts, strangely touched – even appalled. He is amused at her*]. Yes: I'm quite in earnest. Think of how some of our married friends worry one another, tax one another, are jealous of one another, cant bear to let one another out of sight for a day, are more like jailers and slave-owners than lovers. Think of those very same people with their enemies, scrupulous, lofty, self-respecting, determined to be independent of one another, careful of how they speak of one another – pooh! havent you often thought that if they only knew it, they were better friends to their enemies than to their own husbands and wives? Come: depend on it, my dear, you are really fonder of Richard than you are of me, if you only knew it. Eh!

JUDITH Oh, dont say that: dont say that, Tony, even in jest. You dont know what a horrible feeling it gives me.

ANDERSON [*laughing*] Well, well: never mind, pet. He's a bad man; and you hate him as he deserves. And youre going to make the tea, arnt you?

JUDITH [*remorsefully*] Oh yes, I forgot. Ive been keeping you waiting all this time. [*She goes to the fire and puts on the kettle*].

ANDERSON [*going to the press and taking his coat off*] Have you stitched up the shoulder of my old coat?

JUDITH Yes, dear. [*She goes to the table, and sets about putting the tea into the teapot from the caddy*].

ANDERSON [*as he changes his coat for the older one hanging on the press, and replaces it by the one he has just taken off*] Did anyone call when I was out?

JUDITH No, only – [*Someone knocks at the door. With a start which betrays her intense nervousness, she retreats to the further end of the table with the tea caddy and spoon in her hands, exclaiming*] Who's that?

ANDERSON [*going to her and patting her encouragingly on the shoulder*] All right, pet, all right. He wont eat you, whoever he is. [*She tries to smile, and nearly makes herself cry. He goes to the door and opens it. Richard is there, without overcoat or cloak*]. You might have raised the latch and come in, Mr Dudgeon. Nobody stands on much ceremony with us. [*Hospitably*] Come in. [*Richard comes in carelessly and stands at the table, looking round the room with a slight pucker of his nose at the mezzotinted divine on the wall. Judith keeps her eyes on the tea caddy*]. Is it still raining? [*He shuts the door*].

RICHARD Raining like the very [*his eye catches Judith's as she looks quickly and haughtily up*] – I beg your pardon; but [*shewing that his coat is wet*] you see – !

ANDERSON Take it off, sir; and let it hang before the fire a while: my wife will excuse your shirtsleeves. Judith: put in another spoonful of tea for Mr Dudgeon.

RICHARD [*eyeing him cynically*] The magic of property, Pastor! Are even *you* civil to me now that I have succeeded to my father's estate?

Judith throws down the spoon indignantly.

ANDERSON [*quite unruffled, and helping Richard off with his coat*] I think, sir, that since you accept my hospitality, you cannot have so bad an opinion of it. Sit down. [*With the coat in his hand, he points to the railed seat. Richard, in his shirtsleeves, looks at him half quarrelsomely for a*

moment; then, with a nod, acknowledges that the minister has got the better of him, and sits down on the seat. Anderson pushes his cloak into a heap on the seat of the chair at the fire, and hangs Richard's coat on the back in its place].

RICHARD I come, sir, on your own invitation. You left word you had something important to tell me.

ANDERSON I have a warning which it is my duty to give you.

RICHARD [*quickly rising*] You want to preach to me. Excuse me: I prefer a walk in the rain [*he makes for his coat*].

ANDERSON [*stopping him*] Dont be alarmed, sir: I am no great preacher. You are quite safe. [*Richard smiles in spite of himself. His glance softens: he even makes a gesture of excuse. Anderson, seeing that he has tamed him, now addresses him earnestly*]. Mr Dudgeon: you are in danger in this town.

RICHARD What danger?

ANDERSON Your uncle's danger. Major Swindon's gallows.

RICHARD It is you who are in danger. I warned you –

ANDERSON [*interrupting him goodhumoredly but authoritatively*] Yes, yes, Mr Dudgeon; but they do not think so in the town. And even if I were in danger, I have duties here which I must not forsake. But you are a free man. Why should you run any risk?

RICHARD Do you think I should be any great loss, Minister?

ANDERSON I think that a man's life is worth saving, whoever it belongs to. [*Richard makes him an ironical bow. Anderson returns the bow humorously*]. Come: youll have a cup of tea, to prevent you catching cold?

RICHARD I observe that Mrs Anderson is not quite so pressing as you are, Pastor.

JUDITH [*almost stifled with resentment, which she has been expecting her husband to share and express for her at every insult of Richard's*] You are welcome for my husband's sake. [*She brings the teapot to the fireplace and sets it on the hob*].

RICHARD I know I am not welcome for my own, madam. [*He rises*]. But I think I will not break bread here, Minister.

ANDERSON [*cheerily*] Give me a good reason for that.

RICHARD Because there is something in you that I respect, and that makes me desire to have you for my enemy.

ANDERSON Thats well said. On those terms, sir, I will accept your enmity or any man's. Judith: Mr Dudgeon will stay to tea. Sit down: it will take a few minutes to draw by the fire. [*Richard glances at him with a troubled face; then sits down with his head bent, to hide a convulsive swelling of his throat*]. I was just saying to my wife, Mr Dudgeon, that enmity – [*She grasps his hand and looks imploringly at him, doing both with an intensity that checks him at once*]. Well, well, I mustnt tell you, I see; but it was nothing that need leave us worse friend – enemies, I mean. Judith is a great enemy of yours.

RICHARD If all my enemies were like Mrs Anderson, I should be the best Christian in America.

ANDERSON [*gratified, patting her hand*] You hear that, Judith? Mr Dudgeon knows how to turn a compliment.

The latch is lifted from without.

JUDITH [*starting*] Who is that?

Christy comes in.

CHRISTY [*stopping and staring at Richard*] Oh, are *you* here?

RICHARD Yes. Begone, you fool: Mrs Anderson doesnt want the whole family to tea at once.

CHRISTY [*coming further in*] Mother's very ill.

RICHARD Well, does she want to see *me?*

CHRISTY No.

RICHARD I thought not.

CHRISTY She wants to see the minister – at once.

JUDITH [*to Anderson*] Oh, not before youve had some tea.

ANDERSON I shall enjoy it more when I come back, dear. [*He is about to take up his cloak*].

CHRISTY The rain's over.

ANDERSON [*dropping the cloak and picking up his hat from the fender*] Where is your mother, Christy?

CHRISTY At Uncle Titus's.

ANDERSON Have you fetched the doctor?

CHRISTY No: she didnt tell me to.

ANDERSON Go on there at once: I'll overtake you on his doorstep. [*Christy turns to go*]. Wait a moment. Your brother must be anxious to know the particulars.

RICHARD Psha! not I: he doesnt know; and I dont care. [*Violently*] Be off, you oaf. [*Christy runs out. Richard adds, a little shamefacedly*] We shall know soon enough.

ANDERSON Well, perhaps you will let me bring you the news myself. Judith: will you give Mr Dudgeon his tea, and keep him here until I return.

JUDITH [*white and trembling*] Must I –

ANDERSON [*taking her hands and interrupting her to cover her agitation*] My dear: I can depend on you?

JUDITH [*with a piteous effort to be worthy of his trust*] Yes.

ANDERSON [*pressing her hand against his cheek*] You will not mind two old people like us, Mr Dudgeon. [*Going*] I shall not say good evening: you will be here when I come back. [*He goes out*].

They watch him pass the window, and then look at each other dumbly, quite disconcerted. Richard, noting the quiver of her lips, is the first to pull himself together.

RICHARD Mrs Anderson: I am perfectly aware of the nature of your sentiments towards me. I shall not intrude on you. Good evening. [*Again he starts for the fireplace to get his coat*].

JUDITH [*getting between him and the coat*] No, no. Dont go: please dont go.

RICHARD [*roughly*] Why? You dont want me here.

JUDITH Yes, I – [*Wringing her hands in despair*] Oh, if I tell you the truth, you will use it to torment me.

RICHARD [*indignantly*] Torment! What right have you to say that? Do you expect me to stay after that?

JUDITH I want you to stay; but [*suddenly raging at him like an angry child*] it is not because I like you.

RICHARD Indeed!

JUDITH Yes: I had rather you did go than mistake me about that. I hate and dread you; and my husband knows it. If you are not here when he comes back, he will believe that I disobeyed him and drove you away.

RICHARD [*ironically*] Whereas, of course, you have really been so kind and hospitable and charming to me that I only want to go away out of mere contrariness, eh?

Judith, unable to bear it, sinks on the chair and bursts into tears.

RICHARD Stop, stop, stop, I tell you. Dont do that. [*Putting his hand to his breast as if to a wound*] He wrung my heart by being a man. Need you tear it by being a woman? Has he not raised you above my insults, like himself? [*She stops crying, and recovers herself somewhat, looking at him with a scared curiosity*]. There: thats right. [*Sympathetically*] Youre better now, arnt you? [*He puts his hand encouragingly on her shoulder. She instantly rises haughtily, and stares at him defiantly. He at once drops into his usual sardonic tone*]. Ah, thats better. You are yourself again: so is Richard. Well, shall we go to tea like a quiet respectable couple, and wait for your husband's return?

JUDITH [*rather ashamed of herself*] If you please. I – I am sorry to have been so foolish. [*She stoops to take up the plate of toast from the fender*].

RICHARD I am sorry, for your sake, that I am – what I am. Allow me. [*He takes the plate from her and goes with it to the table*].

JUDITH [*following with the teapot*] Will you sit down? [*He sits down at the end of the table nearest the press. There is a plate and knife laid there. The other plate is laid near it; but Judith stays at the opposite end of the table, next the fire, and takes her place there, drawing the tray towards her*]. Do you take sugar?

RICHARD No; but plenty of milk. Let me give you some toast. [*He puts some on the second plate, and hands it to her, with the knife. The action shews quietly how well he knows that she has avoided her usual place so as to be as far from him as possible*].

JUDITH [*consciously*] Thanks. [*She gives him his tea*]. Wont you help yourself?

RICHARD Thanks. [*He puts a piece of toast on his own plate; and she pours out tea for herself*].

JUDITH [*observing that he tastes nothing*] Dont you like it? You are not eating anything.

RICHARD Neither are you.

JUDITH [*nervously*] I never care much for my tea. Please dont mind me.

RICHARD [*looking dreamily round*] I am thinking. It is all so strange to me. I can see the beauty and peace of this home: I think I have never been more at rest in my life than at this moment; and yet I know quite well I could never live here. It's not in my nature, I suppose, to be domesticated. But it's very beautiful: it's almost holy. [*He muses a moment, and then laughs softly*].

JUDITH [*quickly*] Why do you laugh?

RICHARD I was thinking that if any stranger came in here now, he would take us for man and wife.

JUDITH [*taking offence*] You mean, I suppose, that you are more my age than he is.

RICHARD [*staring at this unexpected turn*] I never thought of such a thing. [*Sardonic again*]. I see there is another side to domestic joy.

JUDITH [*angrily*] I would rather have a husband whom everybody respects than – than –

RICHARD Than the devil's disciple. You are right; but I daresay your love helps him to be a good man, just as your hate helps me to be a bad one.

JUDITH My husband has been very good to you. He has forgiven you for insulting him, and is trying to save you. Can you not forgive him for being so much better than you are? How dare you belittle him by putting yourself in his place?

RICHARD Did I?

JUDITH Yes, you did. You said that if anybody came in they would take us for man and – [*She stops, terror-stricken, as a squad of soldiers tramps past the window*]. The English soldiers! Oh, what do they –

RICHARD [*listening*] Sh!

A VOICE [*outside*] Halt! Four outside: two in with me.

Judith half rises, listening and looking with dilated eyes at Richard, who takes up his cup prosaically, and is drinking his tea when the latch goes up with a sharp click, and an English sergeant walks into the room with two privates, who post themselves at the door. He comes promptly to the table between them.

SERGEANT Sorry to disturb you, mum. Duty! Anthony Anderson: I arrest you in King George's name as a rebel.

JUDITH [*pointing at Richard*] But that is not – [*He looks up quickly at her, with a face of iron. She stops her mouth hastily with the hand she has raised to indicate him, and stands staring affrightedly*].

SERGEANT Come, parson: put your coat on and come along.

RICHARD Yes: I'll come. [*He rises and takes a step towards his own coat; then recollects himself, and, with his back to the sergeant, moves his gaze slowly round the room without turning his head until he sees Anderson's black coat hanging up on the press. He goes composedly to it; takes it down; and puts it on. The idea of himself as a parson tickles him: he looks down at the black sleeve on his arm, and then smiles slyly at Judith, whose white face shews him that what she is painfully struggling to grasp is not the humor of the situation but its horror. He turns to the sergeant, who is approaching him with a pair of handcuffs hidden behind him, and says lightly*] Did you ever arrest a man of my cloth before, Sergeant?

SERGEANT [*instinctively respectful, half to the black coat, half to Richard's good breeding*] Well, no sir. At least, only an army chaplain. [*Shewing the handcuffs*] I'm sorry sir; but duty –

RICHARD Just so, Sergeant. Well, I'm not ashamed of them: thank you kindly for the apology. [*He holds out his hands*].

SERGEANT [*not availing himself of the offer*] One gentleman to another, sir. Wouldnt you like to say a word to your missis, sir, before you go?

RICHARD [*smiling*] Oh, we shall meet again before – eh? [*meaning "before you hang me"*].

SERGEANT [*loudly, with ostentatious cheerfulness*] Oh, of course, of course. No call for the lady to distress herself. Still – [*in a lower voice, intended for Richard alone*] your last chance, sir.

They look at one another significantly for a moment. Then Richard exhales a deep breath and turns towards Judith.

RICHARD [*very distinctly*] My love. [*She looks at him, pitiably pale, and tries to answer, but cannot – tries also to come to him, but cannot trust herself to stand without the support of the table*]. This gallant gentleman is good enough to allow us a moment of leavetaking. [*The sergeant retires delicately and joins his men near the door*]. He is trying to spare you the truth; but you had better know it. Are you listening to me? [*She signifies assent*]. Do you understand that I am going to my death? [*She signifies that she understands*]. Remember,

you must find our friend who was with us just now. Do you understand? [*She signifies yes*]. See that you get him safely out of harm's way. Dont for your life let him know of my danger; but if he finds it out, tell him that he cannot save me: they would hang him; and they would not spare me. And tell him that I am steadfast in my religion as he is in his, and that he may depend on me to the death. [*He turns to go, and meets the eye of the sergeant, who looks a little suspicious. He considers a moment, and then, turning roguishly to Judith with something of a smile breaking through his earnestness, says*] And now, my dear, I am afraid the sergeant will not believe that you love me like a wife unless you give one kiss before I go.

He approaches her and holds out his arms. She quits the table and almost falls into them.

JUDITH [*the words choking her*] I ought to – it's murder –

RICHARD No: only a kiss [*softly to her*] for his sake.

JUDITH I cant. *You* must –

RICHARD [*folding her in his arms with an impulse of compassion for her distress*] My poor girl!

Judith, with a sudden effort, throws her arms round him; kisses him; and swoons away, dropping from his arms to the ground as if the kiss had killed her.

RICHARD [*going quickly to the sergeant*] Now, Sergeant: quick, before she comes to. The handcuffs. [*He puts out his hands*].

SERGEANT [*pocketing them*] Never mind, sir: I'll trust you. Youre a game one. You ought to a bin a soldier, sir. Between them two, please. [*The soldiers place themselves one before Richard and one behind him. The sergeant opens the door*].

RICHARD [*taking a last look round him*] Goodbye, wife: goodbye, home. Muffle the drums, and quick march!

The sergeant signs to the leading soldier to march. They file out quickly.

When Anderson returns from Mrs Dudgeon's, he is astonished to find the room apparently empty and almost in darkness except for the glow from the fire; for one of the candles has burnt out, and the other is at its last flicker.

ANDERSON Why, what on earth – ? [*Calling*] Judith, Judith! [*He listens: there is no answer*]. Hm! [*He goes to the cupboard; takes a candle from the drawer; lights it at the flicker of the expiring one on the table; and looks wonderingly at the untasted meal by its light. Then he sticks it in the candlestick; takes off his hat; and scratches his head, much puzzled. This action causes him to look at the floor for the first time; and there he sees Judith lying motionless with her eyes closed. He runs to her and stoops beside her, lifting her head*]. Judith.

JUDITH [*waking; for her swoon has passed into the sleep of exhaustion after suffering*] Yes. Did you call? Whats the matter?

ANDERSON Ive just come in and found you lying here with the candles burnt out and the tea poured out and cold. What has happened?

JUDITH [*still astray*] I dont know. Have I been asleep? I suppose – [*She stops blankly*]. I dont know.

ANDERSON [*groaning*] Heaven forgive me, I left you alone with that scoundrel. [*Judith remembers. With an agonized cry, she clutches his shoulders and drags herself to her feet as he rises with her. He clasps her tenderly in his arms*]. My poor pet!

JUDITH [*frantically clinging to him*] What shall I do? Oh my God, what shall I do?

ANDERSON Never mind, never mind, my dearest dear: it was my fault. Come: youre safe now; and youre not hurt, are you? [*He takes his arms from her to see whether she can stand*]. There: thats right, thats right. If only you are not hurt, nothing else matters.

JUDITH No, no, no: I'm not hurt.

ANDERSON Thank Heaven for that! Come now: [*leading her to the railed seat and making her sit down beside him*] sit down and rest: you can tell me about it to-morrow. Or [*misunderstanding her distress*] you shall not tell me at all if it worries you. There, there! [*Cheerfully*] I'll

make you some fresh tea: that will set you up again. [*He goes to the table, and empties the teapot into the slop bowl*].

JUDITH [*in a strained tone*] Tony.

ANDERSON Yes, dear?

JUDITH Do you think we are only in a dream now?

ANDERSON [*glancing round at her for a moment with a pang of anxiety, though he goes on steadily and cheerfully putting fresh tea into the pot*] Perhaps so, pet. But you may as well dream a cup of tea when youre about it.

JUDITH Oh stop, stop. You dont know – [*Distracted, she buries her face in her knotted hands*].

ANDERSON [*breaking down and coming to her*] My dear, what is it? I cant bear it any longer: you must tell me. It was all my fault: I was mad to trust him.

JUDITH No: dont say that. You mustnt say that. He – oh no, no: I cant. Tony: dont speak to me. Take my hands – both my hands. [*He takes them, wondering*]. Make me think of you, not of him. There's danger, frightful danger; but it is your danger; and I cant keep thinking of it: I cant, I cant: my mind goes back to his danger. He must be saved – no: you must be saved: you, you, you. [*She springs up as if to do something or go somewhere, exclaiming*] Oh, Heaven help me!

ANDERSON [*keeping his seat and holding her hands with resolute composure*] Calmly, calmly, my pet. Youre quite distracted.

JUDITH I may well be. I dont know what to do. I dont know what to do. [*Tearing her hands away*]. I must save him. [*Anderson rises in alarm as she runs wildly to the door. It is opened in her face by Essie, who hurries in full of anxiety. The surprise is so disagreeable to Judith that it brings her to her senses. Her tone is sharp and angry as she demands*] What do you want?

ESSIE I was to come to you.

ANDERSON Who told you to?

ESSIE [*staring at him, as if his presence astonished her*] Are you here?

JUDITH Of course. Dont be foolish, child.

ANDERSON Gently, dearest: youll frighten her. [*Going between them*]. Come here, Essie. [*She comes to him*]. Who sent you?

ESSIE Dick. He sent me word by a soldier. I was to come here at once and do whatever Mrs Anderson told me.

ANDERSON [*enlightened*] A soldier! Ah, I see it all now! They have arrested Richard. [*Judith makes a gesture of despair*].

ESSIE No. I asked the soldier. Dick's safe. But the soldier said you had been taken.

ANDERSON I! [*Bewildered, he turns to Judith for an explanation*].

JUDITH [*coaxingly*] All right, dear: I understand. [*To Essie*] Thank you, Essie, for coming; but I dont need you now. You may go home.

ESSIE [*suspicious*] Are you sure Dick has not been touched? Perhaps he told the soldier to say it was the minister. [*Anxiously*] Mrs Anderson: do you think it can have been that?

ANDERSON Tell her the truth if it is so, Judith. She will learn it from the first neighbor she meets in the street. [*Judith turns away and covers her eyes with her hands*].

ESSIE [*wailing*] But what will they do to him? Oh, what will they do to him? Will they hang him? [*Judith shudders convulsively, and throws herself into the chair in which Richard sat at the tea table*].

ANDERSON [*patting Essie's shoulder and trying to comfort her*] I hope not. I hope not. Perhaps if youre very quiet and patient, we may be able to help him in some way.

ESSIE Yes – help him – yes, yes, yes. I'll be good.

ANDERSON I must go to him at once, Judith.

JUDITH [*springing up*] Oh no. You must go away – far away, to some place of safety.

ANDERSON Pooh!

JUDITH [*passionately*] Do you want to kill me? Do you think I can bear to live for days and days with every knock at the door – every footstep – giving me a spasm of terror? to lie awake for nights and nights in an agony of dread, listening for them to come and arrest you?

ANDERSON Do you think it would be better to know that I had run away from my post at the first sign of danger?

JUDITH [*bitterly*] Oh, you wont go. I know it. Youll stay; and I shall go mad.

ANDERSON My dear, your duty –

JUDITH [*fiercely*] What do I care about my duty?

ANDERSON [*shocked*] Judith!

JUDITH I am doing my duty. I am clinging to my duty. My duty is to get you away, to save you, to leave him to his fate [*Essie utters a cry of distress and sinks on the chair at the fire, sobbing silently*]. My instinct is the same as hers – to save him above all things, though it would be so much better for him to die! so much greater! But I know you will take your own way as he took it. I have no power. [*She sits down sullenly on the railed seat*]. I'm only a woman: I can do nothing but sit here and suffer. Only, tell him I tried to save you – that I did my best to save you.

ANDERSON My dear, I am afraid he will be thinking more of his own danger than of mine.

JUDITH Stop; or I shall hate you.

ANDERSON [*remonstrating*] Come, come, come! How am I to leave you if you talk like this? You are quite out of your senses. [*He turns to Essie*] Essie.

ESSIE [*eagerly rising and drying her eyes*] Yes?

ANDERSON Just wait outside a moment, like a good girl: Mrs Anderson is not well. [*Essie looks doubtful*]. Never fear: I'll come to you presently; and I'll go to Dick.

ESSIE You are sure you will go to him? [*Whispering*] You wont let *her* prevent you?

ANDERSON [*smiling*] No, no: it's all right. All right. [*She goes*]. Thats a good girl. [*He closes the door, and returns to Judith*].

JUDITH [*seated – rigid*] You are going to your death.

ANDERSON [*quaintly*] Then I shall go in my best coat, dear. [*He turns to the press, beginning to take off his coat*]. Where – ? [*He stares at the empty nail for a moment; then looks quickly round to the fire; strides across to it; and lifts Richard's coat*]. Why, my dear, it seems that he has gone in my best coat.

JUDITH [*still motionless*] Yes.

ANDERSON Did the soldiers make a mistake?

JUDITH Yes: they made a mistake.

ANDERSON He might have told them. Poor fellow, he was too upset, I suppose.

JUDITH Yes: he might have told them. So might I.

ANDERSON Well, it's all very puzzling – almost funny. It's curious how these little things strike us even in the most – [*He breaks off and begins putting on Richard's coat*]. I'd better take him his own coat. I know what he'll say – [*imitating Richard's sardonic manner*] "Anxious about my soul, Pastor, and also about your best coat." Eh?

JUDITH Yes, that is just what he will say to you. [*Vacantly*] It doesnt matter: I shall never see either of you again.

ANDERSON [*rallying her*] Oh pooh, pooh, pooh! [*He sits down beside her*]. Is this how you keep your promise that I shant be ashamed of my brave wife?

JUDITH No: this is how I break it. I cannot keep my promises to him: why should I keep my promises to you?

ANDERSON Dont speak so strangely, my love. It sounds insincere to me. [*She looks unutterable reproach at him*]. Yes, dear, nonsense is always insincere; and my dearest is talking nonsense. Just nonsense. [*Her face darkens into dumb obstinacy. She stares straight before her, and does not look at him again, absorbed in Richard's fate. He scans her face; sees that his rallying has produced no effect; and gives it up, making no further effort to conceal his anxiety*]. I wish I knew what has frightened you so. Was there a struggle? Did he fight?

JUDITH No. He smiled.

ANDERSON Did he realize his danger, do you think?

JUDITH He realized yours.

ANDERSON Mine!

JUDITH [*monotonously*] He said, "See that you get him safely out of harm's way." I promised: I cant keep my promise. He said, "Dont for your life let him know of my danger." Ive told you of it. He said that if you found it out, you could not save him – that they will hang him and not spare you.

ANDERSON [*rising in generous indignation*] And you think that I

will let a man with that much good in him die like a dog, when a few words might make him die like a Christian. I'm ashamed of you, Judith.

JUDITH He will be steadfast in his religion as you are in yours; and you may depend on him to the death. He said so.

ANDERSON God forgive him! What else did he say?

JUDITH He said goodbye.

ANDERSON [*fidgeting nervously to and fro in great concern*] Poor fellow, poor fellow! You said goodbye to him in all kindness and charity, Judith, I hope.

JUDITH I kissed him.

ANDERSON What! Judith!

JUDITH Are you angry?

ANDERSON No, no. You were right: you were right. Poor fellow, poor fellow! [*Greatly distressed*] To be hanged like that at his age! And then did they take him away?

JUDITH [*wearily*] Then you were here: thats the next thing I remember. I suppose I fainted. Now bid me goodbye, Tony. Perhaps I shall faint again. I wish I could die.

ANDERSON No, no, my dear: you must pull yourself together and be sensible. I am in no danger – not the least in the world.

JUDITH [*solemnly*] You are going to your death, Tony – your sure death, if God will let innocent men be murdered. They will not let you see him: they will arrest you the moment you give your name. It was for you the soldiers came.

ANDERSON [*thunderstruck*] For me!!! [*His fists clinch; his neck thickens; his face reddens; the fleshy purses under his eyes become injected with hot blood; the man of peace vanishes, transfigured into a choleric and formidable man of war. Still, she does not come out of her absorption to look at him: her eyes are steadfast with a mechanical reflection of Richard's steadfastness*].

JUDITH He took your place: he is dying to save you. That is why he went in your coat. That is why I kissed him.

ANDERSON [*exploding*] Blood an' owns! [*His voice is rough and dominant, his gesture full of brute energy*]. Here! Essie, Essie!

ESSIE [*running in*] Yes.

ANDERSON [*impetuously*] Off with you as hard as you can run, to the inn. Tell them to saddle the fastest and strongest horse they have [*Judith rises breathless, and stares at him incredulously*] – the chestnut mare, if she's fresh – without a moment's delay. Go into the stable yard and tell the black man there that I'll give him a silver dollar if the horse is waiting for me when I come, and that I am close on your heels. Away with you. [*His energy sends Essie flying from the room. He pounces on his riding boots; rushes with them to the chair at the fire; and begins pulling them on*].

JUDITH [*unable to believe such a thing of him*] You are not going to him!

ANDERSON [*busy with the boots*] Going to him! What good would that do? [*Growling to himself as he gets the first boot on with a wrench*] I'll go to them, so I will. [*To Judith peremptorily*] Get me the pistols: I want them. And money, money: I want money – all the money in the house. [*He stoops over the other boot, grumbling*] A great satisfaction it would be to him to have my company on the gallows. [*He pulls on the boot*].

JUDITH You are deserting him, then?

ANDERSON Hold your tongue, woman; and get me the pistols. [*She goes to the press and takes from it a leather belt with two pistols, a powder horn, and a bag of bullets attached to it. She throws it on the table. Then she unlocks a drawer in the press and takes out a purse. Anderson grabs the belt and buckles it on, saying*] If they took him for me in my coat, perhaps theyll take me for him in his. [*Hitching the belt into its place*] Do I look like him?

JUDITH [*turning with the purse in her hand*] Horribly unlike him.

ANDERSON [*snatching the purse from her and emptying it on the table*] Hm! We shall see.

JUDITH [*sitting down helplessly*] Is it of any use to pray, do you think, Tony?

ANDERSON [*counting the money*] Pray! Can we pray Swindon's rope off Richard's neck?

JUDITH God may soften Major Swindon's heart.

ANDERSON [*contemptuously – pocketing a handful of money*] Let him, then. I am not God; and I must go to work another way. [*Judith gasps at the blasphemy. He throws the purse on the table*]. Keep that. Ive taken 25 dollars.

JUDITH Have you forgotten even that you are a minister?

ANDERSON Minister be – faugh! My hat: wheres my hat? [*He snatches up hat and cloak, and puts both on in hot haste*]. Now listen, you. If you can get a word with him by pretending youre his wife, tell him to hold his tongue until morning: that will give me all the start I need.

JUDITH [*solemnly*] You may depend on him to the death.

ANDERSON Youre a fool, a *fool*, Judith. [*For a moment checking the torrent of his haste, and speaking with something of his old quiet and impressive conviction*] You dont know the man youre married to. [*Essie returns. He swoops at her at once*]. Well: is the horse ready?

ESSIE [*breathless*] It will be ready when you come.

ANDERSON Good. [*He makes for the door*].

JUDITH [*rising and stretching out her arms after him involuntarily*] Wont you say goodbye?

ANDERSON And waste another half minute! Psha! [*He rushes out like an avalanche*].

ESSIE [*hurrying to Judith*] He has gone to save Richard, hasnt he?

JUDITH To save Richard! No: Richard has saved him. He has gone to save himself. Richard must die.

Essie screams with terror and falls on her knees, hiding her face. Judith, without heeding her, looks rigidly straight in front of her, at the vision of Richard, dying.

Early next morning the sergeant, at the British headquarters in the Town Hall, unlocks the door of a little empty panelled waiting room, and invites Judith to enter. She has had a bad night, probably a rather delirious one; for even in the reality of the raw morning, her fixed gaze comes back at moments when her attention is not strongly held.

The sergeant considers that her feelings do her credit, and is sympathetic in an encouraging military way. Being a fine figure of a man, vain of his uniform and of his rank, he feels specially qualified, in a respectful way, to console her.

SERGEANT You can have a quiet word with him here, mum.

JUDITH Shall I have long to wait?

SERGEANT No, mum, not a minute. We kep him in the Bridewell for the night; and he's just been brought over here for the court martial. Dont fret, mum: he slep like a child, and has made a rare good breakfast.

JUDITH [*incredulously*] He is in good spirits!

SERGEANT Tip top, mum. The chaplain looked in to see him last night; and he won seventeen shillings off him at spoil five. He spent it among us like the gentleman he is. Duty's duty, mum, of course; but youre among friends here. [*The tramp of a couple of soldiers is heard approaching*]. There: I think he's coming. [*Richard comes in, without a sign of care or captivity in his bearing. The sergeant nods to the two soldiers, and shews them the key of the room in his hand. They withdraw*]. Your good lady, sir.

RICHARD [*going to her*] What! My wife. My adored one. [*He takes her hand and kisses it with a perverse, raffish gallantry*]. How long do you allow a brokenhearted husband for leave-taking, Sergeant?

SERGEANT As long as we can, sir. We shall not disturb you till the court sits.

RICHARD But it has struck the hour.

SERGEANT So it has, sir; but there's a delay. General Burgoyne's just arrived – Gentlemanly Johnny we call him, sir – and he wont have done finding fault with everything this side of half past. I know him, sir: I served with him in Portugal. You may count on twenty minutes, sir; and by your leave I wont waste any more of them. [*He goes out, locking the door. Richard immediately drops his raffish manner and turns to Judith with considerate sincerity*].

RICHARD Mrs Anderson: this visit is very kind of you. And how are you after last night? I had to leave you before you recovered; but I sent word to Essie to go and look after you. Did she understand the message?

JUDITH [*breathless and urgent*] Oh, dont think of me: I havnt come here to talk about myself. Are they going to – to – [*meaning "to hang you"*]?

RICHARD [*whimsically*] At noon, punctually. At least, that was when they disposed of Uncle Peter. [*She shudders*]. Is your husband safe? Is he on the wing?

JUDITH He is no longer my husband.

RICHARD [*opening his eyes wide*] Eh?

JUDITH I disobeyed you. I told him everything. I expected him to come here and save you. I wanted him to come here and save you. He ran away instead.

RICHARD Well, thats what I meant him to do. What good would his staying have done? Theyd only have hanged us both.

JUDITH [*with reproachful earnestness*] Richard Dudgeon: on your honour, what would you have done in his place?

RICHARD Exactly what he has done, of course.

JUDITH Oh, why will you not be simple with me – honest and straightforward? If you are so selfish as that, why did you let them take you last night?

RICHARD [*gaily*] Upon my life, Mrs Anderson, I dont know. Ive been asking myself that question ever since; and I can find no manner of reason for acting as I did.

JUDITH You know you did it for his sake, believing he was a more worthy man than yourself.

RICHARD [*laughing*] Oho! No: thats a very pretty reason, I must say; but I'm not so modest as that. No: it wasnt for his sake.

JUDITH [*after a pause, during which she looks shamefacedly at him, blushing painfully*] Was it for my sake?

RICHARD [*gallantly*] Well, you had a hand in it. It must have been a little for your sake. You let them take me, at all events.

JUDITH Oh, do you think I have not been telling myself that all night? Your death will be at my door. [*Impulsively, she gives him her hand, and adds, with intense earnestness*] If I could save you as you saved him, I would do it, no matter how cruel the death was.

RICHARD [*holding her hand and smiling, but keeping her almost at arms length*] I am very sure I shouldnt let you.

JUDITH Dont you see that I *can* save you?

RICHARD How? By changing clothes with me, eh?

JUDITH [*disengaging her hand to touch his lips with it*] Dont [*meaning "Dont jest"*]. No: by telling the Court who you really are.

RICHARD [*frowning*] No use: they wouldnt spare me; and it would spoil half his chance of escaping. They are determined to cow us by making an example of somebody on that gallows to-day. Well, let us cow them by showing that we can stand by one another to the death. That is the only force that can send Burgoyne

back across the Atlantic and make America a nation.

JUDITH [*impatiently*] Oh, what does all that matter?

RICHARD [*laughing*] True: what does it matter? what does anything matter? You see, men have these strange notions, Mrs Anderson; and women see the folly of them.

JUDITH Women have to lose those they love through them.

RICHARD They can easily get fresh lovers.

JUDITH [*revolted*] Oh! [*Vehemently*] Do you realise that you are going to kill yourself?

RICHARD The only man I have any right to kill, Mrs Anderson. Dont be concerned: no woman will lose her lover through my death. [*Smiling*] Bless you, nobody cares for me. Have you heard that my mother is dead?

JUDITH Dead!

RICHARD Of heart disease – in the night. Her last word to me was her curse: I dont think I could have borne her blessing. My other relatives will not grieve much on my account. Essie will cry for a day or two; but I have provided for her: I made my own will last night.

JUDITH [*stonily, after a moment's silence*] And I!

RICHARD [*surprised*] You?

JUDITH Yes, I. Am I not to care at all?

RICHARD [*gaily and bluntly*] Not a scrap. Oh, you expressed your feelings towards me very frankly yesterday. What happened may have softened you for the moment; but believe me, Mrs Anderson, you dont like a bone in my skin or a hair on my head. I shall be as good a riddance at 12 to-day as I should have been at 12 yesterday.

JUDITH [*her voice trembling*] What can I do to shew you that you are mistaken?

RICHARD Dont trouble. I'll give you credit for liking me a little better than you did. All I say is that my death will not break your heart.

JUDITH [*almost in a whisper*] How do you know? [*She puts her hands on his shoulders and looks intently at him*].

RICHARD [*amazed – divining the truth*] Mrs Anderson! [*The bell of the town clock strikes the quarter. He collects himself, and removes her hands, saying rather coldly*] Ex-

cuse me: they will be here for me presently. It is too late.

JUDITH It is not too late. Call me as witness: they will never kill you when they know how heroically you have acted.

RICHARD [*with some scorn*] Indeed! But if I dont go through with it, where will the heroism be? I shall simply have tricked them; and theyll hang me for that like a dog. Serve me right too!

JUDITH [*wildly*] Oh, I believe you *want* to die.

RICHARD [*obstinately*] No I dont.

JUDITH Then why not try to save yourself? I implore you – listen. You said just now that you saved him for my sake – yes [*clutching him as he recoils with a gesture of denial*] a little for my sake. Well, save yourself for my sake. And I will go with you to the end of the world.

RICHARD [*taking her by the wrists and holding her a little way from him, looking steadily at her*] Judith.

JUDITH [*breathless – delighted at the name*] Yes.

RICHARD If I said – to please you – that I did what I did ever so little for your sake, I lied as men always lie to women. You know how much I have lived with worthless men – aye, and worthless women too. Well, they could all rise to some sort of goodness and kindness when they were in love [*the word love comes from him with true Puritan scorn*]. That has taught me to set very little store by the goodness that only comes out red hot. What I did last night, I did in cold blood, caring not half so much for your husband, or [*ruthlessly*] for you [*she droops, stricken*] as I do for myself. I had no motive and no interest: all I can tell you is that when it came to the point whether I would take my neck out of the noose and put another man's into it, I could not do it. I dont know why not: I see myself as a fool for my pains; but I could not and I cannot. I have been brought up standing by the law of my own nature; and I may not go against it, gallows or no gallows. [*She has slowly raised her head and is now looking full at him*]. I should have done the same for any other man in the town, or any other man's wife. [*Releasing her*] Do you understand that?

JUDITH Yes: you mean that you do not love me.

RICHARD [*revolted – with fierce contempt*] Is that all it means to you?

JUDITH What more – what worse – can it mean to me? [*The sergeant knocks. The blow on the door jars on her heart*]. Oh, one moment more. [*She throws herself on her knees*]. I pray to you –

RICHARD Hush! [*Calling*] Come in. [*The sergeant unlocks the door and opens it. The guard is with him*].

SERGEANT [*coming in*] Time's up, sir.

RICHARD Quite ready, Sergeant. Now, my dear. [*He attempts to raise her*].

JUDITH [*clinging to him*] Only one thing more – I entreat, I implore you. Let me be present in the court. I have seen Major Swindon: he said I should be allowed if you asked it. You will ask it. It is my last request: I shall never ask you anything again. [*She clasps his knee*]. I beg and pray it of you.

RICHARD If I do, will you be silent?

JUDITH Yes.

RICHARD You will keep faith?

JUDITH I will keep – [*She breaks down, sobbing*].

RICHARD [*taking her arm to lift her*] Just – her other arm, Sergeant.

They go out, she sobbing convulsively, supported by the two men.

Meanwhile, the Council Chamber is ready for the court martial. It is a large, lofty room, with a chair of state in the middle under a tall canopy with a gilt crown, and maroon curtains with the royal monogram G. R. In front of the chair is a table, also draped in maroon, with a bell, a heavy inkstand, and writing materials on it. Several chairs are set at the table. The door is at the right hand of the occupant of the chair of state when it has an occupant: at present it is empty. Major Swindon, a pale, sandy-haired, very conscientious looking man of about 45, sits at the end of the table with his back to the door, writing. He is alone until the sergeant announces the General in a subdued manner which suggests that Gentlemanly Johnny has been making his presence felt rather heavily.

SERGEANT The General, sir.

Swindon rises hastily. The general comes in: the sergeant goes out. General Burgoyne is 55, and very well preserved. He is a man of fashion, gallant enough to have made a distinguished marriage by an elopement, witty enough to write successful comedies, aristocratically-connected enough to have had opportunities of high military distinction. His eyes, large, brilliant, apprehensive, and intelligent, are his most remarkable feature: without them his fine nose and small mouth would suggest rather more fastidiousness and less force than go to the making of a first rate general. Just now the eyes are angry and tragic, and the mouth and nostrils tense.

BURGOYNE Major Swindon, I presume.

SWINDON Yes. General Burgoyne, if I mistake not. [*They bow to one another ceremoniously*]. I am glad to have the support of your presence this morning. It is not particularly lively business, hanging this poor devil of a minister.

BURGOYNE [*throwing himself into Swindon's chair*] No, sir, it is not. It is making too much of the fellow to execute him: what more could you have done if he had been a member of the Church of England? Martyrdom, sir, is what these people like: it is the only way in which a man can become famous without ability. However, you have committed us to hanging him; and the sooner he is hanged the better.

SWINDON We have arranged it for 12 o'clock. Nothing remains to be done except to try him.

BURGOYNE [*looking at him with suppressed anger*] Nothing – except to save our own necks, perhaps. Have you heard the news from Springtown?

SWINDON Nothing special. The latest reports are satisfactory.

BURGOYNE [*rising in amazement*] Satisfactory, sir! Satisfactory!! [*He stares at him for a moment, and then adds, with grim intensity*] I am glad you take that view of them.

SWINDON [*puzzled*] Do I understand that in your opinion –

BURGOYNE I do not express my opinion. I never stoop to that habit of profane language which unfortunately coarsens our profession. If I did, sir, perhaps I should be able to express my opinion of the news from Springtown – the news which *you* [*severely*] have apparently not heard. How soon do you get news from your supports here? – in the course of a month, eh?

SWINDON [*turning sulky*] I suppose the reports have been taken to you, sir, instead of to me. Is there anything serious?

BURGOYNE [*taking a report from his pocket and holding it up*] Springtown's in the hands of the rebels. [*He throws the report on the table*].

SWINDON [*aghast*] Since yesterday!

BURGOYNE Since two o'clock this morning. Perhaps we shall be in their hands before two o'clock to-morrow morning. Have you thought of that?

SWINDON [*confidently*] As to that, General, the British soldier will give a good account of himself.

BURGOYNE [*bitterly*] And therefore, I suppose, sir, the British officer need not know his business: the British soldier will get him out of all his blunders with the bayonet. In future, sir, I must ask you to be a little less gener-

ous with the blood of your men, and a little more generous with your own brains.

SWINDON I am sorry I cannot pretend to your intellectual eminence, sir. I can only do my best, and rely on the devotion of my countrymen.

BURGOYNE [*suddenly becoming suavely sarcastic*] May I ask are you writing a melodrama, Major Swindon?

SWINDON [*flushing*] No, sir.

BURGOYNE What a pity! *What* a pity! [*Dropping his sarcastic tone and facing him suddenly and seriously*] Do you at all realize, sir, that we have nothing standing between us and destruction but our own bluff and the sheepishness of these colonists? They are men of the same English stock as ourselves: six to one of us [*repeating it emphatically*] six to one, sir; and nearly half our troops are Hessians, Brunswickers, German dragoons, and Indians with scalping knives. These are the countrymen on whose devotion you rely! Suppose the colonists find a leader! Suppose the news from Springtown should turn out to mean that they have already found a leader! What shall we do then? Eh?

SWINDON [*sullenly*] Our duty, sir, I presume.

BURGOYNE [*again sarcastic – giving him up as a fool*] Quite so, quite so. Thank you, Major Swindon, thank you. Now youve settled the question, sir – thrown a flood of light on the situation. What a comfort to me to feel that I have at my side so devoted and able an officer to support me in this emergency! I think, sir, it will probably relieve both our feelings if we proceed to hang this dissenter without further delay [*he strikes the bell*] especially as I am debarred by my principles from the customary military vent for my feelings. [*The sergeant appears*]. Bring your man in.

SERGEANT Yes, sir.

BURGOYNE And mention to any officer you may meet that the court cannot wait any longer for him.

SWINDON [*keeping his temper with difficulty*] The staff is perfectly ready, sir. They have been waiting your convenience for fully half an hour. *Perfectly* ready, sir.

BURGOYNE [*blandly*] So am I. [*Several officers come in and take*

their seats. One of them sits at the end of the table furthest from the door, and acts throughout as clerk to the court, making notes of the proceedings. The uniforms are those of the 9*th,* 20*th,* 21*st,* 24*th,* 47*th,* 53*rd, and* 62*nd British Infantry. One officer is a Major General of the Royal Artillery. There are also German officers of the Hessian Rifles, and of German dragoon and Brunswicker regiments*]. Oh, good morning, gentlemen. Sorry to disturb you, I am sure. Very good of you to spare us a few moments.

SWINDON Will you preside, sir?

BURGOYNE [*becoming additionally polished, lofty, sarcastic and urbane now that he is in public*] No, sir: I feel my own deficiencies too keenly to presume so far. If you will kindly allow me, I will sit at the feet of Gamaliel. [*He takes the chair at the end of the table next the door, and motions Swindon to the chair of state, waiting for him to be seated before sitting down himself*].

SWINDON [*greatly annoyed*] As you please, sir. I am only trying to do my duty under excessively trying circumstances. [*He takes his place in the chair of state*].

Burgoyne, relaxing his studied demeanor for the moment, sits down and begins to read the report with knitted brows and careworn looks, reflecting on his desperate situation and Swindon's uselessness. Richard is brought in. Judith walks beside him. Two soldiers precede and two follow him, with the sergeant in command. They cross the room to the wall opposite the door; but when Richard has just passed before the chair of state the sergeant stops him with a touch on the arm, and posts himself behind him, at his elbow. Judith stands timidly at the wall. The four soldiers place themselves in a squad near her.

BURGOYNE [*looking up and seeing Judith*] Who is that woman?

SERGEANT Prisoner's wife, sir.

SWINDON [*nervously*] She begged me to allow her to be present; and I thought –

BURGOYNE [*completing the sentence for him ironically*] You thought it would be a pleasure for her. Quite so, quite so. [*Blandly*] Give the lady a chair; and make her thoroughly comfortable.

The sergeant fetches a chair and places it near Richard.

JUDITH *Thank* you, sir. [*She sits down after an awe-stricken curtsy to Burgoyne, which he acknowledges by a dignified bend of his head*].

SWINDON [*to Richard, sharply*] Your name, sir?

RICHARD [*affable, but obstinate*] Come: you dont mean to say that youve brought me here without knowing who I am?

SWINDON As a matter of form, sir, give your name.

RICHARD As a matter of form then, my name is Anthony Anderson, Presbyterian minister in this town.

BURGOYNE [*interested*] Indeed! Pray, Mr Anderson, what do you gentlemen believe?

RICHARD I shall be happy to explain if time is allowed me. I cannot undertake to complete your conversion in less than a fortnight.

SWINDON [*snubbing him*] We are not here to discuss your views.

BURGOYNE [*with an elaborate bow to the unfortunate Swindon*] I stand rebuked.

SWINDON [*embarrassed*] Oh, not you, I as –

BURGOYNE Dont mention it. [*To Richard, very politely*] Any political views, Mr Anderson?

RICHARD I understand that that is just what we are here to find out.

SWINDON [*severely*] Do you mean to deny that you are a rebel?

RICHARD I am an American, sir.

SWINDON What do you expect me to think of that speech, Mr Anderson?

RICHARD I never expect a soldier to think, sir.

Burgoyne is boundlessly delighted by this retort, which almost reconciles him to the loss of America.

SWINDON [*whitening with anger*] I advise you not to be insolent, prisoner.

RICHARD You cant help yourself, General. When you make up your mind to hang a man, you put yourself at a disadvantage with him. Why should I be civil to you? I may as well be hanged for a sheep as a lamb.

SWINDON You have no right to assume that the court has made

up its mind without a fair trial. And you will please not address me as General. I am Major Swindon.

RICHARD A thousand pardons. I thought I had the honor of addressing Gentlemanly Johnny.

Sensation among the officers. The sergeant has a narrow escape from a guffaw.

BURGOYNE [*with extreme suavity*] I believe I am Gentlemanly Johnny, sir, at your service. My more intimate friends call me General Burgoyne. [*Richard bows with perfect politeness*]. You will understand, sir, I hope, since you seem to be a gentleman and a man of some spirit in spite of your calling, that if we should have the misfortune to hang you, we shall do so as a mere matter of political necessity and military duty, without any personal ill-feeling.

RICHARD Oh, quite so. That makes all the difference in the world, of course.

They all smile in spite of themselves; and some of the younger officers burst out laughing.

JUDITH [*her dread and horror deepening at every one of these jests and compliments*] How *can* you?

RICHARD You promised to be silent.

BURGOYNE [*to Judith, with studied courtesy*] Believe me, Madam, your husband is placing us under the greatest obligation by taking this very disagreeable business so thoroughly in the spirit of a gentleman. Sergeant: give Mr Anderson a chair. [*The sergeant does so. Richard sits down*]. Now, Major Swindon: we are waiting for you.

SWINDON You are aware, I presume, Mr Anderson, of your obligations as a subject of His Majesty King George the Third.

RICHARD I am aware, sir, that His Majesty King George the Third is about to hang me because I object to Lord North's robbing me.

SWINDON That is a treasonable speech, sir.

RICHARD [*briefly*] Yes. I meant it to be.

BURGOYNE [*strongly deprecating this line of defence, but still polite*] Dont you think, Mr Anderson, that this is rather – if you will excuse the word – a vulgar line to take? Why

should you cry out robbery because of a stamp duty and a tea duty and so forth? After all, it is the essence of your position as a gentleman that you pay with a good grace.

RICHARD It is not the money, General. But to be swindled by a pig-headed lunatic like King George –

SWINDON [*scandalized*] Chut, sir – silence!

SERGEANT [*in stentorian tones, greatly shocked*] Silence!

BURGOYNE [*unruffled*] Ah, that is another point of view. My position does not allow of my going into that, except in private. But [*shrugging his shoulders*] of course, Mr Anderson, if you are determined to be hanged [*Judith flinches*] there's nothing more to be said. An unusual taste! however [*with a final shrug*] – !

SWINDON [*To Burgoyne*] Shall we call witnesses?

RICHARD What need is there of witnesses? If the townspeople here had listened to me, you would have found the streets barricaded, the houses loopholed, and the people in arms to hold the town against you to the last man. But you arrived, unfortunately, before we had got out of the talking stage; and then it was too late.

SWINDON [*severely*] Well, sir, we shall teach you and your townspeople a lesson they will not forget. Have you anything more to say?

RICHARD I think you might have the decency to treat me as a prisoner of war, and shoot me like a man instead of hanging me like a dog.

BURGOYNE [*sympathetically*] Now there, Mr Anderson, you talk like a civilian, if you will excuse my saying so. Have you any idea of the average marksmanship of the army of His Majesty King George the Third? If we make you up a firing party, what will happen? Half of them will miss you: the rest will make a mess of the business and leave you to the provo-marshal's pistol. Whereas we can hang you in a perfectly workmanlike and agreeable way. [*Kindly*] Let me persuade you to be hanged, Mr Anderson?

JUDITH [*sick with horror*] My God!

RICHARD [*To Judith*] Your promise! [*To Burgoyne*] Thank you, General: that view of the case did not occur to me

before. To oblige you, I withdraw my objection to the rope. Hang me, by all means.

BURGOYNE [*smoothly*] Will 12 o'clock suit you, Mr Anderson?

RICHARD I shall be at your disposal then, General.

BURGOYNE [*rising*] Nothing more to be said, gentlemen. [*They all rise*].

JUDITH [*rushing to the table*] Oh, you are not going to murder a man like that, without a proper trial – without thinking of what you are doing – without – [*she cannot find words*].

RICHARD Is this how you keep your promise?

JUDITH If I am not to speak, you must. Defend yourself: save yourself: tell them the truth.

RICHARD [*worriedly*] I have told them truth enough to hang me ten times over. If you say another word you will risk other lives; but you will not save mine.

BURGOYNE My good lady, our only desire is to save unpleasantness. What satisfaction would it give you to have a solemn fuss made, with my friend Swindon in a black cap and so forth? I am sure we are greatly indebted to the admirable tact and gentlemanly feeling shewn by your husband.

JUDITH [*throwing the words in his face*] Oh, you are mad. Is it nothing to you what wicked thing you do if only you do it like a gentleman? Is it nothing to you whether you are a murderer or not, if only you murder in a red coat? [*Desperately*] You shall not hang him: that man is not my husband.

The officers look at one another, and whisper: some of the Germans asking their neighbors to explain what the woman had said. Burgoyne, who has been visibly shaken by Judith's reproach, recovers himself promptly at this new development. Richard meanwhile raises his voice above the buzz.

RICHARD I appeal to you, gentlemen, to put an end to this. She will not believe that she cannot save me. Break up the court.

BURGOYNE [*in a voice so quiet and firm that it restores silence at once*] One moment, Mr Anderson. One moment, gentlemen. [*He resumes his seat. Swindon and the*

officers follow his example]. Let me understand you clearly, madam. Do you mean that this gentleman is not your husband, or merely – I wish to put this with all delicacy – that you are not his wife?

JUDITH I dont know what you mean. I say that he is not my husband – that my husband has escaped. This man took his place to save him. Ask anyone in the town – send out into the street for the first person you find there, and bring him in as a witness. He will tell you that the prisoner is not Anthony Anderson.

BURGOYNE [*quietly, as before*] Sergeant.

SERGEANT Yes sir.

BURGOYNE Go out into the street and bring in the first townsman you see there.

SERGEANT [*making for the door*] Yes sir.

BURGOYNE [*as the sergeant passes*] The first clean, sober townsman you see.

SERGEANT Yes sir. [*He goes out*].

BURGOYNE Sit down, Mr Anderson – if I may call you so for the present. [*Richard sits down*]. Sit down, madam, whilst we wait. Give the lady a newspaper.

RICHARD [*indignantly*] Shame!

BURGOYNE [*keenly, with a half smile*] If you are not her husband, sir, the case is not a serious one – for *her*. [*Richard bites his lip, silenced*].

JUDITH [*to Richard, as she returns to her seat*] I couldnt help it. [*He shakes his head. She sits down*].

BURGOYNE You will understand of course, Mr Anderson, that you must not build on this little incident. We are bound to make an example of somebody.

RICHARD I quite understand. I suppose there's no use in my explaining.

BURGOYNE I think we should prefer independent testimony, if you dont mind.

The sergeant, with a packet of papers in his hand, returns conducting Christy, who is much scared.

SERGEANT [*giving Burgoyne the packet*] Dispatches, sir. Delivered by a corporal of the 33rd. Dead beat with hard riding, sir.

Burgoyne opens the dispatches, and presently becomes absorbed in them. They are so serious as to take his attention completely from the court martial.

SERGEANT [*to Christy*] Now then. Attention; and take your hat off. [*He posts himself in charge of Christy, who stands on Burgoyne's side of the court*].

RICHARD [*in his usual bullying tone to Christy*] Dont be frightened, you fool: youre only wanted as a witness. Theyre not going to hang *you*.

SWINDON What's your name?

CHRISTY Christy.

RICHARD [*impatiently*] Christopher Dudgeon, you blatant idiot. Give your full name.

SWINDON Be silent, prisoner. You must not prompt the witness.

RICHARD Very well. But I warn you youll get nothing out of him unless you shake it out of him. He has been too well brought up by a pious mother to have any sense or manhood left in him.

BURGOYNE [*springing up and speaking to the sergeant in a startling voice*] Where is the man who brought these?

SERGEANT In the guard-room, sir.

Burgoyne goes out with a haste that sets the officers exchanging looks.

SWINDON [*to Christy*] Do you know Anthony Anderson, the Presbyterian minister?

CHRISTY Of course I do [*implying that Swindon must be an ass not to know it*].

SWINDON Is he here?

CHRISTY [*staring round*] I dont know.

SWINDON Do you see him?

CHRISTY No.

SWINDON You seem to know the prisoner?

CHRISTY Do you mean Dick?

SWINDON Which is Dick?

CHRISTY [*pointing to Richard*] Him.

SWINDON What is his name?

CHRISTY Dick.

RICHARD Answer properly, you jumping jackass. What do they know about Dick?

CHRISTY Well, you *are* Dick, aint you? What am I to say?

SWINDON Address me, sir; and do you, prisoner, be silent. Tell us who the prisoner is.

CHRISTY He's my brother Dick – Richard – Richard Dudgeon.

SWINDON Your brother!

CHRISTY Yes.

SWINDON You are sure he is not Anderson.

CHRISTY Who?

RICHARD [*exasperatedly*] Me, me, me, you –

SWINDON Silence, sir.

SERGEANT [*shouting*] Silence.

RICHARD [*impatiently*] Yah! [*To Christy*] He wants to know am I Minister Anderson. Tell him, and stop grinning like a zany.

CHRISTY [*grinning more than ever*] *You* Pastor Anderson! [*To Swindon*] Why, Mr Anderson's a minister – a very good man; and Dick's a bad character: the respectable people wont speak to him. He's the bad brother: I'm the good one. [*The officers laugh outright. The soldiers grin*].

SWINDON Who arrested this man?

SERGEANT I did, sir. I found him in the minister's house, sitting at tea with the lady with his coat off, quite at home. If he isnt married to her, he ought to be.

SWINDON Did he answer to the minister's name?

SERGEANT Yes sir, but not to a minister's nature. You ask the chaplain, sir.

SWINDON [*to Richard, threateningly*] So, sir, you have attempted to cheat us. And your name is Richard Dudgeon?

RICHARD Youve found it out at last, have you?

SWINDON Dudgeon is a name well known to us, eh?

RICHARD Yes: Peter Dudgeon, whom you murdered, was my uncle.

SWINDON Hm! [*He compresses his lips, and looks at Richard with vindictive gravity*].

CHRISTY Are they going to hang you, Dick?

RICHARD Yes. Get out: theyve done with you.

CHRISTY And I may keep the china peacocks?

RICHARD [*jumping up*] Get out. *Get* out, you blithering baboon, you. [*Christy flies, panicstricken*].

SWINDON [*rising – all rise*] Since you have taken the minister's place, Richard Dudgeon, you shall go through with it. The execution will take place at 12 o'clock as arranged; and unless Anderson surrenders before then, you shall take his place on the gallows. Sergeant: take your man out.

JUDITH [*distracted*] No, no –

SWINDON [*fiercely, dreading a renewal of her entreaties*] Take that woman away.

RICHARD [*springing across the table with a tiger-like bound, and seizing Swindon by the throat*] You infernal scoundrel –

The sergeant rushes to the rescue from one side, the soldiers from the other. They seize Richard and drag him back to his place. Swindon, who has been thrown supine on the table, rises, arranging his stock. He is about to speak, when he is anticipated by Burgoyne, who has just appeared at the door with two papers in his hand: a white letter and a blue dispatch.

BURGOYNE [*advancing to the table, elaborately cool*] What is this? Whats happening? Mr Anderson: I'm astonished at you.

RICHARD I am sorry I disturbed you, General. I merely wanted to strangle your understrapper there. [*Breaking out violently at Swindon*] Why do you raise the devil in me by bullying the woman like that? You oatmeal faced dog, I'd twist your cursed head off with the greatest satisfaction. [*He puts out his hands to the sergeant*] Here: handcuff me, will you; or I'll not undertake to keep my fingers off him.

The sergeant takes out a pair of handcuffs and looks to Burgoyne for instructions.

BURGOYNE Have you addressed profane language to the lady, Major Swindon?

SWINDON [*very angry*] No, sir, certainly not. That question

should not have been put to me. I ordered the woman to be removed, as she was disorderly; and the fellow sprang at me. Put away those handcuffs. I am perfectly able to take care of myself.

RICHARD Now you talk like a man, I have no quarrel with you.

BURGOYNE Mr Anderson –

SWINDON His name is Dudgeon, sir, Richard Dudgeon. He is an impostor.

BURGOYNE [*brusquely*] Nonsense, sir: you hanged Dudgeon at Springtown.

RICHARD It was my uncle, General.

BURGOYNE Oh, your uncle. [*To Swindon, handsomely*] I beg your pardon, Major Swindon. [*Swindon acknowledges the apology stiffly. Burgoyne turns to Richard*]. We are somewhat unfortunate in our relations with your family. Well, Mr Dudgeon, what I wanted to ask you is this. Who is [*reading the name from the letter*] William Maindeck Parshotter?

RICHARD He is the Mayor of Springtown.

BURGOYNE Is William – Maindeck and so on – a man of his word?

RICHARD Is he selling you anything?

BURGOYNE No.

RICHARD Then you may depend on him.

BURGOYNE Thank you, Mr – 'm Dudgeon. By the way, since you are not Mr Anderson, do we still – eh, Major Swindon? [*meaning "do we still hang him?"*]

RICHARD The arrangements are unaltered, General.

BURGOYNE Ah, indeed. I am sorry. Good morning, Mr Dudgeon. Good morning, madam.

RICHARD [*interrupting Judith almost fiercely as she is about to make some wild appeal, and taking her arm resolutely*] Not one word more. Come.

She looks imploringly at him, but is overborne by his determination. They are marched out by the four soldiers: the sergeant, very sulky, walking between Swindon and Richard, whom he watches as if he were a dangerous animal.

BURGOYNE Gentlemen: we need not detain you. Major Swindon:

a word with you. [*The officers go out. Burgoyne waits with unruffled serenity until the last of them disappears. Then he becomes very grave, and addresses Swindon for the first time without his title*]. Swindon: do you know what this is [*shewing him the letter*]?

SWINDON What?

BURGOYNE A demand for a safe-conduct for an officer of their militia to come here and arrange terms with us.

SWINDON Oh, they are giving in.

BURGOYNE They add that they are sending the man who raised Springtown last night and drove us out; so that we may know that we are dealing with an officer of importance.

SWINDON Pooh!

BURGOYNE He will be fully empowered to arrange the terms of – guess what.

SWINDON Their surrender, I hope.

BURGOYNE No: our evacuation of the town. They offer us just six hours to clear out.

SWINDON What monstrous impudence!

BURGOYNE What shall we do, eh?

SWINDON March on Springtown and strike a decisive blow at once.

BURGOYNE [*quietly*] Hm! [*Turning to the door*] Come to the adjutant's office.

SWINDON What for?

BURGOYNE To write out that safe-conduct. [*He puts his hand to the door knob to open it*].

SWINDON [*who has not budged*] General Burgoyne.

BURGOYNE [*returning*] Sir?

SWINDON It is my duty to tell you, sir, that I do not consider the threats of a mob of rebellious tradesmen a sufficient reason for our giving way.

BURGOYNE [*imperturbable*] Suppose I resign my command to you, what will you do?

SWINDON I will undertake to do what we have marched south from Quebec to do, and what General Howe has marched north from New York to do: effect a junc-

tion at Albany and wipe out the rebel army with our united forces.

BURGOYNE [*enigmatically*] And will you wipe out our enemies in London, too?

SWINDON In London! What enemies?

BURGOYNE [*forcibly*] Jobbery and snobbery, incompetence and Red Tape. [*He holds up the dispatch and adds, with despair in his face and voice*] I have just learnt, sir, that General Howe is still in New York.

SWINDON [*thunderstruck*] Good God! He has disobeyed orders!

BURGOYNE [*with sardonic calm*] He has received no orders, sir. Some gentleman in London forgot to dispatch them: he was leaving town for his holiday, I believe. To avoid upsetting his arrangements, England will lose her American colonies; and in a few days you and I will be at Saratoga with 5,000 men to face 18,000 rebels in an impregnable position.

SWINDON [*appalled*] Impossible?

BURGOYNE [*coldly*] I beg your pardon!

SWINDON I cant believe it! What will History say?

BURGOYNE History, sir, will tell lies, as usual. Come: we must send the safe-conduct. [*He goes out*].

SWINDON [*following distractedly*] My God, my God! We shall be wiped out.

As noon approaches there is excitement in the market place. The gallows which hangs there permanently for the terror of evildoers, with such minor advertizers and examples of crime as the pillory, the whipping post, and the stocks, has a new rope attached, with the noose hitched up to one of the uprights, out of reach of the boys. Its ladder, too, has been brought out and placed in position by the town beadle, who stands by to guard it from unauthorized climbing. The Websterbridge townsfolk are present in force, and in high spirits; for the news has spread that it is the devil's disciple and not the minister that King George and his terrible general are about to hang: consequently the execution can be enjoyed without any misgivings as to its righteousness, or to the cowardice of allowing it to take place without a struggle. There is even some fear of a disappointment as midday approaches and the arrival of the beadle with the ladder remains the only sign of preparation. But at last reassuring shouts of Here they come: Here they are, are heard; and a company of soldiers with fixed bayonets, half British infantry, half Hessians, tramp quickly into the middle of the market place, driving the crowd to the sides.

SERGEANT Halt. Front. Dress. [*The soldiers change their column into a square enclosing the gallows, their petty officers, energetically led by the sergeant, hustling the persons who find themselves inside the square out at the corners*]. Now then! Out of it with you: out of it. Some o youll

get strung up yourselves presently. Form that square there, will you, you damned Hoosians. No use talkin German to them: talk to their toes with the butt ends of your muskets: theyll understand that. *Get* out of it, will you. [*He comes upon Judith, standing near the gallows*]. Now then: *youve* no call here.

JUDITH May I not stay? What harm am I doing?

SERGEANT I want none of your argufying. You ought to be ashamed of yourself, running to see a man hanged thats not your husband. And he's no better than yourself. I told my major he was a gentleman; and then he goes and tries to strangle him, and calls his blessed Majesty a lunatic. So out of it with you, double quick.

JUDITH Will you take these two silver dollars and let me stay?

The sergeant, without an instant's hesitation, looks quickly and furtively round as he shoots the money dexterously into his pocket. Then he raises his voice in virtuous indignation.

THE SERGEANT *Me* take money in the execution of my duty! Certainly not. Now I'll tell you what I'll do, to teach you to corrupt the King's officer. I'll put you under arrest until the execution's over. You just stand there; and dont let me see you as much as move from that spot until youre let. [*With a swift wink at her he points to the corner of the square behind the gallows on his right, and turns noisily away, shouting*] Now then, dress up and keep em back, will you.

Cries of Hush and Silence are heard among the towns-

folk; and the sound of a military band, playing the Dead March from Saul, is heard. The crowd becomes quiet at once; and the sergeant and petty officers, hurrying to the back of the square, with a few whispered orders and some stealthy hustling cause it to open and admit the funeral procession, which is protected from the crowd by a double file of soldiers. First come Burgoyne and Swindon, who, on entering the square, glance with distaste at the gallows, and avoid passing under it by wheeling a little to the right and stationing themselves on that side. Then Mr Brudenell, the chaplain, in his surplice, with his prayer book open in his hand, walking beside Richard, who is moody and disorderly. He walks doggedly through the gallows framework, and posts himself a little in front of it. Behind him comes the executioner, a stalwart soldier in his shirtsleeves. Following him, two soldiers haul a light military waggon. Finally comes the band, which posts itself at the back of the square, and finishes the Dead March. Judith, watching Richard painfully, steals down to the gallows, and stands leaning against its right post. During the conversation which follows, the two soldiers place the cart under the gallows, and stand by the shafts, which point backwards. The executioner takes a set of steps from the cart and places it ready for the prisoner to mount. Then he climbs the tall ladder which stands against the gallows, and cuts the string by which the rope is hitched up; so that the noose drops dangling over the cart, into which he steps as he descends.

RICHARD [*with suppressed impatience, to Brudenell*] Look here,

sir: this is no place for a man of your profession. Hadnt you better go away?

SWINDON I appeal to you, prisoner, if you have any sense of decency left, to listen to the ministrations of the chaplain, and pay due heed to the solemnity of the occasion.

THE CHAPLAIN [*gently reproving Richard*] Try to control yourself, and submit to the divine will. [*He lifts his book to proceed with the service*].

RICHARD Answer for your own will, sir, and those of your accomplices here [*indicating Burgoyne and Swindon*]: I see little divinity about them or you. You talk to me of Christianity when you are in the act of hanging your enemies. Was there ever such blasphemous nonsense! [*To Swindon, more rudely*] Youve got up the solemnity of the occasion, as you call it, to impress the people with your own dignity – Handel's music and a clergyman to make murder look like piety! Do you suppose *I* am going to help you? Youve asked me to choose the rope because you dont know your own trade well enough to shoot me properly. Well, hang away and have done with it.

SWINDON [*to the chaplain*] Can you do nothing with him, Mr Brudenell?

CHAPLAIN I will try, sir. [*Beginning to read*] Man that is born of woman hath –

RICHARD [*fixing his eyes on him*] "Thou shalt not kill."

The book drops in Brudenell's hands.

CHAPLAIN [*confessing his embarrassment*] What *am* I to say, Mr Dudgeon?

RICHARD Let me alone, man, cant you?

BURGOYNE [*with extreme urbanity*] I think, Mr Brudenell, that as the usual professional observations seem to strike Mr Dudgeon as incongruous under the circumstances, you had better omit them until – er – until Mr Dudgeon can no longer be inconvenienced by them. [*Brudenell, with a shrug, shuts his book and retires behind the gallows*]. You seem in a hurry, Mr Dudgeon.

RICHARD [*with the horror of death upon him*] Do you think this is

a pleasant sort of thing to be kept waiting for? Youve made up your mind to commit murder: well, do it and have done with it.

BURGOYNE Mr Dudgeon: we are only doing this –

RICHARD Because youre paid to do it.

SWINDON You insolent – [*he swallows his rage*].

BURGOYNE [*with much charm of manner*] Ah, I am really sorry that you should think that, Mr Dudgeon. If you knew what my commission cost me, and what my pay is, you would think better of me. I should be glad to part from you on friendly terms.

RICHARD Hark ye, General Burgoyne. If you think that I like being hanged, youre mistaken. I dont like it; and I dont mean to pretend that I do. And if you think I'm obliged to you for hanging me in a gentlemanly way, youre wrong there too. I take the whole business in devilish bad part; and the only satisfaction I have in it is that youll feel a good deal meaner than I'll look when it's over. [*He turns away, and is striding to the cart when Judith advances and interposes with her arms stretched out to him. Richard, feeling that a very little will upset his self-possession, shrinks from her, crying*] What are you doing here? This is no place for you. [*She makes a gesture as if to touch him. He recoils impatiently*] No: go away, go away: youll unnerve me. Take her away, will you.

JUDITH Wont you bid me good-bye?

RICHARD [*allowing her to take his hand*] Oh good-bye, good-bye. Now go – go – quickly. [*She clings to his hand – will not be put off with so cold a last farewell – at last, as he tries to disengage himself, throws herself on his breast in agony*].

SWINDON [*angrily to the sergeant, who, alarmed at Judith's movement, has come from the back of the square to pull her back, and stopped irresolutely on finding that he is too late*] How is this? Why is she inside the lines?

SERGEANT [*guiltily*] I dunno, sir. She's that artful – cant keep her away.

BURGOYNE You were bribed.

SERGEANT [*protesting*] No, sir –

SWINDON [*severely*] Fall back. [*He obeys*].

RICHARD [*imploringly to those around him, and finally to Burgoyne, as the least stolid of them*] Take her away. Do you think I want a woman near me now?

BURGOYNE [*going to Judith and taking her hand*] *Here*, madam: you had better keep inside the lines; but stand here behind us; and dont look.

Richard, with a great sobbing sigh of relief as she releases him and turns to Burgoyne, flies for refuge to the cart and mounts into it. The executioner takes off his coat and pinions him.

JUDITH [*resisting Burgoyne quietly and drawing her hand away*] No: I must stay. I wont look. [*She goes to the right of the gallows. She tries to look at Richard, but turns away with a frightful shudder, and falls on her knees in prayer. Brudenell comes towards her from the back of the square*].

BURGOYNE [*nodding approvingly as she kneels*] Ah, quite so. Do not disturb her, Mr Brudenell: that will do very nicely. [*Brudenell nods also, and withdraws a little, watching her sympathetically. Burgoyne resumes his former position, and takes out a handsome gold chronometer*]. Now then, are those preparations made? We must not detain Mr Dudgeon.

By this time Richard's hands are bound behind him; and the noose is round his neck. The two soldiers take the shaft of the waggon, ready to pull it away. The executioner, standing in the cart behind Richard, makes a sign to the sergeant.

SERGEANT [*to Burgoyne*] Ready, sir.

BURGOYNE Have you anything more to say, Mr Dudgeon? It wants two minutes of twelve still.

RICHARD [*in the strong voice of a man who has conquered the bitterness of death*] Your watch is two minutes slow by the town clock, which I can see from here, General. [*The town clock strikes the first stroke of twelve. Involuntarily the people flinch at the sound, and a subdued groan breaks from them*]. Amen! my life for the world's future!

ANDERSON [*shouting as he rushes into the market place*] Amen; and stop the execution. [*He bursts through the line of sol-*

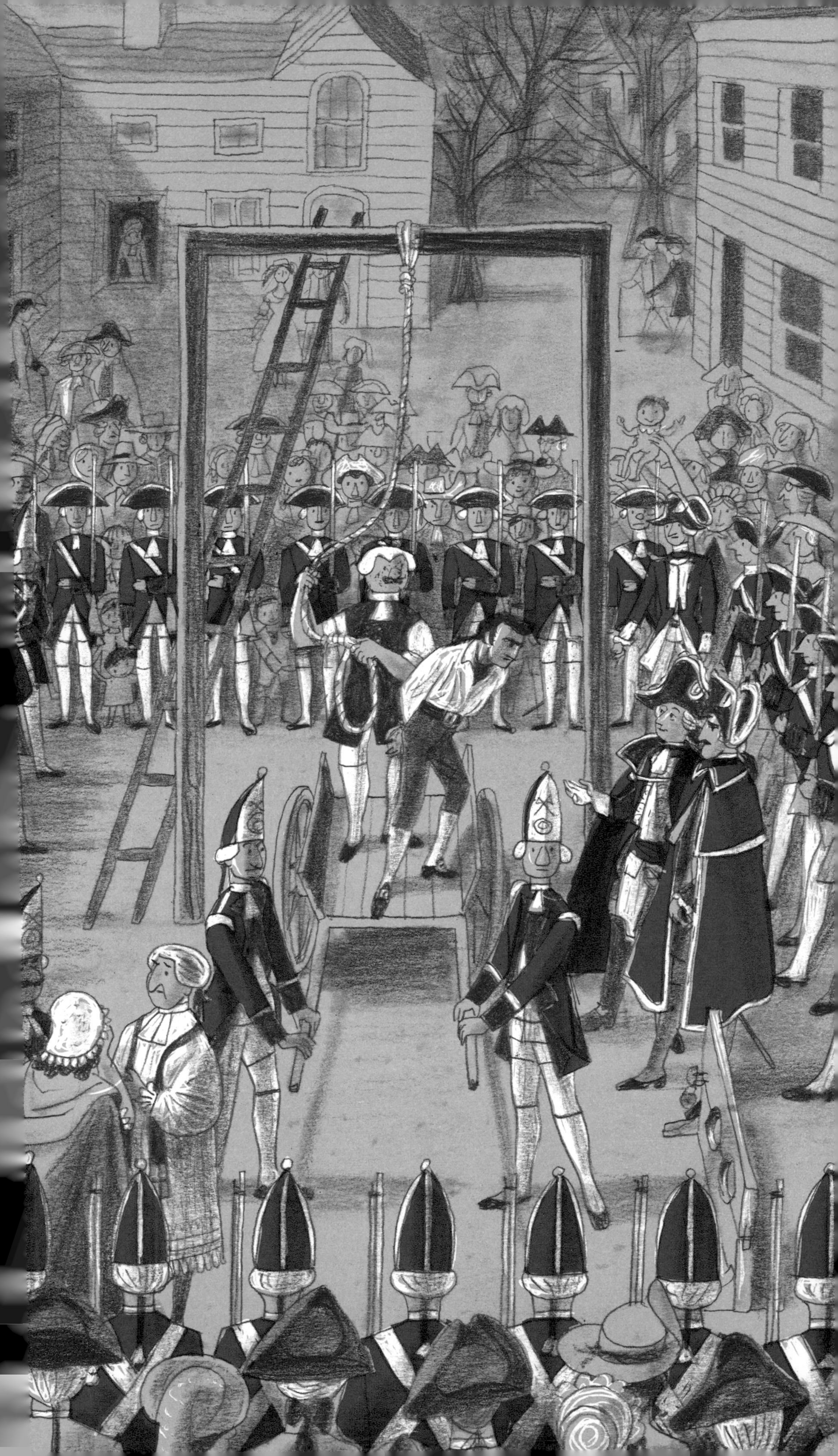

diers opposite Burgoyne, and rushes, panting, to the gallows]. I am Anthony Anderson, the man you want.

The crowd, intensely excited, listens with all its ears. Judith, half rising, stares at him; then lifts her hands like one whose dearest prayer has been granted.

SWINDON Indeed. Then you are just in time to take your place on the gallows. Arrest him.

At a sign from the sergeant, two soldiers come forward to seize Anderson.

ANDERSON [*thrusting a paper under Swindon's nose*] There's my safe-conduct, sir.

SWINDON [*taken aback*] Safe-conduct! Are you – !

ANDERSON [*emphatically*] I am. [*The two soldiers take him by the elbows*]. Tell these men to take their hands off me.

SWINDON [*to the men*] Let him go.

SERGEANT Fall back.

The two men return to their places. The townsfolk raise a cheer; and begin to exchange exultant looks, with a presentiment of triumph as they see their Pastor speaking with their enemies in the gate.

ANDERSON [*exhaling a deep breath of relief, and dabbing his perspiring brow with his handkerchief*] Thank God, I was in time!

BURGOYNE [*calm as ever, and still watch in hand*] Ample time, sir.

Plenty of time. I should never dream of hanging any gentleman by an American clock. [*He puts up his watch*].

ANDERSON Yes: we are some minutes ahead of you already, General. Now tell them to take the rope from the neck of that American citizen.

BURGOYNE [*to the executioner in the cart – very politely*] Kindly undo Mr Dudgeon.

The executioner takes the rope from Richard's neck, unties his hands, and helps him on with his coat.

JUDITH [*stealing timidly to Anderson*] Tony.

ANDERSON [*putting his arm round her shoulders and bantering her affectionately*] Well, what do you think of your husband *now*, eh? – eh?? – eh???

JUDITH I am ashamed – [*she hides her face against his breast*].

BURGOYNE [*to Swindon*] You look disappointed, Major Swindon.

SWINDON You look defeated, General Burgoyne.

BURGOYNE I am, sir; and I am humane enough to be glad of it. [*Richard jumps down from the cart, Brudenell offering his hand to help him, and runs to Anderson, whose left hand he shakes heartily, the right being occupied by Judith*]. By the way, Mr Anderson, I do not quite understand. The safe-conduct was for a commander of the militia. I understand you are a – [*He looks as pointedly as his good manners permit at the riding boots, the pistols, and Richard's coat, and adds*] – a clergyman.

ANDERSON [*between Judith and Richard*] Sir: it is in the hour of trial that a man finds his true profession. This foolish young man [*placing his hand on Richard's shoulder*] boasted himself the Devil's Disciple; but when the hour of trial came to him, he found that it was his destiny to suffer and be faithful to the death. I thought myself a decent minister of the gospel of peace; but when the hour of trial came to me, I found that it was my destiny to be a man of action, and that my place was amid the thunder of the captains and the shouting. So I am starting life at fifty as Captain Anthony Anderson of the Springtown militia; and the Devil's Disciple here will start presently as the Reverend Richard Dudgeon, and wag his pow in my

old pulpit, and give good advice to this silly sentimental little wife of mine [*putting his other hand on her shoulder. She steals a glance at Richard to see how the prospect pleases him*]. Your mother told me, Richard, that I should never have chosen Judith if I'd been born for the ministry. I am afraid she was right; so, by your leave, you may keep my coat and I'll keep yours.

RICHARD Minister – I should say Captain. I have behaved like a fool.

JUDITH Like a hero.

RICHARD Much the same thing, perhaps. [*With some bitterness towards himself*] But no: if I had been any good, I should have done for you what you did for me, instead of making a vain sacrifice.

ANDERSON Not vain, my boy. It takes all sorts to make a world – saints as well as soldiers. [*Turning to Burgoyne*] And now, General, time presses; and America is in a hurry. Have you realized that though you may occupy towns and win battles, you cannot conquer a nation?

BURGOYNE My good sir, without a Conquest you cannot have an aristocracy. Come and settle the matter at my quarters.

ANDERSON At your service, sir. [*To Richard*] See Judith home for me, will you, my boy. [*He hands her over to him*]. Now, General. [*He goes busily up the market place towards the Town Hall, leaving Judith and Richard together. Burgoyne follows him a step or two; then checks himself and turns to Richard*].

BURGOYNE Oh, by the way, Mr Dudgeon, I shall be glad to see you at lunch at half-past one. [*He pauses a moment, and adds, with politely veiled slyness*] Bring Mrs Anderson, if she will be so good. [*To Swindon, who is fuming*] Take it quietly, Major Swindon: your friend the British soldier can stand up to anything except the British War Office. [*He follows Anderson*].

SERGEANT [*to Swindon*] What orders, sir?

SWINDON [*savagely*] Orders! What use are orders now? There's no army. Back to quarters; and be d—— [*He turns on his heel and goes*].

SERGEANT [*pugnacious and patriotic, repudiating the idea of*

defeat] 'Tention. Now then: cock up your chins, and shew em you dont care a damn for em. Slope arms! Fours! Wheel! Quick march!

The drum marks time with a tremendous bang; the band strikes up British Grenadiers; and the sergeant, Brudenell, and the English troops march off defiantly to their quarters. The townsfolk press in behind, and follow them up the market, jeering at them; and the town band, a very primitive affair, brings up the rear, playing Yankee Doodle. Essie, who comes in with them, runs to Richard.

ESSIE Oh, Dick!

RICHARD [*good-humoredly, but wilfully*] Now, now: come, come! I dont mind being hanged; but I will not be cried over.

ESSIE No, I promise. I'll be good. [*She tries to restrain her tears, but cannot*]. I – I want to see where the soldiers are going to. [*She goes a little way up the market, pretending to look after the crowd*].

JUDITH Promise me you will never tell him.

RICHARD Dont be afraid.

They shake hands on it.

ESSIE [*calling to them*] Theyre coming back. They want you.

Jubilation in the market. The townsfolk surge back again in wild enthusiasm with their band, and hoist Richard on their shoulders, cheering him.

CÆSAR AND CLEOPATRA

ACT 1

An October night on the Syrian border of Egypt towards the end of the XXXIII Dynasty, in the year 706 by Roman computation, afterwards reckoned by Christian computation as 48 B.C. *A great radiance of silver fire, the dawn of a moonlit night, is rising in the east. The stars and the cloudless sky are our own contemporaries, nineteen and a half centuries younger than we know them; but you would not guess that from their appearance. Below them are two notable drawbacks of civilization: a palace, and soldiers. The palace, an old, low, Syrian building of whitened mud, is not so ugly as Buckingham Palace; and the officers in the courtyard are more highly civilized than modern English officers: for example, they do not dig up the corpses of their dead enemies and mutilate them, as we dug up Cromwell and the Mahdi. They are in two groups: one intent on the gambling of their captain Belzanor, a warrior of fifty, who, with his spear on the ground beside his knee, is stooping to throw dice with a sly-looking young Persian recruit; the other gathered about a guardsman who has just finished telling a naughty story (still current in English barracks) at which they are laughing uproariously. They are about a dozen in number, all highly aristocratic young Egyptian guardsmen, handsomely equipped with weapons and armor, very unEnglish in*

point of not being ashamed of and uncomfortable in their professional dress; on the contrary, rather ostentatiously and arrogantly warlike, as valuing themselves on their military caste.

Belzanor is a typical veteran, tough and wilful; prompt, capable and crafty where brute force will serve; helpless and boyish when it will not: an effective sergeant, an incompetent general, a deplorable dictator. Would, if influentially connected, be employed in the two last capacities by a modern European State on the strength of his success in the first. Is rather to be pitied just now in view of the fact that Julius Cæsar is invading his country. Not knowing this, is intent on his game with the Persian, whom, as a foreigner, he considers quite capable of cheating him.

His subalterns are mostly handsome young fellows whose interest in the game and the story symbolize with tolerable completeness the main interests in life of which they are conscious. Their spears are leaning against the walls, or lying on the ground ready to their hands. The corner of the courtyard forms a triangle of which one side is the front of the palace, with a doorway, the other a wall with a gateway. The storytellers are on the palace side: the gamblers, on the gateway side. Close to the gateway, against the wall, is a stone block high enough to enable a Nubian sentinel, standing on it, to look over the wall. The yard is lighted by a torch stuck in the wall. As the laughter from the group round the storyteller dies away, the kneeling Persian, winning the throw, snatches up the stake from the ground.

BELZANOR By Apis, Persian, thy gods are good to thee.

THE PERSIAN Try yet again, O captain. Double or quits!

BELZANOR No more. I am not in the vein.

THE SENTINEL [*poising his javelin as he peers over the wall*] Stand. Who goes there?

They all start, listening. A strange voice replies from without.

VOICE The bearer of evil tidings.

BELZANOR [*calling to the sentry*] Pass him.

THE SENTINEL [*grounding his javelin*] Draw near, O bearer of evil tidings.

BELZANOR [*pocketing the dice and picking up his spear*] Let us receive this man with honor. He bears evil tidings.

The guardsmen seize their spears and gather about the gate, leaving a way through for the new comer.

PERSIAN [*rising from his knee*] Are evil tidings, then, so honorable?

BELZANOR O barbarous Persian, hear my instruction. In Egypt the bearer of good tidings is sacrificed to the gods as a thank offering; but no god will accept the blood of the messenger of evil. When we have good tidings, we are careful to send them in the mouth of the cheapest slave we can find. Evil tidings are borne by young noblemen who desire to bring themselves into notice. [*They join the rest at the gate*].

SENTINEL Pass, O young captain; and bow the head in the House of the Queen.

VOICE Go anoint thy javelin with fat of swine, O Blackamoor; for before morning the Romans will make thee eat it to the very butt.

The owner of the voice, a fairhaired dandy, dressed in a different fashion from that affected by the guardsmen, but no less extravagantly, comes through the gateway laughing. He is somewhat battlestained; and his left forearm, bandaged, comes through a torn sleeve. In his right hand he carries a Roman sword in its sheath. He swaggers down the courtyard, the Persian on his right, Belzanor on his left, and the guardsmen crowding down behind him.

BELZANOR Who art thou that laughest in the House of Cleopatra the Queen, and in the teeth of Belzanor, the captain of her guard?

NEW COMER I am Bel Affris, descended from the gods.

BELZANOR [*ceremoniously*] Hail, cousin!

ALL [*except the Persian*] Hail, cousin!

PERSIAN All the Queen's guards are descended from the gods, O stranger, save myself. I am Persian, and descended from many kings.

BEL AFFRIS [*to the guardsmen*] Hail, cousins! [*To the Persian, condescendingly*] Hail, mortal!

BELZANOR You have been in battle, Bel Affris; and you are a

soldier among soldiers. You will not let the Queen's women have the first of your tidings.

BEL AFFRIS I have no tidings, except that we shall have our throats cut presently, women, soldiers, and all.

PERSIAN [*to Belzanor*] I told you so.

THE SENTINEL [*who has been listening*] Woe, alas!

BEL AFFRIS [*calling to him*] Peace, peace, poor Ethiop: destiny is with the gods who painted thee black. [*To Belzanor*] What has this mortal [*indicating the Persian*] told you?

BELZANOR He says that the Roman Julius Cæsar, who has landed on our shores with a handful of followers, will make himself master of Egypt. He is afraid of the Roman soldiers. [*The guardsmen laugh with boisterous scorn*]. Peasants, brought up to scare crows and follow the plough! Sons of smiths and millers and tanners! And we nobles, consecrated to arms, descended from the gods!

PERSIAN Belzanor: the gods are not always good to their poor relations.

BELZANOR [*hotly, to the Persian*] Man to man, are we worse than the slaves of Cæsar?

BEL AFFRIS [*stepping between them*] Listen, cousin. Man to man, we Egyptians are as gods above the Romans.

GUARDSMEN [*exultantly*] Aha!

BEL AFFRIS But this Cæsar does not pit man against man: he throws a legion at you where you are weakest as he throws a stone from a catapult; and that legion is as a man with one head, a thousand arms, and no religion. I have fought against them; and I know.

BELZANOR [*derisively*] Were you frightened, cousin?

The guardsmen roar with laughter, their eyes sparkling at the wit of their captain.

BEL AFFRIS No, cousin; but I was beaten. They were frightened (perhaps); but they scattered us like chaff.

The guardsmen, much damped, utter a growl of contemptuous disgust.

BELZANOR Could you not die?

BEL AFFRIS No: that was too easy to be worthy of a descendant of the gods. Besides, there was no time: all was over in a

moment. The attack came just where we least expected it.

BELZANOR That shews that the Romans are cowards.

BEL AFFRIS They care nothing about cowardice, these Romans: they fight to win. The pride and honor of war are nothing to them.

PERSIAN Tell us the tale of the battle. What befell?

GUARDSMEN [*gathering eagerly round Bel Affris*] Ay: the tale of the battle.

BEL AFFRIS Know then, that I am a novice in the guard of the temple of Ra in Memphis, serving neither Cleopatra nor her brother Ptolemy, but only the high gods. We went a journey to inquire of Ptolemy why he had driven Cleopatra into Syria, and how we of Egypt should deal with the Roman Pompey, newly come to our shores after his defeat by Cæsar at Pharsalia. What, think ye, did we learn? Even that Cæsar is coming also in hot pursuit of his foe, and that Ptolemy has slain Pompey, whose severed head he holds in readiness to present to the conqueror. [*Sensation among the guardsmen*]. Nay, more: we found that Cæsar is already come; for we had not made half a day's journey on our way back when we came upon a city rabble flying from his legions, whose landing they had gone out to withstand.

BELZANOR And ye, the temple guard! did ye not withstand these legions?

BEL AFFRIS What man could, that we did. But there came the sound of a trumpet whose voice was as the cursing of a black mountain. Then saw we a moving wall of shields coming towards us. You know how the heart burns when you charge a fortified wall; but how if the fortified wall were to charge *you?*

THE PERSIAN [*exulting in having told them so*] Did I not say it?

BEL AFFRIS When the wall came nigh, it changed into a line of men – common fellows enough, with helmets, leather tunics, and breastplates. Every man of them flung his javelin: the one that came my way drove through my shield as through a papyrus – lo there! [*he points to the bandage on his left arm*] and would have gone through my neck had I not stooped. They were

charging at the double then, and were upon us with short swords almost as soon as their javelins. When a man is close to you with such a sword, you can do nothing with our weapons: they are all too long.

THE PERSIAN What did you do?

BEL AFFRIS Doubled my fist and smote my Roman on the sharpness of his jaw. He was but mortal after all: he lay down in a stupor; and I took his sword and laid it on. [*Drawing the sword*] Lo! a Roman sword with Roman blood on it!

GUARDSMEN [*approvingly*] Good! [*They take the sword and hand it round, examining it curiously*].

THE PERSIAN And your men?

BEL AFFRIS Fled. Scattered like sheep.

BELZANOR [*furiously*] The cowardly slaves! Leaving the descendants of the gods to be butchered!

BEL AFFRIS [*with acid coolness*] The descendants of the gods did not stay to be butchered, cousin. The battle was not to the strong; but the race was to the swift. The Romans, who have no chariots, sent a cloud of horsemen in pursuit, and slew multitudes. Then our high priest's captain rallied a dozen descendants of the gods and exhorted us to die fighting. I said to myself: surely it is safer to stand than to lose my breath and be stabbed in the back; so I joined our captain and stood. Then the Romans treated us with respect; for no man attacks a lion when the field is full of sheep, except for the pride and honor of war, of which these Romans know nothing. So we escaped with our lives; and I am come to warn you that you must open your gates to Cæsar; for his advance guard is scarce an hour behind me; and not an Egyptian warrior is left standing between you and his legions.

THE SENTINEL Woe, alas! [*He throws down his javelin and flies into the palace*].

BELZANOR Nail him to the door, quick! [*The guardsmen rush for him with their spears; but he is too quick for them*]. Now this news will run through the palace like fire through stubble.

BEL AFFRIS What shall we do to save the women from the Romans?

BELZANOR Why not kill them?

PERSIAN Because we should have to pay blood money for some of them. Better let the Romans kill them: it is cheaper.

BELZANOR [*awestruck at his brain power*] O subtle one! O serpent!

BEL AFFRIS But your Queen?

BELZANOR True: we must carry off Cleopatra.

BEL AFFRIS Will ye not await her command?

BELZANOR Command! a girl of sixteen! Not we. At Memphis ye deem her a Queen: here we know better. I will take her on the crupper of my horse. When we soldiers have carried her out of Cæsar's reach, then the priests and the nurses and the rest of them can pretend she is a queen again, and put their commands into her mouth.

PERSIAN Listen to me, Belzanor.

BELZANOR Speak, O subtle beyond thy years.

THE PERSIAN Cleopatra's brother Ptolemy is at war with her. Let us sell her to him.

GUARDSMEN O subtle one! O serpent!

BELZANOR We dare not. We are descended from the gods; but Cleopatra is descended from the river Nile; and the lands of our fathers will grow no grain if the Nile rises not to water them. Without our father's gifts we should live the lives of dogs.

PERSIAN It is true: the Queen's guard cannot live on its pay. But hear me further, O ye kinsmen of Osiris.

GUARDSMEN Speak, O subtle one. Hear the serpent begotten!

PERSIAN Have I heretofore spoken truly to you of Cæsar, when you thought I mocked you?

GUARDSMEN Truly, truly.

BELZANOR [*reluctantly admitting it*] So Bel Affris says.

PERSIAN Hear more of him, then. This Cæsar is a great lover of women: he makes them his friends and counsellors.

BELZANOR Faugh! This rule of women will be the ruin of Egypt.

THE PERSIAN Let it rather be the ruin of Rome! Cæsar grows old now: he is past fifty and full of labors and battles. He is too old for the young women; and the old women are too wise to worship him.

BEL AFFRIS Take heed, Persian. Cæsar is by this time almost within earshot.

PERSIAN Cleopatra is not yet a woman: neither is she wise. But she already troubles men's wisdom.

BELZANOR Ay: that is because she is descended from the river Nile and a black kitten of the sacred White Cat. What then?

PERSIAN Why, sell her secretly to Ptolemy, and then offer ourselves to Cæsar as volunteers to fight for the overthrow of her brother and the rescue of our Queen, the Great Granddaughter of the Nile.

GUARDSMEN O serpent!

PERSIAN He will listen to us if we come with her picture in our mouths. He will conquer and kill her brother, and reign in Egypt with Cleopatra for his Queen. And we shall be her guard.

GUARDSMEN O subtlest of all the serpents! O admiration! O wisdom!

BEL AFFRIS He will also have arrived before you have done talking, O word spinner.

BELZANOR That is true. [*An affrighted uproar in the palace interrupts him*]. Quick: the flight has begun: guard the door. [*They rush to the door and form a cordon before it with their spears. A mob of women-servants and nurses surges out. Those in front recoil from the spears, screaming to those behind to keep back. Belzanor's voice dominates the disturbance as he shouts*] Back there. In again, unprofitable cattle.

GUARDSMEN Back, unprofitable cattle.

BELZANOR Send us out Ftatateeta, the Queen's chief nurse.

THE WOMEN [*calling into the palace*] Ftatateeta, Ftatateeta. Come, come. Speak to Belzanor.

A WOMAN Oh, keep back. You are thrusting me on the spearheads.

A huge grim woman, her face covered with a network of tiny wrinkles, and her eyes old, large, and wise; sinewy handed, very tall, very strong; with the mouth of a bloodhound and the jaws of a bulldog, appears on the threshold. She is dressed like a person of consequence in the palace, and confronts the guardsmen insolently.

FTATATEETA Make way for the Queen's chief nurse.

BELZANOR [*with solemn arrogance*] Ftatateeta: I am Belzanor, the captain of the Queen's guard, descended from the gods.

FTATATEETA [*retorting his arrogance with interest*] Belzanor: I am Ftatateeta, the Queen's chief nurse; and your divine ancestors were proud to be painted on the wall in the pyramids of the kings whom my fathers served.

The women laugh triumphantly.

BELZANOR [*with grim humor*] Ftatateeta: daughter of a long-tongued, swivel-eyed chameleon, the Romans are at hand. [*A cry of terror from the women: they would fly but for the spears*]. Not even the descendants of the gods can resist them; for they have each man seven arms, each carrying seven spears. The blood in their veins is boiling quicksilver; and their wives become mothers in three hours, and are slain and eaten the next day.

A shudder of horror from the women. Ftatateeta, despising them and scorning the soldiers, pushes her way through the crowd and confronts the spear points undismayed.

FTATATEETA Then fly and save yourselves, O cowardly sons of the cheap clay gods that are sold to fish porters; and leave us to shift for ourselves.

BELZANOR Not until you have first done our bidding, O terror of manhood. Bring out Cleopatra the Queen to us; and then go whither you will.

FTATATEETA [*with a derisive laugh*] Now I know why the gods have taken her out of our hands. [*The guardsmen start and look at one another*]. Know, thou foolish soldier, that the Queen has been missing since an hour past sundown.

BELZANOR [*furiously*] Hag: you have hidden her to sell to Cæsar or her brother. [*He grasps her by the left wrist, and drags her, helped by a few of the guard, to the middle of the courtyard, where, as they fling her on her knees, he draws a murderous looking knife*]. Where is she? Where is she? or – [*he threatens to cut her throat*].

FTATATEETA [*savagely*] Touch me, dog; and the Nile will not rise on your fields for seven times seven years of famine.

BELZANOR [*frightened, but desperate*] I will sacrifice: I will pay. Or stay. [*To the Persian*] You, O subtle one: your father's lands lie far from the Nile. Slay her.

PERSIAN [*threatening her with his knife*] Persia has but one god; yet he loves the blood of old women. Where is Cleopatra?

FTATATEETA Persian: as Osiris lives, I do not know. I chid her for bringing evil days upon us by talking to the sacred cats of the priests, and carrying them in her arms. I told her she would be left alone here when the Romans came as a punishment for her disobedience. And now she is gone – run away – hidden. I speak the truth. I call Osiris to witness –

THE WOMEN [*protesting officiously*] She speaks the truth, Belzanor.

BELZANOR You have frightened the child: she is hiding. Search – quick – into the palace – search every corner.

The guards, led by Belzanor, shoulder their way into the palace through the flying crowd of women, who escape through the courtyard gate.

FTATATEETA [*screaming*] Sacrilege! Men in the Queen's chambers! Sa – [*her voice dies away as the Persian puts his knife to her throat*].

BEL AFFRIS [*laying a hand on Ftatateeta's left shoulder*] Forbear her yet a moment, Persian. [*To Ftatateeta, very sig-*

nificantly] Mother: your gods are asleep or away hunting; and the sword is at your throat. Bring us to where the Queen is hid, and you shall live.

FTATATEETA [*contemptuously*] Who shall stay the sword in the hand of a fool, if the high gods put it there? Listen to me, ye young men without understanding. Cleopatra fears me; but she fears the Romans more. There is but one power greater in her eyes than the wrath of the Queen's nurse and the cruelty of Cæsar; and that is the power of the Sphinx that sits in the desert watching the way to the sea. What she would have it know, she tells into the ears of the sacred cats; and on her birthday she sacrifices to it and decks it with poppies. Go ye therefore into the desert and seek Cleopatra in the shadow of the Sphinx; and on your heads see to it that no harm comes to her.

BEL AFFRIS [*to the Persian*] May we believe this, O subtle one?

PERSIAN Which way come the Romans?

BEL AFFRIS Over the desert, from the sea, by this very Sphinx.

PERSIAN [*to Ftatateeta*] O mother of guile! O aspic's tongue! You have made up this tale so that we two may go into the desert and perish on the spears of the Romans. [*Lifting his knife*] Taste death.

FTATATEETA Not from thee, baby. [*She snatches his ankle from under him and flies stooping along the palace wall, vanishing in the darkness within its precinct. Bel Affris roars with laughter as the Persian tumbles. The guardsmen rush out of the palace with Belzanor and a mob of fugitives, mostly carrying bundles*].

PERSIAN Have you found Cleopatra?

BELZANOR She is gone. We have searched every corner.

THE NUBIAN SENTINEL [*appearing at the door of the palace*] Woe! Alas! Fly, fly!

BELZANOR What is the matter now?

SENTINEL The sacred white cat has been stolen.

ALL Woe! woe! [*General panic. They all fly with cries of consternation. The torch is thrown down and extinguished in the rush. Darkness. The noise of the fugitives dies away. Dead silence. Suspense. Then the blackness and stillness break softly into silver mist and strange airs as the windswept harp of Memnon plays at the dawning of the moon. It rises full over the desert; and a vast horizon comes into relief, broken by a huge shape which soon reveals itself in the spreading radiance as a Sphinx pedestalled on the sands. The light still clears, until the upraised eyes of the image are distinguished looking straight forward and upward in infinite fearless vigil, and a mass of color between its great paws defines itself as a heap of red poppies on which a girl lies motionless, her silken vest heaving gently and regularly with the breathing of a dreamless sleeper, and her braided hair glittering in a shaft of moonlight like a bird's wing.*

Suddenly there comes from afar a vaguely fearful sound (it might be the bellow of a Minotaur softened by great distance) and Memnon's music stops. Silence: then a few faint high-ringing trumpet notes. Then silence again. Then a man comes from the south with stealing steps, ravished by the mystery of the night, all wonder, and halts, lost in contemplation, opposite the left flank of the Sphinx, whose bosom, with its burden, is hidden from him by its massive shoulder.

THE MAN Hail, Sphinx: salutation from Julius Cæsar! I have wandered in many lands, seeking the lost regions from which my birth into this world exiled me, and the company of creatures such as I myself. I have found flocks and pastures, men and cities, but no other Cæsar, no air native to me, no man kindred to me, none who can do my day's deed, and think my night's thought. In the little world yonder, Sphinx, my place is as high as yours in this great desert; only I wander, and you sit still; I conquer, and you endure;

I work and wonder, you watch and wait; I look up and am dazzled, look down and am darkened, look round and am puzzled, whilst your eyes never turn from looking out – out of the world – to the lost region – the home from which we have strayed. Sphinx, you and I, strangers to the race of men, are no strangers to one another: have I not been conscious of you and of this place since I was born? Rome is a madman's dream: this is my Reality. These starry lamps of yours I have seen from afar in Gaul, in Britain, in Spain, in Thessaly, signalling great secrets to some eternal sentinel below, whose post I never could find. And here at last is their sentinel – an image of the constant and immortal part of my life, silent, full of thoughts, alone in the silver desert. Sphinx, Sphinx: I have climbed mountains at night to hear in the distance the stealthy footfall of the winds that chase your sands in forbidden play – our invisible children, O Sphinx, laughing in whispers. My way hither was the way of destiny; for I am he of whose genius you are the symbol: part brute, part woman, and part god – nothing of man in me at all. Have I read your riddle, Sphinx?

THE GIRL [*who has wakened, and peeped cautiously from her nest to see who is speaking*] Old gentleman.

CÆSAR [*starting violently, and clutching his sword*] Immortal gods!

THE GIRL Old gentleman: dont run away.

CÆSAR [*stupefied*] "Old gentleman: dont run away"! ! ! This! to Julius Cæsar!

THE GIRL [*urgently*] Old gentleman.

CÆSAR Sphinx: you presume on your centuries. I am younger than you, though your voice is but a girl's voice as yet.

THE GIRL Climb up here, quickly; or the Romans will come and eat you.

CÆSAR [*running forward past the Sphinx's shoulder, and seeing her*] A child at its breast! a divine child!

THE GIRL Come up quickly. You must get up at its side and creep round.

CÆSAR [*amazed*] Who are you?

THE GIRL Cleopatra, Queen of Egypt.

CÆSAR Queen of the Gypsies, you mean.

CLEOPATRA You must not be disrespectful to me, or the Sphinx will let the Romans eat you. Come up. It is quite cosy here.

CÆSAR [*to himself*] What a dream! What a magnificent dream! Only let me not wake, and I will conquer ten continents to pay for dreaming it out to the end. [*He climbs to the Sphinx's flank, and presently reappears to her on the pedestal, stepping round its right shoulder*].

CLEOPATRA Take care. That's right. Now sit down: you may have its other paw. [*She seats herself comfortably on its left paw*]. It is very powerful and will protect us; but [*shivering, and with plaintive loneliness*] it would not take any notice of me or keep me company. I am glad you have come: I was very lonely. Did you happen to see a white cat anywhere?

CÆSAR [*sitting slowly down on the right paw in extreme wonderment*] Have you lost one?

CLEOPATRA Yes: the sacred white cat: is it not dreadful? I brought him here to sacrifice him to the Sphinx; but when we got a little way from the city a black cat called him, and he jumped out of my arms and ran away to it. Do you think that the black cat can have been my great-great-great-grandmother?

CÆSAR [*staring at her*] Your great-great-great-grandmother! Well, why not? Nothing would surprise me on this night of nights.

CLEOPATRA I think it must have been. My great-grandmother's great-grandmother was a black kitten of the sacred white cat; and the river Nile made her his seventh wife. That is why my hair is so wavy. And I always want to be let do as I like, no matter whether it is the will of the gods or not: that is because my blood is made with Nile water.

CÆSAR What are you doing here at this time of night? Do you live here?

CLEOPATRA Of course not: I am the Queen: and I shall live in the palace at Alexandria when I have killed my brother,

who drove me out of it. When I am old enough I shall do just what I like. I shall be able to poison the slaves and see them wriggle, and pretend to Ftatateeta that she is going to be put into the fiery furnace.

CÆSAR Hm! Meanwhile why are you not at home and in bed?

CLEOPATRA Because the Romans are coming to eat us all. *You* are not at home and in bed either.

CÆSAR [*with conviction*] Yes I am. I live in a tent; and I am now in that tent, fast asleep and dreaming. Do you suppose that I believe you are real, you impossible little dream witch?

CLEOPATRA [*giggling and leaning trustfully towards him*] You are a funny old gentleman. I like you.

CÆSAR Ah, that spoils the dream. Why dont you dream that I am young?

CLEOPATRA I wish you were; only I think I should be more afraid of you. I like men, especially young men with round strong arms; but I am afraid of them. You are old and rather thin and stringy; but you have a nice voice; and I like to have somebody to talk to, though I think you are a little mad. It is the moon that makes you talk to yourself in that silly way.

CÆSAR What! you heard that, did you? I was saying my prayers to the great Sphinx.

CLEOPATRA But this isnt the great Sphinx.

CÆSAR [*much disappointed, looking up at the statue*] What!

CLEOPATRA This is only a dear little kitten of a Sphinx. Why, the great Sphinx is so big that it has a temple between its paws. This is my pet Sphinx. Tell me: do you think the Romans have any sorcerers who could take us away from the Sphinx by magic?

CÆSAR Why? Are you afraid of the Romans?

CLEOPATRA [*very seriously*] Oh, they would eat us if they caught us. They are barbarians. Their chief is called Julius Cæsar. His father was a tiger and his mother a burning mountain; and his nose is like an elephant's trunk. [*Cæsar involuntarily rubs his nose*]. They all have long noses, and ivory tusks, and little tails, and seven arms with a hundred arrows in each; and they live on human flesh.

CÆSAR Would you like me to shew you a real Roman?

CLEOPATRA [*terrified*] No. You are frightening me.

CÆSAR No matter: this is only a dream –

CLEOPATRA [*excitedly*] It is not a dream: it is not a dream. See, see. [*She plucks a pin from her hair and jabs it repeatedly into his arm*].

CÆSAR Ffff – Stop. [*Wrathfully*] How dare you?

CLEOPATRA [*abashed*] You said you were dreaming. [*Whimpering*] I only wanted to shew you –

CÆSAR [*gently*] Come, come: dont cry. A queen mustnt cry. [*He rubs his arm, wondering at the reality of the smart*]. *Am* I awake? [*He strikes his hand against the Sphinx to test its solidity. It feels so real that he begins to be alarmed, and says perplexedly*] Yes, I – [*quite panic-stricken*] no: impossible: madness, madness! [*Desperately*] Back to camp – to camp [*He rises to spring down from the pedestal*].

CLEOPATRA [*flinging her arms in terror round him*] No: you shant leave me. No, no, no: dont go. I'm afraid – afraid of the Romans.

CÆSAR [*as the conviction that he is really awake forces itself on him*] Cleopatra: can you see my face well?

CLEOPATRA Yes. It is so white in the moonlight.

CÆSAR Are you sure it is the moonlight that makes me look whiter than an Egyptian? [*Grimly*] Do you notice that I have a rather long nose?

CLEOPATRA [*recoiling, paralyzed by a terrible suspicion*] Oh!

CÆSAR It is a Roman nose, Cleopatra.

CLEOPATRA Ah! [*With a piercing scream she springs up; darts round the left shoulder of the Sphinx; scrambles down to the sand; and falls on her knees in frantic supplication, shrieking*] Bite him in two, Sphinx: bite him in two. I meant to sacrifice the white cat – I did indeed – I [*Cæsar, who has slipped down from the pedestal, touches her on the shoulder*]. Ah! [*She buries her head in her arms*].

CÆSAR Cleopatra: shall I teach you a way to prevent Cæsar from eating you?

CLEOPATRA [*clinging to him piteously*] Oh, do, do, do. I will steal

Ftatateeta's jewels and give them to you. I will make the river Nile water your lands twice a year.

CÆSAR Peace, peace, my child. Your gods are afraid of the Romans: you see the Sphinx dare not bite me, nor prevent me carrying you off to Julius Cæsar.

CLEOPATRA [*in pleading murmurings*] You wont, you wont. You said you wouldnt.

CÆSAR Cæsar never eats women.

CLEOPATRA [*springing up full of hope*] What!

CÆSAR [*impressively*] But he eats girls [*she relapses*] and cats. Now you are a silly little girl; and you are descended from the black kitten. You are both a girl and a cat.

CLEOPATRA [*trembling*] And will he eat *me?*

CÆSAR Yes; unless you make him believe that you are a woman.

CLEOPATRA Oh, you must get a sorcerer to make a woman of me. Are you a sorcerer?

CÆSAR Perhaps. But it will take a long time; and this very night you must stand face to face with Cæsar in the palace of your fathers.

CLEOPATRA No, no. I darent.

CÆSAR Whatever dread may be in your soul – however terrible Cæsar may be to you – you must confront him as a brave woman and a great queen; and you must feel no fear. If your hand shakes: if your voice quavers; then – night and death! [*She moans*]. But if he thinks you worthy to rule, he will set you on the throne by his side and make you the real ruler of Egypt.

CLEOPATRA [*despairingly*] No: he will find me out: he will find me out.

CÆSAR [*rather mournfully*] He is easily deceived by women. Their eyes dazzle him; and he sees them not as they are, but as he wishes them to appear to him.

CLEOPATRA [*hopefully*] Then we will cheat him. I will put on Ftatateeta's head-dress; and he will think me quite an old woman.

CÆSAR If you do that he will eat you at one mouthful.

CLEOPATRA But I will give him a cake with my magic opal

and seven hairs of the white cat baked in it; and –

CÆSAR [*abruptly*] Pah! you are a little fool. He will eat your cake and you too. [*He turns contemptuously from her*].

CLEOPATRA [*running after him and clinging to him*] Oh please, *please!* I will do whatever you tell me. I will be good. I will be your slave. [*Again the terrible bellowing note sounds across the desert, now closer at hand. It is the bucina, the Roman war trumpet*].

CÆSAR Hark!

CLEOPATRA [*trembling*] What was that?

CÆSAR Cæsar's voice.

CLEOPATRA [*pulling at his hand*] Let us run away. Come. Oh come.

CÆSAR You are safe with me until you stand on your throne to receive Cæsar. Now lead me thither.

CLEOPATRA [*only too glad to get away*] I will, I will. [*Again the bucina*]. Oh come, come, come: the gods are angry. Do you feel the earth shaking?

CÆSAR It is the tread of Cæsar's legions.

CLEOPATRA [*drawing him away*] This way, quickly. And let us look for the white cat as we go. It is he that has turned you into a Roman.

CÆSAR Incorrigible, oh, incorrigible! Away! [*He follows her, the bucina sounding louder as they steal across the desert. The moonlight wanes: the horizon again shows black against the sky, broken only by the fantastic silhouette of the Sphinx. The sky itself vanishes in darkness, from which there is no relief until the gleam of a distant torch falls on great Egyptian pillars supporting the roof of a majestic corridor. At the further end of this corridor a Nubian slave appears carrying the torch. Cæsar, still led by Cleopatra, follows him. They come down the corridor, Cæsar peering keenly about at the strange architecture, and at the pillar shadows between which, as the passing torch makes them hurry noiselessly backwards, figures of men with wings and hawks' heads, and vast black marble cats, seem to flit in and out of ambush. Further along, the wall turns a corner and makes a spacious transept in which Cæsar sees, on his right, a throne, and behind the throne a door. On each side of the throne is a slender pillar with a lamp on it.*

CÆSAR What place is this?

CLEOPATRA This is where I sit on the throne when I am allowed to wear my crown and robes. [*The slave holds his torch to shew the throne*].

CÆSAR Order the slave to light the lamps.

CLEOPATRA [*shyly*] Do you think I may?

CÆSAR Of course. You are the Queen. [*She hesitates*]. Go on.

CLEOPATRA [*timidly, to the slave*] Light all the lamps.

FTATATEETA [*suddenly coming from behind the throne*] Stop. [*The slave stops. She turns sternly to Cleopatra, who quails like a naughty child*]. Who is this you have with you; and how dare you order the lamps to be lighted without my permission? [*Cleopatra is dumb with apprehension*].

CÆSAR Who is she?

CLEOPATRA Ftatateeta.

FTATATEETA [*arrogantly*] Chief nurse to –

CÆSAR [*cutting her short*] I speak to the Queen. Be silent. [*To Cleopatra*] Is this how your servants know their places? Send her away; and do you [*to the slave*] do as the Queen has bidden. [*The slave lights the lamps. Meanwhile Cleopatra stands hesitating, afraid of Ftatateeta*]. You are the Queen: send her away.

CLEOPATRA [*cajoling*] Ftatateeta, dear: you must go away – just for a little.

CÆSAR You are not commanding her to go away: you are begging her. You are no Queen. You will be eaten. Farewell. [*He turns to go*].

CLEOPATRA [*clutching him*] No, no, no. Dont leave me.

CÆSAR A Roman does not stay with queens who are afraid of their slaves.

CLEOPATRA I am not afraid. Indeed I am not afraid.

FTATATEETA We shall see who is afraid here. [*Menacingly*] Cleopatra –

CÆSAR On your knees, woman: am I also a child that you dare trifle with me? [*He points to the floor at Cleopatra's feet. Ftatateeta, half cowed, half savage, hesitates. Cæsar calls to the Nubian*] Slave. [*The Nubian comes to him*]. Can you cut off a head? [*The Nubian*

nods and grins ecstatically, showing all his teeth. Cæsar takes his sword by the scabbard, ready to offer the hilt to the Nubian, and turns again to Ftatateeta, repeating his gesture]. Have you remembered yourself, mistress?

Ftatateeta, crushed, kneels before Cleopatra, who can hardly believe her eyes.

FTATATEETA [*hoarsely*] O Queen, forget not thy servant in the days of thy greatness.

CLEOPATRA [*blazing with excitement*] Go. Begone. Go away. [*Ftatateeta rises with stooped head, and moves backwards towards the door. Cleopatra watches her submission eagerly, almost clapping her hands, which are trembling. Suddenly she cries*] Give me something to beat her with. [*She snatches a snake-skin from the throne and dashes after Ftatateeta, whirling it like a scourge in the air. Cæsar makes a bound and manages to catch her and hold her while Ftatateeta escapes*].

CÆSAR You scratch, kitten, do you?

CLEOPATRA [*breaking from him*] I *will* beat somebody. I will beat him. [*She attacks the slave*]. There, there, there! [*The slave flies for his life up the corridor and vanishes. She throws the snake-skin away and jumps on the step of the throne with her arms waving, crying*] I am a real Queen at last – a real, real Queen! Cleopatra the Queen! [*Cæsar shakes his head dubiously, the advantage of the change seeming open to question from the point of view of the general welfare of Egypt. She turns and looks at him exultantly. Then she jumps down from the step, runs to him, and flings her arms round him rapturously, crying*] Oh, I love you for making me a Queen.

CÆSAR But queens love only kings.

CLEOPATRA I will make all the men I love kings. I will make you a king. I will have many young kings, with round, strong arms; and when I am tired of them I will whip them to death; but you shall always be my king: my nice, kind, wise, good old king.

CÆSAR Oh, my wrinkles, my wrinkles! And my child's heart! You will be the most dangerous of all Cæsar's conquests.

CLEOPATRA [*appalled*] Cæsar! I forgot Cæsar. [*Anxiously*] You will tell him that I am a Queen, will you not? – a real

Queen. Listen! [*stealthily coaxing him*]: let us run away and hide until Cæsar is gone.

CÆSAR If you fear Cæsar, you are no true queen; and though you were to hide beneath a pyramid, he would go straight to it and lift it with one hand. And then – ! [*he chops his teeth together*].

CLEOPATRA [*trembling*] Oh!

CÆSAR Be afraid if you dare. [*The note of the bucina resounds again in the distance. She moans with fear. Cæsar exults in it, exclaiming*] Aha! Cæsar approaches the throne of Cleopatra. Come: take your place. [*He takes her hand and leads her to the throne. She is too downcast to speak*]. Ho, there, Teetatota. How do you call your slaves?

CLEOPATRA [*spiritlessly, as she sinks on the throne and cowers there, shaking*]. Clap your hands.

He claps his hands. Ftatateeta returns.

CÆSAR Bring the Queen's robes, and her crown, and her women; and prepare her.

CLEOPATRA [*eagerly – recovering herself a little*] Yes, the crown, Ftatateeta: I shall wear the crown.

FTATATEETA For whom must the Queen put on her state?

CÆSAR For a citizen of Rome. A king of kings, Totateeta.

CLEOPATRA [*stamping at her*] How dare you ask questions? Go and do as you are told. [*Ftatateeta goes out with a grim smile. Cleopatra goes on eagerly, to Cæsar*] Cæsar will know that I am a Queen when he sees my crown and robes, will he not?

CÆSAR No. How shall he know that you are not a slave dressed up in the Queen's ornaments?

CLEOPATRA You must tell him.

CÆSAR He will not ask me. He will know Cleopatra by her pride, her courage, her majesty, and her beauty. [*She looks very doubtful*]. Are you trembling?

CLEOPATRA [*shivering with dread*] No, I – I – [*in a very sickly voice*] No.

Ftatateeta and three women come in with the regalia.

FTATATEETA Of all the Queen's women, these three alone are left. The rest are fled. [*They begin to deck Cleopatra, who submits, pale and motionless*].

CÆSAR Good, good. Three are enough. Poor Cæsar generally has to dress himself.

FTATATEETA [*contemptuously*] The queen of Egypt is not a Roman barbarian. [*To Cleopatra*] Be brave, my nursling. Hold up your head before this stranger.

CÆSAR [*admiring Cleopatra, and placing the crown on her head*] Is it sweet or bitter to be a Queen, Cleopatra?

CLEOPATRA Bitter.

CÆSAR Cast out fear; and you will conquer Cæsar. Tota: are the Romans at hand?

FTATATEETA They are at hand; and the guard has fled.

THE WOMEN [*wailing subduedly*] Woe to us!

The Nubian comes running down the hall.

NUBIAN The Romans are in the courtyard. [*He bolts through the door. With a shriek, the women fly after him. Ftatateeta's jaw expresses savage resolution: she does not budge. Cleopatra can hardly restrain herself from following them. Cæsar grips her wrist, and looks steadfastly at her. She stands like a martyr*].

CÆSAR The Queen must face Cæsar alone. Answer "So be it."

CLEOPATRA [*white*] So be it.

CÆSAR [*releasing her*] Good.

A tramp and tumult of armed men is heard. Cleopatra's terror increases. The bucina sounds close at hand, followed by a formidable clangor of trumpets. This is too much for Cleopatra: she utters a cry and darts towards the door. Ftatateeta stops her ruthlessly.

FTATATEETA You are my nursling. You have said "So be it"; and if you die for it, you must make the Queen's word good. [*She hands Cleopatra to Cæsar, who takes her back, almost beside herself with apprehension, to the throne*].

CÆSAR Now, if you quail –! [*He seats himself on the throne*].

She stands on the step, all but unconscious, waiting for death. The Roman soldiers troop in tumultuously through the corridor, headed by their ensign with his eagle, and their bucinator, a burly fellow with his instrument coiled round his body, its brazen bell shaped like the head of a howling wolf. When they reach the transept, they stare in amazement at the throne; dress into ordered rank opposite it; draw their swords and lift them in the air with a shout of Hail, Cæsar. *Cleopatra turns and stares wildly at Cæsar; grasps the situation; and, with a great sob of relief, falls into his arms.*

ACT 2

Alexandria. A hall on the first floor of the Palace, ending in a loggia approached by two steps. Through the arches of the loggia the Mediterranean can be seen, bright in the morning sun. The clean lofty walls, painted with a procession of the Egyptian theocracy, presented in profile as flat ornament, and the absence of mirrors, sham perspectives, stuffy upholstery and textiles, make the place handsome, wholesome, simple and cool, or, as a rich English manufacturer would express it, poor, bare, ridiculous and unhomely. For Tottenham Court Road civilization is to this Egyptian civilization as glass bead and tattoo civilization is to Tottenham Court Road.

The young king Ptolemy Dionysus (aged ten) is at the top of the steps, on his way in through the loggia, led by his guardian Pothinus, who has him by the hand. The

court is assembled to receive him. It is made up of men and women (some of the women being officials) of various complexions and races, mostly Egyptian; some of them, comparatively fair, from lower Egypt; some, much darker, from upper Egypt; with a few Greeks and Jews. Prominent in a group on Ptolemy's right hand is Theodotus, Ptolemy's tutor. Another group, on Ptolemy's left, is headed by Achillas, the general of Ptolemy's troops. Theodotus is a little old man, whose features are as cramped and wizened as his limbs, except his tall straight forehead, which occupies more space than all the rest of his face. He maintains an air of magpie keenness and profundity, listening to what the others say with the sarcastic vigilance of a philosopher listening to the exercises of his disciples. Achillas is a tall handsome man of thirty-five, with a fine black beard curled like the coat of a poodle. Apparently not a clever man, but distinguished and dignified. Pothinus is a vigorous man of fifty, a eunuch, passionate, energetic and quick witted, but of common mind and character; impatient and unable to control his temper. He has fine tawny hair, like fur. Ptolemy, the King, looks much older than an English boy of ten; but he has the childish air, the habit of being in leading strings, the mixture of impotence and petulance, the appearance of being excessively washed, combed and dressed by other hands, which is exhibited by court-bred princes of all ages.

All receive the King with reverences. He comes down the steps to a chair of state which stands a little to his right, the only seat in the hall. Taking his place before it, he looks nervously for instructions to Pothinus, who places himself at his left hand.

POTHINUS The king of Egypt has a word to speak.

THEODOTUS [*in a squeak which he makes impressive by sheer self-opinionativeness*] Peace for the King's word!

PTOLEMY [*without any vocal inflexions: he is evidently repeating a lesson*] Take notice of this all of you. I am the first-born son of Auletes the Flute Blower who was your King. My sister Berenice drove him from his throne and reigned in his stead but – but – [*he hesitates*] –

POTHINUS [*stealthily prompting*] – but the gods would not suffer –

PTOLEMY Yes – the gods would not suffer – not suffer – [*He stops; then, crestfallen*] I forget what the gods would not suffer.

THEODOTUS Let Pothinus, the King's guardian, speak for the King.

POTHINUS [*suppressing his impatience with difficulty*] The King wished to say that the gods would not suffer the impiety of his sister to go unpunished.

PTOLEMY [*hastily*] Yes: I remember the rest of it. [*He resumes his monotone*]. Therefore the gods sent a stranger one Mark Antony a Roman captain of horsemen across the sands of the desert and he set my father again upon the throne. And my father took Berenice my sister and struck her head off. And now that my father is dead yet another of his daughters my sister Cleopatra would snatch the kingdom from me and reign in my place. But the gods would not suffer – [*Pothinus coughs admonitorily*] – the gods – the gods would not suffer –

POTHINUS [*prompting*] – will not maintain –

PTOLEMY Oh yes – will not maintain such iniquity they will give her head to the axe even as her sister's. But with the help of the witch Ftatateeta she hath cast a spell on the Roman Julius Cæsar to make him uphold her false pretence to rule in Egypt. Take notice then that I will not suffer – that I will not suffer – [*pettishly, to Pothinus*] What is it that I will not suffer?

POTHINUS [*suddenly exploding with all the force and emphasis of political passion*] The King will not suffer a foreigner to take from him the throne of our Egypt. [*A shout of applause*]. Tell the King, Achillas, how many soldiers and horsemen follow the Roman?

THEODOTUS Let the King's general speak!

ACHILLAS But two Roman legions, O King. Three thousand soldiers and scarce a thousand horsemen.

The court breaks into derisive laughter; and a great chattering begins, amid which Rufio, a Roman officer, appears in the loggia. He is a burly, black-bearded man of middle age, very blunt, prompt and rough, with small clear eyes and plump nose and cheeks, which, however, like the rest of his flesh, are in ironhard condition.

RUFIO [*from the steps*] Peace, ho! [*The laughter and chatter cease abruptly*]. Cæsar approaches.

THEODOTUS [*with much presence of mind*] The King permits the Roman commander to enter!

Cæsar, plainly dressed, but wearing an oak wreath to conceal his baldness, enters from the loggia, attended by Britannus, his secretary, a Briton, about forty, tall, solemn, and already slightly bald, with a heavy, drooping, hazel-colored moustache trained so as to lose its ends in a pair of trim whiskers. He is carefully dressed in blue, with portfolio, inkhorn, and reed pen at his girdle. His serious air and sense of the importance of the business in hand is in marked contrast to the kindly interest of Cæsar, who looks at the scene, which is new to him, with the frank curiosity of a child, and then turns to the king's chair: Britannus and Rufio posting themselves near the steps at the other side.

CÆSAR [*looking at Pothinus and Ptolemy*] Which is the King? the man or the boy?

POTHINUS I am Pothinus, the guardian of my lord the King.

CÆSAR [*patting Ptolemy kindly on the shoulder*] So you are the King. Dull work at your age, eh? [*To Pothinus*] Your servant, Pothinus. [*He turns away unconcernedly and comes slowly along the middle of the hall, looking from side to side at the courtiers until he reaches Achillas*]. And this gentleman?

THEODOTUS Achillas, the King's general.

CÆSAR [*to Achillas, very friendly*] A general, eh? I am a general myself. But I began too old, too old. Health and many victories, Achillas!

ACHILLAS As the gods will, Cæsar.

CÆSAR [*turning to Theodotus*] And you, sir, are – ?

THEODOTUS Theodotus, the King's tutor.

CÆSAR You teach men how to be kings, Theodotus. That is very clever of you. [*Looking at the gods on the walls as he turns away from Theodotus and goes up again to Pothinus*] And this place?

POTHINUS The council chamber of the chancellors of the King's treasury, Cæsar.

CÆSAR Ah! that reminds me. I want some money.

POTHINUS The King's treasury is poor, Cæsar.

CÆSAR Yes: I notice that there is but one chair in it.

RUFIO [*shouting gruffly*] Bring a chair there, some of you, for Cæsar.

PTOLEMY [*rising shyly to offer his chair*] Cæsar –

CÆSAR [*kindly*] No, no, my boy: that is your chair of state. Sit down.

He makes Ptolemy sit down again. Meanwhile Rufio, looking about him, sees in the nearest corner an image of the god Ra, represented as a seated man with the head of a hawk. Before the image is a bronze tripod, about as large as a three-legged stool, with a stick of incense burning on it. Rufio, with Roman resourcefulness and indifference to foreign superstitions, promptly seizes the tripod; shakes off the incense; blows away the ash; and dumps it down behind Cæsar, nearly in the middle of the hall.

RUFIO Sit on that, Cæsar.

A shiver runs through the court, followed by a hissing whisper of Sacrilege!

CÆSAR [*seating himself*] Now, Pothinus, to business. I am badly in want of money.

BRITANNUS [*disapproving of these informal expressions*] My master would say that there is a lawful debt due to Rome by Egypt, contracted by the King's deceased father to the Triumvirate; and that it is Cæsar's duty to his country to require immediate payment.

CÆSAR [*blandly*] Ah, I forgot. I have not made my companions known here. Pothinus: this is Britannus, my secretary. He is an islander from the western end of the world, a day's voyage from Gaul. [*Britannus bows stiffly*]. This gentleman is Rufio, my comrade in arms. [*Rufio nods*]. Pothinus: I want 1,600 talents.

The courtiers, appalled, murmur loudly, and Theodotus and Achillas appeal mutely to one another against so monstrous a demand.

POTHINUS [*aghast*] Forty million sesterces! Impossible. There is not so much money in the King's treasury.

CÆSAR [*encouragingly*] *Only* sixteen hundred talents, Pothi-

nus. Why count it in sesterces? A sestertius is only worth a loaf of bread.

POTHINUS And a talent is worth a racehorse. I say it is impossible. We have been at strife here, because the King's sister Cleopatra falsely claims his throne. The King's taxes have not been collected for a whole year.

CÆSAR Yes they have, Pothinus. My officers have been collecting them all the morning. [*Renewed whisper and sensation, not without some stifled laughter, among the courtiers*].

RUFIO [*bluntly*] You must pay, Pothinus. Why waste words? You are getting off cheaply enough.

POTHINUS [*bitterly*] Is it possible that Cæsar, the conqueror of the world, has time to occupy himself with such a trifle as our taxes?

CÆSAR My friend: taxes are the chief business of a conqueror of the world.

POTHINUS Then take warning, Cæsar. This day, the treasures of the temples and the gold of the King's treasury shall be sent to the mint to be melted down for our ransom in the sight of the people. They shall see us sitting under bare walls and drinking from wooden cups. And their wrath be on your head, Cæsar, if you force us to this sacrilege!

CÆSAR Do not fear, Pothinus: the people know how well wine tastes in wooden cups. In return for your bounty, I will settle this dispute about the throne for you, if you will. What say you?

POTHINUS If I say no, will that hinder you?

RUFIO [*defiantly*] No.

CÆSAR You say the matter has been at issue for a year, Pothinus. May I have ten minutes at it?

POTHINUS You will do your pleasure, doubtless.

CÆSAR Good! But first, let us have Cleopatra here.

THEODOTUS She is not in Alexandria: she is fled into Syria.

CÆSAR I think not. [*To Rufio*] Call Totateeta.

RUFIO [*Calling*] Ho there, Teetatota.

Ftatateeta enters the loggia, and stands arrogantly at the top of the steps.

FTATATEETA Who pronounces the name of Ftatateeta, the Queen's chief nurse?

CÆSAR Nobody can pronounce it, Tota, except yourself. Where is your mistress?

Cleopatra, who is hiding behind Ftatateeta, peeps out at them, laughing. Cæsar rises.

CÆSAR Will the Queen favor us with her presence for a moment?

CLEOPATRA [*pushing Ftatateeta aside and standing haughtily on the brink of the steps*] Am I to behave like a Queen?

CÆSAR Yes.

Cleopatra immediately comes down to the chair of state; seizes Ptolemy; drags him out of his seat; then takes his place in the chair. Ftatateeta seats herself on the step of the loggia, and sits there, watching the scene with sibylline intensity.

PTOLEMY [*mortified, and struggling with his tears*] Cæsar: this is how she treats me always. If I am a king why is she allowed to take everything from me?

CLEOPATRA You are not to be King, you little cry-baby. You are to be eaten by the Romans.

CÆSAR [*touched by Ptolemy's distress*] Come here, my boy, and stand by me.

Ptolemy goes over to Cæsar, who, resuming his seat on the tripod, takes the boy's hand to encourage him. Cleopatra, furiously jealous, rises and glares at them.

CLEOPATRA [*with flaming cheeks*] Take your throne: I dont want it. [*She flings away from the chair, and approaches Ptolemy, who shrinks from her*]. Go this instant and sit down in your place.

CÆSAR Go, Ptolemy. Always take a throne when it is offered to you.

RUFIO I hope you will have the good sense to follow your own advice when we return to Rome, Cæsar.

Ptolemy slowly goes back to the throne, giving Cleopatra a wide berth, in evident fear of her hands. She takes his place beside Cæsar.

CÆSAR Pothinus –

CLEOPATRA [*interrupting him*] Are you not going to speak to me?

CÆSAR Be quiet. Open your mouth again before I give you leave; and you shall be eaten.

CLEOPATRA I am not afraid. A queen must not be afraid. Eat my husband there, if you like: *he* is afraid.

CÆSAR [*starting*] Your husband! What do you mean?

CLEOPATRA [*pointing to Ptolemy*] That little thing.

The two Romans and the Briton stare at one another in amazement.

THEODOTUS Cæsar: you are a stranger here, and not conversant with our laws. The kings and queens of Egypt may not marry except with their own royal blood. Ptolemy and Cleopatra are born king and consort just as they are born brother and sister.

BRITANNUS [*shocked*] Cæsar: this is not proper.

THEODOTUS [*outraged*] How!

CÆSAR [*recovering his self-possession*] Pardon him, Theodotus: he is a barbarian, and thinks that the customs of his tribe and island are the laws of nature.

BRITANNUS On the contrary, Cæsar, it is these Egyptians who are barbarians; and you do wrong to encourage them. I say it is a scandal.

CÆSAR Scandal or not, my friend, it opens the gate of peace. [*He addresses Pothinus seriously*]. Pothinus: hear what I propose.

RUFIO Hear Cæsar there.

CÆSAR Ptolemy and Cleopatra shall reign jointly in Egypt.

ACHILLAS What of the King's younger brother and Cleopatra's younger sister?

RUFIO [*explaining*] There is another little Ptolemy, Cæsar: so they tell me.

CÆSAR Well, the little Ptolemy can marry the other sister; and we will make them both a present of Cyprus.

POTHINUS [*impatiently*] Cyprus is of no use to anybody.

CÆSAR No matter: you shall have it for the sake of peace.

BRITANNUS [*unconsciously anticipating a later statesman*] Peace with honor, Pothinus.

POTHINUS [*mutinously*] Cæsar: be honest. The money you demand is the price of our freedom. Take it; and leave us to settle our own affairs.

EGYPT FOR THE
EGYPTIANS

THE BOLDER COURTIERS [*encouraged by Pothinus's tone and Cæsar's quietness*] Yes, yes. Egypt for the Egyptians!

The conference now becomes an altercation, the Egyptians becoming more and more heated. Cæsar remains unruffled; but Rufio grows fiercer and doggeder, and Britannus haughtily indignant.

RUFIO [*contemptuously*] Egypt for the Egyptians! Do you forget that there is a Roman army of occupation here, left by Aulus Gabinius when he set up your toy king for you?

ACHILLAS [*suddenly asserting himself*] And now under *my* command. *I* am the Roman general here, Cæsar.

CÆSAR [*tickled by the humor of the situation*] And also the Egyptian general, eh?

POTHINUS [*triumphantly*] That is so, Cæsar.

CÆSAR [*to Achillas*] So you can make war on the Egyptians in the name of Rome, and on the Romans – on *me*, if necessary – in the name of Egypt?

ACHILLAS That is so, Cæsar.

CÆSAR And which side are you on at present, if I may presume to ask, general?

ACHILLAS On the side of the right and of the gods.

CÆSAR Hm! How many men have you?

ACHILLAS That will appear when I take the field.

RUFIO [*truculently*] Are your men Romans? If not, it matters not how many there are, provided you are no stronger than 500 to ten.

POTHINUS It is useless to try to bluff us, Rufio. Cæsar has been defeated before and may be defeated again. A few weeks ago Cæsar was flying for his life before Pompey: a few months hence he may be flying for his life before Cato and Juba of Numidia, the African King.

ACHILLAS [*following up Pothinus's speech menacingly*] What can you do with 4,000 men?

THEODOTUS [*following up Achillas's speech with a raucous squeak*] And without money? Away with you.

ALL THE COURTIERS [*shouting fiercely and crowding towards Cæsar*] Away with you. Egypt for the Egyptians! Begone.

Rufio bites his beard, too angry to speak. Cæsar sits as

comfortably as if he were at breakfast, and the cat were clamoring for a piece of Finnan-haddie.

CLEOPATRA Why do you let them talk to you like that, Cæsar? Are you afraid?

CÆSAR Why, my dear, what they say is quite true.

CLEOPATRA But if you go away, I shall not be Queen.

CÆSAR I shall not go away until you are Queen.

POTHINUS Achillas: if you are not a fool, you will take that girl whilst she is under your hand.

RUFIO [*daring them*] Why not take Cæsar as well, Achillas?

POTHINUS [*retorting the defiance with interest*] Well said, Rufio. Why not?

RUFIO Try, Achillas. [*Calling*] Guard there.

The loggia immediately fills with Cæsar's soldiers, who stand, sword in hand, at the top of the steps, waiting the word to charge from their centurion, who carries a cudgel. For a moment the Egyptians face them proudly: then they retire sullenly to their former places.

BRITANNUS You are Cæsar's prisoners, all of you.

CÆSAR [*benevolently*] Oh no, no, no. By no means. Cæsar's guests, gentlemen.

CLEOPATRA Wont you cut their heads off?

CÆSAR What! Cut off your brother's head?

CLEOPATRA Why not? He would cut off mine, if he got the chance. Wouldnt you, Ptolemy?

PTOLEMY [*pale and obstinate*] I would. I will, too, when I grow up.

Cleopatra is rent by a struggle between her newly-acquired dignity as a queen, and a strong impulse to put out her tongue at him. She takes no part in the scene which follows, but watches it with curiosity and wonder, fidgeting with the restlessness of a child, and sitting down on Cæsar's tripod when he rises.

POTHINUS Cæsar: if you attempt to detain us –

RUFIO He will succeed, Egyptian: make up your mind to that. We hold the palace, the beach, and the eastern harbor. The road to Rome is open; and you shall travel it if Cæsar chooses.

CÆSAR [*courteously*] I could do no less, Pothinus, to secure

the retreat of my own soldiers. I am accountable for every life among them. But you are free to go. So are all here, and in the palace.

RUFIO [*aghast at this clemency*] What! Renegades and all?

CÆSAR [*softening the expression*] Roman army of occupation and all, Rufio.

POTHINUS [*desperately*] Then I make a last appeal to Cæsar's justice. I shall call a witness to prove that but for us, the Roman army of occupation, led by the greatest soldier in the world, would now have Cæsar at its mercy. [*Calling through the loggia*] Ho, there, Lucius Septimius [*Cæsar starts, deeply moved*]: if my voice can reach you, come forth and testify before Cæsar.

CÆSAR [*shrinking*] No, no.

THEODOTUS Yes, I say. Let the military tribune bear witness.

Lucius Septimius, a clean shaven, trim athlete of about 40, with symmetrical features, resolute mouth, and handsome, thin Roman nose, in the dress of a Roman officer, comes in through the loggia and confronts Cæsar, who hides his face with his robe for a moment; then, mastering himself, drops it, and confronts the tribune with dignity.

POTHINUS Bear witness, Lucius Septimius. Cæsar came hither in pursuit of his foe. Did we shelter his foe?

LUCIUS As Pompey's foot touched the Egyptian shore, his head fell by the stroke of my sword.

THEODOTUS [*with viperish relish*] Under the eyes of his wife and child! Remember that, Cæsar! They saw it from the ship he had just left. We have given you a full and sweet measure of vengeance.

CÆSAR [*with horror*] Vengeance!

POTHINUS Our first gift to you, as your galley came into the roadstead, was the head of your rival for the empire of the world. Bear witness, Lucius Septimius: is it not so?

LUCIUS It is so. With this hand, that slew Pompey, I placed his head at the feet of Cæsar.

CÆSAR Murderer! So would you have slain Cæsar, had Pompey been victorious at Pharsalia.

LUCIUS Woe to the vanquished, Cæsar! When I served Pompey, I slew as good men as he, only because he conquered them. His turn came at last.

THEODOTUS [*flatteringly*] The deed was not yours, Cæsar, but ours – nay, mine; for it was done by my counsel. Thanks to us, you keep your reputation for clemency, and have your vengeance too.

CÆSAR Vengeance! Vengcance!! Oh, if I could stoop to vengeance, what would I not exact from you as the price of this murdered man's blood? [*They shrink back, appalled and disconcerted*]. Was he not my son-in-law, my ancient friend, for 20 years the master of great Rome, for 30 years the compeller of victory? Did not I, as a Roman, share his glory? Was the Fate that forced us to fight for the mastery of the world, of our making? Am I Julius Cæsar, or am I a wolf, that you fling to me the grey head of the old soldier, the laurelled conqueror, the mighty Roman, treacherously struck down by this callous ruffian, and then claim my gratitude for it! [*To Lucius Septimius*] Begone: you fill me with horror.

LUCIUS [*cold and undaunted*] Pshaw! You have seen severed heads before, Cæsar, and severed right hands too, I think; some thousands of them, in Gaul, after you vanquished Vercingetorix. Did you spare him, with all your clemency? Was that vengeance?

CÆSAR No, by the gods! would that it had been! Vengeance at least is human. No, I say: those severed right hands, and the brave Vercingetorix basely strangled in a vault beneath the Capitol, were [*with shuddering satire*] a wise severity, a necessary protection to the commonwealth, a duty of statesmanship – follies and fictions ten times bloodier than honest vengeance! What a fool was I then! to think that men's lives should be at the mercy of such fools! [*Humbly*] Lucius Septimius, pardon me: why should the slayer of Vercingetorix rebuke the slayer of Pompey? You arc free to go with the rest. Or stay if you will: I will find a place for you in my service.

LUCIUS The odds are against you, Cæsar. I go. [*He turns to go out through the loggia*].

RUFIO [*full of wrath at seeing his prey escaping*] That means that he is a Republican.

LUCIUS [*turning defiantly on the loggia steps*] And what are you?

RUFIO A Cæsarian, like all Cæsar's soldiers.

CÆSAR [*courteously*] Lucius: believe me, Cæsar is no Cæsarian. Were Rome a true republic, then were Cæsar the first of Republicans. But you have made your choice. Farewell.

LUCIUS Farewell. Come, Achillas, whilst there is yet time.

Cæsar, seeing that Rufio's temper threatens to get the worse of him, puts his hand on his shoulder and brings him down the hall out of harm's way, Britannus accompanying them and posting himself on Cæsar's right hand. This movement brings the three in a little group to the place occupied by Achillas, who moves haughtily

away and joins Theodotus on the other side. Lucius Septimius goes out through the soldiers in the loggia. Pothinus, Theodotus and Achillas follow him with the courtiers, very mistrustful of the soldiers, who close up in their rear and go out after them, keeping them moving without much ceremony. The King is left in his chair, piteous, obstinate, with twitching face and fingers. During these movements Rufio maintains an energetic grumbling, as follows:

RUFIO [*as Lucius departs*] Do you suppose he would let us go if he had our heads in his hands?

CÆSAR I have no right to suppose that his ways are any baser than mine.

RUFIO Psha!

CÆSAR Rufio: if I take Lucius Septimius for my model, and become exactly like him, ceasing to be Cæsar, will you serve me still?

BRITANNUS Cæsar: this is not good sense. Your duty to Rome demands that her enemies should be prevented from doing further mischief. [*Cæsar, whose delight in the moral eye-to-business of his British secretary is inexhaustible, smiles indulgently*].

RUFIO It is no use talking to him, Britannus: you may save your breath to cool your porridge. But mark this, Cæsar. Clemency is very well for you; but what is it for your soldiers, who have to fight to-morrow the men you spared yesterday? You may give what orders you please; but I tell you that your next victory will be a massacre, thanks to your clemency. *I*, for one, will take no prisoners. I will kill my enemies in the field: and then you can preach as much clemency as you please: I shall never have to fight them again. And now, with your leave, I will see these gentry off the premises. [*He turns to go*].

CÆSAR [*turning also and seeing Ptolemy*] What! have they left the boy alone! Oh shame, shame!

RUFIO [*taking Ptolemy's hand and making him rise*] Come, your majesty!

PTOLEMY [*to Cæsar, drawing away his hand from Rufio*] Is he turning me out of my palace?

RUFIO [*grimly*] You are welcome to stay if you wish.

CÆSAR [*kindly*] Go, my boy. I will not harm you; but you will be safer away, among your friends. Here you are in the lion's mouth.

PTOLEMY [*turning to go*] It is not the lion I fear, but [*looking at Rufio*] the jackal. [*He goes out through the loggia*].

CÆSAR [*laughing approvingly*] Brave boy!

CLEOPATRA [*jealous of Cæsar's approbation, calling after Ptolemy*] Little silly. You think that very clever.

CÆSAR Britannus: attend the King. Give him in charge to that Pothinus fellow. [*Britannus goes out after Ptolemy*].

RUFIO [*pointing to Cleopatra*] And this piece of goods? What is to be done with *her?* However, I suppose I may leave that to you. [*He goes out through the loggia*].

CLEOPATRA [*flushing suddenly and turning on Cæsar*] Did you mean me to go with the rest?

CÆSAR [*a little preoccupied, goes with a sigh to Ptolemy's chair, whilst she waits for his answer with red cheeks and clenched fists*] You are free to do just as you please, Cleopatra.

CLEOPATRA Then you do not care whether I stay or not?

CÆSAR [*smiling*] Of course I had rather you stayed.

CLEOPATRA Much, *much* rather?

CÆSAR [*nodding*] Much, much rather.

CLEOPATRA Then I consent to stay, because I am asked. But I do not want to, mind.

CÆSAR That is quite understood. [*Calling*] Totateeta.

Ftatateeta, still seated, turns her eyes on him with a sinister expression, but does not move.

CLEOPATRA [*with a splutter of laughter*] Her name is not Totateeta: it is Ftatateeta. [*Calling*] Ftatateeta. [*Ftatateeta instantly rises and comes to Cleopatra*].

CÆSAR [*stumbling over the name*] Tfatafeeta will forgive the erring tongue of a Roman. Tota: the Queen will hold her state here in Alexandria. Engage women to attend upon her; and do all that is needful.

FTATATEETA Am I then the mistress of the Queen's household?

CLEOPATRA [*sharply*] No: *I* am the mistress of the Queen's household. Go and do as you are told, or I will have you

thrown into the Nile this very afternoon, to poison the poor crocodiles.

CÆSAR [*shocked*] Oh no, no.

CLEOPATRA Oh yes, yes. You are very sentimental, Cæsar; but you are clever; and if you do as I tell you, you will soon learn to govern.

Cæsar, quite dumbfounded by this impertinence, turns in his chair and stares at her.

Ftatateeta, smiling grimly, and showing a splendid set of teeth, goes, leaving them alone together.

CÆSAR Cleopatra: I really think I must eat you, after all.

CLEOPATRA [*kneeling beside him and looking at him with eager interest, half real, half affected to shew how intelligent she is*] You must not talk to me now as if I were a child.

CÆSAR You have been growing up since the sphinx introduced us the other night; and you think you know more than I do already.

CLEOPATRA [*taken down, and anxious to justify herself*] No: that would be very silly of me: of course I know that. But – [*suddenly*] are you angry with me?

CÆSAR No.

CLEOPATRA [*only half believing him*] Then why are you so thoughtful?

CÆSAR [*rising*] I have work to do, Cleopatra.

CLEOPATRA [*drawing back*] Work! [*Offended*] You are tired of talking to me; and that is your excuse to get away from me.

CÆSAR [*sitting down again to appease her*] Well, well: another minute. But then – work!

CLEOPATRA Work! what nonsense! You must remember that you are a king now: I have made you one. Kings dont work.

CÆSAR Oh! Who told you that, little kitten? Eh?

CLEOPATRA My father was King of Egypt; and he never worked. But he was a great king, and cut off my sister's head because she rebelled against him and took the throne from him.

CÆSAR Well; and how did he get his throne back again?

CLEOPATRA [*eagerly, her eyes lighting up*] I will tell you. A beautiful young man, with strong round arms, came over the desert with many horsemen, and slew my sister's husband and gave my father back his throne. [*Wistfully*] I was only twelve then. Oh, I wish he would come again, now that I am a queen. I would make him my husband.

CÆSAR It might be managed, perhaps; for it was I who sent that beautiful young man to help your father.

CLEOPATRA [*enraptured*] You know him!

CÆSAR [*nodding*] I do.

CLEOPATRA Has he come with you? [*Cæsar shakes his head: she is cruelly disappointed*]. Oh, I wish he had, I wish he had. If only I were a little older; so that he might not think me a mere kitten, as you do! But perhaps that is because *you* are old. He is many *many* years younger than you, is he not?

CÆSAR [*as if swallowing a pill*] He is somewhat younger.

CLEOPATRA Would he be my husband, do you think, if I asked him?

CÆSAR Very likely.

CLEOPATRA But I should not like to ask him. Could you not persuade him to ask me – without knowing that I wanted him to?

CÆSAR [*touched by her innocence of the beautiful young man's character*] My poor child!

CLEOPATRA Why do you say that as if you were sorry for me? Does he love anyone else?

CÆSAR I am afraid so.

CLEOPATRA [*tearfully*] Then I shall not be his first love.

CÆSAR Not quite the first. He is greatly admired by women.

CLEOPATRA I wish I could be the first. But if he loves me, I will make him kill all the rest. Tell me: is he still beautiful? Do his strong round arms shine in the sun like marble?

CÆSAR He is in excellent condition – considering how much he eats and drinks.

CLEOPATRA Oh, you must not say common, earthly things about him; for I love him. He is a god.

CÆSAR He is a great captain of horsemen, and swifter of foot than any other Roman.

CLEOPATRA What is his real name?

CÆSAR [*puzzled*] His *real* name?

CLEOPATRA Yes. I always call him Horus, because Horus is the most beautiful of our gods. But I want to know his real name.

CÆSAR His name is Mark Antony.

CLEOPATRA [*musically*] Mark Antony, Mark Antony, Mark Antony! What a beautiful name! [*She throws her arms round Cæsar's neck*]. Oh, how I love you for sending him to help my father! Did you love my father very much?

CÆSAR No, my child; but your father, as you say, never worked. I always work. So when he lost his crown he had to promise me 16,000 talents to get it back for him.

CLEOPATRA Did he ever pay you?

CÆSAR Not in full.

CLEOPATRA He was quite right: it was too dear. The whole world is not worth 16,000 talents.

CÆSAR That is perhaps true, Cleopatra. Those Egyptians who work paid as much of it as he could drag from them. The rest is still due. But as I most likely shall not get it, I must go back to my work. So you must run away for a little and send my secretary to me.

CLEOPATRA [*coaxing*] No: I want to stay and hear you talk about Mark Antony.

CÆSAR But if I do not get to work, Pothinus and the rest of them will cut us off from the harbor; and then the way from Rome will be blocked.

CLEOPATRA No matter: I dont want you to go back to Rome.

CÆSAR But you want Mark Antony to come from it.

CLEOPATRA [*springing up*] Oh, yes, yes, yes: I forgot. Go quickly and work, Cæsar; and keep the way over the sea open for my Mark Antony. [*She runs out through the loggia, kissing her hand to Mark Antony across the sea*].

CÆSAR [*going briskly up the middle of the hall to the loggia steps*] Ho, Britannus. [*He is startled by the entry of a*

wounded Roman soldier, who confronts him from the upper step]. What now?

SOLDIER [*pointing to his bandaged head*] This, Cæsar; and two of my comrades killed in the market place.

CÆSAR [*quiet, but attending*] Ay. Why?

SOLDIER There is an army come to Alexandria, calling itself the Roman army.

CÆSAR The Roman army of occupation. Ay?

SOLDIER Commanded by one Achillas.

CÆSAR Well?

SOLDIER The citizens rose against us when the army entered the gates. I was with two others in the market place when the news came. They set upon us. I cut my way out; and here I am.

CÆSAR Good. I am glad to see you alive. [*Rufio enters the loggia hastily, passing behind the soldier to look out through one of the arches at the quay beneath*]. Rufio: we are besieged.

RUFIO What! Already?

CÆSAR Now or to-morrow: what does it matter? We *shall* be besieged.

Britannus runs in.

BRITANNUS Cæsar –

CÆSAR [*anticipating him*] Yes: I know. [*Rufio and Britannus come down the hall from the loggia at opposite sides, past Cæsar, who waits for a moment near the step to say to the soldier*] Comrade: give the word to turn out on the beach and stand by the boats. Get your wound attended to. Go. [*The soldier hurries out. Cæsar comes down the hall between Rufio and Britannus*] Rufio: we have some ships in the west harbor. Burn them.

RUFIO [*staring*] Burn them!!

CÆSAR Take every boat we have in the east harbor, and seize the Pharos – that island with the lighthouse. Leave half our men behind to hold the beach and the quay outside this palace: that is the way home.

RUFIO [*disapproving strongly*] Are we to give up the city?

CÆSAR We have not got it, Rufio. This palace we have; and – what is that building next door?

RUFIO The theatre.

CÆSAR We will have that too: it commands the strand. For the rest, Egypt for the Egyptians!

RUFIO Well, you know best, I suppose. Is that all?

CÆSAR That is all. Are those ships burnt yet?

RUFIO Be easy: I shall waste no more time. [*He runs out*].

BRITANNUS Cæsar: Pothinus demands speech of you. In my opinion he needs a lesson. His manner is most insolent.

CÆSAR Where is he?

BRITANNUS He waits without.

CÆSAR Ho there! admit Pothinus.

Pothinus appears in the loggia, and comes down the hall very haughtily to Cæsar's left hand.

CÆSAR Well, Pothinus?

POTHINUS I have brought you our ultimatum, Cæsar.

CÆSAR Ultimatum! The door was open: you should have gone out through it before you declared war. You are my prisoner now. [*He goes to the chair and loosens his toga*].

POTHINUS [*scornfully*] I *your* prisoner! Do you know that you are in Alexandria, and that King Ptolemy, with an army outnumbering your little troop a hundred to one, is in possession of Alexandria?

CÆSAR [*unconcernedly taking off his toga and throwing it on the chair*] Well, my friend, get out if you can. And tell your friends not to kill any more Romans in the market place. Otherwise my soldiers, who do not share my celebrated clemency, will probably kill you. Britannus: pass the word to the guard; and fetch my armor. [*Britannus runs out. Rufio returns*]. Well?

RUFIO [*pointing from the loggia to a cloud of smoke drifting over the harbor*] See there! [*Pothinus runs eagerly up the steps to look out*].

CÆSAR What, ablaze already! Impossible!

RUFIO Yes, five good ships, and a barge laden with oil grappled to each. But it is not my doing: the Egyptians have saved me the trouble. They have captured the west harbor.

CÆSAR [*anxiously*] And the east harbor? The lighthouse, Rufio?

RUFIO [*with a sudden splutter of raging ill usage, coming down to Cæsar and scolding him*] Can I embark a legion in five minutes? The first cohort is already on the beach. We can do no more. If you want faster work, come and do it yourself.

CÆSAR [*soothing him*] Good, good. Patience, Rufio, patience.

RUFIO Patience! Who is impatient here, you or I? Would I be here, if I could not oversee them from that balcony?

CÆSAR Forgive me, Rufio; and [*anxiously*] hurry them as much as –

He is interrupted by an outcry as of an old man in the extremity of misfortune. It draws near rapidly; and Theodotus rushes in, tearing his hair, and speaking the most lamentable exclamations. Rufio steps back to stare at him, amazed at his frantic condition. Pothinus turns to listen.

THEODOTUS [*on the steps, with uplifted arms*] Horror unspeakable! Woe, alas! Help!

RUFIO What now?

CÆSAR [*frowning*] Who is slain?

THEODOTUS Slain! Oh, worse than the death of ten thousand men! Loss irreparable to mankind!

RUFIO What has happened, man?

THEODOTUS [*rushing down the hall between them*] The fire has spread from your ships. The first of the seven wonders of the world perishes. The library of Alexandria is in flames.

RUFIO Psha! [*Quite relieved, he goes up to the loggia and watches the preparations of the troops on the beach*].

CÆSAR Is that all?

THEODOTUS [*unable to believe his senses*] All! Cæsar: will you go down to posterity as a barbarous soldier too ignorant to know the value of books?

CÆSAR Theodotus: I am an author myself; and I tell you it is better that the Egyptians should live their lives than dream them away with the help of books.

THEODOTUS [*kneeling, with genuine literary emotion: the passion of the pedant*] Cæsar: once in ten generations of men, the world gains an immortal book.

CÆSAR [*inflexible*] If it did not flatter mankind, the common executioner would burn it.

THEODOTUS Without history, death will lay you beside your meanest soldier.

CÆSAR Death will do that in any case. I ask no better grave.

THEODOTUS What is burning there is the memory of mankind.

CÆSAR A shameful memory. Let it burn.

THEODOTUS [*wildly*] Will you destroy the past?

CÆSAR Ay, and build the future with its ruins. [*Theodotus, in despair, strikes himself on the temples with his fists*]. But harken, Theodotus, teacher of kings: you who valued Pompey's head no more than a shepherd values an onion, and who now kneel to me, with tears in your old eyes, to plead for a few sheepskins scrawled with errors. I cannot spare you a man or a bucket of water just now; but you shall pass freely out of the palace. Now, away with you to Achillas; and borrow his legions to put out the fire. [*He hurries him to the steps*].

POTHINUS [*significantly*] You understand, Theodotus: I remain a prisoner.

THEODOTUS A prisoner!

CÆSAR Will you stay to talk whilst the memory of mankind is burning? [*Calling through the loggia*] Ho there! Pass Theodotus out. [*To Theodotus*] Away with you.

THEODOTUS [*To Pothinus*] I must go to save the library. [*He hurries out*].

CÆSAR Follow him to the gate, Pothinus. Bid him urge your people to kill no more of my soldiers, for your sake.

POTHINUS My life will cost you dear if you take it, Cæsar. [*He goes out after Theodotus*].

Rufio, absorbed in watching the embarkation, does not notice the departure of the two Egyptians.

RUFIO [*shouting from the loggia to the beach*] All ready, there?

A CENTURION [*from below*] All ready. We wait for Cæsar.

CÆSAR Tell them Cæsar is coming – the rogues! [*Calling*] Britannicus. [*This magniloquent version of his secretary's name is one of Cæsar's jokes. In later years it would have meant, quite seriously and officially, Conqueror of Britain*].

RUFIO [*calling down*] Push off, all except the longboat. Stand by it to embark, Cæsar's guard there. [*He leaves the balcony and comes down into the hall*]. Where are those Egyptians? Is this more clemency? Have you let them go?

CÆSAR [*chuckling*] I have let Theodotus go to save the library. We must respect literature, Rufio.

RUFIO [*raging*] Folly on folly's head! I believe if you could bring back all the dead of Spain, Gaul and Thessaly to life, you would do it that we might have the trouble of fighting them over again.

CÆSAR Might not the gods destroy the world if their only thought were to be at peace next year? [*Rufio, out of all patience, turns away in anger. Cæsar suddenly grips his sleeve, and adds slyly in his ear*] Besides, my friend: every Egyptian we imprison means imprisoning two Roman soldiers to guard him. Eh?

RUFIO Agh! I might have known there was some fox's trick behind your fine talking. [*He gets away from Cæsar with an ill-humored shrug, and goes to the balcony for another look at the preparations; finally goes out*].

CÆSAR Is Britannus asleep? I sent him for my armor an hour ago. [*Calling*] Britannicus, thou British islander. Britannicus!

Cleopatra runs in through the loggia with Cæsar's helmet and sword, snatched from Britannus, who follows her with a cuirass and greaves. They come down to Cæsar, she to his left hand, Britannus to his right.

CLEOPATRA I am going to dress you, Cæsar. Sit down. [*He obeys*]. These Roman helmets are so becoming! [*She takes off his wreath*]. Oh! [*She bursts out laughing at him*].

CÆSAR What are you laughing at?

CLEOPATRA Youre bald [*beginning with a big B, and ending with a splutter*].

CÆSAR [*almost annoyed*] Cleopatra! [*He rises, for the con-*

venience of Britannus, who puts the cuirass on him].

CLEOPATRA So that is why you wear the wreath – to hide it.

BRITANNUS Peace, Egyptian: they are the bays of the conqueror. [*He buckles the cuirass*].

CLEOPATRA Peace, thou: islander! [*To Cæsar*] You should rub your head with strong spirits of sugar, Cæsar. That will make it grow.

CÆSAR [*with a wry face*] Cleopatra: do you like to be reminded that you are very young?

CLEOPATRA [*pouting*] No.

CÆSAR [*sitting down again, and setting out his leg for Britannus, who kneels to put on his greaves*] Neither do I like to be reminded that I am – middle aged. Let me give you ten of my superfluous years. That will make you 26, and leave me only – no matter. Is it a bargain?

CLEOPATRA Agreed. 26, mind. [*She puts the helmet on him*]. Oh! How nice! You look only about 50 in it!

BRITANNUS [*looking up severely at Cleopatra*] You must not speak in this manner to Cæsar.

CLEOPATRA Is it true that when Cæsar caught you on that island, you were painted all over blue?

BRITANNUS Blue is the color worn by all Britons of good standing. In war we stain our bodies blue; so that though our enemies may strip us of our clothes and our lives, they cannot strip us of our respectability. [*He rises*].

CLEOPATRA [*with Cæsar's sword*] Let me hang this on. Now you look splendid. Have they made any statues of you in Rome?

CÆSAR Yes, many statues.

CLEOPATRA You must send for one and give it to me.

RUFIO [*coming back into the loggia, more impatient than ever*] Now Cæsar: have you done talking? The moment your foot is aboard there will be no holding our men back: the boats will race one another for the lighthouse.

CÆSAR [*drawing his sword and trying the edge*] Is this well set to-day, Britannicus? At Pharsalia it was as blunt as a barrel-hoop.

BRITANNUS It will split one of the Egyptian's hairs to-day, Cæsar. I have set it myself.

CLEOPATRA [*suddenly throwing her arms in terror round Cæsar*] Oh, you are not really going into battle to be killed?

CÆSAR No, Cleopatra. No man goes to battle to be killed.

CLEOPATRA But they do get killed. My sister's husband was killed in battle. You must not go. Let *him* go [*pointing to Rufio. They all laugh at her*]. Oh please, *please* dont go. What will happen to *me* if you never come back?

CÆSAR [*gravely*] Are you afraid?

CLEOPATRA [*shrinking*] No.

CÆSAR [*with quiet authority*] Go to the balcony; and you shall see us take the Pharos. You must learn to look on battles. Go. [*She goes, downcast, and looks out from the balcony*]. That is well. Now, Rufio. March.

CLEOPATRA [*suddenly clapping her hands*] Oh, you will not be able to go!

CÆSAR Why? What now?

CLEOPATRA They are drying up the harbor with buckets – a multitude of soldiers – over there [*pointing out across the sea to her left*] – they are dipping up the water.

RUFIO [*hastening to look*] It is true. The Egyptian army! Crawling over the edge of the west harbor like locusts. [*With sudden anger he strides down to Cæsar*]. This is your accursed clemency, Cæsar. Theodotus has brought them.

CÆSAR [*delighted at his own cleverness*] I meant him to, Rufio. They have come to put out the fire. The library will keep them busy whilst we seize the lighthouse. Eh? [*He rushes out buoyantly through the loggia, followed by Britannus*].

RUFIO [*disgustedly*] *More* foxing! Agh! [*He rushes off. A shout from the soldiers announces the appearance of Cæsar below*].

CENTURION [*below*] All aboard. Give way there. [*Another shout*].

CLEOPATRA [*waving her scarf through the loggia arch*] Goodbye, goodbye, dear Cæsar. Come back safe. Goodbye!

ACT 3

The edge of the quay in front of the palace, looking out west over the east harbor of Alexandria to Pharos island, just off the end of which, and connected with it by a narrow mole, is the famous lighthouse, a gigantic square tower of white marble diminishing in size storey by storey to the top, on which stands a cresset beacon. The island is joined to the main land by the Heptastadium, a great mole or causeway five miles long bounding the harbor on the south.

In the middle of the quay a Roman sentinel stands on guard, pilum in hand, looking out to the lighthouse with strained attention, his left hand shading his eyes. The pilum is a stout wooden shaft 4½ feet long, with an iron spit about three feet long fixed in it. The sentinel is so absorbed that he does not notice the approach from the north end of the quay of four Egyptian market porters carrying rolls of carpet, preceded by Ftatateeta and Apollodorus the Sicilian. Apollodorus is a dashing young man of about 24, handsome and debonair, dressed with deliberate æstheticism in the most delicate purples and dove greys, with ornaments of bronze, oxydized silver, and stones of jade and agate. His sword, designed as carefully as a medieval cross, has a blued blade showing through an openwork scabbard of purple leather and

filagree. The porters, conducted by Ftatateeta, pass along the quay behind the sentinel to the steps of the palace, where they put down their bales and squat on the ground. Apollodorus does not pass along with them: he halts, amused by the preoccupation of the sentinel.

APOLLODORUS [*calling to the sentinel*] Who goes there, eh?

SENTINEL [*starting violently and turning with his pilum at the charge, revealing himself as a small, wiry, sandy-haired, conscientious young man with an elderly face*] Whats this? Stand. Who are you?

APOLLODORUS I am Apollodorus the Sicilian. Why, man, what are you dreaming of? Since I came through the lines beyond the theatre there, I have brought my caravan past three sentinels, all so busy staring at the lighthouse that not one of them challenged me. Is this Roman discipline?

SENTINEL We are not here to watch the land but the sea. Cæsar has just landed on the Pharos. [*Looking at Ftatateeta*] What have you here? Who is this piece of Egyptian crockery?

FTATATEETA Apollodorus: rebuke this Roman dog; and bid him bridle his tongue in the presence of Ftatateeta, the mistress of the Queen's household.

APOLLODORUS My friend: this is a great lady, who stands high with Cæsar.

SENTINEL [*not at all impressed, pointing to the carpets*] And what is all this truck?

APOLLODORUS Carpets for the furnishing of the Queen's apartments in the palace. I have picked them from the best carpets in the world; and the Queen shall choose the best of my choosing.

SENTINEL So you are the carpet merchant?

APOLLODORUS [*hurt*] My friend: I am a patrician.

SENTINEL A patrician! A patrician keeping a shop instead of following arms!

APOLLODORUS I do not keep a shop. Mine is a temple of the arts. I am a worshipper of beauty. My calling is to choose beautiful things for beautiful queens. My motto is Art for Art's sake.

SENTINEL That is not the password.

APOLLODORUS It is a universal password.

SENTINEL I know nothing about universal passwords. Either give me the password for the day or get back to your shop.

Ftatateeta, roused by his hostile tone, steals towards the edge of the quay with the step of a panther, and gets behind him.

APOLLODORUS How if I do neither?

SENTINEL Then I will drive this pilum through you.

APOLLODORUS At your service, my friend. [*He draws his sword, and springs to his guard with unruffled grace*].

FTATATEETA [*suddenly seizing the sentinel's arms from behind*] Thrust your knife into the dog's throat, Apollodorus. [*The chivalrous Apollodorus laughingly shakes his head; breaks ground away from the sentinel towards the palace; and lowers his point*].

SENTINEL [*struggling vainly*] Curse on you! Let me go. Help ho!

FTATATEETA [*lifting him from the ground*] Stab the little Roman reptile. Spit him on your sword.

A couple of Roman soldiers, with a centurion, come running along the edge of the quay from the north end. They rescue their comrade, and throw off Ftatateeta, who is sent reeling away on the left hand of the sentinel.

CENTURION [*an unattractive man of fifty, short in his speech and manners, with a vinewood cudgel in his hand*] How now? What is all this?

FTATATEETA [*to Apollodorus*] Why did you not stab him? There was time!

APOLLODORUS Centurion: I am here by order of the Queen to –

CENTURION [*interrupting him*] The Queen! Yes, yes: [*to the sentinel*] pass him in. Pass all these bazaar people in to the Queen, with their goods. But mind you pass no one out that you have not passed in – not even the Queen herself.

SENTINEL This old woman is dangerous: she is as strong as three men. She wanted the merchant to stab me.

APOLLODORUS Centurion: I am not a merchant. I am a patrician and a votary of art.

CENTURION Is the woman your wife?

APOLLODORUS [*horrified*] No, no! [*Correcting himself politely*] Not that the lady is not a striking figure in her own way. But [*emphatically*] she is *not* my wife.

FTATATEETA [*to the centurion*] Roman: I am Ftatateeta, the mistress of the Queen's household.

CENTURION Keep your hands off our men, mistress; or I will have you pitched into the harbor, though you were as strong as ten men. [*To his men*] To your posts: march! [*He returns with his men the way they came*].

FTATATEETA [*looking malignantly after him*] We shall see whom Isis loves best: her servant Ftatateeta or a dog of a Roman.

SENTINEL [*to Apollodorus, with a wave of his pilum towards the palace*] Pass in there; and keep your distance. [*Turning to Ftatateeta*] Come within a yard of me, you old crocodile; and I will give you this [*the pilum*] in your jaws.

CLEOPATRA [*calling from the palace*] Ftatateeta, Ftatateeta.

FTATATEETA [*looking up, scandalized*] Go from the window, go from the window. There are men here.

CLEOPATRA I am coming down.

FTATATEETA [*distracted*] No, no. What are you dreaming of? O ye gods, ye gods! Apollodorus: bid your men pick up your bales; and in with me quickly.

APOLLODORUS Obey the mistress of the Queen's household.

FTATATEETA [*impatiently, as the porters stoop to lift the bales*] Quick, quick: she will be out upon us. [*Cleopatra comes from the palace and runs across the quay to Ftatateeta*]. Oh that ever I was born!

CLEOPATRA [*eagerly*] Ftatateeta: I have thought of something. I want a boat – at once.

FTATATEETA A boat! No, no: you cannot. Apollodorus: speak to the Queen.

APOLLODORUS [*gallantly*] Beautiful queen: I am Apollodorus the Sicilian, your servant, from the bazaar. I have brought you the three most beautiful Persian carpets in the world to choose from.

CLEOPATRA I have no time for carpets to-day. Get me a boat.

FTATATEETA What whim is this? You cannot go on the water except in the royal barge.

APOLLODORUS Royalty, Ftatateeta, lies not in the barge but in the Queen. [*To Cleopatra*] The touch of your majesty's foot on the gunwale of the meanest boat in the harbor will make it royal. [*He turns to the harbor and calls seaward*] Ho there, boatman! Pull in to the steps.

CLEOPATRA Apollodorus: you are my perfect knight; and I will always buy my carpets through you. [*Apollodorus bows joyously. An oar appears above the quay; and the boatman, a bullet-headed, vivacious, grinning fellow, burnt almost black by the sun, comes up a flight of steps from the water on the sentinel's right, oar in hand, and waits at the top*]. Can you row, Apollodorus?

APOLLODORUS My oars shall be your majesty's wings. Whither shall I row my Queen?

CLEOPATRA To the lighthouse. Come. [*She makes for the steps*].

SENTINEL [*opposing her with his pilum at the charge*] Stand. You cannot pass.

CLEOPATRA [*flushing angrily*] How dare you? Do you know that I am the Queen?

SENTINEL I have my orders. You cannot pass.

CLEOPATRA I will make Cæsar have you killed if you do not obey me.

SENTINEL He will do worse to me if I disobey my officer. Stand back.

CLEOPATRA Ftatateeta: strangle him.

SENTINEL [*alarmed – looking apprehensively at Ftatateeta, and brandishing his pilum*] Keep off, there.

CLEOPATRA [*running to Apollodorus*] Apollodorus: make your slaves help us.

APOLLODORUS I shall not need their help, lady. [*He draws his sword*].

Now, soldier: choose which weapon you will defend yourself with. Shall it be sword against pilum, or sword against sword?

SENTINEL Roman against Sicilian, curse you. Take that. [*He hurls his pilum at Apollodorus, who drops expertly on one knee. The pilum passes whizzing over his head and falls harmless. Apollodorus, with a cry of triumph, springs up and attacks the sentinel, who draws his sword and defends himself, crying*] Ho there, guard. Help!

Cleopatra, half frightened, half delighted, takes refuge near the palace, where the porters are squatting among the bales. The boatman, alarmed, hurries down the steps out of harm's way, but stops, with his head just visible above the edge of the quay, to watch the fight. The sentinel is handicapped by his fear of an attack in the rear from Ftatateeta. His swordsmanship, which is of a rough and ready sort, is heavily taxed, as he has occasionally to strike at her to keep her off between a blow and a guard with Apollodorus. The centurion returns with several soldiers. Apollodorus springs back towards Cleopatra as this reinforcement confronts him.

CENTURION [*coming to the sentinel's right hand*] What is this? What now?

SENTINEL [*panting*] I could do well enough by myself if it werent for the old woman. Keep her off me: that is all the help I need.

CENTURION Make your report, soldier. What has happened?

FTATATEETA Centurion: he would have slain the Queen.

SENTINEL [*bluntly*] I would, sooner than let her pass. She wanted to take boat, and go – so she said – to the lighthouse. I stopped her, as I was ordered to; and she set this fellow on me. [*He goes to pick up his pilum and returns to his place with it*].

CENTURION [*turning to Cleopatra*] Cleopatra: I am loth to offend you; but without Cæsar's express order we dare not let you pass beyond the Roman lines.

APOLLODORUS Well, Centurion; and has not the lighthouse been within the Roman lines since Cæsar landed there?

CLEOPATRA Yes, yes. Answer that, if you can.

CENTURION [*to Apollodorus*] As for you, Apollodorus, you may thank the gods that you are not nailed to the palace door with a pilum for your meddling.

APOLLODORUS [*urbanely*] My military friend, I was not born to be slain by so ugly a weapon. When I fall, it will be [*holding up his sword*] by this white queen of arms, the only weapon fit for an artist. And now that you are convinced that we do not want to go beyond the lines, let me finish killing your sentinel and depart with the Queen.

CENTURION [*as the sentinel makes an angry demonstration*] Peace there. Cleopatra: I must abide by my orders, and not by the subtleties of this Sicilian. You must withdraw into the palace and examine your carpets there.

CLEOPATRA [*pouting*] I will not: I am the Queen. Cæsar does not speak to me as you do. Have Cæsar's centurions changed manners with his scullions?

CENTURION [*sulkily*] I do my duty. That is enough for me.

APOLLODORUS Majesty: when a stupid man is doing something he is ashamed of, he always declares that it is his duty.

CENTURION [*angry*] Apollodorus –

APOLLODORUS [*interrupting him with defiant elegance*] I will make amends for that insult with my sword at fitting time and place. Who says artist, says duellist. [*To Cleopatra*] Hear my counsel, star of the east. Until word comes to these soldiers from Cæsar himself, you are a prisoner. Let me go to him with a message from you, and a present; and before the sun has stooped half way to the arms of the sea, I will bring you back Cæsar's order of release.

CENTURION [*sneering at him*] And you will sell the Queen the present, no doubt.

APOLLODORUS Centurion: the Queen shall have from me, without payment, as the unforced tribute of Sicilian taste to Egyptian beauty, the richest of these carpets for her present to Cæsar.

CLEOPATRA [*exultantly, to the centurion*] Now you see what an ignorant common creature you are!

CENTURION [*curtly*] Well, a fool and his wares are soon parted. [*He turns to his men*]. Two more men to this post here; and see that no one leaves the palace but this man and

his merchandize. If he draws his sword again inside the lines, kill him. To your posts. March.

He goes out, leaving two auxiliary sentinels with the other.

APOLLODORUS [*with polite goodfellowship*] My friends: will you not enter the palace and bury our quarrel in a bowl of wine? [*He takes out his purse, jingling the coins in it*]. The Queen has presents for you all.

SENTINEL [*very sulky*] You heard our orders. Get about your business.

1ST AUXILIARY Yes: you ought to know better. Off with you.

2ND AUXILIARY [*looking longingly at the purse – this sentinel is a hook-nosed man, unlike his comrade, who is squab faced*] Do not tantalize a poor man.

APOLLODORUS [*to Cleopatra*] Pearl of Queens: the centurion is at hand; and the Roman soldier is incorruptible when his officer is looking. I must carry your word to Cæsar.

CLEOPATRA [*who has been meditating among the carpets*] Are these carpets very heavy?

APOLLODORUS It matters not how heavy. There are plenty of porters.

CLEOPATRA How do they put the carpets into boats? Do they throw them down?

APOLLODORUS Not into small boats, majesty. It would sink them.

CLEOPATRA Not into that man's boat, for instance? [*pointing to the boatman*].

APOLLODORUS No. Too small.

CLEOPATRA But you can take a carpet to Cæsar in it if I send one?

APOLLODORUS Assuredly.

CLEOPATRA And you will have it carried gently down the steps and take great care of it?

APOLLODORUS Depend on me.

CLEOPATRA Great, *great* care?

APOLLODORUS More than of my own body.

CLEOPATRA You will promise me not to let the porters drop it or throw it about?

APOLLODORUS Place the most delicate glass goblet in the palace in

the heart of the roll, Queen; and if it be broken, my head shall pay for it.

CLEOPATRA Good. Come, Ftatateeta. [*Ftatateeta comes to her. Apollodorus offers to squire them into the palace*]. No, Apollodorus, you must not come. I will choose a carpet for myself. You must wait here. [*She runs into the palace*].

POLLODORUS [*to the porters*] Follow this lady [*indicating Ftatateeta*]; and obey her.

The porters rise and take up their bales.

FTATATEETA [*addressing the porters as if they were vermin*] This way. And take your shoes off before you put your feet on those stairs.

She goes in, followed by the porters with the carpets. Meanwhile Apollodorus goes to the edge of the quay and looks out over the harbor. The sentinels keep their eyes on him malignantly.

POLLODORUS [*addressing the sentinel*] My friend –

SENTINEL [*rudely*] Silence there.

ST AUXILIARY Shut your muzzle, you.

D AUXILIARY [*in a half whisper, glancing apprehensively towards the north end of the quay*] Cant you wait a bit?

POLLODORUS Patience, worthy three-headed donkey. [*They mutter ferociously; but he is not at all intimidated*]. Listen: were you set here to watch me, or to watch the Egyptians?

SENTINEL We know our duty.

POLLODORUS Then why dont you do it? There is something going on over there [*pointing southwestward to the mole*].

SENTINEL [*sulkily*] I do not need to be told what to do by the like of you.

POLLODORUS Blockhead. [*He begins shouting*] Ho there, Centurion. Hoiho!

SENTINEL Curse your meddling. [*Shouting*] Hoiho! Alarm! Alarm!

1ST AND 2ND AUXILIARIES Alarm! alarm! Hoiho! [*The Centurion comes running in with his guard.*]

CENTURION What now? Has the old woman attacked you again? [*Seeing Apollodorus*] Are *you* here still?

APOLLODORUS [*pointing as before*] See there. The Egyptians are moving. They are going to recapture the Pharos. They will attack by sea and land: by land along the great mole; by sea from the west harbor. Stir yourselves, my military friends: the hunt is up. [*A clangor of trumpets from several points along the quay*]. Aha! I told you so.

CENTURION [*quickly*] The two extra men pass the alarm to the south posts. One man keep guard here. The rest with me – quick.

The two auxiliary sentinels run off to the south. The centurion and his guard run off northward; and immediately afterwards the bucina sounds. The four porters come from the palace carrying a carpet, followed by Ftatateeta.

SENTINEL [*handling his pilum apprehensively*] You again! [*The porters stop*].

FTATATEETA Peace, Roman fellow: you are now singlehanded. Apollodorus: this carpet is Cleopatra's present to Cæsar. It has rolled up in it ten precious goblets of the thinnest Iberian crystal, and a hundred eggs of the sacred blue pigeon. On your honor, let not one of them be broken.

APOLLODORUS On my head be it! [*To the porters*] Into the boat with them carefully.

The porters carry the carpet to the steps.

1ST PORTER [*looking down at the boat*] Beware what you do, sir. Those eggs of which the lady speaks must weigh more than a pound apiece. This boat is too small for such a load.

BOATMAN [*excitedly rushing up the steps*] Oh thou injurious porter! Oh thou unnatural son of a she-camel! [*To Apollodorus*] My boat, sir, hath often carried five men. Shall it not carry your lordship and a bale of pigeons' eggs? [*To the porter*] Thou mangey dromedary, the gods shall punish thee for this envious wickedness.

1ST PORTER [*stolidly*] I cannot quit this bale now to beat thee; but another day I will lie in wait for thee.

APOLLODORUS [*going between them*] Peace there. If the boat were but a single plank, I would get to Cæsar on it.

FTATATEETA [*anxiously*] In the name of the gods, Apollodorus, run no risks with that bale.

APOLLODORUS Fear not, thou venerable grotesque: I guess its great worth. [*To the porters*] Down with it, I say; and gently; or ye shall eat nothing but stick for ten days.

The boatman goes down the steps, followed by the porters with the bale: Ftatateeta and Apollodorus watching from the edge.

APOLLODORUS Gently, my sons, my children – [*with sudden alarm*] gently, ye dogs. Lay it level in the stern – so – tis well.

FTATATEETA [*screaming down at one of the porters*] Do not step on it, do not step on it. Oh thou brute beast!

FIRST PORTER [*ascending*] Be not excited, mistress: all is well.

FTATATEETA [*panting*] All well! Oh, thou hast given my heart a turn! [*She clutches her side, gasping*].

The four porters have now come up and are waiting at the stairhead to be paid.

APOLLODORUS Here, ye hungry ones. [*He gives money to the first porter, who holds it in his hand to shew to the others. They crowd greedily to see how much it is, quite prepared, after the Eastern fashion, to protest to heaven against their patron's stinginess. But his liberality overpowers them*].

1ST PORTER O bounteous prince!

2ND PORTER O lord of the bazaar!

3RD PORTER O favored of the gods!

4TH PORTER O father to all the porters of the market!

SENTINEL [*enviously, threatening them fiercely with his pilum*] Hence, dogs: off. Out of this. [*They fly before him northward along the quay*].

APOLLODORUS Farewell, Ftatateeta. I shall be at the lighthouse before the Egyptians. [*He descends the steps*].

FTATATEETA The gods speed thee and protect my nursling!

The sentry returns from chasing the porters and looks down at the boat, standing near the stairhead lest Ftatateeta should attempt to escape.

APOLLODORUS [*from beneath, as the boat moves off*] Farewell, valiant pilum pitcher.

SENTINEL Farewell, shopkeeper.

APOLLODORUS Ha, ha! Pull, thou brave boatman, pull. Soho-o-o-o-o! [*He begins to sing in barcarolle measure to the rhythm of the oars*]

My heart, my heart, spread out thy wings:
Shake off thy heavy load of love –

Give me the oars, O son of a snail.

SENTINEL [*threatening Ftatateeta*] Now mistress: back to your henhouse. In with you.

FTATATEETA [*falling on her knees and stretching her hands over the waters*] Gods of the seas, bear her safely to the shore!

SENTINEL Bear *who* safely? What do you mean?

FTATATEETA [*looking darkly at him*] Gods of Egypt and of Vengeance, let this Roman fool be beaten like a dog by his captain for suffering her to be taken over the waters.

SENTINEL Accursed one: is she then in the boat? [*He calls over the sea*] Hoiho, there, boatman! Hoiho!

APOLLODORUS [*singing in the distance*]

My heart, my heart, be whole and free:
Love is thine only enemy.

Meanwhile Rufio, the morning's fighting done, sits munching dates on a faggot of brushwood outside the door of the lighthouse, which towers gigantic to the clouds on his left. His helmet, full of dates, is between his knees; and a leathern bottle of wine is by his side. Behind him the great stone pedestal of the lighthouse is

shut in from the open sea by a low stone parapet, with a couple of steps in the middle to the broad coping. A huge chain with a hook hangs down from the lighthouse crane above his head. Faggots like the one he sits on lie beneath it ready to be drawn up to feed the beacon.

Cæsar is standing on the step at the parapet looking out anxiously, evidently ill at ease. Britannus comes out of the lighthouse door.

RUFIO Well, my British islander. Have you been up to the top?

BRITANNUS I have. I reckon it at 200 feet high.

RUFIO Anybody up there?

BRITANNUS One elderly Tyrian to work the crane; and his son, a well conducted youth of 14.

RUFIO [*looking at the chain*] What! An old man and a boy work that! Twenty men, you mean.

BRITANNUS Two only, I assure you. They have counterweights, and a machine with boiling water in it which I do not understand: it is not of British design. They use it to haul up barrels of oil and faggots to burn in the brazier on the roof.

RUFIO But –

BRITANNUS Excuse me: I came down because there are messengers coming along the mole to us from the island. I must see what their business is. [*He hurries out past the lighthouse*].

CÆSAR [*coming away from the parapet, shivering and out of sorts*] Rufio: this has been a mad expedition. We shall be beaten. I wish I knew how our men are getting on with that barricade across the great mole.

RUFIO [*angrily*] Must I leave my food and go starving to bring you a report?

CÆSAR [*soothing him nervously*] No, Rufio, no. Eat, my son, eat. [*He takes another turn, Rufio chewing dates meanwhile*]. The Egyptians cannot be such fools as not to storm the barricade and swoop down on us here before it is finished. It is the first time I have ever run an avoidable risk. I should not have come to Egypt.

RUFIO An hour ago you were all for victory.

CÆSAR [*apologetically*] Yes: I was a fool – rash, Rufio – boyish.

RUFIO Boyish! Not a bit of it. Here [*offering him a handful of dates*].

CÆSAR What are these for?

RUFIO To eat. Thats whats the matter with you. When a man comes to your age, he runs down before his midday meal. Eat and drink; and then have another look at our chances.

CÆSAR [*taking the dates*] My age! [*He shakes his head and bites a date*]. Yes, Rufio: I am an old man – worn out now – true, quite true. [*He gives way to melancholy contemplation, and eats another date*]. Achillas is still in his prime: Ptolemy is a boy. [*He eats another date, and plucks up a little*]. Well, every dog has his day; and I have had mine: I cannot complain. [*With sudden cheerfulness*] These dates are not bad, Rufio. [*Britannus returns, greatly excited, with a leathern bag. Cæsar is himself again in a moment*]. What now?

BRITANNUS [*triumphantly*] Our brave Rhodian mariners have captured a treasure. There! [*He throws the bag down at Cæsar's feet*]. Our enemies are delivered into our hands.

CÆSAR In that bag?

BRITANNUS Wait till you hear, Cæsar. This bag contains all the letters which have passed between Pompey's party and the army of occupation here.

CÆSAR Well?

BRITANNUS [*impatient of Cæsar's slowness to grasp the situation*] Well, we shall now know who your foes are. The name of every man who has plotted against you since you crossed the Rubicon may be in these papers, for all we know.

CÆSAR Put them in the fire.

BRITANNUS Put them – [*he gasps*]! ! ! !

CÆSAR In the fire. Would you have me waste the next three years of my life in proscribing and condemning men who will be my friends when I have proved that my friendship is worth more than Pompey's was – than Cato's is. O incorrigible British islander: am I a bull dog, to seek quarrels merely to shew how stubborn my jaws are?

BRITANNUS But your honor – the honor of Rome –

CÆSAR I do not make human sacrifices to my honor, as your Druids do. Since you will not burn these, at least I can drown them. [*He picks up the bag and throws it over the parapet into the sea*].

BRITANNUS Cæsar: this is mere eccentricity. Are traitors to be allowed to go free for the sake of a paradox?

RUFIO [*rising*] Cæsar: when the islander has finished preaching, call me again. I am going to have a look at the boiling water machine. [*He goes into the lighthouse*].

BRITANNUS [*with genuine feeling*] O Cæsar, my great master, if I could but persuade you to regard life seriously, as men do in my country!

CÆSAR Do they truly do so, Britannus?

BRITANNUS Have you not been there? Have you not seen them? What Briton speaks as you do in your moments of levity? What Briton neglects to attend the services at the sacred grove? What Briton wears clothes of many colors as you do, instead of plain blue, as all solid, well esteemed men should? These are moral questions with us.

CÆSAR Well, well, my friend: some day I shall settle down and have a blue toga, perhaps. Meanwhile, I must get on as best I can in my flippant Roman way. [*Apollodorus comes past the lighthouse*]. What now?

BRITANNUS [*turning quickly, and challenging the stranger with official haughtiness*] What is this? Who are you? How did you come here?

APOLLODORUS Calm yourself, my friend: I am not going to eat you. I have come by boat, from Alexandria, with precious gifts for Cæsar.

CÆSAR From Alexandria!

BRITANNUS [*severely*] That is Cæsar, sir.

RUFIO [*appearing at the lighthouse door*] Whats the matter now?

APOLLODORUS Hail, great Cæsar! I am Apollodorus the Sicilian, an artist.

BRITANNUS An artist! Why have they admitted this vagabond?

CÆSAR Peace, man. Apollodorus is a famous patrician amateur.

BRITANNUS [*disconcerted*] I crave the gentleman's pardon. [*To Cæsar*] I understood him to say that he was a professional. [*Somewhat out of countenance, he allows Apollodorus to approach Cæsar, changing places with him. Rufio, after looking Apollodorus up and down with marked disparagement, goes to the other side of the platform*].

CÆSAR You are welcome, Apollodorus. What is your business?

APOLLODORUS First, to deliver to you a present from the Queen of Queens.

CÆSAR Who is that?

APOLLODORUS Cleopatra of Egypt.

CÆSAR [*taking him into his confidence in his most winning manner*] Apollodorus: this is no time for playing with presents. Pray you, go back to the Queen, and tell her that if all goes well I shall return to the palace this evening.

APOLLODORUS Cæsar: I cannot return. As I approached the lighthouse, some fool threw a great leathern bag into the sea. It broke the nose of my boat; and I had hardly time to get myself and my charge to the shore before the poor little cockleshell sank.

CÆSAR I am sorry, Apollodorus. The fool shall be rebuked. Well, well: what have you brought me? The Queen will be hurt if I do not look at it.

RUFIO Have we time to waste on this trumpery? The Queen is only a child.

CÆSAR Just so: that is why we must not disappoint her. What is the present, Apollodorus?

APOLLODORUS Cæsar: it is a Persian carpet – a beauty! And in it are – so I am told – pigeons' eggs and crystal goblets and fragile precious things. I dare not for my head have it carried up that narrow ladder from the causeway.

RUFIO Swing it up by the crane, then. We will send the eggs to the cook; drink our wine from the goblets; and the carpet will make a bed for Cæsar.

APOLLODORUS The crane! Cæsar: I have sworn to tender this bale of carpet as I tender my own life.

CÆSAR [*cheerfully*] Then let them swing you up at the same

time; and if the chain breaks, you and the pigeons' eggs will perish together. [*He goes to the chain and looks up along it, examining it curiously*].

APOLLODORUS [*to Britannus*] Is Cæsar serious?

BRITANNUS His manner is frivolous because he is an Italian; but he means what he says.

APOLLODORUS Serious or not, he spake well. Give me a squad of soldiers to work the crane.

BRITANNUS Leave the crane to me. Go and await the descent of the chain.

APOLLODORUS Good. You will presently see me there [*turning to them all and pointing with an eloquent gesture to the sky above the parapet*] rising like the sun with my treasure.

He goes back the way he came. Britannus goes into the lighthouse.

RUFIO [*ill-humoredly*] Are you really going to wait here for this foolery, Cæsar?

CÆSAR [*backing away from the crane as it gives signs of working*] Why not?

RUFIO The Egyptians will let you know why not if they have the sense to make a rush from the shore end of the mole before our barricade is finished. And here we are waiting like children to see a carpet full of pigeons' eggs.

The chain rattles, and is drawn up high enough to clear the parapet. It then swings round out of sight behind the lighthouse.

CÆSAR Fear not, my son Rufio. When the first Egyptian takes his first step along the mole, the alarm will sound; and we two will reach the barricade from our end before the Egyptians reach it from their end – we two, Rufio: I, the old man, and you, his biggest boy. And the old man will be there first. So peace; and give me some more dates.

APOLLODORUS [*from the causeway below*] Soho, haul away. So-ho-o-o-o! [*The chain is drawn up and comes round again from behind the lighthouse. Apollodorus is swinging in the air with his bale of carpet at the end of it. He breaks into song as he soars above the parapet*]

Aloft, aloft, behold the blue
That never shone in woman's eyes –

Easy there: stop her. [*He ceases to rise*] Further round! [*The chain comes forward above the platform*].

RUFIO [*calling up*] Lower away there. [*The chain and its load begin to descend*].

APOLLODORUS [*calling up*] Gently – slowly – mind the eggs.

RUFIO [*calling up*] Easy there – slowly – slowly.

Apollodorus and the bale are deposited safely on the flags in the middle of the platform. Rufio and Cæsar help Apollodorus to cast off the chain from the bale.

RUFIO Haul up.

The chain rises clear of their heads with a rattle. Britannus comes from the lighthouse and helps them to uncord the carpet.

APOLLODORUS [*when the cords are loose*] Stand off, my friends: let Cæsar see. [*He throws the carpet open*].

RUFIO Nothing but a heap of shawls. Where are the pigeons' eggs?

APOLLODORUS Approach, Cæsar; and search for them among the shawls.

RUFIO [*drawing his sword*] Ha, treachery! Keep back, Cæsar: I saw the shawl move: there is something alive there.

BRITANNUS [*drawing his sword*] It is a serpent.

APOLLODORUS Dares Cæsar thrust his hand into the sack where the serpent moves?

RUFIO [*turning on him*] Treacherous dog –

CÆSAR Peace. Put up your swords. Apollodorus: your serpent seems to breathe very regularly. [*He thrusts his hand under the shawls and draws out a bare arm*]. This is a pretty little snake.

RUFIO [*drawing out the other arm*] Let us have the rest of you.

They pull Cleopatra up by the wrists into a sitting position. Britannus, scandalized, sheathes his sword with a drive of protest.

CLEOPATRA [*gasping*] Oh, I'm smothered. Oh, Cæsar, a man stood on me in the boat; and a great sack of something fell upon me out of the sky; and then the boat sank; and

then I was swung up into the air and bumped down.

CÆSAR [*petting her as she rises and takes refuge on his breast*] Well, never mind: here you are safe and sound at last.

RUFIO Ay; and now that she *is* here, what are we to do with her?

BRITANNUS She cannot stay here, Cæsar, without the companionship of some matron.

CLEOPATRA [*jealously, to Cæsar, who is obviously perplexed*] Arnt you glad to see me?

CÆSAR Yes, yes; *I* am very glad. But Rufio is very angry; and Britannus is shocked.

CLEOPATRA [*contemptuously*] You can have their heads cut off, can you not?

CÆSAR They would not be so useful with their heads cut off as they are now, my sea bird.

RUFIO [*to Cleopatra*] We shall have to go away presently and cut some of your Egyptians' heads off. How will you like being left here with the chance of being captured by that little brother of yours if we are beaten?

CLEOPATRA But you mustnt leave me alone. Cæsar: you will not leave me alone, will you?

RUFIO What! not when the trumpet sounds and all our lives depend on Cæsar's being at the barricade before the Egyptians reach it? Eh?

CLEOPATRA Let them lose their lives: they are only soldiers.

CÆSAR [*gravely*] Cleopatra: when that trumpet sounds, we must take every man his life in his hand, and throw it in the face of Death. And of my soldiers who have trusted me there is not one whose hand I shall not hold more sacred than your head. [*Cleopatra is overwhelmed. Her eyes fill with tears*]. Apollodorus: you must take her back to the palace.

POLLODORUS Am I a dolphin, Cæsar, to cross the seas with young ladies on my back? My boat is sunk: all yours are either at the barricade or have returned to the city. I will hail one if I can: that is all I can do. [*He goes back to the causeway*].

CLEOPATRA [*struggling with her tears*] It does not matter. I will not go back. Nobody cares for me.

CÆSAR Cleopatra –

CLEOPATRA You want me to be killed.

CÆSAR [*still more gravely*] My poor child: your life matters little here to anyone but yourself. [*She gives way altogether at this, casting herself down on the faggots weeping. Suddenly a great tumult is heard in the distance, bucinas and trumpets sounding through a storm of shouting. Britannus rushes to the parapet and looks along the mole. Cæsar and Rufio turn to one another with quick intelligence*].

CÆSAR Come, Rufio.

CLEOPATRA [*scrambling to her knees and clinging to him*] No, no. Do not leave me, Cæsar. [*He snatches his skirt from her clutch*]. Oh!

BRITANNUS [*from the parapet*] Cæsar: we are cut off. The Egyptians have landed from the west harbor between us and the barricade! ! !

RUFIO [*running to see*] Curses! It is true. We are caught like rats in a trap.

CÆSAR [*ruthfully*] Rufio, Rufio: my men at the barricade are between the sea party and the shore party. I have murdered them.

RUFIO [*coming back from the parapet to Cæsar's right hand*] Ay: that comes of fooling with this girl here.

APOLLODORUS [*coming up quickly from the causeway*] Look over the parapet, Cæsar.

CÆSAR We have looked, my friend. We must defend ourselves here.

APOLLODORUS I have thrown the ladder into the sea. They cannot get in without it.

RUFIO Ay; and we cannot get out. Have you thought of that?

APOLLODORUS Not get out! Why not? You have ships in the east harbor.

BRITANNUS [*hopefully, at the parapet*] The Rhodian galleys are standing in towards us already. [*Cæsar quickly joins Britannus at the parapet*].

RUFIO [*to Apollodorus, impatiently*] And by what road are we to walk to the galleys, pray?

APOLLODORUS [*with gay, defiant rhetoric*] By the road that leads

everywhere – the diamond path of the sun and moon. Have you never seen the child's shadow play of The Broken Bridge? "Ducks and geese with ease get over" – eh? [*He throws away his cloak and cap, and binds his sword on his back*].

RUFIO What are you talking about?

APOLLODORUS I will shew you. [*Calling to Britannus*] How far off is the nearest galley?

BRITANNUS Fifty fathom.

CÆSAR No, no: they are further off than they seem in this clear air to your British eyes. Nearly quarter of a mile, Apollodorus.

APOLLODORUS Good. Defend yourselves here until I send you a boat from that galley.

RUFIO Have you wings, perhaps?

APOLLODORUS Water wings, soldier. Behold!

He runs up the steps between Cæsar and Britannus to the coping of the parapet; springs into the air; and plunges head foremost into the sea.

CÆSAR [*like a schoolboy – wildly excited*] Bravo, bravo! [*Throwing off his cloak*] By Jupiter, I will do that too.

RUFIO [*seizing him*] You are mad. You shall not.

CÆSAR Why not? Can I not swim as well as he?

RUFIO [*frantic*] Can an old fool dive and swim like a young one? He is twenty-five and you are fifty.

CÆSAR [*breaking loose from Rufio*] Old! ! !

BRITANNUS [*shocked*] Rufio: you forget yourself.

CÆSAR I will race you to the galley for a week's pay, father Rufio.

CLEOPATRA But me! me! ! me! ! ! what is to become of *me?*

CÆSAR I will carry you on my back to the galley like a dolphin. Rufio: when you see me rise to the surface, throw her in: I will answer for her. And then in with you after her, both of you.

CLEOPATRA No, no, NO. I shall be drowned.

BRITANNUS Cæsar: I am a man and a Briton, not a fish. I must have a boat. I cannot swim.

CLEOPATRA Neither can I.

CÆSAR [*to Britannus*] Stay here, then, alone, until I recapture the lighthouse: I will not forget you. Now, Rufio.

RUFIO You have made up your mind to this folly?

CÆSAR The Egyptians have made it up for me. What else is there to do? And mind where you jump: I do not want to get your fourteen stone in the small of my back as I come up. [*He runs up the steps and stands on the coping*].

BRITANNUS [*anxiously*] One last word, Cæsar. Do not let yourself be seen in the fashionable part of Alexandria until you have changed your clothes.

CÆSAR [*calling over the sea*] Ho, Apollodorus: [*he points skyward and quotes the barcarolle*]

The white upon the blue above –

APOLLODORUS [*swimming in the distance*]

Is purple on the green below –

CÆSAR [*exultantly*] Aha! [*He plunges into the sea*].

CLEOPATRA [*running excitedly to the steps*] Oh, let me see. He will be drowned [*Rufio seizes her*] – Ah – ah – ah – ah! [*He pitches her screaming into the sea. Rufio and Britannus roar with laughter*].

RUFIO [*looking down after her*] He has got her. [*To Britannus*] Hold the fort, Briton. Cæsar will not forget you. [*He springs off*].

BRITANNUS [*running to the steps to watch them as they swim*] All safe, Rufio?

RUFIO [*swimming*] All safe.

CÆSAR [*swimming further off*] Take refuge up there by the beacon; and pile the fuel on the trap door, Britannus.

BRITANNUS [*calling in reply*] I will first do so, and then commend myself to my country's gods. [*A sound of cheering from the sea. Britannus gives full vent to his excitement*]. The boat has reached him: Hip, hip, hip, hurrah!

ACT 4

Cleopatra's sousing in the east harbor of Alexandria was in October 48 B.C. In March 47 she is passing the afternoon in her boudoir in the palace, among a bevy of her ladies, listening to a slave girl who is playing the harp in the middle of the room. The harpist's master, an old musician, with a lined face, prominent brows, white beard, moustache and eyebrows twisted and horned at the ends, and a consciously keen and pretentious expression, is squatting on the floor close to her on her right, watching her performance. Ftatateeta is in attendance near the door, in front of a group of female slaves. Except the harp player all are seated: Cleopatra in a chair opposite the door on the other side of the room; the rest on the ground. Cleopatra's ladies are all young, the most conspicuous being Charmian and Iras, her favorites. Charmian is a hatchet faced, terra cotta colored little goblin, swift in her movements, and neatly finished at the hands and feet. Iras is a plump, goodnatured creature, rather fatuous, with a profusion of red hair, and a tendency to giggle on the slightest provocation.

CLEOPATRA Can I –

FTATATEETA [*insolently, to the player*] Peace, thou! The Queen speaks. [*The player stops*].

CLEOPATRA [*to the old musician*] I want to learn to play the harp with my own hands. Cæsar loves music. Can you teach me?

MUSICIAN Assuredly I and no one else can teach the queen. Have I not discovered the lost method of the ancient Egyptians, who could make a pyramid tremble by touching a bass string? All the other teachers are quacks: I have exposed them repeatedly.

CLEOPATRA Good: you shall teach me. How long will it take?

MUSICIAN Not very long: only four years. Your Majesty must first become proficient in the philosophy of Pythagoras.

CLEOPATRA Has she [*indicating the slave*] become proficient in the philosophy of Pythagoras?

MUSICIAN Oh, she is but a slave. She learns as a dog learns.

CLEOPATRA Well, then, I will learn as a dog learns; for she plays better than you. You shall give me a lesson every day for a fortnight. [*The musician hastily scrambles to his feet and bows profoundly*]. After that, whenever I strike a false note you shall be flogged; and if I strike so many that there is not time to flog you, you shall be thrown into the Nile to feed the crocodiles. Give the girl a piece of gold; and send them away.

MUSICIAN [*much taken aback*] But true art will not be thus forced.

FTATATEETA [*pushing him out*] What is this? Answering the Queen, forsooth. Out with you.

He is pushed out by Ftatateeta, the girl following with her harp, amid the laughter of the ladies and slaves.

CLEOPATRA Now, can any of you amuse me? Have you any stories or any news?

IRAS Ftatateeta –

CLEOPATRA Oh, Ftatateeta, Ftatateeta, always Ftatateeta. Some new tale to set me against her.

IRAS No: this time Ftatateeta has been virtuous. [*All the*

ladies laugh – not the slaves]. Pothinus has been trying to bribe her to let him speak with you.

CLEOPATRA [*wrathfully*] Ha! you all sell audiences with me, as if I saw whom you please, and not whom I please. I should like to know how much of her gold piece that harp girl will have to give up before she leaves the palace.

IRAS We can easily find out that for you.

The ladies laugh.

CLEOPATRA [*frowning*] You laugh; but take care, take care. I will find out some day how to make myself served as Cæsar is served.

CHARMIAN Old hooknose! [*They laugh again*].

CLEOPATRA [*revolted*] Silence. Charmian: do not you be a silly little Egyptian fool. Do you know why I allow you all to chatter impertinently just as you please, instead of treating you as Ftatateeta would treat you if she were Queen?

CHARMIAN Because you try to imitate Cæsar in everything; and he lets everybody say what they please to him.

CLEOPATRA No; but because I asked him one day why he did so; and he said "Let your women talk; and you will learn something from them." What have I to learn from them? I said. "What they *are*," said he; and oh! you should have seen his eye as he said it. You would have curled up, you shallow things. [*They laugh. She turns fiercely on Iras*]. At whom are you laughing – at me or at Cæsar?

IRAS At Cæsar.

CLEOPATRA If you were not a fool, you would laugh at me; and if you were not a coward you would not be afraid to tell me so. [*Ftatateeta returns*]. Ftatateeta: they tell me that Pothinus has offered you a bribe to admit him to my presence.

FTATATEETA [*protesting*] Now by my father's gods –

CLEOPATRA [*cutting her short despotically*] Have I not told you not to deny things? You would spend the day calling your father's gods to witness to your virtues if I let you. Go take the bribe; and bring in Pothinus. [*Ftatateeta is about to reply*]. Dont answer me. Go.

Ftatateeta goes out; and Cleopatra rises and begins to prowl to and fro between her chair and the door, meditating. All rise and stand.

IRAS [*as she reluctantly rises*] Heigho! I wish Cæsar were back in Rome.

CLEOPATRA [*threateningly*] It will be a bad day for you all when he goes. Oh, if I were not ashamed to let him see that I am as cruel at heart as my father, I would make you repent that speech! Why do you wish him away?

CHARMIAN He makes you so terribly prosy and serious and learned and philosophical. It is worse than being religious, at *our* ages. [*The ladies laugh*].

CLEOPATRA Cease that endless cackling, will you. Hold your tongues.

CHARMIAN [*with mock resignation*] Well, well: we must try to live up to Cæsar.

They laugh again. Cleopatra rages silently as she continues to prowl to and fro. Ftatateeta comes back with Pothinus, who halts on the threshold.

FTATATEETA [*at the door*] Pothinus craves the ear of the –

CLEOPATRA There, there: that will do: let him come in. [*She resumes her seat. All sit down except Pothinus, who advances to the middle of the room. Ftatateeta takes her former place*]. Well, Pothinus: what is the latest news from your rebel friends?

POTHINUS [*haughtily*] I am no friend of rebellion. And a prisoner does not receive news.

CLEOPATRA You are no more a prisoner than I am – than Cæsar is. These six months we have been besieged in this palace by my subjects. You are allowed to walk on the beach among the soldiers. Can I go further myself, or can Cæsar?

POTHINUS You are but a child, Cleopatra, and do not understand these matters.

The ladies laugh. Cleopatra looks inscrutably at him.

CHARMIAN I see you do not know the latest news, Pothinus.

POTHINUS What is that?

CHARMIAN That Cleopatra is no longer a child. Shall I tell you

how to grow much older, and much, *much* wiser in one day?

POTHINUS I should prefer to grow wiser without growing older.

CHARMIAN Well, go up to the top of the lighthouse; and get somebody to take you by the hair and throw you into the sea. [*The ladies laugh*].

CLEOPATRA She is right, Pothinus: you will come to the shore with much conceit washed out of you. [*The ladies laugh. Cleopatra rises impatiently*]. Begone, all of you. I will speak with Pothinus alone. Drive them out, Ftatateeta. [*They run out laughing. Ftatateeta shuts the door on them*]. What are *you* waiting for?

FTATATEETA It is not meet that the Queen remain alone with –

CLEOPATRA [*interrupting her*] Ftatateeta: must I sacrifice you to your father's gods to teach you that *I* am Queen of Egypt, and not you?

FTATATEETA [*indignantly*] You are like the rest of them. You want to be what these Romans call a New Woman. [*She goes out, banging the door*].

CLEOPATRA [*sitting down again*] Now, Pothinus: why did you bribe Ftatateeta to bring you hither?

POTHINUS [*studying her gravely*] Cleopatra: what they tell me is true. You are changed.

CLEOPATRA Do you speak with Cæsar every day for six months: and *you* will be changed.

POTHINUS It is the common talk that you are infatuated with this old man?

CLEOPATRA Infatuated? What does that mean? Made foolish, is it not? Oh no: I wish I were.

POTHINUS You wish you were made foolish! How so?

CLEOPATRA When I was foolish, I did what I liked, except when Ftatateeta beat me; and even then I cheated her and did it by stealth. Now that Cæsar has made me wise, it is no use my liking or disliking: I do what must be done, and have no time to attend to myself. That is not happiness; but it is greatness. If Cæsar were gone, I think I could govern the Egyptians; for what Cæsar is to me, I am to the fools around me.

POTHINUS [*looking hard at her*] Cleopatra: this may be the vanity of youth.

CLEOPATRA No, no: it is not that I am so clever, but that the others are so stupid.

POTHINUS [*musingly*] Truly, that is the great secret.

CLEOPATRA Well, now tell me what you came to say?

POTHINUS [*embarrassed*] I! Nothing.

CLEOPATRA Nothing!

POTHINUS At least – to beg for my liberty: that is all.

CLEOPATRA For that you would have knelt to Cæsar. No, Pothinus: you came with some plan that depended on Cleopatra being a little nursery kitten. Now that Cleopatra is a Queen, the plan is upset.

POTHINUS [*bowing his head submissively*] It is so.

CLEOPATRA [*exultant*] Aha!

POTHINUS [*raising his eyes keenly to hers*] Is Cleopatra then indeed a Queen, and no longer Cæsar's prisoner and slave?

CLEOPATRA Pothinus: we are all Cæsar's slaves – all we in this land of Egypt – whether we will or no. And she who is wise enough to know this will reign when Cæsar departs.

POTHINUS You harp on Cæsar's departure.

CLEOPATRA What if I do?

POTHINUS Does he not love you?

CLEOPATRA Love me! Pothinus: Cæsar loves no one. Who are those we love? Only those whom we do not hate: all people are strangers and enemies to us except those we love. But it is not so with Cæsar. *He* has no hatred in him: he makes friends with everyone as he does with dogs and children. His kindness to me is a wonder: neither mother, father, nor nurse have ever taken so much care for me, or thrown open their thoughts to me so freely.

POTHINUS Well: is not this love?

CLEOPATRA What! when he will do as much for the first girl he meets on his way back to Rome? Ask his slave, Britannus: he has been just as good to him. Nay, ask his very horse! His kindness is not for anything in *me:* it is in his own nature.

POTHINUS But how can you be sure that he does not love you as men love women?

CLEOPATRA Because I cannot make him jealous. I have tried.

POTHINUS Hm! Perhaps I should have asked, then, do *you* love *him?*

CLEOPATRA Can one love a god? Besides, I love another Roman: one whom I saw long before Cæsar – no god, but a man – one who can love and hate – one whom I can hurt and who would hurt me.

POTHINUS Does Cæsar know this?

CLEOPATRA Yes.

POTHINUS And he is not angry?

CLEOPATRA He promises to send him to Egypt to please me!

POTHINUS I do not understand this man.

CLEOPATRA [*with superb contempt*] *You* understand Cæsar! How could you? [*Proudly*] I do – by instinct.

POTHINUS [*deferentially, after a moment's thought*] Your Majesty caused me to be admitted to-day. What message has the Queen for me?

CLEOPATRA This. You think that by making my brother king, you will rule in Egypt, because you are his guardian and he is a little silly.

POTHINUS The Queen is pleased to say so.

CLEOPATRA The Queen is pleased to say this also. That Cæsar will eat up you, and Achillas, and my brother, as a cat eats up mice; and that he will put on this land of Egypt as a shepherd puts on his garment. And when he has done that, he will return to Rome, and leave Cleopatra here as his viceroy.

POTHINUS [*breaking out wrathfully*] That he shall never do. We have a thousand men to his ten; and we will drive him and his beggarly legions into the sea.

CLEOPATRA [*with scorn, getting up to go*] You rant like any common fellow. Go, then, and marshal your thousands; and make haste; for Mithridates of Pergamos is at hand with reinforcements for Cæsar. Cæsar has held you at bay with two legions: we shall see what he will do with twenty.

POTHINUS Cleopatra –

CLEOPATRA Enough, enough: Cæsar has spoiled me for talking to weak things like you. [*She goes out. Pothinus, with a gesture of rage, is following, when Ftatateeta enters and stops him*].

POTHINUS Let me go forth from this hateful place.

FTATATEETA What angers you?

POTHINUS The curse of all the gods of Egypt be upon her! She has sold her country to the Roman, that she may buy it back from him with her kisses.

FTATATEETA Fool: did she not tell you that she would have Cæsar gone?

POTHINUS You listened?

FTATATEETA I took care that some honest woman should be at hand whilst you were with her.

POTHINUS Now by the gods –

FTATATEETA Enough of your gods! Cæsar's gods are all powerful here. It is no use *you* coming to Cleopatra: you are only an Egyptian. She will not listen to any of her own race: she treats us all as children.

POTHINUS May she perish for it!

FTATATEETA [*balefully*] May your tongue wither for that wish! Go! send for Lucius Septimius, the slayer of Pompey. He is a Roman: may be she will listen to him. Begone!

POTHINUS [*darkly*] I know to whom I must go now.

FTATATEETA [*suspiciously*] To whom, then?

POTHINUS To a greater Roman than Lucius. And mark this, mistress. You thought, before Cæsar came, that Egypt should presently be ruled by you and your crew in the name of Cleopatra. I set myself against it –

FTATATEETA [*interrupting him – wrangling*] Ay; that it might be ruled by you and your crew in the name of Ptolemy.

POTHINUS Better me, or even you, than a woman with a Roman heart; and that is what Cleopatra is now become. Whilst I live, she shall never rule. So guide yourself accordingly. [*He goes out*].

It is by this time drawing on to dinner time. The table is laid on the roof of the palace; and thither Rufio is now climbing, ushered by a majestic palace official, wand of office in hand, and followed by a slave carrying an inlaid

stool. After many stairs they emerge at last into a massive colonnade on the roof. Light curtains are drawn between the columns on the north and east to soften the westering sun. The official leads Rufio to one of these shaded sections. A cord for pulling the curtains apart hangs down between the pillars.

THE OFFICIAL [*bowing*] The Roman commander will await Cæsar here.

The slave sets down the stool near the southernmost column, and slips out through the curtains.

RUFIO [*sitting down, a little blown*] Pouf! That was a climb. How high have we come?

THE OFFICIAL We are on the palace roof, O Beloved of Victory!

RUFIO Good! the Beloved of Victory has no more stairs to get up.

A second official enters from the opposite end, walking backwards.

2ND OFFICIAL Cæsar approaches.

Cæsar, fresh from the bath, clad in a new tunic of purple

silk, comes in, beaming and festive, followed by two slaves carrying a light couch, which is hardly more than an elaborately designed bench. They place it near the northmost of the two curtained columns. When this is done they slip out through the curtains; and the two officials, formally bowing, follow them. Rufio rises to receive Cæsar.

CÆSAR [*coming over to him*] Why, Rufio! [*Surveying his dress with an air of admiring astonishment*] A new baldrick! A new golden pommel to your sword! And you have had your hair cut! But not your beard – ? impossible! [*He sniffs at Rufio's beard*]. Yes, perfumed, by Jupiter Olympus!

RUFIO [*growling*] Well: is it to please myself?

CÆSAR [*affectionately*] No, my son Rufio, but to please me – to celebrate my birthday.

RUFIO [*contemptuously*] Your birthday! You always have a birthday when there is a pretty girl to be flattered or an ambassador to be conciliated. We had seven of them in ten months last year.

CÆSAR [*contritely*] It is true, Rufio! I shall never break myself of these petty deceits.

RUFIO Who is to dine with us – besides Cleopatra?

CÆSAR Apollodorus the Sicilian.

RUFIO That popinjay!

CÆSAR Come! the popinjay is an amusing dog – tells a story; sings a song; and saves us the trouble of flattering the Queen. What does she care for old politicians and camp-fed bears like us? No: Apollodorus is good company, Rufio, good company.

RUFIO Well, he can swim a bit and fence a bit: he might be worse, if he only knew how to hold his tongue.

CÆSAR The gods forbid he should ever learn! Oh, this military life! this tedious, brutal life of action! That is the worst of us Romans: we are mere doers and drudgers: a swarm of bees turned into men. Give me a good talker – one with wit and imagination enough to live without continually doing something!

RUFIO Ay! a nice time he would have of it with you when dinner was over! Have you noticed that I am before my time?

CÆSAR Aha! I thought that meant something. What is it?

RUFIO Can we be overheard here?

CÆSAR Our privacy invites eavesdropping. I can remedy that. [*He claps his hands twice. The curtains are drawn, revealing the roof garden with a banqueting table set across in the middle for four persons, one at each end, and two side by side. The side next Cæsar and Rufio is blocked with golden wine vessels and basins. A gorgeous major-domo is superintending the laying of the table by a staff of slaves. The colonnade goes round the garden at both sides to the further end, where a gap in it, like a great gateway, leaves the view open to the sky beyond the western edge of the roof, except in the middle, where a life size image of Ra, seated on a huge plinth, towers up, with hawk head and crown of asp and disk. His altar, which stands at his feet, is a single white stone*]. Now everybody can see us, nobody will think of listening to us. [*He sits down on the bench left by the two slaves*].

RUFIO [*sitting down on his stool*] Pothinus wants to speak to you. I advise you to see him: there is some plotting going on here among the women.

CÆSAR Who is Pothinus?

RUFIO The fellow with hair like squirrel's fur – the little King's bear leader, whom you kept prisoner.

CÆSAR [*annoyed*] And has he not escaped?

RUFIO No.

CÆSAR [*rising imperiously*] Why not? You have been guarding this man instead of watching the enemy. Have I not told you always to let prisoners escape unless there are special orders to the contrary? Are there not enough mouths to be fed without him?

RUFIO Yes; and if you would have a little sense and let me cut his throat, you would save his rations. Anyhow, he *wont* escape. Three sentries have told him they would put a pilum through him if they saw him again. What more can they do? He prefers to stay and spy on us. So would I if I had to do with generals subject to fits of clemency.

CÆSAR [*resuming his seat, argued down*] Hm! And so he wants to see me.

RUFIO Ay. I have brought him with me. He is waiting there [*jerking his thumb over his shoulder*] under guard.

CÆSAR And you want me to see him?

RUFIO [*obstinately*] I dont want anything. I daresay you will do what you like. Dont put it on to me.

CÆSAR [*with an air of doing it expressly to indulge Rufio*] Well, well: let us have him.

RUFIO [*calling*] Ho there, guard! Release your man and send him up. [*Beckoning*]. Come along!

Pothinus enters and stops mistrustfully between the two, looking from one to the other.

CÆSAR [*graciously*] Ah, Pothinus! You are welcome. And what is the news this afternoon?

POTHINUS Cæsar: I come to warn you of a danger, and to make you an offer.

CÆSAR Never mind the danger. Make the offer.

RUFIO Never mind the offer. Whats the danger?

POTHINUS Cæsar: you think that Cleopatra is devoted to you.

CÆSAR [*gravely*] My friend: I already know what I think. Come to your offer.

POTHINUS I will deal plainly. I know not by what strange gods you have been enabled to defend a palace and a few yards of beach against a city and an army. Since we cut you off from Lake Mareotis, and you dug wells in

the salt sea sand and brought up buckets of fresh water from them, we have known that your gods are irresistible, and that you are a worker of miracles. I no longer threaten you –

RUFIO [*sarcastically*] Very handsome of you, indeed.

POTHINUS So be it: you are the master. Our gods sent the north west winds to keep you in our hands; but you have been too strong for them.

CÆSAR [*gently urging him to come to the point*] Yes, yes, my friend. But what then?

RUFIO Spit it out, man. What have you to say?

POTHINUS I have to say that you have a traitress in your camp. Cleopatra –

MAJOR-DOMO [*at the table, announcing*] The Queen! [*Cæsar and Rufio rise*].

RUFIO [*aside to Pothinus*] You should have spat it out sooner, you fool. Now it is too late.

Cleopatra, in gorgeous raiment, enters in state through the gap in the colonnade, and comes down past the image of Ra and past the table to Cæsar. Her retinue, headed by Ftatateeta, joins the staff at the table. Cæsar gives Cleopatra his seat, which she takes.

CLEOPATRA [*quickly, seeing Pothinus*] What is *he* doing here?

CÆSAR [*seating himself beside her, in the most amiable of tempers*] Just going to tell me something about you. You shall hear it. Proceed, Pothinus.

POTHINUS [*disconcerted*] Cæsar – [*he stammers*].

CÆSAR Well, out with it.

POTHINUS What I have to say is for your ear, not for the Queen's.

CLEOPATRA [*with subdued ferocity*] There are means of making you speak. Take care.

POTHINUS [*defiantly*] Cæsar does not employ those means.

CÆSAR My friend: when a man has anything to tell in this world, the difficulty is not to make him tell it, but to prevent him from telling it too often. Let me celebrate my birthday by setting you free. Farewell: we shall not meet again.

CLEOPATRA [*angrily*] Cæsar: this mercy is foolish.

POTHINUS [*to Cæsar*] Will you not give me a private audience? Your life may depend on it. [*Cæsar rises loftily*].

RUFIO [*aside to Pothinus*] Ass! Now we shall have some heroics.

CÆSAR [*oratorically*] Pothinus –

RUFIO [*interrupting him*] Cæsar: the dinner will spoil if you begin preaching your favourite sermon about life and death.

CLEOPATRA [*priggishly*] Peace, Rufio. I desire to hear Cæsar.

RUFIO [*bluntly*] Your Majesty has heard it before. You repeated it to Apollodorus last week; and he thought it was all your own. [*Cæsar's dignity collapses. Much tickled, he sits down again and looks roguishly at Cleopatra, who is furious. Rufio calls as before*] Ho there, guard! Pass the prisoner out. He is released. [*To Pothinus*] Now off with you. You have lost your chance.

POTHINUS [*his temper overcoming his prudence*] I *will* speak.

CÆSAR [*to Cleopatra*] You see. Torture would not have wrung a word from him.

POTHINUS Cæsar: you have taught Cleopatra the arts by which the Romans govern the world.

CÆSAR Alas! they cannot even govern themselves. What then?

POTHINUS What then? Are you so besotted with her beauty that you do not see that she is impatient to reign in Egypt alone, and that her heart is set on your departure?

CLEOPATRA [*rising*] Liar!

CÆSAR [*shocked*] What! Protestations! Contradictions!

CLEOPATRA [*ashamed, but trembling with suppressed rage*] No. I do not deign to contradict. Let him talk. [*She sits down again*].

POTHINUS From her own lips I have heard it. You are to be her catspaw: you are to tear the crown from her brother's head and set it on her own, delivering us all into her hand – delivering yourself also. And then Cæsar can return to Rome, or depart through the gate of death, which is nearer and surer.

CÆSAR [*calmly*] Well, my friend; and is not this very natural?

POTHINUS [*astonished*] Natural! Then you do not resent treachery?

CÆSAR Resent! O thou foolish Egyptian, what have I to do with resentment? Do I resent the wind when it chills me, or the night when it makes me stumble in the darkness? Shall I resent youth when it turns from age, and ambition when it turns from servitude? To tell me such a story as this is but to tell me that the sun will rise to-morrow.

CLEOPATRA [*unable to contain herself*] But it is false – false. I swear it.

CÆSAR It is true, though you swore it a thousand times, and believed all you swore. [*She is convulsed with emotion. To screen her, he rises and takes Pothinus to Rufio, saying*] Come, Rufio: let us see Pothinus past the guard. I have a word to say to him. [*Aside to them*] We must give the Queen a moment to recover herself. [*Aloud*] Come. [*He takes Pothinus and Rufio out with him, conversing with them meanwhile*]. Tell your friends, Pothinus, that they must not think I am opposed to a reasonable settlement of the country's affairs – [*They pass out of hearing*].

CLEOPATRA [*in a stifled whisper*] Ftatateeta, Ftatateeta.

FTATATEETA [*hurrying to her from the table and petting her*] Peace, child: be comforted –

CLEOPATRA [*interrupting her*] Can they hear us?

FTATATEETA No, dear heart, no.

CLEOPATRA Listen to me. If he leaves the Palace alive, never see my face again.

FTATATEETA He? Poth –

CLEOPATRA [*striking her on the mouth*] Strike his life out as I strike his name from your lips. Dash him down from the wall. Break him on the stones. Kill, kill, *kill* him.

FTATATEETA [*shewing all her teeth*] The dog shall perish.

CLEOPATRA Fail in this, and you go out from before me for ever.

FTATATEETA [*resolutely*] So be it. You shall not see my face until his eyes are darkened.

Cæsar comes back, with Apollodorus, exquisitely dressed, and Rufio.

CLEOPATRA [*to Ftatateeta*] Come soon – soon. [*Ftatateeta turns her meaning eyes for a moment on her mistress; then goes grimly away past Ra and out. Cleopatra runs like a gazelle to Cæsar*] So you have come back to me, Cæsar. [*Caressingly*] I thought you were angry. Welcome, Apollodorus. [*She gives him her hand to kiss, with her other arm about Cæsar*].

APOLLODORUS Cleopatra grows more womanly beautiful from week to week.

CLEOPATRA Truth, Apollodorus?

APOLLODORUS Far, far short of the truth! Friend Rufio threw a pearl into the sea: Cæsar fished up a diamond.

CÆSAR Cæsar fished up a touch of rheumatism, my friend. Come: to dinner! to dinner! [*They move towards the table*].

CLEOPATRA [*skipping like a young fawn*] Yes, to dinner. I have ordered *such* a dinner for you, Cæsar!

CÆSAR Ay? What are we to have?

CLEOPATRA Peacocks' brains.

CÆSAR [*as if his mouth watered*] Peacocks' brains, Apollodorus!

APOLLODORUS Not for me. I prefer nightingales' tongues. [*He goes to one of the two covers set side by side*].

CLEOPATRA Roast boar, Rufio!

RUFIO [*gluttonously*] Good! [*He goes to the seat next Apollodorus, on his left*].

CÆSAR [*looking at his seat, which is at the end of the table, to Ra's left hand*] What has become of my leathern cushion?

CLEOPATRA [*at the opposite end*] I have got new ones for you.

MAJOR-DOMO These cushions, Cæsar, are of Maltese gauze, stuffed with rose leaves.

CÆSAR Rose leaves! Am I a caterpillar? [*He throws the cushions away and seats himself on the leather mattress underneath*].

CLEOPATRA What a shame! My new cushions!

MAJOR-DOMO [*at Cæsar's elbow*] What shall we serve to whet Cæsar's appetite?

CÆSAR What have you got?

MAJOR-DOMO Sea hedgehogs, black and white sea acorns, sea nettles, beccaficoes, purple shellfish –

CÆSAR Any oysters?

MAJOR-DOMO Assuredly.

CÆSAR *British* oysters?

MAJOR-DOMO [*assenting*] British oysters, Cæsar.

CÆSAR Oysters, then. [*The Major-Domo signs to a slave at each order; and the slave goes out to execute it*]. I have been in Britain – that western land of romance – the last piece of earth on the edge of the ocean that surrounds the world. I went there in search of its famous pearls. The British pearl was a fable; but in searching for it I found the British oyster.

APOLLODORUS All posterity will bless you for it. [*To the Major-Domo*] Sea hedgehogs for me.

RUFIO Is there nothing solid to begin with?

MAJOR-DOMO Fieldfares with asparagus –

CLEOPATRA [*interrupting*] Fattened fowls! have some fattened fowls, Rufio.

RUFIO Ay, that will do.

CLEOPATRA [*greedily*] Fieldfares for me.

MAJOR-DOMO Cæsar will deign to choose his wine? Sicilian, Lesbian, Chian –

RUFIO [*contemptuously*] All Greek.

APOLLODORUS Who would drink Roman wine when he could get Greek? Try the Lesbian, Cæsar.

CÆSAR Bring me my barley water.

RUFIO [*with intense disgust*] Ugh! Bring *me* my Falernian. [*The Falernian is presently brought to him*].

CLEOPATRA [*pouting*] It is waste of time giving you dinners, Cæsar. My scullions would not condescend to your diet.

CÆSAR [*relenting*] Well, well: let us try the Lesbian. [*The Major-Domo fills Cæsar's goblet; then Cleopatra's and Apollodorus's*]. But when I return to Rome, I will make laws against these extravagances. I will even get the laws carried out.

CLEOPATRA [*coaxingly*] Never mind. To-day you are to be like other people: idle, luxurious, and kind. [*She stretches her hand to him along the table*].

CÆSAR Well, for once I will sacrifice my comfort – [*kissing her hand*] there! [*He takes a draught of wine*]. Now are you satisfied?

CLEOPATRA And you no longer believe that I long for your departure for Rome?

CÆSAR I no longer believe anything. My brains are asleep. Besides, who knows whether I shall return to Rome?

RUFIO [*alarmed*] How? Eh? What?

CÆSAR What has Rome to shew me that I have not seen already? One year of Rome is like another, except that I grow older, whilst the crowd in the Appian Way is always the same age.

APOLLODORUS It is no better here in Egypt. The old men, when they are tired of life, say "We have seen everything except the source of the Nile."

CÆSAR [*his imagination catching fire*] And why not see that? Cleopatra: will you come with me and track the flood to its cradle in the heart of the regions of mystery? Shall we leave Rome behind us – Rome, that has achieved greatness only to learn how greatness destroys nations of men who are not great! Shall I make you a new kingdom, and build you a holy city there in the great unknown?

CLEOPATRA [*rapturously*] Yes, yes. You shall.

RUFIO Ay: now he will conquer Africa with two legions before we come to the roast boar.

APOLLODORUS Come: no scoffing. This is a noble scheme: in it Cæsar is no longer merely the conquering soldier, but the creative poet-artist. Let us name the holy city, and consecrate it with Lesbian wine.

CÆSAR Cleopatra shall name it herself.

CLEOPATRA It shall be called Cæsar's Gift to his Beloved.

APOLLODORUS No, no. Something vaster than that – something universal, like the starry firmament.

CÆSAR [*prosaically*] Why not simply The Cradle of the Nile?

CLEOPATRA No: the Nile is my ancestor; and he is a god. Oh! I have thought of something. The Nile shall name it himself. Let us call upon him. [*To the Major-Domo*] Send for him. [*The three men stare at one another; but the Major-Domo goes out as if he had received the most matter-of-fact order*]. And [*to the retinue*] away with you all.

The retinue withdraws, making obeisance.

A priest enters, carrying a miniature sphinx with a tiny tripod before it. A morsel of incense is smoking in the tripod. The priest comes to the table and places the image in the middle of it. The light begins to change to the

magenta purple of the Egyptian sunset, as if the god had brought a strange colored shadow with him. The three men are determined not to be impressed; but they feel curious in spite of themselves.

CÆSAR What hocus-pocus is this?

CLEOPATRA You shall see. And it is *not* hocus-pocus. To do it properly, we should kill something to please him; but perhaps he will answer Cæsar without that if we spill some wine to him.

APOLLODORUS [*turning his head to look up over his shoulder at Ra*] Why not appeal to our hawkheaded friend here?

CLEOPATRA [*nervously*] Sh! He will hear you and be angry.

RUFIO [*phlegmatically*] The source of the Nile is out of his district, I expect.

CLEOPATRA No: I will have my city named by nobody but my dear little sphinx, because it was in its arms that Cæsar found me asleep. [*She languishes at Cæsar; then turns curtly to the priest*]. Go. I am a priestess, and have power to take your charge from you. [*The priest makes a reverence and goes out*]. Now let us call on the Nile all together. Perhaps he will rap on the table.

CÆSAR What! table rapping! Are such superstitions still believed in this year 707 of the Republic?

CLEOPATRA It is no superstition: our priests learn lots of things from the tables. Is it not so, Apollodorus?

APOLLODORUS Yes: I profess myself a converted man. When Cleopatra is priestess, Apollodorus is devotee. Propose the conjuration.

CLEOPATRA You must say with me "Send us thy voice, Father Nile."

ALL FOUR [*holding their glasses together before the idol*] Send us thy voice, Father Nile.

The death cry of a man in mortal terror and agony answers them. Appalled, the men set down their glasses, and listen. Silence. The purple deepens in the sky. Cæsar, glancing at Cleopatra, catches her pouring out her wine before the god, with gleaming eyes, and mute assurances of gratitude and worship. Apollodorus springs up and runs to the edge of the roof to peer down and listen.

CÆSAR [*looking piercingly at Cleopatra*] What was that?

CLEOPATRA [*petulantly*] Nothing. They are beating some slave.

CÆSAR Nothing!

RUFIO A man with a knife in him, I'll swear.

CÆSAR [*rising*] A murder!

APOLLODORUS [*at the back, waving his hand for silence*] S-sh! Silence. Did you hear that?

CÆSAR Another cry?

APOLLODORUS [*returning to the table*] No, a thud. Something fell on the beach, I think.

RUFIO [*grimly, as he rises*] Something with bones in it, eh?

CÆSAR [*shuddering*] Hush, hush, Rufio. [*He leaves the table and returns to the colonnade: Rufio following at his left elbow, and Apollodorus at the other side*].

CLEOPATRA [*still in her place at the table*] Will you leave me, Cæsar? Apollodorus: are you going?

APOLLODORUS Faith, dearest Queen, my appetite is gone.

CÆSAR Go down to the courtyard, Apollodorus; and find out what has happened.

Apollodorus nods and goes out, making for the staircase by which Rufio ascended.

CLEOPATRA Your soldiers have killed somebody, perhaps. What does it matter?

The murmur of a crowd rises from the beach below. Cæsar and Rufio look at one another.

CÆSAR This must be seen to. [*He is about to follow Apollodorus when Rufio stops him with a hand on his arm as Ftatateeta comes back by the far end of the roof, with dragging steps, a drowsy satiety in her eyes and in the corners of the bloodhound lips. For a moment Cæsar suspects that she is drunk with wine. Not so Rufio: he knows well the red vintage that has inebriated her*].

RUFIO [*in a low tone*] There is some mischief between those two.

FTATATEETA The Queen looks again on the face of her servant.

Cleopatra looks at her for a moment with an exultant reflection of her murderous expression. Then she flings

her arms round her; kisses her repeatedly and savagely; and tears off her jewels and heaps them on her. The two men turn from the spectacle to look at one another. Ftatateeta drags herself sleepily to the altar; kneels before Ra; and remains there in prayer. Cæsar goes to Cleopatra, leaving Rufio in the colonnade.

CÆSAR [*with searching earnestness*] Cleopatra: what has happened?

CLEOPATRA [*in mortal dread of him, but with her utmost cajolery*] Nothing, dearest Cæsar. [*With sickly sweetness, her voice almost failing*] Nothing. I am innocent. [*She approaches him affectionately*]. Dear Cæsar: are you angry with me? Why do you look at me so? I have been here with you all the time. How can I know what has happened?

CÆSAR [*reflectively*] That is true.

CLEOPATRA [*greatly relieved, trying to caress him*] Of course it is true. [*He does not respond to the caress*] *You* know it is true, Rufio.

The murmur without suddenly swells to a roar and subsides.

RUFIO I shall know presently. [*He makes for the altar in the burly trot that serves him for a stride, and touches Ftatateeta on the shoulder*]. Now, mistress: I shall want you. [*He orders her, with a gesture, to go before him*].

FTATATEETA [*rising and glowering at him*] My place is with the Queen.

CLEOPATRA She has done no harm, Rufio.

CÆSAR [*to Rufio*] Let her stay.

RUFIO [*sitting down on the altar*] Very well. Then my place is here too; and you can see what is the matter for yourself. The city is in a pretty uproar, it seems.

CÆSAR [*with grave displeasure*] Rufio: there is a time for obedience.

RUFIO And there is a time for obstinacy. [*He folds his arms doggedly*].

CÆSAR [*to Cleopatra*] Send her away.

CLEOPATRA [*whining in her eagerness to propitiate him*] Yes, I will.

I will do whatever you ask me, Cæsar, always, because I love you. Ftatateeta: go away.

FTATATEETA The Queen's word is my will. I shall be at hand for the Queen's call. [*She goes out past Ra, as she came*].

RUFIO [*following her*] Remember, Cæsar, *your* bodyguard also is within call. [*He follows her out*].

Cleopatra, presuming upon Cæsar's submission to Rufio, leaves the table and sits down on the bench in the colonnade.

CLEOPATRA Why do you allow Rufio to treat you so? You should teach him his place.

CÆSAR Teach him to be my enemy, and to hide his thoughts from me as you are now hiding yours.

CLEOPATRA [*her fears returning*] Why do you say that, Cæsar? Indeed, indeed, I am not hiding anything. You are wrong to treat me like this. [*She stifles a sob*]. I am only a child; and you turn into stone because you think some one has been killed. I cannot bear it. [*She purposely breaks down and weeps. He looks at her with profound sadness and complete coldness. She looks up to see what effect she is producing. Seeing that he is unmoved, she sits up, pretending to struggle with her emotion and to put it bravely away*]. But there: I know you hate tears: you shall not be troubled with them. I know you are not angry, but only sad; only I am so silly, I cannot help being hurt when you speak coldly. Of course you are quite right: it is dreadful to think of anyone being killed or even hurt; and I hope nothing really serious has – [*her voice dies away under his contemptuous penetration*].

CÆSAR What has frightened you into this? What have you done? [*A trumpet sounds on the beach below*]. Aha! that sounds like the answer.

CLEOPATRA [*sinking back trembling on the bench and covering her face with her hands*] I have not betrayed you, Cæsar: I swear it.

CÆSAR I know that. I have not trusted you. [*He turns from her, and is about to go out when Apollodorus and Britannus drag in Lucius Septimius to him. Rufio follows. Cæsar shudders*]. Again, Pompey's murderer!

RUFIO The town has gone mad, I think. They are for tearing

the palace down and driving us into the sea straight away. We laid hold of this renegade in clearing them out of the courtyard.

CÆSAR Release him. [*They let go his arms*]. What has offended the citizens, Lucius Septimius?

LUCIUS What did you expect, Cæsar? Pothinus was a favorite of theirs.

CÆSAR What has happened to Pothinus? I set him free, here, not half an hour ago. Did they not pass him out?

LUCIUS Ay, through the gallery arch sixty feet above ground, with three inches of steel in his ribs. He is as dead as Pompey. We are quits now, as to killing – you and I.

CÆSAR [*shocked*] Assassinated! – our prisoner, our guest! [*He turns reproachfully on Rufio*] Rufio –

RUFIO [*emphatically – anticipating the question*] Whoever did it was a wise man and a friend of yours [*Cleopatra is greatly emboldened*]; but none of *us* had a hand in it. So it is no use to frown at me. [*Cæsar turns and looks at Cleopatra*].

CLEOPATRA [*violently – rising*] He was slain by order of the Queen of Egypt. I am not Julius Cæsar the dreamer, who allows every slave to insult him. Rufio has said I did well: now the others shall judge me too. [*She turns to the others*]. This Pothinus sought to make me conspire with him to betray Cæsar to Achillas and Ptolemy. I refused; and he cursed me and came privily to Cæsar to accuse me of his own treachery. I caught him in the act; and he insulted me – *me*, the Queen! to my face. Cæsar would not avenge me: he spoke him fair and set him free. Was I right to avenge myself? Speak, Lucius.

LUCIUS I do not gainsay it. But you will get little thanks from Cæsar for it.

CLEOPATRA Speak, Apollodorus. Was I wrong?

APOLLODORUS I have only one word of blame, most beautiful. You should have called upon me, your knight; and in fair duel I should have slain the slanderer.

CLEOPATRA [*passionately*] I will be judged by your very slave, Cæsar. Britannus: speak. Was I wrong?

BRITANNUS Were treachery, falsehood, and disloyalty left un-

punished, society must become like an arena full of wild beasts, tearing one another to pieces. Cæsar is in the wrong.

CÆSAR [*with quiet bitterness*] And so the verdict is against me, it seems.

CLEOPATRA [*vehemently*] Listen to me, Cæsar. If one man in all Alexandria can be found to say that I did wrong, I swear to have myself crucified on the door of the palace by my own slaves.

CÆSAR If one man in all the world can be found, now or forever, to *know* that you did wrong, that man will have either to conquer the world as I have, or be crucified by it. [*The uproar in the streets again reaches them*]. Do you hear? These knockers at your gate are also believers in vengeance and in stabbing. You have slain their leader: it is right that they shall slay you. If you doubt it, ask your four counsellors here. And then in the name of that *right* [*he emphasizes the word with great scorn*] shall I not slay them for murdering their Queen, and be slain in my turn by their countrymen as the invader of their fatherland? Can Rome do less then than slay these slayers, too, to shew the world how Rome avenges her sons and her honor. And so, to the end of history, murder shall breed murder, always in the name of right and honor and peace, until the gods are tired of blood and create a race that can understand. [*Fierce uproar. Cleopatra becomes white with terror*]. Hearken, you who must not be insulted. Go near enough to catch their words: you will find them bitterer than the tongue of Pothinus. [*Loftily, wrapping himself up in an impenetrable dignity*] Let the Queen of Egypt now give her orders for vengeance, and take her measures for defence; for she has renounced Cæsar. [*He turns to go*].

CLEOPATRA [*terrified, running to him and falling on her knees*] You will not desert me, Cæsar. You will defend the palace.

CÆSAR You have taken the powers of life and death upon you. I am only a dreamer.

CLEOPATRA But they will kill me.

CÆSAR And why not?

CLEOPATRA In pity –

CÆSAR Pity! What! has it come to this so suddenly, that nothing can save you now but pity? Did it save Pothinus?

She rises, wringing her hands, and goes back to the bench in despair. Apollodorus shews his sympathy with her by quietly posting himself behind the bench. The sky has by this time become the most vivid purple, and soon begins to change to a glowing pale orange, against which the colonnade and the great image shew darklier and darklier.

RUFIO Cæsar: enough of preaching. The enemy is at the gate.

CÆSAR [*turning on him and giving way to his wrath*] Ay; and what has held him baffled at the gate all these months? Was it my folly, as you deem it, or your wisdom? In this Egyptian Red Sea of blood, whose hand has held all your heads above the waves? [*Turning on Cleopatra*] And yet, when Cæsar says to such an one, "Friend, go free," you, clinging for your little life to my sword, dare steal out and stab him in the back? And you, soldiers and gentlemen, and honest servants as you forget that you are, applaud this assassination, and say, "Cæsar is in the wrong." By the gods, I am tempted to open my hand and let you all sink into the flood.

CLEOPATRA [*with a ray of cunning hope*] But, Cæsar, if you do, you will perish yourself.

Cæsar's eyes blaze.

RUFIO [*greatly alarmed*] Now, by great Jove, you filthy little Egyptian rat, that is the very word to make him walk out alone into the city and leave us here to be cut to pieces. [*Desperately, to Cæsar*] Will you desert us because we are a parcel of fools? I mean no harm by killing: I do it as a dog kills a cat, by instinct. We are all dogs at your heels; but we have served you faithfully.

CÆSAR [*relenting*] Alas, Rufio, my son, my son: as dogs we are like to perish now in the streets.

APOLLODORUS [*at his post behind Cleopatra's seat*] Cæsar: what you say has an Olympian ring in it: it must be right; for it is fine art. But I am still on the side of Cleopatra. If we must die, she shall not want the devotion of a man's heart nor the strength of a man's arm.

CLEOPATRA [*sobbing*] But I don't want to die.

CÆSAR [*sadly*] Oh, ignoble, ignoble!

LUCIUS [*coming forward between Cæsar and Cleopatra*] Hearken to me, Cæsar. It may be ignoble; but I also mean to live as long as I can.

CÆSAR Well, my friend, you are likely to outlive Cæsar. Is it any magic of mine, think you, that has kept your army and this whole city at bay for so long? Yesterday, what quarrel had they with me that they should risk their lives against me? But to-day we have flung them down their hero, murdered; and now every man of them is set upon clearing out this nest of assassins – for such we are and no more. Take courage then; and sharpen your sword. Pompey's head has fallen; and Cæsar's head is ripe.

POLLODORUS Does Cæsar despair?

CÆSAR [*with infinite pride*] He who has never hoped can never despair. Cæsar, in good or bad fortune, looks his fate in the face.

LUCIUS Look it in the face, then; and it will smile as it always has on Cæsar.

CÆSAR [*with involuntary haughtiness*] Do you presume to encourage me?

LUCIUS I offer you my services. I will change sides if you will have me.

CÆSAR [*suddenly coming down to earth again, and looking sharply at him, divining that there is something behind the offer*] What! At this point?

LUCIUS [*firmly*] At this point.

RUFIO Do you suppose Cæsar is mad, to trust you?

LUCIUS I do not ask him to trust me until he is victorious. I ask for my life, and for a command in Cæsar's army. And since Cæsar is a fair dealer, I will pay in advance.

CÆSAR Pay! How?

LUCIUS With a piece of good news for you.

Cæsar divines the news in a flash.

RUFIO What news?

CÆSAR [*with an elate and buoyant energy which makes Cleopatra sit up and stare*] What news! What news, did you say, my son Rufio? The relief has arrived: what other news remains for us? Is it not so, Lucius Septimius? Mithridates of Pergamos is on the march.

LUCIUS He has taken Pelusium.

CÆSAR [*delighted*] Lucius Septimius: you are henceforth my officer. Rufio: the Egyptians must have sent every soldier from the city to prevent Mithridates crossing the Nile. There is nothing in the streets now but mob – mob!

LUCIUS It is so. Mithridates is marching by the great road to Memphis to cross above the Delta. Achillas will fight him there.

CÆSAR [*all audacity*] Achillas shall fight Cæsar there. See, Rufio. [*He runs to the table; snatches a napkin; and draws a plan on it with his finger dipped in wine, whilst Rufio and Lucius Septimius crowd about him to watch, all looking closely, for the light is now almost gone*]. Here is the palace [*pointing to his plan*]: here is the theatre. You [*to Rufio*] take twenty men and pretend to go by *that* street [*pointing it out*]; and whilst they are stoning you, out go the cohorts by this and this. My streets are right, are they, Lucius?

LUCIUS Ay, that is the fig market –

CÆSAR [*too much excited to listen to him*] I saw them the day we arrived. Good! [*He throws the napkin on the table,*

and comes down again into the colonnade]. Away, Britannus: tell Petronius that within an hour half our forces must take ship for the western lake. See to my horse and armor. [*Britannus runs out*]. With the rest, *I* shall march round the lake and up the Nile to meet Mithridates. Away, Lucius; and give the word. [*Lucius hurries out after Britannus*]. Apollodorus: lend me your sword and your right arm for this campaign.

POLLODORUS Ay, and my heart and life to boot.

CÆSAR [*grasping his hand*] I accept both. [*Mighty handshake*]. Are you ready for work?

POLLODORUS Ready for Art – the Art of War [*he rushes out after Lucius, totally forgetting Cleopatra*].

RUFIO Come! this is something like business.

CÆSAR [*buoyantly*] Is it not, my only son? [*He claps his hands. The slaves hurry in to the table*]. No more of this mawkish revelling: away with all this stuff: shut it out of my sight and be off with you. [*The slaves begin to remove the table; and the curtains are drawn, shutting in the colonnade*]. You understand about the streets, Rufio?

RUFIO Ay, I think I do. I will get through them, at all events.

The bucina sounds busily in the courtyard beneath.

CÆSAR Come, then: we must talk to the troops and hearten them. You down to the beach: I to the courtyard [*He makes for the staircase*].

CLEOPATRA [*rising from her seat, where she has been quite neglected all this time, and stretching out her hands timidly to him*] Cæsar.

CÆSAR [*turning*] Eh?

CLEOPATRA Have you forgotten me?

CÆSAR [*indulgently*] I am busy now, my child, busy. When I return your affairs shall be settled. Farewell; and be good and patient.

He goes, preoccupied and quite indifferent. She stands with clenched fists, in speechless rage and humiliation.

RUFIO That game is played and lost, Cleopatra. The woman always gets the worst of it.

CLEOPATRA [*haughtily*] Go. Follow your master.

RUFIO [*in her ear, with rough familiarity*] A word first. Tell your executioner that if Pothinus had been properly killed – in the *throat* – he would not have called out. Your man bungled his work.

CLEOPATRA [*enigmatically*] How do you know it was a man?

RUFIO [*startled, and puzzled*] It was not you: you were with us when it happened. [*She turns her back scornfully on him. He shakes his head, and draws the curtains to go out. It is now a magnificent moonlit night. The table has been removed. Ftatateeta is seen in the light of the moon and stars, again in prayer before the white altar-stone of Ra. Rufio starts; closes the curtains again softly; and says in a low voice to Cleopatra*] Was it she? with her own hand?

CLEOPATRA [*threateningly*] Whoever it was, let my enemies beware of her. Look to it, Rufio, you who dare make the Queen of Egypt a fool before Cæsar.

RUFIO [*looking grimly at her*] I will look to it, Cleopatra. [*He nods in confirmation of the promise, and slips out through the curtains, loosening his sword in its sheath as he goes*].

ROMAN SOLDIERS [*in the courtyard below*] Hail, Cæsar! Hail, hail!

Cleopatra listens. The bucina sounds again, followed by several trumpets.

CLEOPATRA [*wringing her hands and calling*] Ftatateeta. Ftatateeta. It is dark; and I am alone. Come to me. [*Silence*] Ftatateeta. [*Louder*] Ftatateeta. [*Silence. In a panic she snatches the cord and pulls the curtains apart*].

Ftatateeta is lying dead on the altar of Ra, with her throat cut. Her blood deluges the white stone.

ACT 5

High noon. Festival and military pageant on the esplanade before the palace. In the east harbor Cæsar's galley, so gorgeously decorated that it seems to be rigged with flowers, is alongside the quay, close to the steps Apollodorus descended when he embarked with the carpet. A Roman guard is posted there in charge of a gangway, whence a red floorcloth is laid down the middle of the esplanade, turning off to the north opposite the central gate in the palace front, which shuts in the esplanade on the south side. The broad steps of the gate, crowded with Cleopatra's ladies, all in their gayest attire, are like a flower garden. The façade is lined by her guard, officered by the same gallants to whom Bel Affris announced the coming of Cæsar six months before in the old palace on the Syrian border. The north side is lined by Roman soldiers, with the townsfolk on tiptoe behind them, peering over their heads at the cleared esplanade, in which the officers stroll about, chatting. Among these are Belzanor and the Persian; also the centurion, vinewood cudgel in hand, battle worn, thick-booted, and much outshone, both socially and decoratively, by the Egyptian officers.

Apollodorus makes his way through the townsfolk and calls to the officers from behind the Roman line.

APOLLODORUS Hullo! May I pass?

CENTURION Pass Apollodorus the Sicilian there! [*The soldiers let him through*].

BELZANOR Is Cæsar at hand?

APOLLODORUS Not yet. He is still in the market place. I could not stand any more of the roaring of the soldiers! After half an hour of the enthusiasm of an army, one feels the need of a little sea air.

PERSIAN Tell us the news. Hath he slain the priests?

APOLLODORUS Not he. They met him in the market place with ashes on their heads and their gods in their hands. They placed the gods at his feet. The only one that was worth looking at was Apis: a miracle of gold and ivory work. By my advice he offered the chief priest two talents for it.

BELZANOR [*appalled*] Apis the all-knowing for two talents! What said the chief Priest?

APOLLODORUS He invoked the mercy of Apis, and asked for five.

BELZANOR There will be famine and tempest in the land for this.

PERSIAN Pooh! Why did not Apis cause Cæsar to be vanquished by Achillas? Any fresh news from the war, Apollodorus?

APOLLODORUS The little King Ptolemy was drowned.

BELZANOR Drowned! How?

APOLLODORUS With the rest of them. Cæsar attacked them from three sides at once and swept them into the Nile. Ptolemy's barge sank.

BELZANOR A marvellous man, this Cæsar! Will he come soon, think you?

APOLLODORUS He was settling the Jewish question when I left.

A flourish of trumpets from the north, and commotion among the townsfolk, announces the approach of Cæsar.

PERSIAN He has made short work of them. Here he comes. [*He hurries to his post in front of the Egyptian lines*].

BELZANOR [*following him*] Ho there! Cæsar comes.

The soldiers stand at attention, and dress their lines. Apollodorus goes to the Egyptian line.

CENTURION [*hurrying to the gangway guard*] Attention there! Cæsar comes.

Cæsar arrives in state with Rufio: Britannus following. The soldiers receive him with enthusiastic shouting.

CÆSAR I see my ship awaits me. The hour of Cæsar's farewell to Egypt has arrived. And now, Rufio, what remains to be done before I go?

RUFIO [*at his left hand*] You have not yet appointed a Roman governor for this province.

CÆSAR [*looking whimsically at him, but speaking with perfect gravity*] What say you to Mithridates of Pergamos, my reliever and rescuer, the great son of Eupator?

RUFIO Why, that you will want him elsewhere. Do you forget that you have some three or four armies to conquer on your way home?

CÆSAR Indeed! Well, what say you to yourself?

RUFIO [*incredulously*] I! I a governor! What are you dreaming of? Do you not know that I am only the son of a freedman?

CÆSAR [*affectionately*] Has not Cæsar called you his son? [*Calling to the whole assembly*] Peace awhile there; and hear me.

THE ROMAN SOLDIERS Hear Cæsar.

CÆSAR Hear the service, quality, rank and name of the Roman governor. By service, Cæsar's shield; by quality, Cæsar's friend; by rank, a Roman soldier. [*The Roman soldiers give a triumphant shout*]. By name, Rufio. [*They shout again*].

RUFIO [*kissing Cæsar's hand*] Ay: I am Cæsar's shield; but of what use shall I be when I am no longer on Cæsar's arm? Well, no matter – [*He becomes husky, and turns away to recover himself*].

CÆSAR Where is that British Islander of mine?

BRITANNUS [*coming forward on Cæsar's right hand*] Here, Cæsar.

CÆSAR Who bade you, pray, thrust yourself into the battle of the Delta, uttering the barbarous cries of your native land, and affirming yourself a match for any four of the Egyptians, to whom you applied unseemly epithets?

BRITANNUS Cæsar: I ask you to excuse the language that escaped me in the heat of the moment.

CÆSAR And how did you, who cannot swim, cross the canal with us when we stormed the camp?

BRITANNUS Cæsar: I clung to the tail of your horse.

CÆSAR These are not the deeds of a slave, Britannicus, but of a free man.

BRITANNUS Cæsar: I was born free.

CÆSAR But they call you Cæsar's slave.

BRITANNUS Only as Cæsar's slave have I found real freedom.

CÆSAR [*moved*] Well said. Ungrateful that I am, I was about to set you free; but now I will not part from you for a million talents. [*He claps him friendly on the shoulder. Britannus, gratified, but a trifle shamefaced, takes his hand and kisses it sheepishly*].

BELZANOR [*to the Persian*] This Roman knows how to make men serve him.

PERSIAN Ay: men too humble to become dangerous rivals to him.

BELZANOR O subtle one! O cynic!

CÆSAR [*seeing Apollodorus in the Egyptian corner, and calling to him*] Apollodorus: I leave the art of Egypt in your charge. Remember: Rome loves art and will encourage it ungrudgingly.

APOLLODORUS I understand, Cæsar. Rome will produce no art itself; but it will buy up and take away whatever the other nations produce.

CÆSAR What! Rome produce no art! Is peace not an art? is war not an art? is government not an art? is civilization not an art? All these we give you in exchange for a few ornaments. You will have the best of the bargain. [*Turning to Rufio*] And now, what else have I to do before I embark? [*Trying to recollect*] There is something I cannot remember: what *can* it be? Well, well: it must remain undone: we must not waste this favorable wind. Farewell, Rufio.

RUFIO Cæsar: I am loth to let you go to Rome without your shield. There are too many daggers there.

CÆSAR It matters not: I shall finish my life's work on my way

back; and then I shall have lived long enough. Besides: I have always disliked the idea of dying: I had rather be killed. Farewell.

RUFIO [*with a sigh, raising his hands and giving Cæsar up as incorrigible*] Farewell. [*They shake hands*].

CÆSAR [*waving his hand to Apollodorus*] Farewell, Apollodorus, and my friends, all of you. Aboard!

The gangway is run out from the quay to the ship. As Cæsar moves towards it, Cleopatra, cold and tragic, cunningly dressed in black, without ornaments or decoration of any kind, and thus making a striking figure among the brilliantly dressed bevy of ladies as she passes through it, comes from the palace and stands on the steps. Cæsar does not see her until she speaks.

CLEOPATRA Has Cleopatra no part in this leavetaking?

CÆSAR [*enlightened*] Ah, I *knew* there was something. [*To Rufio*] How could you let me forget her, Rufio? [*Hastening to her*] Had I gone without seeing you, I should never have forgiven myself. [*He takes her hands, and brings her into the middle of the esplanade. She submits stonily*]. Is this mourning for me?

CLEOPATRA No.

CÆSAR [*remorsefully*] Ah, that was thoughtless of me! It is for your brother.

CLEOPATRA No.

CÆSAR For whom, then?

CLEOPATRA Ask the Roman governor whom you have left us.

CÆSAR Rufio?

CLEOPATRA Yes: Rufio. [*She points at him with deadly scorn*]. He who is to rule here in Cæsar's name, in Cæsar's way, according to Cæsar's boasted laws of life.

CÆSAR [*dubiously*] He is to rule as he can, Cleopatra. He has taken the work upon him, and will do it in his own way.

CLEOPATRA Not in your way, then?

CÆSAR [*puzzled*] What do you mean by my way?

CLEOPATRA Without punishment. Without revenge. Without judgment.

CÆSAR [*approvingly*] Ay: that is the right way, the great way,

the only possible way in the end. [*To Rufio*] Believe it, Rufio, if you can.

RUFIO Why, I believe it, Cæsar. You have convinced me of it long ago. But look you. You are sailing for Numidia to-day. Now tell me: if you meet a hungry lion there, you will not punish it for wanting to eat you?

CÆSAR [*wondering what he is driving at*] No.

RUFIO Nor revenge upon it the blood of those it has already eaten.

CÆSAR No.

RUFIO Nor judge it for its guiltiness.

CÆSAR No.

RUFIO What, then, will you do to save your life from it?

CÆSAR [*promptly*] Kill it, man, without malice, just as it would kill me. What does this parable of the lion mean?

RUFIO Why, Cleopatra had a tigress that killed men at her bidding. I thought she might bid it kill you some day. Well, had I not been Cæsar's pupil, what pious things might I not have done to that tigress! I might have punished it. I might have revenged Pothinus on it.

CÆSAR [*interjects*] Pothinus!

RUFIO [*continuing*] I might have judged it. But I put all these follies behind me; and, without malice, only cut its throat. And that is why Cleopatra comes to you in mourning.

CLEOPATRA [*vehemently*] He has shed the blood of my servant Ftatateeta. On your head be it as upon his, Cæsar, if you hold him free of it.

CÆSAR [*energetically*] On my head be it, then; for it was well done. Rufio: had you set yourself in the seat of the judge, and with hateful ceremonies and appeals to the gods handed that woman over to some hired executioner to be slain before the people in the name of justice, never again would I have touched your hand without a shudder. But this was natural slaying: I feel no horror at it.

Rufio, satisfied, nods at Cleopatra, mutely inviting her to mark that.

CLEOPATRA [*pettish and childish in her impotence*] No: not when a Roman slays an Egyptian. All the world will now see how unjust and corrupt Cæsar is.

CÆSAR [*taking her hands coaxingly*] Come: do not be angry with me. I am sorry for that poor Totateeta. [*She laughs in spite of herself*]. Aha! you are laughing. Does that mean reconciliation?

CLEOPATRA [*angry with herself for laughing*] No, *no*, NO! ! But it is so ridiculous to hear you call her Totateeta.

CÆSAR What! As much a child as ever, Cleopatra! Have I not made a woman of you after all?

CLEOPATRA Oh, it is you who are a great baby: you make me seem silly because you will not behave seriously. But you have treated me badly; and I do not forgive you.

CÆSAR Bid me farewell.

CLEOPATRA I will not.

CÆSAR [*coaxing*] I will send you a beautiful present from Rome.

CLEOPATRA [*proudly*] Beauty from Rome to Egypt indeed! What can Rome give *me* that Egypt cannot give me?

APOLLODORUS That is true, Cæsar. If the present is to be really beautiful, I shall have to buy it for you in Alexandria.

CÆSAR You are forgetting the treasures for which Rome is most famous, my friend. You cannot buy *them* in Alexandria.

APOLLODORUS What are they, Cæsar?

CÆSAR Her sons. Come, Cleopatra: forgive me and bid me farewell; and I will send you a man, Roman from head to heel and Roman of the noblest; not old and ripe for the knife; not lean in the arms and cold in the heart; not hiding a bald head under his conqueror's laurels; not stooped with the weight of the world on his shoulders; but brisk and fresh, strong and young, hoping in the morning, fighting in the day, and revelling in the evening. Will you take such an one in exchange for Cæsar?

CLEOPATRA [*palpitating*] His name, his name?

CÆSAR Shall it be Mark Antony? [*She throws herself into his arms*].

RUFIO You are a bad hand at a bargain, mistress, if you will swop Cæsar for Antony.

CÆSAR So now you are satisfied.

CLEOPATRA You will not forget.

CÆSAR I will not forget. Farewell: I do not think we shall meet again. Farewell. [*He kisses her on the forehead. She is much affected and begins to sniff. He embarks*].

THE ROMAN SOLDIERS [*as he sets his foot on the gangway*] Hail, Cæsar; and farewell!

He reaches the ship and returns Rufio's wave of the hand.

APOLLODORUS [*to Cleopatra*] No tears, dearest Queen: they stab your servant to the heart. He will return some day.

CLEOPATRA I hope not. But I cant help crying, all the same. [*She waves her handkerchief to Cæsar; and the ship begins to move*].

THE ROMAN SOLDIERS [*drawing their swords and raising them in the air*] Hail, Cæsar!

NOTES

Notes to
THE DEVIL'S DISCIPLE

General John Burgoyne, who is presented in this play for the first time (as far as I am aware) on the English stage, is not a conventional stage soldier, but as faithful a portrait as it is in the nature of stage portraits to be. His objection to profane swearing is not borrowed from Mr Gilbert's H.M.S. Pinafore: it is taken from the Code of Instructions drawn up by himself for his officers when he introduced Light Horse into the English army. His opinion that English soldiers should be treated as thinking beings was no doubt as unwelcome to the military authorities of his time, when nothing was thought of ordering a soldier a thousand lashes, as it will be to those modern victims of the flagellation neurosis who are so anxious to revive that discredited sport. His military reports are very clever as criticisms, and are humane and enlightened within certain aristocratic limits, best illustrated perhaps by his declaration, which now sounds so curious, that he should blush to ask for promotion on any other ground than that of family influence. As a parliamentary candidate, Burgoyne took our common expression "fighting an election" so very literally that he led his supporters to the poll at Preston in 1768 with a loaded pistol in each hand, and won the seat, though he was fined £1000, and denounced by Junius, for the pistols.

It is only within quite recent years that any general recognition has become possible for the feeling that led Burgoyne, a professed enemy of oppression in India and elsewhere, to accept his American command when so many other officers threw up their commissions rather than serve in a civil war against the Colonies. His biographer De Fonblanque, writing in 1876, evidently regarded his position as indefensible. Nowadays, it is sufficient to say that Burgoyne was an Imperialist. He sympathized with the colonists; but when they proposed as a remedy the disruption of the Empire, he regarded that as a step backward in civilization. As he put it to the House of Commons, "while we remember that we are contending against brothers and fellow subjects, we must also remember that we are contending in this crisis for the fate of the British Empire." Eighty-four years after his defeat, his republican conquerors themselves engaged in a civil war for the integrity of their Union. In 1885 the Whigs who represented the anti-Burgoyne tradition of American Independence in English politics, abandoned Gladstone and made common cause with their political opponents in defence of the Union between England and Ireland. Only the other day England sent 200,000 men into the field south of the equator to fight out the question whether South Africa should develop as a Federation of British Colonies or as an independent Afrikander United States. In all these cases the Unionists who were detached from their parties were called renegades, as Burgoyne was. That, of course, is only one of the unfortunate consequences of the fact that mankind, being for the most part incapable of politics, accepts vituperation as an easy and congenial substitute. Whether Burgoyne or Washington, Lincoln or Davis, Gladstone or Bright, Mr Chamberlain or Mr Leonard

Courtney was in the right will never be settled, because it will never be possible to prove that the government of the victor has been better for mankind than the government of the vanquished would have been. It is true that the victors have no doubt on the point; but to the dramatist, that certainty of theirs is only part of the human comedy. The American Unionist is often a Separatist as to Ireland; the English Unionist often sympathizes with the Polish Home Ruler; and both English and American Unionists are apt to be Disruptionists as regards that Imperial Ancient of Days, the Empire of China. Both are Unionists concerning Canada, but with a difference as to the precise application to it of the Monroe doctrine. As for me, the dramatist, I smile, and lead the conversation back to Burgoyne.

Burgoyne's surrender at Saratoga made him that occasionally necessary part of our British system, a scapegoat. The explanation of his defeat given in the play (p. 72) is founded on a passage quoted by De Fonblanque from Fitzmaurice's Life of Lord Shelburne, as follows: "Lord George Germain, having among other peculiarities a particular dislike to be put out of his way on any occasion, had arranged to call at his office on his way to the country to sign the dispatches; but as those addressed to Howe had not been fair-copied, and he was not disposed to be balked of his projected visit to Kent, they were not signed then and were forgotten on his return home." These were the dispatches instructing Sir William Howe, who was in New York, to effect a junction at Albany with Burgoyne, who had marched from Quebec for that purpose. Burgoyne got as far as Saratoga, where, failing the expected reinforcement, he was hopelessly outnumbered, and his officers picked off, Boer fashion, by the American farmer-sharpshooters. His own collar was pierced by a bullet. The publicity of his defeat, however, was more than compensated at home by the fact that Lord George's trip to Kent had not been interfered with, and that nobody knew about the oversight of the dispatch. The policy of the English Government and Court for the next two years was simply concealment of Germain's neglect. Burgoyne's demand for an inquiry was defeated in the House of Commons by the court party; and when he at last obtained a committee, the king got rid of it by a prorogation. When Burgoyne realized what had happened about the instructions to Howe (the scene in which I have represented him as learning it before Saratoga is not historical: the truth did not dawn on him until many months afterwards) the king actually took advantage of his being a prisoner of war in England on parole, and ordered him to return to America into captivity. Burgoyne immediately resigned all his appointments; and this practically closed his military career, though he was afterwards made Commander of the Forces in Ireland for the purpose of banishing him from parliament.

The episode illustrates the curious perversion of the English sense of honor when the privileges and prestige of the aristocracy are at stake. Mr Frank Harris said, after the disastrous battle of

Modder River, that the English, having lost America a century ago because they preferred George III, were quite prepared to lose South Africa to-day because they preferred aristocratic commanders to successful ones. Horace Walpole, when the parliamentary recess came at a critical period of the War of Independence, said that the Lords could not be expected to lose their pheasant shooting for the sake of America. In the working class, which, like all classes, has its own official aristocracy, there is the same reluctance to discredit an institution or to "do a man out of his job." At bottom, of course, this apparently shameless sacrifice of great public interests to petty personal ones, is simply the preference of the ordinary man for the things he can feel and understand to the things that are beyond his capacity. It is stupidity, not dishonesty.

Burgoyne fell a victim to this stupidity in two ways. Not only was he thrown over, in spite of his high character and distinguished services, to screen a court favorite who had actually been cashiered for cowardice and misconduct in the field fifteen years before; but his peculiar critical temperament and talent, artistic, satirical, rather histrionic, and his fastidious delicacy of sentiment, his fine spirit and humanity, were just the qualities to make him disliked by stupid people because of their dread of ironic criticism. Long after his death, Thackeray, who had an intense sense of human character, but was typically stupid in valuing and interpreting it, instinctively sneered at him and exulted in his defeat. That sneer represents the common English attitude towards the Burgoyne type. Every instance in which the critical genius is defeated, and the stupid genius (for both temperaments have their genius) "muddles through all right," is popular in England. But Burgoyne's failure was not the work of his own temperament, but of the stupid temperament. What man could do under the circumstances he did, and did handsomely and loftily. He fell, and his ideal empire was dismembered, not through his own misconduct, but because Sir George Germain overestimated the importance of his Kentish holiday, and underestimated the difficulty of conquering those remote and inferior creatures, the colonists. And King George and the rest of the nation agreed, on the whole, with Germain. It is a significant point that in America, where Burgoyne was an enemy and an invader, he was admired and praised. The climate there is no doubt more favorable to intellectual vivacity.

I have described Burgoyne's temperament as rather histrionic; and the reader will have observed that the Burgoyne of the Devil's Disciple is a man who plays his part in life, and makes all its points, in the manner of a born high comedian. If he had been killed at Saratoga, with all his comedies unwritten, and his plan for turning As You Like It into a Beggar's Opera unconceived, I should still have painted the same picture of him on the strength of his reply to the articles of capitulation proposed to him by the victorious Gates (an Englishman). Here they are:

Proposition	Answer
1. General Burgoyne's army being reduced by repeated defeats, by desertion, sickness, etc., their provisions exhausted, their military horses, tents and baggage taken or destroyed, their retreat cut off, and their camp invested, they can only be allowed to surrender as prisoners of war.	*Lieut-General Burgoyne's army, however reduced, will never admit that their retreat is cut off while they have arms in their hands.*
2. The officers and soldiers may keep the baggage belonging to them. The Generals of the United States never permit individuals to be pillaged.	*Noted.*
3. The troops under his Excellency General Burgoyne will be conducted by the most convenient route to New England, marching by easy marches, and sufficiently provided for by the way.	*Agreed.*
4. The officers will be admited on parole and will be treated with the liberality customary in such cases, so long as they, by proper behaviour, continue to deserve it; but those who are apprehended having broke their parole, as some British officers have done, must expect to be close confined.	*There being no officer in this army under, or capable of being under, the description of breaking parole, this article needs no answer.*
5. All public stores, artillery, arms, ammunition, carriages, horses, etc., etc., must be delivered to commissaries appointed to receive them.	*All public stores may be delivered, arms excepted.*
6. These terms being agreed to and signed, the troops under his Excellency's, General Burgoyne's command, may be drawn up in their encampments, where they will be ordered to ground their arms, and may thereupon be marched to the river-side on their way to Bennington.	*This article is inadmissible in any extremity. Sooner than this army will consent to ground their arms in their encampments, they will rush on the enemy determined to take no quarter.*

And, later on, "If General Gates does not mean to recede from the 6th article, the treaty ends at once: the army will to a man proceed to any act of desperation sooner than submit to that article."

Here you have the man at his Burgoynest. Need I add that he had his own way; and that when the actual ceremony of surrender came, he would have played poor General Gates off the stage, had not that commander risen to the occasion by handing him back his sword.

In connection with the reference to Indians with scalping knives, who, with the troops hired from Germany, made up about half Burgoyne's force, I may cite the case of Jane McCrea, betrothed to one of Burgoyne's officers. A Wyandotte chief attached to Burgoyne's force was bringing her to the British camp as a prisoner of war, when another party of Indians, sent by her betrothed, claimed her. The Wyandotte settled the dispute by killing her and bringing her scalp to Burgoyne. Burgoyne let the deed pass. Possibly he feared that a massacre of whites on the Canadian border by the Wyandottes would follow any attempt at punishment. But his own proclamations had threatened just what the savage chief executed.

BRUDENELL

Brudenell is also a real person. At least, an artillery chaplain of that name distinguished himself at Saratoga by reading the burial service over Major Fraser under fire, and by a quite readable adventure, chronicled, with exaggerations, by Burgoyne, concerning Lady Harriet Acland. Others have narrated how Lady Harriet's husband killed himself in a duel, by falling with his head against a pebble; and how Lady Harriet then married the warrior chaplain. All this, however, is a tissue of romantic lies, though it has been repeated in print as authentic history from generation to generation, even to the first edition of this book. As a matter of fact, Major Acland died in his bed of a cold shortly after his return to England; and Lady Harriet remained a widow until her death in 1815.

The rest of the Devil's Disciple may have actually occurred, like most stories invented by dramatists; but I cannot produce any documents. Major Swindon's name is invented; but the man, of course, is real. There are dozens of him extant to this day.

Notes to
CÆSAR AND CLEOPATRA

CLEOPATRA'S CURE FOR BALDNESS

For the sake of conciseness in a hurried situation I have made Cleopatra recommend rum. This, I am afraid, is an anachronism: the only real one in the play. To balance it, I give a couple of the remedies she actually believed in. They are quoted by Galen from Cleopatra's book on Cosmetic.

"For bald patches, powder red sulphuret of arsenic and take it up with oak gum, as much as it will bear. Put on a rag and apply, having soaped the place well first. I have mixed the above with a foam of nitre, and it worked well."

Several other receipts follow, ending with: "The following is the best of all, acting for fallen hairs, when applied with oil or pomatum; acts also for falling off of eyelashes or for people getting bald all over. It is wonderful. Of domestic mice burnt, one part; of vine rag burnt, one part; of horse's teeth burnt, one part; of bear's grease one; of deer's marrow one; of reed bark one. To be pounded when dry, and mixed with plenty of honey til it gets the consistency of honey; then the bear's grease and marrow to be mixed (when melted), the medicine to be put in a brass flask, and the bald part rubbed til it sprouts."

Concerning these ingredients, my fellow-dramatist Gilbert Murray, who, as a Professor of Greek, has applied to classical antiquity the methods of high scholarship (my own method is pure divination), writes to me as follows: "Some of this I dont understand, and possibly Galen did not, as he quotes your heroine's own language. Foam of nitre is, I think, something like soapsuds. Reed bark is an odd expression. It might mean the outside membrane of a reed: I do not know what it ought to be called. In the burnt mice receipt I take it that you first mixed the solid powders with honey, and then added the grease. I expect Cleopatra preferred it because in most of the others you have to lacerate the skin, prick it, or rub it till it bleeds. I do not know what vine rag is. I translate literally."

APPARENT ANACHRONISMS

The only way to write a play which shall convey to the general public an impression of antiquity is to make the characters speak blank verse and abstain from reference to steam, telegraphy, or any of the material conditions of their existence. The more ignorant men are, the more convinced are they that their little parish and their little chapel is an apex to which civilization and philosophy has painfully struggled up the pyramid of time from a desert of savagery. Savagery, they think, became barbarism; barbarism became ancient civilization; ancient civilization became Pauline Christianity; Pauline Christianity became Roman Catholicism; Roman Catholicism became the Dark Ages; and the Dark Ages were finally enlightened by the Protestant instincts of the English race. The whole process is summed up as Progress with a capital P. And any elderly gentleman of Progressive temperament will testify that the improvement since he was a boy is enormous.

Now if we count the generations of Progressive elderly gentlemen since, say, Plato, and add together the successive enormous improvements to which each of them has testified, it will strike us at once as an unaccountable fact that the world, instead of having been improved in 67 generations out of all recognition, presents, on the whole, a rather less dignified appearance in Ibsen's Enemy of the People than in Plato's Republic. And in truth, the period of time covered by history is far too short to allow of any perceptible progress in the popular sense of Evolution of the Human Species. The notion that there has been any such Progress since Cæsar's time (less than 20 centuries) is too absurd for discussion. All the savagery, barbarism, dark ages and the rest of it of which we have any record as existing in the past, exists at the present moment. A British carpenter or stonemason may point out that he gets twice as much money for his labor as his father did in the same trade, and that his suburban house, with its bath, its cottage piano, its drawingroom suite, and its album of photographs, would have shamed the plainness of his grandmother's. But the descendants of feudal barons, living in squalid lodgings on a salary of fifteen shillings a week instead of in castles on princely revenues, do not congratulate the world on the change. Such changes, in fact, are not to the point. It has been known, as far back as our records go, that man running wild in the woods is different from man kennelled in a city slum; that a dog seems to understand a shepherd better than a hewer of wood and drawer of water can understand an astronomer; and that breeding, gentle nurture and luxurious food and shelter will produce a kind of man with whom the common laborer is socially incompatible. The same thing is true of horses and dogs. Now there is clearly room for great changes in the world by increasing the percentage of individuals who are carefully bred and gently nurtured, even to finally making the most of every man and woman born. But that possibility existed in the days of the Hittites as much as it does to-day. It does not give the slightest real support to the common assumption that the civilized contemporaries of the Hittites were unlike their civilized descendants today.

This would appear the tritest commonplace if it were not that the ordinary citizen's ignorance of the past combines with his idealization of the present to mislead and flatter him. Our latest book on the new railway across Asia describes the dulness of the Siberian farmer and the vulgar pursepride of the Siberian man of business without the least consciousness that the string of contemptuous instances given might have been saved by writing simply "Farmers and provincial plutocrats in Siberia are exactly what they are in England." The latest professor descanting on the civilization of the Western Empire in the fifth century feels bound to assume, in the teeth of his own researches, that the Christian was one sort of animal and the Pagan another. It might as well be assumed, as indeed it generally is assumed by implication, that a murder committed with a poisoned arrow

is different from a murder committed with a Mauser rifle. All such notions are illusions. Go back to the first syllable of recorded time, and there you will find your Christian and your Pagan, your yokel and your poet, helot and hero, Don Quixote and Sancho, Tamino and Papageno, Newton and bushman unable to count eleven, all alive and contemporaneous, and all convinced that they are the heirs of all the ages and the privileged recipients of THE truth (all others damnable heresies), just as you have them to-day, flourishing in countries each of which is the bravest and best that ever sprang at Heaven's command from out the azure main.

Again, there is the illusion of "increased command over Nature," meaning that cotton is cheap and that ten miles of country road on a bicycle have replaced four on foot. But even if man's increased command over Nature included any increased command over himself (the only sort of command relevant to his evolution into a higher being), the fact remains that it is only by running away from the increased command over Nature to country places where Nature is still in primitive command over Man that he can recover from the effects of the smoke, the stench, the foul air, the overcrowding, the racket, the ugliness, the dirt which the cheap cotton costs us. If manufacturing activity means Progress, the town must be more advanced than the country; and the field laborers and village artizans of to-day must be much less changed from the servants of Job than the proletariat of modern London from the proletariat of Cæsar's Rome. Yet the cockney proletarian is so inferior to the village laborer that it is only by steady recruiting from the country that London is kept alive. This does not seem as if the change since Job's time were Progress in the popular sense: quite the reverse. The common stock of discoveries in physics has accumulated a little: that is all.

One more illustration. Is the Englishman prepared to admit that the American is his superior as a human being? I ask this question because the scarcity of labor in America relatively to the demand for it has led to a development of machinery there, and a consequent "increase of command over Nature" which makes many of our English methods appear almost medieval to the up-to-date Chicagoan. This means that the American has an advantage over the Englishman of exactly the same nature that the Englishman has over the contemporaries of Cicero. Is the Englishman prepared to draw the same conclusion in both cases? I think not. The American, of course, will draw it cheerfully; but I must then ask him whether, since a modern Negro has a greater "command over Nature" than Washington had, we are also to accept the conclusion, involved in his former one, that humanity has progressed from Washington to the *fin de siècle* Negro.

Finally, I would point out that if life is crowned by its success and devotion in industrial organization and ingenuity, we had better worship the ant and the bee (as moralists urge us to do

in our childhood), and humble ourselves before the arrogance of the birds of Aristophanes.

My reason then for ignoring the popular conception of Progress in Cæsar and Cleopatra is that there is no reason to suppose that any Progress has taken place since their time. But even if I shared the popular delusion, I do not see that I could have made any essential difference in the play. I can only imitate humanity as I know it. Nobody knows whether Shakespear thought that ancient Athenian joiners, weavers, or bellows menders were any different from Elizabethan ones; but it is quite certain that he could not have made them so, unless, indeed, he had played the literary man and made Quince say, not "Is all our company here?" but "Bottom: was not that Socrates that passed us at the Piræus with Glaucon and Polemarchus on his way to the house of Kephalus?" And so on.

CLEOPATRA Cleopatra was only sixteen when Cæsar went to Egypt; but in Egypt sixteen is a riper age than it is in England. The childishness I have ascribed to her, as far as it is childishness of character and not lack of experience, is not a matter of years. It may be observed in our own climate at the present day in many women of fifty. It is a mistake to suppose that the difference between wisdom and folly has anything to do with the difference between physical age and physical youth. Some women are younger at seventy than most women at seventeen.

It must be borne in mind, too, that Cleopatra was a queen, and was therefore not the typical Greek-cultured, educated Egyptian lady of her time. To represent her by any such type would be as absurd as to represent George IV by a type founded on the attainments of Sir Isaac Newton. It is true that an ordinarily well educated Alexandrian girl of her time would no more have believed bogey stories about the Romans than the daughter of a modern Oxford professor would believe them about the Germans (though, by the way, it is possible to talk great nonsense at Oxford about foreigners when we are at war with them). But I do not feel bound to believe that Cleopatra was well educated. Her father, the illustrious Flute Blower, was not at all a parent of the Oxford professor type. And Cleopatra was a chip of the old block.

BRITANNUS I find among those who have read this play in manuscript a strong conviction that an ancient Briton could not possibly have been like a modern one. I see no reason to adopt this curious view. It is true that the Roman and Norman conquests must have for a time disturbed the normal British type produced by the climate. But Britannus, born before these events, represents the unadulterated Briton who fought Cæsar and impressed

Roman observers much as we should expect the ancestors of Mr Podsnap to impress the cultivated Italians of their time.

I am told that it is not scientific to treat national character as a product of climate. This only shews the wide difference between common knowledge and the intellectual game called science. We have men of exactly the same stock, and speaking the same language, growing in Great Britain, in Ireland, and in America. The result is three of the most distinctly marked nationalities under the sun. Racial characteristics are quite another matter. The difference between a Jew and a Gentile has nothing to do with the difference between an Englishman and a German. The characteristics of Britannus are local characteristics, not race characteristics. In an ancient Briton they would, I take it, be exaggerated, since modern Britain, disforested, drained, urbanified and consequently cosmopolized, is presumably less characteristically British than Cæsar's Britain.

And again I ask does anyone who, in the light of a competent knowledge of his own age, has studied history from contemporary documents, believe that 67 generations of promiscuous marriage have made any appreciable difference in the human fauna of these isles? Certainly I do not.

JULIUS CÆSAR

As to Cæsar himself, I have purposely avoided the usual anachronism of going to Cæsar's books, and concluding that the style is the man. That is only true of authors who have the specific literary genius, and have practised long enough to attain complete self-expression in letters. It is not true even on these conditions in an age when literature is conceived as a game of style, and not as a vehicle of self-expression by the author. Now Cæsar was an amateur stylist writing books of travel and campaign histories in a style so impersonal that the authenticity of the later volumes is disputed. They reveal some of his qualities just as the Voyage of a Naturalist Round the World reveals some of Darwin's, without expressing his private personality. An Englishman reading them would say that Cæsar was a man of great common sense and good taste, meaning thereby a man without originality or moral courage.

In exhibiting Cæsar as a much more various person than the historian of the Gallic wars, I hope I have not been too much imposed on by the dramatic illusion to which all great men owe part of their reputation and some the whole of it. I admit that reputations gained in war are specially questionable. Able civilians taking up the profession of arms, like Cæsar and Cromwell, in middle age, have snatched all its laurels from opponent commanders bred to it, apparently because capable persons engaged in military pursuits are so scarce that the existence of two of them at the same time in the same hemisphere is extremely rare. The capacity of any conqueror is therefore more likely than not to be an illusion produced by the incapacity of

his adversary. At all events, Cæsar might have won his battles without being wiser than Charles XII or Nelson or Joan of Arc, who were, like most modern "self-made" millionaires, half-witted geniuses, enjoying the worship accorded by all races to certain forms of insanity. But Cæsar's victories were only advertisements for an eminence that would never have become popular without them. Cæsar is greater off the battle field than on it. Nelson off his quarterdeck was so quaintly out of the question that when his head was injured at the battle of the Nile, and his conduct became for some years openly scandalous, the difference was not important enough to be noticed. It may, however, be said that peace hath her illusory reputations no less than war. And it is certainly true that in civil life mere capacity for work—the power of killing a dozen secretaries under you, so to speak, as a life-or-death courier kills horses—enables men with common ideas and superstitions to distance all competitors in the strife of political ambition. It was this power of work that astonished Cicero as the most prodigious of Cæsar's gifts, as it astonished later observers in Napoleon before it wore him out. How if Cæsar were nothing but a Nelson and a Gladstone combined! a prodigy of vitality without any special quality of mind! nay, with ideas that were worn out before he was born, as Nelson's and Gladstone's were! I have considered that possibility too, and rejected it. I cannot cite all the stories about Cæsar which seem to me to shew that he was genuinely original; but let me at least point out that I have been careful to attribute nothing but originality to him. Originality gives a man an air of frankness, generosity, and magnanimity by enabling him to estimate the value of truth, money, or success in any particular instance quite independently of convention and moral generalization. He therefore will not, in the ordinary Treasury bench fashion, tell a lie which everybody knows to be a lie (and consequently expects him as a matter of good taste to tell). His lies are not found out: they pass for candors. He understands the paradox of money, and gives it away when he can get most for it: in other words, when its value is least, which is just when a common man tries hardest to get it. He knows that the real moment of success is not the moment apparent to the crowd. Hence, in order to produce an impression of complete disinterestedness and magnanimity, he has only to act with entire selfishness; and this is perhaps the only sense in which a man can be said to be *naturally* great. It is in this sense that I have represented Cæsar as great. Having virtue, he has no need of goodness. He is neither forgiving, frank, nor generous, because a man who is too great to resent has nothing to forgive; a man who says things that other people are afraid to say need be no more frank than Bismarck was; and there is no generosity in giving things you do not want to people of whom you intend to make use. This distinction between virtue and goodness is not understood in England: hence the poverty of our drama in heroes. Our stage attempts at them are mere goody-goodies.

Goodness, in its popular British sense of self-denial, implies that man is vicious by nature, and that supreme goodness is supreme martyrdom. Not sharing that pious opinion, I have not given countenance to it in any of my plays. In this I follow the precedent of the ancient myths, which represent the hero as vanquishing his enemies, not in fair fight, but with enchanted sword, superequine horse and magical invulnerability, the possession of which, from the vulgar moralistic point of view, robs his exploits of any merit whatever.

As to Cæsar's sense of humor, there is no more reason to assume that he lacked it than to assume that he was deaf or blind. It is said that on the occasion of his assassination by a conspiracy of moralists (it is always your moralist who makes assassination a duty, on the scaffold or off it), he defended himself until the good Brutus struck him, when he exclaimed "What! you too, Brutus!" and disdained further fight. If this be true, he must have been an incorrigible comedian. But even if we waive this story, or accept the traditional sentimental interpretation of it, there is still abundant evidence of his lightheartedness and adventurousness. Indeed it is clear from his whole history that what has been called his ambition was an instinct for exploration. He had much more of Columbus and Franklin in him than of Henry V.

However, nobody need deny Cæsar a share, at least, of the qualities I have attributed to him. All men, much more Julius Cæsars, possess all qualities in some degree. The really interesting question is whether I am right in assuming that the way to produce an impression of greatness is by exhibiting a man, not as mortifying his nature by doing his duty, in the manner which our system of putting little men into great positions (not having enough great men in our influential families to go round) forces us to inculcate, but as simply doing what he naturally wants to do. For this raises the question whether our world has not been wrong in its moral theory for the last 2,500 years or so. It must be a constant puzzle to many of us that the Christian era, so excellent in its intentions, should have been practically such a very discreditable episode in the history of the race. I doubt if this is altogether due to the vulgar and sanguinary sensationalism of our religious legends, with their substitution of gross physical torments and public executions for the passion of humanity. Islam, substituting voluptuousness for torment (a merely superficial difference, it is true) has done no better. It may have been the failure of Christianity to emancipate itself from expiatory theories of moral responsibility, guilt, innocence, reward, punishment, and the rest of it, that baffled its intention of changing the world. But these are bound up in all philosophies of creation as opposed to cosmism. They may therefore be regarded as the price we pay for popular religion.